T0256362

ADVANCES IN MECHANICS:
THEORETICAL, COMPUTATIONAL AND INTERDISCIPLINARY ISSUES

PROCEEDINGS OF THE 3RD POLISH CONGRESS OF MECHANICS (PCM) & 21ST INTERNATIONAL CONFERENCE ON COMPUTER METHODS IN MECHANICS (CMM) – PCM-CMM-2015 CONGRESS, GDAŃSK, POLAND, 8–11 SEPTEMBER 2015

Advances in Mechanics: Theoretical, Computational and Interdisciplinary Issues

Editors

Michał Kleiber (IFTR PAS)
Tadeusz Burczyński (IFTR PAS)
Krzysztof Wilde (FCEE GUT)
Jarosław Górski (FCEE GUT)
Karol Winkelmann (FCEE GUT)
Łukasz Smakosz (FCEE GUT)

(IFTR PAS) *Institute of Fundamental Technological Research, Polish Academy of Sciences, Warsaw, Poland*
(FCEE GUT) *Faculty of Civil and Environmental Engineering, Gdańsk University of Technology, Gdańsk, Poland*

CRC Press
Taylor & Francis Group
Boca Raton London New York Leiden

CRC Press is an imprint of the
Taylor & Francis Group, an **informa** business
A BALKEMA BOOK

CRC Press/Balkema is an imprint of the Taylor & Francis Group, an informa business

Typeset by V Publishing Solutions Pvt Ltd., Chennai, India

Published by: CRC Press/Balkema
P.O. Box 11320, 2301 EH Leiden, The Netherlands
e-mail: Pub.NL@taylorandfrancis.com
www.crcpress.com – www.taylorandfrancis.com

ISBN: 978-1-138-02906-4 (Hbk)
ISBN: 978-1-315-64506-3 (eBook PDF)

Advances in Mechanics: Theoretical, Computational and Interdisciplinary Issues – Kleiber et al. (Eds)
© 2016 Taylor & Francis Group, London, ISBN 978-1-138-02906-4

Table of contents

Advances in Mechanics: Theoretical, Computational and Interdisciplinary Issues – Kleiber et al. (Eds)
© 2016 Taylor & Francis Group, London, ISBN 978-1-138-02906-4

Preface

The book exposes our great honour and pleasure to let the readers get acquainted with the 132 papers selected for presentation at the PCM-CMM-2015 CONGRESS held on 8–11 September, 2015 in Gdańsk (Poland).

The PCM-CMM-2015 CONGRESS is a joint scientific event of the 3rd Polish Congress of Mechanics (PCM) and the 21st International Conference on Computer Methods in Mechanics (CMM).

The idea of a Polish Congress of Mechanics was firstly suggested in 2005 by the Polish Society of Theoretical and Applied Mechanics. The intended scope was to cover the domain of theoretical, experimental and computational mechanics as well as interdisciplinary issues, including industrial applications.

The 21st International Conference on Computer Methods in Mechanics continues the 44-year series of conferences dedicated to numerical methods and their applications to the mechanics-based problems. The meetings, organized biannually since 1973 provide a forum for presentation and discussion of new ideas referring to the theoretical background and practical applications of computational mechanics.

We would like to express our gratitude to all Authors for their valuable contributions and for their willingness and efforts to share their research and development activities with the international mechanics community. We are particularly grateful to the General Lecturers, Professors: J. Ambrósio (Portugal), M. Geers (The Netherlands), R. Kienzler (Germany), T. Kowalewski (Poland), Z. Kowalewski (Poland), M. Kuczma (Poland), T. Kurtyka (Switzerland), W. Rachowicz (Poland), E. Ramm (Germany), A. Soldati (Italy) and V. Tvergaard (Denmark), for their exceptionally valuable and extensive contributions to this volume. We believe that the presented papers will be warmly greeted by academics, researchers, designers and engineers dealing with various problems of mechanics and computational issues.

Each paper submitted to PCM-CMM-2015 CONGRESS and printed in this book has been reviewed by members of the Scientific Committee and the International Advisory Board, next refined by the Authors according to the referees' comments. We are deeply indebted to all members of the SC and IAB for their help in the creation of the Conference programme and for their important contribution to the publishing process of these volumes. Most of the final texts have been additionally adjusted to technical requirements of the publisher, the linguistic quality of some texts has been slightly refined to better follow the original results. We are grateful to our associates M. Skowronek, A. Mleczek and M. Kujawa for their assistance and help in bringing the volume to its final form.

Tadeusz Burczyński
Gdańsk, December 2015

Organization, Acknowledgements & Media Patronage

ORGANIZATION

Polish Society of Theoretical and Applied Mechanics
Polish Association for Computational Mechanics
Institute of Fundamental Technological Research of the Polish Academy of Sciences
Committee on Mechanics of the Polish Academy of Sciences, Section of Computational Methods and Optimization
Committee on Civil Engineering and Hydroengineering of the Polish Academy of Sciences, Section of Mechanics of Structures and Materials
Committee on Machine Building of the Polish Academy of Sciences
Institute of Fluid-Flow Machinery of the Polish Academy of Sciences
Gdańsk University of Technology, Faculty of Civil and Environmental Engineering, Department of Structural Mechanics

ACKNOWLEDGMENTS

We are grateful for the financial support given by the following institutions:
Ministry of Science and Higher Education of the Republic of Poland
ENERGA Group, Poland
SOFiSTiK AG, Germany

MEDIA PATRONAGE

Acta Energetica, Poland

Advances in Mechanics: Theoretical, Computational and Interdisciplinary Issues – Kleiber et al. (Eds)
© 2016 Taylor & Francis Group, London, ISBN 978-1-138-02906-4

Committees

INTERNATIONAL SCIENTIFIC COMMITTEE

Jorge Ambrósio (Portugal), Klaus Jurgen Bathe (USA), Jiun-Shyan Chen (USA), Rene de Borst (The Netherlands), Leszek Demkowicz (USA), Jüri Engelbrecht (Estonia), Marc Geers (The Netherlands), Dietmar Gross (Germany), Francois Jouve (France), Reinhold Kienzler (Germany), Rimantas Kačianauskas (Lithuania), Pierre Ladevèze (France), Jolanta Lewandowska (France), János Lógó (Hungary), Giulio Maier (Italy), Herbert Mang (Austria), Eugenio Oñate (Spain), Manolis Papadrakakis (Greece), Ekkehard Ramm (Germany), Franz Rammerstorfer (Austria), Bernhard Schrefler (Italy), Paul Steinmann (Germany), João António Teixeira de Freitas (Portugal), Hisaaki Tobushi (Japan), Viggo Tvergaard (Denmark), Wolfgang Wall (Germany).

HONORARY COMMITTEE (POLAND)

Romuald Będziński, Czesław Cempel, Krzysztof Dems, Andrzej Garstecki, Józef Giergiel, Witold Gutkowski, Zbigniew Kączkowski, Józef Kubik, Krzysztof Marchelek, Jarosław Mikielewicz, Zenon Mróz, Janusz Orkisz, Andrzej Styczek, Gwidon Szefer, Eugeniusz Świtoński, Andrzej Tylikowski, Zenon Waszczyszyn, Edmund Wittbrodt.

SCIENTIFIC COMMITTEE (POLAND)

Michał Kleiber – President of the PCM-CMM-2015 Congress
Włodzimierz Kurnik – Vice-President of the PCM-CMM-2015 Congress
Tadeusz Burczyński – Chairman of the PCM-CMM-2015 Congress Scientific Committee
Krzysztof Wilde – Vice-Chairman of the PCM-CMM-2015 Congress Scientific Committee
Arkadiusz Mężyk, Zbigniew Kowalewski – Chairmen of the Permanent Congress Committee
Mieczysław Kuczma – Chairman of the Polish Association for Computational Mechanics
Krzysztof Arczewski, Jan Awrejcewicz, Janusz Badur, Czesław Bajer, Stefan Berczyński, Wojciech Blajer, Roman Bogacz, Ryszard Buczkowski, Witold Cecot, Wojciech Cholewa, Jacek Chróścielewski, Czesław Cichoń, Paweł Dłużewski, Piotr Doerffer, Stanisław Drobniak, Dariusz Gawin, Józef Gawlik, Wojciech Gilewski, Zbigniew Gronostajski, Jan Holnicki-Szulc, Krzysztof Kaliński, Tomasz Kapitaniak, Jan Kiciński, Marian Klasztorny, Paweł Kłosowski, Zbigniew Kołakowski, Piotr Konderla, Witold Kosiński (†), Janusz Kowal, Tomasz Kowalewski, Katarzyna Kowal-Michalska (†), Ireneusz Kreja, Tomasz Krzyżyński, Tomasz Lewiński, Tadeusz Łagoda, Tomasz Łodygowski, Krzysztof Magnucki, Ewa Majchrzak, Bogdan Maruszewski, Bohdan Mochnacki, Wiesław Nagórko, Wiesław Ostachowicz, Jerzy Pamin, Ryszard Parkitny, Henryk Petryk, Ryszard Pęcherski, Wojciech Pietraszkiewicz, Maciej Pietrzyk, Stanisław Radkowski, Wojciech Radomski, Jacek Rokicki, Błażej Skoczeń, Stanisław Stupkiewicz, Andrzej Jacek Tejchman, Jerzy Warmiński.

ORGANIZING COMMITTEE (GDAŃSK, POLAND)

Jarosław Górski – Chairman
Paweł Kłosowski, Jacek Pozorski – Vice-Chairmen
Karol Winkelmann, Łukasz Smakosz – Secretaries
Karol Daszkiewicz, Violetta Konopińska, Alina Kryczałło, Marcin Kujawa, Jacek Lachowicz, Aleksandra Mariak, Anna Mleczek, Magdalena Rucka, Agnieszka Sabik, Marek Skowronek, Mateusz Sondej, Katarzyna Szepietowska, Wojciech Witkowski, Beata Zima.

General lectures

Advances in Mechanics: Theoretical, Computational and Interdisciplinary Issues – Kleiber et al. (Eds)
© 2016 Taylor & Francis Group, London, ISBN 978-1-138-02906-4

Interactions between mechanical systems and continuum mechanical models in the framework of biomechanics and vehicle dynamics

J. Ambrósio
Instituto Superior Técnico, University of Lisbon, Lisbon, Portugal

ABSTRACT: Multibody dynamics approaches provide some of the most general and efficient computational dynamics methodologies to model complex systems in which the relative large overall motion of the components play a major role. Initially addressing only systems made of rigid bodies with relative motion between the components described by perfect kinematic constraints the multibody systems now include the description of the deformation of the components and allow for the joints to be described by contact pairs with local deformations and tribological effects. Here, the traditional construction of multibody systems is first described being alternative formulations for kinematic joints, by perfect kinematic pairs or by contact joints, presented. The possibility for the deformation of the system components is also included in the multibody formulation by using the finite element method to discretize particular components that exhibit deformations that influence the overall performance. The modelling of complex systems by multibody dynamics and finite element methods is further expanded by developing a co-simulation procedure that enables sub-systems to be modelled and analysed using different methods and codes while maintaining the synchronism of their forward time integration. The methods overviewed are applied first to biomechanical models for the human upper and lower limbs, including their detailed musculoskeletal systems, in order to show not only the constructive elements of a traditional multibody model but also the importance of considering imperfect and contact joints instead of the classical kinematic relations. The demonstrative examples are further pursued in the framework of vehicle dynamics in which not only the modelling aspects of realistic mechanical joints are of importance for the system performance but also in which the deformation of the system components play a role. Generalized deformations of sub-systems of the complex multibody system are used in the framework of a satellite deployment for which some of the structural components are made of composite materials and include piezo-electric sensors and actuators for their active control. The use of co-simulation approaches for the simulation of the interaction between structural, or fluid, systems with multibody systems is finally demonstrated by the application to the study of the interaction between pantograph and catenaries for high-speed railway vehicles.

1 INTRODUCTION

Multibody dynamic formulations are the basis for the most efficient computational techniques that deal with large overall motion. Formulations based on nonlinear finite element methods provide powerful and versatile procedures to describe the flexibility of the system components. It is no surprise that many of the most recent formulations on flexible multibody dynamics and on finite element methods with large rotations share common features that are of fundamental importance to the design requirements of advanced mechanical and structural systems and to the real-time simulation of complex systems (Ambrósio & Kleiber 2001).

The framework for a multibody system, shown in Figure 1, describes the general motion of rigid bodies with the Newton-Euler equations of motion. The kinematic relations between the different bodies of the system are included by using Lagrange multipliers (Nikravesh 1988). For components that experience deformations which influence their dynamic performance the assumption of rigidity is not used being their flexibility described with respect to their local reference frames using the finite element method (Geradin & Cardona 2001, Ambrósio 1996). The remaining ingredients for the application of multibody dynamics to realistic complex systems involve the description of the contact between system components or with other systems via the force vector in the equations of motion.

Currently, multipurpose software packages exploit the ease of use of the computational resources available to create virtual prototyping environments (MSC Software 2015, Dassault Systemes 2015). These packages share common geometric modelling features with more or less

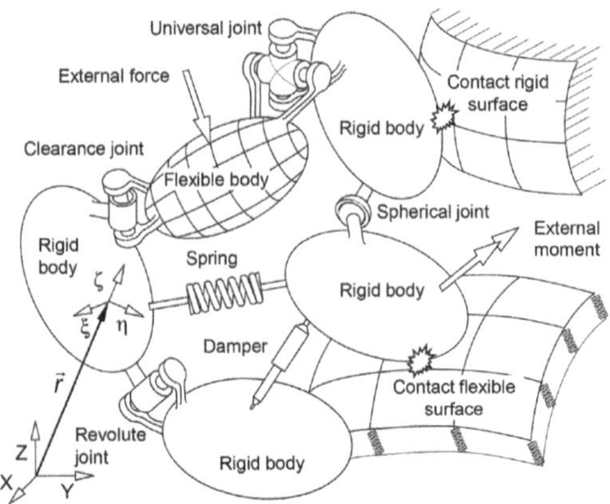

Figure 1. Generic representation of a multibody system.

intuitive interfaces to build multibody or finite element models. Applications to vehicles, deployable structures, space satellites, machines or robot manipulators, which undergo large rigid body motion and material or geometric nonlinear deformations are examples of the application of these approaches.

2 RIGID MULTIBODY SYSTEMS

In a wide number of practical applications the flexibility of the system components does not play a role being enough to represent them as rigid bodies. Vehicle dynamics, biomechanics of human or animal motion or machine dynamics provide extensive examples of this type of systems.

2.1 *Equations of motion and kinematic restrictions*

The equilibrium equations for a multibody system made of rigid bodies is generally represented in the form of (Nikravesh 1988).

$$\begin{bmatrix} \mathbf{M} & \mathbf{\Phi}_{\mathbf{q}}^{T} \\ \mathbf{\Phi}_{\mathbf{q}} & \mathbf{0} \end{bmatrix} \begin{bmatrix} \ddot{\mathbf{q}}_{r} \\ \boldsymbol{\lambda} \end{bmatrix} = \begin{bmatrix} \mathbf{g} \\ \boldsymbol{\gamma} \end{bmatrix} \tag{1}$$

in which $\mathbf{M}$ is the system mass matrix, $\ddot{\mathbf{q}}_r$ is the vector that contains the state accelerations, $\mathbf{g}$ is the generalized force vector, which contains all external forces and moments, except for the vector of the constraint reaction forces described by $\mathbf{g}^{(c)} = -\mathbf{\Phi}_{\mathbf{q}}^{T}\boldsymbol{\lambda}$ and included in left-hand side of the equations. The set of unknown Lagrange multipliers are associated to the intensity of the joint reaction forces. It should be noted the form of equation (1) implies the use of Cartesian (Nikravesh 1988) or of Natural (Garcia de Jalon & Bayo 1993) coordinates to represent the position and orientation of the body fixed coordinate frame associated to each rigid body.

The kinematic restrictions can be viewed as algebraic relations between the rigid body coordinates, as implied in the illustration of Figure 2a or as contact joints between two bodies, as for the clearance joint depicted in Figure 2b.

Any joint that is described as a perfect kinematic joint is included in the equations of motion as a kinematic constraint via the term $\mathbf{g}^{(c)} = -\mathbf{\Phi}_{\mathbf{q}}^{T}\lambda$. However, in many problems of practical importance the mechanical, or biomechanical, joints are not kinematically perfect exhibiting clearances, and eventually local deformable elements such as bushings. In this case clearance contact joints are used being their influence on the equations of motion done via the force vector $\mathbf{g}$ and not by adding the term with the Lagrange multipliers (Flores et al. 2008).

The dynamics of many practical systems in the areas of vehicle dynamics, machines, biomechanics or space vehicles, can be properly addressed with the equations of motion in equation (1) being the large load of the modelling efforts put in the description of the interaction of the system components and external or internal objects. The rail-wheel contact, in railway dynamics, the road-tire contact mechanics in automotive dynamics or the musculoskeletal system actuation, in biomechanics of motion, are examples of cases in which proper constitutive relations that describe the interaction phenomena are required.

2.2 *Biomechanics of the shoulder and shoulder prosthesis*

The dynamics of human and animal motion generally use multibody models of musculoskeletal systems and biological mechanisms, in which the anatomical segments are represented by rigid bodies and the muscles represented either by kinematic

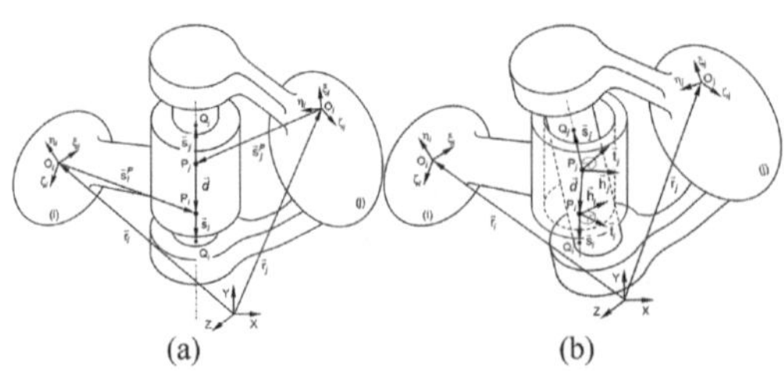

Figure 2. Kinematic restrictions between adjacent bodies: (a) perfect kinematic joint; (b) joint with clearance.

constraints or by actuators. The anatomical joints are generally represented by mechanical joints or, in particular applications, by contact joints. Practical applications focus the identification of the muscle and joint reaction forces developed in the human body to perform a given task or the design of prosthesis that ensure not only a suitable biomechanical functionality but also a long term stability and durability.

A multibody model for the shoulder in which the anatomical joints are described by perfect mechanical joints and the muscles act along straight and curved segments is proposed in (Quental et al. 2012) being some of its modelling features highlighted in Figure 3. The biomechanical problem solved with this model consists on finding the internal muscle and joint reaction forces for prescribed motions acquired in a biomechanical laboratory.

The biomechanical model presented in Figure 3 is further modified to include a reverse shoulder prosthesis, shown in Figure 4a and identifying the muscle forces as well as the loads on the prosthesis ball and cup. The glenohumeral joint with the prosthesis is described by clearance ball joint, seen in Figure 4b, in which the clearance and the contacting surface material properties influence the shoulder dynamics (Quental et al. 2013). Note that due to the techniques used in the surgery part of the deltoid muscle segments are deactivated leading to a difference on the musculoskeletal model with respect to that of the natural shoulder.

The model is further used to find the variation of the shoulder joint reaction forces, i.e., glenohumeral joint reactions, as a function of the different prosthesis geometries. Figure 5 presents the variation of the glenohumeral reaction forces as function of the arm elevation for an unloaded abduction motion in the coronal plane. In the process of the application of the reverse prosthesis shoulder model issues such as the detection of impingement between the scapula and the humerus, the modification of the muscle insertion points or the deactivation of some muscle bundles are addressed, thus providing the surgeon with a powerful design tool to support the surgery procedure adopted.

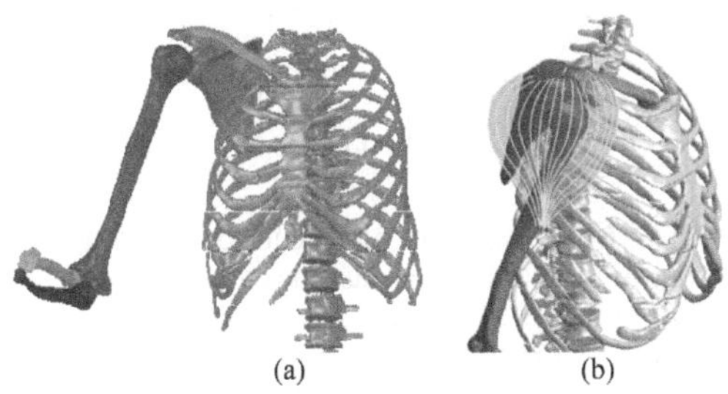

Figure 3. Features of the biomechanical model of the left upper limb using: (a) anatomical segments; (b) deltoid muscle model.

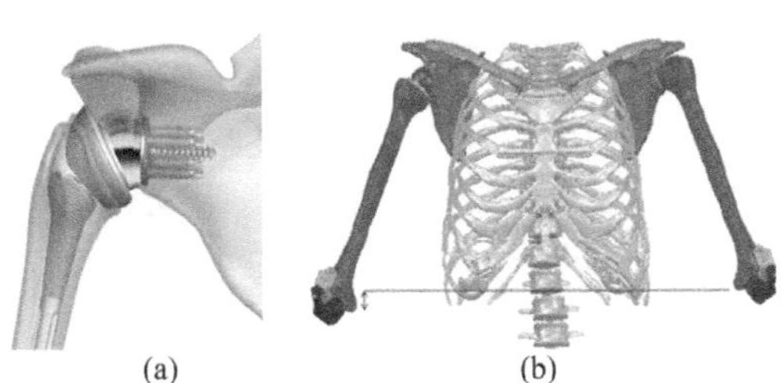

Figure 4. Reverse shoulder prosthesis: (a) prosthesis; (b) model of the left upper limb with a glenohumeral clearance ball joint.

2.3 *Railway vehicle dynamics*

In vehicle dynamics the studies for vehicle stability and manoeuvrability require realistic models for the kinematic joints, without which fundamental interaction phenomena cannot be identified. The study of the virtual approval for operation of a railway vehicle in a mountain track provides an application example in which the detailed modelling of the vehicle with its suspension systems, shown in Figure 4, and of the interaction with the track is of fundamental importance (Magalhães et al. 2015).

The industry specifies a series of limit values on the forces that the vehicle can apply on the rails, on the forces that the vehicle bogies or wheelsets can be subjected to, on the accelerations of the bogies, wheelsets or the carbody and on the dynamic loading that the occupants are subjected to. The

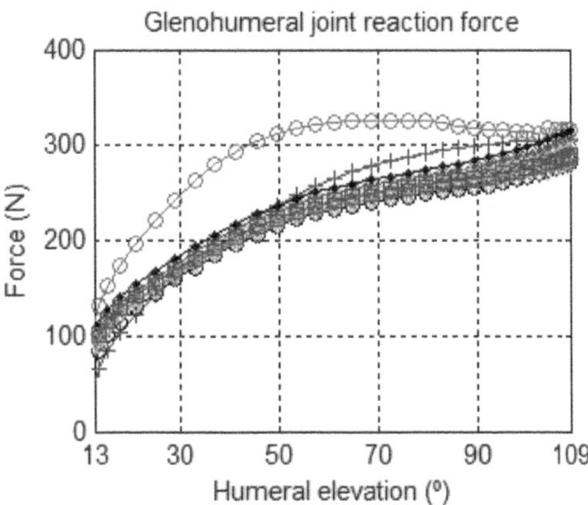

Figure 5. Glenohumeral joint reaction force for diverse reverse shoulder prosthesis (ooo), pathological shoulder (red ooo) and for the natural shoulder (+++).

evaluation of the criteria associated with homologation and acceptance lead to values that must be contained within the thresholds prescribed by the norm (Polach & Evans 2013, Willson et al. 2011).

The quality of the models and their ability to predict a realistic behaviour for the rail-wheel contact. In particular, the modelling decisions on the description of the kinematic joints of the boggies, depicted in Figure 7, influence the vehicle ability to be inserted in the track and to keep the wheels in permanent contact. The analysis show that models in which the realistic joint clearances are not considered lead to premature derailment (Magalhães et al. 2015).

The wheel to rail contact is supposed to occur most commonly between the wheel tread, represented in Figure 8 by the red ball, and the rail and not between the flange and rail, whose closest points for potential contact are represented by two blue balls in Figure 8. Not only the correct rolling contact model used for the wheel-rail contact representation is of crucial importance for the evaluation of the interaction forces, but also the relative kinematics between system components needs to be correctly described to allow for a correct wheel to rail insertion.

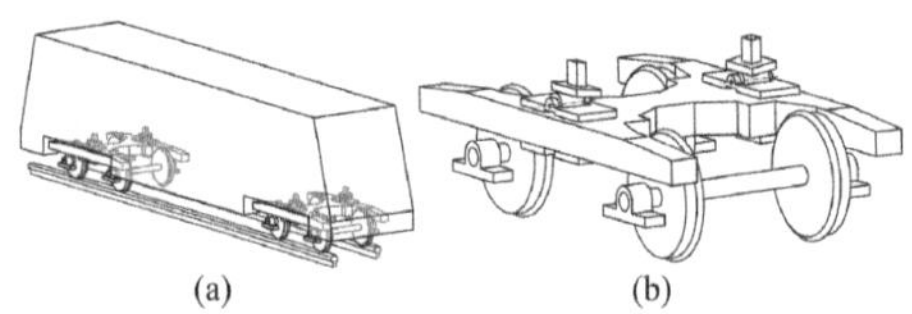

Figure 6. Multibody railway vehicle: (a) vehicle on its tracks; (b) model for one of the boggies.

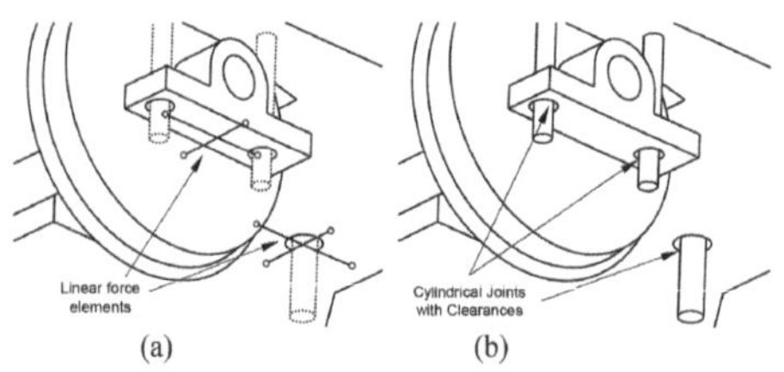

Figure 7. Typical modelling of boggies joints: (a) restrictions by force elements; (b) clearance cylindrical joint.

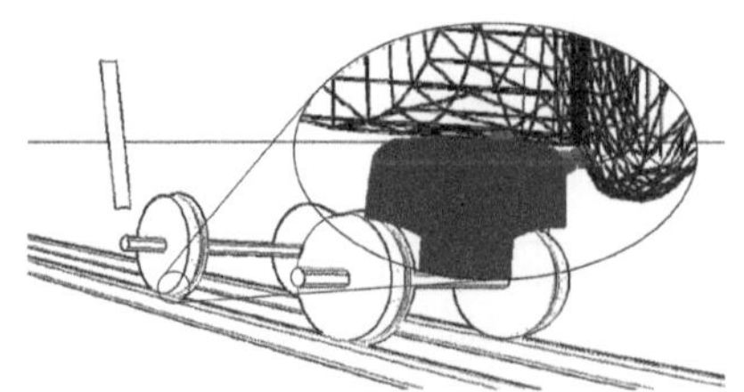

Figure 8. Typical modelling of boggies joints: (a) restrictions by force elements; (b) clearance cylindrical joint.

3 FLEXIBLE MULTIBODY SYSTEMS

The flexibility of the multibody system components plays an important role in the system dynamics response in light weight systems or in classical mechanical systems experiencing particular loading scenarios (Geradin & Cardona 2001). Although alternative procedures have been proposed, such as plastic hinges (Pereira & Ambrósio 1997), finite segments (Huston & Wang 1994) or meshless methods (Fleissner et al. 2007), multibody system that include flexible bodies, as that depicted in Figure 5, include their flexibility described by using the finite element method.

3.1 *Flexible body equations of motion*

In the applications described presented here, the finite elements are described with respect to moving frames associated to the flexible bodies, represented by $\xi\eta\zeta$ in Figure 5 (Ambrósio 1996). The moving frame in this formulation plays the role of the inertial frame XYZ in conventional finite element formulations.

The most general equations of motion for a flexible body that can exhibit nonlinear deformations, due to geometric and material nonlinearities, are written as (Ambrósio 1996)

$$\begin{bmatrix} \mathbf{M}_{rr} & \mathbf{M}_{rf} \\ \mathbf{M}_{fr} & \mathbf{M}_{ff} \end{bmatrix} \begin{bmatrix} \ddot{\mathbf{q}}_r \\ \ddot{\mathbf{u}}' \end{bmatrix} = \begin{bmatrix} \mathbf{g}_r \\ \mathbf{g}'_f \end{bmatrix} - \begin{bmatrix} \mathbf{s}_r \\ \mathbf{s}'_f \end{bmatrix} - \begin{bmatrix} \mathbf{0} \\ \mathbf{f} + (\mathbf{K}_L + \mathbf{K}_{NL})\Delta\mathbf{u}' \end{bmatrix} \quad (2)$$

where $\mathbf{M}_{rr}\ddot{\mathbf{q}}_r = \mathbf{g}_r$ are the rigid body equivalent equations of motion, as in equation (1), and $\mathbf{M}_{ff}\ddot{\mathbf{u}}' = \mathbf{g}'_f - \mathbf{s}'_f - \mathbf{f} + (\mathbf{K}_L + \mathbf{K}_{NL})\Delta\mathbf{u}'$ are the standard nonlinear finite element equations of motion. In equation (2) the nodal coordinates are denote by $\mathbf{u}'$, where $\bullet'$ denotes that the quantity $\bullet$ is expressed in the coordinates of the local frame, and $\mathbf{f}$ is the stress equivalent vector. The mass

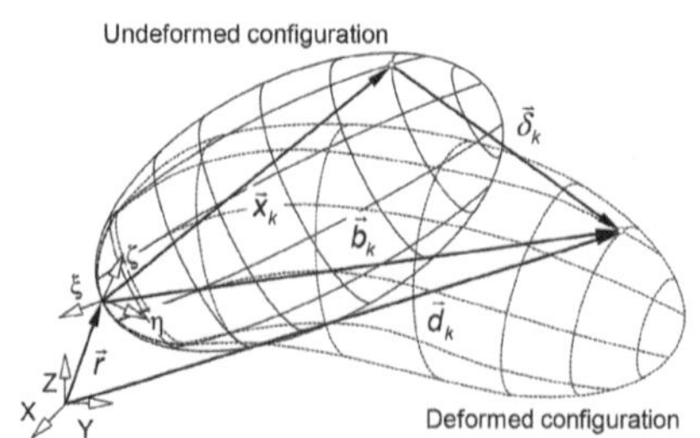

Figure 9. Flexible body including the local reference frame.

matrix sub-matrices $\mathbf{M}_{rf} = \mathbf{M}_{fr}{}^T$ effectively couple the large body motion with its deformations and vectors $\mathbf{s}_r$ and $\mathbf{s}'_f$ are gyroscopic force terms associated to the description of the body flexibility with reference to a moving frame.

The flexible body equation of motion written in the form of equation (2) does not provide a unique description for the displacement field requiring the application of reference conditions, i.e., the equivalent to the boundary conditions used in standard finite elements but with reference to the body fixed frame and not to the inertia frame. The boundary conditions are included in the flexible body equations of motion in the same way the kinematic constraints are included in the rigid multibody system equations, i.e., by using Lagrange multipliers. The equations of motion for a flexible body with the reference conditions included are written as (Ambrósio 2007):

$$\begin{bmatrix} \mathbf{M}_{rr} & \mathbf{M}_{rf} & \mathbf{0} \\ \mathbf{M}_{fr} & \mathbf{M}_{ff} & \mathbf{\Phi}_{\mathbf{u}'}^{(rc)T} \\ \mathbf{0} & \mathbf{\Phi}_{\mathbf{u}'}^{(rc)} & \mathbf{0} \end{bmatrix} \begin{bmatrix} \ddot{\mathbf{q}}_r \\ \ddot{\mathbf{u}}' \\ \boldsymbol{\lambda}^{(rc)} \end{bmatrix} = \begin{bmatrix} \mathbf{g}_r - \mathbf{s}_r \\ \mathbf{g}'_f - \mathbf{s}'_f - \mathbf{f} - (\mathbf{K}_L + \mathbf{K}_{NL})\Delta\mathbf{u}' \\ \boldsymbol{\gamma}^{(rc)} \end{bmatrix} \quad (3)$$

Any set of reference conditions must present at least six independent equations relating the nodal coordinates, i.e, the rigid body motion of the finite element mesh relative to the body reference frame must be removed. The boundary fixed conditions, in which one or more nodes have fixed degrees-of-freedom, are probably the best known and used. However, the mean axis conditions (Cavin & Dusto 1977), in which enforce the kinetic energy associated with the deformation, measured with respect to an observer stationed on the flexible body is minimized, and the principal axis conditions (Nikravesh & Lin 2005), which enforce that the body reference frame is located in the center of mass and that the axis orientation coincides with the principal inertia axis, are important alternatives that must be considered.

The size of the system described by equation (3) and the nonlinearities involved makes the computational solution of a single body already expensive. When used in a multibody system, which may include a moderate or large number of flexible bodies, the use of equation (3) may be computationally prohibitive for analysis with long time periods. Recognizing that in most of the practical applications the deformations of the flexible bodies are elastic and small, the elastodynamic coordinates $\mathbf{u}'$ can be reduced by using a substructuring technique, the mode component synthesis or the Craig-Bampton method (Ambrósio 1996, Neto et al. 2013). For conciseness, let the mode components synthesis be considered here. Let the nodal displacements of the flexible part of the body be described by a weighted sum of a set of assumed modes, which include the modes of vibration associated with the natural frequencies of the flexible body and some selected static modes of deformation

$$\mathbf{u}' = \mathbf{X}\mathbf{w} \quad (4)$$

By substituting the time derivatives of Equation(4) into equation (3) and simplifying, the equation of motion for a flexible body experiencing linear elastic deformations with respect to its local reference frame is

$$\begin{bmatrix} \mathbf{M}_{rr} & \mathbf{M}_{rf} & \mathbf{0} \\ \mathbf{X}^T\mathbf{M}_{fr} & \mathbf{X}^T\mathbf{M}_{ff} & \left(\mathbf{\Phi}_{\mathbf{u}'}^{(rc)}\mathbf{X}\right)^T \\ \mathbf{0} & \mathbf{\Phi}_{\mathbf{u}'}^{(rc)}\mathbf{X} & \mathbf{0} \end{bmatrix} \begin{bmatrix} \ddot{\mathbf{q}}_r \\ \ddot{\mathbf{w}} \\ \boldsymbol{\lambda}^{(rc)} \end{bmatrix} = \begin{bmatrix} \mathbf{g}_r - \mathbf{s}_r \\ \mathbf{X}^T\left(\mathbf{g}'_f - \mathbf{s}'_f\right) - \mathbf{\Lambda}\mathbf{w} \\ \boldsymbol{\gamma}^{(rc)} \end{bmatrix} \quad (5)$$

which represents 6 equations for the large gross motion plus *nm* equations due to the *nm* modes used in the modal composition.

3.2 *Satellite deployment and structural active control*

The use of the flexible multibody methodology is demonstrated in the modelling of the deployment of a satellite with structural components made of composite materials, shown in Figure 10, and in its active control via piezoelectric sensors and actuators (Magalhães et al. 2015). The objective is to find a reliable unfolding mechanism that can ensure not only that no interference exists during the process but also that the loading in all components remains within admissible limits.

All components of the satellite, beams and panels, are made of composite materials being modelled here as independent flexible bodies. The unfolding mechanism is envisaged by implementing two actuators in prescribed joints as seen in Figure 10. When the actuation laws of such actuators is developed with regards only to the kinematics of the different bodies, as if rigid bodies, a very large interference between the unfolding panels, shown in Figure 11, is observed.

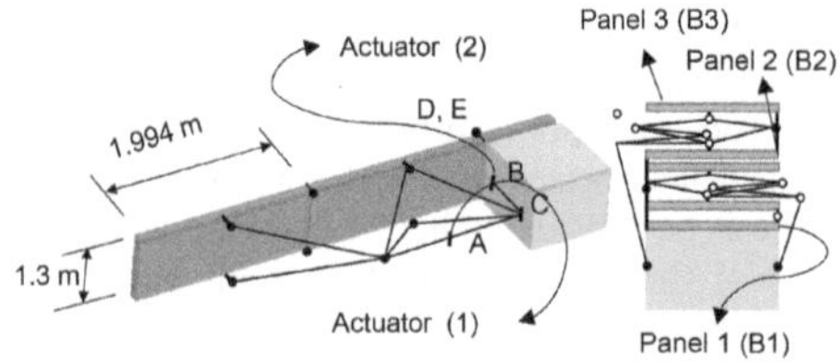

Figure 10. Satellite model with composite materials flexible bodies.

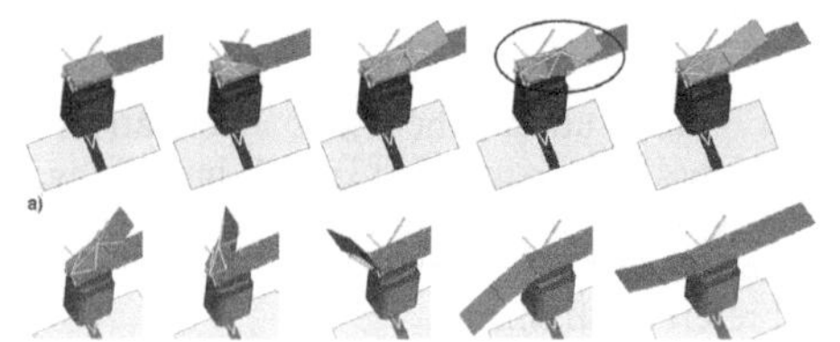

Figure 11. Satellite model with composite materials flexible bodies.

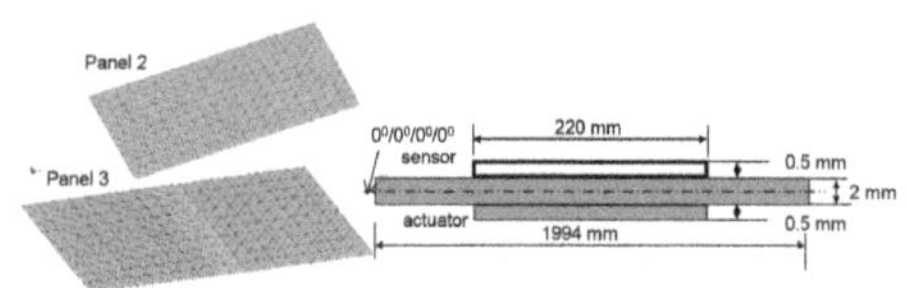

Figure 12. Satellite model with composite materials flexible bodies.

Two types of corrective measures are tested: modification of the panel stiffness by optimization of the composite materials of the panels and beams that minimizes the maximum deformation energy; active control of the deformation of the panels by including two piezoelectric patches, one sensor and other actuator as depicted in Figure 12.

Although not shown here, the optimization of the panel composite material structure minimizes the potential for interference between panels 2 and 3 but does not allow either for the required safety margin or for keeping the joint reaction forces within safe thresholds. The active control of the vibration of the panels, during the unfolding stages enable the complete elimination of the potential for interference while maintaining the joint reaction forces under acceptable limits. The sequence of images during the unfolding process, seen in Figure 13 for the original configuration (top) and active control configuration (bottom) show the separation between the panels.

3.3 *Co-simulation of multibody and finite element models*

Multibody dynamics provides an almost perfect framework for the description of the interaction between systems described by different equilibrium equations and/or analysed with different codes. The interaction between vehicles and infrastructure (Ambrósio et al. 2015), as in the case depicted in Figure 14a for the railway pantograph-catenary contact, or between fluid and structures, as in wind power generator (Moller et al. 2005), as in Figure 14b exemplify cases in which each of the systems are modelled and simulated in different codes that must advance in time synchronously. The co-simulation between multibody and finite element dynamics codes is explored here to address this particular class of interaction between different systems.

The interaction of the pantograph and catenary is achieved through the contact of the pantograph registration strip on the catenary contact wire. The ability of collecting reliable data on the contact

Figure 13. Satellite model with composite materials flexible bodies.

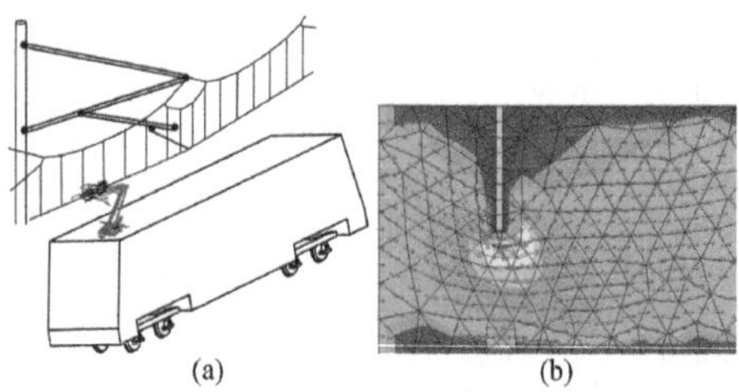

Figure 14. Interaction between systems with co-simulation with different codes: (a) railway pantograph-catenary; (b) fluid-structure.

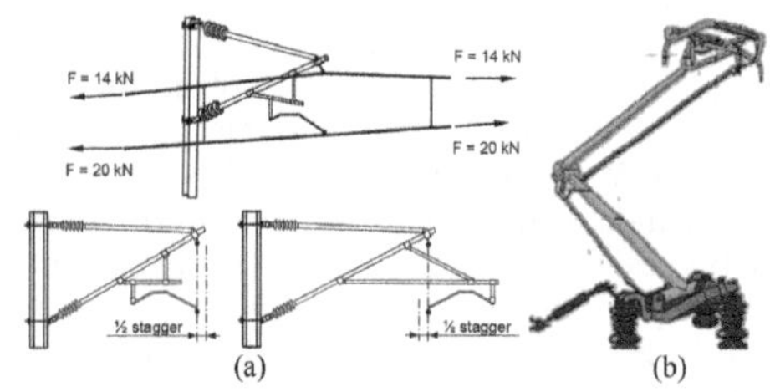

Figure 15. (a) Catenary system for many highspeed railways; (b) typical highspeed pantograph.

forces to allow not only monitoring the operating conditions of the overhead system but also to allow for the validation of numerical models is one of the important key issues of the pantograph/catenary dynamics. The catenary, shown in Figure 15a is a stationary structure mostly made out of beams that exhibits important structural vibrations being those on the contact wire of particular importance. The pantograph, shown in Figure 15b is mechanical system with a large range of motion.

The catenary systems models must be include the proper description of their deformations being the finite element method irreplaceable for the purpose. The length of each section of the catenary, as that represented in Figure 16a are in excess of 1.5 km not being uncommon to require models with two or more sections. As for the pantograph multibody models, the mechanical components can be represented by rigid bodies, as shown in Figure 16b, as their deformations do not play a role in the interaction dynamics.

Being the catenary modelled by complex finite element models it is expected that its dynamics is solved by using suitable finite element codes. By the same token, the multibody pantograph models are handled by multibody dynamics codes. As the integration algorithms of these codes are not only different but also involve diverse time stepping strategies, their co-simulation is required (Ambrósio et al. 2005). The basic idea is that the finite element code supplies the multibody code with the contact forces between the two systems and receives the positions and velocities of the multibody system, as shown in Figure 17.

Typical results of interest for the pantograph-catenary interaction analysis are shown in Figure 18 for a scenario of a highspeed railway vehicle with multiple pantographs.

The variation of the contact force shows a relative periodicity compatible with the pantograph passage under droppers and steady arms but it does not pro-

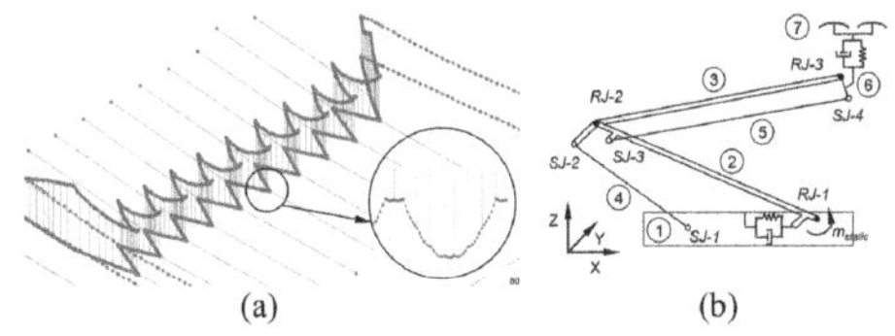

Figure 16. (a) Catenary finite element model; (b) Pantograph multibody model.

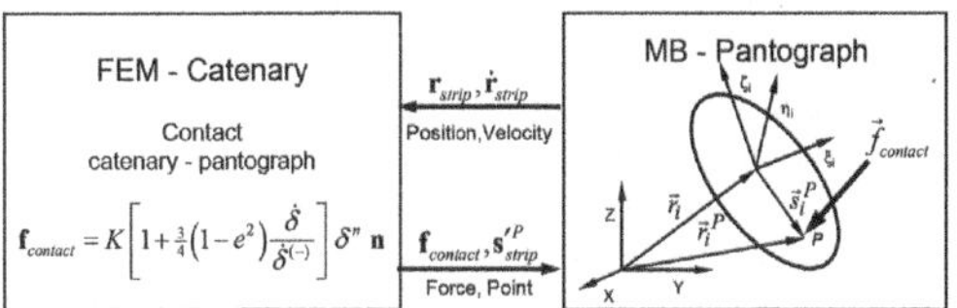

Figure 17. Elements of the co-simulation between a finite element and a multibody code.

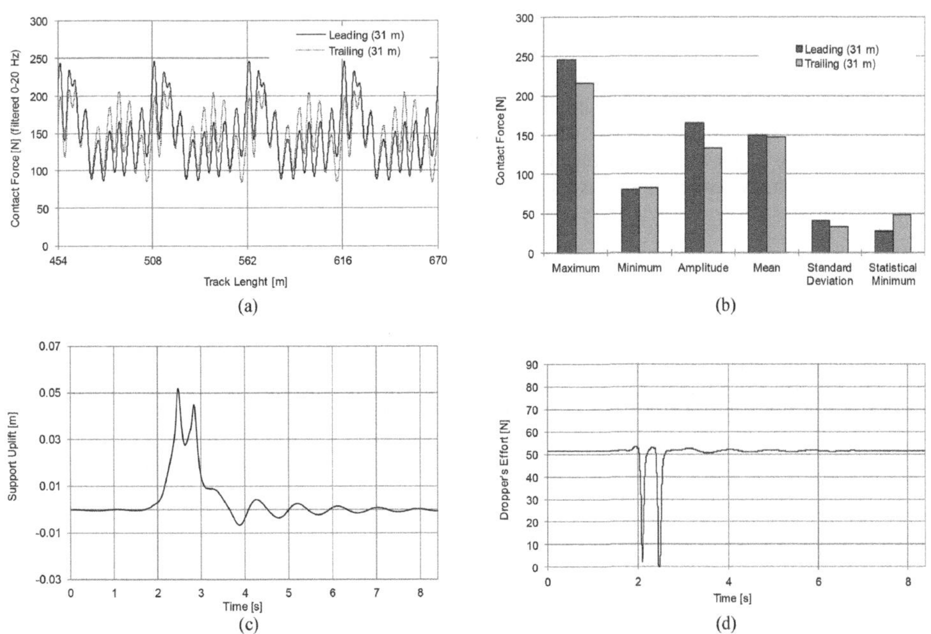

Figure 18. Typical results that characterize the pantograph-catenary contact quality: (a) contact force; (b) statistical measures of the contact quality; (c) uplift of the registration arm; (d) unloading of the droppers.

vide much more useful information. The mean contact force, of 157 N for an operation at 300 km/h, is required for homologation. The standard deviation of the contact force must be lower than 30% of the mean contact force, which is also verified. The uplift of the contact wire under the steady arm is well below the usual maximum of 10 cm allowed for this type of operations. Finally, it is observed that the droppers are almost unloaded when the pantographs run under them, as expected. Therefore, the results collected enable to decide for the acceptance of the compatibility of the pantograph with the catenary for the type of operation analysed here.

4 CONCLUSIONS

The methodologies for the analysis of multibody systems, presented in this work, show a versatility that allows for their use not only to handle complex systems of rigid bodies with restricted relative motions, but also to include their elastodynamics or to be used in co-simulation environments with methods for dynamic analysis typical to other areas. The decisions on the modelling aspects related to the description of the kinematic restrictions, achieved either by kinematic constraints or by contact forces that explore the clearance of the joints between the adjacent bodies are crucial to allow, or prevent, that particular type of analysis are performed. Also, the introduction of the elastodynamics of the different components of the system is naturally done in the framework by using a finite element description. Contrary to the traditional finite element methodologies, the flexible multibody dynamics represents correctly the coupling between the small body deformations and the gross overall motion. The different aspects of the use of multibody dynamics was finally demonstrated in the framework of applications in biomechanics, vehicle dynamics or spacecrafts.

REFERENCES

Ambrósio, J. & Kleiber, M. 2001. *Computational Aspects of Nonlinear Structural Systems with Large Rigid Body Motion*. Amsterdam: IOS Amsterdam.

Ambrósio, J. 1996. Dynamics of Structures Undergoing Gross Motion and Nonlinear Deformations: A Multibody Approach. *Computers and Structures* 59(6): 1001–1012.

Ambrósio, J. 2007. Flexible Multibody Systems with Linear and Nonlinear Deformations. In P. Flores & M. Silva (eds), *Actas da DSM2007—Conferência Nacional de Dinâmica de Sistemas Multicorpo, Guimarães, 6–7 December 2007*: 11–18.

Ambrósio, J., Pombo, J., Antunes, P., Pereira, M. 2015. PantoCat statement of method. *Vehicle System Dynamics* 53(3): 314–328.

Ambrósio, J., Rauter, F., Pombo, J., Pereira, M. 2011. A Flexible Multibody Pantograph Model for the Analysis of the Catenary-Pantograph Contact. In Blajer, W., Fraczek, J., Krzysztof, K. (eds), *Multibody Dynamics: Computational Methods and Applications*: 1–27. Dordrecht: Springer.

Cavin, R.K. & Dusto, A.R. 1977. Hamilton's Principle: Finite Element Method and Flexible Body Dynamics. *AIAA Journal* 15(12): 1684–1690.

Dassault Systemes 2015. Vélizy-Villacoublay, France.

Fleissner, F., Gaugele, T., Eberhard, P. 2007. Applications of the Discrete Element Method in Mechanical Engineering. *Multibody System Dynamics* 18(1): 81–94.

Flores, P., Ambrósio, J., Pimenta Claro, J., Lankarani, H. 2008. *Kinematics and Dynamics of Multibody Systems with Imperfect Joints: Models and Case Studies*. Dordrecht: Springer.

Garcia de Jalon, J. & Bayo, E. 1993. *Kinematic and dynamic simulation of multibody systems*. Heidelberg: Springer-Verlag.

Geradin, M. & Cardona, A. 2001. *Flexible Multibody Dynamics: A Finite Element Approach*. Chichester: John Wiley and Sons.

Huston, R.L. & Wang, Y. 1994. Flexibility effects in multibody systems. In M. Pereira, J. Ambrósio (eds), *Computer aided analysis of rigid and flexible mechanical systems, NATO ASI Series E*. 268: 351–376, Dordrecht: Kluwer Academic Publishers.

Magalhães, H., Ambrósio, J., Pombo, J. 2015. Railway Vehicle Modelling for the Vehicle-Track Interaction Compatibility Analysis. *Proceedings of the Institution of Mechanical Engineers, Part K: Journal of Multibody Dynamics* (accepted).

Moller, H., Lund, E., Ambrósio, J., Gonçalves, J. 2005. Simulation of Fluid Loaded Flexible Multiple Bodies. *Multibody Systems Dynamics* 13(1): 113–128.

MSC Software 2015. Newport Beach, California.

Neto, M.A., Ambrósio, J., Roseiro, L.M., Amaro, A., Vasques, C.M.A. 2013. Active vibration control of spatial flexible multibody systems. *Multibody Systems Dynamics* 30(1): 13–35.

Nikravesh, P. & Lin, Y.-S. 2005. Use of Principal Axes as the Floating Reference Frame for a Moving Deformable Body. *Multibody System Dynamics* 13(2): 211–231.

Nikravesh, P. 1988. *Computer-Aided Analysis of Mechanical Systems*. Englewood Cliffs: Prentice-Hall.

Pereira, M. & Ambrósio, J. 1997. Crashworthiness analysis and design using rigid-flexible multibody dynamics with application to train vehicles. *Int. J. of Nume. Meth. in Engng*. 40: 655–687.

Polach, O. & Evans, J. 2013. Simulations of Running Dynamics for Vehicle Acceptance: Application and Validation. *International Journal of Railway Technology* 2(4): 59–84.

Quental, C., Folgado, J., Ambrósio, J., Monteiro, J. 2012. Dynamic analysis of a multibody system of the upper limb. *Multibody Systems Dynamics* 28(1–2): 83–108.

Quental, C., Folgado, J., Ambrosio, J., Monteiro, J. 2013. Multibody System of the Upper Limb including a Reverse Shoulder Prosthesis. *ASME Journal of Biomechanical Engineering* 135(11), 111005.

Willson, N., Fries, R., Witte, M., et al. 2011. Assessment of safety against derailment using simulation and vehicle acceptance tests: a worldwide comparison of state-of-the-art assessment methods. *Vehicle System Dynamics* 49: 1113–1157.

Advances in Mechanics: Theoretical, Computational and Interdisciplinary Issues – Kleiber et al. (Eds)
© 2016 Taylor & Francis Group, London, ISBN 978-1-138-02906-4

Multiscale mechanics of dynamical metamaterials

M. Geers, V. Kouznetsova, A. Sridhar & A. Krushynska
Department of Mechanical Engineering, Eindhoven University of Technology, Eindhoven, The Netherlands

ABSTRACT: This contribution focuses on the computational multi-scale solution of wave propagation phenomena in dynamic metamaterials. Taking the Bloch-Floquet solution for the standard elastic case as a point of departure, an extended scheme is presented to solve for heterogeneous visco-elastic materials. The physically and geometrically nonlinear case is addressed through a transient computational homogenization scheme. In the particular case of an elastic heterogeneous microstructure, the homogenization scheme can be reduced to the computational analysis of a fluctuation-enriched extended continuum.

1 INTRODUCTION

The mechanical behaviour of materials across the scales often inherits particular properties that are rooted in the intrinsic fine scale level. This is typically the case for problems where the micro-scale reveals special characteristics or pronounced inhomogeneities in the deformation micro-fluctuation fields. A particular class of such materials are acoustic metamaterials, designed to attenuate sound wave propagation for certain frequencies. The unique features of these metamaterials originate from the complex interaction of transient phenomena at the microscopic and macroscopic scales, with local resonance occurring within (one of the) micro-constituents resulting in effective band gaps at the macro-scale. Band-gap materials, like phononic crystals (using Bragg scattering) and acoustic metamaterials (using local resonance), are typically used to control the propagation of elastic waves, where they reveal advanced properties such as negative effective mass density or negative effective elastic moduli. They typically exhibit frequency band gaps (i.e. ranges of frequencies), for which no propagating waves exist.

2 ELASTIC AND VISCO-ELASTIC BAND-GAP METAMATERIALS

Band structures for solid metamaterials are often calculated under the assumption of linear elastic behaviour of the constituting components (Krushynska, Kouznetsova, & Geers 2014). In practice, polymer-based visco-elastic materials are frequently used in these materials, which are in nature dissipative and time-dependent. The influence of the material dissipation on the wave dispersion has been studied for phononic crystals, but the methods used are quite limited when applied to acoustic metamaterials. New methods capable of dealing with a realistic viscoelastic response (e.g. generalized Maxwell) are therefore of great interest. The propagating wave in a deformable continuum complies with the standard equations of motion, linear kinematic strain-displacement relation, and constitutive equations. In the case of a linear viscoelastic material, the constitutive equations can be expressed in terms of Duhamel integrals with time-dependent bulk/shear moduli. For time-harmonic waves, the Bloch-Floquet theorem is generally used to reduce the infinite periodic metamaterial domain to a representative unit cell. The Bloch periodic boundary conditions are applied at the unit cell boundary, whereby the resulting problem is non-linear in the time domain. However, the particular form of the viscoelastic constitutive relations permits a simplified description of all the equations in frequency domain by means of the classical elastic-viscoelastic principle (Hui & Oskay 2013). The transition from the time domain to the frequency domain is achieved by means of the Laplace-Carson transform, which retains the form of the field equations by preserving the physical meaning of the involved coefficients. This allows to extend the existing methods for elastic metamaterials to the visco-elastic regime, whereby the field characteristics will depend on frequency. The considered unit cell of a metamaterial is discretized with a FE-scheme and band structures are calculated using the so-called $k(\omega)$-formulation which has already been successfully applied to phononic crystals with visco-elastic components (Andreassen & Jensen 2013).

3 NONLINEAR DYNAMIC METAMATERIALS

The methodology commonly used for elastic materials, and its extension to visco-elastic materials, cannot be applied for heterogeneous materials that present physical and geometrical nonlinearities. Moreover, scale separation generally does not apply here, and the response of the material is dependent on the structure in which it is embedded. This class of metamaterials is therefore largely unexplored, since it requires a proper methodology to study the transient behaviour accompanying wave propagation.

3.1 *Computational homogenization*

To address the nonlinear response of complex multi-phase solids within a given structure, the Computational Homogenization method is a powerful method (Geers & Kouznetsova 2010). This method is essentially based on the nested solution of two boundary value problems, one at each scale. Though computationally expensive, the procedures developed allow to assess the macroscopic influence of microstructural parameters in a rather straightforward manner. The first-order technique is by now well-established and widely used in the scientific and engineering community. Several extended schemes have been proposed as well: higher-order computational homogenization, continuous-discontinuous homogenization-localization, which aims to incorporate the transition from damage to fracture (via localization) in a multi-scale approach, thermomechanical computational homogenization, substructured thin sheets and shells, multi-scale interfaces or cohesive cracks, multi-physics problems: electro-magnetism, diffusion problems, liquid-phase sintering, heat flow, chemical couplings, contact and friction problems.

3.2 *Computational homogenization for dynamical metamaterials*

The solution in the nonlinear regime is based on a novel transient computational homogenization procedure addressing the evolution in space and in time of materials with a non-steady state microstructure (Pham, Kouznetsova, & Geers 2013). The proposed transient scheme extends the classical concepts used in a first-order computational homogenization framework. The separation of scales hypothesis is relaxed, enabling solutions that go beyond the long wavelength approximation for the microstructural components. It is based on an enriched description of the micro-macro kinematics that allows large spatial fluctuations of the microscopic displacement field (relative to its macroscopic counterpart), originating from local resonance phenomena. An extended Hill-Mandel macro-homogeneity condition is proposed, which couples the coarse scale stress tensor and linear momentum to the solution of the underlying micro-scale problem. Accordingly, the balance of linear momentum is solved at both scales. This novel multi-scale solution method enables innovative designs of finite-size microstructures for locally resonant acoustic metamaterials. Material nonlinearities can be incorporated in a natural manner. Nonlinearities are addressed through the direct solution of an initial boundary value problem, rather than a simplified analysis on an infinite medium, which has mostly been used in existing approximations so far. The transient computational homogenization approach will be illustrated through the multi-scale analysis of a metamaterial subjected to dynamic loading. A direct comparison with Direct Numerical Simulations is made, revealing the consistency of the framework for a large range of loading frequencies and microstructural sizes. Particular emphasis is given to the attenuation of the macroscopic waves at specific loading frequencies. The influence of geometrical and material parameters on the width of the lowest band gap is investigated in detail. The use of different inclusions in the microstructure is assessed by examining the resulting dispersion spectra. As a key engineering result, the proposed method enables to identify the required parameters for the optimal design of locally resonant micro-heterogeneous acoustic metamaterials.

4 FLUCTUATION-ENRICHED EXTENDED CONTINUUM

In the case of a heterogeneous elastic microstructure, the homogenization method allows to eliminate the fine scale, by solving an extended macroscopic continuum. Using the Craig-Bampton reduction technique, internal (fine scale) degrees of freedom can be decomposed into their quasi-static and dynamic parts, and the homogenization method allows to define a (reduced) extra kinematic field η, i.e. a dynamic fluctuation field, to be solved for through a coupled dynamic equation at the level of the representative volume element. This enrichment naturally accounts for the local resonance dispersion phenomena. The working principle of this method will be presented, along with an illustrative example.

REFERENCES

Andreassen, E. & J. Jensen (2013). Analysis of phononic bandgap structures with dissipation. *J. Vib. Acoust. 135.*

Geers, M. & V. Kouznetsova (2010). Multi-scale computational homogenization: trends & challenges. *J. Comp. Math. 234,* 2175–2182.

Hui, T. & C. Oskay (2013). A nonlocal homogenization model for wave dispersion in dissipative composite materials. *Int. J. Sol. Struc. 50,* 38–48.

Krushynska, A., V. Kouznetsova, & M. Geers (2014). Towards optimal design of locally resonant acoustic metamaterials. *J. Mech. Phys. Sol. 71,* 179–196.

Pham, K., V. Kouznetsova, & M. Geers (2013). Transient computational homogenization for heterogeneous materials under dynamic excitation. *J. Mech. Phys. Sol. 61,* 2125–2146.

Advances in Mechanics: Theoretical, Computational and Interdisciplinary Issues – Kleiber et al. (Eds)

Consistent plate theories—a matter still not settled?

R. Kienzler & P. Schneider
University of Bremen, Bremen, Germany

ABSTRACT: Using the uniform-approximation technique in combination with the pseudo-reduction method, a hierarchy of consistent plate theories is derived. After the introduction of non-dimensional quantities, the strain-energy and the dual-energy densities appear as infinite power series in the plate parameter that describes the relative thinness of the structure. The associated Euler-Lagrange equations deliver a countably infinite set of PDEs, where each PDE is an infinite power series with respect to the plate parameter. It is shown that the untruncated set of PDEs is equivalent to the problem of the three-dimensional theory of elasticity. Furthermore, an a-priori error estimation is given for the truncated, finite and therefore tractable PDE system. The error of the Nth-order two-dimensional theory decreases like the $(N+1)$th-power of the characteristic plate parameter, so that a considerable gain of accuracy could be expected for higher-order theories. The resulting equations of a consistent 2nd-order plate theory are used to assess and validate theories established in the literature.

1 INTRODUCTION

Although it is generally accepted that the classical Kirchhoff-plate theory is a consistent first-order approximation of the linear three-dimensional theory of elasticity, a variety of higher-order theories exist, that take shear deformations, cross-sectional warping and normal stresses in thickness direction into account. The variety of existing shear-correction factors (Kaneko 1975) may be regarded as an indication for the problem to be still unsettled.

In the talk we propose to use a theory based on the combination of the uniform-approximation technique (Kienzler 2002) and the pseudo-reduction method (Schneider & Kienzler 2011). In contrary to the direct approach (Altenbach et al. 2010), it belongs to the class of theories, which use series expansions in thickness direction with respect to a proper basis to derive hierarchical sets of approximative theories from the classical equations of the linear three-dimensional theory of elasticity. After introducing non-dimensional quantities, the plate parameter $c^2 = h^2/12a^2$ evolves quite naturally, with h and a denoting the characteristic out-of-plane and in-plane measures, respectively. For plate theories, c^2 is usually assumed to be a small quantity, i.e. $c^2 \ll 1$. The infinite PDE system for the determination of the unknown coefficients of the series expansion, where each PDE appears as an infinite power series with respect to the plate parameter c^2, is exact, i.e., it is equivalent to the equations of the three-dimensional linear theory of elasticity.

The idea of an Nth-order uniformly approximated plate theory is to consider all terms multiplied by c^{2n}, $n \le N$ and neglect terms multiplied by c^{2m}, $m > N$, i.e. terms that are of the order $O(c^{2(N+1)})$. We obtain a finite PDE system of 3 $(N+1)$ equations for 3 $(N+1)$ unknown displacement coefficients. During the reduction of these equations, i.e., elimination of unknowns, terms of the order $O(c^{2(N+1)})$, are also neglected (pseudo-reduction technique, for details Schneider & Kienzler 2011). The procedure results in PDEs, which are derived with recourse to neither a-priory assumption nor shear-correction factors.

By invoking the strain-energy density as well as the dual-energy density, an a-priori error estimate is established for the uniformly truncated series expansion (Schneider 2015). The error of the Nth-order two-dimensional (plate) theory decreases like the $(N+1)$th power of the characteristic plate parameter c.

2 SECOND ORDER THEORY

The derivation of the governing equations by using either monomic polynomials or scaled Legendre-polynomials has already been published (Kienzler 2002, Kienzler 2004, Schneider et al. 2013). In the following, we focus on isotropic material behaviour, and we will use monomic polynomials as basis. In extension of these earlier investigations we introduce here energetic averages $\tilde{w}$ and $\tilde{\psi}_\alpha$ of the transverse displacement w and the slopes ψ_α of the plate middle surface, respectively, as

$$c^2\tilde{w} = c^2 w + \frac{3}{10}\frac{\nu}{1-\nu}c^4\,\Delta w + O(c^6),$$

$$c^2\,\tilde{\psi}_{\alpha} = -c^2 w_{,\alpha} - \frac{3}{10}\frac{8+\nu}{1-\nu}c^4\Delta w_{,\alpha}$$

$$-\frac{6}{5}\varepsilon_{3\alpha\beta}c^2\psi_{,\beta} + O(c^6) \quad (1)$$

where Δ is the two-dimensional Laplace operator, $\Delta() = ()_{,11} + ()_{,22}$ and $\varepsilon_{3\alpha\beta}$ is the completely screw-symmetric permutation tensor. The quantity ψ is defined as $\psi = \psi_{2,1} - \psi_{1,2} \mathrel{\hat{=}} rot\psi_{\alpha}$ and may be regarded as a measure of the transverse-shear deformation. This ψ is next to $\tilde{w}$ one of our two main variables which are governed (after pseudo reduction) by our two main PDEs (Kienzler & Schneider, in prep.)

$$\Delta\Delta\tilde{w} = \frac{a^3}{K}\left(P - \frac{6}{5}\frac{2-\nu}{1-\nu}c^2\Delta P\right) + O(c^6),$$

$$c^2\left(\psi - \frac{6}{5}c^2\Delta\psi\right) = 0 + O(c^6). \quad (2)$$

where K is the classical plate stiffness $K = Eh^3/12(1-\nu^2)$ (Young's modulus E, Poisson's ratio ν) and P is the transverse load applied through the upper and lower plate faces ($x_3 = \pm\, h/2a$).

The stress resultants are calculated as bending moments ($\delta_{\alpha\beta}$—Kronecker's tensor of unity)

$$\tilde{M}_{\alpha\beta} = \frac{K}{a}\left\{\left(1 + \frac{12}{5(1-\nu)}\right)\left((1-\nu)\tilde{w}_{,\alpha\beta} + \nu\tilde{w}_{,\gamma\gamma}\delta_{\alpha\beta}\right)\right.$$

$$\left. + \frac{3}{5}(1-\nu)c^2(\varepsilon_{3\alpha\gamma}\psi_{,\beta} + \varepsilon_{3\beta\gamma}\psi_{,\alpha})_{,\gamma}\right\} \quad (3)$$

$$+ \frac{6}{5}\frac{\nu}{1-\nu}c^2 a^2 P\delta_{\alpha\beta} + O(c^6)$$

and transverse shear forces

$$\tilde{Q}_{\beta} = \frac{K}{a^2}\left\{\Delta\tilde{w}_{,\beta} + \frac{1}{2}(1-\nu)\varepsilon_{3\beta\gamma}\psi_{,\gamma}\right\} +$$

$$-\frac{6}{5}\frac{2-\nu}{1-\nu}c^2 aP_{,\beta} + O(c^6). \quad (4)$$

Higher-order stress resultants are also involved and calculated but turn out to be expressible as linear combinations of the classical stress resultants.

The equilibrium equation deliver.

$$\frac{1}{a}\tilde{Q}_{\beta,\beta} = -P \Rightarrow (2a),$$

$$\frac{1}{a}\tilde{M}_{\alpha\beta,\alpha} = \tilde{Q}_{\beta} \Rightarrow (2b), \quad (5)$$

By the pseudo-reduction method, not all of the displacement-ansatz coefficients are determined. They may be chosen a posterior to fulfil the boundary conditions along the plate faces

$$\sigma_{\alpha 3}\Big|_{\pm\frac{h}{2a}} = 0 + O(c^6),$$

$$\sigma_{33}\Big|_{\pm\frac{h}{2a}} = \pm\frac{1}{2}P + O(c^6) \quad (6)$$

and the local equilibrium conditions (Kienzler & Schneider, in prep.)

$$\sigma_{\alpha\beta,\alpha} + \sigma_{3\beta,3} = 0 + O(c^6),$$

$$\sigma_{\alpha 3,\alpha} + \sigma_{33,3} = 0 + O(c^6). \quad (7)$$

Thus the proposed method leads a well-balanced and contradiction-free second-order plate theory without recourse to any a priori assumptions.

It may be mentioned that for anisotropic plates, i.e., plates of monoclinic materials, the equations for $\tilde{w}$ and ψ are not decoupled anymore and involve 13 material constants (Schneider et al. 2013).

3 DISCUSSION

In the talk, the proposed second-order plate theory will be compared with other theories existing in the literature. We will assess and validate the theories of Reissner/Mindlin, Zhilin, Marguerre, Verkua, Ambartsumyan and Reddy (Schneider & Kienzler 2014).

REFERENCES

Altenbach, J., Altenbach, H. & Eremeyev, V.A. 2010. On generalized Cosserat-type theories of plates and shells; a short review and biography. *Archive of Applied Mechanics* 12: 73–92.

Kaneko, T. 1975. On Timoshenko's correction for shear in vibrating beams. *Journal of Physics D* 8:1927–1936.

Kienzler, R. 2002. On consistent plate theories. *Archive of Applied Mechanics* 72: 229–247.

Kienzler, R. 2004. On consistent second-order plate theories. In R. Kienzler, H. Altenbach & I. Ott (eds), *Theories of plates and shells, critical review and new applications*: 85–96. Berlin: Springer.
Kienzler, R. & Schneider, P. (in prep.). Second-order linear plate theories: Partial differential equations, stress resultants and displacements.
Schneider, P. 2015. On the mathematical justification of the consistent-approximation approach and the derivation of a shear-correction-factor free refined beam theory. Dissertation, University of Bremen. http://nbn-resolving.de/urn:nbn:de:gbv:46-00104458-18.
Schneider, P. & Kienzler, R. 2011. An algorithm for automatisation of pseudo reductions of PDE systems arising from the uniform-approximation technique. In H. Altenbach & V.A. Eremeyev (eds), *Shell-like structures; Non-classical theories and applications*: 377–390. Berlin: Springer.
Schneider, P. & Kienzler, R. 2015. Comparison of various linear plate theories in the light of a consistent second-order approximation. *Mathematics and Mechanics of Solids* 20: 871–882.
Schneider, P., Kienzler, R. & Böhm, M. 2014. Modelling of consistent second-order plate theories for anisotropic materials, *Zeitschrift für Angewandte Mathematik und Mechanik* 94: 21–42.

Advances in Mechanics: Theoretical, Computational and Interdisciplinary Issues – Kleiber et al. (Eds)
© 2016 Taylor & Francis Group, London, ISBN 978-1-138-02906-4

Experimental attempts for creep and fatigue damage analysis of materials—state of the art and new challenges

Z.L. Kowalewski, P. Grzywna & D. Kukla
Institute of Fundamental Technological Research, Warsaw, Poland

ABSTRACT: The development of creep and fatigue damage was investigated using destructive and non-destructive methods in materials commonly applied in power plants or automotive industry. In order to assess such kind of damage the tests for a range of different materials were interrupted for selected time periods (creep) and number of cycles (fatigue). Being destructive methods the standard tension tests were carried out after prestraining. Subsequently, an evolution of the selected tension parameters was taken into account for damage identification. The ultrasonic and magnetic techniques were used as the non-destructive methods for damage evaluation. In the final step of the experimental programme microscopic observations were performed.

1 INTRODUCTION

Many testing techniques commonly used for damage assessment have been developed up till now. Destructive and non-destructive methods may be generally distinguished. Having the parameters of destructive and non-destructive methods for damage development evaluation it seems to be reasonable their further analysis that should provide possible mutual correlations, (Kowalewski et al. 2009). This is because typical destructive investigations, like standard tests, give the macroscopic parameters characterizing the lifetime, strain rate, yield point, ultimate tensile stress, ductility, etc. without a sufficient knowledge concerning microstructural damage development and material microstructure variation. On the other hand, non-destructive methods provide information about damage at a particular time of the entire working period of an element, however, without sufficient knowledge about the microstructure and how it varies with time. Therefore, it seems reasonable to plan damage development investigations in the form of interdisciplinary tests connecting results achieved using destructive and non-destructive methods with microscopic observations in order to find mutual correlations between their parameters. This is the main issue considered in this paper. The microscopic observations of microstructural changes carried out using traditional scanning electron microscopes are possible only after specimen failure, and additionally they suffer on high cost. The non-destructive methods are much more convenient and that is why they are more frequently used in engineering practice for periodic evaluation of the material degradation. The resolution range of particular non-destructive tests covers the entire scope of structural effects and crystallographic defects, Holler & Dobmann (1989), Figure 1.

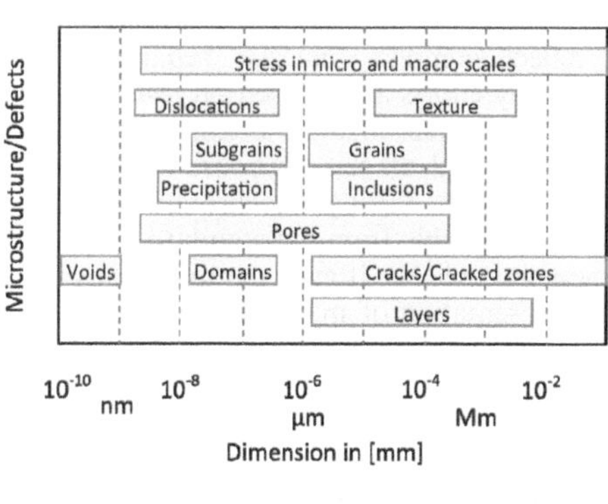

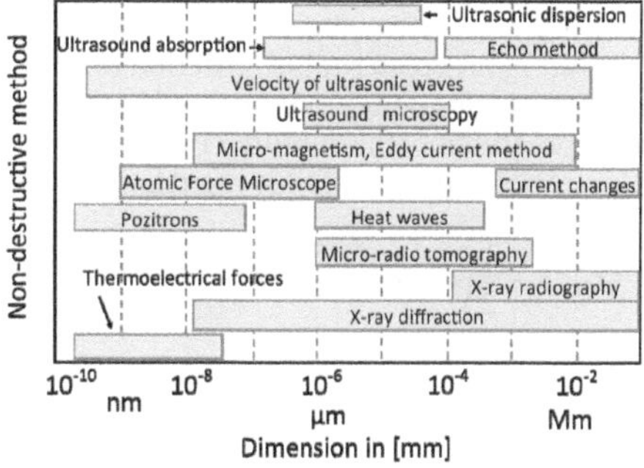

Figure 1. Dimensions of microstructural forms/defects and non-destructive methods suitable for their monitoring.

Considering the detailed conditions and scopes of applying various methods significantly limits the possibilities to use them and makes it difficult to identify and analyse the fatigue damage evolution in the experimental way. This leads to a necessity of the constant improvement of existing non-destructive testing methods and of developing new measurement techniques that would be able to detect and carry out quantitative assessment of the structural failures resulting from the development of processes leading to material fatigue and deterioration of its mechanical properties. The conventional non-destructive techniques (based on the measurement of ultrasonic wave velocities e.g.) are sensitive to material damage in the last stage of material life and do not answer the question when a given element should be withdrawn from the exploitation? Therefore, there is a need to develop a new method based on a single or combination of non-destructive techniques allowing to estimate the microstructural and mechanical degradation observed at different stages of the exploited material.

2 EXPERIMENTAL ATTEMPTS FOR CREEP DAMAGE ANALYSIS

2.1 *Multiaxial creep tests*

The results from uniaxial creep tests are not able to reflect complex material behaviour. Therefore, many efforts are focused on tests carried out under multiaxial loading conditions. The well-known method of creep rupture data from such tests analysis is through isochronous stress-strain curves obtainable from the standard creep curves, Kowalewski (2004). It gives comprehensive graphical representation of material lifetime. The curves of the same time to rupture determined on the basis of experimental programme are compared in Figure 2 with theoretical predictions of the three well known creep rupture hypotheses: (a) the maximum principal stress rupture criterion (1), (b) the Huber-Mises effective stress rupture criterion (2), (c) the Sdobyrev creep rupture criterion (3). For the biaxial stress state conditions, realised in the experimental programme, the rupture criteria mentioned above are defined by the following relations:

$$\sigma_R = \sigma_{\max} = \frac{1}{2}\left(\sigma_{11} + \sqrt{\sigma_{11}^2 + 4\sigma_{12}^2}\right) \tag{1}$$

$$\sigma_R = \sigma_e = \sqrt{\sigma_{11}^2 + 3\sigma_{12}^2} \tag{2}$$

$$\sigma_R = \beta\sigma_{\max} + (1-\beta)\sigma_e \tag{3}$$

As it is clearly seen, the best fit of the aluminium alloy data is obtained using the effective stress rupture criterion. It has to be noted however, that the lifetimes predicted by this criterion are still quite far from experimental data.

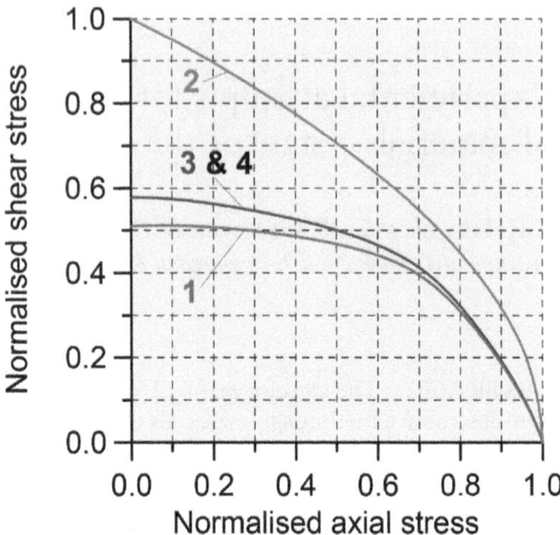

Figure 2. Comparison of the isochronous creep rupture surfaces (t_R = 500 [h]) determined for aluminium alloy (1—experimental results; 2, 3, 4—theoretical predictions using the maximum principal stress criterion; the effective stress criterion; and the Sdobyrev criterion, respectively).

2.2 *New concepts of creep analysis*

To assess damage using destructive method the specimens after different amounts of prestraining were stretched to failure (Kowalewski et al. 2009; Makowska et al. 2014). Afterwards, the selected tension parameters were determined (yield point, ultimate tensile stress) and their variations were used for the identification of damage development. Ultrasonic and magnetic investigations were selected as the non-destructive methods for damage development evaluation. For the ultrasonic method, the acoustic birefringence coefficient was used to identify damage development in the tested steel. In the case of magnetic method a several damage sensitive parameters were identified, e.g. amplitude of Ub or Ua envelopes reflecting the Barkhausen effect (HBE) or Magneto-Acoustic Emission (MAE) variation, respectively, their integrals, and coercivity.

Having selected parameters of destructive and non-destructive techniques, possible relationships between them were evaluated. The representative relationships are illustrated in Figure 3a, b, c. As it is seen, except of the specimen prestrained up to 10.5% due to plastic flow, all results are ordered, and as a consequence, they can be well described

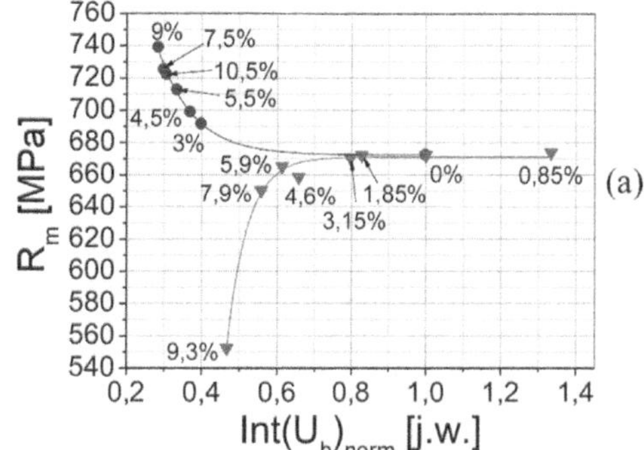

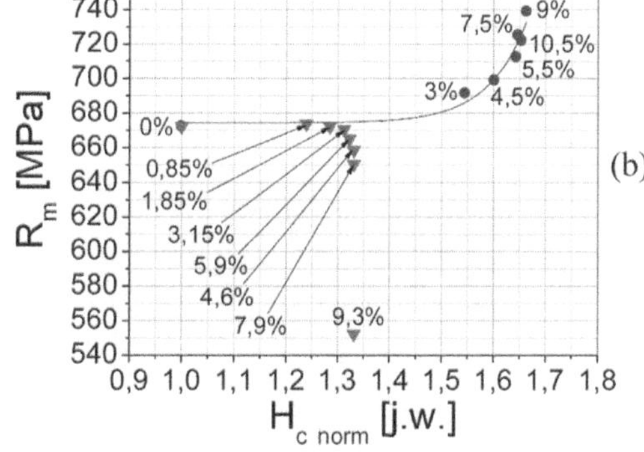

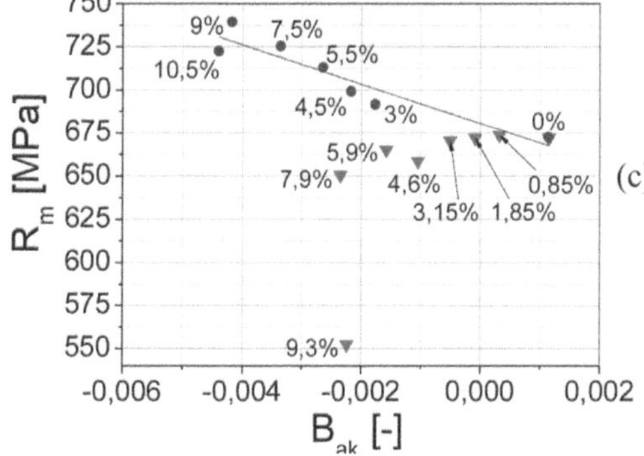

Figure 3. Variation of ultimate tensile stress of the X10CrMoVNb9-1 steel versus: (a) integral of the Ub envelope amplitude; (b) coercivity; (c) birefringence coefficient (triangles—steel after creep; circles—steel after plastic flow).

by adequate functions depending on the type of prior deformation.

Diagram (a) in Figure 3 shows relationship between integral of the Ub envelope amplitude—$Int(U_b)$ and ultimate tensile stress. The magnetic parameter is normalized to the value captured for the non-deformed specimen. Numbers in the figure denote the level of prior deformation. Figure 3a allows concluding that integral $Int(U_b)$ of the Barkhausen noise may be used to estimate a level of the ultimate tensile stress of plastically deformed specimens. It has to be noticed however, that in the case of the material prestrained due to creep the situation is more complicated, since a non-unique relationship between R_m and Int $(Ub)_{norm}$ was found.

The relations in Figure 3a indicate that the steel after plastic deformation, leading to higher values of R_m, can be characterised by lower values of magnetic parameters. This is because the prestrained material contains more dislocation tangles that impede domain walls movement. On the other hand the lower values of magnetic parameters can be attributed to the lower magnitudes of R_m for the steel after creep. The results make evident that the MBE intensity varies significantly due to microstructure modification, however, in different ways depending on prior deformation type. For deformation higher than 2% this intensity decreases after plastic flow and increases after creep. Strongly non-linear character of plots in Figure 3a makes impossible the direct estimation of mechanical parameters when only single magnetic parameter is used. Addressing the issue for practical application of the MBE measurement in assessment of mechanical properties for the damaged steel one can conclude that it is possible only then if at least two magnetic parameters will be taken into account. It can be seen in Figure 3a that relative decrease of the $Int(U_b)$ with prestraining denotes plastic deformation while rapid increase of the $Int(U_b)$ associated with prestrain increase is observed for early stage of creep damage development.

A better correlation was achieved between R_m and coercivity H_c, Figure 3b. As it is seen, except specimen prestrained up to 10.5% due to plastic flow, all results are ordered, and as a consequence, they can be well described by adequate functions depending on the type of prior deformation. The main disadvantage of the relationships between Rm and H_c, is related to the fact that it cannot distinguish a type of prior deformation for small prestrain magnitudes.

Similar remarks can be formulated for the relationships between R_m and acoustic birefringence coefficient B_{ak}, Figure 3c.

The relationships between selected destructive and non-destructive parameters sensitive for damage development show a new feature that may improve damage identification. In order to provide more thorough analysis reflecting physical interpretation of the relationships obtained further investigations are necessary. Programmes of such tests should contain advanced microscopic observations using not only optical techniques, but also SEM and TEM.

3 EXPERIMENTAL ATTEMPTS FOR FATIGUE DAMAGE ANALYSIS

3.1 *Introductory remarks*

The fatigue of structural materials is particularly important for the development of railway and air transport as well as power engineering. Fast introducing of new technologies in those sectors frequently leads to serious crashes. It is enough to mention two crashes of the first jet-propelled passenger aircraft called Comet, crash of the fast German railway in Eschede in 1998 or two jet crashes of the first transoceanic airway line in Poland in 1980 (Chopin Airport in Warsaw) and in 1987 (Kabaty near Warsaw). All those tragic crashes were caused by material fatigue, whose nature has not been examined sufficiently.

Experimental determination of damage development requires on-line recording the material responses to the given cyclic loading during the whole experiment. Difficulty in conducting this task is additionally strengthened by the lack of effective fatigue damage measure describing properly effects of material degradation. Failure mechanics is a relatively new field of studies and the reference fatigue damage indicators described in scientific works should be treated only as suggestions, but not verified and accepted problem solutions. Monitoring of the changes in mechanical properties taking place under cyclic loads by means of recording stress and deformations of the measured sample section in consecutive cycles was an efficient technique enabling not only to define the number of cycles required to damage the specimen, but also to define the fatigue damage indicator and its evolution in the fatigue process.

Behaviour of materials within the range of High Cycle Fatigue (HCF), which means the stress amplitude is below the yield limit of a material, can be divided into two basic types in terms of mechanisms of damage development, Przybyla et al. (2007). Behaviour of the first group of materials is described by the ratcheting generated by local deformation around voids, non-metallic precipitations and other defects of structure. Behaviour of the second group of materials undergoing cyclic loading is described by cyclic plasticity generated by slips on the level of grains and local slip bands. In both cases the changes of measured deformations are a sum of local deformations in the whole volume being measured. Cyclic loading in the range of HCF causes initiation of various mechanisms of damage development.

Initial defects in the form of inclusions initiate a localisation of damage, and as a consequence, its development in consecutive loading cycles. Their imaging (Figs. 4–5) helps not only to define the damage development mechanism, e.g. in the form of debonding or decohesion, but also provides the basis for modelling of their mechanisms development.

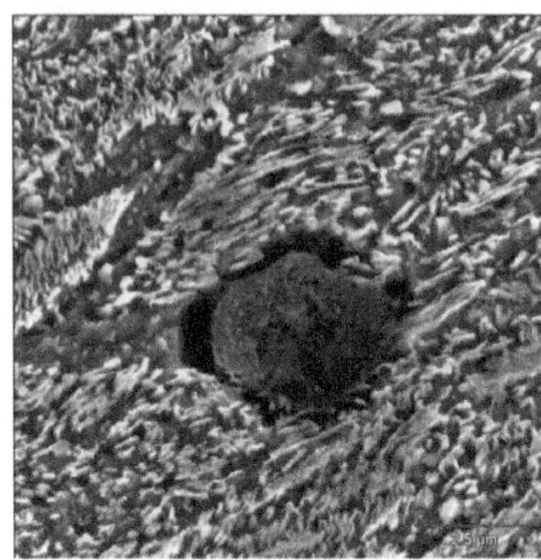

Figure 4. Microscopic image of failure due to fatigue carried out on X10CrMoVNb9-1 steel.

Figure 5. Microscopic image of failure due to fatigue carried out on A356-TiB_2.

The metallographic methods are laborious and problematic due to the specific conditions of sample preparation. Usually they are mainly used after completion of the loading process or after its programmed suspension in the desired phase. Their significant limitations are due to the laborious procedures of sample preparation and the necessity to observe them after load removal. A few years ago, a unique research station has been constructed, with a servo-hydraulic testing machine equipped with digitally controlled signals of load, displacement, total or plastic strain, which can be used for monotonic stretching and compressing of specimens in the range of ±100 kN and for Low-Cycle Fatigue tests (LCF) in temperature range up to 1100°C. The strength testing machine is connected to the set of scanning electron microscopes for examining the sample surface under monotonic load and microscopic observations with the resolution of μm. The second electron microscope enables an examination of the specimen surface under constant load, and microscopic observations with the nano-metric (nm) resolution. The specimen, after fixing and imposing the planned range of fatigue loads, is descended, together with the stress-testing machine frame, into a large vacuum chamber with the capacity of $2m^3$ where the metallographic tests of the entire sample surface are carried out. The device is equipped also with EDX system (Energy Dispersive X-Ray) enabling to carry out local chemical analysis, and with EBSD system (Electron Back Scatter Diffraction) to image the metallographic orientation in the tested field. The first research station of such a type was installed in 2009 at the Institute for Material Science of the Erlangen-Nuernberg University in Germany.

Among many fatigue testing programmes one can distinguish two basic directions: (a) investigations conducted by physicists and metallurgists focusing on trying to learn the mechanisms governing the process of fatigue; (b) theoretical and experimental investigations in order to create a phenomenological theory to allow quantitative description of the phenomenon. Both of these trends are of current, parallel development.

3.2 *Fatigue damage evaluation*

In many cases the process of fatigue damage is controlled by more than one mechanism. The fatigue tests carried out on metallic materials and metal matrix composites have shown that the damage process occurred due to combination of cyclic plasticity and ratcheting mechanisms. Therefore, using adequate damage indicators, Figure 6, damage measure can be defined by the following relationship:

$$\varphi_N(\varepsilon_a^{in}, \varepsilon_m^{in}) = \sum_1^N \left|\varepsilon_a^{in}\right| + \sum_1^N \left|\varepsilon_m^{in}\right| \tag{4}$$

Hence, a definition of damage parameter takes the form:

$$\mathrm{D} = \frac{\varphi_N - (\varphi_N)_{\min}}{(\varphi_N)_{\max} - (\varphi_N)_{\min}} \tag{5}$$

where: φ_N—accumulated strain up to the current loading cycle, $(\varphi_N)_{\min}$—accumulated strain at the first cycle, $(\varphi_N)_{\max}$—accumulated total strain calculated for all cycles, in which damage measure is included in the form of (4).

The results published so far (Rutecka et al. 2011, Socha 2003) confirm the correctness of the adopted methodology for damage analysis of the materials after service loads, that taking into account parameters responsible for cyclic plasticity and ratcheting.

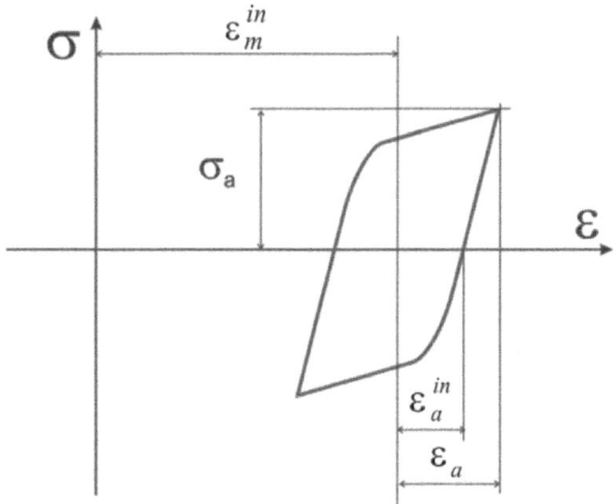

Figure 6. Illustration of strain damage indicators during fatigue conditions.

3.3 *Previous and new concepts of fatigue testing*

In order to assess damage degree due to fatigue of the material in the as-received state and after exploitation the Wöhler diagrams may be elaborated that represent the number of cycles required for failure under selected stress amplitude. The results of such approach are illustrated in Figure 7 for the 13HMF steel. As can be seen, the Wöhler diagrams depending on the state of material differ themselves, thus identifying the fatigue strength reduction due to the applied loading history. Unfortunately, such method of degradation assessment of the material undergoing fatigue suffers on very high cost and additionally it is time-consuming.

Therefore, research is conducted for new solutions that would provide a better assessment of the

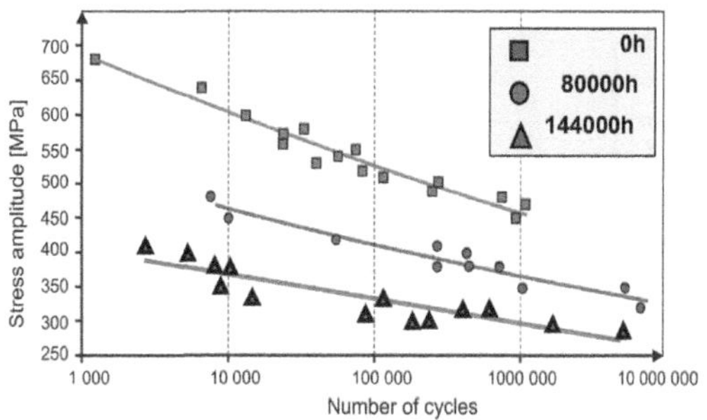

Figure 7. Wöhler diagram for the 13HMF steel before (0h) and after exploitation (80 000h and 144 000h).

fatigue damage development. In order to obtain this effect, an adequate damage parameter must be defined on the basis of the measurable indicators of its development. Selected proposals discussed in section (3.1) have been validated experimentally and the representative results will be presented here.

Force controlled high cycle fatigue tests (20 Hz frequency) were carried out on the servo-hydraulic testing machine MTS 858. During the tests, sine shape symmetric tension-compression cycles were applied to keep constant stress amplitude equal to 70 MPa and 350 MPa for metal matrix composite (Al/SiC) and X10CrMoVNb9-1 steel, respectively. Tests were performed at ambient temperature. Each cylindrical specimen manufactured from both materials was subjected to cyclic loading until fracture. A movement of the subsequent hysteresis loops along the strain axis was observed with an increasing number of cycle (Figs. 8–9). Simultaneously, a width of the subsequent hysteresis loops became almost unchanged for composite and became greater for the steel. Such behaviour identifies the ratcheting effect.

Since ratcheting is the dominant mechanism of the composite deformation, the mean strain was taken into account during a damage parameter calculation in the stable growth period. Hence, the damage parameter can be defined using Equation (4) assuming that its first term is neglected. It is worth noticing that the rate of damage is relatively high at the beginning of the period. Afterwards, it becomes slower, Figure 10. In the case of X10CrMoVNb9-1 steel besides of ratchetting also cyclic plasticity mechanism is responsible (increase of hysteresis loop width) for damage development, Figure 9. Taking into account both mechanisms variation of the damage parameter can be calculated using Equations (4) and (5). It is shown in graphical form in Figure 11. Looking at the diagrams in Figures 10–11 one can say that the linear damage accumulation rule cannot be applied for both tested materials. Moreover, it can be seen that the rate of damage parameter variation for composite depends on the SiC particle content. It increases with an increase of the SiC particle content at the initial stage of fatigue, Figure 10.

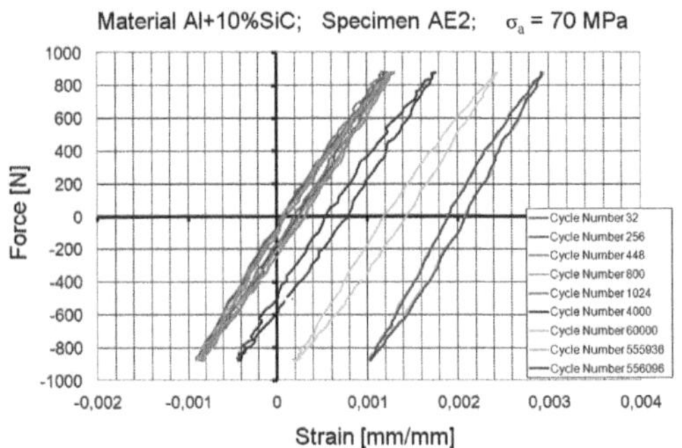

Figure 8. Hysteresis loops for selected cycles—the results for Al/SiC.

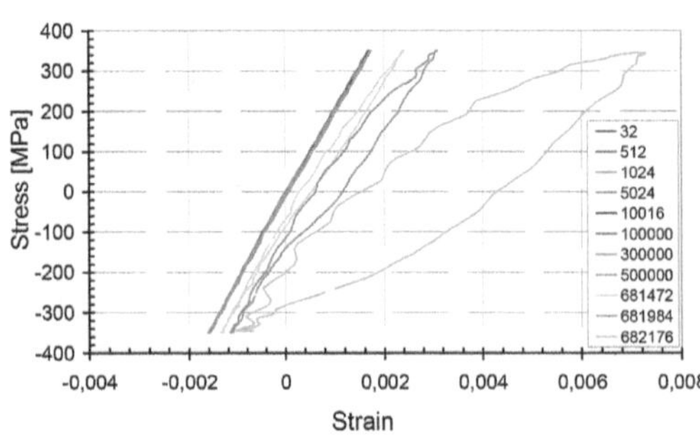

Figure 9. Hysteresis loops for selected cycles—the results for X10CrMoVNb9-1 steel.

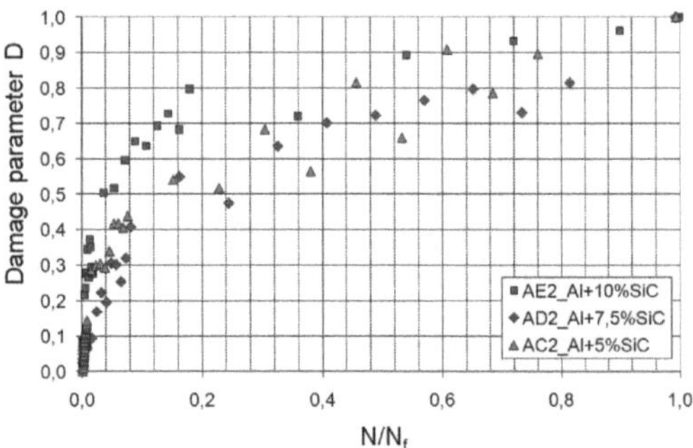

Figure 10. An influence of SiC content on damage parameter variation (stress amplitude 70MPa).

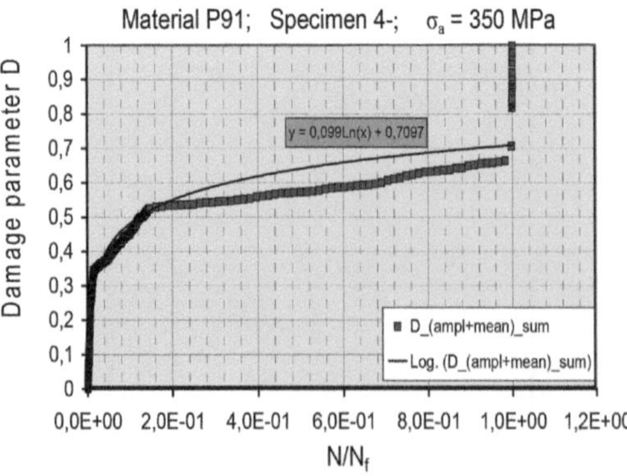

Figure 11. Damage parameter variation for X10CrMoVNb9-1 steel (stress amplitude 350MPa).

The examples of fatigue testing may be considered an alternative method for damage evaluation of materials subjected to cyclic loads. Studies in which a variation of the hysteresis loop width and its movement were recorded for cycles under fixed constant stress amplitude have demonstrated that this procedure gives a possibility to assess safe operation period for materials in question and there is no need to perform so many experiments, as it is required for the Wöhler diagram determination.

The proposed method of assessing fatigue damage evolution makes it possible to determine damage indicators; define damage parameter; assess fatigue and stress levels to find ranges in which an accumulation of damage can be described by the linear law.

Another promising attempts in the fatigue damage analysis are related to relatively new techniques such as Electronic Spackle Pattern Interferometry (ESPI) or Digital Image Correlation (DIC). Some preliminary results of the ESPI usage will be presented here. The cast aluminum alloy AlSi7MgCu0.5 for automotive engine heads was tested after exploitation cycles, Rutecka et al. (2011). An opportunity for strain/stress components measurement of a specimen using ESPI is presented in Figure 12. The specimen was cut from automotive engine head (cast aluminum alloy AlSi7MgCu0.5). Such heads were cast according to the standard procedure that ensures degassing. The ratio of the average porosity was 6%. Observing the strain and stress distributions in Figures 12–13 a serious problem can be noticed, i.e. an averaging of the accuracy of the strain components in the entire volume of the geometrically homogeneous specimen. In this group of materials the development of fatigue damage takes place around various drawbacks, mainly in the form of voids formed in

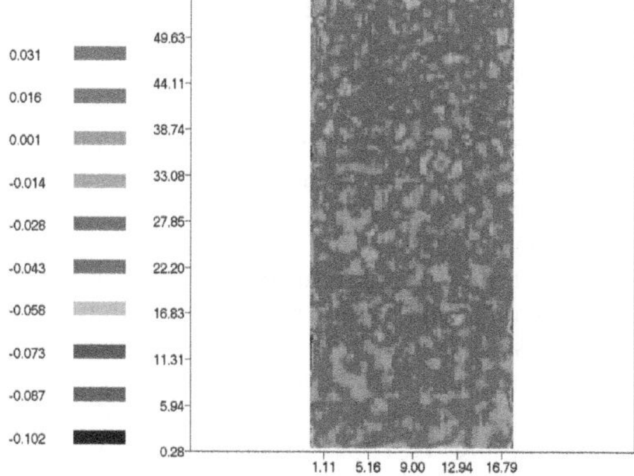

Figure 12. Strain component distribution map for z direction on the plane specimen surface (cross section 18 × 4 *mm*) under load of 1.2 kN.

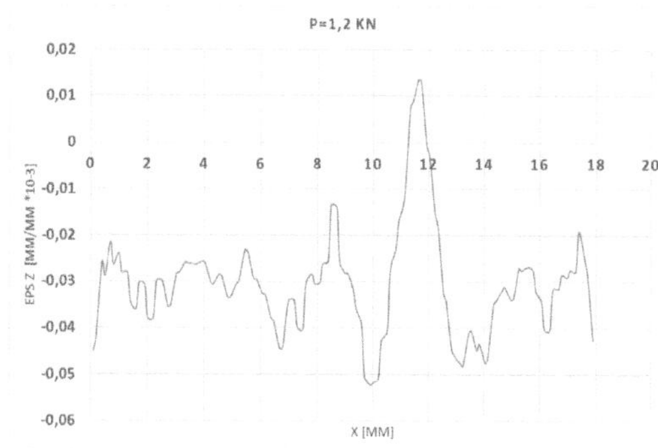

Figure 13. Transversal distribution of the stress component along z axis.

manufacturing processes such as casting. Besides of density and distribution of defects in the volume of tested specimen, the specimen size, and location of individual defects are important factors of the fatigue damage initiation and further development. Development of local strain components around defects leads to ratcheting, i.e. the incremental rise of the strain components being in agreement with acting stress direction. This group includes not only all cast alloys, but also the whole range of modern metal matrix composites.

Solving problems related to development of fatigue damage, degradation of the mechanical properties of structural materials, as well as modeling of mechanical properties necessary to simulate the behaviour of the structure under loading is based mainly on experimental data from tests under uniaxial stress states. However, for such purposes also data coming from complex stress state experiments are required.

The testing technique under complex stress states gives a full information about the mechanical properties of structural materials, necessary for parameters determination during numerical modeling. In this context either ESPI or DIC are very helpful since monitoring of the phenomena related to the fatigue, as the process initiating locally, requires the full-field observations of the displacement brought by cyclic loading. Both methods enable determination of the displacement distribution on the specimen surface, and thus, also strain and stress concentration spots resulting from the defect.

The ESPI method was also applied for evaluation of the fatigue strength of nickel alloys after various surface treatments. On the basis of the displacement field distributions an identification and localization of Thermal Barrier Coating layers was carried out for the material subjected to fatigue tests. Such studies were performed on the

MAR 247 nickel alloy covered by thin films. The aluminizing process was carried out by the Chemical Vapour Deposition (CVD) method using $AlCl_3$ vapours in hydrogen atmosphere as the carrier gas at temperature of 1040°C for the periods of 4 h or 12 h, and the reduced pressure of 150 hPa. Depending on the duration of this process the aluminum layers of 20 μm or 40 μm thickness were obtained on the nickel alloy surface. Tests were performed on plane specimens using the MTS 810 testing machine of the axial force capacity equal to +/− 25 kN at room and elevated temperatures. Figure 14 shows the specimen fixed in the testing machine, which besides of the ESPI system was also equipped with the eddy current defectoscope for non-destructive flaw detection.

All fatigue tests were force controlled under assumption of the zero mean level and constant stress amplitude for a given specimen tested. The frequency during fatigue tests was equal to 20 [Hz], whereas the magnitudes of stress amplitude varied from 500 [MPa] up to 700 [MPa].

Attempts for determination of the initial phase of fracture of layer on the nickel alloy were carried out during fatigue testing using the ESPI camera on the basis of full-field strain distributions. Fatigue tests were interrupted several times for displacement measurements after the fixed number of cycles. In order to keep high accuracy of the subsequent images of phase maps registration was performed using handy operated pump. This is because of the fact that testing machine hydraulics interferes with the ESPI measurement. The measurement methodology shown in Figure 15 was repeated for the fixed number of cycles until initial failure of layer appeared preceding specimen fracture. From five to twelve displacement images

Figure 14. General view of specimen mounted in grips of the testing machine.

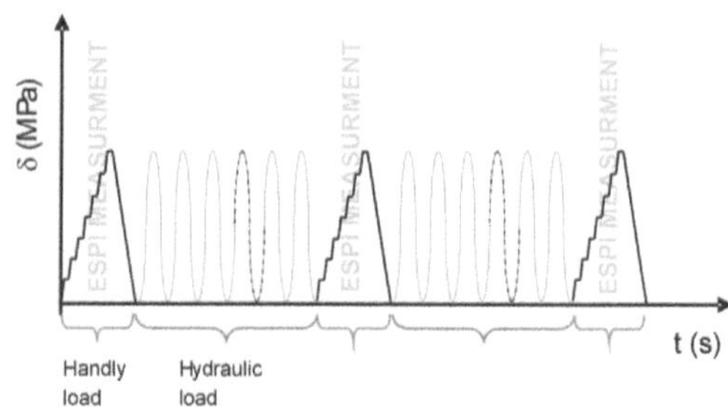

Figure 15. Scheme of the loading program for fatigue tests and ESPI measurements.

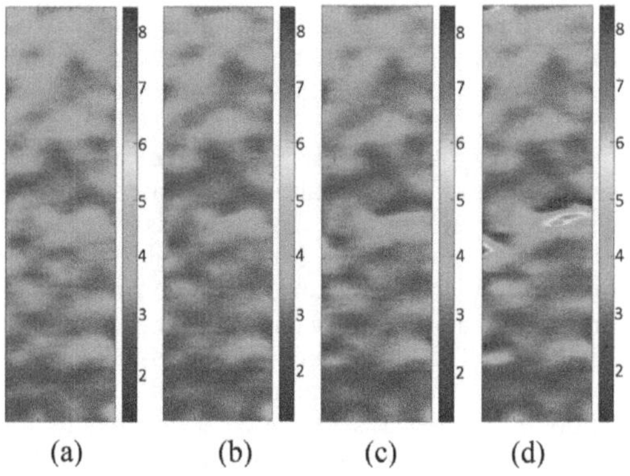

Figure 16. Strain maps captured for specimen subjected to fatigue under strain amplitude of 600MPa after: (a) 1 cycle; (b) 20000 cycles; (c) 40000 cycles; (d) 50000 cycles.

were captured during each stop of the test. It enabled damage development observation on the basis of displacement concentration areas. Strain distribution can be determined using ESPI displacement measurements delivered in the form of interference phase maps capturing from the surface of specimen during handy controlled loading procedure shown in Figure 15.

The representative results are presented in Figure 16 for the specimen subjected to fatigue under 600 MPa stress amplitude.

The maps (a), (b), (c) and (d) illustrate strain distributions at first fatigue cycle and after 20000, 40000 and 50000 cycles, respectively. Figures 16 and 17 show that strain concentrations already take place after 40000 cycles at three regions of the field considered. The last recorded image (d) in Figure 16 shows areas of strong strain concentration. The strain magnitudes in these places may indicate regions where layer failures appeared which process preceding specimen fracture. Simultaneously with the ESPI displacement

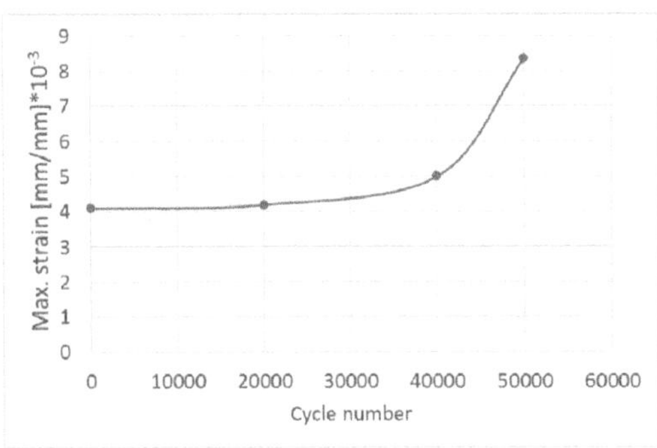

Figure 17. Maximum strain levels calculated on the basis of maps presented in Figure 16.

distribution measurements the eddy current flaw detection was performed in order to identify fracture place. The results enabled fracture identification in the area of greatest deformation, but scarcely after 50000 cycles, i.e. much later than the ESPI system.

4 CONCLUSIONS

The results clearly indicate that selected ultrasonic and magnetic parameters can be good indicators of material degradation and can help to locate the regions where material properties are changed due to prestraining.

Creep or fatigue analysis should be based on the interdisciplinary tests giving a chance to find mutual correlations between parameters assessed by classical macroscopic destructive investigations and parameters coming from the non-destructive experiments. Such relationships should be supported by thorough microscopic tests, giving as a consequence, more complete understanding of the phenomena observed during damage development.

Application of the full-field observation methods (ESPI, DIC) enables localization of the initial spot of the failure resulting from the cyclic stress and monitoring of its development as well.

REFERENCES

Holler, P. & Dobmann, G. 1989. NDT-Techniques for monitoring material degradation. *Mat. Res. Soc. Symp. Proc.* 142: 105–118.

Kowalewski, Z.L. 2004. Isochronous creep rupture loci for metals under biaxial stress. *Journal of Strain Analysis for Engineering Design* 39: 581–593.

Kowalewski, Z.L. et al. 2009. Evaluation of damage development in steels subjected to exploitation loading—destructive and nondestructive techniques. *Journal of Multiscale Modeling* Vol. 1 (3/4): 479–499.

Makowska, K. et al. 2014. Determination of mechanical properties of P91 steel by means of magnetic Barkhausen emission. *Journal of Theoretical and Applied Mechanics* 52: 181–188.

Przybyla, C. et al. 2007. Micromechanical Modeling of High Cycle Fatigue Processes. *ASM/TMS Symp. on Comput. Mat. Design GE Global Research* August 20–21, 2007.

Rutecka, A. et al. 2011. Creep and low cycle fatigue investigations of light aluminium alloys for engine cylinder heads. *Strain—An International Journal of Experimental Mechanics* 47: 374–381.

Socha, G. 2003. Experimental investigation of fatigue cracks nucleation, growth and coalescence in structural steel. *International Journal of Fatigue* 25: 139–147.

Advances in Mechanics: Theoretical, Computational and Interdisciplinary Issues – Kleiber et al. (Eds)
© 2016 Taylor & Francis Group, London, ISBN 978-1-138-02906-4

Micro and nano fluid mechanics

T.A. Kowalewski, P. Nakielski, F. Pierini, K. Zembrzycki & S. Pawłowska
Institute of Fundamental Technological Research, Polish Academy of Science (IPPT PAN), Warsaw, Poland

ABSTRACT: The current task of contemporary fluid mechanics evidently moves from modeling large scale turbulence to lower, molecular scale limit, where assumption of a continuous and deterministic description becomes questionable again. Once the scaling length of flow becomes comparable with structure dimensions, transport phenomena are strongly modulated by molecular interactions and its proper interpretation needs involvement of deeper physics. New experimental tools largely help in understanding transport phenomena at nanoscales. In the following review we give few examples of problems appealing for new theoretical and numerical models embracing continuous flow modeling with molecular scale phenomena.

1 INTRODUCTION

It is generally understood, that the main difference between macro- and nano-scale mechanics originates from rapidly increasing surface to volume ratio along with the decreasing of object size. A total surface of one nanometer particles filling volume of a cubic centimeter is 6000 square meters! Hence, nanoscience is mainly a science of surface forces and surface interactions. It applies particularly to fluids.

The field of microfluidics is characterized by the study and manipulation of fluids at the submillimeter length scale. The fluid phenomena that dominate liquids at this length scale are measurably different from those that dominate at the macroscale. For example, the relative effect of the force produced by gravity at microscale dimensions is greatly reduced compared to its dominance at the macroscale. Conversely, surface tension and capillary forces are dominant at the microscale. These forces can be used for a variety of tasks, such as passively pumping fluids in microchannels, precisely patterning surfaces with user-defined substrates, filtering various analytes, and forming monodisperse droplets in multiphase fluid streams for a variety of applications (Tabeling 2006). Observing biology one obtains plethora of valuable hints how to proceed with such processes (Sasai 2013). One of the fundamental problems in microfluidics is lack of turbulence. Hence, all involved transport processes are based on laminar, creeping flows with very inefficient at such scales diffusion.

Nanoscale fluid mechanics touches completely new area, rediscovered after biologists. Intercellular and cellular transport of molecules, subcellular organelles, and cells immersed in an aqueous environment is based on "stochastic" transport phenomena. Molecular diffusion becomes very efficient at short distances. Whereas typical biomolecule needs hours to be transported by diffusion for distances of centimeter, such transport takes less than 1 s at subcellular distances. Hence, random walk of nano-objects driven by intermolecular interactions initiates all living processes, whereas hydrodynamic interactions play mainly role of a drag force, modifying species mobility.

Physics of nanoscale mechanics can be more or less completely described using quantum mechanics. The methods, called ab initio, are rigorous but limited by present-day computers to systems containing a few hundred atoms at most. To determine the properties of larger ensembles of atoms the Molecular Dynamics method is commonly used, enabling studies of billions of atoms with effective interatomic potentials. Still its practical applications are limited by time and space scales to first nanoseconds of the analyzed phenomena and several nanometers in space. Hence, modelling fluid flow in micro- and nano-scales needs specific technique, which is based on assumption that fluid particle can be represented as a cluster of atoms. Effective clustering can be built using the so-called Voronoi tessellation, describing a special kind of decomposition of the flow domain (Czerwińska 2004). Such coarse grained modelling is useful for general flow description, but needs predefined interactions if we approach molecular distances to interpret specific phenomenon.

Despite of all technical and physical problems with direct experimental analysis of nano/micro scale flows, physical experiment appears to be the only way to validate simplified assumptions, which by definition have to be incorporated to theoretical models. Some experimental examples refer to observing the mother nature, some of them have been recently proposed using completely new for fluid mechanics techniques, like fluorescent microscopy, Atomic Force Microscopy (AFM), and Optical Tweezers (OT).

Most of microfluidic problems concern multiphase flow, suspensions of micro and nanoparticles, cells or macromolecules (proteins, DNA). Understanding and properly interpreting fluid-particle interaction is crucial for interrogating such systems. One of the valuable optical tools is based on Brownian motion (Breuer 2004). The local and bulk mechanical properties of a complex fluid can be obtained by analyzing thermal fluctuations of probe particles embedded within it.

Thermal fluctuations generated by molecules are not only noises, it has been demonstrated that such fluctuations are fundamental to the function of biological systems. Several results are possible with the same probability, in contrast to a mechanical system in which the result is deterministic. In ensemble measurements, the obtained values, which are average values over many molecules, have been usually wrongly interpreted as deterministic values. However, in biological systems, the average values are not necessarily effective, but the values of individual molecules play a decisive role (Kochańczyk et al. 2013, Jaruszewicz et al. 2013).

The biomolecular system may spontaneously fluctuate, and one of the two states occurs alternately. Preferential binding of ligands to one of the spontaneously fluctuating structures of proteins leads to their activation or deactivation. This mechanism appears essential for a long scale evolutionary development of leaving species, and at short time scale to create signaling paths for early immune response of individual cells (Tay et al. 2010). Hence, looking at the "bottom" of our fluid mechanics, there is no place for steady, unique and predictable modeling. Rather, by analogy to quantum mechanics, we have to talk about the most probably evolution of the analyzed system. As an illustration of the difficulties, in the following we cope with two intriguing problems, kinematic boundary conditions in micro and nano scales flow, and mobility of nano-objects suspended in liquids.

2 TO SLIP OR NOT TO SLIP

One of the primary questions, which appeared when fluid mechanists started to play with microfluidics, concerned the interactions between liquids and solid surfaces. From the physical point of view it seems obvious that molecules of liquid cannot be arrested at the solid surface, otherwise the local thermodynamic parameters of liquid should abruptly change. This problem is of fundamental physical interest and has practical consequences in rarefied gas flows. Recently it became rediscovered for small-scale systems, including transport phenomena in biological fluids (Lauga et al. 2005).

The physics of hydrodynamic slip may have different origins. Purely molecular slip is clearly relevant in case of gases. For very small channels formed by Carbon NanoTubes (CNTs) possible drag reduction is expected due to the synchronized slippage of water molecules. It is interrelated with molecular structure of water. For constrained conditions, i.e. tube diameter below 1 nm, water forms a long single molecular chain (Thomas et al. 2009). Thus, its transport properties largely deviate from continuous understanding of fluid flow.

For dense fluids several additional factors appear to play a more or less significant role. One of them involves wetting properties of the solid surface. Molecular scale roughness allows for creation and stabilization of nanobubbles. Such trapped on the wall nanobubbles may effectively work as a gaseous slip layer, responsible for super-hydrophobic surface properties (Borkent et al. 2007).

Identification of the slip appears not a simple task, both numerically as experimentally. Molecular simulations performed up to now neither confirm nor exclude possible deviation from the classical "non-slip" condition formulated by Navier (Navier 1823). Using continuum mechanics we have to combine a grid size that copes with a nanometer while covering enough space to also include a millimeter scale flow. Experimentally, our techniques for looking at the very small objects (e.g., atomic force microscopy) are slow and cannot cope with large areas very well. The attempts to look at both these scales simultaneously show how our intuition about the relationships between nanoscale objects and macroscopic objects can fail badly.

Classical microscopy used for nanoscale observation has resolution limited by the light wavelength of about 500 nm micrometers. Evaluating diffraction disks the measured position of particle coordinates in plane perpendicular to the optical axis can be improved by order of magnitude. However, resolution in depth, along optical axis, remains very low (tenths of micrometer), and is defined solely by focal depth of the microscope lens.

Total Internal Reflection Fluorescence microscopy (TIRF) helps to bypass some of these limitations offering a possibility to locate objects position with resolution of about 20 nm. Laser light illuminating object undergoes total internal reflection at an interface between investigated medium (liquid) and the wall (glass), and part of the light penetrates into the medium parallel to the interface with an intensity that decays exponentially with the normal distance from the interface. This evanescent wave

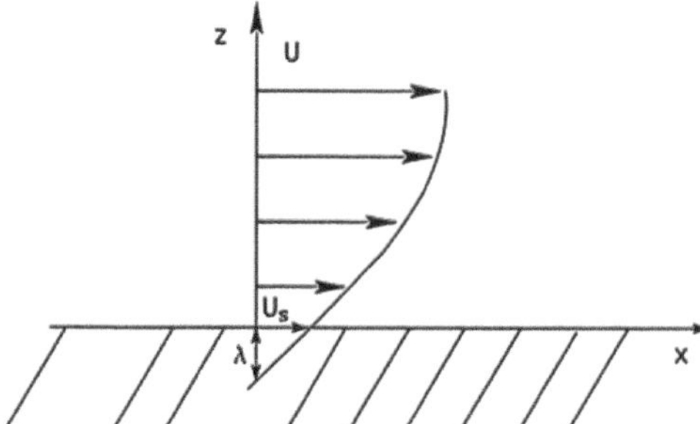

Figure 1. Schematic definition of the slip velocity U_s and the slip length λ for the fluid flow over solid wall.

illumination has been used extensively in the life sciences. Recently it was rediscovered in microfluidics for near wall flow measurements. The main advantage of the method is possibility to reduce the depth of focus of the acquisition system. Hence, it became possible to obtain images of particles, which are in the direct vicinity of the wall.

In our recent study of the Brownian motion of fluorescent particles observed close to the wall (Zembrzycki et al. 2012), the deviation of the particle diffusion rate has been interpreted as an evidence of the slip boundary conditions. According to the theoretical model by Lauga & Squires (2005) the diffusion coefficient of a single colloidal nanoparticle is directly related to the distance from the wall, and the slip velocity. We applied this prediction to determine the slip length from measured and calculated variations of the diffusion coefficient of particles as a function of distance from the wall (Zembrzycki et al. 2012). For this purpose the effect of the wall on the Brownian motion of nanoparticles suspended in water was examined experimentally and compared with numerical simulations performed by Molecular Dynamics approach. The outcome is in the range of uncertainty found in the literature. For a relatively large nanoparticle used in the experiments (300 nm diameter) the evaluated slip length (Fig. 1) measured at 170 nm from the wall appears to be nearly 300 nm. In the numerical analysis, for much smaller particles (24 nm) the evaluated slip length is less than 4 nm. Our difficulties with proper interpretation of available measurements of slip length performed by particle tracking became partly understandable if we look at inherent factors modifying mobility of nanoobjects. We discuss it in the following chapter.

3 MOBILITY OF NANOPARTICLES

Micro and nano scale motion is coupled or sometimes mainly driven by molecular diffusion, direct effect of molecular structure of our environment. Diffusion governed by Brownian motion is an efficient transport mechanism on short time and length scales. Even a highly organized system like a living cell relies on the random Brownian motion of its constituents to fulfill complex functions. A Brownian particle will rapidly explore a heterogeneous environment that in turn strongly alters its trajectory. Thus, detailed information about the environment can be gained by analyzing the particle trajectory. For such an analysis spatial resolution down to the nanometer scale is needed. High resolution is directly connected to the requirement to observe the motion on short time scales. However, at short time scales, the inertia of the particle and of the surrounding fluid can no longer be neglected, and one expects to see a transition from purely diffusive to ballistic motion (Huang et al. 2011). The effect is not negligible for transport phenomena observed in nanoscales, e.g. single-molecule reactions, which are basis for transcription of encoded in DNA information. Thus, for its complete understanding an analysis of Brownian motion at very short time scales is necessary, taking effects of inertia into account.

Biochemical reactions in living systems occur in media of very high molecular concentration. In fact it is difficult to talk about diffusion of molecules at such crowded environment, the interactions between macromolecules hinder their displacements, limiting transport and signaling functions (Długosz & Trylska 2011).

Observation of Brownian motion of micro-objects is a classical basis for particles size measurements, evaluation of liquid properties (viscosity, microrheology), analyzing particle-wall interactions, and many others. Nevertheless even such a seemingly simple problem creates plethora of uncertainties. In all applications it is necessary to maintain the colloid well dispersed and to avoid the formation of aggregates (Chassange & Ibanez 2014). Moreover, it is absolutely necessary to know the fluid-solid interaction in nanoscale and the hydrodynamic properties of the particles.

The equilibrium state and the hydrodynamic properties of many colloid systems in aqueous medium is affected by several environmental parameters. The ionic character of water solutions needs, besides analysis of hydrodynamic friction (famous Stokes formulae), evaluation of surface—liquid interactions. The evaluation of surface charges, ionic streams, creation of the electrostatic double layer theoretically is possible with help of Derjaguin-Landau-Verwey-Overbeek (DLVO) theory (Israelachvili 2011). In practice, we are far from incorporating all necessary molecular and ionic interaction to our macro hydrodynamics, hence commonly used empirical expression called "hydrodynamic diameter" effectively shadows our lack of knowledge. In several cases such simplification can be sufficient for chemical engineering; it becomes unacceptable if size of colloid particles strongly decreases. Any ionic layer, streams of ions attaching particle, steric interactions with suspended molecules, effectively decrease nanoparticle mobility. A proper prediction of such effects is crucial for understanding transport processes at the single cell level.

Figure 2 illustrates our attempts to evaluate size effect for Brownian nano particles (Pawłowska et al. 2014). It is evident that decreasing particle size, the effective (hydrodynamic) diameter affects their diffusion. Moreover, variation of ionic strength of the liquid environment (Fig. 3) strongly modifies diffusion of nanoparticle additionally changing its hydrodynamic diameter.

Detailed experimental analysis of interactions of liquid molecules and surface molecules of individual particle is very difficult. Hence, in practice more or less sophisticated hydrodynamic models are implemented to interpret observed variation of the apparent particle diameter (in fact friction coefficient). Such models used later for measuring and sorting macromolecules are in common use, despite questionable theoretical background given by fluid mechanists.

Recently, a new optical tool called Optical Tweezers (OT) expanded our traditional instrumentation, creating possibility for undisturbed measurements of forces and position in picoNewton and nanometer scales. Detailed evaluation of the forces responsible for particle—particle interactions became possible with help of combined

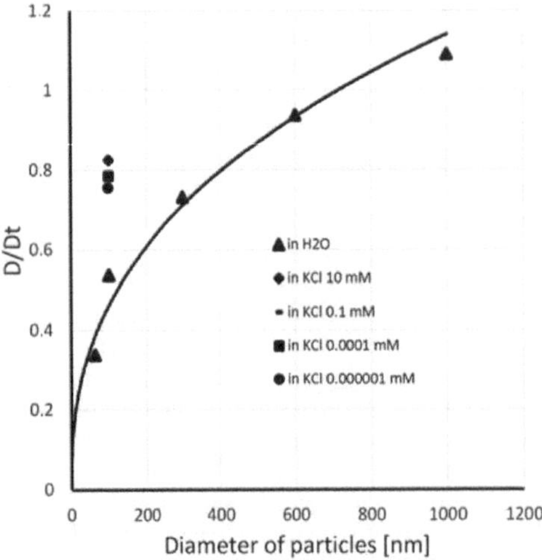

Figure 2. Relative diffusion coefficient measured for polystyrene nano spheres suspended in water; effect of diameter and solution concentration (100 nm particles).

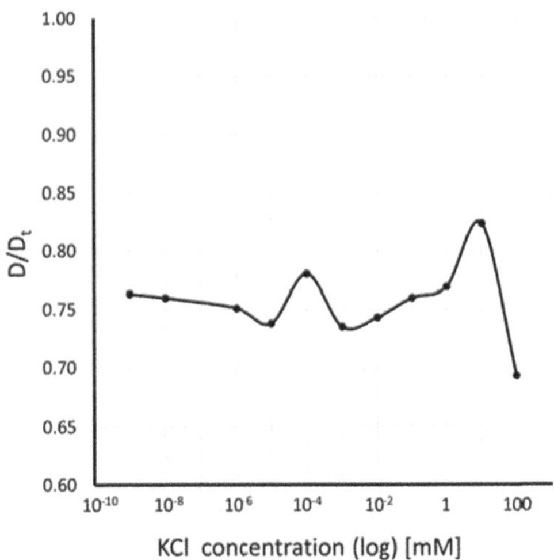

Figure 3. Ionic modulation of diffusion coefficient observed for 100 nm polystyrene spheres suspended in diluted KCl solutions.

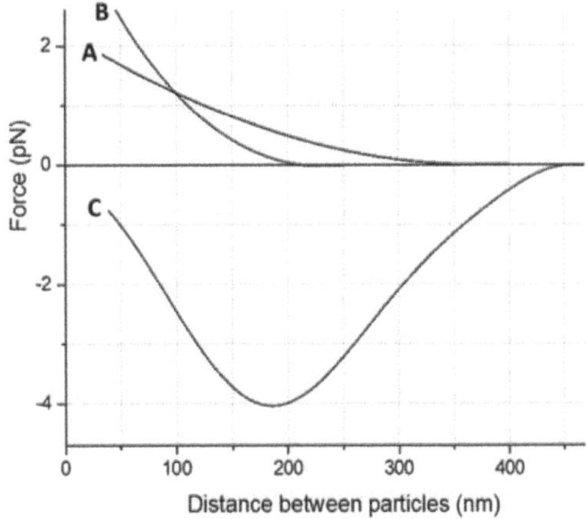

Figure 4. Interaction of two individual colloid particles measured by combined AFM—Optical Tweezer apparatus (Pierini et al. 2015). Data collected for two KCl solutions: 10 3 M (A), 10 5 M (B), and for pure water (C).

AFM-OT apparatus developed at IPPT (Pierini et al. 2015).

Approaching two nanoparticles at nanoscale distances and evaluating intermolecular forces for picoNewton range (Fig. 4) opens attractive possibility to validate existing models of ionic and steric interactions in colloidal suspensions. Dragging, towing single particle trapped by OT allows to perform precise analysis of forces involved by liquid environment, wall interactions, and particle-particle interactions.

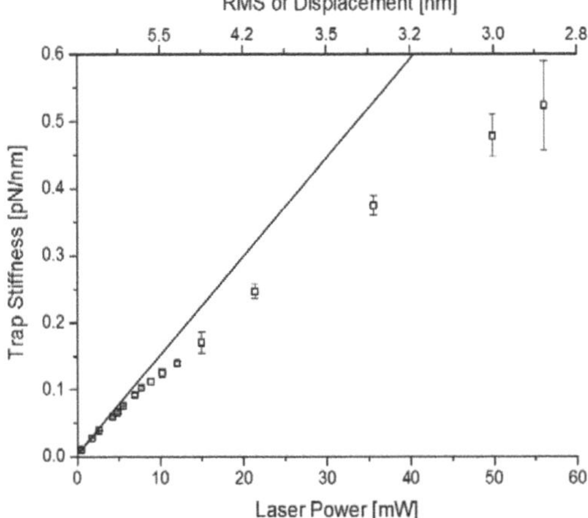

Figure 5. Stiffness of the Optical Tweezers evaluated for 1 μm polystyrene particle suspended in water; particle diffusion increases as the mean Brownian displacement decreases (upper axis). Straight line—large displacements theoretical limit.

An electronic way of signal analysis allows for thermal motion of particle trapped by OT to be evaluated with MHz sampling frequency and displacements below 1 nanometer. Hence, one of the fundamental problems of single particle mobility, namely ballistic regime and effects of inertia creating time dependent recirculation of surrounding liquid molecules, could be proven using OT (Franosch et al. 2011).

In our preliminary study OT developed at IPPT have been used to analyze Brownian motion of trapped polystyrene particle (Zembrzycki et al. 2014). It appears that already at sampling times of 10 kHz diffusion becomes influenced by ballistic regime of molecular interactions (Fig. 5).

4 WORM-LIKE CHAIN (WLC)

The flow of deformable objects (fibers, polymer chains) has a non-Newtonian character, strongly influencing its short time response at microscopic level (Gittes et al. 1993, Jendrejack et al. 2004).

Under the flow these objects are oriented, deformed, and coiled leading to a macroscopic variation of the transport properties.

The flexibility of long biopolymers reflects important aspects of their biological functions. The bending capabilities of DNA have special meaning if we realize that nearly 2 m long human DNA has to be transported through the cell membrane and finally packed into the cell nucleus. The elasticity of cytoskeleton constituents such as actin filaments, microtubules determines cell shape and mechanical response.

The microscopic structures, as well as the macroscopic response of flexible filaments, depend on both the nature of the suspended objects and the flow configuration. Linking mechanical and microscopic properties of the suspended objects to the macroscopic response of the suspension is one of the fundamental scientific challenges of soft matter physics and remains unsolved for a large number of situations typical for intercellular transport of proteins and ligands.

Most of the biomolecules have strongly elongated form, far from idealistic ball like shape. Their penetration through the crowded cellular environment is strongly enhanced by its shape flexibility. The interplay between crowding and thermal bending allows for their controlled mobility.

Deformations of long micro and nano objects are induced not only by thermal fluctuation (Liu et al. 2004), but they are also introduced by non-linear hydrodynamic interactions, leading to some intriguing behavior. For example, it has been predicted that long fibers may perform spectacular windings to form more or less stable knots, phenomenon of fundamental importance for biological macromolecules (Sadlej et al. 2010, Kuei et al. 2015).

Despite recent advances in this field, there is still lack of experimental investigations to validate assumptions of the theoretical and numerical models. Deficiency of experimental studies is mainly due to the absence of good experimental model systems that allow for determining and controlling elasticity and geometry of analyzed flexible objects.

Observation of proteins or DNA is still mostly qualitative, limits of the optical methods permit to find out some predicted characteristics only. Therefore, to systematically investigate the interactions of long molecules with given flow and the resulting macroscopic transport properties, synthetic experimental models of flexible objects can be useful. Hence, we aimed to produce flexible nanofibres to mimic behavior of "worm-like chains". Alas, winding of the objects predicted by simple Stokesian model could not be confirmed in the experiment (Sadlej et al. 2010). It was probably due too high stiffness of nanofibres used in our experiments.

Recently, we have developed new core-shell electrospinning procedure allowing for constructing highly deformable microscopic filaments (Fig. 6), with typical diameter of 100 nm and contour length ranging from single micrometers to millimeters (Nakielski et al. 2015). Introducing such objects directly into microfluidic channels allowed us to observe their deformations due to the flow (Fig. 8) as well as those induced by thermal fluctuations (Table 1). It can be seen that even short

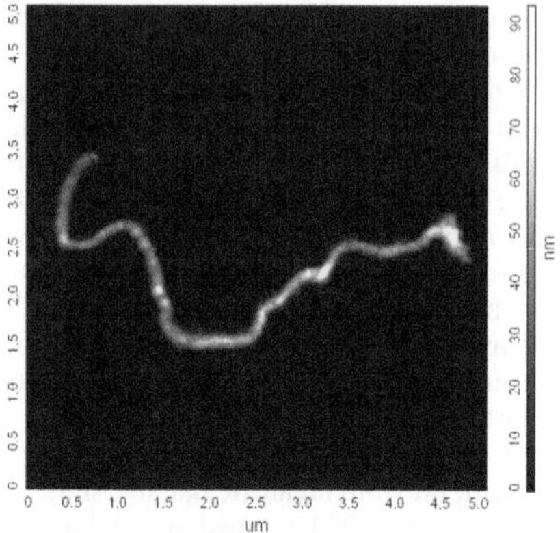

Figure 6. Hydrogel nano-filament observed under an atomic force microscope; contour length 7 μm, diameter 80 nm.

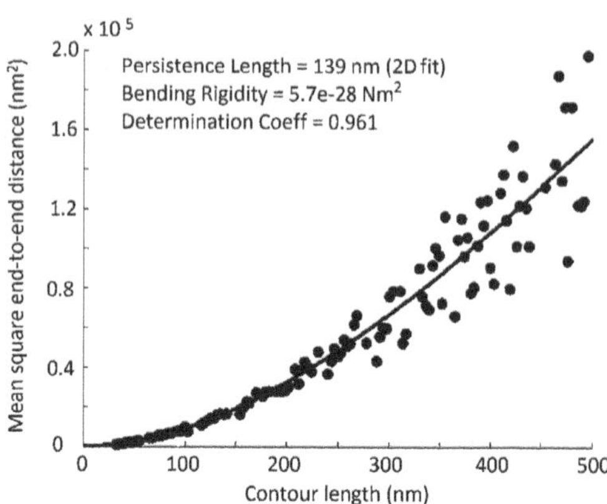

Figure 7. Mean square end to end distance to calculate persistence length of 48 kbp λ-DNA observed with atomic force microscope on mica plate.

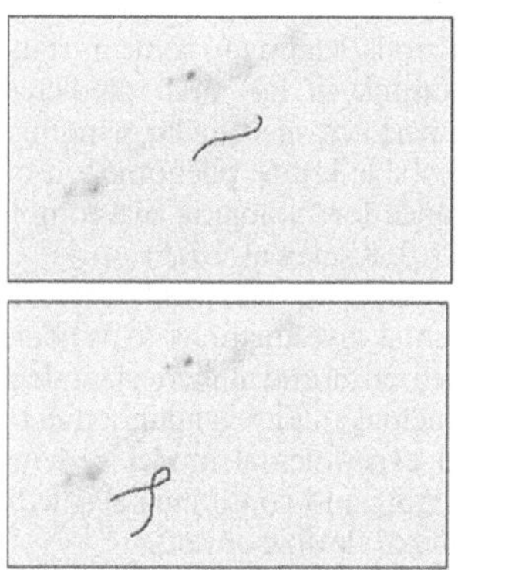

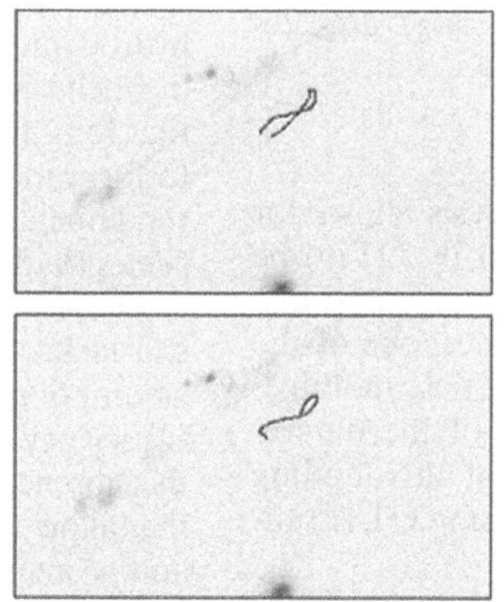

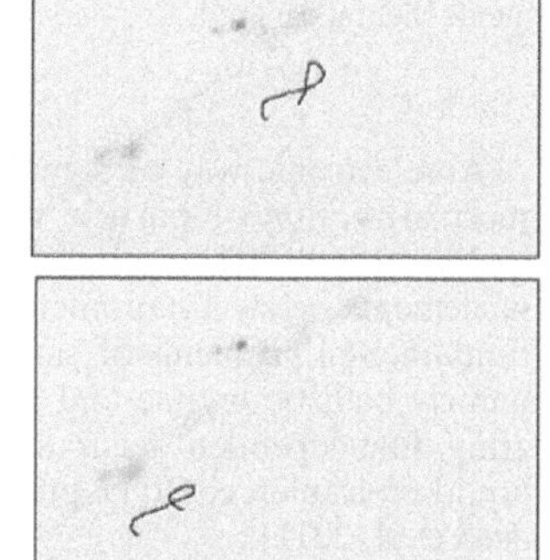

Figure 8. Single highly flexible nanofilament conveyed by Poiseuille flow in 100 μm microchannel. Filament diameter 100 nm, contour length 25 μm. Flow Reynolds number based on the microchannel size equals 4.10–3. Sequence of images (not ordered in time) extracted from the movie recorded under fluorescent microscope and displayed as negatives; image width ~50 μm. There are visible effects of 2-D projection which apparently change contour length of the filament. By selecting images preserving contour length we are able to limit our analysis to in-plane filament deformations.

pieces of filaments exhibit complex mobility, hardly described by the simple theoretical model with translational and rotational diffusion coefficients of prolate ellipsoid (Ortega & Torre 2003).

Additional degrees of freedom created by shape deformation, coiling and uncoiling of longer filaments, change overall redistribution of thermal energy and effective mobility of such objects. The last effect has additional advantage; analysis of thermal fluctuations of flexible objects can be used to evaluate their persistence length (Kratky & Porod 1949), directly correlated with its mechanical properties. Formally the persistence length of long polymers reflects their bending rigidity, parameter describing distance over which the chain of molecules "remembers" its initial position.

For most of biomolecules persistence length varies from nanometers to millimeters (Gittes et al. 1993). Typical values of persistence length obtained for our hydrogel filaments range from 5 μm to 500 μm (Nakielski et al. 2015), being very close to relative values measured for DNA chains (Fig. 7).

Direct observation of shape fluctuations for molecules is presently prohibited by long time illumination necessary to visualize nano objects under optical microscope. Hence, such analysis of DNA persistence length is performed for objects collected onto a functionalized mica substrate, where object-wall interactions may strongly bias statistics of observed deformation variants. In that case collected AFM images of molecules (DNA) are segmented to evaluate local bending curvature (so

Table 1. Translational and rotational diffusion coefficients evaluated from Brownian motion of typical samples of nanofilaments. Experimental data are compared with theoretical diffusion coefficients given for the ideal non-deformable, prolate ellipsoid. Diffusion was measured between two cover slides, about 50 μm apart. Measurements performed in water at temperature T = 30°C, viscosity ηs = 1.0210–3 Pa s, L—filament contour length, R—radius.

	L/2R	D_a (μm²/s)	D_b (μm²/s)	D_r (rad²/s)	L (μm)	Shape
Exper.	43.0	0.052	0.035	0.0006	21.5	
Theor.	43.0	0.144	0.090	0.0019		
Exper.	31.2	0.093	0.067	0.0025	15.6	
Theor.	31.2	0.182	0.116	0.0045		
Exper.	29.8	0.059	0.140	0.0029	14.9	
Theor.	29.8	0.189	0.121	0.0051		
Exper.	47.8	0.231	0.053	0.0005	23.9	
Theor.	47.8	0.133	0.083	0.0014		
Exper.	31.4	0.266	0.070	0.0023	15.6	
Theor.	31.4	0.182	0.116	0.0044		
Exper.	48.4	0.112	0.105	0.0026	24.2	
Theor.	48.4	0.132	0.082	0.0013		
Exper.	104.8	0.054	0.070	0.0005	52.4	
Theor.	104.8	0.072	0.044	0.0002		
Exper.	171.0	0.039	0.150	0.00005	85.5	
Theor.	171.0	0.049	0.029	0.00004		
Exper.	105.6	0.029	0.087	0.0002	52.8	
Theor.	105.6	0.072	0.043	0.0002		

called cosine method), or simply end-to-end length (Fig. 7) is used to build statistical description of molecule deformation.

It is worth to underline that our observations of thermal fluctuations of filaments take place for objects freely suspended in liquid. Hence, despite of 2-D projection limits, we are able to collect statistically relevant sets of data for their thermal deformations. Henceforth, we are able to validate worm-like chain models with material parameterization obtained from the experiment (persistence length). This approach could in the future be used to gain further fundamental understanding of filaments dynamics under flow as a function of their complex mechanical properties as anisotropy or deformability.

It is interesting to note that even for relatively low flow Reynolds number (Fig. 8) we observed typical coiling—uncoiling sequences. It is remarkable similar to WLC modelling performed with Stokesian approach (Sadlej et al. 2010, Kuei et al. 2015). Understanding the link between the microscopic structure of the filaments and the macroscopic flow properties opens the possibility to design nano-objects transported by body fluids for targeted drug release or local tissue regeneration.

5 CONCLUSIONS

Recent development of experimental techniques applicable to fluid mechanics of micro and nano scale permits to have a closer look at applicability of existing mechanical models to small scale phenomena. In the following we have presented few selected problems characterizing nano and micro scale fluid dynamics, namely kinematic boundary condition at solid interfaces and suspension of nano scale intrusions, like nanoparticles and long, deformable filaments. Evidently, similarly to near field fluctuational electrodynamics (Song et al. 2015), *fluctuational fluid mechanics* is needed to describe transport phenomena at nanoscales.

In computational modeling of nanoscale flow problems, similarly to solid mechanics science, there is still unsolved problem of merging atomistic scales with nano, micro and macro systems. In case of fluid mechanics there is additional difficulty. Merging of time stepping for all scales has to be performed for very short time scales. This is challenging, still not available approach. Therefore, we have to cope with simplified models, where solely experimental validation may offer background for tuning and adjustment of crucial model parameters.

ACKNOWLEDGEMENTS

The support of NCN grant no. 2011/03/B/ST8/05481 is acknowledged.

REFERENCES

Borkent, B.M., Dammer, S.M., Schoenherr, H., Vansco, G.J., Lohse, D. 2007. Superstability of surface nanobubbles. *Pysical Review Letters* 98:204502-1-4.

Breuer, K. (ed.) 2004. Microscale Diagnostic Techniques. Berlin:Springer.

Chassange, C., Ibanez, M. 2014. Hydrodynamic size and electrophoretic mobility of latex nanospheres in monovalent and divalent electrolytes. *Colloid and Surfaces A: Physicochemical and Engineering Aspects* 440:208–216.

Czerwińska, J. 2004. Self-diffusion effects in micro scale liquids. Numerical study by a dissipative particle dynamics method. *Bulletin of the Polish Academy of Sciences: Technical Sciences* 55:159–172.

Długosz, M. & Trylska, J. 2011. Diffusion in crowded biological environments: applications of Brownian dynamics. *BMC Biophysics* 4:1–9.

Franosch, T., Grimm, M., Belushkin, M., Mor, F.M., Foffi, G., Forro, L., Jeney, S. 2011. Resonances arising from hydrodynamic memory in Brownian motion. *Nature* 478:85–88.

Gittes, F., Mickey, B., Nettleton, J., Howard, J. 1993. Flexural rigidity of microtubules and actin filaments measured from thermal fluctuations in shape. *The Journal of Cell Biology* 120:923–934.

Huang, R., Chavez, I., Taute, K.M., Lukic, B., Jeney, S., Raizen, M.G., Florin, E.-L. 2011. Direct observation of the full transition from ballistic to diffusive Brownian motion in a liquid. *Nature Physics* 7:576–580.

Israelachvili, J. 2011. Intermolecular and Surface Forces, 3 ed. London:Academic Press.

Jaruszewicz, J., Żuk, P.J., Lipniacki, T. 2013. Type of noise defines global attractors in bistable molecular regulatory systems. *Journal of Theoretical Biology* 317:140–151.

Jendrejack, R.M., Schwartz, D.C., de Pablo, J.J., Graham, M.D. 2004. Shear-induced migration in flowing polymer solutions: simulation of long-chain DNA in microchannels. *Journal of Chemical Physics* 120:2513–2529.

Kochańczyk, M., Jaruszewicz, J., Lipniacki, T. 2013. Stochastic transitions in a bistable reaction system on the membrane. *Journal of the Royal Society Interface* 10:1–12.

Kratky, O. & Porod, G. 1949. Roentgenuntersuchung geloester Fadenmolekuele, *Recueil de travaux Chimiques*. Pays-Bas. 68:1106–1123.

Kuei, S., Słowicka, A.M., Ekiel-Jeżewska, M.L., Wajnryb, E., Stone, H.A. 2015. Dynamics and topology of a flexible chain: knots in steady shear flow. *New Journal of Physics* 17:1–15.

Lauga E. & Squires T.M. 2005. Brownian motion near a partial-slip boundary: A local probe of no-slip condition. *Physics of Fluids* 17:103102-1-16.

Lauga, E., Brenner, M.P. & Stone, H.A. 2005. Microfluidics: The no-slip boundary condition. In Foss J., Tropea C., Yarin A. (eds.), *Handbook of Experimental Fluid Dynamics*, Chap. 15. Berlin:Springer.

Liu, S., Ashok, B., Muthukumar M. 2004. Brownian dynamics simulations of bead-rod-chain in simple shear flow and elongational flow. *Polymer* 45:1383–1389.

Nakielski, P., Pawłowska, S., Pierini, F., Liwinska, V., Hejduk, P., Zembrzycki, K., E. Zabost, E., Kowalewski, T.A. 2015. Hydrogel nanofilaments via core-shell electrospinning. *PLOS ONE* 10(6):e0129816-1-16.

Navier, C. 1823. Memoire sur les lois du mouvement des fluids. *Memoires de Academie Royale des Sciences de Institut de France* 6:389–440

Ortega, A., Garcia de la Torre, J. 2003. Hydrodynamic properties of rodlike and disklike particles in dilute solution. *Journal of Chemical Physics* 119:9914–9919.

Pawłowska, S., Hejduk, P., Nakielski, P., Pierini, F., Zembrzycki, K., Kowalewski, T.A. 2014. Analysis of nanoparticles hydrodynamic diameters in Brownian motion, *Book of Abstracts, XXI Fluid Mechanics Conference, Krakow*, 116–117.

Pierini, F., Zembrzycki, Z., Nakielski, P., Pawłowska, S., Kowalewski, T.A. 2015. Combined atomic force microscopy and optical tweezers (AFM/OT): a hybrid double-probe technique. *Measurement Science and Technology* (submitted).

Sadlej, K., Wajnryb, E., Ekiel-Jezewska, M.L., Lamparska, D., Kowalewski, T.A. 2010. Dynamics of nanofibres conveyed by low Reynolds number flow in a microchannel. *International Journal of Heat and Fluid Flow* 31:996–1004.

Sasai, Y. 2013. Cytosystems dynamics in self-organization of tissue architecture. *Nature* 493:318–326.

Song, B., Ganjeh, Y., Sadat, S., Thompson, D., Fiorino, A., Fernández-Hurtado, V., Feist, J., Garcia-Vidal, J.F., Cuevas, J.C., Reddy, P. & Meyhofer E. 2015. Enhancement of near-field radiative heat transfer using polar dielectric thin films. *Nature Nanotechnology* 10:253–258.

Tabeling, P. 2006. Introduction to Microfluidics, Oxford: Oxford University Press.

Tay, S., Hughey, J.J., Lee, T.K., Lipniacki, T., Quake, S.R., Covert, M.W. 2010. Single-cell NF-kB dynamics reveal digital activation and analogue information processing. *Nature* 466:267–271.

Thomas J.A. & McGaughey, A.J.H. 2009. Water Flow in Carbon Nanotubes: Transition to subcontinuum transport. *Physical Review Lettters* 102:184502-1-4.

Zembrzycki, K., Błoński, S., Kowalewski, T.A. 2012. Analysis of wall effect on the process of diffusion of nanoparticles in a microchannel. *Journal of Physics: Conference Series* 392:012014-1-9.

Zembrzycki, K., Pierini, F., Kowalewski, T.A. 2014. Optical tweezers to interrogate nano-objects in fluid. In Book of Abstracts, 4th National Conference on Nano- and Micromechanics, Wrocław, 25–26.

Advances in Mechanics: Theoretical, Computational and Interdisciplinary Issues – Kleiber et al. (Eds)
© 2016 Taylor & Francis Group, London, ISBN 978-1-138-02906-4

Shape memory materials and structures: Modelling and computational challenges

M. Kuczma
Faculty of Civil and Environmental Engineering, Poznań University of Technology, Poznań, Poland

ABSTRACT: In this contribution we present thermomechanical and mathematical modelling aspects of shape memory materials and structural elements made thereof. Special attention is given to a non-linear physically-based model which takes into account significant hysteresis as well as to the numerical approximation and solution techniques for the corresponding rate boundary value problems. A number of current and future applications of shape memory materials are mentioned. Our aim is to provide a unified constitutive and mathematical frameworks for a class of shape memory materials and structures. Theoretical considerations are illustrated with results of numerical simulations for a unit cell of octet-truss lattice material and a plate, both made from SMAs.

1 INTRODUCTION

Shape memory is the uncanny ability of a material to return to its original configuration from seemingly permanent deformation after removing loading and eventually by heating above a certain characteristic temperature. This intriguing phenomenon is exhibited by some classes of materials, including Shape Memory Alloys (SMAs) and Shape Memory Polymers (SMPs) (Otsuka & Wayman 1998, Meng & Hu 2009). SMAs have been used as active materials in smart structures and various sophisticated devices due to their actuation, damping and sensing properties for more than three decades now, and recently they have also found applications in civil and structural engineering, e.g. (Wilde, Gardoni, & Fujino 2000, Song, Ma, & Lib 2006).

The unusual behaviour of SMAs, including the shape-memory effect and pseudoelasticity, is a result of a crystallographically reversible martensitic phase transformation. It is a first order solid-solid phase transition with displacive nature from a highly ordered parent phase (*austenite*) to a less-ordered product phase (*martensite*) and the reverse one, which can be induced by changes in stress and/or temperature (or magnetic field). Pseudoelasticity (named also superelasticity) of a SMA means that the material is capable to sustain large deformations (8–10%) and to retain to its non-deformed shape upon unloading, at temperatures above A_f (austenitic finish temperature).

A number of classes of SMA models can be distinguished but usually, models aimed at solving realistic boundary value problems are based on continuum thermomechanics and make use of internal variables to account for the microstructural changes due to the martensitic phase transformation, e.g. (Ball & James 1987, Müller & Xu 1991, Bhattacharya & Kohn 1997, Raniecki & Lexcellent 1998, Govindjee, Mielke, & Hall 2003, Roubíček 2004, Bartel & Hackl 2008, Idesman, Levitas, Prestonb, & Cho 2005, Stein & Sagar 2008, Lagoudas 2008, Petryk & Stupkiewicz 2010, Kuczma 2010), to cite a few works.

2 PSEUDOELASTIC HYSTERESIS

A characteristic feature of the pseudoelastic bahaviour of SMAs is a flag-type hysteresis schematically shown in Figure 1. This pertains to the 1D case with austenite A and two variants of martensite M_1 and M_2, and 1D-measures η_1 and η_2 of their phase transformation strains, respectively.

In describing the main loops (bold lines in Fig. 1) we make use of the concept of phase transformation driving forces X_m $(m = 1,2)$ and some threshold functions κ_m^+ and κ_m^- for positive and negative rates of change of volume fractions of martensitic variants c_m, respectively. The phase transformation rules can be expressed as follows

$$\begin{array}{llll} \text{if} & X_m(c_m) = \kappa_m^+(c_m) & \text{then } \dot{c}_m \geq 0, & \\ \text{if} & X_m(c_m) = \kappa_m^-(c_m) & \text{then } \dot{c}_m \leq 0, & (1) \\ \text{if} & \kappa_m^-(c_m) < X_m(c_m) < \kappa_m^+(c_m) & \text{then } \dot{c}_m = 0. & \end{array}$$

Condition $(1)_1$ controls the forward phase transformation of m-th variant of martensite and $(1)_2$—the reverse one, whereas $(1)_3$ means that the

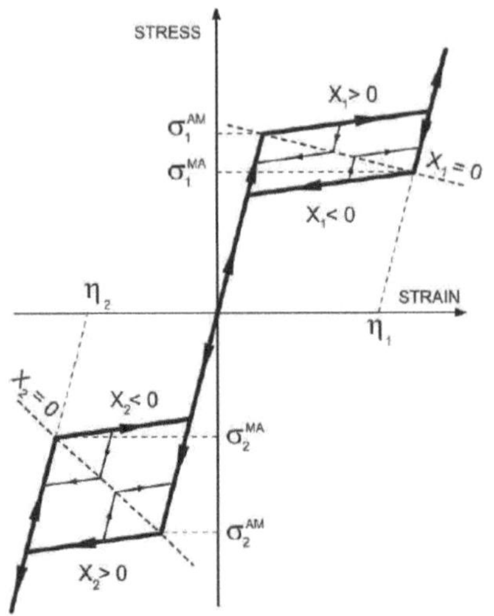

Figure 1. Pseudoelastic hysteresis.

deformation process is purely elastic. The threshold functions κ_m^+ and κ_m^- are measures of the energy dissipated in the course of forward ($\dot{c}_m \geq 0$) and reverse ($\dot{c}_m \leq 0$) phase transformations; for the case shown in Figure 1, κ_m^+ and κ_m^- are linear functions of volume fractions c_m.

3 THERMOMECHANICAL CONSTITUTIVE RELATIONS

We consider a material that may appear in $N+1$ preferred strain states: the parent phase (austenite), indexed with $i = N+1 \equiv a$, and N-variants of martensite ($m = 1, \ldots, N$). The Helmholtz free energy W_i, $i = 1, 2, \ldots, N+1$ is postulated in the form

$$W_i(\epsilon,\theta) = \frac{1}{2}(\epsilon - \mathbf{d}_i)\cdot \mathbf{A}_i[\epsilon - \mathbf{d}_i] + \varpi_i(\theta), \tag{2}$$

where ϵ is the total strain tensor, $\mathbf{A}_i$ is the elasticity tensor and $\mathbf{d}_i$ the transformation strain of i-th phase (variant), and $\varpi_i(\theta)$ is a stress-free part dependent on temperature θ.

The relaxed form of multi-well free energy function with quadratic wells (2) of the austenite-martensite mixture may be defined as

$$\widetilde{W}(\epsilon,\theta,\mathbf{c}) = \frac{1}{2}(\epsilon - \mathbf{d}(\mathbf{c}))\cdot \mathbf{A}[\epsilon - \mathbf{d}(\mathbf{c})] + \varpi(\theta,\mathbf{c}) + \psi_{\text{mix}}(\mathbf{c}), \tag{3}$$

where $\mathbf{c} = \{c_i\}$ is a column matrix that gathers volume fractions of martensitic variants and austenite c_i, ψ_{mix} represents interfacial energy in the mixture, and the effective transformation strain $\mathbf{d}(\mathbf{c})$ is the convex combination,

$$\mathbf{d}(\mathbf{c}) \equiv \sum_{i=1}^{N+1} c_i \mathbf{d}_i, \quad \text{with} \quad c_i \geq 0, \sum_{i=1}^{N+1} c_i = 1. \tag{4}$$

From the reduced form of the dissipation inequality

$$D = -\frac{\partial \widetilde{W}}{\partial \mathbf{c}}\cdot \dot{\mathbf{c}} \equiv \mathbf{X}\cdot\dot{\mathbf{c}} = \sum_{m=1}^{N} X_m \dot{c}_m \geq 0, \tag{5}$$

one obtains components X_m of the phase transformation driving force in the form,

$$X_m = \mathbf{d}_m \cdot \mathbf{A}[\epsilon] - \sum_{i=1}^{N}\left(\mathbf{d}_m \cdot \mathbf{A}[\mathbf{d}_i] + B_{mi}^{\star}\right)c_i - (\varpi_m - \varpi_a) - B_{am}.$$

4 COMPUTATIONAL MODEL

Due to the inequality constraints (1), 4, (5) the considered problem must be solved incrementally in time, and iteratively in space in the case of geometrical nonlinearities.

4.1 *Variational inequality*

Let $\mathbf{u}$ be the displacement vector of a body occupying a region $\Omega \in R^N$ and define finite increments in a typical time step $t_{n-1} \to t_n$ as $\overset{\Delta}{\mathbf{u}}_n \equiv \mathbf{u}_n - \mathbf{u}_{n-1}$ and analogously for the concentration of phases $\overset{\Delta}{\mathbf{c}}_n \equiv \mathbf{c}_n - \mathbf{c}_{n-1}$. Let $V(t_n) \in H^1(\Omega, R^3)$ be the set of kinematically admissible increments of displacements and $K_{\pm}(\mathbf{c}_{n-1})$ stand for sets of increments of volume fractions at time t_{n-1}.

The incremental boundary value problem is governed by the variational inequality:

find $(\overset{\Delta}{\mathbf{u}}_n, \overset{\Delta}{\mathbf{c}}_n) \in V(t_n) \times K(\mathbf{c}_{n-1})$ such that

$$\begin{aligned} &a(\overset{\Delta}{\mathbf{u}}_n, \mathbf{v}) - g(\overset{\Delta}{\mathbf{c}}_n, \mathbf{v}) = f_{n,n-1}(\mathbf{v}), \\ &\mp g(\mathbf{z}_{\pm} - \overset{\Delta}{\mathbf{c}}{}^{\pm}_n, \overset{\Delta}{\mathbf{u}}_n) \pm h(\overset{\Delta}{\mathbf{c}}_n, \mathbf{z}_{\pm} - \overset{\Delta}{\mathbf{c}}{}^{\pm}_n) \\ &\quad \geq \mp b_{n-1}^{\pm}(\mathbf{z}_{\pm} - \overset{\Delta}{\mathbf{c}}{}^{\pm}_n), \end{aligned} \tag{6}$$

for all $\mathbf{v} \in V(t_n)$, $\mathbf{z}_{\pm} \in K_{\pm}(\mathbf{c}_{n-1})$. The conditions (6) reflect the structure of the considered problem: bilinear or linear forms a, g, h, f, b constitute a weak form of the equilibrium equations $(6)_1$ and weak, variational inequality forms of the phase transformation rules (1) defined by two inequalities $(6)_{2\text{-}3}$ for the combinations of upper and lower plus and minus signs in ± and ∓, respectively.

4.2 *Incremental linear complementarity problem*

The finite dimensional counterpart of the variational inequality formulation (6), obtained by the finite element approximation, may be expressed as the following (linear) complementarity problem:

$$\mathbf{D}\mathbf{x}_n + \mathbf{y}_n = \mathbf{b}_{n,n-1},$$
$$\mathbf{x}_n^{'} \geq 0, \mathbf{y}_n^{1} = 0, \mathbf{y}_n \geq 0, \mathbf{x}_n \cdot \mathbf{y}_n = 0, \qquad (7)$$

where $\mathbf{D}$ is a square matrix, $\mathbf{x}_n = (\mathbf{x}_n^1, \mathbf{x}_n^{'})$ is a vector of unknown increments $\mathbf{x}_n^1 = \overset{\Delta}{\mathbf{u}}_n, \mathbf{x}_n^{'} = (\overset{\Delta+}{\mathbf{c}}_n, \overset{\Delta-}{\mathbf{c}}_n)$, (nodal values of the finite element approximations), $\mathbf{y}_n$ denotes a vector of slack variables, and the vector $\mathbf{b}_{n,n-1}$ is known at time t_n.

5 NUMERICAL EXAMPLES

We present here numerical results obtained with author's computer codes and the presented model.

Example 1. The analyzed octahedral unit cell of octet-truss lattice material made from a SMA is shown in Figure 2, following (Deshpande, Fleck, & Ashby 2001) who studied such material but made from an aluminium alloy. Coordinates of nodes $p_1 = (-a,0,0), p_2 = (a,0,0), p_3 = (0,a,0), p_4 = (0,0,b), p_5 = (0,-a,0), p_6 = (0,0,-b)$, with $a = b = 10$ mm. The assumed thermodynamical parameters of the SMA model lead to the following characteristic stresses and unsymmetrical hysteresis loops, Figure 1, for transformation A ↔ M_1: $\sigma_1^{AM}(c_1 = 0) = 100$ MPa, $\sigma_1^{AM}(c_1 = 1) = 101.429$ MPa, $\sigma_1^{MA}(c_1 = 1) = 70$ MPa, $\sigma_1^{MA}(c_1 = 0) = 68.571$ MPa, and $\eta_1 = 0.07$; whereas for transformation A ↔ M_2: $\sigma_2^{AM}(c_2 = 0) = -80$ MPa, $\sigma_2^{AM}(c_2 = 1) = -81.716$ MPa, $\sigma_2^{MA}(c_2 = 1) = -50$ MPa, $\sigma_2^{MA}(c_2 = 0) = -48.339$ MPa, and $\eta_2 = -0.0915$.

Considered here is the case where displacements of node p_6 are fixed, nodes p_1, p_2, p_3, p_5 can move only along z-axis and enforced is horizontal displacement of node p_4, $d \equiv u_4$ with the program by Figure 3. The obtained hysteresis loops are displayed in Figure 4.

Example 2. We have analyzed 1.00 × 1.00 m 3-layer plate of thickness 0.02 m, clamped on its boundary, subjected to uniform loading p = 0,50 MN/m². Its core is made from epoxy resin, SMA layers with $z \in (-0{,}010;-0{,}008)$ and $z \in (0{,}008;0{,}010)$ are made from a Cu-Al-Ni alloy undergoing a cubic → orthorhombic transformation. By assuming the plane state of stress, we have taken into account the 4 transformation strains (martensitic wells):

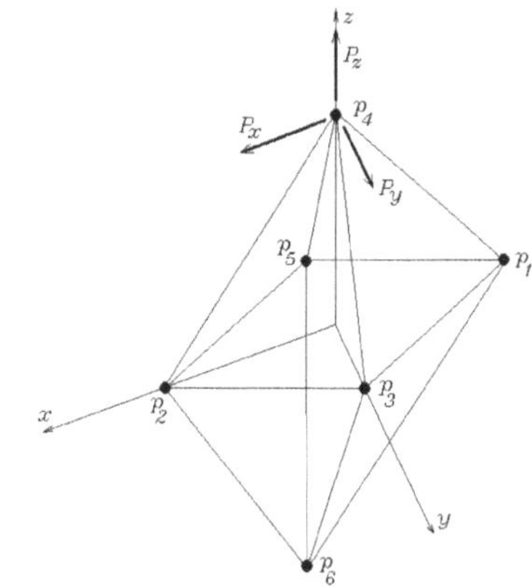

Figure 2. Octahedral cell with the associated coordinate systems.

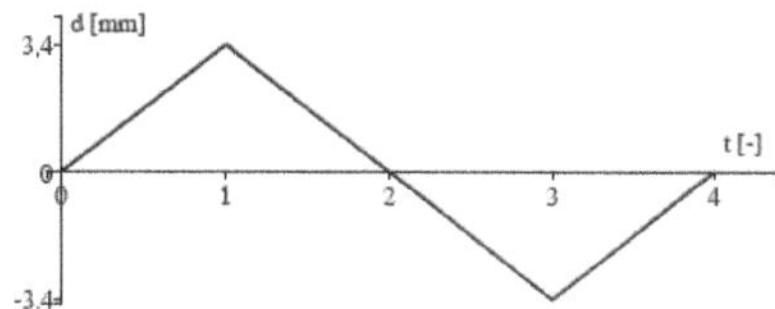

Figure 3. Loading program for displacement $d \equiv u_4$.

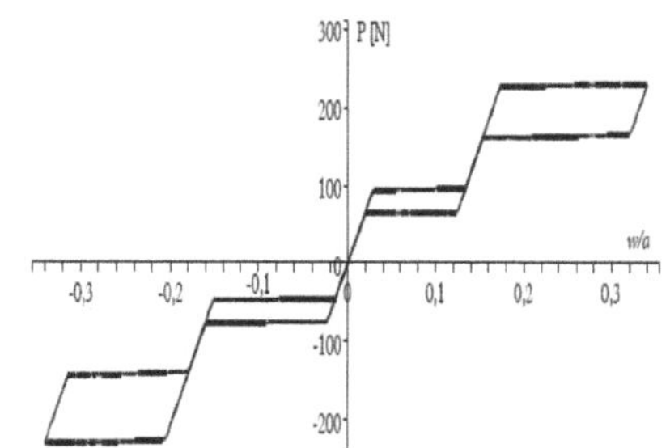

Figure 4. Hysteresis loops $P \equiv P_x$ vs. $d \equiv u_4$ shown in Figure 3.

$$\mathbf{d}_1 = \begin{bmatrix} 0.04245 & 0 \\ 0 & -0.09620 \end{bmatrix},$$

$$\mathbf{d}_2 = \begin{bmatrix} 0.04245 & 0.01945 \\ 0.01945 & 0.04245 \end{bmatrix},$$

$$\mathbf{d}_3 = \begin{bmatrix} 0.04245 & -0.01945 \\ -0.01945 & 0.04245 \end{bmatrix},$$

$$\mathbf{d}_4 = \begin{bmatrix} -0.09620 & 0 \\ 0 & 0.04245 \end{bmatrix}.$$

The obtained distributions of martensitic variants M_1–M_4 and austenite A are displayed in Figures 5–9.

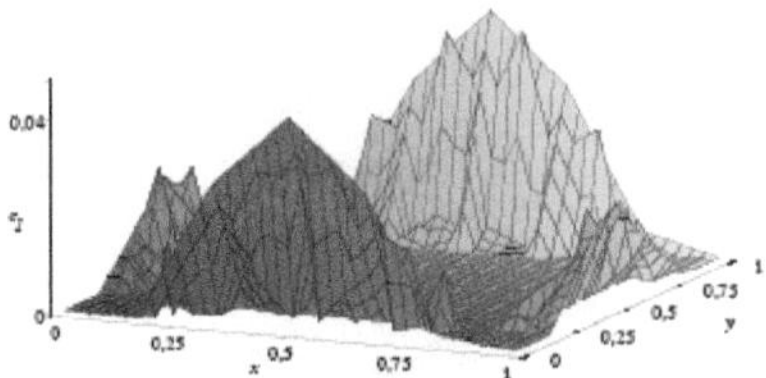

Figure 5. Distribution of martensitic variant M_1.

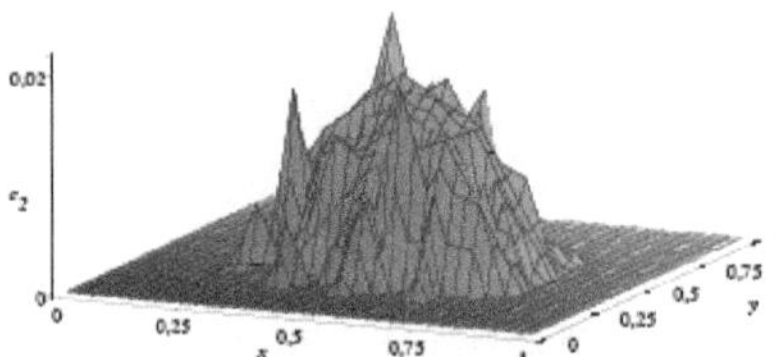

Figure 6. Distribution of martensitic variant M_2.

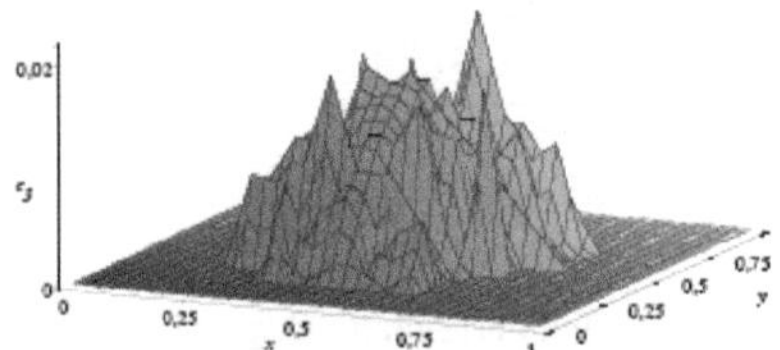

Figure 7. Distribution of martensitic variant M_3.

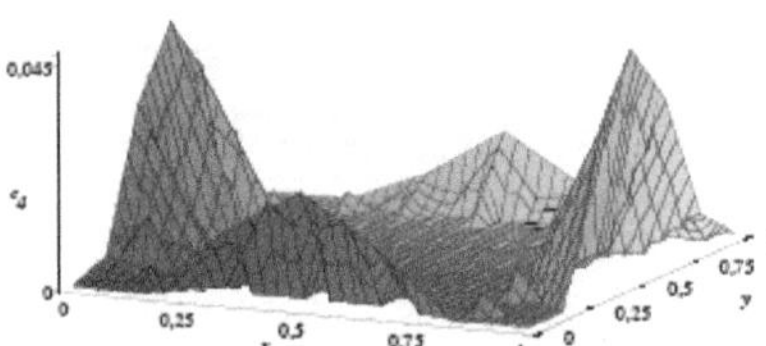

Figure 8. Distribution of martensitic variant M_4.

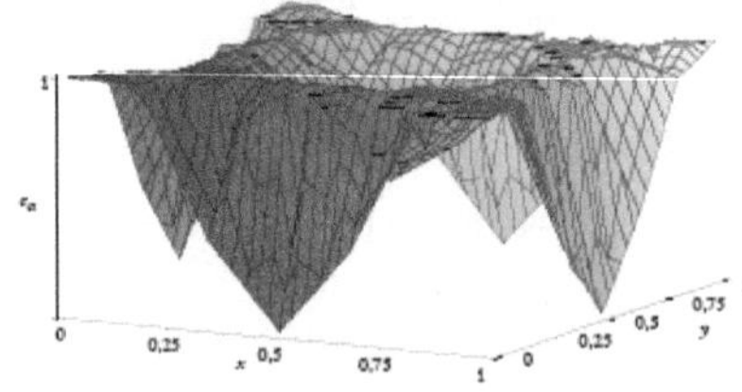

Figure 9. Distribution of austenite A.

6 CONCLUDING REMARKS

The presented variational formulation of the pseudoelastic isothermal deformation process accounts for hysteretic effects. The applied solution algorithm uses the concept of complementarity and automatically determines the moving phase transformation fronts. Some sequential hysteresis loops are obtained for an octahedral cell of octet-truss lattice material. The FEM results for the plate bending problem have demonstrated that the solution of this free BVP is not smooth. The box constraints imposed on the evolution of volume fractions (internal variables) seem to be the most challenging aspect of the presented phase transformation model from the computational viewpoint.

REFERENCES

Ball, J. & R. James (1987). Fine phase mixtures as minimizers of energy. *Arch. Rational Mech. Anal. 100*(1), 13–52.

Bartel, T. & K. Hackl (2008). A novel approach to the modeling of singlecrystalline materials undergoing martensitic phasetransformations. *Mater. Sci. Engng. A*(481–482), 371–375.

Bhattacharya, K. & R. Kohn (1997). Elastic energy minimization and the recoverable strains of polycrystalline shape-memory materials. *Arch. Rational Mech. Anal. 139*, 99–180.

Deshpande, V., N. Fleck, & M. Ashby (2001). Effective properties of the octet-truss lattice material. *J. Mech. Phys. Solids 49*, 1747–1769.

Govindjee, S., A. Mielke, & G. Hall (2003). The free energy of mixing for n-variant martensitic phase transformations using quasi-convex analysis. *J. Mech. Phys. Solids 51*, I–XXVI.

Idesman, A., V. Levitas, D. Prestonb, & J.-Y. Cho (2005). Finite element simulations of martensitic phase transitions and microstructures based on a strain softening model. *J. Mech. Phys. Solids 53*, 495–523.

Kuczma, M. (2010). *Foundations of the mechanics of shape memory structures. Modelling and numerics.* UZ Press, Zielona Góra. (in Polish).

Lagoudas, D. (2008). *Shape memory alloys. Modeling and Engineering applications*. Springer, New York.

Meng, Q. & J. Hu (2009). A review of shape memory polymer composites and blends. *Composites A*(40), 1661–1672.

Müller, I. & H. Xu (1991). On the pseudoelastic hysteresis. *Acta metall. mater. 39*, 263–271.

Otsuka, K. & C. Wayman (1998). *Shape memory materials*. CUP, Cambridge.

Petryk, H. & S. Stupkiewicz (2010). Interfacial energy and dissipation in martensitic phase transformations. *J. Mech. Phys. Solids 58*, 390–408.

Raniecki, B. & C. Lexcellent (1998). Thermodynamics of isotropic pseudoelasticity in shape memory alloys. *Eur. J. Mech. A/Solids 17*, 185–205.

Roubíček, T. (2004). Models for microstructure evolution in shape memory alloys. In P. PonteCastaneda, J. Telega, and B. Gambin (Eds.), *Nonlinear Homogenization and its Applications to Composites, Polycrystals and Smart Materials*, Dodrecht, pp. 269–304. Kluwer Academic Publishers.

Song, G., N. Ma, & H.-N. Lib (2006). Applications of shape memory alloys in civil structures. *Eng. Struct. 28*, 1266–1274.

Stein, E. & G. Sagar (2008). Theory and finite element computation of cyclic martensitic phase transformation at finite strain. *Int. J. Numer. Meth. Eng. 74*(1), 1–31.

Wilde, K., P. Gardoni, & Y. Fujino (2000). Base isolation system with shape memory alloy device for elevated highway bridges. *Engineering Structures 22*, 222–229.

Advances in Mechanics: Theoretical, Computational and Interdisciplinary Issues – Kleiber et al. (Eds)
© 2016 Taylor & Francis Group, London, ISBN 978-1-138-02906-4

Advanced mechanics in High Energy Physics experiments

T. Kurtyka
CERN—European Organization for Nuclear Research, Geneva, Switzerland

ABSTRACT: Research in High Energy Physics (HEP) is now based on a world-wide collaboration. The challenges of construction of giant instruments of this research, like the Large Hadron Collider, require advances in other domains of science and technology, also including several fields of mechanics. Large HEP laboratories and CERN in particular, rely here on collaboration with a network of external research centers capable of providing such an advanced mechanics expertise. This network includes already some Polish partners; future projects of HEP will require a still wider collaboration in various fields of mechanics. The paper gives an overview of some of these fields, and may be treated as an invitation for broadening this network.

1 INTRODUCTION

The construction of the Large Hadron Collider (LHC), the world's largest and most powerful particle collider, has required an extensive collaboration and development work in various fields of mechanics, encompassing new theoretical developments, novel experimental techniques and more efficient computational tools. Some of them are briefly described in the paper which also signals the domains where further collaboration would be highly desirable.

2 THERMOMECHANICS OF INTERACTION OF HIGH ENERGY PARTICLE BEAMS WITH MATTER

Highly energetic particle beams interact with matter producing a shower of secondary particles and depositing heat. In the case of short beam pulses and high energy densities this leads to thermally induced, dynamic wave-type phenomena in the impacted solid materials. Historically, this effects were first studied in the context of target-type physics experiments, where deposited beam energies were low and analytical solutions possible (Bertarelli et al. 2008).

In the collider-type experiments the beam energies are much higher, reaching unprecedented level in the case of the LHC; its two opposite colliding beams of protons carry the energy of up to 724 MJ (equivalent to 173 kg TNT), which must be safely handled by *beam intercepting devices*, including a system of *collimators* removing a small part of this energy during normal operation, but designed against higher accidental beam losses, up to the system of *beam-dumps* used for discharging the total beam energy.

The design challenges of the beam intercepting devices for the LHC, exposed to potentially destructive beam impacts, have driven at CERN a vigorous development of the methods of mechanical analysis of such phenomena, supported by an intensive collaboration with external research centers specialized in impact studies and in computational mechanics. At present, a considerable expertise has been gathered allowing analysis of a wide range of material and structural behavior, from the elasto-plastic up to the shock wave regimes, and with respective computational tools ranging from implicit up to explicit FEM codes and "Hydrocodes", respectively.

Effective use of these computational tools requires an extensive experimental calibration of dynamic yield conditions, failure criteria, and Equations of State (EOS). Here, CERN relies on external collaborators and available databases, quite scarce in the case of the EOS. An important element has recently been added to this experimental base, namely an unique, dedicated HiRadMat test facility at CERN (Bertarelli et al. 2013), where structures and materials can be tested under real high energy proton beam impacts, and where the results of computational simulations could be verified by tests involving destructive levels of beam energies, Figure 1.

This new test facility is essential for the current design work devoted to the next generation of collimators for the LHC and for the future HEP projects where the stored beam energies may be much higher. These new projects call again for collaboration in the development of new computational tools, novel materials, their experimental characterization and constitutive modelling which, apart from complex dynamic characteristics, should also account for the effects of the *radiation induced damage*.

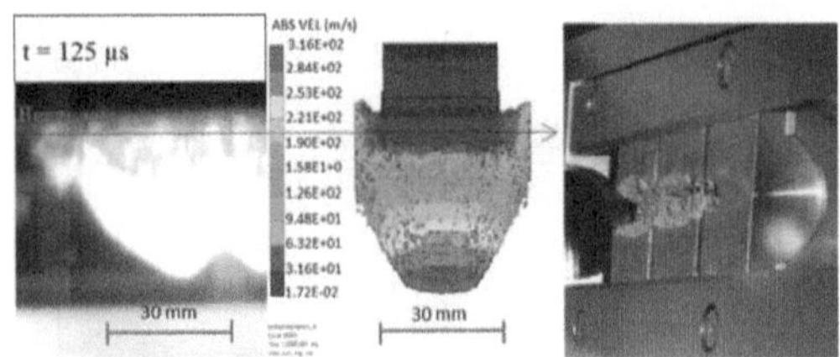

Figure 1. Flow of material fragments from a specimen under proton beam impact; fast-camera measurements, numerical simulation and *post mortem* inspection (Bertarelli et al. 2013).

3 MECHANICS OF MATERIALS

The construction of the LHC required extensive testing of structural materials for high temperature applications, as discussed above, but also for the cryogenic environment of the LHC superconducting magnets, cooled by superfluid (1.9 K) or liquid (4.2 K) helium. Mechanical testing of materials at liquid He temperatures has been intensively developed, giving a new insight into material behavior and driving in turn development of new constitutive material models. Such phenomena as discontinuous plastic flow ("serrated yielding"), plastic strain induced phase changes and damage effects have found new formulations, both by extending the classical phenomenological framework of plasticity, but also by demonstrating that for their adequate description a more deep physical, *ab initio* approach (e.g. based on phonon mechanism), is needed (Skoczen et al. 2014).

This need is particularly visible in the description of the so called *strain induced degradation* of superconductors. This effect is now intensively studied, mainly for the Nb_3Sn superconductors foreseen for the next generation of superconducting magnets. The phenomenon, known for a wide class of intermetallics (e.g. A15-phase compounds), is observed as a marked decrease of superconducting properties under mechanical strain. In the current engineering practice this effect is accounted for by various empirical *scaling laws* (Bordini et al. 2013), however there are now several attempts to describe this effect on the basis of more physical arguments, e.g. again on phonon mechanism, as initiated by Markiewicz, 2008.

4 OTHER DOMAINS OF MECHANICS

4.1 *Dynamics, vibrations*

Analysis of vibrations is now routinely performed for accelerator and detector elements, usually requiring extremely high accuracy of positioning and stabilization, and therefore sensitive to mechanical and "white noise" type seismic disturbances. Future analysis of linear colliders is a particularly challenging task, with a necessity of stabilizing some of its elements to the accuracy of nanometers with *active vibration isolation systems* (Janssens et al. 2011).

4.2 *Mechanics of fluids*

Mechanics of fluids, as applied in the context of HEP experiments, encompasses a wide range of problems, including thermo-hydraulics of liquid and superfluid He, which would require a separate specialist overview. Let us signal here only some analyses and tests, essential for the safety of experimental facilities at CERN, including the problem of sudden bursts and propagation of gases in confined spaces (Chorowski et al. 2006), or the analysis of pressure waves following a break of large vacuum systems. Let us also note that for several future projects of physics, liquid metal beam targets impacted by proton beams are considered as the only feasible solution to handle extremely high power loads necessary to produce the secondary particles for experiments in neutrino and muon physics or for testing the concepts of Accelerator Driven Systems for future sub-critical nuclear reactors. This extends the problems discussed above in Par. 2 to the field of interaction of particle beams with liquids.

REFERENCES

Bertarelli, A., Dallocchio, A., & Kurtyka, T. 2008. Dynamic response of rapidly heated cylindrical rods: longitudinal and flexural behaviour. *Journal of Applied Mechanics, ASME* 75(3), 031010.

Bertarelli, A., et al. 2013. An experiment to test advanced materials impacted by intense proton pulses at CERN HiRadMat facility. *Nuclear Instruments and Methods in Physics Research B* 308: 88–99.

Bordini, A., et al. 2013. An exponential scaling law for the strain dependence of the Nb_3Sn critical current density. *Supercond. Sci. Technol.* 26.

Chorowski, M., et al. 2006. Safety oriented analysis of cold helium-air mixture formation and stratification. *Cryogenics* 46: 262–272.

Janssens, S., et al. 2011. Stabilization and positioning of CLIC quadrupole magnets with sub-nanometre resolution. *Proc. ICALEPCS2011, Grenoble,* MOMMU005: 74–77.

Markiewicz, W.D. 2004. Elastic stiffness model for the critical temperature Tc of Nb_3Sn including strain dependence. *Cryogenics* 44: 767–782.

Skoczeń, B., Bielski, J., & Tabin, J. 2014. Multiaxial constitutive model of discontinuous plastic flow at cryogenic temperatures. *International Journal of Plasticity* 55: 198–218.

Advances in Mechanics: Theoretical, Computational and Interdisciplinary Issues – Kleiber et al. (Eds)
© 2016 Taylor & Francis Group, London, ISBN 978-1-138-02906-4

Hierarchic isogeometric analyses of beams and shells

B. Oesterle, M. Bischoff & E. Ramm
Institute of Structural Mechanics, University of Stuttgart, Stuttgart, Germany

ABSTRACT: The higher inter-element continuity of the Isogeometric Analysis (IGA) applying NURBS functions for geometry as well as mechanics opens up new possibilities in the analysis of thin-walled structures, i.e. beams, plates and shells. The contribution addresses the straightforward implementation of classical theories requiring C^1-continuity, such as the Euler-Bernoulli beam and Kirchhoff-Love shell theory. Based on these "simplest" models shear deformable theories, introducing Timoshenko and Reissner-Mindlin kinematics, are formulated in a hierarchic manner. In contrast to the usual Finite Element concept using total rotations the present model picks up traditional formulations introducing incremental rotations as primary variables. Furthermore an alternative version is discussed with a split of the displacements into bending and transverse shear parts. Both hierarchic concepts can be easily extended to 3D–shell models. The key aspect of this alternative parameterization is the complete a-priori removal of the transverse shear locking and curvature thickness locking (in the case of 3D-shells).

1 INTRODUCTION

From the different concepts for dimensional reduction of thin-walled structures, in the present study a derived approach is applied utilizing a-priori assumptions across the thickness for selected mechanical parameters. In the early days Finite Elements were mostly derived for transverse shear free formulations, e.g. a Kirchhoff-Love model for plates and shells. It turned out that the C^1-continuity requirement was a severe obstacle leading to sophisticated discretization schemes. Consequently later on most formulations applied shear deformable theories requiring only C^0-continuity, e.g. applying Reissner-Mindlin kinematics. This in turn opened the field of locking problems for equal low order interpolation, such as transverse shear, membrane and thickness locking, resulting in an almost never-ending discussion on appropriate unlocking schemes up to the present time, see review article Bischoff et al. (2004).

This situation leads to the objective of the present study: find a modified parameterization for a primal formulation, leading to satisfying stress resultants, as hierarchic model with equal (low) order interpolation free from locking avoiding any numerical unlocking scheme. The hierarchic concept means that starting from an Euler-Bernoulli or a Kirchhoff-Love model the inclusion of transverse shear deformations is simply added in an incremental way. This in turn allows extending the model towards thickness changes or even other 3D higher order effects, again in an additive manner. It is obvious that such a hierarchical family of formulations eases model adaptation, but also leads to a unified implementation.

It is found that the Isogeometric Analysis (IGA) (Hughes et al. 2009; Echter et al. 2010) using NURBS functions as shape functions for geometrical and mechanical variables is an elegant basis for the raised task. B-Splines and NURBS elements of order p have a C^{p-1}-continuity in contrast to the standard Lagrange Finite Elements with C^0-continuity; thus quadratic order functions $p = 2$ are C^1-continuous as required for Euler-Bernoulli and Kirchhoff-Love formulations. These low order NURBS elements with C^1-continuity even show an outstanding performance when applied to transverse shear deformable shells (Echter et al. 2013). On the other hand it has to be mentioned that NURBS elements show similar locking phenomena than Lagrange elements and need unlocking schemes.

2 HIERARCHIC FORMULATION OF BEAMS AND SHELLS

2.1 *Model case of straight beam*

For simplicity the hierarchic concept is explained for straight Timoshenko beams:

$$\kappa = \varphi' \tag{1}$$

$$\gamma = v' + \varphi \tag{2}$$

Equation (1) describes the curvature κ and (2) denotes the shear strain γ. Instead following the standard way and discretizing the displacement v and the total rotation φ, displacement v and the shear angle γ itself can be introduced as primal parameters:

$$\kappa = -v'' + \gamma \tag{3}$$

$$\gamma = \gamma \tag{4}$$

It is apparent that this alternative discretization yields the second derivative of v, which requires C^1-continuity of the trial functions. Equations (3) and (4) can be identified as the kinematic equations of the Euler-Bernoulli beam enriched by a shear strain γ; it plays the role of a hierarchic rotation superimposed to the shear-free beam. In the thin limit γ vanishes what can be easily satisfied by (4). Thus this formulation is a priori free from shear locking.

It should be noted though that the unbalance in derivatives in Equation (2) is now shifted to κ. The second derivative in the displacement v leads to oscillations in the shear force which are in particular pronounced for quadratic shape functions. This disadvantage may be remedied resorting to a different parameterization decomposing v into a bending part v_b and shear part v_s

$$v = v_b + v_s \tag{5}$$

so that bending and shear are completely decoupled:

$$\kappa = -v_b'' \tag{6}$$

$$\gamma = v_s' \tag{7}$$

v_s plays the role of an incremental or hierarchic displacement. In this version again shear locking does not occur per definition.

The three versions with different parameterization are tested for the elementary model case of a simply supported straight Timoshenko beam under uniform transverse load q; it has a length of 10, a Young's modulus E = 1000 and Poisson's ratio ν = 0. Load q is scaled with t^3, where t denotes the thickness of the beam. Thus the exact displacements are independent of t in the thin limit. 10 elements with quadratic B-spline shape functions, having a maximum inter-element continuity of C^1, are used for the three pairs of kinematic variables v/φ, v/γ, and v_b/v_s, respectively. Two cases with different slenderness are investigated, a very thick and a very thin beam with L/t = 10 and L/t = 1000, respectively. Figure 1 shows bending moments and shear forces.

For both cases applying the three different parameterizations. Using the total rotations bending moments are well approximated for the thick

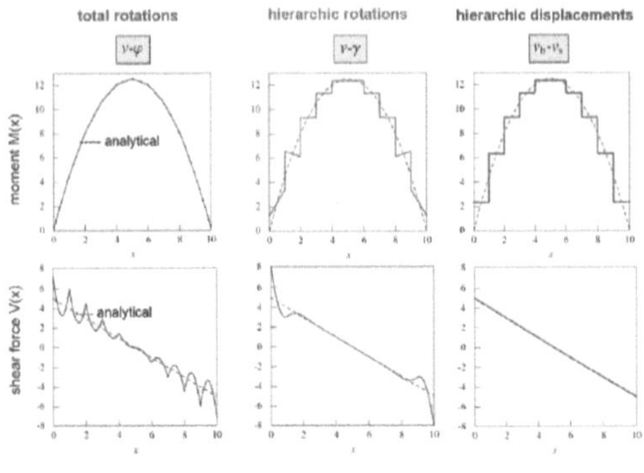

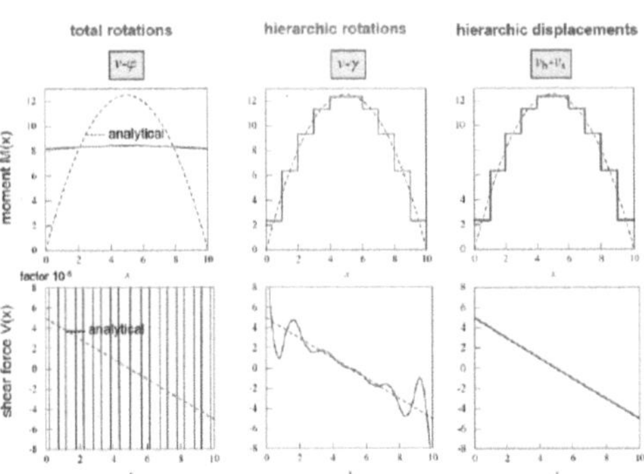

Figure 1. Simply supported beam, bending moment and shear force for thick beam L/t = 10 (above) and very thin beam L/t = 1000 (below).

beam however are severely underestimated for the thin beam. The plots for the shear forces illustrate the oscillating behavior which completely deteriorates for the thin case. Both hierarchic formulations lead to reasonable bending moments independent of the thickness; thus these concepts are a-priori free of transverse shear locking. The quality suffers a bit from the reduced approximation for the curvature resulting from the second derivative of v. Shear forces still show oscillations for the hierarchic rotation formulation however are exact for the hierarchic displacement version independent of the slenderness.

2.2 *Curved beams and shells*

The extension to curved beams and shells can be done in a similar way. For a detailed description it is referred to Echter et al. (2013) and Oesterle et al. (2015). The starting point is again a transverse shear-free solution (Euler-Bernoulli or Kirchhoff-Love), here denoted as 3-Parameter formulation referring to the three midsurface displacements **v** as primary parameters. In Figure 2,

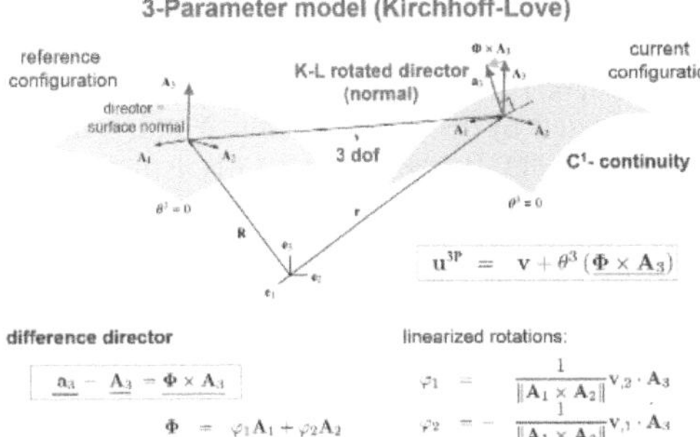

Figure 2. 3-Parameter formulation (Kirchhoff-Love).

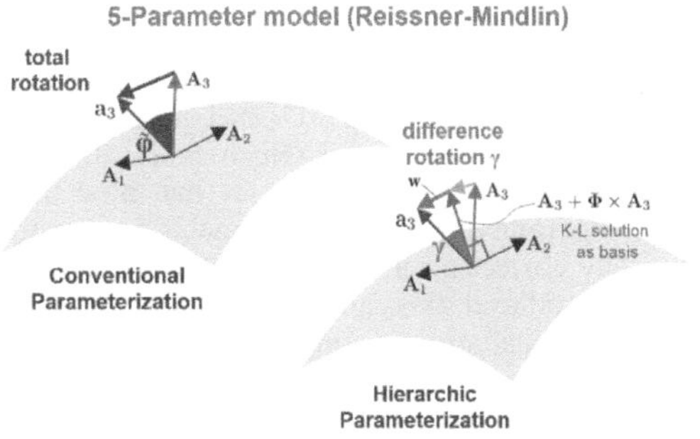

Figure 3. 5-Parameter formulation (Reissner-Mindlin), conventional vs. hierarchic parametrization.

main features are given such as convective coordinates θ^i, basis vectors $\mathbf{A}_i$ and $\mathbf{a}_i$ in the reference and current configuration, respectively, and rotation tensor $\mathbf{\Phi}$ with its linearized components φ_α expressing the rotated director following the usual kinematic constraint.

The C^1-continuity is easily satisfied by the NURBS discretization within the IGA concept, even for quadratic trial functions (Kiendl et al. 2009; Echter et al. 2013).

The extension to a 5-Parameter formulation including transverse shear deformations in the IGA context is straightforward (Benson et al. 2010). Figure 3 shows the conventional parameterization using total rotations $\tilde{\mathbf{\Phi}}$ and the hierarchic version applied here, introducing two difference rotations $\boldsymbol{\gamma}$ superimposed to the already determined Kirchhoff-Love rotations. This version utilizes a traditional concept, and was proposed in Long et al. (2012) applying subdivision surfaces and in Echter et al. (2013) in the IGA context. Alternatively the displacement $\mathbf{w}$ at the shell surface can be introduced instead of $\boldsymbol{\gamma}$.

Figure 4 indicates that the difference of the current and reference director $\mathbf{a}_3 - \mathbf{A}_3$ contains the two parameters $\mathbf{w}$ in addition to the Kirchhoff-Love term leading to five primary variables. It can be observed that the transverse shear terms ε_{13} and ε_{23} depend only on $\mathbf{w}$ and thus are completely decoupled from $\mathbf{v}$. This is the reason that transverse shear vanishes in the thin limit leading per definition to a shear locking free formulation.

In order to diminish oscillating transverse shear forces an alternative 5-Parameter version can be introduced (Oesterle et al. 2015). In this case the transverse displacement $\mathbf{v}$ is decomposed into a bending part $\mathbf{v}_b$ and the shear parts $\mathbf{v}_{s1}$ and $\mathbf{v}_{s2}$. These two values bring in the influence of the two shear angles γ_1 and γ_2 for both directions.

In Figure 5 both versions are contrasted. For the hierarchic displacements the difference director depends on the total rotations $\tilde{\mathbf{\Phi}}$; their

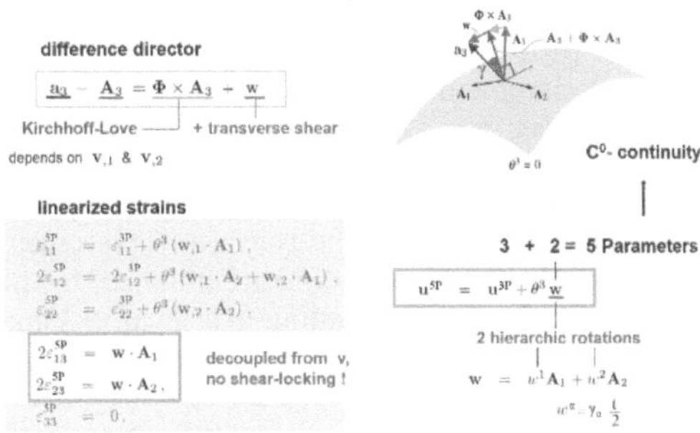

Figure 4. 5-Parameter formulation (Reissner-Mindlin), kinematics with hierarchic rotation.

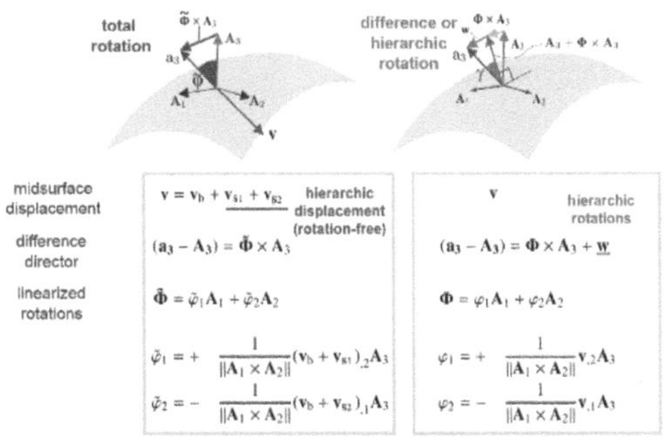

Figure 5. 5-Parameter formulation (Reissner-Mindlin), kinematics with hierarchic displacements vs. hierarchic rotations.

components contain the derivatives of the bending displacement $\mathbf{v}_b$ and the respective shear part $\mathbf{v}_{s\alpha}$. On the other hand the hierarchic rotations lead to the Kirchhoff-Love rotations supplemented by the incremental shear rotations, here represented by the difference displacements $\mathbf{w}$ of the director at the shell surface.

2.3 *3D-shells*

The hierarchic Reissner-Mindlin formulations can easily be expanded to a 3D-shell model including the thickness change (Büchter et al. 1994; Bischoff & Ramm 1997), i.e. **w** has now three primal parameters. Since this model is not asymptotically correct an additional 7th parameter has to be introduced; this can be the linear component of the normal strain ε_{33} in thickness direction or alternatively a quadratic mode of the transversal displacement. This model has been denoted as 7-Parameter formulation.

2.4 *IGA discretization*

The described formulations are free of any transverse shear and curvature locking. Figure 6 summarizes their discretization in the sense of IGA. Initial position and base vectors, the displacements of the midsurface as well as the hierarchic rotations are discretized by NURBS shape functions $N^i(\xi, \eta)$, defined in the parameter space ξ and η. The three components of the midsurface displacement (3-Parameter K-L formulation) are enhanced by the two hierarchic rotations leading to a shear locking-free 5-Parameter formulation with Reissner-Mindlin (R-M) kinematics (Echter et al. 2013).

As mentioned above an alternative parameterization could be carried out splitting the midsurface displacement field into the bending part and two shear parts in each direction, which in turn are discretized in the IGA format (Oesterle et al. 2015).

Figure 6 also refers to the discretization of the two extra kinematic variables w_3 and $\bar{\mathbf{W}}$ for the 7-Parameter formulation.

2.5 *Numerical examples*

Two conceptual examples should underline the main differences between the conventional and the hierarchic parameterization. In both cases NURBS functions of order p = 2 with C^1-continuity have been applied. The plate under a scaled load of $q_z = 1.0\ t^3$ is discretized by 10 × 10 elements; Young's modulus is E = 1000; Poisson's ratio 0.3 is assumed. The "locking diagram", Figure 7, shows the maximum displacements plotted versus the slenderness parameter L/t; the analytical reference solution is 0.4429.

The 3-Parameter model (K-L) is naturally locking free, whereas the 5-Parameter model (R-M) applying the conventional parameterization without unlocking schemes shows a distinct shear locking in the slender regime; in contrast its hierarchic counterpart is per definition free of transverse shear locking.

The shear forces q_{xz} determined for a thin and a very thin plate are depicted in Figure 8. The solution for the conventional parameterization illustrates unacceptable oscillations for both slenderness values. For the very thin case the solution scaled to 10^{-7} completely deteriorates. For further insight into the described effects the shear force results for all formulations scaled to 10^{-8} are shown for the very thin case in more detail.

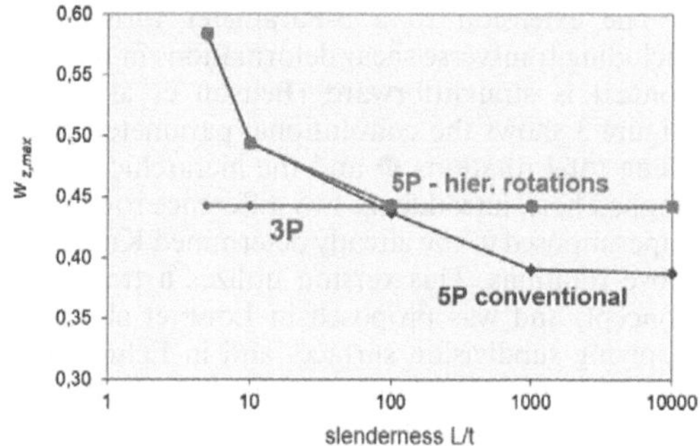

Figure 7. Simply supported plate under uniform load, locking diagram.

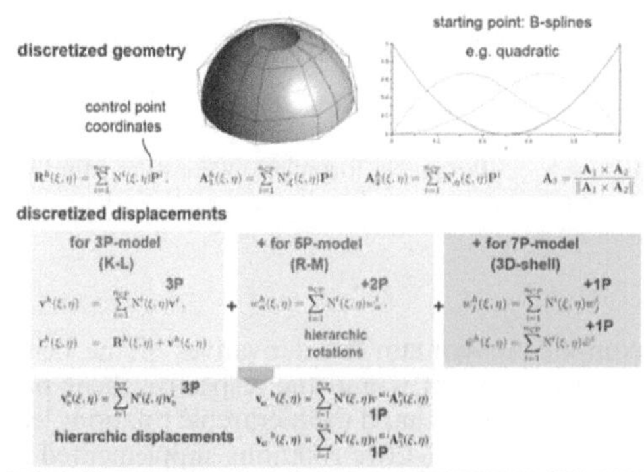

Figure 6. NURBS discretization for Isogeometric Analysis (IGA).

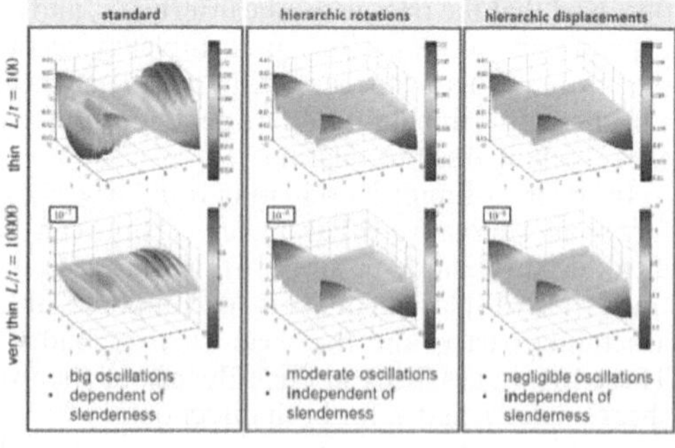

Figure 8. Simply supported plate under uniform load, shear forces q_{xz}.

Figure 9 shows the stress resultant q_{xz} in a section at y = 5.0. The parameterization with hierarchic rotations still results in oscillations mainly at the boundaries; in contrast these are not present when using hierarchic displacements. The results for the conventional formulation appear as almost vertical lines due to severe oscillatory behavior.

In Figure 10 the Scordelis–Lo roof often used as benchmark example is investigated applying the 5P-formulation with hierarchic rotations. The rather poor convergence behaviour for the conventional as well as the hierarchic displacement parameterization is due to a severe membrane locking, still present in this formulation. However if a Hybrid Stress (HS) approach is applied using linear shape functions for all membrane forces

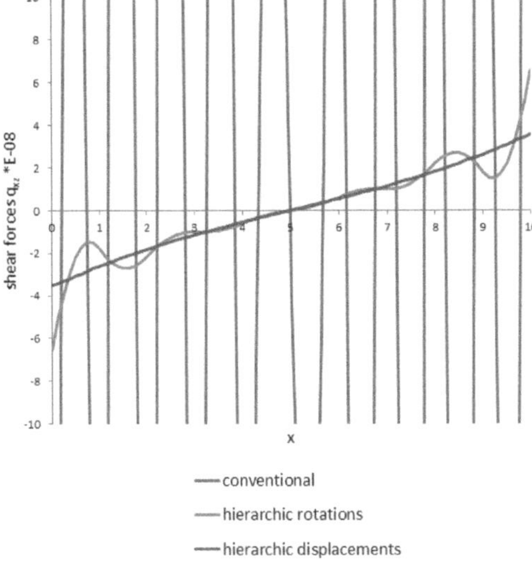

Figure 9. Simply supported plate under uniform load, shear forces q_{xz}, section at y = 5.

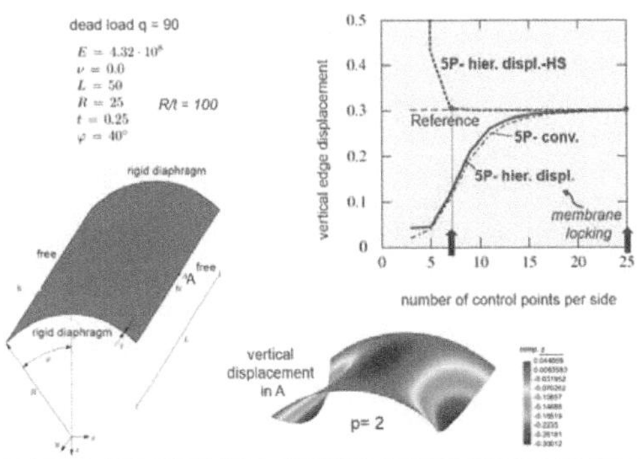

Figure 10. Scordelis-Lo roof.

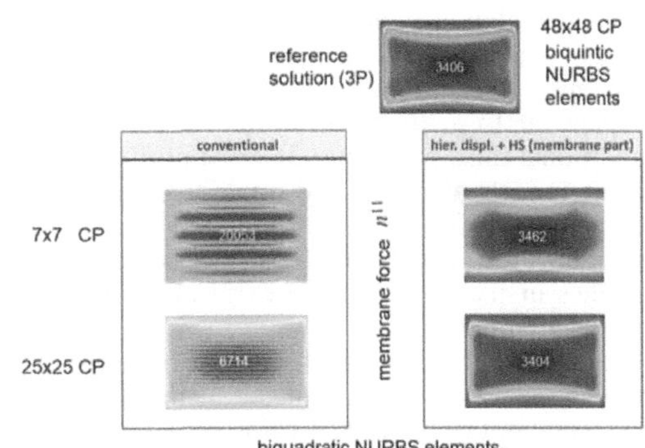

Figure 11. Scordelis-Lo roof, membrane forces n^{11} (hoop forces).

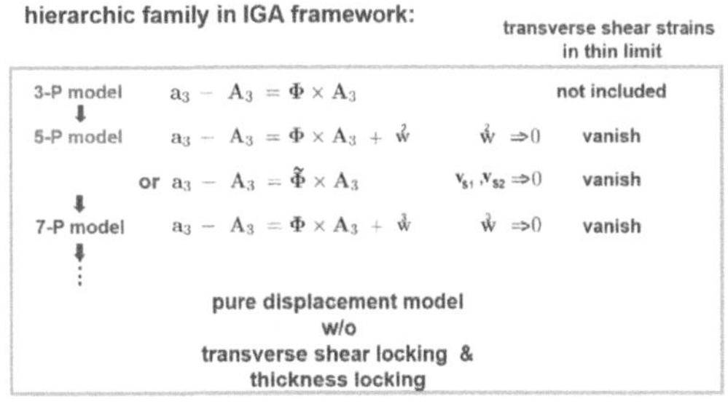

Figure 12. Hierarchic family of shell formulations, kinematics of director difference.

both locking phenomena are removed. It is mentioned that alternatively also the Discrete Strain Gap (DSG) method could successfully be utilized to omit membrane locking; references for HS and DSG methods are given in Echter et al. (2013).

The quality of the membrane forces n^{11} in hoop direction is investigated in Figure 11; the results of a very fine mesh with 48 × 48 Control Points (CP) applying biquintic shape functions are used as reference with a maximum value of 3406. Two meshes with 7 × 7 and 25 × 25 CP (marked in Figure 10 by big arrows) applying biquadratic NURBS functions are analyzed for the conventional and the hierarchic displacement parameterization; the latter one is enhanced by HS for the membrane part. Although the displacements in the fine mesh for the conventional formulation converged, the membrane forces are too big and oscillate. In contrast the 'hier.displ.+HS' formulation shows acceptable results even for the coarse mesh.

3 CONCLUSIONS

Figure 12 summarizes the kinematics of the director difference $\mathbf{a}_3 - \mathbf{A}_3$ for the hierarchic family of shell formulations. The K-L formulation, its extensions to a hierarchic R-M model with the two incremental rotations or displacements and the expansion to a 3D-model where **w** contains an additional third parameter describe pure displacement models avoiding any transverse shear and thickness locking on the formulation level before discretization. The family constitutes beam or shell models in the IGA framework ideally suited for model adaptation.

The ideal objectives as defined in the introduction are satisfied except that the membrane locking could not be removed a-priori in the sense of a primal formulation. However this goal is worth to be further investigated. Additional research is also necessary for an adequate formulation of boundary conditions and the extension to geometrical nonlinear response.

REFERENCES

Bischoff, M., Wall, W.A., Bletzinger, K.-U. & Ramm, E. 2004. Models and Finite Elements for Thin-Walled Structures. In E. Stein et al. (eds), *Encyclopedia of Computational Mechanics,* Vol. 2: *Solids and Structures*, 59–137. Chichester: John Wiley & Sons, Ltd.

Benson, D.J., Bazilevs, Y., Hsu, M.C. & Hughes, T.J.R. 2010. Isogeometric shell analysis: The Reissner-Mindlin shell. *Comput. Methods Appl. Mech. Engng.* 199: 276–289.

Echter, R. & Bischoff, M. 2010. Numerical efficiency, locking and unlocking of NURBS finite elements. *Comput. Methods Appl. Mech. Engng.* 199: 374–382.

Echter, R., Oesterle, B. & Bischoff, M. 2013. A hierarchic family of isogeometric shell finite elements. *Comput. Methods Appl. Mech. Engng.* 254: 170–180.

Hughes, T.J.R., Cottrell, J.A., & Bazilevs, Y. 2009. *Isogeometric Analysis: Toward Integration of CAD and FEA.* John Wiley & Sons, Ltd.

Kiendl, J., Bletzinger, K.-U., Linhard, J. & Wüchner, R. 2009. Isogeometric shell analysis with Kirchhoff-Love elements. *Comput. Methods Appl. Mech. Engng.* 198: 3902–3914.

Long, Q., Bornemann, P.B. & Cirak, F. 2012. Shear-flexible subdivision shells. *Int. J. Numer. Meth. Engng.* 90: 1549–1577.

Oesterle, B., Ramm, E. & Bischoff, M. 2015. A shear deformable, rotation-free isogeometric shell formulation. *Comput. Methods Appl. Mech. Engng.* Submitted.

Advances in Mechanics: Theoretical, Computational and Interdisciplinary Issues – Kleiber et al. (Eds)
© 2016 Taylor & Francis Group, London, ISBN 978-1-138-02906-4

Finite Element Method simulations of linear and nonlinear elasticity problems with error control and mesh adaptation

W. Rachowicz
Institute of Computer Science, Cracow University of Technology, Cracow, Poland

A. Zdunek
Aeronautics and System Integration, Swedish Defense Research Agency FOI, Stockholm, Sweden

W. Cecot
Institute for Computational Civil Engineering, Cracow University of Technology, Cracow, Poland

ABSTRACT: The paper presents the main concepts of error estimation and adaptivity for linear and non-linear elasticity problems solved with the Finite Element Method. We discuss possibility of simulations of boundary-value problems in classical linear elasticity and in finite elasticity with nearly incompressible and (possibly) nearly inextensible materials. We present the most popular a posteriori residual error estimation techniques, the goal-oriented error estimation and the rules of adaptivity of meshes.

1 INTRODUCTION: LINEAR ELASTICITY AS A MODEL ELLIPTIC BOUNDARY-VALUE PROBLEM

The classical statement of linear elasticity includes the equilibrium equation, the definition of strains, the Hooke's law and boundary conditions:

$$\begin{cases} -\nabla\cdot\sigma = f \quad \text{in } \Omega, \\ \epsilon = \frac{1}{2}(\nabla u + \nabla^T u), \\ \sigma = 2\mu\epsilon + \lambda \mathrm{tr}(\epsilon) I, \\ u = \hat{u} \quad \text{on } \Gamma^D, \\ \sigma n = \hat{t} \quad \text{on } \Gamma^N, \end{cases} \tag{1}$$

where σ denotes the Cauchy stress tensor, ϵ is the infinitesimal strain tensor, u denotes the field of displacements. Parameters μ and λ are the Lamé constants. The problem is driven by the volume forces f, the Dirichlet data $\hat{u}$, and the Neumann data $\hat{t}$. The Dirichlet and Neumannn boundary conditions are specified on separate parts of the boundary Γ_D and Γ_N. By multiplying the equilibrium equation $(1)_1$ by a smooth test function v, integrating over domain Ω, integrating by parts and using the Neumann boundary condition one obtains the equivalent variational formulation:

$$\begin{cases} \text{Find } u \in V + \hat{u}: \\ a(u,v) = l(v) \quad \forall v \in V, \\ a(u,v) = \int_\Omega [\mu\,\epsilon(u) : \epsilon(v) + \lambda \mathrm{tr}\,\epsilon(u) \mathrm{tr}\,\epsilon(v)] dx, \\ l(v) = \int_\Omega f\cdot v\, dx + \int_{\Gamma^N} \hat{t}\cdot n\, dS, \\ V = \{v \in [H^1(\Omega)]^3 : v = 0 \text{ on } \Gamma^D\}. \end{cases} \tag{2}$$

Under customary assumptions concerning parameters μ and λ and smoothness of the loads one finds that bilinear form $a(u, v)$ and linear functional $l(v)$ are continuous, and that a is V-coercive with respect to the H^1-norm:

$$\begin{aligned} &\exists M > 0: \ |a(u,v)| \le M \|u\|_{1,\Omega} \|v\|_{1,\Omega} \quad \forall u,v \in V, \\ &\exists \alpha > 0: \ a(v,v) \ge \alpha \|v\|^2_{1,\Omega} \quad \forall v \in V \subset V, \\ &\exists c > 0: \ |l(v)| \le c \|v\|_{1,\Omega} \quad \forall v \in V. \end{aligned} \tag{3}$$

2 FINITE ELEMENT APPROXIMATION

We assume that the bounded domain Ω is subdivided into elements K (being hexahedral, tetrahedral or triangular prisms) forming a regular mesh (i.e. adjacent elements share their common faces, edges and vertices). We define a finite element space $V_h \subset V$ which consists of continuous piecewise polynomials on elements. Finite element approximation is defined as a solution of variational formulation (2) in V_h:

Find $u_h \in V_h + \hat{u}$ such that

$$a(u_h, v_h) = l(v_h) \ \ \forall v_h \in V_h. \tag{4}$$

We define the energy norm as $\| v \|_E = a(v,v)^{1/2}$ and the associated dual norm of linear functionals $r(v)$ on V:

$$\| r \|_* := \sup_{v \in V} \frac{|r(v)|}{\| v \|_E}. \tag{5}$$

It is well-known that the dual norm of the residual functional $r(v) := a(u_h, v) - l(v)$ is equal to the energy norm of the error

$$\| u - u_h \|_E = \| r \|_*, \tag{6}$$

and that the error is the solution of the variational problem:

$$a(u - u_h, v) = r(v) \ \ \forall v \in V. \tag{7}$$

3 RESIDUAL ERROR ESTIMATION TECHNIQUES

Relation (6) is the basis of the so-called residual error estimation methods. Obviously an exact solution of the optimization problem (5) for the residual functional is very expensive so that the residual methods are based on approximate evaluation of $\| r \|_*$. Among these methods we find an explicit approach and a few implicit methods which we discuss next.

3.1 *An explicit residual method*

In this technique (Babuška & Rheinboldt 1979) we express the dual norm of the residual (and at the same time the energy norm of the error) as a sum of error indicators, $\| u - u_h \|_E = (\Sigma_K \eta_K^2)^{1/2}$, defined as follows:

$$\eta_K^2 = C\Big\{ h_K^2 \, \| f + \nabla \cdot \sigma \|_{0,K}^2 + \frac{h_K}{4} \| [| t |] \|_{0,\partial K \setminus \partial\Omega}^2 + h_K \| \hat{t} - t_h \|_{0,\partial K \cap \Gamma^N}^2 \Big\}^6 \tag{8}$$

where h_K denotes the size of element K $[|t|] = (\sigma_h^K - \sigma_h^L)n$ is a jump of the flux between neighboring elements K and L, n is the normal to their common face. As we can see the error is expressed by a sum of adequately scaled element L^2-norms of the element residuals: in the equilibrium of volume forces, surface fluxes between elements and on Γ_N.

3.2 *A subdomain residual method*

In this approach we consider a set of patches Ω_i of elements which are supports of vertex shape functions ϕ_i, $\Omega_i = \mathrm{supp}(\phi_i)$, (Babuška & Rheinboldt 1979). For a regular mesh they are found as elements sharing a common i-th vertex. In this technique we solve a sequence of local Dirichlet boundary-value problems on subdomains Ω_i:

Find $\phi_i \in H^1(\Omega_i)$ such that $\phi_i = u_h$ on $\partial\Omega_i$ and

$$a(\phi_i, v) = l(v) \ \ \forall v \in H_0^1(\Omega_i). \tag{9}$$

Then the total error is expressed as follows:

$$\| u - u_h \|_E \le C \left(\sum_{i=1}^{N} \| \phi_i - u_h \|_{E,K}^2 \right)^{1/2}. \tag{10}$$

Since in general exact solution of the continuous problem (9) is not available we solve it in finite element space of the patch Ω_i enriched by raising the order of approximation from p to $p + 1$.

3.3 *An element residual method*

This technique estimates the error considering solutions of local boundary-value problems on separate elements. Moreover, these are problems defined on discrete spaces. The method can be formulated as follows (Bank & Weiser 1985).

For every element K we consider its original finite element space $V_{hp}(K)$ of order p and an auxiliary enriched space of order $p + 1$, $V_{h,p+1}(K)$. We take into account an interpolation operator Π_{hp} corresponding to $V_{hp}(K)$ (this might be, for instance, the Lagrangian interpolation operator). We define its kernel in $V_{h,p+1}(K)$:

$$M_K = \{ v \in V_{h,p+1}(K) : \ \Pi_{hp} v = 0 \}. \tag{11}$$

Next we solve the following discrete boundary-value problem for every element K:

Find $\phi_K \in M_K$ such that

$$a(\phi_K, v) = r_K(v) \ \ \forall v \in M_K, \tag{12}$$

where

$$r_K(v) = \int_K (f + \nabla \cdot \sigma_h) \cdot v \, dx + \int_{\partial K \setminus \partial\Omega} \frac{1}{2} [| \sigma_h n |] \cdot v \, dS + \int_{\partial K \cap \Gamma_N} (\hat{t} - \sigma_h n) \cdot v \, dS \tag{13}$$

is an element contribution to the residual functional. The final estimate is expressed by a sum of energy norms of functions ϕ_K:

$$\| u-u_h \|_E \leq C\left(\sum_K \| \phi_k \|_{E,K}^2 \right)^{1/2}. \tag{14}$$

An essential advantage of this method is that the dimensions of problems (12) grow as $O(p^2)$ as opposed to the growth $O(p^3)$ for majority other methods which is essential especially for higher order approximations. Constant C shows only very moderate growth with p by a factor of 1.2 for $p \leq 10$.

3.4 *A residual method with self-equilibrated fluxes*

This technique was proposed by Ainsworth and Oden (Ainsworth & Oden 1993). The method is considered the most reliable, moreover, it does not use any generic constants. We begin description of the procedure by defining a special split of the residual functional into element contributions:

$$\begin{aligned} r(v) &= a(u_h,v) - l(v) = \sum\nolimits_K [a_K(u_h,v) - l_K(v)] \\ &= \sum\nolimits_K \underbrace{[a_K(u_h,v) - l_K(v) - \lambda_K(v)]}_{r_K(v)} = \sum\nolimits_K r_K(v). \end{aligned} \tag{15}$$

In this representation λ_k denote a specially designed linear functionals which satisfy the consistency condition $\sum_K \lambda_K = 0$. In addition, we assume that they vanish on the trilinear shape functions ψ_K^n of element K.

$$r_K(\psi_K^n) = 0, \; n = 1, \ldots, 3n_v, \tag{16}$$

where n_v is the number of vertices of the element. Since infinitesimal rigid motions belong to $\text{span}_n \psi_K^n$ the condition above guarantees that such motions are in the kernel of r_K. This allows us to solve the local Neumann boundary-value problem:

Find $\phi_K \in V(K)$ such that

$$a(\phi_K, v) = r_K(v) \;\; \forall v \in V_K, \tag{17}$$

(vanishing of r on rigid motions can be interpreted as self-equilibration of residual forces). The definition of ϕ_K allows one to estimate the error as follows:

$$\| u-u_h \| \leq \left(\sum_K \| \phi_K \|_{E,K}^2 \right)^{1/2}. \tag{18}$$

Like in the subdomain residual method in practice we solve problems (17) in the space of enriched element shape functions. The major difficulty in this method is construction of functionals λ_K. The consistency condition implies that they act exclusively on fluxes on interelement boundaries, and they can be expressed by specially designed average values of stresses $t_h = \sigma_h n$ on both sides of the interelement boundary. Such averaging procedures are proposed in (Ladeveze & Mounder 1996).

3.5 *A goal oriented error estimation*

An important issue of engineering practice is to know the accuracy of some parameters characterising a finite element solution. As an example we can mention a displacement or a stress at a selected point or an average value of such parameters in some small subdomain. It turns out that in many cases accuracy of such parameters, often called quantities of interst, q.o.i., can be established based on the residual techniques (Becker & Rannacher 1996). Let us assume that the quantity of interest is a linear and continuous (in the H^1-norm) functional $q(u)$. We define a generalized Green's function $G \in V$ as a solution of the auxiliary boundary-value problem:

Find $G \in V$ such that

$$a(w,G) = q(w) \;\; \forall w \in V. \tag{19}$$

Let us assume that we also find the finite element approximation $G_h \in V_h$ of the Green function G. Then accuracy of q.o.i. can be found as follows:

$$\begin{aligned} &| q(u-u_h) | \\ &= | a(u-u_h,G) | = | a(u-u_h,G-G_h) | \\ &= \sum\nolimits_K | a_K(u-u_h,G) | \\ &\leq \sum\nolimits_K a(u-u_h,u-u_h)^{1/2} a(G-G_h,G-G_h)^{1/2} \\ &= \sum\nolimits_K \| u-u_h \|_{E,K} \| G-G_h \|_{E,K} \approx \sum\nolimits_K \eta_K^u \eta_K^G \end{aligned}$$

where η_K^u, η_K^G are element error indicators for u_h and G_h. In the first step above we used the definition of G, next the orthogonality condition $a(u-u_h,G_h) = 0$ of finite element solutions. Finally, we applied the Cauchy-Schwarz inequality and we replaced the exact energy errors by their estimates.

4 ADAPTIVITY

Solutions of boundary-value problems are frequently irregular which appears by high gradients in some small subdomains. Such concentration of stresses requires adequately fine approximation. At the same time covering the whole domain with very fine mesh is unnecessary and it would be prohibitevely expensive. The idea behind the mesh adaptation is to detect automatically areas with large errors and refine the mesh in such subdomains. Two kinds of local improvements of approximation are available: h-adaptivity based on dividing elements into smaller so-called "elements-sons", and p-adaptivity consisting in local raising of order of approximation. In addition, both operations can be anisotropic and they can be combined together leading to hp-adaptivity.

Presence of singularities degrades the optimal order $O(N^{-p/d})$ of convergence of FEM (N is the number of degrees-of-freedom while d the spatial dimension). It turns out that application of h-adaptivity for singularities recovers the optimal rate. Moreover, appropriate hp-adaptivity leads to exponential convergence $O(e^{-\beta N^{1/3}})$, $\beta > 0$ even for singular solutions. These factors made adaptivity an essential tool in simulations of strongly irregular solutions.

Adaptivity of the h-type was first investigated in a systematic way by (Babuška and Rheinboldt 1978). Their idea is based on the conclusion from Cea's lemma that the approximation error is bounded (within to a constant M/α from (3)) by the interpolation error:

$$\| u - u_h \|_{1,\Omega} \leq \frac{M}{\alpha} \| u - u_h \|_{1,\Omega}. \tag{20}$$

This indicates that constructing an optimal mesh from the point of view of interpolation error allows to obtain a small approximation error. On the other hand optimization from the point of view of interpolation can be relatively easily solved due to the local character of this procedure. Babuška and Rheinboldt proposed to define the optimal mesh in a sense of minimization of the interpolation error with a constraint that the number of elements is fixed:

Find distribution of element sizes $h(x)$ such that

$$\begin{cases} \int_\Omega h^{2p} w(x) dx \longrightarrow \min, \\ \int_\Omega \frac{dx}{h^d} = N, \end{cases} \tag{21}$$

where $w(x)$ corresponds to a sum of derivatives of u being involved in the *a priori* interpolation estimate $|u_I - u|^2_{1,K} \simeq h_K^{2p} \| u \|^2_{p+1,K}$. Simple calculation reveals that optimal mesh is charaterized by equidistribution of errors i.e. ideally errors corresponding to all elements should be equal. This prompts the following feed-back adaptive strategy:

1. Solve the boundary-value problem on the current mesh.
2. Find *a posteriori* error indicators η_K.
3. If $(\Sigma_K \eta_K^2)^{1/2} \leq \mathrm{TOL}$ (or if the resources of the computer were exhausted)—STOP.
4. Otherwise: subdivide the elements whose errors exceed some prescribed percentage α of the maximum error: $\eta_K \geq \alpha \max_L \eta_L$.
5. GO TO 1.

The limit α is often set so that some fixed fraction of elements are refined.

Combined h- and p-adaptivity was a subject of investigation of Ivo Babuška and his co-workers in the 80 s. One of their findings was construction of hp-meshes resulting with exponential convergence of solutions with pointwise singularities (for instance caused by the re-entrant corners). According to this recipe the mesh should consist of layers of elements surrounding the center of singularity, with their size being reduced by a factor of 0.15 when approaching the center, and with the orders p growing by 1 with the increasing number of the layer.

An approach designing the hp-meshes for any kind of irregularities in 1,2 and 3D was proposed by Demkowicz and his co-workers in (Demkowicz, Kurtz, Pardo, Paszynski, Rachowicz, & Zdunek 2008). It is based on investigation of solutions on two meshes: the actual mesh referred to as the solution coarse mesh u_{hp}, and solution $u_{h/2,p+1}$ on the fine mesh obtained from the coarse one by breaking all its elements and raising its orders by one. The feed-back adaptive algorithm is based on the fact that for optimal refinements the reduction of the square of the interpolation error per one newly introduced degree-of-freedom should be the largest possible. A practical implementation of the procedure is as follows:

1. Investigate interpolation error of $u_{h/2,p+1} := u$ on the elements of the coarse mesh with various hypothetical trial refinements.
2. Select refinements resulting in the largest reduction of the square of the interpolation error $\| u - \Pi_{hp} u \|^2_E$ per 1 new dof. (Π_{hp} is an hp interpolation operator).
3. Perform selected refinements on the coarse mesh.
4. Obtain the new fine mesh.
5. Solve the problem on fine mesh.
6. Go to 1.

The "final product" of the algorithm is the solution on the fine mesh so that the essential expense of obtaining this solution is by no means futile.

4.1 *An example*

A hyperboloidal shell is considered whose middle surface is defined as follows:

$$\frac{x^2+y^2}{a^2}-\frac{z^2}{c^2}=1, \qquad (22)$$

with $a=1, c=2$ and $z\in(-3,1)$. The thickness of the shell is $t = 0.05$. The external surface is loaded by a constant horizontal stress $t_x = 1$. To avoid the numeriacal locking due to thin elements the initial order of approximation in the tangential direction is assumed $p = 3$ and for the fine mesh $p = 4$ as the latter value guarantees avoiding numerical locking due to presence of very thin elements.

Figures 1 and 2 present the coarse and the fine mesh after 6 levels of *hp* refinements, and selected components of the solution.

4.2 *Goal-oriented adaptivity*

In some applications one might be interested in obtaining high accuracy of a selected parameter characterizing the solution. Assuming that the parameter is given as a linear and continuous functional, and accepting a heuristic assumption that the element approximation errors are reduced with the same rate as the element interpolation errors we may glean from the goal oriented estimate that possibly fast reduction of the error of q.o.i. is obtained if the elements with the largest contributions $\eta_K^u\eta_K^G$ are refined.

Considering the *hp*-adaptive algorithm one finds that in order to make it goal-oriented, i.e. focused on fast reduction of the error of q.o.i., we must decide to perform the refinements which instead of reducing the square of the error $|u-\Pi_{hp}u|_{1,K}^2$ reduce the product $|u-\Pi_{hp}u|_{1,K}|G-\Pi_{hp}G|_{1,K}$ with the fastest rate (Demkowicz, Kurtz, Pardo, Paszynski, Rachowicz, & Zdunek 2008). The remaining principles of the *hp*-adaptive algorithm remain unchanged.

4.3 *A numerical example*

We consider a hyperboloidal shell as before. We define q.o.i. as average values of selected internal forces of the shell:

$$q(u)=\int_{D(\Delta S)} w\sigma_{ij}\zeta^k dx, \qquad (23)$$

where *i, j* denumerate the local shell directions, ΔS is a small part of the middle surface and $D(\Delta S)$ the corresponding 3D part of the shell. Parameter ζ is the distance of the current point from the middle surface, $w(\xi)$ is a weight function of shell coordinates ξ such that $\int_{D(\Delta S)} wdx=1$. We can observe that selection of *i, j* and k = 1 corresponds to appropriate component of the bending or twisting moment while the choice of $k = 0$ leads to a membrane force. In our test we choose:

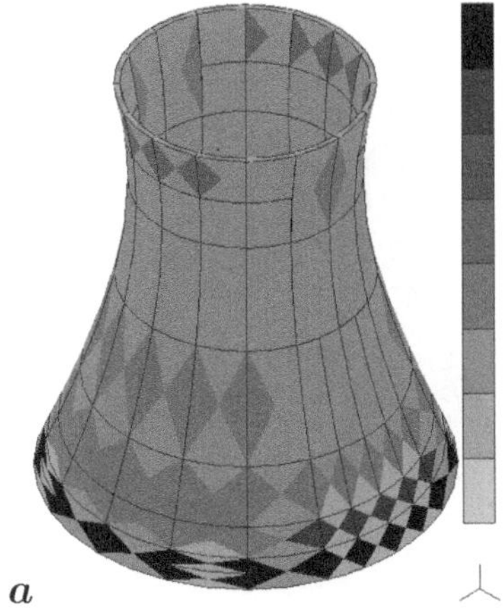

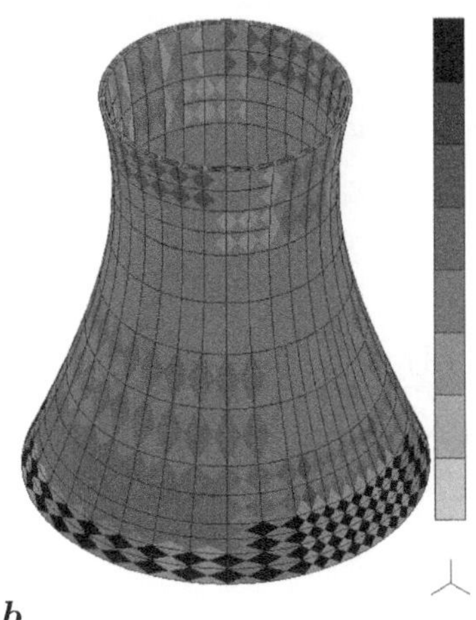

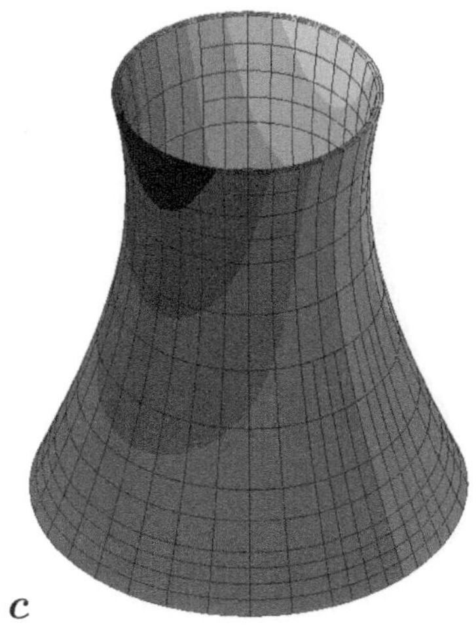

Figure 1. Energy based adaptivity: a) coarse and b) fine *hp*-adaptive meshes, c) displacements u_r.

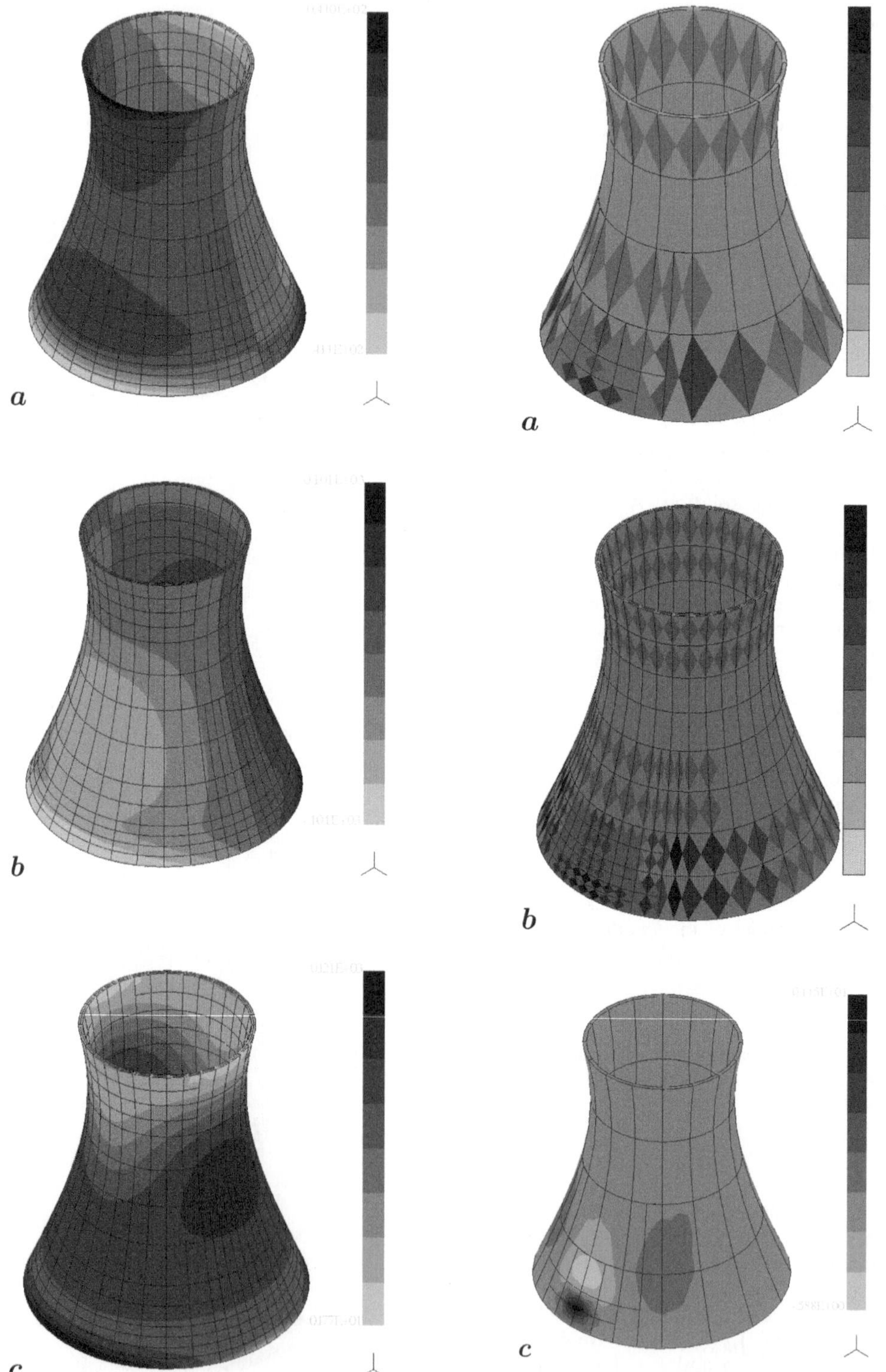

Figure 2. Energy based adaptivity. Contour plots of solution: a) stresses $\sigma_{\phi\phi}$, b) stresses σ_{zz}, c) effective stress σ_0.

Figure 3. Goal-oriented adaptivity for bending moment m_{11}. a) coarse mesh, b) fine mesh, c) displacements u_r of the Green function.

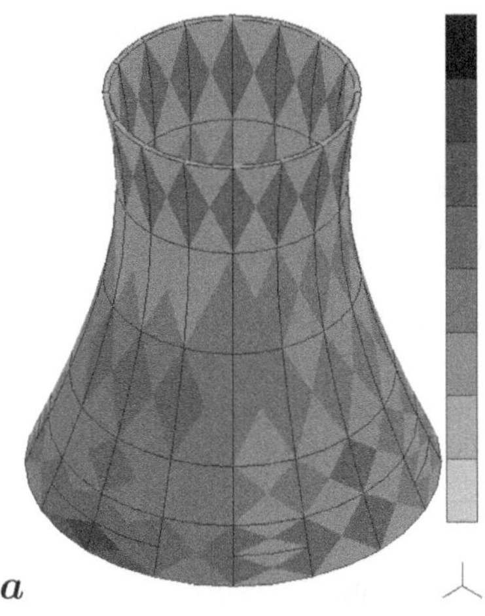

a

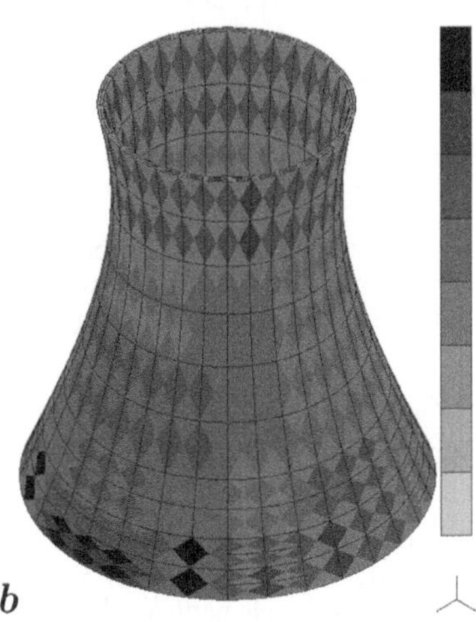

b

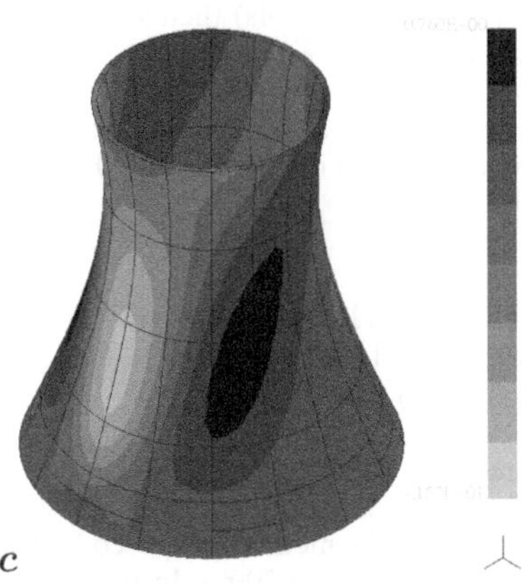

c

Figure 4. Goal-oriented adaptivity for membrane force n_{11}. a) coarse mesh, b) fine mesh, c) displacements u_r of the Green function.

$i, j = 1,\ k = 1:\ q(u) = (m_{11})_{av}$, bending moment
$i, j = 1,\ k = 0:\ q(u) = (n_{11})_{av}$, membrane force.

(index $i = 1$ corresponds to meridional while $i = 2$ to circumferential direction). Figures 3 and 4 shows the coarse and fine meshes and selected characteristics of the Green functions. The values of q.o.i. are $m_{11} = -0.151 \pm 0.7\%$ and $n_{11} = -5.03 \pm 0.06\%$.

5 FORMULATION OF FINITE ELASTICITY BOUNDARY-VALUE PROBLEM

We introduce the standard notions of Lagrange description. Let X denote location of a point in the reference configuration and $x(X, t)$ its location in the deformed body at time instant t. We define the deformation gradient F, the right Cauchy-Green tensor C and the volume ratio J:

$$F = \frac{\partial x}{\partial X},\ C = F^T F,\ J = \det(F). \tag{24}$$

We consider a possibility of reinforcing the material by up to two families of fibres with material directions $M_i, i = 1, 2, |M_i| = 1$ which define the corresponding structural tensors $A_i = M_i \otimes M_i$. A multiplicative decomposition of deformation gradient and Cauchy-Grren tensor is introduced:

$$\bar{F} J^{-1/3} F,\ \bar{C} = J^{-2/3} C. \tag{25}$$

We consider a set of invariants of C and $\bar{C}$ and auxiliary invariants associated with A_i:

$$\begin{aligned} I_1 &= tr(C), & \bar{I}_1 &= J^{-2/3} I_1 \\ I_2 &= \frac{1}{2}[(tr(C)^2 - tr(C)^2], & \bar{I}_2 &= J^{-1/3} I_2, \\ I_{4i} &= C : A_i, & \bar{I}_{4i} &= J^{-2/3} I_{4i}. \end{aligned} \tag{26}$$

The strain energy function per unit volume is expressed as follows:

$$\Psi = \Psi_{vol}(J) + \Psi_{iso}(\bar{I}_1, \bar{I}_2) + \sum_{i=1}^{2} \Psi_{fi}(\bar{I}_{4i}), \tag{27}$$

where the first component corresponds to volumetric response, the remaining ones to isochoric response, including the last term reflecting possible anisotropy associated with the presence of reinforcing fibres. A particular form of these components that is used for instance in biomechanics of soft tissues is as follows:

$$\Psi_{vol} = \frac{K}{4}(J-1)^2,$$
$$\Psi_{iso} = \frac{\mu}{2}(\bar{I}_1 - 3),$$
$$\Psi_{fi} = \frac{k_{1i}}{2k_{2i}}\left\{\exp[k_{2i}(\bar{I}_{4i}-1)^2]-1\right\}. \quad (28)$$

The form of the strain energy implies the constitutive relation expressing the second Piola-Kirchhoff stress tensor:

$$S = 2\frac{\partial \Psi}{\partial C} = -pJC^{-1} + 2J^{-2/3}\{\Psi_{iso,1}Dev[I] + \Psi_{iso,2}Dev[\bar{C}^{-2}] + \sum_{i=1}^{2} W_{fi,4}Dev[A_i]\}. \quad (29)$$

where $p = \Psi'_{vol}$ is the pressure. We focus our attention on nearly incompressible materials which is manifested by large values of the ratio K/μ, typically of the order of 10^3. In such situation it is useful to apply a mixed approximation of the solid mechanics problem with displacement $u = x - X$ and pressure p being separate variables. The formulation involves the principle of virtual displacements expressing the equilibrium and the weak enforcement of the constitutive relation for pressure:

$$p = \frac{K}{2}(J-1), \quad (30)$$

and it can be stated as follows:

Find $u \in V + \hat{u}$ and $p \in Q$ such that

$$\begin{cases} \int_{\Omega_0} S : E'(\phi,v)dX = \int_{\Omega_0} b \cdot v\, dX + \int_{\Gamma_0^N} (F^T v)\cdot \hat{t}\, dA, \forall v \in V, \\ \int_{\Omega_0} -\left(\frac{2}{K}p - (J-1)\right) q\, dX = 0, \forall q \in Q, \end{cases} \quad (31)$$

where $E'(\phi,v) = 1/2(F^T \nabla_X v + \nabla_X^T v F)$ and $\hat{u}$ is the data of the Dirichlet boundary condition on Γ_D while $\hat{t}$ is the prescribed load on the Neumann part of the boundary Γ_N, b denotes the force per unit volume in the reference configuration. The functional spaces for the solution and test functions are taken as

$$V = \{v \in (H^1\Omega)^3 : v = 0 \text{ on } \Gamma_D\},$$
$$Q = L^2(\Omega). \quad (32)$$

The formulation above is nonlinear and it is solved with the Newton-Raphson method. The linearization of (31) takes the following form:

$$\begin{cases} \int_{\Omega} [E'(\phi,v) : IC : E(\phi,u) + \nabla_X^T v S \nabla_X^T u]dX - \int_{\Omega} \nabla_X \cdot vq\, dX = -R_1(\phi,p;v) \\ \int_{\Omega_0} \pi[-J\nabla_X \cdot u + 2/Kq]dX = -R_2(\phi,p,\pi) \end{cases} \quad (33)$$

where we introduced elasticities

$$IC = 2\frac{\partial S}{\partial C} \quad (34)$$

and the residuals:

$$R_1(\phi,p,;v) = \int_{\Omega_0} S : E'(\phi,v)dX - \int_{\Omega_0} Jb \cdot v dX - \int_{\Gamma_0^N} (F^T v)\cdot SN\, dA,$$
$$R_2(\phi,p;\pi) = \int_{\Omega_0} -\left(\frac{2}{K}p - J\right)\pi\, dX. \quad (35)$$

To discretize the saddle point problem (33) we use the Q^p/P^{p-1} family of hexahedral elements following Simo and Taylor (Simo & Taylor 1991) that is

$$u \simeq u_h \in V_h \subset H^1(\Omega), \quad V_h(K) = Q^p(K),$$
$$p \simeq p_h \in Q_h \subset L^2(\Omega), \quad Q_h(K) = P^{p-1}(K). \quad (36)$$

where p is the order of approximation. Identifications above are valid for elements K being affine images of the pattern element $\hat{K} = [0,1]^3$. If isoparametric mapping $T_K : x(\xi) = \sum_{i=1}^{ndof} \hat{\psi}(\xi)x^i$ is used (where $\hat{\psi}_i$ are the shape functions on master element, x^i geometrical dofs) then spaces $Q^p(K)$ are replaced by $T_K^{-1}(Q^p(\hat{K}))$.

It should be mentioned that since approximation of pressure is discontinuous we can perform static condensation of the pressure dofs on the element level.

6 ERROR ESTIMATION FOR FE APPROXIMATION OF FINITE ELASTICITY

Residual error estimation techniques are also applicable for nonlinear formulation of finite elasticity. We outine here the method proposed by Ruter and Stein (Rüter and Stein 2000). In this approach we define an error function $\psi \in V$ corresponding to the residual functional of the principle of virtual displacements:

$$\int_{\Omega} \nabla \psi \cdot \nabla v \, dx = R(v), \tag{37}$$

where

$$R(v) = \int_{\Omega_0} b \cdot v \, dX + \int_{\Gamma_0^N} (F^T v) \cdot \hat{t} \, dA - \int_{\Omega_0} S : E'(\phi_h, v) dX. \tag{38}$$

According to findings of Rüter and Stein the total error of FE approximation is estimated as follows:

$$|u - u_h|_{1,\Omega} + \| p - p_h \|_{0,\Omega} \leq \left(|\psi|_{1,\Omega} + \| (J-1) - \frac{2}{K} p_h \|_{0,\Omega} \right), \tag{39}$$

Estimation of the H^1-seminorm of ψ is possible without solution of (37). It can be done using one of the residual techniques discussed before since these methods estimate the dual norm of the residual and, at the same time, the energy norm (here H^1-seminorm) of the solution corresponding to $R(v)$. In this work we use the element residual method of Bank and Weiser to estimate $|\psi|_{1,\Omega}$.

7 A NUMERICAL EXAMPLE: PRESSURIZATION OF A TUBE

We test the error estimate and the h-adaptive procedure on a benchmark problem proposed by Yosibash and Priel. We consider a cylindrical tube with a structure resembling a typical human artery. It consists of two layers: internal media and external adventitia. The height of the tube is $h = 10$, the radii $r_1 = 3.317$, $r_2 = 3.8103$, $r_3 = 4.052$. The parameters characterising the materials of the layers are collected in Table 1. The tube is clamped at the bottom and subject to vertical no-penetration condition at the top. It is pressurized with the internal pressure $p = 13.3$. We consider only 1/4 of the tube exploiting the fact that the solution is axisymmetric. Seven levels of h-adaptation were applied to obtain the final solution which is depicted in Figure 5. We also show the convergence history of the adaptive process comparing it with convergence of the smooth solution: pressurization of the tube with no-penetration boundary conditions at the bottom and at the top. It shows that both rates of convergence are equal which is an experimental confirmation of the fact known from the linear analysis that h-adaptivity recovers the rate of convergence of smooth solutions.

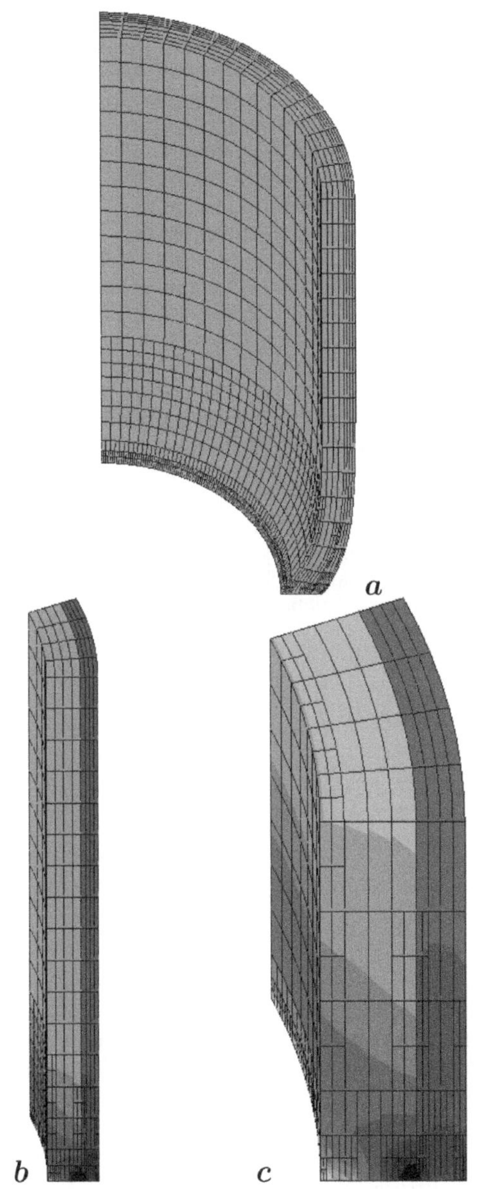

Figure 5. Pressurization of tube: a) displacements u_r on deformed configuration, b) pressure, c) pressure—detail at the bottom.

8 A NUMERICAL EXAMPLE: ANGIOPLASTY

As the next example we present the numerical simulation of the medical procedure of angioplasty with stent. This procedure is aimed at the treatment of the medical condition of artherosclerosis. This vascular decease consists in narrowing of the lumen of the artery due to plaque deposits which slow down the blood flow. The procedure of angioplasty consists in mechanical broadening of the lumen in the deceased area by a balloon (catheder) inserted into the artery and pressurized with fluid filling the balloon. In addition, to make the increase of the lumen permanent, the balloon is surrounded by a stent—a structure of thin metallic struts which deforms increasing its diameter according to deformation of the balloon. Large widening of the stent causes its plastic deformation which keeps the increased radius of the device and so it keeps widened the artery after removing the pressure and the balloon. The idea is illustrated in Figure... We consider a simplified model of the deceased artery as a cylidnrical tube of hight $h = 10$ and internal and external radii $r_1 = 1.5$ and $r_2 = 3.0$ [mm], respectively. The narrowing is modeled as a toroidal bump of the radial dimension $g = 0.28$ and vertical size $l = 3.0$ [mm]. In our simple model we assume that the artery is made of a homogeneous hyperelasic material characterized by the following strain energy function

$$\Psi = U(J) + \bar{\Psi}(\bar{I}_1)$$

$$U = \frac{1}{2}\kappa(J-1)^2$$

$$\bar{\Psi} = \frac{\mu}{2}(\bar{I}_1 - 3) + \frac{k_1}{2k_2}\left\{\exp\left[k_2\left(\bar{I}_1 - 3\right)^2\right] - 1\right\}$$

$$\kappa = 3.0\ MPa$$

$$\mu = 0.066\ MPa$$

$$k_1 = 2.140\ MPa,\ k_2 = 277.6[-] \tag{40}$$

The balloon is a tube of internal and external radii $r_1 = 0.8$ and $r_2 = 1.0$ and height $h = 12$ [mm]. It is made of a homogeneous anisotropic material enforced with circumferential and longitudinal fibres with the strain energy:

$$\bar{\Psi}(\bar{C}, A_1, A_2) = \bar{\Psi}_{iso}(\bar{I}_1) + \bar{\Psi}_{aniso}(\bar{I}_{4,i}),\ i = 1, 2$$

(1—**vertical**, 2 circumferential)

$$\bar{\Psi}_{iso} = \frac{\mu}{2}(\bar{I}_1 - 3)$$

$$\bar{\Psi}_{aniso} = \sum_{i=1,2} \frac{d_{1,i}}{n}(\bar{I}_{4,i} - d_{2,i})^2$$

$$\mu = 100\ N/mm^2,\ d_{2,1} = 1,$$

$$d_{2,2} = 2.25,\ d_{1,i} = 500\ N/mm^2,\ n = 3 \tag{41}$$

Finally, in our test we modeled the stent neglecting its plastic properties, assuming that it is made of a hyperelastic material with parameters

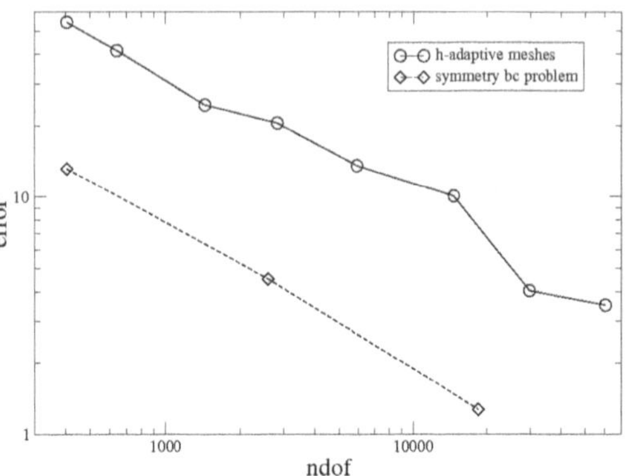

Figure 6. Convergence of estimated error on adaptive and uniform meshes (rates 0.54 ÷ 0.60 and 0.5, resp.); comparison with smooth solution of a tube with a sliding bottom (rate = 0.62).

$K = 167000, \mu = 1546$ [MPa]. The structure of struts of the stent is shown in Figure 7. It consists of 7 typical cells in circumferential direction and 4 cells in vertical direction. The thickness of struts is $t = 0.1$ mm.

The problem under consideration obviously involves the phenomenon of contact: between the balloon and stent, stent and artery, and possibly the balloon and artery. We use a mortar, Segment-To-Segment (STS) technique (Puso & Laursen 2004) to enforce no-penetration condition between the bodies in contact. This technique has an essential advantage over popular Node-To-Segment (NTS) approach in the context of the adaptive FEM. Namely, enforcing no-penetration of nodes of one surface into the second surface in case it is covered by essentially finer mesh than the first mesh resembles enforcing pointwise Dirichlet boundary condition and it results in artificial large gradients on the surface with fine mesh, thus it causes problems with numerical treatment. Unlike, the mortal method is using a weak condition of no-penetration with smooth approximation of contact forces which practically does not impose any restrictions concerning densities of meshes. A disadvantage of the method is a complex integration procedure over pairs of elements of surfaces in contact.

We also use the Primal-Dual Active Set Strategy (PDASS) advocated by Popp et al. (Popp, Gitterle, Gee, & Wall 2010), to search for nodes being in contact. This approach allows one to express the standard inequality constraints describing no-penetration by an equivalent equality constraint thus making it possible to include this condition in the regular Newton-Raphson solution procedure.

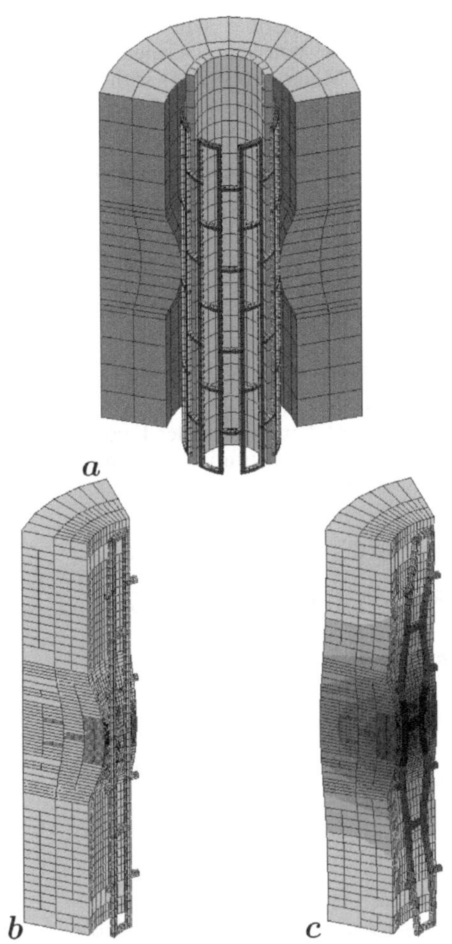

Figure 7. An idea of angioplasty. a) artery, balloon and stent, b) artery and stent at initial configuration, and c) at deformed configuration.

The error estimation method discussed in Sec. 3.3 was generalized for contact problems. The modification consists in using the contact forces as additional exterior loads acting on surfaces of the body in contact. This includes also the complex mortar integrating procedure over non-matching sides of elements of both surfaces.

We performed the simulation of angioplasty using the internal pressure $p = 8 \cdot 10^{-4}$ [kPa]. We started from the initial mesh as shown in Figure 7. Due to symmetry of the solution we ran computation only for 1 of 7 circumferential sections of the structure adding appropriate no-penetration conditions on symmetry faces. Three levels of h-refinements were performed. We present the selected characteristics of the solution, contour maps on the deformed configuration in Figure 8 Location of the internal surface of the artery before and after deforming the balloon is presented in Figure 9.

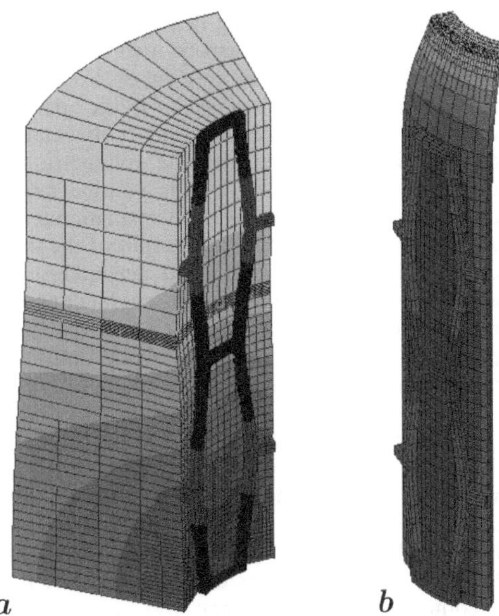

Figure 8. Angioplasty: a) contours of u_r on deformed configuration of artery, b) u_r on ballon.

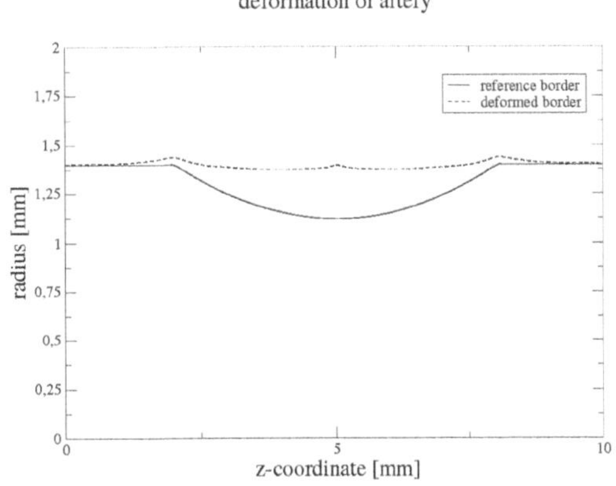

Figure 9. Deformation of artery wall for 4 kinds of artery bumps and available balloon expansions.

9 NEARLY INCOMPRESIBLE AND NEARLY INEXTENSIBLE MATERIAL

Hyperelastic materials reinforced with fibres may becom almost inextensible in directions of the fibres because of the exponential growth of their stiffness. This can cause difficulties in approximation similar as one encounters modeling nearly incompressible materials. A remedy that can be used to deal with this problem was proposed in (Zdunek, Rachowicz, & Eriksson 2014). and it is similar to the approach for almost incompressible materials. In that case we decompose deformation into the volumetric and isochoric parts expressed by the volume ratio J and the unimodular part of the deformation tensor $\bar{F}$. Then we postulate an additive split of the strain energy function into components depending on J and $\bar{C} = \bar{F}^T\bar{F}$, and we introduce separate approximation for pressure p and J. In an analogy to this procedure one can propose to further decompose $\bar{F}$ into unimodular stretch along direction of fibre M:

$$\bar{f}_{ext}(\bar{\lambda},a) := \bar{\lambda}^{-1/2}(I-a)+\bar{\lambda}a$$

where $\bar{\lambda} := (\bar{M}\cdot \bar{F}^T\bar{F}M)^{1/2}$ is the isochoric fibre stretch, $a = m\otimes m$ the spatial structural tensor, and $m = \bar{\lambda}^{-1}FM$ denotes spatial direction of fibres, and the remaining stretch-less deformation. The final decomosition of F takes form:

$$\tilde{F}(\phi,\tilde{J},\tilde{\bar{\lambda}}) = \tilde{J}I\circ \bar{f}_{ext}(\tilde{\bar{\lambda}},a)\circ f_{ext}^{-1}(\bar{\lambda},a)\circ \bar{F}$$

where we introduced a possibility of separate approximation of variables $\tilde{J} \doteq J$ and $\tilde{\bar{\lambda}} \doteq \bar{\lambda}$. We denote $F = \bar{f}_{ext}^{-1}(\bar{\lambda},a)\bar{F}$. The split implies the following definition of the right Cauchy-Green tensor

$$\tilde{C}(\bar{C},\tilde{J},\tilde{\bar{\lambda}}) := \tilde{F}^T\tilde{F} = \tilde{J}^{2/3}q^{-1}[\bar{C}-(1-q^3)\bar{\lambda}^{-1}\bar{C}A\bar{C}],$$
$$q = \tilde{\bar{\lambda}}/\bar{\lambda},$$

and its "no-stretch" version:

$$\overset{\nrightarrow}{C} := \overset{\nrightarrow}{F}^T\overset{\nrightarrow}{F} = \bar{\lambda}[\bar{C}-(1-\bar{\lambda}^{-3})\bar{\lambda}^{-2}\bar{C}A\bar{C}].$$

One can show that the representation of kinematics introduced above leads to the definition of internal forces conjugate to $\tilde{J}$ and $\bar{\lambda}$ which are the pressure and a quantity ϱ proportional to the Kirchhoff fibre stress τ^+:

$$\varrho = \frac{3}{2}J^{1/3}\lambda^{-3}Dev[S]CAC =: \frac{3}{2}\bar{\lambda}\tau^+$$

In addition, the following constitutive relations can be obtained:

$$\begin{cases} S = -\bar{p}JC^{-1} + J^{-3/2}Dev\left[\varrho\bar{\lambda}^{-1}A + \tilde{S}\dfrac{\partial\tilde{C}}{\partial\bar{C}}\right], \\ p = -\dfrac{\partial\Psi(\tilde{C})}{\partial\tilde{J}} \\ \varrho = \dfrac{\partial\Psi(\tilde{C})}{\partial\tilde{\bar{\lambda}}} \end{cases}$$

As an example of a material falling into this description we consider a Generalized Standard Reinforced Material (GSRM) with the following strain energy function:

$$\Psi_{\text{GSRM}}(\tilde{J},\tilde{\bar{\lambda}},\overset{\nrightarrow}{C};A) = \kappa\bar{\Psi}_{vol}(\tilde{J}) + \mu\Gamma\bar{\Psi}_{ext}(\tilde{I}_4(\tilde{J},\tilde{\bar{\lambda}})) + \mu\bar{\Psi}_{iso}(\overset{\nrightarrow}{I}_1),$$
$$\bar{\Psi}_{vol}(\tilde{J}) = \tilde{J}-1-\ln\tilde{J},$$
$$\bar{\Psi}_{ext}(\tilde{I}_4) = \tfrac{1}{2}(\tilde{I}_4-1)^2,$$
$$\bar{\Psi}_{iso}(\overset{\nrightarrow}{I}_1) = \tfrac{1}{2}(\overset{\nrightarrow}{I}_1-3).$$

where $\overset{\nrightarrow}{I}_1 = tr[\overset{\nrightarrow}{C}]$ and parameter Γ controls level of anisotropy. The constitutive relations corresponding to Ψ are as follows:

$$\begin{cases} S = -\bar{p}JC^{-1} + J^{-3/2}Dev[\varrho\bar{\lambda}^{-2}A + \bar{S}_{iso}] \\ \bar{p} = -\kappa\bar{\Psi}'_{vol}(\tilde{J}) - \mu\Gamma\dfrac{2}{3}\tilde{J}^{-1}\tilde{I}_4\bar{\Psi}'_{ext}(\tilde{I}_4), \\ \varrho = 2\mu\Gamma\tilde{J}^{2/3}\tilde{\bar{\lambda}}^2\,\bar{\Psi}'_{ext}(\tilde{I}_4), \end{cases}$$

where

$$\bar{S}_{iso} = \bar{\lambda}[\overset{\nrightarrow}{S} - \iota A + 2(1-\bar{\lambda}^{-3})(\bar{\lambda}^{-4}\bar{C}A\bar{C}:\overset{\nrightarrow}{S})A - \tfrac{1}{2}\bar{\lambda}^{-2}(\overset{\nrightarrow}{S}\bar{C}A + A\bar{C}\overset{\nrightarrow}{S}))]$$

and

$$\iota = \tfrac{3}{2}\bar{\lambda}^{-4}(\bar{C}Dev[A]\bar{C}:\overset{\nrightarrow}{S}),\quad \overset{\nrightarrow}{S} := 2\frac{\partial\Psi_{iso}(\overset{\nrightarrow}{I}_1)}{\partial\overset{\nrightarrow}{C}}/ = \mu I$$

A 5-field Hu-Washizu mixed formulation for this material was developed. It involves:

Equilibrium:

$$G_U(\phi)[\delta U] = \int_{\Omega_0} E'(\phi,\delta U):S\,dV + \int_{\Omega_0}\delta U\cdot\rho_0 B\,dV + \int_{\Gamma_0^N}\delta U\cdot\bar{P}\,dA = 0,$$
$$\forall\delta U\in V.$$

Definition of $\tilde{J}$ and $\tilde{\bar{\lambda}}$:

$$G_{\tilde{J}}(U,\tilde{J})[\delta\pi] = \int_{\Omega_0}(J-\tilde{J})\delta\pi\,dV = 0,\quad \forall\delta\pi\in Q,$$

$$G_{\tilde{\bar{\lambda}}}(U,\tilde{\bar{\lambda}})[\delta\varrho] = \int_{\Omega_0}(\bar{\lambda}-\tilde{\bar{\lambda}})\delta\varrho\,dV = 0,\quad \forall\delta\varrho\in Q,$$

Constitutive relations:

$$G_\pi(U,\tilde{\bar{\lambda}},\tilde{J},p)[\delta\tilde{J}] = \int_{\Omega_0}\left(p + \frac{\partial\Psi(\tilde{C})}{\partial\tilde{J}}\right)\delta\tilde{J}\,dV = 0,$$

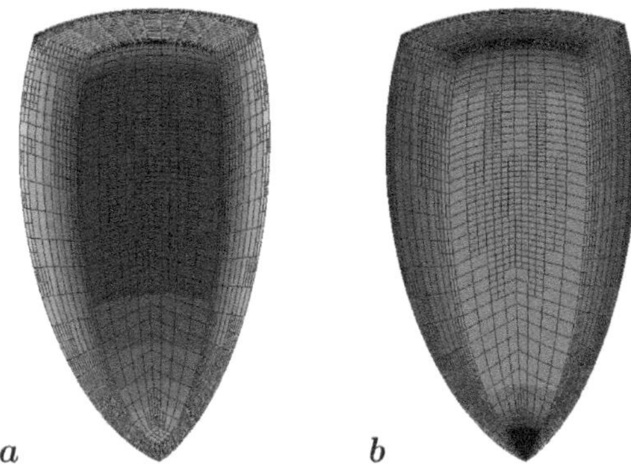

Figure 10. Anisotropic material: a) u_r, b) pressure p.

$$G_{\varrho}(U,\bar{\tilde{\lambda}},\tilde{J},\varrho)[\delta\bar{\tilde{\lambda}}] = \int_{\Omega_0} \left(\varrho - \frac{\partial\Psi(\tilde{C})}{\partial\bar{\tilde{\lambda}})} \right) \delta\bar{\tilde{\lambda}} \, dV = 0,$$

The formulation is subject to linearization.

As a simple example we present pressurization of an ellipsoidal body whose top part was sliced and clamped. The semi-axes of the internal and external boundary are assumed as $a_1 = 18.3$, $b_1 = 6.9$, $a_2 = 19.3$, $b_2 = 12.7$, [mm]. Parameters defining the strain energy: $\Gamma = 10$, $\mu = 0.1$ [MPa], $\kappa = 1000\mu$. We chose circumferential direction of fibres. Internal pressure is $p = 13.3$ [kPa]. Figure 10 presents the solution obtained after performing 6 levels of mesh adaptation.

ACKNOWLEDGEMENTS

W. Rachowicz and W. Cecot acknowledge the financial support via grant No. UMO-2011/01/B/ST6/07306 received from the Polish National Center of Science.

REFERENCES

Ainsworth, M. & J. Oden (1993). A unified approach to a posteriori error estimation using element residual methods. *Numer. Math 65*, 23–50.

Babuška, I. & W. Rheinboldt (1978). A posteriori error estimates for the finite element method. *Int. J. Numer. Meth. Engrg. 12*, 1597–1615.

Babuška, I. & W. Rheinboldt (1979). Analysis of optimal finite element meshes in r^1. *Math. Comp. 33*, 435–463.

Bank, R. & A. Weiser (1985). Some a posteriori error estimates for elliptic partial differential equations. *Math. Comp. 44*, 283–301.

Becker, R. & R. Rannacher (1996). A feedback approach to error control in finite element methods: Basic analysis and examples. *EAST-West J. of Numer. Math. 4*, 237–264.

Demkowicz, L., J. Kurtz, D. Pardo, M. Paszynski, W. Rachowicz, & A. Zdunek (2008). *Computing with hp-adaptive finite elements, vol 2, Frontiers: Three Dimensional Elliptic and Maxwell Problems with Applications*. Boca Raton, London, New York: Chapman and Hall/CRC.

Ladeveze, P. & E. Mounder (1996). A general method for recovering equilibrating element tractions. *Comput. Meth. Appl. Mech. Engng. 137*, 111–151.

Popp, A., M. Gitterle, M. Gee, & W. Wall (2010). A dual mortar approach for 3d finite deformation contact with consistent linearization. *Int. J. Numer. Meth. Engng. 83*, 1428–1465.

Puso, M. & T. Laursen (2004). A mortar segment-to-segment contact method for large deformation solid mechanics. *Comp. Meth. Appl. Mech. Engng. 193*, 601–629.

Rüter, M. & E. Stein (2000). Analysis, finite element computation and error estimation in transversely isotropic nearly incompressible finite elasticity. *Comput. Meth. Appl. Mech. Engng. 190*, 519–541.

Simo, J. & R. Taylor (1991). Quasi-incompressible finite elasticity in principal stretches. *Comput. Meth. Appl. Mech. Engng. 85*, 273–310.

Zdunek, A., W. Rachowicz, & T. Eriksson (2014). A novel computational formulation for nearly incompressible and nearly inextensible finite hyperelasticity. *Comput. Meth. Appl. Mech. Engng. 281*, 220–241.

Advances in Mechanics: Theoretical, Computational and Interdisciplinary Issues – Kleiber et al. (Eds)
© 2016 Taylor & Francis Group, London, ISBN 978-1-138-02906-4

Physics and computations of turbulent dispersed flows: Macro-consequences from micro-interactions

A. Soldati
Department of Electrical, Management and Mechanical Engineering, University of Udine, Udine, Italy

ABSTRACT: We use Direct Numerical Simulations of turbulence and Lagrangian Particle Tracking to elucidate the physics of inertial particles motion in different turbulence instances and we provide insights to model and simulate turbulent dispersed flows in industrial, environmental and geophysical applications.

1 INTRODUCTION

Turbulent fluids and small particles or droplets or bubbles are common to a number of key processes in energy production, product industry and environmental phenomena. In modelling these processes, the dispersed phase is usually assumed uniformly distributed. Indeed, it is not true. Dispersed phases can be focused by turbulence structures and can have a time-space distribution which barely resembles prediction of simplified averaged modelling. Preferential distribution controls the rate at which sedimentation and re-entrainment occur, reaction rates in burners or reactors and can also determine raindrop formation and, through plankton, bubble and droplet dynamics, the rate of oxygen-carbon dioxide exchange at the ocean-atmosphere interface.

In the talk, we review a number of physical phenomena in which particle segregation in turbulence is a crucial effect describing the physics by means of Direct Numerical Simulation of turbulence. We elucidate concepts and modeling ideas derived from a systematic numerical study of the turbulent flow field coupled with Lagrangian tracking of particles under different modeling assumptions. We underline the presence of the strong shear which flavors wall turbulence with a unique multiscale aspect and adds intricacy to the role of inertia, gravity and buoyancy in influencing particle motion. We describe the role of free surface turbulence in dispersing and clustering the light particles such as plancton and the role of the distribution of dissipation in non-homogeneous turbulence to control breakage rates of brittle and ductile aggregates. Through a number of physical examples of practical interest such as boundary layers, free-surface and stratified flows, we show that a sound rendering of turbulence mechanisms is required to produce a physical understanding of particle trapping, segregation and ultimately macroscopic flows such as surfacing, settling and re-entrainment.

We initially focus on the role of inertia on a single particle in a vortex to discuss the effects of the wall vortices present in a turbulenyt boundary layer on particle deposition and re-entrainment. Specifically, we will give precise identification of coherent structures responsible for particle sedimentation and reentrainment (Marchioli & Soldati 2002) as presented in Figure 1.

However, turbulence features change according with the geometric features of the flow. Some significant environmental problems are relative to free surface turbulence. The free surface turbulence, albeit constrained onto a two-dimensional space, exhibits features which barely resemble predictions of simplified two-dimensional modelling. In particular, in a three dimensional open channel flow, surface turbulence is characterized by upscale energy transfer which controls the long term evolution of the larger scales. This can be demonstrated by associating downscale and upscale energy transfer at the surface with the trace of the velocity gradient tensor as shown in Figure 2. The presence

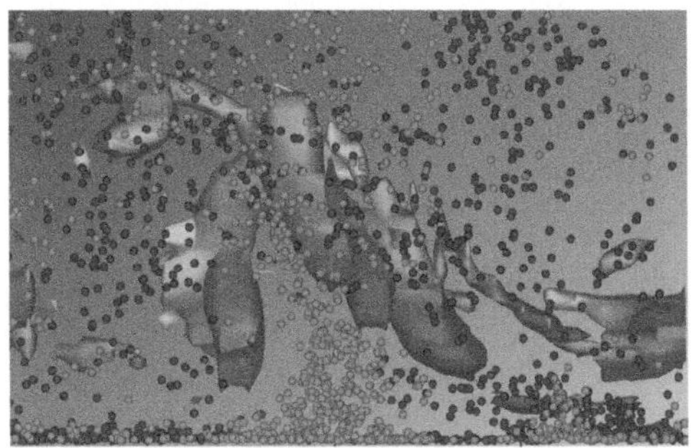

Figure 1. Vortices and inertial particles in a boundary layer. Different color for the vortices indicate clockwise or counter-clockwise rotation in the streamwise direction. Blue particles are directed away from the wall; Purple particles are directed towards the wall.

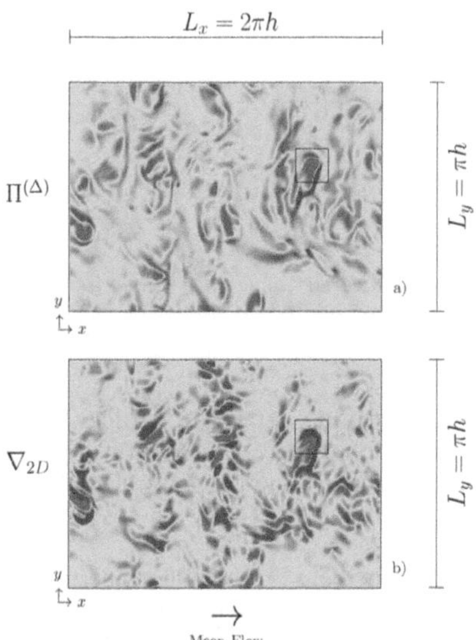

Figure 2. Contour maps of the energy flux (panel a) and of the two-dimensional surface divergence (panel b) computed at the free surface for Re = 509.

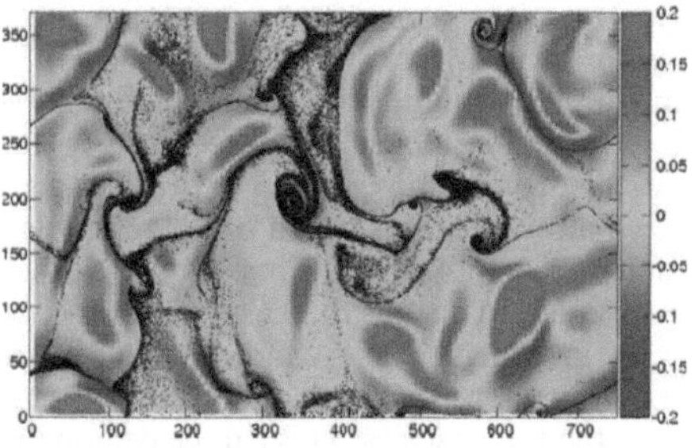

Figure 3. Light particles floating on a flat shear-free surface of a turbulent open water. This configuration mimics the motion of buoyant matter (e.g. phytoplankton, pollutants or nutrients). Correlation between floater clusters and surface divergence ∇_{2D}. Floaters segregatein $\nabla_{2D} < 0$ regions (in blue, footprint of sub-surface downwelings) avoiding footprint of sub-surface upwelings). Particle buoyancy induces clusters that evolve towards a long-term fractal distribution in a time much longer than the Lagrangian integral fluid time scale, indicating that such clusters overlive the surface turbulent structures which produced them.

of the inverse energy cascade at the free-surface is crucial in determining the pattern evolution of floaters and planctonc species. In particular it is possible to demonstrate that that particle buoyancy induces clusters that evolve towards a long-term fractal distribution in a time much longer than the Lagrangian integral fluid time scale, indicating that such clusters overlive the surface turbulent structures which produced them (Lovecchio, Zonta & Soldati 2015) as presented in Figure 3.

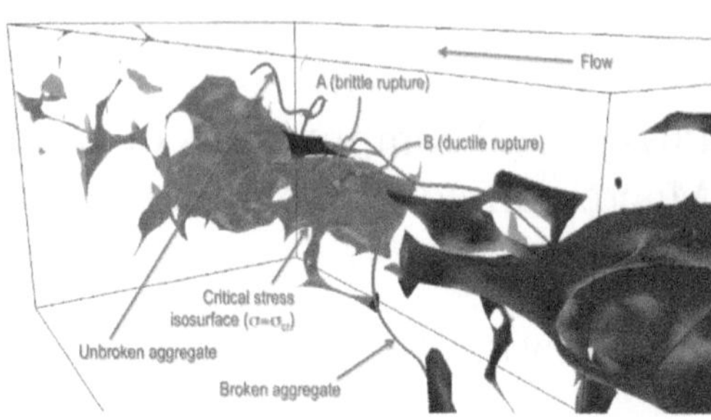

Figure 4. Rendering of brittle and ductile rupture in turbulent flow. Trajectories of two different aggregates are shown superimposed onto the isosurface of the critical stress $\sigma = \sigma_{cr}$ required to produce brittle rupture or activate ductile rupture. The aggregate trespasses the σ_{cr} isosurface at point A (potential brittle rupture) and undergoes ductile rupture at point B (where breakage condition $E > E_{cr}$ is met). The unbroken aggregate avoids all regions where $\sigma > \sigma_{cr}$ and does not break within the time window considered. Critical stress isosurface is taken at the time of ductile rupture. Aggregate trajectories are tracked several time steps backward from this time.

We also discuss the effects of thermal stratification (Zonta, Onorato, & Soldati 2012), (Lovecchio, Zonta, & Soldati 2014) on the distribution of passive and active planctonic species and swimmers.

A final issue which is addressed in this talk is the local shearing action induced by turbulence on the rupture of aggregates. Brittle and ductile aggregates will be examined and physics and statistical features of the rupture will be discussed (Babler, Biferale, Brandt, Feudel, Guseva, Lanotte, Marchioli, Picano, Sardina, Soldati, & Toschi 2015, (Marchioli & Sol dati 2015) as presented in Figure 4.

REFERENCES

Babler, M., L. Biferale, L. Brandt, U. Feudel, K. Guseva, A. Lanotte, C. Marchioli, F. Picano, G. Sardina, A. Soldati, & F. Toschi (2015). Numerical simulations of aggregate breakup in bounded and unbounded turbulent flows. *Journal of Fluid Mechanics* 766, 104–128.

Lovecchio, S., F. Zonta, & A. Soldati (2014). Influence of thermal stratification on the surfacing and clustering of floaters in free surface turbulence. *Advances in Water Resources* 72, 22–31.

Lovecchio, S., F. Zonta, & A. Soldati (2015). Upscale energy transfer and flow topology in free surface turbulence. *Physical Review E* 91.

Marchioli, C. & A. Soldati (2002). Mechanisms for particle transfer and segregation in turbulent boundary layer. *Journal of Fluid Mechanics* 468, 283–315.

Marchioli, C. & A. Soldati (2015). Turbulent breakage of ductile aggregates. *Physical Review E* 91.

Zonta, F., M. Onorato, & A. Soldati (2012). Turbulence and internal waves in stably-stratified channel flow with temperature-dependent fluid properties. *Journal of Fluid Mechanics* 697, 175–203.

Advances in Mechanics: Theoretical, Computational and Interdisciplinary Issues – Kleiber et al. (Eds)
© 2016 Taylor & Francis Group, London, ISBN 978-1-138-02906-4

Finite strain analyses of deformations in polymer specimens

V. Tvergaard
Department of Mechanical Engineering, Solid Mechanics, Technical University of Denmark, Denmark

ABSTRACT: Analyses of the stress and strain state in test specimens or structural components made of polymer are discussed. This includes the Izod impact test, based on full 3D transient analyses. Also a long thin polymer tube under internal pressure has been studied, where instabilities develop, such as bulging or necking. An axisymmetric bulge develops on the tube followed by necking in the bulge, and neck propagation is observed in both the circumferential and the axial directions. Analyses of indentation tests have been carried out, with focus on the effect of the material parameters characterizing viscoplastic flow on the indentation response. The ability of the simpler expanding spherical cavity model to reproduce the trends from the 3D finite element solutions has been assessed too.

1 INTRODUCTION

The analyses to be discussed here focus on determining the stress and strain state in test specimens or structural components made of polymer. When polymers are compressed or stretched to large strains, the plastic straining usually initiates at a stress peak and subsequently the stress level decays during large straining until increased network stiffness gives very high stresses, as network locking essentially stops plastic deformation. Constitutive models for this type of material behaviour have been developed by (Argon 1973, Boyce et al. 1973, Boyce & Arruda 1990, Wu & van der Giessen 1996). Including models for the viscoelasticity effect has been discussed by Anand & Ames (2006). These material models have been implemented in the analyses to be discussed here, carried out in collaboration with A. Needleman.

2 IZOD TEST

The Izod pendulum impact test is frequently used to measure the impact resistance of plastics. The test specimen is somewhat similar to the Charpy V-notch specimens often applied to test the brittle-ductile transition in structural steels, but in the Izod test the notched specimen stands clamped in a vertical position, and the pendulum strikes the clamped specimen near the free end. The specimen has standard length, depth and notch dimensions, but tests are carried out with different specimen widths, ranging from a width equal to the specimen depth (square specimen cross-section) to a width about a quarter of that value. The small width specimens give deformations under conditions near plane stress while the square cross-section specimens give higher constraint on plastic flow.

Full 3D numerical analyses have been carried out in Tvergaard & Needleman (2008a) using material parameters that qualitatively represent a polycarbonate. Subsequently the effect of various deviations from these material parameters has been studied in Tvergaard & Needleman (2008b). Regarding the mean stress, the maximum value is directly ahead of the notch, at the notch root. It is worth contrasting this with the distribution in metal plasticity, where the maximum mean stress occurs some distance into the material, directly ahead of the notch as a result of the increased stress triaxiality. But in the polymer the high stress at the notch results from the network stiffening, which prevents further straining but gives very large stresses. Thus, for the polymer the maximum strain with corresponding high stresses occurs first at the notch root when the limit stretch is reached, and then this region of high mean stress gradually spreads into the material from the notch root, as more material is affected by the network stiffening.

3 TUBE UNDER INTERNAL PRESSURE

For a polymer tube loaded by internal pressure the expansion of the tube wall leads to necking. This has been studied, first under the idealized assumption of a plane strain analysis, where the necks cover the full length of the tube (Lindgreen et al. 2008a). The tube is assumed to have an initial thickness imperfection with a number of slightly thinner regions around the circumference. At each of these thin points a neck starts to develop, and each of these necks grow into the neck propagation mode that is characteristic for polymers. For a metal the force transferred through a neck keeps reducing so that the localized necking typically ends as a fracture, but in a polymer neck the reduction of the force stops when network stiffening sets in and therefore the neck becomes longer and longer, as more material is pulled through the softening region to reach the limit stretch where network stiffening occurs.

A more realistic full 3D analysis of a long polymer tube has been carried out in (Lindgreen et al. 2008b). In addition to the small initial thickness imperfection with a number of slightly thinner points around the circumference another initial imperfection specified a slightly larger radius at one end of the tube, because it is known that tubes made of a softening material will tend to develop a bulge instability. Indeed it turns out that a bulge forms at one end of the tube and subsequently necking of the tube wall starts to develop in the bulge.

4 INDENTATION

The indentation test is frequently used to measure the hardness of materials and to get some information on mechanical properties such as the yield strength and the elastic modulus. For polymers full 3D analyses of indentation have been carried out in Tvergaard & Needleman (2011), considering conical and pyramidal indenters. Also the spherical cavity model was considered, as suggested for metals in (Bishop et al. 1945) and discussed in Johnson (1970). It was observed that the subsurface displacements produced by a blunt indenter are approximately radial from the point of first contact, with roughly hemispherical contours of equal strain. This model requires only the 1D analysis of the expansion of a spherical void.

The analyses in Tvergaard & Needleman (2011) use material parameters with a large limit stretch that give a reasonable approximation of tensile tests for a high-density polyethylene. In Tvergaard & Needleman (2012) the same type of computations are carried out for a number of different sets of material parameters all giving the up-down-up tensile or compressive law characteristic for polymers, some of them with a smaller limit stretch. All the analyses in (Tvergaard & Needleman 2011, 2012) show that also for the polymers the hardnesses predicted by the full 3D numerical analyses are rather well approximated by the spherical cavity model where the analyses are 1D with spherical symmetry.

It is well known for metals that their hardness is about three times the yield strength Johnson (1970). But if the initial peak stress on the uniaxial tensile or compressive curve is considered the effective yield strength of the polymer, the hardness is much lower than a factor three times this yield strength. The effect of elastically soft or plastically compressible solids on the hardness has recently been investigated (Needleman et al. 2015). The plastic deformations in the computations (Tvergaard & Needleman 2011, 2012) for polymers were taken to be incompressible, but the ratio of Young's modulus to the effective yield strength for these materials was only 19, much lower than the value of this ratio for metals. Therefore, the hardness found for the polymers is only about 1.05 times the effective yield strength.

REFERENCES

Anand, L. & Ames, N.M. 2006. On modeling the micro-indentation response of an amorphous polymer. *Int. J. Plasticity* 22: 1123–1170.

Argon, A.S. 1973. A theory for the low-temperature plastic deformation of glassy polymers. *Polymer* 44: 6013–6032.

Bishop, R.F., Hill, R. & Mott, N.F. 1945. The theory of indentation and hardness tests. *Proc. Phys. Society* 57: 147–159.

Boyce, M.C. & Arruda, E.M. 1990. An experimental and analytical investigation of the large strain compressive and tensile response of glassy polymers. *Polym. Eng. Sci.* 30: 1288–1298.

Boyce, M.C., Parks, D.M. & Argon, A.S. 1988. Large inelastic deformation of glassy polymers. Part I: Rate dependent constitutive model. *Mech. Mater.* 19: 193–212.

Johnson, K.L. 1970. The correlation of indentation experiments. *J. Mech. Phys. Solids* 18: 115–126.

Lindgreen, B., Tvergaard, V. & Needleman, A. 2008a. Dynamic neck development in a polymer tube under internal pressure loading. *Int. J. Solids Structures* 45: 580–592.

Lindgreen, B., Tvergaard, V. & Needleman, A. 2008b. Bulge formation and necking in a polymer tube under dynamic expansion. *Modelling and Simulation in Materials Science and Engineering* 16: 085003.

Needleman, A., Tvergaard, V. & van der Giessen, E. 2015. Indentation of elastically soft and plastically compressible solids. *Acta Mech. Sin.* 31, 473–480.

Tvergaard, V. & Needleman, A. 2008a. An analysis of thickness effects in the Izod test. *Int. J. Solids Structures* 45: 3951–3966.

Tvergaard, V. & Needleman, A. 2008b. Effect of material parameters in the Izod test for polymers. *Proceedings of the IUTAM Symposium on theoretical, computational and modelling aspects of inelastic media*: 297–306. Springer.

Tvergaard, V. & Needleman, A. 2011. Polymer indentation: Numerical analysis and comparison with a spherical cavity model. *J. Mech. Phys. Solids* 59: 1669–1684.

Tvergaard, V. & Needleman, A. 2012. Effect of viscoplastic material parameters on polymer indentation. *Modelling and Simulation in Materials Science and Engineering* 20: 065002.

Wu, P.D. & van der Giessen, E. 1996. Computational aspects of localized deformations in amorphous glassy polymers. *Eur. J. Mech. A/Solids* 15: 799–823.

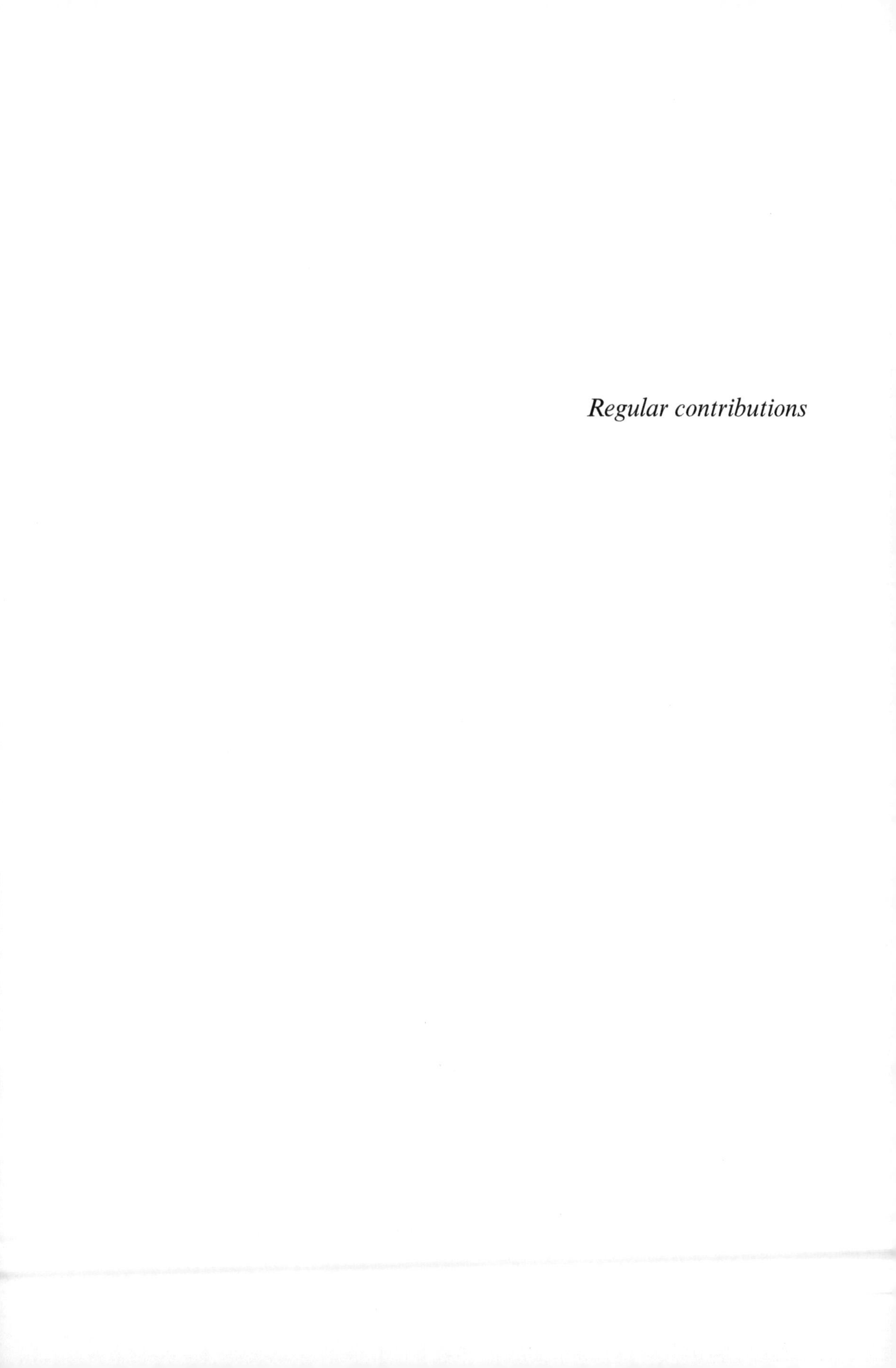

Regular contributions

Advances in Mechanics: Theoretical, Computational and Interdisciplinary Issues – Kleiber et al. (Eds)

Experimental tests for the determination of mechanical properties of PVC foil

A. Ambroziak
Gdańsk University of Technology, Gdańsk, Poland

ABSTRACT: The paper presents an application of the PVC film in civil engineering. Mechanical behaviour of the PVC film applied for suspended ceilings, in the form of a stretch ceiling is investigated under uniaxial and biaxial tensile tests. The study is focused on the determination of mechanical properties from experimental data (uniaxial and biaxial tensile tests). The uniaxial cyclic tests are performed in order to observe a variation of immediate mechanical properties under cyclic load. This paper is suggested to be an introduction to a comprehensive investigation on building engineering application of the PVC foils.

1 INTRODUCTION

The applications of plastics are widespread in various branches of industry. Sorts of PolyVinyl Chloride (PVC) films are used in stretch ceilings. The application of flexible PVC films in stretched ceilings is a modern form of design and interior decoration. Stretched film ceiling installation consists of a lightweight, highly durable film or a decorative fabric to a framing construction intended to hold it stretched in place. The architectural applications of flexible PVC films are still being developed.

Recent developments in membrane building materials for lightweight structures were described by Saxe 2000. The advantages of lightweight tensioned coated fabrics and foils applied to the existing building sector in order to improve the insulating/shading performance of the external building envelopes was reviewed by Baccarelli & Chilton 2013. Another types of foils made of ETFE are used in civil engineering for tensile structures e.g. roofs and claddings. Mechanical behaviour of ETFE foils under uniaxial and biaxial loading was investigated by Galliot & Luchsinger 2011.

The building site supervision makes it possible to recognize new technologies and applications of new materials. The author performs construction site supervision on the Alchemia building site in Gdansk where the PVC foil is applied for stretched ceilings. The Alchemia building is a modern multipurpose complex in Gdansk, Poland. It covers office spaces and supplies a sports and recreation facility offering a 25-meter swimming pool, a sports hall, fun climbing zone, a gym. The swimming pool is covered by a stretched ceiling. The PVC foil is suspended on a steel-aluminium structure (Fig. 1). The investigation is performed to confirm that the materials, all equipment and systems, function well and conform to the quality standards well.

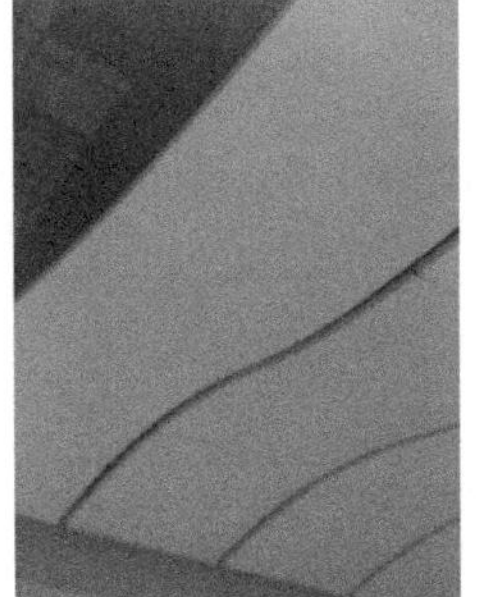

Figure 1. Stretching ceiling—under erection.

2 LABORATORY TESTS

Uniaxial and biaxial tensile tests are chosen from a large group of experimental tests (see e.g. Ambroziak & Kłosowski 2013) to model the material behaviour of a PVC foil. The aim of the laboratory tests is to determine mechanical properties of a polymer, named DPS stretch ceiling. According to the technical data specified by the manufacturer (http://www.grupadps.com/en/) the following properties are defined: weight 240 g/m^2, total thickness 0.17 mm, tensile strength for longitudinal and transverse directions 17.1 N/mm^2 and 18.7 N/mm^2, respectively. The tensile strength is established according to PN-EN ISO 527-1. Material parameters necessary to analyze a real structure were not found in technical specification. In order to perform a relevant assessment of a material, laboratory tests were carried out.

2.1 *Uniaxial tensile tests*

Uniaxial cyclic tensile laboratory tests were made using, computer-controlled Zwick testing machine. Specimens from the same batch of foil in two perpendicular directions T (Transverse) and L (Longitudinal) were firstly prepared. Their dimensions were: width 50 mm, length 300 mm. The specimens were subjected to tension with a displacement rate of grip equal to 50 mm/min. The tests were controlled by a video extensometer with a constant base about 50 mm grip separation of 200 mm. Laboratory tests were performed at room temperature (about 20°C). Three stress ranges (A, B, C) were assumed. In each cyclic test 20 load cycles are carried out up to one of load levels specified in Table 1 and full unloading. Finally, after this sequence the specimen was loaded up to rupture.

Comparison of the results obtained from the different stress ranges of uniaxial cyclic tests are given in Figure 2 for the longitudinal direction and in Figure 3 for the transverse direction, respectively. It can be observed that the stress-strain curves exhibit a hysteresis effect. The values of the breaking strain (ε_R) and the Ultimate Tensile Strength (UTS) for Uniaxial And Biaxial Tensile tests are specified in Table 2 and Table 3. The difference (7–10%) between longitudinal and transverse direction are observed.

For all cyclic tests in the assumed load range the stress-strain curves for immediate state and cycles $n = 1$, $n = 10$ and $n = 20$ (where n is a number of the cycle) are shown in Figures 4–6. It can be shown that for $n = 0$ (first tension of a film) the results for T and L directions converge. Next cycles give similar stress-strain curves for T and L directions, but for the A stress range only. For B and C stress range a difference is observed between stress-strain curves for T and L directions.

Table 1. Stress range—cyclic tests.

Tests type	Stress values (N/m)
A	200
B	400
C	800

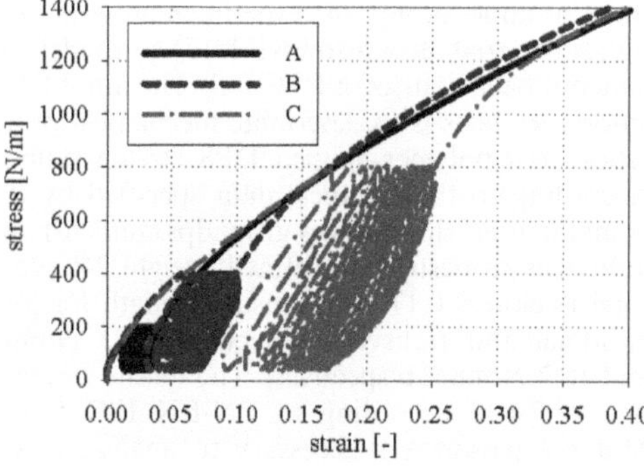

Figure 2. Uniaxial cyclic tensile tests—longitudinal direction.

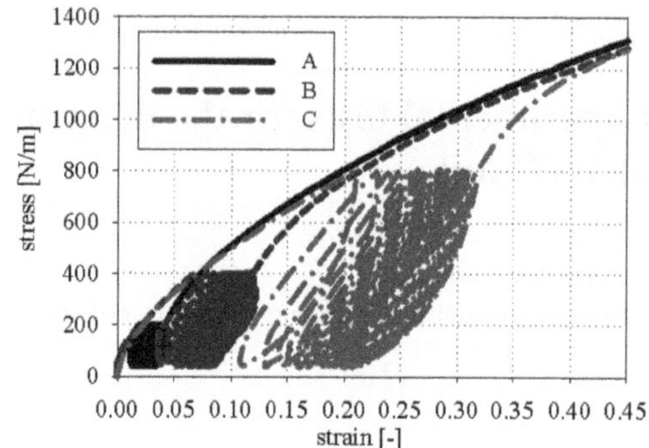

Figure 3. Uniaxial cyclic tensile tests—transverse direction.

Table 2. Values of strain at break—uniaxial tests.

Tests type	Transverse (–)	Longitudinal (–)
A	0.873	0.842
B	0.886	0.854
C	0.857	0.754
Mean	0.872	0.816

Table 3. Ultimate tensile strength—uniaxial tests.

Tests type	Transverse (kN/m)	Longitudinal (kN/m)
A	1.89	2.16
B	1.87	2.22
C	1.82	1.83
Mean	1.86	2.07

For a detailed comparison of the uniaxial cyclic tests results, the residual strains $\varepsilon_I^{n=1}$, $\varepsilon_I^{n=10}$, $\varepsilon_I^{n=20}$ (where n is a number of the cycle) and longitudinal stiffness values $F_n = 1$, $F_n = 10$, $F_n = 20$ have been assigned. It should be noted that the longitudinal stiffness parameter F (called tension stiffness) is specified in [N/m], in the case uniaxial tensile tests corresponding to Young's modulus. The unit [N/m] is accepted in accordance with the units generally used for fabrics. The results of the identification are collected in Table 4. The values of F [N/m] parameters are specified for stress ranges >100 N/m (>5% of

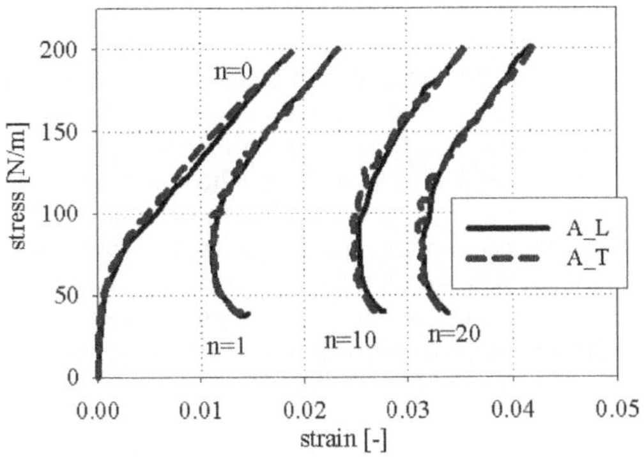

Figure 4. Uniaxial cyclic tensile tests—A stress range.

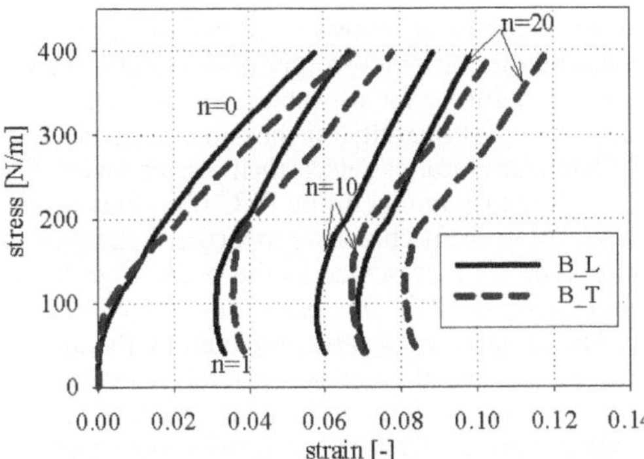

Figure 5. Uniaxial cyclic tensile tests—B stress range.

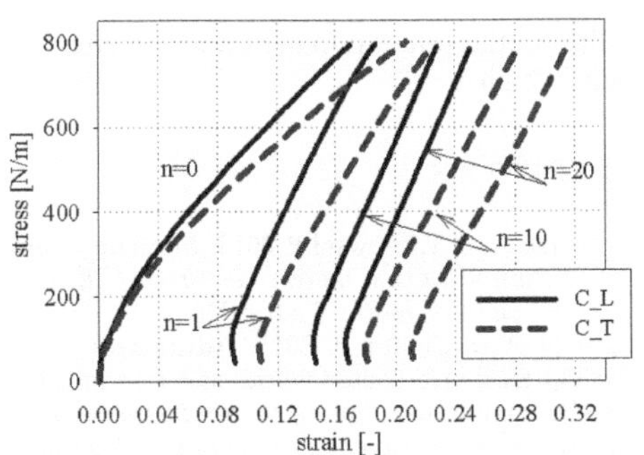

Figure 6. Uniaxial cyclic tensile tests—C stress range.

the Ultimate Tensile Strength—UTS). Observation for each stress level yields that the increment of cycle number grows makes the longitudinal stiffness increase ($F_n = 20 > F_n = 10 > F_n = 1 > Fn = 0$). The immediate mechanical parameters ($F_n = 0$) change during the cyclic tests in the analysed range of loads. The values of residual strains $\varepsilon_l^{n=20}$ range between 3% and 20%.

Table 4. Results of identification—cyclic tests.

	Transverse			Longitudinal		
	A	B	C	A	B	C
$F_n = 0$ [N/m]	6990	4500	3280	7570	5480	3970
$F_n = 1$ [N/m]	7840	5960	5670	8080	7780	6890
$\varepsilon_l^{n=1}$ [–]	0.011	0.035	0.108	0.011	0.031	0.089
$F_n = 10$ [N/m]	8310	6670	6510	9570	8970	7905
$\varepsilon_l^{n=10}$ [–]	0.025	0.067	0.180	0.025	0.058	0.144
$F_n = 20$ [N/m]	8730	6800	6440	9170	9230	7950
$\varepsilon_l^{n=20}$ [–]	0.031	0.081	0.211	0.031	0.069	0.167

Figure 7. Biaxial laboratory tests stand.

It should be noted that the immediate properties (n = 0) specified in Table 4 refer to the initial stress state during the erection process. These parameters are indispensable in the film cutting pattern process too. The concern is how the PVC film behaves in each direction at the initial stress state. Additionally, the behaviour of PVC film under cyclic loads should be specified. These tests show the variation of immediate mechanical properties under cyclic loads.

2.2 *Biaxial tensile tests*

The biaxial tensile tests performed on the Zwick system with a video extensometer attached (Fig. 7), were made for the cross-shaped specimens. The arm width was 100 mm, therefore on the testing area of 100 × 100 mm the gage length of about 50 mm in both directions was indicated. Due to all the tests initial grip separation of 300 mm was selected. The specimens were subjected to cyclic tension (constant force rate) in longitudinal and transverse directions with load (stress) ratios 1:1, 1:2, 2:1 (σ_L: σ_T, where σ_L and σ_T are stresses in the longitudinal and transverse directions, respectively, see Fig. 8).

In the beginning, the stress values at brake of specimens are established, see Table 5. The mean

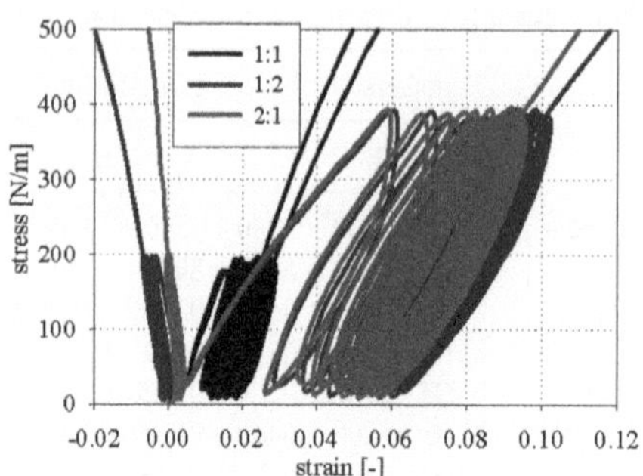

Figure 8. Biaxial tensile tests—comparison of tests.

Table 5. Stress at brake of specimens—biaxial tests.

Tests type	Transverse (kN/m)	Longitudinal (kN/m)
1:1	1.29	1.28
1:2	0.76	1.49
2:1	1.52	0.78

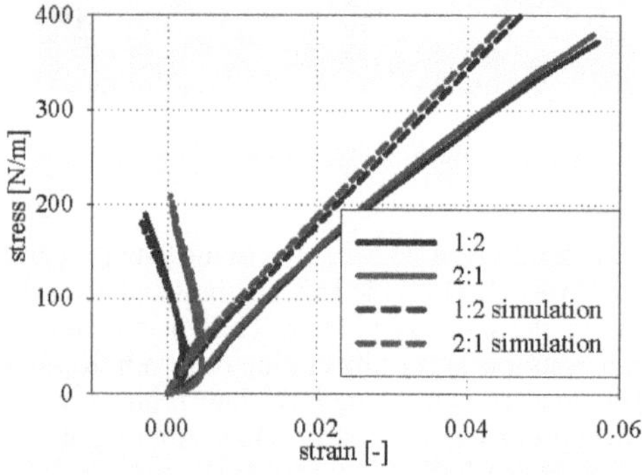

Figure 9. Comparison of numerical simulations with tests results for 2:1 and 1:2.

stress value at brake is about 1.43 kN/m and is about 70% of the mean uniaxial tensile strength. This result can be explained by means of the failure mode of a specimen, see Figure 7. Generally, the damage of specimens begins in the corner zone, next the cross arm is torn.

For the sake of engineering calculations it is possible to assume the isotropic film theory. This theory yields the following:

$$\sigma_L = \frac{F}{1-\nu^2}(\varepsilon_L + \nu\varepsilon_T)$$
$$\sigma_T = \frac{F}{1-\nu^2}(\varepsilon_T + \nu\varepsilon_L) \quad (1)$$

where ν is the Poisson's ratio; $F = 0.5(F_L + F_T)$; ε_T, ε_L are strains in transverse and longitudinal directions, respectively. An isotropic model is frequently used for the analysis of suspension structures made of plastic film, rubber-like materials, etc. Assumed $F = 7280$ N/m (the mean value of $F_n = 0$ for A range, see Table 4) the value $\nu = 0.4$ according to Equation 1 is specified. Comparison of numerical simulations for loading curves with tests results for 2:1 and 1:2 are given in Figure 9. A good correlation is observed between numerical simulations and tests results.

3 CONCLUSION

The study presents test methods to investigate the mechanical properties of PVC foils. A modern laboratory equipment allows for different variants of tests and the computer storage of the results, important for a future identification process. The comparison of tests results on the same material from different laboratories detects their discrepancies. The procedure of manufacturing PVC films may results in variation of mechanical properties. Examples of such tests were presented in the paper in order to understand the PVC film behaviour better.

The research program completed by the author for the construction of new stretched ceilings of the Alchemia building in Gdansk presents various applications modern testing devices. Such tests are necessary in all stages—before, during and after construction. The tests results may be used in the solution of problems inherent to the roof assembly and its service. The investigation confirms that the quality of the foil, equipment and systems is sufficiently high.

REFERENCES

Ambroziak A. & Kłosowski P. 2013. Mechanical testing of technical woven fabrics. *Journal of Reinforced Plastics and Composites* 32(10): 726–739.

Baccarelli P. & Chilton J. 2013. Advantages of lightweight tensioned coated fabrics and foils façades for the building sector. *Proc. of the 6th International Conference on Textile Composites and Inflatable Structures, Structures Membranes*, 200–211.

Galliot C. & Luchsinger R.H. 2011. Uniaxial and biaxial mechanical properties of ETFE foils. *Polymer Testing* 30(4): 356–365.

Saxe K. 2000. Recent development in membrane building materials for lightweight structures. *Asian Textile Journal.* 9(8): 34–36.

Advances in Mechanics: Theoretical, Computational and Interdisciplinary Issues – Kleiber et al. (Eds)
© 2016 Taylor & Francis Group, London, ISBN 978-1-138-02906-4

Asymptotic solution for free vibration of weakly nonlinear oscillator with two serially connected springs

J. Awrejcewicz
Department of Automation, Biomechanics and Mechatronics, Lodz University of Technology, Łódź, Poland
Institute of Vehicles, Warsaw University of Technology, Warsaw, Poland

R. Starosta & G. Sypniewska-Kamińska
Institute of Applied Mechanics, Poznan University of Technology, Poznań, Poland

ABSTRACT: The subject of the paper is analysis of the dynamical regular behaviour of the oscillator with two serially connected springs exhibiting the nonlinearity of cubic type. The motion of the system is described by a set of Differential-Algebraic Equations (DAEs). The problem has been solved using the Multiple Scales Method (MSM) in time domain, which is an asymptotic one. In order to apply this method for solving the governing equations of such a type the classical approach that is used for solving the differential equations has been appropriately modified. The approximate analytical solutions have been verified by comparison with the results obtained in the numerical manner. The analytical form of the solution allows, among others, to formulate an explicit relationship between the period and the amplitude of the oscillations of the system.

1 INTRODUCTION

The linear simplification is sometimes too rough to describe the behaviour of physical objects enough precisely. Therefore, models of nonlinear oscillators have been widely considered in physics and engineering. Nonlinear oscillators with serially connected springs were investigated, mainly numerically, by many authors. Most of the papers concern the case, when one of the springs is linear and the second one is nonlinear (Telli & Kopmaz 2006, Lai & Lim, 2007). Bayat et al. (2014) proposed an analytical approach for determining the equivalent stiffness of an oscillator with two nonlinear springs.

In the paper an oscillator with two nonlinear springs is analyzed using multiple scales method in time domain. Telli & Kopmaz (2006) showed that the vibrations of a body mounted via two springs in series are described by a set of differential-algebraic equations. Similar situation occurring in our investigations implies a need of a modification of the standard form of MSM.

2 FORMULATION OF THE PROBLEM

Let us consider one-dimensional motion of a body of mass m attached by two massless nonlinear springs to an immovable support. The studied system is shown in Figure 1. The restoring force of the spring with the nonlinearity of the cubic type is

$$F_i = k_i(\Delta l_i + \Lambda_i \Delta l_i^3), \quad i = 1,2 \tag{1}$$

where Δl_i is the elongation of the i-th spring, k_i is its linear stiffness, and Λ_i is the nonlinearity parameter. Lengths of the untensioned springs are L_{10} and L_{20}.

The spring characteristics is called "hard" for $\Lambda_i > 0$, while for $\Lambda_i < 0$ it is called "soft". In further considerations we take into account only the case when $\Lambda_i > 0$ and the nonlinearity is weak.

The governing equations for the considered system are as follows:

$$m\ddot{X}_2 + k_2(X_2 - X_1 + \Lambda_2(X_2 - X_1)^3) = 0, \tag{2}$$

$$k_2(X_2 - X_1 + \Lambda_2(X_2 - X_1)^3) - k_1(X_1 + \Lambda_1 X_1^2) = 0. \tag{3}$$

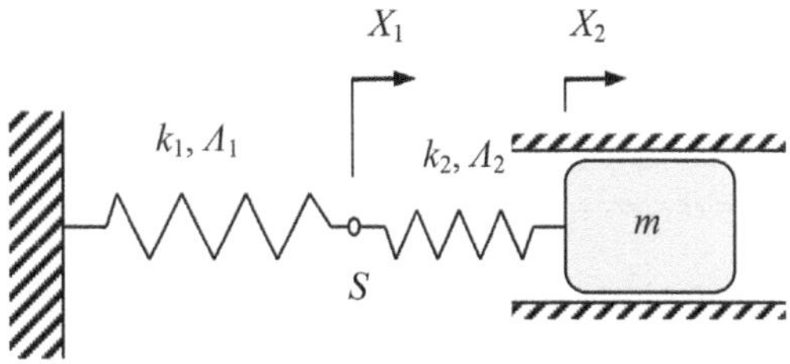

Figure 1. Series connection of two nonlinear springs.

The initial conditions referring to Equation (2) are

$$X_2(0) = X_0, \dot{X}_2(0) = V_0, \tag{4}$$

where X_0 and V_0 are known.

Let $L = L_{10} + L_{20}$ be the characteristic dimension. Assuming $k_e = k_1k_2/(k_1 + k_2)$ is the effective stiffness of the analogous linear system we introduce the dimensionless variables and parameters in the following manner: $x_i = X_i/L$, $\alpha_i = \Lambda_i/L^2$ for $i = 1, 2$, $\lambda = k_2/k_1$. Moreover, $\tau = \omega_0 t$, where $\omega_0^2 = k_e/m$, is the dimensionless time.

It is convenient to define the functions representing the dimensionless elongations of the both springs

$$u(\tau) = x_1(\tau), w(\tau) = x_2(\tau) - x_1(\tau). \tag{5}$$

All these arrangements allow us to rewrite Equations (2)–(4) as follows:

$$\ddot{w} + \ddot{u} + (1+\lambda)(1+\alpha_2 w^2)w = 0, \tag{6}$$

$$\lambda(1+\alpha_2 w^2)w - (1+\alpha_1 u^2)u = 0, \tag{7}$$

$$w(0) + u(0) = x_0, \dot{w}(0) + \dot{u}(0) = v_0, \tag{8}$$

where $x_0 = X_0/L$, $v_0 = V_0/(\omega_0 L)$.

3 METHOD OF SOLUTION

The problem given by Equations (6)–(8) can be solved analytically using the multiple scales method in time domain which is detail presented for instance in Awrejcewicz & Krysko (2006). However, the applying this method for DAEs requires significant modification.

The necessary relations between the functions $w(\tau)$ and $u(\tau)$ and their first and second derivatives are obtained by differentiating Equation (7)

$$(1+3\alpha_1 u^2)\dot{u} - \lambda(1+3\alpha_2 w^2)\dot{w} = 0, \tag{9}$$

$$6\alpha_1 u\dot{u}^2 + \left(1+3\alpha_1 u^2\right)\ddot{u} - 6\lambda\alpha_2 w\dot{w}^2 - \lambda\left(1+3\alpha_2 w^2\right)\ddot{w} = 0. \tag{10}$$

The assumptions concerning smallness of the nonlinearity parameters are proposed in the form

$$\alpha_1 = \tilde{\alpha}_1\varepsilon, \alpha_2 = \tilde{\alpha}_2\varepsilon, \tag{11}$$

where ε is the small parameter.

The functions $w(\tau)$ and $u(\tau)$ are seek in the form of the power series with respect to the small parameter. Assuming two time scales and taking into account only two first terms of the power series we can write

$$w(\tau,\varepsilon) = \sum_{k=0}^{k=1} \varepsilon^k w_k(\tau_0,\tau_1),\ u(\tau,\varepsilon) = \sum_{k=0}^{k=1} \varepsilon^k u_k(\tau_0,\tau_1), \tag{12}$$

where $\tau_0 = \tau$ is the "slow" time scale, while $\tau_1 = \varepsilon\,\tau$ is the "fast" time scale.

The time derivative is then replaced by the partial differential operator as follows

$$\frac{d}{d\tau} = \frac{\partial}{\partial\tau_0} + \varepsilon\frac{\partial}{\partial\tau_1}. \tag{13}$$

Introducing the assumptions given by (11)–(13) into Equations (6)–(7) we get the differential equation and algebraic one in which the small parameter ε occurs in various nonnegative powers. Each of them should be satisfied for any value of ε. Therefore, after rearranging of the terms in these equations according to the powers of the small parameter ε, the original governing equations (6)–(8) can be approximated by the system consisting of the partial linear differential equations and algebraic ones. Taking into account only the equations corresponding to the constant terms and the terms linear with respect to ε we obtain the equations of the first order approximation (i.e. involving all the terms of degree zero)

$$(1+\lambda)w_0 + \frac{\partial^2 u_0}{\partial\tau_0^2} + \frac{\partial^2 w_0}{\partial\tau_0^2} = 0, \tag{14}$$

$$u_0 - \lambda w_0 = 0, \tag{15}$$

and the equations of the second order approximation (i.e. involving all the terms of degree one with respect to ε)

$$(1+\lambda)\left(\tilde{\alpha}_2 w_0^3 + w_1\right) + \frac{\partial^2 u_1}{\partial\tau_0^2} + \frac{\partial^2 w_1}{\partial\tau_0^2} + 2\left(\frac{\partial^2 u_0}{\partial\tau_0\partial\tau_1} + \frac{\partial^2 w_0}{\partial\tau_0\partial\tau_1}\right) = 0, \tag{16}$$

$$u_1 - \lambda w_1 + \tilde{\alpha}_1 u_0^3 - \lambda\tilde{\alpha}_2 w_0^3 = 0. \tag{17}$$

The relationships between partial derivatives of the unknown functions u_0, u_1 and w_0, w_1 with respect to variable τ_0 should be determined in order to solve the system consisting of Equations (14)–(17). They can be obtained from Equation (10) in the similar way as above, i.e. by introducing Equations (11)–(13) into Equation (10) and taking into account only the terms of degrees zero and one. The relation of the first order approximation has very simple form

$$\frac{\partial^2 u_0}{\partial \tau_0^2} - \lambda\frac{\partial^2 w_0}{\partial \tau_0^2} = 0, \tag{18}$$

and the relationship corresponding to the second order approximation is as follows:

$$\frac{\partial^2 u_1}{\partial \tau_0^2} - \lambda\frac{\partial^2 w_1}{\partial \tau_0^2} + 2\left(\frac{\partial^2 u_0}{\partial \tau_0 \partial \tau_1} - \lambda\frac{\partial^2 w_0}{\partial \tau_0 \partial \tau_1}\right)$$
$$6\tilde{\alpha}_1 u_0\left(\frac{\partial u_0}{\partial \tau_0}\right)^2 - 6\lambda\tilde{\alpha}_2 w_0\left(\frac{\partial w_0}{\partial \tau_0}\right)^2$$
$$+3\tilde{\alpha}_1 u_0^2\frac{\partial^2 u_0}{\partial \tau_0^2} - 3\tilde{\alpha}_2\lambda w_0^2\frac{\partial^2 w_0}{\partial \tau_0^2} = 0. \tag{19}$$

The system (14)–(17) is solved recursively, starting from the equations of the first order approximation. Solving Equation (18) with respect to the partial derivative of u_0, and then introducing this solution into Equation (14) we obtain the linear, homogenous differential equation

$$w_0 + \frac{\partial^2 w_0}{\partial \tau_0^2} = 0. \tag{20}$$

Its general solution may be written in the form

$$w_0 = A(\tau_1)\exp(i\tau_0) + \bar{A}(\tau_1)\exp(-i\tau_0), \tag{21}$$

where $A_1(\tau_1)$ is the unknown complex-valued function of the slow time, and by upper bar one denotes its complex conjugate.

From Equation (15) we obtain

$$u_0 = \lambda\left(A(\tau_1)\exp(i\tau_0) + \bar{A}(\tau_1)\exp(-i\tau_0)\right). \tag{22}$$

Afterwards, the solutions (21)–(22) as well as the formula for the second order derivative of u_1 derived from (19) are introduced into equations of the second order approximation (16)–(17). The equation obtained in this way from (16) contains the so-called secular terms that are responsible for the appearance of the unbounded solutions that in turn are not acceptable in autonomous systems. Elimination of such terms is equivalent to the fulfillment of the following solvability condition

$$3\lambda^3 A^2\bar{A}\tilde{\alpha}_1 + 3A^2\bar{A}\tilde{\alpha}_2 + 2i\dot{A}(1+\lambda) = 0. \tag{23}$$

After eliminating secular terms, the solution of the equations of the second order approximation are as follows:

$$u_1 = \frac{9\lambda^3\tilde{\alpha}_1 + \tilde{\alpha}_2 - 8\lambda\tilde{\alpha}_2}{8(1+\lambda)}A^3\exp(3i\tau_0) + CC, \tag{24}$$

$$w_1 = -3\lambda A^2\bar{A}\left(\lambda^2\tilde{\alpha}_1 - \tilde{\alpha}_2\right)\exp(i\tau_0)$$
$$+\frac{\lambda^3(\lambda-8)\tilde{\alpha}_1 + 9\lambda\tilde{\alpha}_2}{8(1+\lambda)}A^3\exp(3i\tau_0) + CC, \tag{25}$$

where symbol CC denotes complex conjugates to the parts of solutions that are written explicitly.

Let us write the function $A_1(\tau_1)$ in the exponential form

$$A(\tau_1) = \frac{1}{2}b(\tau_1)\exp(i\psi(\tau_1)), \tag{26}$$

where real-valued functions $b(\tau_1)$ and $\psi(\tau_1)$ are the amplitude and the phase of vibration, respectively.

Introducing substitution (26) into solvability condition (23) we obtain the modulation equations

$$\dot{b}(\tau_1) = 0, \quad \dot{\psi}(\tau_1) = \frac{3b(\tau_1)^3(\lambda^3\tilde{\alpha}_1 + \tilde{\alpha}_2)}{8(1+\lambda)}. \tag{27}$$

The initial conditions complementing the modulation equations are

$$b(0) = b_0, \quad \psi(0) = \psi_0, \tag{28}$$

where the initial values b_0 and ψ_0 are strictly related to the initial values x_0 and v_0.

Taking advantage of (26) and then introducing the solutions of the modulation problem (27)–(28)

$$b(\tau_1) = b_0, \quad \psi(\tau_1) = \frac{3b_0^3(\lambda^3\tilde{\alpha}_1 + \tilde{\alpha}_2)}{8(1+\lambda)}\tau_1 + \psi_0 \tag{29}$$

into Equations (24)–(25) allow us to write the approximate solution for functions $w(\tau)$ and $u(\tau)$ in the analytical form

$$w(\tau) = b_0\cos((1+\Gamma)\tau + \psi_0)$$
$$+\frac{b_0^3(\alpha_2 - 8\alpha_2\lambda + 9\alpha_1\lambda^3)}{32(1+\lambda)}\cos(3(1+\Gamma)\tau + 3\psi_0), \tag{30}$$

$$u(\tau) = \lambda b_0\left(1 + \frac{3b_0^2(\alpha_2 - \alpha_1\lambda^2)}{4}\right)\cos((1+\Gamma)\tau + \psi_0)$$
$$+\frac{\lambda b_0^3(9\alpha_2 + \alpha_1(\lambda-8)\lambda^2)}{32(1+\lambda)}\cos(3(1+\Gamma)\tau + 3\psi_0), \tag{31}$$

where $\Gamma = \dfrac{3b_0^2(\alpha_2 + \alpha_1\lambda^3)}{8(1+\lambda)}$.

The assumptions (11) and the relation between the time τ and the slow time τ_1 have been taken into account in the above formulae.

4 RESULTS

Time histories of the oscillator motion obtained analytically according to (30)–(31) are presented in Figure 2.

The comparison of the numerical and analytical solutions confirms correctness of the asymptotic calculations.

The explicit form of the solution allows for deeper analysis of the motion. It is easily to observe that the period of the vibration is

$$T = \frac{2\pi}{\Gamma + 1} = \frac{16\pi(\lambda+1)}{3b_0^2(\alpha_2 + \alpha_1\lambda^3) + 8(\lambda+1)}. \tag{32}$$

Expression (32) involving the nonlinearities α_1 and α_2 as well as the parameter λ quantitatively describes the dependence of the period with respect to b_0 which is the first approximation of the amplitude of the function $w(\tau)$. For hardening springs the period decreases with b_0 what is shown in Figure 3.

The influence of the parameter λ on the period and the whole amplitude of $w(\tau)$, according to (30), is presented in Figures 4 and 5, respectively.

In Figure 6 the value of period of the vibration versus nonlinearity parameters α_1 and α_2 is shown.

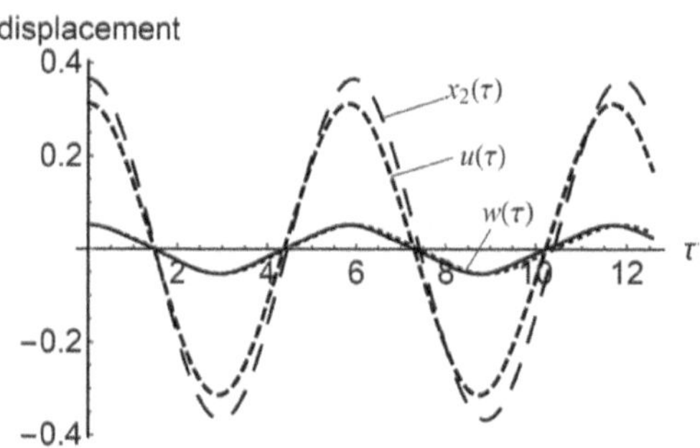

Figure 2. Time history of the system motion for $\alpha_1 = \alpha_2 = 1.8$, $\lambda = 7$, $b_0 = 0.05$, $\psi_0 = 0$; dotted line—numerical solution.

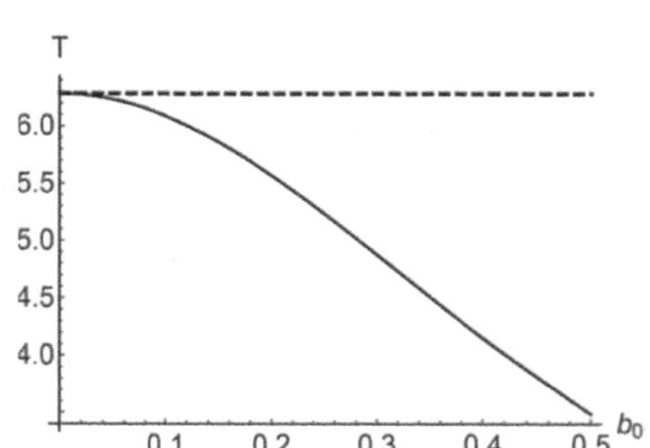

Figure 3. Vibration period versus b_0 for $\alpha_1 = \alpha_2 = 1.8$, $\lambda = 2.5$, $\psi_0 = 0$.

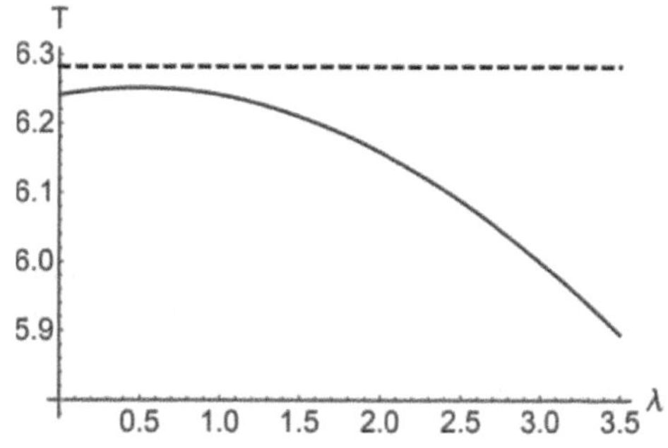

Figure 4. Vibration period versus λ for $\alpha_1 = \alpha_2 = 1.8$, $b_0 = 0.1$, $\psi_0 = 0$.

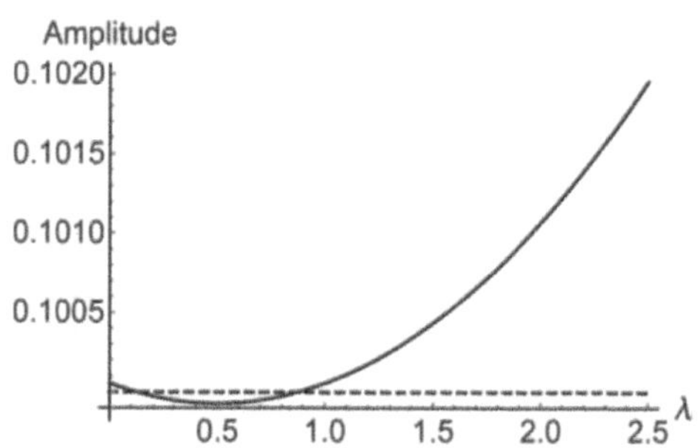

Figure 5. Amplitude of $w(\tau)$ versus λ for $\alpha_1 = \alpha_2 = 1.8$, $b_0 = 0.1$, $\psi_0 = 0$.

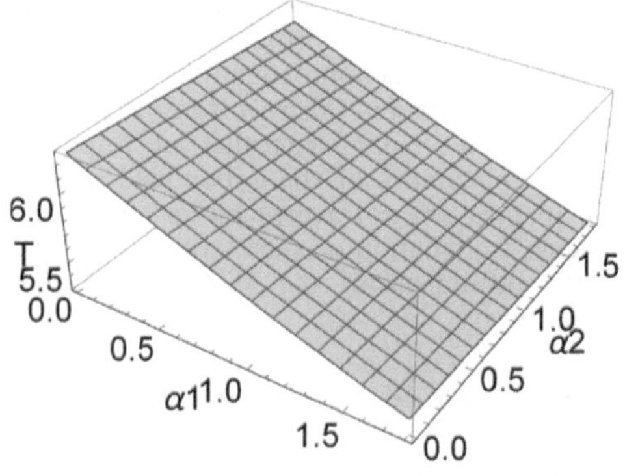

Figure 6. Vibration period versus α_1 and α_2, for $\lambda = 2$, $b_0 = 0.1$, $\psi_0 = 0$.

Generally, greater influence on the period comes from the nonlinearity of that spring, which has smaller linear stiffness coefficient k_i.

5 CONCLUSIONS

The analytical solution of the dynamical response of the oscillator with two serially connected nonlinear springs has been obtained. The properly modified multiple scale method in time domain allows to solve efficiently the differential—algebraic system of equations, which describe motion of the oscillator.

The analytical solution allows to study directly the influence of the parameters on the dependence between period and amplitude of the free system vibration.

ACKNOWLEDGEMENTS

This paper was financially supported by the National Science Centre of Poland under the grant MAESTRO 2, No. 2012/04/A/ST8/00738, for years 2013–2016 and grant 02/21/DSPB/3463.

REFERENCES

Awrejcewicz, J. & Krysko, V.A. 2006. *Introduction to Asymptotic Methods.* Boca Raton: Chapman and Hall.

Bayat, M., Pakar, I., Cveticanin L. 2014. Nonlinear free vibration of systems with inertia and static type cubic nonlinearities: An analytical approach. *Mechanism and Machine Theory* 77: 50–58.

Lai, S.K. & Lim, C.W. 2007. Accurate approximate analytical solutions for nonlinear free vibration of systems with serial linear and nonlinear stiffness, *Journal of Sound and Vibration* 307: 72–736.

Telli, S. & Kopmaz, O. 2006. Free vibrations of a mass grounded by linear and nonlinear springs in series. *Journal of Sound and Vibration* 289: 689–710.

Advances in Mechanics: Theoretical, Computational and Interdisciplinary Issues – Kleiber et al. (Eds)
© 2016 Taylor & Francis Group, London, ISBN 978-1-138-02906-4

Strain hardening effect on elastic-plastic contact of a rigid sphere against a deformable flat

Ł. Bąk, F. Stachowicz & T. Trzepieciński
Rzeszow University of Technology, Rzeszów, Poland

S. Bosiakov & S. Rogosin
Belarusian State University, Minsk, Belarus

ABSTRACT: This paper considers the effect of strain hardening on the elastic-plastic contact of a rigid ball with an elastic-plastic flat using experiments and finite-element software Abaqus. The strain hardening is an increase in the strength and hardness of the metal due to a mechanical deformation in the microstructure of the metal. Flat tensile samples with different values of strain are considered, in order to study the effect of strain hardening. It was found that the strain hardening phenomenon and anisotropy of both friction and material have a great influence on the ball indentation value and the maximum indentation force. The anisotropy of both material and friction conditions influenced the non-uniformity of the stress and strain distribution around the pin axis.

1 INTRODUCTION

Accurate calculation of the contact area is a major problem in the field of tribology, and leads to improved understanding of friction and wear. In metal forming operations, rough surfaces consist of asperities having different radii and height, so it is a difficult task to evaluate the contact area and contact load. The problem is simplified by Hertz (1882). The assumption of surfaces having asperities of spherical shape is adopted to simplify the contact problems. Timoshenko & Goodier (1970) stated that the results of normal loading under friction differ from the frictionless Hertzian contact problem. Later, Johnson (1985) found that friction can increase the total load required to produce a contact of given size by at most 5% compared to Hertz.

Based on a model of rigid anisotropic asperities, a theoretical investigation on friction limit surfaces and sliding rules has been carried out by Mróz & Stupkiewicz (1994). The effect of strain hardening on the elastic-plastic contact of a deformable sphere against a rigid flat under full-stick contact conditions was studied by Chatterje & Saho (2012). Chang et al. (1987) proposed an elastic-plastic contact where the sphere remains in elastic contact until a critical interference is reached, above which the volume conservation of the sphere tip is imposed. In the microcontact model for the contact between two nominally flat surfaces analysed by Zhao et al. (2000) the transition from elastic deformation to fully plastic flow of the contacting asperity is modelled, based on contact-mechanics theories.

The Finite Element (FE) method can be used to calculate accurately the contact parameters like contact load, contact area, and pressure, and so forth, removing some of the assumptions made in the earlier analytical theories regarding asperity interaction (e.g. Sahoo et al. 2010a). One of the first works that provided an accurate result of elastic-plastic contact of a hemisphere and a rigid flat using FE was the study of Kogut & Etsion (2002). They concluded a negligible effect of strain hardening for the frictionless and the nonadhesive contact. Jackson & Green (2003) observed the effect of the deformed geometry on the effective hardness and presented some empirical relations of contact area and contact load. Sahoo et al. (2010b) found that the interfacial parameters like contact load and real contact area during loading are dependent on the material properties of the deformable sphere.

Most of friction models are completely defined by the friction conditions, which specify a set of admissible contact forces, and the sliding rule, which stipulates the allowed directions of sliding (Hol et al. 2012). The limit surface is usually assumed to be isotropic predicting a frictional behaviour independent of the sliding direction. For many industrial applications, this assumption seems to be unrealistic and many experimental studies show that the frictional behaviour can change drastically with the sliding direction, requiring an anisotropic model (Trzepieciński 2010).

Deep drawing sheets usually exhibit directional surface topography. Furthermore, the initial surface topography constantly changes with an increase of strain. This paper considers the effect of strain hardening on an elastic-plastic contact of a rigid ball with an elastic-plastic flat using experiments and the finite element method. To fully understand the mechanism of asperity deformation, in our study we analysed the change in surface topography of pre-strained sheet specimens cut along 3 directions with respect to the rolling direction. The anisotropic friction model that corresponded to experimental results was implemented into a Finite Element (FE) model built using the ABAQUS software.

2 EXPERIMENTS

2.1 *Material*

Experiments were carried out for DC04 steel sheet metal. The mechanical properties of the sheet metal listed in Table 1 have been determined through uniaxial tensile tests along three directions with respect to the rolling direction.

2.2 *Methods*

The friction properties of the deep drawing quality steel sheets used in the experiments were determined by using the pin-on-disc tribometer T01-M. In this study, the friction coefficient as a function of angular position with respect to the rolling direction of the sheet metal was measured.

Continuous indentation tests were performed on DC04 steel sheets of a 2 mm thickness.

Table 1. Mechanical properties of DC04 steel sheet.

	Orientation			
Parameter	0	45	90	Mean
Yield stress $R_{p0,2}$, MPa	182.1	196	190	189.9
Ultimate strength R_m, MPa	322.5	336.2	320.9	326.5
Elongation A_{50}, %	45.8	41.6	45.6	44.3
Hardening coefficient C, MPa	549.3	564.9	541.6	555.2
Strain hardening exponent n	0.214	0.205	0.209	0.208
Lankford's coefficient r	1.751	1.124	1.846	1.461

The samples for indentation tests were cut into 3 directions: along the rolling direction, at 45° to the rolling direction and transverse to the rolling direction. The flat samples of 20 mm width and 200 mm length were straightened using uniaxial tensile test to receive different strain values ε_1: 5, 10, 15, 20, 25 and 30%. Then the surface of pre-strained samples was grinded.

The indentation tests were performed using a modified Zwick Roell Z030 operated in the compression mode. The applied load versus the cross-head displacement or depth of the indentation were continuously recorded throughout the tests. A maximum load of 60, 80, 100, 125, 150, 200, 250, 300 N was applied for each sample. In our analyses, the indenter made of bearing steel has a 6 mm diameter.

3 NUMERICAL MODELING

The symmetry of the process was assumed, in order to reduce computational time. Only one quarter of blank and the ball with symmetry boundary conditions was modelled. Geometry of the ball was assumed discrete rigid. An elastic-plastic material model of sheet was implemented. The elastic behaviour is specified in numerical simulations by the value of Young's modulus, E = 210000 MPa, and of Poisson's ratio $\nu = 0.3$. In the numerical model, the anisotropy of the material has been established using the Hill yield criterion (Hill 1948). For the blank and hemisphere meshing the 3-dimensional 8-node brick elements were used.

The experimental results of friction tests show that the friction coefficient depends on the measured angle from the rolling direction, and corresponds to the surface topography. The anisotropic elliptic friction model (Trzepieciński 2010) was implemented by specifying different friction coefficients in two orthogonal directions on the contact surface. These orthogonal directions coincide with the defined slip directions. To use an anisotropic friction model, 2 friction coefficients (i.e. $\mu = 0.142$ and $\mu = 0.157$) were specified, where the first is the coefficient of friction in the first slip direction along the rolling direction and second is the coefficient of friction in the perpendicular slip direction.

4 RESULTS AND DISCUSSION

4.1 *Surface roughness*

The value of the Ra and Rz parameters increased with the tensile strain level (Table 2). This dependence is nearly linear. The linear character of this relation is confirmed by Stachowicz (2015). The increase in surface roughness of the sheet is the

Table 2. Effect of tensile strain on change of the Rz parameter value of sheet surface.

	Longitudinal to rolling direction	Perpendicular to rolling direction		
Strain, %	Ra, μm	Rz, μm	Ra, μm	Rz, μm
0	0.965	5.043	1.002	5.597
5	1.144	5.848	1.105	5.677
10	1.182	6.416	1.187	6.457
15	1.441	7.913	1.375	8.426
20	1.546	8.287	1.501	9.663
25	1.583	8.378	1.608	9.870
30	1.586	8.584	1.646	10.77

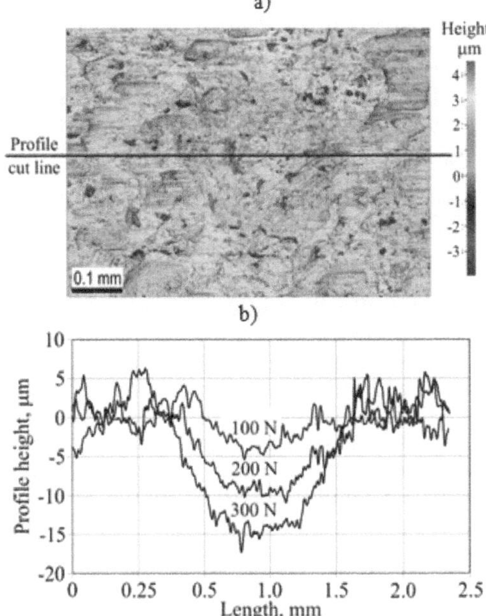

Figure 1. Optical micrograph of indentation for normal load 60 N (a) and influence of normal load on indentation depth (b).

result of reorientation and fragmentation of the individual grains, mainly in the subsurface. The ball indentation with load up to 60 N caused a visible effect on the surface profile (Fig. 1a). The values of indentation depth during testing non-prestrained sheets for normal loads 100 N and 200 N (Fig. 1b) are smaller than the value of Rz parameter.

This fact leads to the conclusion that only the asperities of the surface were deformed. Increasing the normal load to 300 N causes the plastic deformation occurring in the subsurface, some distance below the roughness profile. Evidently, the observed relationship between impression depth and the Rz parameter is different to pre-strained sheets due to the strain hardening effect, and will be interrogated in next part of the paper.

4.2 *Normal load*

The depth of impression h is measured between the average line of the profile curve and average line of the indentation profile (Fig. 2). The linear dependence between the normal load and impression depth is observed (Figs. 3 and 4). Furthermore, it is evident that increasing the normal load allows an increase in the depth. It was found that the value of penetration depth for specific force value decreases non-linearly with the increase of sample strain. This finding is a result of increasing

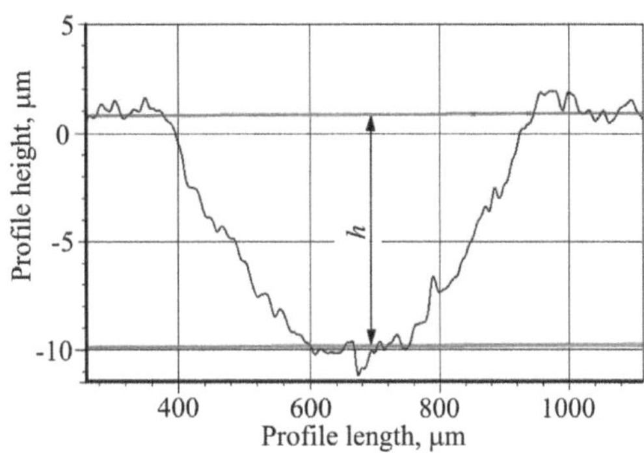

Figure 2. The scheme of penetration depth measurement.

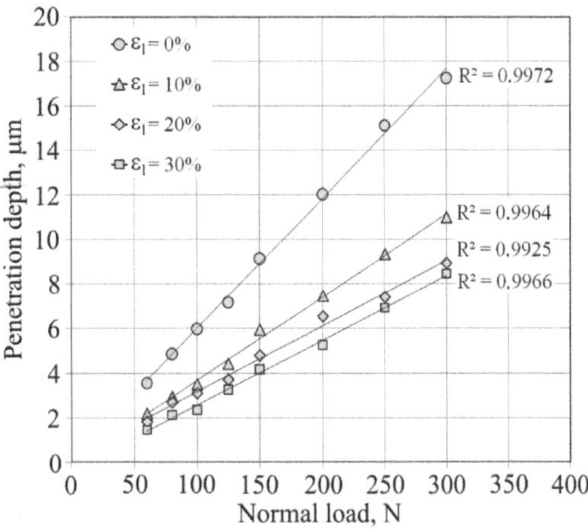

Figure 3. Effect of sample strain on normal load value for samples cut along the rolling direction.

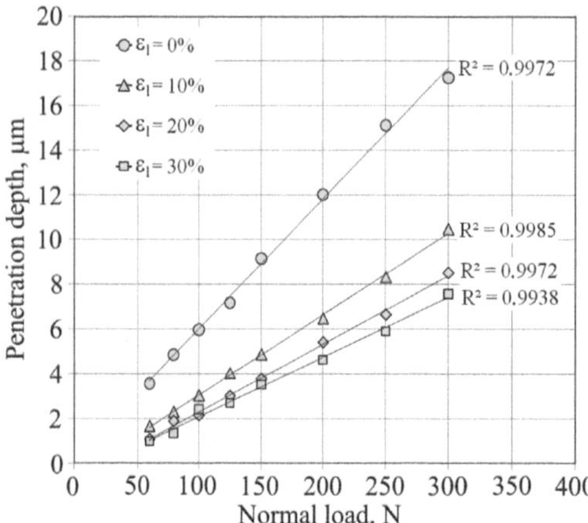

Figure 4. Effect of sample strain on normal load value for samples cut transverse to the rolling direction.

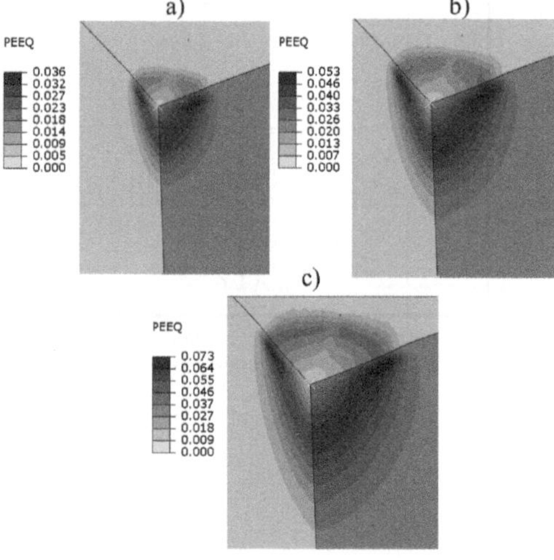

Figure 5. Distribution of equivalent plastic strain in pre-strained sheet (ε_1 = 30%) at indentation depths: a) 0.005 µm, b) 0.010 µm and c) 0.015 µm.

deformation resistance due to the strain hardening phenomenon.

Pre-strained samples cut transverse to the rolling direction exhibit higher deformation resistance than samples cut along the rolling direction. For all pre-strained samples cut at 0° (Fig. 3) the indentation depths are smaller than in case of samples cut transverse to the rolling direction (Fig. 4). It is an effect of the anisotropic properties of the tested sheet metals.

Despite the linear relationship between the normal load and penetration depth, the real contact area between the ball and sheet increases non-linearly with the penetration depth.

The evaluation of normal pressure in case of small normal loads is a problematic issue because only the asperities of the roughness undergo a deformation (Fig. 1b).

4.3 *Strain distribution*

The maximum value of equivalent plastic strain is found at the sub-surface, some distance below the centre of the contact region (Fig. 5). It was observed for all sample pre-strains. The anisotropy of both material and friction conditions influenced the non-uniformity of the strain distribution around the line of axial symmetry of the model. It was clearly visible for higher values of ball indentations.

5 CONCLUSIONS

Tensile strain of the sheets causes strain hardening of the material and a change in the surface topography of the sheet. The increase in surface roughness of the sheet is the result of a reorientation and fragmentation of the individual grains, mainly in the subsurface. Furthermore, the initial directionality of surface topography constantly changes under the influence of deformation. The relationship between sample strain and roughness parameters (Ra, Rz) is nearly linear.

The ball indentation test allows us to investigate the surface topography change of the sheets subjected to earlier pre-strain. The strain hardening effect on the surface parameters of the steel sheet plate was observed, which caused a different contact area for each strain level of a specimen. The indentation load of 60 N deformed only the asperities of the material on the surface. The linear dependence between the normal load and the impression depth is observed. It was found that the value of penetration depth for specific force values decreases non-linearly with increasing material hardening.

The simulations of ball indentation show that as the load increases, the plastic zone continues to grow until the edge of the plastic zone reaches the surface near the edge of the contact radius. Strain hardening of a material causes a decrease of maximum values of the equivalent plastic strain for specific indentation depths.

ACKNOWLEDGEMENTS

This work was supported by the European Research Agency—FP7-PEOPLE-2013-IRSES Marie Curie Action "International Research Staff Exchange Scheme", grant agreement No. 610547.

REFERENCES

Chang, WR. & Etsion, I. & Bogy, D.B. 1987. An elastic-plastic model for the contact of rough surfaces. *ASME Journal of Tribology* 109: 257–263.

Chatterje, B. & Sahoo, P. 2012. Effect of strain hardening on elastic-plastic contact of a deformable sphere against a rigid flat under full stick contact condition. *Advances in Tribology* 2012: 1–8.

Hertz, H. 1882. Über die berührung fester elastischer köper, *Journal für die Reine und Angewandte Mathematik* 92: 156–171.

Hill, R. 1948. A theory of the yielding and plastic flow of anisotropic metals. *Proceedings of the Royal Society of London* 193: 281–297.

Hol, J. & Cid Alfaro M.V. & de Rooij M.B. & Meinders T. 2012. Advanced friction modeling for sheet metal forming. *Wear*, 286–287: 66–78.

Jackson R.L. & Green, I. 2003. A finite element study of elastoplastic hemispherical contact against a rigid flat. *ASME Journal of Tribology* 46: 383–390.

Johnson, K.L. 1985. *Contact Mechanics*. Mass: Cambridge University Press.

Kogut, L. & Etsion, I. 2002. Elastic-plastic contact analysis of a sphere and a rigid flat. *ASME Journal of Applied Mechanics* 69: 657–662.

Mróz, Z. & Stupkiewicz, S. 1994. An anisotropic friction and wear model. *International Journal of Solids and Structures* 31: 1113–1131.

Sahoo, P. & Adhikary, D. & Saha, K. 2010a. Finite element based elastic-plastic contact of fractal surfaces considering strain hardening. *Journal of Tribology and Surface Engineering* 1: 39–56.

Sahoo, P. & Chatterjee, B. & Adhikary, D. 2010b. Finite element based elastic-plastic contact behavior of a sphere against a rigid flat—effect of strain hardening. *International Journal of Engineering, Science and Technology* 2: 1–6.

Stachowicz, F. 1991. On the mechanical and geometric inhomogenity and formability of aluminium and aluminium alloy sheets. *Archives of Metallurgy* 41: 61–75.

Timoshenko, S.P. & Goodier, J.N. 1970. *Theory of Elasticity*. New York: McGraw-Hill.

Trzepieciński, T. 2010. 3D elasto-plastic FEM analysis of the sheet drawing of anisotropic steel sheet. *Archives of Civil and Mechanical Engineering* 10: 95–106.

Zhao, Y. & Maietta, D.M. & Chang, L. 2000. An asperity microcontact model incorporating the transition from elastic deformation to fully plastic flow. *ASME Journal of Tribology* 122: 86–93.

Advances in Mechanics: Theoretical, Computational and Interdisciplinary Issues – Kleiber et al. (Eds)
© 2016 Taylor & Francis Group, London, ISBN 978-1-138-02906-4

Multiscale evolutionary optimization of Functionally Graded Materials

W. Beluch & M. Hatłas
Institute of Computational Mechanics and Engineering, Silesian University of Technology, Gliwice, Poland

ABSTRACT: The paper is devoted to the multiobjective and multiscale optimization of composite structures made of Functionally Graded Materials (FGMs). The aim of the optimization is to design structures made of FGM by modifying the microstructure parameters in order to satisfy contradictory criteria on a macro scale. Numerical homogenization methods are used to perform a multiscale analysis. Finite element method is employed to solve a boundary-value problem in both scales. Computational intelligence method in form of a multiobjective evolutionary algorithm is used to perform the optimization procedure. Numerical example presenting the efficiency of the proposed attitude is attached.

1 INTRODUCTION

Composites are structural materials made of at least two different constituents. Typically, they consist of a continuous phase (matrix) and a reinforcement. Due to their properties, especially the high stiffness-to-weight ratio, composites are an interesting alternative to traditional, usually isotropic, structural materials, like metals or their alloys.

The aim of the paper is to perform the optimization of the structures made of a particular group of composites—functionally graded materials. As more than one criterion is considered during the optimization and the criteria are contradictory, the multiobjective optimization is performed. To avoid the premature convergence to local optima as well as problems with the calculation of the fitness functions gradient, an evolutionary algorithm is employed to perform the multiobjective optimization.

2 FUNCTIONALLY GRADED MATERIALS

Composites are materials whose properties can be designed by a proper selection of the constituent materials, the volume fraction of constituents and the shape and location of the reinforcement. An increasingly important group of composites state Functionally Graded Materials (FGMs).

FGMs are characterized by the gradual variation in composition and structure along at least one specific direction (Koizumi 1997). FGMs are readily used for applications related to severe thermal gradients and different structures under thermal or combined thermo-mechanical loads. The main application areas for FGMs state: aerospace industry machining tools, machine parts (e.g. ceramic-metal composites for gears) and the applications as biomaterials (bones, interfaces between implants and bones).

The performance of FGMs is not only affected by properties of relative amounts of constituents, but is also related to the optimal utilization of the constituent materials. Designing of FGMs allows obtaining products well fit for the foreseeable conditions of its operation simultaneously reducing as well the material cost as the costs associated with the operation and maintenance of the product.

In order to obtain the structure made of FGMs best suited for a particular application, it is necessary to use optimization methods. Computational intelligence methods like evolutionary algorithms or artificial immune systems belong to global optimization methods and may be used as the optimization procedures in considered case.

3 NUMERICAL HOMOGENIZATION OF HETEROGENOUS MATERIALS

Some materials, like composites or porous materials are heterogeneous on a certain level of observation and their structure should be considered to model them with proper accuracy. The direct application of more than one scale (macro, meso, micro, nano) for the calculations performed by means of Finite Element Method (FEM) or Boundary Element Method (BEM) leads to extremely large equation systems. To overcome this problem, numerical homogenization methods may be used (Kouznetsova 2002).

Assuming local or global periodicity of the considered structure, numerical homogenization allows determining the parameters of the equivalent, macroscopically homogeneous material. Typically, the numerical homogenization makes use of Representative Volume Element (RVE) and consists in the determination of the constitutive relation between averaged field variables, like stresses and strains.

The RVE fully represents part of the medium (assuming local periodicity) or the whole medium (global periodicity). The RVE is usually the smallest possible volume representing the entire medium (or its part). The conditions that RVE has to satisfy are:

1. its size l_{RVE} is significantly larger than the microstructure characteristic dimensions l_{micro} and considerably smaller than the characteristic dimensions l_{macro} in the macro scale:

$$l_{micro} << l_{RVE} << l_{makro} \tag{1}$$

2. Hill condition for the equality of the average energy density in the micro scale and the macroscopic energy density at the point of macrostructure corresponding to the RVE:

$$\langle \sigma_{ij}\varepsilon_{ij} \rangle = \langle \sigma_{ij} \rangle \langle \varepsilon_{ij} \rangle \tag{2}$$

where: σ_{ij} and ε_{ij}—stress and strain tensors in the micro scale; $\langle \cdot \rangle$—the averaged value of the field;

3. proper boundary conditions, e.g. periodic boundary conditions imposed on the opposite faces of RVE and strain boundary conditions taken from the higher scale.

The averaged stresses obtained in the micro scale are transferred to the macro scale which allows calculating the equivalent material parameter values.

4 NUMERICAL EXAMPLES

Two numerical examples of FGM optimization are presented—i) three-point bending of a beam and ii) a helicopter rotor.

4.1 *Three-point bended beam*

A composite beam of $b \times h \times l = 10 \times 10 \times 250$ mm dimensions, made of epoxy resin reinforced with longitudinally placed carbon fibers in form of FGM is optimized. The beam is divided into $n = 8$ sections of the same thickness h_i (Fig. 1a) and the FEM model consists of 610 3-D finite elements. The fibers in each section have the same diameter but fiber diameters may vary between the sections. Each section is represented by RVE or the same external dimensions $a = 10 \times 10 \times 10$ µm. Reinforcement diameter values d_i state the design variables of the optimization.

A 3-dimensional RVE with centrally placed fibre (Fig. 1b) and divided into 648 finite elements has been used to calculate the equivalent material parameters for each section.

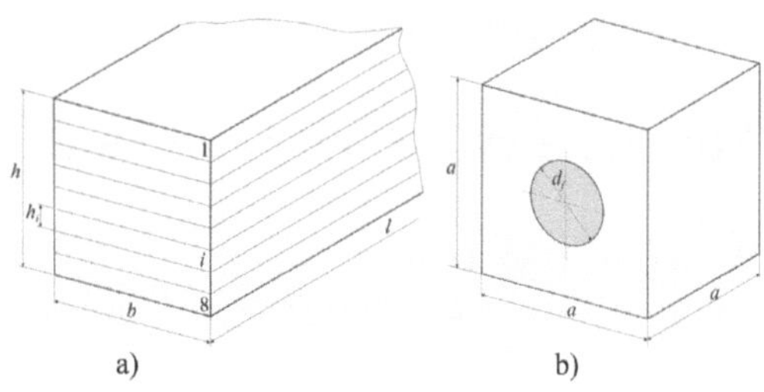

Figure 1. a) Sections of the beam and b) exemplary RVE.

4.1.1 *Formulation of the optimization problem*

The aim of the optimization is to obtain the desired properties of considered structure made of FGM. The structure is divided into n sections of the same volume fraction of the reinforcement. The aim of the optimization is to simultaneously minimize two following contradictory objectives:

1. the maximum deflection of the beam:

$$\arg\min\{f_1(x);\, x \in D\};\; f_1(x) = \max(u_i) \tag{3}$$

2. the total cost of the structure, represented by the total volume fraction of the reinforcement:

$$\arg\min\{f_2(x);\, x \in D\};\; f_2(x) = \sum_{i=1}^{n} V_r^i \Big/ V^i \tag{4}$$

where: x—a vector of the design variables, D—a set of admissible solutions being a subset of design space $\mathbf{X}$, u_i—the vertical component of displacement, V_r^i—the reinforcement volume in the section i of FGM, V^i—the volume of the section i.

4.1.2 *Calculations*

To solve the multiobjective optimization problem two multiobjective Evolutionary Algorithms (EA) based on NSGA-II algorithm have been used. The first is available in Matlab environment (Deb et al. 2002), the second one (MOOPTIM) is an in-house code (Długosz & Burczyński 2012). MSC Patran/Nastran FEM software has been employed to solve boundary-value problems in both scales.

The parameters of the EA are: the number of chromosomes $n_{ch} = 40$; the number of generations: $n_g = 50$; the arithmetical crossover probability: $p_{ac} = 0.8$; the uniform mutation probability: $p_{um} = 0.01$; the fibers diameter range $d_f = 1 \div 9$ µm.

The results of the multiobjective optimization in form of the Pareto frontier are presented in Figure 2. The values of the objective functions were normalized to the range $0 \div 10$ and presented as dimensionless.

The values of the design variables (fibre dimensions, from top to bottom sections) and the values of the objective functions for distinctive points A, B and C (Fig. 2) are collected in Table 1.

4.1.3 *Conclusions*

The multiobjective and multiscale optimization of the beam made of FGM has been performed by means of the evolutionary computations. The results presented in Figure 2 and Table 1 show that the considered objective functions are contradictory. Points A and C represent solutions with one of the criteria being dominating, which is proved by the obtained results. The values of design variables for the point B have high values in the outer sections and much lower inside the beam, which results in obtaining a functionally I-beam structure.

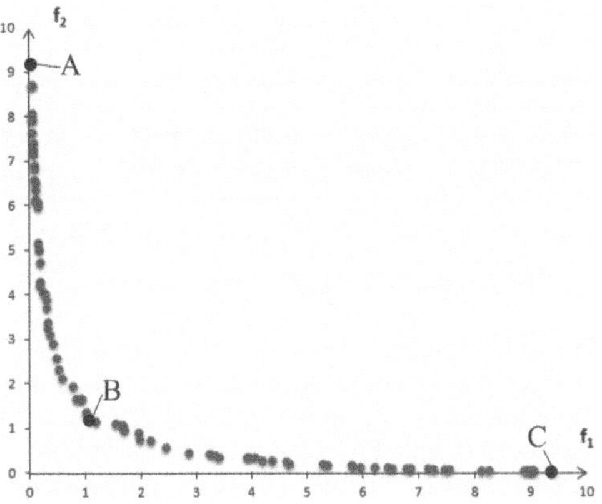

Figure 2. Beam optimization results—Pareto frontier.

4.2 *Helicopter rotor blade*

A composite blade of $h \times b \times l = 150 \times 600 \times 6700$ mm dimensions, made of epoxy resin reinforced with steel particles in form of FGM is optimized.

The blade is divided into $n = 14$ sections (Fig. 3b). The reinforcements volume fraction each section is the same, but may vary between the sections. Each section is represented by RVE or the same external dimensions a = 10 × 10 × 10 μm. Reinforcement volume fraction V_i state the design variables of the optimization. The blade model is loaded by rotor angular velocity ω = 15 rad/s.

A 3-dimensional RVE with 27 uniformly placed particles (Fig. 4) and divided into 7228 finite elements has been used to calculate the equivalent material parameters for each section.

4.2.1 *Formulation of the optimization problem*

The aim of the optimization is to simultaneously minimize three contradictory objectives:

1. the maximum deflection of the blade:

$$\arg\min\{f_1(x);\, x \in D\};\; f_1(x) = \max(u_i) \quad (5)$$

2. the maximum von Mises stress:

$$\arg\min\{f_2(x);\, x \in D\};\; f_2(x) = \max(\sigma_{red}) \quad (6)$$

Table 1. Design variables and objective function in specific points.

	Design variables (fiber diameters) [μm]								Objective f. values	
	d_{f_1}	d_{f_2}	d_{f_3}	d_{f_4}	d_{f_5}	d_{f_6}	d_{f_7}	d_{f_8}	f_1	f_2
Point A	8.8890	8.4215	8.0346	8.5714	8.7735	8.9586	8.8338	8.4805	0.0352	9.1737
Point B	5.5992	1.4117	1.3231	1.1295	1.3762	2.4043	1.6212	6.1826	1.0797	1.2008
Point C	1.0000	1.3339	1.0000	1.0000	1.0000	1.0000	1.2500	1.0000	9.3946	0.0202

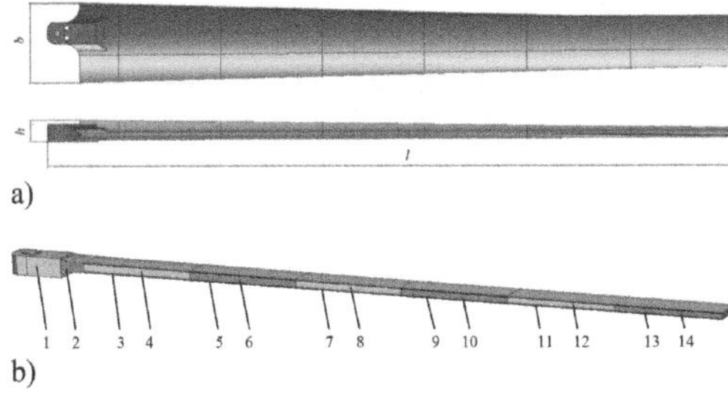

Figure 3. Helicopter rotor blade: a) geometry b) sections.

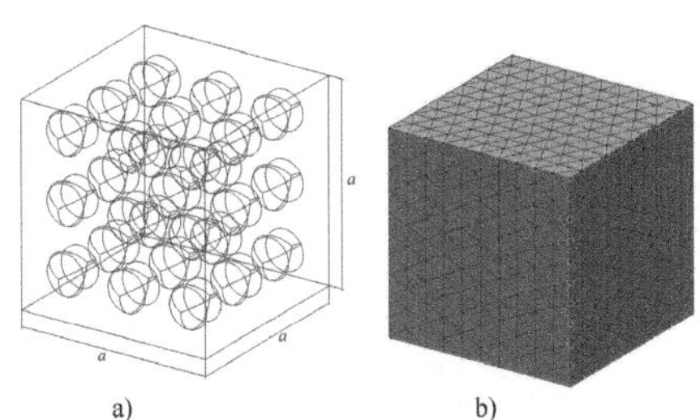

Figure 4. a) Geometry of exemplary RVE and b) RVE mesh.

3. the total cost of the structure, represented by the total volume fraction of the reinforcement:

$$\arg\min\{f_3(x); x \in D\};\ f_3(x) = \sum_{i=1}^{n} V_r^i \Big/ V^i \quad (7)$$

4.2.2 *Calculations*

The parameters of the EA are: the number of chromosomes n_{ch} = 40; the number of generations: n_g = 30; the arithmetical crossover probability: p_{ac} = 0.8; the uniform mutation probability: p_{um} = 0.01; the volume fraction range V_f = 5÷50%.

The results of the optimization is 3-dimentional Pareto frontier presented in Figure 5. Exemplary design variables and objective function values are presented in Table 2 and Figure 6 (better results are in bold type). To compare results obtained from NSGA-II and MOOPTIM algorithms, optimization is performed six times for each algorithm and hypervolume indicator value is calculated.

The results of algorithms comparison for hypervolume indicator values are presented in Table 3.

4.2.3 *Conclusions*

The multiobjective and multiscale optimization of the helicopter rotor blade made of FGM has been performed by means of the evolutionary computations. The results presented in Figure 6 show that the considered objective functions are contradictory. The values of design variables for the point closest to the ideal have the highest values in the outer sections placed close to rotor fixing.

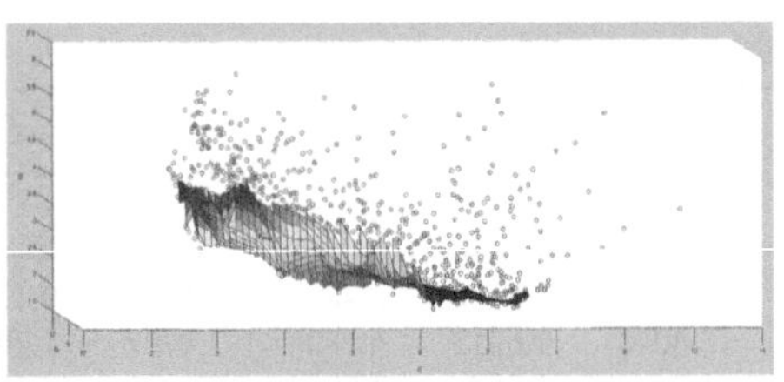

Figure 5. Exemplary Pareto frontier obtained from MOOPTIM.

Table 2. Exemplary objective function values.

	NSGA-II			MOOPTIM		
Point	f_1	f_2	f_3	f_1	f_2	f_3
Min (f_1)	2.4666	4.3142	5.5027	**1.6348**	4.2029	4.6722
Min (f_2)	6.5151	1.6245	2.8367	6.7606	**1.5745**	2.9641
Min (f_3)	8.1499	1.9612	**2.4378**	8.4025	1.7483	2.7627
Closest to ideal	4.1983	4.1872	**2.3519**	**2.5106**	**3.3847**	3.9125

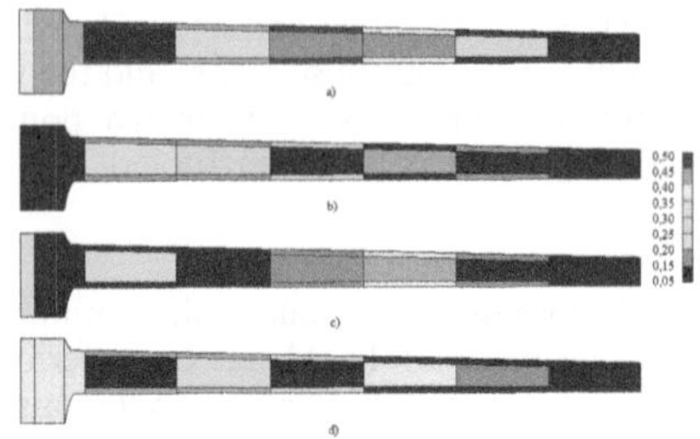

Figure 6. Exemplary design variables obtained from MOOPTIM: a) min (f_1), b) min (f_2), c) min (f_3), d) closest to the "ideal".

Table 3. Values of the hypervolume indicator.

	Hypervolume indicator			
Algorithm	Min	Max	Avg.	St. dev.
NSGA II	0.396	0.436	0.418	**0.015**
MOOPTIM	**0.404**	**0.520**	**0.4645**	0.043

5 FINAL CONCLUSIONS

The multiscale and multi-objective optimization of FGMs has been performed. Numerical homogenization methods, multiobjective EAs and FEM have been coupled to solve presented problems.

The proposed attitude allows the proper designing of structures made of FGMs for certain purposes with a different number of different optimization criteria.

The choice of the one of the results at the Pareto frontier should state the next step and this is out of the scope of the present paper.

ACKNOWLEDGMENTS

The research is partially financed from the Polish science budget resources as the research projects no. 10/990/BK_15/0013 and 10/040/BK 15/0006.

REFERENCES

Deb, K. et al. 2002. A fast and elitist multi-objective genetic algorithm: NSGA-II. *IEEE Transaction on Evolutionary Computation* 6 (2): 181–197.

Długosz, A. & Burczyński, T. 2012. Multiobjective shape of selected coupled problems by means of evolutionary algorithms. *Bull. Pol. Acad. Sci., Tech. Sci.* 60(2): 215–222.

Koizumi, M. 1997. FGM activities in Japan. *Compos. Part B: Engineering*, 28(1–2): 1–4.

Kouznetsova, V.G. 2002. *Computational homogenization for the multi-scale analysis of multiphase materials.* Ph.D. thesis TU Eindhoven.

Advances in Mechanics: Theoretical, Computational and Interdisciplinary Issues – Kleiber et al. (Eds)
© 2016 Taylor & Francis Group, London, ISBN 978-1-138-02906-4

Optimization of structural topology using unstructured Cellular Automata

B. Bochenek & K. Tajs-Zielińska
Institute of Applied Mechanics, Cracow University of Technology, Cracow, Poland

ABSTRACT: In recent years the Cellular Automata concept has been successfully applied to structural topology optimization problems. The majority of results that have been obtained so far were based on regular lattices of cells. Practical engineering analysis and design require however using, in many cases, highly irregular meshes for complicated geometries and/or stress concentration regions. The aim of the paper is to extend the concept of Cellular Automata towards implementation of unstructured grid of cells related to non-regular mesh of finite elements. Introducing irregular lattice of cells allows to reduce number of design variables without losing accuracy of results and without excessive increase of number of elements caused by using fine mesh for a whole structure. It is worth noting that the non-uniform density of finite elements can be, but not necessary is, directly related to design variables which are related to cells of Cellular Automaton. The implementation of non-uniform cells of Cellular Automaton requires a reformulation of standard local rules, for which the influence of neighborhood on current cell is independent of sizes of neighboring cells.

1 INTRODUCTION

For a few decades topology optimization has been one of the most important aspects of structural design. Since the early paper by Bendsoe & Kikuchi (1988) one can find in the literature numerous approaches to generating optimal topologies based both on optimality criteria and evolutionary methods. A general overview as well as a broad discussion on topology optimization concepts are provided by many survey papers e.g. (Eschenauer & Olhoff 2001, Rozvany 2008, Sigmund & Maute 2013, Deaton & Grandhi 2014). At the same time hundreds of paperspresent numerous solutions including classic Michell examples as well as complicated spatial engineering structures, implementing specific methods ranging from gradient based approaches to evolutionary structural optimization, biologically inspired algorithms, material cloud method, spline based topology optimization and level set method. It is a permanently developing area and one of the most important issues stimulating this progress nowadays is implementation of efficient and versatile methods to generation of optimal topologies for engineering structural elements.

In recent years the Cellular Automata paradigm has been successfully applied to topology optimization problems. In engineering implementation of Cellular Automaton the design domain is decomposed into a lattice of cells, and a particular cell together with cells to which it is connected form neighborhood. It is assumed that the interaction between cells takes place only within the neighborhood, and the states of cells are updated synchronously according to some local rules. The first application of Cellular Automata to the optimization of structures was proposed in the mid-1990s and the basic idea was described by Inou et al. 1994. The majority of papers dealing with that subject were published during the last two decades. The discussion on the application of Cellular Automata to structural optimization can be found, e.g. in research papers by (Kita & Toyoda 2000, Missoum et al. 2005, Tovar et al. 2006, Sanaei & Babaei 2011, Bochenek & Tajs-Zieli´nska 2012, Du et al. 2013).

The majority of structural topology optimization results that have been obtained so far were based on regular lattices of cells. Practical engineering analysis and design require however using, in many cases, highly irregular meshes for complicated geometries and/or stress concentration regions. The aim of the present paper is to extend the concept of Cellular Automata towards implementation of unstructured grid of cells related to non-regular mesh of finite elements. Introducing irregular lattice of cells allows to reduce number of design variables without losing accuracy of results and without excessive increase of number of elements caused by using fine mesh for the whole structure. It is worth noting that the non-uniform density of finite elements can be, but not necessary is, directly related to design variables which are represented by cells of Cellular Automaton. The implementation of non-uniform cells of Cellular Automaton requires a reformulation of standard local rules, for which the influence of neighborhood on current cell is independent of sizes of neighboring cells. This paper proposes therefore new local update rules dedicated to unstructured Cellular Automata.

2 UNSTRUCTURED CELLULAR AUTOMATA

The most of applications of Cellular Automata in structural optimization up till now are conventionally based on regularly spaced, structured meshes. On the other hand using unstructured computational meshes provides more flexibility for fitting complicated geometries and allows local mesh refinement. Some attempts to implement unstructured Cellular Automata have been already reported in the literature e.g. (O Sullivan 2001, Lin et al. 2011), but application to topology optimization is rather incidental (see Talischi et al. 2012).

In this paper the concept of topology generator based on Cellular Automata rules is extended to unstructured meshes. Similar to structured (regular) Cellular Automata, several neighborhood schemes can be identified. The two most common ones are the von Neumann and the Moore types. As can be seen in the Figure 1 in case of the von Neumann configuration only three immediate neighbors are taken into account. These neighboring cells share common edges with the central cell. In the Moore type neighborhood (see Fig. 2) any triangle that has common edge or common vertices with the central cell can be considered as a neighbor of the central triangle. It is worth noting that this type of neighborhood considers more neighbors around the central cell, as well as that the number of neighbors can vary since this depends on the particular unstructured mesh arrangement.

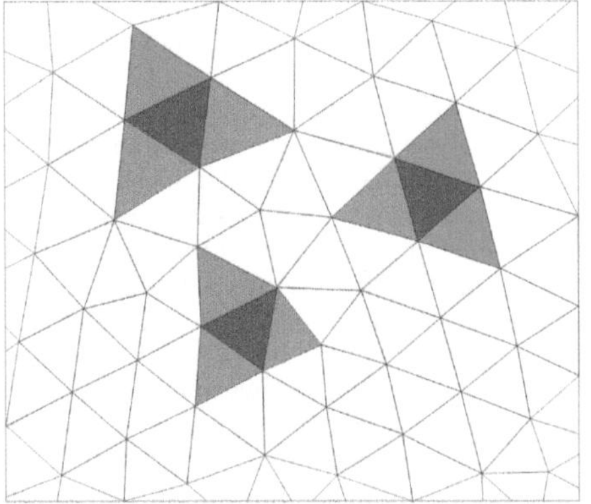

Figure 1. Unstructured triangular mesh. The von Neuman type neighborhood.

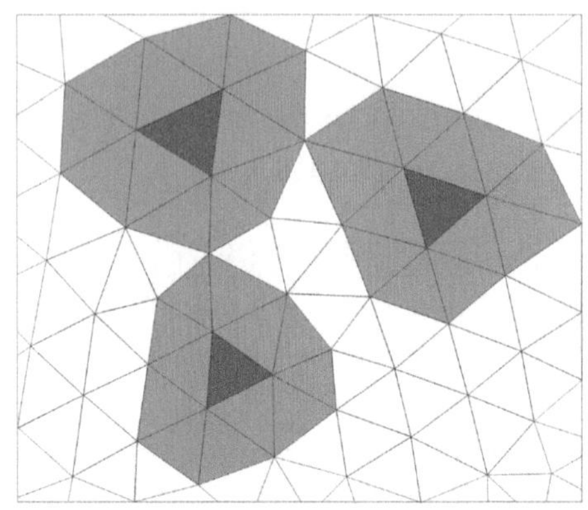

Figure 2. Unstructured triangular mesh. The Moore type neighborhood.

3 THE ALGORITHM

The performance of Cellular Automata algorithms, reported in literature, is often based on heuristic local rules. Similarly, in the present paper the efficient heuristic algorithm, being extension of the one introduced by Bochenek & Tajs-Zielińska (2012), (2013), has been implemented. The power law approach defining Solid Isotropic Material with Penalization (SIMP) with design variables being relative densities of a material has been utilized. The elastic modulus of each cell element is modelled as a function of relative density d_i using power law, according to (1). This power p penalizes intermediate densities and drives design to a solid/void structure.

$$E_i = d_i^p E_0, \; d_{min} \le d_i \le 1 \tag{1}$$

The local update rule applied to design variables d_i associated with central cells is now constructed based on the information gathered from adjacent cells forming the Moore or the von Neumann type neighborhood.

$$d_i^{(t+1)} = d_i^{(t)} + \delta d_i,$$
$$\delta d_i = \left(\alpha_0 + \sum_{k=1}^{N} \alpha_k \right) m = \tilde{\alpha} m \tag{2}$$

It is set up as a sum of design variables corrections (2), the values of which are influenced by the states of the neighborhood surrounding each cell.

$$\alpha_0 = \begin{cases} -C_{\alpha 0} & \text{if} \quad U_i \le U^* \\ C_{\alpha 0} & \text{if} \quad U_i > U^* \end{cases} \tag{3}$$

$$\alpha_k = \begin{cases} -C_\alpha & \text{if} \quad U_{ik} \dfrac{A_{ik}}{A_i} \le U^* \\ C_\alpha & \text{if} \quad U_{ik} \dfrac{A_{ik}}{A_i} > U^* \end{cases} \tag{4}$$

The compliance values calculated for central cell U_i and N neighboring cells U_{ik} are compared to a selected threshold value U^*. The quantities A_i and A_{ik} stand for areas of central and neighboring cells, respectively. Based on relations (3) and (4) specially selected positive or negative parameters C_{α_0} for central cell and C_α for surrounding cells are transferred to the design variable update. The move limit m implemented in the above algorithm controls the allowable changes of the design variables values. The numerical algorithm has been build in order to implement the above proposed design rule. As to the optimization procedure the sequential approach, has been adapted, meaning that for each iteration, the structural analysis performed for the optimized element is followed by the local updating process. Simultaneously a global volume constraint can be applied for specified volume fraction. If so the generated optimal topology preserves a specified volume fraction of a solid material.

4 GENERATION OF OPTIMAL TOPOLOGIES

Selected examples of compliance-based topologies generated using the approach being presented in this article are discussed in this section. The first example it is the L-shaped structure shown in Figure 3. The unstructured mesh that consists of triangular elements/cells has been applied. The more dense mesh surrounds two corners within stress concentration regions as shown in Figure 4. The topology optimization has been performed and the obtained results are presented in Figure 5, with final compliance 4.1 10^{-6} Nm for irregular lattice of 11396 cells and in Figure 6 with compliance 4.0 10^{-6} Nm for regular lattice of 36490 cells, respectively. Calculating the Huber-Mises-Hencky equivalent stress gives maximal value of 38.5 kPa for both unstructured and regular meshes. It is worth noting that in order to obtain comparable results over 3 times more cells for regular mesh were required. The second example it is the knee type structure shown in Figure 7. As for the previous case, the irregular/unstructured mesh that consists of triangular elements/cells has been applied. The more dense mesh surrounds region of loading application as presented in Figure 8. The minimal compliance topologies have been found for

Figure 4. L-shaped structure, irregular lattice of triangular cells.

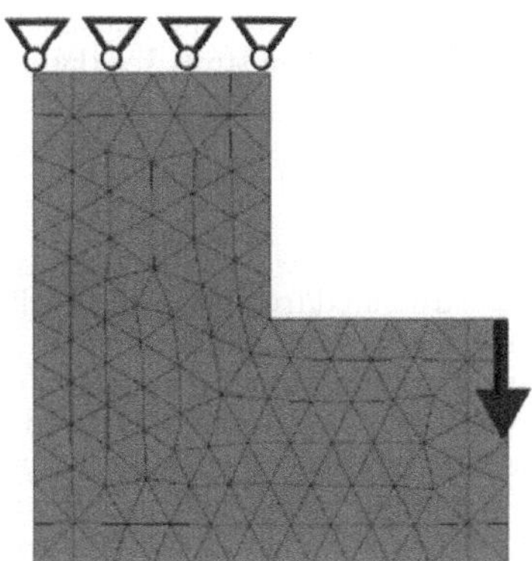

Figure 3. L-shaped structure, loading application, support scheme.

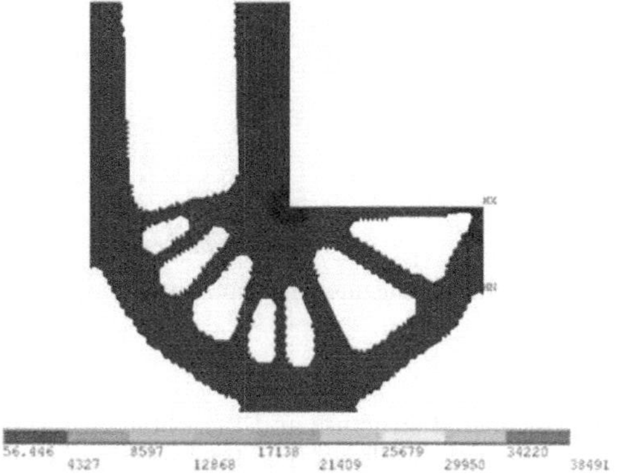

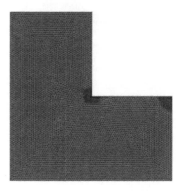

Figure 5. Final topology obtained for unstructured lattice of 11396 cells. Distribution of equivalent stress.

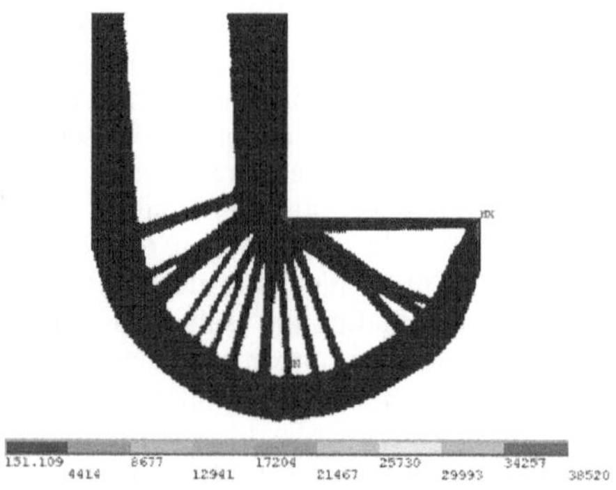

Figure 6. Final topology obtained for regular lattice of 36490 cells. Distribution of equivalent stress.

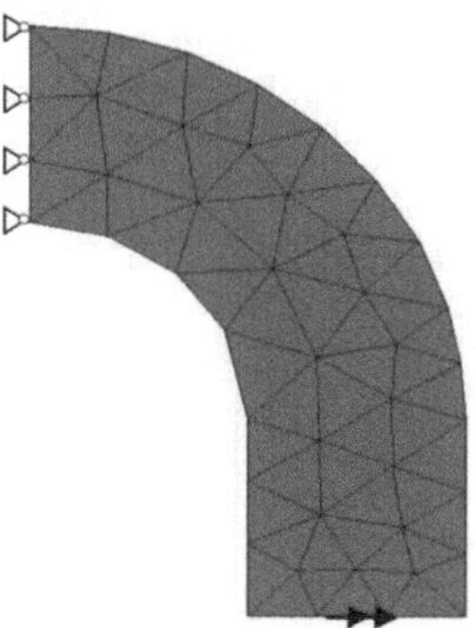

Figure 7. Knee structure, loading application, support scheme.

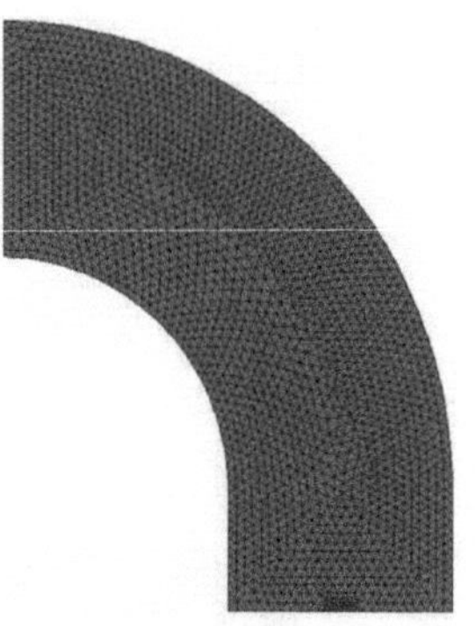

Figure 8. Knee structure, irregular lattice of triangular cells.

the considered structure and the obtained results are presented as follows. The final topology given in Figure 9 has been found for irregular mesh of 3811 cells, whereas the one shown in Figure 10 refers to regular lattice of 16155 cells. That many

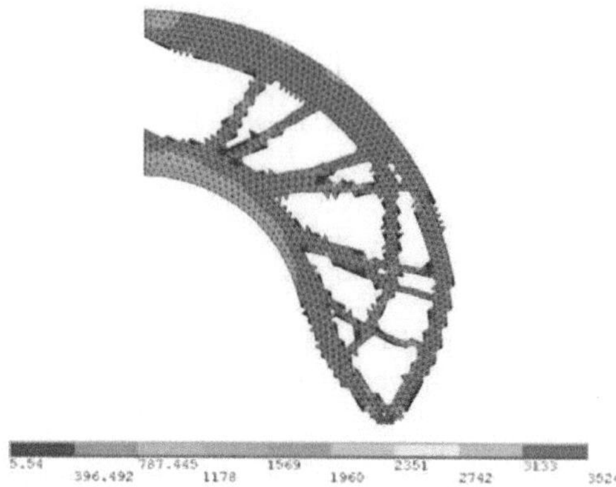

Figure 9. Final topology obtained for unstructured lattice of 3811 cells. Distribution of equivalent stress.

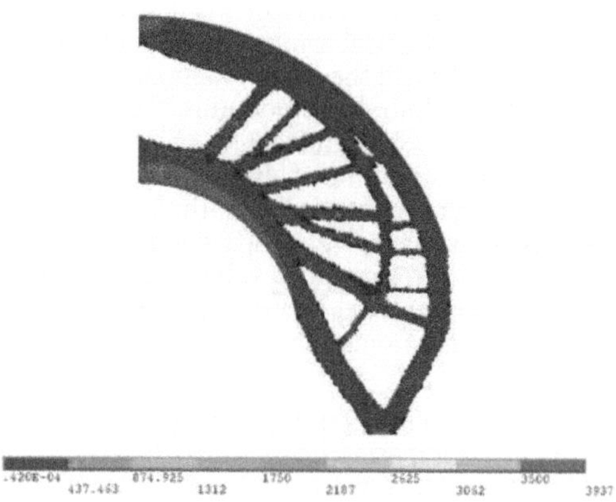

Figure 10. Final topology obtained for regular lattice of 16155 cells. Distribution of equivalent stress.

elements were necessary in order to end up with the same compliance value 1.2 10^{-7} Nm, maximal displacement 0.9 10^{-8} m, and maximal Huber-Mises-Hencky equivalent stress 3.9 kPa.

5 CONCLUDING REMARKS

The extension of Cellular Automata concept towards unstructured/irregular grid of cells related to non-regular mesh of finite elements has been presented. The obtained results confirm that local update rules proposed for unstructured Cellular Automata can be efficiently used for generation of minimal compliance topologies and irregular mesh suited for structural analysis can be directly used in optimization process. Since it is not necessary to use a very fine regular mesh for whole structure therefore number of elements/design variables can be reduced. Although the number of cells is limited, because of local mesh refinement, information about stresses and displacements can still be correct. Implementation of irregular lattice does not influence significantly value of compliance of final structure. The approach presented in this paper demonstrates a significant potential of application to problems which cannot be adequately represented by regular grids. The use of unstructured meshes may be helpful while modelling a domain geometry, accurately specify design loads or supports and compute structure response.

REFERENCES

Bendsoe M.P. & Kikuchi N. (1988). Generating optimal topologies in optimal design using a homogenization method. *Compututational Methods and Applications in Mechanical Engineering* 71, 197–224.

Bochenek B. & Tajs-Zielińska K. (2012). Novel local rules of Cellular Automata applied to topology and size optimization. *Engineering Optimization* 44, 1, 23–35.

Bochenek B. & Tajs-Zielińska K. (2013). Topology optimization with efficient rules of cellular automata. *Engineering Computations* 30, 8, 1086–1106.

Deaton J.D. & Grandhi R.V. (2014). A survey of structural and multidisciplinary continuum topology optimization: post 2000. *Structural and Multidisciplinary Optimization* 49, 1–38.

Du Y., Chen D., Xiang X., Tian Q. & Zhang Y. (2013). Topological design of structures using a cellular automata method. *Computer Modeling in Engineering and Sciences* 94, 1, 53–75.

Eschenauer H.A. & Olhoff N. (2001). Topology optimization of continuum structures: a review. *Applied Mechanics Review* 54, 331–390.

Inou N., Shimotai N. & Uesugi T. (1994). A cellular automaton generating topological structures. *Proceedings of the 2nd European Conference on Smart Structures and Materials* 2361, 47–50.

Kita E. & Toyoda T. (2000). Structural design using cellular automata. *Structural and Multidisciplinary Optimization* 19, 64–73.

Lin Y., Mynett A.E. & Li H. (2011). Unstructured Cellular Automata for modelling macrophyte dynamics. *Journal of River Basin Management* 9, 3–4, 205–220.

Missoum S., Gurdal Z. & Setoodeh S. (2005). Study of a new local update scheme for cellular automata in structural design. *Structural and Multidisciplinary Optimization* 29, 103–112.

O Sullivan D. (2001). Exploring spatial process dynamics using irregular cellular automaton models. *Geographical Analysis* 33, 1, 1–18.

Rozvany G.I.N. (2008). A critical review of established methods of structural topology optimization. *Structural and Multidisciplinary Optimization* 37, 217–237.

Sanaei E. & Babaei M. (2011). Cellular Automata in topology optimization of continuum structures. *International Journal of Engineering Science and Technology* 3, 4, 27–41.

Sigmund O. & Maute K. (2013). Topology optimization approaches. *Structural and Multidisciplinary Optimization* 48, 1031–1055.

Talischi C., Paulino G.H., Pereira A. & Menezes I.F.M. (2012). PolyTop: a Matlab implementation of a general topology optimization framework using unstructured polygonal finite element meshes. *Structural and Multidisciplinary Optimization* 45, 329–357.

Tovar A., Patel N.M., Niebur G.L., Sen M. & Renaud J.E. (2006). Topology optimization using a hybrid cellular automaton method with local control rules. *Journal of Mechanical Design* 128, 1205–1216.

Advances in Mechanics: Theoretical, Computational and Interdisciplinary Issues – Kleiber et al. (Eds)
© 2016 Taylor & Francis Group, London, ISBN 978-1-138-02906-4

A mesh-free particle model for simulation of trimming of aluminum alloy sheet

L. Bohdal & R. Patyk
Koszalin University of Technology, Koszalin, Poland

ABSTRACT: In the paper, the applications of mesh-free SPH (Smoothed Particle Hydrodynamics) continuum method to the simulation and analysis of trimming process of aluminum alloy sheet is presented. In dealing with trimming simulations existing literatures Finite Element Method (FEM) is applied to analyse this process. Approach presented in this work and its application to trimming of aluminum alloy sheet allows for a complex analysis of physical phenomena occurring during the process without significant deterioration in the quality of the finite element mesh during large deformation. This allows for accurate representation of the loss of cohesion of the material under the influence of cutting tools. Analysis of state of stress, strain and fracture mechanisms of the material is presented. In experimental studies, advanced vision-based technology based on Digital Image Correlation (DIC) for monitoring the cutting process is applied.

1 INTRODUCTION

Shearing plays a very important role in mechanical manufacturing area, the nature of trimming being concerned with many related subjects of technology and industry (Ilinich et al. 2011). However, trimming is a very complicated process. There are several factors, such as nonlinearities (geometrical, physical and thermal), large deformation, friction, large strain rates of material and material characteristics that have a direct influence on the process and are sensitive to each other. The knowledge of the trimming process is based mainly on experimental methods, which are often expensive and unable to be extrapolated to other cutting configurations. Trimming modelling is becoming an increasingly important tool in the improvement understanding and developing this process. At the moment the trimming numerical models are based on Lagrangian or Arbitrary Lagrangian Eulerian (ALE) Finite Element Methods (FEM) (Golovashchenko 2006, 2007). Hence, the use of grid/mesh can lead to various difficulties in dealing with problems with free surface, deformable boundary, moving interface, and extremely large deformation and crack propagation. There are significant difficulties in modeling process of material failure, as well as a significant sheet mass loss during deletion of mesh elements in the shearing area (Hilditch & Hodgson 2005). Element deletion leads not only to mass loss, but also results in significant decrease in predicted trimming force, residual stresses, workpiece strain and strain state. Moreover, for the problems with complicated geometry, generation of a quality mesh has become a difficult, time-consuming and costly process.

The mentioned disadvantages of the finite element models can be eliminated using the following mesh-free methods: Smoothed Particle Hydrodynamics (SPH), element-free Galerkin method, discrete element method, amongst others.

In the paper, the applications of mesh-free SPH methodology to the simulation and analysis of 3-D trimming process is presented. The developed model is used to analyse residual stresses in workpiece during and after process under different conditions. Next, the model is validated with experimental research using the vision-based solutions.

2 SPH METHOD

SPH is total Langrangian and is a truly mesh-free technique initially developed by Gingold & Monaghan (1977) and others for the analysis and simulation of astrophysics problems. The technique was later extended to model of discontinuous flows with large deformations and to analysis and simulate for large strain solid mechanics problems (Bagci 2011, Das & Cleary 2007, Heisel et al. 2011). SPH combines the advantages of mesh-free, particle methods and Langrangian. The advantage of the mesh-free method is its capability to solve problems that cannot be influencely solved using other numerical approaches.

The main differences between FEM and SPH are the absence of a grid and in the discretisation of continuum. It does not suffer from the mesh distortion problems that limit Lagrangian methods based on structured mesh when simulating large deformations. The idea of this method is to divide a continuum into discrete elements, called particles which are placed at some distance *d* from each other. This distance is called particle density *d*. The smoothing of field variables is performed in the area with radius *h*, called the smoothing length, over which the variables are smoothed by a kernel function. This means that the value of a variable in any spatial point can be obtained by adding the relevant values of the variables within two smoothed lengths.

The SPH approximation of the equation for continuum mechanics uses the following approaches. A function $f(x)$ is substituted by its approximation $A_f(x,h)$, characterising a body condition. For example, the velocities of body points in a particular area are approximated with the following expression (Bagci 2011, Heisel et al. 2011):

$$A_f(x, h) = \int f(y) \cdot W(x, h) dy, \tag{1}$$

where $W(x, h)$ is a smoothed kernel function. The size of the smoothing kernel is defined by the function of Θ:

$$W(x, h) = \left(\frac{1}{h(x)^p} \right) \cdot \Theta(x), \tag{2}$$

where p is the dimension of space.

The function $W(x, h)$ has to be symmetrical in relation to the vertex. The majority of the smoothing kernels used in the SPH method is represented as cubic B-spline, determining the selection of the function Θ as follows:

$$\Theta(x) = C \cdot \begin{cases} 1 - \frac{3}{2} \cdot x^2 + \frac{3}{4} \cdot x^3, & \text{if} |x| \le 1 \\ \frac{1}{4} \cdot (2-x)^3, & \text{if } 1 \le |x| \le 2, \\ 0 & \text{if } 2 \le |x| \end{cases} \tag{3}$$

where C is the normalisation constant. Integration time step can be determined by following equation:

$$\Delta\tau = C_{\Delta\tau} \cdot \min_i \left(\frac{h_i}{C_i + v_i} \right) \tag{4}$$

where i is the particle number; $C_{\Delta\tau}$ is the time step increase coefficient; v_i is the velocity of particle i. It is important to notice that coefficient $C_{\Delta\tau}$ directly influences the integration time step (Das & Cleary 2007).

A quadratic approximation of the particle motion is mainly used for the SPH method. A motion of the particles can be described here with the following equation:

$$\frac{\partial v_i^\alpha}{\partial t} = \sum_{j=1}^{N} m_j \cdot \left(\frac{\sigma_i^{\alpha\beta}}{d_i^2} + \frac{\sigma_j^{\alpha\beta}}{d_j^2} + A_{ij} \right) \cdot \frac{\partial W_{ij}}{\partial x_i^\beta}, \tag{5}$$

where j is particle number; N is the number of neighbouring particles; $v_i^\alpha = dx_i^\alpha / dt$ is the velocity of particle i; m_j is the mass of particle j, $\sigma_i^{\alpha\beta}$, $\sigma_j^{\alpha\beta}$ are the stress tensors of i and j particles respectively; d_i and d_j are the densities of i and j particles respectively; A_{ij} are the specific external forces; $W_{ij} = W(x_i - x_j, h)$ is the smoothing kernel.

3 SPH MODEL OF TRIMMING

In the suggested approach to the modeling of the trimming process, a coupling of the FEM model and the model based on hydrodynamic particles (the SPH method) has been proposed. A three-dimensional model of trimming is built in LS—Dyna solver. Numerical calculations are performed for the 3D state of strain and 3D state of stress in this model.

An AA6111-T4 aluminum alloy is used as the material to be cut in numerical and experimental studies, with thickness of $t = 1$ mm. A velocity of $v = 100$ mm/s is applied to the upper knife in the y direction. The contact between ideally rigid tools and the deformable sheet metal is described using Coulomb's friction model, and constant coefficients of static friction $\mu_s = 0.08$ and kinetic friction $\mu_d = 0.009$ are accepted. In trimming models, accurate and reliable flow stress models are considered as highly necessary to represent workpiece material constitutive behaviour, the constitutive material model reported by Johnson and Cook is employed in this study, it is often used for ductile materials in cases where strain rate vary over a large range and where adiabatic temperature increase due to plastic heating cause material softening.

A series of simulations are carried out to determine the optimal parameters of the solver, ensuring a minimum prediction error of trimming variables and minimal simulation cost. The computer simulations are executed for different initial particle densities: d_1 = 0,04 mm, d_2 = 0,05 mm, d_3 = 0,06 mm, d_4 = 0,07 mm, d_5 = 0,08 mm, d_6 = 0,09 mm, d_7 = 0,1 mm.

A initial particle density of d_2 = 0,05 mm is selected as a optimal particle density for used model dimensions. This initial particle density is selected so as to have a reasonable number of particles at the thickness of material. Larger particle density not illustrate the material flow features and stress distributions appropriately (Fig. 1), and a smaller initial particle density strongly increase the computing time.

The obtained results make it clear that initial particle density has an effect on force-tool displacement (penetration) curves (Fig. 2). Stabilization of

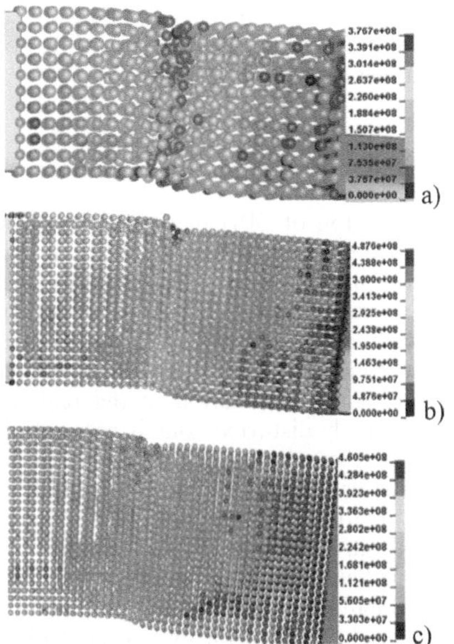

Figure 1. Stress intensity distribution in plastic flow phase of the process for different initial particle density: a) $d_7 = 0,1$ mm (simulation time: 40 min), b) $d_2 = 0,05$ mm (simulation time: 164 hours, 37 min), c) $d_1 = 0,04$ mm (simulation time: 391 hours, 11 min).

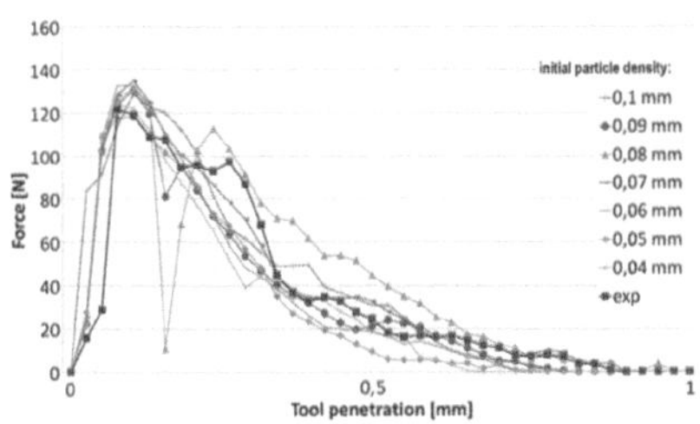

Figure 2. Force-tool penetration curves for analyzed initial particle density.

the force-tool displacement curves is reached when initial particle density carry out d ≤ 0,07 mm.

4 EXAMPLES OF NUMERICAL AND EXPERIMENTAL RESULTS

Sample results of numerical and experimental investigations at various stages of the trimming process are shown in Figures 3–5. At the beginning of the cutting process during the elastic-plastic phase the plasticised zone occurs only at the highest concentration of stresses, that is in direct contact with the cutting edges of tools. During the second phase the intensive plastic flow of the material in the surroundings of the cutting surface can be observed (Fig. 3). A characteristic distortion of SPH particles in this areas can be seen. In experimental studies, an advanced vision-based technology based on Digital Image Correlation (DIC) for monitoring the cutting process is used (Fig. 3b). In two-dimensional digital image correlation, displacements of material are directly detected from digital images of the surface of an object. Then, the images on the surface of the object, one before and another after deformation are recorded, digitized and stored in a computer as digital images. These images are compared to detect displacements by searching a salient features from one image to another. Using this method it is possible to determine the areas of strong nonlinearities and deformation of material structure.

Both in numerical model and experimental investigations the plastic strain localization zones propagate much faster from the moving (upper knive) blade than from the bottom stationary blade (Fig. 4). The deformation zone is non-symmetric with respect to the top and bottom blades. Since the crack will initiate and propagate through the localization zone, the cut surface will be curved and the burr will form because the crack will not run to the bottom blade tip. This trend is evident in both the numerical model as well as experimental studies. In mechanically affected zone a characteristic material flow along the crack path can be observed. To accurately measure the material flow on the surface of an object, a texture patterns need to be related to small group areas. Obtained results are in agreement with these experimental observations

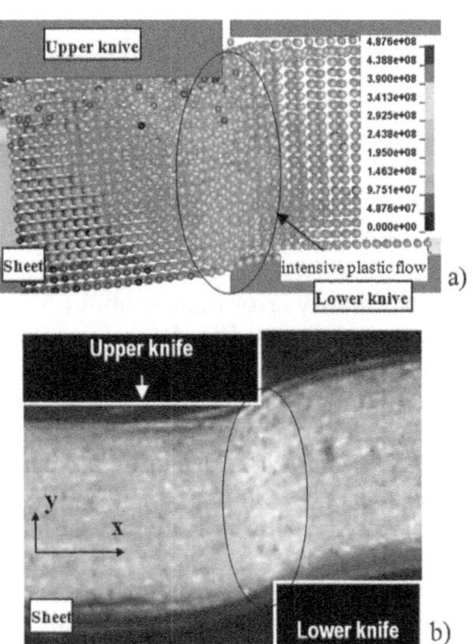

Figure 3. Comparison of the numerical and experimental results for upper knive penetration w = 0,25 mm: a) FEM+SPH model (equivalent stress distribution), b) recorded image.

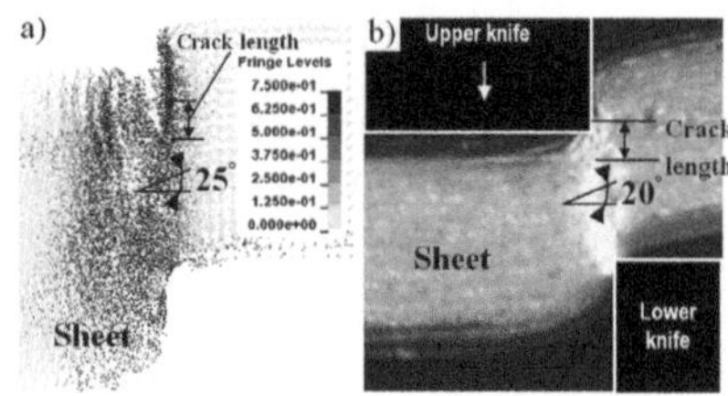

Figure 4. Comparison of the numerical and experimental results for upper knive penetration w = 0,45 mm: a) FEM+SPH model (equivalent strain distribution), b) recorded image from a high-speed camera.

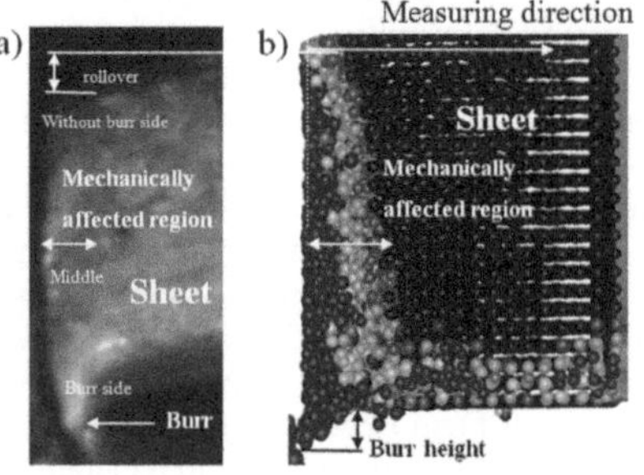

Figure 5. Quality of cut surface measurement: a) experiment, b) FEM+SPH model.

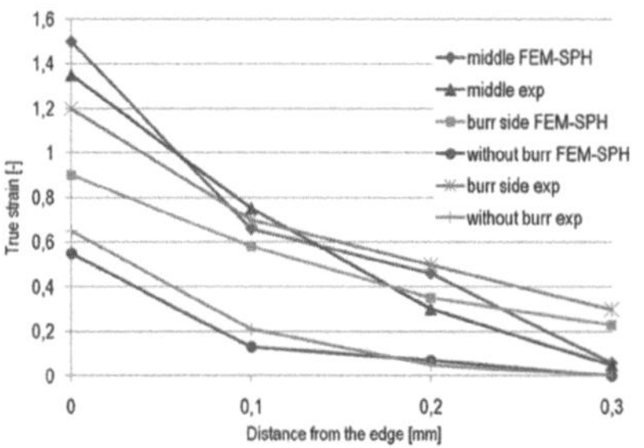

Figure 6. Results of the true strain measured in sheared edge.

with approximately error margin about 5° in flowing angle measurement (Fig. 4).

Analysis of strain state in the cut surface shows that the maximum value of plastic strain occurs at the edge and then gradually decreases with increasing distance from the cut edge until it reaches a low constant value (Figs. 5 and 6).

The measured results indicate that the maximum true strain has reached a value of 1,5 at the middle of cut surface, and it exponentially decays away from the edge. The lowest values of strain occur in the upper part of the sample. The width of the deformed area is small and carry out approximately 0.1 mm (Fig. 6). Comparison of the characteristic features of cut surface obtained from numerical model and experiment showed good agreement in the depth angle and the burr height. Some differences occurred in the measurement of the rollover of the upper edge of workpiece.

5 CONCLUSIONS

The application of SPH method is fairly new in metal forming simulations. Its features are not fully understood and the most effective means to exploit it are still being discovered. A new hybrid approach that used the SPH formulation in the shearing zone with high material distortion and the Lagrangian formulation in the region away from the highly distorted zone is very robust and reliable. It reduces the simulation time needed to obtain the results. The FEM-SPH method is validated using the advanced vision based system and DIC method. Based on the experimental data, the model has been shown to be able to provide adequate estimation of the deformation state of material and might be used for process control as well as optimizing the trimming parameters. Additionally, the next paper will be focused on examination and analysis of influence of process technological parameters for example: the clearance, tool geometry, cutting velocity on residual stresses and quality of cut surface using the SPH method.

REFERENCES

Bagci, E. 2011. 3-D numerical analysis of orthogonal cutting process via mesh-free method, *International Journal of Physical Sciences* 6: 1267–1282.

Das, R. & Cleary, PW. 2007. Modeling plastic deformation and thermal response in welding using smoothed particle hydrodynamics, *16th Australasian fluid mechanics conference*, 2–7 December 2007.

Gingold, R.A. & Monaghan, J.J. 1977. Smooth particle hydrodynamics: theory and application to non-spherical stars, *Monthly Notices of the Royal Astronomical Society* 181, 375–389.

Golovashchenko, S.F. 2006. A study on trimming of aluminum autobody sheet and development of a new robust process eliminating burrs and slivers, *International Journal of Mechanical Sciences* 48, 1384–1400.

Golovashchenko, S.F. 2007. Analysis of Trimming of Aluminum Closure Panels, *JMEPEG* 16, 213–219.

Heisel, U. Zaloga, W. Krivoruchko, D. Storchak, M. & Goloborodko, L. 2013. Modelling of orthogonal cutting processes with the method of smoothed particle hydrodynamics, *Production Engineering Research and Development* 7, 639–645.

Hilditch, T.B. & Hodgson, P.D. 2005. Development of the sheared edge in the trimming of steel and light metal sheet, Part 2—Mechanisms and modeling. *Journal of Materials Processing Technology*, 169, 192–198.

Ilinich, A.M. Golovashchenko, S.F. Smith, L.M. 2011. Material anisotropy and trimming method effects on total elongation in DP500 sheet steel, *Journal of Materials Processing Technology* 211, 441–449.

Advances in Mechanics: Theoretical, Computational and Interdisciplinary Issues – Kleiber et al. (Eds)
© 2016 Taylor & Francis Group, London, ISBN 978-1-138-02906-4

New method of generating Strut and Tie models using truss topology optimization

K. Bołbotowski & T. Sokół
Faculty of Civil Engineering, Warsaw University of Technology, Poland

ABSTRACT: The paper deals with automatic generating of Strut and Tie models via truss topology optimization. A method is developed to modify the classic ground structure approach. In order to rationalize the solution a penalty parameter, that represents cost of the nodes, is introduced. Stresses in concrete are controlled by an iterative construction of the ground structure, hence the design code requirements are met. The linear programming problem is handled with the use of the adaptive ground structure by means of a solution of both primal and dual formulations of the optimization problem. Algorithms allowing generating the ground structure in arbitrary domains bounded by polygons assure versatility of the method.

1 INTRODUCTION

Strut and Tie (ST) models are used to approximate complex stress fields in a reinforced concrete structures by trusses in which bars in compression (struts) map the stresses in the concrete, whilst the bars in tension (ties) correspond to the steel reinforcement. ST models are incorporated in design codes, such as Eurocode 2 (BS-EN 1992-1-1:2004), as a tool for the ultimate limit state design of the reinforcement in two dimensional plate-like elements of the reinforced concrete structure, namely the bearing capacity of the truss is a lower bound of the capacity of the structure by virtue of the limit load theorem known from plasticity. The codes, however, do not offer any indication on how such models should be constructed for arbitrary structures, hence leaving this problem for designers. The only restriction the codes set on the models is that directions of the bars must not differ significantly from the directions of principal stresses in the corresponding solid, elastic body in order to satisfy the ductility requirements and to avoid excessive cracking.

Among several examples of ST models made available in Eurocode 2 there is a model for a corner of the frame in which the depth of the beam is more than twice the width of the column. In this model the directions of bars do not coincide with directions of principal stresses in the corresponding solid body; moreover, the amount of steel required with the use of this model is alarmingly high.

The example of the frame corner model exposes issues with creating ST models for nonstandard reinforced concrete structures. It implies a need for developing a versatile method allowing to automatically construct ST models for various reinforced concrete members. Literature review shows, that topology optimization can be applied for this purpose; e.g. Victoria et al. 2011 obtain truss-resembling layouts by means of solving compliance minimization problem for 2D solid body using two materials.

In the paper a topology optimization problem is formulated for a structure that is a truss by definition. The method used is a modification of the classic ground structure approach where truss volume is minimized subject to stress constraints (see Sokół 2014) and its first version was published in Bołbotowski et al. (2013). Here, an algorithm based on the adaptive ground structure approach was incorporated in the method increasing its efficiency.

2 FORMULATION OF TRUSS TOPOLOGY OPTIMIZATION PROBLEM

2.1 *Definition of ground structure "T" and "C"*

Let us consider a feasible domain bounded by arbitrary set of polygons. Regular mesh of points is stretched over a rectangle circumscribed about the domain, the points contained in the domain being called *truss nodes*. Each node is connected with each other by a line segment; the segments contained in the domain become *bars* of dense truss called *initial ground structure*. Dedicated, versatile algorithms were developed to determine whether given node or bar is contained in the feasible domain.

Two substructures "T" and "C" are extracted from the initial ground structure, the construction materials being assigned to them—steel and concrete respectively. The ground structure "T" does not contain, in comparison to the initial one, bars that are placed too close to the boundary of the domain due to concrete cover requirements. Construction of the structure "C" is more complex and will be discussed in section 3, at this point let us assume that the structure "C" coincides with the initial ground structure. The support conditions are implied, as well as the loads determined by vector $\boldsymbol{P}_{[s]}$, where s denotes the number of degrees of freedom of the initial ground structure. For each "T" and "C" trusses the following quantities are determined: number of bars *nt*, *nc*, vectors of bar lengths $\boldsymbol{L}_{T\,[nt]}$, $\boldsymbol{L}_{C\,[nc]}$ and geometric matrices $\boldsymbol{B}_{T\,[nt\,x\,s]}$, $\boldsymbol{B}_{C\,[nc\,x\,s]}$. Vectors of cross section areas of bars $\boldsymbol{A}_{T\,[nt]}$, $\boldsymbol{A}_{C\,[nc]}$ are treated as design variables in the optimization process.

Figure 1 presents a ground structure in a problem of short cantilever, with the feasible domain being rectangular. A sparse mesh of 6×5 nodes is used (for the sake of clarity), which generates 435 bar ground structure.

2.2 *Modified problem of truss volume minimization. Primal and dual formulation*

In order to effectively design the reinforcement of the structure with the use of ST models one could pose an optimization problem as follows: find vectors $\boldsymbol{A}_T$, $\boldsymbol{A}_C$ that minimize volume of the bars in tension and for which ultimate limit state under the given load $\boldsymbol{P}$ is not exceeded, namely under the stress limits (σ_T and σ_C are yield stresses for steel and concrete respectively). By solving such a problem we arrive at a highly dense truss that approximate a Michell structure (see Fig. 1b in Lewiński et al. 2013) which is not applicable in practice.

Given the above we formulate, after eliminating the design variables $\boldsymbol{A}_T$, $\boldsymbol{A}_C$ in the first place, the following linear programming problem

$$V_{T\,opt} = \min\left\{ V_T = \frac{1}{\sigma_T}\tilde{\boldsymbol{L}}_T \cdot \mathbf{T} \,\middle|\, \mathbf{B}_T^{\mathrm{T}}\mathbf{T} - \mathbf{B}_C^{\mathrm{T}}\mathbf{C} = \mathbf{P};\ \mathbf{T}, \mathbf{C} \geq 0 \right\} \tag{1}$$

which is a modified problem of truss volume minimization. The modification is introduced by the modified length vector $\tilde{L}_{T,k} = L_{T,k} + p$ for $k = 1, \ldots, nt$; p is a *penalty parameter* representing the cost of the truss nodes and it is equal for all the bars in the truss "T". The larger the parameter p the less complex the solution, since efficiency of short bars decreases with its value. The parameter p is to be set by the designer.

Treating problem $V_{T\,opt}$ as its primal formulation one can pose a dual problem arriving at

$$V_{T\,opt,dual} = \max\left\{ \boldsymbol{P} \cdot \boldsymbol{v} \,\middle|\, \frac{1}{\sigma_T}\tilde{\boldsymbol{L}}_T \geq \boldsymbol{\Delta}_T, \boldsymbol{\Delta}_C \geq 0 \right\} \tag{2}$$

where: $\boldsymbol{v}_{[s]}$ is a vector of *adjoint displacements* of the truss nodes; $\boldsymbol{\Delta}_{T\,[nT\,x\,s]} = \boldsymbol{B}_T\,\boldsymbol{v}$, $\boldsymbol{\Delta}_{C\,[nC\,x\,s]} = \boldsymbol{B}_C\,\boldsymbol{v}$ are vectors of *adjoint elongations* of bars of the ground structures "T", "C".

Figure 1 includes a solution of the problem $V_{T\,opt}$ formulated for the ground structure presented there (with parameter p being equal to zero).

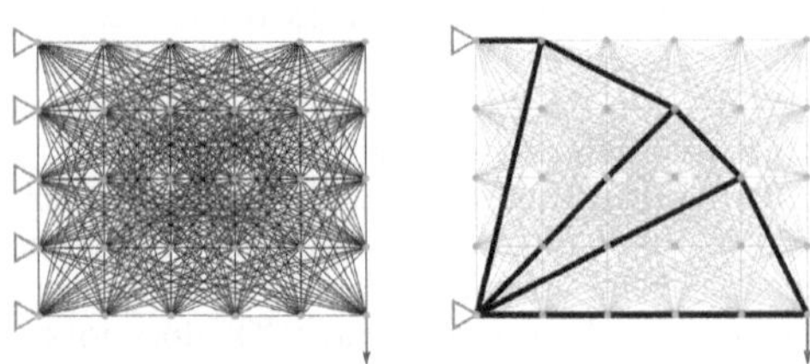

Figure 1. A 435 bar ground structure spread on 6×5 mesh of nodes (on the left) and solution of problem $V_{T\,opt}$ (on the right).

3 ALGORITHMS OF THE METHOD

3.1 *Control of the stresses in concrete*

According to the design codes the stresses in the bars in compression cannot exceed limit stresses for concrete. A bar of given compressive force must thus have properly large cross section area, which is, on the other hand, bounded by geometry of the reinforced concrete structure. This limitation cannot be introduced directly as a constraint of the problem $V_{T\,opt}$ with its linearity being preserved.

A solution for the ground structure "C" being equal to the initial one contains, in general, bars in compression that do not fit in the feasible domain. This particular solution is treated as first iteration, namely $V_{T\,opt} = V^{(1)}_{T\,opt}$. By means of elimination of the bars intersecting the boundary, a new ground structure "C" is determined and used in the second iteration $V^{(2)}_{T\,opt}$. This procedure is continued until iteration *i*, for which all the compressed bars of solution are contained in the domain. The algorithm for concrete stress control is demonstrated in Figure 2.

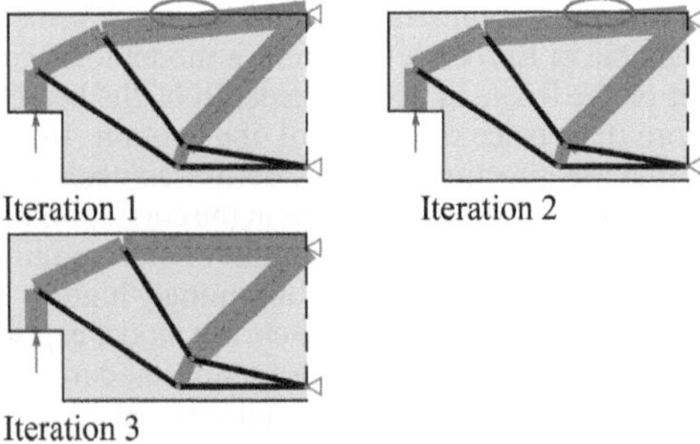

Figure 2. Illustration of the algorithm allowing control of the stresses in concrete.

3.2 *New algorithm based on the adaptive ground structure approach*

In the paper by Bołbotowski et al. (2013) a Strut and Tie model for the frame corner problem was generated via topology optimization with the use of ground structure consisting of over half of million bars, while computing time reached almost 10 minutes. A study of other problems shows, that for more complex feasible domains (i.e. domains with openings) a very dense ground structure (even 10 million bars) is required, if more detailed design is desired. For such a ground structure the problem $V_{T\ opt}$ becomes, despite its linearity, impossible to solve directly while using a personal computer.

In order to cope with multimillion ground structures and to increase the effectiveness of the method for smaller ones, each problem $V^{(i)}_{T\ opt}$ is solved with the approach of the adaptive ground structure. Instead of handling the problem $V^{(i)}_{T\ opt}$ for the fully connected ground structure straightaway, a sequence of subproblems $V^{(i,j)}_{T\ opt}$ is solved for far smaller structures (each iteration $V^{(i)}_{T\ opt}$ is solved in one level deeper iterative manner). The algorithm starts with first subiteration $V^{(i,1)}_{T\ opt}$ formulated for ground structure consisting of bars of length 1 only (in the sense of maximum norm), that is bars connecting neighbouring nodes (node mesh is kept full). Such a construction yields number of bars merely 4 times the number of nodes, which even for very dense meshes does not exceed 100 thousand. A solution $\boldsymbol{T} = \boldsymbol{T}^{(i,1)}$, $\boldsymbol{C} = \boldsymbol{C}^{(i,1)}$ of the primal problem is then obtained, as well as a solution $\boldsymbol{v} = \boldsymbol{v}^{(i,1)}$ of the dual one. Let $\boldsymbol{\Delta}_T = \boldsymbol{\Delta}^{(i,1)}{}_T$ and $\boldsymbol{\Delta}_C = \boldsymbol{\Delta}^{(i,1)}{}_C$ denote vectors of adjoint elongations determined for fully connected ground structure and for adjoint displacements $\boldsymbol{v}^{(i,1)}$. In the second subiteration ground structures "T" and "C" are constructed by adding bars of length 2 and of indices kt and kc, for which elongations $\Delta^{(i,1)}{}_{T,kt}$ and $\Delta^{(i,1)}{}_{C,kc}$ violate dual problem constraints – in general such bars do exist, since displacements $\boldsymbol{v}^{(i,1)}$ guarantee satisfying these constraints only for bars of the ground structure from the previous subiteration. These steps are repeated until subiteration j for which all the bar elongations $\boldsymbol{\Delta}^{(i,j)}{}_T$, $\boldsymbol{\Delta}^{(i,j)}{}_C$ in full ground structure satisfy the dual problem constraints. At this point solution $\boldsymbol{v}^{(i,j)}$ of dual problem $V^{(i,j)}_{T\ opt,dual}$ is at the same time the solution $\boldsymbol{v}^{(i)}$ of the dual problem $V^{(i)}_{T\ opt,dual}$, which implies, that the solution $\boldsymbol{T}^{(i,j)}$, $\boldsymbol{C}^{(i,j)}$ of the primal problem $V^{(i,j)}_{T\ opt}$ is, in fact, the solution $\boldsymbol{T}^{(i)}$, $\boldsymbol{C}^{(i)}$ of the problem $V^{(i)}_{T\ opt}$, which was sought in the first place.

Figure 3 illustrates iterations of the algorithm described above for problem stated in Figure 1. The reader should notice, that throughout all the iterations the problem (1) is being solved for a ground structure nearly twice smaller than the full one. This profit increases drastically with the density of node mesh, what will be confirmed in section 4.

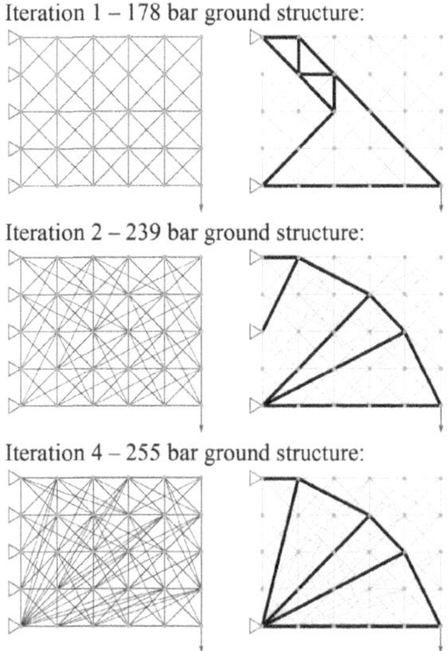

Figure 3. Solution of the optimization problem stated in Figure 1 with the adaptive ground structure approach. Ground structure and solution in each iteration are respectively on the left and right-hand side.

4 EXAMPLES OF GENERATION OF STRUT AND TIE MODELS USING THE NEW METHOD

In the paper Bołbotowski et al. (2013) present authors carried out a precise analysis of generation of Strut and Tie model for the frame corner mentioned in the introduction. A method used there was quite the same as described here, except each iteration $V^{(i)}_{T\ opt}$ was solved directly, namely without using the algorithm of the adaptive ground structure.

In the paper the solution of the very frame corner problem is obtained using new method incorporating the adaptive ground structure approach. The method was implemented as a computer program in Wolfram Mathematica environment. The mesh used in computations is the same as in the paper Bołbotowski et al. (2013): 32 × 40 nodes which generates 1 253 353 bars contained in both ground structures "T" and "C". The computation time in the case of the method with and without the adaptive algorithm was 69 sec and 572 sec, respectively, hence the new method turned out to be over 8 times faster.

The Strut and Tie model generated for the frame corner is presented in Figure 4 (identical for both methods). It is worth mentioning here, that the reinforcement designed with the use of the model requires about 45% less steel than with the one proposed in Eurocode 2. Two more examples of generated ST models are showed in Figure 4—for a cantilever of medium length and short cantilever

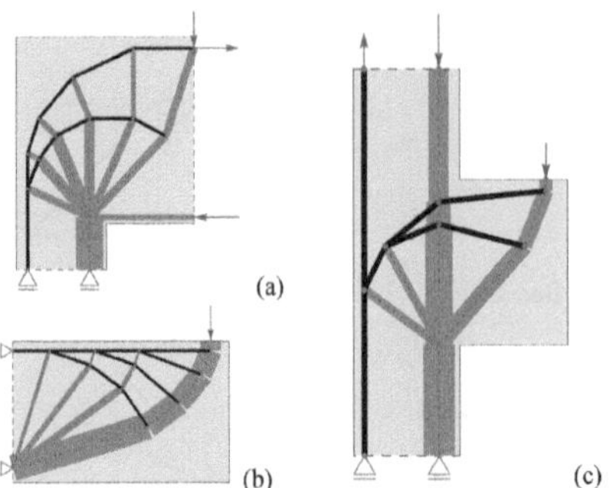

Figure 4. Examples of Strut and Tie models generated for different problems: (a) the frame corner, (b) cantilever of medium length, (c) short cantilever fixed in a column.

Table 1. Efficiency comparison of the method with and without the use of the adaptive ground structure: F.C.—frame corner, M.C.—medium-long cantilever, S.C.—short cantilever.

Problem	Ground structure	Without adapt. CPU time (sec)	With adapt. CPU time (sec)
F.C.	1 172 147	572	69
M.C.	1 656 977	920	108
S.C.	186 997	106	14

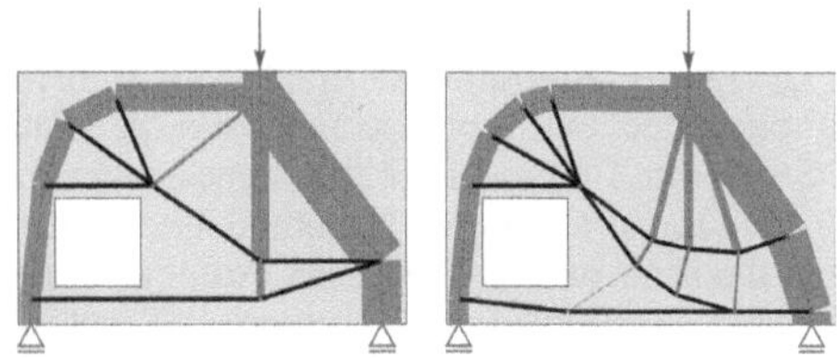

Figure 5. Comparison of Strut and Tie models generated with the use of old (left-hand side) and new method (right-hand side).

fixed in a column. Table 1 deals with efficiency comparison of two methods based on computation time needed to generate each model from Figure 4.

On the left side of Figure 5a Strut and Tie model for deep beam with opening is presented. This model was first showed in Bołbotowski et al. 2013, where a mesh of 32 × 20 nodes was used, arriving at 229 540 bar ground structure. Due to a complex feasible domain and a low resolution of the mesh, the penalty parameter had to be set relatively high, namely p = 5 cm (for lower parameters the resulting model was far from useful in practice). As a consequence the model obtained may be considered too simplistic, lacking detail.

The new method allows to solve this problem with 3 times denser mesh (96 × 60 nodes), which produces a ground structure of 16 759 168 bars. Much lower parameter p = 0.6 cm was used which resulted in a model demonstrated on the right hand side of Figure 5. The model offers more detailed design of reinforcement and, what is more, requires about 5,5% less steel.

5 CONCLUDING REMARKS

Due to the algorithm based on the adaptive ground structure approach, the new method allows far more efficient generation of Strut and Tie models via truss topology optimization. Except for almost 10 times faster execution of the computer program, it paves the way for finding more detailed and accurate reinforcement layouts, which, in the end, result in lower steel consumption.

ACKNOWLEDGEMENTS

The paper was prepared within the Research Grant no 2013/11/B/ST8/04436 financed by the National Science Centre (Poland), entitled: Topology optimization of engineering structures. An approach synthesizing the methods of: free material design, composite design and Michell-like trusses.

REFERENCES

Bołbotowski K., Knauff M., Sokół T. 2013. Applications of truss topology optimization in design of reinforced concrete structures using "Strut and Tie" models (in polish). *Budownictwo i Architektura*, Vol. 12(1), pp. 91–98, PolitechnikaLubelska, Lublin, 2013. ISSN 1899-0665.

BS-EN 1992-1-1:2004 Design of concrete structures. General rules and rules for buildings, *European Committee for Standardization*, 2010.

He L., Gilbert M. 2015. Rationalization of trusses generated via layout optimization. *Struct Multidisc Optim*, 52(4):677–694.

Lewiński T., Rozvany G.I.N., Sokół T., Bołbotowski K. 2013. Exact analytical solutions for some popular benchmarks problems in topology optimization III: L-shaped domains revisited. *Struct Multidisc Optim*, 47(6):937–942.

Sokół T. 2014. Multi-load truss topology optimization using the adaptive ground structure approach, Chapter 2, pp. 9–16, In T. Łodygowski, J. Rakowski, P. Litewka (Eds.), *Recent Advances in Computational Mechanics*, CRC Press, London.

Victoria M., Querin O.M., Martí P. 2011. Generation of strut-and-tie models by topology design using different material properties in tension and compression, *Struct Multidisc Optim*, 44(2):247–258.

Zegard T., Paulino G.H. 2014. GRAND—Ground structure based topology optimization for arbitrary 2D domains using MATLAB. *Struct Multidisc Optim*, 50(5):861–882.

Advances in Mechanics: Theoretical, Computational and Interdisciplinary Issues – Kleiber et al. (Eds)
 ISBN 978-1-138-02906-4

The influence of imperfect interfaces on the overall mechanical response of metal-matrix composites

G. Bolzon & P. Pandi
Politecnico di Milano, Milano, Italy

ABSTRACT: A systematic study of the overall mechanical response of strongly or weakly bonded periodic composites was performed. Results have been recovered in the extreme conditions of perfect adhesion and pure friction. The aim of the investigation is to evaluate the sensibility of the overall material behavior to the interface characteristics and to evidence the load transfer mechanisms between the matrix and the inclusions. The problem is particularly relevant for metal-ceramic composites, due to low chemical affinity and difficult production processes that can prevent the desirable coupling between the components.

1 INTRODUCTION

Metal-ceramic composites have been developed for their potentially interesting thermal, wear and corrosion overall properties. The most promising coupling is often designed on the basis of simple bi-phase mixture laws, possibly grading the composition in order to meet the desired functionality (Bolzon 2011). However, recent experiences have shown that the actual macroscopic characteristics of the resulting composite may not be well predicted even in the simple case of linear elastic response. On the contrary, the measured effective moduli can even fall outside the classical homogenization bounds (Bolzon et al. 2010, Bruck & Rabin 1999).

These apparent inconsistencies have been attributed to the poor adhesion existing between the reinforcement and the matrix due to both the low chemical compatibility and the difficult production processes of metal-ceramics composites. However, noteworthy experimental observation evidence separation at the interface between the matrix and the ceramic inclusions even in affine systems like Al-Al_2O_3 or SiC-C_f (Bruck & Rabin 1999, Babout et al. 2004a, Park et al. 2008, Padmavathi et al. 2012).

Imperfect interfaces have been often conceived as thin interphases between the matrix and its inclusions, still ensuring perfect material continuity between the components (Miloh & Benveniste 2001, Hashin 2002). The relative stiffness of the inclusion coating layer has been shown to influence the overall characteristics of the composite but this material model is not capable to interpret the extreme results detected in some experiments.

The hypothesis of material continuity has been released and replaced by different conjectures in recent research contributions.

In the most common approaches, which follow quite consolidated assumptions in different application fields (Needleman 1990), a cohesive damaging behavior is attributed to the interface among the matrix and its inclusions (Zhang et al. 2005, Yuan et al. 2012, Xu & Lu 2013). An alternative proposal suggests that the failure mechanism of reinforced metal-matrix composites consists of plastic strain localization and void formation in a narrow band around the particle inclusions (Babout et al. 2004b, Balasivanandha Prabu & Karunamoorthy 2008, Sabirov & Kolednik 2010). On the other hand, Bonora & Ruggiero (2006a, b) assume that the connection between the constituents is purely mechanical, with no chemical bonding. Thus, load transfer mechanisms are ensured only by local contact stresses that develop during the manufacturing process due to the mismatch of the thermal expansion coefficients.

In most cases, the actual mechanical behavior can be identified only indirectly, on the basis of the macroscopic system response. The present study is carried out with this peculiar feature in mind, with the aim of evaluating the sensibility of the overall material behavior to the interface characteristics.

Results concerning the extreme conditions of perfect adhesion and frictional coupling in simple representative volume elements are briefly summarized in this contribution, which evidences the spectrum of responses that originates even from this simple material coupling model.

2 THE PROBLEM

The presented study considers simple periodic representative volume elements, visualized for instance in Figure 1, derived from a square array of cylindrical ceramic inclusions embedded in a metal matrix. The purpose is to simulate the transversal response of fiber-reinforced composites with relative ceramic volume content varying between 0.1 and 0.7.

The focus of this investigation is on the initial material response. Thus, the bulk constitutive law is assumed linear elastic and isotropic and the components are characterized by the values of their Young modulus (E_m, E_c) and Poisson ratio (ν_m, ν_c).

The results presented in this contribution refer to Al-SiC coupling, defined by the parameters suggested by Xu & Lu (2013), namely: $E_m = 72.4$ GPa and $\nu_m = 0.33$ for the matrix; $E_c = 420$ GPa and $\nu_c = 0.25$ for the ceramic.

The only considered non-linear phenomenon consists of the purely frictional response of the interface, characterized by Coulomb coefficient $\mu = 0.01, 0.5, 1$. The lowest value of this series has been selected on the basis of numerical stability studies.

Comparison is made with the hypothesis of perfectly bonded components. The void inclusion case is also considered, consistently with the observation reported by Bruck & Rabin (1999) and by Bolzon et al. (2010) on their experimental results, which suggested that detached or damaged material portions behave as equivalent porous inclusions in the remaining material system.

Macroscopically uniaxial tensile and compressive tests have been simulated by introducing the due symmetry and periodic boundary conditions to the discretized version of the domains represented in Figure 1 with the aid of a Python routine (Phyton 2.7 2010). The overall strain and stress components have been calculated from the boundary displacement values and from the recovered reaction forces, respectively, since classical volume-averaging schemes lead to misleading results in this context, where discontinuities arise. Plane stress analyses have been performed by a popular finite element code (Abaqus/Standard 2013). Some representative results are summarized in the following.

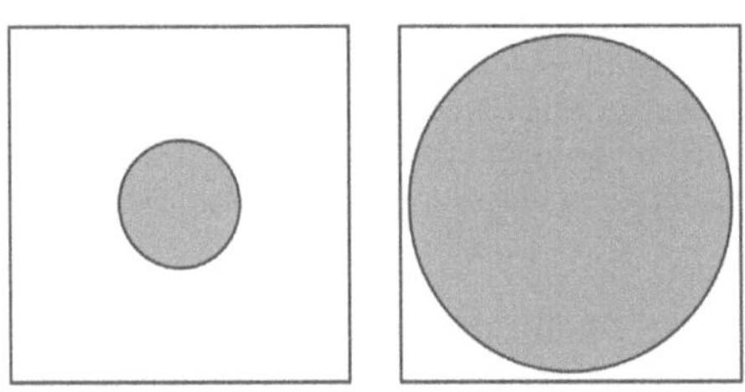

Figure 1. Representative volume elements for 0.1 (left) and 0.7 (right) relative volume content of the inclusion.

3 SELECTED RESULTS

The simulated overall material response to uniaxial compression test of the considered material samples is represented in Figures 2 and 3; the graphed results refer to 70% ceramic volume content.

The relationship between the macroscopic stress and the corresponding strain in the loading direction looks linear, as visualized by Figure 2. On the contrary, the results reported in Figure 3 show that the longitudinal and transversal macroscopic strains do not grow proportionally during the deformation process. In fact, sliding at the interface is initiated as soon as the sample experiences the load, while the position of the kink points in the graphs of Figure 3 depends on the strain level imposed in the initial step of the performed numerical analysis.

Sliding along the interface is accompanied by opening displacements on the interface portions orthogonal to the loading direction. This occurrence is clearly reflected by the transversal expansion at the macroscopic scale, which becomes even larger than the longitudinal strain.

The non-linear phenomena are enhanced by the reduction of the friction coefficient but occur even in the case of a large μ value.

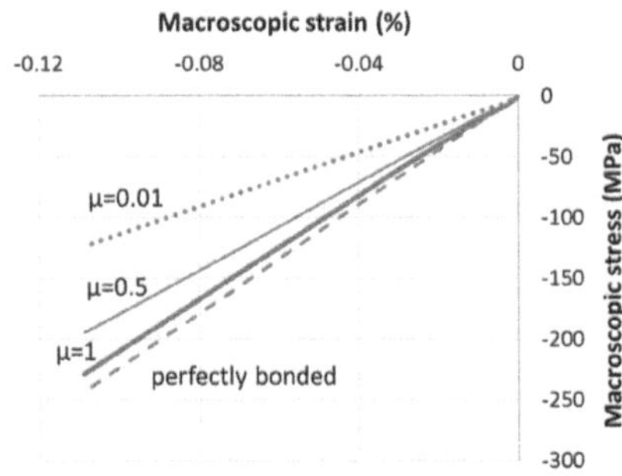

Figure 2. Overall stress versus strain response for 70% ceramic volume content.

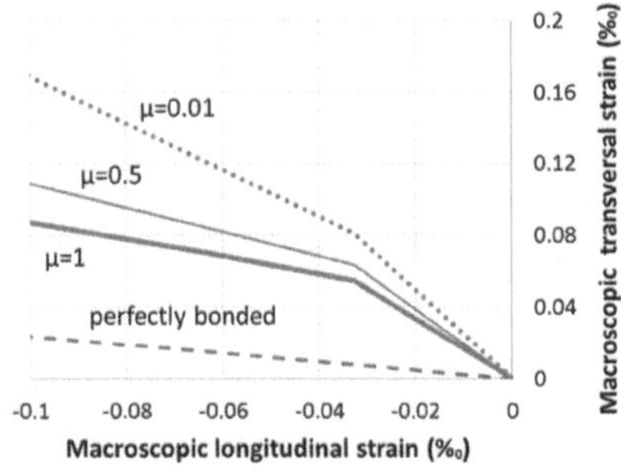

Figure 3. Overall strain components in uniaxial compression 70% ceramic volume content.

Similar macroscopic response is observed for all investigated ceramic content although the effects of the interface characteristics are magnified for large volume fraction of the inclusion.

The deformed configuration and the local stress distribution of the considered representative volume elements at 0.1% macroscopic strain is visualized in Figures 4 and 5, respectively, for 10% and 70% particle volume content. Displacements are amplified by a factor 100. The represented results correspond to the extreme values of the considered friction coefficient, namely 0.01 and 1. The contour plots refer to the stress components parallel to the macroscopic loading direction. The detached portion of the interface can be clearly identified, especially in Figure 5. The significant contribution of the displacement discontinuities to the overall strain values is also evidenced. The influence of the interface characteristics on the mechanical response of the considered composites under compression is summarized by the graphs drawn in Figure 6.

The curves represent the slope of the almost linear curves visualized in Figure 2 for 70% ceramic

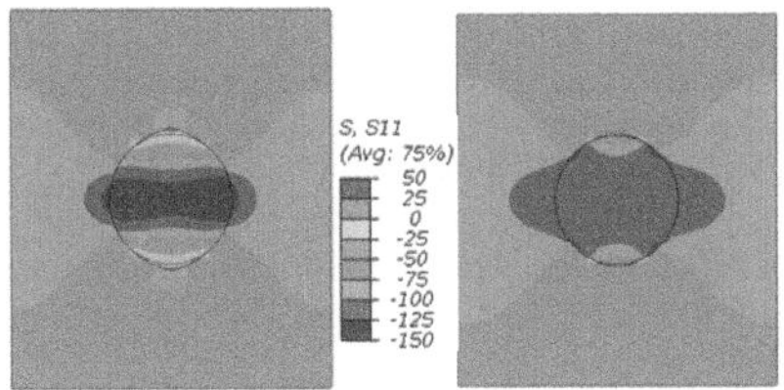

Figure 4. The local value distribution of stresses parallel to the loading direction in the simulated case of uniaxial compression for: 10% ceramic volume content; macroscopic strain 0.1%; friction coefficient $\mu = 0.01$ (left) and $\mu = 1$ (right). The support of the contour plots represents the deformed configuration of the representative volume element; displacements are amplified by a factor 100.

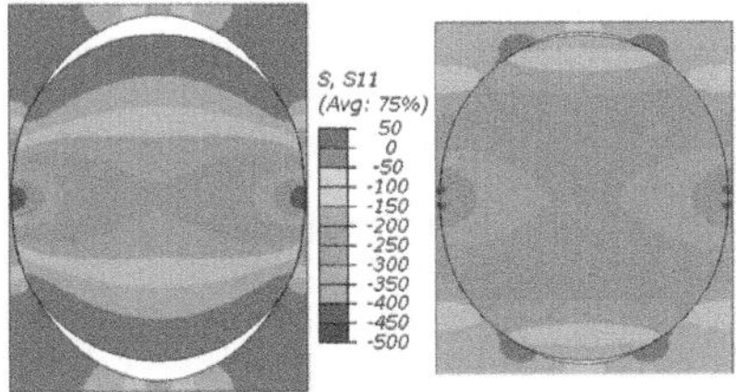

Figure 5. The local value distribution of stresses parallel to the loading direction in the simulated case of uniaxial compression for: 70% ceramic volume content; macroscopic strain 0.1%; friction coefficient $\mu = 0.01$ (left) and $\mu = 1$ (right). The support of the contour plots represents the deformed configuration of the representative volume element; displacements are amplified by a factor 100.

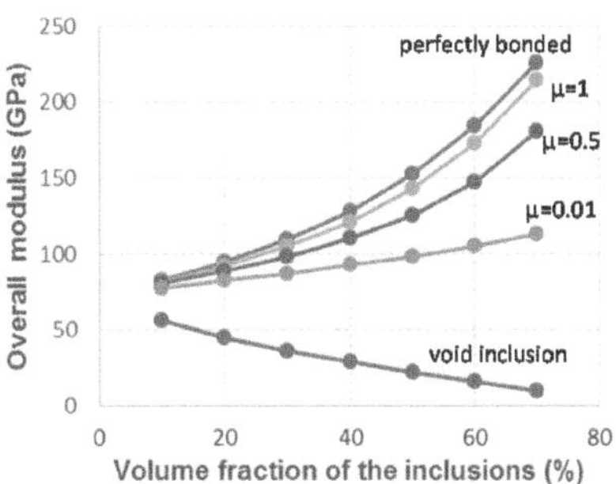

Figure 6. The overall modulus relating the macroscopic stress and strain component in the loading direction in the case of uniaxial compression.

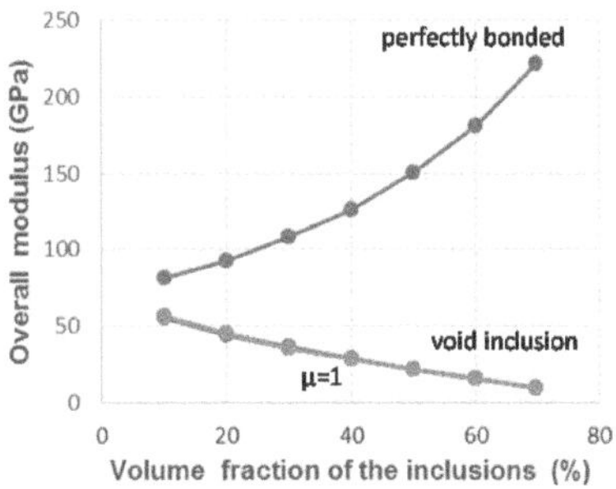

Figure 7. The overall modulus relating the macroscopic stress and strain component in the loading direction in the case of uniaxial tension.

content, and of the analogous ones for different volume fraction of the components.

A significant drop of the overall modulus is produced as the ceramic fraction is increased and Coulomb coefficient is reduced. This result reflects the load transfer mechanisms that are activated at the microstructure level, as a consequence of allowed or controlled interface sliding.

The effects of an imperfect interface are even more pronounced in the case of the simulated tensile tests, as shown by the graphs drawn in Figure 7. While the overall modulus estimated from truly elastic solutions (corresponding to either perfect adhesion or void inclusion) coincides with the compressive one, as expected, the resulting effective values are otherwise insensitive to the presence of the reinforcement. Under tension, any friction value and ceramic content lead to results, which essentially coincide with those of void inclusion, consistently with the experimental observation reported by Bruck & Rabin (1999), Bolzon et al. (2010).

This peculiar response can be attributed to the poor contribution of the interface portions orthogonal to the loading direction, which are under compression during overall uniaxial tensile tests.

4 CLOSING REMARKS

The overall response to uniaxial tests of metal-ceramic composites with frictional interfaces has been computationally evaluated. The obtained results are consistent with the output former experimental observations. The analysis of the microscopic material response in the investigated cases shows that stresses and strains are not purely proportional due to the initial relative sliding of the materials in contact. The deviation from linearity at the macroscopic scale is however light and prone to be likely confused with the initial settlement of the equipment in real experiments.

On the contrary, the interface characteristics have a strong influence on the transversal expansion of the material sample, with increasing sensitivity to the friction coefficient with the increase of the ceramic content. This peculiar feature suggests a simple and effective operative procedure for the quality assessment of the coupling resulting from the production processes of the considered metal-matrix composites. The above results concern periodic representative volume elements derived from a square array of cylindrical ceramic inclusions embedded in a metal matrix. Physical and modelling assumptions alternative to the presented ones may be introduced, considering different material coupling or geometrical arrangement (for instance, hexagonal fiber packing), with results comparable only to some extent. The shape of the inclusions (e.g., particles instead of fibers) may also lead to different conclusions at least in quantitative terms.

Due to the difficult workability of specimens, the class of metal-matrix composites is often characterized by means of experiments like indentation tests, which induce biaxial stress states that influence the macroscopic material response.

The presence of micro-cracks or flaws, which do not alter significantly the relative volume content, may also have dramatic consequence on the results.

These aspects are discussed in forthcoming contributions.

ACKNOWLEDGEMENTS

P. Pandi gratefully acknowledges the financial support received from Heritage Erasmus Mundus Project.

REFERENCES

Abaqus/Standard 2013. Release 6.13-1. Dassault Systèmes Simulia Corp., Providence, RI, USA.

Babout, L., Brechet, Y. & Fougères, R. 2004a. Damage initiation in model metallic materials: X-ray tomography and modelling. *Acta Materialia* 52: 2475–2487.

Babout, L., Brechet, Y., Maire, E. & Fougères, R. 2004b. On the competition between particle fracture and particle decohesion in metal matrix composites. *Acta Materialia* 52: 4517–4525.

Balasivanandha Prabu, S. & Karunamoorthy, L. 2008. Micro-structure-based finite element analysis of failure prediction in particle-reinforced metal-matrix composites. *Journal of Materials Processing Technology* 207: 53–62.

Bolzon, G. 2011. The mechanical response of metal-ceramic functionally graded materials: models and experiences. In N.J. Reynolds (ed.), *Functionally Graded Materials*: 181–192, Hauppauge NY: Nova Science Publishers.

Bolzon, G., Chiarullo, E.J., Egizabal, P. & Estournes, C. 2010. Constitutive modelling and mechanical characterization of aluminium-based metal matrix composites produced by spark plasma sintering. *Mech. of Materials* 42: 548–558.

Bonora, N. & Ruggiero, A. 2006a. Micromechanical modeling of composites with mechanical interface—Part I: Unit cell model development and manufacturing process effects. *Composites Science and Technology* 66: 314–322.

Bonora, N. & Ruggiero, A. 2006b. Micromechanical modeling of composites with mechanical interface—Part II: Damage mechanics assessment. *Composites Science and Technology* 66: 323–332.

Bruck, H.A. & Rabin, R.H. 1999. Evaluating microstructural and damage effects in rule-of-mixtures predictions of the mechanical properties of Ni–Al_2O_3 composites. *Journal of Materials Science* 34: 2241–2251.

Hashin, Z. 2002. Thin interphase/imperfect interface in elasticity with application to coated fiber composites. *Journal of the Mechanics and Physics of Solids* 50: 2509–2537.

Miloh, T. & Benveniste, Y. 2001. Imperfect soft and stiff interfaces in two-dimensional elasticity. *Mechanics of Materials* 33: 309–323.

Needleman, A. 1990. An analysis of decohesion along an imperfect interface. *Inter. Journal of Fracture* 42: 21–40.

Padmavathi, N., Ghosal, P., Eswara Prasad, N. Subramanyam J. & Ray, J.J. 2012. Synthesis of carbon fibre-reinforced, silicon carbide composites by soft-solution approach. *Sadhana* 37: 493–502.

Park, B.G., Crosky A.G. & Hellier, A.K. 2008. Fracture toughness of microsphere Al_2O_3-Al particulate metal matrix composites. *Composites Part B: Engineering* 39: 1270–1279.

Python 2.7 2010. Python Software Foundation, Beaverton, OR, USA.

Sabirov, I. & Kolednik, O. 2010. Local and global measures of the fracture toughness of metal matrix composites. *Materials Science and Engineering* A 527: 3100–3110.

Xu, Q. & Lu, Z. 2013. An elastic-plastic cohesive zone model for metal-ceramic interfaces at finite deformations. *International Journal of Plasticity* 41: 147–164.

Yuan, M.N., Yang, Y.Q., Li, C., Heng, P.Y. & Li, L.Z. 2012. Numerical analysis of the stress–strain distribution in the particle reinforced metal matrix composite SiC/6064 Al. *Materials & Design* 38: 1–6.

Zhang, H., Ramesh, K.T. & Chin, E.S.C. 2005. Effects of interfacial debonding on the rate-dependent response of metal matrix composites. *Acta Materialia* 53: 4687–4700.

Advances in Mechanics: Theoretical, Computational and Interdisciplinary Issues – Kleiber et al. (Eds)

The effect of technological parameters of shot peening on the surface roughness of 51CrV4 steel

M. Bucior, L. Gałda, F. Stachowicz & W. Zielecki
Faculty of Mechanical Engineering and Aeronautics, Rzeszow University of Technology, Rzeszow, Poland

ABSTRACT: In the paper the effect of selected technological parameters of shot peening on the surface roughness of 51CrV4 steel is presented. The experiments were conducted according to statistical Hartley's plan PS/DS-P:Ha_3. In the article the results of investigations, which were carried out on stylus and optical profilometer, were compared. Technological parameters were changed in the range: 1÷3 min shot peening time t, 1.5÷2.5 mm diameter of the balls db and 0.2÷0.4 MPa pressure p. For these parameters the analysis of experiment reproducibility, impact significance and adequacy of equations were made. As the result of investigations the adequate equations were obtained describing amplitude parameters such as: Rv, Rt, Rq. The recommendation to use the stylus profilometer to measure the surface roughness after shot peening was established.

1 INTRODUCTION

The increasing requirements for machine parts force the engineers to apply special treatments to increase the strength and durability of these parts as well as to use sophisticated measurement techniques to control them (Chmielik et al. 2013). The surface roughness has an important effect on the machine parts performance therefore it is essential to choose the appropriate method of measurement. There are some techniques which could be employed such as the contact method with mechanical stylus usage and the non-contact methods: interferometry, confocal microscopy, focus variation or electron microscopy.

Shot peening has a great impact on the machine parts strength increase. This process is widely used to improve the fatigue strength of the material by the creation of compressive residual stresses in surface layers of materials (Tekeli 2002, Zielecki 1987). The fatigue strength of peened austenitic chromium-nickel steel demonstrates higher values compared to untreated one (Śledź et al. 2013, Śledź et al. 2015). There are several technological parameters which influence the process and should be controlled. Dorr et al. (1999) and Obata et al. (1993) discussed the increase in surface hardness and surface roughness with the increase in shot size and the peening intensity. As per the guidelines given by Champaine (1989), the exposure time is an important factor to achieve desired peening coverage for the material.

One of the statistical techniques for improving the product design and solving production problems is Design of Experiments (DoE). The DoE enables the test realization, the results could be impossible to get in any other way or could determine high costs. It is possible to find the appropriate design of the experiment allowing the rational application (Korzyński 2006). The statistical Hartley's plan PS/DS-P:Ha_i is strongly recommended for the surface layer tests (Korzyński 2006).

2 METHOD

The effect of selected technological parameters of shot peening on the surface roughness was examined with 51CrV4 usage. Chemical composition of 51CrV4 steel is presented in Table 1. The experiments were conducted according to statistical Hartley's plan PS/DS-P:Ha_3. All tests were conducted with threefold repetition. The significance level $\alpha = 0.05$ was adopted for calculations. Pneumatic shot peening process was realized applying the device shown in Figure 1.

Table 1. Chemical composition of 51CrV4 steel.

Chemical composition [%]							
Steel 51CrV4							
C	Mn	Si	S	P	Cr	Ni	V
0.46–0.54	0.5–0.8	0.15–0.4	0.03–max	0.03–max	0.8–1.1	0.4–max	0.1–0.2

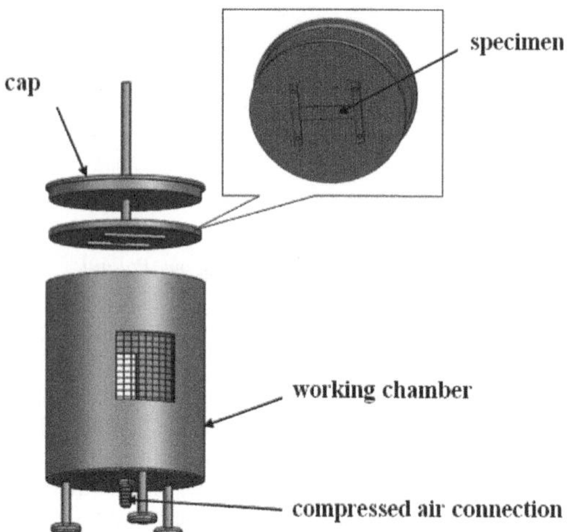

Figure 1. Device for shot peening process.

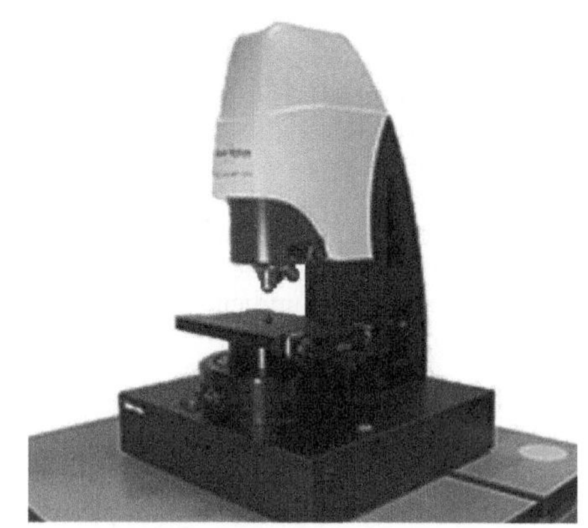

Figure 2. Optical profilometer Talysurf CCI Lite.

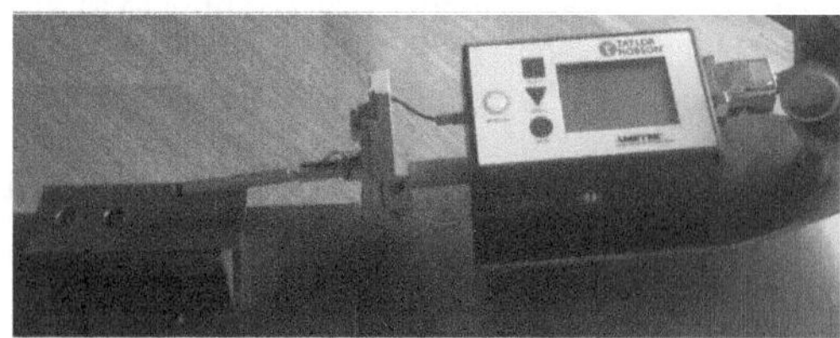

Figure 3. Stylus profilometer Surtronic 25 Taylor Hobson.

The surface roughness was examined with two different instruments usage: optical profilometer Talysurf CCI Lite (Fig. 2) based on non-contact interferometry method of measurement and stylus profilometer Surtronic 25 Taylor Hobson (Fig. 3).

Table 2. Matrix of plan PS/DS-P:Ha_3.

No	x_1	x_2	x_3	x_1^2	x_2^2	x_3^2	x_1x_2	x_1x_3	x_2x_3
1	–	–	+	+	+	+	+	–	–
2	+	–	–	+	+	+	–	–	+
3	–	+	–	+	+	+	–	+	–
4	+	+	+	+	+	+	+	+	+
5	–	0	0	+	0	0	0	0	0
6	+	0	0	+	0	0	0	0	0
7	0	–	0	0	+	0	0	0	0
8	0	+	0	0	+	0	0	0	0
9	0	0	–	0	0	+	0	0	0
10	0	0	+	0	0	+	0	0	0
11	0	0	0	0	0	0	0	0	0

Where: x_1—coded value of time t, x_2—coded value of ball diameter d_b, x_3—coded value of pressure p.

The plan of the experiment implies the adoption of three levels of controlling factors: the minimum (–), the central (0) and the maximum (+). Technological parameters were changed in the range: shot peening time t = 1–3 min, ball diameter d_b = 1.5–2.5 mm and pressure p = 0.2–0.4 MPa. The matrix of the Hartley's plan with three input factors is presented in Table 2.

3 RESULTS AND DISCUSSION

The selected topography maps of specimens surface measured using the optical profilometer Talysurf CCI Lite before and after shot peening are presented in Figure 4. In Table 3 there are average values of selected surface roughness parameters obtained after shot peening process at differential technological parameters according to Hartley's plan.

After shot peening process the values of height roughness parameters increased considerably in comparison to those of non-peened surfaces.

Several adequate equations (Eqs. 1÷4) were obtained as the results of conducted experiments. According to the methodology of Hartley's plan the significance of technological parameters was assessed.

For surface roughness parameter such as Rv the adequate mathematical models were obtained independently of the surface measurement method. Additionally, there were found mathematical descriptions for Rt and Rq roughness parameters which values were measured with the usage of a stylus profilometer.

The roughness parameter Rv representing maximum depth of valleys depends mainly on two technological parameters the pressure and ball diameter. When the pressure was greater, the

a) before shot peening

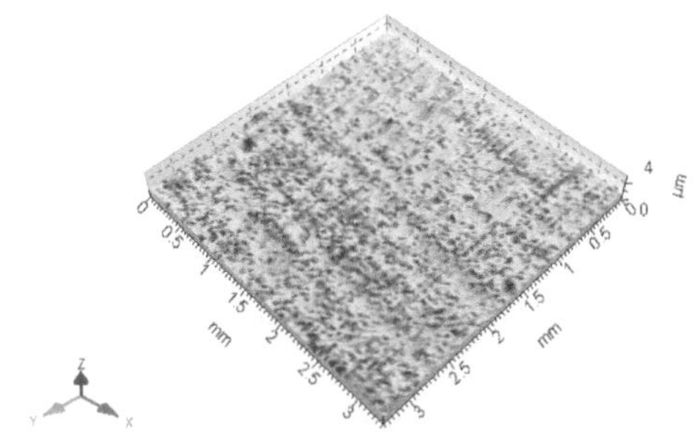

b) after shot peening

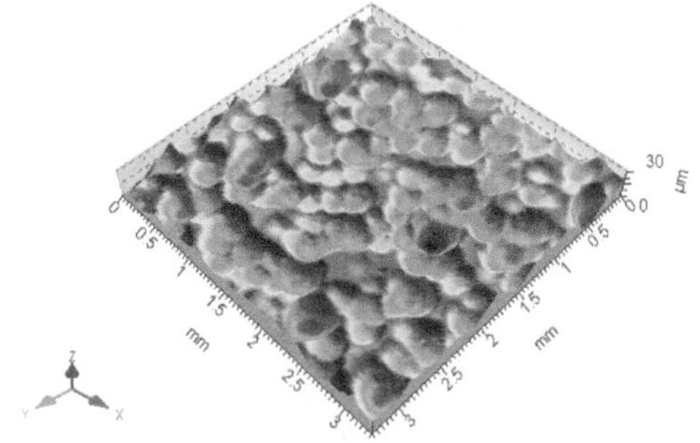

Figure 4. Selected surface topography maps of 51CrV4 steel sheets before (a) and after (b) pneumatic shot peening obtained from optical profilometer (Bucior et al. 2015).

Table 3. Average values of selected surface roughness parameters obtained after shot peening process at differential technological parameters according to Hartley's plan (S—Surtronic 25, T—Talysurf CCI Lite instruments).

	Input factors			Output factors			
	t	d_b	p	Rv S	Rv T	Rt S	Rq S
No	min	mm	MPa	µm	µm	µm	µm
0	–	–	–	0.851	0.873	1.75	0.299
1	1	1.5	0.4	5.10	6.18	14.80	2.680
2	3	1.5	0.2	4.233	4.550	13.833	2.150
3	1	2.5	0.2	2.600	2.393	8.167	1.164
4	3	2.5	0.4	4.327	4.920	15.167	2.890
5	1	2	0.3	3.767	3.880	11.300	1.963
6	3	2	0.3	3.987	4.167	12.833	2.397
7	2	1.5	0.3	4.827	5.407	16.233	2.483
8	2	2.5	0.3	3.837	3.227	11.000	2.203
9	2	2	0.2	3.367	3.093	9.010	1.810
10	2	2	0.4	4.447	4.943	13.133	2.760
11	2	2	0.3	3.857	4.227	12.833	2.330

Where: 0—before shot peening.

bigger values of Rv parameter were achieved. However, with a larger ball diameter the depth values were smaller (Table 3 and Fig. 5). The time of the process duration has no significant effect on the Rv parameter except the interaction influence with the pressure and with the ball diameter. Similar adequate models (Eq. 1) and (Eq. 2) with the same tendency were found comparing two different measurement methods:

$$\text{Rv S} = 5.9814 - 0.7881t - 3.7254d_b + 18.156p + 1.2966td_b - 6.017tp \quad (1)$$

and:

$$\text{Rv T} = 7.934 - 1.5933t - 6.0176d_b + 27.071p + 2.0766td_b - 8.533tp \quad (2)$$

In comparison of the adequateness variance concerned Rv parameter the mechanical contact method of measurement is better for such applications.

Maximum value 16.233 µm of Rt roughness parameter was achieved at medium pressure but at minimum ball diameter of 1.5 mm and medium process time duration of 2 min (Table 3). Similar values were obtained in first and fourth series and taking into account the efficiency of the process to get the maximum height of surface irregularities the minimum process time is the most appropriate option that could be set with minimum ball diameter and maximum pressure (first series). The Rt parameter depends mainly on two technological parameters the ball diameter, pressure and interaction influence with the both parameters (Eqs. 3).

$$\text{Rt S} = 36.721 - 2.731t - 20.528d_b - 13.682p + 3.983td_b - 13.25tp + 30.17d_bp \quad (3)$$

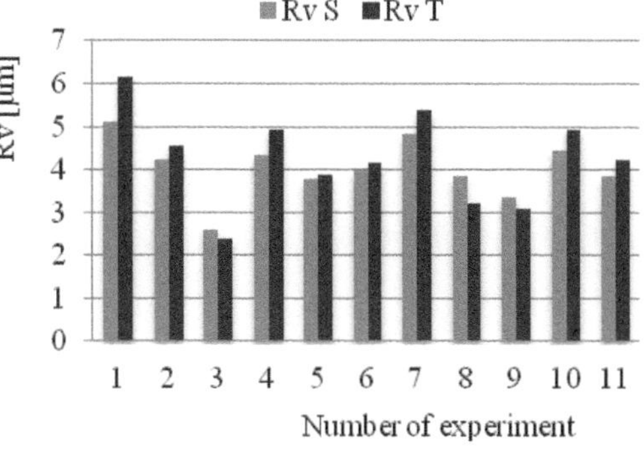

Figure 5. Averaged values of Rv obtained with Surtronic (S) and Talysurf CCI Lite (T) instruments (Bucior et al. 2015).

At the maximum pressure of 0.4 MPa the greatest values of Rq parameter were obtained. In this case also the first set of technological parameters is recommended.

$$Rq\ S = 8.663 - 1.9882t - 4.0484d_b - 6.579p + 1.1296td_b + 5.964d_bp \quad (4)$$

In comparison of the adequateness variance concerned the analysed roughness parameters and also taking into account the fact that more mathematical models were obtained from the results received from the stylus profilometer the mechanical contact method of measurement is better for such applications. It is recommended to use the stylus profilometer to measure the surface roughness after shot peening. The values measured with the optical instrument were usually bigger than those from the stylus profilometer.

4 CONCLUSIONS

The significant effect on the surface roughness is detected with respect to pressure and the ball diameter. Substantial increase in height parameters values was obtained after shot peening process in the whole range of applied technological parameters. For surface roughness parameter such as Rv the adequate mathematical models were obtained independently of the surface measurement method. Additionally, mathematical description was found for Rt and Rq roughness parameters which values were measured with the usage of the stylus profilometer. The stylus profilometer is recommended to measure the surface roughness after shot peening.

REFERENCES

Bucior, M. & Gałda, L. & Stachowicz, F. & Zielecki, W. 2015. *The effect of technological parameters of shot peening on surface roughness of 51CrV4 steel.* PCM-CMM-2015 Congress Book of Short Papers.

Champaigne, J. 1989. *Controlled shot peening, 2nd edition.* Mishawaka: The shot peener.

Chmielik, I.P. & Czarnecki, H. & Tomasik, J. 2013. Comparative analysis of surface roughness measurements in the 3D system by the contact and optical methods. *Mechanics 7: 544–547.*

Dorr, T. & Hilpert, M. & Beckmerhagan, P. & Kiefer, A. & Wagner, L. 1999. *Influence of shot peening on fatigue performance of high-strength aluminum- and magnesium alloys.* Warsaw: Proceedings of the ICSP-7 conference.

Korzyński, M. 2006. *Metodyka eksperymentu, planowanie, realizacja i statystyczne opracowanie wyników eksperymentów technologicznych.* Warszawa: WNT.

Obata, M. & Sudo, A. 1993. *Effect of shot peening on residual stress and stress corrosion cracking for cold worked austenitic stainless steel.* Oxford: Proceedings of the ICSP-5 conference.

Śledź, M. & Bąk, Ł. & Stachowicz, F. & Zielecki, W. 2013. Analysis of the effect of shot peening on mechanical properties of steel sheets used as screener sieve materials. *J. Physics: Conf. Series, 451, 012029.*

Śledź, M. & Stachowicz, F. & Zielecki, W. 2015. *The effect of shot peening on the fatigue strength of steel sheets.* Kovove Mater. 53: 91–95.

Tekeli, S. 2002. Enhancement of fatigue strength of SAE 9245 steel by shot peening. *Mater. Letters 57: 604–608.*

Zielecki, W. 1987. Wpływ intensywności pneumokulowania na wytrzymałość zmęczeniową i stan powierzchni stali 50 HF. *ZN PRz, Mechanika 14: 35–40.*

Advances in Mechanics: Theoretical, Computational and Interdisciplinary Issues – Kleiber et al. (Eds)

Analysis approach for diffusor augmented small wind turbine rotor

J. Bukała, K. Damaziak, M. Krzeszowiec, J. Małachowski & M. Tomaszewski
Department of Mechanics and Applied Computer Sciences, Military University of Technology, Warsaw, Poland

K. Kroszczyński
Department of Applied Geomatics, Military University of Technology, Warsaw, Poland

K. Sobczak
Institute of Turbomachinery, Łódź University of Technology, Łódź, Poland

ABSTRACT: The paper proposes a workflow for initial design and simulation of a small diffusor augmented wind turbine rotor using modern computer methods. The authors describe the methodology behind the process and cover subjects of numerical wind data acquisition, establishing blade geometry using a 2D porous body approach and simulating the considered geometry in a series of 3D, 2-way fluid-structure interaction CFD simulations to evaluate its mechanical response to typical work conditions.

1 INTRODUCTION

Increasing interest in renewable energy sources, thanks to country level tax incentives and rising oil prices, is driving research towards developing a range of advanced new energy generation devices and unit solutions. Wind power is a good example of this. As a direct result of rising understanding of climate change and negative aspects of green gas emission there is an increasing number of people taking interest in setting up privately owned small wind power plants. As opposed to large wind turbines which are dominated by the classic three-bladed horizontal axis setup (ABB SACE 2011), the small turbine market offers a number of different solutions, often advertised as more efficient than the popular three-blade design. There has been some in-depth analysis of the European market of small wind turbines performed by various authors, with limited insight on the specifics of the turbine design (Simic et al. 2012, Bortolini et al. 2014). Based on the global market data the number of small wind turbines should steadily increasing (Fig. 1). An interesting design solution for a small wind turbine with a horizontal axis of rotation is the Diffusor Augmented Wind Turbine (DAWT). The design by itself was first patented and built over 150 years ago by Ernest Bolée. Because of decades of stagnancy in the wind turbine market, the design was somewhat forgotten and has seen a major interest increase in recent years with many academic (Kosasih et al. 2012, Ohya et al. 2012) and industrial centers proposing various solutions to the DAWT concept. The main possible advantage of such a turbine is the option to use smaller size rotor blade, which in turn greatly reduces the overall turbine cost, as the blades are some of the most expensive parts (ABB SACE 2011). Because of the possible cost reduction, the authors have chosen the DAWT concept as a design candidate for a new small wind turbine with the intent of private ownership and micro energy generation.

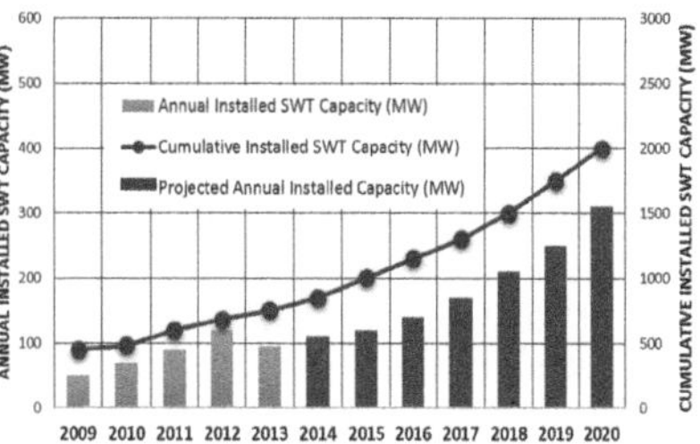

Figure 1. Small Wind Turbine installed capacity world market forecast 2009–2020.

2 WIND STATISTIC

The initial task of establishing the turbine parameters was based on a broad statistical study. In order for the analysis to be meaningful, accurate wind data with a high enough sample rate was required. Obtaining archive wind data covering various regions has proved to be difficult. Therefore, it was decided to obtain the data using the non-hydrostatic mesoscale Weather Research & Forecasting (WRF) model. The simulated period covered one calendar year with a 24-hour forecast-

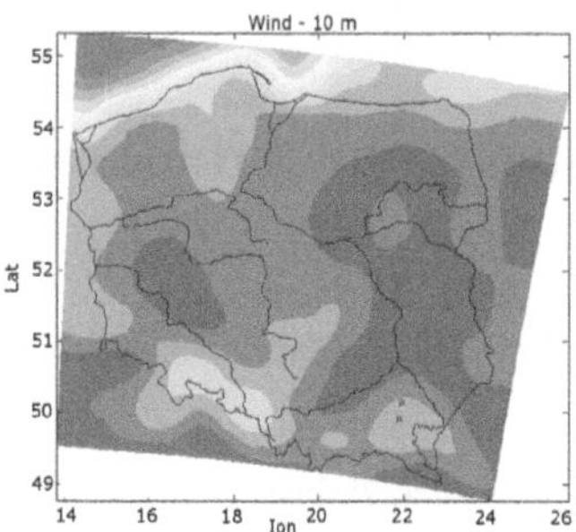

Figure 2. Wind speed contour lines (Bukała et al. 2014).

ing cycle and a hourly sampling rate. Each sample point was a five-dimensional field of prognostic parameters spanning over the whole area of Poland. These parameters included, amongst others: pressure values, geopotential, temperature and three-dimensional wind fields (Fig. 2).

Wind speed data averaged over a 60 minute period has been proved useful for long term energy yield predictions for small wind turbines (Elliott et al. 2012). Data analysis was mainly focused on pinpointing the vital characteristics of the planned new small wind turbine. Wind turbine parameters taken into account were: electrical and aerodynamic efficiency, rotor size and tower height. A more in-depth look into the methods used can be found in the authors' previous work (Bukala et al. 2014). The direct result of this statistical analysis was a set of geometric and systemic goals for the device.

3 BLADE AND DIFFUSOR DESIGN

After establishing the planned rotor size and target power output, the next phase was focused on creating the blade and diffusor proposed geometry. The initial work was to perform a series of 2D simulations using CFX software. The model was based on a method proposed and described in (Ohya & Karasudani 2012) and used a parametric porous body domain, which simulated the pressure drop caused by the operating rotor. The 2D model schematics can be seen in Figure 3.

The mathematical description of the porous body accounted for a given, specific airfoil profile and chord length at the blade root and tip in return providing estimated power output and optimal geometric twist angle values along the blade for increased aerodynamic efficiency. After obtaining the main geometrical characteristics of the blade, it was modelled as a solid (Fig. 4) and used in preparation of the 3D CFD simulations. In order to minimize the number of elements and thus simulation time, symmetry of the structure was taken into account.

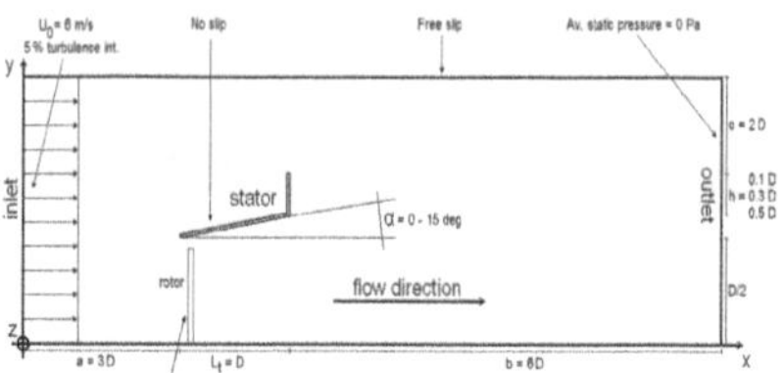

Figure 3. Schematic for the 2D simulation (Lipian et al. 2014).

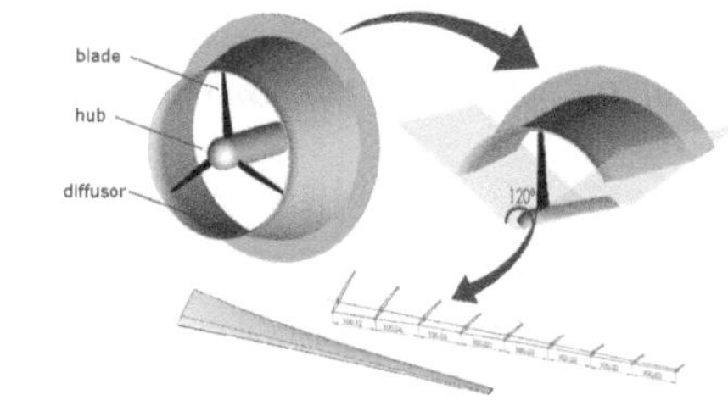

Figure 4. DAWT geometry.

4 FSI ANALYSIS OF THE BLADE

Based on the established blade geometry, a much more sophisticated 3D flow model was created. All the numerical simulations were prepared using Ansys commercial software: ANSYS Mechanical (FEM solver) and Ansys Fluent (CFD solver). The CFD software relies on the basic Euler equations for the fluid dynamics or Navier—Stokes equations for fluid dynamics with the viscosity effect. During the analysing of the flow in macroscopic scale the impact of molecular structure is ignored. Description of the liquid is based only on the macroscopic properties and includes: pressure, speed, density, temperature or derivatives of these values over time. The equation of mass and momentum conservation is applied separately for each discrete cell in the model. The sum of all cell gives a global continuity equation for the calculated model. The first step to prepare analysis was to build a flow domain. Flow domain is a kind of virtual wind tunnel which allows to obtain all necessary data such as pressure and stress values in a structure. Mesh for the domain was built using ICEM CFD preprocessor. The total number of discrete elements was 10513503. The considered fluid domain was composed out of three zones. Each of those was described using a different element topology (Fig. 5). Boundary conditions were described according to Figure 3. Boundary layer effect was taken into account using grid transition. In the close distance to structure the narrow elements have been created, which become larger further away from the examined structure.

The blade model was also used to prepare an initial, steady-state strength FEM analysis with regard

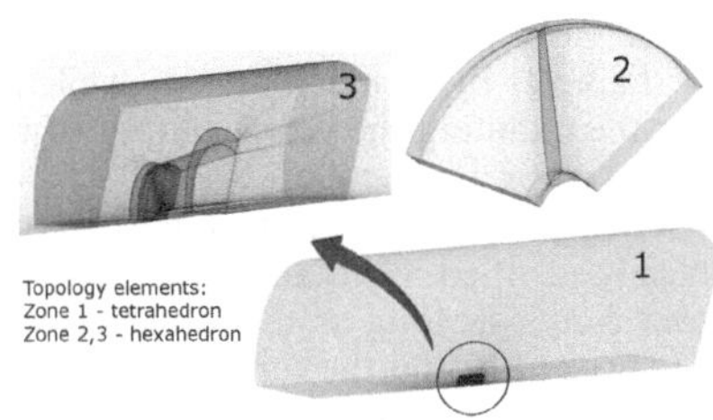

Figure 5. Flow domain.

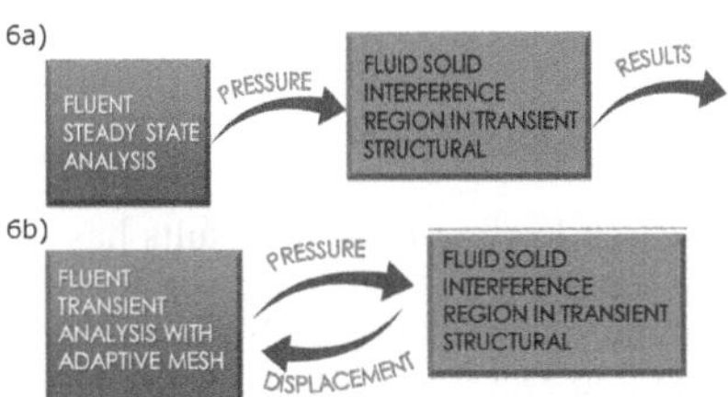

Figure 6. Schematic view of FSI methods: a) one way Fluid Structure Interaction, b) two way FSI.

only to inertia loads. Main emphasis was put on the comparison of fluid—structure interaction techniques (one way and two way FSI). One way fluid structure interaction takes place when one way connection between two solvers is performed (Fig. 6a). In this case resultant pressure from CFD steady state analysis is applied to a FE model in transient structural as a load (boundary or initial). To prepare such analysis the area of data exchange must be defined. Two way fluid structure interaction (Fig. 6b) is the extended version of the one way FSI. In this method the data from one solver is sent to the second solver and vice versa. During the analysis pressure from CFD analysis causes the blade displacement. Displacement values obtained from the structural analysis are being sent back to the CFD solver, where the flow is calculated for the new geometry position. This algorithm is repeated to achieve convergence between the two programs.

5 RESULTS

The results for a series of simulations confirmed the power augmenting possibilities of the planned diffusor shroud. Results predict an increase of wind velocity as high as 50% in some areas, in regard to the initial 12 m/s, free flow wind speed. This speed-up effect can be seen in Figure 7, which presents the wind velocity as a function of the distance from the hub in the radial direction, inside the diffusor and before the flow stream interacts with the rotating blade for two different method.

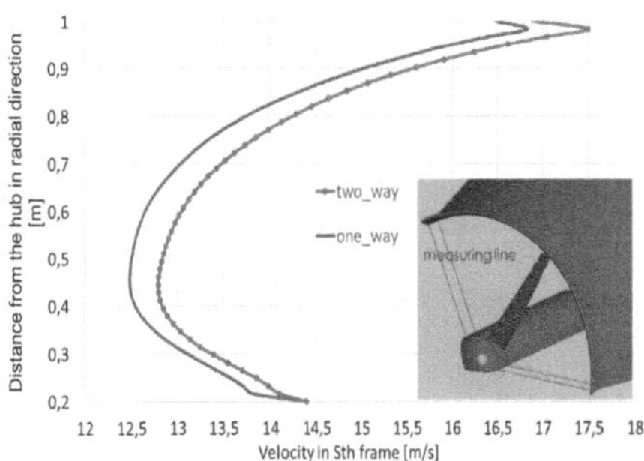

Figure 7. Sample result presenting flow acceleration from the 3D flow simulation of the shrouded rotor.

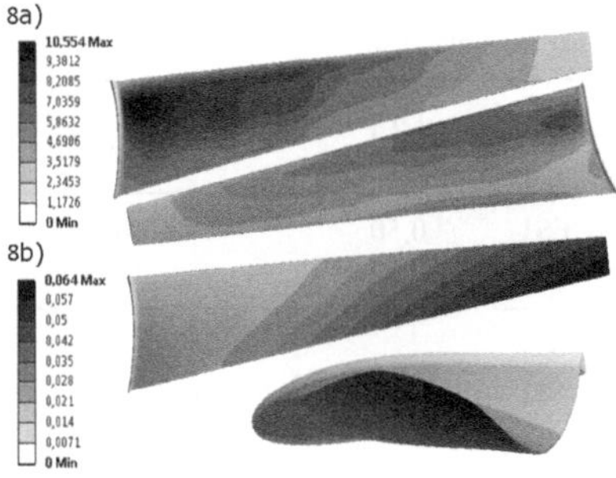

Figure 8. Sample result presenting: a) Von Mises Stress, b) displacement maps for steel blade (inertia load only).

Table 1. Results from initial steady—state strength FEM analysis (inertia loads).

Material	Displacement [mm]	Von mises stress [MPa]
Steel	0.06	10.50
Aluminium	0.58	8.94
Glass fibre composite	2.23	9.49
Carbon fibre composite	1.30	8.73

5.1 *Initial steady—state FEM analysis*

As a result of static FEM analysis, stress (Fig. 8a) and displacement (Fig. 8b) fringes were obtained for a turbine blade made of different materials (Table 1). Based on the stress fringe, the incoming wind influence can be observed. For a turbine blade made of steel and loaded only by the inertia forces, the maximum equivalent stress (von Mises) in equal to 10.55 [MPa] (Fig. 8a). In case where the pressure is included (which is the result of incoming wind influence) the considered value rises to 18.8 [MPa] for two way FSI (Fig. 9a) and 17.8 [MPa] for one way FSI which corresponds to an increase of 78% and 69% respectively. The greatest displacement

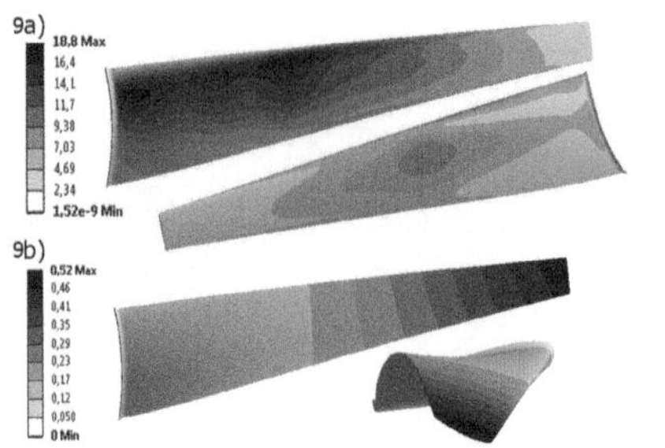

Figure 9. Sample result presenting: a) Von Mises Stress, b) displacement maps for steel blade (inertia and pressure loads).

Table 2. Comparison between one way and two way FSI method for steel blade.

Method	Displacement [mm]	Von mises stress [MPa]
One way FSI	0.50	10.50
Two way FSI	0.52	18.80

(deflection) can be found at the tip of a blade (Fig. 8b).

5.2 *One way and two way FSI*

According to the performed analyses for one way and two way fluid structure it can be observed that obtained results are similar for both FSI algorithms (Table 2), however the model preparation and analysis time for two way FSI is much longer (6h45′ for one way FSI and 16h30′ for two way FSI).

Contrary to initial steady—state FEM analysis it can be observed that stress concentration area in FSI analyses is located in the middle part of the blade (Fig. 9a). The greatest displacement (deflection) can still be found at the tip of a blade (Fig. 9b).

6 CONCLUSIONS

The methods presented in the paper provides a robust and very efficient approach to designing new small wind turbines. It is based on a natural synthesis of various advanced computer methods such as weather forecasting, statistical analysis, CFD and FEM simulations. Such a broad range of tools, allows for early identification of the most important turbine parameters and incorporating numerous findings into the proposed design with relative ease.

The work is still in progress, but as of now the authors concluded the following:

- Weather prediction modeling can be used with success for turbine parameter response analysis.
- The use of shrouded turbine can reduce the rotor diameter by as much as 60% while at the same time retaining the power output of non—shrouded rotor.
- A well designed small wind turbine, can be economically justified at in land mean wind speed of over 5 m/s, as long as the initial investment is not very costly and the supporting tower is at least 15 m high (preferably higher).
- At this stage of research, glass fiber composite materials for the blade of small wind turbine appear to be the most versatile and effective.

ACKNOWLEDGEMENTS

The research leading to these results has received funding from the Polish-Norwegian Research Programme operated by the National Centre for Research and Development under the Norwegian Financial Mechanism 2009e2014 in the frame of Project Contract No. Pol-Nor/200957/47/2013.

REFERENCES

ABB SACE—ABB S.p.A. L.V. Breakers. 2011. Wind power plants, *Technical Application Papers*.

Bortolini M., Gamberi M., Graziani A., Manzini R. & Pilati F. 2014. Performance and viability analysis of small wind turbines in the European Union, *Renewable Energy, Volume 62: 629–639*.

Bukała J., Damaziak K., Karimi H.R., Kroszczyński K., Krzeszowiec M. & Małachowski J. 2015. Modern small wind turbine design solution comparison in terms of estimated cost to energy output ratio, *Renewable Energy: 83: 1166–1173*.

Elliott D. & Infield D. 2012. An assessment of the Impact of Reduced Averaging Time on Small Wind Turbine Power Curves, Energy Capture Predictions and Turbulence Intensity Measurements, *Wind Energy, Volume 17, Issue 2: 337–342*.

Kosasih B. & Tondelli A. 2012. Experimental study of shrouded micro wind turbine, *Procedia Engineering, no. 49: 92–98*.

Lipian M. & Olasek K. & Karczewski M. 2014. Sensitivity study of diffuser angle and brim height parameters for the design of 3 kW Diffuser Augmented Wind Turbine, *11th International Symposium on Compressor and Turbine Flow Systems in Lodz, Turbomachinery no. 145: 79–80*.

Ohya Y., Karasudani T. & Matsuura C.T. 2011. A highly efficient wind turbine with wind-lens shroud, in 13th International Conference on Wind Engineering. Amsterdam.

Ohya Y. & Karasudani Y. 2012. A shrouded wind turbine generating high output power with wind-lens technology, *Energies: 634–649*.

Simic Z. & Havelka J.G. & Vrhovcak M.B. 2012. Small wind turbines—A unique segment of the wind power market, *Renewable Energy, Volume 50: 1072–1036*.

Advances in Mechanics: Theoretical, Computational and Interdisciplinary Issues – Kleiber et al. (Eds)
© 2016 Taylor & Francis Group, London, ISBN 978-1-138-02906-4

Geometric analysis of a 1-dof, six-link feeder

J. Buśkiewicz
Poznań University of Technology, Poznań, Poland

ABSTRACT: The paper deals with the geometric analysis of an one degree of freedom feeder built up of six links. The proposed structure does not require any extra drive of the gripper as the jaws of the gripper are driven by the same active link which drives the whole feeder. The jaws catch and transport the product to other work stand where the gripper releases the product and moves back to its initial position. There exist various different links assemblies realizing the motion when the gripper is closed and the gripper is open but they result in mechanisms differing in kinematic parameters. The analysis is carried out to ensure that the active link makes full revolution and that that the area enclosing the trajectory of the gripper (workspace) has a reasonable size.

1 INTRODUCTION

The automatic devices for transporting loads, products and half-products to work stands or magazines are widely used in manufacturing process. The feeders take the part for assembly from the pod, transport it and place it into the working space. The industrial robots are the state-of-the-art solutions in the processes of self-operating transport between production zones. Electronically controlled gripper of the feeder carries the elements or tools on machine tools in automatic production lines. The gripper equipped with electromagnet is used in steelworks, strap yards or steel mills for transporting ferromagnetic materials. The gripper in excavators for lifting and transporting powdery materials consists of two hinged buckets controlled by hydraulic system. In all these structures the gripper requires the drive independent of the other feeder links, and requires the system which controls the drive. For many reasons one degree of freedom kinematic systems built of possibly small number of links are used to perform specific industrial tasks. There are known the grippers which fail to require an additional, independent mechanical drive or an electronic steering. In these feeders a load is pressed by the arm attached to the pull spring in the phase of carrying. The spring is tightened when the arm meets the resistance of the immoveable pin, and the load is let out (Fig. 1a). The feeders based on the mechanisms without any pull strings are presented in Erdman et al. (2001). Mechanism of 1 Degree of Freedom (DOF) may transport the load between two points by means of a properly shaped gripper. The load is thrown out by tilting the gripper (Fig. 1b) or the load is pushed out by the other link of the feeder.

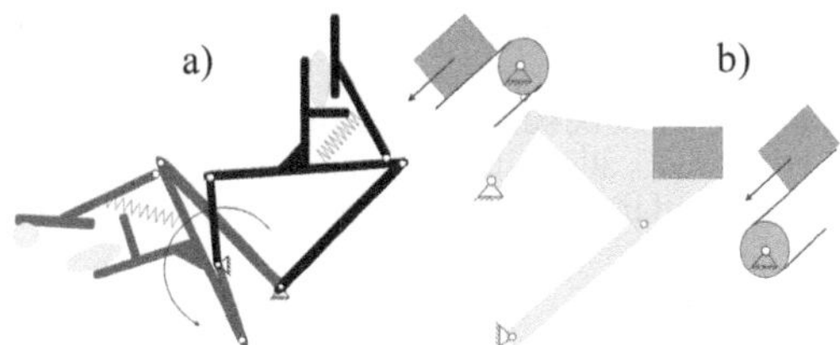

Figure 1. The double crank mechanism in which the arm of the gripper keeps the load by applying the pull spring—the arm opens when meets the immoveable pin (two positions are drawn) (a). The feeder based on a four-bar linkage for transporting the load between two lines, in which the load is thrown out by tilting the coupler (b).

In order to design an one degree of freedom feeder built up of six links connected by means of revolute joints, the techniques for optimal path/motion synthesis of four link planar mechanisms outlined in Buśkiewicz (2014a, b) were employed. Compared to the works Buśkiewicz (2014a, b) this paper presents the detailed analysis carried out to ensure that the active link makes full revolution and to estimate the working space of the mechanism.

The proposed structure of a feeder is based on the kinematic chain in which the phase of motion exists when two links rotate at the same angular velocity. This effect can be obtained by locking the coupler of the four-bar linkage in the dwell mechanism (Radovan et al. 2013, Mohan 2011, Pennock & Isra 2009, Kota 1991, Shiakolas et al. 2005). Then the frame of the dwell mechanism becomes a moving coupler, and the other coupler is the output rotating link in the dwell mechanism. The advantage of the proposed solution is that the motion is transmitted mechanically from the driving link to the gripper and therefore the jaws do not require any additional independent control, steering and other tightening elements as other known feeders do Erdman et al. 2001.

2 GEOMETRY OF 1 DOF SIX-LINK FEEDER

The main chain of the feeder is the four-bar linkage O_1ABO_2 with point D tracing the trajectory a part of which is the circular arc of radius R, centred at point O_3 and spanned by a given angle (Fig. 2a). A crank-rocker linkage is accepted only in order to simplify the drive of the mechanism. Let the mechanism be set in the initial position, in which point D is located at the beginning point of the circular arc. We fix additional link O_3N to the ground at the pivot O_3. The length of the link is $l_7 = R$. We choose an arbitrary point M on coupler AB and we connect the coupler to link MN by means of the revolute joint at this point. The links MN and NO_3 are connected by revolute joint N. The location of point M must be such that the full revolution of the active link is allowed. The adequate conditions are formulated in order that links O_3N and NM do not approach to the position in which they are collinear. Then, the following inequality must be met: $|O_3N| + |NM| > |O_3M|$, for any θ_1. The position of joint N is coincident with the position of point D only when D moves along circular arc. Then link MN and coupler AB rotate with approximately the same angular velocity. One jaw of the gripper is attached to coupler AB, the other one is attached to link MN (Fig. 2b). When point D starts drawing the arc, the jaws of the gripper catch the product, shut and keep closed as long as point D moves along the circular arc. Subsequently, the gripper opens, the product is taken away and point D moves back to its initial position along the remaining part of its coupler curve.

The last step is to choose the location of the characteristic point G_1 of the first jaw in the local system Ax'y' (Fig. 2a) attached to coupler AB. The location of this point is very important when the path, along which the load is carried, is also prescribed. Theoretically the location of point G_1 can be chosen arbitrarily since the feeder can be translated, scaled and rotated so that the load moves between preset pickup and destination points. Nonetheless, the location of G_1 influences on the size of the working area of the feeder and determines whether it is possible to fix the feeder into existing system of machines. Further, one computes the coordinates of the point of the jaw on coupler NM (Fig. 2b). The point is denoted by G_2 and its coordinates are unequivocally defined in terms of the coordinates of G_1. For the sake of computing G_2, the local system Nx"y" attached to coupler NM with origin at N is introduced. When the jaws of the gripper are closed, the coordinates of both points G_1 and G_2 in the global immoveable reference frame O_1xy are equal to each other. This is expressed by means of the linear equation from which the coordinates of G_2 are computed.

It is obvious that there exist different assemblies of links MN and O_3N providing the phase when the couplers do not rotate with respect to each other, but they give mechanisms differing in kinematic parameters depending on the location of pivot M and jaws G. The analysis has to be carried out to ensure that the position of point M allows for full revolution of active link O_1A and that the working space of the gripper has an acceptable size.

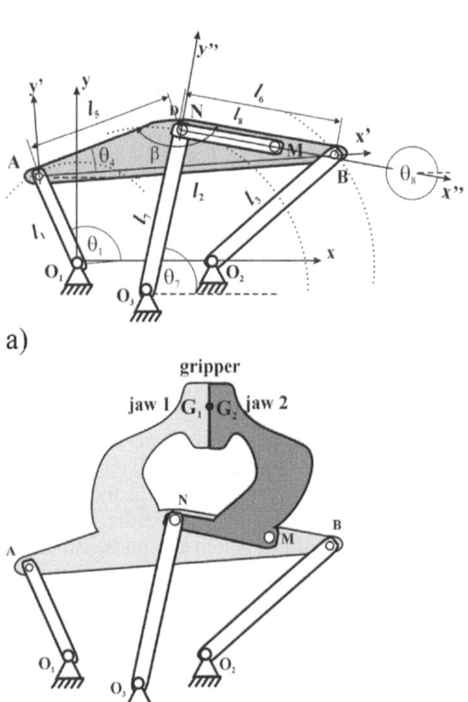

Figure 2. The geometric scheme (a), the general concept (b).

3 NUMERICAL ANALYSIS

The example presents how the locations of points M and G affect the working space of the mechanisms. A following six-bar mechanism realizes motion with the phase when the couplers do not rotate with respect to each other: $O_1(-5, 0)$, $O_2(5.78, 0)$, $O_3(0, 0)$, $l_1 = 2.785$, $l_3 = 3.348$, $l_5 = 5.03$, $l_6 = 5.83$, $\beta = 3.12$ rad,

$\theta_4 = 0.009$ rad. Lengths are nondimensional. The length of link O_3N (pinned at O_3) is 3. The four bar linkage approximating the circular with the centre at O_3 and radius O_3N can be obtained using any synthesis method, e.g. those presented in Buśkiewicz (2014a, b). The allowed positions of joint M in prescribed area are presented in Figure 3.

From this area the following coordinates of M in the local coordinate Ax'y' system are taken: (–4.346, 2.173). The length of link MN is 7.267. The following coordinates are taken for G_1 on the first jaw: (1.63, 16.297). Having these coordinates, the coordinates of the point G_2 on the other jaw in the reference frame attached to coupler MN are derived: (6.911, –15.096). Applicability of the solution can be assessed on the basis of the space occupied by the moving feeder. Therefore, the paths of all the joints which influence the size of the housing are presented. Figure 4 shows the instant when the gripper is closed. The difference between the angular positions of the both links supporting the jaws is less than 0.0005 rad as shown in Figure 5. The transmission angle of the four-bar linkage O_1ABO_2 is presented in Figure 6. To provide the motion without any abrupt changes in velocities (high accelerations) this angle cannot be less than 30 and higher than 150 degrees. This condition is checked during mechanism synthesis.

Let us change the location of joint M on coupler AB, e.g. M(–4.35, 2.17). The length of link MN is then 9.62. The positions of G_1 and G_2 are the same. The kinematic scheme of the feeder obtained along with the trajectories of joints is presented in Figure 7.

A designer may be interested in a mechanism occupying smaller area compared to the area encircled by the load path. Not only can the positions of points M and G be changed for this purpose. Search for most optimal mechanisms may be carried out by changing the length of link NO_3 and the position of the pivot O_3. In this case the locations of ground pivot O_3 as well as the length of link O_3N are changed. Link O_3N is pinned at point $O_3(0, -1.5)$ and its length is equal to $R = 2$. The coordinates of joints of the four-bar linkage which approximates the arc are as follows: $O_1(-5, 0)$, $O_2(5.4152, 0)$. Links dimensions are: $l_1 = 1.888$, $l_3 = 2.11$, $l_5 = 5.157$, $l_6 = 5.636$, $\beta = 3.7$, $\theta_4 = -0.29$ rad.

The area of allowed locations of joint M on coupler AB is shown in Figure 8. From this area

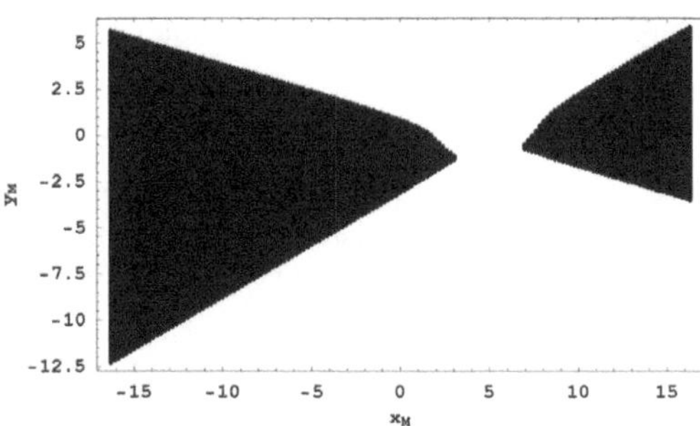

Figure 3. Area of allowed locations of joint M in Ax'y'.

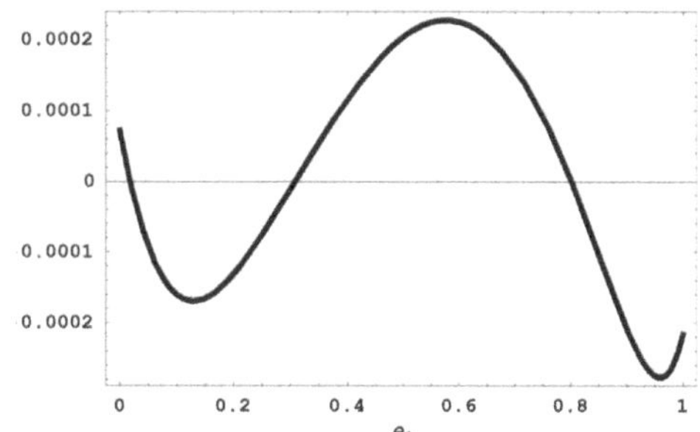

Figure 5. The relative angular motion of the jaws when the load is transported.

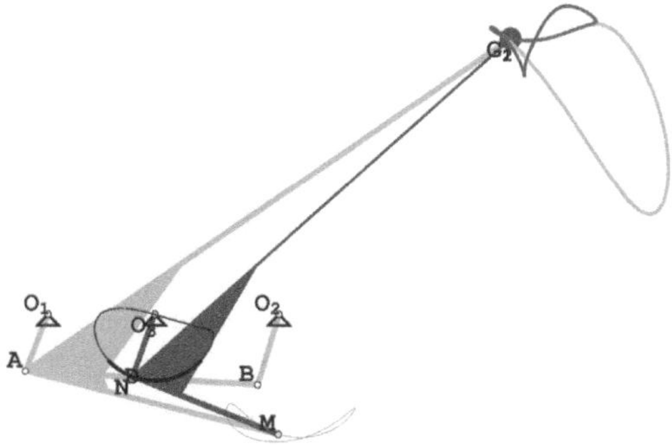

Figure 4. A position of the mechanism while the gripper is closed (the trajectories of G_1 and G_2 overlaps).

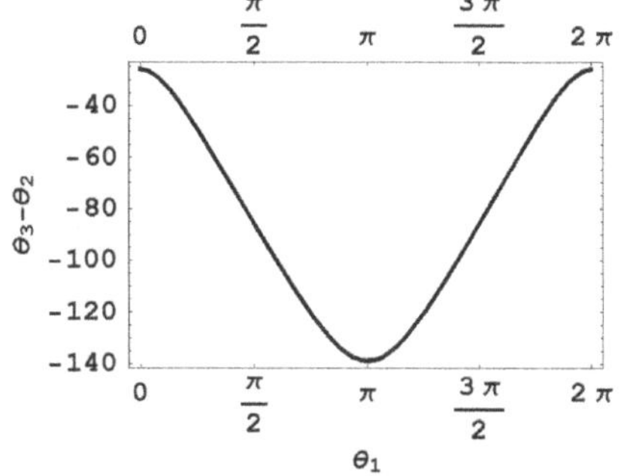

Figure 6. Transmission angle of the four bar linkage O_1ABO_2.

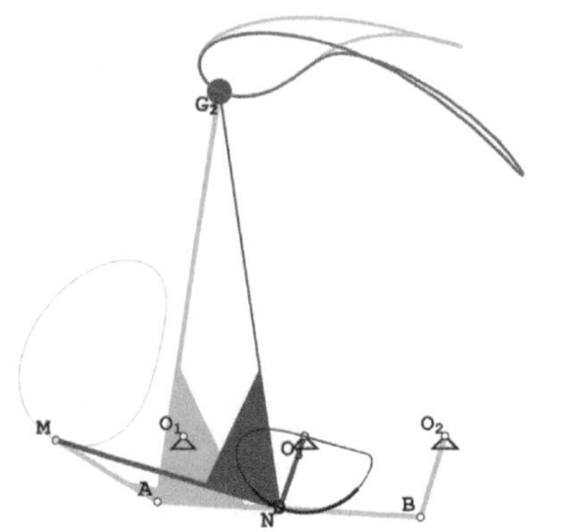

Figure 7. A position of the feeder for changed location of M.

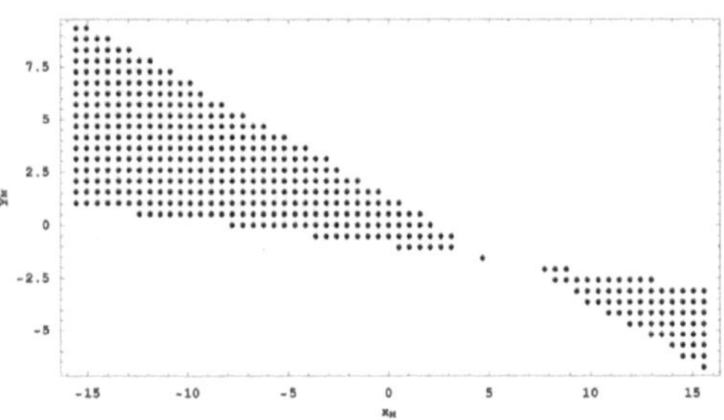

Figure 8. The area of allowed locations of joint M in the local coordinate system attached to coupler AB (the case with changed length of link O_3N).

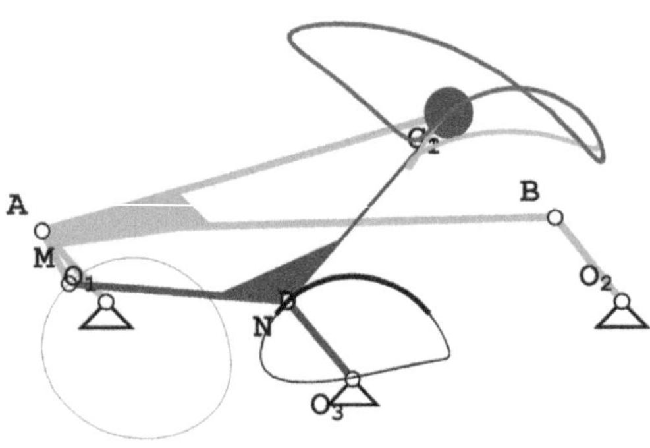

Figure 9. A position of the mechanism when the load is transported. The trajectories of the chosen joints (the case with changed length of link O_3N).

the following point is taken: M(0.519, −1.037). The length |MN| = 4.441. The coordinates of the middle point of the first jaw are: G_1(8.297, 2.074). The coordinates of the second jaw are: G_2(−2.974, −3.899). A position of this mechanism with the paths of the chosen joints are shown in Figure 9.

4 CONCLUSIONS

A countless number of feeders may be generated by means of the method presented here. Nonetheless the examples show that feeders occupy relatively large working space to transport a load between two points. After finding the positions of pivot M ensuring the full revolution of the active link, one can analyze how the positions of jaws and pivot M affect the size of the working space in searching for a satisfactory solution. The design process is not fully deterministic as the evolutionary algorithm is employed to search four-bar linkage meeting required geometry. The choices of the locations of points M and G_1 are also made by hand. These choices in particular industrial application are conditioned by concrete design requirements. Similarly the parameters of the locations of the ground pivot O_3 and the length of link O_3N (defining the synthesized circular arc)—very important in the first stage of the feeder designing—are chosen manually. It follows that many simulations are carried to find a satisfactory solution. For this reason the problem is open with respect to formulation of mathematical formulae enabling to find the optimal feeder dimensions. The aim of further studies is to make the method more automatic in searching for feeders optimally satisfying the prescribed requirements.

REFERENCES

Buśkiewicz, J. 2014a. Design of a feeder with the use of the path synthesis method. New Advances in Mechanisms, Transmissions and Applications, *Mechanisms and Machine Science* 17: 251–259.

Buśkiewicz, J. 2014b. Specific problem of mechanism synthesis. *International Journal of Applied Mechanics and Engineering* 19(3): 513–522.

Erdman, A.G., Sandor, G.N. & Kota, S.S. 2001. *Mechanism design: Analysis & Synthesis. 4th Ed. (Web Enhanced)*, Volume I: Prentice-Hall.

Kota, S. 1991. Generic models for designing dwell mechanisms: A novel kinematic design of Stirling engines as an example. *J. Mechanical Design, Transaction of the ASME* 113: 446–450.

Mohan, J. 2011. Optimisation Design of Six-Bar Double Dwell Mechanisms: A New Approach. *Applied Mechanics and Materials* 110–116: 5216–5222.

Pennock, G.R. & Isra, A. 2009. Kinematic analysis and synthesis of an adjustable six-bar linkage. *Mechanism and Machine Theory* 44: 306–323.

Radovan, R., Bulatović, R.R., Dordević, S.R. & Dordević, V.S. 2013. Search algorithm: A metaheuristic approach to solving the problem of optimum synthesis of a six-bar double dwell linkage. *Mechanism and Machine Theory* 61: 1–13.

Shiakolas, P.S., Koladiya, D. & Kebrle, J. 2005. On the optimum synthesis of six-bar linkages using differential evolution and the geometric centroid of precision positions technique. *Mechanism and Machine Theory* 40: 319–335.

Advances in Mechanics: Theoretical, Computational and Interdisciplinary Issues – Kleiber et al. (Eds)
© 2016 Taylor & Francis Group, London, ISBN 978-1-138-02906-4

Modal approach in the fluid-structure interaction in aerospace

W. Chajec
Materials and Structures Research Center, Institute of Aviation, Warszaw, Poland

A. Dziubiński
Center of New Technologies, Institute of Aviation, Warsaw, Poland

ABSTRACT: The paper contains examples of computational analyses based on modal modelling of the structural motion. This approach is typical for the investigation of aircraft aeroelastic phenomena. The flutter computation in the frequency domain, using simple unsteady aerodynamic models and specialized software is nowadays an aerospace standard. The selected normal modes, control surface modes and rigid body modes should be here taken into account during design. The mechanical properties of structure are represented by modal parameters as mode shape, generalized mass, normal frequency and damping coefficient. For selected configuration of structure they can be determined only once, by computation or ground vibration test. Similar approach can be used for more detailed simulations in the time domain, in which a more credible unsteady aerodynamic models can be applied. Due to this approach, the "structural" part of problem is very simple. The resulting database (structural modes) is small and unvarying, so the whole time-domain computation can be provided only in a flow computation system, without data exchange between structure's and flow computation systems.

1 INTRODUCTION

Modal modelling of the structural motion is a time and cost efficient way of dynamic analysis. The selected normal modes, control surface modes (including control system modes) and rigid body modes (for unsupported structures as flying aircraft) should be here taken into account.

The rigid body modes of unsupported structures can be treated as a simple extension of normal modes. The ordinary normal modes (vibration modes) are orthogonal to rigid body modes due to constant momentum equations. The rigid body modes are mutually orthogonal, if they were selected as displacements along and rotation about the main central axes of the structural inertia.

The mechanical properties of structure, which should be close to linear, are represented by modal parameters as mode shape, generalized mass (unity for normalized mode shape), normal frequency (zero for rigid body modes) and damping coefficient.

For a selected configuration of structure, the modes can be determined only once, by computation or ground vibration test. Some alterations of this configuration (as fuel or payload mass distribution) can be taken into account also by modal approach (modal mass matrix alteration or rapid new normal modes re-computation, (Chajec & Seibert 2011).

Similar approach can be used for more detailed time-domain simulations, in which a more credible unsteady aerodynamic models can be applied. Thanks to this approach, the number of DOF is reduced and the "structural" part of the problem is very simple. The equations of motion are coupled only by aerodynamic, gyroscopic or other external forces.

The limitation of modal approach are: nonlinearities, large deformations of very flexible structures, etc.

2 MODAL APPROACH

It is assumed that the structural movement (a vector function of the both, time and localization) is done by the following expression:

$$\boldsymbol{u}(t,\boldsymbol{r}) = \sum\nolimits_{i=1}^{n} q_i(t)\, \boldsymbol{\varphi}_i(\boldsymbol{r}), \tag{1}$$

where **n** is the number of modes, $q_i(t)$ is an unknown scalar function of time, $\boldsymbol{r}$ is the vector that identifies the point of structure and $\varphi_i(\boldsymbol{r})$ is i-th mode shape. The mode shapes, of normal modes as well as control surface modes and rigid body modes, and its frequencies can be pre-determined using the numerical FEM model of structure, or by Ground Vibration Test (GVT) of real structure. They are fixed for each separate configuration.

Using modal approach, the equations of motion can be written as (each mode shape is normalized with respect to unit generalized mass):

$$\ddot{q}_i + c_i\dot{q}_i + \omega_i^2 qi = f_{i,} \quad i = 1, \ldots, n, \tag{2}$$

where i is the mode number (selected normal modes, control surface modes and rigid body modes are taken into account), $q_i(t)$ is the contribution of i-th normal mode in the resultant motion (modal coordinate), c_i is damping coefficient (no cross mode couplings due to damping is assumed), ω_i is circular frequency, zero for rigid body modes—the term ω_i^2 is the generalized stiffness, and f_i are generalized aerodynamic forces, determined by fluid dynamics, gyroscopic forces and other forces treated as external ones.

The generalized forces can be determined using formula:

$$f_i = \int \boldsymbol{\varphi}_i(\boldsymbol{r}) \cdot \boldsymbol{F}(\boldsymbol{r})\, dV \quad i = 1, \ldots, n. \tag{3}$$

where $\boldsymbol{F}(\boldsymbol{r})$ is the external forces distribution.

The solution of Eqs (2) is a column matrix of modal coordinates $q_i(t)$, $i = 1, \ldots, n$.

3 TYPICAL ADDITIONAL ASSUMPTIONS

Harmonic motion and linear model of unsteady aerodynamic are typical assumptions for investigation of aircraft aeroelastic phenomena, such as: aerodynamic flutter, airfoil divergence, control reversal, aeroelastic dynamic response (with external excitation) as well as whirl-flutter of propeller power plants (with gyroscopic forces and aerodynamic forces on rotating propeller).

The flutter computation in the frequency domain using simple unsteady aerodynamic models (DLM or even strip model) and home (as the Polish JG2, (Nowak & Potkański 1976) or commercial software (as the ZONA software, NASTRAN or Martin Hollmann's SAF) is nowadays an aerospace industrial standard. The results are usually presented as g(V) and f(V) diagrams, like in Figures 1, 2.

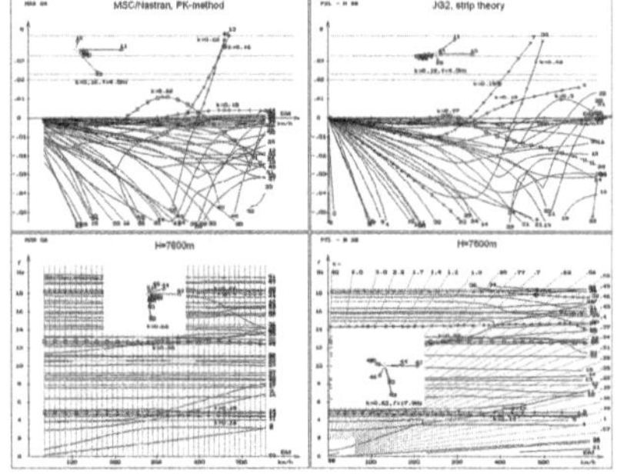

Figure 1. Flutter calculation results for the same computed normal modes set obtained by two analytical methods (Chajec 2007).

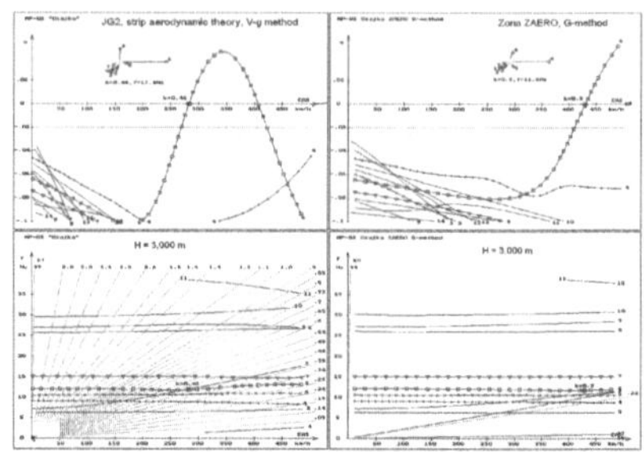

Figure 2. Flutter calculation results based on measured modes using JG2 and Zona ZAERO [Chajec 2015].

4 COMPARISON OF RESULTS OBTAINED BY A FEW COMPUTATION METHODS

In this paragraph selected results of flutter frequency-domain computations obtained by software used in the Institute of Aviation, Warsaw, i.e. JG2, MSC/NASTRAN as well as ZONA ZAERO are compared.

Figure 1 presents the flutter computation results of PZL M28-06 aircraft that is a designed, but not finalized modification with longer fuselage of M28 that is a Polish version of two turboprop Antonov An-28. The normal modes were obtained by MSC/Nastran using a detailed FEM model of both: right and left sides of the aircraft. The flutter calculation were provided using both MSC/Nastran and JG2.

In this case, no modal structural damping coefficients were introduced to flutter computation. The structural damping were taken into account separately for each flutter mode on g(V) diagram (horizontal lines above the g = 0 - line.

Another interesting example of flutter computation based on computed normal modes, verified by a comparison with GVT-results is presented by Chajec (2013).

In the author's opinion, the use of GVT-measured modes is a better way of flutter analysis that can be provided directly, using measured generalized masses of resonant modes, or indirectly, using theoretical mass model of aircraft. In the second case, the measure of generalized masses is not necessary. In addition, the resonant modes orthogonalization can be provided as well as some extra-cases, as control systems' disconnection and the mass alterations can be analyzed (Chajec & Seibert 2011).

Figure 2 presents the comparison of antisymmetric flutter computation results for MP-02 "Czajka" ultralight airplane, case of flutter analysis with additional masses on pedals and after increasing on rudder mass balace. During GVT the aircraft had the origin rudder mass balance. Some resonant modes are measured twice, before and after additional masses addition to stick and pedals. The masses are introduced to simulate of pilots hand/

foots masses. The flutter computation are provided using JG2 home software with strip aerodynamic theory and, for comparison, using Zona ZAERO commercial aeroelastic software. The flutter critical speeds determined by both methods were different, but the flutter modes were similar. It is the fuselage and fin bending and torsional flutter with rudder deflection.

5 RECTANGULAR WING WITH CONTROL SURFACE

For comparison of several numeric flutter analysis method the simple object was used, that was tested by Lorenc (2009) and numerically analyzed by Chajec (2010) for a few parameters of model, as spring stiffnesses. For most cases the calculating results were conservative, but for one parameters' set, identified as case 13, the wind tunnel test critical speed was 17.7 m/s, but in the MSC.Nastran flutter computation no flutter occurred, but the results strongly depended on the a wing/rudder division parameter in aerodynamic model. For a = 72.9%, expressed in percent of chord, the flatter occurs and its critical speed is ca 20 m/s. In the case 13 the normal modes are: 1. 3.35 Hz, wing yawing; 2. 4.44 Hz, wing translation; 3. 5.37 Hz, rudder flapping.

The flutter calculations of case 13 were repeated (Chajec & Dziubiński 2015) using Zona ZAERO software. In the case of ZAERO v.8.5 or higher a small correction of a parameter is sufficient. For a = 73.5% the critical flutter speed is ca 17 m/s.

6 TIME-DOMAIN COMPUTATION USING ADVANCED MODELS OF FLOW

The Equations 2 can be solved directly, in the time domain. This idea was used by several authors, i.a. in the TAURUS and CESAR European projects, (Roszak et al. 2011). To determine the aerodynamic forces, the normal modes $\boldsymbol{\varphi}_i(\boldsymbol{r})$ should be known in the selected points. It is possible due to numeric interpolation procedure.

A simple example of such approach has been done in (Chajec & Dziubiński 2015) for model, case 13 described on previous paragraph. An advanced fluid dynamics method to solve Reynolds Averaged Navier Stokes (RANS) equations has been used to simulate the movement of a two-body object. An airfoil with control surface, here named also a flap, with an ability to move in three degrees of freedom: vertical movement, the whole body rotation and a flap rotation, has been simulated in time domain. In this way the object is prepared to represent a flutter phenomena on a wingtip.

Since this movement is well-defined in two-dimensional computational space, a tetrahedral mesh representing 2D geometry has been created and flow has been simulated on widely recognized commercial FLUENT code. The software has an ability to be extended by parts of the code written by the user (UDF), so movement of both parts of a body has been described using such a code utilizing a modal approach mentioned above.

The movement of a body inside the mesh causes the mesh either to be deflected or to be partially rebuilt. Of course, this process could cause a mesh to degenerate, and for cases where the movement of body parts has been large in comparison to the mesh density, it did. Unexpectedly, this degeneration did not influence the flow solution quality in a similar way that it affected the mesh. It is due to very effective interpolation procedures which the mentioned above commercial code uses. The results are corresponding with the well-known behavior of the simulated phenomena, so there is a boundary velocity below which the oscillations tend to be dampened, but the above simulation led to very rapid movements of a model.

The simulation was a recreation of the case 13 described in the previous paragraph, so a computational domain had to be bounded by the walls at least on the upper and lower surface, but authors decided to use a condition, which was supposed to be safer in terms of mesh rebuild thus avoiding possibility of wall collision. The walls of domain were sufficiently far from the model to use far field boundary condition, as in free flight simulations.

Also a complication of a fluid model has been investigated. Omitting the viscosity forces at different levels of model (mechanic, fluid) has been tested and caused almost no effect if only the shear force has been neglected during integration over the model as on inviscid surface, but clearly increased the oscillation frequency when inviscid fluid model has been used.

Examples of damped and undamped flutter cases have been shown in Figures 3, 4 and 5.

The flow field in Figures 4 and 5 is visualized by means of velocity magnitude colour map. In places where velocity is low, colour gets darker while high velocity areas are lighter. The wake and separation area appearing at the high deflection are clearly visible. This kind of phenomena limits calculations using simple models, but using finite volume method and RANS solver area of oscillations where detachment appears and interaction with shedded vortexes is achievable. It is worth adding that a very similar mechanism appears during flight of small insects and humming-birds which use shedded vortex energy and energy stored in flexible wings deflection to minimize their energy cost during hover. This area is currently widely explored by UAV designers, so proper modelling of new design examples is crucial.

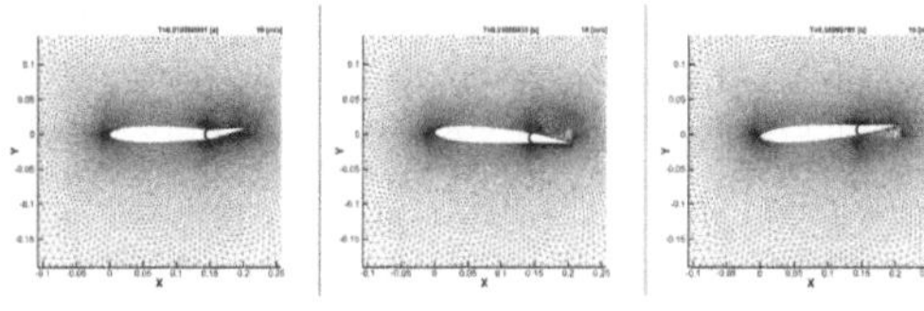

Figure 3. Airfoil movement and mesh accommodation in three time steps: 0.0, 0.25 and 0.5 s for low supercritical velocity 10 m/s.

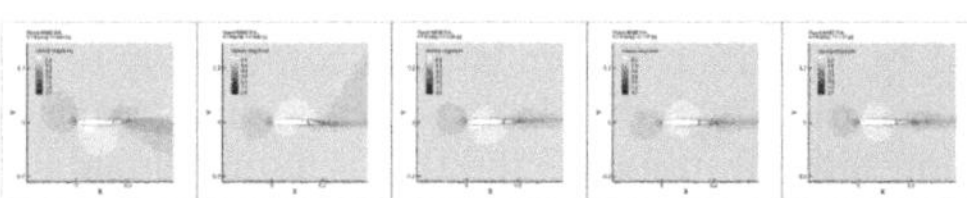

Figure 4. Airfoil movement and velocity magnitude in five time steps for the subcritical velocity 6 m/s.

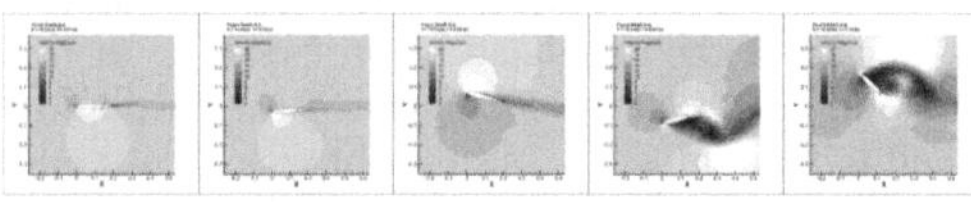

Figure 5. Airfoil movement and velocity magnitude in five time steps for high supercritical velocity 16 m/s.

The simplifications of omitting the tangential forces on two levels of abstraction have been tested. First, the tangential force has been omitted during forces on airfoil calculation. Next, the inviscid fluid model has been assumed. As it could be predicted, neglecting shear forces during calculation of airfoil (and rudder) loading does not introduce much change in oscillation frequency and amplitude. The inviscid fluid model assumption increases the frequency of rotational oscillation and also increases the amplitude. What is more interesting, the oscillations where the inertia of fluid is important, both models predict similar frequency, but using inviscid method a slightly greater amplitude is obtained.

In the Ansys time domain computation, the resulting movement contains mixed 3 complex modes with different frequencies. Here existed some troubles with the flow mesh accommodation for big displacements (see Fig. 3), but critical flutter speed can be determined properly—it is between 6 and 10 m/s.

7 CONCLUSION

Flutter computation should be conservative, because the flight flutter tests are dangerous. The sensitive data, like the *a* parameter, which for some control surface geometry, like the Friese-aileron, is hard to idealise, but also control surfaces balancing masses—should be varied in the analysis. The typical outcome of flutter analysis is the determination of optimal control surface mass balance.

Frequency domain flutter computation are rapid and relatively inexpensive.

Time domain flutter computation is very time consuming and expensive. The modal approach can simplify it.

After some improvements, the time domain Ansys (FMV) flutter computation is useful, especially for a verification of selected frequency domain computation results.

REFERENCES

Chajec, W. 2007. Obliczenia flatteru samolotu M 28 06 alternatywną metodą z wykorzystaniem pasowej teorii aerodynamicznej (in Polish, Flutter analysis of M 28-06 by an alternative method using the strip aerodynamic theory), an internal document No. M2806/PRG/Of/344/07 of PZL-Mielec.

Chajec, W. 2010. Analiza flatteru w systemie Nastran zamocowanego podatnie płata ze sterem o trzech stopniach swobody (in Polish, Flutter analysis in Nastran of elastic suported airfoil with control surface with three degrees of freedom), an internal document No. 5/BU/2010 of Institute of Aviation.

Chajec, W.C. & Seibert, T.R. 2011. Flutter calculation based on GVT-results and theoretical mass model, *Proceedings of the International Forum of Aeroelasticity and Structural Dynamics*, IFASD-2011-186, Paris.

Chajec, W.C. 2013. Aeroelastic Calculation of Innovative Non-conventional Aircraft with Swinging Canard Surface, Proceedings of the International Forum of Aeroelasticity and Structural Dynamics, IFASD-2013-11C, Bristol 2013 (published in *ASD Journal, Politechnico di Milano*, www.asdjournal.org/index.php/ASD/issue/view/5).

Chajec, W. & Dziubiński, A. 2015. Obliczenia flatteru prostokątnego płata ze sterem za pomocą MSC. NASTRAN, ZONA ZAERO i ANSYS FLUENT—porównanie z wynikami badań w tunelu aerodynamicznym (in Polish, MSC.NASTRAN, ZONA ZAERO and ANSYS FLUENT flutter computation of rectangular wing with control surface—comparison with flutter wind tunnel results), *Prace Instytutu Lotnictwa/Transaction of Institute of Aviation*, Warsaw.

Chajec, W. 2015. Instrukcja wykorzystania programu ZAERO do wykonania obliczeń flatteru na podstawie wyników badań rezonansowych (in Polish, Manual of ZAERO using for flutter calculation based on GVT results), internal document No. CBMK/CBMS/23018.01/01/2015 of Institute of Aviation.

Lorenc, Z. 2009. Badania w tunelu aerodynamicznym małej turbulencji modelu flatterowego o trzech stopniach swobody (in Polish, The low-turbulence wind tunnel tests of a flutter model with tree degrees of freedom), internal report No. 19/BU/2009 of Institute of Aviation, Warsaw.

Nowak, M. & Potkański W. 1976. Metodyka analizy flatteru samolotów lekkich (in Polish, Flutter analysis methodology of light aeroplanes), *Prace Instytutu Lotnictwa*, No. 65, Warszawa.

Roszak, R. et al. 2011. Fluid Structure Interaction for Symmetric Manoeuvre Base on Ultra Light Plane, American Institute of Physics, Melville, New York.

Advances in Mechanics: Theoretical, Computational and Interdisciplinary Issues – Kleiber et al. (Eds)
© 2016 Taylor & Francis Group, London, ISBN 978-1-138-02906-4

Investigations on the chip shape and its upsetting and coverage ratios in partial symmetric face milling process of aluminium alloy AW-7075 and the simulation of the process with the use of FEM

J. Chodór & L. Żurawski
Koszalin University of Technology, Koszalin, Poland

ABSTRACT: The paper presents the results of experimental research of milling aluminium alloy AW-7075 in low temperature. Shapes of chips and their ratios of upsetting and coverage after process are shown. The investigations were carried out using the Sandvik CoroMill R245-125Q40-12M milling head and exchangeable R245-12T3E-ALH10 carbide plates with superfinishing surface. Numerical analysis of milling (cutting) process were conducted in ANSYS/LS-PrePost programme. Technological and geometrical parameters of the tool and the workpiece material were identical to those in the experimental tests. Chip shapes and the ratios of upsetting that were obtained experimentally and numerically were compared.

1 INTRODUCTION

One of the problems of modern technology is the fulfilment of growing requirements connected with the operation of plant and machinery, concerning an increase of their durability and reliability. This is determined by an appropriate formation of the top layer. Its condition is essential because almost all tribological and wear processes occur on the object surface. The properties of the top layer after milling depend chiefly from the variant and realization conditions of the process.

In order to analyze the milling processin a comprehensive manner it is necessary to develop an adequate mathematical model and numerical methods in relation to its solution. In connection with this, the present paper makes use of the physical and mathematical models of the milling process, developed in the research team.

The algorithms of the solutions of the discrete systems obtained of equations including appropriate initial and boundary conditions were developed on their basis. The milling process was considered a geometrically and physically non-linear boundary and initial problem with an assumption of the occurrence of non-linear moving and boundary conditions that are variable in time and space, whereas these conditions are unknown in the contact area between the head and the workpiece material. The algorithms that had been prepared in this way were implemented in the ANSYS software.

The main purpose of the study was to develop an application in the ANSYS system to allow one to observe those phenomena that occur in extremely small areas and that take place with high velocities, which last very short time, and yet they determine the results of the milling process. Similar problems occurs in other works (Chodor 2007÷2014). These phenomena include among others the chip formation phenomenon, which in the conditions of an experiment are hard to observe due to the dynamics of the process. A numerical analysis allows a detailed observation of the chip formation phenomenon and also an observation of the distribution of the intensity of stresses, the intensity of strains and distribution of temperature in any time and at any place of the process (Kaldunski 2014, Kukielka 2007, 2015).

The research problem of the present study was the determination of the chip shape as well as the determination of the upsetting ratio λ_h and chip coverage ratio λ_b in the process of cylindrical, face, symmetrical and incomplete milling in low temperature. The shapes of chips and the values of upsetting ratios from the experimental tests were compared with the chip shapes and the values of upsetting ratio obtained in computer simulations. In computer simulations, the technological parameters, the tool geometry and the parameters of the workpiece material were identical with those from the experimental tests. On the background of a comparison of chips, the correctness was to be found of the applications developed and of the preparation of numerical analyses.

2 EXPERIMENTAL RESEARCHES

The investigations related to the chip shape and its ratios of upsetting and coverage were carried out with the use of the following devices and tools:

- Sandvik CoroMill R245-125Q40-12M milling head, diameter $D_c = 125$ [mm] with the number of blades $z = 8$,
- exchangeable carbide plates, type R245-12T3E-ALH10.
- Geometry of the machining blade:
- the main set angle and the support set angle: $\kappa = 45^{\circ}$, $\kappa' = 45^{\circ}$,
- the blade rounding radius $r_{\varepsilon} = 1.5$ [mm],
- the plates possessed superfinishing surface $b_s = 2.3$ [mm].

The research covered the performance of cylindrical and face milling of the aluminium alloy AW-7075 with the feed velocity of the milling machine table $v_f = 307$ and 505 [mm/min]. In the experiment, the versatile FWD32U vertical milling machine was used (Zurawski 2012). Once the cutter had covered the specified machining path L_s, the measurements were performed of the thickness, width and length of chips. Partial symmetric face milling process in low temperature −180°C was discontinuous process. Milling head had machined thin layer of material, then returned to initial position and later machined next layer of material. After every pass of milling head chips were collected. They were assigned to length of cut path L_s. Form of chip, upsetting ratio, and chip coverage ratio were determined. The results were averaged and presented graphically. Their digital images were made.

The dependence of the chip upsetting ratio λ_h and the chip coverage ratio λ_b from the length of the machining path L_s for feed $v_f = 307$ [mm/min] is presented in Figures 1a, b.

Figure 2 shows the shapes of chips documented after milling with feed speed $v_f = 307$ [mm/min] at cryogenic treatment. On the whole length of cut Ls from 100 to 601 [m] obtained chips were in the form of: tubular short, ribbon short, and tubular with two or three curls and changed to the form of ribbon chips (Fig. 2).

The dependence of the chip upsetting ratio λ_h and the chip coverage ratio λ_b from the length of the machining path L_s for feed $v_f = 505$ [mm/min] is presented in Figure 3a, b.

Figure 4 shows the shapes of chips documented after milling with feed speed $v_f = 505$ [mm/min] at cryogenic treatment. On the whole length of cut Ls from 60 to 609 [m] obtained chips were in the form of: tubular short and tubular from three to five curls (Figure 2).

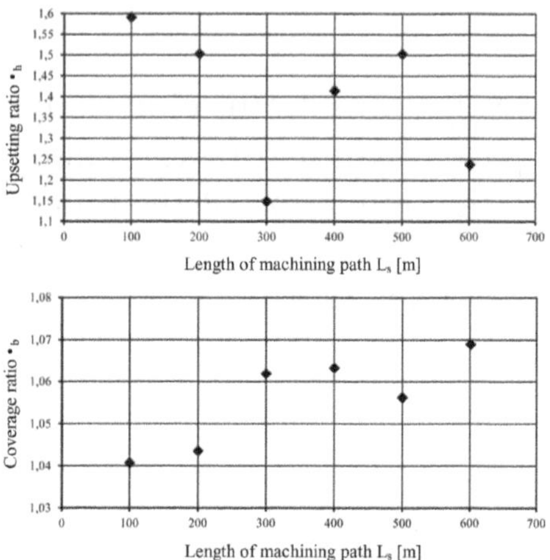

Figure 1. Diagram of the dependence of the following ratios: a) upsetting λ_h, b) coverage λ_b from the machining path L_s for machining with feed $f_b = 0.032$ [mm/blade], cutting velocity $v_c = 471$ [m/min] during cooling with liquid nitrogen.

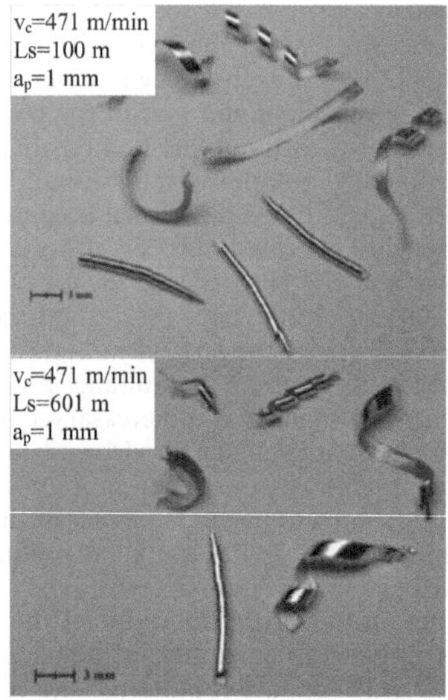

Figure 2. Chosen form of chip for milling with the feed $f_b = 0.032$ [mm/blade] and cutting velocity $v_c = 471$ [m/min] for various values of the machining path L_s.

Additional research shows the results of temperature measurements during milling with cooling with liquid nitrogen (Fig. 5). From the obtained results for the cutting feed speed $v_f = 307$ [mm/min] can be seen that the function is linear, and the temperature is in the range of 22÷29°C.

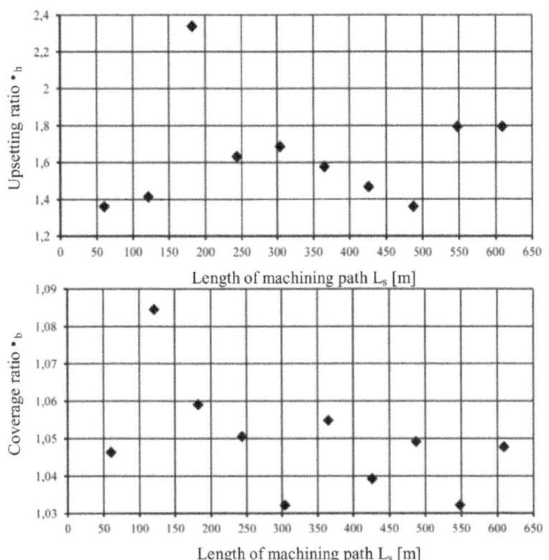

Figure 3. Diagram of the dependence of the following ratios: a) upsetting λ_h, b) coverage λ_b from the machining path L_s for machining with feed $f_b = 0.052$ [mm/blade], cutting velocity $v_c = 471$ [m/min] during cooling with liquid nitrogen.

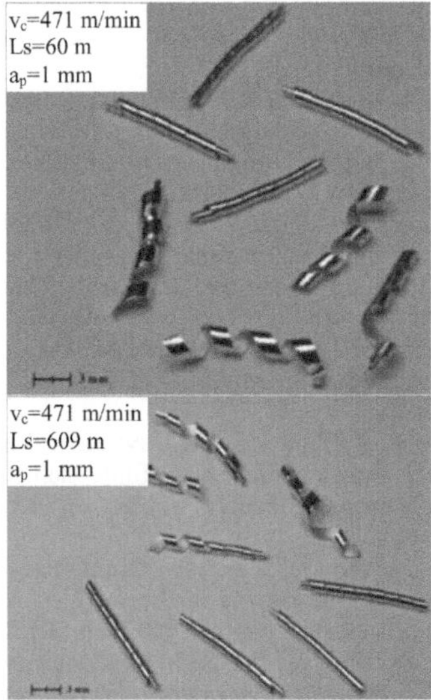

Figure 4. Form of chip for milling with the feed $f_b = 0.052$ [mm/blade] and cutting velocity $v_c = 471$ [m/min] for various values of the machining path L_s.

Other results obtained for milling with speed feed $v_f = 505$ [mm/min], because the temperature is in the range of 12÷30°C. At the beginning of length of cut L_s obtained parabolic function but after 180 [m] of the L_s the temperature function is linear.

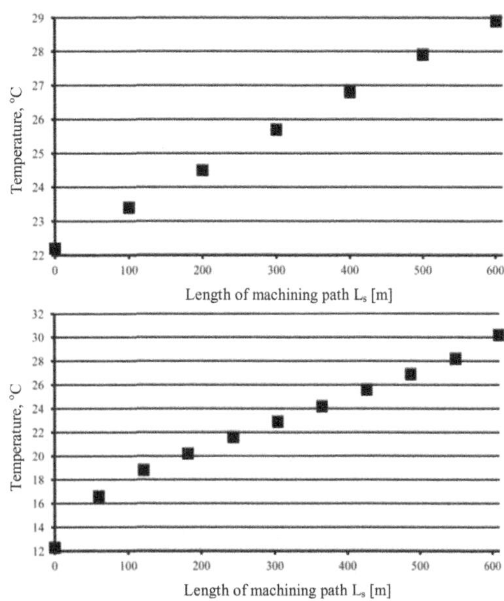

Figure 5. Diagram of the dependence of the temperature for milling cutting feed speed: a) $v_f = 307$ [mm/min], b) $v_f = 505$ [mm/min] for cutting velocity $v_c = 471$ [m/min] during cooling with liquid nitrogen.

3 NUMERICAL ANALYSIS

Milling is considered a geometrically and physically nonlinear initial and boundary problem. The analytical solution of this problem like: determination of states of deformations and stresses in the any moment of duration of the process is impossible. Therefore this problem by Finite Element Method (FEM) was solved.

Application was developed in the Ansys/LS-PrePost programme, which makes possible a complex time analysis of the states of deformations (displacements and strain) and stress in the surface layer of object at/after milling. It was accepted in the simulations that the tool is a non–deformable body, but it can be elastic body as well for precise calculations, while the object is an elastic/visco–plastic body described with the aid of Cowper–Symonds model and has material parameters adequate for aluminium alloy AW-7075. Huber–Mises–Hencky's plasticity model is used together with the associated flow rule. Below (Fig. 6) are shown exemplary chips received during numerical simulations.

The numerical algorithms implemented in the ANSYS system can be used in an evaluation of the impact of the technological conditions of machining on the product quality, chip shape etc. They permit a better understanding of those

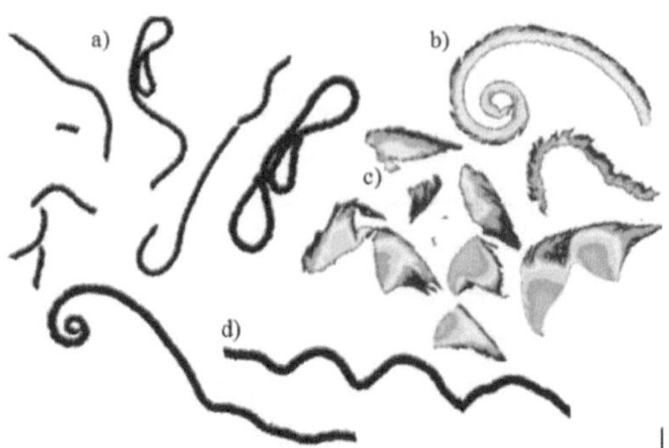

Figure 6. Forms of chips received during numerical simulations of milling Aluminum alloy AW-7075.

phenomena that occur in contact areas; hence, they may constitute the basis for the development of the guidelines for the selection of machining conditions in relation to the technological quality of the product as required.

Chips shapes in numerical simulations are similar with experimental chips received after milling process. Upsetting ratio, chip coverage ratio also were on the similar level. That confirms the justifiability of the use of computer simulations and their reliability.

4 CONCLUSIONS

The values of upsetting and coverage ratios of chip after cooling with liquid nitrogen were different for whole machining path L_s during milling with cutting feed speed $v_f = 307$ and 505 [mm/min], the forms of the chips were: short tubular or short ribbon.

Such shapes of chips shows that effect of cooling with liquid nitrogen has huge influence on the cutting process.

At a greater value of feed speed v_f obtained lower values of cutting temperature depending on the initial machining path L_s. After the experiment, the cutting temperatures measured were similar for smaller and higher cutting speeds.

When analysing the shapes of chips obtained in simulation tests (realized in the same conditions as in the case of experiments) with those chips that were produced in the experimental tests, it can be observed that the chip shapes are similar. The values of upsetting ratios are also similar. Several chip shapes were produced. The chip shape depends from the machining path. The upsetting and coverage ratios presented in the diagram accept a saw course, which depends chiefly from the feed value.

REFERENCES

Chodor J. & Kukielka L. 2007. Numerical analysis of the influence of abrasive grain geometry and cutting angle on states of strain and stress in the surface layer of object, *Surface/Contact Conference, Wessex Institute, Ashurst Lodge, Ashurst, Southampton, UK.*

Chodor J. & Kukielka L. 2008. Numerical analysis of chip formation during machining for different value of failure strain. *Journal PAMM*, Volume 7, Issue 1: 4030031–4030032.

Chodor J. & Kukielka L. 2008. Numerical analysis of micromachining of C45 steel with single abrasive grain. *GAMM79th Annual Meeting of the International Association of Applied Mathematics and Mechanics*, Bremen.

Chodor J., Kukielka L., Storch B. 2009. New method of determination of tool rake angle on the basis of crack angle of specimen in tensile test and numerical simulations. *Ninth International Conference on Surface Effects and Contact Mechanics Computational Methods and Experiments, Algarve, Portugal.*

Chodor J., Forysiewicz M., Kukielka L. 2011. Numerical analysis of flash and chip creating for elasto/visco-plastic body in the process of wedge movement, *XXXIV Scientific School of Abrasive Machining*, Gdansk. (In Polish)

Chodor J. & Kaldunski P. 2014. Experimental researches of influence of sliding burnishing technological conditions for surface roughness of 41Cr4 steel product, *IBEN*, Gorzów Wielkopolski.

Forysiewicz M., Chodor J., Kukielka L. 2009. Discrete modeling and numerical analysis of the process of cutting with a single abrasive grain using finite element method. *Numerically controlled machine tools and techniques programming in manufacturing operations.* Printing in Radom University of Technology, Radom (In Polish)

Kaldunski P. & Chodor J. 2014. Numerical analysis of the deep drawing process by finite element method for anisotropic model of object, *IBEN*, Gorzów Wielkopolski.

Kukielka L. & Kukielka K. 2007. Numerical analysis of the physical phenomena in the working zone in the rolling process of the round thread, in: J.T.M. de Hosson, C.A. Brebbia, S-I Nishida, *Computer Methods and Experimental Measurements for Surface Effects and Contact Mechanics VIII*, WIT*PRESS*, Southampton-Boston: 125–124.

Kukielka L. & Kukielka K. 2015. Modelling and analysis of the technological processes using finite element method, *Mechanik*.

Zurawski L., Palka T., Zawada-Tomkiewicz A. 2012. Improving the efficiency of milling of flat surfaces. *School of Machining—Effective Production*, Wroclaw. (In Polish)

Advances in Mechanics: Theoretical, Computational and Interdisciplinary Issues – Kleiber et al. (Eds)
© 2016 Taylor & Francis Group, London, ISBN 978-1-138-02906-4

Sensitivity analysis of behavior of sandwich plate with PU foam core with respect to boundary conditions and material model

M. Chuda-Kowalska & M. Malendowski
Institute of Structural Engineering, Poznań University of Technology, Poznań, Poland

ABSTRACT: Sandwich panels consisting of thin metal sheets and a closed-cell structure polyurethane foam core are considered in the paper. The problem of the influence of two main factors on the response of modeled plate is analyzed, mainly: material model of the PU foam and support conditions of plate. The main attention is focused on applications and limitations of the most commonly used for PU foam isotropic material model, since the foam is definitely an orthotropic material. The orthotropic model of the PU foam is proposed and recommended, alternatively. The results obtained from FE analyses are compared with the experimental results carried out by the authors and proper conclusions are drawn.

1 INTRODUCTION

Sandwich panels made up of two external thin and stiff metal facings separated by a thick, lightweight core are considered in the paper. In the literature, it is possible to find many papers focused on sandwich structures, their applications, designing and testing procedures which take into account the influence of the soft core on the behavior of the layered structure (Davies 2008, Zenkert 1995). The vast majority of them assumed that the foam core is isotropic, linear-elastic and homogeneous material (Gibson 2011, Studzinski et al. 2015). Then the shear modulus of the core material plays a crucial role for structural performance of a sandwich panel. Therefore, establishing the reliable experimental methods for determining this parameter is important and still under consideration (Juntikka and Hallstrom 2007). Unfortunately, foams can exhibit highly anisotropic properties depending on the direction of measurement (Caliri Junior et al. 2012, Chen and Fatt 2013, Chuda-Kowalska et al. 2015). Additionally, very short elastic range and the fact that the same material may exhibit brittle, perfect plastic or hardened responses, depending on the direction along which the load is applied, complicate the determination of mechanical response of analyzed material. Therefore, using porous materials in the structural applications (e.g. as a core in three-layered panels), a knowledge about their mechanical behavior on both a micro- and a macroscopic scale is necessary (Gibson & Ashby 1997, Mills 2007) and more advanced analyses are required. In that case, the set of parameters required by material model reflect the need for numerous experimental tests. Thus, simplifying the material model of foam to be elastic and isotropic is very attractive and useful for engineers and designers as well as for scientists. Nevertheless, degree of anisotropy of PU foam should be always controlled and taken into account if it is needed.

2 FORMULATION OF THE PROBLEM

In a design process engineers often assume an elastic range of loading and therefore focus on investigating the elastic properties only, namely Young's modulus E_i and Poisson's ratio ν_{ij}, using conventional procedures based on tension test with a standardized geometry. For analyzed sandwich plates in accordance with code EN 14509 foam core is treated as an isotropic and homogeneous material. Then, only two parameters of the core material are needed, because the relation (1) must obligatory hold in this classical model. For analyzed sandwich plates methods to determine Young's modulus E_C and shear modulus G_C are presented in code 14509.

$$G = \frac{E}{2(1+\nu)} \tag{1}$$

In order to accurately identify the behavior of the foam, a series of tests (tension, compression, bending) were carried out using standard procedures and a Digital Image Correlation (DIC) technique, named Aramis. Methods and samples were presented in detail by Chuda-Kowalska and Urbaniak (2015) (in print). Obtained results were summarized in Table 1 and revealed pronounced anisotropy of analyzed foam.

Table 1. Material parameters of analyzed PU foam.

Young's moduli [MPa]			Shear moduli [MPa]			Poisson's ratios [–]		
E_x	E_y	E_z	G_{xy}	G_{zx}	G_{yz}	ν_{xy}	ν_{zx}	ν_{yz}
13.45	4.41	3.65	2.51	3.0	2.3	0.55	0.27	0.2

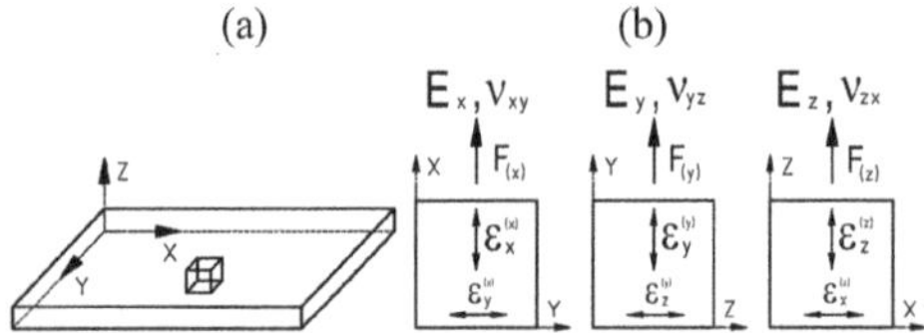

Figure 1. (a) Adopted material directions, (b) Nomenclature.

Material directions and coordinate system adopted in the paper are chosen based on the plate's dimensions and shown in Figure 1.

The main goal of the work is to demonstrate the sensitivity of sandwich plates on boundary conditions and material model. Based on numerical and experimental studies, two material models are compared. The authors reveal, that despite the large anisotropy of the foam, the application of the simplified isotropic model can be sufficiently accurate if certain conditions are met. Nevertheless, the cases when orthotropic model is more appropriate are also presented.

3 NUMERICAL ANALYSIS

Numerical models were created in Abaqus system environment. Facings were modeled as four node, doubly curved, thin shell, finite membrane strains elements S4 (2D element). The core was modeled using eight node linear brick elements C3D8 (3D element). The "tie" interaction has been used between the layers. The panel is supported by basing plates modeled as rigid bodies. In numerical analyses it was assumed that the facings are flat and made of steel with Young's modulus $E_F = 210$ GPa and Poisson's ratio $\nu_F = 0.3$. All numerical analyses were done in linear-elastic range. A linear elasticity in an orthotropic material is defined according to equation (3), by giving the engineering constants summarized in Table 1. In general, relationship between the Poisson's ratios ν_{ij} and ν_{ji} is defined by (2).

$$\frac{\nu_{ij}}{E_i} = \frac{\nu_{ji}}{E_j} \tag{2}$$

3.1 *Example 1: four-point bending test*

3.1.1 *Sensitivity analysis with respect to material model—beam specimen*

The first numerical example refers to structure investigated during laboratory experiments. In this case the authors analyzed elastic behavior of sandwich panel loaded by two line forces. The scheme of the test is shown in Figure 2. This kind of test is used to identify the shear modulus of the core material G_C.

For both supporting base plates (b = 0.1 m), only the rotation about the axis perpendicular to the beam axis is free, in order to obtain simply supported boundary conditions. The connection between the plate and the support was defined as a contact "surface to surface" with hard normal contact and 0.3 friction coefficient. Shell element S4 has dimension 2 × 2 cm, while brick element C3D8—2 × 2 × 2 cm. The geometric parameters of the plate are: width B = 0.1 m, total length L = 1.0 m, length of the span L_0 = 0.9 m, thickness of the core d_C = 94.36 mm, and the thickness of steel facings t = 0.35 mm. The deflection of the midpoint of the span w is presented in Figure 3 (No 0) as a function of the load F, where $w = f—w_0$ (Fig. 2). In this figure, the another lines represent the numerical solutions and are drawn for: 1) isotropic model with real Young's modulus E_z = 3.65 MPa and ν_{zx} = 0.27, 1a) isotropic model with real shear modulus G_{zx} = 3.0 MPa and $\nu_{zx} = 0.27 \Rightarrow E_{fiction}$ = 7.62 MPa according to (1) and the last case 1b) orthotropic model.

The obtained results show that isotropic model with real value of Young's modulus E_z and Poisson's ratio ν_{zx} (Table 1) is an incorrect approach

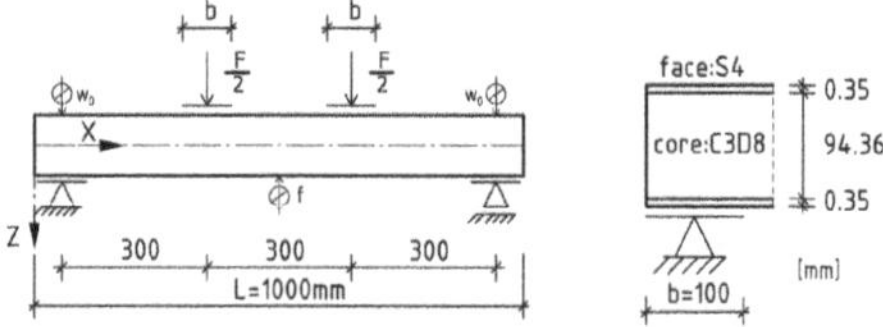

Figure 2. (a) The scheme of the tested sample, (b) Close-up of the support area.

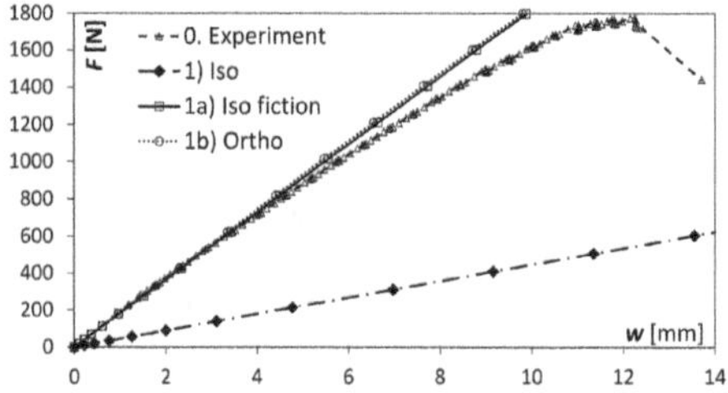

Figure 3. Four-point bending test—beam.

(plot No 1 in Fig. 3). Assuming that G_{zx} and ν_{zx} (Table 1) are the most important parameters, the isotropic model (iso-fiction No 1a) and orthotropic model (No 1b) give similar responses, comparable with the real behavior.

$$\begin{Bmatrix} \varepsilon_x \\ \varepsilon_y \\ \varepsilon_z \\ \gamma_{xy} \\ \gamma_{zx} \\ \gamma_{yz} \end{Bmatrix} = \begin{pmatrix} 1/E_x & -\nu_{yx}/E_y & -\nu_{zx}/E_z & 0 & 0 & 0 \\ -\nu_{xy}/E_x & 1/E_y & -\nu_{zy}/E_z & 0 & 0 & 0 \\ -\nu_{xz}/E_x & -\nu_{yz}/E_y & 1/E_z & 0 & 0 & 0 \\ 0 & 0 & 0 & 1/G_{xy} & 0 & 0 \\ 0 & 0 & 0 & 0 & 1/G_{zx} & 0 \\ 0 & 0 & 0 & 0 & 0 & 1/G_{yz} \end{pmatrix} \begin{Bmatrix} \sigma_x \\ \sigma_y \\ \sigma_z \\ \sigma_{xy} \\ \sigma_{zx} \\ \sigma_{yz} \end{Bmatrix} \tag{3}$$

3.1.2 *Sensitivity analysis with respect to material model—plate specimen*

The second example differs only in the width of the sample, and now B = 1.0 m. This example shows that if the panel has dimensional proportion L/B = 1 but the other conditions (load and support) provide that it works as a beam element then simplified isotropic model (with G_{zx} as a most important parameter) can be used with reasonable accuracy. The obtained relations between load and deflection of the plate are presented in Figure 4.

3.1.3 *Sensitivity analysis with respect to boundary conditions*

The next study is focused on the sensitivity of the sandwich plate for numerical approach used in definition of support conditions. Therefore, the "tie" and "contact" interactions, used between the plate and the supports, have been compared between each other. However, no changes were observed in the behavior of the analyzed models (B = 0.1 m and B = 1.0 m).

3.2 *Example 2: sandwich panel supported on three edges loaded by force*

The next two cases concern the plate supported on three edges with asymmetrically applied downward concentrated force (Fig. 5). A load is applied on the free edge at mid-span through a rigid 100 mm × 100 mm block.

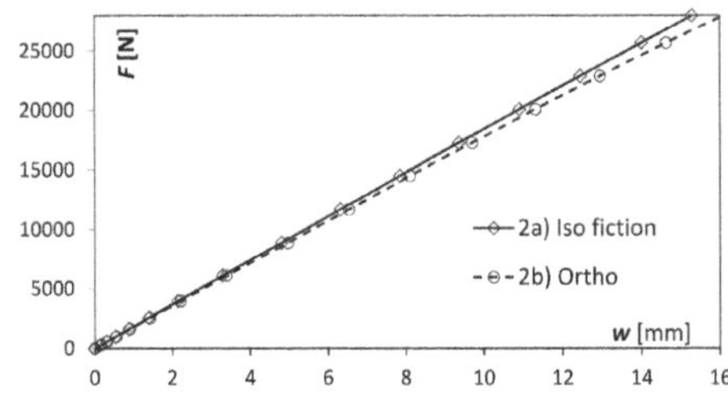

Figure 4. Four-point bending test—plate.

3.2.1 *"Contact" interaction*

In the first case, the connection between the plate and the support was defined as a "contact" with the same manner like in *Example* 1. In this case numerical results were compared to experimental data carried out and presented in detail by Studzinski et al. 2015. Therefore, the same geometric parameters of the plate were used (B = 1.0 m, L = 3.0 m, d_C = 79.0 mm, t = 0.42 mm). The width and length of the supports in FE analysis were also in agreement with the dimensions of the test bed supports. Shell element S4 has dimension 3 × 3 cm, while brick element C3D8—3 × 3 × 3 cm. The distribution of the displacements field for orthotropic model is presented in Figure 6a.

3.2.2 *"Tie" interaction*

In the second case plate is tied to the support. Comparing to Figure 6a and b, the distribution of the displacements field and behavior of the plate, must be, are completely different.

The comparison of obtained results, which is presented in Figure 7, shows that for plate supported at three edges, for both analyzed cases, the isotropic and orthotropic models give different results. Comparative analysis was performed for load F = 3.5 kN. In case with "contact" interaction (paths No 1 and 2) the difference between "iso" and "ortho" approach is 15%. But in case with "tie" interaction the scatter in the results is even higher, and equal nearly to 20%.

For plate with "contact" interaction experimental verification was done. Orthotropic model (No 2—dash-dot line in Fig. 7) gives similar response as real experiment (No 0—continuous line). Therefore, we can say, that for *Example* 2 (both analyzed cases), orthotropic model is definitely more accurate.

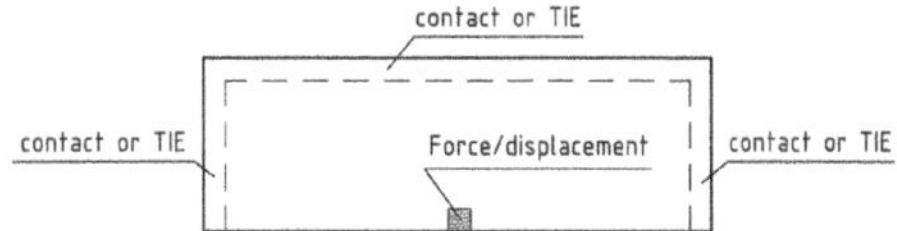

Figure 5. Scheme of boundary conditions.

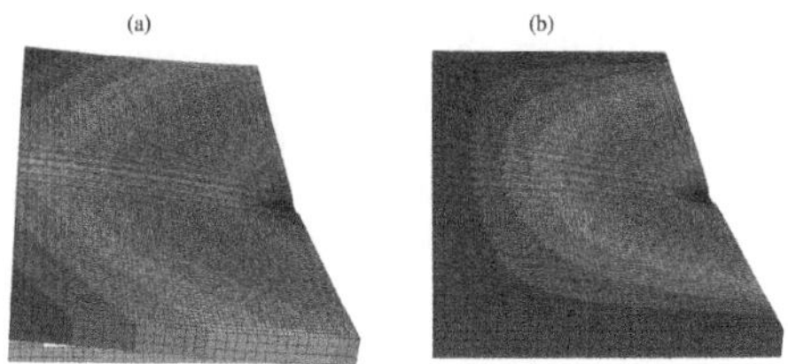

Figure 6. Distribution of the vertical displacement in plate supported on three edges: (a) "contact" interaction, (b) "tie" interaction.

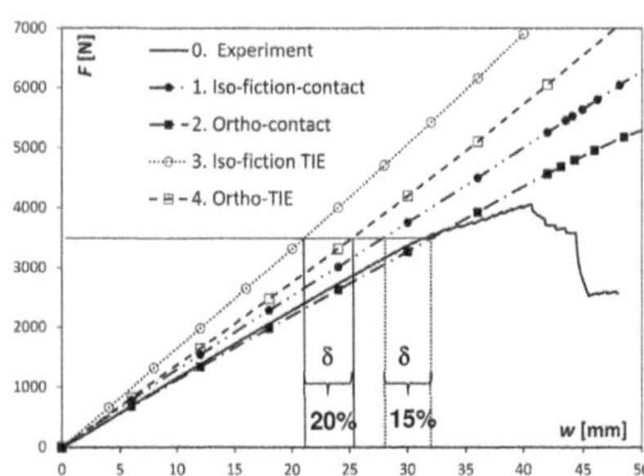

Figure 7. Plate supported on three edges loaded by force: FEM solutions No 1–4 versus experimental plot No 0.

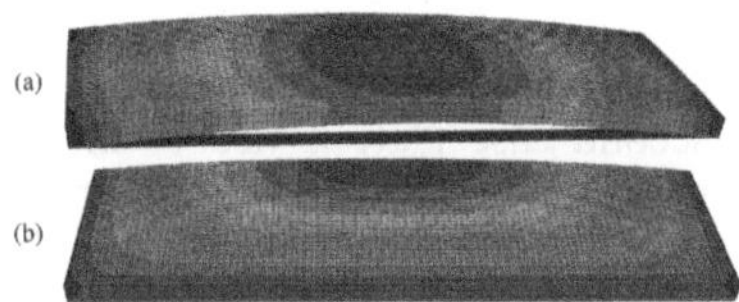

Figure 8. Plate supported on three edges with temperature load: (a) "contact" interaction, (b) "tie" interaction.

Table 2. Temperature load—results.

Interaction	ISO w [mm]	ORTHO w [mm]	δ [%]
"Contact"	5.201	5.196	0.10
"Tie"	2.567	2.906	13.21

3.3 *Example 3: sandwich panel supported on three edges with temperature load*

For sandwich plates, climatic effects (e.g. due to a temperature difference between the faces of a panel) shall be taken into account in the design. Therefore, the influence of the temperature gradients resulting from the difference between the outside temperature $T_1 = +55^\circ C$ (for very light colors accordance with code 14509) and the inside temperature $T_2 = +20^\circ C$ are examined. Similar like in *Example* 2, two cases of boundary conditions are considered. The geometric parameters of the plate are: width $B = 1.0$ m, total length $L = 3.0$ m, thickness of the core $d_C = 99$ mm, thickness of external steel facing $t_1 = 0.46$ mm and thickness of internal steel facing $t_2 = 0.36$ mm. Shell element S4 has dimension 3×3 cm, while brick element C3D8—$3 \times 3 \times 3$ cm. The distribution of the displacements field for orthotropic models is presented in Figure 8. The results obtained by this analysis, summarized in Table 2, show that in "contact" interaction the model behaves like a simple supported plate. Then, isotropic model with real value of G can be used with sufficient accuracy. However, in case with "tie" interaction orthotropic model should be used.

4 CONCLUSIONS

Different effects act on the behavior of modeled sandwich panels. In the article, the main attention has been focused on the study of the structural sensitivity with respect to the material model, geometric parameters, as well as the type of boundary conditions (load excitations, support). The range of applicability of classical isotropic PU material model has been discussed base on numerical examples. It has been proved, that in some special cases, where sandwich panel starts to act as plate, in contrast to beam behavior, introduction of amore complex orthotropic material model may be particularly useful. Due to reliable FE model, effort of evaluation of the structural response of panel subjected to the difficult to realize experimentally actions can be significantly minimized. That means mainly the number of laboratory tests and subsequently the costs of testing procedure.

REFERENCES

Caliri Junior, M., G.P. Soares, R. Angelico, R. Bresciani Canto, & V. Tita (2012). Study of anisotropic polymeric cellular material under compression loading. *Journal of Materials Research 15(3)*, 359–364.

Chen, L. & M.S.H. Fatt (2013). Transversely isotropic mechanical properties of pvc foam under cyclic loading. *Journal of Materials Science 48(19)*, 6786–6796.

Chuda-Kowalska, M., T. Gajewski, & T. Garbowski (2015). Mechanical characterization of orthotropic elastic parameters of a foam by the mixed experimentalnumerical analysis. *Journal of Theoretical and Applied Mechanics 53(2)*, 383–394.

Chuda-Kowalska, M. & M. Urbaniak (2015). *Continuous Media with Microstructure*, Chapter Orthotropic Parameters of PU Foam Used in Sandwich Panels. Springer. inprint.

Davies, J.M. (2008). *Lightweight sandwich construction*. John Wiley & Sons.

Gibson, L. & M. Ashby (1997). *Cellular solids: structure and properties*. Cambridge University Press.

Gibson, R. (2011). A simplified analysis of deflections in sheardeformable composite sandwich beams. *Journal of Sandwich Structures and Materials 13(5)*, 579–588.

Juntikka, R. & S. Hallstrom (2007). Shear characterization of sandwich core materials using four-point bending. *Journal of Sandwich Structures and Materials 9*, 67–94.

Mills, N. (2007). *Polymer foams handbook. Engineering and Biomechanics Applications and Design Guide.* Butterworth-Heinemann. ISBN-10: 0750680695, ISBN-13: 978-0750680691.

Studzinski, R., Z. Pozorski, & A. Garstecki (2015). Structural behavior of sandwich panels with asymmetrical boundary conditions. *Journal of Constructional Steel Research 104*, 227–234.

Zenkert, D. (1995). *An Introduction to Sandwich Construction*. EAMS.

Advances in Mechanics: Theoretical, Computational and Interdisciplinary Issues – Kleiber et al. (Eds)
© 2016 Taylor & Francis Group, London, ISBN 978-1-138-02906-4

Distribution of damages in unconventionally reinforced concrete slabs subjected to impact loads

K. Cichocki & M. Ruchwa
Koszalin University of Technology, Koszalin, Poland

ABSTRACT: The study concerns numerical analysis performed in order to evaluate the final distribution of damages developed in circular plates made of fibre reinforced concrete with brick rubble aggregate subjected to impact loads. Adequate experimental tests have been carried out on entire set of slabs produced with various types and amount of fibres. The development of damages, its distribution and intensity was observed and documented for each slab defining the mechanism of this phenomenon. Numerical discrete models were built and applied to simulate the entire problem using an advanced finite element method approach.

1 INTRODUCTION

One of the goals of this study is to investigate the possibility to reuse the waste material (sorted brick rubble—see Figs. 1, 2) as aggregate in the production of concrete elements (Cichocki, K., Domski, J., Katzer, J. & Ruchwa, M. 2014, Katzer, J. & Domski, J. 2015). Application of steel fibres (unconventional reinforcement—see Fig. 3) increased considerably the resistance and durability of such elements for impact loads. This is a very important characteristic for structures exposed to this load: elements of bridges, roadways, protective structures applied in order to minimize the effects of explosions and impacts, etc. (Cichocki & Ruchwa 2013a).

The necessity to improve the impact resistance for concrete structures used in many areas of civil engineering was the main reason to undertake this research (Cichocki & Ruchwa 2013b). Various factors and phenomena have been studied during experimental tests performed on series of circular plates, both qualitative and quantitative, as the

Figure 1. Ceramic rubble.

Figure 2. Section of concrete sample produced with the application of waste aggregate.

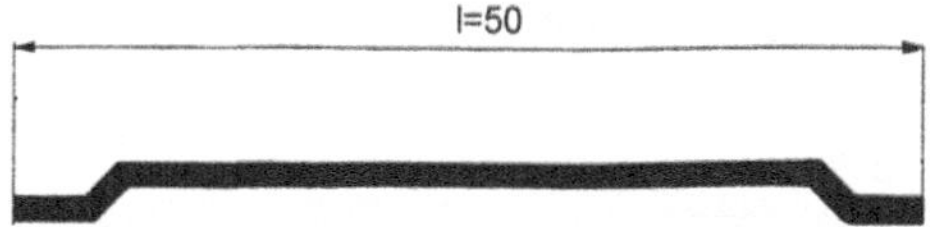

Figure 3. Schematic view of a steel fibre (length in mm).

failure mode of various types of plates, the number and dimensions of cracks propagated through the material, development of damage, etc.

Experimental tests carried out during the research served as a basis to create the discrete numerical model of the Finite Element Method in ABAQUS computer code environment (Belytschko et al. 2000, Dassault Systemes SIMULIA

2013a). Thus it was made possible to calibrate the parameters of applied numerical material model for concrete, describe the nature and characteristics of impact loads, define the applied algorithms of contact between the impacting objects and concrete element under investigation, as well as other parameters of numerical analysis in order to obtain the adequate and reliable numerical discrete model (Smadi & Bani Yasin 2008, Taqieddin & Voyiadjis 2009, Veselý & Frantík 2011).

2 EXPERIMENTAL TESTS

A special experimental stand (Fig. 4) was build in order to perform the impact tests. All tests were carried out on circular slabs of 1 m diameter, thickness 0.1 m, with various percentage of steel fibres. A detailed description of the experiment is provided in (Cichocki et al. 2014), the results are also discussed.

The examples of final distribution of damages for the slabs are given in Figures 5 and 6. For the last slab the pattern of cracks traversing the entire thickness was obtained after 7 impacts of 40 kg weight (free fall from a 1 m height).

Figure 4. Experimental stand.

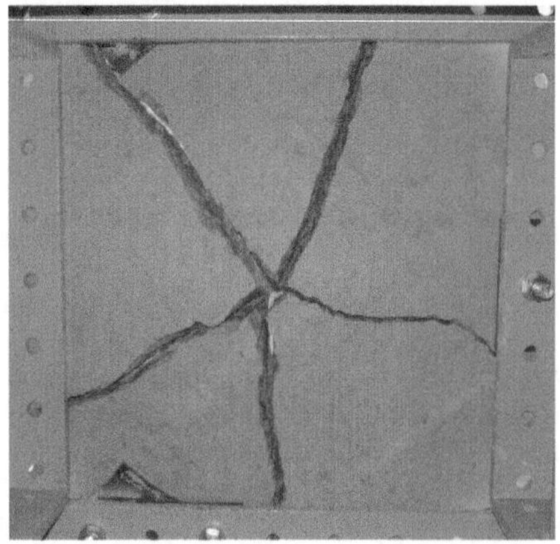

Figure 5. Plate without reinforcement (M0) after first impact, bottom view.

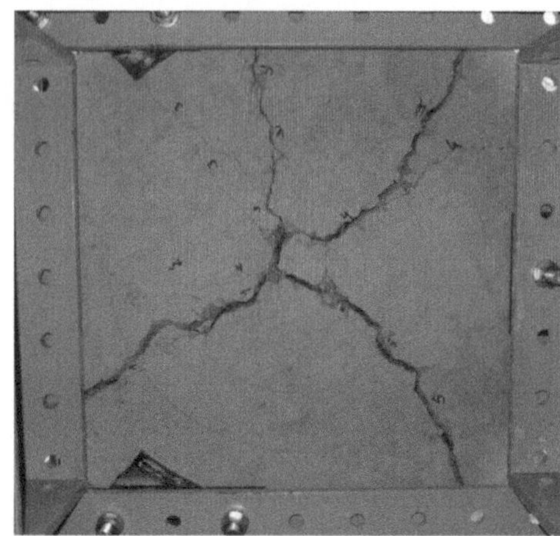

Figure 6. Reinforced plate (E2-10) after several impacts, bottom view.

Table 1. Type of tested plates and number of impacts necessary to damage the slab (selected slabs).

Slab	Reinforcement	Number of impacts
M0	Without fibres	1
E2-05	0.5%	8
E2-10	1.0%	11
E2-15	1.5%	19

Table 2. Basic characteristics of applied concrete (selected slabs).

Slab	Young modulus (GPa)	Compressive/ Tensile strength (MPa)	
M0	22	39.14	3.12
E2-05	22	44.10	3.38
E2-10	23	50.54	4.02
E2-15	25	51.89	4.56

In Tables 1 and 2, M0 means the slab without reinforcement, E2-05/10/15 is the slab with 0.5%/1.0%/1.5% Ecomet fibres reinforcement (length 50 mm, thickness 1 mm), etc.

For other reinforcement configurations the results were different, especially for the highest percentage of reinforcement (1.5%) and good quality fibres the pattern presented numerous cracks developed around the contact zone with the impacting weight.

3 NUMERICAL ANALYSIS

Adequate numerical models for the tested plates were created, based on Finite Element Method (FEM) approach (Belytschko et al. 2000). Due to an impulsive character of the applied load, the explicit procedure of integration for equations of motion was used, with the application of ABAQUS/Explicit (Dassault Systemes SIMULIA 2013a). The main assumptions concerning all variants of the slabs were identical, the only difference concerned the description of material characteristics for an analysed group of slabs. For each model identical slab dimensions were considered constant (diameter 1 m, thickness 0.10 m), the same type of supports (modeled as rigid, nondeformable surfaces) and the same type of impacting load, in order to obtain the most realistic scenario for the tests. Concrete slabs and impactor were modeled as three dimensional deformable solid bodies, the supports were discretized with quadrilateral rigid elements. The master-slave type of contact between impactor and slab, as well as between slab and supports has been introduced.

The Concrete Damage Plasticity (CDP) material model implemented in a computer code ABAQUS (Dassault Systemes SIMULIA 2013a) has been applied in this study. This constitutive model takes into account the elasto-plastic behaviour of concrete in compression and elasto-brittle characteristics response in tension. Two independent scalar parameters for damage are used in this formulation (De Souza Neto et al. 2008). Application of CDP material model allowed to achieve the satisfactory results on numerical analyses, both for static and dynamic loads (Głodkowska & Ruchwa 2010, Cichocki & Ruchwa 2013b).

For the fibres the von Mises elasto-plastic material model with isotropic hardening has been assumed (De Souza Neto et al. 2008).

The final response of slabs depends on characteristics of constituent materials and the quantity and distribution of fibres in the volume, the adequate proprietary code (*script*) to create the numerical model was prepared by authors using *Python programming language* (Dassault Systemes SIMULIA 2013b), included in ABAQUS computer code environment. As the result, the script creates the entire discrete model of the problem (input file for FEM computer code ABAQUS/Explicit), with all geometric, material and other necessary data, where the distribution of fibres was randomly defined by pseudo-random number generator. The random position of each fibre was defined by coordinates of its centroid and adequate angles of inclination with respect to assumed reference system axes. Additionally the information about spatial distribution of fibres in such kinds of structural elements was taken into account (e.g., contribution of fibres with various inclination angles, distribution in thickness, etc.).

All fibres were modelled using three-dimensional truss finite elements. Adequate interaction between concrete matrix and steel fibres was made possible by means of insertion of randomly distributed "cloud of finite elements" of fibres into concrete volume of solid elements and application of embedded elements definition, which allows for the interaction of two different meshes of finite elements, one inserted into another. Schematic view of the numerical model is shown in Figure 7.

On the basis of independent drawing several computational models have been defined for each of considered slabs, and adequate different (in certain limits) distributions of damages in slabs were obtained in numerical analyses.

An example of numerical results is given in Figure 8, with well visible zones of totally damaged material, very close to distributions obtained in experimental tests.

Similar numerical results are given in Figure 9, where the presence of steel fibres reduced the extend

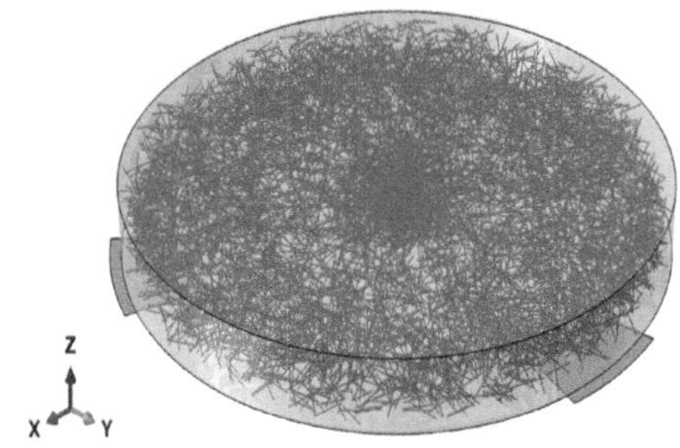

Figure 7. Schematic view of the numerical model.

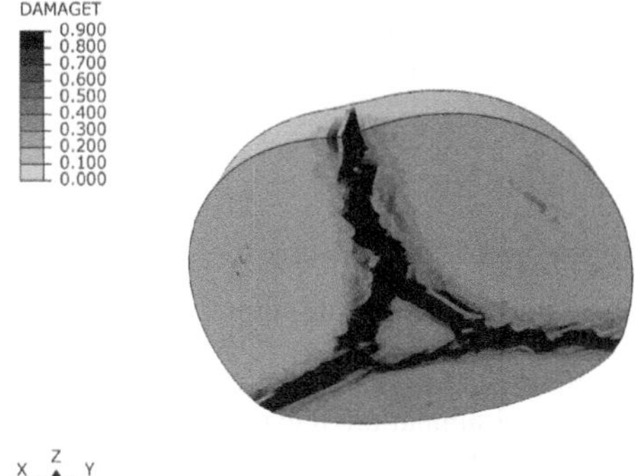

Figure 8. Distribution of damages. Plate without reinforcement (M0), after first impact, bottom view.

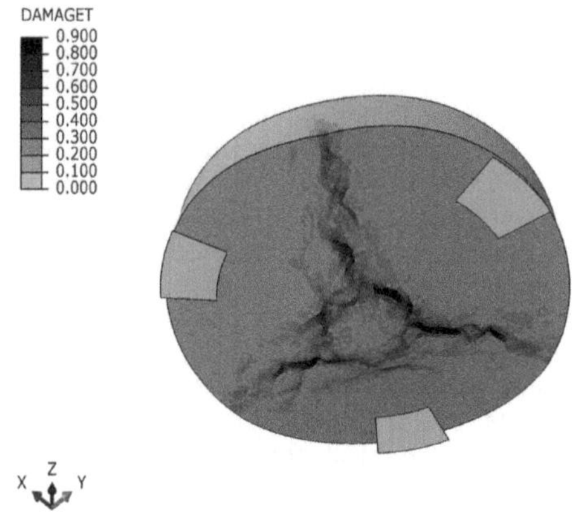

Figure 9. Distribution of damages. Plate with reinforcement (E2-10), after three impacts, bottom view.

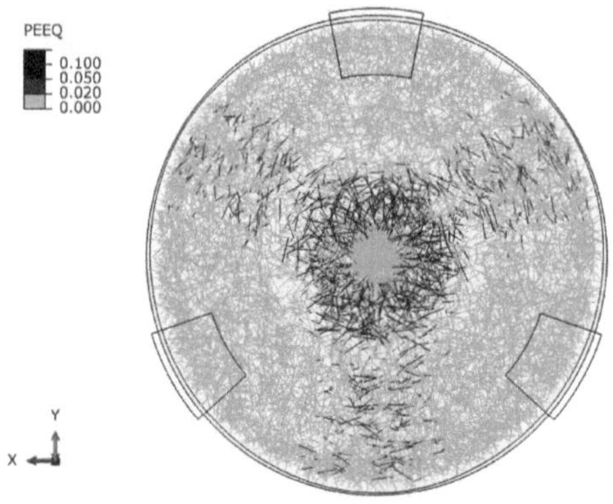

Figure 10. Distribution of equivalent plastic strains in fibres. Plate with reinforcement (E2-10), after three impacts, upper view.

of damages in concrete matrix. Distribution on equivalent plastic strains for this case in presented in Figure 10, with well-visible areas of high values concentrated in damaged concrete volume.

4 SUMMARY

Two main goals were the object of this study:

- experimental investigation on dynamic response of slabs subjected to repetitive impact loading up to total damage of material;
- numerical simulation of experimental tests, comparison of results, etc.

The experiments allowed to describe the development of damage patterns in a function of number of impacts. Additionally the secondary effects revealed during the entire phenomena have been registered and documented.

Numerical simulations performed using the advanced features of FEM code ABAQUS show the adequate correspondence between experimental and numerical results, very useful for future practical applications.

The experimental and numerical results obtained in the study allowed for a better understanding of the specific behaviour of investigated structural elements subjected to impact loads. A detailed description of final conclusions and remarks is given in (Cichocki et al. 2014) together with recommendations concerning the practical use of such elements.

REFERENCES

Belytschko, T., Liu W.K. & Moran, B. 2000. *Nonlinear Finite Elements for Continua and Structures*. John Wiley & Sons.

Cichocki, K., Domski, J., Katzer, J. & Ruchwa, M. 2014. Impact resistant concrete elements with nonconventional reinforcement. *Rocznik Ochrona Srodowiska—Annual Set The Environment Protection*, Monograph No. 1, 16, Part 2: 1–99.

Cichocki, K. & Ruchwa, M. 2013a. Numerical analysis of fibre reinforced slabs under impact loads. In: *20th Intern. Conf. on Computer Methods in Mechanics, CMM-2013. Short Papers. Poznań (Poland), 27–31 August 2013*. Poznań: Poznań University of Technology.

Cichocki, K. & Ruchwa, M. 2013b. Integrity analysis for multistorey buildings. *Ovidius University Annals—Constantza. Series: Civil Engineering* 15: 45–52.

Dassault Systemes SIMULIA 2013a. *ABAQUS Analysis User's Manual*. Providence.

Dassault Systemes SIMULIA 2013b. *ABAQUS Scripting User's Guide*. Providence.

De Souza Neto, E.A., Perić, D. & Owen, D.R.J. 2008. *Computational methods for plasticity. Theory and applications*. John Wiley & Sons.

Głodkowska, W. & Ruchwa, M. 2010. Static analysis of reinforced concrete beams strengthened with CFRP composites. *Archives of Civil Engineering* LVI (2): 111–122.

Katzer, J. & Domski, J. 2015. Static and dynamic characteristics of fibre reinforced WCA concrete. In: *Proc. 8th Intern. Conf. FIBRE CONCRETE 2015. Prague (Czech Republic), 10–11 September 2015*, Prague: Czech Technical University in Prague.

Smadi, M.M. & Bani Yasin, I.S. 2008. Behavior of high-strength fibrous concrete slab–column connections under gravity and lateral load. *Construction and Building Materials* 22(8): 1863–1873.

Taqieddin, Z.N. & Voyiadjis, G.Z. 2009. Elastic Plastic and Damage Model for Concrete Materials: Part II—Implementation and Application to Concrete and Reinforced Concrete. *International Journal of Structural Changes in Solids* 1(1): 187–209.

Veselý, V. & Frantík, P. 2011. Reconstruction of fracture process zone during tensile failure of quasi-brittle materials. *Applied and Computational Mechanics* 4: 237–250.

Advances in Mechanics: Theoretical, Computational and Interdisciplinary Issues – Kleiber et al. (Eds)
 ISBN 978-1-138-02906-4

Experimental verification of the relationships between Young's modulus and bone density using DIC

Ł. Cyganik, M. Binkowski, P. Popik & Z. Wróbel
Institute of Computer Science, University of Silesia, Sosnowiec, Poland

G. Kokot & A. John
Faculty of Mechanical Engineering, Silesian University of Technology, Gliwice, Poland

T. Rusin
Dantec Dynamics GmbH, Ulm, Germany

F. Bolechała
Medical College, Jagiellonian University, Kraków, Poland

R. Nowak
Department of Orthopedics, Medical University of Silesia, Katowice, Poland

ABSTRACT: The paper presents the methodology of prediction of mechanical properties of trabecular bone based on the X-ray microtomography and the relationships between bone density and mechanical properties. Five relationships between bone density and Young's modulus available in the literature were selected for the tests (Carter and Hayes, Ciarelli et al., Kaneko et al., Keller and Li and Aspden). Twelve cubic specimens of the trabecular bone from femoral heads were cut out and X-ray Microtomography (XMT) scanned. The mechanical compression tests were performed on these specimens with the displacement and strain full field measurements using Digital Image Correlation (DIC) method. On the base of the XMT scans numerical models of tested specimens were generated and numerically tested by the simulation of the compression test using the finite element method. The results of the experimental tests and results of numerical analyses were compared.

1 INTRODUCTION

According to the literature, there are many mathematical relationships between bone density and mechanical properties, which were determined experimentally. Application of these relationships in the Finite Element Analyses (FEA) may result in different accuracy of numerical simulations of the bone mechanical behaviour (Cyganik et al. 2014) especially in the case of the trabecular bone structure. The aim of the presented study was to indicate the relationship that assures the best accuracy of the numerical simulation in comparison with the experiment, using a methodology consisting of X-ray Microtomography (XMT) scanning, compression tests with Digital Image Correlation (DIC) measurements, nanoindentation tests and FEA. The mechanical compression tests with the DIC were used to acquire the experimental displacements and strains fields of trabecular bone samples and compared with the fields acquired from finite element analysis for the FEA accuracy estimation. The material data for finite element analysis were calculated on the base of mineral bone density using selected relationships between bone density and its mechanical properties.

2 MATERIALS AND METHOD

Twelve cubic trabecular bone specimens (edge length-10 mm) were cut out from the heads of human femur and scanned by means of X-ray microtomography scanner v|tome|x s, GE Sensing & Inspection Technologies, Phoenix|x-ray, Wunstorf, Germany. A hydroxyapatite phantom was used during scanning the specimens to convert the relative Hounsfield Units into real equivalent bone density values (Cyganik et al. 2014).

The images obtained from XMT scans were reconstructed using a manufacturer's software (Datos 2.0). Then, the trabecular bone structures were segmented and 3D surface models of bone samples were created using Mimics software. The Mimics software was also used for the generation of finite element meshes using ten-noded tetrahedral elements (Fig. 1, Table 1). The material properties for each numerical model were calculated on the base of the one of five selected relationships: proposed by Carter & Hayes (1977), Ciarelli et al. (2000), Kaneko et al. (2004), Keller (1994) and Li & Aspden (1997). The greyscale density calculated from XMT scans using a hydroxyapatite phantom was used as the basis for calculation of ash density ρ_{ash}, apparent density ρ_{app} and bone volume fraction BV/TV being variables in selected relationships.

The created FE models were virtually placed between two plates and the glue contact was applied between plates and finite elements. The contact tolerance of 0.01 mm was used. The load was applied to the top plate, which was free to move. The bottom plate was fixed. The applied load values are presented in Table 1. If the specimen's mesh had a relief on the top surface caused by the inaccuracies in the cutting process of the specimen, the initial contact was applied to the highest point of the mesh. Then, the load was applied linearly in 5 steps of calculation and the contact surface was increasing with each step. In the case of flat top surface on the specimen's mesh, the initial contact was applied to the whole surface, the load was applied in a single step of calculation.

Figure 1. Exemplary FE mesh of trabecular bone cubic specimen with material properties assigned using relationship given by Carter & Hayes (1977).

Table 1. Numerical models data.

Specimen no	Number of elements (–)	Load (N)
1	1,114,130	1,200
2	769,950	600
3	853,516	1,200
4	1,119,803	1,020
5	1,126,503	1,000
6	1,069,141	1,000
7	1,242,083	1,000
8	751,447	700
9	935,911	900
10	1,087,268	1,000
11	870,917	627
12	802,708	1,200

Numerical models with five different materials property sets were prepared for each of twelve specimens. In total, 60 numerical models were generated and the compression test was simulated for each of them. Calculations were performed using MSC. Marc & Mentat 2012 software (MSC Software Corporation, Santa Ana, California, USA). A symmetric multi-frontal sparse solver was used. The calculations were run on a PC (Dell T7500 Intel® Xeon® X5650, 2 processors 2.67 GHz, 96 GB RAM).

The results of numerical analyses were compared with the results of the experiments. The experiments consisted of mechanical compression tests (Fig. 2) performed on the MTS machine (MTS Insight 2, Eden Prairie, MN, USA) with the full field measurements of displacements and strains by means of Digital Image Correlation system (model Q-400, Dantec Dynamics GmbH, Ulm, Germany).

Additionally, the nanoindentation measurement of Young's modulus of single trabeculae was conducted after the compression tests. The mechanical compression tests were performed at room temperature (~25°C) and the displacement rate was 0.5 mm/min. The load range was set up from 0 to 2000 N (or till the specimen's failure). The cameras of the DIC system were taking pictures every 100 N of the load increment. The exemplary displacements fields obtained from numerical calculation and experiments are presented in Figure 3. It can be seen that

Figure 2. The mechanical compression test with the digital image correlation measurements.

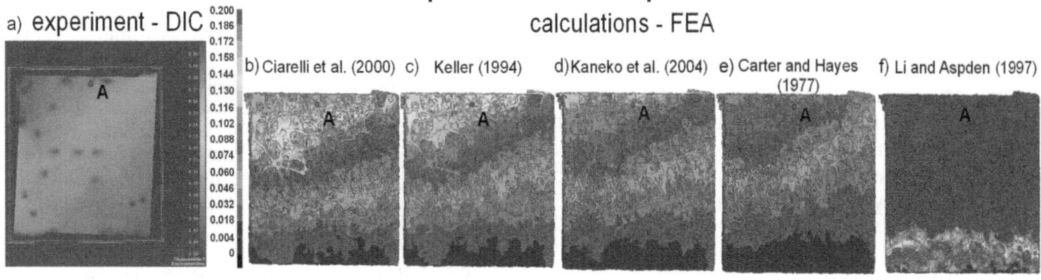

Figure 3. Displacement distributions obtained from a) digital image correlation and b—f) numerical analyses by means of different relationships. It is seem that the displacement field in numerical solution where the relationship given by Ciarelli et al. (b) was used is the most similar to the field acquired from DIC (a).

the displacement field in numerical solution where the relationship given by Ciarelli et al. (Fig. 3b) was used is the most similar to the field acquired in experiment with the DIC method (Fig. 3a). Three quantities were compared between results acquired from DIC and FEA: the maximum displacement $\overline{\delta}_{max}$, the displacement in control point A $\overline{\delta}_A$ (Fig. 3), and the average strain value in the selected area $\overline{\delta}_{\varepsilon}$. From these quantities, the average relative errors of numerical solution were calculated and compared to the results of experiments.

The ranges of Young's modulus values calculated using each relationship were additionally compared with the nanoindentation measurements of Young's modulus on specimen's trabeculae.

3 RESULTS

The best accuracy of the numerical analysis was obtained for the relationship given by Ciarelli et al. (2000) ($\overline{\delta}_{\max} = 14.11\%$, $\overline{\delta}_A = 17.87\%$, $\overline{\delta}_{\varepsilon} = 50.94\%$) as compared to others: Keller (1994) ($\overline{\delta}_{max} = 23.20\%$, $\overline{\delta}_A = 25.88\%$, $\overline{\delta}_{\varepsilon} = 60.27\%$), Kaneko et al. (2004) ($\overline{\delta}_{max} = 29.42\%$, $\overline{\delta}_A = 37.26\%$, $\overline{\delta}_{\varepsilon} = 62.29\%$), Carter and Hayes (1977) ($\overline{\delta}_{max} = 35.47\%$, $\overline{\delta}_A = 45.18\%$, $\overline{\delta}_{\varepsilon} = 67.99\%$), Li and Aspden (1997) ($\overline{\delta}_{max} = 328.81\%$, $\overline{\delta}_A = 364.74\%$, $\overline{\delta}_{\varepsilon} = 216.62\%$).

The relationship given by Ciarelli et al. (2000) allows obtaining the displacement distribution in the numerical analysis with a relative error of 17.87%. The relative error for the strain distribution is much higher and equal 50.94%.

This relationship ensures that the calculated and assigned to the numerical models range of Young's modulus (1.461–11.829 GPa) is the most similar to the ranges of Young's modulus measured in nanoindentation tests (2.4–15.6 GPa). The dispersion of the Young's modulus values is natural for the trabecular bone tissue and similar results were several times reported in the literature e.g. 3.27–10.58 GPa (Choi et al. 1990), 11.4 ± 5.6 GPa (Zysset et al. 1999).

4 CONCLUSIONS

The conducted research shown, that not all of relationships between bone mineral density and mechanical properties can be used for mapping mechanical properties of trabecular bone. However, using the relationship given by Ciarelli et al. (2000) for material properties assignment in the numerical models it is possible to obtain the displacement distribution with a relative error of 17.87%. Using this relationship it is not possible to obtain the same accuracy for strain distribution (relative error for the strain distribution is about 50.94%). Therefore, the application of the relationship given by Ciarelli et al. (2000) is limited only to the determination of displacements in linear simulations and cannot be used in nonlinear simulations such as cracking, where the accuracy of determined stain and stress concentrations should be very high.

However, the proposed methodology of research that include: XMT scanning, experimental compression tests, nanoindentation measurements and FE analysis is a remarkable way for estimation/verification of relationships between material properties. Although, that in our experiments we used this methodology for experimental verification of existing relationships, the presented methodology can be successfully applied for research and estimation of new relationships and dependences. The XMT scanning allow for accurate acquisition of material structure data, compression tests combined with DIC measurements deliver displacement and strain fields under a certain load that are necessary to validate the numerical results. In addition, the nanoindetation tests allow to verify the correctness of calculated Young's modulus values from proposed relationship and FEA finally verify the correctness of relationship by the compatibility of simulation with experiments.

Therefore, using the presented method it is necessary to conduct further research to formulate new relationship between Young's modulus and bone density that will allow to achieve improved accuracy between experimental and numerical results especially in the strain field.

ACKNOWLEDGMENTS

The research is partially funded from the projects Silesian University of Technology, Faculty of Mechanical Engineering 10/990/BK_15/0013, 10/040/BK_15/0006.

REFERENCES

Carter, D.R., & Hayes, W.C. 1977. The Compressive Behavior of Bone as a Two-phase Porous Structure. *The Journal of Bone & Joint Surgery*, 59(7), 954–962.

Choi, K., Kuhn, J.L., Ciarelli, M.J., Goldstein, S.A. 1990. The Elastic Moduli of Human Subchondral, Trabecular, and Cortical Bone Tissue and the Size-dependency of Cortical Bone Modulus. *Journal of Biomechanics*, 23(11), 1103–13.

Ciarelli, T.E., Fyhrie, D.P., Schaffler, M.B., Goldstein, S.A. 2000. Variations in Three-dimensional Cancellous Bone Architecture of the Proximal Femur in Female Hip Fractures and in Controls. *Journal of Bone and Mineral Research*, 15(1), 32–40.

Cyganik, Ł., Binkowski, M., Kokot, G. et al. 2014. Prediction of Young's modulus of trabeculae in microscale using macro-scale's relationships between bone density and mechanical properties, *Journal of the Mechanical Behavior of Biomedical Materials*, 36, 120–134.

Kaneko, T.S., Bell, J.S., Pejcic, M.R., Tehranzdeh, J., Keyak, J.H. 2004. Mechanical Properties, Density and Quantitative CT Scan Data of Trabecular Bone with and Without Metastases. *Journal of Biomechanics*, 37(4), 523–30.

Keller, T.S, 1994. Predicting the Compressive Mechanical Behavior of Bone. *Journal of Biomechanics*, 27(9), 1159–68.

Li, B. & Aspden, R.M. 1997. Composition and Mechanical Properties of Cancellous Bone from the Femoral Head of Patients with Osteoporosis or Osteoarthritis. *Journal of Bone and Mineral Research*, 12(4), 641–51.

Zysset, P.K., Guo, X.E., Hoffler, C.E., Moore, K.E., Goldstein, S.A. 1999. Elastic Modulus and Hardness of Cortical and Trabecular Bone Lamellae Measured by Nanoindentation in the Human Femur. *Journal of Biomechanics*, 32(10), 1005–1012.

Advances in Mechanics: Theoretical, Computational and Interdisciplinary Issues – Kleiber et al. (Eds)
 ISBN 978-1-138-02906-4

Construction of stress trajectories in optimal, non-homogeneous elastic bodies

S. Czarnecki & P. Wawruch
Warsaw University of Technology, Warsaw, Poland

ABSTRACT: Trajectories of principal stresses deliver a key information on the response of the structure due to a given load. The paper is aimed at construction of stress trajectories in 3D structures of minimal compliance in which elastic moduli are optimally distributed assuring the least compliant design. The trajectories are constructed by the Runge-Kutta algorithm implemented into the topology optimization procedures of the stress-based Free Material (Isotropic Material) Design (FMD/IMD) methods. The method proposed allows for construction of stress trajectories in optimal three-dimensional non-homogeneous anisotropic (or isotropic, in case of the IMD design) elastic bodies.

1 INTRODUCTION

Structural engineers often use the term *flow lines of forces* which may be understood as a visualization of the way in which the loads are transmitted to the supports. The primary and the most reliable source of information about the *flow lines of forces* are trajectories of principal stresses. Determination of the stress trajectories is relatively simple in a homogeneous and isotropic elastic body Ω with a constant in the whole area Young's modulus E and Poisson's ratio ν. However, the present technology allows also to design the properties and various characteristics of the material of the structure. Typical examples are now widely used in the design of laminated plates and shells made of composite materials in which (depending on the technology) there are designed, inter alia, the optimal directions of the fibers. Thus, in addition to searching for the optimal shape of the body, it is possible to control the optimal distribution of material characterized by Young's modulus E and Poisson's ratio ν or many other moduli. The progress in the field of material technology allows to assume that it will soon be possible to design all components of non-homogeneous Hooke elasticity tensor not only in the case of isotropic or orthotropic bodies, but also in the case of fully anisotropic materials. It follows that in the design process it is worthwhile to analyze the trajectories of principal stresses in heterogeneous bodies with optimal distribution of Hooke elasticity tensor. The main aim of the work is to develop a set of tools that allows to design and visualize stress trajectories in three-dimensional homogeneous isotropic bodies, as well as in non-homogeneous, optimal isotropic or anisotropic elastic bodies. Optimal layouts of bulk and shear moduli k and μ or Kelvin modulus λ were found in accordance with stress-based version of IMD or FMD method, respectively.

2 BRIEF INTRODUCTION TO THE STRESS-BASED APPROACH IN FREE MATERIAL DESIGN

2.1 *Basic notions and definitions*

The purpose of this section is to present a brief review of the basic notions and definitions that are necessary for the formulation of the optimization problem. Mathematical proofs of basic concepts and issues are in detail presented in Czarnecki & Lewiński 2014 and Czarnecki 2015. Consider a three-dimensional elastic body Ω. The domain Ω is parameterized by the Cartesian coordinates x_1, x_2, x_3. The body is fixed at the boundary surface $\Gamma_2 \subset \partial\Omega$. The segment of the boundary $\Gamma_1 \subset \partial\Omega$ is subject to the tractions of intensity $\mathbf{t}$. The unknown displacement field $\mathbf{u} = (u_i)$ and the virtual displacement field $\mathbf{v} = (v_i)$ vanish on Γ_2. The strains constitute a tensor $\boldsymbol{\varepsilon} = (\varepsilon_{ij})$ of components $\varepsilon_{ij} = (u_{i,j} + u_{j,i})/2$ determined by the displacement vector $\mathbf{u}$. The stresses $\boldsymbol{\sigma} = (\sigma_{ij})$ are linked with strains by the Hooke law $\boldsymbol{\sigma}(\mathbf{u}) = \mathbf{C}\,\boldsymbol{\varepsilon}(\mathbf{u})$ where $\mathbf{C} = (C_{ijkl})$ is the 4th rank, non-homogeneous, anisotropic or isotropic Hooke tensor. Almost everywhere in Ω, the components $C_{ijkl} \in L^\infty(\Omega)$ of tensor $\mathbf{C}$ satisfy well known symmetry conditions and the positivity condition $C_{ijkl}\varepsilon_{ij}\varepsilon_{kl} \geq c\varepsilon_{ij}\varepsilon_{ij}$ for some $c > 0$ (summation convention over indices i, j, k, l adopted). In the case of a non-homogeneous, isotropic elastic body, Hooke tensor $\mathbf{C}$ additionally satisfies isotropy condition at all points $x \in \Omega$. Tensor fields $\mathbf{C}$ (anisotropic or isotropic) satisfying all above conditions constitute a set $H(\Omega)$. Let $\Lambda = E_0|\Omega|$, where E_0 be a given elastic modulus and $|\Omega|$ denotes volume of the body Ω. The set $H_\Lambda(\Omega) = \{\mathbf{C} \in H(\Omega) | \int \mathrm{tr}\mathbf{C}\, dx = \Lambda\}$ of admissible (anisotropic or isotropic) Hooke's tensors denotes the domain of design parameters (integration is performed over the whole domain Ω). The stress field $\boldsymbol{\tau} = (\tau_{ij})$ is said to be statically admissible if the following variational equation holds:

$$\forall \mathbf{v} \in V(\Omega) \quad \int_\Omega \boldsymbol{\tau} \cdot \boldsymbol{\varepsilon}(\mathbf{v})\, dx = \int_{\Gamma_1} \mathbf{t} \cdot \mathbf{v}\, da \tag{1}$$

where $V(\Omega)$ represents the space of kinematically admissible displacements. The set of statically admissible stresses is denoted by $\Sigma(\Omega)$.

2.2 *Stress-based formulation of FMD and IMD method*

Let us remind in brief the original formulation of displacement-based FMD or IMD problem: find an admissible, non-homogeneous, anisotropic or isotropic Hooke tensor field $\mathbf{C}^* \in H_\Lambda(\Omega)$ such that:

$$f^* = \min_{\mathbf{C} \in H_\Lambda(\Omega)} f(\mathbf{C}) \tag{2}$$

where:

$$f = f(\mathbf{C}) = \int_{\Gamma_1} \mathbf{t} \cdot \mathbf{u}(\mathbf{C})\, da \tag{3}$$

is the compliance of the body Ω and $\mathbf{u} = \mathbf{u}(x)$ is solution of the equilibrium problem:

$$\forall \mathbf{v} \in V(\Omega) \quad \int_\Omega C_{ijkl}\, \varepsilon_{kl}(\mathbf{u})\, \varepsilon_{ij}(\mathbf{v})\, dx = \int_{\Gamma_1} t_i v_i\, da \tag{4}$$

In Czarnecki & Lewiński 2014 and Czarnecki 2015 it was shown that the displacement-based problem can be re-formulated to its stress-based version: find statically admissible stress tensor field $\boldsymbol{\tau}^* \in \Sigma(\Omega)$ such that:

$$Z^* = \min_{\boldsymbol{\tau} \in \Sigma(\Omega)} \int_\Omega |||\boldsymbol{\tau}|||\, dx \tag{5}$$

where:

$$Z^* = \int_\Omega |||\boldsymbol{\tau}^*|||\, dx \tag{6}$$

and norm $|||\cdot|||$ definitions are:

$$|||\boldsymbol{\tau}||| = \sqrt{\boldsymbol{\tau} \cdot \boldsymbol{\tau}} \quad \text{or} \quad |||\boldsymbol{\tau}||| = \alpha|\mathrm{tr}\, \boldsymbol{\tau}| + \beta\|\mathrm{dev}\, \boldsymbol{\tau}\| \tag{7}$$

for anisotropic or isotropic optimization, respectively. For three-dimensional bodies: $\alpha = 10^{1/2}$, $\beta = 5 \times 6^{1/2}$. It was also shown (Czarnecki & Lewiński 2014, Czarnecki 2015) that the solution Z^*, $\boldsymbol{\tau}^*$ of the stress-based version (5) determines explicitly the optimal compliance f^*, optimal distribution of Kelvin moduli λ_i^* and eigenstates $\boldsymbol{\omega}_i^*$ ($i = 1, 2, \ldots, 6$) or bulk k^* and shear μ^* moduli for anisotropic or isotropic optimization, respectively. These formulas allow to calculate all components C^*_{ijkl} of optimal, non-homogeneous Hooke tensor $\mathbf{C}^*$ at any point $x \in \Omega$ (in particular, at Gaussian integration points) and finally find a solution $\mathbf{u}^*$ of the static problem:

$$\forall \mathbf{v} \in V(\Omega) \quad \int_\Omega C^*_{ijkl}\, \varepsilon_{kl}(\mathbf{u})\, \varepsilon_{ij}(\mathbf{v})\, dx = \int_{\Gamma_1} t_i v_i\, da \tag{8}$$

based on the Hooke law $\boldsymbol{\sigma}(x) = \mathbf{C}^*(x)\, \boldsymbol{\varepsilon}(x)$, $x \in \Omega$. The performed in Czarnecki & Lewiński (2014) and Czarnecki & Wawruch (2015) numerical tests have shown that the FMD (anisotropic) designs are 5 or even 6 times stiffer than their IMD (isotropic) counterparts, the latter being about 2 times stiffer than the non-optimal homogeneous isotropic designs. On the other hand, the FMD designs are appropriate only for the given load, since the optimal Hooke tensor $\mathbf{C}^*$ becomes degenerated; only one (the biggest) optimal Kelvin moduli λ_1^* is positive and appropriate tools are necessary to deal with the conditionally stable numerical scheme while finding the solution $\mathbf{u}^*$ of the problem (8) for optimal, non-homogeneous, anisotropic Hooke tensor $\mathbf{C}^*$ found by FMD method. This is due to the fact that the global stiffness matrix of FEM becomes singular or numerically very close to singular. To overcome the difficulties, inter alia, the conventional LU (or other) decomposition was replaced by the numerically stable SVD method to achieve reliable values of nodal displacements. The above, briefly mentioned numerical problems is less pronounced in optimal, non-homogeneous, isotropic Hooke tensor $\mathbf{C}^*$ found by IMD method, but only under the condition, that the optimal field of Young modulus E^* does not reach zero (at least numerically). Unfortunately, a common feature of the both optimal solutions (found by FMD and IMD method) is that in large subdomains, the optimal values of elastic moduli reach numerical values close to zero, which simply means the need to remove material from all these sub-domains and design new shape of the body. In other words, topology optimization developed implies the geometry optimization or shape optimization of the body. Keeping the material in all these sub-domains, which on the basis of the optimal solutions should be removed, generates (very briefly described above) numerical problems, including the construction of stress trajectories. To simplify (significantly) the numerical calculations, the material has not been removed while solving the problem (8). This made it possible to maintain the original finite element mesh, but on the other hand, confirmed the above-mentioned numerical problems. It is worth noting that in special cases of several independently applied external loadings, only Pareto optimal solutions guarantee a non-singularity of the (Pareto optimal) stiffness matrix (Czarnecki & Lewiński 2014).

3 CONSTRUCTION OF STRESS TRAJECTORIES IN ELASTIC BODIES

3.1 *Basic notions and definitions*

Construction of stress trajectories requires finding the family of solutions

$$[0,\pm\infty) \ni s \rightarrow \mathbf{r}^{*}(s) \in \mathrm{R}^{3} \tag{9}$$

of the system of (three in 3D) ordinary differential equations:

$$\frac{d\mathbf{r}}{ds} = \boldsymbol{\sigma}_{k}[s,\mathbf{r}(s)], \quad \mathbf{r}(0) = \mathbf{r}_{0}^{*} \tag{10}$$

where s = denotes the natural parameter of the sought curve $\mathbf{r}(s)$ and $\boldsymbol{\sigma}_k[\mathbf{s},\mathbf{r}(s)]$ (k = 1, 2 or 3) = is one of the three eigenvectors of the stress tensor $\boldsymbol{\sigma}[\mathbf{s},\mathbf{r}(s)]$ sorted with respect to its eigenvalues $\sigma_k[\mathbf{s},\mathbf{r}(s)]$. Assuming that it is possible to calculate all components $\sigma_{ij}(x)$ of the stress tensor $\boldsymbol{\sigma}(x)$ at any point $x\in\Omega$ of non-homogeneous elastic body Ω, searching for a solution (9) of ordinary differential equations (10) requires implementation of any numerical method, e.g. Runge-Kutta method used here. Determining the next point $\mathbf{r}_{i+1}$ lying on sought trajectory requires a calculation of all components of the stress tensor $\boldsymbol{\sigma}(\mathbf{r}_i)$ at the current point $\mathbf{r}_i$ so that eventually select the appropriate vector of the three eigenvectors $\boldsymbol{\sigma}_k(\mathbf{r}_i)$ associated with the sorted eigenvalues $\sigma_k(\mathbf{r}_i)$, k = 1, 2, 3. This task is trivial in the case of the displacement approximation or interpolation performed on real domain Ω, e.g. in most meshless/meshfree methods, because shape functions in such methods are very often defined on real domain Ω. In classical FEM, shape functions (for obvious reasons) are almost never defined on the real finite elements $\Omega_e \subset \Omega$, but on master element ω. Additionally, we have always to assume that each finite element Ω_e is the image $x^e(\omega)$ of the master element ω for given geometric mapping $x^e{:}\omega\rightarrow\Omega_e$ defining, usually only approximate, shape of the body Ω. It follows that the calculation of the value of any field (displacements, stresses, etc.) at a given point $x\in\Omega$ must be preceded by determining the number e-th of finite element Ω_e in which this point lies, next determining the values of nodal parameters in Ω_e and finally finding its pre-image $\xi = (x^e)^{-1}(x)\in\omega$ in master element ω to calculate the value of the investigated field (stress in our case). It follows that it is necessary to perform very fast and accurate calculations to find pre-image $\xi\in\omega$ of an arbitrary point $x\in\Omega$, preferably based on the analytical formulae $(x^e)^{-1}$: $\Omega_e\rightarrow\omega$, which are generally not known in FEM, even for the simplest finite elements. Only in few cases of simple geometry of Ω, for example if the final shapes of elements Ω_e are defined by translation, rotation, and scaling of master element ω, then the analytical formulae for inverse mappings $(x^e)^{-1}$ are generally trivial.

3.2 *Case studies*

The aim of the section is to construct a few families of stress trajectories in non-optimal, homogeneous and optimal, non-homogeneous elastic bodies for a visual comparison. In view of the difficulties mentioned above, the shape of the bodies has been limited to one or a number of "glued together" cubic blocks. The numerical approach starts from division of the design domain Ω into 8-node, cubic, conforming, isoparametric C3D8 finite elements. Due to the necessity of finding the optimal distribution of elastic modules based on stress-based approach in FMD/IMD method, the number of finite elements did not exceed a few thousand (Czarnecki & Lewiński 2014, Czarnecki 2015).

Example 1. The first example (Fig. 1a) concerns a rectangular cantilever of width l and height $2l$. The unit, vertical force is applied to the middle point on upper horizontal side.

The course of stress trajectories in non-optimal, homogeneous and in optimal, non-homogeneous, anisotropic body clearly differs (compare Fig. 1b, c). Particularly evident is the narrowing (close to two non-sliding supports and place where the vertical force is applied) of the vertical curves of principal compressive stresses confirming the optimal distribution of material similar to the characteristic shape of a bell. Emerging irregularities in the course of the horizontal (blue) trajectories (especially visible in the upper and lower middle subdomains) are primarily caused by the degeneration of the optimal Hooke elasticity tensor.

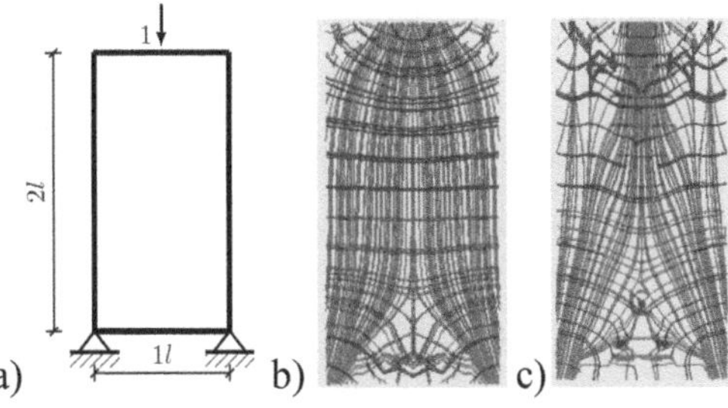

Figure 1. Rectangular cantilever loaded with unit vertical force (a). Stress trajectories in non-optimal, homogeneous (b) and in optimal, non homogeneous, anisotropic body (c).

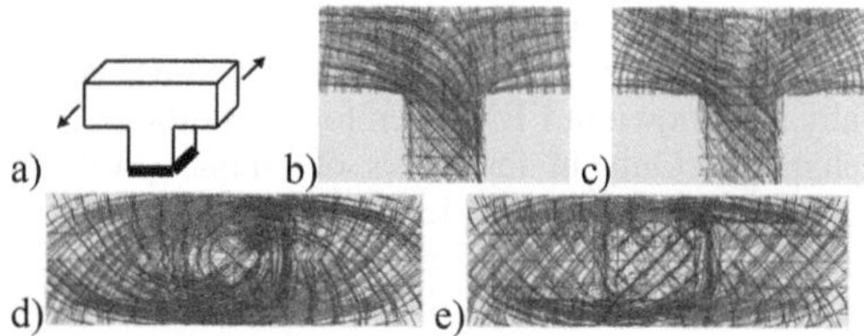

Figure 2. Clamped at bottom side and twisted T-shape frame cantilever (a). Stress trajectories in non-optimal, homogeneous (b), (d) and in optimal, non-homogeneous, isotropic body (c), (e).

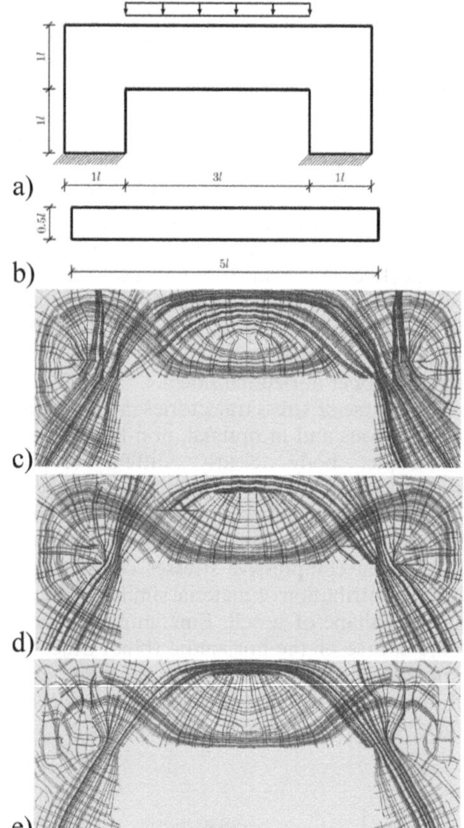

Figure 3. Three-dimensional frame structure clamped at bottom sides of vertical plates—side (a) and top view (b). Stress trajectories in non-optimal, homogeneous (c), in optimal, non-homogeneous, isotropic (d) and in optimal, non-homogeneous, anisotropic body (e).

Example 2. The second example concerns the visualization of the stress trajectories in non-optimal, homogeneous and in optimal, non-homogeneous, isotropic, three-dimensional, clamped at bottom side T-shape frame which is twisted (Fig. 2a).

As in the previous example, the course of stress trajectories in non-optimal, homogeneous and in optimal, non-homogeneous, isotropic body clearly differs, especially in the core of a twisted column (compare top views in Fig. 2d and e). Irregularities in the course of the trajectories are in this case much less noticeable due to the fact that the optimal body in this case is isotropic—not anisotropic.

Example 3. The third example (Fig. 3a) concerns the three-dimensional plate that has a shape of frame composed of three rods, two of which (vertical ones) are clamped. The vertical unit pressure is applied to the upper side of the middle, horizontal plate.

As in the previous example, the course of stress trajectories in non-optimal, homogeneous in optimal, non-homogeneous, isotropic and anisotropic body clearly differs, especially in the subareas of the two corners and rafter framework (compare Fig. 3c, d, e).

3.3 *Conclusions*

Knowledge of principal stress trajectories can be a valuable suggestion for engineers designing e.g. the concrete structures in selecting the shape of reinforcing bars in Strut & Tie model recommended by European Committee for Standardization. The geometry of principal stress curves in the optimal non-homogeneous elastic body delivers a number of important information on the state of stresses, indispensable for the design of the reinforcement.

ACKNOWLEDGEMENTS

The paper was prepared within the Research Grant no 2013/11/B/ST8/04436 financed by the National Science Centre (Poland), entitled: Topology optimization of engineering structures. An approach synthesizing the methods of: free material design, composite design and Michell-like trusses.

REFERENCES

Czarnecki, S. & Lewiński, T. 2014. The Free Material Design in Linear Elasticity, *CISM Courses and Lectures, Topology Optimization in Structural and Continuum Mechanics, CISM International Centre for Mechanical Sciences, Courses and Lectures.* In G. Rozvany & T. Lewiński (eds), 549(9): 213–257. Wien, New York: Springer.

Czarnecki, S. 2015. Isotropic Material Design. *Computational Methods in Science and Technology* 21(2): 49–64.

Czarnecki, S. & Wawruch, P. 2015. The emergence of auxetic material as a result of optimal isotropic design. *Phys. Status Solidi B* 252(7): 1620–1630.

Advances in Mechanics: Theoretical, Computational and Interdisciplinary Issues – Kleiber et al. (Eds)
 ISBN 978-1-138-02906-4

Multiobjective optimization of electrothermal microactuators by means of Immune Game Theory Multiobjective Algorithm

A. Długosz
Institute of Computational Mechanics and Engineering, Faculty of Mechanical Engineering, Silesian University of Technology, Gliwice, Poland

P. Jarosz
Institute of Computer Science, Faculty of Physics, Mathematics and Computer Science, Cracow University of Technology, Cracow, Poland

ABSTRACT: The paper presents an application of the IMGAMO (Immune Game Theory Multiobective Algorithm) in multiobjective optimization of electrothermal actuators. Different types of functionals, which depend on equivalent stresses, deflection of the actuators, total heat generated by the actuator and volume of the actuators, have been formulated. The values of the functionals are calculated on the basis of the results obtained from numerical simulations. Boundary-value problem of electro-thermo-mechanical analysis is solved by means of Finite Element Method (FEM). Numerical example of multiobjective optimization of electrothermal actuator is presented. The comparison of the proposed method with the NSGAII algorithm has been presented.

1 INTRODUCTION

Electrothermal microactuators have proved to be very useful for generating forces or displacements in MEMS. Comparing to the other types of mictroactuators (e.g. electrostatic, piezoelectric), electrothermal microactuators can generate larger force per unit volume. Low electric potential used for control of such structures is another advantage of the electrothermal actuators. The deflection of the electrothermal actuator is produced when the electrical potential difference is applied to electrical pads. Due to the material properties—high electrical resistivity and thermal expansion—arms of the actuator can elongate. Two types of the electrothermal actuators are mainly used in MEMS: U-beam or V-beam ones. U-beam microactuator consist of beams (arms) which have different cross-section area. It produces different amount of Joule heating which causes the bending of the actuator. For the V-beam actuator the displacement is produced due to the pre-bending angle of beams (arms). Microactuators composed of greater number of pairs of beams are named chevron type (see Fig. 2). The problems of optimal design of electrothermal actuators have been considered by many researches, e.g. (Henneken V. 2008, Jain N. 2012). In order to efficiently optimize such structures, proper optimization techniques have to be applied, especially when more than one criterion is taken into account. Multiobjective optimization by means of evolutionary algorithms of U-beam actuators has been solved by authors in previous papers (Dlugosz 2010). The present work proposes methodology of multiobjective optimization of actuators which is more effective, comparing to the previous work, if more then three criteria are considered. It was obtained by application an in-house implementation of algorithm based on Artificial Immune System and the Game Theory. The proposed algorithm has been tested on mathematical benchmark problems, showing its superiority comparing to known multiobjective optimization techniques.

2 FORMULATION OF THE MULTIOBJECTIVE OPTIMIZATION PROBLEM

In the multiobjective optimization, solution of the problem is represented by more than one objective function. In such problems all of the objective functions cannot be simultaneously improved, moreover they are usually in conflict with each other, so the term *optimize* means finding such a solution which gives the values of all objective functions acceptable to the designer. Instead of one optimal solution, the set of optimal solutions can be received. Such problems usually are performed using the Pareto concept (Coello 1999).

The goal of the multiobjective optimization is to find vector of design variables (e.g.: geometry of the structure, material properties, boundary conditions) for which defined objective functions achieve extrema. An optimization task can be formulated as minimization or maximization problem. For the minimization problem, the vector of design variables $\mathbf{x} = [x_1, x_2, \dots x_n]$ which minimizes the vector of k objective functions, is searched.

$$f(\mathbf{x}) = [f_1(\mathbf{x}), f_2(\mathbf{x}), \dots, f_k(\mathbf{x})] \tag{1}$$

Generally, p equality constrains ($h_i(\mathbf{x}) = 0$ $i = 1, 2, \dots, p$) and m inequality constrains ($g_i(\mathbf{x}) \geq 0 \quad i = 1, 2, \dots, m$) are imposed on the optimization problem.

Thus, the decision domain is n-dimensional and the objective domain is k dimensional in multiobjective optimization. Pareto concept requires the definitions of such terms as: dominated solutions, non-dominated solutions, Pareto-optimality and Pareto front. Considering two vectors **x** and **y** in the searching domain: solution **x** strongly dominates **y**, if:

$$\forall i \in \{1, 2, \dots, k\}: \quad f_i(\mathbf{x}) < f_i(\mathbf{y}) \tag{2}$$

solution **x** weakly dominates **y**, if:

$$\begin{aligned} &\forall i \in \{1, 2, \dots, k\}: \quad f_i(\mathbf{x}) \leq f_i(\mathbf{y}) \quad \wedge \\ &\exists j \in \{1, 2, \dots, k\}: \quad f_j(\mathbf{x}) < f_j(\mathbf{y}) \end{aligned} \tag{3}$$

solution **x** is neutral (incomparable) to the **y**, if:

$$\begin{aligned} &\exists i, j \in \{1, 2, \dots, k\}: \\ &f_i(\mathbf{x}) < f_i(\mathbf{y}) \quad \wedge \quad f_j(\mathbf{y}) > f_j(\mathbf{x}) \end{aligned} \tag{4}$$

The set of non-dominated solutions creates Pareto front (Fig. 1). For three criteria the set of Pareto-optimal solutions creates surface, whereas for higher number of dimensions an additional technique is needed for representation of the Pareto front (Chiba K. 2007, Cui Q. 2006).

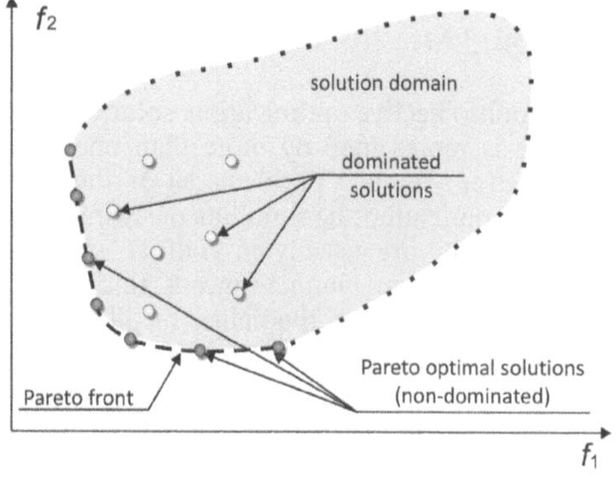

Figure 1. Representation of the Pareto front for the bi-objective case.

3 MULTIOBJECTIVE OPTIMIZATION ALGORITHM

The metaphors of the game theory and immunology are used to solve the problems of multiobjective optimization using IMGAMO (Immune Game Theory Multi-Objective Algorithm) (Jarosz 2010, Jarosz 2011). Each player has its own objective (a payoff function in the Nash equilibrium). The strategy for a particular player is the optimum solution for this player's problem remembering that other players also play their best strategies. The solution of the optimized problem consists of several parameters, each of which is assigned to one of the players. Each player optimizes only its parameters (its strategy) taking the rest of them as constant. The rest of the parameters are set by taking the best solutions from the other players. Solutions from all players should establish the solution of the problem. Then all players use the immune algorithm to optimize their objectives. IMGAMO algorithm was tested on several benchmark problems (ZDT2, ZDT3, ZDT4, ZDT6, DTLZ1, DTLZ2, DTLZ3, DTLZ4, DTLZ5, DTLZ6, DTLZ7). The results of multiobjective optimization are compared to the results obtained by NSGA2 nad SPEA2 algorithms. The metrices of General Distance (GD) and spread of the Pareto optimal solutions (S) are used for comparison between algorithms. For the bi-objective optimization results obtained by IMGAMO and NSGA2 or SPEA2 algorithms are comparable, whereas for higher number of criteria (up to 7) IMGAMO gives significantly better results. Details of the algorithm have been described in (Jarosz 2011). The most important assumptions to the IMGAMO algorithm are as follows: (i) each player has its own fitness function (payoff function), (ii) each player has assigned the part of parameters of the solution (strategy of this player) (this is changed at each iteration), the rest of parameters are set as constants and taken from the best solutions from the other players (this is done at each iteration), (iii) all solutions are coded with real values, (iv) the results of the algorithm (the determined Pareto frontier) are stored in the result population, (v) each player optimize using the immune algorithm to optimize its own objective (each player has its own population).

4 FORMULATION OF THE PROBLEM

Multiobjective optimization of the chevron-type electrothermal actuator is considered (Fig. 2). The deflection or force generated by the microactuator

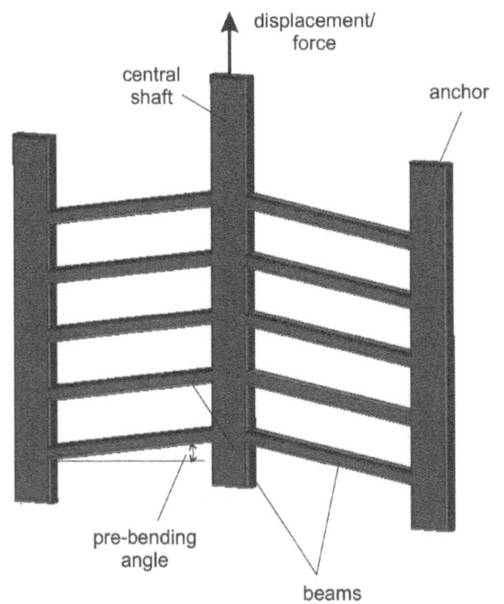

Figure 2. Geometry of the chevron-type microactuator.

can be calculated analytically for the straight shape arms only. For the more complicated shapes and if the stress field should be examined, a numerical model has to be created. In order to simulate numerically electrothermal actuators, a coupled electrical, thermal, and mechanical analysis has to be solved. Such a problem is described by the appropriate partial differential equations (Beer 1983, Kleiber 1998, Zienkiewicz & Taylor 2000). The equations with arbitrary geometries and boundary conditions are solved by Finite Element Method. This problem is weakly coupled and it requires solving electrical, thermal and mechanical analysis separately. The coupling is carried out by transferring loads between the considered analyses and by means of staggered procedures. Matrix equations of electrical, thermal and mechanical problem can be expressed as follows:

$$\begin{aligned} \mathbf{K_E V} &= \mathbf{I} \\ \mathbf{K_T T} &= \mathbf{Q} + \mathbf{Q_E} \\ \mathbf{K_M u} &= \mathbf{F} + \mathbf{F_T} \end{aligned} \tag{5}$$

where: $\mathbf{K_E}$ is the electrical conductivity matrix, $\mathbf{K_T}$ is the thermal conductivity matrix, $\mathbf{K_M}$ is the stiffness matrix, $\mathbf{Q_E}$ is the heat generation vector due to current flow, $\mathbf{F_T}$ is the force due to thermal strain vector, **V**, **T**, **u** are the nodal vector of voltage, temperature and displacements, respectively, **I**, **Q**, **F**, are the nodal vectors of current, heat fluxes and applied forces, respectively. The thermal and mechanical problems are coupled through thermal strain loads $\mathbf{F_T}$. Coupling between the electrical and thermal problems is done by heat generation due to the electrical flow $\mathbf{Q_E}$.

The FEM commercial software MSC Mentat/Marc is adapted to solve the coupled boundary-value problem.

The following functionals are defined:

- the minimization the volume of the structure

$$\min_x \quad f_1 \overset{def}{=} \int_\Omega d\Omega \tag{6}$$

- the minimization of the maximum value of the equivalent stress

$$\min_x \quad f_2 \overset{def}{=} \max\left(\sigma_{eq}\right) \tag{7}$$

- the maximization the deflection of the actuator

$$\max_x \quad f_3 \overset{def}{=} u_0 \tag{8}$$

- the minimization of the total heat generated by the actuator

$$\max_x \quad f_4 \overset{def}{=} \int_\Omega (q)\, d\Omega \tag{9}$$

5 NUMERICAL EXAMPLE

The model of the structure, which contains one pair of the beams of microelectrothermal actuator is considered. The geometry and parametrization of the model are shown in Figure 3. Four design variables are responsible for the shape of the beam. The beam is modeled by means of NURBS curve which consist of 4 control points (a symmetry is assumed along centerline of the beam). Fifth, sixth and seventh design variable are pre-bending angle of the beam, radius *R*1 and radius *R*2, respectively. Box constraints imposed on the design variables are as follows (in μm): P1–P4 (2.5 ÷ 20.0), α (0.0 ÷ 12.0), R1, R2 (1.0 ÷ 20.0). The actuator is modeled as plane stress structure fabricated from polycrystalline silicon. The structure is subjected to the proper electrical, thermal and mechanical boundary conditions. The multi-objective optimization concerns determining the 7 design variables which minimize 4 previously defined functionals are performed by means of IMGAMO algorithm. Following values of IMGAMO parametres are assumed: size of the population—10, number of clones—10, number of iterations—60.

The results of the optimization are shown in Figure 4. Figure presents set of Pareto optimal solutions in 3-dimentional space, whereas 2-nd functional depending on of the maximal value of the equivalent stress is presented as a color map (in the paper in the gray scale).

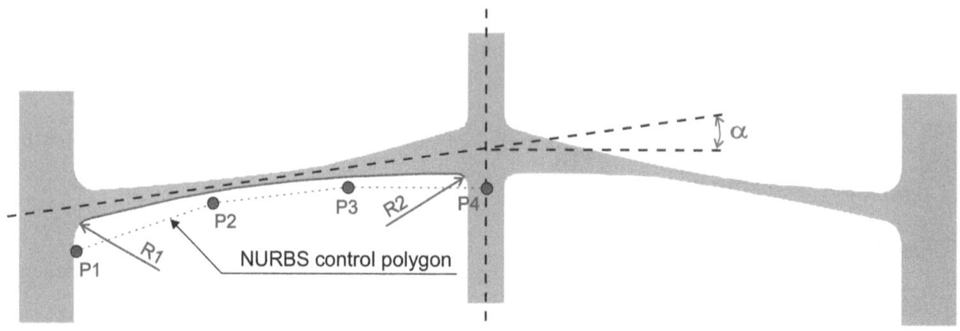

Figure 3. Geometry and parametrization of the chevron-type microactuator.

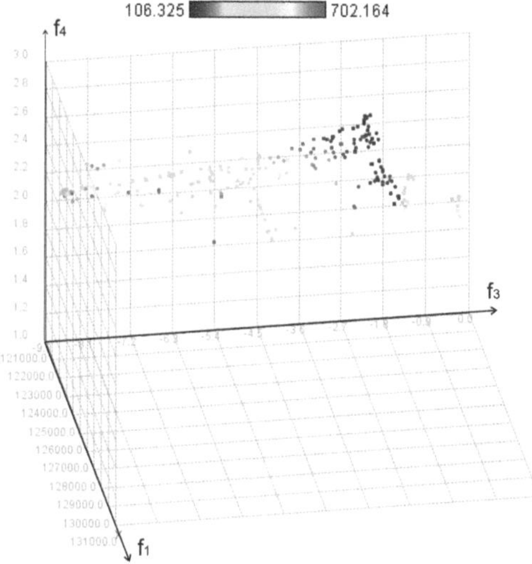

Figure 4. The set of Pareto optimal solutions for simultaneously optimization of 4 functionals.

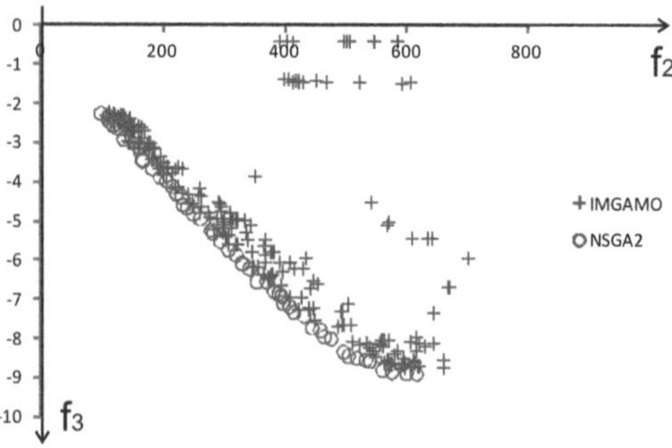

Figure 5. Comparison between 4-objective optimization in the plane of functional (7)—f_2 and (8)—f_3 (IMGAMO) and 2-objective optimization (NSGA2).

Results of optimization were compared with optimization performed by means of NSGA2 algorithm for each combination of defined functionals. The total number of combination between all functionals is equal to 6 (6 graphs are needed for representation of 4 functional). Results obtained by IMGAMO algorithm and NSGA2 algorithm are comparable for all pairs of functionals. An example of the comparison is presented in Figure 5. Figure depicts the non-dominated individuals of 4-objective optimization in the plane of functional (7)—f_2 and (8)—f_3, together with the results to 2-objective optimization found by NSGA2 algorithm.

6 CONCLUSIONS

The multiobjective shape optimization of thermal microactuator was presented. The problem was solved for different criteria based on thermal, mechanical and other quantities. Another type of functionals can be easy formulated. IMGAMO algorithm allows to obtain multidimensional set of Pareto-optimal solution as a run of only one optimization task, whereas for multiobjective algorithms, which works efficiently for small number of criteria more optimization tasks have to be performed (ex. NSGA2 algorithm). 4-dimensional objective space of Pareto optimal solutions may be presented in different form, e.g. colored $3D$ graphs or projection to the planes. Projection of the results, obtained by IMGAMO are compared to the results obtained by NSGA2 algoritm, for all 6 planes. The results of 4-objective optimization include the Pareto fronts of each 2-objective optimization (see Fig. 5). Analysis of the 4-objective optimization results in the consecutive planes gives more optimal choices for the designer. The analysis of other results (both for objective space and design variable space) may lead to the following conclusions: (i) for all obtained

Pareto optimal solution beams are thin (values of the position of the control points are close to box constrains), (ii) sensitivity of changing design parameters on functional f_4 (dissipation of the heat flux) is very low (see Fig. 4), (iii) in order to maximize deflection of the microactuator pre-bending angle should be low (between 1.5–1.8 μm), (iv) in order to minimize equivalent stress, the value of the pre-bending angle and value of the second radius ($R2$) should be greater.

ACKNOWLEDGMENTS

The research is partially financed by research project no. 10/040/BK15/0006.

REFERENCES

Beer, G. (1983). Finite element, boundary element and coupled analysis of unbounded problems in elastostastics. *Int. J. Numer. Meth. Eng. 19*, 567–580.

Chiba K., O.S. (2007). Data mining for multidisciplinary design space of regional-jet wing. *Journal of Aerospace Computing, Information and Communication 4 (11)*, 1019–1036.

Coello, C. (1999). A comprehensive survey of evolutionary based multi-objective optimization techniques. *Knowledge and Information Systems 1 (3)*, 129–156.

Cui Q., Ward M., R.E. (2006). Enhancing scatterplot matrices for data with ordering or spatial attributes. In *Proc. SPIE 6060, Visualization and Data Analysis.*

Dlugosz, A. (2010). Multiobjective evolutionary optimization of mems structures. *Computer Assisted Mechanics and Engineering Sciences 17 (1)*, 41–50.

Henneken V., Tichem M., S.P. (2008). Improved thermal u-beam actuators for micro-assembly. *Sensors and Actuators 142*, 298–305.

Jain N., Sharma A., K.S. (2012). Design and simulation of bidirectional in-plane chevron beam microtweezer. *International Journal of Emerging Technologies in Computational and Applied Sciences*, 43–49.

Jarosz, P., B.T. (2010). Coupling of immune algorithms and game theory in multiobjective optimization. *Artifical Intelligence and Soft Computing*, 500–507.

Jarosz, P., B.T. (2011). Artificial immune system based on clonal selection and game theory priniciples for multiobjective optimization. *Lecture Notes in Computer Science*, 321–333.

Kleiber, M. (1998). *Handbook of Computational Solid Mechanicsd.* Berlin: Springer-Verlag.

Zienkiewicz, O. & R. Taylor (2000). *The Finite Element Method*. Oxford: Butterworth-Heinemann.

Advances in Mechanics: Theoretical, Computational and Interdisciplinary Issues – Kleiber et al. (Eds)
© 2016 Taylor & Francis Group, London, ISBN 978-1-138-02906-4

Detection of damages in a riveted plate

Ł. Doliński, M. Krawczuk, M. Palacz & A. Żak
Gdańsk University of Technology, Gdańsk, Poland

ABSTRACT: The paper presents the results of damage detection in a riveted aluminium plate. The detection method was based on Lamb wave propagation. The plate has been analysed numerically and experimentally. Numerical calculations have been carried out by the use of the time-domain spectral finite element method, while for the experimental analysis Laser Scanning Doppler Vibrometry (LSDV) has been utilised. The panel has been excited by a 5-pulse sinusoidal wave packet of a 35 kHz carrier frequency. A comparison of the experimental and analytical result indicates a good quality of the proposed damage detection technique.

1 INTRODUCTION

Elastic waves are a type of mechanical waves propagating in an elastic medium as an effect of forces associated with its volume deformation (compression and extension) as well as shape deformation (shear) (Ostachowicz et al. 2012).

Lamb waves is a type of elastic waves, that propagate in infinite media bounded by two surfaces and arise as a result of superposition of multiple reflections of longitudinal waves and shear waves from the bounding surfaces. In the case of these waves medium particle oscillations are very complex in character. Various anomalies in wave propagation patterns resulting from wave-damage interaction can be observed, interpreted and next employed for damage assessment. It is well known that the presence of damage results in reflection and scattering of propagating elastic waves and such features are commonly used for damage detection purposes. Wave propagation may be analysed in two ways—experimentally and numerically. For a practical wave analysis laser vibrometry may be depicted as one of the most efficient tools.

In the presented example of damage detection a comparison of two approaches, numerical and experimental, was made.

2 GETTING STARTED

For numerical modelling the time-domain spectral finite element method (Patera 1984) is recommended due to its simplicity of application. The method originates from the application of spectral series for solution of partial differential equations, while at the same time its base ideas are analogous to the classical finite element (Zienkiewicz 1989) approach.

Its main assumption is the application of orthogonal Lobatto polynomials as approximation functions defined at appropriate Gauss-Lobatto-Legendre integration points. As a consequence of that the inertia matrix obtained in this spectral approach is diagonal making the total cost of numerical calculations much less demanding. Additionally, thanks to the orthogonality of the approximation polynomials the spectral finite element method is characterised by exponential convergence (Ostachowicz et al. 2012).

In the current formulation of a spectral shell elements according to the time-domain SFEM are used. As an example the 5th order polynomials were chosen. The resulting grid of element nodes in this case is presented in Figure 1.

For the presented isotropic shell element the Lobatto node distribution was used based on the 6-th order complete Lobatto polynomial (Ostachowicz et al. 2012). In the normalised (curvilinear) coordinate system $\xi\eta\zeta$ of the element

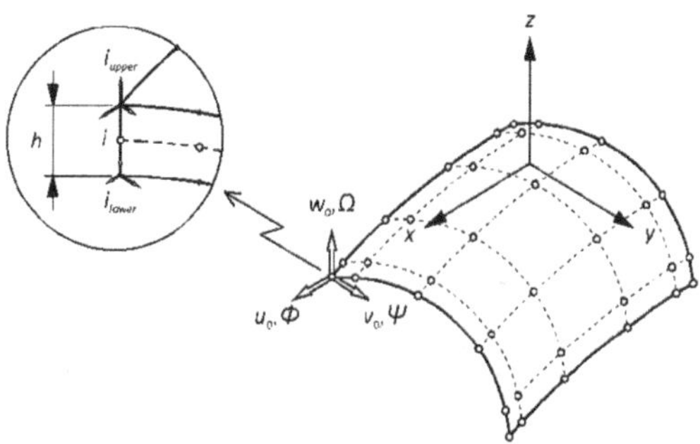

Figure 1. Displacement components in a 36-node spectral shell finite element.

the coordinates of the element nodes ξ_i ($i = 1, \ldots, 6$) and η_j ($j = 1, \ldots, 6$) were assumed as the roots of:

$$
\begin{aligned}
L_6^c(\xi_i) &= 0, \qquad i = 1, \ldots, 6 \\
L_6^c(\eta_j) &= 0, \qquad j = 1, \ldots, 6
\end{aligned}
\tag{1}
$$

where the complete Lobatto polynomial is defined in the following manner (Żak & Krawczuk 2012):

$$
L_6^c(x) = (1 - x^2)L_4(x) = (1 - x^2)P_5'(x) \tag{2}
$$

where L is the 4-th order Lobatto polynomial, P is the 5-th order Legendre polynomial and the symbol $'$ denotes differentiation in respect of x.

In the present definition of the spectral shell finite element an extended form of the displacement field is employed (Żak 2009). The element has six degrees of freedom per node including two in-plane displacement components u_0 and v_0 and four out-of-plane displacement components w_0, Φ and Ψ as well as Ω, as shown in Figure 1. The reader is kindly encouraged to see (Żak 2009, Ostachowicz et al. 2012) for similar solution technique used for element definition.

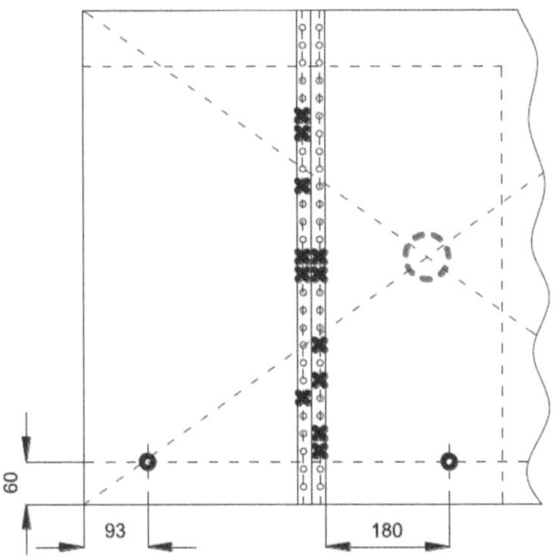

Figure 2. The assumed damage scenario: --- measured region, (o)—additional mass, (x)—removed rivet.

3 LAYOUT OF TEXT

For numerical and experimental test an aluminium panel was prepared (E = 72.7 GPa, ν = 0.33, ρ = 2700 kg/m^3) with two T-shaped stiffeners. The dimensions of the plate were: $1 \times 0.7 \times 0.001$ m and of the stiffeners: $0.04 \times 0.04 \times 0.004$ m. The stiffeners were fixed with 54 rivets located in 2 parallel rows. The assumed additional mass was not bigger than 1% of the total riveted plate.

As damages in both cases were tested some removed rivets as well as additional masses attached to the plate surface.

For the numerical modelling of the described element a 6-node spectral element was used, based on 3-mode theory of higher order, shortly given in previous paragraph. The damages were modelled by modulation the magnitudes of mass and stiffness in adequate elements. The model built for numerical simulations consisted of 54761 nodes in total. The calculation time was 0.5 ms for the analysis of 8000 time steps.

Figure 3 shows 2 different calculation results of wave propagation caught in 3 different time snapshots. Figures marked with (a), (b) and (c) show numerically obtained energy of the signal (calculated as RMS value of the signal velocity) recorded in 78 μs, 350 μs and 505 μs respectively for a plate without damage. Pictures marked with (d), (e) and (f) represent signal RMS values for the same snap shots but calculated for a damaged plate. It may be noticed, that excitation area as well as lines of rivets have been identified very well.

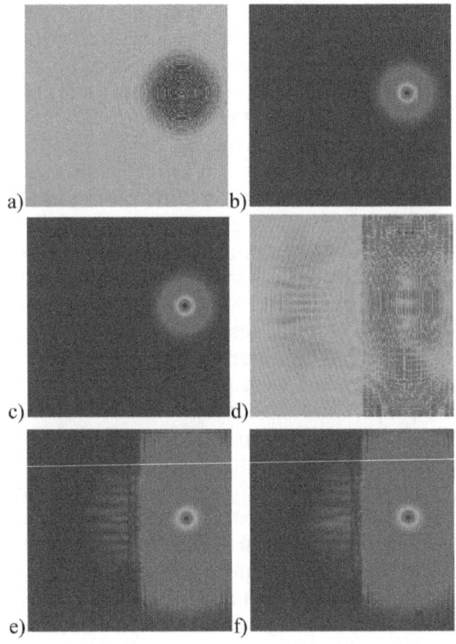

Figure 3. Comparison of wave propagating in element with and without damage. Calculation result for panel without damage at (a) 78 μs, (b) 350 μs and (c) 505 μs and result for panel with damage at (a) 78 μs, (b) 350 μs and (c) 505 μs.

For laser measurements the specimen was isolated from the influence of external vibration source by locating it on polyurethane foam. For excitation, a PZT actuator (T216-A4 NO-273X) was used with resonant frequency of 7.3 kHz. All data measurement was done using a Polytec PSV-400 SLDV. As Lamb wave source there have been used sinusoidal signal of 35 kHz frequency multiplied by Hanning window, covering 5 periods of a sinusoid. All signals were generated via digital generator of ± 10V output. Next, the signals were amplified to ± 100V.

During the measurements the velocities were taken at every measurement point in direction perpendicular to the plate surface. The signal was quantified to 1024 time samples and every measurement has been repeated 10 times and the average value was calculated in order to increase of the signal–noise ratio. The time between every excitation has been long enough to dump previously activated wave. Also the release time was determined for every periodic frequency by registering in a sample point the time necessary to attenuate the wave totally.

The Figure 4 illustrates a comparison of the numerical and experimental results of wave propagating in an analysed riveted damaged panel. Picture located on left hand side shows numerical results, whereas right hand side stands for experiments. It may be seen that both, numerical and experimental approaches, are suitable for location an additional mass, moreover—experimental approach allows also location of damaged rivets. This conclusion is very important for practical application of described methodology as the proposed technique requires LSDV for measurements with all its benefits. It might be successfully used for more complicated geometries.

a) 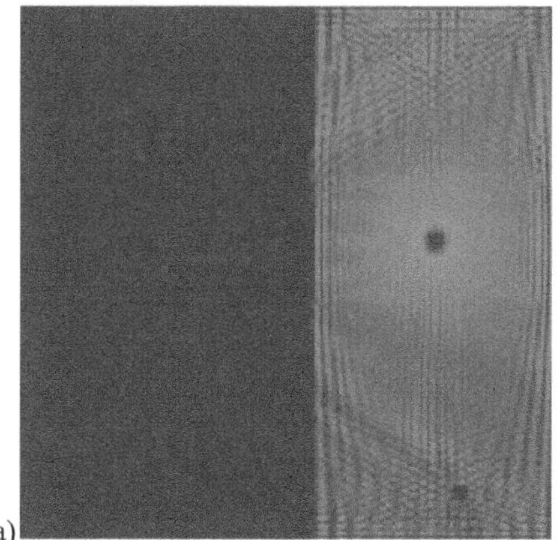

b) 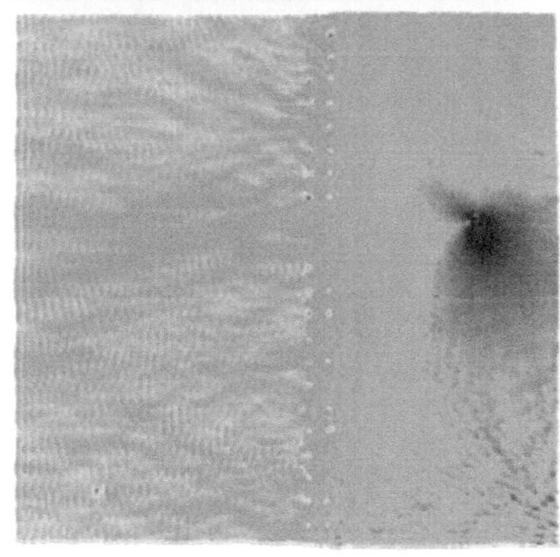

Figure 4. Comparison of numerical (a) and experimental (b) results.

4 CONCLUSIONS

The paper presents a comparison of the results obtained during numerical calculations and experimental measurements. It has been shown that the proposed numerical model based on higher modes theory is suitable for modelling real riveted panel for analysis of such a complicated phenomenon like elastic wave propagation. From the practical point of view it is worth to emphasize that numerical simulation of complicated processes may allow significant cost saving of different prototypes. That is why the development of efficient numerical tools is still important for the mechanical applications.

It has been also shown that utilisation of wave propagation for damage detection makes possible identification of isolated damaged rivets. On the bases of performed research results it may be concluded that adequate selection of analysis parameters for wave based damage detection techniques guarantees a very sensitive damage diagnostics tool.

ACKNOWLEDGEMENTS

The authors of this work would like to gratefully acknowledge the support for their research provided by the Academic Computer Centre in Gdańsk. All results presented in this paper have been obtained by the use of the software available at the Academic Computer Centre in Gdańsk. The authors also would like to thank M. Radzieński, MSc., for his important input in measurements.

REFERENCES

Chang, F.K. 2011. *Structural Health Monitoring*. Stanford.

Giurgiutiu, V. & Bao, J.J. 2004. Embedded-ultrasonics Structural Radar for In Situ Structural Health Monitoring of Thin-wall Structures. *Struc. Health Monitoring* 3: 121.

Inman, D.J., Farrar, C.R., Lopes, Jr V. & Steffen, Jr V. 2005. *Damage Prognosis. For Aerospace, Civil and Mechanical Systems*. West Sussex: Wiley & Sons.

Lakshmanan, K.A. & Pines, D.J. 1997. Detecting crack size and location in composite rotorcraft flex beams. *SPIE Smart Structures and Materials* 3041: 408.

Ostachowicz, W., Kudela, P., Krawczuk, M. & Żak, A. 2012. *Guided Waves in Structures for SHM*. West Sussex: Wiley & Sons.

Palacz, M. & Krawczuk, M. 2002. Analysis of longitudinal wave propagation in a cracked rod by the spectral element method. *Computers and Structures* 80: 1809.

Patera, A.T. 1984. A spectral element method for fluid dynamics, laminar flow in a channel expansion. *Journal of Computational Physics* 54: 468.

Patera, A.T. 1994. A spectral element method for fluid dynamics, laminar flow in a channel expansion. *Journal of Computational Physics* 54: 468.

Radzieński, M. et al. 2011. Application of RMS for damage detection by guided elastic waves. *Journal of Physics Conference Series* 07/2011, 305 (1): 012085.

Radzieński, M., Doliński, Ł., Krawczuk, M. & Palacz, M., 2013. Damage localisation in a stiffened plate structure using a propagating wave. *Mechanical Systems and Signal Processing* 39 (1–2): 388.

Wilcox, P.D. 2003. Omni-directional guided wave transducer arrays for the rapid inspection of large areas of plate structures. *IEEE Transactions of Ultrasonic for Frequency Conference* 50 (6): 699.

Worlton, D.C. 1961. Experimental confirmation of Lamb waves at megacycle frequencies. *Journal of Applied Physics* 32: 967.

Żak, A. & Krawczuk, M. 2012. Static and dynamic analysis of isotropic shell structures by the spectral finite element method. *Journal of Physics: Conference Series* 382.

Żak, A. 2009. A novel formulation of a spectral plate element for wave propagation in isotopic structures. *Finite Elements in Analysis and Design* 45: 650.

Żak, A. 2009. A novel formulation of a spectral plate element for wave propagation in isotopic structures. *Finite Element in Analysis and Design* 45: 650.

Zienkiewicz, O.C. 1989. *The finite element method.* London: McGraw-Hill.

Zumpano, G. & Meo, M. 2006. A new damage detection technique based on wave propagation for rails. *International Journal of Solids and Structures* 43 (5): 1023.

Advances in Mechanics: Theoretical, Computational and Interdisciplinary Issues – Kleiber et al. (Eds)
 ISBN 978-1-138-02906-4

Nonlinear dynamic response of periodically inhomogeneous Rayleigh beams

Ł. Domagalski & J. Jędrysiak
Lodz University of Technology, Łódź, Poland

ABSTRACT: The paper deals with large amplitude vibrations of slender periodic beams with inhomogeneity period much smaller than their span. A beam with geometrical and physical properties varying periodically along its axis is considered. The beam can interact with visco-elastic subsoil. Replacement model equations with constant coefficients are obtained via application of the tolerance modelling technique. The proposed model can serve as a tool in parametric analysis of large amplitude vibrations of considered structures.

1 INTRODUCTION

Flexural periodic structures can exhibit desirable mechanical properties, such as band-pass filtering, Olhoff et al. (2012). Their dynamical behaviour in the small amplitude regime is well recognized, especially for infinite systems, and has been studied in numerous papers. Most of the research effort is devoted into analysis of wave propagation. Amongst them, to name only few, the Floquet-Bloch theory (Chen, 2013) and the transfer matrix method (Yu et al., 2012) are applied. The most general approach in analysis of periodic microstructured media is based on the theory of asymptotic homogenization, cf. Bensoussan et al. (1978).

The main objective of this contribution is modelling and analysis of geometrically nonlinear dynamic problems of beams with periodic structure. In the most general case considered beams have periodically variable stiffness and mass density and interact with visco-elastic subsoil. An example fragment of structure of this type is depicted in Figure 1.

We propose an averaged mathematical model of this type of structure. A simplified version of this model was presented in Domagalski & Jędrysiak (2014). The equivalent system of differential equations is based on the tolerance modelling approach, cf. Woźniak et al. (2010), a nonasymptotic technique of averaging differential operators. This method was applied in analysis of numerous thermomechanics problems of microheterogeneous structures, such as linear vibrations of plates with periodic microstructure (Mazur-Śniady et al., 2004, Marczak & Jędrysiak, 2015). The static analysis problems of thin periodic plates subjected to moderately large deflections were undertaken by Domagalski & Jędrysiak (2012). In particular, the linear vibrations of periodic beams were analysed by Mazur-Śniady & Śniady (2001).

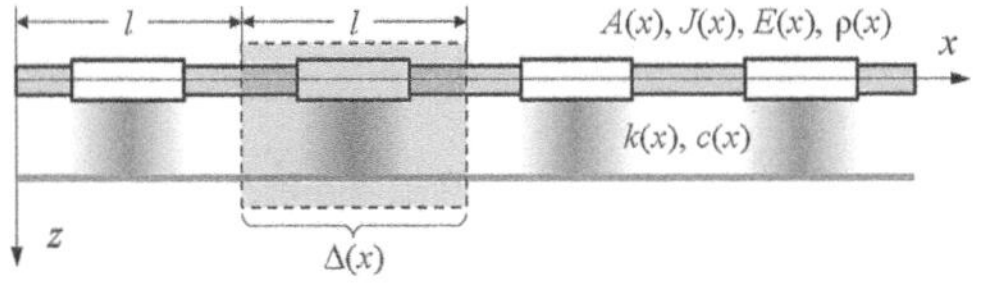

Figure 1. A fragment of a periodically inhomogeneous beam.

2 FUNDAMENTAL EQUATIONS

The dynamics of considered beams is governed by the following system of coupled nonlinear differential equations:

$$
\begin{aligned}
&\partial\left[EA\left(\partial u_0+\tfrac{1}{2}\partial w\partial w\right)\right]=0,\\
&\mu\ddot{w}-\partial\left(\vartheta\partial\ddot{w}\right)+c\dot{w}+kw+\partial^2\left(EJ\partial^2 w\right)+\\
&\quad-\partial\left[EA\left(\partial u_0+\tfrac{1}{2}\partial w\partial w\right)\partial w\right]=q,
\end{aligned}
\tag{1}
$$

where $E = E(x)$ stands for the Young's modulus, $A = A(x)$ and $J = J(x)$ are the cross-section's area and moment of inertia, $\mu = \mu(x)$ and $\vartheta = \vartheta(x)$ are the mass density and the rotational inertia per unit length, $k = k(x)$ and $c = c(x)$ the stiffness and damping coefficients of the subsoil. The transverse deflection and longitudinal displacements are denoted by $w = w(x,t)$, $u_0 = u_0(x,t)$.

In a general case all the coefficients of the equations above are discontinuous, highly oscillating periodic functions in x, which lead to calculational difficulties.

3 THE AVERAGED DESCRIPTION

Since, all the fundamental assumptions, lemmas and assertions of the tolerance modelling approach are described and discussed in Woźniak et al. (2010), only the crucial concepts are given below.

Let $\Delta(x) = x + \Delta$ be a periodicity cell with center at $x \in \Pi_\Delta$, $\Pi_\Delta = \{x \in \Pi: \Delta(x) \in \Pi\}$, $\Pi = [0, L]$. The averaging operator for an arbitrary integrable function f is defined by

$$<f>(x) = l^{-1}\int_{\Delta(x)} f(y)dy, \tag{2}$$

where l is the periodicity cell length, called the microstructure parameter.

It is assumed that the response of a periodic medium has a tolerance-periodic (TP) character. Thus, the unknown displacements are decomposed into their averaged and fluctuating parts as follows:

$$\begin{aligned} & w(x,t) = W(x,t) + h^A(x)V^A(x,t), \quad A = 1, \ldots, N, \\ & W(\cdot),\ V^A(\cdot) \in SV_d^2(\Pi,\Delta), \quad h^A(\cdot) \in FS_d^2(\Pi,\Delta); \\ & u_0(x,t) = U(x,t) + g^K(x)T^K(x,t), \quad K = 1, \ldots, M, \\ & U(\cdot),\ T^K \in SV_d^1(\Pi,\Delta), \quad g^K(\cdot) \in FS_d^1(\Pi,\Delta). \end{aligned} \tag{3}$$

In the above relations new basic kinematic unknowns $W(\cdot)$ and $U(\cdot)$ are called the macrodeflection and the in-plane macrodisplacements, respectively; $V^A(\cdot)$, $T^K(\cdot)$ are additional kinematic unknowns called the fluctuation amplitudes. All these functions are Slowly-Varying (SV) in x. The Fluctuation Shape (FS) functions $h^A(\cdot)$ and $g^K(\cdot)$ describe the disturbance of the displacements in a periodicity cell. The tolerance parameter is denoted by d.

Let us introduce the following basic denotations:

$$\begin{aligned} & \xi = x/L, \quad \tau = \Omega t, \quad \lambda = l/L, \quad \eta = r_0/L, \\ & \bar{w} = W/r_0, \quad \bar{v}^A = l^2 V^A/r_0, \quad \bar{u} = U/r_0, \\ & \Omega = \pi^2\sqrt{E_0 A_0 r_0^2/\mu_0 L^4}, \\ & \delta_0 = \eta^{-1}\int_0^1 u_{,\xi} d\xi = \eta^{-1}(u(1) - u(0)), \end{aligned} \tag{4}$$

where r_0, A_0, E_0 μ_0 are the radius of gyration, area, Young's modulus and mass density per unit length of a reference cross section, L is the beam length, Ω stands for the reference angular frequency and δ_0 is the initial elongation of the beam axis.

Making use of decomposition (Eq. 3) and averaging operation (Eq. 2) we obtain the following system of dimensionless differential and integral equations:

$$\begin{aligned} & D\bar{w}_{,\xi\xi\xi\xi} + K\bar{w} + C\bar{w}_{,\tau} + M\bar{w}_{,\tau\tau} - \eta^2 J\bar{w}_{,\xi\xi\tau\tau} \\ & - \bar{N}\bar{w}_{,\xi\xi} + D^A\bar{v}^A_{,\xi\xi} + \lambda^2\left(K^A\bar{v}^A + C^A\bar{v}^A_{,\tau} + M^A\bar{v}^A_{,\tau\tau}\right) \\ & - \eta^2\lambda J^A\bar{v}^A_{,\xi\tau\tau} - \lambda\tilde{N}^A\bar{v}^A_{,\xi} - Q = 0, \end{aligned} \tag{5}$$

$$\begin{aligned} & D^{AB}\bar{v}^B + \lambda^4 K^{AB}\bar{v}^B + \lambda^4 C^{AB}\bar{v}^B_{,\tau} + \lambda^4 M^{AB}\bar{v}^B_{,\tau\tau} \\ & + \lambda^2 C^A\bar{w}_{,\tau} + \lambda^2\eta^2 J^{AB}\bar{v}^B_{,\tau\tau} + \lambda\tilde{N}^A\bar{w}_{,\xi} + \lambda^2\tilde{N}^{AB}\bar{v}^B \\ & + D^A\bar{w}_{,\xi\xi} + \lambda^2 K^A\bar{w} + \lambda^2 M^A\bar{w}_{,\tau\tau} + \lambda\eta^2 J^A\bar{w}_{,\xi\tau\tau} + \\ & -\lambda^2 Q^A = 0, \end{aligned} \tag{6}$$

$$\begin{Bmatrix} \bar{N} \\ \tilde{N}^A \\ \tilde{N}^{AB} \end{Bmatrix} = \begin{bmatrix} \bar{B}_0 & \lambda\bar{B}_0{}^C & \lambda^2\bar{B}_0{}^{CD} \\ \bar{B}_0{}^A & \lambda\tilde{B}_0{}^{AC} & \lambda^2\bar{B}_0{}^{ACD} \\ \bar{B}_0{}^{AB} & \lambda\bar{B}_0{}^{ABC} & \lambda^2\bar{B}_0{}^{ABCD} \end{bmatrix} \times\left[\int_0^1 \begin{Bmatrix} \frac{1}{2}\bar{w}_{,\xi}\bar{w}_{,\xi} \\ \bar{v}^C\bar{w}_{,\xi} \\ \frac{1}{2}\bar{v}^C\bar{v}^D \end{Bmatrix} d\xi + \begin{Bmatrix} \delta_0 \\ 0 \\ 0 \end{Bmatrix}\right]. \tag{7}$$

with constant averaged coefficients (denoted by capital letters) that represent the periodically varying beam properties averaged with appropriate derivatives of fluctuation shape functions. It should be emphasized that the solutions to this model have physical significance only when the unknown functions are slowly-varying.

4 SOLUTION PROCEDURE

The obtained partial differential equations are transformed into a finite system of ordinary differential equations by using the Bubnov-Galerkin method. The solutions to Equations 5–7 are sought for in the form of truncated series of orthogonal functions:

$$\begin{aligned} & \bar{w}(\xi,\tau) = \sum_{m=1}^{M_w} w_m(\tau)X_m(\xi), \\ & \bar{v}^A(\xi,\tau) = \sum_{m=1}^{M_V^A} v_m^A(\tau)Y_m^A(\xi), \quad A = 1, \ldots, N. \end{aligned} \tag{8}$$

The resulting equations can be written in convenient matrix form as follows:

$$\mathbf{K}\mathbf{y} + \mathbf{C}\dot{\mathbf{y}} + \mathbf{M}\ddot{\mathbf{y}} = \mathbf{q}, \quad \mathbf{K} = \mathbf{K}_0 + \mathbf{K}_{NL}(\mathbf{y}). \tag{9}$$

The linearized equations ($\mathbf{K} = \mathbf{K}_0$) are solved for linear eigenvalues. Independently, Equations 9 are rewritten as a system of the first order equations

$$\begin{cases} \dot{\mathbf{y}} = \mathbf{v}, \\ \dot{\mathbf{v}} = \mathbf{M}^{-1}(\mathbf{q} - \mathbf{K}\mathbf{y} - \mathbf{C}\mathbf{v}), \end{cases} \quad \begin{cases} \mathbf{y}(0) = \mathbf{y}_0, \\ \mathbf{v}(0) = \mathbf{v}_0, \end{cases} \tag{10}$$

and numerically integrated by means of the Runge-Kutta-Fehlberg method.

5 APPLICATIONS

5.1 *Linear vibrations modes and frequencies*

The obtained model is applied in analysis of a pinned-pinned beam of length $L = 1$ m and uniform cross-section $b \times h = 0.005 \times 0.0025$ m^2 carrying a system of periodically distributed masses $M_1 = 10\ \mu_0 l$ and $M_2 = \alpha M_1$ with mass moments of inertia I_1 and $I_2 = \alpha I_1$, $\lambda = 1/10$, $\mu_0 = b \times h \times 7850$ kgm^{-3}. The beam and its fragment are depicted in Figure 2.

The fluctuation shape functions $h_1(y)$, $h_2(y)$, $h_3(y)$ and $g_1(y)$ can be found by means of finite element discretization of a periodicity cell as the periodic eigenforms of the cell vibrations, e.g. as depicted in Figure 3.

The solutions to the tolerance model Equations 5–7 are assumed in the form:

$$X_m(\xi) = \sin m\pi\xi,$$
$$Y_m^A(\xi) = \begin{cases} \sin m\pi\xi & \text{for} \quad A = 1, \\ \cos(m-1)\pi\xi & \text{for} \quad A = 2,3, \end{cases} \tag{11}$$

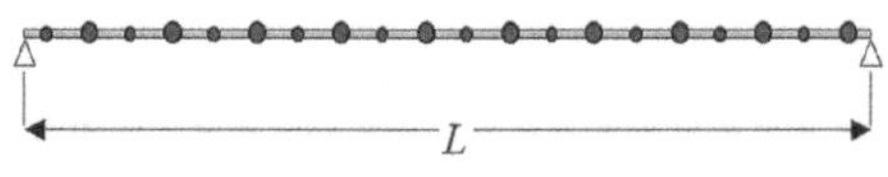

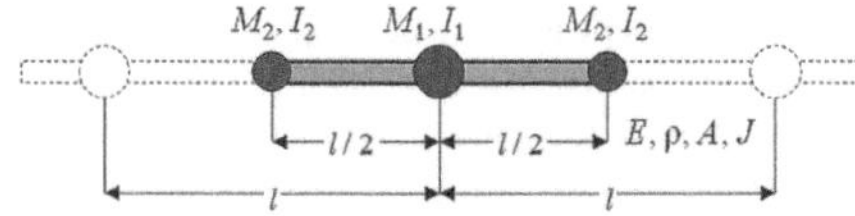

Figure 2. Considered beam and its periodicity cell.

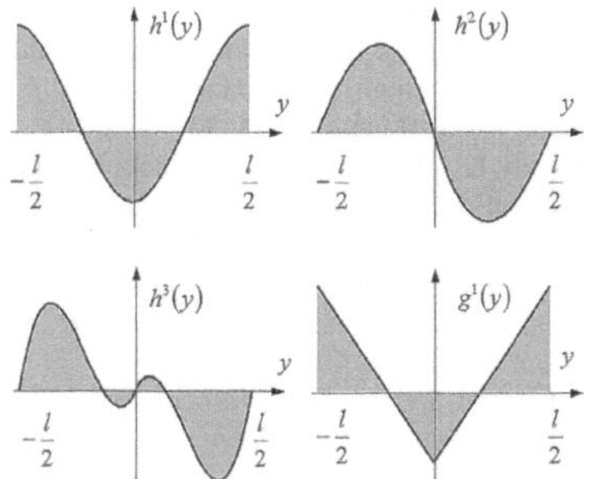

Figure 3. Plots of fluctuation shape functions h^1, h^2, h^3 and g^1.

that satisfy the boundary conditions for a simply-supported beam.

Figure 4 depicts the linear eigenfrequencies versus parameter α for a 20-element 40-degree-of-freedom finite element (circles) and a 4-degree-of-freedom ($m = 1$) tolerance (solid lines) models of considered beam. It can be seen that the proposed model is able to determine the stop- and pass-band boundaries, which correspond to the higher-order vibration modes.

5.2 *Nonlinear vibrations frequencies*

The nonlinear eigenfrequencies are calculated as $\omega_{NLIN} = 2\pi/T_0$, where T_0 is the first return period obtained from time series with appropriate initial conditions. The first lower order and the first higher order frequency are taken into account.

Figures 5 and 6 depict the time series of total deflection and corresponding discrete Fourier

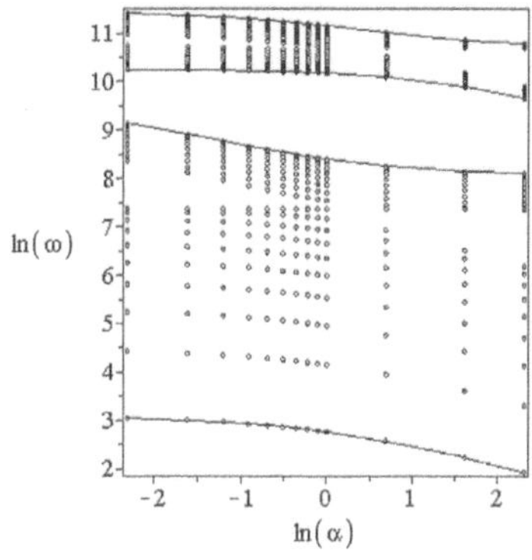

Figure 4. Eigenfrequencies of the considered beam.

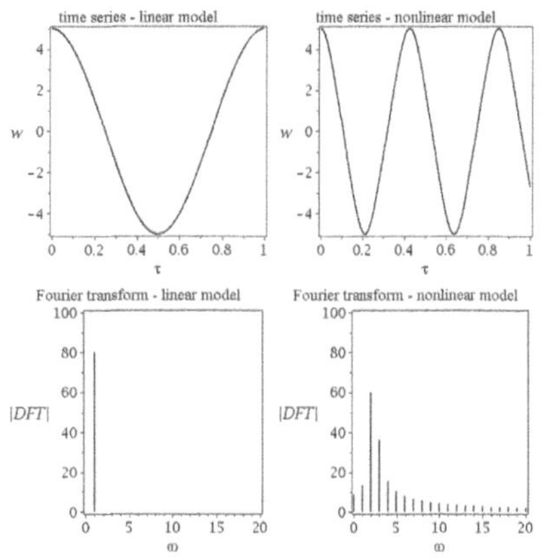

Figure 5. Time series and discrete Fourier transform of free undamped macro-mode vibrations calculated by linear and nonlinear model.

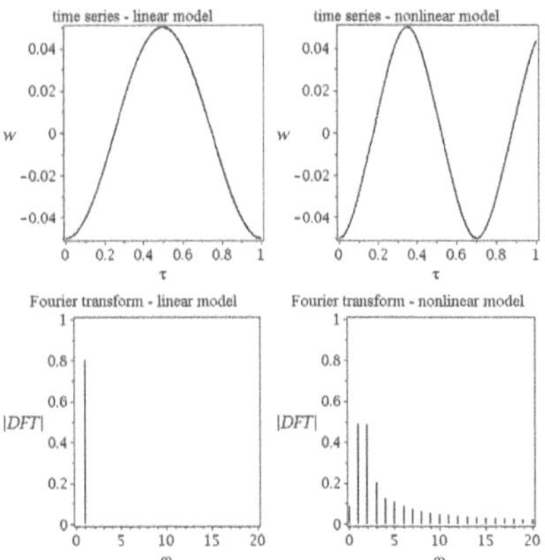

Figure 6. Time series and discrete Fourier transform of free undamped micro-mode vibrations calculated by linear and nonlinear model.

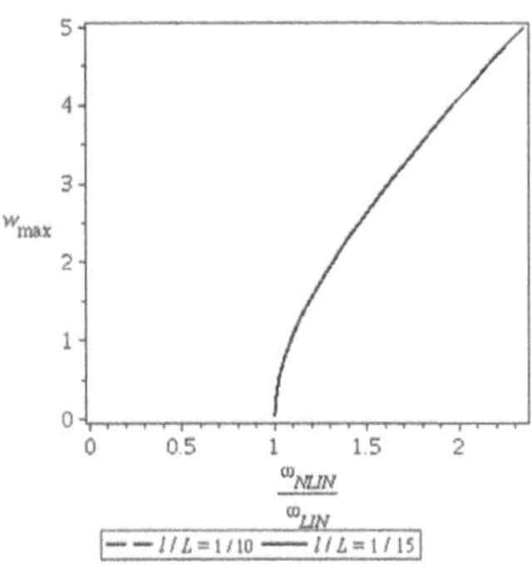

Figure 7. Backbone curves in the low order frequency range and their dependence of microstructure parameter.

transform obtained in the framework of linearized and nonlinear models for initial conditions:

$$\begin{aligned} &w_1(0)=4,\ \dot{w}_1(0)=0,\ v_1^1(0)=0,\ \dot{v}_1^1(0)=0, \\ &w_1(0)=0,\ \dot{w}_1(0)=0,\ v_1^1(0)=4,\ \dot{v}_1^1(0)=0, \end{aligned} \tag{12}$$

respectively.

The described procedure is applied to determine the nonlinear natural vibrations frequencies. The dependence of the nonlinear frequencies of the vibration amplitudes (the backbone curves) in the lower- and higer-order frequency range are shown in Figures 7 and 8 for two different values of the microstructure parameter λ. It can be seen that the backbone curves lean towards the higher frequency values with increasing amplitude, which is typical for nonlinear oscillators with hardening type cubic nonlinearity. Although, only the higher order frequencies are dependent on the microstructure size in the considered case.

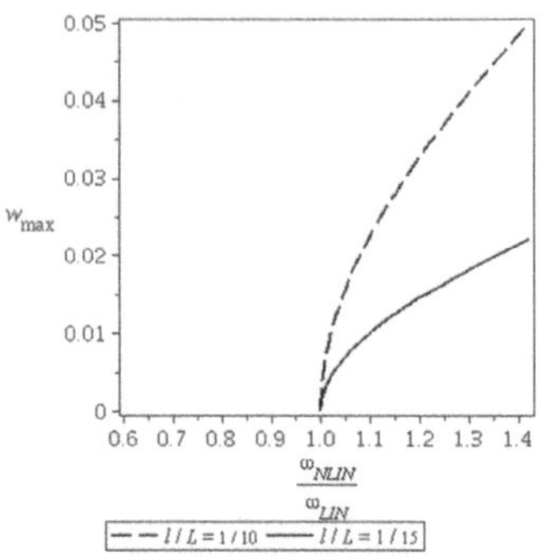

Figure 8. Backbone curves in the high order frequency range and their dependence of microstructure parameter.

6 CONCLUSIONS

The tolerance modelling approach was applied to obtain a geometrically nonlinear 1-D model of locally resonant one-dimensional flexural periodic structures. The original description was transformed into a system of differential and integral equations with constant coefficients, some of which depend on the length of the periodicity cell. The proposed model can be applied in qualitative parametric analysis of large amplitude vibrations of periodic beams.

REFERENCES

Bensoussan A., Lions J.L., Papanicolaou, G. 1978. *Asymptotic analysis for periodic structures*. Amsterdam: North-Holland.

Chen, T. 2013. Investigations on flexural wave propagation of a periodic beam using multi-reflection method. *Archive of Applied Mechanics* 83: 315–329.

Domagalski, Ł. & Jędrysiak J. 2014. Nonlinear vibrations of periodic beams. *Vibrations in Physical Systems* 26: 73–78.

Marczak, J. & Jędrysiak, J. 2015. Tolerance modelling of vibrations of periodic three-layered plates with inert core. *Composite Structures* 134: 854–861.

Mazur-Śniady, K. & Śniady, P. 2001. Dynamic response of a micro-periodic beam under moving load—deterministic and stochastic approach. *Journal of Theoretical and Applied Mechanics* 39: 323–338.

Mazur-Śniady K., Woźniak C., Wierzbicki E. 2004. On the modelling of dynamic problems for plates with a periodic structure. *Archive of Applied Mechanics* 74: 179–90.

Olhoff N., Niu B., Cheng G. 2012. Optimum design of band-gap beam structures. *International Journal of Solids and Structures* 49: 3158–3169.

Woźniak, C. et al. (eds). 2010. *Mathematical modeling and analysis in continuum mechanics of microstructured media*. Gliwice: Silesian University of Technology Press.

Yu D., Wena J., Shen H., Xiao Y., Wen X. 2012. Propagation of flexural wave in periodic beam on elastic foundations. *Physics Letters A* 376: 626–630.

Advances in Mechanics: Theoretical, Computational and Interdisciplinary Issues – Kleiber et al. (Eds)
© 2016 Taylor & Francis Group, London, ISBN 978-1-138-02906-4

Investigation of Turbulent Boundary Layers at moderate Reynolds number in the vicinity of separation

A. Dróżdż, W. Elsner & A. Kępiński
Częstochowa University of Technology, Częstochowa, Poland

ABSTRACT: The paper deals with the experimental analysis of turbulent boundary layer at the flat plate developed under the influence of strong adverse pressure gradient. The special design of the test section equipped with perforated, movable upper wall allow to generate on the bottom flat plate the turbulent boundary layer, which is at the verge of separation. The emphasis is on the analysis of the streamwise fluctuating velocity and mean velocity profiles as well as on some criterion, which are able to predict the detachment position. The change of mean velocity profiles and fluctuating velocity fields confirmed the presence of specific conditions for the separation of the turbulent layer. The increasing of APG strength cause the decrease in mean velocity below the log law region. The effect of APG is especially visible in significant modification of the distribution of Reynolds stresses.

1 INTRODUCTION

Among various types of near wall flows the Turbulent Boundary Layers (TBLs) subjected to an Adverse Pressure Gradient (APG) are in the spotlight. It is because TBLs under the influence of the Adverse Pressure Gradient (APG) are frequently encountered in many engineering applications, such as diffusers, compressor and turbine blades, and the trailing edges of airfoils. The performance of such flow devices is significantly affected by the presence of the adverse pressure gradient. It is due to the inner part of the boundary layer nearest to the wall, which is crucial in determining the skin friction drag. If a turbulent boundary layer flow encounters a strong APG, the flow becomes unstable and, if the APG is sufficiently large, it separates from the surface. The existence of separation involves an increase of energy losses connected sometimes with pressure and velocity fluctuations. The evidence of the latter phenomenon was given by Cherry et al. (1984), who investigated the unsteady structure of a separated and reattaching flow. Unstable location of turbulent separation results from the impact of large-scale outer region vortex structures that fall into the area of separation, causing a temporary increase in momentum. Downstream the incipient of separation (1% instantaneous back flow) these large-scale structures grow rapidly and agglomerate with one another. It is also a region where the displacement thickness of the boundary layer begins to increase strongly. An extensive phenomenological description of the flow separation distinguishing various stages of separation was given by Simpson (1989). The flow separation is more or less an unsteady phenomenon covering the region from the so-called incipient separation up to the total separation defined by the zero mean value of the wall shear stress.

The turbulent boundary layer that is maintained on the verge of separation has already been studied experimentally (Krogstad & Skare 1995). Authors claim that the flow close to separation exhibited a definite non-equilibrium character, indicated by the different scales required for collapse of the mean velocity and turbulence intensity profiles. Castillo et al. (2004) showed, however that outer part of turbulent boundary layer under strong adverse pressure gradient and even near and past the separation tends to remain in equilibrium state.

Pressure gradient and especially APG has a complex effect on the mean velocity profile and so on the shape factor $H = \delta^*/\theta$, which has been shown to increase in the presence of an adverse pressure gradient (Krogstad & Skare 1995). The effect of APGs is also visible in significant modification of the distribution of Reynolds stresses. This effect was reported by many authors for different pressure gradients, as for example by Krogstad & Skare (1995) and Nagano & Houra (2002). Similar observations were published by Materny et al. (2208), who demonstrated the appearance, independently of the first peek in the inner layer, of the second maximum in profiles of velocity fluctuations. They observed that the second maximum moves away from the wall with pressure gradient but the underlying mechanism of this effect is still not well understood. It is believed that it is associated with the production and breakdown of organized flow structures in the same sense as it is proposed for the transition process to turbulence in wall-bounded flows and based on streak instabilities.

Many researches try to investigate turbulent separation and properly detect the detachment point. The paper fits well into these studies. It presents experimental investigation of turbulent boundary layer at the verge of separation at the flat plate, where the adverse pressure gradient is controlled by curvature of the upper wall and additionally by flow suction at that same region. The main focus of the paper is on detection of the separation zone and the analysis of mean and fluctuating velocity field.

2 EXPERIMENTAL SETUP

The experiments were performed in an open-circuit wind tunnel, where the turbulent boundary layer developed along the flat plate, which is 7 m long. The newly developed test section located at the end of the wind-tunnel (see Fig. 1) is equipped with perforated, movable upper wall enabling to generate flow conditions for the separation of turbulent boundary layer at the bottom flat plate. The profile of the upper wall and rate of suction were adjusted to have a constant gradient of pressure coefficient along the test section. In order to accelerate the transition to turbulent boundary layer the tripping wire followed by the strip of coarse-grained sandpaper was used. Flow parameters determined in core flow at the inlet plane to test section (i.e. 5000 mm downstream the leading edge), located in the zero pressure gradient area, are the mean velocity $U = 15$ m/s and turbulence intensity $Tu < 1\%$. The value of Reynolds number, based on the momentum loss thickness θ, equal $Re_\theta \approx 8300$.

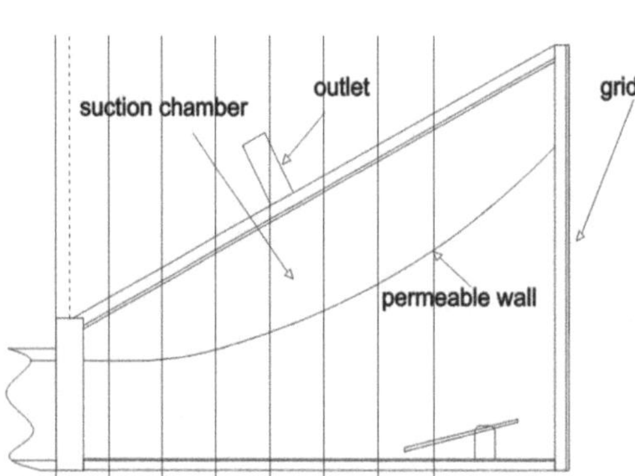

Figure 1. Test section geometry.

The measurements were performed with hot-wire anemometry CCC developed by Polish Academy of Science in Krakow. A single hot-wire probe of a diameter $d = 3$ μm and length $l = 0.4$ mm was used. Acquisition was maintained at frequency 50 kHz with 30 s sampling records. For the assumed sampling frequency the non-dimensional inner scale representation was $f^+ \approx 1$, what is consistent with the assumption of Hutchins et al. (2009).

3 RESULTS

Measurements were performed for three pressure gradient conditions controlled by means of upper wall position and by flow suction. The first case (Case 1) was the reference case where no suction was applied. For the next one (Case 2) external conditions were set in order to achieve turbulent boundary layer on the verge of separation at the end of test section and for the last one (Case 3) higher pressure gradient were applied to shift the separation point upstream. The flow conditions are characterised by pressure coefficient $Cp = 1 - (U_\infty/U_{\infty 0})^2$, where U_∞ is free stream velocity, while subscript 0 denotes the inlet plane value.

Figure 3 presents the downstream evolution of boundary layer thickness δ and displacement thickness δ^*. Initially, the increase in the value of these parameters is slow, but after entering in an area of strong pressure gradient both the boundary layer thickness and the displacement thickness begins to increase rapidly, going into the area of separation.

There are number of separation criteria available in the literature and their interesting review was published by Castillo et al. (2004). For the purposes of this study it was decided to use the criterion proposed by Sandborn & Kline (1961). They showed that the shape factor defined as:

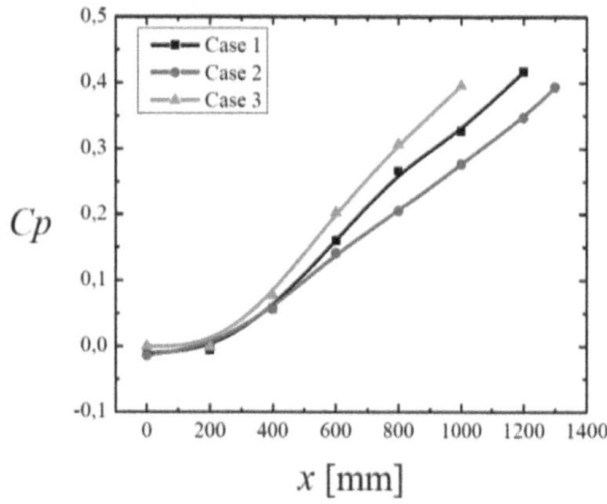

Figure 2. Distribution of pressure coefficient Cp.

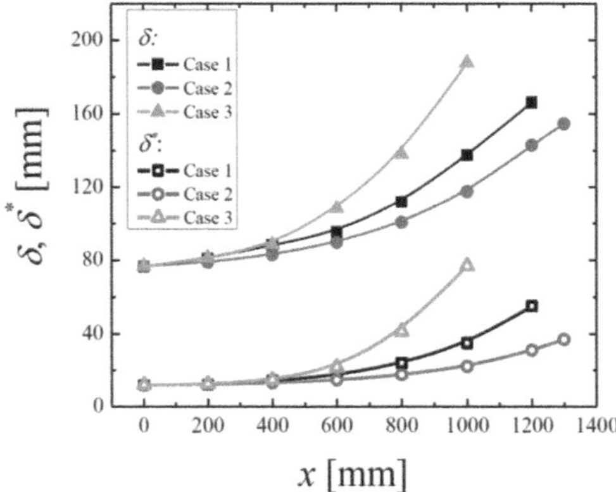

Figure 3. Distribution of the boundary layer thickness δ and displacement thickness δ^*.

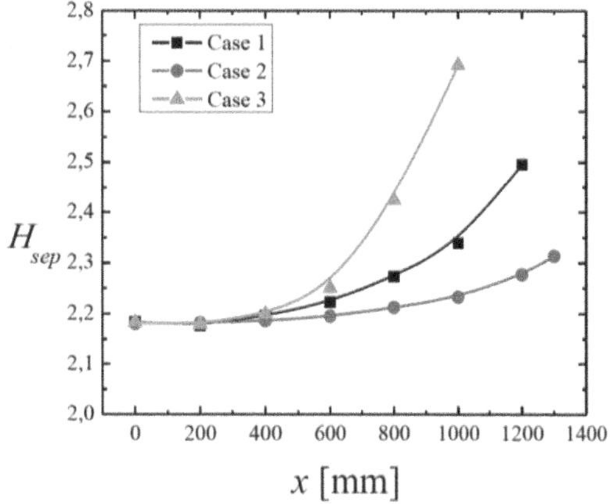

Figure 4. Distribution of shape factor H_{sep} at separation.

$$H_{sep} = 1 + \frac{1}{1 - \delta^*/\delta} \qquad (1)$$

has a value 2.7 at the Intermittent Transitory Detachment (ITD) position. According to definition of Simpson (1989) the Intermittent Transitory Detachment point is when the reverse flow occurs of about 20% of the time. This point is a bit ahead of detachment point, where the time averaged wall shear stress is zero. Figure 4 presents the downstream evolution of H_{sep}. For the reference Case 1 the level of H_{sep} parameter is maintained at a constant level until $x \approx 600$ mm and then rise reaching the level above 2.5.

In other cases the changed conditions forced the earlier boundary layer separation, what is confirmed by the significant change of H_{sep} distribution. It is due to the abrupt rise of the displacement thickness. It can be concluded that for all cases case the ITD point is almost reached, but for various downstream locations.

Figures 5–6 show the development of the mean streamwise velocity profiles a) and streamwise Reynolds stresses b) for the two chosen cases. The velocity is normalized by the local freestream velocity U_∞ and Reynolds stress by outer scale velocity $U_0 = 2(U_\infty - U_{y=0.5\delta})$, while the y coordinate is scaled by the boundary layer thickness δ.

Only the Cases 1 and 3, characterized by medium and strong pressure gradients, were chosen for comparison. In Figure 5a in downstream direction a strong deformation of velocity profile and the disappearance of near-wall maximum of streamwise Reynolds stress are seen Figure 5b. The stronger decrease is however, observed for Case 3 (Fig. 6), where near the wall the extremely low values of U remains until $y/\delta = 0.1$. In the same time the outer peak of streamwise Reynolds stress appears in the

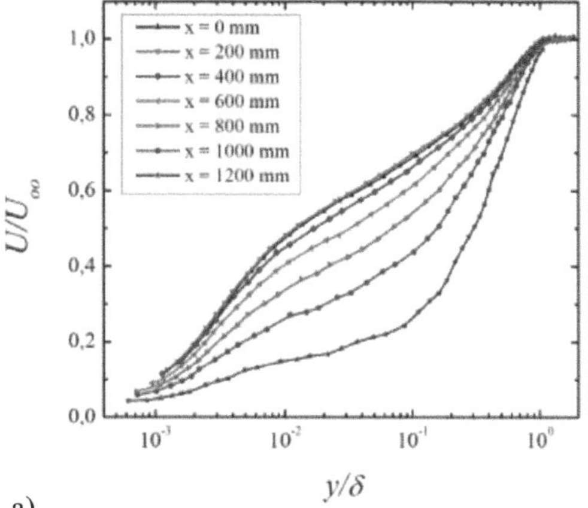

a)

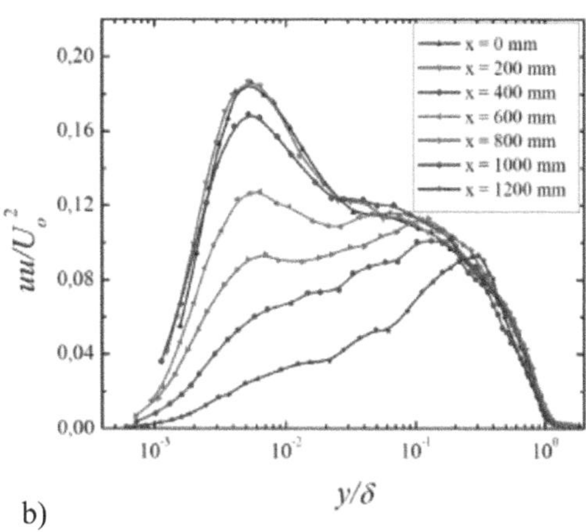

b)

Figure 5. Case 1 streamwise statistics: a) mean velocity normalized with U_∞, b) Reynolds normal stress normalized with U_0.

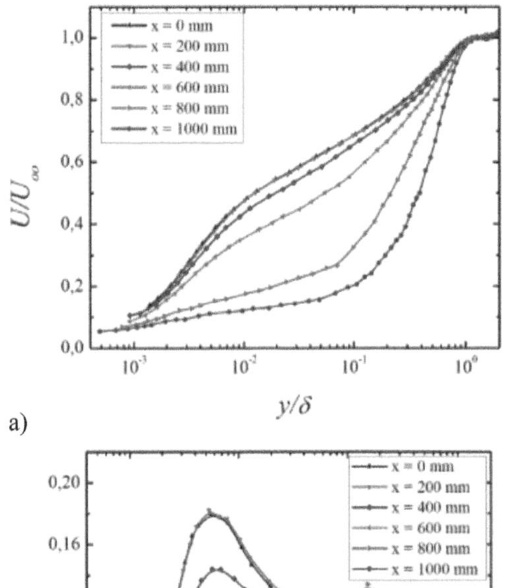

a)

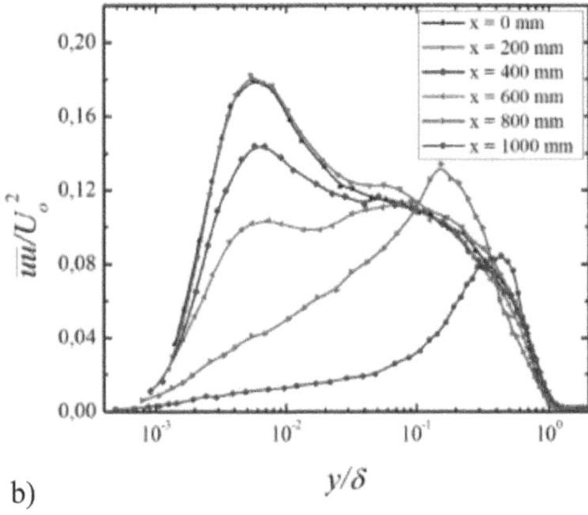

b)

Figure 6. Case 3 streamwise statistics: a) mean velocity normalized with U_e, b) Reynolds normal stress normalized with U_0.

outer zone. However, for the Case 3 (Fig. 5b) it is much more distinct than for the Case 1 (Fig. 6b) indicating the more advanced separation stage. In Figure 6b at x > 600 the overshot of the maximum in the outer zone can be noticed. The behavior is consistent with investigation of Gongur et al. 2014 for near wall flow with shape factor $H < 2$ very strong pressure gradient that is also a case in the current paper.

4 CONCLUSIONS

The presented results confirm that the turbulent boundary layer is on the verge of separation for all pressure gradient conditions, but clearly for the different locations. The increasing of APG strength cause the decrease in mean velocity below the log law region. The effect of APG is especially visible in significant modification of the distribution of Reynolds stress. In downstream direction the disappearance of near-wall maximum and the increase in intensity of the outer region are clearly seen. The changing behavior of mean velocity profiles can be observed in the boundary later thickness and the shape factor.

ACKNOWLEDGMENTS

The investigations presented in this paper have been obtained with funding from Polish National Centre of Science within the grant DEC 2011/03/B/ST8/06401 as well as statutory funds BS-03-301/98.

REFERENCES

Castillo, L., Wang, X., & George, W.K. 2004. Separation Criterion for Turbulent Boundary Layers Via Similarity Analysis. *J. Fluids Eng.* 126: 297.

Cherry, N.J., Hillier, R. & Latour, M.E.M.P. 1984. Unsteady measurements in a separated and reattaching flow. *J. Fluid Mech.* 144: 13–46.

Gungor, A.G., Maciel, Y., Simens, M.P. & Soria, J. 2014. Analysis of a Turbulent Boundary Layer Subjected to a Strong Adverse Pressure Gradient. *J. Phys. Conf. Ser.* 506: 012007.

Hutchins, N., Nickels, T.B., Marusic, I. & Chong, M.S. 2009. Hot-wire spatial resolution issues in wall-bounded turbulence. *J. Fluid Mech.* 635: 103.

Krogstad, P.-A. & Skare, E. 1995. Influence of a strong adverse pressure gradient on the turbulent structure in a boundary layer. *Phys. Fluids* 7: 2014–2024.

Materny, M., Dróżdż, A., Drobniak, S. & Elsner, W. 2008. Experimental analysis of turbulent boundary layer under the influence of adverse pressure gradient. *Arch. Mech.* 60: 449–466.

Nagano, Y. & Houra, T. 2002. Higher-order moments and spectra of velocity fluctuations in adverse-pressure-gradient turbulent boundary layer. *Exp. Fluids* 33: 22–30.

Sandborn, V.A. & Kline, S.J. 1961. Flow models in boundary-layer stall inception. *J. Fluids Eng.* 83: 317–327.

Simpson, R.L. 1989. Turbulent Boundary-Layer Separation. *Annu. Rev. Fluid Mech.* 21: 205–234.

Advances in Mechanics: Theoretical, Computational and Interdisciplinary Issues – Kleiber et al. (Eds)
© 2016 Taylor & Francis Group, London, ISBN 978-1-138-02906-4

Design of fuzzy logic controller for the unloading system in mechatronic device for gait reeducation

S. Duda, D. Gąsiorek, G. Gembalczyk, S. Kciuk & A. Mężyk
Institute of Mechanical Engineering, Silesian University of Technology, Gliwice, Poland

ABSTRACT: The article presents a structural solution and a control system of a device responsible for patient unloading during therapeutic tasks related to the walk re-education process. The device is incorporated in a complex mechatronic system which constitutes a rehabilitation device. Unlike traditional devices of this type which often operate as classic hoists, the applied system is equipped with two independent motors. One of the motors is responsible for the dynamic compensation of the patient's body weight and acts as the so-called "elastic series actuator"—a drive applied in walking machines and human-assisting robots. The other one is responsible for the operation of the winding reel. Due to the need to ensure proper cooperation between the two engines, the selection of the optimal control system of the device constitutes an interesting engineering problem. The work presents a solution using an implemented algorithm based on fuzzy logic. The selection of parameters was conducted based on numerical studies in which the walk of a patient through a step was simulated.

1 INTRODUCTION

Mechatronic devices equipped with intelligent control systems are starting to take on tasks that used to be conducted by humans only. This may be exemplified by modern rehabilitation devices which take the place of physical therapists or at least significantly ease their work Behrman & Harkema (2000). As far as rehabilitation equipment is concerned, the device supporting the process of walk re-education designed at the Institute of Theoretical and Applied Mechanics is one of such solutions. It operates as a travelling crane in which the unloading system performs a movement following the moving patient while unloading the patient with a pre-set force value. The patient may thus move at the entire area, limited only by the structure of the crane. This type of solution allows to perform various types of exercises under the rehabilitator's supervision, e.g. climbing stairs, squats or standing up. Another matter that is essential to the rehabilitation process is also to ensure the patient a possibility to move in the most natural manner. The unloading devices used until now have only allowed for a movement in a single direction. This limits the patient's freedom of movement and largely hinders the natural moves of the rehabilitated person. As the experimental studies (Duda et al. 2015) have proven, in which the displacement of the centre of gravity of the unloaded person during walk was analyzed using the videogrametric method, the application of the follow-up movement in all directions (both in the sagittal and frontal planes of the patient) allows to walk in a near-natural manner.

2 CONSTRUCTION OF THE DEVICE

From the structural point of view, the device resembles a travelling crane in which the girder travels on tracks located in the top part of the support frame of the device. The structure is made of aluminium profiles. In one direction, the follow-up movement of the crane keeping up with the patient (horizontal plane movement) is conducted using a belt transmission and using a screw transmission in the other. The electrical part of the device is comprised of BLDC motors controlled by servo inverters operating in speed mode (Duda et al. 2015).

The operation of the unloading system should guarantee a rapid reaction to the changes of the unloading force values as well as to allow the patient to move along a track with elements of a different height. To allow the fulfilment of both the criteria, two separately operating motors have been installed in the device (Pratt et al. 2002). One, using a planetary gear, drives the reel of the hoist (Z1 drive), which allows to control the height of the sling. The second engine has been coupled with a screw drive which is responsible for the movement of the travelling element which is separated from the pulley with elastic elements (Z2 drive).

A cable (1) is stretched on the pulleys system (2)—one end the cable is fixed to a drum (3) and the other to the orthopaedic harness (4) in which

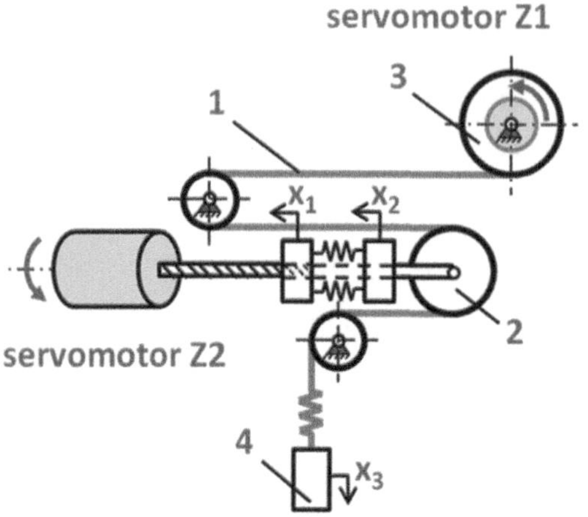

Figure 1. Scheme of unloading system.

the patient is located. The diagram of the developed drive has been presented in Figure 1.

Moreover, the device is equipped with designed measurement systems: sensor of the rope deflection angle and the sensor for the measurement of the unloading force.

3 CONTROL SYSTEM

The control of the device is conducted in a real-time mode, using RT-DAC4/PCI cards connected to a servomotors control cabinet by means of a dedicated signal conditioning interface.

The algorithm responsible for the follow-up movement utilizes PD regulators operating in a feedback loop. The misalignment signal is equal to the deflection angle of the rope along a given gait direction (Duda et al. 2015).

In case of the patient's movement along a non-uniform track (climbing stairs, inclined planes), the control unit of the unloading system also manages the operation of both the drives—Z1 and Z2.

Due to the fact that the control unit of the unloading system is a multi-input and multi-output system, the control algorithm (in opposition to the control of the drives responsible for the follow-up movement) has been based on fuzzy logic (Marchal-Crespo et al. 2009), Smoczek J. (2014), (Tu et al. 2000).

The conception of the proposed control system has been presented as a block diagram in Figure 2.

In the presented configuration, the fuzzy controller, on the basis of information about the error value of set relieving force and movement of actuator element in Z2 drive, determines the value of signals controlling the work of Z1 and Z2 engine.

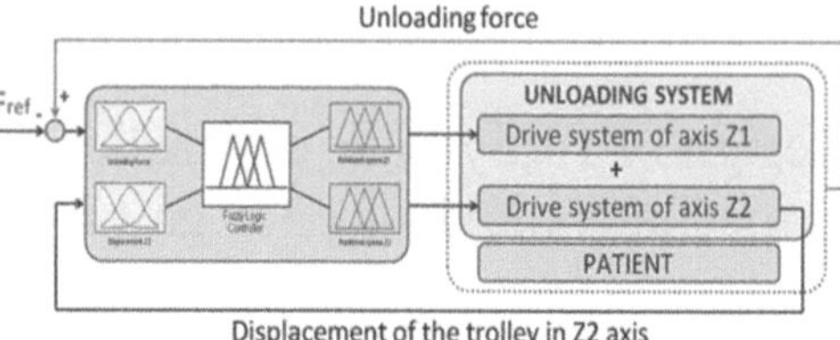

Figure 2. Scheme of control system for unloading unit.

The cooperation of these two servo drives should ensure realization of two basic assumptions. First, maintain a constant value of a relieving force. Second, keep the drive Z2 actuators within the allowable range between limit switches.

4 OPTIMIZATION OF A CONTROL ALGORITHM

The operation of classic fuzzy controller is divided into three stages:

- Fuzzification—input signals (analog or digital) are converted from scalar values to values described by membership functions,
- Set of rules—based on fuzzy input signals created in the process of fuzzification, a block with a defined set of rules generates fuzzy output signals
- Defuzzification—fuzzy output signals are converted into scalar values.

For the controlled which is being designed, the set of rules has been prepared on the basis of the desired behavior of the system at the given configurations of input signals. The controlling rules have not undergone further modification.

4.1 *The variables in the optimization process*

In order to optimize the operation of the fuzzy controller, for each input and output signal, membership functions have been defined by the equation:

$$Mf(x,a,b)=\begin{pmatrix} 0, & x<a \\ \dfrac{2(x-a)}{b}, & a<x<a+\dfrac{b}{2} \\ \dfrac{a+b-x}{a+\dfrac{b}{2}}, & a+\dfrac{b}{2}<x<a+b \\ 0, & a+b<x \end{pmatrix} \tag{1}$$

where x = the value of input signal; a = the starting point of membership function; i b = range of membership functions.

For each of the signals, seven membership functions have been defined. Because of the fact that the ranges of input and output signals were symmetrical with respect to zero, it has been assumed that the membership functions will also be arranged symmetrically with respect to zero point. This allowed to reduce the number of optimized variables twice. In the case of input functions, a fixed value of optimal range has been adopted—describing the values considered as accepted. Another simplification was to define the fixed ranges of membership functions describing the output signals (rotational speeds of Z1 and Z2 engines). The distribution of these functions resulted from the maximum rotational speeds of engines.

In connection with the presented simplifications, optimized parameters related only to membership functions describing input signals. The number of optimized variables was 12.

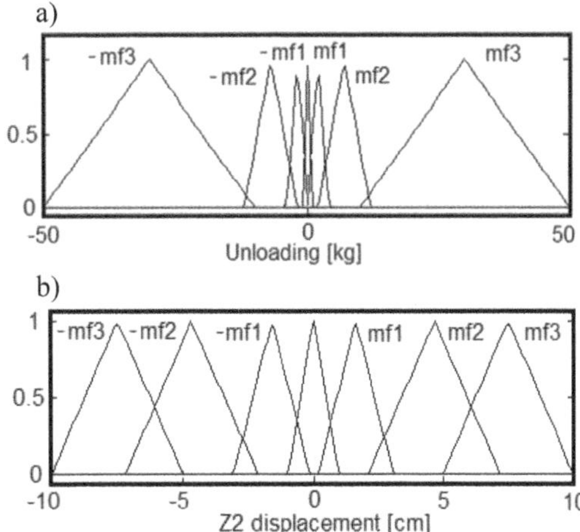

Figure 3. Membership functions for the fuzzification of input signals: a) the relieving force error, b) movement of actuator element of Z2 drive.

4.2 *The optimization process*

Due to a large scope of the search for the best variables, to optimize the values of described parameters of the fuzzy controller, a genetic algorithm has been selected.

After drawing initial population, each individual generated file *.fis with implemented variable values of the given individual, and activated a numerical model of the device. Based on the results of numerical simulation it was possible to calculate the adaptation function, defined as:

$$\min(y \in \Omega) \quad f_c(y) = w_i \cdot |\Delta F_t| + w_k \cdot |Z2_t| \tag{2}$$

where Ω = the area of possible solutions in the space of input signals; y = the vector of optimized variables; w = weight matrix of criterion functions, such as $0 < w_{i,k} < 1$, and $w_i + w_k = 1$; ΔF_t = the error value vector of the relieving force in every computational step of the simulation; and $Z2_t$ = the movement of the actuator element of Z2 drive in each computational step of the simulation.

The numerical model of the device has been developed in Matlab/Simulink environment. The model parameters have been estimated on the basis of experimental investigations.

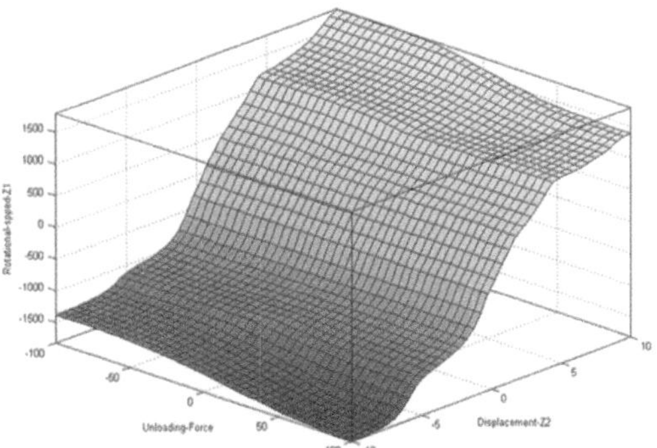

Figure 4. The output signal controlling the Z1 engine speed presented as a surface.

4.3 *The results of the optimization*

The carried out optimization calculations allowed to formulate the membership function of input signals, which defined small (membership function mf1), medium (mf2) and large (mf3) deviation from the back-off value.

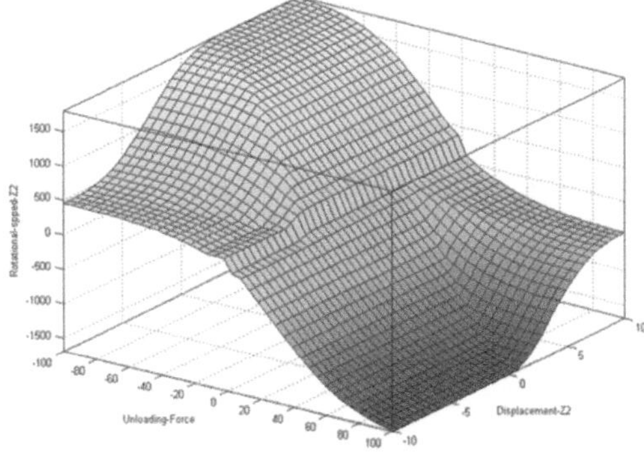

Figure 5. The output signal controlling the Z1 engine speed presented as a surface.

The obtained values of control signals, after the process of defuzzification, have been shown as surfaces in the function of input signals.

The obtained setting values of the control system with the fuzzy controller have been implemented to the numerical model of the device. During the simulation tests, the analysis of the patient's passage through an eight-centimeter obstacle has been performed. The following results of the simulation tests show: set enforcement in the vertical direction (Fig. 6), the value of the relieving force acting on the patient (Fig. 7) and the movement of actuator element in a dynamic compensation system Z2 (Fig. 8). During carried out numerical simulations, the set value of the relieving force was 300 N.

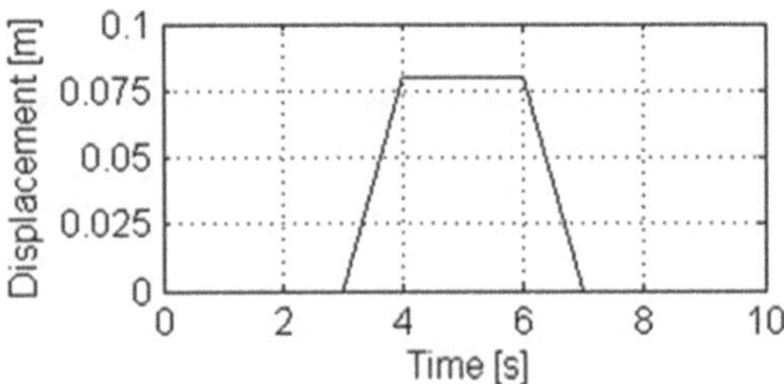

Figure 6. The set enforcement in the vertical direction as a function of time.

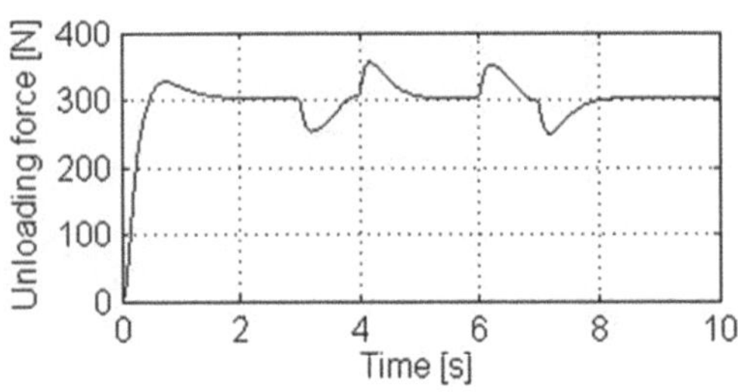

Figure 7. Change of unloading force values as a function of time.

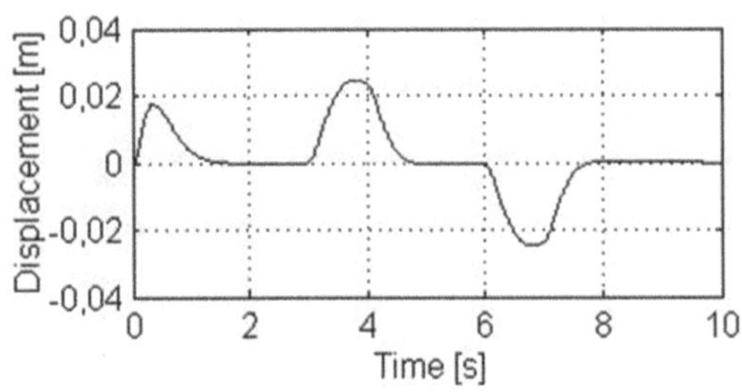

Figure 8. Displacement of the pulley in the Z2 axis as a function of time.

5 CONCLUSIONS

The article presents the system controlling the relief system, installed in mechatronic device for the gait re-education, based on fuzzy controller. The results of the optimization of membership functions for input signals, and the results of numerical calculations presenting the operation of the device have been illustrated. Simulation tests show that the relief system together with converted control, properly implements the basic assumptions—it maintains a constant value of the relieving force, and the actuator element of drive moves within the acceptable range of motion and regardless of the height at which the patient is, it returns to the starting position. In relation to the previously developed control of relief system with one active drive (Duda et al., in press), ensuring cooperation between both engines included in the relief system, increased the cumulative error of the relieving force, but it did not change the maximum error value. The relieving force of the patient fluctuates within ± 40 N in relation to the set value. Therefore, the proposed control system allows for the performance of exercises in relieving, which require considerable movement in the vertical direction from the patient. This is particularly important e.g. in climbing stairs training—one of the basic functions of a person's everyday life.

REFERENCES

Behrman, A.L. & Harkema, S.J. 2000. Locomotor training after human spinal cord injury: a series of case studies. *Physical Therapy* 80(7): 688–700.

Duda S., Gasiorek D., Gembalczyk G., Kciuk S., Mezyk A. Mechatronic device for loco motor training. *Acta Mechanica et Automatica* (in press).

Duda, S., Gembalczyk G., Jurkojć J., Kciuk S., Michnik R. 2015. Mechatronic solution of components cooperation in the device for gait reeducation. In Josep M. Font-Llagunes (ed.), *Multibody dynamics 2015; ECCOMAS Thematic Conference, Barcelona, 29 June—July 2, 2015.*

Duda, S., Kawlewski, K., Gembalczyk G. 2015. Concept of the System for Control over Keeping Up the Movement of a Crane. *Solid State Phenomena* 220/221: 339–344.

Marchal-Crespo, L. & Reinkensmeyer, D.J. 2009. Review of control strategies for robotic movement training after neurologic injury. *Journal of neuroengineering and rehabilitation* 6(1): 20.

Pratt, J., Krupp, B., Morse. Ch. 2002. Series elastic actuators for high fidelity force control. *Industrial Robot: An International Journal* 29: 234–241.

Smoczek, J. 2014. Fuzzy crane control with sensorless payload deflection feedback for vibration reduction. *Mechanical Systems and Signal Processing* 46: 70–81.

Tu, K.Y., Lee, T.T., Wang, W.J. 2000. Design of a multilayer fuzzy logic controller for multi-input multi-output systems. *Fuzzy sets and systems* 111: 199–214.

Advances in Mechanics: Theoretical, Computational and Interdisciplinary Issues – Kleiber et al. (Eds)
© 2016 Taylor & Francis Group, London, ISBN 978-1-138-02906-4

The reliability assessment for steel industrial building

A. Dudzik & U. Radoń
Faculty of Civil Engineering and Architecture, Kielce University of Technology, Kielce, Poland

ABSTRACT: The study presents a probabilistic approach to the problems of static and stability analyses of a steel industrial building. Structural design parameters were defined as deterministic values and random variables. The latter were not correlated. The Hasofer-Lind index was used as a reliability measure. The FORM method was employed as a primary research tool. In order to validate the correctness of the computations, SORM, Monte Carlo and Importance Sampling methods were used. The sensitivity of the reliability index to random variables was defined. In the examples of reliability analysis the STAND program was used.

1 THE DEVELOPMENT OF RELIABILITY

Many of the current approaches to achieving structural safety evolved over many centuries. As time passed, by the laws of nature and structural behavior became better understood. The actual load and design parameters often are irregular, accidental, in other words—random. Mathematical theories of material and structural behavior evolved, providing a more rational basis for structural design. In turn, these theories furnished the necessary framework in which probabilistic methods could be applied to quantify structural safety and reliability.

The first mathematical formulation of the structural safety problem can be attributed to Mayer (1926), Streletzkii (1947) & Wierzbicki (1936). They recognized that load and resistance parameters are random variables and therefore, for each structure, there is a finite probability of failure. Their concepts were further developed by Freudenthal (1956). The formulations involved convolution functions that were too difficult to evaluate by hand.

The work of Hasofer & Lind (1974) is recognized as the first important step towards the contemporary methods that make it possible to effectively and accurately estimate structural safety. Hasofer and Lind developed the concept of localization of the so-called "design point", i.e. such realization of random variables from the failure surface which corresponds to the greatest value of the probability density function. With the linearization of the limit state function at the design point, it is possible to obtain a measure of reliability which is invariant due to the equivalent formulations of the boundary condition, i.e. the so-called Hasofer-Lind reliability index. The lack of invariance was the basic shortcoming of the previously applied Cornell index. It was difficult to use this measure of reliability for comparison of the degrees of safety of different structures. The concept of Hasofer-Lind index was used in 1978 by Rackwitz & Fiessler (1978). Additionally, they applied transformations of independent random variables with arbitrary probability distributions to standard normal variable and they proposed an algorithm of the approximation of the design point. Hohenbichler & Rackwitz (1981) proposed the utilization of Rosenblatt's transformation (1952) to the transformation of dependent random variables to standard space. Rosenblatt's transformation and Nataf's transformation, applied for the first time by Der Kiureghian & Liu (1986), are most frequently used at present. Those make it possible to analyze reliability problems both when the joint probability density function of random variables as well as the marginal probability density functions of variables and correlation matrix is known. Approximation of the limit state function at the design point by the function of the first degree leads to the method of first-order reliability analysis (FORM). Probabilistic methods allow the quantitative assessment of structural reliability. Depending on the manner of computing the failure probability, the methods can be classified as approximation and simulation ones by Melchers (2007).

2 AIM OF PAPER

The study presents a probabilistic approach to the problems of static and stability analyses of a steel industrial hall. Structural design parameters were defined as deterministic values and random variables. The latter were not correlated. The Hasofer-Lind index was used as a reliability measure. The FORM method, which is one of the approximation methods, was a basic tool employed in the study. In order to validate the correctness of the computations, SORM, Monte Carlo and Importance Sampling methods were used. The reliability index sensitivity to random variables was defined. In the examples presented below, the STAND software

(2009) was applied to the reliability analysis. That software was developed at the Institute of Fundamental Technological Research of the Polish Academy of Sciences by Stocki et al. (2009).

3 NUMERICAL EXAMPLES

3.1 *Probabilistic approach to the static analysis of steel frame*

The reliability analysis was carried out for a steel industrial hall located in Poland. The basic supporting structure of the building consists of steel frames (Fig. 1). All members are made of S235 JR steel, the yield point of which is $f_y = 235$ MPa and modulus of elasticity is $E = 210$ GPa. Columns are made from HEB 450 I-shaped beams, and girts from HEA 700 I-shaped beams. The reliability analysis was performed for the most unfavourable load configuration, i.e. permanent load and uniform snow load (Fig. 1).

The limit functions imposed on the load bearing structure are displacement constraints related to the serviceability limit state and stress constraints related to the ultimate limit state. The initial analysis of the displacement state and internal forces, performed with the *Autodesk Robot Structural Analysis Professional 2014* software, made it possible to locate the sites at which their extreme values occur. Then, using the Finite Element Method, the formulas for the maximum vertical displacement and the maximum bending moment were determined in module for symbolic calculations with the *Mathematica* software. Limit functions were defined as functions of random variables grouped in vector: $\boldsymbol{X} = \{X_1, X_2, X_3, X_4, X_5, X_6, X_7\}$.

Below, random quantities are specified:

X_1—permanent load from the covering self-weight from 1 m of the roof plane,
X_2—permanent load from the cladding self-weight from 1 m of the curtain wall,
X_3—permanent load from a fragment of the roof and wall at the eave reduced to a focused force acting on the structure column,
X_4—variable load from uniform snow load imposed on the roof plane,
X_5—variable load from a fragment of the roof and wall at the eave reduced to a focused force acting on the structure column,

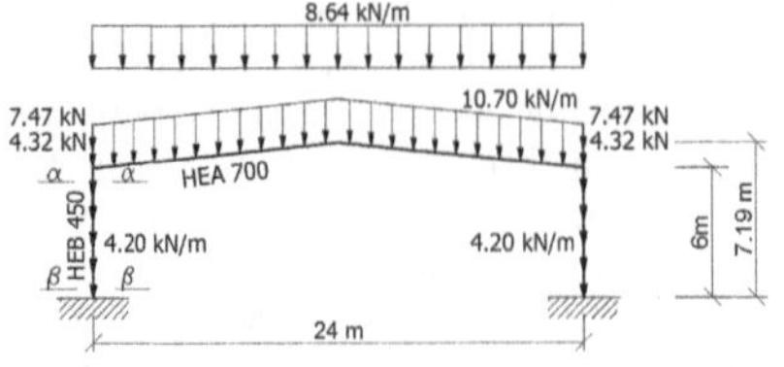

Figure 1. Supporting structure of a steel industrial building.

X_6—elastic modulus for S235 JR steel,
X_7—yield point for S235 JR steel.

The values of coefficients of variation were selected on the basis of statistical studies of the building strength and buildings materials and products from work of Gwóźdź & Machowski (2011). For variables with load range coefficients of variation were adopted in accordance with recommendations of Probabilistic Model Code (2001).

Member section areas, member inertia moments, and lengths of columns and girts were assumed to be deterministic quantities. Limit state functions were formulated. These describe the following:

- serviceability limit state

$$g_1(\boldsymbol{X}) = 1 - \frac{|q_{\max}}{q_{dop}} \tag{1}$$

where $q_{dop} = L/250$ ($q_{dop} = 0.096$ m) denotes permissible displacement in accordance with EN standard;

$$q_{\max} = (634878X_1 + 825.69X_2 + 275.23X_3 + 631377X_4 + 275.23X_5)/X_6$$

- ultimate limit state

$$g_2(\boldsymbol{X}) = 1 - \frac{\left|\dfrac{M_{\max}}{W_s}\right|}{f_y} \tag{2}$$

where W_s = denotes the elastic index of the column strength.

For the HEB 450 profile, value $W_s = 0.00624$ m³.

$$M_{\max} = 25.69X_1 - 0.11X_2 - 0.04X_3 + 25.53X_4 - 0.04X_5$$

The values of the Hasofer-Lind reliability index were determined with the FORM method, and for the sake of comparison, with other methods, i.e. SORM, Monte Carlo and Importance Sampling. The STAND software was used. The results are presented in Table 2. Relative error of the Hasofer-Lind reliability index was estimated assuming that the reference is the Monte Carlo method (Table 3).

In addition, graphs that show the sensitivity of the reliability index to random variables for both limit functions were plotted (Fig. 2).

3.2 *Probabilistic approach to the stability analysis of steel frame*

The frame analyzed in Example 3.2. was made of the material having the same characteristics as this in Example 3.1. The external loads, to which the frame

Table 1. Description of the random variables.

X_i	Probability density function	μ_{Xi}	σ_{Xi}	v_{Xi} [%]
X_1	Normal	10.70 kN/m	1.07 kN/m	10
X_2	Normal	4.20 kN/m	0.42 kN/m	10
X_3	Normal	7.47 kN	0.745 kN	10
X_4	Normal	8.64 kN/m	0.864 kN/m	10
X_5	Normal	4.32 kN	4.32 kN	10
X_6	Normal	$210 \cdot 10^6$ kN/m²	$84 \cdot 10^5$ kN/m²	4
X_7	Normal	$235 \cdot 10^3$ kN/m²	$18.8 \cdot 10^3$ kN/m²	8

Table 2. Comparative analysis of reliability index computations.

	Type of analysis	
The reliability methods	Serviceability limit state	Ultimate limit state
FORM	1.952	3.794
SORM	1.949	3.702
Monte Carlo	1.948	3.719
Importance Sampling	1.922	3.816

Table 3. The relative error of determining the reliability index as regards the Monte Carlo method.

	The relative error [%]	
The reliability methods	Serviceability limit state	Ultimate limit state
FORM	0.2	2.0
SORM	0.1	0.5
Importance Sampling	1.3	2.6

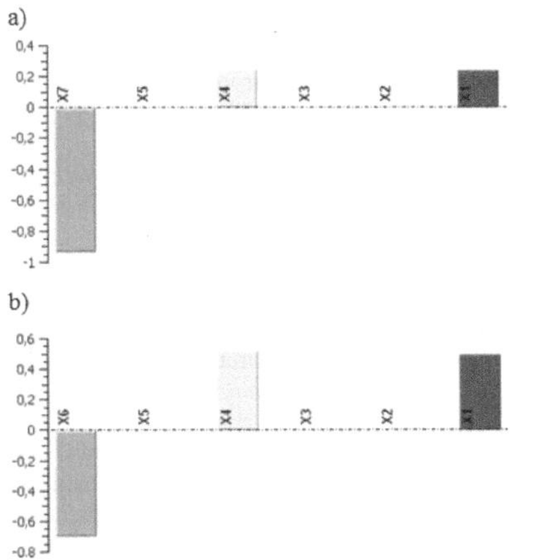

Figure 2. Sensitivity of the reliability index β to random variables for a) ultimate limit state b) serviceability limit state.

was subjected, were also the same, but the columns were additionally loaded with a compressive force S (Fig. 3). The latter can be a reaction, transferred from the external structure above the building roof. The value of the critical load was determined using the Finite Element Method with the *Mathematica* software. In the computations, the geometric stiffness matrices for compression elements were used. The value of critical force obtained was $S_{kr} = 3.13*10^4$ kN.

The probabilistic approach to the stability analysis involved the determination of the reliability index value for columns loaded with the compressive force, which constituted a certain percentage value of the critical load. In the reliability analysis, deterministic and probabilistic variables were used. These were defined in Example 3.1. When the structure was loaded with successive compressive forces, the new limit functions were defined. The general formula for the limit function can be represented as follows:

- for the serviceability limit state, in accordance with Equation (1), where q_{max} describes the vertical displacement of roof ridge node,
- for the ultimate limit state, in accordance with Equation (2), where M_{max} describes the bending moment at the column-to-foundation connection (cross-section β-β).

Tables 4 and 5 present the values of the Hasofer-Lind reliability index which were determined with the FORM, SORM, Monte Carlo and Important Sampling methods for the serviceability and ultimate limit state, respectively. The bending moment, for which the static analysis of reliability was performed, is located in the column-to-beam connection (cross-section α-α)., and it is the highest value of the bending moment in the analyzed structure. As regards the stability analysis, it is important to take into account the effect of compressive forces on the magnitude and distribution of cross-sectional forces. In frame structures, the greatest impact of these forces on the values of the moments in column-to-foundation connection (cross-section β-β) is observed. The value of this bending moment is twice smaller than that for which the static analysis was performed. Therefore, the stability analysis for the ultimate limit state is restricted to presenting only the extreme cases—the lack of compressive

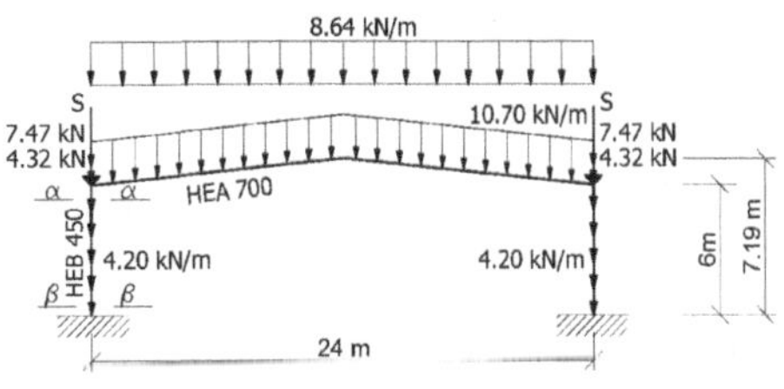

Figure 3. Supporting structure of a steel industrial building in stability analysis.

Table 4. Comparative analysis of the reliability index computations for different values of the critical load for the serviceability limit state.

Value of the compressive force	Reliability method			
	FORM	SORM	Monte Carlo	Importance Sampling
0% S_{kr}	1.952	1.949	1.948	1.922
70% S_{kr}	1.515	1.514	1.518	1.511
80% S_{kr}	1.446	1.445	1.446	1.452
95% S_{kr}	1.339	1.339	1.321	1.332

Table 5. Comparative analysis of the reliability index computations for different values of the critical load for ultimate limit state.

Value of the compressive force	Reliability method			
	FORM	SORM	Monte Carlo	Importance Sampling
0% S_{kr}	4.937	4.930	5.020	4.999
95% S_{kr}	4.784	4.776	5.002	4.835

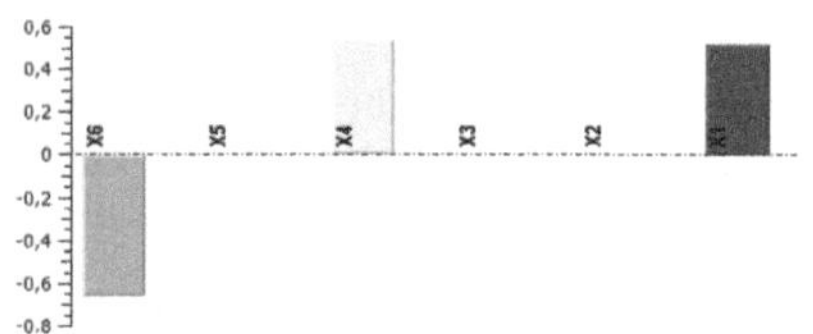

Figure 4. Sensitivity of the reliability index β to random variables for serviceability limit state.

forces and force load of 95% S_{kr}. A graph was plotted to show the sensitivity of the reliability index to random variables for limit functions describing the serviceability limit state was plotted (Fig. 4).

4 CONCLUSIONS

The analysis of the results demonstrates that the FORM method is good enough and definitely much simpler to apply. The relative error of the reliability index was estimated for individual methods at the assumption that the Monte Carlo (where sample size is 10000) served as the reference method. In Example 3.1. concerning the ultimate limit state, the error amounted to: for the FORM method—2%, for the SORM method—0.5%, for the Importance Sampling method—2.6%. As regards the serviceability limit state, the error values were as follows: for the FORM method— 0.2%, for the SORM method—0.1%, and for the Importance Sampling method—1.3%. Another important component of the study was to investigate the sensitivity of the reliability index to changes in the probabilistic characteristics of the random variables under consideration. If the reliability index sensitivity due to the random variable X_i is low when compared with other variables, it can be stated that the impact of this variable on failure probability is small. Therefore, in successive computations it can be treated as a deterministic parameter.

The probabilistic approach to the stability analysis shows the effect of large compressive forces on the reliability index value. In the serviceability limit state, between the extreme cases (the lack of compressive forces and force load of 95% S_{kr}), the reliability index was observed to decrease to 31% but large compressive forces doesn't effect on sensitivity analysis (Figs. 2b, 4). The FORM method is good enough and definitely much simpler to apply. However, it has to be remembered at all times that the FORM method yields the best results when only one design point exists, the limit state function is not strongly nonlinear and it is differentiable.

REFERENCES

Der Kiureghian, A. & Pei-Ling Liu 1986. Structural reliability under incomplete probability information. *Journal of Engineering Mechanics*. ASCE. 112 (1): 85–104.

EN 1993: Eurocode 3: *Design of steel structures*—Part 1–1: *General rules and rules for buildings.*

Gwóźdź, M. & Machowski, A. 2011. *Wybrane badania i obliczenia konstrukcji budowlanych metodami probabilistycznymi*. Kraków: Wydawnictwo Politechniki Krakowskiej.

Hasofer, A.M. & Lind, N.C. 1974. Exact and invariant second moment code format. *Journal of the Engineering Mechanics Division*. ASCE. 100: 111–121.

Hohenbichler, M. & Rackwitz, R. 1981. Non-normal dependent vectors in structural safety. *Journal of the Engineering Mechanics Division*. ASCE. 107: 1227–1238.

JCSS, 2001. *Probabilistic Model Code.* Joint Committee of Structural Safety.

Knabel, J. Kolanek, K. Nguyen Hoang, V. Stocki, R. Tauzowski, P. 2008. Structural reliability analysis using object oriented environment STAND. *In proc. of the 36th Solid Mechanics Conference. 9–12 September 2008*. Gdansk. Poland.

Mayer, M. 1926. *Die Sicherheit der Bauwerke und ihre Berechnung nach Grenzkraften statt nach zulassigen Spannungen.* Berlin: Springer-Verlag.

Rackwitz, R. & Fiessler, B. 1978. Structural reliability under combined random load sequences. *Computers & Structures.* 9 (5): 489–494.

Rosenblatt, M. 1952. Remarks on a Multivariate Transformation. *The Annals of Mathematical Statistic* 23 (3): 470–472.

Stocki, R. Kolanek, K. Knabel, J. Tauzowski, P. 2009. FE based structural reliability analysis using STAND environment. *Computer Assisted Mechanics and Engineering Sciences*. 16: 35–58.

Streletzkii, N.S. 1947. *Statistical basis for evaluation of the structural safety factor*. Moscow: State Publishing House for Buildings. (in Russian).

Advances in Mechanics: Theoretical, Computational and Interdisciplinary Issues – Kleiber et al. (Eds)

Optimal design of eigenfrequencies for functionally graded piezoelectric plate by two-scale model and harmony search

G. Dziatkiewicz
Silesian University of Technology, Gliwice, Poland

ABSTRACT: The optimization problem of eigenfrequencies for the functionally graded piezoelectric plate is considered. The eigenvalue problem of the plate is modeled by the state-space approach. The discrete multi-layer model of the plate is considered. For the discrete model the transfer matrix of the state-space approach is given in a closed form by the exponential matrix function. To determine the effective properties of the FGM piezocomposite the micromechanical model is applied. The effective properties are determined by the Mori-Tanaka approach. The optimization problem of the plate eigenfrequencies is formulated across the micro- and macroscale and the harmony search method is applied to solve it.

1 PROBLEM FORMULATION

Materials with a Functionally Graded Microstructure (FGM) belong to the modern and rapidly growing class of smart structures (Taya 2005). However, the modelling problem of the FGM piezoelectric materials requires the special treatment due to the anisotropy of the properties and the coupling of the electric and mechanical fields (Chen et al. 2004, Fedelinski et al. 2014, Taya 2005). Also the influence of the microstructure must be taken into account by using the micromechanical models, hence the eigenvalue problem of the FGM piezoelectric plate can be treated as a two-scale model. To model the plate at the macroscale the state-space approach is applied (Chen et al. 2004). The Mori-Tanaka micromechanical model (Gross & Seelig 2011) is used to obtain the effective properties of plate at the microscale. In the present work the soft computing method, namely the harmony search method (Lee & Geem 2005) is used to optimize the eigenfrequency function of the FGM piezoelectric plate. The simply-supported piezoelectric FGM plate is considered. The following unconstrained optimization problem is formulated:

$$\max_{\mathbf{x}\in D} \omega_{111}^{OC}(\mathbf{x}),\ \mathbf{x}^L \leq \mathbf{x} \leq \mathbf{x}^U, \tag{1}$$

where ω_{111} is the non-dimensional first eigenfrequency for the mode (1,1) obtained with the Open-Circuit (OC) electrical boundary conditions for the plate. The design variables vector $\mathbf{x}$ contains the parameters of the structure related to the macro- or microscale. On the vector $\mathbf{x}$ the box constraints are imposed as shown in (1). The optimization problem requires the macroscale model for the eigenvalue problem of the plate and also the microscale model which is able to take into account the microstructure influence on the macroscale effective properties of the plate.

2 MACROSCALE MODEL

The linear constitutive equations of piezoelectric orthotropic material can be expressed by the generalized quantities (Chen et al. 2004):

$$\mathbf{\Sigma} = \mathbf{C}(\mathbf{U}). \tag{2}$$

The generalized stress vector $\mathbf{\Sigma} = [\sigma_{11}\ \sigma_{22}\ \sigma_{33}\ \sigma_{23}\ \sigma_{13}\ \sigma_{12}\ D_1\ D_2\ D_3]^T$ contains the mechanical stress tensor components σ_{ij} and the electric displacement vector components D_i; the generalized displacements $\mathbf{U} = [u_1\ u_2\ u_3\ \phi]^T$ are the displacement vector components u_i and the electric potential ϕ. The operator $\mathbf{C}(\cdot)$ has various form depending on the anisotropy. The physical properties of the plate are described by the elastic, piezoelectric and dielectric tensor components. The mentioned physical properties are dependent on the microstructure of the plate and the spatial coordinates. For plate problems usually the spatial dependency is restricted to the coordinate x_3 measured through the thickness of the plate. The application of the equation of motion allows to obtain the so-called state equation of the plate (Chen et al. 2004):

$$\partial\mathbf{V}/\partial x_3 = \mathbf{M}(\mathbf{V}), \tag{3}$$

where the state vector $\mathbf{V} = [u_1\ u_2\ \sigma_{33}\ D_3\ \sigma_{13}\ \sigma_{23}\ u_3\ \phi]^T$ and the differential operator $\mathbf{M}(\cdot)$ can be found in (Chen et al. 2004). The solution (3) is known from the control theory for the linear time-varying system (Logemann & Ryan 2014):

$$\mathbf{V}(x_3) = \mathbf{T}(x_3, 0)\mathbf{V}(0), \tag{4}$$

where the state transition operator $\mathbf{T}(x_3, 0)$ can be expressed by the Peano-Baker series (Zhong & Yu 2006):

$$\mathbf{T}(x_3,0)=\mathbf{I}+\int_0^{x_3}\mathbf{M}(s_1)ds_1+\int_0^{x_3}\mathbf{M}(s_1)\int_0^{s_1}\mathbf{M}(s_2)ds_2ds_1+\ldots$$
$$+\int_0^{x_3}\mathbf{M}(s_1)\int_0^{s_1}\mathbf{M}(s_2)..\int_0^{s_{j-1}}\mathbf{M}(s_j)ds_j..ds_2ds_1+\ldots \tag{5}$$

If the operator $\mathbf{M}(\cdot)$ does not depend on the coordinate x_3, the series in (5) can be summed and the state transition operator $\mathbf{T}(x_3, 0)$ is expressed by the exponential matrix function (Chen et al. 2004, Logemann & Ryan 2014):

$$\mathbf{T}(x_3,0) = e^{\mathbf{M}x_3}. \tag{6}$$

To use the solution (6) the discrete model of the plate is considered as shown in (Fig. 1).

It is assumed that the physical properties are constant in each layer of the plate. In this case, the state equation are given by (Chen et al. 2004):

$$\frac{d\bar{\mathbf{V}}(\zeta)}{d\zeta} = \bar{\mathbf{M}}\bar{\mathbf{V}}(\zeta). \tag{7}$$

In (7) the non-dimensional quantities are applied and the operator $\mathbf{M}(\cdot)$ takes the form of the algebraic operator and it can be found in (Chen et al. 2004). Equation (7) is valid for each layer, hence the solution for the whole plate can be obtained by using the global state transition matrix (Chen et al. 2004):

$$\mathbf{T} = \prod_{j=N}^{1}\mathbf{Q}_j,\ \mathbf{Q}_j = \exp\left[\bar{\mathbf{M}}\left(\zeta_j - \zeta_{j-1}\right)\right]. \tag{8}$$

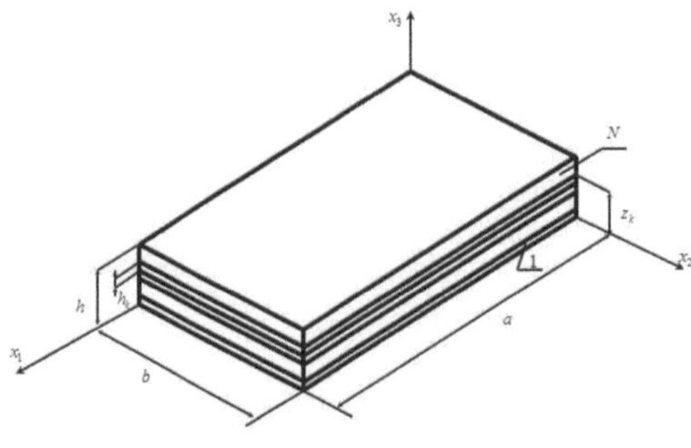

Figure 1. Discrete model of the FGM piezoelectric plate.

where: ζ_j = is the non-dimensional coordinate of the j-th layer. The ansatz which fullfils the simply supported boundary conditions can be found in (Chen et al. 2004). For the eigenvalue problem the up and the bottom surface of the plate are traction-free, hence introducing such boundary conditions and taking into account the open-circuit electrical boundary conditions one can obtain the transcendental equation for the each mode shape of non-dimensional eigenvalues (Chen et al. 2004). To solve the mentioned equation the bisection method with the automatic detection of the isolation interval is applied.

3 MICROSCALE MODEL

The physical properties expressed by the appropriate tensor components in the operator $\mathbf{M}(\cdot)$ are strongly dependent on the microstructure of the material of the plate. To assure the properties of the FGM material the gradation of physical properties is obtained by changing the parameters of the microstructure (Fedelinski et al. 2014, Gross & Seelig 2011, Taya 2005), e.g. the volume fractions of the constituents and the morphology of the constituents.

To calculate the effective properties of the composite the generalized Mori-Tanaka approach is used. For the electroelastic composite the effective moduli tensor $\mathbf{L}^{eff}$ is given by (Gross & Seelig 2011, Taya 2005):

$$\mathbf{L}^{eff} = \mathbf{L}_0 + \sum_{r=1}^{N} f_r(\mathbf{L}_r - \mathbf{L}_0) : \mathbf{A}_r, \tag{9}$$

where $\mathbf{L}_0$ and $\mathbf{L}_r$ = denote the moduli of the composite matrix and the r-th filler, respectively. The volume fraction of the r-th filler is denoted by f_r. The generalized global strain concentration tensor $\mathbf{A}_r$ can be calculated by the Eshelby approach, in this case it is called the local generalized strain concentration tensor $\mathbf{T}_r$ (Gross & Seelig 2011):

$$\mathbf{T}_r = \left[\mathbf{I} + \mathbf{S}_r : \mathbf{L}_0^{-1} : (\mathbf{L}_r - \mathbf{L}_0)\right]^{-1}, \tag{10}$$

where $\mathbf{I}$ = is the unit tensor and $\mathbf{S}_r$ = is the famous Eshelby tensor, it is worth to note that the presented formulation is valid for the ellipsoidal inclusion. The Mori-Tanaka approach takes into account the interaction between the particles inside the matrix, hence the effective properties can be obtained by the following equation (Gross & Seelig 2011):

$$\mathbf{L}^{eff} = \mathbf{L}_0 + \sum_{r=1}^{N} f_r(\mathbf{L}_r - \mathbf{L}_0) : \mathbf{T}_r : \left[\sum_{n=1}^{N} f_n\mathbf{T}_n\right]^{-1}. \tag{11}$$

In (11) the $\mathbf{S}_r$ is the most important quantity, but for the general piezoelectric medium the electroelastic Eshelby tensor cannot be calculated analytically. In the present work, the numerical method is applied and the components of the electroelastic Eshelby tensor are given by the following integral formulation (Huang et al. 1998):

$$S_{MnAb} = \frac{L^0_{iJAb}}{4\pi}\begin{cases}\frac{1}{2}\int_{-1}^{1}\int_{0}^{2\pi}[J_{mJin}(\mathbf{z})+J_{nJim}(\mathbf{z})]\,d\theta d\xi_3 \\ \int_{-1}^{1}\int_{0}^{2\pi} J_{4Jin}(\mathbf{z})\,d\theta d\xi_3,\ M=4\end{cases} \tag{12}$$

details are given in (Huang et al. 1998). To calculate the integrals in (12) the 10-point Gauss integration quadrature is applied. Manufacturing process requirements establish the certain way of gradation the physical properties of the FGM piezocomposites. In the present work the discrete version of the power gradation law is used. The value of the property P_i in the i-th layer of the plate can be calculated by the following equation (Taya 2005):

$$P_i = (P_b - P_a)\left(\frac{i-1}{N_s-1}\right)^m + P_a. \tag{13}$$

For the rainbow-type gradation P_a and P_b denote the value of the property P in the first and the last layer of the plate, respectively. In this case N_s is the number of layers. For the symmetric gradation with respect to the middle layer P_a and P_b denote the value of the property P in the first and the middle layer of the plate, respectively. The character of the gradation is described by the exponent m. The property P changed through the thickness of the plate may be related to the micro—or macroscale parameter of the plate model, e.g. the porosity, the volume fractions of composite constituents or the thickness of the layers of the plate.

4 OPTIMIZATION METHOD

To solve the problem (1) the harmony search method is applied. The harmony search method (HS) was developed on the basis of a improvisation process observed in music, especially in jazz and blues (Lee & Geem 2005). The effort to search and to find the harmony in music is analogous to searching and to finding the optimum in an optimization process. The HS algorithm takes a few elements from the jazz improvisation process: usage of harmony memory, pitch adjusting and randomization (Lee & Geem 2005). The unknown optimum is the searched best harmony. The harmonies contain design variables and change during the optimization process. The new harmonies are created from harmony memory and by using the pitch adjustment and randomization process. For the new harmonies the objective function is evaluated and better ones exchange harmony memory. The main advantage of the HS method is the fact that this approach does not need any information about the gradient of the objective function and gives a great probability of finding the global optimum (Lee & Geem 2005). In the present work the self-adaptive global variant of the harmony search method (GS-HS) (Wang & Huang 2010) was applied.

5 NUMERICAL EXAMPLES

The maximization of the fundamental eigenfrequency for the porous $BaTiO_3$ plate is considered. The $BaTiO_3$ ceramic material has controlled porosity as in (Taya 2005). The plate is $a = 100$ mm long and $b = 20$ mm wide. It is assumed uniform thickness of the layers and the total thickness of the plate is equal to $h = a/5$. The FGM bimorph structure is considered with the symmetrical pattern of the porosity gradation. The design variables are: the porosity in the first layer, the porosity in the middle layer and the exponent of the gradation law. The lower and upper bound of the porosity are equal to 0.02 and 0.25, respectively. The exponent m takes values from 0 to 10. The voids are spheroidal. The harmony memory size of the GS-HS algorithm is equal to 20, the harmony memory accepting rate is equal to 0.95. The number of iteration set to 250, and the number of replication of the algorithm is equal to 30. Models with different number of layers were considered. The models with 5, 9, 13, 17 and 21 layers were considered. In Figure 2 the mean of the best solutions with the standard deviation versus the number of layers of the plate is shown.

It can be seen that the best solution is statistically insensitive with respect to the number of layers in

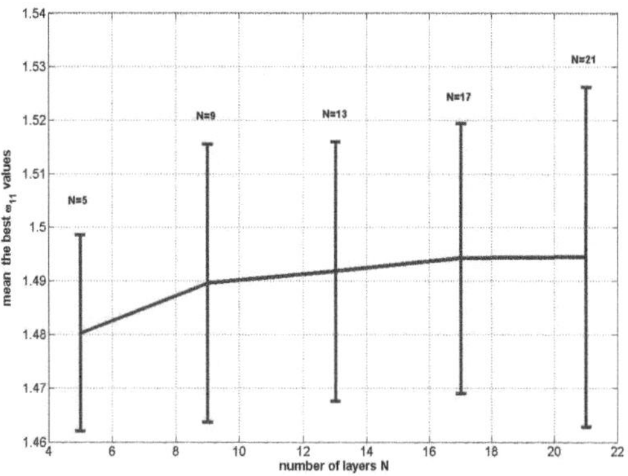

Figure 2. The best solutions, influence of the number of layers.

the plate model. The best obtained optimal solutions are shown in Figure 3. The best solutions tend to the sandwich-type plate. The outermost layers are highly porous, the values of porosity are almost equal to the upper bounds, but the inner layers are of low-porosity almost equal to the lower bounds of the appropriate design variables. The optimal values of the gradation exponent are low. To investigate the influence of the exponent m, the several calculations were performed for the best obtained models. It can be observed that the value of the objective function is practically insensitive with respect to the gradation exponent if m is smaller than 10^{-4}.

In the next example the free gradation pattern of the porosity is considered. The number of layer is equal to $N = 7$, hence 7 design variables denote the porosity in each layer of the plate. The porosity bounds are the same as in the previous examples. In this case the vector of design variables for the best solution is equal to $\boldsymbol{x}^{best} = [0.243\ 0.054\ 0.043\ 0.025\ 0.054\ 0.037\ 0.031]$; in the first layer the porosity is almost on the upper bound, the porosity for the rest of the layers is close to the lower bound of the design variables. Although the porosity pattern is different than in the previous example, the value of the objective function is close to the previous one. The comparison between global optimization methods is the challenging task. To compare the GS-HS method with other global optimization methods the several tests were performed with the Simulated Annealing algorithm and the Genetic Algorithm. The basic performance metrics can be compared (Poteralski et al. 2013), however in the present work the nonparametric statistical test was applied (Derrac et al. 2011), namely the Wilcoxon signed ranks test. The differences were not detected between the compared algorithms.

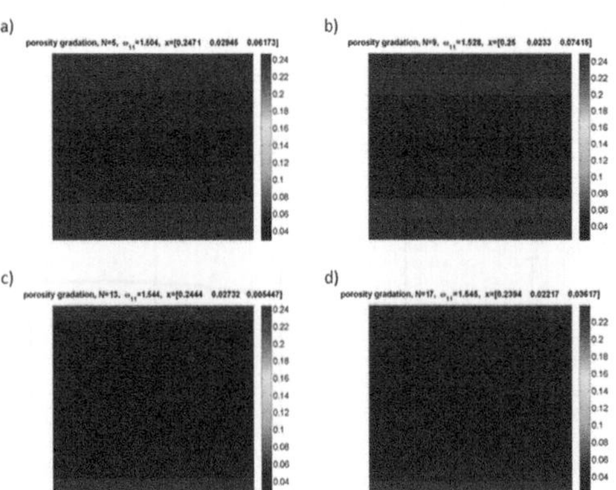

Figure 3. The best optimal solutions for different number of layers N: a) $N = 5$, b) $N = 9$, c) $N = 13$, d) $N = 17$.

6 CONCLUSIONS

The two-scale model was applied to solve the optimization problem of the FGM piezoelectric plate. The state space approach with the harmony search optimization method seems to be robust computational tool for optimization problems solving. The HS optimal solutions of the fundamental eigenfrequency optimization problem for porous $BaTiO_3$ piezoelectric material lead to the sandwich type FGM plate. The best obtained solutions are insensitive with respect to the number of layers. The solution for the free-gradation problem shows the asymmetric pattern of the porosity gradation.

REFERENCES

Chen, W.Q., Cai, J.B., Ye, G.R. & Wang, Y.F. 2004. Exact three-dimensional solutions of laminated orthotropic piezoelectric rectangular plates featuring interlaminar bonding imperfections modeled by a general spring layer. *International Journal of Solids and Structures* 41: 5247–5263.

Derrac, J., Garcia, S., Molina, D. & Herrera, F. 2011. A practical tutorial on the use of nonparametric statistical tests as a methodology for comparing evolutionary and swarm intelligence algorithms. *Swarm and Evolutionary Computation* 1: 3–18.

Fedelinski, P., Gorski, R., Czyz, T., Dziatkiewicz, G. & Ptaszny, J. 2014. Analysis of effective properties of materials by using the boundary element method. *Archives of Mechanics* 66: 19–35.

Gross, D. & Seelig, T. 2011. *Fracture mechanics with an introduction to micromechanics*. Heidelberg: Springer.

Huang, J.H., Chiu, Y.-H. & Liu, K.-H. 1998. Magneto-electro-elastic Eshelby tensors for a piezoelectric–piezomagnetic composite reinforced by ellipsoidal inclusions. *Journal of Applied Physics* 83: 5364–5370.

Lee, K.S. & Geem, Z.W. 2005. A new meta-heuristic algorithm for continuous engineering optimization: harmony search theory and practice. *Computer Methods in Applied Mechanics and Eng*ineering 194: 3902–3933.

Logemann, H. & Ryan, E.P. 2014. *Ordinary differential equations. Analysis, qualitative theory and control*. London: Springer-Verlag.

Poteralski, A., Szczepanik, M., Dziatkiewicz, G., Kuś, W. & Burczyński, T. 2013. Comparison between PSO and AIS on the basis of identification of material constants in piezoelectrics. In L. Rutkowski, M. Korytkowski, R. Sherer, R. Tadeusiewicz, L.A. Zadeh & J.M. Zurada (eds.), *Artificial Intelligence and Soft Computing*; *Lecture Notes in Computer Science* 7895: 569–581, Berlin—Heidelberg: Springer.

Taya, M. 2005. *Electronic composites*. New York: Cambridge University Press.

Wang, C. & Huang, Y. 2010. Self-adaptive harmony search algorithm for optimization. *Expert Systems with Applications* 37: 2826–2837.

Zhong, Z. & Yu, T. 2006. Vibration of a simply supported functionally graded piezoelectric rectangular plate. *Smart Materials Structures* 15: 1404–1412.

Advances in Mechanics: Theoretical, Computational and Interdisciplinary Issues – Kleiber et al. (Eds)
© 2016 Taylor & Francis Group, London, ISBN 978-1-138-02906-4

Reliability analysis of reinforced concrete components: A comparative study

P. Evangeliou, A. Roy & M.A.N. Hendriks
Department of Structural Engineering, Delft University of Technology, Delft, The Netherlands

R. Steenbergen
TNO, Delft, The Netherlands

A. de Boer
Rijkswaterstaat, Utrecht, The Netherlands

ABSTRACT: Full probabilistic assessment of the structural reliability of reinforced concrete structures holds the key to proper calibration of the existing safety formats. The scope of the present work is to compare different level II reliability methods that take into account variability and perform accurately in the left side tail of the probability density function of the resistance, in the framework of nonlinear finite element analysis. Alongside, the effectiveness and accuracy of these probabilistic methods are investigated and, through a comparative study, the most reliable method is determined. As a numerical example, structural reliability of a reinforced concrete beam is assessed through Probabilistic Nonlinear Finite Element Analysis (PNLFEA). The methods investigated in the paper are: Directional Adaptive Response Surface, Directional Simulation, Response Surface combined with Crude Monte Carlo Simulation and First Order Reliability Method. The software facilitated for the implementation of the aforementioned methods is the probabilistic module of DIANA finite element software and Matlab.

1 INTRODUCTION

Studies concerning the reliability assessment of existing reinforced concrete structures, which makes up the bulk of infrastructures worldwide, are of paramount importance. The deterministic analysis of RC structures is carried out by Nonlinear Finite Element Analysis (NLFEA). The source of non-linearity includes cracking, crushing, shearing and de-bonding of reinforcement bars. However, the geometry, loading pattern and material properties vary significantly not only from one structure to another and but also within the same structure. As a result, the quest of assessing the safety of RC structures can be best attempted within the framework of probability and statistics.

The basic philosophy is to generate the distribution of the structural resistance output based on the input distributions of the model, loading and geometry uncertainties. This NLFEA output distribution of the resistance, particularly around the left sidetail region, leads to the structural reliability computation. Nonetheless, the high computational cost associated with each NLFEA makes the Monte Carlo Simulation, which requires a very high number of simulations, a non viable option. In the wake of this, one has to resort to level II reliability methods. With the level II methods a full probabilistic analysis can be carried out relatively cheaply. However, in this process, accuracy is partially sacrificed. The current study aims to compare several of these level II methods in conjunction with the random variable input model.

2 BRIEF SUMMARY OF METHODS

The level II reliability methods like Directional Adaptive Response Surface (DARS), Directional Simulation (DS), First Order Reliability Method (FORM) and Response Surface (RS) methods are considered in this study. The stochastic input is simulated with the random variable model. In this model the stochastic parameters vary from a sample to another, but for each particular sample they are considered to be fixed in the entire domain. In other words, the input parameters are homogeneous in the random variable model. The advantage of this method is the relative ease with which the stochasticity can be included in the FE formulation. However, the obvious disadvantage is that this approach ignores the inhomogeneity of the input parameters.

2.1 *First Order Reliability Method*

The principle of the First Order Reliability Method (FORM) is to compute the reliability index β through a Taylor series expansion truncated after the first order term in the neighborhood of the design point. The design point is calculated through an iterative procedure which requires information about the limit state function and its gradient. Once β is computed, the probability of failure can be calculated as,

$$P_f = \Phi(-\beta) \tag{1}$$

where Φ is the joint probability density function of the standard-normal distribution. In NLFEA, the gradient of the limit state function structure can be obtained from the nonlinear FE system of equations, as illustrated in (Haldar & Mahadevan 2000). Moreover, implementing the gradient of the limit state function requires less computational effort in the standard normal space. Consequently, it becomes essential to formulate the limit state function $g(\xi)$ in the standard normal space, as following,

$$g(\xi(\mathbf{u})) = G(\mathbf{u}) \tag{2}$$

The FORM is very efficient if a structure has one predominant failure mode, and therefore, more suitable for relatively simple structural components.

2.2 *Directional Sampling*

The philosophy behind Directional Simulation (DS) is similar to Monte Carlo Simulation (MCS). The key difference between MCS and DS is that in DS, instead of sampling random numbers in the entire stochastic domain and then determining whether they lie in the failure domain or not, random directions are generated and the failure probability content in each of these directions is evaluated. The failure probability is finally calculated as a mean of all those individual contributions.

2.3 *Response surface*

This approach uses FE analysis to generate a response surface on which computationally cheap reliability calculations can be performed. If $\hat{g}(\xi)$ is an approximate reliability surface, then

$$\hat{G}(\mathbf{u}) = \sum_{i=0}^{N-1} a_i \psi_i(\mathbf{u}) \tag{3}$$

represents the response surface, where $\psi_i(\mathbf{u})$ are the Hermite polynomials and a_i are the coefficients to be determined through minimization of the error,

$$error(\mathbf{a}) = \sum_{k=1}^{NF} (G(\mathbf{u}^k) - \hat{G}(\mathbf{u}^k))^2 \tag{4}$$

Here k denotes the sample points. Hermite polynomials are chosen to represent the RS because they are characterized by being orthogonal to each other, this way providing independence of cross terms, $\psi_0(\mathbf{u}) = 1$, $<\psi_i(\mathbf{u})> = 0$, $<\psi_i(\mathbf{u})\psi_j(u)> = 0$, $i \neq j$ where $< >$ denote expectation operation. Once the response surface is constructed, the failure probability can be computed by a straightforward MCS. The accuracy of this method depends on the proper selection of the sample points.

2.4 *Directional Adaptive Response Surface*

The Directional Adaptive Response Surface (DARS) is such a method of selecting proper sample points (Grooteman 2011). DARS is a refinement of DS in the sense that sampling occurs as in DS, but instead of performing a complete calculation for each sample point, calculations are only carried out for sampling points that are nearly critical. For non-critical sample points the structural response is approximated. This method is more efficient than DS on the grounds that the number of the time consuming computations that have to be performed is reduced.

3 NUMERICAL EXAMPLE

3.1 *Model*

The model evaluated is a 4 m span, 4-point bending reinforced concrete beam without shear reinforcement. The deterministic and stochastic material properties are presented in Table 1. The beam thickness is 200 mm and the reinforcement area 622 mm^2. Due to symmetry, only half of the beam is modeled, by applying horizontal deflection constraints at the midsection nodes. The meshed model is illustrated in Figure 1.

3.2 *Deterministic NLFEA*

The deterministic NLFEA is carried out with the DIANA FE software considering the 2D model of the beam. Based on the beam theory main assumptions, the concrete area is meshed with 2D plane stress, 8-node quadrilateral elements with exact Gauss integration and quadratic interpolation. Smeared cracking is utilized to capture the crack propagation and the bi-axial stress state by the use of the rotating total crack strain model. The tension and compression softening concrete behavior is entailed by a linear and a quasi-exponential softening model, respectively. Additionally, the

Table 1. Deterministic and stochastic input of material properties.

Material	Concrete					Reinforcement steel		
Parameter	Compressive strength (f_{cm})	Young's modulus (E_{cm})	Tensile strength (f_{ct})	Fracture energy (G_f)	Poisson's ratio (v_c)	Yield strength (f_{ym})	Young's modulus (E_{ym})	Poisson's ratio (v_y)
Units	MPa	MPa	MPa	N/mm	–	MPa	MPa	–
Mean	25	29180	2.5	0.1	0.15	440	210000	0.2
Distribution	Lognormal	Lognormal	–	–	–	Lognormal	–	–
COV	10%	8%	–	–	–	10%	–	–

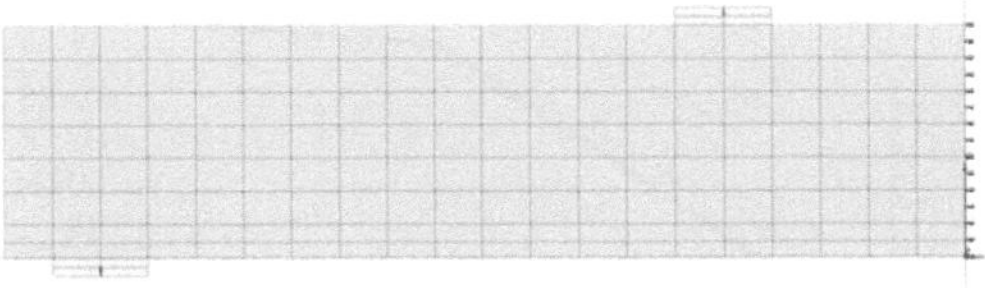

Figure 1. Reinforced concrete beam mesh, symmetry model.

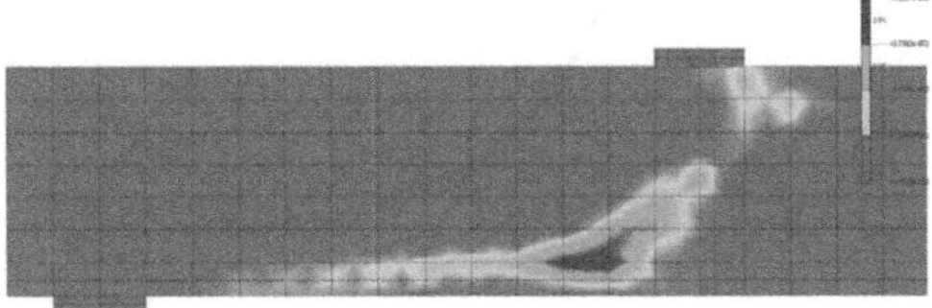

Figure 2. Principle strain ϵ_1 contour plot at shear failure.

reinforcement is modeled with the V. Mises constitutive model and meshed as embedded.

The load is imposed incrementally with a displacement control via loading plates. For the iterative procedure the Secant-Newton scheme is implemented, combined with the force convergence norm. Finally, a line search technique is applied for enhancement ofthe convergence rate.

Initially, a deterministic NLFEA is carried out for the mean values of the material properties. The contour plot of the principal strain ϵ_1 for the mean values of material properties is illustrated in Figure 2. Note that shear failure mode is governing in this case. The reaction force-deflection curves plotted in Figure 3 are the output of 3 independent deterministic NLFEA's for 3 different realizations of the concrete compressive strength f_c. By Figure 3, it is concluded that for different realizations of f_c different failure modes are induced, namely shear failure for high and bending failure for low values. Thus, material uncertainty may cause modal failure differentiation.

3.3 *Reliability analysis*

The stochastic input is comprised of 3 random variables, namely the concrete compressive strength f_c, the concrete Young's modulus E_c and the reinforcement yield strength f_y. Their mean values and coefficients of variation are exhibited in Table 1. Both f_c and E_c are perfectly correlated whereas a correlation coefficient of 0.5 is assumed for f_y with respect to the other two variables. The Limit State Function (LSF) is generally formulated as:

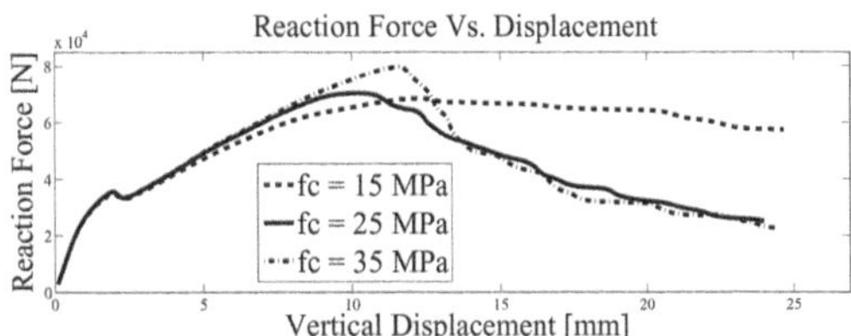

Figure 3. Reaction force vs vertical deflection; varying f_c.

$$G = R - E \quad (5)$$

with R the resistance and E the action effects. The safety requirement is: $G \geq 0$. In the current study two LSF formulations are presented. The first, eq. 6, LSF formulation is based on the ultimate reaction force:

$$G = R_{critical} - R(f_c, E_c, f_y) \quad (6)$$

where $R_{critical}$ is the critical value of the reaction force and $R(f_c, E_c, f_y)$ is the reaction force output for each realization of the stochastic input. Note that in the context of concrete reliability literature G should have been defined as the additive inverse of G in 6. The LSF defined in equation 6 is due to the limitations of the current development version of the software. In the second, eq. 7 LSF formulation, the limit state is based on the maximum vertical deflection of the beam:

$$G = disp_{critical} - disp(f_c, E_c, f_y) \quad (7)$$

Table 2. Reliability results for both the reaction force and the vertical deflection limit state function formulations.

Reaction force	Reliability results				Error (%)	Vertical deflection	Reliability results				Error (%)
Critical (N)	DARS (–)	DS (–)	FORM (–)	RS (–)	DARS vs RS	Critical (mm)	DARS (–)	DS (–)	FORM (–)	RS (–)	DARS vs RS
74788.78	0.9746	0.9737	0.9703	0.9705	–0.4225	–1.9484	0.9816	0.9822	0.9839	0.9863	0.4848
77368.48	0.9968	0.9967	0.9963	0.9969	0.01304	–2.0343	0.9966	0.9967	0.9970	0.9969	0.0288
79800.04	0.9997	0.9997	0.9997	0.9999	0.02630	–2.1172	0.9996	0.9996	0.9997	0.9992	–0.0403

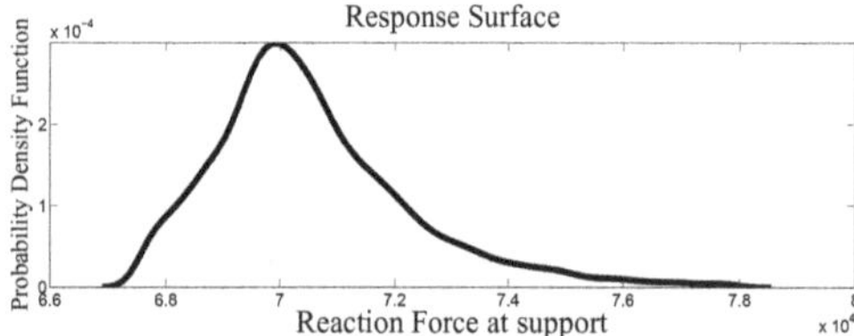

Figure 4. Probability density function of reaction force output.

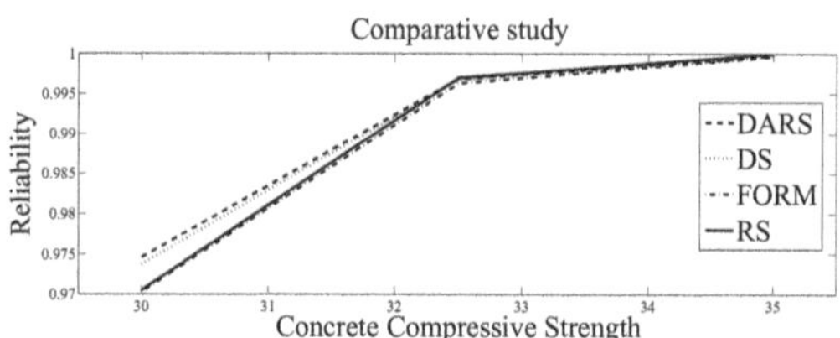

Figure 5. Reliability comparison for varying f_c.

where $disp_{critical}$ is the critical value of the maximum vertical deflection and $disp(f_c, E_c, f_y)$ is the vertical deflection output for each realization of the stochastic input.

The PNLFEA is carried out with the probabilistic module of DIANA software. The probabilistic methods applied are DS, DARS and FORM. The number of samples used for each PNLFEA is kept to minimum, testing the efficiency of the methods. The polynomial terms used to fit the response surface are 15.

The reliability results for the two, eq.'s 6 and 7, LSF formulations are exhibited on Table 2, for various values of $R_{critical}$ and $disp_{critical}$ for each case, respectively. Critical is the fact the error between the PNLFEA methods and the RS method is significantly low for both the LSF formulations. Note also that the reliability increases with the increase of the critical value $R_{critical}$ as well as with the increase of the critical absolute value $disp_{critical}$. Moreover, applying a minimum number of 40 samples for each PNLFEA is proven to lead to accurate results, highlighting the efficiency of the methods.

The structural reliability is verified by comparison to results generated by a response surface approach. In the RS the Probability Density Function (PDF) of the response is computed in the space of Hermite polynomials. Then, a crude Monte Carlo Simulation is performed to calculate the structural reliability.

In this study, 14 Hermite polynomials are employed. Furthermore, the PDF of the response is approximated by 40 sample points. Therefore, 40 independent deterministic NLFEA's have to be conducted for 40 realizations of the stochastic input. These specific realizations are uniformly sampled from a 10^6-sample generated with Matlab, taking into account the correlation between the variables. Then, the crude MCS is implemented for the same sample and the structural reliability is computed. The PDF of the response, for the first, eq. 6, LSF formulation, is displayed in Figure 4.

Successively, a comparative study between the probabilistic methods is carried out. For simplicity, the stochastic input is limited to one random variable, the concrete compressive strength f_c. Only the first, eq. 6, LSF formulation is applied. As demonstrated in Figure 4, the convergence error between the methods is decreasing for higher reliability values. Nevertheless, the error, even in the case of decreasing reliability, remains very low.

4 CONCLUSIONS AND FUTURE SCOPE

The full probabilistic methods DS, DARS and FORM have been utilized to compute the structural reliability of reinforced concrete beams and have proved to be efficient and accurate for this case. This is highlighted by the high accuracy achieved with minimum number of samples. Moreover, all methods result in reliability results with a negligible error between them. For a higher degree of uncertainty the agreement between these methods is reducing, not significantly however.

The probabilistic module of the DIANA FE software has been verified against reliability analysis of reinforced concrete beams, by comparison to reliability results from a robust and accurate response surface method generated with Matlab.

The Hermite polynomials have been applied for the construction of a robust response surface. Their orthogonality feature constitutes them as a suitable candidate for generating a response surface.

Of all the implemented methods, DARS is more likely to produce the most accurate results. Its accuracy can be further enhanced by the use of Hermite polynomials for fitting response surface generated in DARS in the neighborhood of the design point.

Future studies will address the inhomogeneity of the random input parameters in reinforced concrete structures.

REFERENCES

Grooteman, F. (2011). An adaptive directional importance sampling method for structural reliability. *J. Prob. Eng. Mech. 26*, 134–141.

Haldar, A. & S. Mahadevan (2000). *Reliability assessment using stochastic finite element analysis*. John Wiley and Sons.

Advances in Mechanics: Theoretical, Computational and Interdisciplinary Issues – Kleiber et al. (Eds)
© 2016 Taylor & Francis Group, London, ISBN 978-1-138-02906-4

Nonlinear model of spacecraft relative motion in an elliptical orbit

P. Felisiak & W. Wróblewski
Department of Cryogenic, Aviation and Process Engineering, Wrocław University of Technology, Wrocław, Poland

K. Sibilski
Institute of Aeronautics and Applied Mechanics, Warsaw University of Technology, Warsaw, Poland

ABSTRACT: Modern control strategies, such as model predictive control, requires accurate models of the process in time domain. This paper presents an formulation of nonlinear, time-variant model of spacecraft motion relative to body orbiting in an elliptical orbit. Since the mass of expelled propellant has significant influence on the system dynamics, the model assumes variable mass of the controlled spacecraft. The model is described by the state-space representation, while the parameters of the model are dependent on time, state and control signal only. The presented model can be applied for control of spacecraft relative motion, for example control of spacecraft during orbital rendezvous, one of the key space technology. The formulation of the model is suitable for state observers, especially where some of the states are unmeasured and the model-based estimation is needed as accurate as possible. Simulations of control system where controller was based on the proposed model indicates a relevant improvement of accuracy relative to the classical models in cases where separation between the satellites was relatively large.

1 INTRODUCTION

The reason of investigation presented in this paper is the need for model formulation suitable for model predictive control algorithm, wherein one of the design guidelines was the ability to provide a spacecraft control in case where reference point moves in an elliptical orbit.

The spacecraft relative motion presented in this paper describes the motion of the so-called deputy satellite, frequently referred as chaser satellite, relative to the chief satellite, often called target satellite. It is assumed that the deputy satellites controllable, while the chief satellite is uncontrolled and moves along a known elliptical orbit. The chief satellite determines a reference point, furthermore chief satellite does not have to be a physical object—it can be only a selected reference point.

In most cases, the models of spacecraft relative motion are the results of linearization of nonlinear equations. In the case of circular orbit problem, the example is the well known Hill-Clohessy-Wiltshire model (Hill 1878), (Clohessy & Wiltshire 1960). Several models were derived for relative motion in elliptical orbit, such as Tschauner-Hempel (Tschauner & Hempel 1965) equations. However, these linearized models are valid only for small separations between the spacecraft and reference point. Their applicability is limited to the initial distance of separation between chief and deputy satellites less than 1 km.

In order to improve the precision of relative dynamics equations, some investigations considered the second-order terms of relative positions and velocities. Such approach was presented in (London 1963). A system of equations was derived with quadratic terms, its approximate solution was based on the solution of the Hill-Clohessy-Wiltshire equations. (Kechichian 1992) describes an application of the second-order equations set to a rendezvous trajectory with an initial relative distance of 2000 km.

The model presented in the paper is an augmentation of the exact nonlinear model. One of the requirements to the model was the description in the time domain, while this kind of models are usually formulated in domain of so-called true anomaly, the angular position in the orbit. The problem was solved by numerical solution of the Kepler's equation. Since it is assumed that the spacecraft uses propulsion which operates using expulsion of the significant amount of mass, the principal model of relative motion was augmented using a mass model.

2 REFERENCE FRAME

In order to describe the deputy spacecraft motion relative to the chief satellite, Cartesian Local-Vertical Local-Horizontal (LVLH) coordinate

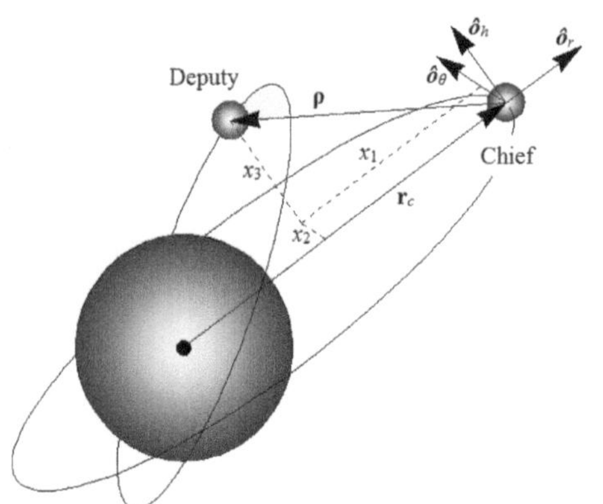

Figure 1. Local-Vertical Local-Horizontal (LVLH) coordinate frame.

frame is used. This reference frame sometimes is referred as Hill frame. The frame is attached to the chief satellite and rotates with the chief's radius vector $\mathbf{r}_c$ as shown in Figure 1.

Orientation of the LVLH frame is determined by the unit vector triad $\{\hat{\mathbf{o}}_r, \hat{\mathbf{o}}_\theta, \hat{\mathbf{o}}_h\}$ where vector $\hat{\mathbf{o}}_r$ lies in the chief's radial direction, $\hat{\mathbf{o}}_h$ is parallel to the orbit angular momentum vector, and $\hat{\mathbf{o}}_\theta$ completes the right-handed orthogonal triad. Then, the position of the deputy satellite relative to the chief satellite can be expressed by Cartesian coordinate vector ρ:

$$\rho = x_1\hat{\mathbf{o}}_r + x_2\hat{\mathbf{o}}_\theta + x_3\hat{\mathbf{o}}_h \tag{1}$$

3 PRINCIPAL MODEL

Derivation of the exact nonlinear equations of relative motion in LVLH frame can be found in (Schaub & Junkins 2003).

In this investigation the impact of the external control force has been included and the equations are transformed into state-space representation.

The set of time-variant nonlinear equations is formulated as follows:

$$\ddot{x}_1 - 2\dot{f}\left(\dot{x}_2 - x_2\frac{\dot{r}_c}{r_c}\right) - x_1\dot{f}^2 - \frac{\mu}{r_c^2} = -\frac{\mu}{r_d^3}(r_c + x_1) + \frac{u_{d1}}{m_d} \tag{2}$$

$$\ddot{x}_2 + 2\dot{f}\left(\dot{x}_1 - x_1\frac{\dot{r}_c}{r_c}\right) - x_2\dot{f}^2 = -\frac{\mu}{r_d^3}x_2 + \frac{u_{d2}}{m_d} \tag{3}$$

$$\ddot{x}_3 = -\frac{\mu}{r_d^3}x_3 + \frac{u_{d3}}{m_d} \tag{4}$$

wherein x_1, x_2 and x_3 are components of the deputy satellite relative position vector in LVLH frame, according to Figure 1, f denotes true anomaly (angular position in the orbit) of the chief satellite, μ is the standard gravitational parameter, r_c is the current radius of the chief satellite orbit and r_d is the current radius of the deputy satellite orbit. u_{d1}, u_{d2} and u_{d3} are forces acting on the deputy satellite in the radial, in-track and cross-track directions respectively. Finally, m_d denotes the current deputy spacecraft mass.

3.1 *State-space representation*

Modern model predictive control applications utilize models represented in state-space. Here we will transform equations 2, 3 and 4 into a form suitable for such applications.

First, let x_4, x_5 and x_6 be components of a relative velocity vector, according to the frame depicted in Figure 1 and define the state vector:

$$\mathbf{x}_{rm} = \begin{bmatrix} x_1 & x_2 & x_3 & x_4 & x_5 & x_6 \end{bmatrix}^T \tag{5}$$

Furthermore, the following control force vector is assumed:

$$\mathbf{u} = \begin{bmatrix} u_1 & u_2 & u_3 \end{bmatrix}^T \tag{6}$$

where the components u_1, u_2 and u_3 are in the radial, in-track and cross-track directions respectively. Then, the state-space representation of the principal model of relative motion takes the form:

$$\dot{\mathbf{x}}_{rm} = \mathbf{A}_{rm}\mathbf{x}_{rm} + \mathbf{B}_{rm}\mathbf{u} + \mathbf{V}_{rm} \tag{7}$$

wherein the state matrix is defined as:

$$\mathbf{A}_{rm} = \begin{bmatrix} 0 & 0 & 0 & 1 & 0 & 0 \\ 0 & 0 & 0 & 0 & 1 & 0 \\ 0 & 0 & 0 & 0 & 0 & 1 \\ \dot{f}^2 - \frac{\mu}{r_d^3} & -2\dot{f}\frac{\dot{r}_c}{r_c} & 0 & 0 & 2\dot{f} & 0 \\ 2\dot{f}\frac{\dot{r}_c}{r_c} & \dot{f}^2 - \frac{\mu}{r_d^3} & 0 & -2\dot{f} & 0 & 0 \\ 0 & 0 & -\frac{\mu}{r_d^3} & 0 & 0 & 0 \end{bmatrix} \tag{8}$$

The input matrix has the form:

$$\mathbf{B}_{rm} = \begin{bmatrix} 0 & 0 & 0 \\ 0 & 0 & 0 \\ 0 & 0 & 0 \\ \frac{1}{m_d} & 0 & 0 \\ 0 & \frac{1}{m_d} & 0 \\ 0 & 0 & \frac{1}{m_d} \end{bmatrix} \tag{9}$$

Finally, the matrix of the nonlinear term can be written as:

$$\mathbf{V}_{rm} = \begin{bmatrix} 0 & 0 & 0 & \mu\left(\frac{1}{r_c^2} - \frac{r_c}{r_d^3}\right) & 0 & 0 \end{bmatrix}^T \tag{10}$$

3.2 *Model parameters calculation*

The principal model of relative motion represented by equations 2, 3 and 4 as well as equation 7 have time-variant parameters $\dot{f}$, r_c, $\dot{r}_c$, r_d and m_d. The procedures of their calculation are treated as a part of the complete model. Character of the model predictive control application entails the necessity of their calculation in time domain.

First, let us consider the true anomaly rate $\dot{f}$. Its calculation requires knowledge of the current value of true anomaly f, which in turn can be calculated using the eccentric anomaly E:

$$\tan\frac{f}{2} = \sqrt{\frac{1+e}{1-e}}\tan\frac{E}{2} \tag{11}$$

where e is the chief's orbit eccentricity.

The eccentric anomaly E can be obtained by solving of the Kepler's Equation (Curtis 2005):

$$M = E - e\sin E \tag{12}$$

However, Kepler's Equation needs to be solved using a numerical method, for example Newton's method.

The mean anomaly M in equation 12 can be found using the following definition:

$$M = M_0 + n(t - t_0) \tag{13}$$

wherein $0 \le M \le 2\pi$ and the mean angular motion n:

$$n = \sqrt{\frac{\mu}{a^3}} \tag{14}$$

where a is the semi-major axis of the chief satellite orbit.

Equations 11, 12 and 13 enables us to find the true anomaly f as a function of time t. Then, the true anomaly rate $\dot{f}$ can be obtained using:

$$\dot{f} = \sqrt{\frac{\mu p}{r_c^4}} \tag{15}$$

wherein p is a parameter known as *semilatus rectum*:

$$p = a\left(1 - e^2\right) \tag{16}$$

The current radius of the chief satellite orbit r_c can be found as the function of the true anomaly f:

$$r_c = \frac{p}{1 + e\cos f} \tag{17}$$

The radial velocity $\dot{r}_c$ of the chief satellite is calculated using:

$$\dot{r}_c = \frac{\mu e}{h}\sin f \tag{18}$$

wherein h is massless angular momentum:

$$h = \sqrt{\mu p} \tag{19}$$

The current radius of the deputy satellite orbit $\dot{r}_d$ can be obtained by:

$$r_d = \sqrt{(r_c + x_1)^2 + x_2^2 + x_3^2} \tag{20}$$

Finally, calculation of the current mass of the deputy satellite m_d is described in section 4.

4 MASS MODEL

The mass model describes flow rate of a propellant mass, what enables to calculate the current mass of the deputy satellite m_d. The general concept of the model is based on the relationship between mass flow rate and the thrusters' parameter known as the specific impulse (Sidi 1997). The relationship is given below:

$$\dot{m} = \frac{F_{thrust}}{I_{sp}g} \tag{21}$$

where $\dot{m}$ denotes the mass flow rate, F_{thrust} is the force obtained from the thruster, g is the acceleration at the Earth's surface and I_{sp} denotes the specific impulse. Assume that $\mathbf{x}_{ep}$ having dimensions $\dim[\mathbf{x}_{ep}] = 1 \times 1$ denotes mass of the propellant expelled by the deputy satellite thrusters.

Then, the state-space model of the expelled mass can be expressed as:

$$\dot{\mathbf{x}}_{ep} = \mathbf{A}_{ep}\mathbf{x}_{ep} + \mathbf{B}_{ep}\mathbf{u} \tag{22}$$

Since the expelled mass $\mathbf{x}_{ep}$ does not affect the mass flow rate, the state matrix $\mathbf{A}_{ep} = \mathbf{0}_{1,1}$.

The input matrix $\mathbf{B}_{ep}$ having dimensions $\dim\left[\mathbf{B}_{ep}\right] = 1 \times 3$ contains components which correspond to mass flow rate in the radial, in-track and cross-track directions respectively. The components are formulated based on equation 21, however the sign of each component depends on the sign of corresponding component in the control vector $\mathbf{u}$. In the case where the expelled mass

$x_{ep} = \|\mathbf{x}_{ep}\|$ is less than the initial propellant mass m_{p0} available to the deputy satellite, i-th component ($i = 1, 2, 3$) of the $\mathbf{B}_{ep}$ row matrix is defined as:

$$\forall x_{ep} < m_{p0} \left\{ u_i \geq 0 \Rightarrow B_{ep1,i} = \frac{1}{I_{sp}g} \right\} \quad (23)$$

$$\forall x_{ep} < m_{p0} \left\{ u_i < 0 \Rightarrow B_{ep1,i} = -\frac{1}{I_{sp}g} \right\} \quad (24)$$

In the case where the propellant mass available to the deputy satellite is fully expelled, the $\mathbf{B}_{ep}$ input matrix is a zero matrix:

$$\exists x_{ep} = m_{p0} \left\{ \mathbf{B}_{ep} = \begin{bmatrix} 0 & 0 & 0 \end{bmatrix} \right\} \quad (25)$$

Finally, the deputy spacecraft mass m_d, which is a parameter in the NERM model given by equation 7, can be found using the following formula:

$$m_d = m_{dry} + m_{p0} - x_{ep} \quad (26)$$

wherein m_{dry} denotes mass of the deputy satellite without available propellant.

5 AUGMENTED MODEL

The final continuous model is formulated by an augmentation of the principal relative motion model given by equation 7 using the mass model described by equation 22.

Assuming an augmented state vector:

$$\mathbf{x}_m = \begin{bmatrix} \mathbf{x}_{rm} \\ \mathbf{x}_{ep} \end{bmatrix} \quad (27)$$

the augmented, continuous state-space model of relative motion is given by:

$$\dot{\mathbf{x}}_m(t) = \overbrace{\begin{bmatrix} \mathbf{A}_{rm} & \mathbf{0}_{6,1} \\ \mathbf{0}_{1,6} & \mathbf{A}_{ep} \end{bmatrix}}^{\mathbf{A}_{cm}} \mathbf{x}_m(t) + \overbrace{\begin{bmatrix} \mathbf{B}_{rm} \\ \mathbf{B}_{ep} \end{bmatrix}}^{\mathbf{B}_{cm}} \mathbf{u}(t) + \overbrace{\begin{bmatrix} \mathbf{V}_{rm} \\ \mathbf{V}_{ep} \end{bmatrix}}^{\mathbf{V}_{cm}} \quad (28)$$

wherein $\mathbf{V}_{ep} = \mathbf{0}_{1,1}$.

6 DISCRETE EQUIVALENT

Design of digital controller requires conversion of continuous state-space model into discrete equivalent. This transformation was performed with assumption that matrices $\mathbf{A}_{cm}$, $\mathbf{B}_{cm}$ and $\mathbf{V}_{cm}$ from equation 28 are invariant within time range $[t_0, t]$.

Define discrete state matrix:

$$\mathbf{A}_m = e^{\mathbf{A}_{cm}T} \quad (29)$$

where T is a sampling period, discrete input matrix:

$$\mathbf{B}_m = \int_0^T e^{\mathbf{A}_{cm}\eta} \mathbf{B}_{cm} d\eta \quad (30)$$

where η is an integration variable:

$$\eta = (k+1)T - \tau \quad (31)$$

wherein k denotes the current sampling instant, and discrete nonlinear term matrix:

$$\mathbf{V}_m = \mathbf{V}_{cm}T \quad (32)$$

Then, the discrete state-space representation can be written as:

$$\mathbf{x}_m(k+1) = \mathbf{A}_m \mathbf{x}_m(k) + \mathbf{B}_m \mathbf{u}(k) + \mathbf{V}_m \quad (33)$$

$$\mathbf{y}_m(k) = \mathbf{C}_m \mathbf{x}_m(k) + \mathbf{D}_m \mathbf{u}(k) \quad (34)$$

where $\mathbf{C}_m = \mathbf{C}_{cm}$ and $\mathbf{D}_m = \mathbf{D}_{cm}$.

Note that we can write:

$$\mathbf{A}_m = \mathbf{I} + \mathbf{A}_{cm} T \Psi \quad (35)$$

and:

$$\mathbf{B}_m = \sum_{i=0}^{\infty} \frac{\mathbf{A}_{cm}^i T^i}{(i+1)!} T \mathbf{B}_{cm} = \Psi T \mathbf{B}_{cm} \quad (36)$$

wherein:

$$\Psi = \mathbf{I} + \frac{\mathbf{A}_{cm}T}{2!} + \frac{\mathbf{A}_{cm}^2 T^2}{3!} + \ldots \quad (37)$$

7 CONCLUSION

In order to prove the usefulness and validity of the presented model, a numerical experiment with participation of control system was performed. The control system included presented nonlinear continuous model of relative motion as a control object and model predictive controller equipped with the proposed discrete model. Simulations of the control system shown that a rendezvous maneuver can be completed successfully with the initial separation above 36000 km. Such result is unachievable using the linearized models.

REFERENCES

Clohessy, W.H. & P.S. Wiltshire (1960, September). Terminal guidance system for satellite rendezvous. *Journal of Aerospace Sciences 27*, 653–658.

Curtis, H.D. (2005). *Orbital Mechanics for Engineering Students*. Elsevier Aerospace Engineering Series. Butterworth-Heinemann.

Hill, G.W. (1878). Researches in the lunar theory. *American Journal of Mathematics 1*, 526,129147,24260.

Kechichian, J.A. (1992). Techniques of accurate analytic terminal rendezvous in near-circular orbit. *Acta Astronautica 26*(6), 377–394.

London, H.S. (1963). Second approximation to the solution of the rendezvous equations. *AIAA Journal 1*(7), 1691–1693.

Schaub, H. & J.L. Junkins (2003, October). *Analytical Mechanics of Space Systems*. Reston, Virginia, US: AIAA Education Series.

Sidi, M.J. (1997). *Spacecraft Dynamics and Control: A Practical Engineering Approach*. Cambridge Aerospace Series. Cambridge University Press.

Tschauner, J. & P. Hempel (1965). Rendezvous zu einem in elliptischer bahn umlaufenden ziel. *Astronautica Acta 11*(2), 104–109.

Advances in Mechanics: Theoretical, Computational and Interdisciplinary Issues – Kleiber et al. (Eds)
 ISBN 978-1-138-02906-4

Reliable mechanical characterization of layered pavement structures

T. Garbowski
Institute of Structural Engineering, Poznan University of Technology, Poznan, Poland

A. Venier
Faculty of Civil and Environmental Engineering, Politecnico di Milano, Milan, Italy

ABSTRACT: Non-destructive testing is often used in pavement engineering to determine the structural status of pavements. One of the most used devices in this field is the Falling Weight Deflectometer (FWD), which produces a set of recorded surface deflections later used for parameter identification purposes. It is then important to create a proper model to interpret FWD data. In literature many authors have proposed different approaches for analysing the behavior of pavement structures. Some of them are simpler, other are more rigorous. Most commercial programs employ quasi-static analyses due to its simplicity and performance. This choice, however, may be in conflict with the purely dynamic response of a real structure. A better alternative might contemplate dynamic finite element models, which unfortunately are very costly and therefore impractical in real life applications. In this work however, the dynamic behavior of the system is described through the Spectral Element Method (SEM). Its high-performance and precision represent a valuable element in back-analysis. Besides, the introduction of different minimization algorithms (Powell's method, Levenberg-Marquardt algorithm, Extended Kalman Filter) combined with multiple starting points result in a more confident and stable solution. In conclusion, the study reports a back-calculation example for a 3-layered pavement system. The intention is to show the importance of lower frequencies and the influence of resonance phenomena in system identification.

1 INTRODUCTION

1.1 *Maintenance and rehabilitation*

Deterioration of the pavement is an inevitable process and takes place from the day after construction. In such a scenario, it is mandatory to schedule a precise maintenance plan in order to ensure the full functionality of the road during its design life (Fig. 1). It is thus imperative to intervene in the right moment, when rehabilitation can still be conducted with lower resources. A posterior operation, dealing with structure in worse conditions, might require a different amount of finances. Fortunately, with just a limited quantity of new materials and by reemploying the existing ones, flexible pavements may be easily recycled. Nevertheless, it is still an expensive operation and, in order to plan a proper rehabilitation strategy for the whole network, it is necessary to regularly monitor the status of the roads. Such process is called pavement management and contemplates the use of specific devices for assessing the conditions of the structure by means of Nondestructive Testing and Evaluation (NDT and NDE) methods. A typical machine used for such purpose is, in fact, the Falling Weight Deflectometer (FWD) (Uzan 1994, Zaghloul, White, Drnevich, & Coree 1998).

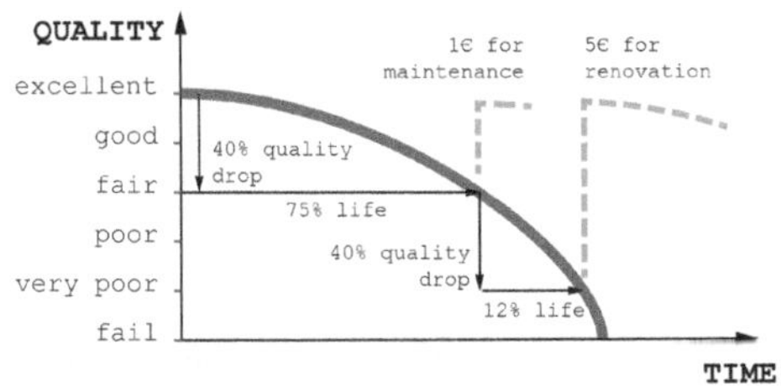

Figure 1. Pavement life management.

1.2 *The falling weight deflectometer*

The FWD represents a precious instrument for the structural evaluation of existing flexible pavements: it produces fast and easy-repeatable tests. The basic concept behind the FWD is that the magnitude of the surface deflections caused by a given load are indirectly indicating the structural health of the system. The mechanism attempts, in fact, to replicate the effect of single heavy moving wheel acting on the pavement by applying a dynamic load. Such impulse is produced by a considerable weight and delivered to the surface by a circular plate (typically 300 mm in diameter). A rubber base at the bottom permits a uniform load pressure. The instrument is also equipped with a load-measuring cell that accounts for the exact pressure at each time interval. Several geophones (7 or 9) are located at different distances from the loading device and are responsible for rigorous displacement recording. Indeed, the sensor precision ranges between $\pm 1.0\,\mu m$ (or even $\pm 0.1\,\mu m$): accurate recording is imperative when small variations in the deflection have considerable influence on the structural response. Such deflections are then employed in a back-calculation procedure in order to assess the pavement mechanical properties. Typical tests contemplate the use of four different load levels (the entire procedure is normally completed in less than two minutes) and they are conducted every 150 meters, in order to ascertain unexpected shifts of soil and pavement properties. FWD tests are often preferred over destructive methods because they are much more rapid and do not entail the removal of pavement materials. This is a great advantage since the sought information may be available soon after the test is performed, thus saving expensive laboratory tests.

2 SPECTRAL BASED FORWARD MODEL

The proposed here procedure uses a Spectral Element Method (SEM) instead of dynamic FE models. The main advantage, beside a substantial decrease of the computational time, is a significant increase of an experimental data to be incorporated into the inverse analysis. This is mainly because one can now freely sample the dynamic response not only in space but also in time. Moreover, since the spectral element technique makes use of the exact solution of the wave problem, a very high level of precision is guaranteed.

2.1 *Governing differential equations*

A vertical impulse load acting on a homogeneous isotropic half-space generates axisymmetric perturbations. Therefore, by adopting a cylindrical reference system, it is possible to exploit Navier's equations and the Helmholtz potential decomposition in order to obtain the wave motion equations. Denoting the potentials (ϕ and ψ) and the vertical and radial coordinates (z and r), the governing differential equation reads:

$$\frac{\partial^2\phi}{\partial r^2}+\frac{1}{r}\frac{\partial\phi}{\partial r}+\frac{\partial^2\phi}{\partial z^2}=\frac{1}{c_p^2}\frac{\partial^2\phi}{\partial t^2}$$
$$\frac{\partial^2\psi}{\partial r^2}+\frac{1}{r}\frac{\partial\psi}{\partial r}+\frac{\partial^2\psi}{\partial z^2}-\frac{\psi}{r^2}=\frac{1}{c_s^2}\frac{\partial^2\psi}{\partial t^2} \quad (1)$$

where:

$$c_p=\left(\frac{\lambda+2\mu}{\rho}\right)^{1/2};\quad c_s=\left(\frac{\mu}{\rho}\right)^{1/2} \quad (2)$$

while the displacement field respectively in radial and vertical downward direction is:

$$u=\frac{\partial\phi}{\partial r}-\frac{\partial\psi}{\partial z};\quad w=\frac{\partial\phi}{\partial z}+\frac{1}{r}\frac{\partial(r\psi)}{\partial r} \quad (3)$$

A way of solving the differential equations (1) is through Fourier transform. The shifting from time domain to frequency domain reduces the equations to simple Bessel equations: its solution is the Bessel's function J_0. In order to discretize the domain it is necessary to impose zero amplitude at distance $r = R$ far enough from the origin.

$$J_0(kR)=0 \quad (4)$$

Choosing the constants (wavenumbers) as $k_m = \alpha_m/R$, where α_m represents the m-th root of J_0, the solution of equation (4) is achieved. The potentials in the frequency domain are consequently defined as:

$$\hat{\phi}_{mn}(r,z)=A_{mn}e^{-ik_{pzmn}z}J_0(k_m r)$$
$$\hat{\psi}_{mn}(r,z)=B_{mn}e^{-ik_{szmn}z}J_1(k_m r) \quad (5)$$

where:

$$k_{pzmn}=\left(\frac{\omega_n^2}{c_p^2}-k_m^2\right)^{0.5};\quad k_{szmn}=\left(\frac{\omega_n^2}{c_s^2}-k_m^2\right)^{0.5} \quad (6)$$

being k_{pzmn} the wavenumber in vertical direction, k_{szmn} the wavenumber in radial direction, A_{mn} and B_{mn} constants defined by boundary conditions, J_1 the Bessel's function of the first kind.

2.2 *Discrete spectral solution*

Due to the linearity and homogeneity of the governing equations it is possible to use the principle of superposition. This means that by double summation over M wavenumbers and N angular frequencies ω_n one can reconstruct the entire oscillating system.

$$\phi_{mn}(r,z,t)=\sum_n\sum_m A_{mn}e^{-ik_{pzmn}z}J_0(k_m r)e^{-i\omega_n t}$$
$$\psi_{mn}(r,z,t)=\sum_n\sum_m B_{mn}e^{-ik_{szmn}z}J_1(k_m r)e^{-i\omega_n t} \quad (7)$$

At this point, it is straightforward to build the stiffness matrix for the layer elements from the nodal displacement obtained by (3) and the applied boundary tractions. The process needs to be done for every wavenumber and every frequency while the global stiffness matrix is assembled in the same way as in the finite element method. The final displacement field is the following:

$$u(r,z,t)=\sum_{n}\sum_{m}\hat{u}_{mn}(z,k_m,\omega_n)\hat{F}_m J_1(k_m r)\hat{F}_n e^{i\omega_n t}$$
$$w(r,z,t)=\sum_{n}\sum_{m}\hat{w}_{mn}(z,k_m,\omega_n)\hat{F}_m J_0(k_m r)\hat{F}_n e^{i\omega_n t} \tag{8}$$

being $\hat{u}_{mn}$ and $\hat{w}_{mn}$ the displacements for a unit load condition, while F_m and F_n represent the Fourier-Bessel spatial coefficients and fast Fourier time coefficients of the load. Differently from FEM, however, thanks to the exactness of the found solution, one element per layer is sufficient to represent adequately the behavior of the system. More details can be found in (Al-Khoury, Scarpas, Kasbergen, & Blaauwendraad 2001).

3 PARAMETER IDENTIFICATION STUDY

3.1 *Back-calculation analysis*

Back-calculation analysis with a particular application to constitutive model calibration is a tool widely used by many researchers (see e.g. (Maier, Bolzon, Buljak, Garbowski, & Miller 2010, Garbowski, Maier, & Novati 2011, Garbowski, Maier, & Novati 2012, Buljak 2012, Maier, Buljak, Garbowski, Cocchetti, & Novati 2014)). In general, it involves the minimization of the discrepancy between the numerically computed $\mathbf{U}_{\mathrm{NUM}}$ and measurable quantities $\mathbf{U}_{\mathrm{EXP}}$. A vector of residuals $\mathbf{R}$ in frequency domain f can be constructed in the following way:

$$\mathbf{R}^f(\mathbf{x}) = \mathbf{U}^f_{\mathrm{EXP}} - \mathbf{U}^f_{\mathrm{NUM}}(\mathbf{x}) \tag{9}$$

By adjusting the constitutive parameters (encapsulated in the vector $\mathbf{x}$) embedded in the numerical model, which in turn mimic the experimental setup, an iterative convergence towards the required solution can be achieved. The minimization of the objective function ω (within the least square frame) takes the form:

$$min_{\mathbf{x}}\ \omega(\mathbf{x})^f = \left\|\mathbf{R}^f(\mathbf{x})\right\|_2^2 \tag{10}$$

and it is updated through the use of gradient-based algorithms (Levenberg-Marquardt (LM), Extended Kalman FIlter (EKF)) and gradient-less algorithms (Powell (P)). More details can be found in (Nocedal & Wright 2006). Fortunately, the profound differences between each of these optimization techniques constitute the backbone of the entire program. Diverse solutions may be found by each algorithm and, among those, only the one that shows the lowest error norm is selected.

3.2 *Multistart approach*

It is known that optimization problems may be more or less dependent on the user provided initial guesses. This is due to the very nature of the objective function which is in general non-convex: the algorithm might get trapped inside a local minimum and fail to reach the optimum. There is not much to do on the algorithm side. However, as the spectral implementation is relatively not demanding, the program can afford to contemplate the coexistence of multiple initial values. By selecting the best estimate related to the lowest error norm, it is possible to eradicate the dependency on the starting guess. Thus, few more seconds of computations can produce a massive enhancement of code stability. Figure 2 documents the improvement achieved.

3.3 *Example: 3-layered pavement structure with rigid subbase on shallow bedrock*

The current study considers the presence of a stabilized subbase in a 3-layered system (the properties are indicated in Table 1). This scenario represents an important practical case because a stiffer layer is typically required for pavements constructed above weak soil. Moreover, it is known that when the stiffness ratio between layers is significant, some back-calculation instabilities may arise. The here presented study is based on computer-generated deflection data.

As speculated the program strives to find a correct solution (represented by the dashed line). It can be noticed, however, that the lower frequencies (0–10 Hz) are able to correctly find the stiffness values for all layers. Analyzing the following frequency range (10–20 Hz), one can notice a sudden dispersion peak which leads to highly incorrect results. This behavior is caused by resonance phenomena due to the shallow bedrock. Therefore, in this interval, found parameters are not to be trusted. In addition, this example shows that stiffer layers are the ones which are better identified, regardless of the depth of the layer.

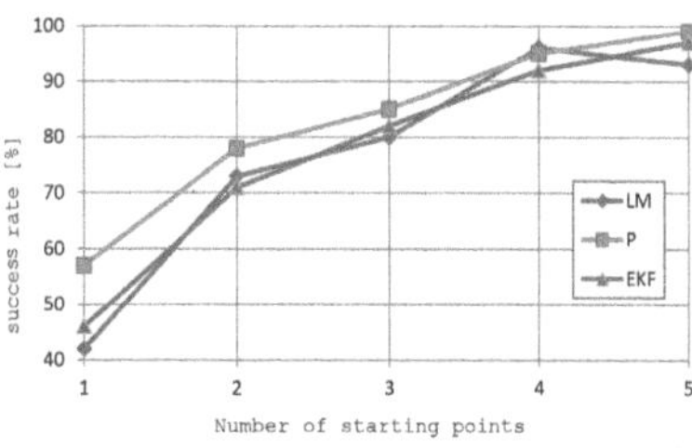

Figure 2. Success rate for different amounts of starting points.

Table 1. Material properties.

Layer	E [MPa]	h [cm]	ν	ρ [Kg/m^3]
1	2000	15	0.35	2300
2	5000	25	0.35	2000
3	25	430	0.35	1500

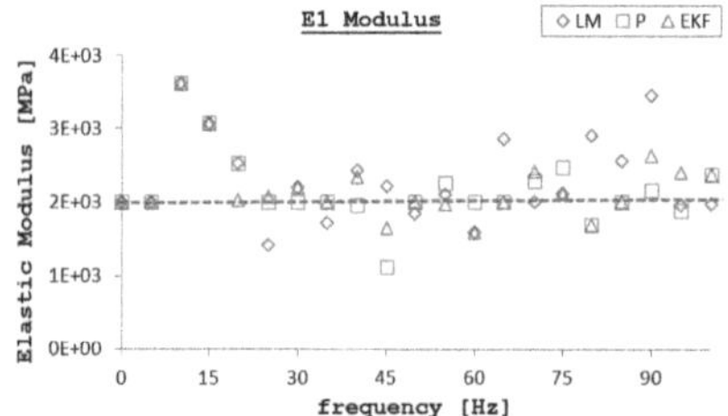

Figure 3. Dispersion representation of the first layer elastic modulus.

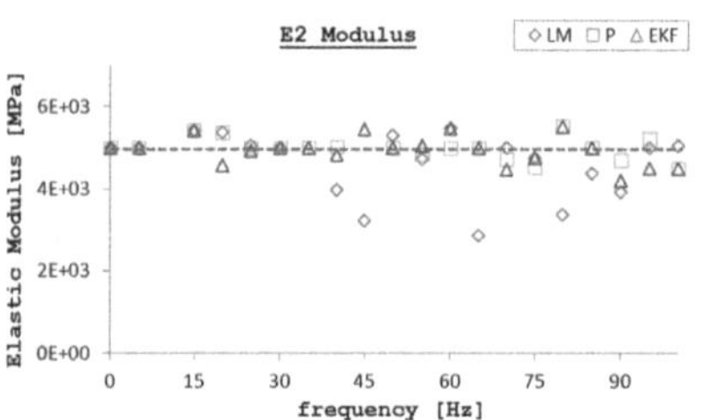

Figure 4. Dispersion representation of the second layer elastic modulus.

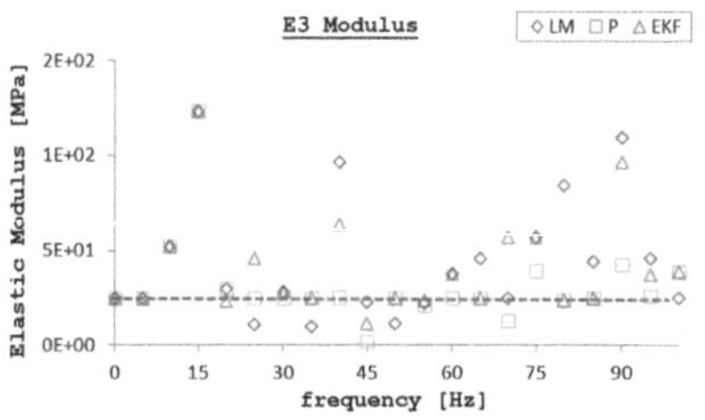

Figure 5. Dispersion representation of the third layer elastic modulus.

4 CONCLUSIONS

Herein a procedure based on the discrete spectral solution has been implemented and its verification has been properly obtained. One of the most powerful features of SEM is the double summation approach, which avoids the typical complications rising with numerical integration between zero and infinity. Also, it was discovered that one finite spectral element is representative of the entire layer. This results in a dramatic reduction of the model complexity if compared, for instance, to FEM. Therefore, it is possible to assert that the spectral element method represents a robust and highly efficient tool for recreating the dynamic nature of the FWD test. The association of the numerical model with three distinct minimization algorithms yielded to an exhaustive back-calculation program. The program, in fact, selects only the solution with the lowest error norm, thus ensuring a better estimate. It is known, however, that search techniques are strongly conditioned by the choice of the initial guess. Here, a simple consideration, contemplating the coexistence of multiple random starting points, was able to dramatically improve the stability of the code. Later, a 3-layered example with rigid subbase and shallow bedrock was studied. It was noticed that lower frequencies usually permit more successful parameter identification. Besides, great prudence is required in the eventuality of shallow bedrock (or seasonal stiff layers). Resonance phenomena may in fact contaminate measurement and conduce to erroneous results.

REFERENCES

Al-Khoury, R., A. Scarpas, C. Kasbergen, & J. Blaauwendraad (2001). Spectral element technique for efficient parameter identification of layered media. i. forward calculation. *International Journal of Solids and Structures 38*, 1605–1623.

Buljak, V. (2012). *Inverse analisyes with model reduction—Proper orthogonal decomposition in structural mechanics.* Springer.

Garbowski, T., G. Maier, & G. Novati (2011). Diagnosis of concrete dams by flat-jack tests and inverse analyses based on proper orthogonal decomposition. *Journal of Mechanics of Materials and Structures 6*(1–4), 181–202.

Garbowski, T., G. Maier, & G. Novati (2012). On calibration of orthotropic elastic-plastic constitutive models for paper foils by biaxial tests and inverse analyses. *Structural and Multidisciplinary Optimization 46*(1), 111–128.

Maier, G., G. Bolzon, V. Buljak, T. Garbowski, & B. Miller (2010). Synergistic combinations of computational methods and experiments for structural diagnosis. In M. Kuczma and K. Wilmanski (Eds.), *Advanced Structured Materials—Proceedings of the 18th International Conference on Computer Methods in Mechanics, CMM 2009*, pp. 453–476. Springer-Varlag Berlin Heidelberg.

Maier, G., V. Buljak, T. Garbowski, G. Cocchetti, & G. Novati (2014). Mechanical characterization of materials and diagnosis of structures by inverse analyses: some innovative procedures and applications. *International Journal of Computational Methods 11*(3).

Nocedal, J. & S. Wright (2006). *Numerical Optimization.* Springer-Verlag New York.

Uzan, J. (1994). Testing of pavements and back calculation of modulus. *American Society for Testing and Materials*, 3–37.

Zaghloul, S., T. White, V. Drnevich, & B. Coree (1998). Dynamic analysis of fwd loading and pavement response using a three dimensional dynamic finite element program. *American Society for Testing and Materials*, 125–138.

Advances in Mechanics: Theoretical, Computational and Interdisciplinary Issues – Kleiber et al. (Eds)
 ISBN 978-1-138-02906-4

The impact of fire situation on the static and stability response of the bearing steel structure

A. Garstecki
Polytechnic Institute, Stanislaw Staszic University of Applied Sciences in Pila, Piła, Poland

K. Rzeszut, Ł. Polus, M. Klój & M. Terech
Faculty of Civil and Environmental Engineering, Poznan University of Technology, Poznań, Poland

ABSTRACT: In the article, the behavior of multi-bay portal frame steel structures in fire conditions are analyzed. The analyses are carried out for fire accidental action using the standard ISO-fire curve. The iterative procedure to investigate critical temperature is used. At each increment of temperature the ultimate limit state is checked using the EC recommendation. Special attention is focused on failure analysis. First, using the public domain program LUCA the structure is checked if it does not collapse towards the outside of the building in case of fire occurring in one of the building compartments. The criteria which guarantee the safety of occupants and firemen are considered. Secondly, using the same program it is checked if the localized failure of the unprotected compartment leads to the collapse of the protected compartment. At this stage of analysis the additional tensile forces and lateral displacement resulting from the collapse of unprotected compartment are calculated. The analyzed single-story building has two bays and one fire wall which separates the bays. Only one of the bays has a fire protection. The conducted analyses lead to the improvement of structural safety. It must be underlined here that the impact of fire situation can be significant, enforcing important changes in the project. Therefore in design of steel structures fire protection and the behavior of the structure during fire must be seriously taken into account.

1 INTRODUCTION

One of the most important tasks of single-storey steel buildings design is to ensure the safety of people and materials in fire conditions. A broad discussion on this topic is conducted by Biegus (2014). Single-storey buildings can be designed traditionally or according to the principles of fire safety engineering. In traditional procedures the fire resistance of the building is expressed as the time in minutes, before the structure loses its load capacity. Design Manuals (2008) present methods which allow the designer to evaluate the mechanical behavior of steel structure exposed to high temperature. In engineering practice the most widespread simplified methods are those based on the tabulated data or critical temperature concept. Some interesting analysis of boundary conditions influence on the thermal response of selected steel members subjected to fire loads based on simplified method are performed in Kucz et al. (2013). In the paper simply supported and restrained steel beams subjected to fire accidental action are analyzed using the standard ISO fire curve and the recommendations implemented in Eurocode 0 (2004), Eurocode 1 (2002), Eurocode 3 (2005) and DIFISEK (2008). The authors found out that the beam boundary conditions can modify structural response and result in reduction of critical temperature and fire resistance time. Another interesting issue is analysis of steel structures made of members of fourth class cross-sections. Due to local and distortional buckling, which result in decrease in load bearing capacity of the member this issue is very complicated as presented in Franssen et al. (2009, 2010). This problem is discussed in Rzeszut & Polus (2013) too, where the influence of fire load on class of cross-section of steel structural elements is presented. By means of several examples the authors show that calculating procedures proposed in Annex E of Eurocode 3 for members with Class 3 in normal situation and Class 4 in fire case, lead to critical temperature higher than 350°C and result in savings in costs of fire protection. Nevertheless, the relatively simple conventional methods are usually used to design buildings that require a low level of fire resistance and only allow for analysis of separate structural elements. Nowadays, fire designing of steel multi-bay single-storey and industrial buildings with the firewalls (Fig. 1) requires an analysis of complete bearing structure, taking into account the different fire scenarios. For economic reasons, fire protection is often used only in specially separated fire zones.

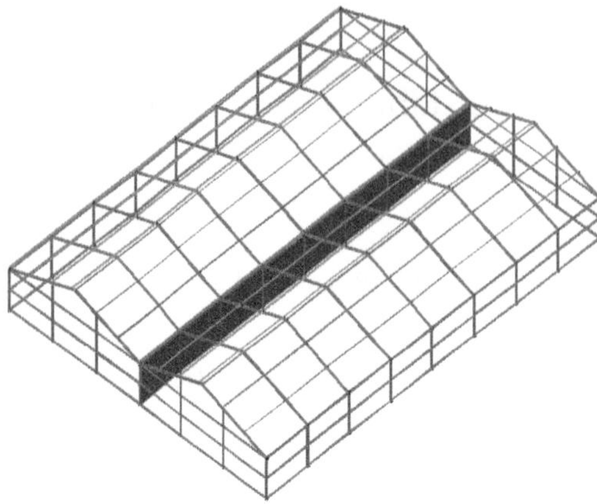

Figure 1. Industrial hall.

This solution may cause the collapse mechanisms in fire event of unsecured bay which may pulling together secured parts of single-storey building. The analysis of described above mechanism is very complex, therefore very fast progress is observed in the simplified analysis of this phenomenon which may improve the design process routine.

However, there is still lack of examples which present application of the simplified methods with proper conscience of real situations and need of using more sophisticated analyses. One of the earliest European regulations including fundamental design rules for single-storey steel buildings accounting for fire conditions are provided in Design Manuals (2008). Some interesting fire resistance analyses of simple steel frame were presented in Maślak & Tkaczyk (2011). Instead of method relying on a single failure mechanism, the two most common failure mechanisms for pitched portal frames under fire conditions are considered in Song et al. (2009) to predict the critical temperatures of the frame.

In the paper a single-storage two-bay building constructed with a steel structure in the fire situation is analyzed using several simplified design methods. In the analyzed example only one bay of the building has passive fire protection, while the second bay remains unprotected. The main aim of the research is to prove that absence in one bay of passive fire protection do not cause the risk of collapse of protected part and provide the safety of occupants and firemen. The analysis of collapse mechanism is carried out using the public domain program LUCA.

2 PROBLEM FORMULATION

The most common way to protect single-storey steel buildings against fire is the use of firewalls and fire protection of the steel elements. Firewalls divide buildings into fire compartments limiting damage only to the area where the fire occurred (Fig. 2).

The safety of occupants and firemen is guaranteed by the following criteria: no collapse towards the outside (Fig. 3) and no progressive collapse (Fig. 4).

The localized failure of the unprotected compartment in fire situation may leads to the collapse of the neighboring bays Design Manuals (2008). The most dangerous situation for firemen is when the structure collapses towards the outside. For this reason, it is necessary to check if it can occur. However, the main purpose of the paper is to analyse the localized failure of the unprotected compartment in fire condition and its influence on the other parts of the building. This influence may be evaluated using simplified methods implemented in public domain programme Luca. It makes possible

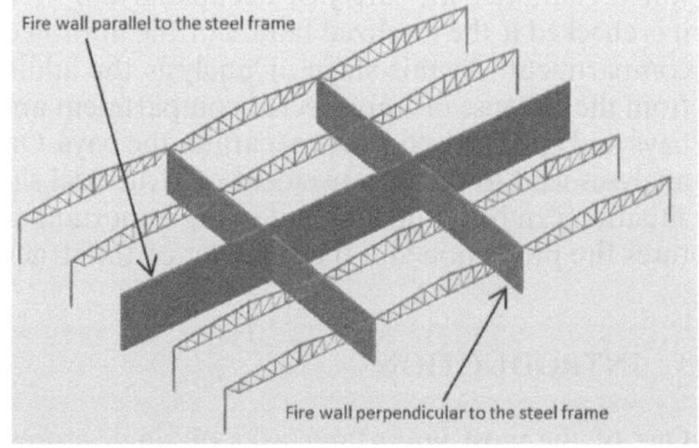

Figure 2. Partition of the single-storey building into fire compartments.

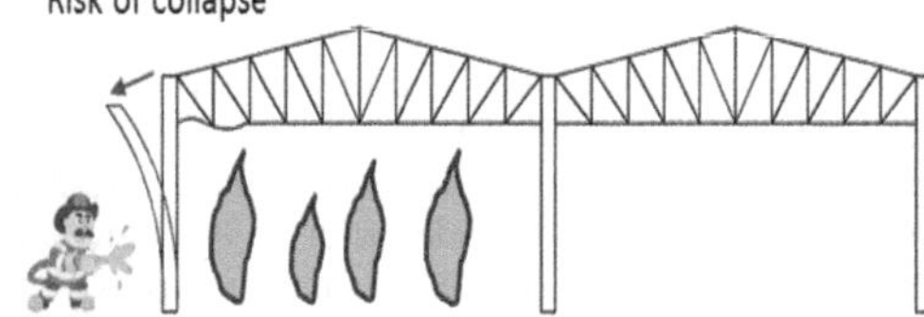

Figure 3. Criteria—no collapse towards the outside.

Figure 4. Criteria—no progressive collapse.

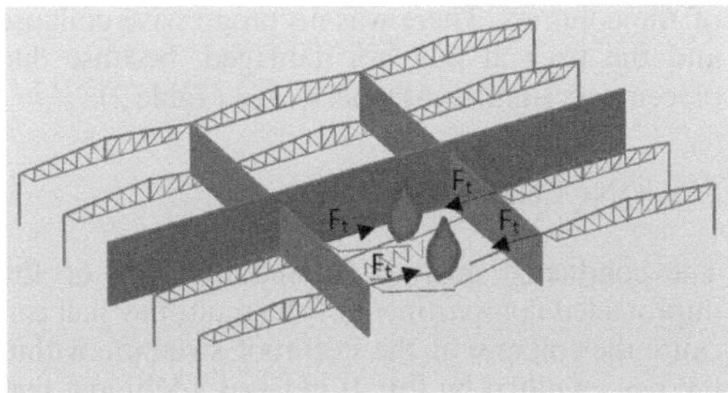

Figure 5. Forces caused during collapsing of the structures of the compartment in fire.

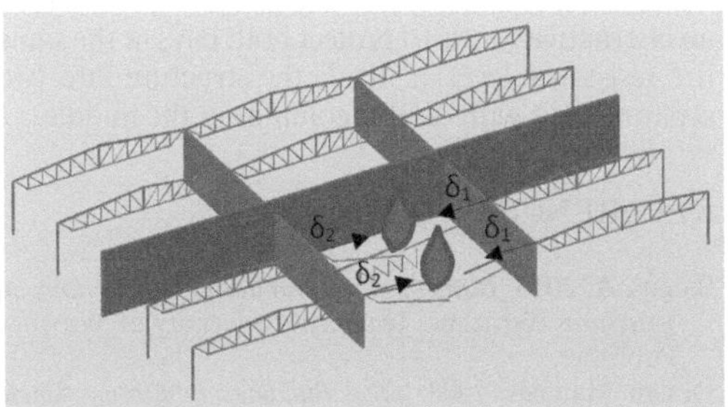

Figure 6. Displacements during collapsing of the structures of the compartment in fire.

the determination of forces which are generated by the collapsing structure in the compartment area covered by the fire (Fig. 5).

These forces are used as additional horizontal load during the structural analysis at the room temperature. Moreover, using simplified methods, it is possible to evaluate maximum horizontal displacements and check whether fire walls or building facades will be destroyed or not (Fig. 6).

3 TENSILE FORCES AND LATERAL DISPLACEMENT AT THE FIRE COMPARTMENT

The horizontal tensile force at the fire compartment boundaries may be determined using the recommendations in Design Manuals (2008) and the formula:

$$F = c_p n_{eff} q l \tag{1}$$

where c_p = an empirical coefficient depending on the slope of the roof and the type of steel structure; n_{eff} = a coefficient related to the total number of heated bays in the fire compartment; q = the linear load on roof applied on the beam in the fire situation [N/m]; l = the span of heated bay connected to the column [m].

The forces are used to check resistance of the frame remaining after fire. Tensile forces may lead to collapse of the cold frames.

Maximum lateral displacements at the top of columns δ_i (i = 1, 2) located at the compartment boundary can be obtained from:

$$\delta_i = \begin{cases} \dfrac{K_t}{K_i} c_{th} n l & \text{fire at the end of the building} \\ Max\left\{\dfrac{K_t}{K_i} c_{th} n l; \dfrac{F}{K_i}\right\} & \text{fire in the middle of the building} \end{cases} \tag{2}$$

where n = is the number of heated bays; K_i = is the equivalent lateral stiffness of the considered part i of the structure [N/m]; K_t = the equivalent stiffness depending on equivalent stiffness K_1 and K_2 [N/m]; l = is the span of the heated bay connected to the column; F = the tensile force [N]; c_{th} = an empirical coefficient dependent on the slope of the roof and type of steel structure.

Displacements are used to check resistance of the fire wall. When they are too great, fire wall may be destroyed.

4 CALCULATION EXAMPLE

In the paper a numerical example is presented. The structure consists of two bays steel frame. Only one of them had fire protection (Fig. 7).

First, frames were designed in normal conditions, then in fire conditions where critical temperatures for elements were obtained. Fire resistance time was obtained using ISO curve. Next, the mechanical behavior of steel structure exposed to high temperature was evaluated. The fire wall was located in the middle of the hall and fire burst up only in one of the compartments, where there was not fire protection. The roof of this compartment was collapsed and additional tensile forced were created. The forces and displacements were obtained using LUCA program.

The computed values of force and lateral displacement at the boundary of fire compartment are presented in the Table 1.

Next, this force was used as additional horizontal load in structural analysis in fire load condition at the room temperature. Additional horizontal forces led to significant lateral displacement and firewall destruction. For this reason, the overall constructional stiffness was improved and the stiffness of the columns was increased. The construction of the frame was strengthened using greater cross-sections

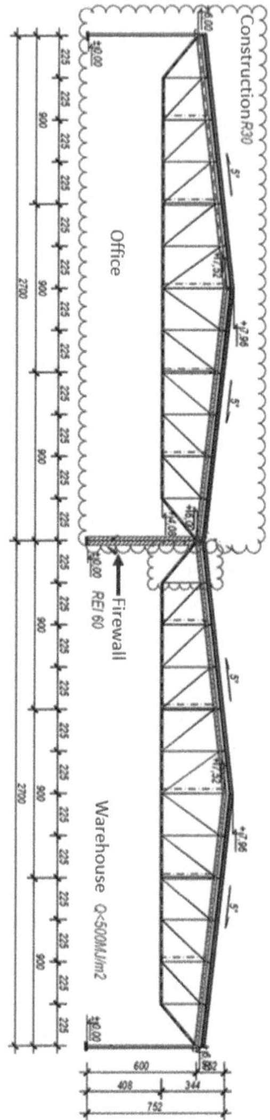

Figure 7. Analysed example.

Table 1. Numerical results.

Force (kN)	Lateral displacement (mm)
100.2	952

Table 2. Numerical results.

Force (kN)	Lateral displacement (mm)
100.2	84

of the columns. There was no progressive collapse and the firewall was not damaged, because displacements significantly decreased (Table 2).

5 CONCLUSIONS

The conducted analysis showed that fire of the unprotected compartment of the multi-bay hall can cause the collapse of the steel roof structure within the bay engulfed by fire. It induced additional tensile force which affected the neighboring bays. As a result, large displacements occurred at the top of the columns and damaged the firewalls. In order to avoid it the stiffeners of the column should be increased but in some cases it can be uneconomical. Therefore, an alternative way is to protect both bays at the same fire resistance level or divide the structure into two separate bays with double columns in the middle.

REFERENCES

Biegus, A. 2014. *Basis of structural design and actions on structures*. Wrocław: Technical University of Wrocław Publisher.

Design Manuals. 2008. *Steel Buildings in Europe Single storey buildings Part 7 Fire engineering*, Arcelor Mittal.

DIFISEK. 2008. *Dissemination of Structural Fire Safety Engineering Knowledge*, Research project ECSC nr RFS-C2–03048.

Eurocode 0. 2004. EN 1990, Basic of structural design. European Committee for Standardization.

Eurocode 1. 2002. EN 1991-1-2, Actions on Structures Part 1–2 General Actions—Actions on structures exposed to fire. European Committee for Standardization.

Eurocode 3. 2005. EN 1993-1-1, Design of steel structures Part 1–1 General rules and rules for buildings. European Committee for Standardization.

Eurocode 3. 2005. EN 1993-1-2, Design of steel structures Part 1–2 Structural fire design. European Committee for Standardization.

Franssen, J.M. & Vila Real, P. 2010. *Fire design of Steel Structures, Eurocode 1: Actions on structures Part 1–2:Actions on structures exposed to fire, Eurocode 3:Design of steel structures Part 1–2 Structural fire design*. European Convention for Constructional Steelwork.

Franssen, J.M. et al. 2009. *Designing steel structures for fire safety*. Taylor & Francis Group.

Kucz, M. et al. 2013. Influence of Boundary Conditions on the Thermal Response of Selected Steel Members. *Procedia Engineering* 57: 977–985.

Maślak, M. & Tkaczyk, A. 2011. Fire resistance of simple steel frame—kinematic approach to evaluation, *Eurosteel 2011: Vol. B: Fire: Proceedings of 6th European Conference on Steel and Composite Structures*, Brussels: ECCS.

Rzeszut, K. & Polus, Ł. 2013, Classes of Cross-Sections of Steel Structural Elements in the Fire Situation, *Procedia Engineering* 57: 967–976.

Song, Y. et al. 2009. A new design method for industrial portal frames in fire. *Application of Structural Fire Engineering, Prague, Czech Republic.*

Advances in Mechanics: Theoretical, Computational and Interdisciplinary Issues – Kleiber et al. (Eds)
© 2016 Taylor & Francis Group, London, ISBN 978-1-138-02906-4

Form finding of tensegrity structures *via* Singular Value Decomposition of compatibility matrix

W. Gilewski
Warsaw University of Technology, Warsaw, Poland

J. Kłosowska & P. Obara
Kielce University of Technology, Kielce, Poland

ABSTRACT: Application of Singular Value Decomposition (SVD) of a compatibility matrix to form-finding of pin-joined structures is presented in the paper. This decomposition allows to identify the mechanism of geometrical variation of a truss, its preliminary classification as finite or infinitesimal and determination of longitudinal forces which respond to the self-stress state. The computational program for tensegrity form-finding and qualitative verification of truss properties was based on the finite element analysis within the *Mathematica* environment.

1 INTRODUCTION

The idea of tensegrity was first described about 50 years ago. According to Emmerich (1988) the first tensegrity structure was introduced in Russia in the early 1900s. The definition of tensegrity systems is not unique. A widely accepted version is proposed by Pugh (1976). The concept concerns specific trusses which consist of compression and tensile components which stabilize each other despite the fact that there are mechanisms in the structures. A detailed history of tensegrity systems is given by Motro (1992).

There are many benefits of tensegrity structures (Skelton et al. 2002), let us take some:

- tension stabilizes the structure,
- tensegrity structures can be more reliably modeled,
- tensegrity structures facilitate high precision control,
- tensegrity structures are efficient,
- tensegrity structures are deployable,
- tensegrity structures are easily tunable.

Tensegrity structures are built of bars and strings. The bars can resist compressive forces, the strings cannot do so. The rigidity of a tensegrity is a result of a self-stressed equilibrium between bars and strings. A compressive element loses stiffness under a load whereas a tensile element gains stiffness. Hence, in tensegrity, a large stiffness-to-mass ratio can be achieved by increasing the use of tensile elements.

All elements of a tensegrity structures are axially loaded. None of individual elements of these structures experience bending moments. Models for axially loaded elements can be expected more reliable compared to models for elements in bending. Structures that can be more precisely modeled can be more precisely controlled.

Efficiency of a structure increases with the minimal mass design for a given set of stiffness properties. Tensegrity structures use longitudinal elements arranged in a very unusual pattern to achieve maximum strength with small mass.

Since the compressive elements of tensegrity structures are either disjoint or connected with ball joints, large displacement, deployability and stowage in a compact volume are possible in tensegrity structures. This feature offers operational and portability advantages. Deployable structures can save transport costs by reducing the mass required, or by eliminating the requirement of humans for assembly. The deployment technique can also make small adjustments for fine tuning of the loaded structures, or adjustment of a damaged structure.

Nowadays the concept of tensegrity structures is understood in many ways. Key features of tensegrity structures can be presented in the following sections:

- the structure is a truss (K),
- there are self-stress states (S),
- there are infinitesimal mechanisms and they are stabilized by these self-stress states (M),
- the system of compression elements is discontinuous (N),
- compression members are inside the tensile elements (W),
- tensile elements have no rigidity (C)

The presence of all of these characteristics allows qualifying a structure as a tensegrity structure. In the case if three characteristics like K, S, C and one of M, N or W for a structure will be determined, then this is a structure with tensegrity features.

A key step in the design of tensegrity structures is the determination of their geometrical configuration, known as form-finding. The form-finding process determines a possible pre-stress distribution and geometry for a tensegrity (Fuller 1962, Snelson 1965).

The spectral analysis of trusses matrix is presented for form-finding of tensegrity structures in this paper. It was made using the decomposition matrix method which describes the elongations in a truss according to Singular Values (SVD) (Klema 1980, Stewart 1998). This method allows us to identify whether the structure is geometrically variable and whether there are self-stress states.

In the SVD decomposition a given matrix is presented in the form of the product of the unitary square matrix, the rectangular diagonal matrix with non-negative real coefficients and the Hermitian conjugation of unitary square matrix. Coefficients of the diagonal matrix are called singular values of the analyzed matrix. When the given matrix has real coefficients the unitary matrices become orthogonal matrices and the Hermitian conjugation becomes a transposition.

The paper presents the essence of singular value decomposition of the matrix and an example, explaining what information about the structure can be obtained according to this distribution. SVD of compatibility matrix of truss theory has been hinted at in literature, but usually for statically determinate structures.

2 GOVERNING EQUATIONS

The subject of the analysis is N-membered, unloaded and supported truss with following characteristics: material constants E_e, cross-sectional areas A_e and element lengths L_e $(e = 1, \ldots, N)$. Its mechanical properties are described by three linearized equations: compatibility, material properties and equilibrium with boundary conditions included:

$$\mathbf{\Delta} = \mathbf{Bq} \tag{1}$$

$$\mathbf{S} = \mathbf{E\Delta} \tag{2}$$

$$\mathbf{B}^T\mathbf{S} = \mathbf{P} \tag{3}$$

where $\mathbf{q}$ is M-membered displacement vector, $\mathbf{B}$ is compatibility matrix, $\mathbf{\Delta}$ is extension vector, $\mathbf{S}$ is longitudinal forces, $\mathbf{E}$ is elasticity matrix and $\mathbf{P}$ is load vector.

The equilibrium equations for analyzed, unloaded and supported, trusses can be presented in the form of displacements:

$$\mathbf{K}_L\mathbf{q} = \mathbf{0} \tag{4}$$

where

$$\mathbf{K}_L = \mathbf{B}^T\mathbf{EB} \tag{5}$$

is the linear stiffness matrix, or in the form of stresses (3) after a symmetrisation:

$$\mathbf{DS} = \mathbf{0} \tag{6}$$

where

$$\mathbf{D} = \mathbf{BB}^T \tag{7}$$

The compatibility matrix $\mathbf{B}$ of an analyzed truss can be determined directly or using the formalism of the finite element method (Gilewski et al. 2012, Zienkiewicz 2000).

3 SINGULAR VALUE DECOMPOSITION

The analysis of the compatibility matrix $\mathbf{B}$ is allows determining whether a structure has the characteristics of tensegrity. Making the Singular Value Decomposition (SVD) of this matrix we can identify whether the structure is geometrically variable and whether there are self-stress states.

The singular value decomposition of an $N \times M$ real matrix $\mathbf{B}$ is a factorization of the form:

$$\mathbf{B} = \mathbf{YNX}^T \tag{8}$$

where $\mathbf{Y}$ is an $N \times N$ real orthogonal matrix, $\mathbf{X}$ is an $M \times M$ real orthogonal matrix and $\mathbf{N}$ is an $N \times M$ rectangular diagonal matrix. Let us consider two eigenproblems:

$$\left(\mathbf{BB}^T - \mu\mathbf{I}\right)\mathbf{y} = 0 \tag{9}$$

and

$$\left(\mathbf{B}^T\mathbf{B} - \lambda\mathbf{I}\right)\mathbf{x} = 0 \tag{10}$$

with the solutions in the form of eigenvalues and eigenvectors:

$$\mu_1, \mathbf{y}_1;\ \mu_2, \mathbf{y}_2;\ \ldots;\ \mu_N, \mathbf{y}_N \tag{11}$$

and

$$\lambda_1, \mathbf{x}_1;\ \lambda_2, \mathbf{x}_2;\ \ldots;\ \lambda_M, \mathbf{x}_M \tag{12}$$

Full solutions of the above eigenproblems can be expressed in the condensed form

$$\mathbf{BB}^T = \mathbf{YMY}^T \tag{13}$$

$$\mathbf{B}^T\mathbf{B} = \mathbf{XLX}^T \tag{14}$$

where

$$\mathbf{M} = diag[\mu_1, \mu_2, ..., \mu_M] \tag{15}$$

$$\mathbf{L} = diag[\lambda_1, \lambda_2, ..., \lambda_M] \tag{16}$$

$$\mathbf{Y} = [\mathbf{y}_1\ \mathbf{y}_2 \cdots \mathbf{y}_N] \tag{17}$$

$$\mathbf{X} = [\mathbf{x}_1\ \mathbf{x}_2 \ldots \mathbf{x}_M] \tag{18}$$

The matrices (17) and (18) are orthogonal, it means that:

$$\mathbf{X}^T\mathbf{X} = \mathbf{I} \text{ and } \mathbf{Y}^T\mathbf{Y} = \mathbf{I} \tag{19}$$

One can notice that the product $\mathbf{BB}^T$ (Eq. 13) can be considered a matrix of symmetrised equations of equilibrium with non-negative eigenvalues (Eq. 6). If all the eigenvalues (Eq. 15) are positive there are no self-stress states in the truss. In the tensegrity structures, one or more of the eigenvalues are zero. Zero eigenvalues (if any) are related to the non-zero solution of homogeneous equations named self-stress. The self-stress can be considered as an eigenvector related to zero eigenvalue.

In a similar way the product $\mathbf{B}^T\mathbf{B}$ (Eq. 14) can be considered a particular form of linear stiffness matrix (Eq. 5) with unit elasticity matrix. Eigenvalues (Eq. 16) describe the energy states of the module, while the eigenvectors describe the form of deformation. There are no movements in the case when all the eigenvalues are greater than zero. If we analyses the tensegrity structure one or more of the eigenvalues are zero. Zero eigenvalues are related to the finite or infinitesimal mechanisms, but in general the information from the null-space analysis alone does not suffice to establish the difference. The mechanism can be considered an eigenvector related to zero eigenvalue. To establish if the mechanism is infinitesimal it is necessary to apply the nonlinear analysis with the use of geometric stiffness matrix, which is possible if the self-stress exist. Lack of self-stress means that the mechanism is finite.

Based on the above two eigenproblems (Eq. 9) and (10) it is easy to proof the singular value decomposition of the compatibility matrix **B**:

$$\mathbf{BB}^T = \mathbf{YNX}^T\mathbf{XN}^T\mathbf{Y}^T = \mathbf{YNN}^T\mathbf{Y}^T = \mathbf{YMY}^T \tag{20}$$

$$\mathbf{B}^T\mathbf{B} = \mathbf{XN}^T\mathbf{Y}^T\mathbf{YNX}^T = \mathbf{XN}^T\mathbf{NX}^T = \mathbf{XLX}^T \tag{21}$$

with the following relations included:

$$\mathbf{M} = \mathbf{NN}^T \tag{22}$$

$$\mathbf{L} = \mathbf{N}^T\mathbf{N} \tag{23}$$

4 EXAMPLES

This paper explains what kind of information about a structure can be obtained from the singular value decomposition of the compatibility matrix **B** (Eq. 8). In order to demonstrate the method, the two-dimensional and three-dimensional truss will be analyzed. Calculations were made in the *Mathematica* environment.

4.1 *2D truss*

The two-dimensional truss shown in Figure 1 is presented. The vectors of nonzero generalized coordinates and longitudinal forces are:

$$\mathbf{q} = \{q_3\ q_4\ q_5\ q_6\ q_7\ q_8\ q_9\ q_{10}\ q_{13}\ q_{14}\ q_{15}\ q_{16}\ q_{17}\ q_{18}\ q_{19}\ q_{20}\} \tag{24}$$

and

$$\mathbf{S} = \{S_1\ S_2 \ldots S_{15}\} \tag{25}$$

In this case number of elements N equals 15 but number of nonzero generalized coordinates $M = 16$. For this **B** is the 15×16 real matrix. The roots of $\mathbf{BB}^T$ matrices are found on the diagonal entries of:

$$\mathbf{M} = diag\{3.85\ 3.59\ 3.43\ 3.02\ 2.69\ 2.12\ 2.0\ 1.48\ 1.23\ 1.07\ 0.71\ 0.49\ 0.18\ 0.12\ 0\} \tag{26}$$

and the roots of $\mathbf{B}^T\mathbf{B}$—on the diagonal entries of:

$$\mathbf{L} = diag\{3.85\ 3.59\ 3.43\ 3.02\ 2.69\ 2.12\ 2.0\ 1.48\ 1.23\ 1.07\ 0.71\ 0.49\ 0.18\ 0.12\ 0\ 0\} \tag{27}$$

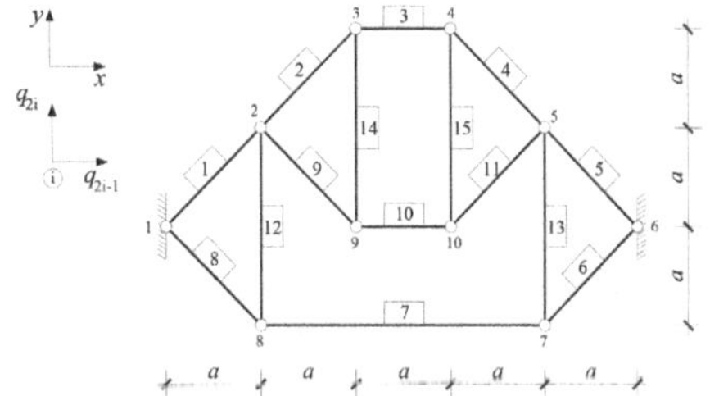

Figure 1. Scheme of the 2D truss.

Zero eigenvalue presented in the matrix (Eq. 26) is responsible for existence of the one self-stress state defined by the eigenvector of $\mathbf{BB}^T$:

$$\mathbf{y}_{15} = \{0.38\ 0.19\ 0.13\ 0.19\ 0.38\ 0.38\ 0.26\ 0.38\ 0.19\ 0.13\ 0.19 - 0.16\ -0.26\ -013\ -0.13\} \tag{28}$$

Two zero eigenvalues in the matrix (Eq. 27) are responsible for existence of two mechanisms defined by the two eigenvectors of $\mathbf{B}^T\mathbf{B}$ corresponding to those values:

$$\mathbf{x}_{15} = \{0.21\ -0.21\ 0.32\ -0.32\ 0.32\ 0.32\ 0.21\ 0.21\ -0.21\ 0.21\ -0.21\ -0.21\ 0.11\ -0.32\ 0.11\ 0.32\} \tag{29}$$

and

$$\mathbf{x}_{16} = \{0.16\ -0.16\ -0.2\ 0.2\ -0.2\ -0.2\ 0.16\ 0.16\ -0.16\ 0.16\ -0.16\ -0.16\ 0.52\ 0.2\ 0.52\ -0.2\} \tag{30}$$

Analysis of the compatibility matrix $\mathbf{B}$ is allows determining that in this case there is one self-stress and two mechanisms. The self-stress observed in the structure with the normalized forces (Eq. 28). It means that four elements on this truss (12,13,14,15) are compressed and other are tensile. The mechanisms (Eq. 29, 30) are presented in Figure 2.

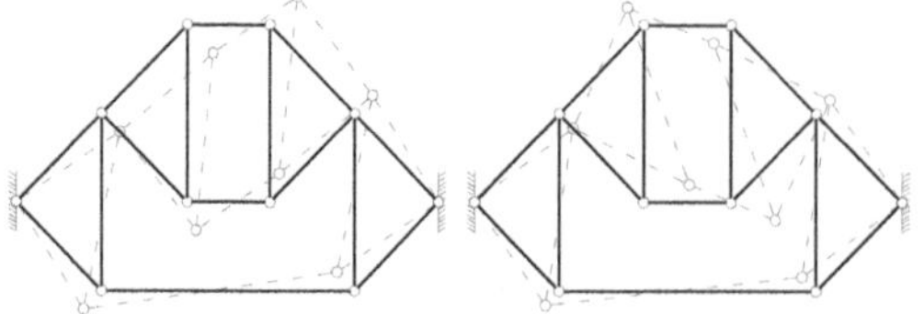

Figure 2. Mechanisms of the 2D truss.

4.2 *3D truss*

The three-dimensional truss named Simplex shown in Figure 3 is presented. The vectors of nonzero generalized coordinates and longitudinal forces are:

$$\mathbf{q} = \{q_2\ q_4\ q_7\ q_{10}\ q_{11}\ q_{12}\ q_{13}\ q_{14}\ q_{15}\ q_{16}\ q_{17}\ q_{18}\} \tag{31}$$

and

$$\mathbf{S} = \{S_1, S_2, ..., S_{12}\} \tag{32}$$

In this case number of elements N is equal to number of nonzero generalized coordinates M:

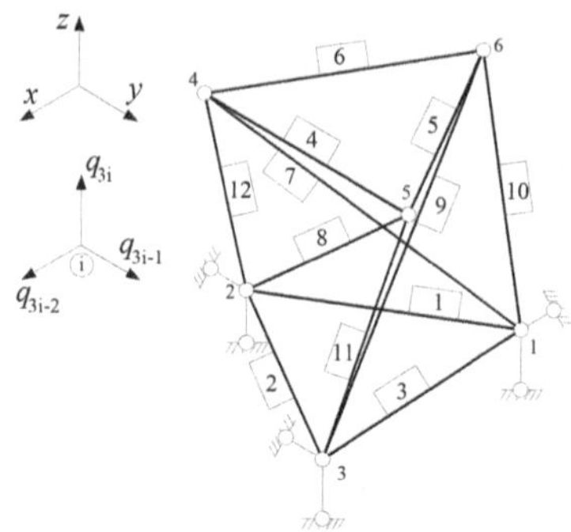

Figure 3. Scheme of the 3D tensegrity truss—the Simplex.

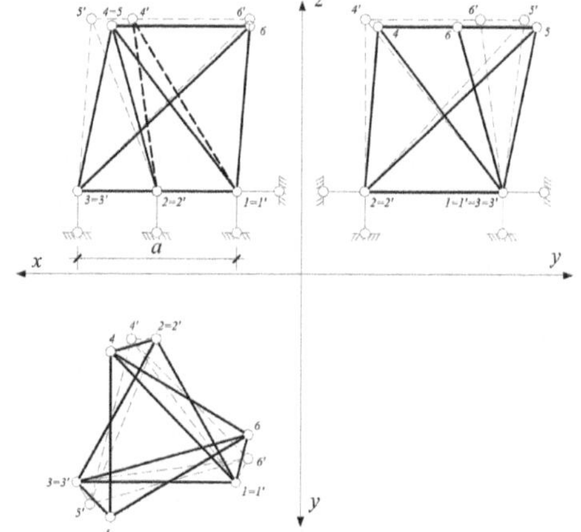

Figure 4. Mechanism of the Simplex.

$N = M = 12$. For this $\mathbf{B}$ is the 12×12 real matrix. The roots of both $\mathbf{BB}^T$ and $\mathbf{B}^T\mathbf{B}$ matrices are found on the diagonal entries of:

$$\mathbf{M} = \mathbf{L} = diag\{3.81\ 2.54\ 2.31\ 1.77\ 1.51\ 1.27\ 1.07\ 0.86\ 0.21\ 0.11\ 0.06\ 0\} \tag{33}$$

The zero eigenvalue presented in matrix (Eq. 33) is responsible for existence of the self-stress state defined by the eigenvector of $\mathbf{BB}^T$:

$$\mathbf{y}_{12} = \{0.17\ 0.17\ 0.17\ 0.17\ 0.17\ 0.17\ -0.43 - 0.43 - 0.43\ 0.3\ 0.3\ 0.3\} \tag{34}$$

and for existence of the mechanism defined by the eigenvector of $\mathbf{B}^T\mathbf{B}$ corresponding to this value too:

$$\mathbf{x}_{12} = \{0\ 0\ 0\ -0.48\ -0.28\ 0.16\ 0.48\ -0.28\ 0.16\ 0\ 0.55\ 0.16\} \tag{35}$$

The analysis of the compatibility matrix **B** is allows determining that in this case there is one self-stress and one mechanisms. The self-stress observed in the structure with the normalized forces (Eq. 34). It means that in truss elements (number shown in Fig. 3) are the following forces: 0.17 in top and bottom cables (1,2,3,4,5,6), 0.3 in vertical cables (10,11,12) and –0.43 in vertical struts (7,8,9). The mechanism (Eq. 35) is presented in Figure 4.

5 CONCLUSIONS

The paper discusses the use of singular value decomposition for the qualitative analysis of trusses, including tensegrity structures. The analysis allows one to draw the following conclusions:

- those structures have all of tensegrity characteristics (K+S+M+N+W+C) what allow qualifying as tensegrity structures,
- the knowledge of the compatibility matrix **B** is sufficient to analyze the qualitative properties of the truss, including the configuration of nodes and rods. Therefore only the geometric description is needful. It is not necessary to know the cross bars and the material properties,
- singular value decomposition allows to identify the mechanism of geometrical variation of a truss and its preliminary classification as finite or infinitesimal,
- singular value decomposition allows determination of the normal forces which respond to the self-stress state, if there are any in the analyzed truss,
- the knowledge of the normal forces in elements of a truss allows to determine a geometric stiffness matrix. Spectral analysis taking into account a geometric stiffness matrix should be carried out to determine whether the mechanism is infinitesimal.

The considerations set out in the work can be used to analyze both two- and three-dimensional trusses.

REFERENCES

Emmerich, D.G. 1988. Structures tendues et autotendantes, Monographies de Ge´ometrie Constructive, *E´ditions de l'E´ cole d'Architecture de Paris La Villette*, Paris.

Fuller, R.B. 1962. *Tensile-integrity structures*, United States Patent 3.063.521.

Gilewski, W. & Kasprzak, A. 2012. Introduction to tensegrity modules (in Polish), *Theoretical Foundations of Civil Engineering*, Vol. I. Mechanics of Materials and Structures, 83–94. Warszawa.

Klema, V.C. 1980. The singular value decomposition: it's computation and some applications, *IEEE Transactions on Automatic Control* AC-25, 2. 164–176.

Motro, R. 1992. Tensegrity systems: the state of the art, *International Journal of Space Structures* 7(2). 75–83.

Pugh, A. 1976. *An Introduction to Tensegrity*, first ed., Berkeley: University of California Press, Los Angeles, California.

Skelton, R.E., Helton, J.W., Adhikari, R., Pinaud, J.P., Chan, W. 2002. *An Introduction to the Mechanics of Tensegrity Structures*, CRC Press LLC.

Snelson, K.D. 1965. *Continuous tension, discontinuous compression structures*, United States Patent 3.169.611.

Stewart, G.W. 1998. *Matrix algorithms: basic decompositions*, SIAM, Philadelphia.

Zienkiewicz, O.C. & Taylor, R.L., 2000. *The Finite Element Method*, Vol. 1: The Basis, fifth ed., Butterworth-Heinemann, Oxford.

Advances in Mechanics: Theoretical, Computational and Interdisciplinary Issues – Kleiber et al. (Eds)
 ISBN 978-1-138-02906-4

Technical coefficients for continuum models of orthotropic tensegrity modules

W. Gilewski & A. Al Sabouni-Zawadzka
Faculty of Civil Engineering, Warsaw University of Technology, Warsaw, Poland

ABSTRACT: The paper focuses on the application of continuum models for orthotropic tensegrity modules and the determination of their technical coefficients. Tensegrities are cable-strut systems with a special node configuration, which ensures the occurrence of infinitesimal mechanisms balanced with self-stress states. The continuum model of the module is built assuming that the strain energy of the unsupported tensegrity structure is equivalent to the strain energy of the cube. After the proper validation, the proposed model can be used to determine and interpret physical properties of the module. From the obtained elasticity matrix and the inverse matrix, technical coefficients such as: Young's moduli, Poisson's ratios and shear moduli can be computed.

1 INTRODUCTION

The term "tensegrity" was first introduced by Buckminster Fuller (see Skelton & de Oliveira 2005 for historical details). Several definitions of this concept can be found in the literature (see Motro 2003). For the purpose of this paper, a tensegrity structure is defined as a pin-jointed system with a particular configuration of cables and struts that form a statically indeterminate structure in a stable equilibrium. Tensegrities consist of a discontinuous set of compressed elements inside a continuous set of tensioned members, which have no compressive stiffness. Infinitesimal mechanisms, which occur in tensegrity structures, are balanced with self-stress states. To the major advantages of tensegrity systems belong: large stiffness-to-mass ratio, deployability, reliability and controllability.

The continuum model of tensegrity modules should allow the following:

- estimate properties of the module with typical deformation modes (tension, shear),
- evaluate the influence of self-stress for the defined deformation,
- evaluate the influence of cables and struts for the properties of the module,
- compare the elastic properties of typical tensegrity modules,
- determine the technical coefficients,
- find a physical interpretation for the technical coefficients,
- find conditions limiting the values of technical coefficients.

2 3D CONTINUUM MODEL

Discrete models of tensegrity modules are mathematically described with the use of the Finite Element Method (Zienkiewicz 2000). The strain energy is a quadratic form of nodal displacements $\mathbf{q}$:

$$E_s^{FEM} = \frac{1}{2}\mathbf{q}^T\mathbf{K}\mathbf{q} \tag{1}$$

with the global linear and geometric stiffness matrix $\mathbf{K}=\mathbf{K}_L+\mathbf{K}_G$ as a kernel. The self-stress of the module (proportional to the tension force S) is represented by the geometric stiffness matrix.

In order to build a continuum model of tensegrity modules (Gilewski & Kasprzak 2014, Green & Zema 2002, Kebiche et al. 2008), the symmetric linear 3D elasticity theory is considered. According to this theory, the strain energy can be expressed as:

$$E_s^{LES} = \frac{1}{2}\int_V \boldsymbol{\varepsilon}^T\mathbf{E}\boldsymbol{\varepsilon}\, dV \tag{2}$$

where: $\boldsymbol{\varepsilon}$—the strain vector, $\mathbf{E}$—the elasticity matrix.

In the proposed concept it is assumed that the strain energy of the unsupported tensegrity structure is equivalent to the strain energy of the cube of edge length a (Fig. 1) and that the strain energy of the cube is constant in its whole volume.

To compare the energies and build the equivalent matrix $\mathbf{E}$, the nodal displacements are expressed

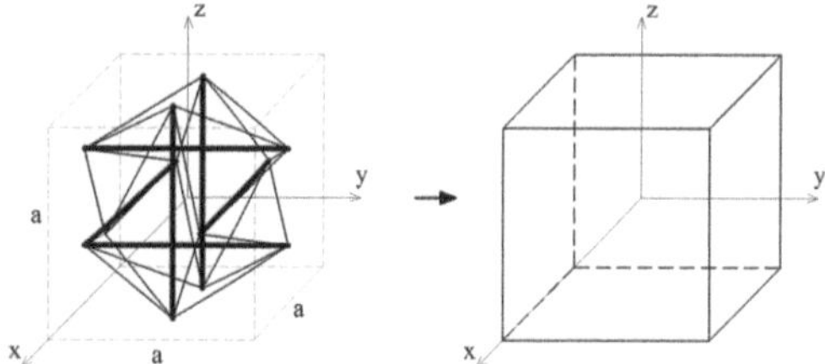

Figure 1. Tensegrity and continuum.

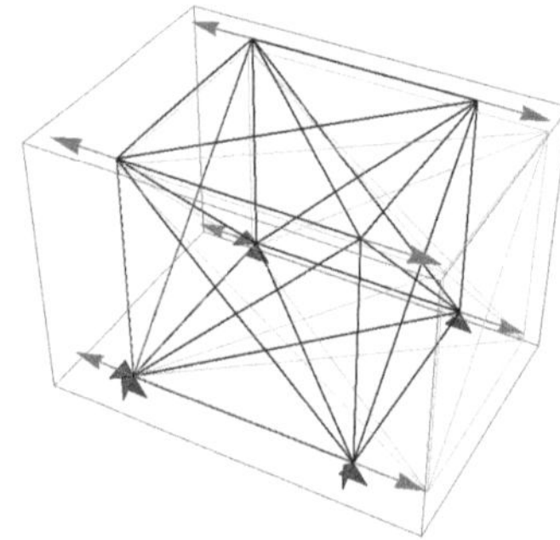

Figure 2. Determination of E and ν in one direction.

by the average mid-values of displacements and their derivatives with the use of Taylor series expansion. Coordinates of nodal points $\{\alpha_{xi}\cdot a, \alpha_{yi}\cdot a, \alpha_{zi}\cdot a\}$ are expressed in Taylor series by the edge length a with the increments: $\Delta x = \alpha_{xi}\cdot a$, $\Delta y = \alpha_{yi}\cdot a$, $\Delta z = \alpha_{zi}\cdot a$.

The obtained elasticity tensor can be expressed in Voight's form $\mathbf{E} = [e_{ij}]\ i, j = 1,2, \ldots 6$. There are 21 independent coefficients for anisotropy. However, the analyses presented in this paper have been limited to orthotropic cases with 9 independent coefficients.

3 MODEL VALIDATION

The proposed continuum model has been validated by comparing several results obtained from this model with the corresponding results obtained from the truss analysis. The validation was performed on an orthotropic space truss. The truss was tested for the technical coefficients: Young's modulus E, Poisson's ratio ν and shear modulus G. Nine independent coefficients were determined and compared. Figure 2 shows the test which was carried out in order to determine the Young's modulus and the Poisson's ratios for one direction. The same tests were performed for the other two directions.

Similar tests were performed to determine shear moduli in three directions.

4 EXAMPLE – TENSEGRITY

The example presented in the study concerns a continuum model of the orthotropic tensegrity module—an expanded octahedron inscribed into a cube of edge length a (Fig. 3). The geometry is described by three parameters: $\alpha_1 = 0.65$, $\alpha_2 = 0.30$, $\alpha_3 = 0.56$, which are ratios between struts spacings and length a in three directions.

After the process described above, the following equivalent elasticity matrix for the expanded octahedron was obtained:

$$\mathbf{E} = \begin{bmatrix} e_{11} & e_{12} & e_{13} & 0 & 0 & 0 \\ & e_{22} & e_{23} & 0 & 0 & 0 \\ & & e_{33} & 0 & 0 & 0 \\ & & & e_{12} & 0 & 0 \\ & & & & e_{13} & 0 \\ sym. & & & & & e_{23} \end{bmatrix} \tag{3}$$

with the following components depending on the physical properties and the self-stress of the module:

$$e_{11} = \frac{EA}{a^2}(3.04649 + 2\cdot k + 0.25845\cdot\sigma) \tag{4}$$

$$e_{22} = \frac{EA}{a^2}(2.71823 + 2\cdot k + 0.274056\cdot\sigma) \tag{5}$$

$$e_{33} = \frac{EA}{a^2}(1.84239 + 2\cdot k + 0.32202\cdot\sigma) \tag{6}$$

$$e_{12} = e_{44} = \frac{EA}{a^2}(0.845615 - 0.105243\cdot\sigma) \tag{7}$$

$$e_{13} = e_{55} = \frac{EA}{a^2}(1.26604 - 0.153207\cdot\sigma) \tag{8}$$

$$e_{23} = e_{66} = \frac{EA}{a^2}(1.51283 - 0.168813\cdot\sigma) \tag{9}$$

$$e_{14} = e_{15} = e_{16} = e_{24} = e_{25} = e_{26} = e_{34} = e_{35} = e_{36} = e_{45} = e_{46} = e_{56} = 0 \tag{10}$$

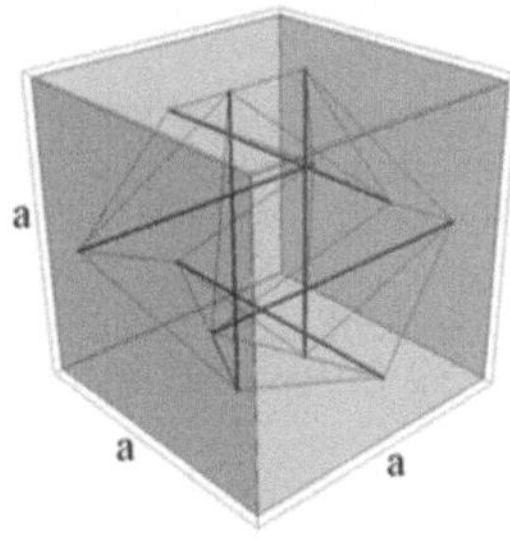

Figure 3. Expanded octahedron inscribed into a cube.

where:

$$k=\frac{(EA)_{strut}}{(EA)_{cable}},(EA)_{cable}=EA,\sigma=\frac{S}{EA} \tag{11}$$

In order to determine technical coefficients of the module, several operations have to performed. Shear moduli can be computed directly from the components of the elasticity matrix. Young's moduli and Poisson's ratios, on the other hand, can be determined from the components of the inverse elasticity matrix $\mathbf{E}^{-1}$:

$$\mathbf{E}^{-1}=\begin{bmatrix} \frac{1}{E_1} & -\frac{\nu_{21}}{E_2} & -\frac{\nu_{31}}{E_3} & 0 & 0 & 0 \\ -\frac{\nu_{12}}{E_1} & \frac{1}{E_2} & -\frac{\nu_{32}}{E_3} & 0 & 0 & 0 \\ -\frac{\nu_{13}}{E_1} & -\frac{\nu_{23}}{E_2} & \frac{1}{E_3} & 0 & 0 & 0 \\ 0 & 0 & 0 & \frac{1}{G_1} & 0 & 0 \\ 0 & 0 & 0 & 0 & \frac{1}{G_2} & 0 \\ 0 & 0 & 0 & 0 & 0 & \frac{1}{G_3} \end{bmatrix} \tag{12}$$

The following values of technical coefficients were obtained for the expanded octahedron:

$$E_1=\frac{e_{13}^2e_{22}-2e_{12}e_{13}e_{23}+e_{11}e_{23}^2+e_{12}^2e_{33}-e_{11}e_{22}e_{33}}{e_{23}^2-e_{22}e_{33}} \tag{13}$$

$$E_2=\frac{e_{13}^2e_{22}-2e_{12}e_{13}e_{23}+e_{11}e_{23}^2+e_{12}^2e_{33}-e_{11}e_{22}e_{33}}{e_{13}^2-e_{11}e_{33}} \tag{14}$$

$$E_3=\frac{e_{13}^2e_{22}-2e_{12}e_{13}e_{23}+e_{11}e_{23}^2+e_{12}^2e_{33}-e_{11}e_{22}e_{33}}{e_{12}^2-e_{11}e_{22}} \tag{15}$$

$$G_1=e_{12},\quad G_2=e_{13},\quad G_3=e_{23} \tag{16}$$

$$\nu_{12}=\frac{e_{13}e_{23}-e_{12}e_{33}}{e_{23}^2-e_{22}e_{33}},\quad \nu_{21}=\frac{e_{13}e_{23}-e_{12}e_{33}}{e_{13}^2-e_{11}e_{33}} \tag{17}$$

$$\nu_{13}=\frac{-e_{13}e_{22}+e_{12}e_{23}}{e_{23}^2-e_{22}e_{33}},\quad \nu_{31}=\frac{-e_{13}e_{22}+e_{12}e_{23}}{e_{12}^2-e_{11}e_{22}} \tag{18}$$

$$\nu_{23}=\frac{e_{12}e_{13}-e_{11}e_{23}}{e_{13}^2-e_{11}e_{33}},\quad \nu_{32}=\frac{e_{12}e_{13}-e_{11}e_{23}}{e_{12}^2-e_{11}e_{22}} \tag{19}$$

with the symmetry conditions:

$$\frac{\nu_{12}}{E_1}=\frac{\nu_{21}}{E_2},\quad \frac{\nu_{13}}{E_1}=\frac{\nu_{31}}{E_3},\quad \frac{\nu_{23}}{E_2}=\frac{\nu_{32}}{E_3} \tag{20}$$

Matrices $\mathbf{E}$ and $\mathbf{E}^{-1}$ must be positive definite, so that the elastic strain energy is non-negative definite. The strain energy is equal to zero only when there are no deformations. If the deformations appear, the strain energy has to be positive.

In case of the positive definite matrices, their principal minors are positive. This imposes limitations on the obtained technical coefficients:

$$E_1>0,\quad E_2>0,\quad E_3>0 \tag{21}$$

$$G_1>0,\quad G_2>0,\quad G_3>0 \tag{22}$$

$$\nu_{12}\nu_{21}<1,\quad \nu_{13}\nu_{31}<1,\quad \nu_{23}\nu_{32}<1 \tag{23}$$

$$\nu_{12}\nu_{21}+\nu_{13}\nu_{31}+\nu_{23}\nu_{32}+\nu_{12}\nu_{31}\nu_{23}+\nu_{21}\nu_{13}\nu_{32}<1 \tag{24}$$

The last condition is global and comes from the positive determinants of the submatrix 3×3.

Dependence of the selected technical coefficients on the cable to strut ratio k and the effect of self-stress σ is presented in Figures 4–8 with the limitation conditions included. Following these graphs it is possible to define a range of the cable to strut ratio k and the self-stress ratio σ for the correctly defined tensegrity module.

The above graphs present how the Young's modulus E_1 depends on two parameters: k and σ. Taking into consideration the given ranges of the parameters: $k\in\{0;1\}$ and $\sigma\in\{0;3\}$, the

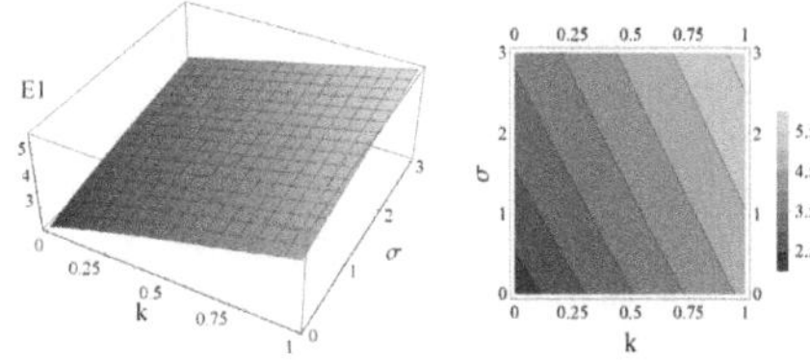

Figure 4. Young's modulus E_1 multiplied by the factor: EA/a^2 with the limitation condition: $E_1>0$.

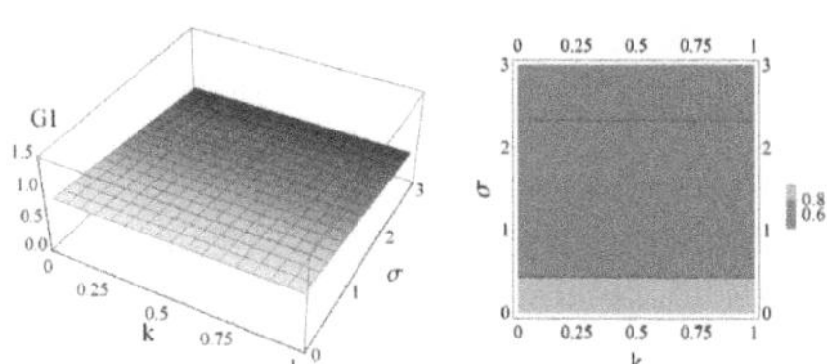

Figure 5. Shear modulus G_1 multiplied by the factor: EA/a^2 with the limitation condition: $G_1>0$.

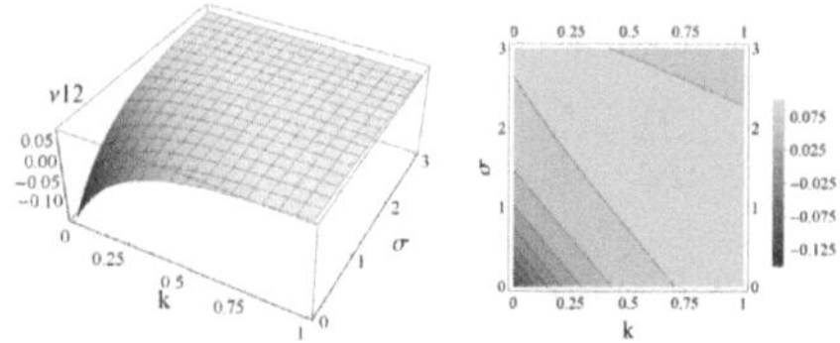

Figure 6. Poisson's ratio ν_{12} multiplied by the factor: EA/a^2.

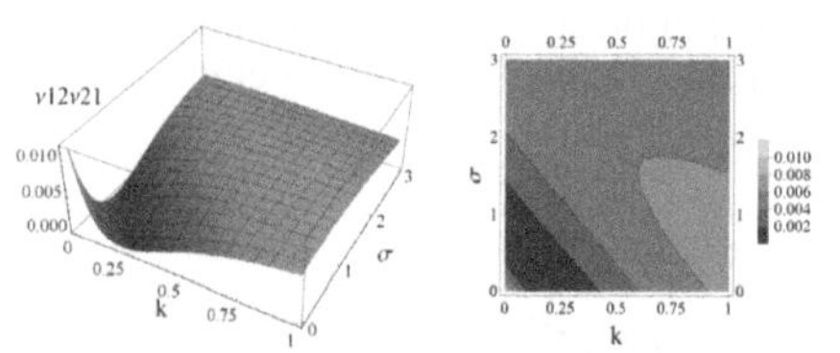

Figure 7. Product of Poisson's ratios ν_{12} and ν_{21} with the limitation condition: $\nu_{12}\ \nu_{21} < 1$.

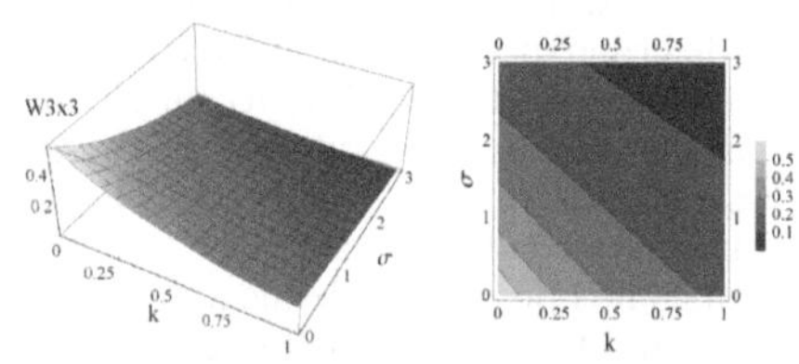

Figure 8. Condition obtained from the positive determinant of the submatrix 3×3.

value of Young's modulus E_1 multiplied by the factor EA/a^2 increases from 2.15 to 5.66, which is a ~163% increase. It might be noticed that the limitation condition $E_1 > 0$ is fulfilled for the whole given range of the parameters k and σ.

Similar graphs can be obtained for the other two Young's moduli: E_2 and E_3. The qualitative character of all these graphs is similar because of the orthotropic properties of the analysed module. The only thing that differs the three parameters is the increase range: E_2 increases by ~264%, E_3 by ~490%.

The similar situation can be noticed in case of the shear moduli. For the given ranges of the parameters k and σ, the value of shear modulus G_1 multiplied by the factor EA/a^2 decreases from 0.85 to 0.53, which is a ~38% decrease. The limitation condition $G_1 > 0$ is fulfilled for the whole given range of the parameters k and σ.

Similar graphs can be obtained for the other two shear moduli: G_2 and G_3. As in the case of Young's moduli, the qualitative character of all these graphs is similar because of the orthotropic properties of the analysed module. The only thing that differs the three parameters is the decrease range: G_2 decreases by ~36%, G_3 by ~33%.

In case of Poisson's ratio ν_{12}, the graphs have slightly different character. For the small values of the parameters k and σ, the value ν_{12} changes rapidly and after reaching a certain level, its increase slows down. In the given range of the parameters k and σ, the value ν_{12} increases by ~154%.

The Poisson's ratio ν_{21} changes in the same way as ν_{12}. The other four: ν_{13}, ν_{31}, ν_{23}, ν_{32}, on the other hand, decrease in the given range of the parameters k and σ.

The product of Poisson's ratios $\nu_{12}\ \nu_{21} < 1$ is in fact a limitation condition obtained from the positive determinant of the submatrix 2×2. It can be noticed that in case of the analysed module, the condition is fulfilled in the whole given range of the parameters k and σ.

The above graphs present the global limitation condition obtained from the positive determinant of the submatrix 3×3. This condition is also fulfilled in the whole given range of the parameters k and σ.

5 CONCLUSIONS

Tensegrity structures are complicated regarding both their geometry and mechanics. In order to understand their properties and identify technical coefficients, a continuum model is suggested.

The continuum model of the orthotropic tensegrity module—an expanded octahedron—was built by assuming that the strain energy of the unsupported tensegrity structure is equivalent to the strain energy of the cube. After the proper validation, the proposed model was used to determine and interpret physical properties of the module. From the obtained elasticity matrix and the inverse matrix, the technical coefficients such as: Young's moduli, Poisson's ratios and shear moduli were computed.

Similar analyses can be performed for other types of symmetry and for anisotropic systems.

Qualitative and quantitative evaluations of tensegrity modules are possible with the use of the proposed technique.

REFERENCES

Gilewski, W. & Kasprzak, A. 2014. 3D continuum models of tensegrity modules with the effect of self-stress, *WCCM XI*, Barcelona.

Green, A.E. & Zerna, W. 2002. *Theoretical elasticity*, Couver Dover Publications.

Kebiche, K., Kazi Aoual, M.N. & Motro, R. 2008. Continuum models for systems with a self-stress state, *Int. Journ. of Space Structures*, 23, pp. 103–115.

Motro, R. 2003. *Tensegrity structural systems for the future*, Kogan Page, London.

Skelton, R.E. & de Oliveira, M.C. 2009. *Tensegrity systems*, Springer, London.

Zienkiewicz, O.C. & Taylor, R.L. 2000. *The finite element method*, Vol. 1: *The Basis*, Prentice Hall, New Jersey.

Advances in Mechanics: Theoretical, Computational and Interdisciplinary Issues – Kleiber et al. (Eds)
© 2016 Taylor & Francis Group, London, ISBN 978-1-138-02906-4

Truss model of *origami* inspired folded structures

W. Gilewski & J. Pełczyński
Faculty of Civil Engineering, Warsaw University of Technology, Warsaw, Poland

ABSTRACT: *Origami* is an old art of paper folding developed in Japan. Initially it was limited only to religious purposes. However nowadays architects and engineers takes an inspiration from *origami* while creating structures. From mechanical point of view *origami* can be defined as a folded structure and should be described by a six parameter shell theory. The paper shows an attempt to build a simplified model of *origami*-inspired folded structures based on the pin-joined space bars. The model allows to analyse the influence of the introduced geometry on the global properties of the structure. The results for two basic *origami* modules Eggbox and Miura Ori are presented and discussed.

1 INTRODUCTION

Folded plates are attractive solutions for engineers. There are a lot of interesting engineering structures of this kind, some of them are based on the concept of origami—an old art of paper folding dated for the 7th century (Buri & Weinand 2000) developed in Japan with origins in China. The term is a combination of two Japanese words: "oru"—to fold and "kami"—a paper. There are several typical folds in the *origami* art: a "mountain fold", a "valley fold" and a "swivel fold". After changing the direction of the mountain fold one can receive an "inside reverse fold" or an "outside reverse fold". The other possibilities are a "squash fold" and a "sink fold". A detailed description of *origami* folds and patterns can be found in (Duresseix 2012). An example of the *origami* shape is presented in Figure 1.

Origami is an inspiration for engineers in the fields of civil engineering, architecture, biotechnology, medicine, space engineering and other technical applications (Chudoba, van der Woerd, Schmerl, & Hegger 2014, Lebee 2015, Nishiyama & Miura 2012, Schenk 2011, Sekularac, Sekularac, & Tovarovic 2012). The most attractive areas of civil engineering applications are:

- deployable structures,
- stiff structures with minimal expense of weight,
- shock absorbing devices.

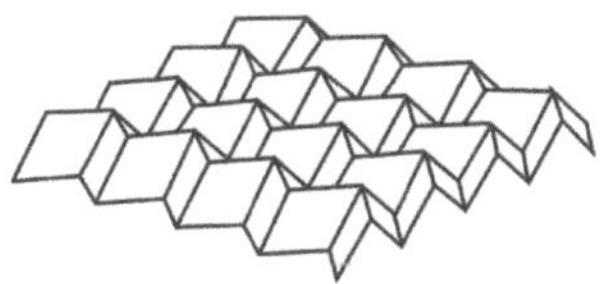

Figure 1. Miura Ori *origami* shape.

The folded structures can be capable of changing their shape to accommodate to the new requirements, whilst maintaining a continuous external surface. The process can be done automatically while the new conditions occurs or in advance to prepare the structure for new demands.

2 MATHEMATICAL MODEL

From mechanical point of view the correct theory to describe folded plates is a six parameter shell theory (Chróścielewski, Makowski, & Pietraszkiewicz 2004, Gilewski, Pełczyński, & Stawarz 2014) with three displacements and three rotations in the displacement field (1).

$$u_\alpha,\ w,\ \varphi_\alpha,\ \psi \tag{1}$$

The third rotation ψ is necessary due to the folding of the structure. Each fold consists of flat surfaces, so the equations of the theory can be simplified by the assumption of $b_{\alpha\beta} = 0$ and given by: strain-displacement equations

$$\begin{gathered}\gamma_{\alpha\beta} = u_{\alpha,\beta} - \in_{\alpha\beta}\psi \quad \gamma_{\alpha 3} = u_\alpha - w_{,\alpha},\\ \kappa_{\alpha\beta} = \varphi_{\alpha,\beta} \quad \kappa_{\alpha 3} = \psi_{,\alpha},\\ \gamma_{33} = \psi,\end{gathered} \tag{2}$$

constitutive equations

$$\begin{gathered}N_{\alpha\beta} = B^0_{\alpha\beta\lambda\mu} \qquad N_{\alpha 3} = k^2 B^0_{\alpha 3\beta 3}\gamma_{\beta 3},\\ M_{\alpha\beta} = \frac{h^2}{12} B^0_{\alpha\beta\lambda\mu} \qquad M_{\alpha 3} = \frac{h^2}{12} l^2 B^0_{\alpha 3\beta 3}\kappa_{\beta 3},\\ B^0_{\alpha\beta\lambda\mu} = Gh\left(a_{\alpha\lambda}a_{\beta\mu} + a_{\alpha\mu}a_{\beta\lambda} + \lambda a_{\alpha\beta}a_{\lambda\mu}\right),\end{gathered} \tag{3}$$

$$B^0_{\alpha 3\beta 3} = Gha_{\alpha\beta},$$

where

$$G = \frac{E}{2(1-\nu)}, \quad \lambda = \frac{2\nu}{1-\nu}, \quad k^2 = \frac{5}{6}, \quad l^2 = \frac{7}{10}$$

and equilibrium equations

$$\begin{aligned} &N_{\alpha\beta,\alpha} + f_\beta = 0, \\ &N_{\alpha 3,\alpha} + f_3 = 0, \\ &M_{\alpha\beta,\alpha} - N_{\beta 3} + m_\beta = 0, \\ &M_{\alpha 3,\alpha} + \in_{\alpha\beta} N_{\alpha\beta} + m_3 = 0. \end{aligned} \tag{4}$$

In equations (1–4) Einstein summation notation is used with $\alpha, \beta, \lambda, \mu = 1, 2$.

For numerical analysis it is necessary to use the finite element method, because of the complex character of the structures. Finite elements with six d.o.f. per node are to be used. However, shell models of folded structures provide the models with a lot of degrees of freedom and need application of professional software. The most challenging task is to develop an effective technique for efficient computation of structures with a lot of folds (Gilewski, Pełczyński, & Stawarz 2014).

One of the initial issues is the type of the connections between folds. They can be connected fixedly (see Fig. 2a). The other possibility is modelling of linear hinges on each edge (see Fig. 2b) or even hinges only between the folds corners. One can also take into account only the edges and build a truss (see Fig. 2c).

The paper is an attempt to create a simplified model of *origami*-inspired folded structures based on the pin-joined space bars. The authors are not interested in the minutiae of the stress distribution. The aim of the model is to obtain different information—to analyze the effect of the introduced geometry on the global properties of the sheet. Dominant mechanisms and self-stress states—the internal normal forces states, which satisfy the homogeneous equilibrium equations—are considered.

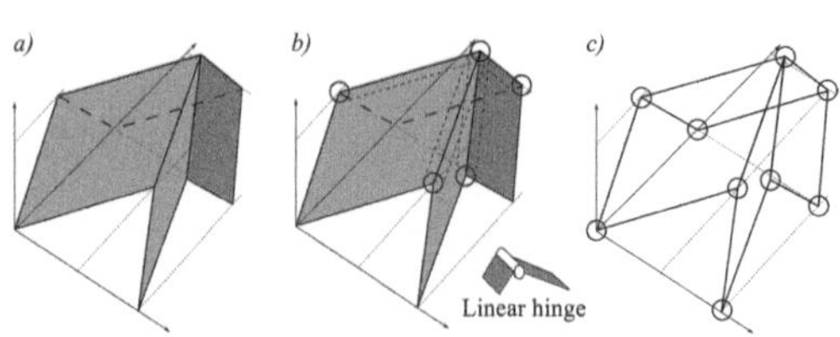

Figure 2. Connection types in *origami* modules: fixed (a), linear hinge (b), *origami*-inspired truss structure (c).

3 SIMPLIFIED MECHANICAL MODEL

The paper makes the use of a pin-jointed bar framework to represent the *origami* folding. The geometry of the truss is defined by edges of the folding. Its mechanical properties are described by three linearized equations: compatibility (5), material properties (6) and equilibrium (7):

$$\Delta = \mathbf{Bq}, \tag{5}$$

$$\mathbf{S} = \mathbf{E}\Delta, \tag{6}$$

$$\mathbf{B}^T\mathbf{S} = \mathbf{P}, \tag{7}$$

where $\mathbf{q}$ is a displacement vector, $\mathbf{B}$ is a compatibility matrix, Δ is an extension vector, $\mathbf{S}$ is an internal bar tensions vector, $\mathbf{E}$ is an elasticity matrix and $\mathbf{P}$ is a load vector. Combining equations (5–7) one can obtain the truss solving equation (8)

$$\mathbf{K}_L\mathbf{q} = \mathbf{P}, \qquad \mathbf{K}_L = \mathbf{B}^T\mathbf{EB}, \tag{8}$$

where $\mathbf{K}_L$ is the linear stiffness matrix.

Following the singular value decomposition of the compatibility matrix $\mathbf{B}$ it is possible to define the finite or infinitesimal modes of the structure as well as the self-stress states. It gives the qualitative information about the global stability of the structure.

4 ANALYSIS

For further analysis two truss structures, inspired by Miura Ori and Eggbox *origami* schemes (see Fig. 3), are prepared. Both are presented in Figure 4. It is assumed that the bars are characterized by the Young's modulus E and cross-sectional area A. The supports are realized by bars that are simply supported at one end so as to maintain the properties of single module taken from bigger structure.

For each truss the matrices $\mathbf{B}$ and $\mathbf{E}$ are built. Thereby, by finding the eigensolution of the matrix $\mathbf{B}^T\mathbf{B}$, it is determined that four infinitesimal modes

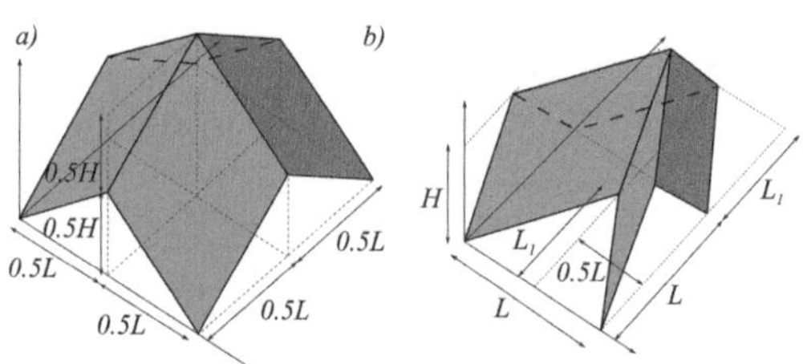

Figure 3. Modules of Eggbox (a) and Miura Ori (b) structures.

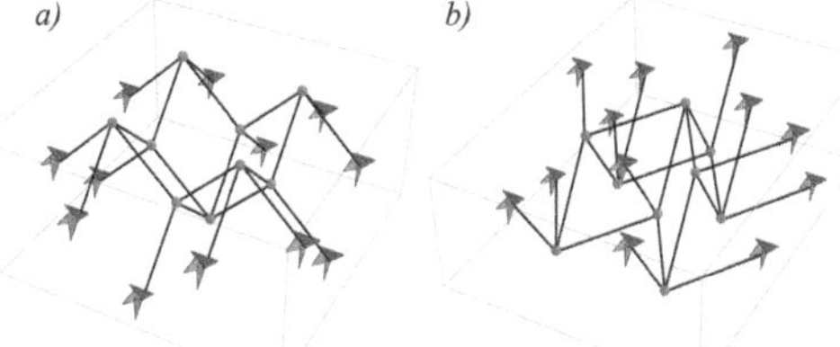

Figure 4. Trusses inspired by modules of Eggbox (a) and Miura Ori (b) with supports.

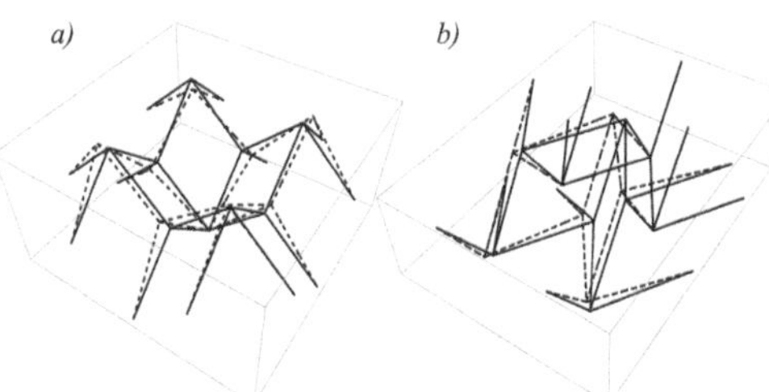

Figure 5. Examples of rigid movements for Eggbox (a) and Miura Ori (b) trusses.

occur in the analyzed structures. Vectors corresponding to the zero eigenvalues indicates the form of rigid movements. The examples are shown in Figure 5. Also the eigensolution of the matrix $\mathbf{BB}^T$, indicating the number of self-stresses occurring in the truss, and the corresponding eigenvectors (see Table 1), which specify the distribution of the self-stress force over bars, are found. Both trusses are characterized by one self-stress state. Figure 6 shows bars under tension and compression in the self-stress state.

A linear stiffness matrix (8) is calculated. In order to eliminate four rigid movements the self-stress state is introduced to the structure. For this purpose the geometrical stiffness matrix $\mathbf{K}_G(\sigma)$ is developed following a standard FEM procedure (Zienkiewicz & Taylor 2000) by the aggregation of the single finite element geometrical stiffness matrices (9):

$$\mathbf{K}_G^e = S\frac{s_e}{L_e}\begin{pmatrix} 0 & 0 & 0 & 0 & 0 & 0 \\ 0 & 1 & 0 & 0 & -1 & 0 \\ 0 & 0 & 1 & 0 & 0 & -1 \\ 0 & 0 & 0 & 0 & 0 & 0 \\ 0 & -1 & 0 & 0 & 1 & 0 \\ 0 & 0 & -1 & 0 & 0 & 1 \end{pmatrix}, \tag{9}$$

where S is the reference self-stress force, L_e is the length of e-th finite element, s_e and $S \cdot s_e$ are self-stress vector component and self-stress force

Table 1. Eigenvalues of $\mathbf{BB}^T$ and vector corresponding to zero eigenvalues for Eggbox and Miura Ori modules.

Eggbox		Miura Ori	
Eigenvalues	Eigenvector "0"	Eigenvalues	Eigenvector "0"
3.62	1	3.59	1
3.10	−1	3.38	−1
3.10	1	2.83	1
2.71	−1	2.83	1.6
2.37	−1	2.77	−1.6
2.37	1	2.60	1
2.16	−1	2.25	1.6
2.16	−1	2.09	−1
1.71	−1	2.00	−1.6
1.57	1	2.00	1
1.57	1	1.28	−1.6
1.38	1	1.28	1.6
1.32	1	1.24	1
1.32	1	1.20	−1.6
1.29	−1	0.847	−1
1.00	1	0.837	1.6
0.664	1	0.768	1
0.664	−1	0.725	1.6
0.634	−1	0.477	−1.6
0.634	1	0.339	1
0.293	−1	0.322	1.6
0.175	−1	0.283	−1
0.175	−1	0.0629	−1.6
0	1	0	1

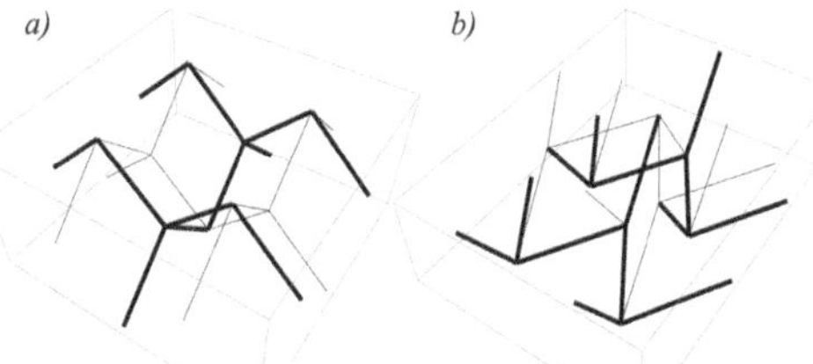

Figure 6. Bars under self-stress for modules of Eggbox (a) and Miura Ori (b). Thin—tension, thick—compression.

corresponding to the e-th finite element respectively and parameter σ is given by (10):

$$\sigma = \frac{S}{EA}. \tag{10}$$

Afterwards matrix $\mathbf{K}_G$ is added to $\mathbf{K}_L$ forming stiffness matrix $\mathbf{K}$. Thereby the equation (8) is modified to form (11):

$$\mathbf{Kq} = \mathbf{P}, \quad \mathbf{K} = \mathbf{K}_L + \mathbf{K}_G. \tag{11}$$

Eigenvalues of matrix **K** without the addition of $\mathbf{K}_G$ ($\sigma = 0$) are shown in Table 1 (Eggbox module, Miura Ori module). Further analysis ($\sigma \neq 0$) of the stiffness matrix **K** shows that all zero eigenvalues change to positive or negative values (see Table 2 and Table 3). It should be mentioned that elimination of zero eigenvalues and hence the infinitesimal modes is possible only by appropriate choice of parameter σ. For $\sigma = 1$ in the Eggbox module one zero eigenvalue remained (Table 2). Although the matrix **K** is not positive definite it is possible to solve the $\mathbf{Kq} = \mathbf{P}$ system and obtain reasonable results.

For the truss composed of more than one single module some results are obtained. The analysis shows that this kind of structure possess respectively more infinitesimal modes. What is very important the number of self-stress states does not depend on the size of the structure and always equals one. This fact means that it is possible to create smart structures, which stiffness can be controlled only with pretension of one of the cables. Example results are presented in Figure 7.

Table 2. Eigenvalues of stiffness matrix **K** for Eggbox module.

$\sigma = 0$	$\sigma = 1/100$	$\sigma = 1/10$	$\sigma = 1$
5.12	5.12	5.16	7.47
4.39	4.40	4.53	7.28
4.39	4.38	4.29	7.18
3.83	3.83	3.84	6.82
3.35	3.36	3.42	6.79
3.35	3.34	3.28	6.29
3.06	3.06	3.14	−4.41
3.06	3.05	3.03	−4.34
2.41	2.41	2.49	4.22
2.23	2.25	2.48	−3.97
2.23	2.21	2.17	3.46
1.95	1.95	1.95	−3.46
1.87	1.88	1.94	3.29
1.87	1.86	1.84	2.83
1.83	1.83	1.78	2.83
1.41	1.42	1.67	2.83
0.939	0.941	0.935	2.60
0.939	0.936	0.909	2.43
0.897	0.904	0.872	2.05
0.897	0.885	0.641	−1.84
0.414	0.414	0.397	1.52
0.248	0.256	0.248	−1.17
0.248	0.239	0.140	−0.636
0	−0.00470	−0.0783	0.561
0	0.00310	−0.0759	0.352
0	−0.000591	−0.0566	−0.0559
0	0.0000602	−0.0391	0

Table 3. Eigenvalues of stiffness matrix **K** for Miura Ori module.

$\sigma = 0$	$\sigma = 1/100$	$\sigma = 1/10$	$\sigma = 1$
3.80	3.81	3.86	6.37
3.58	3.58	3.63	6.09
3.00	3.01	3.11	5.86
2.99	2.99	3.06	5.67
2.93	2.93	2.86	5.28
2.75	2.75	2.71	4.81
2.39	2.40	2.59	4.77
2.22	2.22	2.32	3.78
2.12	2.13	2.15	3.62
2.12	2.12	2.13	−3.51
1.36	1.37	1.69	−3.38
1.36	1.36	1.40	3.18
1.31	1.34	1.38	3.12
1.28	1.29	1.33	−2.57
0.897	0.924	1.18	−2.30
0.887	0.905	1.11	2.21
0.814	0.813	0.844	2.13
0.769	0.757	0.648	2.12
0.506	0.515	0.604	1.63
0.358	0.357	0.380	1.59
0.343	0.341	0.277	−1.45
0.301	0.308	0.252	−1.39
0.0667	0.0698	−0.131	0.955
0	0.00767	0.120	0.774
0	0.00690	−0.110	0.741
0	−0.00557	0.0333	0.621
0	−0.00429	0.0111	0.165

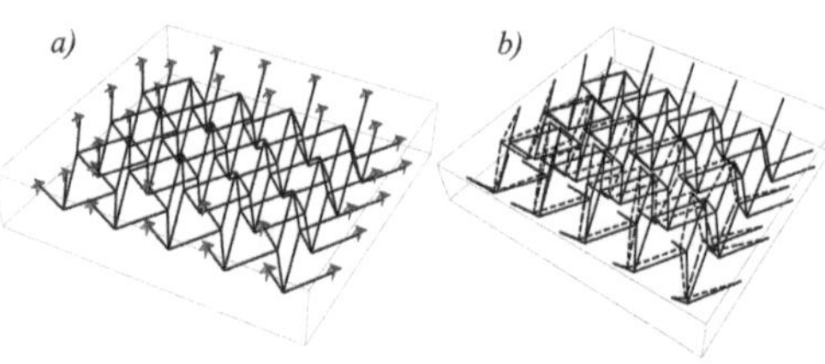

Figure 7. Truss consisting of 12 Miura Ori modules. Geometry (a) and example of rigid movement (b).

5 CONCLUSIONS

Approaches which simplify the process of obtaining information about the global properties of *origami*-inspired folded structures are proposed. Spectral analysis of the matrices **B** and **K** is performed. Thereby infinitesimal modes of the structure are found, which afterwards are eliminated by addition of self-stress states. It should be noted that in structures composed of more than one module many infinitesimal modes occur, however all can be eliminated by single self-stress state.

The qualitative analysis of structures with self-stress included is the subject of further analysis, because of non-elliptic form of the strain energy.

The proposed approach can lead to the development of engineering structures such as deployable forms, very light but stiff enough structures or shock absorbing devices.

REFERENCES

Buri, H. & Y. Weinand (2000). *ORIGAMI—Folded plate structure*. Architecture. EPF Lausanne.

Chróścielewski, J., J. Makowski, & W. Pietraszkiewicz (2004). *Statics and dynamics of folded shells*. IPPT PAN.

Chudoba, R., J. van der Woerd, M. Schmerl, & J. Hegger (2014). Modeling support for design and manufacturing of folded concrete structures. *Advances in engineering software 72*, 119–127.

Duresseix, D. (2012). An overview of mechanisms and patterns with origami. *International Journal of Space Structures* 27, 1–14.

Gilewski, W., J. Pełczyński, & P. Stawarz (2014). A comparative study of origami inspired folded plates. *Procedia Engineering* 91, 220–225.

Lebee, A. (2015). From folds to structures, a review. *International Journal of Space Structures* 30(2), 55–74.

Nishiyama, Y. & Y. Miura (2012). Folding: applying origami to space exploration. *International Journal of Pure and Applied Mathematics* 79, 269–279.

Schenk, M. (2011). *Folded shell structures*. Ph. D. thesis, University of Cambridge.

Sekularac, N., I. Sekularac, & J. Tovarovic (2012). Folded structures in modern architecture. *Facta Universitatis, Architectura and Civil Engineering* 10, 1–16.

Zienkiewicz, O. & R. Taylor (2000). *The finite element method*. Prentice Hall, New Yersey.

Advances in Mechanics: Theoretical, Computational and Interdisciplinary Issues – Kleiber et al. (Eds)
 ISBN 978-1-138-02906-4

Parametric analysis of Istanbul's ring road viaduct for three levels of seismic load

K. Grębowski, M. Hirsz & K. Wilde
Gdańsk University of Technology, Gdańsk, Poland

A. Nadolny
Gdansk Bridges Ltd., Gdańsk, Poland

ABSTRACT: The paper presents numerical analysis of the Istanbul's ring road viaduct that is currently under construction within the Northern Marmara Highway project. The structure, due to its location on seismic prone areas is exposed to seismic loads of different magnitude and different return periods. The study is focused on concrete bridge supports that are designed to work in a nonlinear range. The study, conducted in SOFISTIK environment, aim at global responses of the whole viaduct. The detailed nonlinear analysis of concrete hinge development at the bridge support is performed in ABAQUS. The obtained results confirmed the correctness of the assumptions stated at the beginning of the design process.

1 INTRODUCTION

Istanbul is located near the North Anatolian Fault which includes a place where two tectonic plates, African and Eurasian, meets. As a result, the entire region is one of the most seismically active on the Earth. In 1999, in Kocaeli province, in the location several dozen kilometers away from Istanbul, the 7.6 Richter scale earthquake took place. It caused the death of almost 18 000 people and led to the destruction of many buildings and bridges that played a very important role in the region's development.

In 2010, due to the development of towns and suburbs of the Black Sea, an idea of building a third ring road of Istanbul within the framework of the Northern Marmara Highway project arose. In 2013 the project started to be implemented. The route includes 67 viaducts, and the length of the designed section is 260 kilometers.

2 EARTHQUAKE LOAD OF KOCAELI AND DESCRIPTION OF THE V6 VIADUCT

The map (Fig. 1) of seismicity of Turkey shows earthquake prone regions with red and brown colors. Two black lines denote the fault lines that divide the area into regions particularly often haunted by earthquakes. The line that stretches from east to west along the Black Sea is the North Anatolian Fault, whereas the line between Georgia and Syria is the East Anatolian Fault. The whole western Turkey, the area where the third ring road of Istanbul is built, is very active seismically (Fig. 1).

The Turkish norm DLH 2008 includes three earthquake intensity levels for instances of earthquakes:

- **D1**: return period: 72 years (50% in 100 years);
- **D2**: return period; 475 years (10% in 100 years);
- **D3**: return period: 2475 years (2% in 100 years).

The design of the viaduct structure assumed that the object exposed to seismic loads, corresponding

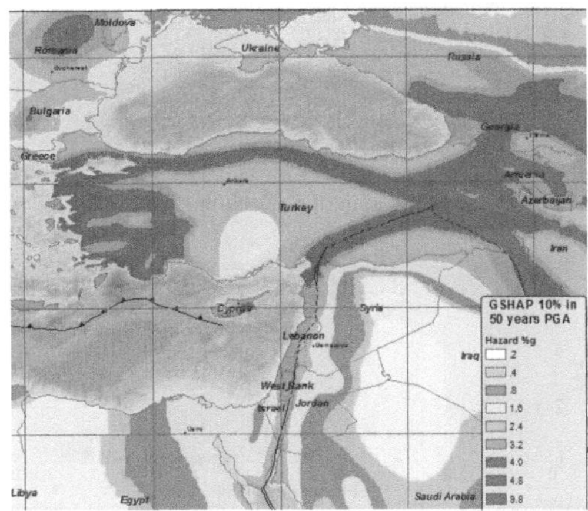

Figure 1. Seismicity map of Turkey.

to an occurrence of earthquake of level D1 will avoid damaging. For earthquakes of level D2 the viaduct structural elements will develop some local material deterioration zones, but the damage will occur in controlled areas which might be repaired within a few months period with little traffic restrictions (level D2). Controlled damage is intended to be in the upper part of the supports by developing plastic hinges. The deck should not fall off from the supports at level D3 earthquake. For numerical simulations the seismic load that was recorded during the quake of 17 August 1999, near Istanbul's province Kocaeli is adopted (Figs. 2–3).

The analyzed V6 object is located in the western part of the northern ring road. The object consists of two parallel viaducts, each of which has a four lane road. Its total length is 640 m (Figs. 4–5).

The viaduct's deck has a box cross-section made of transversely and longitudinally prestressed concrete. The concrete corresponds to the C45/50 class (according to EC). The width of the deck is 22 m and it consists of: 2 m wide pedestrian path, 19 m wide road and the 1 m cantilever for safety barriers (Fig. 6). The road consists of 4 lanes of 3.75 m each and 3 m wide emergency lane. The distance between the decks of two parallel viaducts equals 3 m.

The cross-section of the box is variable along the length of the object. The base trapezoidal box section consists of a 30 cm thick top plate, 25 cm thick bottom plate, and 45 cm thick inclined webs (Fig. 6). The section of the support area (Fig. 7) has an enlarged 55 cm thick web.

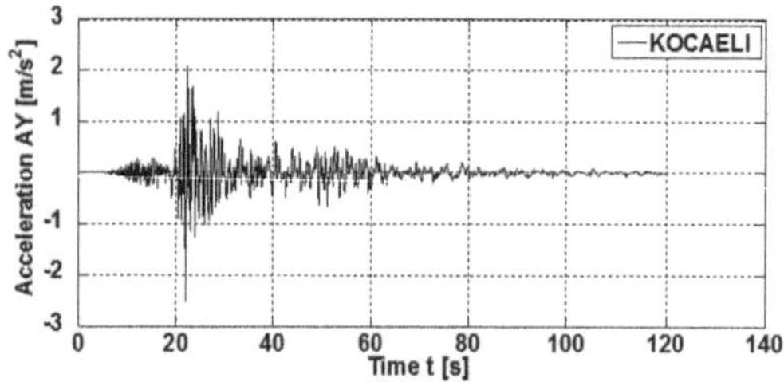

Figure 2. Seismic load of Kocaeli in 1999.

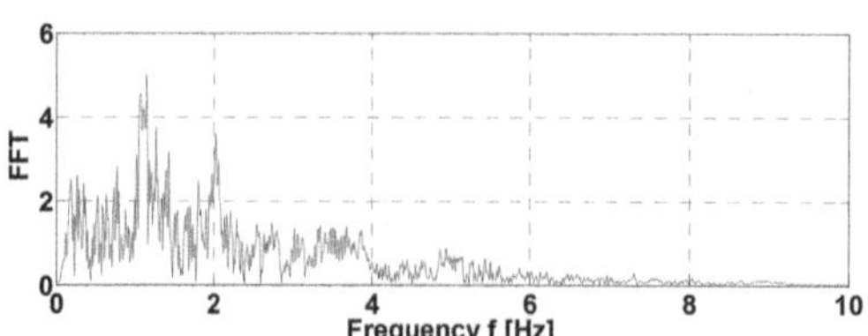

Figure 3. The Fourier transform's load of Kocaeli in 1999.

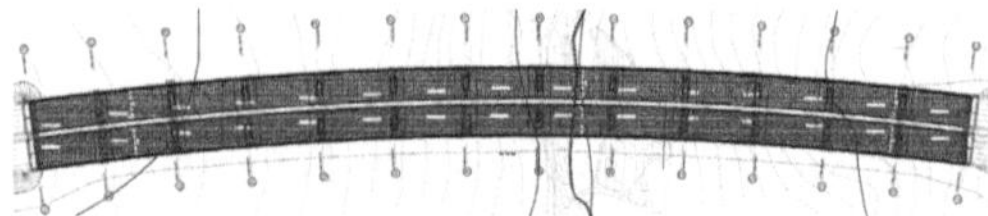

Figure 4. V6 Viaduct—view from the top.

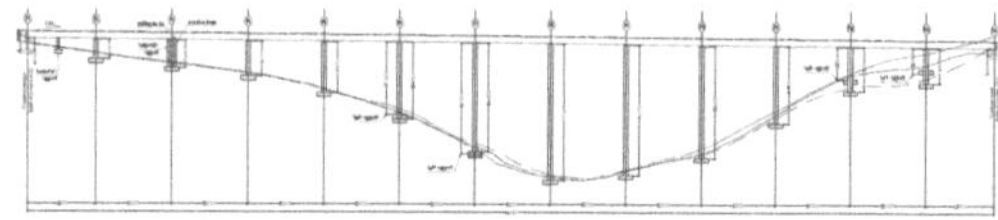

Figure 5. Side view of the V6 Viaduct.

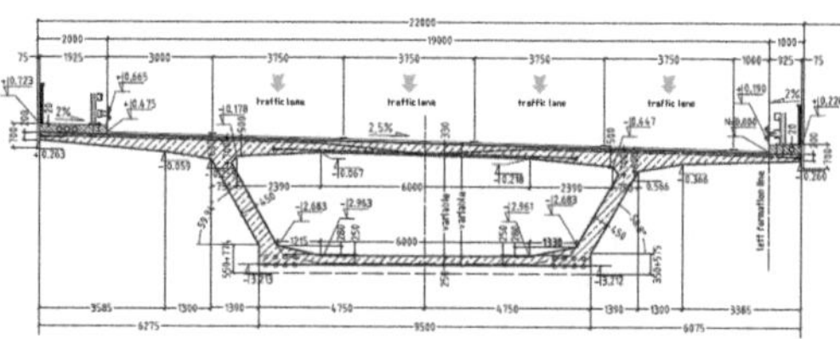

Figure 6. A typical mid-span cross-section of the deck.

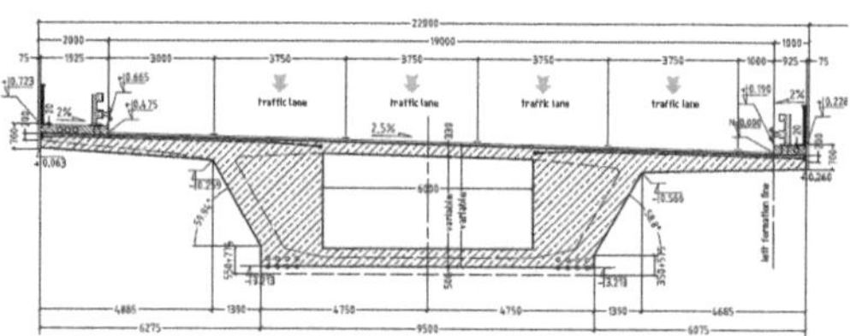

Figure 7. A typical cross-section of the deck near supports.

The aforementioned part of the section is located from the support wall face to 9 m from the support axis, on each side.

The section directly above the intermediate supports has a full reinforced concrete membrane, along with a rectangular 5.5 × 2.1 m hole intended for bridge bearing inspection. The section above the abutments has a square hole with 2.1 m for the bridge bearing inspection.

Intermediate supports are designed as variable stiffness structures to limit the internal forces of the lowest supports during strong earthquakes. The intermediate supports are divided into two parts. The upper part consists of a rafter plate with length of 12 m, width of 3 m and height of 2 m. The rafter plate is equipped with a reinforced concrete buffer that absorbs platform's impact in case of a high-intensity earthquake. The rafter plate is connected with two slender reinforced concrete

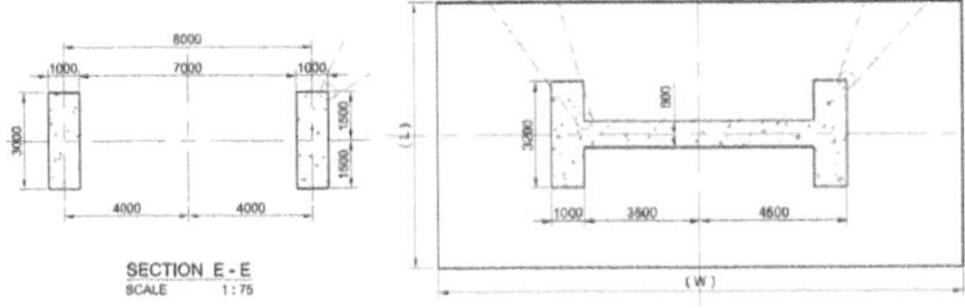

Figure 8. The cross-section of the slender and stiff parts of the supports.

Figure 9. Picture taken at the construction site on Sept. 2015.

columns with dimensions: 3 × 1 m and 19 m high. The geometry of the supports' upper parts has a fixed height for the entire object, in case when their total height does not exceed 21 m. The upper part of the support is shown in Figure 8 (section E-E) and Figure 9.

When a support height exceeds 21 m, the bottom part consists of a stiff I-section topped with a reinforced concrete slab. The web's thickness of the I-section is fixed at 80 cm with the height of 7 m. The width of the section's flange is dependent on the total height of the intermediate support (Fig. 8).

Pot bearings are used in the object. The sliding in the longitudinal direction is blocked on one of the abutments.

3 ANALYSIS OF THE V6 VIADUCT IN TIME AND FREQUENCY DOMAIN

The FEM model of the V6 viaduct has been developed in the SOFISTIK environment. The model consists of beam elements for description of the deck and intermediate supports. The intermediate supports are restrained at the base. In order to render the longitudinal friction forces, elastic bonds with suitable susceptibility were introduced between the nodes of the deck and the rafter plate.

The first three eigenvalues of the bridge are: 1.32 Hz, 1.67 Hz and 2.11 Hz. The first three eigenmodes have dominant displacement component in vertical, transvers direction associated with bending of the upper part of the intermediate supports. The sixth mode shape is a vertical bending mode and has frequency of 2,47 Hz (Fig. 10).

The results of the lateral, vertical displacement in time of the top of the highest intermediate support is shown in Figure 11. The proportional Rayleigh damping is assumed and the Kocaeli earthquake of peak ground acceleration 2.50 m/s^2 is used as an excitation. The support time lateral displacement is shown in Figure 11. The maximum displacement of the top of the support is 1.5 m. The deformation of the whole structure at the maximum displacement at 22nd second of the response is given in Figure 12.

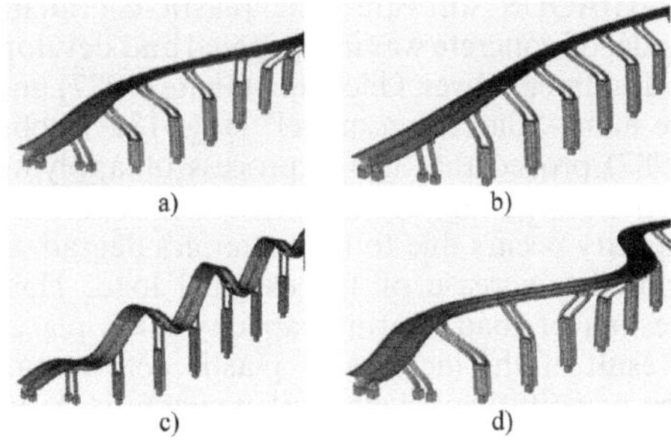

Figure 10. Mode shapes of the V6 viaduct: a) first form—1.321 [Hz], b) second form—1.674 [Hz] c) third form—2.111 [Hz], d) sixth form—2.472 [Hz].

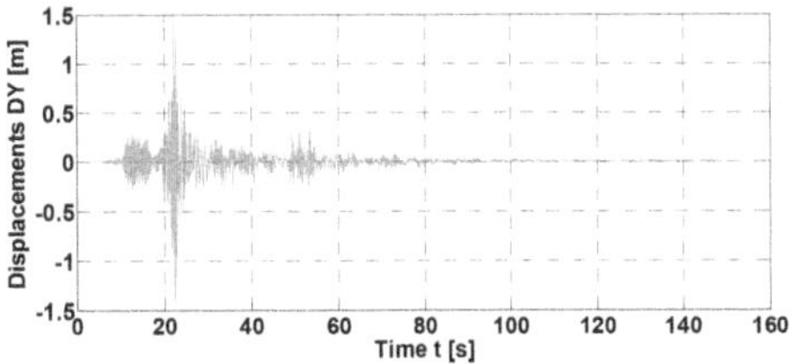

Figure 11. Horizontal displacement in the span center of the viaduct for the D1 load level.

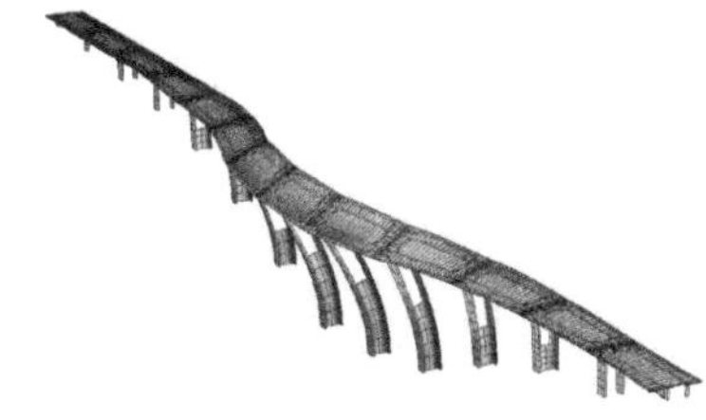

Figure 12. The viaduct's response at t = 22 s; the maximum lateral displacement under Kocaeli earthquake.

The zones with largest values of stresses develop plastic hinges in the locations where stiff, I-section of the lower part connects with the slender columns of the upper part of the support and in the location where the slender columns connect with the rafter plate. The results obtained through the FEM model developed in SOFISTIK cannot describe the assumed plastic hinge formation during the bridge response induced by earthquake.

4 PLASTIC-DEGRADATION MODEL OF CONCRETE

Numerical analysis of the plastic hinge formation in a local sense is conducted with use of the plastic-degradation model of concrete implemented in ABAQUS software. The plastic-degradation model of concrete was investigated and developed by Lubliner, Oliver, Oiler, and Onate (1987) under the name "Barcelona model" (Fig. 13). Lubliner (1987) proved that in the process of applying a load on concrete, the exhaustion of load-beaing capacity occurs due to the material's degradation from the increase of the external load. Hence, the loss of load bearing capacity takes place as a result of the increase of plastic deformations. The constitutive behavior of concrete is defined by introducing a scalar variable D which quantifies the influence of microcracking. The change of the stress due to variable D is given by the following formula:

$$\sigma = (1 - D) \cdot \sigma' \tag{1}$$

were σ' denotes "real" stress damage.

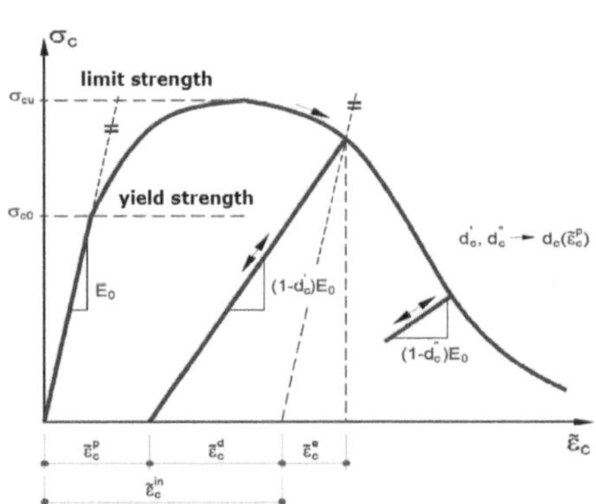

Figure 13. Concrete damaged-plasticity—Barcelona model.

5 NUMERICAL ANALYSIS OF THE VIADUCT'S INTERMEDIATE SUPPORT DYNAMIC RESPONSE

The detailed model of the intermediate support at place were stiff bottom part connects with the slender column is shown in Figure 14.

The analysis of plastic deformation of the hinges has been performed by a non-linear analysis in ABAQUS software with use of the Ductility Demand-Capacity concept. The conducted analysis is consistent with the approach defined in Turkish norm (DLH 2008) that takes into account three levels of earthquakes D1, D2 and D3 by multiplying the reference record by appropriate factors. In these simulations a Kocaeli earthquake is used as the reference one.

The results of the non-linear analysis of the intermediate support qusi-dynamic response are given in Figure 15. The maximum displacements

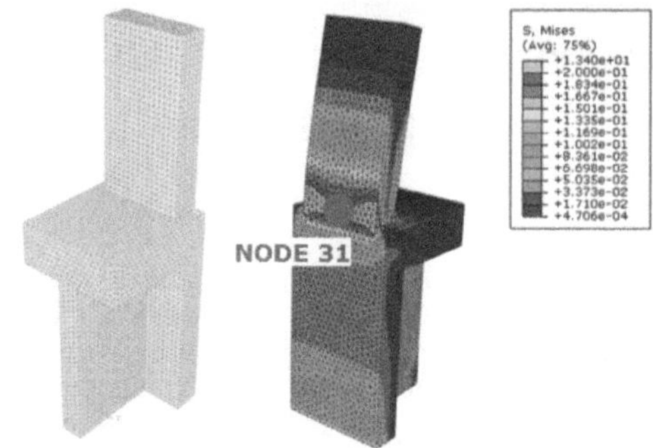

Figure 14. The numerical model of the connection between stiff bottom part with slender upper column.

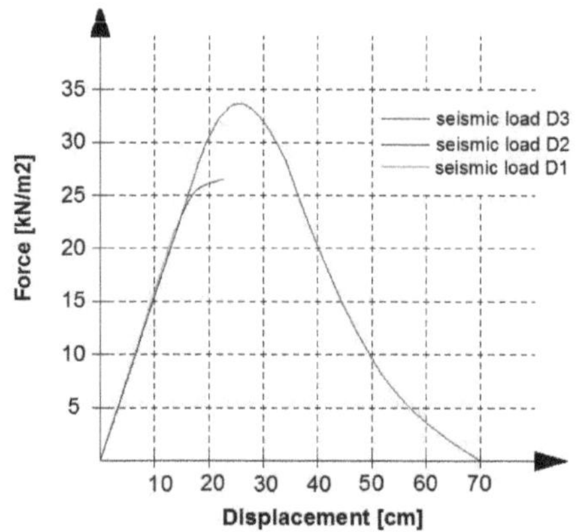

Figure 15. The resultant lateral force-displacement diagram for D1, D2 and D3 seismic load.

of the top surface of the local model for earthquakes D1, D2 and D3 have been calculated with use of the global model. Then, the obtained displacements are used for the load functions for all three levels of earthquake intensity for calculation with the local model. Finally, the resultant lateral force corresponding to each load case is computed. For earthquake level D1 the intermediate support work in linear range, for level D2 at some points the function force-displacement exhibits the nonlinear behavior. For excitation level D3 concrete deteriorates in the bottom part of the slender column which is clearly visible for example at the node 31 (Fig. 14).

6 CONCLUDING REMARKS

The numerical analysis, of V6 viaduct of a third ring road of Istanbul within the framework of the Northern Marmara Highway project, shows that dividing the supports into flexible upper part and bottom, stiff part allows the relatively equal distribution of seismic load on all columns during earthquakes of different intensities. The inelastic design of the viaduct assumes plastic hinge formation in the slender columns of the intermediate supports. Nevertheless, the obtained results do not provide clear answer about the structure state after the occurrence of earthquake level D3. Further studies on combined global—detailed models are under investigation.

REFERENCES

AASHTO 2002. *Standard Specifications for Highway Bridges.*

DLH 2008. Earthquake norm—Turkish.

Erdik, M. 2008. *Report on 1999 Kocaeli and Düzce earthquakes.* Bogazici University, Dept. of Earthquake Engineering.

Eurocode 8—Design of structures for earthquake resistance—Part 2: Bridges.

Soyluk, A., & Harmankaya, Z.Y. 2012. The History of Development in Turkish Seismic Design Codes. *International Journal of Civil & Environmental Engineering* 12(1), Ankara.

Advances in Mechanics: Theoretical, Computational and Interdisciplinary Issues – Kleiber et al. (Eds)
 ISBN 978-1-138-02906-4

New method of modelling nonlinear multi-bolted systems

R. Grzejda
Faculty of Mechanical Engineering and Mechatronics, West Pomeranian University of Technology, Szczecin, Poland

ABSTRACT: Multi-bolted connections are systems of many bodies in a contact. These systems are composed of: fasteners, joined elements and contact joints between them. There are known methods for modelling and calculation of parts of typical multi-bolted connections, which are geometrically symmetrical or symmetrically loaded. A novelty is a treatment of the multi-bolted connection as a system consisting of components, which can be considered, modelled and calculated as separate subsystems using methods adequate to their properties. The aim of the paper is to present the concept of a new modelling method of multi-bolted connections dealing with as multi-bolted systems. The model assumptions and modelling bases of separate subsystems of a physical model of the multi-bolted connection are given. The outcome of actions described in this paper is establishment of the equilibrium equations, which can be applied for any condition of the multi-bolted system. Calculations results of a selected multi-bolted connection are presented.

1 INTRODUCTION

Reliability of the designed structure of a machine or device depends primarily on reliability of joints used in this structure. This applies particularly to systems created with bolted connections. Quality of the execution of these connections immensely determines the safety of the machine users. Meanwhile, according to the literature review (Witek et al. 2011), the comprehensive solution of problems occurring in bolted connections has not yet been carried out. Therefore, the task of modelling bolted connections is still valid and important.

One of the essential features of multi-bolted connections is that they are usually composed of many bodies being in a contact. Therefore, these connections should be treated in calculations as nonlinear systems. The source of this nonlinearity are, for instance, gaskets or washers often used as additional intermediate elements in this type of connections (Over et al. 2003, Schaaf & Bartonicek 2003) but also all of contact connections existing between joined elements. There are many additional phenomena related thereto, such as: self-loosening (Yokoyama et al. 2012), leakage (Magnucki & Sekulski 2005) or connection opening (Saberi et al. 2014).

Another important feature of multi-bolted connections is that they occur in two states of loading and deformation:

- the assembly state—when the joint is mounted,
- the operational state—when the preloaded joint is loaded by an external force.

Therefore, the calculations of these connections should be carried out respectively in two stages. Meanwhile, in the literature, the assembly state of multi-bolted connections is commonly omitted. The authors assume a constant force in all of the bolts and uniform clamping of joined elements after the preloading process (Mathan & Siva Prasad 2011, Zerres & Guérout 2004). However, the final tension of the joint at the end of its assembly operation depends on the way of bolt tightening (Abid & Khan 2013, Kumakura & Saito 2003, Sawa et al. 2003, Takaki & Fukuoka 2003).

Most of publications dealing with bolted connections concerns modelling single-bolted joints (Bhonge et al. 2011, Esmaeili et al. 2014, Ferjaoui et al. 2015, Li W. et al. 2015, Nechache & Bouzid 2008, Żyliński & Buczkowski 2010). A second group of papers is related to modelling typical multi-bolted connections as follows:

- beam-to-column connections (Brunesi et al. 2015, Hu et al. 2014, Saberi et al. 2014, Zapico-Valle et al. 2012),
- double lap connections (Ascione 2010, Sallam et al. 2011),
- flange connections (Jakubowski & Schmidt 2004, Mourya et al. 2015, Semke et al. 2006, Wang et al. 2013).

In any of these publications, a systemic approach to modelling, calculation and analyzing bolted connections is not undertaken. They concern mainly typical structures with bolted connections.

The studies broached in this paper deal with geometrically asymmetric and nonlinear

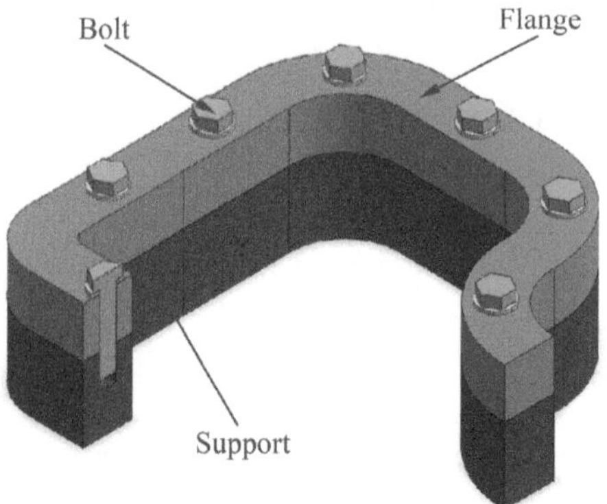

Figure 1. Multi-bolted connection.

multi-bolted connections (Fig. 1). The connections are preloaded and can be externally loaded by an arbitrary force. A physical model of the connection is proposed, which is treated as a system composed of the following subsystems: a pair of flexible joined elements (the flange and the support), a contact layer between the joined elements and a set of bolts screwed directly in the support.

The model of the multi-bolted connection is developed based on the model of a joint of a flexible flange element with a rigid support, which modelling and calculations were presented at the 2nd Polish Congress of Mechanics (Witek et al. 2011). The new generalized model of the multi-bolted connection is described by the equilibrium equations, which can be used in calculations both for the assembly and operational condition of the connection.

2 STRUCTURE OF THE MULTI-BOLTED CONNECTION

In general, the structure of the multi-bolted connection model arise from the concept presented in Witek et al. (2011).

The model is built with a flexible flange element fastened to a flexible support by means of a set of bolts. A conventional contact layer is introduced between joined elements (Fig. 2).

The equation of system equilibrium can be written in the form:

$$\boldsymbol{K}\cdot\boldsymbol{q}=\boldsymbol{p} \tag{1}$$

where $\boldsymbol{K}$ = stiffness matrix; $\boldsymbol{q}$ = displacements vector; and $\boldsymbol{p}$ = loads vector.

It can be assigned the following four different subsystems in the discussed multi-bolted connection:

- subsystem **B**, built with the bolts,
- subsystem **F**, which is the flexible flange element,
- subsystem **C**, related to the conventional contact layer,
- subsystem **S**, which is the flexible support.

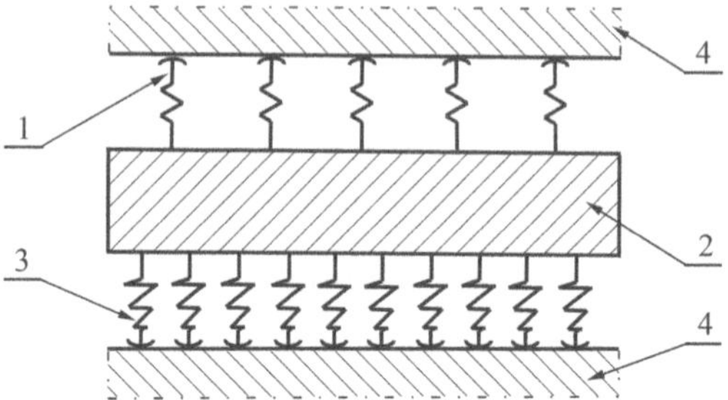

Figure 2. Division of the multi-bolted connection into subsystems (1—subsystem **B**, 2—subsystem **F**, 3—subsystem **C**, 4—subsystem **S**).

Table 1. FE-models of parts of the multi-bolted connection.

Subsystem	Types of FE-models
B	No-bolt model, rigid cylindrical body, plane model, beam model, coupled bolt model, rigid body bolt model, spider bolt model, spatial bolt model
F	Plane model, spatial model
C	Plane or spatial conventional contact layer
S	Plane model, spatial model

The most common method applied for modelling components of bolted connections is the finite element method—FEM. Table 1 shows models of the individual parts of subsystems which can be created using FEM (Cho et al. 2014, Grzejda 2015, Khurshid 2014, Kim et al. 2007, Li Y.-F. et al. 2015, Montgomery 2006, Schaumann et al. 2001, Williams et al. 2009).

For modelling bolted connections the following methods are also used:

- the Boundary Element Method—BEM (Zhang 2001, Zografos & Dini 2009),
- the Rigid Finite Element Method—RFEM (Witek & Grzejda 2005).

Adopting the division of the connection into four subsystems, the vectors $\boldsymbol{q}$ and $\boldsymbol{p}$ are defined as:

$$\boldsymbol{q}=\text{col}\,[\boldsymbol{q}_B,\boldsymbol{q}_F,\boldsymbol{q}_C,\boldsymbol{q}_S] \tag{2}$$

$$\boldsymbol{p}=\text{col}\,[\boldsymbol{p}_B,\boldsymbol{p}_F,\boldsymbol{p}_C,\boldsymbol{p}_S] \tag{3}$$

where $\boldsymbol{q}_i$ = displacements vector of the i-th subsystem; $\boldsymbol{p}_i$ = loads vector of the i-th subsystem; and i = symbol of the subsystem, $i \in \{B, F, C, S\}$.

The stiffness matrix can be represented in the block form. Then, Eqn (1) is rearranged to the formula:

$$\begin{bmatrix} K_{BB} & K_{BF} & 0 & K_{BS} \\ K_{FB} & K_{FF} & K_{FC} & 0 \\ 0 & K_{CF} & K_{CC} & K_{CS} \\ K_{SB} & 0 & K_{SC} & K_{SS} \end{bmatrix} \cdot \begin{bmatrix} q_B \\ q_F \\ q_C \\ q_S \end{bmatrix} = \begin{bmatrix} p_B \\ p_F \\ p_C \\ p_S \end{bmatrix} \tag{4}$$

where $\boldsymbol{K}_{BB}$, $\boldsymbol{K}_{FF}$, $\boldsymbol{K}_{CC}$, $\boldsymbol{K}_{SS}$ = stiffness matrices of separate subsystems; and $\boldsymbol{K}_{BF}$, $\boldsymbol{K}_{FB}$, $\boldsymbol{K}_{BS}$, $\boldsymbol{K}_{SB}$, $\boldsymbol{K}_{FC}$, $\boldsymbol{K}_{CF}$, $\boldsymbol{K}_{CS}$, $\boldsymbol{K}_{SC}$ = matrices of elastic couplings among separate subsystems.

In view of the introduced division of the multi-bolted system, each subsystem can be separately analyzed with varying degrees of simplification.

3 NUMERICAL RESULTS OF CALCULATIONS

Sample calculations are performed for a selected asymmetrical multi-bolted connection, which model is shown in Figure 3.

The thickness of the flange is equal to 30 mm and the thickness of the support is equal to 45 mm. The flange is fastened to the support by means of eleven bolts M10 made in the mechanical property class 10.9. The preload of the bolt F_m is equal to 17,2 kN and it was set down on the base of Polish Standard PN-EN 1591-1. The preloaded multi-bolted connection is subjected by external force F equal to 30 kN and applied at the middle of the arc between the bolts No. 7 and No. 8 (Fig. 3).

In the connection the bolts were replaced by the rigid body bolt models consisted of a flexible plain part of the bolt and a rigid bolt head (Grzejda 2015). For modelling of the contact joint between the joined elements the general type of the contact joint available in the Midas NFX 2014 program are used.

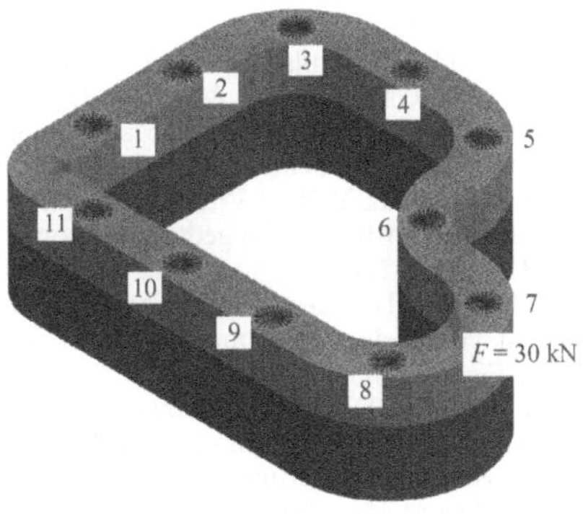

Figure 3. FEM-based model of the multi-bolted connection.

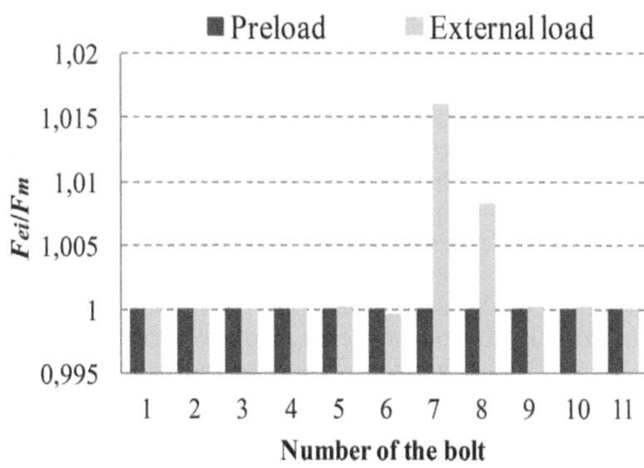

Figure 4. Bolt force values in the multi-bolted connection loaded externally.

The received distribution of operational bolt forces F_{ei} referenced to the preload F_m is illustrated in Figure 4. From results of calculations it can be noticed, that the bolts which are closest to the location of external force applying are loaded most.

4 CONCLUSIONS

The paper presents a general systemic approach to modelling and calculations of arbitrary multi-bolted connections. It is based on the consideration in the process of modelling the real structure this types of systems, which are composed of: fasteners, joined elements and contact joints between them.

The method can be implemented in both the assembly and operational state of the connection. After the adoption of appropriate assumptions, it can also be used for calculations in the case of typical multi-bolted connections.

REFERENCES

Abid, M. & Khan, Y.M. 2013. The effect of bolt tightening methods and sequence on the performance of gasketed bolted flange joint assembly. *Struct. Eng. Mech.* 46(6): 843–852.

Ascione, F. 2010. A preliminary numerical and experimental investigation on the shear stress distribution on multi-row bolted FRP joints. *Mech. Res. Commun.* 37(2): 164–168.

Bhonge, P.S., Foster, B.D. & Lankarani, H.M. 2011. Finite element modeling and analysis of structural joints using nuts and bolts. *IMECE 2011: Vol. 3: Design and Manufacturing, Proceedings of the ASME 2011 International Mechanical Engineering Congress and Exposition*. Denver: ASME, 73–83.

Brunesi, E., Nascimbene, R. & Rassati, G.A. 2015. Seismic response of MRFs with partially-restrained bolted beam-to-column connections through FE analyses. *J. Constr. Steel Res.* 107: 37–49.

Cho, Y.H., Kim, S.H. & Kim, T.S. 2014. Structural behaviors and curling influence of single shear bolted connections with aluminum alloys (7075-T6). *Int. J. Precis. Eng. Man.* 15(1): 183–187.

Esmaeili, F., Chakherlou, T.N. & Zehsaz, M. 2014. Investigation of bolt clamping force on the fatigue life of double lap simple bolted and hybrid (bolted/bonded) joints via experimental and numerical analysis. *Eng. Fail. Anal.* 45: 406–420.

Ferjaoui, A., Yue, T., Wahab, M.A. & Hojjati-Talemi, R. 2015. Prediction of fretting fatigue crack initiation in double lap bolted joint using Continuum Damage Mechanics. *Int. J. Fatigue*, 73: 66–76.

Grzejda, R. 2015. Modelling of bolts in multi-bolted connections using MIDAS NFX. *Technical Sciences* 18(1): 61–68.

Hu, F., Shi, G., Bai, Y. & Shi, Y. 2014. Seismic performance of prefabricated steel beam-to-column connections. *J. Constr. Steel Res.* 102: 204–216.

Jakubowski, A. & Schmidt, H. 2004. Numerical investigations on the fatigue-relevant bolt stresses in preloaded ring flange connections with imperfections (in German). *Stahlbau* 73(7): 517–524.

Khurshid, H. 2014. Development of layout factor for mechanical shear joints using finite element analysis. *AU Journal of Technology* 17(3): 120–128.

Kim, J., Yoon, J.-Ch. & Kang, B.-S. 2007. Finite element analysis and modeling of structure with bolted joints. *Appl. Math. Model.* 31(5): 895–911.

Kumakura, S. & Saito, K. 2003. Tightening sequence for bolted flange joint assembly. *Analysis of bolted joints, Proceedings of the 2003 ASME Pressure Vessels and Piping Conference*, Cleveland: ASME, 9–16.

Li, W., Cai, H., Li, C., Wang, K. & Fang, L. 2015. Micromechanics of failure for fatigue strength prediction of bolted joint structures of carbon fiber reinforced polymer composite. *Compos. Struct.* 124: 345–356.

Li, Y.-F., Jobe, O., Yu, Ch.-Ch. & Chiu, Y.-T. 2015. Experiment and analysis of bolted GFRP beam-beam connections. *Compos. Struct.* 127: 480–493.

Magnucki, K. & Sekulski, Z. 2005. Effective design of a bolted flange joint with a flat ring gasket. *The Archive of Mechanical Engineering* 52(3): 267–281.

Mathan, G. & Siva Prasad, N. 2011. Studies on gasked flange joint under bendig with anisotropic Hill plasticity model for gasket. *Int. J. Pres. Ves. Pip.* 88(11–12): 495–500.

Montgomery, J. 2006. Modeling multi-bolted systems. *Proc. of the ANSYS 2006 User's Conference*. Pittsburgh.

Mourya, R.K., Banerjee, A. & Sreedhar, B.K. 2015. Effect of creep on the failure probability of bolted flange joints. *Eng. Fail. Anal.* 50: 71–87.

Nechache, A. & Bouzid, A.-H. 2008. On the use of plate theory to evaluate the load relaxation in bolted flanged joints subjected to creep. *Int. J. Pres. Ves. Pip.* 85(7): 486–497.

Over, H.-H., Igelmund, A., Zerres, H., Guérout, Y. & Kockelmann, H. 2003. Web-enabled database for gasket parameters. *Analysis of bolted joints, Proceedings of the 2003 ASME Pressure Vessels and Piping Conference*, Cleveland: ASME, 97–109.

PN-EN 1591-1: 2014–04. Flanges and their joints. Design rules for gasketed circular flange connections. Part 1: Calculation.

Saberi, V., Gerami, M. & Kheyroddin, A. 2014. Comparison of bolted end plate and T-stub connection sensitivity to component thickness. *J. Constr. Steel Res.* 98: 134–145.

Sallam, H.E.M., El-Sisi, A.E.A., Matar, E.B. & El-Hussieny, O.M. 2011. Effect of clamping force and friction coefficient on stress intensity factor of cracked lapped joints. *Eng. Fail. Anal.* 18(6): 1550–1558.

Sawa, T., Matsumoto, M. & Nagata, S. 2003. Effects of scatter in bolt preload of pipe flange connections with gaskets on sealing performance. *Analysis of bolted joints, Proceedings of the 2003 ASME Pressure Vessels and Piping Conference*, Cleveland: ASME, 65–75.

Schaaf, M. & Bartonicek, J. 2003. Calculation of bolted flange connections of floating and metal-to-metal contact type. *Analysis of bolted joints, Proceedings of the 2003 ASME Pressure Vessels and Piping Conference*, Cleveland: ASME, 59–64.

Schaumann, P., Kleineidam, P. & Seidel, M. 2001. FE-modelling of connections with bolts in tension (in German). *Stahlbau* 70(2): 73–84.

Semke, W.H., Bibel, G.D., Jerath, S., Gurav, S.B. & Webster, A.L. 2006. Efficient dynamic structural response modelling of bolted flange piping systems. *Int. J. Pres. Ves. Pip.* 83(10): 767–776.

Takaki, T. & Fukuoka, T. 2003. Methodical guideline for bolt-up operation of pipe flange connections (A case using sheet gasket and spiral wound gasket). *Analysis of bolted joints, Proceedings of the 2003 ASME Pressure Vessels and Piping Conference*, Cleveland: ASME, 23–30.

Wang, Y.Q., Zong, L. & Shi, Y.J. 2013. Bending behavior and design model of bolted flange-plate connection. *J. Constr. Steel Res.* 84: 1–16.

Williams, J.G., Anley, R.E., Nash, D.H. & Gray, T.G.F. 2009. Analysis of externally loaded bolted joints: Analytical, computational and experimental study. *Int. J. Pres. Ves. Pip.* 86(7): 420–427.

Witek, A. & Grzejda, R. 2005. Analysis of a nonlinear multi-bolted joint loaded by a normal force. *Archives of Mechanical Technology and Automation* 25(2): 211–219.

Witek, A., Grzejda, R., Grudziński, P. & Konowalski, K. 2011. Modelling and numerical calculations of asymmetrical multi-bolted connections (in Polish). *Proc. of the 2nd Polish Congress of Mechanics*. Poznan: Poznan University of Technology.

Yokoyama, T., Olsson, M., Izumi, S. & Sakai, S. 2012. Investigation into the self-loosening behavior of bolted joint subjected to rotational loading. *Eng. Fail. Anal.* 23: 35–43.

Zapico-Valle, J.L, Abad-Blasco, J., González-Martínez, M.P., Franco-Gimeno, J.M. & García-Diéguez, M. 2012. Modelling and calibration of beam-column joint based on modal data. *Comput. Struct.* 108–109: 31–41.

Zerres, H. & Guérout, Y. 2004. Present calculation methods dedicated to bolted flanged connections. *Int. J. Pres. Ves. Pip.* 81(2): 211–216.

Zhang, J.M. 2001. Design and analysis of mechanically fastened composite joints and repairs. *Eng. Anal. Bound. Elem.* 25(6): 431–441.

Zografos, A. & Dini, D. 2009. A combined BEM/Contact Asymptotics (BEM-CA) semi-analytical formulation for the assessment of fretting damage in bolted joints. *Procedia Engineering* 1(1): 201–204.

Żyliński, B. & Buczkowski, R. 2010. Analysis of bolt joint using the finite element method. *Archive of Mechanical Engineering* 57(3): 275–292.

Advances in Mechanics: Theoretical, Computational and Interdisciplinary Issues – Kleiber et al. (Eds)
© 2016 Taylor & Francis Group, London, ISBN 978-1-138-02906-4

Fretting wear simulation in model studies

S. Guzowski & M. Michnej
Institute of Rail Vehicles, Faculty of Mechanical Engineering, Cracow University of Technology, Cracow, Poland

ABSTRACT: Fretting wear occurs on the surfaces pressed to each other and loaded by variable force of elements. An example of its occurrence is a connection a wheel—a wheelset axle with an automatic change of a wheel track (SUW2000). The mechanism of wear in such connections has not yet been fully studied. Studying it on a real object is significantly difficult, therefore a model study was proposed of a connection of running fit. The received results can be transferred onto a real object on condition that mechanical similarity criteria are satisfied. In the article these criteria have been laid down as well as possibilities have been given of using them in fretting wear studies of a connection a wheel—a wheelset axle with an automatic change of a wheel track.

1 INTRODUCTION

The notion of fretting comprises the effects of complex phenomena actions on the surfaces pressed against each other and loaded by a variable force of the elements. At present mechanical engineers agree that fretting is a phenomenon of a very complex wear mechanism, in which there overlap or follow in succession: adhesive wear, surface fatigue, exfoliation, oxidation, irregularities apex wear and loose products wear (Furmanik 2014). Particular researchers differ in their opinions, what is the result of taking on one of these processes as the initiating one of the fretting wear development. The studies of fretting wear were carried out mainly for matching concentrated or flat contacts and the proposed wear models also referred to them (Shinde 2006). Those studies were mainly carried out in the 50's – 80's of the past century. The present query of periodicals dealing with the above subject, especially Wear or International Tribology, shows that the fretting wear problem is mainly considered in the aspect of the influence of material selection or technological process on the initiation of wear in the tribological node. There are no publications on the studies of fretting wear development mechanism in actual nodes of machines or vehicles.

The wheelset with an automatic change of a wheel track (SUW2000) consists of wheel sets with the automatic gauge changing system and a rail gauge change stand which enables the change of the wheel gauge in a rail vehicle. The wheel set of the SUW 2000 system shown in Figure 1 has a far more complicated construction than a standard wheel set. It consists of the following parts:

- the axle sub-assembly (1),
- the wheel (2),
- the locking mechanisms (3),
- the axle bearings with casings (4),
- the external cover (5),
- the internal cover (6),
- the stopper rings (7),
- the self-locking nuts (8).

Wheel sets can adapt to three different combinations of rail gauges (1435/1520 mm, 1435/1668 mm, and 1435/1520/1668 mm). Wheel gauge change takes place when a train moves along the rail gauge change stand where the wheels are automatically adjusted to the required track gauge.

The initial exploitation of the set showed that already after a low mileage (of about 1.5 thousand km) there occurred problems in the wheel track

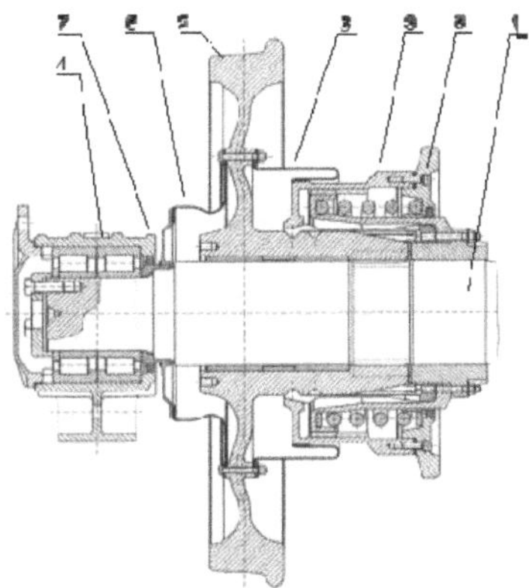

Figure 1. Basic elements of the SUW 2000 wheel set.

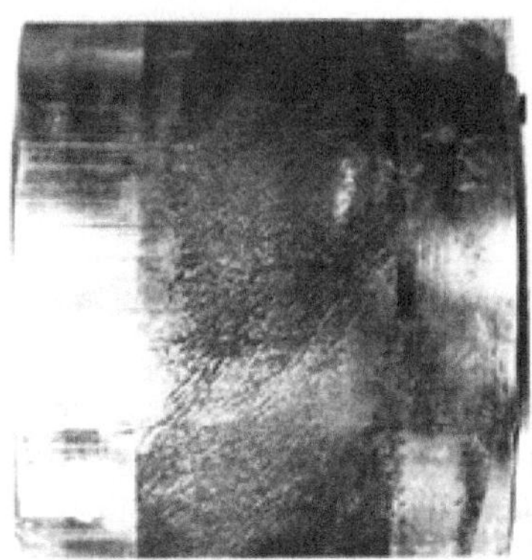

Figure 2. Fretting wear at the edge of a wheel seat of a wheelset axle with an automatic change of a wheel track.

change (Suwalski 2006). Observations of the surface of the axle wheel seat after dismantling the wheel set presented in Figure 2, showed fretting damages in the area of contact with the wheel hub, causing locking of a wheel on the axle when changing its track.

Carrying out tests on a real object such as a wheelset is very difficult. Fretting wear studies of a connection a wheel—a wheelset axle, in agreement with Heinke's classification of experimental tests, can be carried out as model tests on simple specimens when simulating real exploitation conditions (Szücs 1977). The advantages of such studies are: a simple construction of a sample, easy selection of a typical stand for fretting wear tests, a significant shortening of the testing time, the possibility of detailed evaluation of wear in the connection area by making use of, among others, metallographic and metrological tests, lowering the costs of making the connection model and the tests themselves. On the other hand the disadvantage is the necessity of the achieved results verification with those carried out on a real object.

2 ESTABLISHING MECHANICAL SIMILARITY CRITERIA

In experimental tests carried out on models it is very important to satisfy the mechanical similarity criteria which would allow transferring the results of these tests onto real objects. Modelling on a test stand only the fundamental parameters of the process simplifies and facilitates the experiment, whereas reproduction of all the parameters is usually impossible due to the complexity and scope of the experiment. The difficulty in fretting wear modelling results from the fact that accepting the same materials, contact and load conditions and the like does not represent all fretting wear phenomena that can be encountered in practice. In such studies similarity theory may prove useful, whose mathematical tool is dimensional analysis (Waterhouse 1981). In the present article elements of this theory were used to establish mechanical similarity criteria accepted for testing the model of running fit connection in a wheel—a wheelset axle with an automatic change of a wheel track in relation to the real object.

Measures of fatigue strength of a wheel axle running fit connection are the following:

- pressure on the wheel—axle contact surface: $p\,[\mathrm{kg \cdot m^{-1} \cdot s^{-2}}]$
- bending moment that can transfer the connection: $Mg\,[\mathrm{N \cdot mm}]$

Fatigue strength can be influenced by the following factors treated as independent variables:

1. substitute modulus of elasticity:
2. $E\,[\mathrm{kg \cdot m^{-1} \cdot s^{-2}}]$
3. connection length of frictional pair elements: l [mm]
4. axle wheel seat diameter: D [mm]
5. clearance gap: h [mm]
6. bearing length of the effective roughness profile coefficient: t_p [%]
7. load frequency: n [Hz]
8. axle wheel seat surface roughness: Ra_1 [μm]
9. surface roughness of the wheel hub opening: Ra_2 [μm]
10. normal load: P [N]

Normal contact stress can be presented as a function:

$$p = F(t_p, P, h, Ra_1, Ra_2, l, D, E, n) \tag{1}$$

In accordance with denotations, dimensional matrix of the quantities in dependencies (1) will have the form (Table 1).

Function F gains its form on the basis of experimental studies. All the assumptions of matrix method application are fulfilled. Equation 1 is equivalent to the dimensionless Equation 2 in agreement with Buckingham's theorem which states that the number of dimensionless moduli is equal to the number of independent physical parameters reduced by the number of basic dimensions.

$$\pi_0 = F(\pi_1, \ldots \pi_i) \tag{2}$$

where: π_i ($i = 0,1,\ldots 6$) form a complete system of variable criteria numbers: $p \ldots n$.

From the above matrix the following dimensional system of Equations 3–5 results, which was solved for the unknown p_7, p_8 and p:

Table 1. Dimensional matrix.

	p_0	p_1	p_2	p_3	p_4	p_5	p_6	p_7	p_8	p_9
	p	t_p,	P	h	Ra_1	Ra_2	l	D	E	n
L	−1	0	1	1	1	1	1	1	−1	0
M	1	0	1	0	0	0	0	0	1	0
T	−2	0	−2	0	0	0	0	0	−2	−1

where: **L**[m]—length, **M**[kg]—mass, **T**[s]—time.

Table 2. Solution matrix.

	p_0	p_1	p_2	p_3	p_4	p_5	p_6	p_7	p_8	p_9
π	p	t_p,	P	h	Ra_1	Ra_2	l	D	E	n
π_0	1	0	0	0	0	0	0	0	−1	0
π_1	0	1	0	0	0	0	0	0	0	0
π_2	0	0	1	0	0	0	0	−2	−1	0
π_3	0	0	0	1	0	0	0	−1	0	0
π_4	0	0	0	0	1	0	0	−1	0	0
π_5	0	0	0	0	0	1	0	−1	0	0
π_6	0	0	0	0	0	0	1	−1	0	0

Table 3. Numbers of calculated values of chosen characteristic.

		Real object	Model	
π_0	p [MPa]	69,2–103,9	37,4–110,8	Ob ≈ M
π_2	P/D^2 [N·mm²]	2,01	1,09	Ob ≠ M
π_3	h/D [mm]	0,001	0,001	Ob = M
$\pi_{4,5}$	Ra_1/Ra_2	0,7–0,2	0,7–0,2	Ob ≈ M
π_6	l/D	1,84	1,84	Ob ≈ M

$$-p_0 + p_2 + p_3 + p_4 + p_5 + p_6 + p_7 - p_8 = 0 \quad (3)$$

$$p_0 + p_2 + p_8 = 0 \quad (4)$$

$$-2p_0 - 2p_2 - 2p_8 - p_9 = 0 \quad (5)$$

Considering free variables p_i $(i = 0, 1, \ldots 6)$ and special sequence numbers the solution matrix was received (Table 2).

From Table 2 matrix after transformations the following set of characteristic numbers was achieved.

Table 3 shows examples of calculated values of chosen characteristic numbers according to the above method.

From the comparison of the values of characteristic numbers it appears that first of all the conditions of geometric similarity of the connection itself and the stress state were satisfied. However, the condition of external load was not satisfied, but it was the result of fatigue testing machine limitations as well as the connection model itself. Failures generated in the connection model at a smaller range of external loads may suggest their greater intensiveness on the real object.

Thus, preserving mechanical similarity criteria allows to transfer the test results achieved on a connection model of running fit onto the real connection.

3 RESEARCH METHODOLOGY

What was taken into consideration when choosing a sample modelling the connection sliding bush—wheel set axle was to preserve dimensional similarity in the connection area. In order to achieve this, a proportion of the connection length and the axle diameter and matching was kept. Fatigue testing machine of MUJ type was used, what allowed receiving temporarily variable load together with bending the rotating sample thus simulating actual work conditions of the wheel set.

The selection of the modeling sample of the slide sleeve—wheel set axle joint was determined by the dimension similarity in the contact area. Therefore the proportion of the length of the joint and the diameter of the axle, as well as the fit was maintained.

The sleeve was fixed on the shaft using a headless screw which provided protection against axial and circumferential movements and enabled the sleeve radial movement during rotations.

For the preliminary tests, a shaft was made from steel 45 while the sleeve from steel 36HNM. Such a choice of materials provided similar material properties to those of which the prototype of the real object was built. This model was to be a point of reference for other alternatives of the axle-sleeve joints which should reduce or eliminate wear in the joint and simultaneously ensure the lowest shift force in the axial direction.

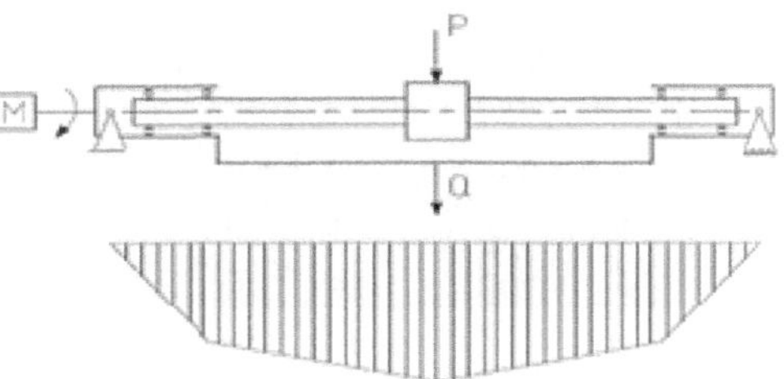

Figure 3. A sample scheme and referring to it bending moment distribution.

4 WEAR TESTS OF A WHEEL—AN AXLE CONNECTION MODEL

Testing the base sample aimed at verification of the choice of model by achieving a similar wear image as in the case of a real object and as a reference to further studies. When choosing a sample modelling bushing—wheelset axle connection attention was paid to preserve dimensional similarity in the area of connection. For this the length proportion and the axle diameter and fit were preserved. In wear tests a fatigue testing machine was used what allowed obtaining periodically variable load with a simultaneous bending of the rotating sample in this way simulating real work conditions of the wheelset.

Figure 4 presents a characteristic fretting wear image occurring on the real axle surface of the wheelset in comparison to the wear registered on the sample modelling the axle in simulation tests. Failures on the shaft surface in fatigue tests both in respect of the image and the place of failure are similar to the failures on the axle surface. This proves that the choice of the model and the conditions of its work were correct.

Preserving equality of the characteristic numbers for the real object and the connection model proves that the choice was correct. It is a very important statement as it allows transferring the results of fretting wear tests obtained on a connection model directly onto a real object. Studying fretting wear phenomena on the proposed connection model allows to make a number of observations and measurements which would not have been possible on a real object. Especially important is to examine the contact surface of both elements before the tests as well as after fretting wear tests. The above studies and observations of the contact surface are possible in the case of such a big object as a wheelset only on the accepted connection model. Dimensional analysis method in determining mechanical similarity proves to be a convenient tool in the case of fretting wear studies on big objects.

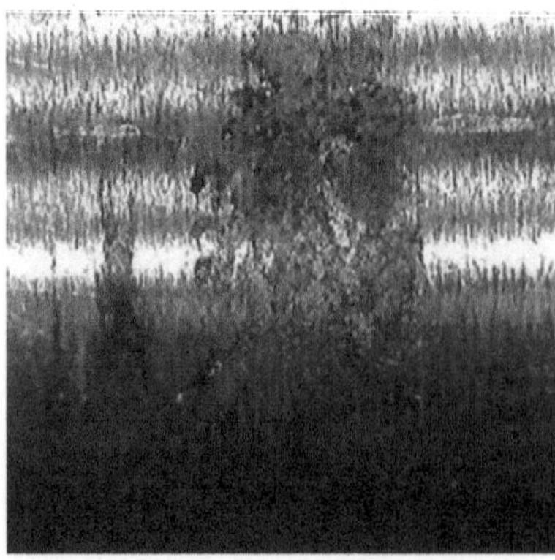

Figure 4. Fretting wear: a) at the edge of a wheel seat of a wheelset axle with an automatic change of a wheel track [8], b) at the edge of a shaft modelling a wheel seat of the wheelset axle.

5 CONCLUSIONS

Preserving equality of the characteristic numbers for the real object and the connection model proves that the choice was correct. It is a very important statement as it allows transferring the results of fretting wear tests obtained on a connection model directly onto a real object. Studying fretting wear phenomena on the proposed connection model allows making a number of observations and measurements which would not have been possible on a real object. Especially important is to examine the contact surface of both elements before the tests as well as after fretting wear tests. The above studies and observations of the contact surface are possible in the case of such a great object as a wheelset only on the accepted connection model. Dimensional analysis method in determining mechanical similarity proves to be a convenient tool in the case of fretting wear studies on great objects.

REFERENCES

Furmanik K., Oleksiak Z. 2002. Analiza wymiarowa w modelowaniu tarcia i zużycia. *Problemy Eksploatacji, nr 1.*

Furmanik, K.: 2014. Możliwości wykorzystania analizy wymiarowej w badaniach oporu przeginania taśmy na bębnach przenośnika. *Mining Science—Fundamental Problems of Conveyor Transport, vol. 21(2).*

Guzowski S. 2003. A model of fretting wear in wheel set clamped joint. *Archives of Transport, wyd. PAN, vol. 15.*

Hoeppner, D.W. 2006:Fretting fatigue case studies of engineering components. *Tribology International, Vol. 39, Issue: 10.*

Neyman, A. 2003. Fretting w elementach maszyn. *Wyd. Politechniki Gdańskiej, Gdańsk.*

Nowak, Z. 1969. Ogólna metoda wyznaczania zupełnego układu iloczynów bezwymiarowych. *Czasopismo Techniczne z.6, Wyd. Politechniki Krakowskiej, Kraków.*

Shinde S., Hoeppner D.W. 2006. Observations on fretting damage transition to cracking; state of the art and preliminary observations. *Tribology International, Vol: 39 Issue: 10.*

Suwalski R.M. 2006. System samoczynnej zmiany rozstawu kół pojazdów szynowych. *Monografia AGH, Kraków.*

Szczerek, M. 1997. Metodologiczne problemy systematyzacji eksperymentalnych badań tribologicznych. *Wyd. ITE, Radom.*

Szücs, E. 1977. Modelowanie matematyczne w fizyce i technice. *WNT Warszawa.*

Waterhouse, R.B. 1981. Fretting fatigue. Applied Science Publishers Ltd., *London.*

Advances in Mechanics: Theoretical, Computational and Interdisciplinary Issues – Kleiber et al. (Eds)
 ISBN 978-1-138-02906-4

Dynamics analysis of the RUSP linkage with joint friction modelled by the Stribeck effect

A. Harlecki & A. Urbaś
Department of Mechanics, University of Bielsko-Biala, Bielsko-Biała, Poland

ABSTRACT: The dynamics analysis of a one-dof spatial RUSP linkage is presented in the paper. It was assumed that friction occurs in the revolute joints and in the prismatic joint of the mechanism, whereas the spherical joint was treated as an ideal one. The Stribeck effect was taken into account in the process of friction modelling. The linkage was divided by using the cut joint technique in the place of the spherical joint into two open-loop kinematic chains, and joint forces were introduced. The equations of motion of the chains were derived using the Lagrange equations. Cut joint constraints were introduced to complete the equations of motion. In order to determine the values of friction torques in the revolute joints and the friction force in the prismatic joint in each integration step of the equations of motion the values of joint forces and torques in these joints were determined using the recursive Newton-Euler algorithm.

NOMENCLATURE

c —symbol of a chain
(c,p) —symbol of link (joint) p in chain c
g —acceleration of gravity
$n_l^{(c)}$ —number of links in chain c
$l^{(c,p)}$ —length of link p in chain c
$m^{(c,p)}$ —mass of link p in chain c
$\tilde{\mathbf{r}}_\alpha^{(c,p)}\Big|_{\alpha\in\{C^{(p)},S\}}$ —vector of position of point $C^{(c,p)}$ or S defined in the local coordinate system $\{c,p\}$ of link (c,p)
$\mathbf{H}^{(c,p)}$ —inertial matrix 4×4 of link (c,p)
$\tilde{\mathbf{T}}^{(c,p)}$ —transformation matrix 4×4 from the local coordinate system $\{c,p\}$ of link (c,p) to the system $\{c,p-1\}$ of link $(c,p-1)$
$\mathbf{T}^{(c,p)}$ —transformation matrix 4×4 from the local coordinate system of link $\{c,p\}$ to the global reference system $\{1,0\}$

$$\mathbf{T}_i^{(c,p)} = \frac{\partial \mathbf{T}^{(c,p)}}{\partial q_i^{(c,p)}}, \mathbf{T}_{i,j}^{(c,p)} = \frac{\partial^2 \mathbf{T}^{(c,p)}}{\partial q_i^{(c,p)} \partial q_j^{(c,p)}}$$

dof —degree(s)-of-freedom

Friction parameters:
$\mu_\alpha^{(c,p)}\Big|_{\alpha\in\{A,B,C\}}, \mu^{(c,p)}$ —instantaneous friction coefficient in revolute joint (c,p) and prismatic joint (c,p), respectively
$\mu_{s,\alpha}^{(c,p)}\Big|_{\alpha\in\{A,B,C\}}, \mu_s^{(c,p)}$ —static (limiting) friction coefficient
$\gamma_\alpha^{(c,p)}\Big|_{\alpha\in\{A,B,C\}}, \gamma^{(c,p)}$ —parameter of the Stribeck curve
$\varepsilon_\alpha^{(c,p)}\Big|_{\alpha\in\{A,B,C\}}, \varepsilon^{(c,p)}$ —parameter of the Stribeck curve

1 MATHEMATICAL MODEL OF THE SYSTEM

The one-dof RUSP linkage analysed here is presented in Figure 1. Its links are connected by means of joints—revolute R, universal U (which is a system of two revolute joints R), spherical S and prismatic P. The mechanism contains neither passive constraints nor redundant degrees of freedom. The driving link is loaded by driving torque $\mathbf{t}_{dr}^{(1,1)}$ and reduced resistance torque $\mathbf{t}_{res}^{(1,1)}$.

For the requirements of the analysis, the linkage was cut in the place of spherical joint S and, as a result, open-loop kinematic chains connected with the fixed base were obtained (Fig. 2): 1 is formed by links (1,1), (1,2), (1,3), 2 is formed only by link (2,1) which is a slider in prismatic joint P.

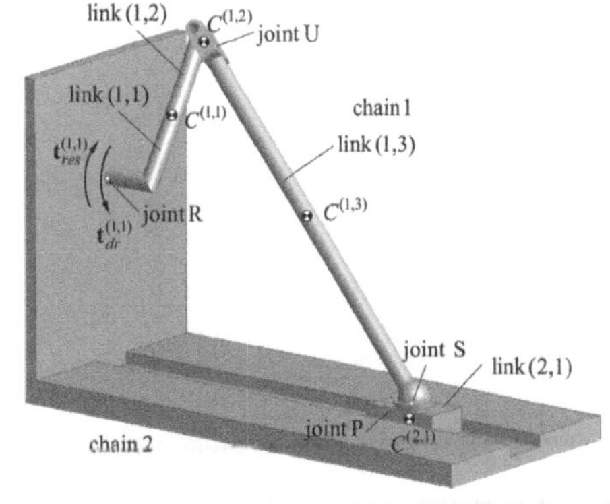

Figure 1. Spatial RUSP linkage.

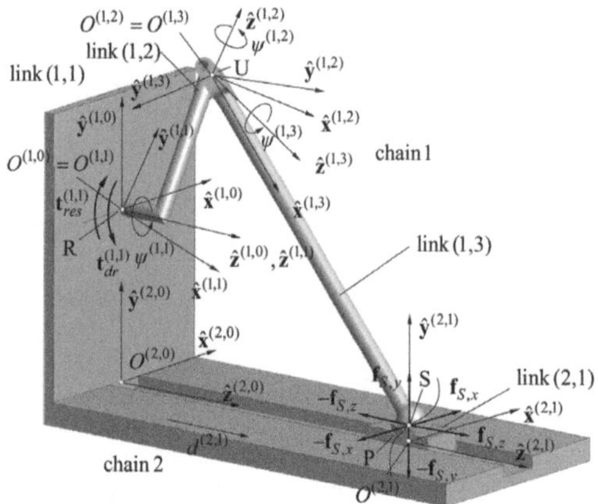

Figure 2. Joint forces acting in cut-joint S.

The local coordinate systems were attached to particular links according to the Denavit-Hartenberg notation (Craig, 1989).

The motion of both chains was described by the vectors of joint coordinates:

$$\mathbf{q}^{(c,n_l^{(c)})} = \left(q_j^{(c,n_l^{(c)})}\right)\Big|_{\substack{c=1,n_l^{(1)}=3,\, j=1,2,3\\ c=2,n_l^{(2)}=1,\, j=1}}, \tag{1}$$

where: $\mathbf{q}^{(1,3)} = [\psi^{(1,1)}\ \psi^{(1,2)}\ \psi^{(1,3)}]^T$, $\mathbf{q}^{(2,1)} = [d^{(2,1)}]$.

The homogeneous transformation matrices from the local coordinate systems $\{c,p\}$ of links (c,p) to the global reference system $\{1,0\}$ were determined according to the relationship:

$$\mathbf{T}^{(c,p)} = \mathbf{T}^{(c,p-1)}\tilde{\mathbf{T}}^{(c,p)}, \tag{2}$$

where $\mathbf{T}^{(c,0)} = \mathbf{I}$,

$$\tilde{\mathbf{T}}^{(1,1)} = \begin{bmatrix} c\,\psi^{(1,1)} & -s\,\psi^{(1,1)} & 0 & 0 \\ s\,\psi^{(1,1)} & c\,\psi^{(1,1)} & 0 & 0 \\ 0 & 0 & 1 & 0 \\ 0 & 0 & 0 & 1 \end{bmatrix},$$

$$\tilde{\mathbf{T}}^{(1,2)} = \begin{bmatrix} c\,\psi^{(1,2)} & -s\,\psi^{(1,2)} & 0 & 0 \\ 0 & 0 & 1 & l^{(1,1)} \\ -s\,\psi^{(1,2)} & -c\,\psi^{(1,2)} & 0 & 0 \\ 0 & 0 & 0 & 1 \end{bmatrix},$$

$$\tilde{\mathbf{T}}^{(1,3)} = \begin{bmatrix} c\,\psi^{(1,3)} & -s\,\psi^{(1,3)} & 0 & 0 \\ 0 & 0 & 1 & 0 \\ -s\,\psi^{(1,3)} & -c\,\psi^{(1,3)} & 0 & 0 \\ 0 & 0 & 0 & 1 \end{bmatrix},$$

$$\tilde{\mathbf{T}}^{(2,1)} = \begin{bmatrix} 1 & 0 & 0 & 0 \\ 0 & 1 & 0 & 0 \\ 0 & 0 & 1 & d^{(2,1)} \\ 0 & 0 & 0 & 1 \end{bmatrix},$$

$$c\,\psi^{(\alpha,\beta)}\Big|_{\substack{\alpha\in\{1\}\\ \beta\in\{1,2,3\}}} = \cos\psi^{(\alpha,\beta)}\Big|_{\substack{\alpha\in\{1\}\\ \beta\in\{1,2,3\}}},$$

$$s\,\psi^{(\alpha,\beta)}\Big|_{\substack{\alpha\in\{1\}\\ \beta\in\{1,2,3\}}} = \sin\psi^{(\alpha,\beta)}\Big|_{\substack{\alpha\in\{1\}\\ \beta\in\{1,2,3\}}}.$$

Joint forces $\mathbf{f}_{S,x}, \mathbf{f}_{S,y}, \mathbf{f}_{S,z}$ and $-\mathbf{f}_{S,x}, -\mathbf{f}_{S,y}, -\mathbf{f}_{S,z}$ acting on chains 1 and 2, respectively, in accordance with the versor directions of the global reference system {1,0} were applied in cut-joint S (precisely—at the centre *S* of this joint).

2 EQUATIONS OF MOTION

The equations of motion of both chains were formulated on the basis of Lagrange equations by using the algorithm presented in a monograph by Jurevič (1986):

$$\begin{bmatrix} \mathbf{A}^{(1,3)} & \mathbf{0} & -\mathbf{D}^{(1,3)} \\ \mathbf{0} & \mathbf{A}^{(2,1)} & \mathbf{D}^{(2,1)} \\ \mathbf{D}^{(1,3)^T} & -\mathbf{D}^{(2,1)^T} & \mathbf{0} \end{bmatrix} \begin{bmatrix} \ddot{\mathbf{q}}^{(1,3)} \\ \ddot{\mathbf{q}}^{(2,1)} \\ \mathbf{f}_S \end{bmatrix} = \begin{bmatrix} \mathbf{e}^{(1,3)} + \mathbf{t}_{dr}^{(1,3)} - \mathbf{t}_{res}^{(1,3)} - \mathbf{s}_f^{(1,3)} \\ \mathbf{e}^{(2,1)} - \mathbf{s}_f^{(2,1)} \\ \mathbf{c}^{(1,2)} \end{bmatrix}, \tag{3}$$

where:

$$\mathbf{A}^{(c,n_l^{(c)})} = \left(\left(\mathbf{A}^{(c,n_l^{(c)})}\right)_{i,j}\right)_{i,j=1,\ldots,n_l^{(c)}},$$

$$\left(\mathbf{A}^{(c,n_l^{(c)})}\right)_{i,j} = \sum_{p=\max\{i,j\}}^{n_l^{(c)}} \mathrm{tr}\left\{\mathbf{T}_i^{(c,p)}\mathbf{H}^{(c,p)}\mathbf{T}_j^{(c,p)^T}\right\},$$

$$\mathbf{e}^{(c,n_l^{(c)})} = \left(\left(\mathbf{e}^{(c,n_l^{(c)})}\right)_k\right)_{k=1,\ldots,n_l^{(c)}},$$

$$\left(\mathbf{e}^{(c,n_l^{(c)})}\right)_k = \sum_{p=k}^{n_l^{(c)}}\left[-h_k^{(c,p)} - g_k^{(c,p)}\right],$$

$$h_k^{(c,p)} = \sum_{i,j=1}^{p} \mathrm{tr}\left\{\mathbf{T}_k^{(c,p)}\mathbf{H}^{(c,p)}\mathbf{T}_{i,j}^{(c,p)^T}\right\}\dot{q}_i^{(c,p)}\dot{q}_j^{(c,p)},$$

$$g_k^{(c,p)} = m^{(c,p)} g\, \mathbf{j}_2 \mathbf{T}_k^{(c,p)} \tilde{\mathbf{r}}_{C^{(c,p)}}^{(c,p)},$$

$$\mathbf{t}_{dr}^{(1,3)} = \left[t_{dr}^{(1,1)}\ \ 0\ \ 0\right],\ \mathbf{t}_{res}^{(1,3)} = \left[t_{res}^{(1,1)}\ \ 0\ \ 0\right],$$

$$\mathbf{s}_f^{(1,3)} = \left[t_f^{(1,1)} \quad t_f^{(1,2)} \quad t_f^{(1,3)} \right]^T, \mathbf{s}_f^{(2,1)} = \left[f_f^{(2,1)} \right],$$

$$\mathbf{f}_S = \left[f_{S,x} \quad f_{S,y} \quad f_{S,z} \right]^T,$$

$$\mathbf{D}^{(c,n_l^{(c)})^T} = \mathbf{J} \left[\mathbf{T}_1^{(c,n_l^{(c)})} \tilde{\mathbf{r}}_S^{(c,n_l^{(c)})} \quad \cdots \quad \mathbf{T}_{n_l^{(c)}}^{(c,n_l^{(c)})} \tilde{\mathbf{r}}_S^{(c,n_l^{(c)})} \right],$$

$$\mathbf{c}^{(1,2)} = \mathbf{J} \left[\left(\sum_{i,j=1}^{n_l^{(2)}} \mathbf{T}_{i,j}^{(2,1)} \dot{q}_i^{(2,1)} \dot{q}_j^{(2,1)} \right) \tilde{\mathbf{r}}_S^{(2,1)} - \left(\sum_{i,j=1}^{n_l^{(1)}} \mathbf{T}_{i,j}^{(1,3)} \dot{q}_i^{(1,3)} \dot{q}_j^{(1,3)} \right) \tilde{\mathbf{r}}_S^{(1,3)} \right],$$

$$\mathbf{J} = \begin{bmatrix} 1 & 0 & 0 & 0 \\ \hdashline 0 & 1 & 0 & 0 \\ \hdashline 0 & 0 & 1 & 0 \end{bmatrix} = \begin{bmatrix} \mathbf{j}_1 \\ \mathbf{j}_2 \\ \mathbf{j}_3 \end{bmatrix}.$$

The equations presented here were solved using the Newmark method with the iteration procedure.

In order to determine the values $t_f^{(1,1)}$, $t_f^{(1,2)}$, $t_f^{(1,3)}$ and $f_f^{(2,1)}$ of the friction torques in revolute joints R and the friction force in prismatic joint P, respectively, in each integration step of the equations of motion the values $\tilde{f}_{O^{(c,p)}}^{(c,p)}$ of the joint forces and the values $\tilde{n}_{O^{(c,p)}}^{(c,p)}$ of the joint torques in these joints were calculated by using the recursive Newton-Euler algorithm (Craig, 1989).

For the purposes of the method, models of revolute joints R (Fig. 3) and prismatic joint P (Fig. 4) were worked out (Harlecki & Urbaś).

According to the Coulomb formula the values of friction torques in revolute joints R and the value of the friction force in prismatic joint P depend on the values of the normal reaction forces acting on the frictional surfaces of these joints and, respectively, on friction coefficients:

$$\mu_\alpha^{(c,p)} \Big|_{\alpha \in \{A,B,C\}}$$

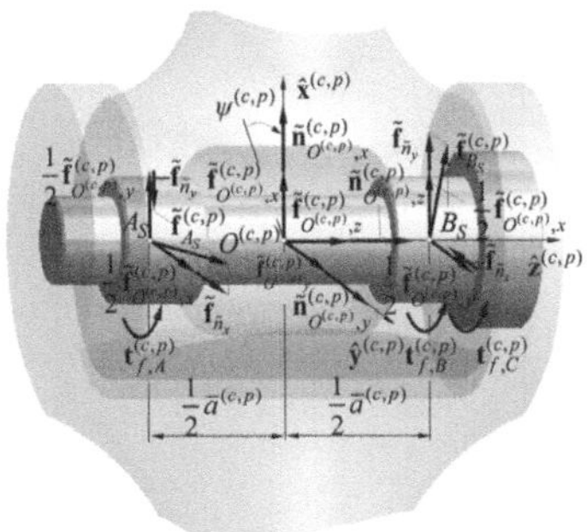

Figure 3. Model of the revolute joint.

on the frictional surfaces of the revolute joints and friction coefficient $\mu^{(c,p)}$ on the frictional surfaces of the prismatic joint. In the case of revolute joints there are three frictional surfaces—rotational A, B and face C.

It was assumed that the friction coefficients:

$$\mu_\alpha^{(c,p)} \Big|_{\alpha \in \{A,B,C\}}$$

and $\mu^{(c,p)}$ change along with the change of relative (generalised) velocities $\dot{q}_j^{(c,n_l^{(c)})}$ in joints (c,p). The courses of these coefficients (Fig. 5), defined in the paper as the friction characteristics, were determined in the continuous form based on the formula given in an article by Van de Vrande et al. (1999) as:

$$\mu \Big|_{\substack{\mu=\mu_\alpha^{(c,p)}|_{\alpha\in\{A,B,C\}} \\ \mu=\mu^{(c,p)}}} = \mu_s \frac{2}{\pi} \frac{\arctan\left(\varepsilon \dot{q}_j^{(c,n_l^{(c)})}\right)}{1+\gamma \left|\dot{q}_j^{(c,n_l^{(c)})}\right|} \Bigg|_{\substack{\mu_s=\mu_{s,\alpha}^{(c,p)},\, \varepsilon_\alpha=\varepsilon_\alpha^{(c,p)},\, \gamma_\alpha=\gamma_\alpha^{(c,p)}|_{\alpha\in\{A,B,C\}} \\ \mu_s=\mu^{(c,p)},\, \varepsilon=\varepsilon^{(c,p)},\, \gamma=\gamma^{(c,p)}}} \quad (4)$$

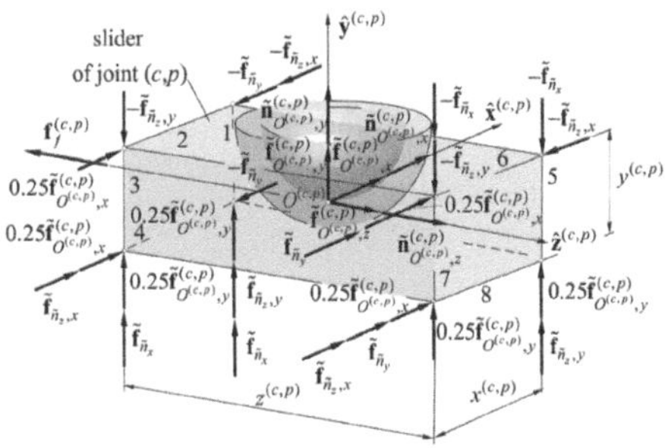

Figure 4. Model of the prismatic joint.

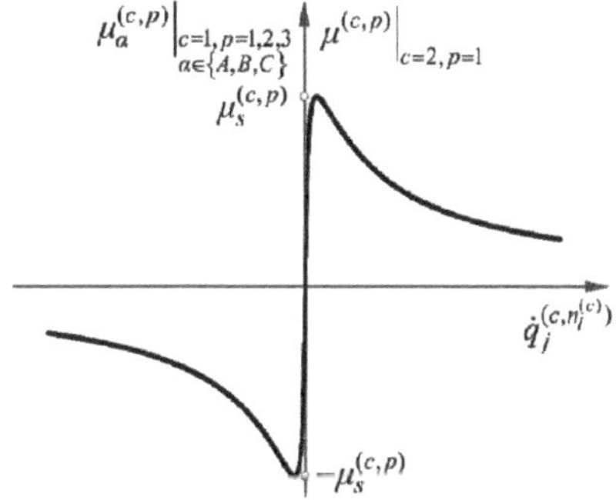

Figure 5. Friction characteristics.

As can be observed, phases of static friction in the joints were omitted in the analysis, but the Stribeck effect (Stribeck, 1902) in the form of the descending part of the friction characteristic was taken into account. Such a procedure is allowed because, in practice, the stiffness of the drive of the linkages and the stiffness of their links are sufficiently large, therefore these phases are negligibly short.

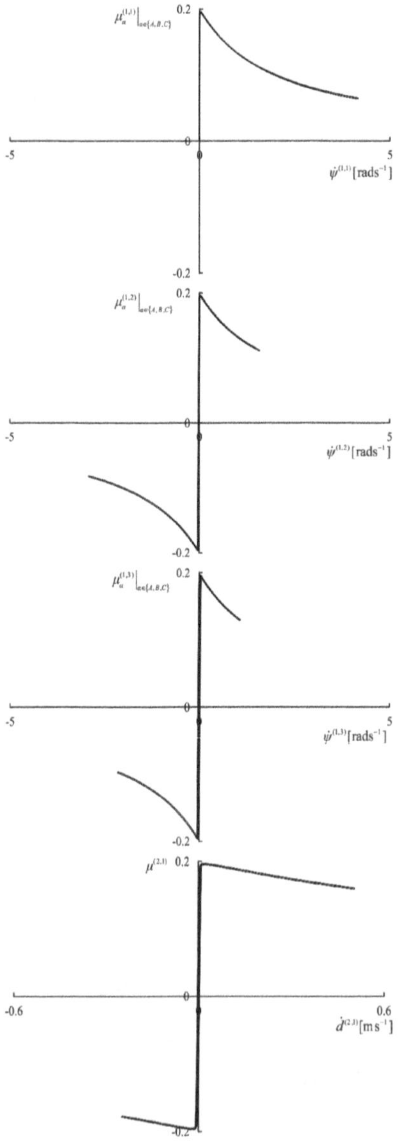

Figure 6. Friction characteristics in the joints of the linkage.

The continuous friction characteristics were also considered in other papers devoted to friction modelling in problems of machine dynamics, but their authors took into account only the constant values of these coefficients in the phase of gross-sliding, thus they omitted the Stribeck effect (Bernard, 1980, Ronney & Deravi, 1982, Threlfall, 1978).

3 RESULTS OF CALCULATIONS

The subsequent figures present examples of the results of numerical calculations. The friction characteristics, omitting the static friction phases and taking into account the Stribeck effect in the joints of the linkage considered here, are presented in Figure 6.

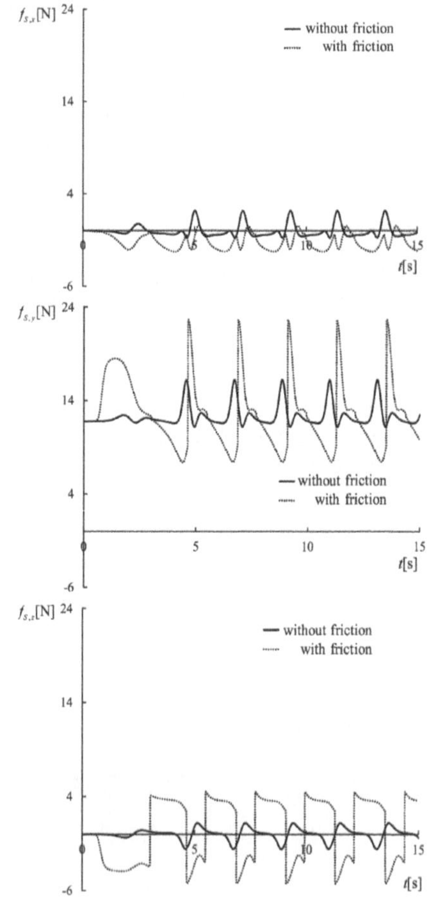

Figure 7. Courses of the components of joint forces in cut-joint S.

The courses of components of joint forces in cut-joint S, expressed as a function of time, are shown in Figure 7 in cases when friction was either taken into account or omitted.

The characteristics obtained when the friction in the joints was considered are significantly different from the characteristics determined in the case of ideal joints. This is a result of taking into account the Stribeck effect in the process of friction modelling. The numerical experiments that were carried out showed that the differences between both kinds of courses would not be so significant if the constant values of friction coefficients in the phase of gross-sliding were taken into account, i.e. if the Stribeck effect was omitted.

4 CONCLUSIONS

In the author's opinion the method proposed for modelling friction in the joints of the linkage considered here can be successfully used in dynamics analysis of other linkages with the similar structure, i.e. RSRRR and RSUP mechanisms. In the analysis, the static friction phases in the joints are omitted but, as was already stated, this is permissible in the case of such systems because the approach sufficiently reflects the real behaviour of these linkages. It is important to note that the friction characteristics that are assumed take into account the Stribeck effect which occurs in real lubricated joints of linkages. In the procedure considered these characteristics are expressed by a single formula.

REFERENCES

Bernard, J.E. 1980. The simulation of Coulomb friction in mechanical systems, Simulation, January.

Craig, J.J. 1989. *Introduction to robotics. Mechanics and Control*, Addison-Wesley Publishing Company, Inc.

Harlecki, A. & Urbaś, A. On the dynamics analysis of spatial linkages with friction in joints, *Journal of Theoretical and Applied Mechanics*, in review.

Jurevič, E.I. (ed.) 1984. *Dynamics of robot control*, Nauka, Moscow, in Russian.

Ronney, G.T. & Deravi, P. 1982. Coulomb friction in mechanisms sliding joints, *Mechanism and Machines Theory* 17(1).

Stribeck, R. 1902. Die wesentlichen Eigenschaften der Gleit- und Rollenlager, *Zeitschrift des Vereines Deutscher Ingenieure* 46(36).

Threlfall, D.C. 1978. The inclusion of Coulomb friction in mechanisms programs with particular reference to DRAM, *Mechanism and Machines Theory* 13(4).

Van de Vrande, B.L., van Campen, D.H. de Kraker, A. 1999. An approximate analysis of dry-friction-induced stick-slip vibrations by a smoothing procedure, *Nonlinear Dyn.* 19: 157–169.

Advances in Mechanics: Theoretical, Computational and Interdisciplinary Issues – Kleiber et al. (Eds)
 ISBN 978-1-138-02906-4

Dynamic response of a sandwich beam with a periodic core due to fuzzy stochastic moving load

R. Idzikowski & P. Śniady
Wrocław University of Environmental and Life Sciences, Wrocław, Poland

K. Misiurek
Wrocław University of Technology, Wrocław, Poland

ABSTRACT: We consider the problem of dynamic response of a finite, simply supported sandwich beam with periodic structure of the core with uncertain parameters under a moving fuzzy random load. The solution of the problem was found thanks by introducing the random dynamic moving influence function and applying average tolerance approach method. The tolerance approach method allows to pass from differential equations with periodic variable coefficients to differential equations with constant coefficients.

1 INTRODUCTION

Over the past decades sandwich beams had widespread applications in the fields of aerospace, civil and mechanical engineering as structural members with high strength to weight ratios. Dynamics response of a sandwich beams has been studied by many authors in the recent decades. Very often the cores used in load carrying sandwich beam have a cellular, periodic nature such as honeycomb or corrugated sheet. The problem of a dynamic response of a structure subjected to moving loads is interesting and important. This problem occurs in dynamics of bridges, roadways, railways and runways as well as missiles and aircrafts. The structural engineering is significantly influenced by uncertainty which originates from human mistakes and errors in the manufacturing process, from the use and maintenance of the construction and from a lack of information. Therefore, in recent years the uncertainty modelling in computational mechanics has received particular attention. In the dynamic analysis of structures the uncertainty can be divided into two types, namely the randomness (stochastic variability) described by the use of probability theory and the fuzziness (imprecision) described by the use of fuzzy sets. In the case when sufficient statistical data is not available a fuzzy function (fuzzy process) or a fuzzy random variable (fuzzy stochastic process) could be employed for modeling purposes. In the paper the problem of dynamic response of a finite, simply supported sandwich beam with a periodic structure of the core with uncertain parameters under a moving fuzzy random load is considered. The solution of the problem was found by introducing the random dynamic moving influence function and applying the average tolerance approach method. The average tolerance approach method allows to pass from differential equations with periodic variable coefficients to differential equations with constant coefficients. The tolerance averaging method (Woźniak 1993), Woźniak & Wierzbicki 2000) has several advantages and may be used as an alternative to the well-known homogenization method.

2 GETTING STARTED

Let us consider a sandwich beam with a rectangular cross-section consisting of two thin, stiff, elastic sheets and a thick, periodic core layer. The following general assumptions are made when developing the differential equations of motions of a sandwich beam under load excitation:

- the theory of linear elasticity applies,
- transverse direct strains in the face sheets and core are negligible so that transverse displacements are the same for all points in a normal section,
- the face sheets carry only axial forces,
- the periodic core carries only shear,
- there is no slippage or delamination between the face sheets and the core,
- only transverse inertia is taken into account,
- the damped vibrations are considered.

The forces and displacement are shown on Figures 1–2.

The set of differential equations which describe vibrations of the sandwich beam are presented in the form (Misiurek & Śniady 2013):

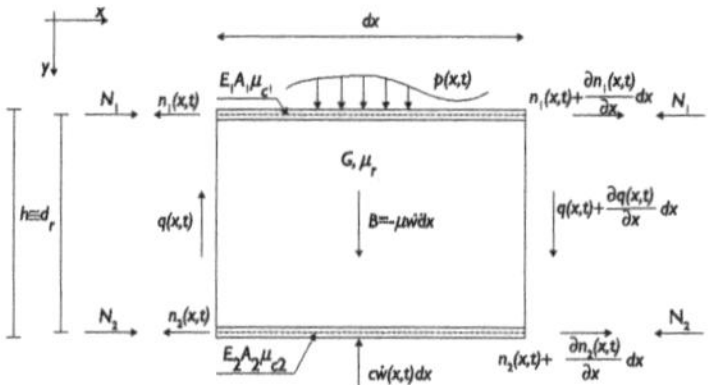

Figure 1. The internal forces in the beam section.

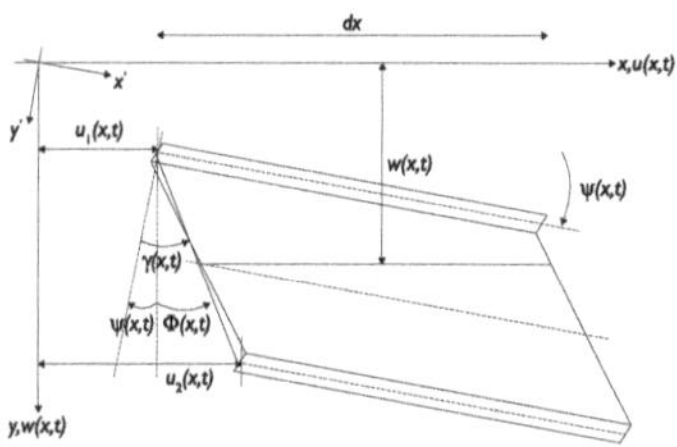

Figure 2. Displacements of the beam.

$$-bd\frac{\partial}{\partial x}\left\{G_\alpha(x)\left[\frac{\partial w_\alpha(x,t)}{\partial x}+\Phi_\alpha(x,t)\right]\right\}$$
$$+c(x)\frac{\partial w_\alpha(x,t)}{\partial t}+\mu(x)\frac{\partial^2 w_\alpha(x,t)}{\partial t^2}=p_\alpha(x,t), \quad (1)$$

$$\frac{\partial^2\Phi_\alpha(x,t)}{\partial x^2}-\frac{G_\alpha(x)b}{\varsigma d}\Phi_\alpha(x,t)$$
$$-\frac{G_\alpha(x)b}{\varsigma d}\frac{\partial w_\alpha(x,t)}{\partial x}=0, \quad (2)$$

where $\Phi_\alpha(x,t)=[u_{2\alpha}(x,t)-u_{1\alpha}(x,t)]/d$, $\varsigma=E_1A_1E_2A_2/(E_1A_1+E_2A_2)$ and where E_1, E_2 = are the Young's moduli of the upper and the lower sheets, respectively; $G_\alpha(x)$ = is the shear modulus of the beam core; A_1, A_2 = are the areas of the sheets cross-sections; b is the beam width and $\mu=\mu_1+\mu_2+\mu_c$ is the mass of the beam; μ_1, μ_2 = are the masses of the sheets; μ_c = is the mass of the beam core. The quantity 2ς denotes the harmonic average of the axial stiffnesses E_1A_1 and E_2A_2 of the sheets. The symbol α denotes the fuzziness of the parameters or functions. The axial force in the sheets is equal to:

$$n_\alpha(x)=\left(\frac{E_1A_1E_2A_2}{E_1A_1+E_2A_2}\right)d\frac{\partial\Phi_\alpha(x)}{\partial x}=\varsigma d\frac{\partial\Phi_\alpha(x)}{\partial x}. \quad (3)$$

The boundary conditions for a simply supported beam have forms:

$$w_\alpha(0,t)=w_\alpha(L,t)=0, \quad (4)$$

$$\left.\frac{\partial\Phi_\alpha(x,t)}{\partial x}\right|_{x=0}=\left.\frac{\partial\Phi_\alpha(x,t)}{\partial x}\right|_{x=L}=0, \quad (5)$$

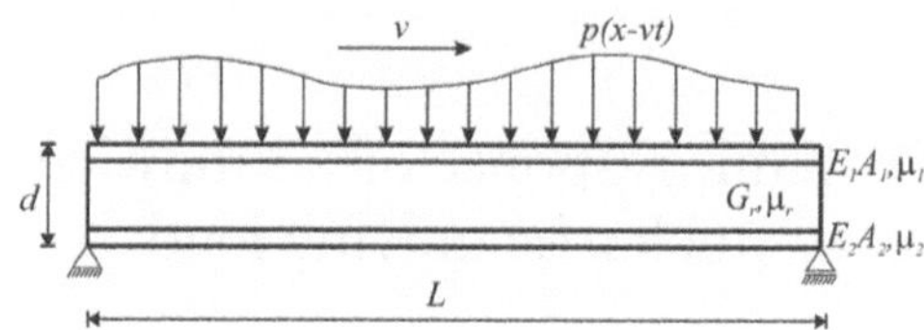

Figure 3. Beam loaded by a randomly distributed load moving with a constant velocity.

where L = is the beam length.

The standard methods of analyzing the beam dynamics are effective only if the coefficients in (7–8) are deterministic and either constant or slowly varying. The shear modulus $G_\alpha(x)$ in this study is modeled as a fuzzy, random periodic field and is a rapidly varying l-periodic function:

$$G_\alpha(x)=a_\alpha\bar{G}(x)=G_\alpha(x+l)=a_\alpha\bar{G}(x+l) \quad (6)$$

The length l is small compared to the length L of the beam ($l << L$). The function $\bar{G}(x)$ = is deterministic and a_α = is a fuzzy random variable. It is assumed that the expected value $E[a_\alpha]$ and the variance $\sigma^2_{a\alpha}=E[a_\alpha^2]-E^2[a_\alpha]$ are known.

Let the vibrations of the beam be caused by a randomly distributed load $p(x-vt)$ moving with a constant velocity v (Fig. 3).

Let us notice that the load process $p(x-vt)=p(\tau)$ is a weak space-time stationary stochastic process and can be assumed a sum of a deterministic, and a random part:

$$p(x-vt)=\bar{p}+\tilde{p}(x-vt)=\bar{p}+\tilde{p}(\tau), \quad (7)$$

where $\bar{p}$ = const., $E[\tilde{p}(x-vt)]=0$ and the symbol $E[\,]$ denotes expectation. The solution for the deterministic part of the load $\bar{p}$ = const. has been omitted in this paper and for this reason we consider only, vibration of the beam due to the stochastic part.

Let us assume that the covariance function of the moving load process is known:

$$C_{pp}(\tau_1,\tau_2)=E[\tilde{p}(\tau_1)\tilde{p}(\tau_2)]=C_{pp}[(\tau_1-\tau_2)]. \quad (8)$$

3 THE TOLERANCE AVERAGING APPROXIMATION METHOD

It is difficult to find the solution to (1–2) as the coefficients are strongly periodic. Thus, we solve (1–2) based on the tolerance-averaged model (Woźniak 1993, Woźniak & Wierzbicki 2000). Using this procedure it is possible to transform (1–2) to the form of a system of averaged differential equations with constant coefficients. This approximation describes the effect of the structural length parameter of the beam. The periodic functions will be averaged by means of the formula:

$$< g(x,t) > = \frac{1}{l} \int_{x-l/2}^{x+l/2} g(\xi,t) d\xi, \quad x \in \Omega \tag{9}$$

where $g(x,t)$ is an arbitrary function defined on $\Omega = (0, L)$. We base on Conformability Assumption (Woźniak 1993) that the functions $w_\alpha(x, t)$ and $\Phi_\alpha(x, t)$ conform to the l-periodic structure of the beam, and together with all their derivatives are periodic-like. Let us introduce the following decomposition of these functions:

$$w_\alpha(x,) = W_\alpha(x,t) + g^A(x) V_\alpha^A(x,t), \tag{10}$$

$$\Phi_\alpha(x,) = \Psi_\alpha(x,t) + g^A(x) \varphi_\alpha^A(x,t), \tag{11}$$

where $W_\alpha(x, t)$, $\Psi_\alpha(x, t)$ are the averaged parts of the functions $w_\alpha(x, t)$, $\Phi_\alpha(x, t)$ and $V_\alpha^A(x, t)$, $\varphi_\alpha^A(x, t)$ will be referred to as the fluctuating parts of the functions $w_\alpha(x, t)$, $\Phi_\alpha(x, t)$ (the summation convention over $A = 1, 2, \ldots$ holds). The functions $g^A(x)$ are a priori known oscillating l-periodic-like functions and the amplitudes $V_\alpha^A(x, t)$, $\varphi_\alpha^A(x, t)$ are sufficiently regular and slowly varying functions.

The functions $g^A(x)$ should satisfy conditions:

$$< g^A(x) > = \frac{1}{l} \int_{x-l/2}^{x+l/2} g^A(\xi) d\xi = 0, \tag{12}$$

$$< \mu(x) g^A(x) > = \frac{1}{l} \int_{x-l/2}^{x+l/2} \mu(\xi) g^A(\xi) d\xi = 0. \tag{13}$$

Because the functions $V_\alpha^A(x, t)$ and $\varphi_\alpha^A(x, t)$ are slowly varying in contrast to the oscillating functions $g^A(x)$, it follows that the derivatives with respect to the x variable are given by:

$$\frac{\partial w_\alpha(x,t)}{\partial x} = \frac{\partial W_\alpha(x,t)}{\partial x} + \frac{dg^A(x)}{dx} V_\alpha^A(x,t), \tag{14}$$

$$< \mu(x) g^A(x) > = \frac{1}{l} \int_{x-l/2}^{x+l/2} \mu(\xi) g^A(\xi) d\xi = 0. \tag{15}$$

The virtual work principle for the considered sandwich beam vibrations has the form:

$$\begin{aligned} &\int_0^L q_{c\alpha}(x,t) \delta \frac{\partial w_\alpha(x)}{\partial x} dx \\ &= \int_0^L \left[p_\alpha(x,t) + -c(x) \frac{\partial w_\alpha(x,t)}{\partial t} \right. \\ &\left. -\mu(x) \frac{\partial^2 w_\alpha(x,t)}{\partial t^2} \right] \delta w_\alpha(x) dx, \end{aligned} \tag{16}$$

where shear force in the core is equal to:

$$\begin{aligned} q_{c\alpha}(x,t) &= db\tau_{c\alpha}(x,t) \\ &= dbG_\alpha(x) \left[\frac{\partial w_\alpha(x,t)}{\partial x} + \Phi_\alpha(x,t) \right]. \end{aligned} \tag{17}$$

By averaging the left-hand-side of (16) and taking into account (10, 11, 14) and (15) we obtain (Mazur-Śniady et al. 2009):

$$\begin{aligned} &-a_\alpha db \left\{ < \bar{G}_\alpha > \left[\frac{\partial^2 W_\alpha(x)}{\partial x^2} + \frac{\partial \Psi_\alpha(x)}{\partial x} \right] + \right. \\ &\quad - < \bar{G}_\alpha \frac{dg^A(x)}{dx} > \frac{\partial V_\alpha^A(x)}{\partial x} + \\ &\quad \left. - < \bar{G}_\alpha g^A(x) > \frac{\partial \varphi_\alpha^A(x)}{\partial x} \right\} + < c > \frac{\partial W_\alpha(x)}{\partial x} \\ &\quad + < \mu > \frac{\partial^2 W_\alpha(x)}{\partial x^2} = < p(x,t) >, \end{aligned} \tag{18}$$

$$\begin{aligned} &a_\alpha db \left\{ < \bar{G}_\alpha \frac{dg^B(x)}{dx} > \left[\frac{\partial W_\alpha(x,t)}{\partial x} + \Psi_\alpha(x,t) \right] \right. \\ &\quad + < \bar{G}_\alpha \frac{dg^A(x)}{dx} \frac{dg^B(x)}{dx} > V_\alpha^A(x,t) \\ &\quad \left. + < \bar{G}_\alpha g^A(x) \frac{dg^B(x)}{dx} > \varphi_\alpha^A(x,t) \right\} \\ &= < p(x,t) g^B(x) >. \end{aligned} \tag{19}$$

After averaging (2) and taking into account the relationships (10–11) and (14–15) one obtains:

$$\begin{aligned} &\frac{\partial^2 \Psi_\alpha(x,t)}{\partial x^2} - \frac{a_\alpha b}{\varsigma d} < \bar{G}_\alpha > \Psi_\alpha(x,t) + \\ &\quad - \frac{a_\alpha b}{\varsigma d} < \bar{G}_\alpha g^A(x) > \varphi_\alpha^A(x) \\ &\quad - \frac{a_\alpha b}{\varsigma d} < \bar{G}_\alpha > \frac{\partial W_\alpha(x,t)}{\partial x} \\ &\quad - \frac{a_\alpha b}{\varsigma d} < \bar{G}_\alpha \frac{dg^A(x)}{dx} > V_\alpha^A(x,t) = 0, \end{aligned} \tag{20}$$

$$\begin{aligned} &-\frac{a_\alpha b}{\varsigma d} < \bar{G}_\alpha g^B > \Psi_\alpha(x,t) - \left[\frac{a_\alpha b}{\varsigma d} < \bar{G}_\alpha g^A(x) g^B > + \right. \\ &\quad \left. - < \frac{d^2 g^A(x)}{dx^2} g^B > \right] \varphi_\alpha^A(x,t) - \\ &\quad + \frac{a_\alpha b}{\varsigma d} < \bar{G}_\alpha g^B > \frac{\partial W_\alpha(x,t)}{\partial x} + \\ &\quad - \frac{a_\alpha b}{\varsigma d} < \bar{G}_\alpha \frac{dg^A(x)}{dx} > V_\alpha^A(x,t) = 0. \end{aligned} \tag{21}$$

4 FUZZY RANDOM DYNAMIC INFLUENCE FUNCTION

For a moving load modeled by a space-time stationary stochastic process it is difficult to find the solution using direct sine and cosine transformation which can be done in some other stochastic excitations. In order to solve the problem stochastic vibrations of the beam due to a moving random load let us introduce the Fuzzy Random Dynamic Influence Functions (FRDIF) $H_w(x, t)$ and $H_\varphi(x, t)$ which are the responses of the sandwich beam loaded by a moving force equal to one. In this case the load process has the form $p(x-vt) = \delta(x-vt)$, where δ denotes the Dirac delta function.

5 GENERAL SOLUTION

If the Fuzzy Random Dynamic Influence Function (FRDIF) are known then the response of the sandwich beam under space-time stochastic excitation can be presented in the following form:

$$w_\alpha(x,t) = \int_{t_0}^{t} H_{W\alpha}(x, t-\tau) p_\alpha(\tau) d\tau + g^A(x) \int_{t_0}^{t} H_\alpha^A(x, t-\tau) p_\alpha(\tau) d\tau, \tag{22}$$

$$\varphi_\alpha(x,t) = \int_{t_0}^{t} H_{\varphi\alpha}(x,t-\tau) p_\alpha(\tau) d\tau + g^A(x) \int_{t_0}^{t} H_{\varphi\alpha}^A(x,t-\tau) p_\alpha(\tau) d\tau, \tag{23}$$

where if $t_0 = 0$ then one considers transition vibrations and for $t_0 = -\infty$ one considers steady-state vibration case.

Thus, in order to determine the probabilistic characteristics of the displacement of the sandwich beam, one can apply the expectancy operator to (23–24) and consequently obtain the variance functions of the sandwich beam displacement:

$$\sigma_{w\alpha}^2(x,t) = \int_{t_0}^{t}\int_{t_0}^{t} E\left[H_{W\alpha}(x,t-\tau_1) H_{W\alpha}(x,t-\tau_2)\right] \cdot C_{PP\alpha}(\tau_1-\tau_2) d\tau_1 d\tau_2 + \int_{t_0}^{t}\int_{t_0}^{t} C_{(H_w H_w)_\alpha}(x,x,t-\tau_1,t-\tau_2) \cdot E\left[p(t-\tau_1)\right] E\left[p(t-\tau_2)\right] d\tau_1 d\tau_2 \tag{24}$$

$$\sigma_{\Phi\alpha}^{w}(x,t) = \int_{t_0}^{t}\int_{t_0}^{t} E\left[H_{\varphi\alpha}(x,t-\tau_1) H_{\varphi\alpha}(x,t-\tau_2)\right] \cdot C_{PP\alpha}(\tau_1-\tau_2) d\tau_1 d\tau_2 + \int_{t_0}^{t}\int_{t_0}^{t} C_{(H_\varphi H_\varphi)_\alpha}(x,x,t-\tau_1,t-\tau_2) \cdot E\left[p(t-\tau_1)\right] E\left[p(t-\tau_2)\right] d\tau_1 d\tau_2 \tag{25}$$

where:

$$C_{(H_w H_w)_\alpha}(x,x,t-\tau_1,t-\tau_2) = E\left[H_{w\alpha}(x_1,t_1) H_{w\alpha}(x_2,t_2)\right]_\alpha + -E\left[H_{w\alpha}(x_1,t_1)\right] E\left[H_{w\alpha}(x_2,t_2)\right]_\alpha \tag{26}$$

$$C_{(H_\Phi H_\Phi)_\alpha}(x,x,t-\tau_1,t-\tau_2) = E\left[H_{\Phi\alpha}(x_1,t_1) H_{\Phi\alpha}(x_2,t_2)\right]_\alpha + -E\left[H_{\Phi\alpha}(x_1,t_1)\right] E\left[H_{\Phi\alpha}(x_2,t_2)\right]_\alpha \tag{27}$$

Using α-level optimization procedure (Möller & Graf 2000) for arbitrary $\alpha = \alpha_k \in [0, 1]$, the smallest and the largest expected values and variance at an established point x and time t can be found and since the membership functions of the displacement of the sandwich beam.

6 CONCLUSIONS

The average tolerance approach was used to pass from differential equations with periodic coefficients to differential equations with constant coefficients. The presented computational algorithm based on the fuzzy random influence function is quite universal and can be applied in the vibration analysis of various types of beams and plates.

REFERENCES

Misiurek, K. & Śniady, P. 2013. Vibrations of sandwich beam due to a moving load. *Composite Structures* 104: 85–93.

Mazur-Śniady, K., Śniady, P. & Zielichowski-Haber, W. 2009. Dynamic response of micro-periodic composite rods with uncertain parameters under moving random load. *Journal of Sound and Vibration* 320: 273–288.

Woźniak, Cz. 1993. A macro-dynamics of elastic and visco-elastic microperiodic composites. *Journal of Theoretical and Applied Mechanics* 39: 763–770.

Woźniak Cz. & Wierzbicki E. 2000. *Averaging techniques in thermomechanics of composite solids*. Wydawnictwo Politechniki Świętokrzyskiej.

Möller B. & Graf G. & Beer M. 2000. Fuzzy structural analysis using α-level optimization. *Computational Mechanics* 26: 547–565.

Advances in Mechanics: Theoretical, Computational and Interdisciplinary Issues – Kleiber et al. (Eds)
© 2016 Taylor & Francis Group, London, ISBN 978-1-138-02906-4

Numerical methods for the assessment of bridge safety barriers

K. Jamroz, S. Burzyński, W. Witkowski & K. Wilde
Faculty of Civil and Environmental Engineering, Gdańsk University of Technology, Gdańsk, Poland

G. Bagiński
SafeRoad RRS Polska Sp. z o. o., Poland

ABSTRACT: The paper presents a numerical study of the bus crash test for bridge safety barriers. The analysis is conducted with the use of explicit code LS-DYNA. The problem of hourglass effect in shell elements is discussed. The simulations of bus impact against the barrier show that the analyzed barrier withstands the impact forces and restrains the bus trajectory to the bridge road.

1 INTRODUCTION

Road accidents continue to tragically affect many families in Poland. Many of the crashes are run-off road accidents whose consequences include rollovers or hitting roadside objects. Despite significant progress in hazard reduction over the last decade, these types of accidents have claimed more than 6 300 lives on Polish roads, as Budzyński M., Jamroz K. (2014) reported. Road and bridge safety barriers, crash cushions and passive support structures are road safety devices that reduce the probability of a run-off-road accident and its consequences. With high traffic volumes and high vehicle speeds, national roads are most frequently provided with road safety devices. National roads account for about 19 000 km of roads and have 6 900 bridges of the total length of 367 km. This represents almost 2% of national roads. While bridges are the scene of about 0.5% of all accidents, the severity is much worse than on other road sections.

The operation and performance of safety barriers and supports is set out in the standards which are systematically improved and published. Standard PN-EN 1317 defines the requirements for "road restraint systems" and tests that should be conducted before the structures are considered fit for purpose. The standards describe classes of safety barrier performance. Barriers are characterized based on their functional features (containment levels, barrier deformation and impact of dynamic forces on car occupants) that are achieved in crash tests. Safety barrier performance classes as defined in the standard, depend on the speed, weight and angle impact as confirmed in crash tests. The technical parameters of barriers are defined in road authority requirements. In Poland in 2010 the General Directorate for National Roads and Motorways issued guidelines for using safety barriers on the national roads network. Bridge safety barriers and the way they should be used are not regulated in Polish design and construction practice. The safety barrier guidelines, including the barriers on of bridge structures, are to be updated in 2016 and the recommendations of particular solutions require extensive set of experimental, expensive crash tests. The presented study is focused on the development of procedures for analysis of bridge barriers safety features based on numerical simulations supported by limited number of experimental crash tests.

2 BRIDGE SAFETY BARRIERS

A number of publications addresses topics such as experimental tests, modelling, simulation, validation and experimental verification of road crash tests, e.g. Ren & Vesenjak (2005), Borovinsek et al. (2007). In particular, bridge safety barriers were studied by Barnas & Edl (2006) on the performance of concrete barriers upon heavy impact, Thai (2010) on the performance of steel and composite bridge safety barriers and Atahan and Cansiz (2005) on the performance of different road and bridge transition structures. There is no research, however, on the distribution of forces in anchors that fix bridge safety barriers and the reinforcement of the bridge structural elements.

The basic problems that are still unresolved include: selecting the right barrier containment class for the level of risk and modal split, the effects of additional elements of bridge barriers and curb height on changes in functionality, linking rigid bridge barriers with deformable safety barriers and defining the strength of forces on bridge support structures for the different containment levels and types of engineering structure design (steel vs. concrete).

3 FEM ANALYSIS OF BRIDGE SAFETY BARRIERS

Physically a crash is a phenomenon of a dynamic violent nature, lasting, as a rule, not more than 1 s. However, during this short time, large deformations occur that are accompanied usually with extensive damage. Usually, the materials exhibit rate dependence. From the FEM modelling viewpoint, to account for these observable effects it is necessary to apply FEM codes that use explicit time integration schemes to compute the evolution of internal variables. Since these schemes are conditionally stable, it is very important to apply extremely short time increments for the integration of equations of motion. This necessitates the use of high-end computers (or computer clusters) to run the calculations. Therefore, substantial experience in FEM and structural dynamics is required to properly interpret the results. One of the most popular explicit codes, targeted at crash test simulations is LS-DYNA. In this paper, the analysis of a bus which hits a bridge safety barrier wall, is calculated and discussed. Figure 1 shows the initial configuration and conditions of the crash test, with considered support post.

The bus FEM model has been developed by the National Crash Analysis Center (NCAC) of the George Washington University under a contract with the FHWA and NHTSA of the US DOT. The barrier geometry (Fig. 2) is based on barrier manufactured by Orsta Stal A/S. The whole model consists of 63930 nodes and 595582 finite elements. The material properties of the barrier correspond to steel S235 and S355 grades. In simulations both materials are treated as elastoplastic with isotropic hardening. The barrier is connected to the support by four bolts with 20 mm diameter made of S355 elasto-plastic steel. The bridge

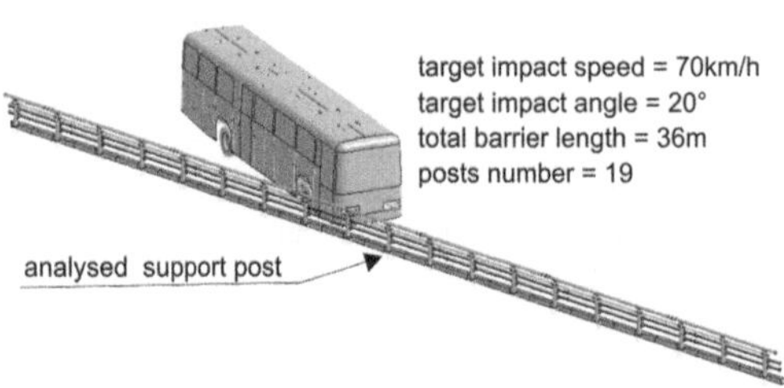

Figure 1. FEM model of the bus and barrier, initial configuration of the crash test.

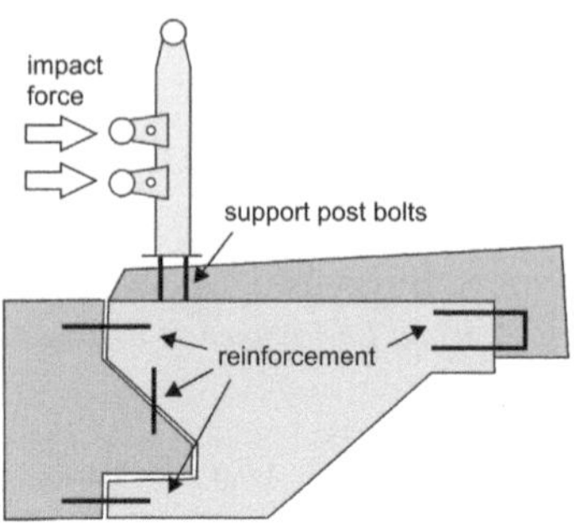

Figure 2. Bridge barrier on a concrete bridge—cross section.

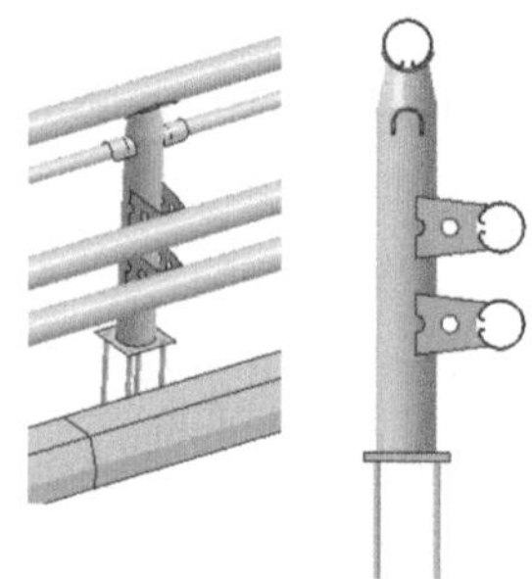

Figure 3. A detailed view of support post.

structure is considered to be solid, homogenous, non-deformable structure. The normal and shear forces in support post bolts, their magnitude are used to predict barrier slipping or extraction from bridge structure.

The simulation time is 2.0 s. The wallclock time of simulation on a cluster (360 processors) amounted to approx. 19 hours.

3.1 *Hourglass and energy analysis*

The FEM discretization of the barrier and the bus consists mainly of shell elements. Due to efficiency of calculation time, reduced integration Belytschko-Tsay shell elements have been used. The Reduced Integration (RI) technique yields substantial speed-up of calculations in comparison with Full Integration (FI) of element matrices. However, RI must be always used with care since it is accompanying always by the spurious zero-energy modes (hourglass modes), unless special techniques are applied. These spurious forms result in non-physical deformations of the analyzed structure. Therefore, various approaches are used to minimize the influence of spurious forms on the overall results. In the present case the viscous form (IHQ = 1) and the stiffness form (Flangan-Belytschko, IHQ = 4) was used with parameter qh = 0.03, accessible in LS-Dyna.

The latter technique was selected as the one which removes hourglass effect significantly comparing to other methods available in the code. By comparing the energy balance during simulation span some of the methods yielded non-physical jump of the internal energy. In Figure 4 we present the energy balance as obtained using IHQ = 1 and IHQ = 4. In the latter case, it can be noticed that the contribution of the hourglass energy in negligible as compared to the physical components.

3.2 *Barrier deformation analysis*

The impact forces due to the bus crash into the barrier lead to local damage of the barrier and can also affect the bridge secondary elements like concrete pavement cover and steel reinforcement. That may lead to the barrier slipping from the bridge, even though the barrier seems to remain functional. Figure 5 depicts the deformed configuration of the bus-barrier system at 4 time instances. The simulations showed that the barrier restrains the trajectory of the moving bus and withstands the impact loads. The part of the barrier that is exposed to impact undergoes large dynamic deformations and extensive permanent displacements characterize the final state of the structure.

Figures 6 and 7 show detailed deformation of the barrier. It is visible that the assumed geometric parameters allow for substantial irreversible deformations of large portions of the barriers. The horizontal rails, made of circular cross sections, deform into band-like shape efficiently dissipating the crash energy.

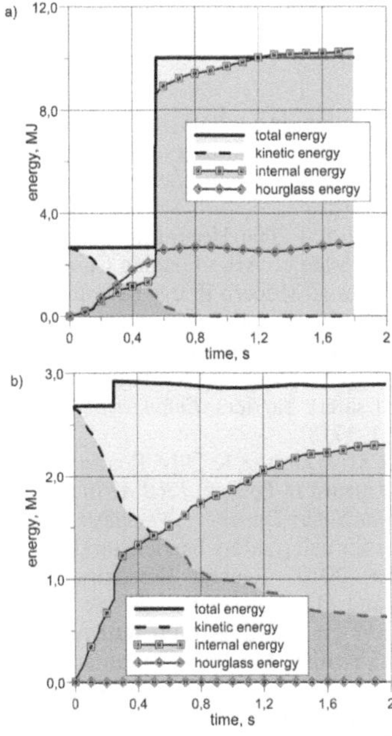

Figure 4. Energy balance during analysis a) viscous form (IHQ = 1) b) stiffness form (IHQ = 4).

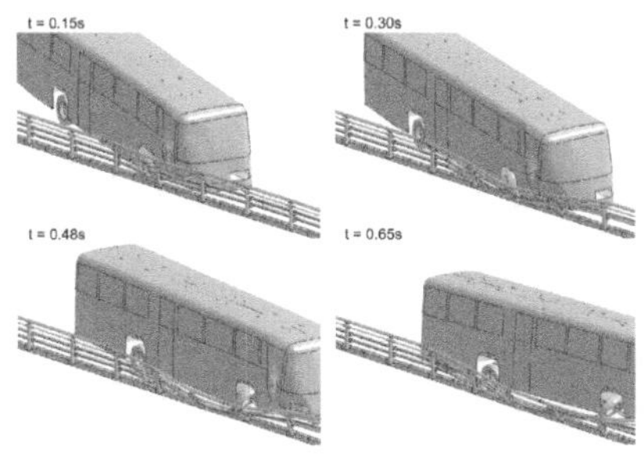

Figure 5. Snapshot of bus crash at 4 time instances.

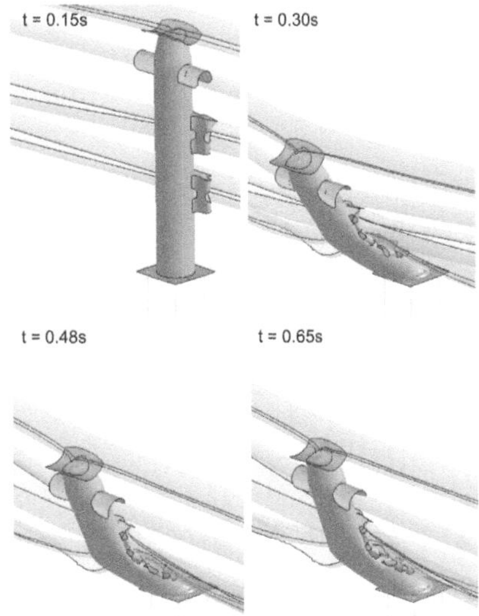

Figure 6. The support post deformation at 4 time instances.

3.3 *Impact forces on post bolts*

The barrier support posts are connected to the concrete pavement cover by four bolts (Fig. 2). The bus impact induces, in all four bolts the time dependent forces. The axial force, resultant shear force and resultant bending moment acting on the second bolt in the first raw (looking form the side of the road) are given in Figure 8. The maximum axial force reaches 120 kN at t = 0.18 s and it is a tensile force. The maximum bending moment is 0.53 kNm and the resultant shear force do not exceed 25 kN. Plastic deformation of the bolt starts at the time instance of about t = 0.11 s.

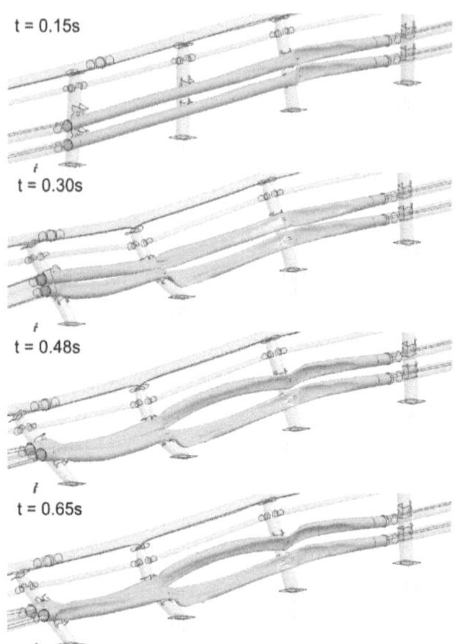

Figure 7. Safety rails deformations at 4 time instances.

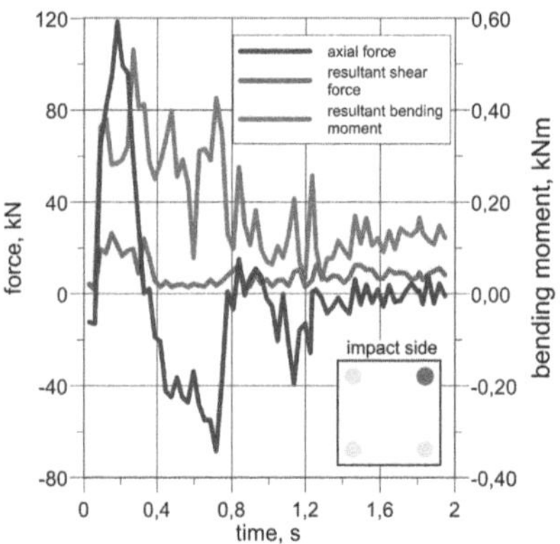

Figure 8. Forces in the bolt in the support post base plate at the impact side.

The axial force, resultant shear force and resultant bending moment acting on the second bolt in the second raw (looking form the side of the road) are given in Figure 9. This bolt is compressed by the axial force of the maximum magnitude of about 130 kN. The maximum bending reaches 0.50 kNm.

4 FINAL REMARKS

The paper shows the results of the numerical crash tests of the bridge barriers manufactured by Orsta Stal A/S exposed to impact of the bus moving with velocity of 70 km/h and impact angle of 20 degrees. The impact forces induce plastic deformations of the

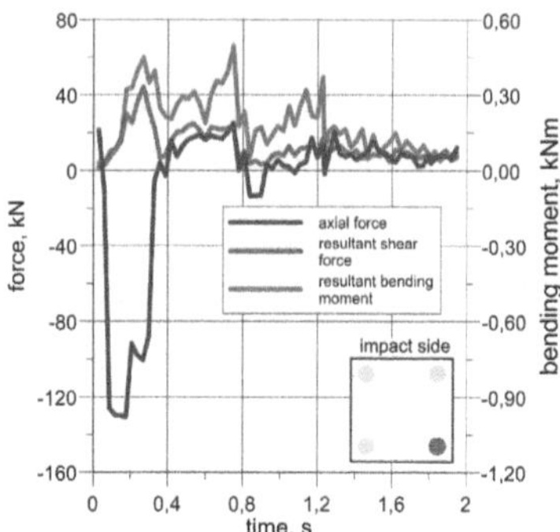

Figure 9. Forces in the bolt in the support post base plate at the side opposite to the impact.

barrier posts, the horizontal rails, made of circular cross sections and bolts connecting the post to the bridge pavement cover. The horizontal rails, made of circular cross sections, deform into ribbons distributing the impact energy along the barrier. The impact causes damage in 6 posts of the barrier section directly affected by the moving bus. The barrier withstands the impact from the bus keeping it at the bridge road.

ACKNOWLEDGEMENTS

The calculations have been carried out at the Academic Computer Center in Gdańsk, Gdańsk University of Technology.

REFERENCES

Atahan, A. & Cansiz, O. 2005. Impact analysis of vertical flared back bridge rail-to-guardial transition structure using simulation. *Finite Elements in Analysis Design,* 41: 371–396.

Barnas, A. & Edl, T. 2006. Heavy Impact on Bridge Restraint Systems—State of Art—Loading Cases and analysis—Advantages of Modern Bridge Restraint Systems. *International Conference on Bridges*, Dubrownik, Croatia.

Borovinsek, M., Vesenjak, M., Ulbin, M., Ren, Z. 2007. Simulation of crash test for high containment levels of road safety barriers. *Engineering Failure Analysis*, 14: 1711–1718.

Budzyński, M. & Jamroz, K. 2014. Roadside Trees as a Road Safety Hazard in Poland. *TRB AFB(20) 2 Meeting on Road Side Safety Design—5 November 2014*, Brussel.

Guidelines for using safety barriers on the national roads network. 2010. General Directorate for National Roads and Motorways, Warsaw.

PN-EN 1317-1/8, Road restraint systems.—Parts 1–8.

Ren, Z. & Vesenjak, M. 2005. Computational and experimental crash analysis of the road safety barrier. *Engineering Failure Analysis*, 12: 963–973.

Thai, H. 2010. Performance evaluation of steel and composite bridge safety barriers by vehicles crash simulation. *Interaction and Multiscale Mechanics, vol. 3*, No 4: 405–414.

Advances in Mechanics: Theoretical, Computational and Interdisciplinary Issues – Kleiber et al. (Eds)
© 2016 Taylor & Francis Group, London, ISBN 978-1-138-02906-4

XFEM analysis of intermediate crack debonding of FRP strengthened RC beams

I. Jankowiak
Poznan University of Technology, Poznań, Poland

ABSTRACT: In the paper a comparative analysis of concrete beams strengthened by CFRP strips using the Finite Element Method (FEM) and the eXtended Finite Element Method (XFEM) is presented. The good agreement between results obtained by means of FEM and the laboratory tests was observed and has been reported (Jankowiak 2010, 2012). Application of FEA has proved to be effective way to assessment of load carrying capacity of RC beams strengthened by CFRP strips. The applied Concrete Damage Plasticity (CDP) material model for RC beams has allowed to satisfactory prediction of both layout and propagation of cracks which lead to intermediate crack debonding failure. The preliminary study of application of XFEM to model a mechanism of intermediate crack debonding failure of concrete beams strengthened by CFRP strips is presented and first results are demonstrated.

1 INTRODUCTION

Strengthening of reinforced concrete elements by bonding to their tension and/or shear zones composite strips or mats of carbon fibers (CFRP) is nowadays a well-known technique as an alternative to traditional ones. Main failure modes of such systems debonding along the FRP-concrete interface can be categorized into two basic modes (Teng et al. 2002) (Fig. 1):

- Plate-End (PE) debonding which initiates at the end of FRP strip and propagates in the direction of increasing moment results from the high interface stresses at end discontinuities.
- Intermediate Crack-induced (IC) debonding which initiates in the beam span at a flexural or shear/flexural crack in the concrete and propagates towards the strip end in the direction of decreasing moment.

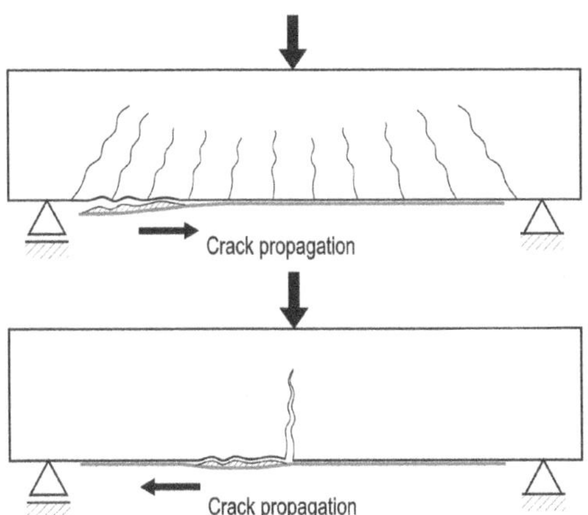

Figure 1. Two main modes of CFRP debonding: Plate-End (PE) debonding (up) and Intermediate Crack (IC) debonding (down).

The second failure mode is initiated at the location at which the interface is subjected to mixed-mode loading. This mode of failure was observed in series of experiments performed on the RC simply supported beams in real scale strengthened by CFRP strips in their flexural zones (Jankowiak 2010). The numerical analyses of considered RC beams were performed using Abaqus, the Finite Element Method commercial code (Abaqus 2014). Good compatibility of FEM analyses with experiment results was obtained for different cases of CFRP strengthening (Jankowiak 2012, Jankowiak & Madaj 2013).

The presence of cracks in concrete elements causes that conventional methods of modeling, including FEM, turn out to be inadequate and not enough efficient to represent some aspects of real behavior of concrete before failure. In the case of the CFRP strengthened beams it concerns for example the way of initiation and propagation of main cracks during the process of CFRP debonding. In the paper the eXtended Finite Element Method is applied to model CFRP intermediate crack debonding failure which is initiated at tip of flexural crack in concrete beam.

2 FEM ANALYSIS

In the previous author's study (Jankowiak 2010, 2012, Jankowiak & Madaj 2013) numerical models of RC beams have been created and carried out in the environment of the Abaqus code. The Concrete Damage Plasticity material model (CDP) was used to describe behavior of concrete (Lee & Fenves 1998). It is a continuum, plasticity-based, damage model for concrete. It can be used for plain concrete, even though it is intended primarily for analysis of RC structures.

In the analyses it was assumed that reinforced concrete can work in tensile zones even after cracks occur ('tension stiffening' effect). In order to represent the behavior of concrete after cracking the description of concrete expressed by a function of the fracture energy G_f released in the process of formation of cracks was used. The strain-softening behavior of concrete in tension was given as the stress-displacement relationship (σ-w) according to the concept of fictitious cracks model of Hillerborg (Hillerborg et al. 1976). Thus, tension stiffening effect was presented by applying the fracture energy cracking criterion. Additionally, this approach enabled to account for effects of reinforcement interaction with concrete to be simulated in a simple manner. The reinforcement rebars were modeled in a discrete manner as truss elements embedded in 2D or 3D solid elements of concrete.

Numerical model validation and verification procedure of numerical modeling based on the results of laboratory tests have been completed successfully (Jankowiak 2010, 2012). The FEM has proved to be useful in determination of load carrying capacity as well as the stiffness of strengthened elements at every level of loading. The identification of crack arrangement for all tested beams was made on the basis of maps of damage parameter which only imitated cracks. The modes of failure of beams (including propagation of delamination of strips) were also difficult to determine. The applied approach allowed, after introduction of failure criteria, to follow the whole RC beam behavior up to collapse. However, the intermediate crack debonding failure observed during the experiments was not possible to obtain using the standard FEA modeling. The Extended Finite Element Method is employed then to simulate intermediate CFRP debonding along the CFRP–concrete interface.

3 THE XFEM ANALYSIS

The XFEM was proposed first in context of fracture by Belytschko & Black (1999) and is nowadays successfully used for analysis of phenomena of concrete cracking elements as well as debonding of FRP from concrete surface (Harries et al. 2012, Mohammadi et al. 2013).

The XFEM is an alternative method to the FEM which extends the allowable basis functions known as partition of unity methods. When applied to crack propagation problems, XFEM introduces two extra sets of functions in addition to standard FEM nodal basis functions: a step function H(x) to capture the discontinuity in displacement across a crack and a set of functions $F_i(x)$, typically expressed in polar coordinates centered on the crack tip, that capture stress singularity in that region. The extra basis functions are defined as the product of these enrichment functions with the standard nodal basis function in order to ensure that the basic functions remain mesh-based and local to the enriched nodes.

This method does not require the mesh to match the geometry of discontinuities. It can be used to simulate initiation and propagation of a discrete crack by using fracture energy criterion along an arbitrary, solution-dependent path in the bulk material without the requirement of remeshing (crack propagation is not tied to the element boundaries in a mesh).

The XFEM allows the presence of discontinuities (cracks) in the elements of a finite element analysis by enriching the element with additional degrees of freedom. This technique can model crack opening and increase accuracy of approximation near the crack tip. The degree of freedom enrichment is done with special displacement function consisted of usual nodal shape function as well as the associated discontinuous jump function across the crack (Heaviside distribution) and the asymptotic crack tip functions:

$$u = \sum_i N_i(x)\left[u_i + a_i H(x) + \sum_{j=1}^{4} b_i^j F_j(x)\right] \tag{1}$$

where $N_i(x)$ = the usual nodal shape function; u_i = the usual nodal displacement; a_i = enrichment degrees of freedom associated with crack separation away from the tip; $H(x)$ = the associated discontinuous jump function across the crack (Heaviside distribution); b_i^j = the nodal enriched degrees of freedom associated with near-tip displacement; $F_j(x)$ = the asymptotic crack tip functions.

In (1) the first term on the right-hand side is valid for all nodes in the model; the second term is used for nodes whose shape function is cut by the crack interior (the part between the old and new crack tips) and the third term is applicable only for nodes whose shape function is cut by the new crack tip.

The failure including degradation and eventual separation between two surfaces consists of two components: a damage initiation criterion and a damage evolution law. The damage is initiated (an additional crack is introduced or the crack length of an existing crack is extended after an equilibrium increment) when the fracture criterion, f, reaches the value 1,0 within a given tolerance (for example when the contact stresses and/or contact separations satisfy the damage initiation criterion). The damage evolution law describes the rate at which the cohesive stiffness is degraded after initiation criterion has been reached.

In the paper two damage initiation criterions were considered:

- The maximum principal stress criterion:

$$f = \left\{ \frac{\langle \sigma_{\max} \rangle}{\sigma^0_{\max}} \right\} \tag{2}$$

where $\sigma^0_{\max}$ represents the maximum allowable principal stress.

- The quadratic nominal stress criterion:

$$f = \left\{ \frac{\langle t_n \rangle}{t_n^0} \right\}^2 + \left\{ \frac{t_s}{t_s^0} \right\}^2 + \left\{ \frac{t_t}{t_t^0} \right\}^2 \tag{3}$$

where t_n^0, t_s^0, t_t^0 represent the peak values of nominal stress as components of the nominal traction stress vector, t (t_n is the component normal to the likely cracked surface and t_s and t_t are two shear components on the likely cracked surface).

The symbol $\langle \ \rangle$ is used to signify that a purely compressive stress does not initiate damage (i.e. $\langle \sigma_{\max} \rangle = 0$ if $\sigma_{\max} < 0$ and $\langle \sigma_{\max} \rangle = \sigma_{\max}$ if $\sigma_{\max} \geq 0$).

In the following analyzed beams as a damage evolution law the fracture criterion described by a value of fracture energy G_f was adopted.

4 PRELIMINARY NUMERICAL RESULTS

Before applying XFEM to analysis of intermediate crack debonding of FRP strengthened beams the validation of model was performed based on the experimental results obtained in (Jankowiak 2010) for concrete beams only. The four-point-bending case of concrete beams was studied (Fig. 2).

In the series of experiments the mean ultimate force was measured as 31.0 ± 2.45 kN whereas in the XFEM analyses it was read as 35.9 kN. The mode of failure was determined in both studies as a single crack that occurs close to the middle of beam (Fig. 3). The concrete beam was modeled as 2D plane strain elements. The elastic material model was used to describe concrete behavior with

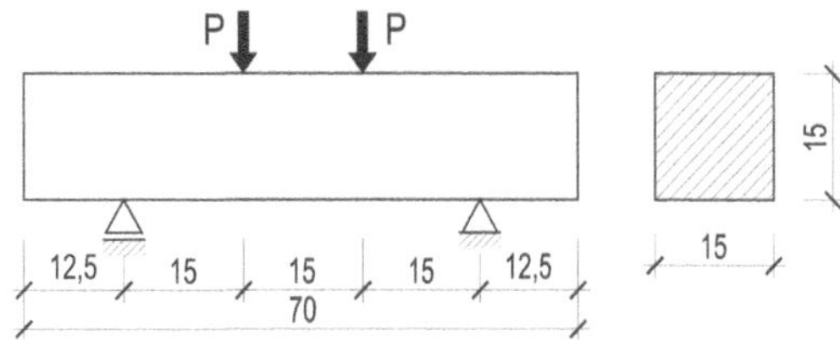

Figure 2. Concrete beam used for XFEM analysis [cm].

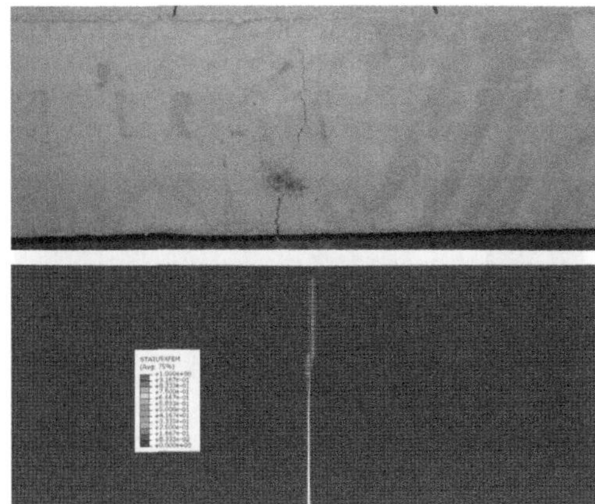

Figure 3. The crack layout in concrete beam obtained in the experiment (up) and in XFEM analysis (down) (map of status of cracked elements).

criterion of crack initiation as the maximum principal stress (eq. 2). In the numerical simulations there was not necessary to predefine in layout of mesh an onset of crack for further propagation.

5 XFEM ANALYSES OF CONCRETE BEAMS WITH CFRP STRIP

In the XFEM analyses of concrete beams strengthened with CFRP strips the same assumptions have been adopted as for analyzing plain concrete beams under four-point bending. Additionally the CFRP strip and adhesive layer were modeled as 2D plane strain elements and perfect bond between CFRP/adhesive layer and adhesive layer/beam was assumed.

Two different cases of concrete beams have been studied (Fig. 4): beam without notch and with notch. The first beam without notch failed after developing a vertical crack which occurred close to the middle of beam. In the second beam with notch a horizontal crack has started to develop causing the failure along the CFRP strip as observed in the laboratory tests.

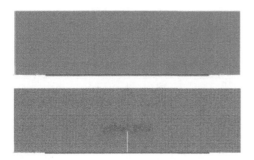

Figure 4. Numerical model of analyzed beam strengthened using CFRP strips without (up) and with notch (down).

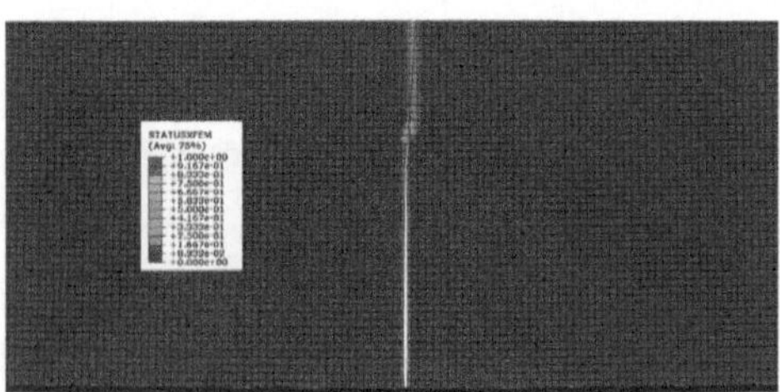

Figure 5. Crack opening in the middle of beam without notch strengthened using CFRP strip for P = 57.7 kN (map of status of cracked elements).

Figure 6. IC debonding of CFRP strip from concrete surface for P = 76.4 kN (map of status of cracked elements).

In the real behavior of beam a horizontal crack starts to develop from the vertical crack which occurs due to flexure (first case). Since such mechanism is not possible to reproduce now in the used FEM code (crack bifurcation) a notch was employed in the study of intermediate crack debonding failure in order to simulate the previously formed vertical crack.

In the model without notch the maximum principal stress (2) and in the model with notch the quadratic nominal stress criterion (3) were assumed as criteria of crack initiation. In both analyses as a damage evolution law the criterion described by a value of fracture energy G_f was adopted.

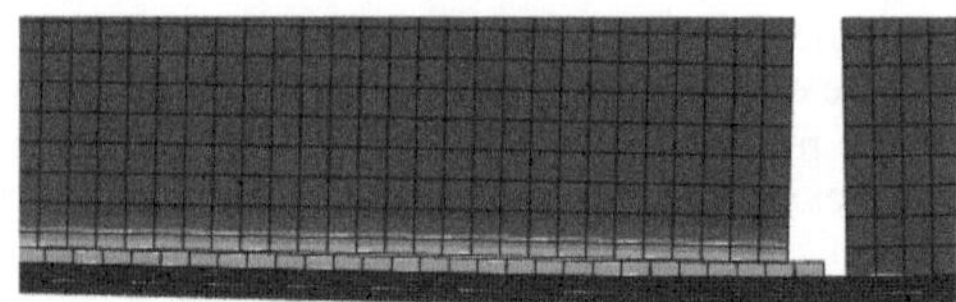

Figure 7. IC debonding near notch at failure moment for P = 76.4 kN (map of status of cracked elements).

Figure 8. IC debonding at the end of CFRP strip at failure moment for P = 76.4 kN (map of status of cracked elements).

During the analysis in the first model (without notch) the crack was initiated in the middle of beam when P = 26.87 kN and afterwards propagated vertically until total failure of beam has been obtained for P = 57.7 kN (Fig. 5). In the second analysis (with notch) debonding started in the first element of the damage band adjacent to the concrete/FRP interface at the tip of notch (Fig. 6). When element stresses meet the initiation criterion (for P = 5.4 kN), the debonding horizontal crack started to propagate towards the support (Figs. 7–8).

6 CONCLUSIONS

The eXtended Finite Element Method was successfully used for analysis of concrete cracking elements taking into account local issues such as intermediate crack debonding. It was shown by means by preliminary study that in the case of concrete beams strengthened by CFRP strips it is a good alternative for standard FEM. The presented technique can take into account modeling of intermediate crack debonding leading to failure and increasing in such way the accuracy of simulation.

Further studies are required to investigate e.g. influence of adhesive and its stiffness on failure of intermediate crack debonding phenomena. Laboratory tests are planned to be carried out in the near future as well.

REFERENCES

Abaqus 6.14, *User's Manual*, Dassault Systeme, USA, 2014.

Belytschko, T. & Black, T. 1990. Elastic crack growth in finite elements with minimal remeshing, *International Journal for Numerical Methods in Engineering* 45: 601–620.

Harries, K., Hamilton, H.R., Kasan, J. & Tatar, J. 2012. Development of standard bond capacity test for FRP bonded to concrete. *6th international conference on Fiber Reinforced Polymer (FRP) Composites in Civil Engineering (CICE).*

Hillerborg, A., Modéer, M. & Petersson, P.E. 1976. Analysis of Crack Formation and Crack Growth in Concrete by Means of Fracture Mechanics and Finite Elements. *Cement and Concrete Research* 6: 773–782.

Jankowiak, I. 2010. Effectiveness of strengthening of RC bridge beams using composite materials, *PhD thesis (in Polish), Poznan University of Technology*, Poznan, Poland.

Jankowiak, I. 2012. Analysis of RC beams strengthened by CFRP strips—Experimental and FEA study. *Archives of Civil and Mechanical Engineering* 12: 376–388.

Jankowiak, I. & Madaj, A. 2013. Load carrying capacity of RC beams strengthened by pre-tensioned CFRP strip. *Recent Advances in Computational Mechanics: proceedings of the 20th International Conference on Computer Methods in Mechanics (CMM 2013), 27–31 August, 2013*, Poznań, Poland.

Lee, J. & Fenves, G.L. 1998. Plastic-damage model for cyclic loading of concrete structures. *Journal of Engineering Mechanics* 124 (8): 892–900.

Mohammadi, T., Wan, B. & Harries, K. 2013. Intermediate crack debonding model of FRP-strengthened concrete beams using XFEM. *2013 SIMULIA Community Conference*, Providence, USA.

Teng, J.G., Chen, J.F., Smith, S.T. & Lam, L. 2002. *FRP-strengthened RC Structures*. John Wiley & Sons LTD: Chichester, England.

Advances in Mechanics: Theoretical, Computational and Interdisciplinary Issues – Kleiber et al. (Eds)
© 2016 Taylor & Francis Group, London, ISBN 978-1-138-02906-4

Performance of the parallel FEAP in calculations of effective material properties using RVE

P. Jarzębski & K. Wiśniewski
Institute of Fundamental Technological Research, Polish Academy of Sciences, Warsaw, Poland

ABSTRACT: In this paper, we consider speedup of computations of large 3D models of advanced materials using the parallelized FE code for machines with shared memory. We parallelize the loop over elements in the research code FEAP using OpenMP, which required several modifications of the code and a specific method of synchronization for assembling, see (Jarzebski, Wisniewski, & Taylor 2015). Besides, the interface to the parallel solver HSL MA86 (Hogg & Scott 2010), with various re-ordering methods, is implemented.

We demonstrate performance of the parallelized FEAP, designated as ompFEAP, in calculations of effective material properties using the RVE method. Two large RVE examples are computed for: a heterogenous metal-ceramic composite and a ceramic foam with a complicated micro-structure. We conclude that ompFEAP provides a very good speedup and efficiency at a small increase of the memory usage. Additionally, we verify that the HSL MA86 solver performs best for the multi-thread reordering by mtMETIS of (LaSalle and Karypis 2015).

1 INTRODUCTION

The Representative Volume Element (RVE) method is a computational technique developed to determine average properties of composite, porous and cellular materials. In the RVE computations, typically, 3D finite elements are used, models have the size of millions of degrees of freedom and must be repeated many times when a database of material properties is built. These properties can be subsequently used in computation of shells (Wisniewski 2010).

Therefore, to improve efficiency of an FE code, concurrent capabilities of multi-core processors and multi-machine clusters should be used, which requires several advanced computational techniques. The minimum is to use one of the available parallel solvers, such as PARDISO, MUMPS, PaStiX, HSL and others. The next step is parallelization of computations of the FE matrices and vectors, e.g. using OpenMP. In the third, the so-called hybrid approach, additionally the domain decomposition is performed, e.g. by METIS, and computations for subdomains are scattered over a cluster of computers, which requires, e.g. MPI. Implementation of these techniques in a large and complicated existing serial FE code requires a significant programming effort.

In this paper, we consider a parallelization of the research code FEAP (Taylor 2014) using OpenMP to enable parallel computations on a multi-core machine with shared memory. The parallelized FEAP is designated as ompFEAP. We assess the efficiency of parallelization by computing effective properties of two advanced materials: (1) a metal-ceramic composite and (2) a ceramic foam. We show that the applied parallelization significantly reduces the time of execution and causes only a small increase in memory usage.

2 PARALLELIZATION OF FEAP

FEAP is a research FE code used at many universities, developed by Prof. R.L. Taylor at the UCB (Taylor 2014). A serial and a cluster (MPI) version of FEAP exists, but not the version for one shared memory machine with multi-core processors so we undertook the task of developing it. Our parallelization of the serial version of FEAP consists of: (1) implementing a parallel loop over elements, which includes also the assembling of the global matrix in the sparse format, see (Jarzebski et al. 2015), and (2) adding the parallel solver.

The loop over elements was parallelized using OpenMP, which is a library for developing parallel codes for multi-core machines with the shared memory. It implements the 'fork-join' paradigm of parallel programming, i.e. the main program branches off ('fork') in parallel at some point in the program and all branches meet ('join') at a subsequent point. Then the program resumes in a sequential fashion. Figure 1 explains this concept for a loop and the PARALLEL DO directive.

2.1 *Assembling of stiffness matrix*

The main difficulty in parallelization of the loop over elements in FEAP is caused by the existing architecture of this code, which has been developed over many years. Several changes in FEAP were needed, for details see (Jarzebski, Wisniewski, & Taylor 2015), but the overall architecture of the code remained unmodified. The main features of the implementation are as follows:

1. the assembling of elemental matrices is performed immediately after they have been generated, without an additional storage which in some codes is used prior to assembling,
2. The crucial part of the loop is the assembling of elemental matrices into a global matrix, in some papers designated as the 'reduction', as there is the possibility of the race condition. To prevent it, we implemented and tested all the available directives for the mutual exclusion synchronization of OpenMP; the directive ATOMIC provided the best scalability in our tests. Using this directive, we have many critical sections, but they are very small and their execution is supported by hardware.

2.2 *Direct solution of system of linear equations*

The process of direct solving system of linear equations is divided into three phases: (1) reordering and analysis, (2) numerical factorization, and (3) forward and back substitution.

The first phase computes a permutation of the coefficient matrix to obtain the properties suitable for parallel factorization, such as: (1) a low number of non-zeros that show up after factorization ('fill-in'), (2) the as even as possible load balance. The best known ordering heuristic in this class is called the *nested dissection*. We used the serial METIS (Karypis 2014) and the multi-thread mtMETIS (LaSalle and Karypis 2015), which implement such an algorithm.

The second phase is based on computing $\mathbf{A} = \mathbf{LU}$, where $\mathbf{L}$ and $\mathbf{U}$ are lower and upper triangular parts, respectively. If $\mathbf{A}$ is symmetric then a factorization of the form $\mathbf{A} = \mathbf{LL}^T$ or $\mathbf{A} = \mathbf{LDL}^T$ is computed via *Cholesky factorization*, where

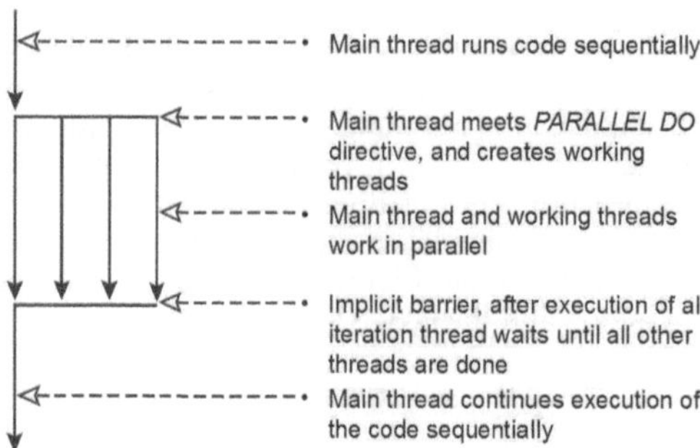

Figure 1. Loop parallelism in OpenMP.

$\mathbf{D}$ is a diagonal matrix. In FE codes, the system of linear equations is sparse, so this phase must be adapted for sparse matrices, in which a large fraction of entries are zero. The parallelization strategy for factorization is based on the Direct Acyclic Graphs (DAG), where vertices denote tasks and edges specify dependencies among the tasks. Each thread takes the task from the graphs as long as there are any. The third phase is done in a similar way to the second one.

For executing the latter two phases, we implemented in ompFEAP the interface to the parallel solver HSL MA86, see (Hogg & Scott 2010). This solver divides the sparse factorization into tasks, each of which alters a single block or a block column. The three different types of tasks are referred to as: the factorize column task, the update internal task, and the update between task. The tasks are partially ordered; for example, the updating of a block from a block column of $\mathbf{L}$ has to wait until all the rows that it needs from the block column have been calculated. As soon as all the data that a task needs are available, the task is placed in a pool of tasks for execution by any thread.

3 DEFINITIONS OF MODELS OF MATERIALS

Two advanced materials are considered: (1) a heterogenous metal-ceramic composite, and (2) a ceramic foam with a complicated micro-structure. Here we provide only a short description of them, for details see (Weglewski et al. 2013) and (Nowak et al. 2013), respectively.

3.1 *Metal-ceramic composite*

The sample of chromium—alumina—rhenium composite was manufactured by Spark Plasma Sintering (SPS) method, see (Weglewski et al. 2013). The following composition of powders (given in vol.%) was used: 75% Cr + 25% Al_2O_3 + 5% Re. The average grain size was: 2–5 μm for Cr, 5 μm for Al_2O_3, and 80 μm for Re. (In the cited paper, other compositions of powders are considered as well.)

To obtain an FE mesh for the so-produced composite, the test specimens were scanned using the X-ray micro-computed tomography with the voxel resolution of 0.9 μm. The scanned specimens were cubes of the size 1 × 1 × 1 mm, to ensure a good resolution and well represent the composite's microstructure.

To transform the micro-CT scans into an FE mesh, a commercial software ScanIP/FE was used. This is a fully automatic tool for volume meshing, which can be performed by choosing between two algorithms: (i) the traditional image-based meshing creating mixture

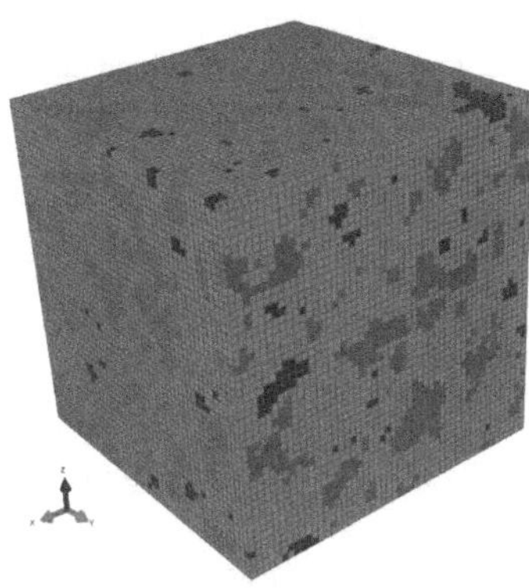

Figure 2. Finite element mesh for metal-ceramic composite.

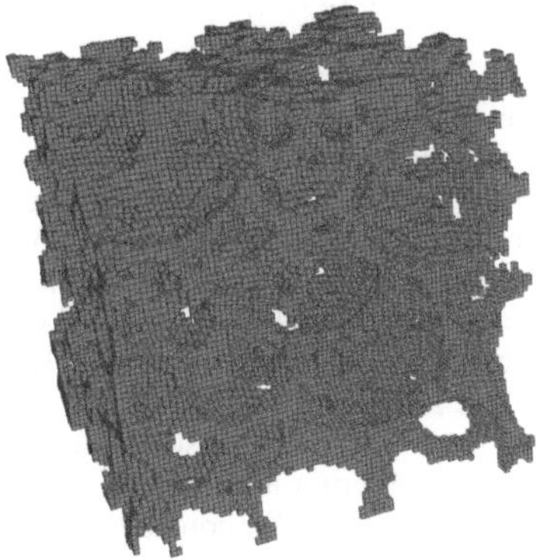

Figure 3. Finite element mesh for ceramic foam.

of hexagonal and tetragonal elements (FE-Grid), and (ii) a new algorithm, which allows for mesh adaptation based on features which can significantly reduce mesh size, (FE-Free). The mesh obtained by the first algorithm is shown in Fig. 2.

The FE mesh is by courtesy of the authors of (Weglewski et al. 2013), which we gratefully acknowledge. We additionally checked this mesh using VERDE, and its high quality is confirmed. Material properties used in numerical simulations are given in the cited paper. The computational FE model involves about 0.4 million dofs.

3.2 *Ceramic foam*

The ceramic foam is an open-cell alumina foam manufactured by gelcasting, which is a method for molding ceramic powder. The foam with the walls made of Al_2O_3, has 86% porosity and is composed of approximately spherical cells interconnected by circular windows, see (Nowak et al. 2013). The specimen was a cuboid of the size 11 × 8 × 8 mm. The CT-scans are by courtesy of the authors of (Nowak et al. 2013), which we gratefully acknowledge.

To transform the micro-CT scans into a FE mesh a commercial software ScanIP/FE was used. The traditional image-based meshing creating hexagonal elements (FE-Grid) was selected; the obtained mesh is shown in Fig. 3. Additionally, we confirmed high quality of this mesh using VERDE.

In numerical calculations, the cell wall material is isotropic, the Young's modulus is 370 GPa and Poisson's ratio is 0.3. The computational FE model of the foam involves 3.8 million dofs.

4 NUMERICAL RESULTS

In order to assess the effects of parallelization of FEAP, we perform the computations for the RVEs described in the preceding section. To determine the effective Young's modulus, the uniaxial tension test is simulated numerically, which amounts to generating and solving the system of linear equation for one right-hand side. The RVEs are modeled by the 3D 8-node standard finite element.

The computations are performed on the machine with 2 processors Xeon X5650 2.66GHz (6 cores each) and 24GB DDR3 1333 MHz RAM. The Intel Compiler ver. 14.0.0, optimization flag 2 is used, and the correctness of implementation is checked by Intel Inspector 2015.

A. To verify the correctness of parallelization, we compare the computed displacements and effective material parameters to these yielded by the serial code; they are identical regardless of the number of threads used.
B. The speedup ratio is calculated by dividing the calculation time for 1 thread by that for 12 threads. Two solution phases are distinguished: (1) the loop over elements, which includes the matrix generation/assembling, and (2) the factorization by the parallel solver HSL MA86. The scalability for both FE models is shown in Figures 4 and 5. For the bigger model, i.e. the ceramic foam, the speedup ratios are as follows: for the loop over elements ~11.27, for the factorization ~9.05, and for both steps together ~9.18. The peak memory usage increased by 12% for 12 threads compared to this for 1 thread, from 15.45 to 17.68GB.

Note that the speedup ratio for the loop over elements reported in (Jarzebski et al. 2015) is ~10.7. It is obtained for the unit cube test and the regular mesh of $N \times N \times N$ elements ($N = 64$, ~0.81 million dofs). We see that similar or slightly higher speedup ratios are obtained for irregular RVE meshes.

We also checked how the other re-ordering methods included in the HSL package perform: a) *cMark*—Markovitz Criterium Method, b) *MD*—Minimum Degree Method, c) *AMD*—Approximate Minimum Degree Order-

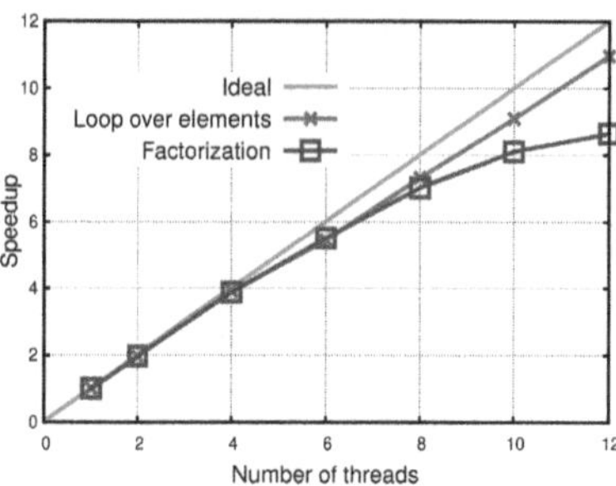

Figure 4. Scalability of the parallel parts of the code for metal-ceramic composite. 0.4 million dofs. Re-ordering by METIS.

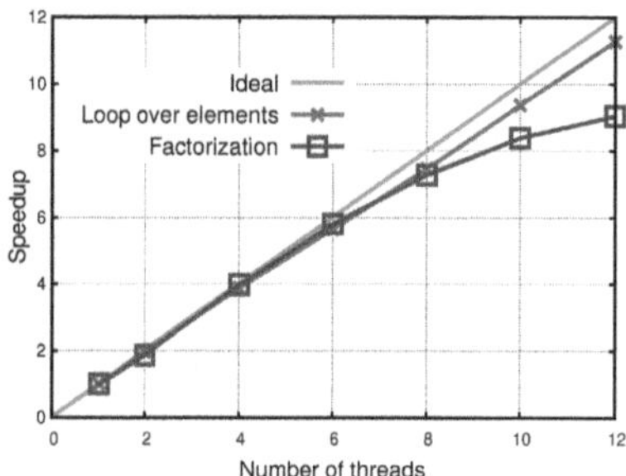

Figure 5. Scalability of the parallel parts of the code for ceramic foam. 3.8 million dofs. Re-ordering by METIS.

Table 1. Time and speedup for various re-ordering algorithms for metal-ceramic composite. 0.4 million dofs.

Threads	Phase	AMD	METIS	mtMETIS
Time [secs]				
1	Re-order	0.53	8.33	8.33
	Factor	1316.06	318.67	318.67
	Total	1316.59	327.00	327.00
12	Re-order	0.53	8.33	1.70
	Factor	147.85	36.88	50.20
	Total	148.38	45.21	51.90
Speedup ratio				
	Factor	8.90	8.64	6.51
	Total	8.87	7.23	6.30

ing Algorithm, d) *METIS*—Nested Bisection Algorithm. Additionally, we prepared the interface to the multi-thread *mtMETIS* of (LaSalle and Karypis 2015). In Tables 1 and 2, the time of re-ordering and factorization is shown for the three fastest methods; the results for *cMark* and *MD* are not reported as they are worse than by *AMD*.

We see that for the bigger model, i.e. the ceramic foam, the reordering provided by mtMETIS yields the lowest time of factorization, the lowest total time, and the best speedup.

Table 2. Time and speedup for various re-ordering algorithms for ceramic foam. 3.8 million dofs.

Threads	Phase	AMD	METIS	mtMETIS
Time [secs]				
1	Re-order	7.74	73.98	73.98
	Factor	1813.78	421.08	421.08
	Total	1821.52	495.06	495.06
12	Re-order	7.74	73.98	9.00
	Factor	279.18	46.55	45.09
	Total	286.92	120.53	**54.09**
Speedup ratio				
	Factor	6.49	9.05	9.34
	Total	6.35	4.11	9.15

5 CONCLUSIONS

In this paper, we describe the main features of ompFEAP, which is the parallelized OpenMP version of the research FE code FEAP, and we verify this code on two examples of the RVE for advanced materials, the larger one involving 3.8 million dofs. We test the parallelization of the loop over elements, which includes also the assembling of the global matrix in the sparse format, and the parallel solver HSL MA86. Both yield a significant reduction of the execution time and good scalability. Besides, we test several methods of re-ordering and, for the bigger model, mtMETIS performs best. We conclude that ompFEAP is a reliable and highly efficient tool, which can be applied to large models of advanced materials.

REFERENCES

Hogg, J. & J. Scott (2010). An indefinite sparse direct solver for multicore machines. *Technical Report TRRAL-2010-011*.

Jarzebski, P., K. Wisniewski, & R.L. Taylor (2015). On paralelization of the loop over elements in feap. *Computational Mechanics 56*(1), 77–86.

Karypis, G. (2014). Metis. *http://glaros.dtc.umn.edu/gkhome/views/metis*.

LaSalle, D. & G. Karypis (2015). Efficient nested dissection for multicore architectures. *EuroPar*.

Nowak, M., Z. Nowak, R. Pecherski, M. Potoczek, & R. Sliwa (2013). On the reconstruction method of ceramic foam structures and the methodology of young modulus determination. *Archive of Metallurgy and Materials 58*, 1219–1222.

Taylor, R.L. (2014). *FEAP Ver. 8.4*.

Weglewski, W., K. Bochenek, M. Basista, T. Schubert, U. Jehring, J. Litniewski, & S. Mackiewcz (2013). Comparative assessment of youngs modulus measurements of metalceramic composites using mechanical and nondestructive tests and micro-ct based computational modeling. *Computational Materials Science 77*, 19–30.

Wisniewski, K. (2010). *Finite Rotation Shells*. Springer.

Advances in Mechanics: Theoretical, Computational and Interdisciplinary Issues – Kleiber et al. (Eds)

Receptance coupling for turning with a follower rest

M. Jasiewicz & B. Powałka

Faculty of Mechanical Engineering and Mechatronics, West Pomeranian University of Technology Szczecin, Szczecin, Poland

ABSTRACT: The paper presents an issue of modelling dynamics of turned parts with a follower rest for machining stability prediction. Dynamic properties of a spindle, tailstock and the follower rest are assumed to be constant and can be determined experimentally based on results of impact test. Hence, the variable of system "machine—handle—part—tool" is the machined part and follower rest setting, which can be modelled analytically. The method of receptance coupling enables a synthesis of experimental (spindle, tailstock, follower rest) and analytical (machined part) models, so impact testing of the entire system becomes unnecessary. The paper presents methodology of synthesis analytical model of the machined parts with the spindle, tailstock and follower rest experimental models. In the summary experimental verification of the calculations is presented.

1 INTRODUCTION

Machining of compliant parts is difficult to carry out due to occurring high—amplitude vibrations. An example of such machining may be slender rods turning, usually implemented using steady or follower rest (Fig. 1) to provide stiffening of the workpiece so that the turning process remains stable.

In order to assure stable cutting conditions, selection of the cutting depth and spindle speed can be carried out using so-called stability lobes (Marchelek et al. 2002), which depend on the dynamic properties of the system "machine—handle—part—tool". Considering machining with the follower rest moving along with a tool post, it should be noted that besides the geometry and material properties of the workpiece, the rest location will have a significant influence on these properties.

Experimental determination of the dynamic properties of the system for each object mounted on a lathe at different follower rest locations is too laborious, and requires access to the appropriate measuring equipment.

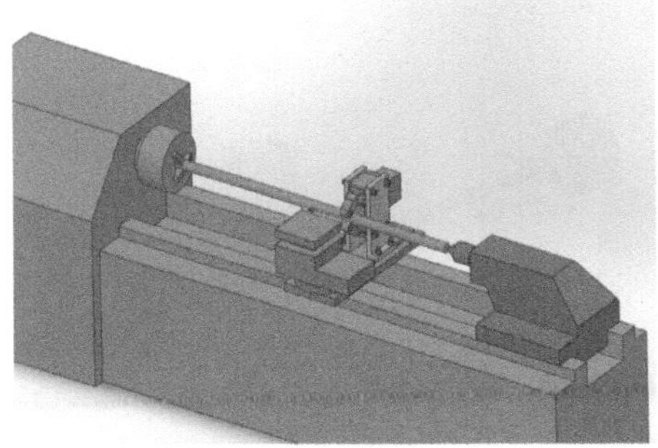

Figure 1. Turning with a follower rest.

The elements of the system, where the dynamic properties can be considered constant during the machining are spindle, tailstock and follower rest and can be determined experimentally (eg. impact test), or based on the FEM modelling. Furthermore the workpiece can be considered as a circular cross section beam and its dynamic properties could be determined analytically. Having transfer functions of all system components a synthesis using the method of receptance coupling (Matuszak & Powałka 2002) can be carried out.

In the references (Erturk et al. 2006, Matuszak & Powałka 2010, Park et al. 2003, Schmitz & Duncan 2009) the synthesis of the transfer function is made for the purpose of determining the dynamic properties of milling spindle—tool assembly. In the paper, the method of receptance coupling used for determining the dynamic properties of lathe is presented, where the workpiece is modelled analytically and the spindle, tailstock and follower rest—experimentally.

2 SYSTEM COMPONENTS

In this part the methodology of determining the transfer function of the "machine—handle—part—tool" system components is presented.

2.1 *Spindle*

The dynamic properties of the spindle does not change during machining and may be determined experimentally (impact test).

Apart from determining the transfer function in x direction it is also necessary to identify Rotational Degrees of Freedom (RDOF) (Park et al. 2003), given in Figure 2 because chuck fixture causes that rotation angles of the spindle and workpiece at the point of attachment are the same.

Spindle dynamics could be expressed as:

$$\begin{bmatrix} x_{s1} \\ \varphi_{s1} \end{bmatrix} = \begin{bmatrix} H_{s11} & H_{s12} \\ H_{s21} & H_{s22} \end{bmatrix} \cdot \begin{bmatrix} F_{s1} \\ M_{s1} \end{bmatrix} \tag{1}$$

where: x_s—translational displacement, φ_s rotation angle, F_s—force, Ms c torque, H—transfer functions.

RDOF transfer functions can be determined indirectly based on impact testing using the first order finite difference method (Duarte & Ewins 2000, Varoto et al. 2006).

2.2 *Workpiece*

Another element of the system is a workpiece. It could be modelled as a free-free Timoshenko beam with circular cross section. The beam is considered to be free, because the boundary conditions are applied to the system during its synthesis and arise from the spindle, tailstock and follower rest properties.

Point and transfer Frequency Response Functions (FRFs) of two beam points can be expressed as:

$$H_{z1z2}(\omega) = \sum_{m=1}^{\infty} \frac{\omega_m^2 + \tau j \omega \omega_m - \omega^2}{\rho A \gamma_m^2} X_m(z_1) X_m(z_2) \tag{2}$$

$$\gamma_m^2 = \int_0^l X_m^2(z) dz \tag{3}$$

where: ω_m—mth mode natural frequency, τ—beam damping loss factor, ρ—beam density, A—cross-sectional area, X_m—mth mode eigenfunction of the beam.

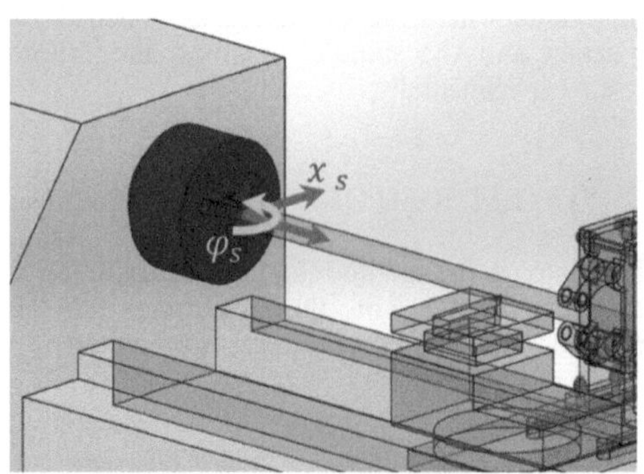

Figure 2. Lathe spindle with the chuck.

2.3 *Follower rest and tailstock*

Since rotational motion of the workpiece is independent of rotation of the tailstock and follower rest it is sufficient to determine only translational response functions on the basis of the results of impact tests.

3 RECEPTANCE COUPLING

Having all required transfer functions of the system components (Fig. 3) synthesis using receptance coupling could be performed.

Dynamics of separated (not connected) components of the system can be expressed by following transfer matrix relationships:

Spindle:

$$\begin{bmatrix} x_{s1} \\ \varphi_{s1} \end{bmatrix} = \begin{bmatrix} H_{s11} & H_{s12} \\ H_{s21} & H_{s22} \end{bmatrix} \cdot \begin{bmatrix} F_{s1} \\ M_{s1} \end{bmatrix} \tag{4}$$

Workpiece:

$$\begin{bmatrix} x_{B1} \\ \varphi_{B1} \\ x_{B2} \\ x_{B3} \\ x_{B4} \end{bmatrix} = \begin{bmatrix} H_{B11} & H_{B12} & H_{B13} & H_{B14} & H_{B15} \\ H_{B21} & H_{B22} & H_{B23} & H_{B24} & H_{B25} \\ H_{B31} & H_{B32} & H_{B33} & H_{B34} & H_{B35} \\ H_{B41} & H_{B42} & H_{B43} & H_{B44} & H_{B45} \\ H_{B51} & H_{B52} & H_{B53} & H_{B54} & H_{B55} \end{bmatrix} \cdot \begin{bmatrix} F_{B1} \\ M_{B1} \\ F_{B2} \\ F_{B3} \\ F_{B4} \end{bmatrix} \tag{5}$$

Follower rest:

$$[x_{FR3}] = [H_{FR3}] \cdot [F_{FR3}] \tag{6}$$

Tailstock:

$$[x_{T4}] = [H_{T4}] \cdot [F_{T4}] \tag{7}$$

Imposing the boundary conditions (Eq. 8) and the equilibrium of forces (Eq. 9) between the isolated coordinate systems of components and assigning them to a coordinate of a merged system its transfer function could be synthesised.

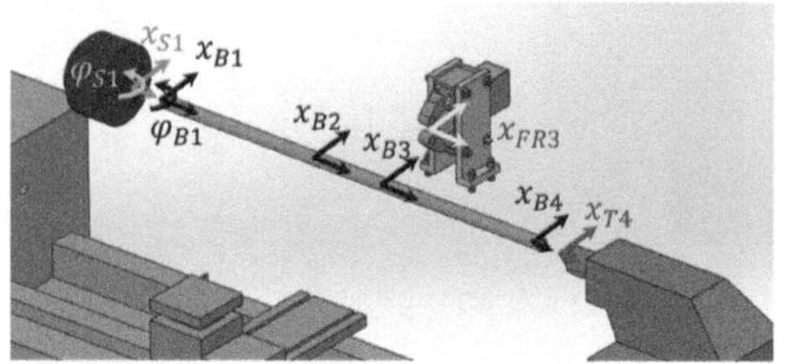

Figure 3. System components.

Boundary conditions:

$$\begin{cases} x_{S1} = x_{B1} = x_1 \\ \varphi_{S1} = \varphi_{B1} = \varphi_1 \\ x_{B2} = x_{FR2} = x_2 \end{cases} \tag{8}$$

Equilibrium of forces:

$$\begin{cases} F_{w1} + F_{B1} = F_1 \\ M_{w1} + M_{B1} = M_1 \end{cases} \tag{9}$$

Coupled system FRFs matrix can be expressed as product of inverse Transformation matrix $\mathbb{T}$ and beam matrix:

$$\begin{bmatrix} x_1 \\ \varphi_1 \\ x_{B2} \\ x_{B3} \\ x_4 \end{bmatrix} = \mathbb{T}^{-1} \cdot \begin{bmatrix} H_{B11} & H_{B12} & H_{B13} & H_{B14} & H_{B15} \\ H_{B21} & H_{B22} & H_{B23} & H_{B24} & H_{B25} \\ H_{B31} & H_{B32} & H_{B33} & H_{B34} & H_{B35} \\ H_{B41} & H_{B42} & H_{B43} & H_{B44} & H_{B45} \\ H_{B51} & H_{B52} & H_{B53} & H_{B54} & H_{B55} \end{bmatrix} \cdot \begin{bmatrix} F_1 \\ M_1 \\ F_2 \\ F_3 \\ F_4 \end{bmatrix} \tag{10}$$

The Transformation matrix $\mathbb{T}$ contains spindle, beam, follower rest and tailstock FRFs, and could be expressed as:

Figure 4. Testing object.

Figure 5. FRFs verification point.

$$\mathbb{T} = \begin{bmatrix} 1+\dfrac{H_{B11}H_{S22} - H_{B12}H_{S12}}{H_{S11}H_{S22} - H_{S12}H_{S21}} & \dfrac{H_{B12}H_{S11} - H_{B11}H_{S12}}{H_{S11}H_{S22} - H_{S12}H_{S21}} & \dfrac{H_{B13}}{H_{FR4}} & 0 & \dfrac{H_{B15}}{H_{T4}} \\ \dfrac{H_{B21}H_{S22} - H_{B22}H_{S12}}{H_{S11}H_{S22} - H_{S12}H_{S21}} & 1+\dfrac{H_{B22}H_{S11} - H_{B21}H_{S12}}{H_{S11}H_{S22} - H_{S12}H_{S21}} & \dfrac{H_{B23}}{H_{FR4}} & 0 & \dfrac{H_{B25}}{H_{T4}} \\ \dfrac{H_{B41}H_{S22} - H_{B42}H_{S12}}{H_{S11}H_{S22} - H_{S12}H_{S21}} & \dfrac{H_{B32}H_{S11} - H_{B31}H_{S12}}{H_{S11}H_{S22} - H_{S12}H_{S21}} & 1+\dfrac{H_{B33}}{H_{FR2}} & 0 & \dfrac{H_{B35}}{H_{T4}} \\ \dfrac{H_{B41}H_{S22} - H_{B42}H_{S12}}{H_{S11}H_{S22} - H_{S12}H_{S21}} & \dfrac{H_{B42}H_{S11} - H_{B41}H_{S12}}{H_{S11}H_{S22} - H_{S12}H_{S21}} & \dfrac{H_{B43}}{H_{P2}} & 1 & \dfrac{H_{B45}}{H_{T4}} \\ \dfrac{H_{B51}H_{S22} - H_{B52}H_{S12}}{H_{S11}H_{S22} - H_{S12}H_{S21}} & \dfrac{H_{B52}H_{S11} - H_{B51}H_{S12}}{H_{S11}H_{S22} - H_{S12}H_{S21}} & \dfrac{H_{B43}}{H_{FR2}} & 0 & 1+\dfrac{H_{B55}}{H_{T4}} \end{bmatrix} \tag{11}$$

4 EXPERIMENTAL VERIFICATION

The next step of the research was to conduct experimental verification. The testing object (presented in the Figure 4) was a lathe with a hydraulic follower rest SMW Autoblock SLU-X1 mounted on the saddle and the workpiece—shaft of 1300 mm length and 35 mm diameter. The verification was carried out for three follower rest locations: 300 mm, 600 mm and 800 mm from the spindle.

The identification of the dynamics of the spindle, follower rest and tailstock were based on the results of impact tests. Using the obtained transfer function and determining the geometry of the workpiece in the analytical model the transfer function of merged system was synthesized.

For all follower rest locations impacts test were performed at equally spaced points on the workpiece (dist. 100 mm) and mode shapes were identified. Experimental and synthetized FRFs were compared at the tool point (100 mm from the follower rest on the side of the spindle, x—direction), marked with the arrow in the Figure 5.

5 SUMMARY AND CONCLUSIONS

In this section the comparison of selected mode shapes and FRFs determined analytically using receptance coupling method and identified experimentally are presented (Figs. 6–8) and the results are discussed.

For all follower rest locations presented receptance coupling method allows for accurate mode shapes identification (Figs. 6–7). Inaccuracies in FRFs estimation (example—Fig. 8) may result from difficulties with obtaining accurate rotational transfer functions of the spindle and disregarding contacts between the system components—these issues are the subject of further work.

The presented method allows for quick and computationally efficient determination of the dynamics of a complex system, because the fundamental variable is determined analytically. Moreover, the method may be useful for industrial applications, where on the particular lathe are machined objects with different geometry and material properties. This approach also allows to define the dynamic properties of any point of the system what gives development opportunities, such as active follower rests application.

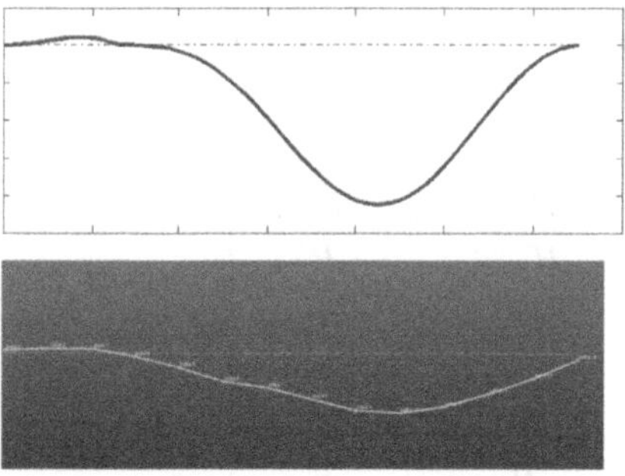

Figure 6. First mode shape for 300 mm follower rest location: upper—determined analytically, lower—identified experimentally.

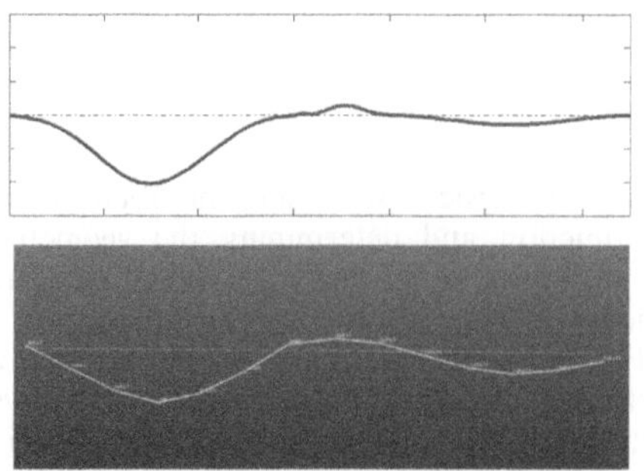

Figure 7. Second mode shape for 600 mm follower rest location: upper—determined analytically, lower—identified experimentally.

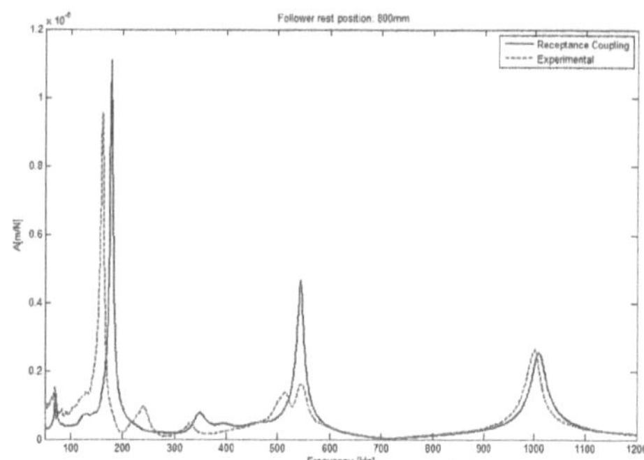

Figure 8. FRFs comparison for 800 mm follower rest location.

ACKNOWLEDGEMENTS

The researches were carried out within the project INNOTECH-K3/IN3/19/226994/NCBR/14 financed by NCBiR.

REFERENCES

Duarte, M.L.M.; Ewins, D.J., 2000, Rotational degrees of freedom for structural coupling analysis via finite-difference technique with residual compensation. Mechanical Systems and Signal Processing, 14.2: 205–227.

Erturk A., Ozguven H.N., Budak E., 2006, Analytical modeling of spindle-tool dynamics on machine tools using Timoshenko beam model and receptance coupling for the prediction of tool point FRF, International Journal of Machine Tools & Manufacture 46 1901–1912.

Marchelek K., Pajor, M., Powałka, B., 2002, Vibrostability of the milling process described by the time-variable parameter model. Journal of Vibration and Control, 8.4: 467–479

Matuszak M., Powałka B., 2010, Wybrane problemy badawcze właściwości dynamicznych obrabiarki do mikroskrawania, Modelowanie Inżynierskie 39, s. 151–158, ISSN 1896–771X, Gliwice.

Park, S. S., Altintas, Y., Movahhedy M., 2003, Receptance coupling for end mills. International Journal of Machine Tools and Manufacture, 43.9: 889–896.

Schmitz T.L., Duncan G.S., 2009, Three-Component Receptance Coupling Substructure Analysis for Tool Point Dynamics Prediction, International Journal of Machine Tools and Manufacture, 49/12–13/ 947–957.

Varoto, P.S., Lofarno, M., Cicogna, T.R., Oliveira, L.P.R., 2006, Moment Mobility FRF Measurement Techniques, IMAC-XXIV Conference & Exposition on Structural Dynamics. St. Louis, Missouri.

Advances in Mechanics: Theoretical, Computational and Interdisciplinary Issues – Kleiber et al. (Eds)
© 2016 Taylor & Francis Group, London, ISBN 978-1-138-02906-4

Analysis of the influence of differences in strength parameters of steel S235 on passive safety of lighting columns

T.I. Jedliński
Research and Development Centre, Europoles sp. z o. o., Poland

J. Buśkiewicz
Faculty of Mechanical Engineering and Management, Poznan University of Technology, Poznan, Poland

ABSTRACT: Available lighting columns have different effects on road safety. Properly designed and manufactured column can reduce the range of injuries sustained in a road accident, and may even save life. The article describes the impact of the strength parameters of the steel used in the production of the whole structure on passive safety. The material S235 has partially defined strength boundaries, however, due to their very large extent one can get much harder material. In this case, the developed products have functional properties different from the expected ones.

1 INTRODUCTION

Nowadays, the main objective in the design of new roads and their surroundings is to ensure maximum safety of users. As a result of such an approach, the guidelines of the design of modern lighting columns are described in European standard EN 12767 "Passive safety of support structures for road equipment. Requirements, classification and test methods". Therefore, it is required that a pole, which is tested in terms of the abovementioned standard, is produced from the same material and maintains its unchanged structure. It is difficult to satisfy the second condition as the mechanical properties of S235 provided by iron works are very different.

Review of the literature on the general issues of the strength of materials was related mainly to the books (Rudnik 1997). However, the research part was based on scientific studies and articles in specialized journals (Janszen 2007, Vila'n et al. 2006) due to the relatively narrow industry. In these articles the chosen test methods are verified by comparing with the results obtained in crash tests. The tests conditions, assembly methods, methodology of calculations are based on European standards of lightning poles tests worked out by European Committee for Standardization.

2 EQUIPMENT—MEASURING APPARATUS

2.1 *Apparatus to static test*

Equipment required is different in two parts of this project:

- In the first draft, the column was calculated using static software of Europoles Company. The software allows to calculate the static load of columns with a constant loads suspended on and exerted by wind pressure.
- Produced columns were then tested on two test posts:
  - Mechanical properties of steel are determined using sampling machine. During the test, the elongation of samples was measured by the extensometer, and the strength with the strain gauge force sensor placed on the hydraulic cylinder.
  - The stand for the static and dynamic tests of the complete lighting columns enables mounting the columns to a rigid base in the horizontal position and loading them anywhere with the use of a hydraulic cylinder. The actuator is controlled automatically and the preset load program is carried out by the driver. During the test, deflection and bending moment of the column were measured.

2.2 *Crash test apparatus*

Crash track for static safety testing consists of a specially designed truck imitating the behavior of a conventional vehicle on which the measuring apparatus is installed (Figs. 1 and 2). During the test overloads are measured in three axes in the range of ± 600 g and the rotational speed of the vehicle up to 50 rad/s. All data is stored on a computer with a sampling rate of 10 kHz. In addition, an area of 6 m from and 12 m behind the point of the impact is recorded by two high-speed cameras (500 frames/sec and a resolution of 800 × 600 pixels).

Figure 1. The truck imitating the behavior of a conventional vehicle.

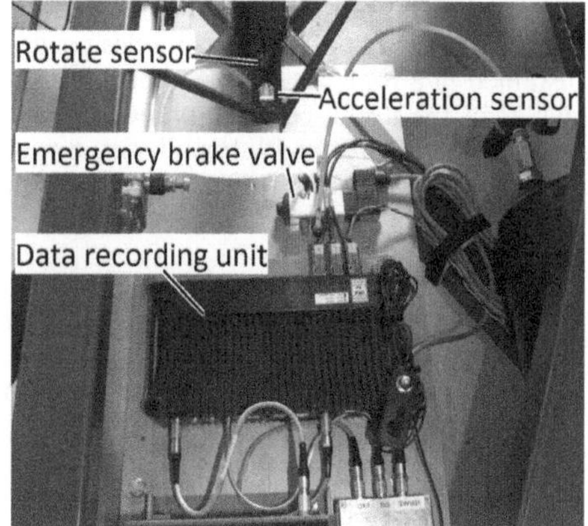

Figure 2. Measuring apparatus on the vehicle.

Recording of all data is triggered by contact sensor placed on the surface of the column.

3 CRASH TEST—MEASUREMENT PARAMETERS

All tests were conducted in accordance with the assumptions of the standard EN 12767 "Passive safety of support structures for road equipment. Requirements, classification and test methods". Columns are divided with respect to the absorption of energy of the vehicle as shown in Figure 3 and Table 1. Within each category the driver's safety is defined from 1 (lowest safety level) to 4 (highest safety level) (Table 2) depending on Acceleration Severity Index (ASI) and the Theoretical Head Impact Velocity (THIV) (Vila'n et al. 2006, European Committee for Standardization 2009).

Both indicators are calculated in accordance with EN 12767. ASI depends on the g-force accelerations in all axes (Eq. 1), while THIV is the theoretical speed of the driver's head when hitting the steering wheel, and is calculated from the negative g-forces and vehicle speed (European Committee for Standardization 2010).

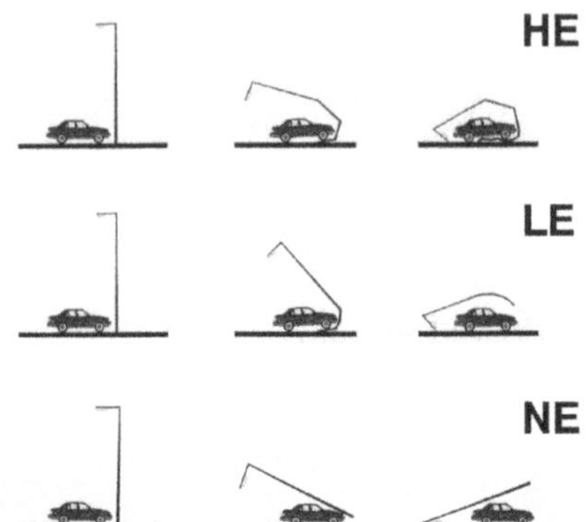

Figure 3. Non-Energy-absorbing (NE), Low Energy-absorbing (LE) and High Energy-absorbing (HE) lighting columns (Vila'n et al. 2006).

Table 1. Energy absorption level for roadside (Vila'n et al. 2006).

Energy absorption level	Impact velocity Vb [km/h] Velocity after an impact Ve [km/h]		
	Vb = 50	Vb = 70	Vb = 100
HE	0	0 < Ve ≤ 5	0 < Ve ≤ 50
LE	0 < Ve ≤ 5	5 < Ve ≤ 30	50 < Ve ≤ 70
NE	5 < Ve ≤ 50	30 < Ve ≤ 70	70 < Ve ≤ 100
"0"	–	–	–

Table 2. Occupant safety performance class (Vila'n et al. 2006).

Energy absorption level	Security level for the occupants	Impact at speed 35 km/h		Impact at speed 50, 70 and 100 km/h	
		ASI	THIV [km/h]	ASI	THIV [km/h]
HE	1	1.0	27	1.4	44
HE	2	1.0	27	1.2	33
HE	3	1.0	27	1.0	27
LE	1	1.0	27	1.4	44
LE	2	1.0	27	1.2	33
LE	3	1.0	27	1.0	27
NE	1	1.0	27	1.2	33
NE	2	1.0	27	1.0	27
NE	3	1.0	11	0.8	11
NE	4	-	-	-	3

ASI is a function of time defined as:

$$\mathrm{ASI}(t) = [(A_x/12)^2 + (A_y/9)^2 + (A_z/10)^2]^{1/2} \tag{1}$$

where A_x, A_y, A_z are the filtered components of vehicle acceleration.

4 LIGHTING POLE—DESCRIPTION AND TESTING

The tested lighting column was sunken into concrete foundation 0.5 m deep and attached to it by a clamp. The column had a height of 4.5 m, a peak diameter of 76 mm, the convergence of 14 mm/m. The column was produced of S235 steel 3 mm thick. 12 kg weight lighting lamp was mounted at the top of the column. The total weight of the lighting columns was 47 kg. The columns were made of the same grade of steel having different strength properties. The first set of columns wa made of S235 steel with a yield stress of 248 MPa and tensile strength of 369 MPa. The other set of columns was made of theoretically the same steel, but with the yield stress of 328 MPa and tensile stress of 408 MPa. Both materials are steel S235JR+N according to the standard EN 10025-2 and received certificates (European Committee for Standardization 2010). All the poles were sited in the ground, using the same backfill, with the same level of soil compaction. Moreover the screws in all the poles were twisted with the same torque.

To determine strength parameters of the steel, nine samples from both the pole and the steel coils used in its production were taken. The material of various samples taken from a single coil indicated a low variability of parameters. The minimal limit of yield strength at the 235 MPa level and tensile strength between 360 and 510 MPa is defined in the norm (European Committee for Standardization 2010).

The next step was to calculate static strength of columns, and to confirm the calculation experimentally. The main parameter was assumed maximum bending moment equal respectively 6.47 kNm for the first column and 8.39 kNm for the second column. Static tests were performed to verify the analytical calculations. The pole was fixed in the horizontal position in the test stand, and the load was exerted by a hydraulic actuator. During the test the poles were damaged in the area of the door hole for the bending moment similar to that obtained in the calculation which confirmed the correctness of the analytical calculations.

The last step was to determine the effect of the difference in strength parameters on the behavior of steel columns during the crash tests and to determine the category to which the columns are assigned. For this purpose crash tests were performed maintaining the same conditions and removing the maximum number of variables. The tests were conducted at speed of 70 km/h because of the small size of the columns. Figure 4 presents the frames taken 100, 150 and 200 ms after the collision. It may be seen that the pole made of the material with higher strength remains longer in the vertical position. Such large differences in deformation of columns during the collision have been reflected on readings from sensors mounted on the vehicle. The most significant changes have been observed for g-force acceleration in the X-axis, which resulted in the values of ASI and THIV (Figs. 5–7). The most visible changes occurred 100 ms after the impact. The acceleration increased from 3.8 *g* to 5.8 *g* (Fig. 6) and the time of energy absorption lengthened about 70 ms which reduced exit speed from 40 to 23.4 km/h. The poles differ in the level of vehicle energy absorption and the security class of a passenger. The column made of the material with lower strength parameters is assigned to category 70NE2 whereas the other column is assigned to 70LE3.

Figure 4. Pictures 100, 150 and 200 *ms* after the crash.

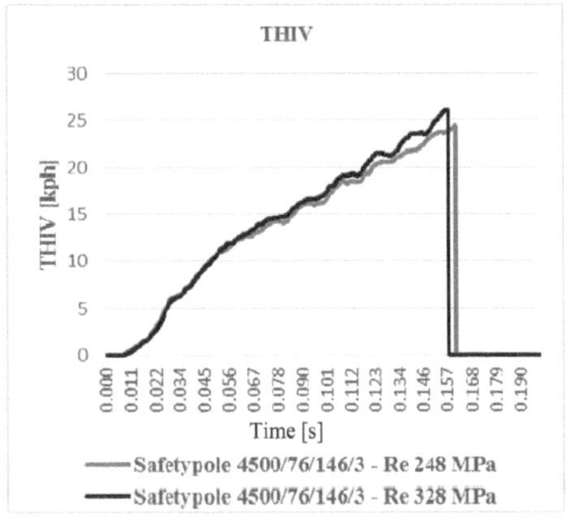

Figure 5. Comparison of THIV value for both poles.

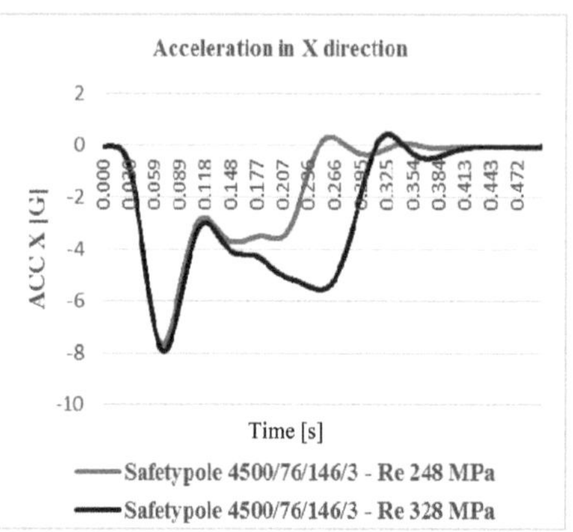

Figure 6. Comparison of acceleration in X direction for both poles.

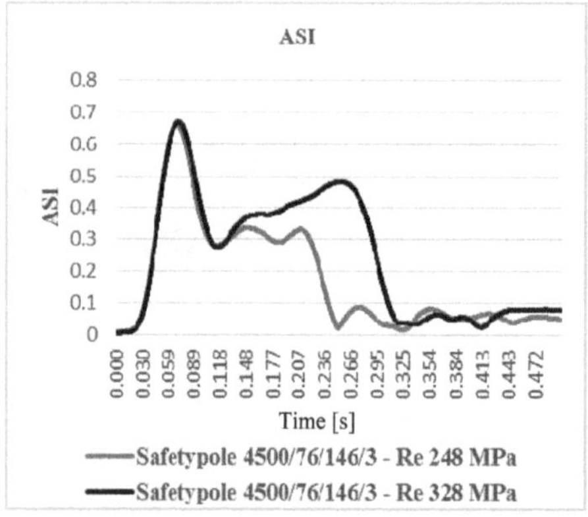

Figure 7. Comparison of ASI value for both poles.

5 CONCLUSIONS

The differences in the results of the crash tests were so significant that the columns must have been assigned to different categories of energy absorption and levels of driver's safety. A very wide range of strength parameters of steel S235 and the lack of restrictions as to their controls before crash testing undermines their credibility and meaning of the performance. The standard EN 12767 ought to impose the control of the strength parameters of steel. The other solution might be the modification of the standard EN 10025-2 in the range of acceptable values of Re and Rm.

At present, it is difficult to ensure the same pole behaviour at the instant of collision with vehicle. It is an opportunity for dishonest producers to manipulate the test results to achieve a desired effect.

REFERENCES

European Committee for Standardization (CEN). 2007. Technical delivery conditions for non-alloy structural steels. *European Standard EN 10025-2*. Warsaw: CEN.

European Committee for Standardization (CEN). 2009. Passive Safety of Support Structures for Road Equipment—Requirements and Test Methods. *European Standard EN 12767*. Warsaw: CEN.

European Committee for Standardization (CEN). 2010. Metallic materials—Tensile testing—Part 1: Method of test at room temperature. *European Standard EN ISO 6892-1*. Warsaw: CEN.

European Committee for Standardization (CEN). 2010. Road restraint systems. Terminology and general criteria for test methods. *European Standard EN 1317-1*. Warsaw: CEN.

Janszen, G. 2007. Vehicle crash test against a light ing pole: experimental analysis and numerical simulation. *WIT Transactions on The Built Environment* Vol 94: 347–355.

Rudnik. S. 1997. *Metallography*. Warsaw: PWN.

Vilán, J.A., Segade, A. & Casqueiro, C. 2006. Development and testing of a non-energy-absorbing anchorage system for roadside poles. *International Journal of Crashworthiness* Vol 11: 143–152.

Advances in Mechanics: Theoretical, Computational and Interdisciplinary Issues – Kleiber et al. (Eds)
© 2016 Taylor & Francis Group, London, ISBN 978-1-138-02906-4

Tolerance modelling of dynamics of microstructured functionally graded plates

J. Jędrysiak
Łódź University of Technology, Łódź, Poland

ABSTRACT: Microstructured functionally graded plates are considered. The size of the microstructure is of an order of the plate thickness. In order to take into account the effect of the microstructure on dynamic behaviour of these plates *the tolerance modelling* is applied. Using this method model equations are derived with smooth functional coefficients involving terms dependent of the microstructure size.

1 INTRODUCTION

In this note plates with constant thickness d made of two materials are analysed. On the microlevel plate components are tolerance-periodic distributed along one direction $x \equiv x_1$ (Fig. 1). But averaged plate properties are slowly varying along this direction, normal to interfaces between materials. Hence, such plates consist of many small elements on the micro-level, being treated as plate bands with span l (called *the microstructure parameter*). These plates are called *functionally graded plates* (Woźniak et al. 2008, Jędrysiak 2010), as made of a *Functionally Graded Material (FGM)*, Suresh & Mortensen (1998). Thickness d is the same order of the microstructure parameter l, $d \sim l$ (Mazur-Śniady et al. 2004, Baron 2006, Jędrysiak 2012, 2013), material properties are tolerance-periodic functions in x and independent of x_2. These plates can be investigated using various kinematic assumptions. Here, two various model equations are presented—one based on the thin plates theory relations and the other based on the medium thickness plates theory. Because the governing equations have highly oscillating, tolerance-periodic, non-continuous functional coefficients in x, they are not well-suited tools to analyse various problems of these plates. To investigate thermomechanical problems of FGM-type structures (also for plates) modelling methods for periodic structures can be used (Suresh & Mortensen 1998, Woźniak et al. 2008), but the effect of the microstructure size is usually neglected in these models equations. Composite plates as parts of more complicated structures are analysed by Kołakowski (2012), the three layered composite plates with foam cores regarded by Magnucka-Blandzi (2011). The list of papers, in which theoretical and numerical results of various problems of functionally graded structures are presented, is large. The third-order shear deformation theory is used to formulation finite element models for static and dynamic analysis of functionally graded plates by Reddy (2000). The higher order shear and normal deformable plate theory is applied to modelling such plates by Batra (2007). Natural frequencies can be considered successfully using also meshless methods, e.g. for functionally graded plates by Ferreira et al. (2006). Dynamics of functionally graded shells is also investigated by Tornabene & Viola (2009). A new low-order shell element is used to analyse shell structures with functionally graded properties in three directions by Kugler et al. (2013).

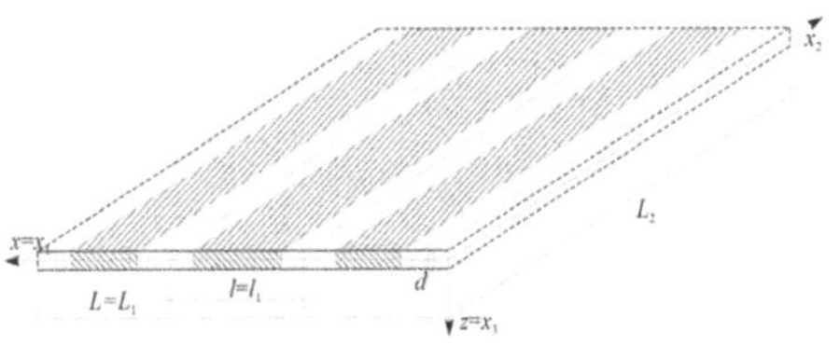

Figure 1. Functionally graded plate with microstructure.

However, the model equations describing the effect of the microstructure size can be obtained using *the tolerance modelling* (Woźniak et al. 2008), which replaces differential equations with highly oscillating, tolerance-periodic, non-continuous, functional coefficients by equations of new non-asymptotic models of the plates with smooth continuous coefficients. This method is a general one, useful to analyse problems of microstructured media.

Certain applications of this approach for periodic structures are shown in a series of papers, e.g. periodic wavy-plates by Michalak (2001), periodic shells by Tomczyk (2007), thin periodic plates by Jędrysiak (2000, 2003, 2009) and by Domagalski & Jędrysiak (2012), nonlinear periodic beams resting on a visco-elastic foundation by Domagalski &

Jędrysiak (2014) and for functionally graded media, e.g. thin functionally graded plates with the microstructure size of an order of the plate thickness by Jędrysiak (2010, 2012, 2013, 2014), thin functionally graded plates with the plate thickness small in comparing to the microstructure size by Kaźmierczak & Jędrysiak (2011, 2013, 2014), Michalak & Wirowski (2012), functionally graded laminated plates by Jędrysiak et al. (2005), thin functionally graded shells by Jędrysiak & Woźniak (2010). The list of papers can be found in the book Woźniak et al. (2008) and Jędrysiak (2010).

The aim of this note is to present averaged equations of *tolerance models of functionally graded thin/medium thickness plates*, with slowly-varying coefficients in x. These equations derived using *the tolerance modelling* (Woźniak et al. 2008, Jędrysiak 2010), involve terms describing the effect of the microstructure size. Similar tolerance model for thin transversally graded plates under the condition $d << l$ is shown in Woźniak et al. (2008), Jędrysiak (2010), Kaźmierczak & Jędrysiak (2011, 2013).

2 FUNDAMENTAL RELATIONS

Let us denote $\mathbf{x} \equiv (x_1, x_2)$, $x = x_1$, $z \equiv x_3$; and $\Omega \equiv \{(\mathbf{x}, z): -d/2 \leq z \leq d/2, \mathbf{x} \in \Pi\}$ as the region of undeformed plate, with midplane Π and the constant plate thickness d. Let $\Delta \equiv [-l/2, l/2] \times \{0\}$ be the "basic cell" on Ox_1x_2, with cell length dimension along the x-axis equal l, which satisfies conditions $d \sim l$ and $l << \min(L_1, L_2)$, and is called *the microstructure parameter*. Denote by ∂_α derivatives of x_α, and also $\partial_{\alpha \cdots \delta} \equiv \partial_\alpha .. \partial_\delta$. Properties of the plate material, as mass density ρ and elastic moduli a_{ijkl}, are assumed to be tolerance-periodic functions in x, even functions in z and independent of x_2. Denote also $c_{\alpha\beta\gamma\delta} \equiv a_{\alpha\beta\gamma\delta} - a_{\alpha\beta33}a_{\gamma\delta33}(a_{3333})^{-1}$.

A mass density per unit area μ, a rotational inertia ϑ and stiffnesses $b_{\alpha\beta\gamma\delta}$, $d_{\alpha\beta}$ of the plate are defined:

$$\mu \equiv \int_{-d/2}^{d/2} \rho dz, \quad \vartheta \equiv \int_{-d/2}^{d/2} \rho z^2 dz,$$
$$b_{\alpha\beta\gamma\delta} \equiv \int_{-d/2}^{d/2} c_{\alpha\beta\gamma\delta} z^2 dz, \quad d_{\alpha\beta} \equiv \frac{5}{6}\int_{-d/2}^{d/2} c_{\alpha 3\beta 3} dz, \tag{1}$$

and are tolerance-periodic functions in x. Let u_i $(i = 1, 2, 3)$ be plate displacements, p—loads normal to the midplane. Using the kinematic assumptions of the thin plates theory, dynamics of a functionally graded plates is described by:

$$\partial_{\alpha\beta}(b_{\alpha\beta\gamma\delta}\partial_{\gamma\delta}u) + \mu\ddot{u} - \partial_\alpha(\vartheta\partial_\beta\ddot{u})\delta_{\alpha\beta} = p, \tag{2}$$

for the plate deflection $u = u_3$. From the kinematic assumptions of the medium thickness plate theory the system of three equations is derived:

$$\partial_\beta(b_{\alpha\beta\gamma\delta}\partial_\delta\varphi_\gamma) - d_{\alpha\beta}(\partial_\beta w + \varphi_\beta) - \vartheta\ddot{\varphi}_\alpha = 0,$$
$$\partial_\alpha[d_{\alpha\beta}(\partial_\beta u + \varphi_\beta)] - \mu\ddot{u} = -p, \tag{3}$$

for the plate deflection u and rotations ϕ_α, $\alpha = 1, 2$.

The governing equations (2) and (3) have highly oscillating, tolerance-periodic, non-continuous functional coefficients in x. They are not relevant tools to analyse various special problems of these plates. To derive model equations with smooth continuous coefficients the tolerance modelling is applied.

3 TOLERANCE MODELLING

Basic concepts and assumptions of *the tolerance modelling* are presented in Woźniak et al. (2008), Jędrysiak (2010, 2012, 2013). There are: an averaging operator, a slowly-varying function $SV^\alpha_\delta(\Pi, \Delta)$, a tolerance-periodic function $TP^\alpha_\delta(\Pi, \Delta)$, a highly oscillating function $HO^\alpha_\delta(\Pi, \Delta)$, a fluctuation shape function $FS^\alpha_\delta(\Pi, \Delta)$. Here, only two of them are mentioned. Let $\Delta(\mathbf{x}) \equiv \mathbf{x} + \Delta$ be a cell at $\mathbf{x} \in \Pi_\Delta$; $\Pi_\Delta = \{\mathbf{x} \in \Pi: \Delta(\mathbf{x}) \subset \Pi\}$. *The averaging operator* for an arbitrary integrable function f is defined by:

$$< f > (x) = \frac{1}{l} \int_{\Delta(x)} f(y)dy. \tag{4}$$

For the tolerance-periodic in x function f its averaged value by (4) is a slowly-varying function in x. The important concept of this method is also *the fluctuation shape function*, which is assumed in the form of a saw-type function of x (Jędrysiak 2012, 2013). Using the above concepts the fundamental assumptions of the tolerance modelling for the considered plates can be formulated. The main assumption is *the micro-macro decomposition* of basic unknown fields in the problem under consideration. Below, there are introduced two independent assumptions, which lead to various tolerance models of functionally graded plates.

The micro-macro decomposition for the thin plates, in which it is assumed that the plate midplane displacements u_i $(i = 1, 2, 3)$ can be decomposed as for periodic plates (Mazur-Śniady et al. 2004):

$$u_3(\mathbf{x}, z, t) = u(\mathbf{x}, t) = w(\mathbf{x}, t),$$
$$u_\alpha(\mathbf{x}, z, t) = -z[\partial_\alpha w(\mathbf{x}, t) + h(x)v_\alpha(\mathbf{x}, t)], \tag{5}$$

with new unknowns: *macrodeflection* w and *fluctuation variables* v_α $(\alpha = 1, 2)$, being slowly-varying functions in x; and the known fluctuation shape function h, assumed in the form of a saw-type function of x.

The micro-macro decomposition for the medium thickness plates, in which displacements of the plate midplane u_i (i = 1, 2, 3) are assumed as for periodic plates (Baron 2006):

$$u_3(\mathbf{x},z,t) = u(\mathbf{x},t) = w(\mathbf{x},t),$$
$$u_\alpha(\mathbf{x},z,t) = z[\phi_\alpha(\mathbf{x},t) + g(x)\theta_\alpha(\mathbf{x},t)], \tag{6}$$

with new unknowns: *macrodeflection* w, *macrorotations* φ_α (α = 1, 2), and *fluctuation variables* θ_α (α = 1, 2), being slowly-varying functions in x; and the known fluctuation shape function g, which has the form of a saw-type function of x.

The second assumption is *the tolerance averaging approximation*, in which it is assumed that in the modelling procedure terms of an order of tolerance parameter δ are negligibly small, e.g. in formulas:

$$(i) \quad <\varphi>(\mathbf{x}) = <\bar{\varphi}>(\mathbf{x}) + O(\delta),$$
$$(ii) \quad <\varphi F>(\mathbf{x}) = <\varphi>(\mathbf{x})F(\mathbf{x}) + O(\delta),$$
$$(iii) \quad <\varphi\partial_\gamma(gF)>(\mathbf{x}) = <\varphi\partial_\gamma g>(\mathbf{x})F(\mathbf{x}) + O(\delta),$$
$$\mathrm{x} \in \Pi;\ \gamma = 1,\alpha;\ \alpha = 1,2;\ 0 < \delta << 1;$$
$$\varphi \in TP_\delta^\alpha(\Pi,\Delta),\ F \in SV_\delta^\alpha(\Pi,\Delta),\ g \in FS_\delta^\alpha(\Pi,\Delta); \tag{7}$$

they can be neglected. The above concepts and assumptions can be applied in various tolerance modelling procedures (Woźniak et al. 2008, Jędrysiak 2010). Here, the procedure from Jędrysiak (2012) is used, which can be divided on four steps. The first of them is to substitute micro-macro decompositions (5) or (6) to equations (2) or (3), respectively. In the second step these equations are averaged by using the averaging operator (4). In the third step the problem to find the fluctuation variables is formulated. To find these unknowns the orthogonal method is used, i.e. the governing equations (2) or (3) are multiplied by the fluctuation shape function and then averaged by formula (4). The fourth step is to substitute micro-macro decompositions (4) or (5) into obtained equations.

4 TOLERANCE MODEL EQUATIONS

From the above tolerance modelling procedure governing equations of two various tolerance models are derived.

4.1 *Tolerance model equations of the thin functionally graded plates*

The tolerance modelling replaces thin plates equation (2) by the following governing equations:

$$\partial_{\alpha\beta}(<b_{\alpha\beta\gamma\delta}>\partial_{\gamma\delta}w + <b_{\alpha\beta\gamma 1}\partial_1 h> v_\gamma)$$
$$+ <\mu>\ddot{w} - <\vartheta>\partial_{\alpha\beta}\ddot{w}\delta_{\alpha\beta} = <p>,$$
$$<b_{\alpha 1\gamma\delta}\partial_1 h>\partial_{\gamma\delta}w \tag{8}$$
$$= \underline{<b_{\alpha 2\gamma 2}h^2>}\partial_{22}v_\gamma - \underline{<\vartheta h^2>}\ddot{v}_\alpha$$
$$- <b_{\alpha 1\gamma 1}(\partial_1 h)^2> v_\gamma.$$

Equations (8) are differential equations: *macrodeflection* w and *fluctuation variables* v_α.

4.2 *Tolerance model equations of the medium thickness functionally graded plates*

The tolerance modelling leads from medium thickness plates equations (3) to the following governing equations:

$$\partial_\beta(<b_{\alpha\beta\gamma\delta}>\partial_\delta\phi_\gamma) + \partial_\beta(<b_{\alpha\beta\gamma 1}\partial_1 g>\theta_\gamma)$$
$$- <d_{\alpha\beta}>(\partial_\beta w + \phi_\beta) - <\vartheta>\ddot{\phi}_\alpha = 0,$$
$$\partial_\alpha[<d_{\alpha\beta}>(\partial_\beta w + \phi_\beta)] - <\mu>\ddot{w} = -<p>, \tag{9}$$
$$<b_{\alpha 1\gamma\delta}\partial_1 g>\partial_\delta\phi_\gamma + [<b_{\alpha 1\beta 1}(\partial_1 g)^2>$$
$$+ \underline{<d_{\alpha\beta}g^2>}]\theta_\beta - \underline{<b_{\alpha 2\gamma 2}g^2>}\partial_{22}\theta_\gamma$$
$$+ \underline{<\vartheta g^2>}\ddot{\theta}_\alpha = 0.$$

It is a system of differential equations—two for *macrorotations* φ_α, one for *macrodeflection* w and two equations for *fluctuation variables* θ_α.

In the above equations (8) or (9) the effect of the microstructure size is described by the underlined terms involving the microstructure parameter l. Equations (8) or (9) have physical sense for unknowns w, v_α or w, φ_α, θ_α, respectively, being slowly-varying functions in x, for every t. It is a certain *a posteriori* criterion of physical reliability for these models. Boundary conditions have to be formulated only for macrofunctions: *macrodeflection* w and *macrorotations* φ_α on all edges. However, boundary conditions for *fluctuation variables* v_α and θ_α can be defined only on edges $x_2 = 0, L_2$.

5 ASYMPTOTIC MODEL EQUATIONS

In order to evaluate obtained results asymptotic model equations are introduced. These governing equations can be obtained using the formal asymptotic modelling procedure (Kaźmierczak & Jędrysiak 2013), or from equations (8) or (9) after neglecting terms involving the microstructure parameter l (the underlined terms). Hence, the governing equations with continuous coefficients of two various asymptotic models are derived.

5.1 *Asymptotic model equations of the thin functionally graded plates*

From tolerance model equations (8) the following governing equations are obtained:

$$\begin{aligned}
&\partial_{\alpha\beta}(<b_{\alpha\beta\gamma\delta}>\partial_{\gamma\delta}w+<b_{\alpha\beta\gamma 1}\partial_1 h>v_\gamma)\\
&\quad+<\mu>\ddot{w}-<\vartheta>\partial_{\alpha\beta}\ddot{w}\delta_{\alpha\beta}=<p>, \qquad (10)\\
&<b_{\alpha 1\gamma\delta}\partial_1 h>\partial_{\gamma\delta}w=-<b_{\alpha 1\gamma 1}(\partial_1 h)^2>v_\gamma.
\end{aligned}$$

Equations (10) form a system of one differential equation for *macrodeflection* w and two algebraic equations for *fluctuation variables* v_α.

5.2 *Asymptotic model equations of the medium thickness functionally graded plates*

Simplifying tolerance model equations (9) we arrive at the following governing equations:

$$\begin{aligned}
&\partial_\beta(<b_{\alpha\beta\gamma\delta}>\partial_\delta\phi_\gamma)+\partial_\beta(<b_{\alpha\beta\gamma 1}\partial_1 g>\theta_\gamma)\\
&\quad-<d_{\alpha\beta}>(\partial_\beta w+\phi_\beta)-<\vartheta>\ddot{\phi}_\alpha=0,\\
&\partial_\alpha[<d_{\alpha\beta}>(\partial_\beta w+\phi_\beta)]-<\mu>\ddot{w}=-<p>, \qquad (11)\\
&<b_{\alpha 1\gamma\delta}\partial_1 g>\partial_\delta\phi_\gamma+<b_{\alpha 1\beta 1}(\partial_1 g)^2>\theta_\beta=0.
\end{aligned}$$

It is a system of three differential equations—two for *macrorotations* φ_α, one for *macrodeflection* w and two algebraic equations for *fluctuation variables* θ_α, which are basic unknowns.

In equations (10) or (11) the effect of the microstructure size is neglected. Equations (10) or (11) describe the behaviour of these microstructured plates only on the macro-level. These equations have physical sense for unknowns w, v_α or w, φ_α, θ_α, respectively, being slowly-varying functions in x.

6 REMARKS

The tolerance modelling used for governing equations of thin or medium thickness functionally graded plates with a microstructure makes it possible to pass from the equations with tolerance-periodic, non-continuous functional coefficients to equations with averaged, slowly-varying functional coefficients.

The characteristic feature of these tolerance equations is that they have some terms, depending on the microstructure parameter l. Hence, *the tolerance model of the microstructured plates* makes it possible to analyse dispersion phenomena of these plates.

REFERENCES

Baron, E. 2006. On modelling of periodic plates having the inhomogeneity period of an order of the plate thickness. *Journal of Theoretical and Applied Mechanics* 44: 3–18.

Batra, R.C. 2007. Higher-order shear and normal deformable theory for functionally graded incompressible linear elastic plates. *Thin-Walled Structures* 45: 974–982.

Domagalski, Ł. & Jędrysiak, J. 2012. On the elastostatics of thin periodic plates with large deflections. *Meccanica* 41: 1659–1671.

Domagalski, Ł. & Jędrysiak, J. 2014. Nonlinear vibrations of periodic beams. *Vibrations in Physical Systems* 26: 73–78.

Ferreira, A.J.M., Batra, R.C., Roque, C.M.C., Qian, L.F. & Jorge, R.M.N. 2006. Natural frequencies of functionally graded plates by a meshless method. *Composite Structures* 75: 593–600.

Jędrysiak, J. 2000. On the stability of thin periodic plates. *European Journal of Mechanics—A Solids* 19: 487–502.

Jędrysiak, J. 2003. The length-scale effect in the buckling of thin periodic plates resting on a periodic Winkler foundation. *Meccanica* 38: 435–451.

Jędrysiak, J. 2009. Higher order vibrations of thin periodic plates. *Thin-Walled Structures* 47: 890–901.

Jędrysiak, J. 2010. *Thermomechanics of laminates, plates and shells with functionally graded properties*. Łódź: Wyd. Politechniki Łódzkiej, (in Polish).

Jędrysiak, J. 2012. Vibrations of microstructured functionally graded plates. *Vibrations in Physical Systems* 25: 193–198.

Jędrysiak, J. 2013. Modelling of dynamic behaviour of microstructured thin functionally graded plates. *Thin-Walled Structures* 71: 102–107.

Jędrysiak, J. 2014. On the tolerance modeling of thin microstructured plates. In W. Pietraszkiewicz & J. Górski (eds), *SSTA*. Vol. 3: 101–104. London—Leiden: Balkema.

Jędrysiak, J. & Pazera, E. 2014. Free vibrations of thin microstructured plates. *Vibrations in Physical Systems* 26: 93–98.

Jędrysiak, J., Rychlewska, J. & Woźniak, C. 2005. Microstructural 2D-models of functionally graded laminated plates. In W. Pietraszkiewicz & C. Szymczak (eds), *SSTA*. Vol. 1: 119–123. London—Leiden: Balkema.

Jędrysiak, J. & Woźniak, C. 2010. Modelling of thin functionally graded shells. In W. Pietraszkiewicz & I. Kreja (eds), *SSTA*. Vol. 2: 67–70. London—Leiden: Balkema.

Kaźmierczak, M. & Jędrysiak, J. 2011. Tolerance modelling of vibrations of thin functionally graded plates. *Thin-Walled Structures* 49: 1295–1303.

Kaźmierczak, M. & Jędrysiak, J. 2013. A new combined asymptotic-tolerance model of vibrations of thin transversally graded plates. *Engineering Structures* 46: 322–331.

Kaźmierczak-Sobińska, M. & Jędrysiak, J. 2014. Some problems of modelling stability of thin functionally graded shells. In W. Pietraszkiewicz & J. Górski (eds), *SSTA*. Vol. 3: 105–107. London—Leiden: Balkema.

Kołakowski, Z. 2012. Static and dynamic interactive buckling of composite columns. *Journal of Theoretical and Applied Mechanics* 47: 177–192.

Kugler, S., Fotiu, P.A. & Murin, J. 2013. The numerical analysis of FGM shells with enhanced finite elements. *Engineering Structures* 49: 920–935.

Magnucka-Blandzi, E. 2011. Mathematical modelling of a rectangular sandwich plate with a metal foam core. *Journal of Theoretical and Applied Mechanics* 49: 439–455.

Mazur-Śniady, K., Woźniak, C. & Wierzbicki, E., On the modelling of dynamic problems for plates with a periodic structure, *Archive of Applied Mechanics* 74, 179–190, 2004.

Michalak, B. 2001. *The meso-shape functions for the meso-structural models of wavy-plates*. ZAMM 81: 639–641.

Michalak, B. & Wirowski, A. 2012. Dynamic modelling of thin plate made of certain functionally graded materials. *Meccanica* 47: 1487–1498.

Reddy, J.N. 2000. Analysis of functionally graded plates. *Inter. Jour. for Num. Methods in Engineering* 47: 663–684.

Suresh, S. & Mortensen, A. 1998. *Fundamentals of functionally graded materials*. Cambridge: The University Press.

Tomczyk, B. 2007. A non-asymptotic model for the stability analysis of thin biperiodic cylindrical shells. *Thin-Walled Structures* 45: 941–944.

Tornabene, F. & Viola, E. 2009. Free vibration analysis of functionally graded panels and shells of revolution. *Meccanica* 44: 255–281.

Woźniak, C., Michalak, B. & Jędrysiak, J. 2008. *Thermomechanics of microheterogeneous solids and structures*. Łódź: Wyd. Politechniki Łódzkiej.

Advances in Mechanics: Theoretical, Computational and Interdisciplinary Issues – Kleiber et al. (Eds)
© 2016 Taylor & Francis Group, London, ISBN 978-1-138-02906-4

Tolerance modeling of vibration of visco-elastic thin periodic plates with moderately large deflections

J. Jędrysiak
Łódź University of Technology, Łódź, Poland

ABSTRACT: Visco-elastic periodic plates with moderately large deflections are considered. Geometrically nonlinear vibrations of these plates are investigated. In order to take into account the effect of the microstructure on dynamics of these plates *the tolerance modelling* is applied. This method makes it possible to derive model equations with constant coefficients with terms dependent of the microstructure size.

1 INTRODUCTION

Thin visco-elastic plates with a periodic structure in planes parallel to the plate midplane are considered. These plates consist of many identical small elements, called *periodicity cells*. It is assumed that the plates have deflections of the order of their thickness and rest on a periodic visco-elastic foundation. Hence, plates of this kind can be called *thin visco-elastic plates with moderately large deflections*. A fragment of this plate is shown in Figure 1. Dynamic behaviour of these plates can be described by nonlinear partial differential equations with coefficients being highly oscillating, periodic and non-continuous functions of x_1, x_2, which are not good tools to analyse specific problems of plates. In order to obtain equations with constant coefficients simplified approaches are proposed, introducing effective plate properties, e.g. those based on the asymptotic homogenization, Kohn & Vogelius (1984).

Thin plates with moderately large deflections can be described by known geometrically nonlinear equations, e.g. Levy (1942), Timoshenko & Woinowsky-Krieger (1959), Woźniak (2001). The Von Kármán-type plate theory equations can be derived from equations of the three-dimensional nonlinear continuum mechanics, Meenen & Altenbach (2001). Bending problems of such plates are analysed using different methods, Levy (1942), Timoshenko & Woinowsky-Krieger (1959), but other new methods are also proposed. The influence of damping on vibrations of nonlinear periodic plates is shown in Reinhall & Miles (1989). Large amplitude vibrations of composite plates using von Karman's assumptions and Galerkin's method are analysed in Singha & Rupesh Daripa (2009). The governing equations of the proposed models neglect usually the effect of the microstructure size on the plate behaviour.

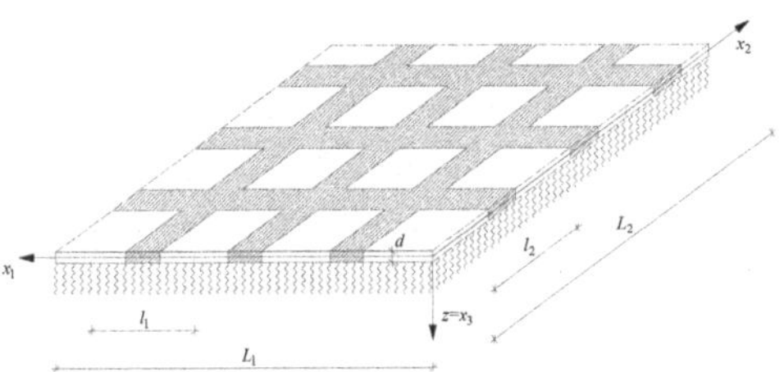

Figure 1. A fragment of visco-elastic thin periodic plate resting on a foundation.

The model equations describing this effect can be obtained using *the tolerance modelling*, Woźniak et al. (2008), which leads from differential equations with highly oscillating, non-continuous, functional coefficients to equations of new non-asymptotic models of the plates. This method is very general and is useful to analyse problems of microstructured media. Certain applications of this approach for periodic structures are shown in a series of papers, e.g. Michalak (2001), Tomczyk (2007), Jędrysiak (1999, 2000, 2003, 2009), Nagórko & Woźniak (2002) and for functionally graded media, e.g. Jędrysiak (2013), Kaźmierczak & Jędrysiak (2013). Moreover, the tolerance method is applied to investigate nonlinear bending of thin periodic plates (Domagalski & Jędrysiak 2012) and nonlinear vibrations of periodic beams resting on a visco-elastic foundation (Domagalski & Jędrysiak 2014). The list of papers can be found in the book Woźniak et al. (2008).

The aim of this note is to present averaged equations of *the tolerance model of thin visco-elastic periodic plates with moderately large deflections resting on an elastic foundation with damping*, with constant coefficients.

2 FUNDAMENTAL RELATIONS

Let us denote the orthogonal Cartesian co-ordinate system in the physical space by $0x_1x_2x_3$ and the time co-ordinate by t; and also $x \equiv (x_1, x_2)$, $z \equiv x_3$; and by $\Omega \equiv \{(\mathbf{x}, z): -d(\mathbf{x})/2 \le z \le d(\mathbf{x})/2, x \in \Pi\}$ the region of undeformed plate, with midplane Π and the plate thickness $d(\mathbf{x})$. These plates are assumed to have a periodic structure along the x_1- and x_2-axis directions with periods l_1, l_2, respectively, in planes parallel to the plate midplane. The periodicity basic cell on $0x_1x_2$ plane is denoted by $\Delta \equiv [-l_1/2, l_1/2] \times [-l_2/2, l_2/2]$. The cell size is described by parameter $l \equiv [(l_1)^2 + (l_2)^2]^{1/2}$, satisfying the condition $\max(d) << l << \min(L_1,L_2)$, and is called *the microstructure parameter*. Subscripts $\alpha, \beta,\ldots(i,j,\ldots)$ run over 1,2 (1, 2, 3) and indices $A,B,\ldots$ $(a, b,\ldots)$ run over $1, \ldots, N$ $(1, \ldots, n)$. The partial derivatives with respect to space coordinates are denoted by $(\cdot)_{,\alpha}$. Denote displacements, strains and stresses by u_i, e_{ij} and s_{ij}, respectively, virtual displacements and virtual strains by $\bar{u}_i$ and $\bar{e}_{ij}$. Let $w(\mathbf{x})$ be the plate midplane deflection, $u^0_\alpha(\mathbf{x})$ be the in-plane displacements along the x_α-axes, $p(\mathbf{x})$ be the total loadings in the z-axis; $x \in \Pi$. Thickness d can be a periodic function in $\mathbf{x}$, elastic moduli a_{ijkl} and mass density ρ can be also periodic functions in $\mathbf{x}$, but even functions in z. The non-zero components of the elastic moduli tensor are $a_{\alpha\beta\gamma\delta}$, $a_{\alpha\beta33}$, a_{3333}. Let $c_{\alpha\beta\gamma\delta} \equiv a_{\alpha\beta\gamma\delta} - a_{\alpha\beta33}a_{\gamma\delta33}(a_{3333})^{-1}$, but proper visco-elastic moduli be denoted by $\tilde{c}_{\alpha\beta\gamma\delta}$.

It is also assumed that these periodic plates interact with a visco-elastic foundation, periodic in planes parallel to the plate midplane, resting on a undeformable base. Hence, foundation properties: a mass density per unit area $\hat{\mu}$, Winkler's coefficient k, damping parameter c, can be periodic functions in $\mathbf{x}$. Moreover, the plate cannot be torn off from the foundation. Assumptions of the nonlinear theory of thin plates (Levy 1942, Timoshenko & Woinowsky-Krieger 1959, Woźniak 2001) are used.

- The thin plates kinematic assumptions

$$u_\alpha(\mathbf{x},z,t) = u^0_\alpha(\mathbf{x},t) - zw_{,\alpha}(\mathbf{x},t),$$
$$u_3(\mathbf{x},z,t) = w(\mathbf{x},t), \tag{1}$$

where $w(\mathbf{x},t)$ = is the deflection of the midplane, $u^0_\alpha(\mathbf{x},t)$ = is the in-plane displacements; and similarly for virtual displacements:

$$\bar{u}_\alpha(\mathbf{x},z) = \bar{u}^0_\alpha(\mathbf{x}) - z\bar{w}_{,\alpha}(\mathbf{x}),$$
$$\bar{u}_3(\mathbf{x},z) = \bar{w}(\mathbf{x}). \tag{2}$$

- The strain-displacement relation

$$e_{\alpha\beta} = u_{(\alpha,\beta)} + \frac{1}{2}u_{3,\alpha}u_{3,\beta}. \tag{3}$$

- The stress-strain relations (it is assumed that the plane of elastic symmetry is parallel to the plane $z = 0$).

$$s_{\alpha\beta} = c_{\alpha\beta\gamma\delta}e_{\gamma\delta} + \tilde{c}_{\alpha\beta\gamma\delta}\dot{e}_{\gamma\delta}. \tag{4}$$

where:

$$\begin{aligned}
c_{\alpha\beta\gamma\delta} &= a_{\alpha\beta\gamma\delta} - a_{\alpha\beta33}a_{33\gamma\delta} / a_{3333},\\
c_{\alpha3\gamma3} &= a_{\alpha3\gamma3} - a_{\alpha333}a_{33\gamma3} / a_{3333},\\
\tilde{c}_{\alpha\beta\gamma\delta} &= \tilde{a}_{\alpha\beta\gamma\delta} - \tilde{a}_{\alpha\beta33}\tilde{a}_{33\gamma\delta} / \tilde{a}_{3333},\\
\tilde{c}_{\alpha3\gamma3} &= \tilde{a}_{\alpha3\gamma3} - \tilde{a}_{\alpha333}\tilde{a}_{33\gamma3} / \tilde{a}_{3333},
\end{aligned} \tag{5}$$

- The virtual work equation

$$\begin{aligned}
&\int_\Pi\int_{-d/2}^{d/2}\rho\ddot{u}_i\bar{u}_i dz da + \int_\Pi\int_{-d/2}^{d/2}s_{\alpha\beta}\bar{e}_{\alpha\beta}dz da\\
&= \int_\Pi p\bar{u}_3(\mathbf{x},d/2)da\\
&\quad -\int_\Pi(ku_3 + \hat{\mu}\ddot{u}_3 + c\dot{u}_3)\bar{u}_3(\mathbf{x},-d/2)da,
\end{aligned} \tag{6}$$

which is satisfied for arbitrary virtual displacements (Equation 2), neglecting on the plate boundary and being sufficiently regular, independent functions; moreover: $da = dx_1dx_2$.

The plate properties, i.e. stiffness tensors: $b_{\alpha\beta\gamma\delta}$, $d_{\alpha\beta\gamma\delta}$, visco-elastic tensors: $\tilde{b}_{\alpha\beta\gamma\delta}, \tilde{d}_{\alpha\beta\gamma\delta}$, inertia properties: μ, j, are defined as periodic functions in $\mathbf{x}$:

$$\begin{aligned}
b_{\alpha\beta\gamma\delta}(\mathbf{x}) &= \int_{-d/2}^{d/2}c_{\alpha\beta\gamma\delta}(\mathbf{x},z)dz,\\
d_{\alpha\beta\gamma\delta}(\mathbf{x}) &= \int_{-d/2}^{d/2}c_{\alpha\beta\gamma\delta}(\mathbf{x},z)z^2dz,\\
\tilde{b}_{\alpha\beta\gamma\delta}(\mathbf{x}) &= \int_{-d/2}^{d/2}\tilde{c}_{\alpha\beta\gamma\delta}(\mathbf{x},z)dz,\\
\tilde{d}_{\alpha\beta\gamma\delta}(\mathbf{x}) &= \int_{-d/2}^{d/2}\tilde{c}_{\alpha\beta\gamma\delta}(\mathbf{x},z)z^2dz,\\
\mu(\mathbf{x}) &= \int_{-d/2}^{d/2}\rho(\mathbf{x},z)dz, \quad j(\mathbf{x}) = \int_{-d/2}^{d/2}\rho(\mathbf{x},z)z^2dz,
\end{aligned} \tag{7}$$

Using the above known assumptions of the nonlinear thin plate theory, dynamics of *thin visco-elastic plates resting on an elastic foundation with damping* is described by the following governing equations:

- the constitutive equations:

$$\begin{aligned}
m_{\alpha\beta} &= -h_{\alpha\beta\gamma\delta}u_{\gamma,\delta} + d_{\alpha\beta\gamma\delta}w_{,\gamma\delta} - 1/2h_{\alpha\beta\gamma\delta}w_{,\gamma}w_{,\delta}\\
&\quad - \tilde{h}_{\alpha\beta\gamma\delta}\dot{u}_{\gamma,\delta} + \tilde{d}_{\alpha\beta\gamma\delta}\dot{w}_{,\gamma\delta} + 1/2\tilde{h}_{\alpha\beta\gamma\delta}(\dot{w}_{,\gamma}w_{,\delta} + w_{,\gamma}\dot{w}_{,\delta}),\\
n_{\alpha\beta} &= b_{\alpha\beta\gamma\delta}u_{\gamma,\delta} - h_{\alpha\beta\gamma\delta}w_{,\gamma\delta} + 1/2b_{\alpha\beta\gamma\delta}w_{,\gamma}w_{,\delta}\\
&\quad + \tilde{b}_{\alpha\beta\gamma\delta}\dot{u}_{\gamma,\delta} - \tilde{h}_{\alpha\beta\gamma\delta}\dot{w}_{,\gamma\delta} + 1/2\tilde{b}_{\alpha\beta\gamma\delta}(\dot{w}_{,\gamma}w_{,\delta} + w_{,\gamma}\dot{w}_{,\delta}),
\end{aligned} \tag{8}$$

– the equilibrium equations:

$$m_{\alpha\beta,\alpha\beta} - (n_{\alpha\beta}w_{,\alpha})_{,\beta} + \mu\ddot{w} - j\ddot{w}_{,\alpha\alpha} + kw + \hat{\mu}\ddot{w} + c\dot{w} = p,$$
$$-n_{\alpha\beta,\beta} + \mu\ddot{u}_{\alpha} = 0. \tag{9}$$

The governing equations (8–9) can have highly oscillating, periodic, non-continuous functional coefficients in **x**.

3 LAYOUT OF TEXT

The tolerance modelling is based on some concepts and assumptions shown in Woźniak et al. (2008), e.g.: an averaging operator, a slowly-varying function $SV^{\alpha}_{\delta}(\Pi, \Delta)$, a tolerance-periodic function $TP^{\alpha}_{\delta}(\Pi, \Delta)$, a highly-oscillating function $HO^{\alpha}_{\delta}(\Pi, \Delta)$, a fluctuation shape function $FS^{\alpha}_{\delta}(\Pi, \Delta)$. These concepts are used in the fundamental assumptions.

The first assumption is *the micro-macro decomposition*, where it is assumed that deflections w and in-plane displacements u^0_{α}, $\alpha = 1,2$, can be decomposed as ($A = 1, \ldots, N$; $a = 1, \ldots, m$):

$$w(\mathbf{x},t) = W(\mathbf{x},t) + g^A(\mathbf{x})V^A(\mathbf{x},t),$$
$$u^0_{\alpha}(\mathbf{x},t) = U_{\alpha}(\mathbf{x}) + f^a(\mathbf{x})T^a_{\alpha}(\mathbf{x},t), \tag{10}$$

with new unknowns: *macrodeflection* W, *in-plane macrodisplacements* U_{α}, *fluctuation amplitudes of the deflection* V^A and *the in-plane displacements* T^a_{α}, being slowly-varying functions in **x**; and the known fluctuation shape functions g^A, f^a, assumed usually in the form of trigonometric functions. Similar micro-macro decomposition is also posed on virtual displacements $\bar{w}(\cdot)$, $\bar{u}^0_{\alpha}(\cdot)$:

$$\bar{w}(\mathbf{x}) = \bar{W}(\mathbf{x}) + g^A(\mathbf{x})\bar{V}^A(\mathbf{x}),$$
$$\bar{u}^0_{\alpha}(\mathbf{x}) = \bar{U}_{\alpha}(\mathbf{x}) + f^a(\mathbf{x})\bar{T}^a_{\alpha}(\mathbf{x}), \tag{11}$$

with slowly-varying functions $\bar{W}(\cdot), \bar{V}^A(\cdot), \bar{U}_{\alpha}(\cdot), \bar{T}^a_{\alpha}(\cdot)$.

The second assumption is *the tolerance averaging approximation*, in which it is assumed that in the modelling terms of an order of tolerance parameter δ are negligibly small, e.g. in formulas:

$$(i)\quad <\varphi>(\mathbf{x}) = <\bar{\varphi}>(\mathbf{x}) + O(\delta),$$
$$(ii)\quad <\varphi F>(\mathbf{x}) = <\varphi>(\mathbf{x})F(\mathbf{x}) + O(\delta),$$
$$(iii)\quad <\varphi\partial_{\gamma}(gF)>(\mathbf{x}) = <\varphi\partial_{\gamma}g>(\mathbf{x})F(\mathbf{x}) + O(\delta),$$
$$\mathrm{x} \in \Pi;\ \gamma = 1, \alpha;\ \alpha = 1, 2;\ 0 < \delta << 1;$$
$$\varphi \in TP^{\alpha}_{\delta}(\Pi,\Delta),\ F \in SV^{\alpha}_{\delta}(\Pi,\Delta),\ g \in FS^{\alpha}_{\delta}(\Pi,\Delta); \tag{12}$$

they can be neglected.

In the modelling procedure the above concepts and fundamental assumptions are used. This procedure can be divided into four steps. In the first step micro-macro decompositions (10) are substituted into the virtual work equation of such plate resting on a foundation (6). Then, in the next step the resulting equations are averaged by the averaging operation over a periodicity cell (Jędrysiak 2003). In the third step, after using formulas of *the tolerance averaging approximation* (Jędrysiak 2003) we arrive at the tolerance averaged virtual work equation. Then, after some manipulations governing equations of the proposed approximate model can be derived.

4 TOLERANCE MODEL EQUATIONS

From the tolerance modelling procedure, Woźniak et al. (2008), the governing equations with constant coefficients are derived:

– the constitutive equations:

$$M_{\alpha\beta} = D_{\alpha\beta\gamma\delta}W_{,\gamma\delta} + D^A_{\alpha\beta}V^A + \tilde{D}_{\alpha\beta\gamma\delta}\dot{W}_{,\gamma\delta} + \tilde{D}^A_{\alpha\beta}\dot{V}^A,$$

$$M^A = D^A_{\alpha\beta}W_{,\gamma\delta} + D^A V^B + \tilde{D}^A_{\alpha\beta}\dot{W}_{,\gamma\delta} + \tilde{D}^{AB}\dot{V}^B,$$

$$\begin{aligned} N_{\alpha\beta} = {} & B_{\alpha\beta\gamma\delta}(U_{\gamma,\delta} + 1/2W_{,\gamma}W_{,\delta}) + B^a_{\alpha\beta\gamma}T^a_{\gamma} \\ & + lF^A_{\alpha\beta\gamma}W_{,\gamma}V^A + 1/2l^2F^{AB}_{\alpha\beta}V^AV^B + \tilde{B}^a_{\alpha\beta\gamma}\dot{T}^a_{\gamma} \\ & + l\tilde{F}^A_{\alpha\beta\gamma}(\dot{W}_{,\gamma}V^A + W_{,\gamma}\dot{V}^A) \\ & + \tilde{B}_{\alpha\beta\gamma\delta}[\dot{U}_{\gamma,\delta} + 1/2(\dot{W}_{,\gamma}W_{,\delta} + W_{,\gamma}\dot{W}_{,\delta})] \\ & + 1/2l^2\tilde{F}^{AB}_{\alpha\beta}(\dot{V}^AV^B + V^A\dot{V}^B), \end{aligned}$$

$$\begin{aligned} N^a_a = {} & B^a_{\alpha\gamma\delta}(U_{\gamma,\delta} + 1/2W_{,\gamma}W_{,\delta}) \\ & + B^{ab}_{\alpha\gamma}T^b_{\gamma} + lF^{aB}_{\alpha\gamma}W_{,\gamma}V^B + 1/2l^2F^{aBC}_a V^BV^C \\ & + \tilde{B}^a_{\alpha\gamma\delta}[\dot{U}_{\gamma,\delta} + 1/2(\dot{W}_{,\gamma}W_{,\delta} + W_{,\gamma}\dot{W}_{,\delta})] \\ & + \tilde{B}^{ab}_{\alpha\gamma}\dot{T}^b_{\gamma} + l\tilde{F}^{aB}_{\alpha\gamma}(\dot{W}_{,\gamma}V^B + W_{,\gamma}\dot{V}^B) \\ & + 1/2l^2\tilde{F}^{aBC}_a(\dot{V}^BV^C + V^B\dot{V}^C), \end{aligned}$$

$$\begin{aligned} Q^A_a = {} & lF^A_{\alpha\gamma\delta}(U_{\gamma,\delta} + 1/2W_{,\gamma}W_{,\delta}) + lF^{aA}_{a\gamma}T^a_{\gamma} \\ & + l^2F^{AB}_{a\gamma}W_{,\gamma}V^B + l\tilde{F}^A_{\alpha\gamma\delta}\dot{U}_{\gamma,\delta} + 1/2l^3F^{ABC}_{\alpha}V^BV^C \\ & + l\tilde{F}^{aA}_{\alpha\gamma}[\dot{T}^a_{\gamma} + 1/2(\dot{W}_{,\gamma}W_{,\delta} + W_{,\gamma}\dot{W}_{,\delta})] \\ & + l^2\tilde{F}^{AB}_{\alpha\gamma}(\dot{W}_{,\gamma}V^B + W_{,\gamma}\dot{V}^B) \\ & + 1/2l^3\tilde{F}^{ABC}_{\alpha}(\dot{V}^BV^C + V^B\dot{V}^C), \end{aligned}$$

$$R^{AB}=l^2F_{\gamma\delta}^{AB}(U_{\gamma,\delta}+1/2W_{,\gamma}W_{,\delta})+1/2l^4F^{ABCD}V^CV^D$$
$$+l^2F_{\gamma}^{aAB}T_{\gamma}^a+l^3F_{\gamma}^{ABC}W_{,\gamma}V^C$$
$$+l^2\tilde{F}_{\gamma\delta}^{AB}[\dot{U}_{\gamma,\delta}+1/2(\dot{W}_{,\gamma}W_{,\delta}+W_{,\gamma}\dot{W}_{,\delta})]$$
$$+l^2\tilde{F}_{\gamma}^{aAB}\dot{T}_{\gamma}^a+l^3\tilde{F}_{\gamma}^{ABC}(\dot{W}_{,\gamma}V^C+W_{,\gamma}\dot{V}^C)$$
$$+1/2l^4\tilde{F}^{ABCD}(\dot{V}^CV^D+V^C\dot{V}^D), \tag{13}$$

– the equilibrium equations:

$$M_{\alpha\beta,\alpha\beta}-(N_{\alpha\beta}W_{,\alpha}+Q_{\beta}^AV^A)_{,\beta}+(m+\hat{m})\ddot{W}$$
$$+l^2(m^A+\hat{m}^A)\ddot{V}^A-\vartheta\ddot{W}_{,\alpha\alpha}-l\vartheta_{\alpha}^A\ddot{V}_{,\alpha}^A+KW$$
$$+l^2K^AV^A+C\dot{W}+l^2C^A\dot{V}^A=P,$$
$$M^A+Q_{\alpha}^AW_{,\alpha}+R^{AB}V^B+l^2(m^A+\hat{m}^A)\ddot{W}$$
$$+l^2(K^AW+C^A\dot{W})+l\vartheta_{\alpha}^A\ddot{W}_{,\alpha}+l^2[l^2(m^{AB}+\hat{m}^{AB})$$
$$+\vartheta_{\alpha\beta}^{AB}]\ddot{V}^B+l^4(K^{AB}V^B+C^{AB}\dot{V}^B)=l^2P^A,$$
$$-N_{\alpha\beta,\beta}+m\ddot{U}_{\alpha}+l\overline{m}^a\ddot{T}_{\alpha}^a=0,$$
$$N_{\alpha}^a+l\overline{m}^a\ddot{U}_a+l^2\overline{m}^{ab}\ddot{T}_a^b=0. \tag{14}$$

Equations (13–14) and (10) constitute *the nonlinear tolerance model of thin visco-elastic periodic plates resting on an elastic foundation with damping.* This model describes the effect of the microstructure size on the plate behaviour in terms of the microstructure parameter l. There have to be formulated boundary conditions only for macrodeflection W and in-plane macrodisplacements U_{α}. Moreover, the basic unknowns of equations (13–14) have to be slowly-varying functions in **x**.

5 ASYMPTOTIC MODEL EQUATIONS

Using the formal asymptotic modelling procedure (Kaźmierczak & Jędrysiak 2013) or in equations (13–14) neglecting terms involving the microstructure parameter l, the following governing equations with constant coefficients are obtained:

– the constitutive equations:

$$M_{\alpha\beta}=D_{\alpha\beta\gamma\delta}W_{,\gamma\delta}+D_{\alpha\beta}^AV^A+\tilde{D}_{\alpha\beta\gamma\delta}\dot{W}_{,\gamma\delta}+\tilde{D}_{\alpha\beta}^A\dot{V}^A,$$
$$M^A=D_{\alpha\beta}^AW_{,\gamma\delta}+D^{AB}V^B+\tilde{D}_{\alpha\beta}^A\dot{W}_{,\gamma\delta}+\tilde{D}^{AB}\dot{V}^B,$$
$$N_{\alpha\beta}=B_{\alpha\beta\gamma\delta}(U_{\gamma,\delta}+1/2W_{,\gamma}W_{,\delta})+B_{\alpha\beta\gamma}^aT_{\gamma}^a$$
$$+\tilde{B}_{\alpha\beta\gamma\delta}[\dot{U}_{\gamma,\delta}+1/2(\dot{W}_{,\gamma}W_{,\delta}+W_{,\gamma}\dot{W}_{,\delta})]+\tilde{B}_{\alpha\beta\gamma}^a\dot{T}_{\gamma}^a,$$
$$N_{\alpha}^a=B_{\alpha\gamma\delta}^a(U_{\gamma,\delta}+1/2W_{,\gamma}W_{,\delta})+B_{\alpha\gamma}^{ab}T_{\gamma}^b$$
$$+\tilde{B}_{\alpha\gamma\delta}^a[\dot{U}_{\gamma,\delta}+1/2(\dot{W}_{,\gamma}W_{,\delta}+W_{,\gamma}\dot{W}_{,\delta})]+\tilde{B}_{\alpha\gamma}^{ab}\dot{T}_{\gamma}^b,$$
$$Q_{\alpha}^A=0,\quad R^{AB}=0 \tag{15}$$

– the equilibrium equations:

$$M_{\alpha\beta,\alpha\beta}-(N_{\alpha\beta}W_{,\alpha})_{,\beta}+(m+\hat{m})\ddot{W}$$
$$-\vartheta\ddot{W}_{,\alpha\alpha}+KW+C\dot{W}=P,$$
$$-N_{\alpha\beta,\beta}+m\ddot{U}_{\alpha}=0,$$
$$M^A=0,\quad N_{\alpha}^a=0. \tag{16}$$

Equations (15–16) and (10) constitute *the nonlinear asymptotic model of thin visco-elastic periodic plates resting on an elastic foundation with damping.* This model neglects the effect of the microstructure size on the plate behaviour. Hence, it describes the behaviour of these plates only on the macro-level. In this model similar boundary conditions have to be formulated for the tolerance model. Moreover, the basic unknowns of equations (15–16) are also slowly-varying functions in **x**.

6 FINAL REMARKS

The tolerance modelling replaces governing equations of *thin visco-elastic periodic plates with moderately large deflections resting on a visco-elastic foundation*, which have highly oscillating, periodic, non-continuous functional coefficients by *tolerance model equations* with constant coefficients. The characteristic feature of these derived tolerance equations is that they involve some terms, which depend on the microstructure parameter l. Hence, *the nonlinear tolerance model of these plates* makes it possible to analyse dispersion phenomena of these plates. The proposed model is a certain generalization of tolerance models of various periodic structures, i.e. of nonlinear bending of thin periodic plates by Domagalski & Jędrysiak (2012), of nonlinear vibrations of periodic beams by Domagalski & Jędrysiak (2014), of damping vibrations of thin plates with periodically distributed concentrated masses by Marczak & Jędrysiak (2014). Some different examples of dynamic problems for such visco-elastic periodic plates will be investigated in the future in the framework of the proposed tolerance model.

REFERENCES

Domagalski, Ł. & Jędrysiak, J. 2012. On the elastostatics of thin periodic plates with large deflections. *Meccanica* 41: 1659–1671.

Domagalski, Ł. & Jędrysiak, J. 2014. Nonlinear vibrations of periodic beams. *Vibrations in Physical Systems* 26: 73–78.

Jędrysiak, J. 1999. Dynamics of thin periodic plates resting on a periodically inhomogeneous Winkler foundation. *Archive of Applied Mechanics* 69: 345–356.

Jędrysiak, J. 2000. On the stability of thin periodic plates. *European Journal of Mechanics—A Solids* 19: 487–502.

Jędrysiak, J. 2003. Free vibrations of thin periodic plates interacting with an elastic periodic foundation. *International Journal Mechanical Sciences* 45(8): 1411–1428.

Jędrysiak, J. 2009. Higher order vibrations of thin periodic plates. *Thin-Walled Structures* 47: 890–901.

Jędrysiak, J. 2013. Modelling of dynamic behaviour of microstructured thin functionally graded plates. *Thin Walled Structures* 71: 102–107.

Kaźmierczak, M. & Jędrysiak, J. 2013. A new combined asymptotic-tolerance model of vibrations of thin transversally graded plates. *Engineering Structures* 46: 322–331.

Kohn, R.V. & Vogelius, M. 1984. A new model of thin plates with rapidly varying thickness. *International Journal of Solids and Structures* 20: 333–350.

Levy, S. 1942. *Bending of rectangular plates with large deflections*. NACA Rep. 737; NACA Tech Note 846.

Maczak, J. & Jędrysiak, J. 2014. Analysis of vibrations of plate strip with concentrated masses using tolerance averaging technique. *Vibrations in Physical Systems* 26: 161–168.

Meenen, J. & Altenbach, H. 2001. A consistent deduction of von Kármán-type plate theories from three-dimensional nonlinear continuum mechanics. *Acta Mech.* 147: 1–17.

Michalak, B. 2001. The meso-shape functions for the meso-structural models of wavy-plates. *ZAMM* 81: 639–641.

Nagórko, W. & Woźniak, C. 2002. Nonasymptotic modelling of thin plates reinforced by a system of stiffeners. *Electronic Journal of Polish Agricultural Universities—Civil Engineering* 5(2): www.ejpau.media.pl.

Reinhall, P.G. & Miles, R.N. 1989. Effect of damping and stiffness on the random vibration of non-linear periodic plates. *Journal of Sound and Vibration* 132: 33–42.

Singha, M.K. & Rupesh, Daripa. 2009. Nonlinear vibration and dynamic stability analysis of composite plates. *Journal of Sound and Vibration* 328: 541–554.

Timoshenko, S. & Woinowsky-Krieger, S. 1959. *Theory of plates and shells*. New York: McGraw-Hill.

Tomczyk, B. 2007. A non-asymptotic model for the stability analysis of thin biperiodic cylindrical shells. *Thin-Walled Structures* 45: 941–944.

Woźniak, C. 2001. *Mechanics of elastic plates and shells*. Warszawa: PWN (in Polish).

Woźniak, C., Michalak, B. & Jędrysiak, J. 2008. *Thermomechanics of microheterogeneous solids and structures*. Łódź: Wyd. Politechniki Łódzkiej.

Advances in Mechanics: Theoretical, Computational and Interdisciplinary Issues – Kleiber et al. (Eds)
© 2016 Taylor & Francis Group, London, ISBN 978-1-138-02906-4

Thermoelastic phenomena in functionally graded laminates

J. Jędrysiak & E. Pazera
Faculty of Civil Engineering, Architecture and Environmental Engineering, Łódź University of Technology, Łódź, Poland

ABSTRACT: In the paper the thermoelasticity problems in laminates are presented. These laminates are made of two components non-periodically distributed as micro-laminas along one direction x_1. In this note, three models are presented: the tolerance and the asymptotic-tolerance model, taking into account the effect of the microstructure size, and the asymptotic model, in which this effect is omitted. The equations of these models are obtained using the tolerance averaging technique.

1 INTRODUCTION

In the note we deal with the problem of thermoelasticity in two phase laminates. The cells of considered laminates are composed of two sublayers made of different materials. The thickness of the cell is constant and denoted by l. This thickness is treated as the size of the microstructure and is called *the microstructure parameter*. The macroscopic properties of this composite are changing continuously along one direction, perpendicular to the laminas (Fig. 1). This type of composites can be called functionally graded laminates, Suresh & Mortensen (1998). The overall behaviour of these laminates can be described by adopted methods, which are used for macroscopically homogeneous composites and in the analysis of various issues concerning these laminates, the same approaches are often used as for composites with periodic structures, e.g. the asymptotic homogenization method, Bensoussan et al. (1978), the homogenization based on the microlocal parameters, Matysiak & Perkowski (2015).

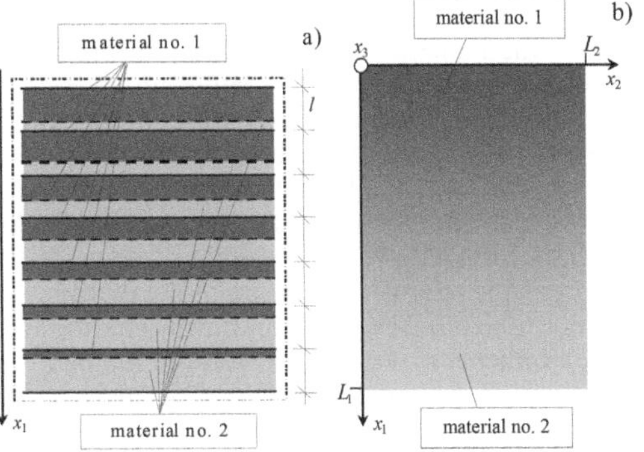

Figure 1. The cross-section of considered laminates: a) microstructure, b) macrostructure.

Unfortunately, most of the known equations models, do not take into account the effect of the microstructure size.

In order to obtain the averaged equations taking into account this effect, the tolerance averaging technique is applied (Woźniak et al. 2008). This technique was applied in many studies to obtain equations describing the various issues of periodic structures (Woźniak et al. 2008, 2010), e.g. thermoelastic problems (Wierzbicki et al. 1997, Woźniak C. et al. 2002, Woźniak M. et al. 2002, Baczyński 2003, Ignaczak 2004). Recently, this technique was modified and adopted to analyse cases of composites made of materials with functional gradation of properties. The examples of applications of this method can be found in (Woźniak et al. 2008, Jędrysiak 2011, 2013, Ostrowski & Michalak 2011, 2015, Jędrysiak & Radzikowska 2012, Kaźmierczak & Jędrysiak 2013, Marczak & Jędrysiak 2015, Pazera & Jędrysiak 2015, Rabenda & Michalak 2015). By using the tolerance modelling it is possible to obtain a system of differential equations with slowly-varying or constant coefficients instead equations with functional, highly-oscillating, tolerance-periodic or periodic and non-continuous coefficients. The introductory concepts related to this technique apply a slowly-varying function, a tolerance-periodic function and an averaging operation (Woźniak et al. 2008, 2010, Jędrysiak 2011).

In the paper the equations of three models are obtained: the tolerance, the asymptotic-tolerance and the asymptotic model. The equations of the tolerance model involve terms describing the effect of the microstructure size on the overall behaviour of the laminates under consideration. The asymptotic-tolerance model equations involve also terms with the microstructure parameter. Equations of the asymptotic model neglect the effect of the microstructure size.

2 THE MODELLING PROCEDURES

The starting point of the modelling are governing equations describing thermoelasticity problems of composites:

$$\partial_j(c_{ijkl}\partial_l u_k) - \rho\ddot{u}_i = \partial b_{ij}\theta + b_{ij}\partial_j\theta,$$
$$\partial_j(k_{ij}\partial_i\theta) = c\dot{\theta} + T_0 b_{ij}\partial_j\dot{u}_i, \tag{1}$$

where c_{ijkl}, ρ, b_{ij}, k_{ij}, c are highly-oscillating, non-continuous functional coefficients of x_1 for displacements u_i ($i, j, k, l = 1, 2, 3$) and temperature θ.

Using the tolerance or the asymptotic-tolerance modelling it is possible to obtain equations with slowly-varying coefficients, describing the effect of the microstructure size.

The tolerance modelling is based on two assumptions. The first is the micro-macro decomposition assumed in the following form:

$$u_i(x,\mathbf{x},t) = w_i(x,\mathbf{x},t) + h(x)v_i(x,\mathbf{x},t),$$
$$\theta(x,\mathbf{x},t) = \vartheta(x,\mathbf{x},t) + g(x)\psi(x,\mathbf{x},t), \tag{2}$$

where $x \equiv x_1$, $\mathbf{x} \equiv (x_1,x_2)$; w_i, v_i, ϑ, ψ are slowly-varying functions in x; functions w_i and θ are the basic unknowns, called the macrodisplacements and the macrotemperature, respectively; v_i and ψ are additional basic unknowns, called the fluctuation amplitudes of displacements and temperature, respectively; $h(x)$, $g(x)$ are the fluctuation shape functions, assumed as saw-like functions. The second assumption is the tolerance averaging approximation in which terms of an order of $O(\delta)$ are negligibly small. Substituting micro-macro decomposition (2) to governing equations (1), by doing appropriate averaging and transformations the final equations of the tolerance model can be obtained.

The asymptotic-tolerance modelling procedure can be carried out in two steps, shown in (Woźniak et al. 2010, Jędrysiak 2013, Kaźmierczak & Jędrysiak 2013). In the first step, the asymptotic model solution is obtained in the following form:

$$u_{0i}(x,\mathbf{x},t) = w_i(x,\mathbf{x},t) + h(x)v_i(x,\mathbf{x},t),$$
$$\theta_0(x,\mathbf{x},t) = \vartheta(x,\mathbf{x},t) + g(x)\psi(x,\mathbf{x},t). \tag{3}$$

As a result, the system of equations is derived only for the macrodisplacements and the macrotemperature. The second step is application of the additional micro-macro decomposition to equations, with the known functions w_i, v_i, ϑ, ψ (from the asymptotic model solution):

$$u_i(x,\mathbf{x},t) = u_{0i}(x,\mathbf{x},t) + f(x)r_i(x,\mathbf{x},t),$$
$$\theta(x,\mathbf{x},t) = \theta_0(x,\mathbf{x},t) + d(x)\chi(x,\mathbf{x},t), \tag{4}$$

where r_i and χ are new unknowns being slowly-varying functions; $f(x)$, $d(x)$ are new additional fluctuation shape functions similar to $h(x)$ and $g(x)$.

3 THE MODEL EQUATIONS

From the tolerance modelling procedure (Woźniak et al. 2010, Jędrysiak 2011), the equations of the tolerance model of functionally graded laminates are obtained:

$$\partial_j(<c_{ijkl}>\partial_l w_k + <c_{ijk1}\partial h> v_k) - <\rho>\ddot{w}_i$$
$$= \partial <b_{i1}> \vartheta + <b_{ij}>\partial_j\vartheta,$$
$$-\underline{c_{i\alpha k\beta}hh>\partial_\alpha\partial_\beta v_k} + <c_{i1k1}\partial h\partial h> v_k + <c_{i1kl}\partial h>\partial_l w_k$$
$$+\underline{<\rho hh>\ddot{v}_i} = -<b_{i1}\partial g>\vartheta + \underline{<b_{i\beta}gh>\partial_\beta\psi},$$
$$\partial_j(<k_{ij}>\partial_i\vartheta + <k_{1j}\partial g>\psi)$$
$$= <c>\dot{\vartheta} + <T_0 b_{ij}>\partial_j\dot{w}_i + <T_0 b_{i1}\partial h>\dot{v}_i,$$
$$\underline{<k_{\alpha\beta}gg>\partial_\alpha\partial_\beta\psi} - <k_{i1}\partial g>\partial_i\vartheta - <k_{11}\partial g\partial g>\psi$$
$$= \underline{<cgg>\dot{\psi}} + \underline{<T_0 b_{i\beta}hg>\partial_\beta\dot{v}_i}, \tag{5}$$

where the underlined terms depend on the microstructure parameter l; the coefficients in <·> are slowly-varying functions in x, in contrast to the main equations of thermoelasticity problem, where the coefficients are highly-oscillating, tolerance-periodic and non-continuous. The equations of the tolerance model give the possibility to analyse the effect of the microstructure size in the thermoelasticity problems. The basic unknowns are averaged displacements w_i, $i = 1, 2, 3$, averaged temperature ϑ, fluctuation amplitudes of displacement v_i and of temperature ψ, being slowly-varying functions in x.

Equations of the asymptotic-tolerance model of functionally graded laminates are derived by the procedure presented in (Woźniak et al. 2010) in the following form:

$$\partial_j(<c_{ijkl}>\partial_l w_k+<c_{ijk1}\partial h>v_k)-<\rho>\ddot{w}_i$$
$$=\partial<b_{i1}>\vartheta+<b_{ij}>\partial_j\vartheta,$$
$$<c_{i1k1}\partial f\partial f>r_k-\underline{<c_{i\alpha k\beta}ff>\partial_\alpha\partial_\beta r_k}+\underline{<\rho ff>\ddot{r}_i}$$
$$=-<c_{i1k1}\partial h\partial f>v_k-<c_{i1kl}\partial f>\partial_l w_k-<b_{i1}\partial f>\vartheta,$$
$$\partial_j(<k_{ij}>\partial_i\vartheta+<k_{1j}\partial g>\psi)$$
$$=<c>\dot{\vartheta}+<T_0 b_{ij}>\partial_j\dot{w}_i+<T_0 b_{i1}\partial h>\dot{v}_i,$$
$$\underline{<k_{\alpha\beta}dd>\partial_\alpha\partial_\beta\chi}-<k_{11}\partial d\partial d>\chi=<k_{i1}\partial d>\partial_i\vartheta$$
$$+<k_{11}\partial g\partial d>\psi+\underline{<cdd>\dot{\chi}}+\underline{<T_0 b_{i\beta}fd>\partial_\beta\dot{r}_i},$$

$$<c_{i1k1}\partial h\partial h>v_k=-<c_{i1kl}\partial h>\partial_l w_k-<b_{i1}\partial g>\vartheta,$$

$$<k_{11}\partial g\partial g>\psi=-<k_{i1}\partial g>\partial_i\vartheta. \tag{6}$$

The above equations describe the thermoelastic problems in laminates with the effect of the microstructure size by the underlined terms.

Equations of the asymptotic model of functionally graded laminates can be obtained by the steps shown in books (Woźniak et al. 2010, Jędrysiak 2011) or directly from equations of the tolerance model (6), neglecting underlined terms (involving the microstructure parameter l):

$$\partial_j(<c_{ijkl}>\partial_l w_k+<c_{ijk1}\partial h>v_k)-<\rho>\ddot{w}_i$$
$$=\partial<b_{i1}>\vartheta+<b_{ij}>\partial_j\vartheta,$$
$$\partial_j(<k_{ij}>\partial_i\vartheta+<k_{1j}\partial g>\psi)$$
$$=<c>\dot{\vartheta}+<T_0 b_{ij}>\partial_j\dot{w}_i+<T_0 b_{i1}\partial h>\dot{v}_i,$$
$$<c_{i1k1}\partial h\partial h>v_k+<c_{i1kl}\partial h>\partial_l w_k=-<b_{i1}\partial g>\vartheta,$$
$$-<k_{i1}\partial g>\partial_i\vartheta-<k_{11}\partial g\partial g>\psi=0. \tag{7}$$

The above equations describe the thermoelasticity in laminates only on the macro-level.

4 THE EXAMPLE

The considered problem is to analyse stationary problem along two directions in a laminated layer made of two components. These components are different materials. It is assumed that the materials are isotropic and homogeneous. Moreover, in the macro-scale the material properties are varying continuously along one direction, perpendicular to the laminas.

The fluctuation shape functions $h(y)$, $g(y)$ can be given in the following form:

$$h(y)=g(y)=-l\sqrt{3}\frac{\xi}{\xi_1}\left(\frac{2y}{l}+\xi_2\right)$$
$$\text{for } y\in\left(-\frac{l}{2},-\frac{l}{2}+\xi_1\right),$$
$$h(y)=g(y)=l\sqrt{3}\frac{\xi}{\xi_2}\left(\frac{2y}{l}-\xi_2\right)$$
$$\text{for } y\in\left(-\frac{l}{2}+\xi_1,\frac{l}{2}\right), \tag{8}$$

where $\xi=(\xi_1\xi_2)^{1/2}$ is non-homogeneity ratio; ξ_1 and ξ_2 are the share of the first and the second material in the cell, respectively. All these ratios $\xi(\cdot)$, $\xi_1(\cdot)$, $\xi_2(\cdot)$ are slowly-varying functions in x.

In this case the fluctuation shape functions $f(\cdot)$, $d(\cdot)$ are assumed as the same as $h(\cdot)$ and $g(\cdot)$.

Taking into account above denotations and assumptions, the equations for considered laminates can be obtained, for the tolerance, asymptotic-tolerance and the asymptotic model.

Next the six different distribution functions of material properties are considered: the periodic, linear, quadratic, cubic, exponential and logarithmic function and adopted proportion of structure and material properties between the first and the second material (Young's modulus, Poisson's ratio, mass density, expansion coefficient and heat conduction). Based on these properties the Lame's and thermoelastic constants are counted (for the first and the second sublayers).

The load in this issue is assumed as a temperature: for $x_1=0$ and $x_2\in[0,L_2]$ is equal 90°C, for $x_1\in[0,L_1]$ and $x_2=L_2$ is equal 90°C, for $x_1\in[0,L_1]$ and $x_2=0$ is changing linearly from 90°C to 10°C and for $x_1=L_1$ and $x_2\in[L_2,0]$ is changing linearly from 90°C to 10°C. All displacements at the edges are assumed zero.

Including the load and the boundary conditions and by using the finite difference method, the equations describing the considered issue can be solved.

Below, the plots are shown for the temperature (Fig. 2) and for the displacements in both directions (Figs. 3–4).

In the above Figures 2–4 some exemplary results obtained for the asymptotic-tolerance model and quadratic distribution function of material properties are presented.

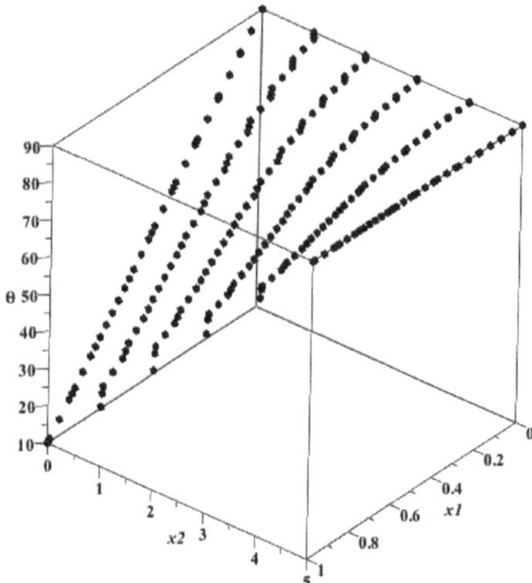

Figure 2. Temperature θ.

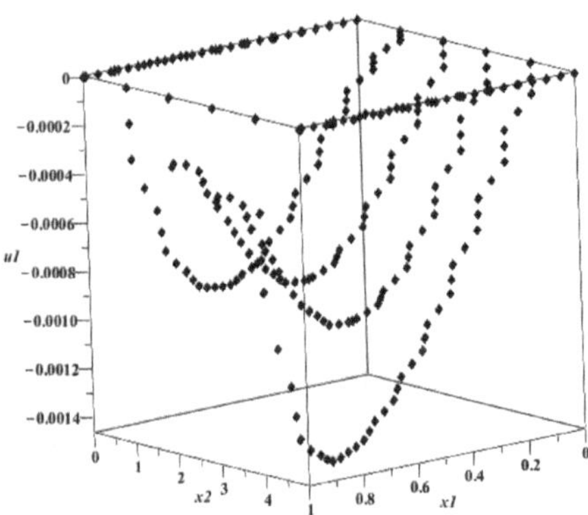

Figure 3. Displacements u_1.

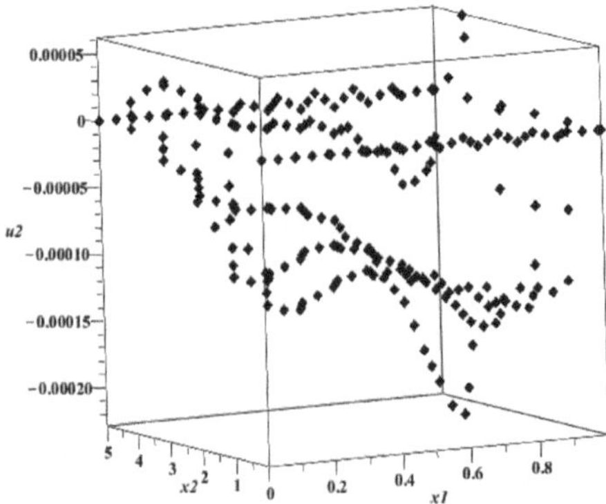

Figure 4. Displacements u_2.

5 REMARKS

Using the tolerance modelling it is possible to replace the differential equations of thermoelasticity for functionally graded laminates with highly-oscillating, non-continuous coefficients by the differential equations with smooth, slowly-varying coefficients.

The proposed modelling procedures lead to obtain the governing equations of two models—the tolerance and the asymptotic-tolerance, describing the effect of the microstructure size in thermoelasticity problems of the considered laminates, and of one model—the asymptotic, in which this effect is neglected.

The equations for these models, tolerance, asymptotic and asymptotic-tolerance, can be applied in the analysis of some specific cases, namely where distribution of the ingredients is functional but non-periodic.

Distributions of the temperature and the displacements depend on the distribution functions of material properties.

Results obtained by the asymptotic–tolerance and the asymptotic model are overlayed.

REFERENCES

Baczyński, Z.F. 2003. Dynamic thermoelastic process in periodic composites. *Journal of Thermal Stresses* 26: 55–66.

Bensoussan A., Lions J.L., Papanicolaou G. 1978. *Asymptotic analysis for periodic structures*. Amsterdam: North-Holland.

Ignaczak, J. 2004. Plane harmonic waves in a microperiodic layered thermoelastic solid revisited. *Journal of Thermal Stresses* 27: 779–793.

Jędrysiak, J. 2011. On the tolerance modelling of thermoelasticity problems for transversally graded laminates. *Archives of Civil and Mechanical Engineering* 11: 61–74.

Jędrysiak, J. 2013. Modelling of dynamic behaviour of microstructured thin functionally graded plates. *Thin Walled Structures* 71: 102–107.

Jędrysiak, J. & Radzikowska, A. 2012. Tolerance averaging of heat conduction in transversally graded laminates. *Meccanica* 47: 95–107.

Kaźmierczak, M. & Jędrysiak, J. 2013. A new combined asymptotic-tolerance model of vibrations of thin transversally graded plates. *Engineering Structures* 46: 322–331.

Marczak, J. & Jędrysiak, J. 2015. Tolerance modelling of vibrations of periodic three-layered plates with inert core. *Composite Structures* 134: 854–861.

Matysiak, S.J. & Perkowski, D.M. 2015. On heat conduction in periodically stratified composites with slant layering to boundaries. *Thermal Science* 19: 83–94.

Ostrowski, P. & Michalak, B. 2011. Non-stationary heat transfer in hollow cylinder with functionally graded material properties. *Journal of Theoretical and Applied Mechanics* 49(2): 385–397.

Ostrowski, P. & Michalak, B. 2015. The combined asymptotic-tolerance model of heat conduction in a skeletal micro-heterogeneous hollow cylinder. *Composite Structures* 134: 343–352.

Pazera, E. & Jędrysiak, J. 2015. Thermoelastic phenomena in functionally graded laminates. *Composite Structures* 134: 663–671.

Rabenda, M. & Michalak, B. 2015. Natural vibrations of prestressed thin functionally graded plates with dense system of ribs in two directions. *Composite Structures* 133: 1016–1023.

Suresh, S. & Mortensen, A. 1998. *Fundamentals of functionally graded materials*. Cambridge: The University Press.

Wierzbicki, E., Woźniak, C. & Woźniak, M. 1997. Thermal stresses in elastodynamic of composite materials. *International Journal of Engineering Science* 35: 187–196.

Woźniak, C., Wierzbicki, E. & Woźniak, M. 2002. A macroscopic model for the heat propagation in the microperiodic composite solids. *Journal of Thermal Stresses* 25: 283–293.

Woźniak, M., Wierzbicki, E. & Woźniak, C. 2002. A macroscopic model of the diffusion and heat transfer processes in a periodically micro-stratified solid layer. *Acta Mechanica* 157: 175–185.

Woźniak, C., Michalak, B. & Jędrysiak, J. (eds). 2008. Thermomechanics of microheterogeneous solids and structures. Tolerance averaging approach. Łódź: Publishing House of Łódź University of Technology.

Woźniak, C. et al. (eds). 2010. *Mathematical modelling and analysis in continuum mechanics of microstructured media.* Gliwice: Publishing House of Silesian University of Technology.

Advances in Mechanics: Theoretical, Computational and Interdisciplinary Issues – Kleiber et al. (Eds)
© 2016 Taylor & Francis Group, London, ISBN 978-1-138-02906-4

Inverted joined-wing multidisciplinary optimization

M. Kalinowski
Engineering Design Center, Institute of Aviation, Warsaw, Poland

ABSTRACT: The joined-wing system is an unconventional way to connect the lifting surfaces that is becoming more popular with designers to use in the prototypes of new aircrafts. However, due to numerous constraints during aircraft design a large number of iterations is necessary to obtain satisfactory results. The most apparent solution is to prepare an automatic algorithm that shall manage the whole design process taking into consideration specific behaviour of the joined-wing. The goal of the algorithm is a globally optimized structure with specific assumptions. Thus, a multi-field optimization process is proposed for the preliminary design of the joined-wing. Based on an automatic geometry generator, a FEM solver and the aerodynamic panel method a modular algorithm is developed. According to this methodology a computer program is coded to test the presented optimization algorithm and its performance. The article presents a way of multidisciplinary optimization of the inverted joined-wing that can be useful during its preliminary design and the results of a performed case study.

1 INTRODUCTION

The inverted joined wing configuration is an unconventional proposal for the future airplanes. It consists of two similar size lifting surfaces. The first one is attached to the forward section of the fuselage and acts as a classical wing. The second is attached to the aft end of the fuselage instead of a horizontal stabilizer (see Fig. 1). Usually, the forward lifting surface is located in front of the center of gravity, while the second is behind it. Moreover both surfaces are connected by wing tip surfaces. Therefore, the lifting surfaces create a closed circuit–box wing.

This configuration was proposed for the first time by Prandtl in 1924. It has many possible advantages like a low induced drag caused by the closed wing loop or a lower weight due to a higher stiffness of the closed wing. As a result it could have a higher lift to drag ratio in comparison to the classical configuration aircrafts with the same weight (Wolkovitch 1986). Thus, the associated range could be higher. To adequately avail all of the joined-wing advantages, the aircraft should be

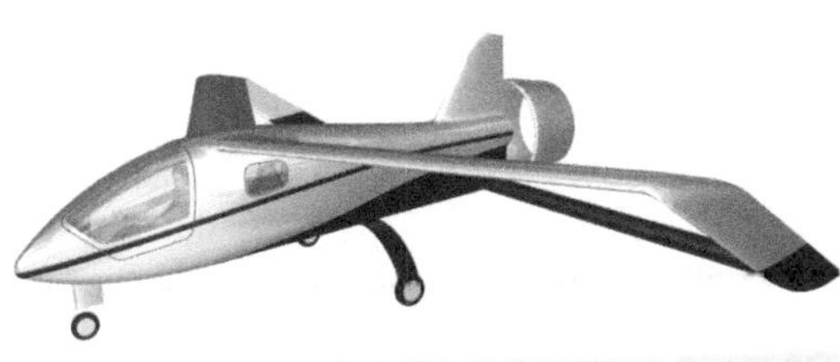

Figure 1. Reversed joined-wing configuration aircraft.

well optimized during the design process. This is not a simple task due to the strong aerodynamic interference of the lifting surfaces and the static indeterminacy of the structure. The joined-wing is not like the classical configuration a cantilever structure; rather a closed frame. Searching for the optimal solution is highly limited by the specific phenomena for this configuration like the global and local buckling of the wings-fuselage system. That kind of constraint is not as important for the classic configuration. For the joined-wing, however, it cannot be omitted if the minimum weight is a target of the design process. Aeroelasticity is another phenomenon highly important for the closed wing configuration that should be taken into consideration.

2 OPTIMIZATION

Numerous constraints during the design of a joined-wing aircraft enforce a large number of iterations to obtain satisfactory results for many design variables. Taking into account the mentioned problems, the apparent solution is to perform the optimization of this kind of an aircraft during the design process. The most efficient way is to prepare a multidisciplinary optimization of an entire aircraft with an automatic algorithm that shall manage the whole design process taking into consideration specific behaviour of the joined-wing. The goal of the algorithm is a globally optimized structure for specified assumptions. A multi-field optimization process is proposed for the preliminary design of the joined-wing. Two disciplines are considered:

aerodynamics (aerodynamic performance) and structural strength. The modular algorithm that consists of an automatic geometry generator, a FEM solver and the aerodynamic panel method is created. The whole process is optimized to decrease the huge computation cost to the minimum.

The optimization algorithm was developed for a small, electric, general aviation aircraft and was presented in detail beforehand (Galiński et al. 2013). Due to a very low efficiency in the task solution a thorough reconstruction of the algorithm was performed (Fig. 2). The general objective is the maximization of range of the aircraft for an assumed mission, as low range is a major disadvantage of electric aircrafts. The payload and battery capacity are fixed. A very small reserve of electric energy and high consumption during the climb and descent leads to the calculation of the total range as a sum of ranges during climb, cruise and descent. During all of the phases of flight the aircraft has a constant mass that is unlike all other engine types. Only the batteries discharge during the mission.

The global geometry and local structural parameters are selected as the design variables. The proposed optimization approach has been designed to allow the global optimization with a large number of geometric design variables. Generally optimization is divided into two separate optimization tasks: performance and structural. This separation is possible due to the fact that for a specified global geometry of the aircraft, maximum range is obtained for minimum mass. Thus, both optimizations can be performed using different methods, the most efficient for the corresponding disciplines.

The algorithm (Fig. 2) starts with the generation of a number of sets of global geometry variables using Latin hypercube method. Then, for thus obtained set the structural optimization is performed and maximum ranges for the current

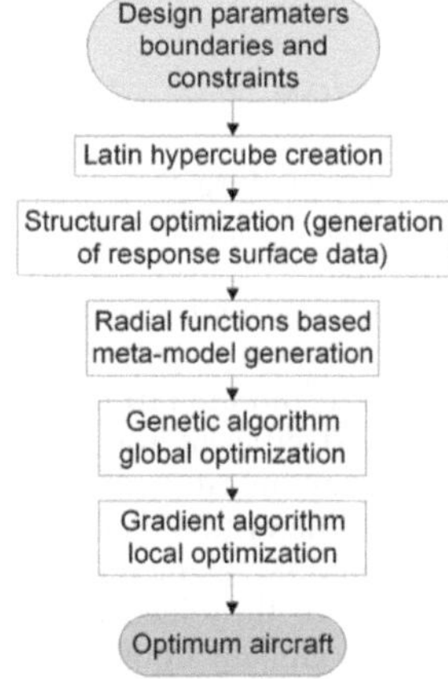

Figure 2. Multidisciplinary optimization algorithm.

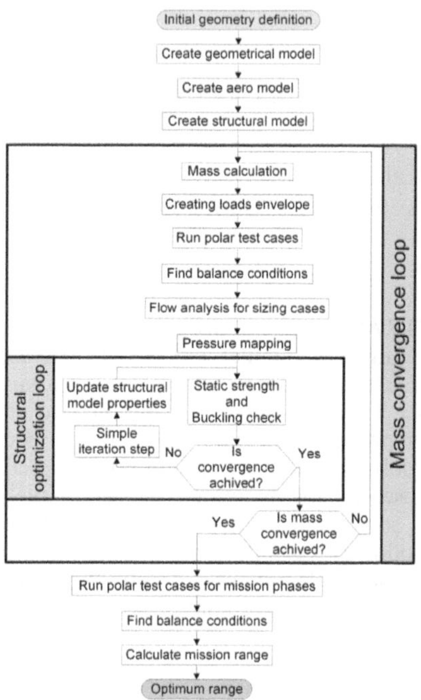

Figure 3. Structural optimization (generation of response surface data).

parameters are calculated. During the structural optimization (Fig. 3) only the structural variables are changed. These optimizations are performed with the use of a simple iteration method. After it is finished, the set of response surface points (the set of aircraft ranges) is collected. In the following stage the set is used for the creation of a response surface point meta-model. After that, the two-step optimization is performed. The first step driven by a genetic algorithm is an initial step that searches for a global optimum. The next step is a final gradient method based on optimization that allows obtaining better estimation of an optimum configuration generated from the previous step.

During structural optimization the strength is checked for a few sizing load cases obtained from the load envelope. Only the structural parameters vary in this loop. For the aerodynamic optimization the objective function (the range) is optimized by changing the geometrical parameters only.

3 GEOMETRY GENERATION

The geometry is generated automatically using a predefined parametric model. This model has a fixed topology (the wings connected to the fuselage, the vertical stabilizer, the internal structure) but can change in some ways defined by the parameters.

It consists of B-spline surfaces that are created by defined cross sections of the fuselage, wings, side wing and vertical stabilizer. The parameters that define geometry of these sections are global geometry design variables, e.g.: the points defining airfoils, wing span, aircraft length, angles of skew, dihedral, twist, chords and the position of lifting surfaces.

4 AERODYNAMIC CALCULATIONS

Aerodynamic analyses are the most common in the proposed algorithm. They are used in the generation of the load data for the strength calculation and for the estimation of optimum range of the aircraft. The panel method program PanAir (Magnus & Epton 1980) is a solver used for the flow calculations. As only symmetric loads are considered in the optimization a panel model is created on the geometry of a half of the aircraft. Based on this model a set of blows at a certain range of velocities and angles of attack is run to obtain the aerodynamic characteristics of the whole aircraft. Although, the computation cost of the panel method solver is much lower than Navier-Stokes codes, the friction drag is omitted. The reason of this is solving a potential flow by PanAir. Hence an additional correction of drag results need to be done. This is estimated using simple engineering formulas (Bertin & Cummings 2009).

Final characteristics are validated by a wind tunnel tests performed by Lis et. al in 2014. Based on the characteristics, the balance calculations are performed and balanced characteristics are generated. Finally this enables a load envelope creation for specified flight parameters and generation of the sizing pressure distributions for the characteristic load envelope points.

5 STRUCTURE CALCULATIONS

For the purpose of strength calculations another model is created. This finite element model is more complex than the aerodynamic one as it

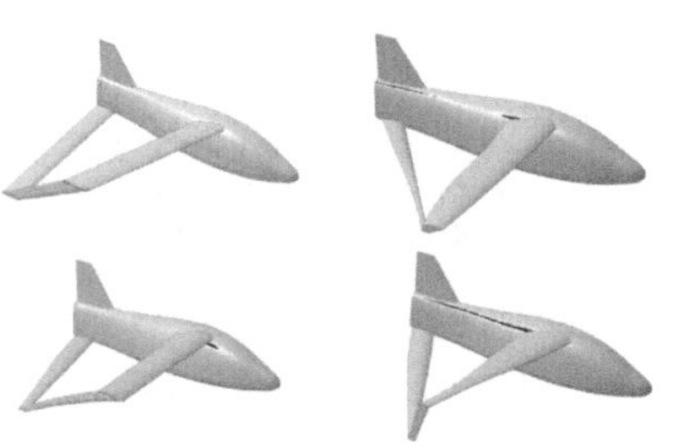

Figure 4. Different configurations of joined-wing generated by parametric model.

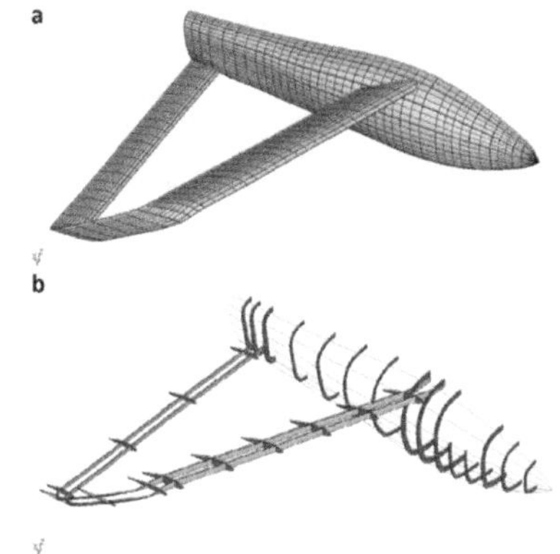

Figure 5. Structural model: a—external, b—internal structure.

consists of internal and external structure representation (Fig. 5). It is also generated on the half of the aircraft. Notably, the vertical stabilizer was omitted as it is sized by non-symmetrical loads that are not considered. In general, the aerodynamic and structural meshes do not match each other. This leads to aerodynamic pressures (which are a loads to strength checks) mapping onto the structural model. This is done using radial basis functions.

The integrity of the aircraft is checked against two failure modes: static strength and local buckling analyses. Based on the reserve factor level the optimization of the structure is performed using a simple iteration method (see the internal loop in Fig. 3). The mass changing during optimization proves problematic. As the aircraft loads depend on the mass of the aircraft, there is one more loop with loads recalculation that is needed to get a convergence of the mass. After the strength calculations an optimized structure for the specified global geometry design variables is estimated.

6 CASE STUDY

To check the performance of the algorithm, a case study was proposed. The aim was to maximize the range of the aircraft carrying 12.3 kg of payload and batteries of 9.6 Ah capacity. Optimization was performed for 18 global geometry (parameters of wings geometry only) and 232 structural variables (whole aircraft structure). The parameters were limited by minimum and maximum boundaries. Additional assumption was a structure made of aluminum alloy 2024 with corresponding strength allowable and moduli.

Table 1. Comparison of obtained optimum ranges (in meters).

No	Population	Generations number	Model	Genetic opt. range (initial)	Gradient opt. range (final)
1	60	100	Meta-model	99779	102422
			Exact	99632	100030
2	100	200	Meta-model	101093	102537
			Exact	101783	101817
3	200	200	Meta-model	101286	102212
			Exact	101066	101022

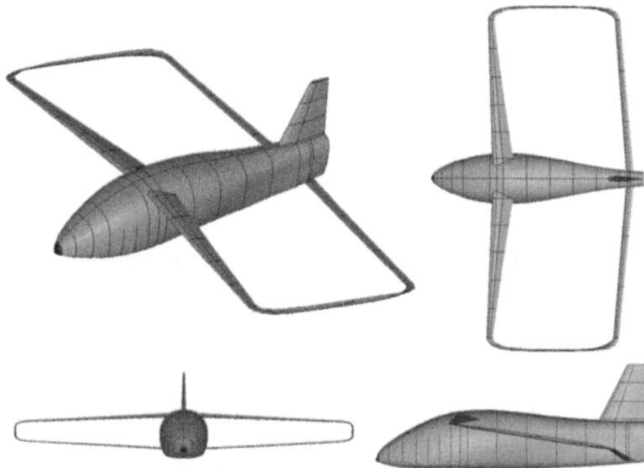

Figure 6. Optimum configuration geometry (set No. 2).

A Latin hypercube of 500 sets of global design variables was generated and based on this a meta-model was crated. To validate it 50 randomly generated seeds were used. A determination coefficient calculated for this meta-model was $R^2 = 0.61$. This is not the best result but a sufficient value for an algorithm testing purpose.

Three different optimization tasks were executed on the generated meta-model. These differed by population quantity and number of generations used in the initial genetic algorithm optimization. The purpose was to check if the meta-model has a local optimum and find out what are the sufficient parameters for future optimizations. Results are shown in Table 1. Final optimum ranges calculated using the meta-model are compared on results generated by the structural optimization algorithm (Fig. 3). The concurrence of results is very good (maximum error approx. 2%). Moreover, the optimum range for all three sets of optimization is very similar. Also the geometries of the optimum configurations are almost the same.

The optimization duration was relatively small thanks to the use of the meta-model (5 min for No. 1, 35 min for No. 3). In comparison, the generation of the meta-model takes about 310 hours and was performed on a 32 processors of 2.6 GHz clocks computer.

7 CONCLUSIONS

The automatic multidisciplinary optimization algorithm for the preliminary optimization of the joined-wing aircraft was proposed and successfully tested. The major advantage is the possibility of optimizing the whole aircraft at once without dividing it into parts and making simplifying assumptions. The algorithm considers two disciplines but can be easily expanded thanks to its modular form.

Very little influence of parameters of genetic optimization has been noticed. Optimization is very quick in comparison to the generation of the meta-model. However, the meta-model makes the algorithm foolproof and allows for an effective global optimization.

REFERENCES

Bertin J., R. Cummings. 2009. *Aerodynamics for engineers*. United States Air Force Academy.

Galinski C., J. Hajduk, M. Kalinowski, K. Seneńko. 2013. The Concept of the Joined Wing Scaled Demonstrator Programme. *Proceedings of the CEAS'2013 conference*, Linkoping, Sweden, 244–253.

Lis M., A. Dziubiński, C. Galiński, G. Krysztofiak, P. Ruchała, K. Surmacz. 2014. Predicted flight characteristics of the, Proceedings of 29th Congress of the International Council of the Aeronautical Sciences. *ICAS*, Petersburg, Russia.

Magnus A.E., M.A. Epton. 1980. PAN AIR—A Computer Program for Predicting Subsonic or Supersonic Linear Potential Flows About Arbitrary Configurations Using A Higher Order Panel Method, Vol. I. Theory Document (Version 1.0). *NASA Contractor Report 3251*.

Prandtl, L. 1924. Induced drag of multiplanes. *NACA TN 182*.

Wolkovitch, J. 1986. The joined wing—An overview. *Journal of Aircraft*, Vol. 23 (No. 3), 161–178.

Advances in Mechanics: Theoretical, Computational and Interdisciplinary Issues – Kleiber et al. (Eds)
© 2016 Taylor & Francis Group, London, ISBN 978-1-138-02906-4

Plane anisotropy parameters identification based on Barlat's model

P. Kałduński
Department of Technical Mechanics and Strength of Materials, Koszalin University of Technology, Koszalin, Poland

ABSTRACT: The paper presents the method to determine anisotropy parameters of flat sheet, according to Hayer's method, using anisotropic Barlat's model. Hayer's method is based on the principle of a constant material volume. A numerically tensile test of three flat specimens has been carried out. The first sample according to the sheet rolling direction, the second at 45 degrees and a third perpendicular to the rolling direction. Barlat's model parameters were changed individually. Then measurements vary the width of the measurement section as a result of the elongation of 20% has been performed, according to the standard. The paper also presents the results of a drawing process simulation, using a designation of anisotropy model parameters. In addition the differences in the amount of the resulting ears on the periphery and their location relative to the rolling direction of the sheet have been compared.

1 INTRODUCTION

In the modelling of the deep drawing process isotropic materials are usually applied. This assumption is true only in the case of shaping high quality metal sheet for deep drawing not showing anisotropy phenomena. However, in most cases, drawing sheet or deep drawing sheet is used, characterized by different properties relative to the direction of rolling so there is a need, to determine the properties of the planar anisotropy of the drawing sheet.

In the paper, the method of determining the plane anisotropy parameters, based on the adopted Barlat's model has been presented (Barlat et al. 1991, 1997). Parameters can be also determined by linearization of the material function (Barlat et al. 2005).

2 NUMERICAL RESULTS OF TENSILE TEST

In order to determine the Hayer's anisotropy parameters, the flat tensile specimen model has been built. The total length of the samples was 250 mm. Whereby the measuring section was exactly 50 mm. A specimen has been numerically stretched in accordance with the x axis, that is compatible with the sheet rolling direction in Figure 1. A second sample has been at 45 degrees to the rolling direction stretched Figure 2a and the last perpendicular to the rolling direction Figure 2b. The samples have been stretched to strain equal 0,2 of the measured section, which gives the final length 60 mm. Anisotropy value is determined by the Hayer's formula (Eq. 1). Depending on the direction of sampling, anisotropy can be determined for 0°, 45° i 90° relative to the rolling direction.

$$r = \frac{\ln \frac{b_0}{b}}{\ln \frac{b \cdot l}{b_0 \cdot l_0}} \qquad (1)$$

In the Ansys/Ls-Dyna program anisotropic properties have been as the Barlat's model implemented (Barlat et al. 1991). This model is based on the six anisotropy factors: a, b, c, f, g, h. The first three are

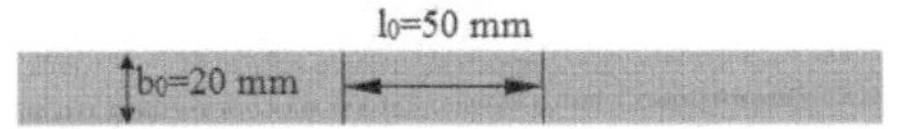

Figure 1. Sample according to the rolling direction.

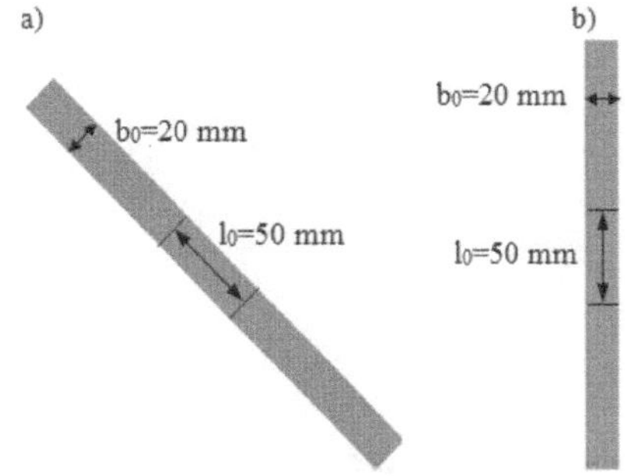

Figure 2. Samples: a) at 45 degrees, b) perpendicular to the rolling direction.

most important, related to the main directions of the sheet, the last three the tangents. The yield function Φ is defined as Equation 2 (Barlat et al. 1991):

$$\Phi = \left|S_{xx} - S_{yy}\right|^m + \left|S_{yy} - S_{zz}\right|^m + \left|S_{zz} - S_{xx}\right|^m = 2\bar{\sigma}^m, \tag{2}$$

where σ is the effective stress and $S_{i = xx,yy,zz}$ are the principal values of the symmetric matrix S (Eq. 3):

$$S = \begin{bmatrix} S_{xx} & S_{xy} & S_{xz} \\ S_{yx} & S_{yy} & S_{yz} \\ S_{zx} & S_{zy} & S_{zz} \end{bmatrix} = \begin{bmatrix} \frac{cC - bB}{3} & hH & gG \\ hH & \frac{aA - cC}{3} & fF \\ gG & fF & \frac{bB - aA}{3} \end{bmatrix}, \tag{3}$$

where:

$$S_{xx} = \left[c\left(\sigma_{xx} - \sigma_{yy}\right) - b\left(\sigma_{zz} - \sigma_{xx}\right)\right]/3,$$

$$S_{yy} = \left[a\left(\sigma_{yy} - \sigma_{zz}\right) - c\left(\sigma_{xx} - \sigma_{yy}\right)\right]/3,$$

$$S_{zz} = \left[b\left(\sigma_{zz} - \sigma_{xx}\right) - a\left(\sigma_{yy} - \sigma_{zz}\right)\right]/3,$$

$$S_{yz} = f\sigma_{yz},$$

$$S_{xz} = g\sigma_{xz},$$

$$S_{xy} = h\sigma_{xy}.$$

The material constants **a**, **b**, **c**, **f**, **g** and **h** represents anisotropic properties. When all equals 1 the material is isotropic. The yield strength of the material is (4):

$$\sigma_y = k\left(\varepsilon^p + \varepsilon_0\right)^n, \tag{4}$$

where ε_0 is the strain corresponding to the initial yield stress and ε^p is the plastic strain. The first stage of the research is to determine the impact of changes individual factors on the anisotropy according to Hayer. The parameters in the Barlat's model have been individually changed, whereas the value 1 specifies isotropic material. The results of numerical calculations are presented in Table 1. The last line in Table 1 shows the normal anisotropy value (5), which in the case of drawing process, should be as large as possible.

$$\bar{r} = \frac{1}{4}\left(r_0 + 2r_{45} + r_{90}\right) \tag{5}$$

Moreover, the change of the single Barlat's model parameters on the Hayer's anisotropy has been tested. Figure 3 shows the influence of the

Table 1. Numerical results tensile test.

a	1	0.5	1	1	1	1	1	1.5	1	1	1	1	1
b	1	1	0.5	1	1	1	1	1	1.5	1	1	1	1
c	1	1	1	0.5	1	1	1	1	1	1.5	1	1	1
f	1	1	1	1	0.5	1	1	1	1	1	1.5	1	1
g	1	1	1	1	1	0.5	1	1	1	1	1	1.5	1
h	1	1	1	1	1	1	0.5	1	1	1	1	1	1.5
r_0	0.994	2.028	1.398	0.169	0.994	0.995	0.997	0.547	0.716	2,125	0.996	0.996	0.993
r_{45}	0.992	2.287	2.279	0.989	0.992	0.99	−0.12	0.488	0.488	0.989	0.992	0.992	2.723
r_{90}	0.992	1.398	2.033	0.169	0.993	0.995	0.993	0.715	0.546	2.101	0.993	0.99	0.988
r	0.993	2.000	1.997	0.579	0.993	0.993	0.438	0.560	0.560	1.551	0.993	0.993	1.857

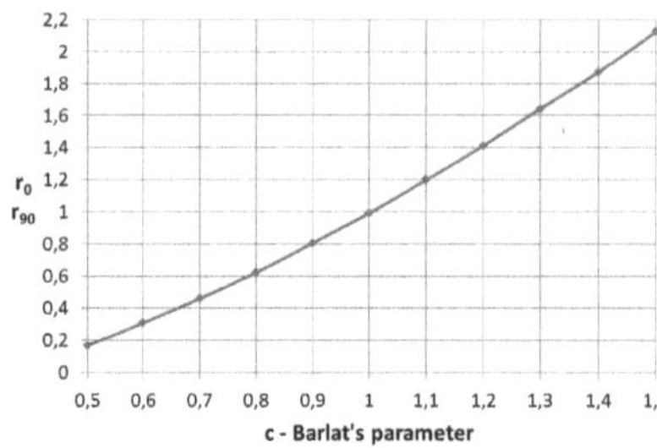

Figure 3. Influence of the c—Barlat's parameter on the Hayer.

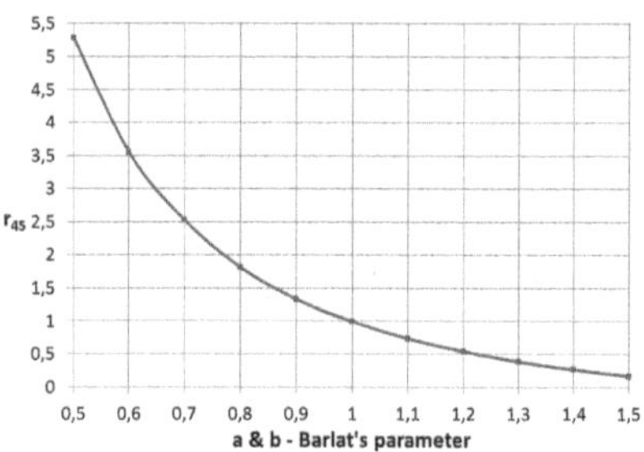

Figure 4. Influence of the a & b—Barlat's parameter on the Hayer.

c—Barlat's parameter, on the Hayer anisotropy at 0° and 90° to the sheet rolling direction presented. The **c**—Barlat's parameter has an affects only at $\mathbf{r_0}$ and $\mathbf{r_{90}}$, because in symmetric matrix S: c $(\sigma_{xx} - \sigma_{yy})$. This influence is identical at the rolling direction and at the perpendicular direction.

Figure 4 shows the symmetrical change influence of the **a** and **b** Barlat's parameters on the Hayers anisotropy at 45° to the rolling direction. Symmetrical change has affect only at $\mathbf{r_{45}}$ because in symmetric matrix S: a $(\sigma_{yy} - \sigma_{zz})$ and b $(\sigma_{zz} - \sigma_{xx})$. In both cases the Barlat's parameters were changed between 0.5 and 1.5. This range is sufficient because the minimum value of the anisotropy in Figure 4 is equal 0.25 and maximum is 5.25.

3 NUMERICAL RESULTS OF DEEP DRAWING PROCESS

Figure 5 shows in cross-sectional a discrete model of the drawing process.

The following conditions for the sheet metal forming were admitted:

$D_0 = 70$—initial diameter of a disk [mm],
$g_0 = 2$—disk thickness [mm],
$d_d = 40$—die-block hole diameter [mm],
$d_p = 35$—punch diameter [mm],
$r_p = 5$—punch edge rounding diameter [mm],
$r_d = 18$—die-block edge rounding diameter [mm].

The process was considered isothermal and quasi-static. Thus the strain rate and temperature affect neither stress nor strain values. The tools used in the process of sheet metal forming, i.e. a die block and a punch were assumed non-deformable bodies. Coulomb friction model has been adopted according to the Equation 6:

$$\mu_c = FD + (FS - FD)e^{-DC|v_{rel}|}, \tag{6}$$

where: FD—dynamic friction coefficient, FS—static friction coefficient, DC—exponential decay coefficient, e—Euler's number, v_{rel}—relative velocity of the surfaces in contact.

The following coefficients of friction in the contact zone between the punch and the sheet has been adopted:

$\mu_s = 0{,}2$—static friction coefficient
$\mu_d = 0{,}1$—dynamic friction coefficient.

The following coefficients of friction in the contact zone between the die block and the sheet has been adopted:

$\mu_s = 0{,}1$—static friction coefficient
$\mu_d = 0{,}01$—dynamic friction coefficient.

The DC exponential decay coefficient equal 10. The solid line on the graph (Fig. 6) shows dependence of the friction coefficient between the sheet and the punch. While the dotted line shows the dependence of the friction coefficient between the sheet and the die block.

The drawing process analysis has been performed using the Barlat's model. The problem was approached using the central-difference method, also called as the explicit-integration method. It includes a larger group of methods for the direct integration of dynamic equations of motion. The central-difference method does not require reversing the stiffness matrix, which its great advantage, especially at diagonal mass and attenuation matrixes.

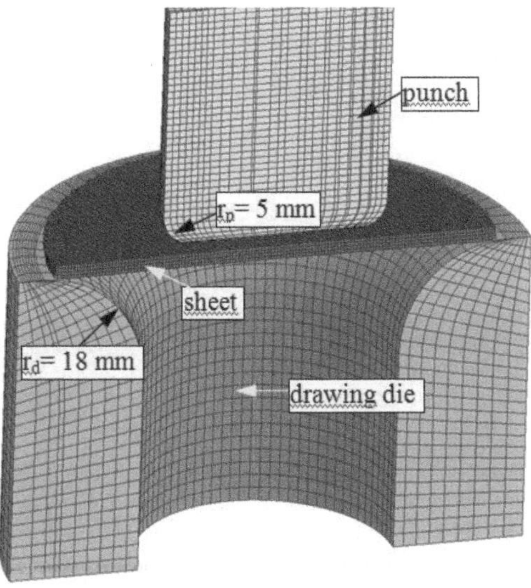

Figure 5. Discrete model of the drawing process.

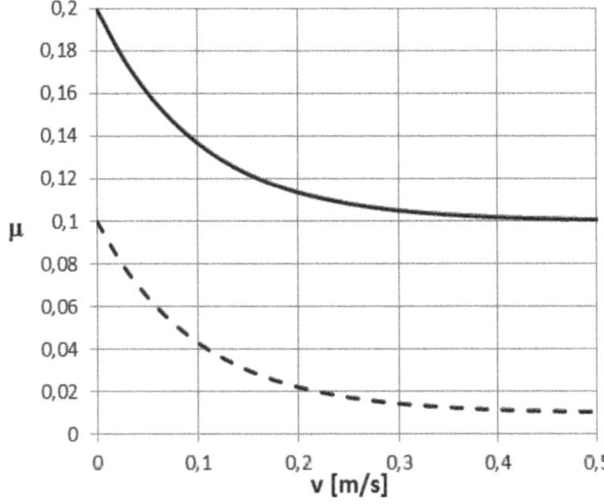

Figure 6. Friction model.

However, the fundamental disadvantage is a lack of the unconditional algorithm stability, which requires the selection of the step length after time Δt in such a way so that it was smaller than the critical time Δt_{kr} depending on properties of the whole system. The Ansys/Ls-Dyna program has an option of automatic length selection for each step, which makes it more users friendly.

The result for an isotropic material shown in Figure 7 where all the parameters in the model have a value equal to 1. The maximum value of von Mises stress is about 538 MPa. Figures 8 and 9a and 9b shows the drawpiece with adopted anisotropy parameter a = 0.5 in Barlat's model (bolded column in the Table 1). The lowest value of the anisotropy has been for the r_{90} = 1.398 obtained (Fig. 9b). Therefore, in this direction the largest recess has been achieved. The highest value obtained for r_{45} = 2.287 has given rise to, so-called, ear in this direction. In this case maximum value of von Mises stress is about 836 MPa. This value is located on the drawpiece edge between the "ears" (Fig. 8).

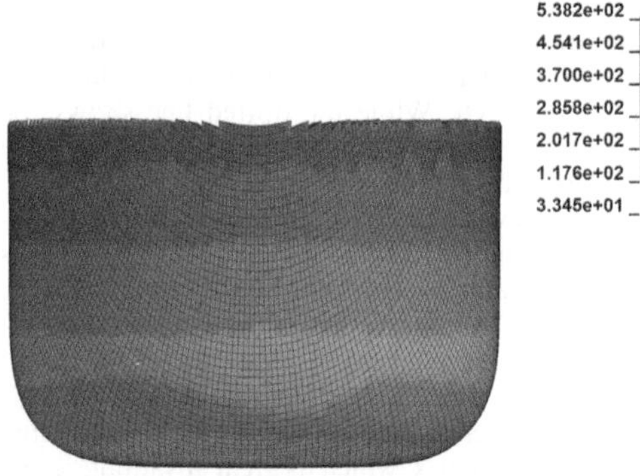

Figure 7. Drawpiece obtained with isotropic model.

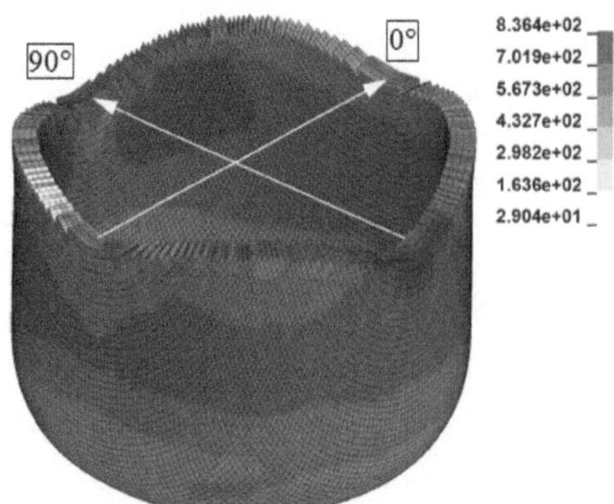

Figure 8. Drawpiece obtained with anisotropic model.

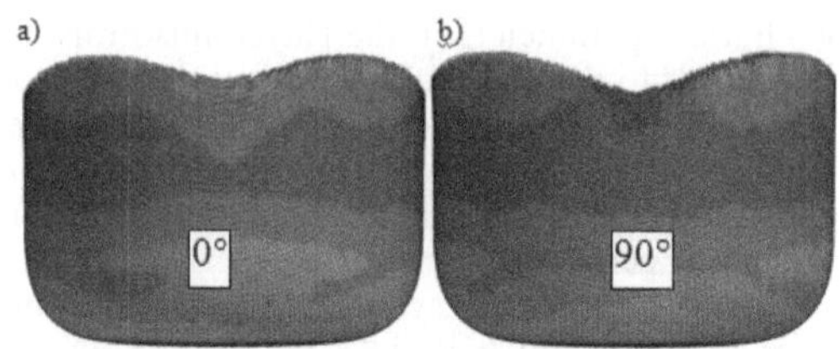

Figure 9. Drawpiece obtained with anisotropic model: a) rolling direction view, b) perpendicular direction view.

4 CONCLUSIONS

The paper presents a numerical method for determining the standard planar anisotropy parameters according to Hayer, using the numerical anisotropic model. Barlat's model parameters are correlated so it's impossible to provide a simple functional dependence between Hayer's values. To calculate the Hayer's anisotropic parameters, a numerical tensile tests have been conducted. After this calculation the basic anisotropic parameters in the plane of the sheet have been known. It also have been the deep drawing analysis with anisotropic Barlat's model presented. The plane anisotropy has a large influence on the drawpiece shape and the best option is when it is minimum. In the following research is planned to conduct the inverse problem. That is the calculation of six Barlat's model parameters: **a**, **b**, **c**, **f**, **g** and **h** based on the specified anisotropic parameters r_0, r_{45} and r_{90}. This will allow, based on the results of the experiment, to develop an adequate model to simulate the anisotropic phenomena occurring in the deep drawing process. So the Hayer's anisotropic parameters is much easier to determine from experiment than Barlat's parameters.

REFERENCES

Barlat, F., Aretz, H., Yoon, J.W., Karabin, M.E., Brem, J.C., Dick, R.E. 2005. Linear transformation-based anisotropic yield functions, *International Journal of Plasticity*, Vol. 21: 1009–1039.

Barlat, F., Becker, R.C., Hayashida, Y., Maeda, Y., Yanagawa, M., Chung, K., Brem, J.C., Lege, D.J., Matsui, K., Murtha, S.J., Hattori, S. 1997. Yielding description for solution strengthened aluminium alloys, *International Journal of Plasticity*, Vol. 13, No. 4: 385–401.

Barlat, F., Lege, D.J., Brem, J. 1991. A six-component yield function for anisotropic materials, *International Journal of Plasticity*, Vol. 7: 693–712.

Advances in Mechanics: Theoretical, Computational and Interdisciplinary Issues – Kleiber et al. (Eds)
© 2016 Taylor & Francis Group, London, ISBN 978-1-138-02906-4

Reliability of some axisymmetric shell structure by the response function method and the generalized stochastic perturbation technique

M. Kamiński & B. Pokusiński
Department of Structural Mechanics, Faculty of Civil Engineering, Architecture and Environmental Engineering, Łódź University of Technology, Łódź, Poland

ABSTRACT: The main idea of the paper is to demonstrate an application of numerical analysis based on the generalized stochastic perturbation method and the response function technique to determine the reliability indices for different performance functions concerning eigenfrequencies, stresses or vertical deflections for a large scale shell structure. Computational analysis has been carried out for the example of the steel skeletal structure supporting axisymmetric large span shell exhibiting Gaussian uncertainty in the wall thickness of all structural members. The resulting values of the reliability indices can be further employed in reliability assessment by comparing the minimal one with its limit given in Eurocode and depends upon reliability class, reference period and limit state type. Further extension of this model towards the Second Order Reliability Method application would be accompanied with the stochastic forced vibration analysis.

1 INTRODUCTION

Computer analysis of civil engineering structures with random parameters has a remarkably increasing influence on structural design process, optimization and reliability modeling, at least because of a variety of uncertainty sources. One of the most important parameters in shell structures is their wall thickness, whose uncertainty may arise during the manufacturing processes or as the effect of their non-uniform corrosion. The solution of such a problem is accompanied by the reliability index verification according to the rules given by European standards. There is a lot of methods leading to a determination of such indices, we propose the generalized stochastic perturbation-based method originating from the second order approach (Kleiber & Hien 1992). The basic features of the generalized version used here are: relatively short computational time, no need for massive computers and relatively easy implementation into different FEM computer systems (Kamiński 2007, Kamiński 2013). Computational implementation is carried out for static problems using the FEM system ROBOT and polynomial response functions approximated in the computer algebra system MAPLE from the series of original problem solutions obtained for the wall thicknesses varying about its mean value. Such a form of response enables determination of all its partial derivatives used further to obtain the basic probabilistic characteristics and the reliability indices.

2 THE GENERALIZED STOCHASTIC PERTURBATION TECHNIQUE

The basic idea of the stochastic perturbation approach is an expansion of all random functions of a given problem via Taylor series of the required order about their expectations. It is done using the perturbation parameter $\varepsilon = 1$. In case of some real function $f(b)$ of the stationary random variable b, the following expression is employed:

$$f(b) = f^0(b^0) + \sum_{n=1}^{\infty} \frac{\varepsilon^n}{n!} \frac{\partial^n f(b)}{\partial b^n}\bigg|_{b=b^0} (\Delta b)^n \quad (1)$$

Having the input random variable with symmetric probability density function, one can show that the first two probabilistic moments of the structural response function yield in the second order perturbation approach are represented as:

$$E(f(b);b) \approx f(b^0) + \frac{1}{2}\varepsilon^2 \frac{\partial^2 f(b)}{\partial b^2}\bigg|_{b=b^0} \mu_2(b^0) \quad (2)$$

and:

$$Var\left(f(b);b\right) \approx \varepsilon^2 \mu_2(b^0) \left(\frac{\partial f(b)}{\partial b}\right)^2 \bigg|_{b=b^0} \quad (3)$$

correspondingly, where $\mu_2(b)$ denotes second central probabilistic moment of the quantity b.

3 RESPONSE FUNCTION METHOD

As shown above, one of the crucial issues during application of the generalized perturbation based approach is numerical determination of partial derivatives of the structural response function of up to the nth order with respect to the randomized parameter. In order to complete this task it is necessary to determine this function by a multiple solution of the boundary value problem about the mean of our random parameter: $b = [b^0 - \Delta b, b^0 + \Delta b]$. The unknown response function is approximated by the following polynomial form:

$$f(b) = D_1 \cdot b^{n-1} + D_2 \cdot b^{n-2} + \cdots + D_n \cdot b^0 \quad (4)$$

and the coefficients D_i for $i = 1, \ldots, n$ are computed using the Least Square Method. It makes it possible to calculate up to nth order derivatives of the random function $f(b)$ with respect to the Gaussian variable b at its mean value as:

$$\frac{\partial^k f(b)}{\partial b^k} = \prod_{i=1}^{k}(n-i) \cdot D_1 \cdot b^{n-k} + \prod_{i=2}^{k}(n-i) \cdot D_2 \cdot b^{n-(k+1)} + \cdots + D_{n-k} \quad (5)$$

4 FINITE ELEMENT ANALYSIS

Numerical tests have been performed on the example of the BGŻ Arena steel roof structure presented in Figure 1. Two independent computational models have been created—the first one (Model I) includes the very detailed mesh of the entire shell, while Model II is based upon surrogate spatial discretization: one half of the structure with bonds in plane of symmetry foreclosing antisymmetric deformation. Therefore, the total number of the 3D frame finite elements is reduced from 5785 to 2923 with parallel reduction of the nodal points—from 2120 to 1121, correspondingly (6 degrees of freedom at each node).

All structural members have been assigned the properties of the stainless steel S235. These chords have been made from rectangular and the rest from circular hollow sections (Fig. 1).

The supports along the perimeter have been defined as completely rigid in Model I, while Model II includes the additional symmetry boundary conditions avoiding any antisymmetric deformation. The rigid connections have been used in modeling of the trusses and predefined load cases have been connected with wind action, permanent and snow loads.

Modal analysis has been performed in Model I by a verification of the first eigenmode only (Fig. 2), while global deformation of this structure in the Serviceability Limit State (SLS, Fig. 3) and the efficiency ratio of its members in the Ultimate Limit State (ULS, Fig. 4) have been computed both in Model II.

The fundamental purpose of the computational FEM analysis of this structure is to determine its fundamental eigenfrequency, the extreme reduced

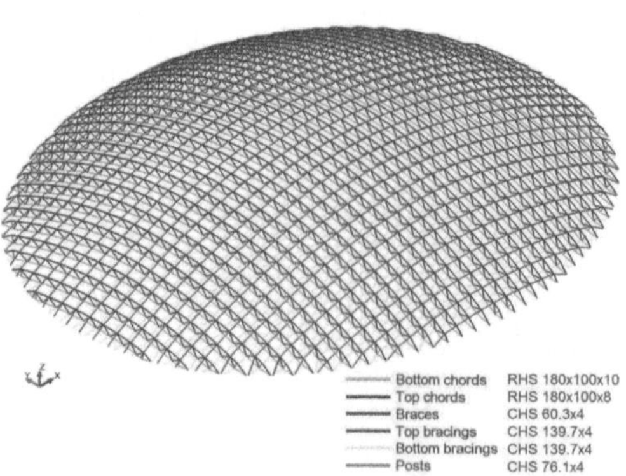

Figure 1. BGŻ Arena steel roof structure.

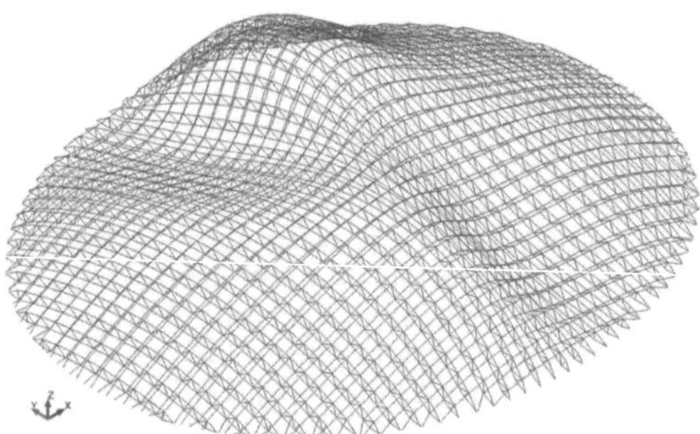

Figure 2. The first eigenmode of the roof shell.

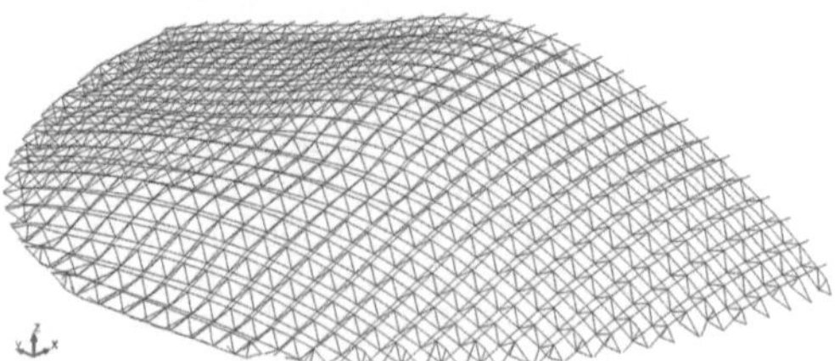

Figure 3. Deformation of the structure.

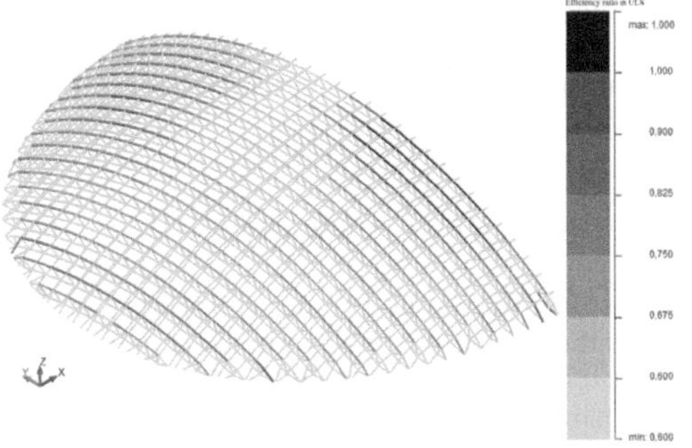

Figure 4. Efficiency ratio for the principal structural elements.

stresses and the vertical deflections in characteristic and frequent external load combinations. The first extreme has been determined with the use of Model I, whereas the rest result from the Model II.

5 COMPUTATIONAL RELIABILITY ANALYSIS

The linear response functions (see Fig. 5) of the reduced stresses, vertical deflections and the first eigenfrequency in addition to the randomized element thickness have been adopted. It is done after the initial determination of the correlation coefficients of the approximating polynomial of various order to the set of trials results coming from the deterministic FEM analysis. They are approximately equal for the linear polynomials to 0.99957 for the reduced stresses to 0.99994 for the fundamental eigenfrequency.

The above mentioned form enables to obtain constant function of expected values (Fig. 6) and quadratic function of the variance: $Var(f) = C \cdot \alpha^2$ (Fig. 7), on the assumption that wall thickness coefficient is the Gaussian input random variables defined by its expected value and coefficient of variation (α). In this case values of skewness and kurtosis equal 0.

The Cornell reliability indicators have been calculated using of the limit function g according to the following formula:

$$\beta = \frac{E[g]}{\sigma(g)} = \frac{E[R-E]}{\sigma(R-E)} = \frac{E[R]-E[E]}{\sqrt{Var(R-E)}} = \frac{E[R]-E[E]}{\sqrt{Var(R)+Var(E)}} \tag{6}$$

where R denotes structural resistance, E stands for the effect of all actions upon the given structure

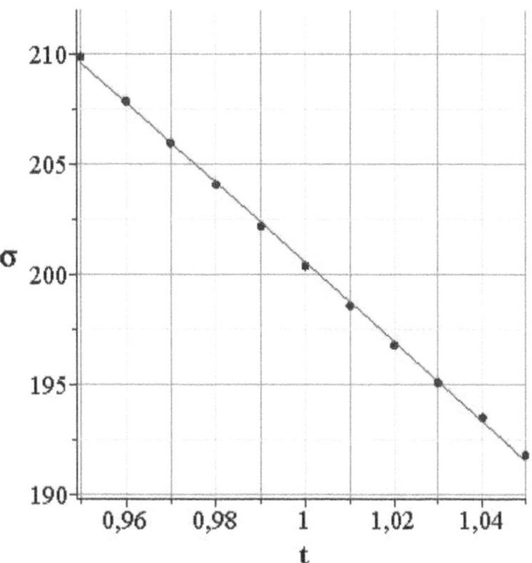

Figure 5. Linear variation of the maximum reduced stress response function with respect to the wall thickness coefficient.

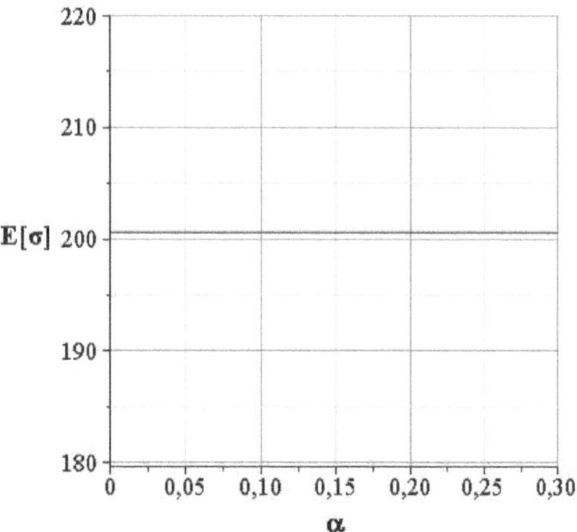

Figure 6. Constant functions of expected values on the example of maximum of reduced stresses.

and their first two probabilistic moments have been determined using the stochastic perturbation-based numerical approach.

All the reliability indices have been also computed with the following assumptions: (a) the vertical deflections are limited to 1/250 of the entire span, (b) admissible stress equals 235 MPa and (c) the difference between frequency of forced vibration and fundamental eigenfrequency should be bigger than 25%. Their variations upon the input coefficient of variation of wall thickness (α) have been presented in Figure 8a–8d.

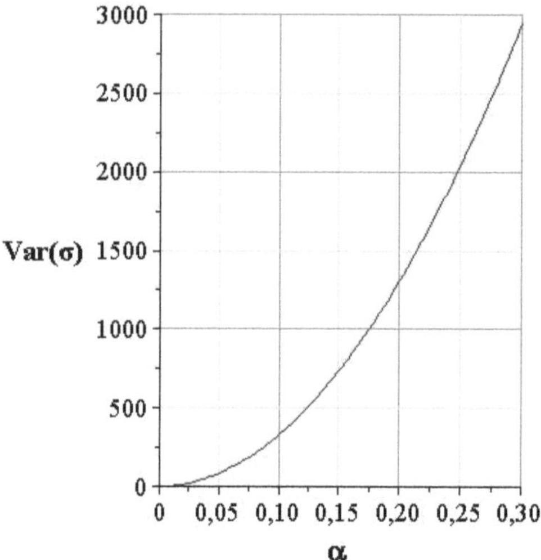

Figure 7. Quadratic functions of variance on the example of the extreme reduced stresses.

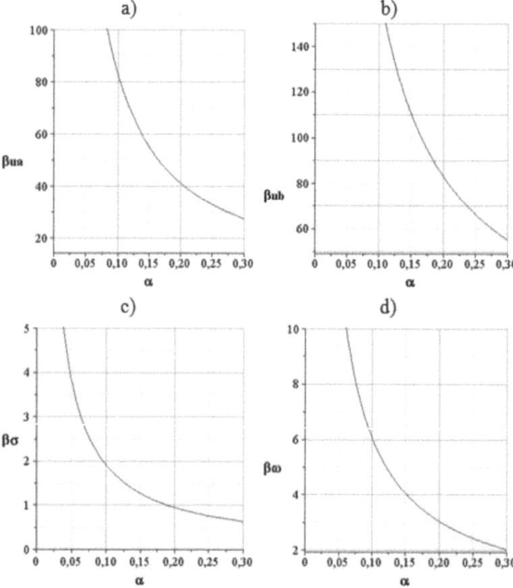

Figure 8. The reliability index of a) maximum of vertical deflections (characteristic combinations) b) maximum of vertical deflections (frequent combinations) c) maximum of reduced stresses d) fundamental frequency of the structure.

All of them show that reliability indicates nonlinearly (hyperbolically) decrease together with an increase of the uncertainty of wall thickness, thereby the probability of failure raise, which is compliant with the engineering intuition in this matter. The final values are significantly sensitive to the input coefficient α and even its small variation results in the remarkable change of these indices.

6 CONCLUDING REMARKS

The stochastic method of determination the reliability indices presented above appears to be an efficient tool to determinate the β values for different performance functions. It is confirmed directly by the computations that the reliability indices of the analyzed shell structure always decrease together with an additional increase of the uncertainty in the wall thickness.

It needs to be mentioned that the BGŻ Arena steel structure roof designed without any initial probabilistic safety analysis appears to be above any reliability limits while computed according to Eurocode 0 procedures.

The values of the index β for different limit states can be further used in reliability assessment by comparing the minimal one (connected in this case with stresses) with its limit dependent on reliability class, reference period and type of limit states, given in the European standards appendices (Eurocode 2002).

Further computational models in this area should consider higher order response functions polynomials and more extended stochastic perturbation approach to get higher order statistics, more accurate results and higher order reliability models. This study may be also extended to include uncertainty of some other, perhaps turning out as more decisive, parameters as well as stochastic fire reliability of such a large scale shell structure.

REFERENCES

Eurocode 2002. Basis of structural design. *European Committee for standardization*. Brussels.

Kamiński, M. 2007. Generalized perturbation-based stochastic finite element method in elastostatics. *Computers and Structures* 85(10): 586–594.

Kamiński, M. 2013. The Stochastic Perturbation Method for Computational Mechanics. Chichester. Wiley.

Kleiber, M., Hien T.D. 1992. The Stochastic Finite Element Method. Chichester: Wiley.

Advances in Mechanics: Theoretical, Computational and Interdisciplinary Issues – Kleiber et al. (Eds)
 ISBN 978-1-138-02906-4

A method of estimating the stability coefficient of a considerably degraded cooling tower

T. Kasprzak, P. Konderla, R. Kutyłowski & G. Waśniewski
Wrocław University of Technology, Wrocław, Poland

ABSTRACT: The aim of this research was to develop a method of estimating the reduction in the coefficient of stability of a cooling tower caused by wear. As a result of material deterioration Young's modulus of the concrete and the thickness of the cooling tower shell change, contributing to changes in the stiffness of the shell. This paper presents an algorithm for determining the function of structural stability coefficient sensitivity in the cooling tower shell to changes in shell stiffness, whereby a map of the sensitivity of the cooling tower shell structure is obtained. Having data on the actual degree of degradation of the shell one can estimate the current stability coefficient of the analyzed cooling tower.

1 INTRODUCTION

In their design form cooling towers (Fig. 1) meet the stability condition usually with a certain margin. Over years of service the materials of the cooling tower structure undergo degradation, whereby their physical parameters, especially those of the concrete, change. As a result of the surface degradation of the cooling tower shell the effective thickness of the latter decreases. Consequently, the cross-sectional stiffness of the cooling tower shell also decreases, lowering the cooling tower structure's stability coefficient (Altmeyer & Scharf 2012). Although routine tests supply information about the parameters of the material and the structure, the number of local parameter determinations is small because of the unit costs (considering the large cooling tower surface) of such tests. The aim of this research was to develop a methodology for determining the shell safety level with regard to stability loss on the basis of a small set of in situ (Fig. 2) and laboratory (Fig. 3) test results documenting the degree of cooling tower shell degradation.

Figure 1. Cooling tower.

A structural safety level with regard to stability is determined by solving the initial stability

Figure 2. In situ thickness examination.

Figure 3. Core specimens for laboratory tests.

problem for the structure. In particular, the critical load multiplier, called a stability coefficient (λ) (Ledwoń & Golczyk 1967), is calculated. For a cooling tower the coefficient should meet the condition $\lambda \geq 5$ (IASS 1977). The critical load level directly depends on the structure's stiffness and the latter is the function of two parameters: the Young's modulus of concrete and the thickness of the shell wall.

The proposed method of stability analysis was tested on a cooling tower investigated in situ by the authors (Kasprzak et al. 2012). The cooling tower had been in service for several decades and its shell showed considerable degradation.

One of the results of the analysis is a function of the sensitivity of the cooling tower to a change in the stability coefficient due to the degradation of the particular cooling tower areas. This function can used in a similar way as the areas of influence of, e.g., the internal forces. Having such a function and appropriate material test results in discrete points of the cooling tower shell one can determine the degree of safety of the cooling tower with regard to stability loss.

2 THEORETICAL FOUNDATIONS

It is assumed that in a newly built cooling tower the distribution of shell flexural stiffness is axially symmetric ($K_0(z)$). For homogenous concrete a change in shell stiffness along the height results from a change in shell thickness. Over time as the material undergoes degradation the cooling tower stiffness parameters change. Let $K(\mathbf{x})$ represent the current distribution of shell thickness under bending, where $\mathbf{x} = (z, \varphi)$. In order to identify function $K(\mathbf{x})$ tests were carried out in N points of the shell by measuring the thickness of the shell and the Young modulus of the concrete and on this basis the values of function $K(\mathbf{x}_i)$ in points $(\mathbf{x}_1, \mathbf{x}_2, \ldots, \mathbf{x}_N)$ were calculated. The reduction in stiffness in the i-the point is:

$$D_i = K_0(z_i) - K(\mathbf{x}_i). \tag{1}$$

By analogy to the division into finite element in FEM, shell surface V is divided into subareas V_i whereby changes in stiffness can be approximated in the form of the series:

$$D(\mathbf{x}) = \sum_i D_i N_i(\mathbf{x}) \tag{2}$$

where $N_i(\mathbf{x})$ performs the role of base (shape) functions while D_i performs the role of a nodal point of stiffness change approximation. For a proper selection of base functions function $D(\mathbf{x})$ is continuous on V.

Let λ_i be a stability coefficient of the cooling tower if changes in stiffness in V_i, in the form:

$$K_i(\mathbf{x}) = K_0(z) - \Delta K \cdot N_i(\mathbf{x}) \tag{3}$$

where ΔK is a fixed small quantity relative to $\min\{K_0(z)\}$, are superimposed on the ideal geometry of the shell.

The difference $\lambda_0 - \lambda_i$ is a measure of stability coefficient sensitivity to shell stiffness function in area V_i, with distribution $\Delta K \cdot N_i(\mathbf{x})$. The notion of sensitivity $\eta(\mathbf{x})$, understood as the intensity of stability coefficient change as a result of a change in shell stiffness in a unit surface area by a unit value, is introduced. This function can be presented in base $\{N_i(\mathbf{x})\}$:

$$\eta(\mathbf{x}) = \sum_i \eta_i \cdot N_i(\mathbf{x}). \tag{4}$$

Then the nodal values in Equation 4 are equal to:

$$\eta_i = \frac{\lambda_0 - \lambda_i}{\Delta K}\left[\int_V N_i(\mathbf{x})\mathrm{d}V\right]^{-1}. \tag{5}$$

For a set field of change in cooling tower shell stiffness, the current stability coefficient at arbitrary time instant T can be calculated from the Equation 6:

$$\lambda_T = \lambda_0 - \int_V D(\mathbf{x}) \cdot \eta(\mathbf{x})\mathrm{d}V. \tag{6}$$

3 CALCULATION EXAMPLE

A cooling tower the static diagram of which is shown in Figure 4a was analyzed. The shell

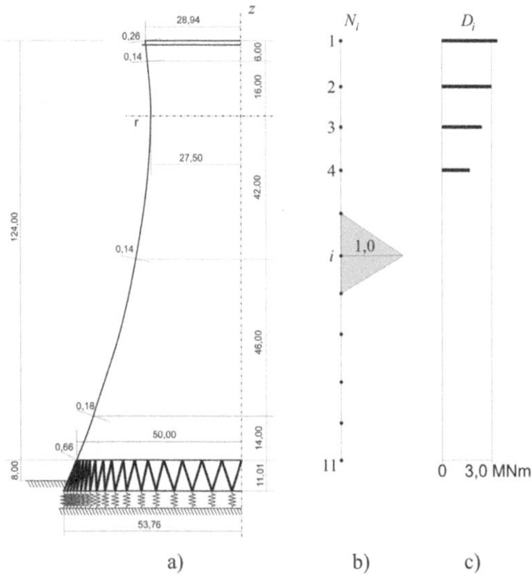

Figure 4. Static diagram of cooling tower a), base function b), distribution of shell stiffness change c).

Table 1. Values by which stiffness parameters decreased.

Level	ΔE [MPa]	Δh [m]
1	300	0.025
2	2000	0.020
3	1500	0.015
4	1000	0.010

parameters were tested on 11 latitudinal levels. A calculation test was carried out for cooling tower concrete with Young's modulus of 31.4 GPa. Table 1 shows how much the Young's modulus and shell thickness values decreased on four successive levels of the cooling tower (counting from its top down) as a result of material degradation. On the other levels the Young modulus and shell thickness values remained unchanged. Assuming a constant change in stiffness relative to the design stiffness, discrete distribution $D_i(z)$ of shell change as shown in Figure 4c, was obtained. The base functions were defined as bilinear along the generating line (Fig. 4b) and constant along the circumference, whereby a continuous distribution of function $D(\mathbf{x})$ within cooling tower shell area V is obtained.

By determining stability coefficient λ_0 for the cooling tower design parameters and then cooling tower stability coefficients $\lambda_1, \lambda_2, \ldots, \lambda_{11}$ the sensitivity function shown in Figure 5 was obtained for the next stiffness fluctuations according to stiffness distribution (Eqn. 2).

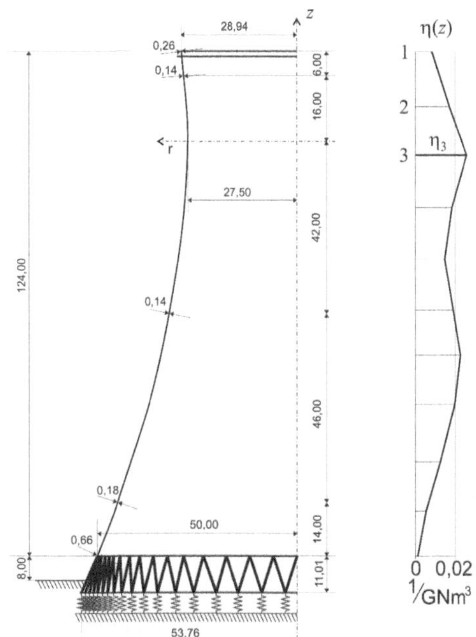

Figure 5. Sensitivity function.

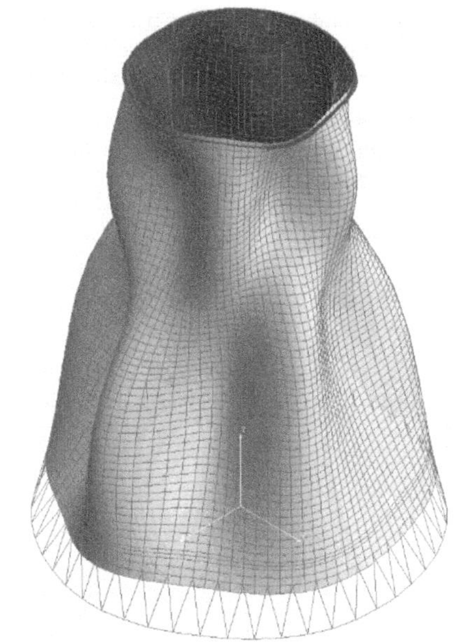

Figure 6. Buckling mode for coefficient $\lambda_T' = 5.897$.

In order to determine the influence of the measured stiffness parameters of the analyzed cooling tower on the lowering of the stability coefficient one should use expression (Eqn. 6). For the considered case, after transformations one gets:

$$\Delta\lambda = \int_V D(\mathbf{x}) \cdot \eta(\mathbf{x}) \mathrm{d}V = \sum_i \sum_j D_i \eta_j \int_V N_i(\mathbf{x}) \cdot N_j(\mathbf{x}) \mathrm{d}V = 0.352. \quad (7)$$

It appears from the above that due to shell material degradation the stability coefficient from the initial value $\lambda_0 = 6.260$ decreased to $\lambda_T = 5.908$.

The obtained result was verified by determining stability coefficient λ_T' on the basis of the measured changes. Performed FEM analysis gave in this case coefficient value: $\lambda_T' = 5.897 \approx \lambda_T$ (Fig. 6), which is very close to value obtained by presented algorithm. That indirectly validating the proposed methodology.

4 CONCLUSION

Having proper sets of laboratory-determined cooling tower shell Young's modulus values and in-situ cooling tower shell thickness test results and using the proposed algorithm one can determine the actual degree of safety of the cooling tower shell with regard to stability loss. The proposed methodology opens up possibilities for the synthetic analysis of the stability problem, in particular for:

1. determining the shell areas the degradation of which significantly contributes to the lowering of the stability coefficient,
2. estimating the stability coefficient as a random variable (treating the test results as random variables).

Considering the geometrical similarity between a large number of cooling towers built in the 1970s and 1980s, the sensitivity functions for the structures will be qualitatively very similar.

REFERENCES

Altmeyer, F. & Scharf, H. 2012. Safety analysis and rehabilitation of natural draft cooling tower at the power plant of Niederaussem, R. Harte & R. Meiswinkel (eds), *6th International Symposium on Cooling Towers.* Wuppertal: Departament of Civil Engineering, Bergische Universität.

IASS. 1977. *Recommendations for the design of hyperbolic or rather similarity shaped cooling towers.* Brussels: IASS.

Kasprzak, T., Konderla, P., Kutyłowski, R. & Waśniewski, G. 2012. Ocena skutków rozbudowy elektrowni na stan bezpieczeństwa istniejącej chłodni kominowej. *Przegląd Budowlany* 5: 36–39.

Ledwoń, J. & Golczyk, M. 1967. *Chłodnie kominowe i wentylatorowe.* Warszawa: Arkady.

Advances in Mechanics: Theoretical, Computational and Interdisciplinary Issues – Kleiber et al. (Eds)

Numerical analysis of contact between 3-D beams with deformable circular cross-section

O. Kawa & P. Litewka
Poznan University of Technology, Poznań, Poland

ABSTRACT: In the paper frictionless contact between three-dimensional elastic beams with deformations at the contact zone is analysed. It is assumed that the analysed beams undergo large displacements, the strains remain small and the beams cross-section are deformed. In order to include deformation of the cross-sections the classical analytical result from Hertzian theory of contact between two elastic cylinders is used. The physical law for contacting bodies is introduced so high precision contact is considered. The appropriate kinematic variables are defined and discretised using the finite element methodology.

1 INTRODUCTION

The main purpose of computational contact mechanics is to provide numerical tool to properly describe physical behaviour of bodies coming in contact, and especially the deformation and forces acting in the vicinity of the contact interface. There are several contributions which are related to beam-to-beam contact, e.g. Wriggers & Zavarise (1997), Zavarise & Wriggers (2000), Litewka (2010), Kawa & Litewka (2014).

In order to include the cross-section deformation the classical analytical result from Hertzian contact between two elastic cylinders representing the contacting beams has been proposed in (Kawa & Litewka 2014). In the paper the authors show a different way to include such cross-section deformations in the contact zone. In the new element the deformation is introduced by replacing the penalty parameter with the contact stiffness.

2 CONTACT FORMULATION

2.1 *Introduction*

We consider two beams with circular cross-sections coming into contact. In order to detect contact the penetration function is defined, which for two beams with circular cross-sections can be written as:

$$g_N = d_n - r_m - r_s \tag{1}$$

where d_n = is the minimum distance between the beams, r_m, r_s = are the radii of the of beams cross-sections.

The function of penetration can be used as the criterion of contact. The contact condition can be defined as:

$$g_N = d_n - r_m - r_s \leq 0 \tag{2}$$

In order to evaluate g_N two closest points are found lying on two curves representing the beam axes (m and s). After finding the position vectors $\mathbf{x}_{mn}$ and $\mathbf{x}_{sn}$ of the points C_{mn} and C_{sn} the minimum distance is calculated between the beams, d_n, so the penetration function is defined.

2.2 *Contact points*

Location of the points on the curves in the 3D space is defined by local curvilinear co-ordinates: ξ_m for first beam and ξ_s for the second one (Fig. 1). In the global Cartesian system each point on the curve can be expressed as:

$$\begin{aligned} \mathbf{x}_m &= \mathbf{X}_m + \mathbf{u}_m \\ \mathbf{x}_s &= \mathbf{X}_s + \mathbf{u}_s \end{aligned} \tag{3}$$

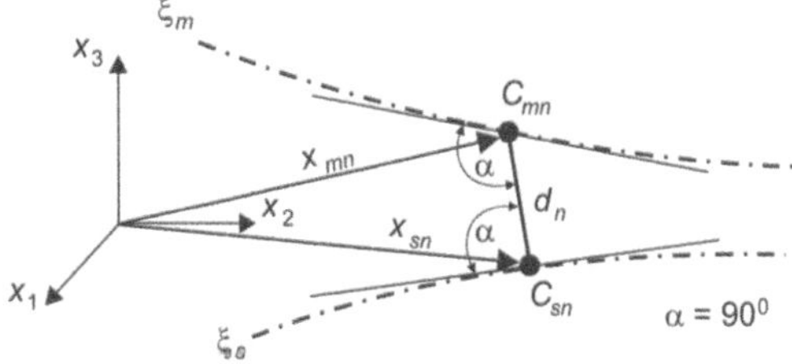

Figure 1. The closest points on two curves.

where $\mathbf{X}_m$, $\mathbf{X}_s$ = the position vectors for points at the initial configuration; $\mathbf{u}_m$, $\mathbf{u}_s$ = the displacement vectors.

For the closest points C_{mn} and C_{sn} the position vectors $\mathbf{x}_{mn}$ and $\mathbf{x}_{sn}$ must fulfil simultaneously the orthogonality conditions:

$$\begin{aligned}(\mathbf{x}_{mn} - \mathbf{x}_{sn}) \circ \mathbf{x}_{mn,m} = 0 \\ (\mathbf{x}_{mn} - \mathbf{x}_{sn}) \circ \mathbf{x}_{sn,s} = 0\end{aligned} \tag{4}$$

where:

$$\mathbf{x}_{mn,m} = \frac{\partial \mathbf{x}_{mn}}{\partial \xi_m}, \quad \mathbf{x}_{sn,s} = \frac{\partial \mathbf{x}_{sn}}{\partial \xi_s} \tag{5}$$

The solution, which defines the location of the points C_{mn} and C_{sn}, can be obtained using the Newton-Raphson method. Linearization leads to a set of two equations, which allow for calculation of co-ordinate increments $\Delta\xi_s$ and $\Delta\xi_m$. After finding the position vectors $\mathbf{x}_{mn}$ and $\mathbf{x}_{sn}$ of the points C_{mn} and C_{sn} one can calculate the minimum distance between the beams:

$$d_n = \|\mathbf{x}_{mn} - \mathbf{x}_{sn}\| \tag{6}$$

2.3 *Hertzian contact*

To include the deformations at the contact zone the classical analytical result from Hertzian contact between two elastic cylinders is used. One starts with the contact analysis between a rigid sphere and an elastic half-space. The case of two identical cylinders with perpendicular axes representing two contacting beams leads to the same result as for a pair of spheres with radii r_s and r_m (Fig. 2).

The normal force at the contact zone is defined by (Popov 2010):

$$F_N = \frac{4}{3} \cdot E \cdot R^{1/2} \cdot d^{3/2} \tag{7}$$

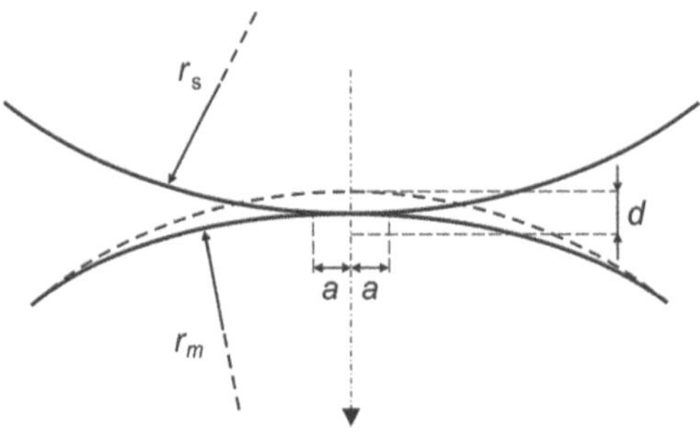

Figure 2. The contraction of radii of contacting spheres.

where R = the value of the effective radius; E = the mean Young's modulus; d = the change of r_m and r_s due to the deformation of cross-sections.

The Hertz equations are given for the circular contact region, so some approximation has to be done. In equation (7) the value of the effective radius R is used:

$$R = (r_s \cdot r_m)^{1/2} \tag{8}$$

For two elastic bodies the following expression for the mean Young's modulus E must be used:

$$\frac{1}{E} = \frac{1 - v_s^2}{E_s} + \frac{1 - v_m^2}{E_m} \tag{9}$$

so the value d present in (7) can be written as:

$$d = \sqrt[3]{\frac{9}{16}} \cdot \frac{F_N^{2/3}}{E^{2/3} \cdot R^{1/3}} \tag{10}$$

The normal gap used to define contact can be regarded as the cross-sections deformation, thus:

$$g_N = d \tag{11}$$

The normal force can be expressed as:

$$F_N = \frac{4}{3} \cdot E \cdot R^{1/2} \cdot g_N^{\ 3/2} \tag{12}$$

Contrary to this, in the standard penalty method, which is a way to regularize the Signorini contact conditions, the normal force is:

$$F_N = \varepsilon_N \cdot g_N \tag{13}$$

Therefore, in fact now the penalty parameter is replaced by:

$$\frac{4}{3} \cdot E \cdot R^{1/2} \cdot \frac{3}{2} \cdot g_N^{1/2} \tag{14}$$

The cross-section deformation is introduced by replacing the penalty parameter with the Hertzian contact stiffness. In the iterative solution procedure the radii change d in the current step can be evaluated using the normal force and the normal gap g_N from the current step, too.

2.4 *Weak form and its linearisation*

The solution of a frictional contact problem for two bodies involves finding a minimum of the potential energy functional Π. The contact contribution takes the form:

$$\min \Pi = \min(\Pi_1 + \Pi_2 + \Pi_c) \tag{15}$$

where Π_1, Π_2 = the potential energy values of first and second body; Π_c = the part of energy resulting from contact.

The problem of contact leads to a solution of a functional minimization with inequality constraints. To solve this problem the concept of an active set is used. This procedure involves choosing the contact pairs which fulfil the condition $g_N < 0$. The inequality constraints can be replaced by the equality constraints (11). The problem takes a form of a functional minimization with equality constraints.

The contact contribution to the corresponding weak form is:

$$\delta\Pi_c = F_N \delta g_N \tag{16}$$

Substituting (12) into (16) gives:

$$\delta\Pi_c = \frac{4}{3} \cdot E \cdot R^{1/2} \cdot g_N^{\ 3/2} \delta g_N \tag{17}$$

The related linearization required for the Newton solution scheme for the non-linear contact problem reads:

$$\Delta\delta\Pi_c = \left(\frac{4}{3} \cdot E \cdot R^{1/2} \cdot g_N^{\ 3/2}\right) \Delta g_N \delta g_N + \left(\frac{4}{3} \cdot E \cdot R^{1/2} \cdot g_N^{\ 3/2}\right) \Delta \delta g_N \tag{18}$$

The variation, linearization and second variation of the penetration function g_N are computed in the same way as in the analysis without the cross-section deformation. The finite element discretisation follows the same lines, too.

3 NUMERICAL EXAMPLES

3.1 *Introduction*

In this section two examples of beam-to-beam contact are presented. Each beam is discretized using ten identical co-rotational finite elements proposed by Crisfield (1990). The curves representing the beam axes are defined using Hermite's polynomials (Litewka 2010). A comparison of results from the new and earlier approach (Kawa & Litewka 2014) is presented.

3.2 *Example 1*

The contact between two clamped-clamped beams is considered. The initial configuration of beam axes is shown in Figure 3. The beams have circular cross-sections with radius $r = 0.1$, length 6.0 and are initially spaced at 0.001. They are made of a material with Young's modulus $E = 250 \cdot 10^5$ and Poisson's ratio $\nu = 0.3$. The imposed displacements $\Delta = 0.5$ are shown in Figure 3. They are applied simultaneously in 60 equal increments.

The values of penetration and final normal forces F_N are presented in Table 1. In the first column the values of penetration without the cross-section

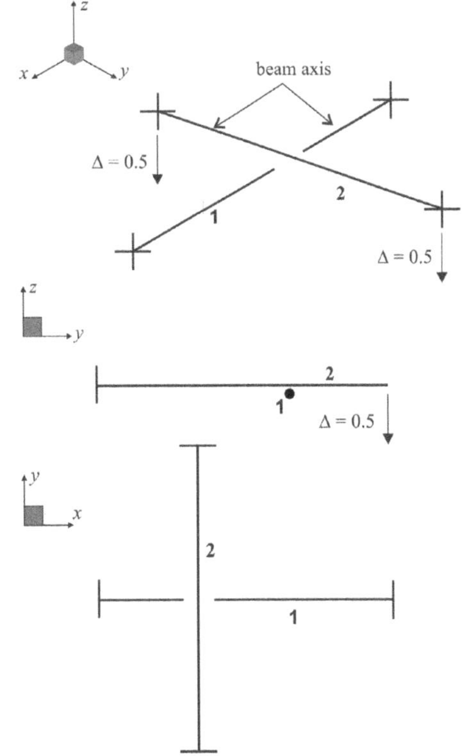

Figure 3. The initial configuration of beams axes in Example 1.

Table 1. Values of penetration in Example 1.

Increment	Without deformation	Earlier approach	New approach
10	0.0000930788	0.0000923594	0.0000067839
20	0.0002109870	0.0002092683	0.0000153834
30	0.0003739436	0.0003703912	0.0000272815
40	0.0006007104	0.0005939721	0.0000438612
50	0.0009081203	0.0008962469	0.0000663729
60	0.0013116440	0.0012919988	0.0000959776
F_N	1311	1291	1318

deformation are presented. In the second column the values of the penetration with cross-section deformations from the approach which has been proposed in (Kawa & Litewka 2014) are given. In both of these cases the penalty method is used. The penalty parameter used in the analysis is $\varepsilon_N = 1 \cdot 10^6$. In the last column the values of penetration from the new approach which represent the cross-section deformation are shown.

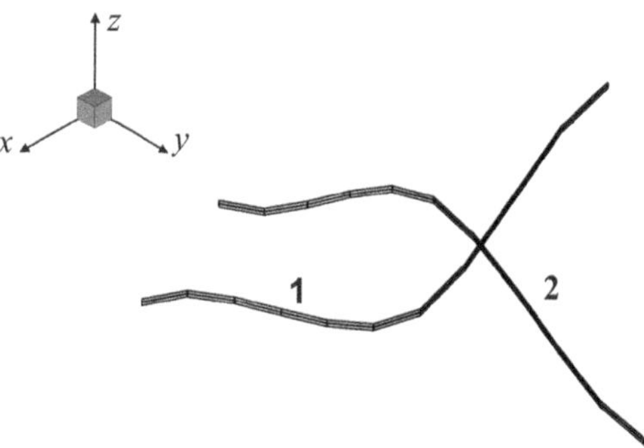

Figure 4. Deformed configuration of beams axes in Example 1.

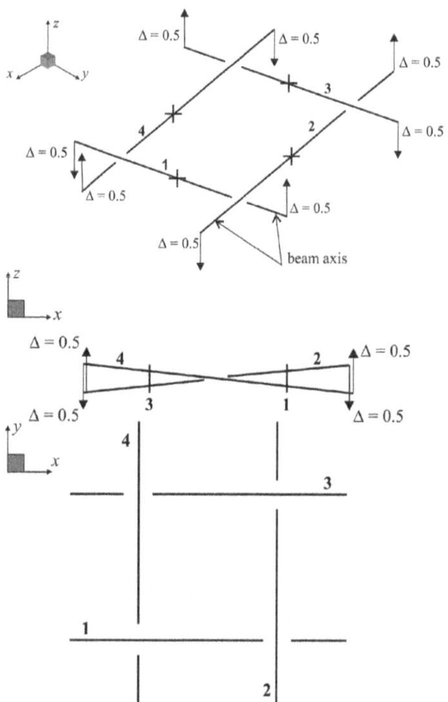

Figure 5. The initial configuration of beams axes in Example 2.

The deformed configuration of the axes of beams is presented in Figure 4.

3.3 *Example 2*

In this example contact between four beams forming a symmetric assembly is considered. The beam axes in the initial configuration are presented in Figure 5. The beams have circular cross-sections with radius $r = 0.1$, length 8.0 and are initially spaced at 0.001. They are made of the same material with $E = 250 \cdot 10^5$ and $\nu = 0.3$. Each of the beams has almost fully constrained centre points, except for the freedom of rotation about the axes lying in the plane perpendicular to the beams. The imposed displacements $\Delta = 0.5$ are applied in 30 equal increments at the free ends of the beams.

The values of penetration of beams for three different approaches are presented in Table 2. As previously the values in first and second column refer to the cases where the penalty method is used. The penalty parameter used in this analysis is $\varepsilon_N = 2 \cdot 10^6$. In Figure 6 the deformed configuration of beams axes is presented.

Table 2. Values of penetration in Example 2.

Increment	Without deformation	Earlier approach	New approach
5	0.0000212855	0.0000211680	0.0000037805
10	0.0000492265	0.0000489932	0.0000087430
15	0.0000774648	0.0000771360	0.0000137584
20	0.0001061011	0.0001056835	0.0000188445
25	0.0001352435	0.0001347382	0.0000240206
30	0.0001650178	0.0001644230	0.0000293090
F_N	330	328	330

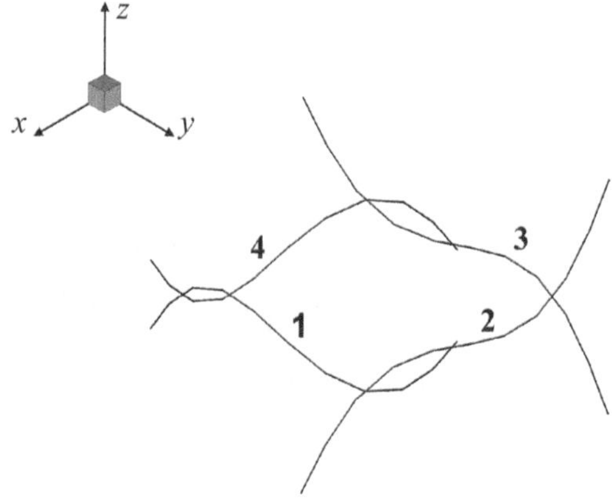

Figure 6. Deformed configuration of beams axes in Example 2.

4 SUMMARY

An alternative method to include the cross-section deformation in the analysis of beam-to-beam contact has been presented. The method uses a classical Hertz solution for two contacting cylinders. The physical law for contacting bodies is introduced so in this new approach high precision contact is considered. In the new approach the gap function is the real value of the deformation of the cross-sections, while in the penalty approach it actually represents the error of the method in the fulfillment of non-penetrability conditions.

An important thing from the computational practice, is that no noticeable differences in the computer time or convergence speed have been observed, when compared with the low precision formulation.

The future work will include an application of more general models to determine the cross-sections change, e.g. for beams contacting at any angle.

ACKNOWLEDGEMENTS

This research is done under the internal university grants 01/11/DSMK/0509 and 01/11/DSPB/400.

REFERENCES

Crisfield M. A. 1990. A consistent co-rotational formulation for non-linear, three-dimensional beam-elements. *Comput. Meth. Appl. Mech. Eng.* 81: 131–150.

Kawa O. & Litewka P., 2014. Contact between 3-D beams with deformable circular cross sections. *Recent Advances in Computational Mechanics, T. Łodygowski, J. Rakowski, P. Litewka (eds.),* CRC Press/Balkema, Taylor & Francis Group, London: 183–190.

Litewka P. 2010. *Finite Element Analysis of Beam to Beam Contact*. Berlin Heidelberg: Springer.

Popov Valentin L. 2010. *Contact Mechanics and Friction*. Springer—Verlag Berlin Heidelberg: 55–64.

Wriggers P. & Zavarise G. 1997. On contact between three-dimensional beams undergoing large deflections. *Communications in numerical method in engineering,* 13: 429–438.

Zavarise G. & Wriggers P. 2000. Contact with friction between beams in 3-D space. *International Journal for Numerical Methods in Engineering Int. J. Numer. Meth. Engng.* 49: 977–1006.

Advances in Mechanics: Theoretical, Computational and Interdisciplinary Issues – Kleiber et al. (Eds)
© 2016 Taylor & Francis Group, London, ISBN 978-1-138-02906-4

Analysis of wave propagation in a bolted steel joint

R. Kędra & M. Rucka
Faculty of Civil and Environmental Engineering, Gdańsk University of Technology, Gdańsk, Poland

ABSTRACT: The paper presents results of numerical and experimental investigations of elastic wave propagation in a bolted single-lap joint. The research was carried out for both symmetric and antisymmetric Lamb wave modes. In experimental studies the PZT transducers were used to excite and register wave propagation signals. The controlled value of the torque during measurements was ensured by the use of high precision torque wrenches. Numerical model of the joint was developed by the finite element method using commercial software Abaqus. The analysis took into account varying contact conditions between the connected parts of the joint depending on the value of the bolt torque. Both experimental and numerical results showed the influence of the value of the pretension force on the amplitude and phase shifts of registered signals.

1 INTRODUCTION

Prestressed bolted joints due to their high load capacity, durability and easy installation are very popular and often used in many industrial sectors, including numerous applications in construction industry. A disadvantage of pretensioned connection is the possibility of the decrease in the load capacity with time due to self-loosening and rheological effects. This results in the need for continuous monitoring of prestressed bolted joints crucial for the overall structural stability. Recently, many diagnostic techniques have been developed for non-destructive testing and evaluation of such type of connection. One of the most efficient strategies of diagnostics of bolted connections is associated with the use of elastic wave propagation (e.g. Amerini & Meo 2011, Kędra and Rucka, 2014, 2015).

Qualitative and quantitative indicators based on the variability of recorded signals can be used in the assessment of the condition state of bolted joints. In many cases, the research on those diagnostic methods was connected with the development of analytical models describing wave propagation. The main issue lies in the solution of the problem of the contact between connected elements. Different modelling approaches were used to describe propagation at the interface surface. A detailed analysis of the effect of the pressure on the contact condition of composite panels in micro-scale using Hertz contact theory allowed Yang & Chang (1989a, b) to confirm the hypothesis of the existence of the relationship between a nominal and a true contact area and the value of dissipated energy at a contact zone. Rhee et al. (2012) created a relatively simple model of a multi-bolt lap joint. They analyzed two models of bolts, namely solid and rigid, and took into account contact between bolts and plates by a constrained lateral surface of a bolt stud with edges of holes in plates. In addition, they created an accurate model of piezoelectric transducers. Finally, the authors were able to reproduce accurately the shape of experimentally recorded signals. They concluded a low impact of prestressing force on the variability of recorded waveforms. Huda et al. (2013) used a multi-step approach. They assumed a circular shape of a contact area and determined its radius by means of static analysis. Afterwards, in the second step, they assumed constant contact conditions between the bolt and plates. Furthermore, they introduced the stress and strain results obtained in static analysis as initial conditions in wave propagation analysis. As a result, the authors obtained a very good agreement of dynamic parameters of the developed model and experimental data for the cases of one- and multi-bolt joints.

The models described above are relatively simple and they are intended to reproduce results of a single experiment. They do not include the full complexity of the phenomenon of wave propagation. Therefore, there is still a need to develop a more accurate model, able to cover all physical aspects related to wave propagation. This study is a first step towards developing of such complex FEM model. The paper includes numerical studies performed in parallel with experimental tests. The obtained results are compared and evaluated in terms of a possible application in diagnostics of bolted lap joints.

2 STUDIES OF SINGLE LAP BOLTED JOINTS

2.1 *Model description*

The research was conducted for a laboratory model of a single lap bolted joint. It was made of two steel plates, which were assembled together by a single bolt of a nominal diameter of 12 mm and a length of 49.5 mm, nut and two stainless steel washers. The dimensions of plates were 6.4 mm × 40.4 mm × 440.5 mm. The geometry of the connection is illustrated in Figure 1.

2.2 *Experimental tests*

Experimental set-up is presented in Figure 2. During experimental investigations the model of a bolted joint was placed in a steel frame to ensure a constant position of steel plates during the process of bolt tightening (Fig. 2a). The excitation and acquisition of signals of propagated elastic waves were carried out by the device PAQ-16000D and piezoelectric transducers Noliac NAC2002 with dimensions of 3 mm × 3 mm × 2 mm. The actuator and sensors were attached to the bolted joint with the use of wax at selected points located along symmetry plane. Actuator was located on the upper surface of the top plate at a distance of 40 mm from the axis of the bolt. Three sensors were attached to the bottom plate at a distance of 40 mm (S1), 60 mm (S2) and 80 mm (S3) from the axis of the fastener (Fig. 2b). The excitation signal was a 5-cycle sine function with a frequency of 100 kHz modulated by the Hanning window. The sampling frequency during the measurements was set as 2 MHz. In order to reduce influence of noise, each measurement was repeated 100 times and averaged.

2.3 *Numerical analysis*

Numerical analysis of wave propagation at the joint with different values of prestressed force was performed by the Finite Element Method (FEM) in Abaqus environment. In order to reduce the computational cost, the symmetry of the joint was utilized, i.e. only a half of the actual joint geometry was modeled and constraint displacement in the direction perpendicular to in the plane of symmetry was defined (Fig. 3). All parts of the joint and sensors were discretized using solid brick elements (six-node C3D6 and eight-node C3D8) with second order accuracy, enhanced hourglass control and maximum edge size equal to 2 mm. Linear elastic isotropic material model was established, described by the following parameters: Young's modulus $E = 200$ GPa, Poisson's ratio $\nu = 0.3$, density $\rho = 7850$ kg/m^3 and friction coefficient $\mu = 0.2$. The fixed-fixed boundary conditions of the bolted joint were assumed. Bolt load was introduced using 1 DoF pretension nodes section.

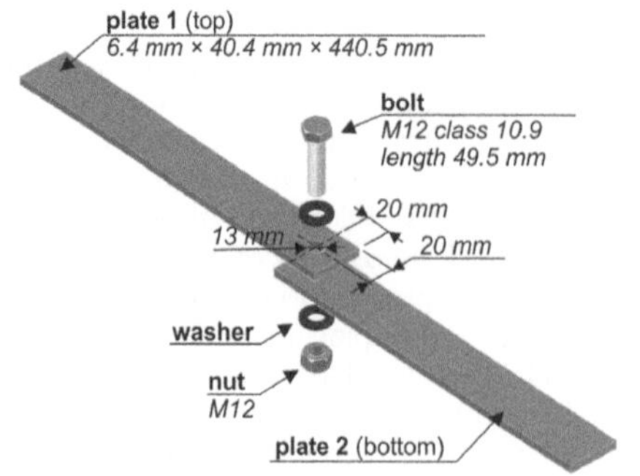

Figure 1. Scheme of analyzed bolted connection.

a)

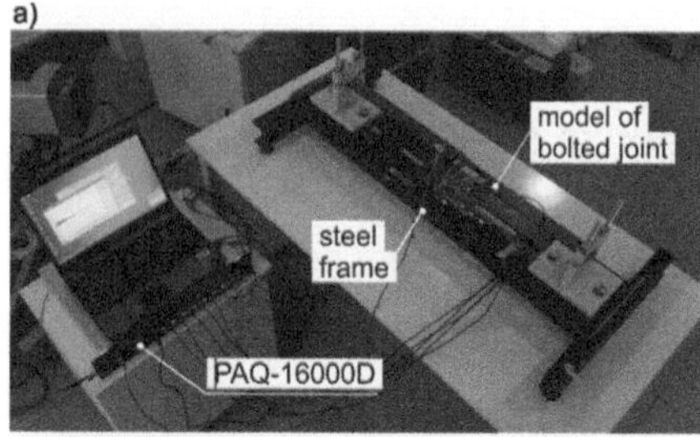

b)

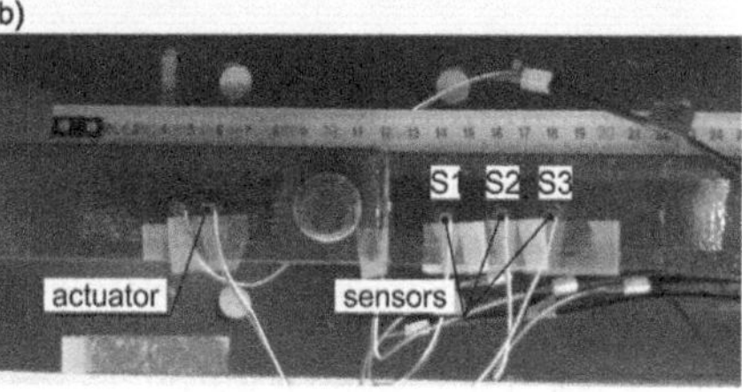

Figure 2. Laboratory tests: a) experimental set-up; b) actuator and sensors arrangement.

Contact between elements of the connection was modelled introducing a thin layer (0.1 mm) of brick elements. For simplicity, constant contact stresses in all contact zones (between plates, plates and washers, washer and nut, washer and bolt head) and conical shape of interaction region with a slope of 30° according to Osgood was assumed (Budynas & Nisbett 2006). A value of Young's modulus of the contact layer was selected in order to ensure suitable value of reflection coefficient. Other parameters of the layer were the same as joint parts.

The elastic waves in the considered prestressed bolted joint were induced applying an electric field to finite elements constituting a piezoelectric transducer. Discretization of sensor element was carried

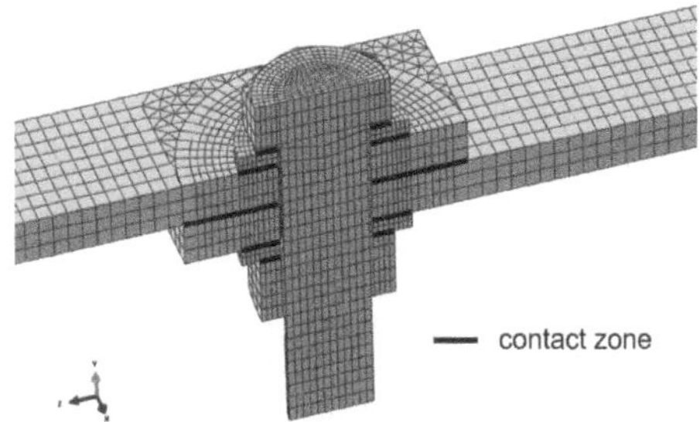

Figure 3. Visualization of finite element mesh in lap region.

out based on structural mesh using hex elements (C3D8E) with the maximum edge size 1 mm. The actuator was attached to joint by elastic layer of finite elements with relative small stiffness. Apart from a number of limitations, including the effect of temperature and non-linearity between the various parameters for low electric field (small strain), the constitutive equations for the actuator material can be written as:

$$\sigma_{ij} = D^E_{ijkl}(\varepsilon_{kl} - d_{mkl}E_m) \tag{1a}$$

$$q_i = e_{ijk}\varepsilon_{jk} + D^{\varphi(\varepsilon)}_{ij}E_j \tag{1b}$$

where: σ_{ij}—(Cauchy) stress tensor, ε_{kl}—strain tensor, q_i—electric displacement (electric flux) vector, D^E_{ijkl}—material elastic stiffness tensor, d_{mkl}—piezoelectric strain (charge) constants tensor, e_{ijk}—piezoelectric stress (voltage) constants tensor and $D^{\phi(\varepsilon)}_{ij}$—material dielectric prosperities tensor (Abaqus 2012). For piezoelectric transducers, which are used in the research, the numerical values of material components are given below:

$$D^E_{ijkl} = \begin{bmatrix} 1.34 & 8.89 & 9.09 & 0 & 0 & 0 \\ & 1.34 & 9.09 & 0 & 0 & 0 \\ & & 12.1 & 0 & 0 & 0 \\ & & & 2.24 & 0 & 0 \\ & symm. & & & 2.05 & 0 \\ & & & & & 2.05 \end{bmatrix} \times 10^{10}\left[\frac{N}{m^2}\right],$$

$$d_{mkl} = \begin{bmatrix} 0 & 0 & 0 & 0 & 0 & 669 \\ 0 & 0 & 0 & 0 & 669 & 0 \\ -208 & -208 & 443 & 0 & 0 & 0 \end{bmatrix} \times 10^{-12}\left[\frac{C}{N}\right],$$

$$e_{mkl} = \begin{bmatrix} 0 & 0 & 0 & 0 & 0 & 13.7 \\ 0 & 0 & 0 & 0 & 13.7 & 0 \\ -6.06 & -6.06 & 17.2 & 0 & 0 & 0 \end{bmatrix} \times 10^{-3}\left[\frac{C}{m^2}\right],$$

$$D^{\varphi(\varepsilon)}_{ij} = \begin{bmatrix} 1.72 & 0 & 0 \\ 0 & 1.72 & 0 \\ 0 & 0 & 1.68 \end{bmatrix} \times 10^{-8}\left[\frac{F}{m}\right].$$

Numerical calculations were carried out in two stages. Initially, the geometrical nonlinear static analysis was made to introduce pretension force in the bolt. In the second step, wave propagation was simulated. The time-variation of excitation was identical to the value from experimental studies. The time step of numerical implicit integration was assumed as 2×10^{-7} s.

3 RESULTS

The results of conducted experimental and numerical analyses were presented as voltage and acceleration signals, respectively, in the form of A-scans in the time domain. Additionally, in the case of numerical calculations, wave propagation fields were shown in the form of two-dimensional acceleration maps.

The experimental results indicated an influence of a bolt torque value on the registered voltage signals. With the increase of pretension value, the phase shifts and growth of the amplitude in the initial part of signals can be observed (Fig. 4).

Figure 5 shows the numerical waveforms at point S1 for various values of the Young's modulus: 125 MPa, 317 MPa, 564 MPa and 1140 MPa, which correspond to the following values of the transmission coefficient: 0.1, 0.2, 0.3 and 0.4. It can

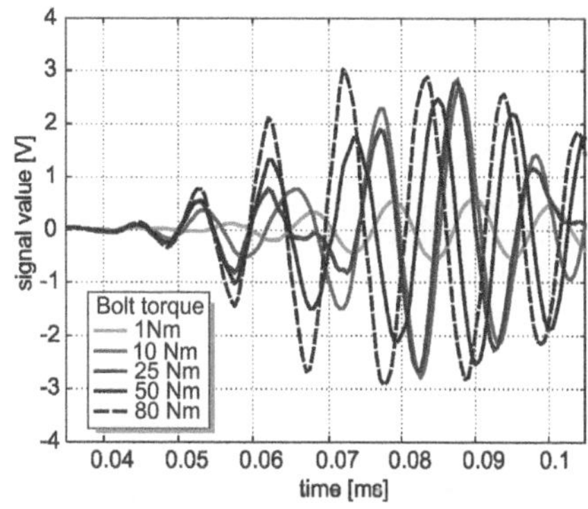

Figure 4. Comparison of experimental signals registered for different values of tightening torque.

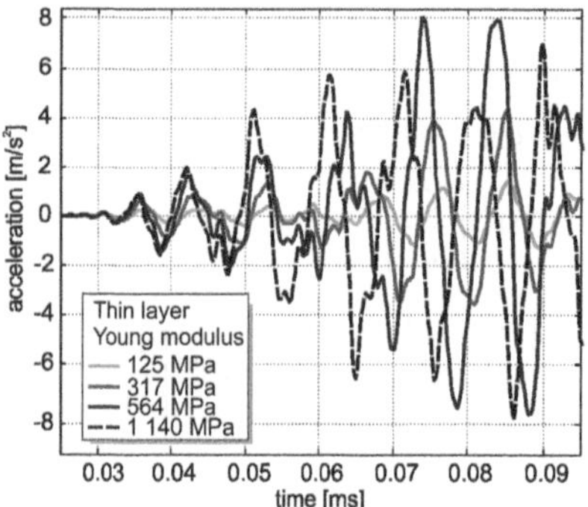

Figure 5. Comparison of numerical signals computed for different values of Young's modulus of contact layer.

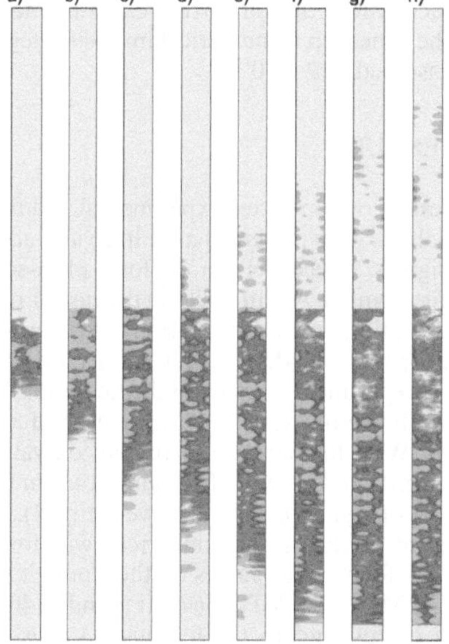

Figure 6. Numerical maps of acceleration field in bolted joint during wave propagation analysis at selected time instances: a) 0.02 ms; b) 0.04 ms; c) 0.06 ms; d) 0.08 ms; e) 0.1 ms; f) 0.12 ms, g) 0.14 ms; h) 0.16 ms.

be seen that with the gradual increase of the value of the modulus of elasticity of the contact layer, an increase of signal amplitudes and phase shifts can be observed. The difference between the beginning of the wave registered in experimental and numerical investigations (Figs. 4, 5) results from a delay of signal trigger of device PAQ-16000D in relation to start of recording.

Figure 6 illustrates results of numerical analysis for contact layer with a Young's modulus of 125 MPa. It can be seen that the amplitude of the disturbances at the lower plate are relatively small and only a small part of the energy is transferred to the opposite side of the connection. This is due to big value of reflection coefficient, which results that most of the wave is reflected on the contact zone.

4 CONCLUSIONS

In the paper experimental and numerical investigations of elastic wave propagation in a laboratory model of the single lap bolted joint were presented. The obtained results showed the possibility of using quantitative characteristics of signals for the purpose of diagnostics of bolted lap joints.

Experimental results indicated an influence of a bolt torque value and hence a level of normal contact pressure on the registered voltage signals. The increase of the bolt torque value resulted in the phase shifts and growth of the amplitude at the initial part of registered waveforms. It has been confirmed that similar results can be achieved numerically using a thin contact layer of solid elements with appropriate material parameters and the relationship between the reflection coefficient and the normal contact stress.

ACKNOWLEDGMENTS

Calculations were carried out at the Academic Computer Center in Gdańsk.

REFERENCES

Abaqus 2012. *Abaqus 6.12 Documentation*. Providence: Dassault Systèmes Simulia Corp.

Amerini, F. & Meo, M. 2011. Structural health monitoring of bolted joints using linear and nonlinear acoustic/ultrasound methods. *Structural Health Monitoring* 10: 659–372.

Budynas, R.G. & Nisbett, J.K. 2006. *Shigley's Mechanical Engineering Design*. NewYork: McGraw-Hill.

Huda, F., Kajiwara, I., Hosoya N. & Kawamura, S. 2013. Bolt loosening analysis and diagnosis by noncontact laser excitation vibration tests. *Mechanical Systems and Signal Processing* 40: 589–604.

Kędra R. & Rucka M. 2014. Diagnostics of bolted lap joint using guided wave propagation. *Diagnostyka*, 15(4): 35–40.

Kędra R. & Rucka M. 2015. Research on assessment of bolted joint state using elastic wave propagation. 2015. *Journal of Physics: Conference Series* 628: 012025.

Rhee, I., Choi, E. & Roh., Y-S. 2012. Guided wave propagation introduced by piezoelectric actuator in bolted thin steel members, *KSCCE Journal of Civil Engineering* 16(3): 398–406.

Yang, J. & Chang, F.-K. 2006a. Detection of bolt loosening in C-C composite thermal protection panels: I. Diagnostic principle. *Smart Materials and Structures* 15: 581–590.

Yang, J. & Chang, F.-K. 2006b. Detection of bolt loosening in C-C composite thermal protection panels: II. Experimental verification. *Smart Materials and Structures* 15: 591–599.

Advances in Mechanics: Theoretical, Computational and Interdisciplinary Issues – Kleiber et al. (Eds)
 ISBN 978-1-138-02906-4

An application of the Discrete Wavelet Transform to defect localization in plates

A. Knitter-Piątkowska & M. Guminiak
Poznan University of Technology, Poznan, Poland

ABSTRACT: The paper presents the defect detection in plates considering the influence of external static and dynamic loads. The aim if this work is to detect the localization of defect provided that damage exists in the considered plate structure. The Kirchhoff plate bending is described and solved by the Boundary Element Method. Plates have rectangular shape, different boundary conditions and additionally, they rest on the internal column supports. The analysis of a structural response is conducted with the use of signal processing tool namely Discrete Wavelet Transformation (DWT). Defects in plates are modeled as slots near the plate boundary.

1 INTRODUCTION

Defect detection in engineering structures is significant for monitoring of a structural behaviour. There are different non-destructive techniques which enable the identification of defective part of structure. This problem is extensively investigated by scientists and some approaches based on e.g. optimization of loads (Mróz & Garstecki 2005), information of natural frequencies (Dems & Mróz 2001), heat transfer (Ziopaja et al. 2011), inverse analysis (Garbowski et al. 2011), soft computing methods such as evolutionary algorithms (Burczyński et al. 2004) or Artificial Neural Networks (ANN) (Rucka & Wilde 2010, Waszczyszyn & Ziemiański 2001) are applied. Defect (damage) can be effectively detected using a relatively new method of signal analysis called Wavelet Transformation (WT) (Wang & Deng 1999), also in its discrete form (Knitter-Piątkowska et al. 2006, Knitter-Piątkowska et al. 2014). Combining this method with, earlier mentioned, ANN or inverse analysis (Knitter-Piątkowska et al. 2012) one can precisely identify defect (damage) details. The paper presents the issue of defect detection in thin plates excited by external static loads. The influence of a kind of structural response signal on obtained results is considered too. In order to carry out the numerical investigation of the structural behavior the Finite Element Method (FEM) is usually used (Knitter-Piątkowska et al. 2006). In the work the Boundary Element Method (BEM) is applied to carry out the numerical investigation of plate bending. Numerical examples are presented in the paper.

2 BASIS OF WAVELET TRANSFORMATION

Any signal can be portrayed as a sum of sinusoidal signals. It becomes a basis for the widely used Fourier analysis. In the paper we implement another kind of analysis, namely wavelet transform, in which for the representation of the signal $f(t)$ a linear combination of wavelet functions is used. In contrast to Fourier transform, wavelets are localized in time and frequency domain. Therefore they are well-suited for dealing with signals which detect discontinuities. A basis of wavelet transformation was widely described in the work Knitter-Piątkowska et al. (2014).

An important role in engineering applications is played by Discrete Wavelet Transform (DWT) due to the numerical effectiveness of the procedure. In practice, DWT requires neither integration, nor explicit knowledge of scaling (father) and wavelet (mother) function. The wavelet family can be defined in the following form (Knitter-Piątkowska et al. 2014):

$$\psi_{j,k}(t) = 2^{(j/2)} \cdot \psi\left(2^j \cdot t - k\right), \tag{1}$$

where integers k and j = are scale and translation parameters, respectively. The meaning of these parameters can be easily explained for the simplest Haar wavelet (Knitter-Piątkowska et al. 2014).

A discrete signal decomposition can be written in the form (Mallat 1998):

$$f_J = S_J + D_J + \cdots + D_n + \cdots + D_1, \quad n = J - j, \tag{2}$$

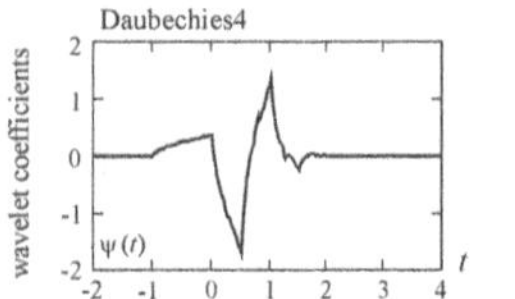

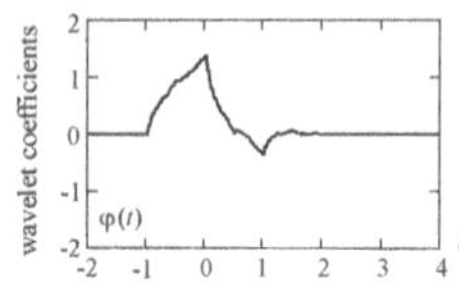

Figure 1. Basis function (mother) and scaling function (father): Daubechies 4 wavelet.

where each component in signal representation is associated with a specific range of frequency and provides information at the scale level $j = 1, \ldots, J$. S_J = is smooth signal representation, D_n, S_n = are details and rough parts of a signal. D_1 = corresponds with the most detailed representation of the signal. The above method of signal representation is called a Multiresolution Analysis (MRA), where J = is the level of MRA. Equation (2) describes the Mallat pyramid algorithm (Mallat 1998, Knitter-Piątkowska et al. 2014). In order to fulfill the dyadic requirements of DWT the function f_J must be approximated by $N = 2^J$ discrete values.

In the paper Daubechies wavelet family is applied. This is the first discovered family of orthogonal continuous wavelets with compact support. These wavelets are asymmetrical, compactly supported, with sharp edges. They require a small number of coefficients therefore are widely used in solving a broad range of problems, e.g. image analysis. Their order is between 2 and 20 (always even numbers). A second order Daubechies (1992) wavelet corresponds to the simplest Haar wavelet. Basis and scaling functions of Daubechies 4 wavelet are presented in the Figure 1.

3 DEFECT DETECTION

3.1 *Problem formulation*

The aim of the presented work is to detect the localization of defects provided that they exist in the considered plate structure. Numerical investigation is conducted based on signal analysis of structural static and dynamic response. Plate bending problem is described and solved by the BEMethod. The boundary integral equations are derived according to the non-singular approach (Guminiak & Sygulski 2007). Rectangular plates supported on boundary and resting on the internal column supports under static external loading are considered. Internal supports are introduced using Bèzine approach (Bèzine 1978). The analysis of a structural response is carried out with the use of DWT. The multiresolution signal analysis using Mallat pyramid algorithm (Mallat 1998) is applied too. Defects in plates are modelled as the slots near the plate boundary.

The considered plates are loaded by a single concentrated force P moving along the indicated line, for example 1, 2 or 3. The force P can have static character. At the selected point D in equal intervals of time, there are measured different structural responses such as deflection w, angle of rotation in arbitrary direction φ, curvatures κ and internal forces: bending and twisting moments or shear forces. The measured response parameters have the character of the influence lines in its discrete form. The structural response signal defined in this way is processed using Discrete Wavelet Transformation (DWT) which basis are described in Section 2.

The plate bending is described using the Boundary Element Method according to the simplified approach of formulation the boundary-domain integral equations in which the Kirchhoff corner forces are avoided (Guminiak & Sygulski 2007). The boundary-domain integral equations have the form:

$$
\begin{aligned}
& c(\mathbf{x}) \cdot w(\mathbf{x}) + \int_{\Gamma} \left[T_n^*(\mathbf{y},\mathbf{x}) \cdot w(\mathbf{y}) - M_{ns}^*(\mathbf{y},\mathbf{x}) \cdot \frac{dw(\mathbf{y})}{ds} + \right. \\
& \qquad \left. - M_n^*(\mathbf{y},\mathbf{x}) \cdot \phi_n(\mathbf{y}) \right] \cdot d\Gamma(\mathbf{y}) \\
& = \int_{\Gamma} \left[\tilde{T}_n(\mathbf{y}) \cdot w^*(\mathbf{y},\mathbf{x}) - M_n(\mathbf{y}) \cdot \phi_n^*(\mathbf{y},\mathbf{x}) \right] \cdot d\Gamma(\mathbf{y}) + \\
& \quad - \int_{\Omega_r} q_r \cdot w^*(r,\mathbf{x}) \cdot d\Omega_l + \int_{\Omega} p \cdot w^*(\mathbf{y},\mathbf{x}) \cdot d\Omega(\mathbf{y})
\end{aligned}
\tag{3}
$$

$$
\begin{aligned}
& c(\mathbf{x}) \cdot \phi_n(\mathbf{x}) + \int_{\Gamma} \left[\bar{T}_n^*(\mathbf{y},\mathbf{x}) \cdot w(\mathbf{y}) - \bar{M}_{ns}^*(\mathbf{y},\mathbf{x}) \cdot \frac{dw(\mathbf{y})}{ds} + \right. \\
& \qquad \left. - \bar{M}_n^*(\mathbf{y},\mathbf{x}) \cdot \phi_n(\mathbf{y}) \right] \cdot d\Gamma(\mathbf{y}) \\
& = \int_{\Gamma} \left[\tilde{T}_n(\mathbf{y}) \cdot \bar{w}^*(\mathbf{y},\mathbf{x}) - M_n(\mathbf{y}) \cdot \bar{\phi}_n^*(\mathbf{y},\mathbf{x}) \right] \cdot d\Gamma(\mathbf{y}) + \\
& \quad - \int_{\Omega_r} q_r \cdot \bar{w}^*(r,\mathbf{x}) \cdot d\Omega_l + \int_{\Omega} p \cdot \bar{w}^*(\mathbf{y},\mathbf{x}) \cdot d\Omega(\mathbf{y})
\end{aligned}
\tag{4}
$$

where the fundamental solution of the biharmonic equation:

$$\nabla^4 w^*(\mathbf{y},\mathbf{x}) = \frac{1}{D} \cdot \delta(\mathbf{y},\mathbf{x}) \tag{5}$$

is given as Green function:

$$w^*(\mathbf{y},\mathbf{x}) = \frac{1}{8\pi D} \cdot r^2 \cdot \ln(r) \tag{6}$$

for a thin isotropic plate, $r = |\mathbf{y} - \mathbf{x}|$, δ = is the Dirac delta; $\mathbf{x}$ = is the source point and $\mathbf{y}$ = is a field point; $D = Eh^3/12(1-\nu^2)$ is the plate stiffness;

h = is a plate thickness; E and ν = are the Young's modulus and the Poisson's ratio respectively; p = is external distributed loading and q = is the reaction distributed over the cross-section of the column (internal support). The coefficient $c(\mathbf{x})$ is taken as: $c(\mathbf{x}) = 1$, when $\mathbf{x}$ is located inside the plate domain; $c(\mathbf{x}) = 0.5$, when $\mathbf{x}$ is located on the smooth boundary; $c(\mathbf{x}) = 0$, when $\mathbf{x}$ is located outside the plate domain. The set of algebraic equations will have the form proposed by Guminiak & Sygulski (2006).

4 NUMERICAL EXAMPLES

Rectangular plates, supported on the boundary and resting on the internal column supports are considered. Defects are introduced by the additional edges forming a crack in relation to the basic plate domain. The static external concentrated load P is imposed at selected points along direction parallel to the one edge of the plate. Measured variables are: deflections and curvatures. The data are gathered at one measurement point, located near and slightly distal position in relation to the loss of plate basic domain, in equal time intervals. Decomposition of the obtained signal is carried out using DWT and Daubechies 4 wavelet. The plate properties are: $E = 205$ GPa, $\nu = 0.3$; plate thickness $h = 0.05$ m; the dimensions of the slot are: $d = 0.005$ m, $e = 0.25$ m; external load $P = 10$ kN; the internal column supports have the square cross-section with dimensions 0.08 m × 0.08 m and they are treated as additional sub-domains of the constant type; all of plate edges are divided into 30 boundary elements of the constant type.

4.1 *Example 1*

The plate with all edges free and resting on eight column supports shown is considered (Fig. 2). The response signals of the structure as the plate deflection and curvatures κ_x and κ_y are measured at the D point (2.35 m; 0.35 m). The results of calculations are shown in Figures 3–5.

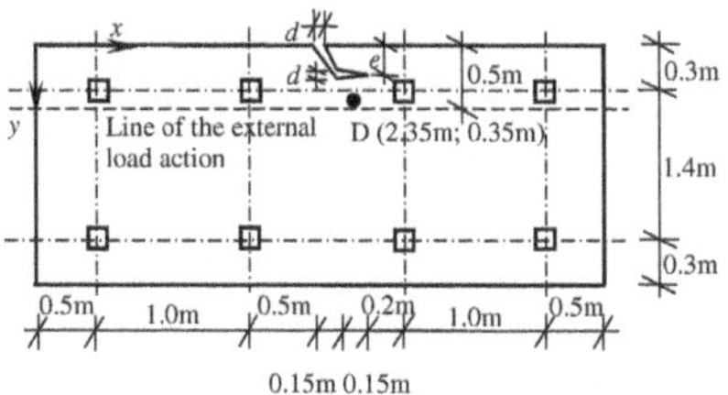

Figure 2. The plate structure resting on eight column supports with all edges free.

4.2 *Example 2*

The plate resting on four column supports with two opposite edges simply-supported and two edges free is considered (Fig. 6). The response signals (deflection and curvatures κ_x and κ_y) of the structure are measured at the D point (3.6 m; 0.55 m). The results of calculations are shown in Figures 7–9.

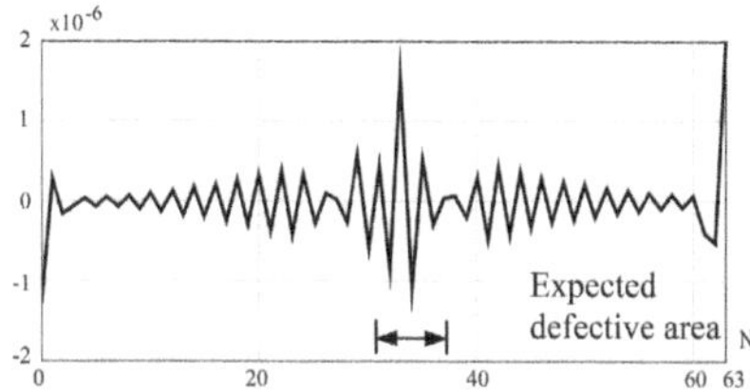

Figure 3. DWT (Daubechies4, detail 1) signal: vertical displacements measured at D point, N—number of measurements.

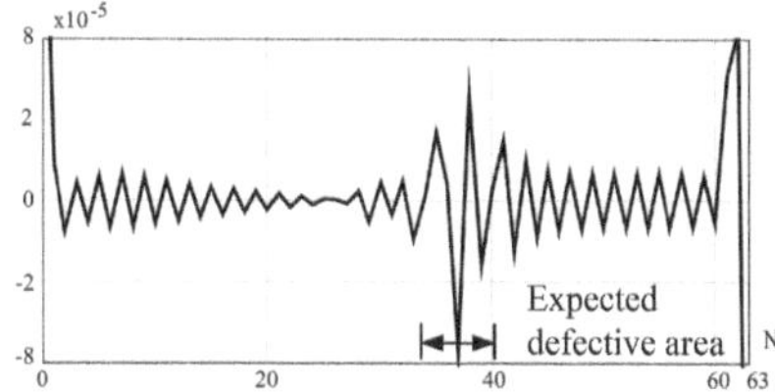

Figure 4. DWT (Daubechies4, detail 1) signal: curvature κ_x measured at D point, N—number of measurements.

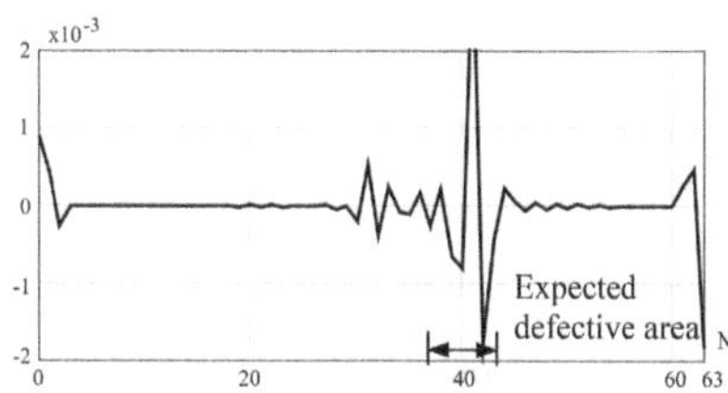

Figure 5. DWT (Daubechies4, detail 1) signal: curvature κ_y measured at D point, N—number of measurements.

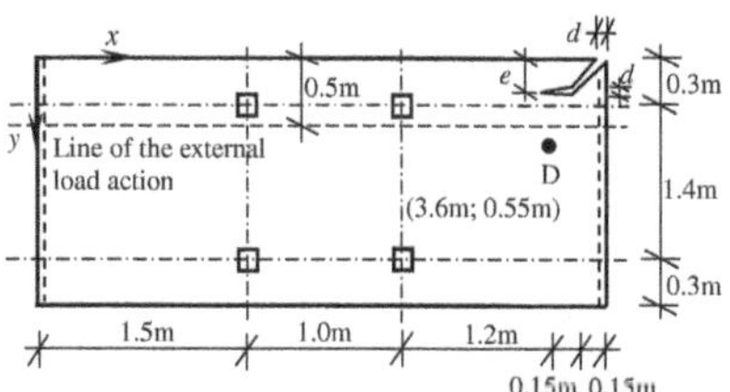

Figure 6. The plate resting on four column supports with two opposite edges simply-supported.

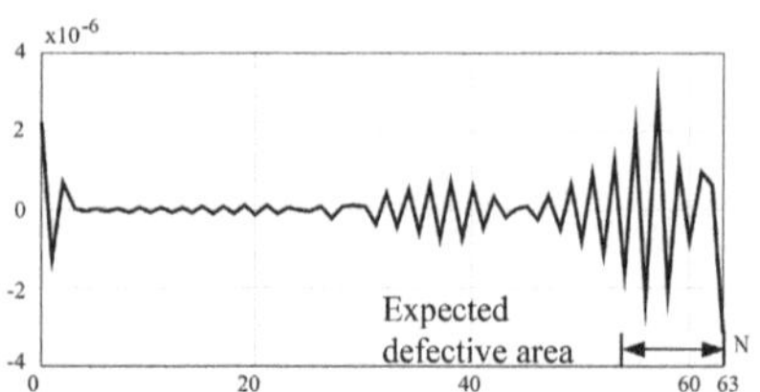

Figure 7. DWT (Daubechies4, detail 1) signal: vertical displacements measured at D point, N—number of measurements.

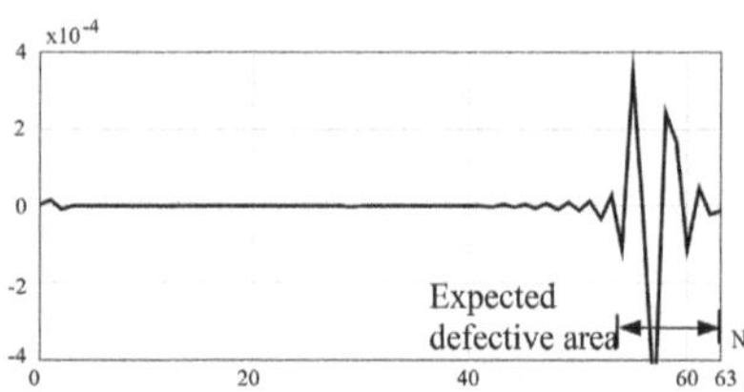

Figure 8. DWT (Daubechies4, detail 1) signal: curvature κ_x measured at D point, N—number of measurements.

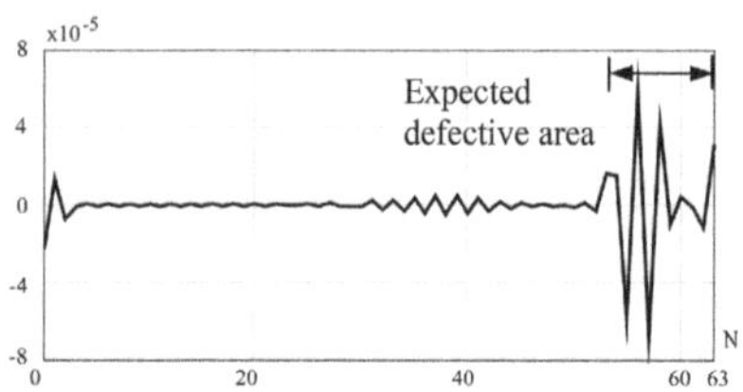

Figure 9. DWT (Daubechies4, detail 1) signal: curvature κ_y measured at D point, N—number of measurements.

5 CONCLUDING REMARKS

The implementation of DWT to identification of signal discontinuity in the analysis of plates is presented in the paper. The Boundary Element Method was used to describe the thin plate bending. Although the considered issue is two-dimensional from the point of view of deformation description, applied one-dimensional DWT leads to efficient results in defect detection. It discovers small disturbances in response signal of defective structure and does not require the reference to a signal from undamaged structure. The convenience of the present approach is that data are gathered in one measurement point in equal time intervals. The distance of the measurement point from damaged area is crucial for proper defect localization. The analysis is carried out without any signal noise reduction. The considered examples, exhibiting an evident disturbances of the transformed signal, identify quite correctly the presence and position of defects (Figs. 5, 8–9). The existence of additional column supports does not substantially impair the response signal. The presented examples identify quite correctly the presence and position of a defect.

REFERENCES

Burczyński, T., Kuś, W., Długosz, A. & Orantek, P. 2004. Optimization and defect identification using distributed evolutionary algorithms. *Engineering Applications of Artificial Intelligence* 17: 337–344.

Bèzine, G. & Gamby, D.A. 1978. A new integral equations formulation for plate bending problems. *Advances in Boundary Element Method*, Pentech Press, London.

Daubechies, I. 1992. *Ten lectures on wavelets*. Society for Industrial and Applied Mathematics, Philadelphia.

Dems, K. & Mróz, Z. 2001. Identification of damage in beam and plate structures using parameter dependent frequency changes. *Engineering Computation* 18 (1/2): 96–120.

Garbowski, T., Maier, G. & Novati, G. 2011. Diagnosis of concrete dams by flat-jack tests and inverse analysis based on proper othogonal decomposition. *Journal of Mechanics of Materials and Structures* 6 (1–4): 181–202.

Guminiak, M. & Sygulski, R. 2007. The analysis of internally supported thin plates by the Boundary Element Method. Part 1—Static analysis. *Foundations of Civil and Environmental Engineering* 9: 17–41.

Knitter-Piatkowska, A., Garbowski, T. & Garstecki, A. 2012. Damage detection through wavelet transform and inverse analysis. *Proc. 8th European Solid Mechanics Conference, ESMC 2012, 9–13 July, Graz, Austria.*

Knitter-Piątkowska, A., Pozorski, Z. & Garstecki A. 2006. Application of discrete wavelet transformation in damage detection. Part I: Static and dynamic experiments. *Computer Assisted Mechanics and Engineering Sciences* 13: 21–38.

Knitter-Piątkowska A., Guminiak M. & Przychodzki M. 2014. Damage detection in truss structure being the part of historic railway viaduct using wavelet transformation. In. T. Łodygowski et al. (eds), *Recent Advances in Computational Mechanics*, CRC Press/ Balkema: 157–163.

Mróz, Z. & Garstecki, A. 2005. Optimal loading conditions in design and identification of structures. Part 1: Discrete formulation. *Int. J. Struct. and Multidisciplinary Optim.* 29: 11–18.

Mallat, S.G. 1998. A theory for multiresolution signal decomposition: The wavelet representation. *IEE Trans. on Pattern Anal. and Machine Int.* 11(7): 674–693. San Diego.

Rucka, M. & Wilde, K. 2010. Neuro-wavelet damage detection technique in beam, plate and shell structures with experimental validation. *Journal of Theoretical and Applied Mechanics* 48(3): 579–604.

Wang, Q. & Deng, X. 1999. Damage detection with spatial wavelets. *Journal of Solids and Structures* 36: 3443–3468.

Waszczyszyn, Z. & Ziemiański L. 2001. Neural networks in mechanics of structures and materials—new results and prospects of application. *Computers and Structures* 79 (22–25): 2261–2276.

Ziopaja, K., Pozorski, Z. & Garstecki, A. 2011. Damage detection using thermal experiments and wavelet transformation. *Inverse Problems in Science and Engineering* 19 (1): 127–153.

Advances in Mechanics: Theoretical, Computational and Interdisciplinary Issues – Kleiber et al. (Eds)
© 2016 Taylor & Francis Group, London, ISBN 978-1-138-02906-4

Optimally tuned fuzzy control for smart, possibly damaged piezocomposites

P. Koutsianitis, G.A. Drosopoulos, G.K. Tairidis & G.E. Stavroulakis
Technical University of Crete, Chania, Greece

G. Foutsitzi
Technological Educational Institute of Epirus, Preveza, Greece

ABSTRACT: A mathematical model of smart composite beams and plates with piezoelectric layers that take into account the adhesive between the layers has been created. The finite element method has been used for the numerical solution of related static and dynamic problems. Furthermore, delamination between various layers has been modelled, in order to estimate one of the most commonly appearing damage in composites. Finally, active control is applied by means of optimally tuned fuzzy inference rules. Tuning has been achieved using genetic optimization. The results demonstrate a viable way of modelling and optimal design of active smart piezocomposites.

1 INTRODUCTION

Finite element modelling of the dynamic response of smart piezocomposites with adhesive and piezoelectric layers is first outlined. Smart composites may suffer from fatigue, damage and delamination, so that the response of the nominal model of the structure changes. Robust controllers are required, in order to work even in the presence of these deviations. Fuzzy controllers are generally believed to have robust properties. Therefore a fuzzy controller is used in order to study the effect of delamination on the control of a smart piezoelectric plate.

2 ELECTROMECHANICAL MODEL

Consider a plate structure consisting of three layers: The upper and lower layers simulate the behavior of a "smart" piezoelectric material, while the middle illustrates the behavior of elastic material. Between the interfaces of these layers, two new thin layers of adhesive material have been placed. Each adhesive layer is assumed to be thin enough that its peel and shear strains can be assumed constant through its thickness and other strains are negligible. For the non-adhesive layers, the theory of Mindlin have been considered. A finite element model based on the layerwise displacement theory which incorporates the electro-mechanical coupling effects and the adhesive layer flexibility has been used. The full derivation and parameters are explained in Foutsitzi et al. (2006). The equations of motion and charge equilibrium of the system can be written as:

$$[M]\{\ddot{d}\}+[K_{uu}]\{d\}+[K_{u\phi}]\{\phi\}=\{F_m\}$$
$$[K_{\phi u}]\{d\}+[K_{\phi\phi}]\{\phi\}=\{F_q\} \tag{1}$$

where $\{d\}$ and $\{\phi\}$ = are the global mechanical and electrical Degrees of Freedom (DoFs) vectors which include the corresponding DoFs of each layer. $[M]$ = is the global mass matrix, $[K_{uu}]$, $[K_{u\phi}]$ = $[K_{u\phi}]^{-1}$ and $[K_{\phi\phi}]$ = are the global mechanical stiffness, mechanical–electrical coupling stiffness and dielectric stiffness matrices respectively. $\{F_m\}$and $\{F_q\}$ = are the respective global mechanical and electrical loads vectors. It should be noted, that due to the adhesive layer, global mass and stiffness matrices contain coupling terms between the layers (Foutsitzi et al. 2006, 2010).

3 DELAMINATION MODELING

The purpose of this part of the article is the insertion of non-linear delamination law, in the existing composite piezoelectric finite element model of the plate described above. The finite element model, with the non-linear delamination law, includes the possibility of applying both mechanical and electrical load and the possibility of simulating static and dynamic problems.

For the simulation of delamination, a non-linear delamination law is considered in the previously outlined electromechanical model as shown in Figure 1.

The above law leads to appropriate modification of the existing stiffness matrix of the structure in order to take into account the differences in the behavior before and after the delamination.

For each finite element, the element average strain ε_{zz} is calculated as follows:

$$\varepsilon_{zz}^{(a_i)} = \frac{w^{(i+1)} - w^{(i)}}{h^{(a_i)}} \tag{2}$$

where w = is the vertical displacement of layers, whose behavior in delamination is investigated; $h^{(a)}$ = is the thickness of the adhesive layer. If the strain is less than ε_o; no delamination appears in the element. Otherwise, delamination takes place. The delamination can occur between the lower-middle and the middle-upper plate layer.

The iterative solution follows classical Newton-Raphson techniques and specialized algorithms, see (Mistakidis & Stavroulakis 1998).

It is worth noting that the existing composite piezoelectric model of the plate, fully facilitate the above iterative process of finding the tangential stiffness of the structure, by appropriately removing members from the full, initial, stiffness matrix.

Here, delamination between the middle and the upper plate layer have been considered. The above mentioned numerical scheme is applied to a composite plate with dimensions 100 × 100 mm, fixed on the left side. Each element consists of 12 nodes: 4 nodes belong to the lower layer, 4 nodes to the middle and 4 nodes to the upper layer. Every node has 5 degrees of freedom (three translational and two rotational). Therefore, the structure has 363 nodes and 1815 degrees of freedom.

When the mechanical load is increased, the behavior of the structure changes thus, delamination and non-linear behavior appear. According to the shape of the displacement which follows, the delamination occurs across the upper layer, Figures 2–3, as it is expected (Frémont 2002).

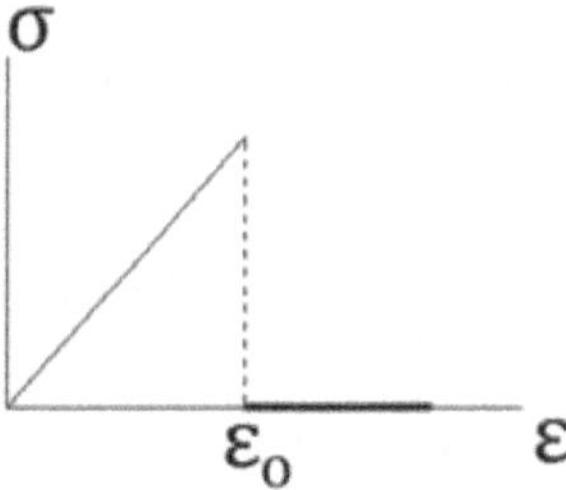

Figure 1. Stress-strain behavior in the adhesive.

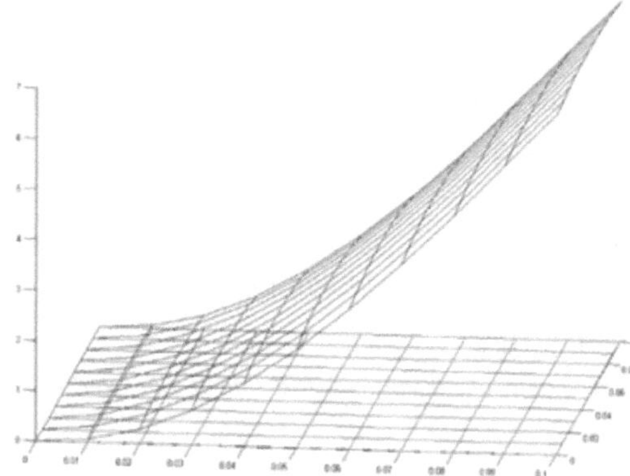

Figure 2. Displacement of the structure, when delamination in the whole middle-upper layer interface appears.

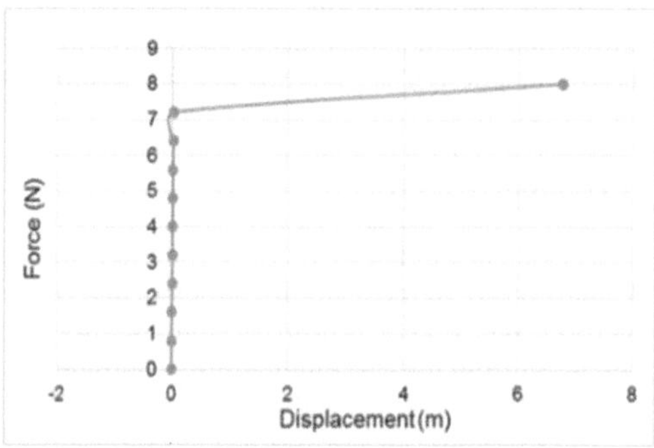

Figure 3. Force—displacement diagram (load control).

At the end of the analysis, this operates completely independently of the underlying layers. As a result, large displacements occur in this layer.

Since the softening behavior cannot be depicted by a load control procedure, a displacement controlled procedure is used to capture partial delamination effects, see Figures 4–5.

The change of the slope of the load-displacement diagram shown below, which is nonlinear, expressing the phenomenon of delamination, which is accompanied by a decrease in the stiffness of the system.

As already mentioned above, the delamination in the last example happens simultaneously to all elements of the interface. The gradual appearance of delamination only to certain parts of the interface, cannot be illustrated by a force mechanical load (load control), due to the numerical instability which is introduced by the iterative Newton-Raphson procedure up to the point of failure. Consequently, the softening behavior cannot be depicted by the load control procedure.

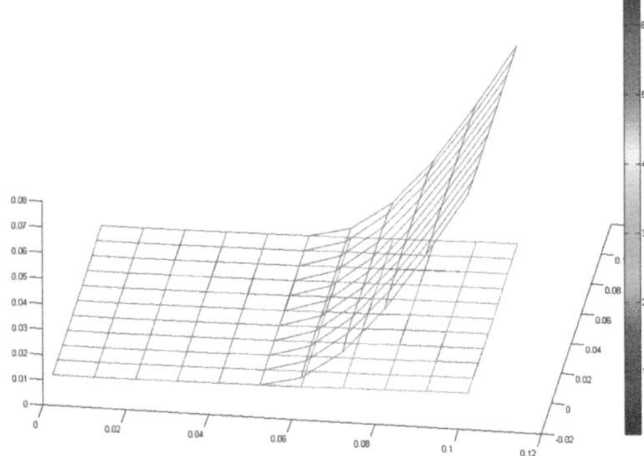

Figure 4. Displacement of the structure, when partial delamination in the whole middle-upper layer interface appears.

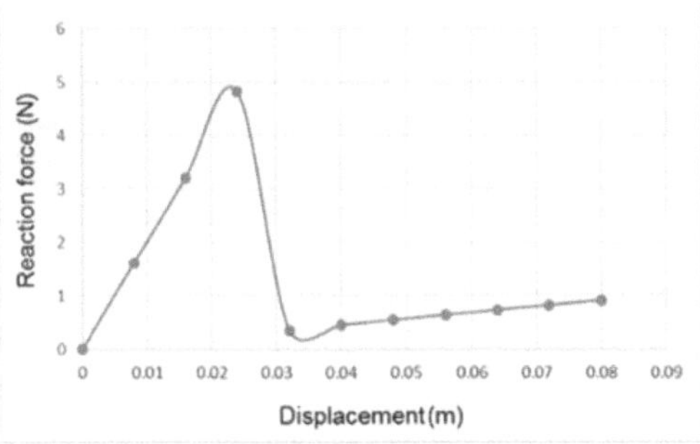

Figure 5. Force—displacement diagram for the structure with delamination (displacement control).

For this reason, the code has been changed, so that the applied mechanical force is a displacement control. As seen from the figures that follow, it is possible then to display gradual delamination. At the same time, the load-displacement diagram shows a softening branch similar to the stress-strain diagram.

4 OPTIMALLY TUNED FUZZY CONTROLLER

A Mamdani-type fuzzy logic controller is developed, using Matlab, with two inputs (displacement, velocity) and one output (control force), collocated at the free end of the plate. The fuzzy inference system is shown in Figure 6.

Trapezoidal and triangular membership functions are used as shown in Figures 7–9.

The decision is based on a set of 15 if-then rules, the AND operator and the centroid defuzzification method, as shown in Table 1.

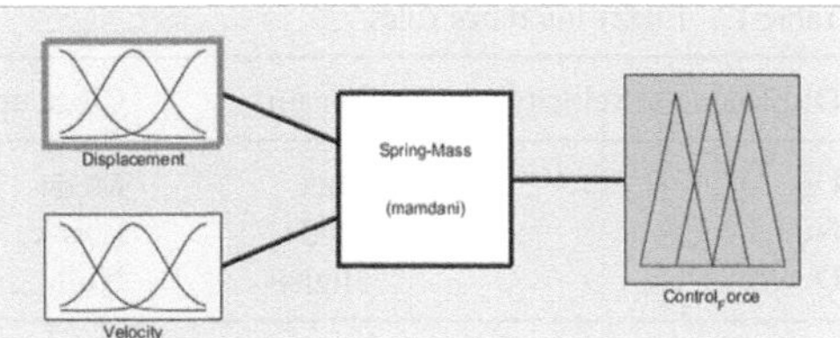

Figure 6. Mamdani fuzzy inference system.

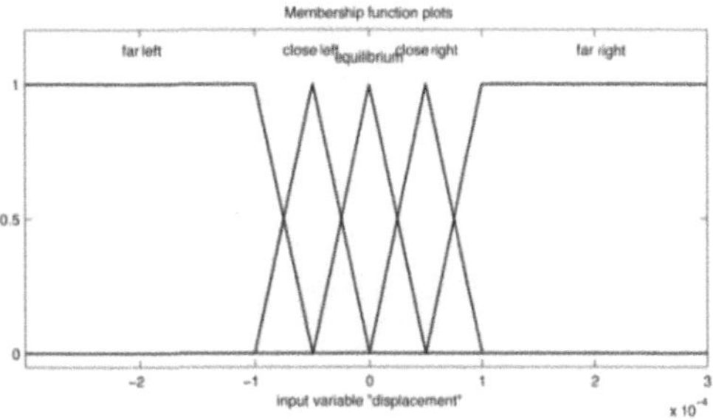

Figure 7. Membership function of displacement (input 1).

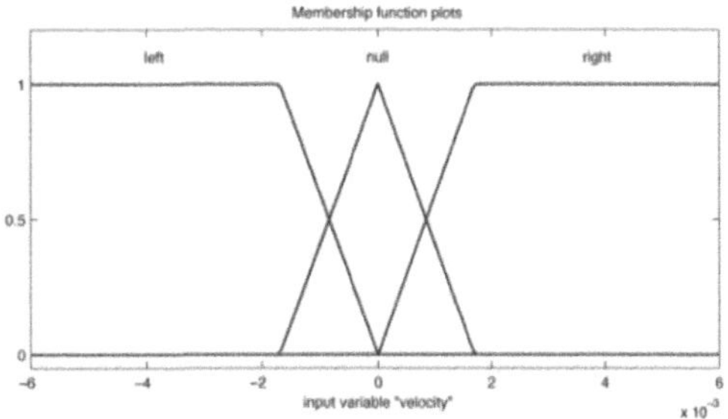

Figure 8. Membership function of velocity (input 2).

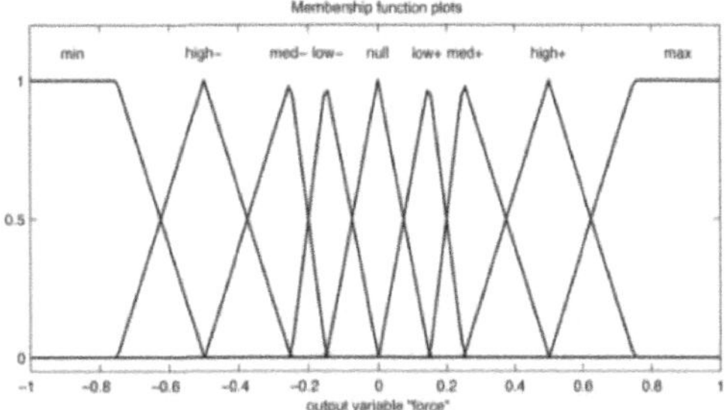

Figure 9. Membership function of control force (output).

Table 1. Fuzzy inference rules.

Displacement velocity	Far up	Close up	Equilibrium	Close down	Far down
Up	Max	Med+	Low+	Null	Low–
Null	Med+	Low+	Null	Low–	Med–
Down	High+	Null	Low+	Med–	Min

* e.g. if displacement is far up and velocity is up then the control force is max.

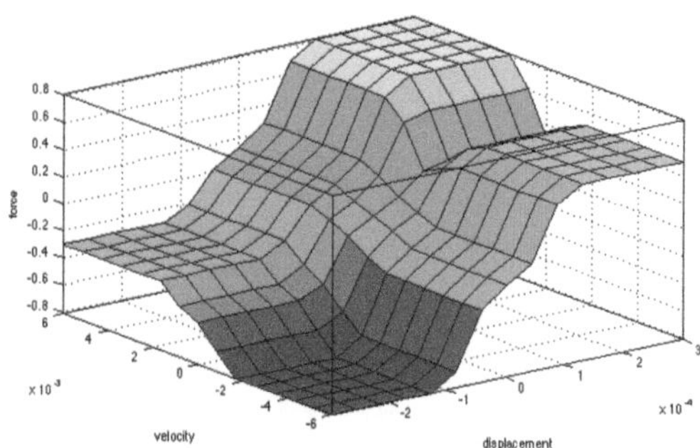

Figure 10. The fuzzy surface of the nonlinear fuzzy controller.

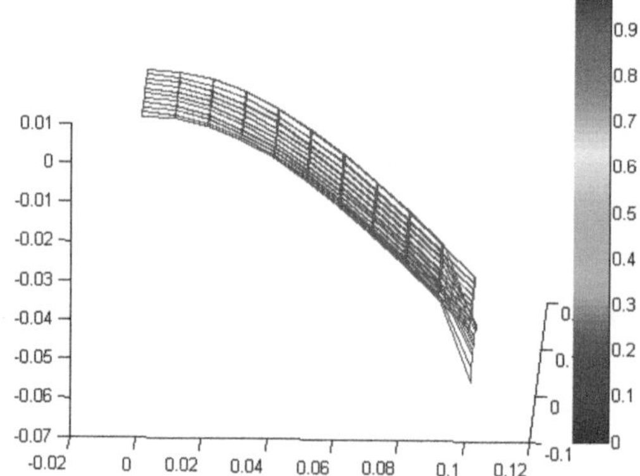

Figure 11. Plate with 10% delamination.

As defuzzification method the centroid method has been chosen.

Genetic algorithms can be used for the optimal tuning of the parameters (Stavroulakis et al. 2011).

Alternatively, fine tuning can be performed using particle swarm optimization (see the application on Smart beams in Marinaki et al. (2010). Smoothing of velocities and accelerations can be done using ANFIS techniques (Stavroulakis et al. 2011).

The nonlinear control feedback resulting from the optimized fuzzy logic controller is shown in Figure 10.

5 NUMERICAL EXAMPLE

The fuzzy controller, initially created for the control of the non-delaminated structure, has been applied on a structure with 10% and 50% delamination of the sensor layer, with satisfactory results.

In the first case, the plate presents 10% loss of the piezoelectric material (Fig. 11).

The results are shown in Figure 12.

From these diagrams, it can be observed that a significant vibration suppression in terms of displacement, velocity and acceleration is achieved.

This means that the behavior of the structure and the operation of the controller are very satisfactory.

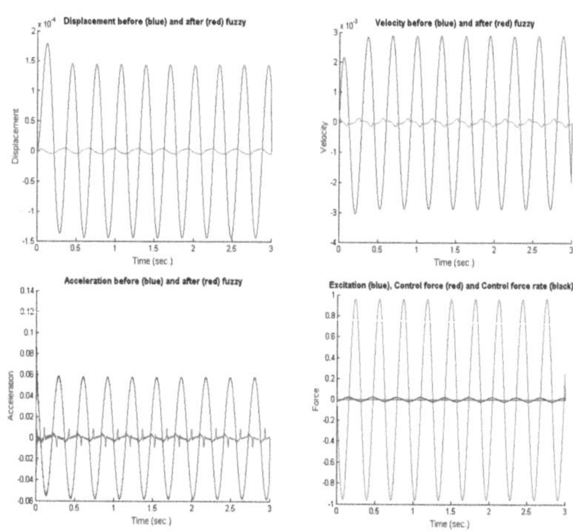

Figure 12. Displacement, velocity, acceleration and forces of the plate with 10% delamination.

A disadvantage of the present investigation is that a large control force is needed for the vibration suppression. A model with delamination equal to the 50% of the piezoelectric material of the sensor is shown in Figure 13.

The results for this case are shown in Figure 14. The vibration suppression in terms of

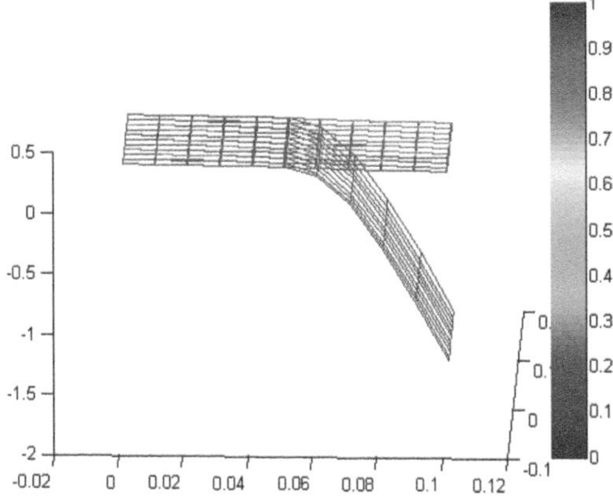

Figure 13. Plate with 50% delamination.

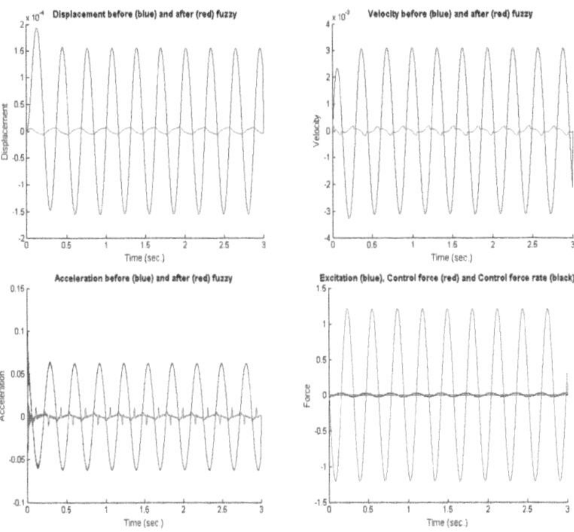

Figure 14. Displacement, velocity, acceleration and forces of the plate with 50% delamination.

displacement, velocity and acceleration is again very satisfactory, while the disadvantage of the large control force needed remains. It is concluded, that the functionality of the controller in an extended delamination is quite satisfying.

ACKNOWLEDGEMENTS

This research has been co-financed by the European Union (European Social Fund—ESF) and Greek national funds through the Operational Program "Education and Lifelong Learning" of the National Strategic Reference Framework (NSFR) —Research Funding Program: ARCHIMEDES III.—Investing in knowledge society through the European Social Fund. The authors greatfully acknowledge this support.

REFERENCES

Foutsitzi, G.A., Stavroulakis, G.E. & Hadjigeorgiou, E. 2010. Modelling and Simulation of Smart Structures, *Proceedings of the 9th International Congress on Mechanics, 9th HSTAM,* Limassol, Cyprus, 12–14 July, 2010.

Foutsitzi, G.A., Marinova, D., Stavroulakis, G.E. & Hadjigeorgiou, E. 2006. Vibration Control Analysis of smart Piezoelectric Composite Plates, *2nd IC-SCCE: Proceedings of the 2nd International Conference "From Scientific Computing to Computational Engineering,* Athens, 5–8 July 2006: 1436–1443.

Frémont, M. 2002. *Non-smooth thermomechanics,* Springer.

Marinaki M., Marinakis, Y. & Stavroulakis G.E. 2010. Fuzzy control optimized by PSO for vibration suppression of beams. *Control Engineering Practice* 18 (6): 618–629.

Mistakidis, E.S. & Stavroulakis, G.E. 1998. *Nonconvex optimization in mechanics*, Springer.

Stavroulakis, G.E., Papachristou, I., Salonikidis, S., Papalaios, I. & Tairidis, G.K. 2011. Neurofuzzy Control for Smart Structures, *Soft Computing Methods for Civil and Structural Engineering*, In. Tsompanakis Y. & Topping B.H.V. (eds,) Saxe-Coburg Publications, Stirlingshire, UK, Chapter 7: 149–172.

Advances in Mechanics: Theoretical, Computational and Interdisciplinary Issues – Kleiber et al. (Eds)
ISBN 978-1-138-02906-4

Analysis of sloping brace stiffness influence on stability and load bearing capacity of a truss

M. Krajewski, P. Iwicki, N. Sałyk & A. Łukowicz
Faculty of Civil and Environmental Engineering, Gdansk University of Technology, Gdansk, Poland

ABSTRACT: The paper is focused on the numerical study of stability and load bearing capacity of a truss with side elastic braces. The structure is made in reality. The rotational and sliding brace stiffnesses were taken into account. Linear buckling analysis and non-linear static analysis with geometric and material nonlinearity were performed for the beam and shell model of the truss with respect to the angle of sloping braces. As a result the buckling load and limit load in relation to the truss bracing stiffness was obtained. The threshold bracing condition necessary to provide maximum buckling resistance of the structure was proposed. The truss bearing capacity due to Eurocode requirements was calculated.

1 INTRODUCTION

Steel trusses have a much greater strength and stiffness in-plane than out-of-plane, therefore should be braced against lateral deflection and twisting. The problem of bracing requirements necessary to provide lateral stability of compressed members is present in a design codes PN (1990) and EC3 (2006). The out-of-plane buckling of trusses was studied in numerical research by Iwicki (2010), Biegus & Wojczyszyn (2011) or Biegus (2014) and also in experimental and numerical analysis conducted by Kołodziej & Jankowska-Sandberg (2013) and Iwicki & Krajewski (2013, 2014).

The present research is focused on lateral buckling of truss with sloping linear elastic side supports. The translational and rotational stiffness of braces was taken into account. In the paper a full bracing condition was investigated. This condition describes the required bracing stiffness that ensures the out-of-plane truss buckling is prevented which means that buckling occurs between braces or in the truss plane. At the threshold bracing stiffness other truss elements, such as compressed diagonals, or vertical members, or the truss top chord may buckle in the truss plane. The full bracing condition may also be defined as the bracing stiffness necessary to obtain the maximal buckling load of the truss, or when an increase in bracing stiffness does not cause any further increase of the buckling load.

In the paper linear buckling analysis and nonlinear analysis of the imperfect truss model were carried out. Both geometric and material non-linearity was considered. The buckling load (LBA) and limit load (GMNIA) of the truss due to stiffness and angle of side supports are calculated. The results are compared with Eurocode 3 requirements.

2 DESCRIPTION OF THE TRUSS

In the present research the model of a steel truss (Fig. 1, 2) braced by four elastic side supports was taken into consideration. The truss length was $L_t = 28.0$ m its height was $h = 1.8$ m. The distance between the braces was 5.6 m. The top chord consist of 2 × L160 × 15 rolled profiles, and the bottom chord of 2 × L150 × 15. The diagonals and

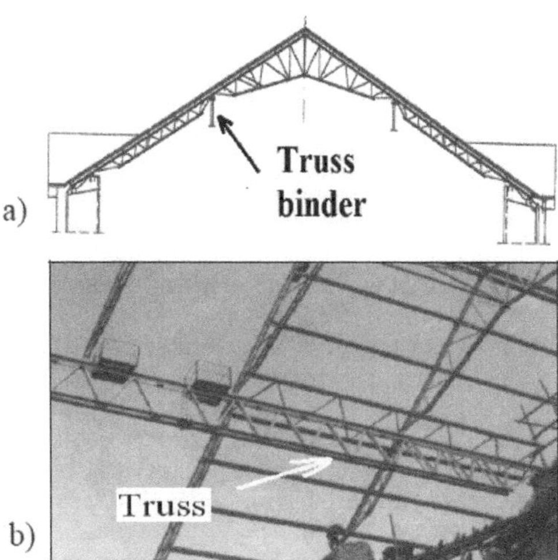

Figure 1. a) roof construction, b) truss binder with sloping braces.

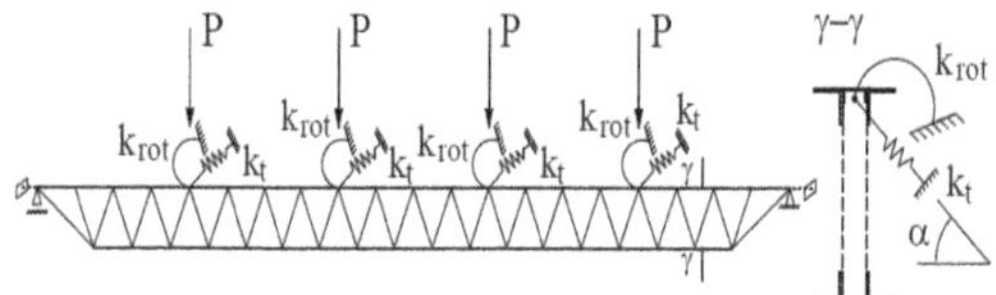

Figure 2. Static schema of the truss with elastic braces.

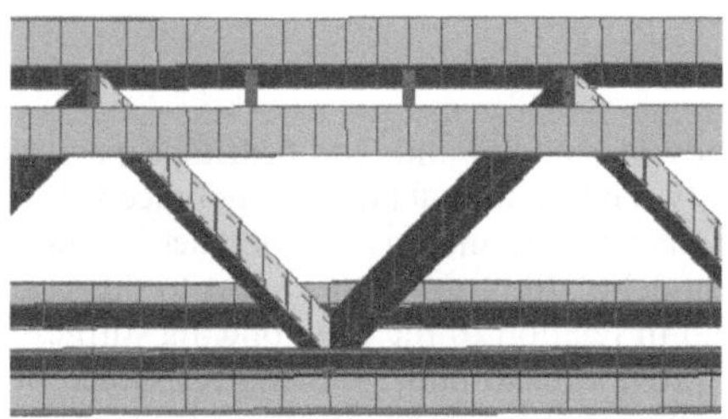

Figure 3. Model of the truss—beam model detail.

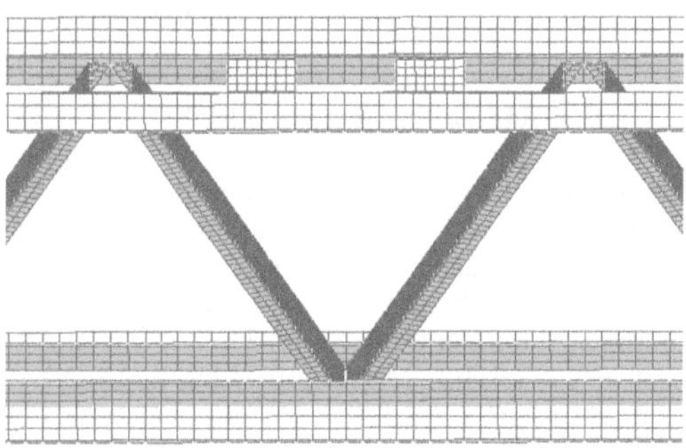

Figure 4. Model of the truss—shell model detail.

battens (at the top chord) were made of C140 profile except three most compressed members near the supports (at each side) which were made of 2 × L90 × 10 rolled profiles. The yield strength of the truss steel was 235 MPa.

In numerical analysis it was assumed that the load is applied in the form of point forces at four top chord braced joints. The truss was modelled by beam elements (Robot Structural Analysis 2010) or shell elements (Femap with NX Nastran 2009). The beam model of the truss was made by means of standard 1D elements with 6 degrees of freedom at node. Each member of top and bottom chord (each angle bar) was modelled separately (Fig. 3). In the shell model (3D), 4-node shell elements QUAD4 (with 6 degrees of freedom in node) were used (Fig. 4). The truss members were connected by means of RIGID links. The lateral sloping supports (sloped at angles α = 0°, 15°, 30°, 45° measured from the horizontal line) were modelled by means of ROD elements of axial stiffness only (k_t [kN/m]). The DOF-spring elements were used as the elastic supports with rotational stiffness (k_{rot} [kNm/deg]).

3 RESULTS OF NUMERICAL ANALYSIS

3.1 *Linear buckling analysis*

Linear buckling analysis confirmed that buckling load depends on brace stiffness and an angle α (Figs. 5–8). There was a threshold (minimum) stiffness (k_t [kN/m]—translational stiffness and k_{rot} [kNm/deg]—rotational stiffness) of braces which ensures that buckling load does not increase (or the load increase is less than 5%) (Table 1). The threshold rotational stiffness $k_{rot\text{-}lim}$ was found for the models with translational stiffness equal to

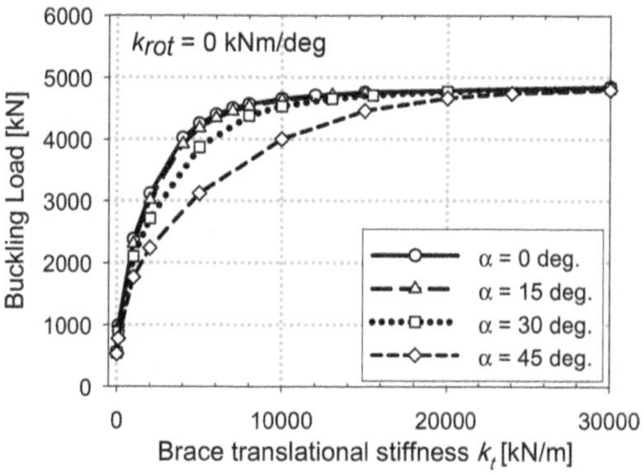

Figure 5. Comparison between the buckling load of the truss with respect to the angle and translational stiffness of braces for beam model.

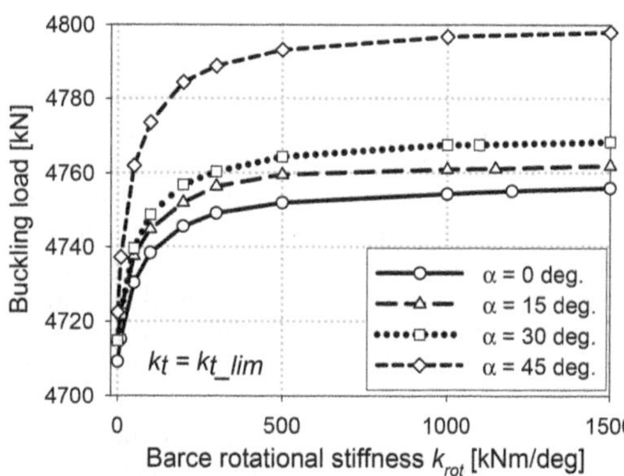

Figure 6. Comparison between the buckling load of the truss with respect to the angle and rotational stiffness of braces for beam model.

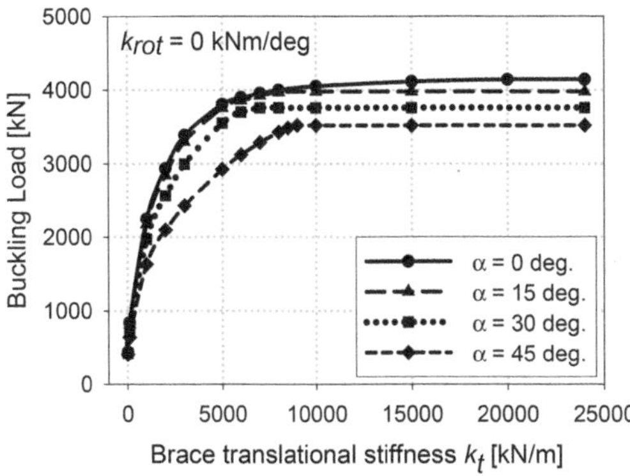

Figure 7. Comparison between the buckling load of the truss with respect to the angle and translational stiffness of braces for shell model.

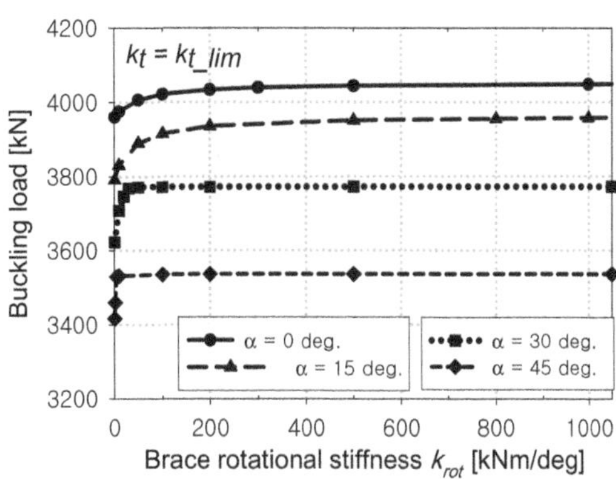

Figure 8. Comparison between the buckling load of the truss with respect to the angle and rotational stiffness of braces for shell model.

Table 1. Threshold stiffness of the braces $k_{s\text{-}lim}$ [kN/m], $k_{rot\text{-}lim}$ [kNm/deg].

	Beam model		Shell model	
	$k_{s\text{-}lim}$	$k_{rot\text{-}lim}$	$k_{s\text{-}lim}$	$k_{rot\text{-}lim}$
$\alpha = 0^\circ$	12000	1200	7000	1000
	$P_{cr} = 4700$ kN	$P_{cr} = 4756$ kN	$P_{cr} = 3956$ kN	$P_{cr} = 4048$ kN
$\alpha = 15^\circ$	13000	1150	5200	800
	$P_{cr} = 4704$ kN	$P_{cr} = 4760$ kN	$P_{cr} = 3796$ kN	$P_{cr} = 3956$ kN
$\alpha = 30^\circ$	15500	1100	5300	30
	$P_{cr} = 4712$ kN	$P_{cr} = 4768$ kN	$P_{cr} = 3588$ kN	$P_{cr} = 3768$ kN
$\alpha = 45^\circ$	22500	1000	7500	10
	$P_{cr} = 4720$ kN	$P_{cr} = 4796$ kN	$P_{cr} = 3356$ kN	$P_{cr} = 3532$ kN

$k_{s\text{-}lim}$. The differences between the magnitudes of buckling load (with respect to the brace angle) for the truss with rigid lateral braces ($k_t = 10^6$ kN/m, $k_{rot} = 0$ kNm/deg) were less than 1% for the beam model and about 15% for the shell model. The differences between magnitudes of buckling load for 1D and 3D models were up to 30% ($k_{rot} = 0$ kNm/deg). The possible explanation of the above described differences might be the built-up cross section layout of the truss members (influence of battens). The reason could be also different performance of the elastic braces in 1D and 3D models. In the beam model several members of the truss (top chord, diagonals) and elastic support are joined at one node. In the shell model braces were combined only with the nodes placed on the walls of the top chord cross-section.

The magnitude of buckling load raised up to 5% due to increase of brace rotational stiffness (Figs. 6 and 8). The buckled shape of the truss (beam and shell models) is presented in Figure 9.

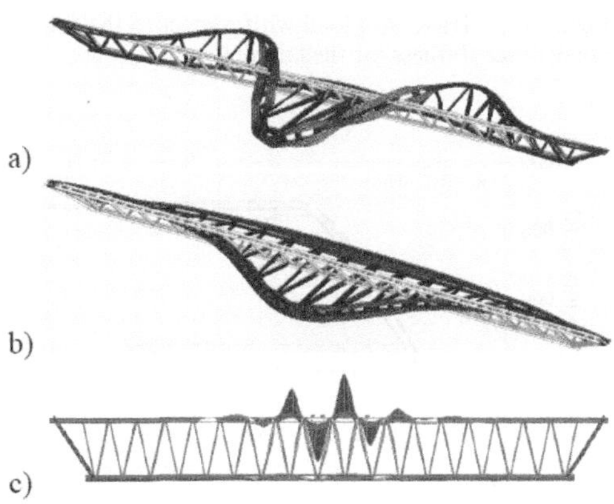

Figure 9. Buckled shape of the truss with braces of stiffness ($k_{rot} = 0$ kNm/deg): a) $\alpha = 0^\circ$, $k_t = k_{t\_lim}$ – shell model and $\alpha = 0^\circ, 15^\circ, 30^\circ, 45^\circ$ $k_t = k_{t\_lim}$ – beam model, b) $\alpha = 45^\circ$, $k_t = k_{t\_lim}$ – shell model, c) $\alpha = 0^\circ, 15^\circ, 30^\circ, 45^\circ$ $k_t = 10^6$ kN/m – shell model.

3.2 *Non-linear analysis*

Variants of non-linear static analysis (shell model) with material and geometric non-linearity for the imperfect truss model were carried out. The initial imperfection (Fig. 10) was implemented in the form of arc curvature at the top chord with an amplitude $L_t/500$ according to the code requirements (EC3 2006). As a result the magnitudes of limit load were obtained (depended on brace translational stiffness Fig. 11). For the truss with horizontal side

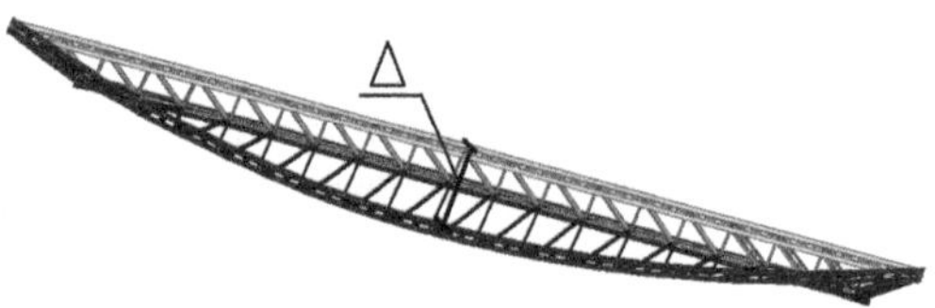

Figure 10. Initial geometric imperfection of the truss: $\Delta = L_t/500$.

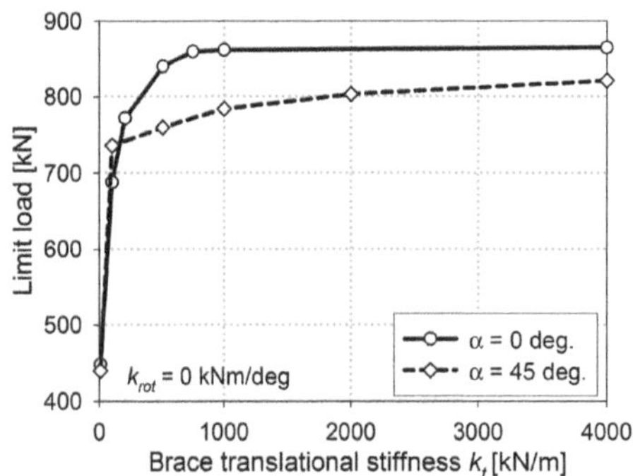

Figure 11. Truss limit load with respect to the translational brace stiffness for shell model.

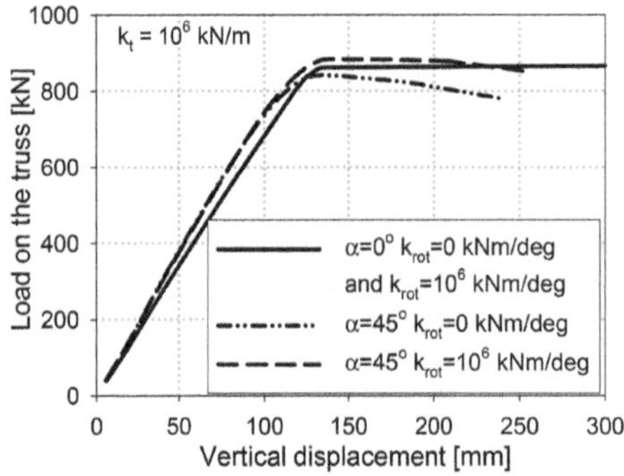

Figure 12. Truss loading due to the vertical displacement (at the midspan) for different stiffnesses of rotational side-supports.

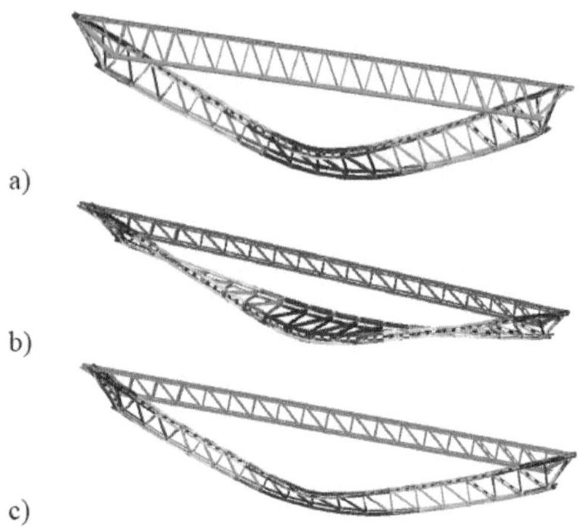

Figure 13. Post-buckling deformation of the truss (at the limit state—shell model) with braces of stiffness: a) $\alpha = 0°$, $k_t = 750$ kN/m ($k_{rot} = 0$ kNm/deg or 106 kNm/deg), b) $\alpha = 45°$, $k_t = 4000$ kN/m ($k_{rot} = 0$ kNm/deg), c) $\alpha = 45°$, $k_t = 106$ kN/m and $k_{rot} = 106$ kNm/deg.

Table 2. Truss bearing capacity according to EC3 with respect to the brace angle, (shell model), N_{brd} [kN], P_{max} [kN] (all top chord joints).

	$\alpha = 0°$	$\alpha = 15°$	$\alpha = 30°$	$\alpha = 45°$
N_{brd}	1956	1947	1935	1919
P_{max}	782	779	774	768

supports ($\alpha = 0^0$) the rotational stiffness had no influence on the magnitude of limit load (Fig. 12). In the case of truss with braces situated at angle equal to 45° the magnitude of limit load increased about 5% due to the increase of brace rotational stiffness. The post-buckling deformation of the structure is presented in Figure 13.

4 TRUSS BEARING CAPACITY ACCORDING TO EC3

On the basis of LBA results (buckling load, critical normal force at the middle of the span-top chord) for the structure with rigid braces the bearing capacity N_{brd} [kN] due to EC3 (2006) requirements was calculated. The maximum magnitude of truss loading P_{max} [kN] (referred to N_{brd}) is presented in Table 2.

5 CONCLUSIONS

The linear buckling analysis results confirmed that buckling load depends on translational and rotational brace stiffness and angle.

The threshold translational stiffness of braces raises with the increase of brace angle (except $\alpha = 0^{\circ}$ - 3D model).

The threshold rotational stiffness of braces decreases with the increase of brace angle.

For the shell model of the structure with rigid braces, the top chord of the truss buckled locally in the truss plane (bending and torsion of the top chord cross section). In this case the influence of battens (neglected in 1D beam model) on stiffness of the built-up cross section was very important.

The results of the non-linear static analysis (GMNIA) for the shell model of the truss showed that rotational stiffness of braces may cause an increase in bearing capacity (up to 5%).

Load bearing capacity of the truss calculated due to EC3 (2006) requirements was up to 13% lower than the results obtained from GMNIA.

REFERENCES

Autodesk Robot Structural Analysis Professional 2010, Autodesk Inc., 2010.

Biegus, A. 2014. Trapezoidal sheet as a bracing preventing flat trusses from out-of-plane buckling, *Archives of Civil and Mechanical Engineering*, http://dx.doi.org/10.1016/j.acme.2014.08.007.

Biegus, A. Wojczyszyn, D. 2011. Studies on buckling lengths of chords for out-of-plane instability, *Archives of Civil and Mechanical Engineering*, Vol. XI, No. 3:507–517.

Femap with NX Nastran 2009. Version 10.1.1. Siemens Product Lifecyde Management Software Inc.

Iwicki, P. 2010. Sensitivity analysis of critical forces of trusses with side bracing. *Journal of Constructional Steel Research* 66: 923–930.

Iwicki, P. Krajewski, M. 2013. 3D buckling analysis of a truss with horizontal braces, *Mechanics and Mechanical Engineering*, Vol. 17, No. 1: 49–58.

Jankowska-Sandberg, J. Kołodziej, J. 2013. Experimental study of steel truss lateral-torsional buckling. *Engineering Structures* 46: 165–172.

Krajewski, M. Iwicki. P. 2014. Stability and load bearing capacity of a truss with elastic braces, *Recent Advances in Computational Mechanics*, CRC Press Balkema, Taylor & Francis, London: 17–22.

Krajewski, M. Iwicki. P. 2014. Stability analysis of a truss with sloping elastic braces, *Proceedings of EUROSTEEL: 7th European Conference on Steel and Composites Structures*, Book of abstracts and electronic version, Neapol: 150–152.

PN-90 B-03200 Konstrukcje stalowe. Obliczenia statyczne i projektowanie.

PN-EN 1993-1-1 2006. Eurocode 3: Projektowanie konstrukcji stalowych. Część 1–1: Reguły ogólne i reguły dla budynków.

Advances in Mechanics: Theoretical, Computational and Interdisciplinary Issues – Kleiber et al. (Eds)
© 2016 Taylor & Francis Group, London, ISBN 978-1-138-02906-4

Experimental investigation of the innovative flow control blowing devices

A. Krzysiak
Institute of Aviation, Warsaw, Poland

ABSTRACT: Research of the use of new flow control methods and their implementation on real objects, especially on the aircrafts, is one of the priority directions in the development of new technologies. In the case of aircraft, the interest in new flow control methods are associated with potential opportunities to improve their aerodynamics and to reduce operating costs. The paper presents the results of experimental tests related to the application of the innovative flow control methods on aircraft lift surfaces or helicopter blades through the use of air blowing. Three different applications of the flow control blowing devices are presented. The experiments were carried out in Institute of Aviation low speed wind tunnels T-1, T-3 and trisonic wind tunnel N-3.

1 INTRODUCTION

Active flow control, regarded as a possibility of active or passive manipulation of flowfield, became widely used in many fields of science and technology. Many different flow control techniques are currently the subject of intense experimental and numerical studies in a number of research centers, (Gad-el-Had 2000, 2001). Improvement of the efficiency of currently used aircraft control systems or replace them by unconventional flow control methods, can be a source of measurable benefits. These benefits can still be significantly enhanced by the use of flow control operating in the Close Loop Control (CLC) System (Alam et al. 2006, Bright et al. 2005).

One of the methods of active flow control, described in the literature (Wilson et al. 2013, Ikeda et al. 2008), is an additional blowing on a wetted surface. In this method, properly targeted additional air jets increase energy of the flow. Boundary layer supplied with additional energy becomes less susceptible to separation, even at angles of attack higher than the critical (for the condition without blowing). Postponed flow separation may contributed to the increase of maximum lift and simultaneously to drag decrease. This in turn, may improve e.g. the airplane aerodynamic performance.

In the paper three different applications of the flow control blowing devices on the aircraft wings and helicopter blades are shown. The presented blowing devices were used not only to delay the flow separation, but also to control the load on the aircraft wing. Experimental studies are related to:

- flow control using conventional and self-supplying air jet vortex generators installed on the segment of the airfoil,
- separation control on the wing flap controlled by close loop system,
- wing load control using blowing fluidic devices.

2 WIND TUNNELS

The experiments were carried out at the Institute of Aviation low speed wind tunnels T-1, T-3 and trisonic wind tunnel N-3. The low speed wind tunnel T-1 is closed-circuit, continuous-flow wind tunnel with 1.5 meter diameter open test section. The range of freestream velocity is 15÷40 m/s.

The T-3 low speed wind tunnel is an atmospheric, closed-circuit tunnel with an open test section of 5 meter diameter, which can reach velocity of 90 m/s.

The N-3 wind tunnel is a blow-down type with partial re-circulation of the flow. It can operate in subsonic, transonic and supersonic flow regimes at Mach numbers $M = 0.2 \div 1.2$, 1.5 and 2.3. The closed test section is equipped with perforated top and bottom walls for tests at subsonic and transonic flow velocities and solid walls—at supersonic range.

3 EXPERIMENTAL RESULTS AND ANALYSIS

3.1 *Flow control using self-supplying air jet vortex generators*

Experimental studies of flow control using air jet vortex generators (both conventional and proposed self-supplying) were conducted in the Institute of Aviation (IoA) low speed wind tunnels T-1, in the range of Mach numbers M = 0.05 ÷ 0.1. Air jet vortex generators, which are an alternative to traditional vane vortex generators consist of a number of small streams outgoing from the wetted surface and properly oriented with respect to the undisturbed flow. The interaction between the air jets and undisturbed flow generates a well-organized vortex structures, Figure 1.

The vortices are able to withstand the adverse pressure gradient appearing on the upper surface of the airfoil, at higher angles of attack. As a result the flow separation is delayed. Proposed self-supplying air jet vortex generators in comparison with conventional ones use the airfoil overpressure regions, as a source of the air for them, Figure 2.

Experimental and numerical study of conventional and proposed self-supplying air jet vortex generators allowed to make an analyze of the impact of the generated vortex structures on the flow separation on the upper surface of the NACA 0012 airfoil. The optimal design parameters of air jet vortex generators were obtained.

In Figure 3 the influence of conventional and proposed self-supplying air jet vortex generators on lift coefficient and critical angle of attack is presented. The lift coefficient C_L was calculated by integrating pressure distribution along the airfoil chord. Experimental studies showed, that the proposed self-supplying air jet vortex generators produce so strong turbulence areas, that their effectiveness is only slightly lower than optimal designed conventional ones and they can become an alternative to the previously used vortex generators. An important advantage of the use of self-supplying air jet vortex generators, instead of conventional ones, would be a significant simplification of their design, as they would not need to use an external air supply source (i.e. compressor), but would be supplied with air from the overpressure areas.

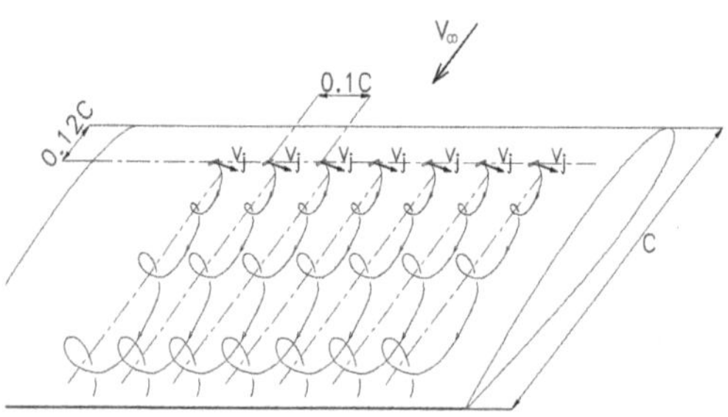

Figure 1. Air jet vortex generators.

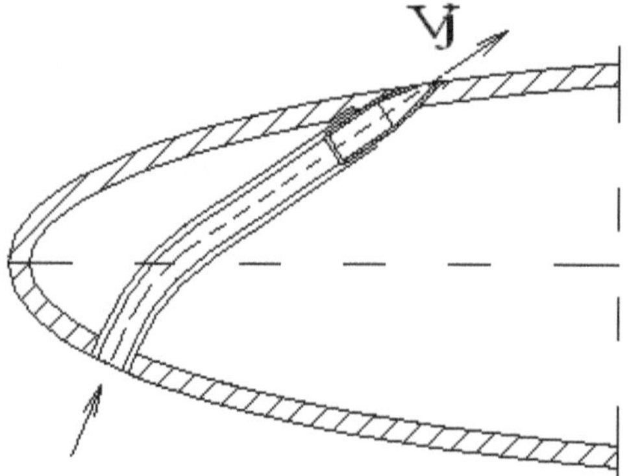

Figure 2. Proposed self-supplying air jet vortex generators.

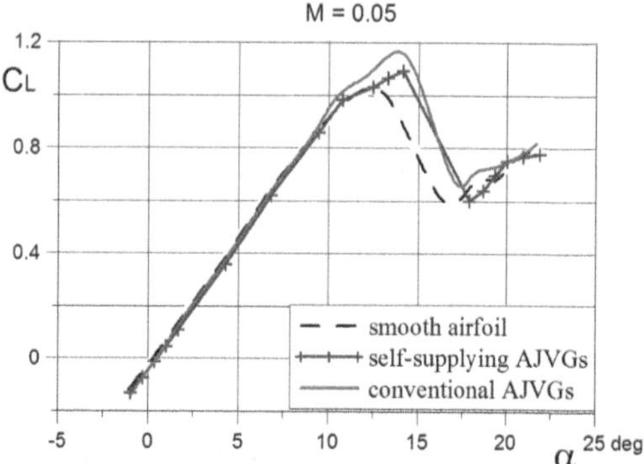

Figure 3. Lift coefficient versus angle of attack for conventional and proposed self-supplying air jet vortex generators.

3.2 *Separation control on the wing flap controlled by close loop system*

Wind tunnel tests of flow control, using an additional blowing on the airfoil movable flap, were performed in the IoA low speed wind tunnels T-1 in the range of Mach numbers M = 0.05 ÷ 0.1. Blowing was realized through the set of nozzles located on the trailing edge of the main body of the airfoil. Air flow through the nozzles was controlled by a set of the electromagnetic valves located inside the model.

Minimization of air flow rate necessary to maintain the desired state of the boundary layer on the upper flap surface, required the implementation of such process in a control loop. The main task of the Close Loop Control (CLC) system used for fluidic active flow control was to keep flow attached, during flap movement, in the range from $\delta = 0^0$ to $\delta = 45^0$. This process was based on "on-line" analysis of the boundary layer. Pressure, measured by the sensors mounted on the upper flap surface (close to its trailing edge), created a feedback signal for a CLC system. This signal was analyzed by control unit. In the case of attached flow the coefficient C_p of measured pressure had a positive value and the valves were closed. In the case of a detached flow, the coefficient C_p of measured pressure had a negative value and the valves were opened. In Figure 4, general idea of closed loop control system for fluidic active flow control is presented.

Experimental studies showed, that the air blowing on the airfoil flap eliminates the flow separation even on the strongly deflected flap. The result of flow attachment to the upper flap surface is significant increase in the airfoil lift coefficient. In Figure 5 the influence of steady air blowing on the airfoil flap (deflected at $\delta = 400$), with volume mass flow rate VFR = 120 m3/h, on the lift coefficient versus airfoil angle of attack is presented.

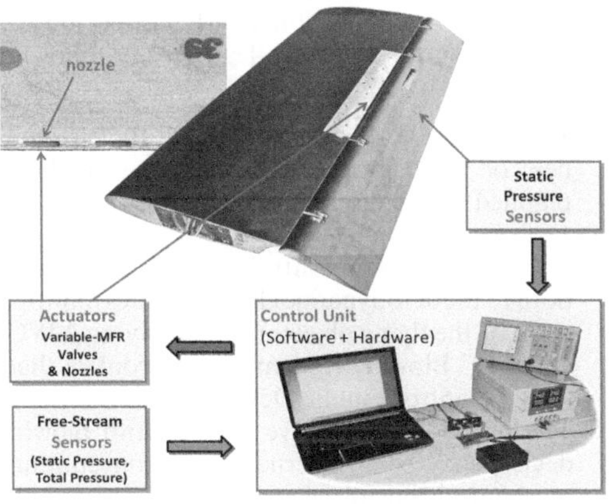

Figure 4. The general concept of CLC-system.

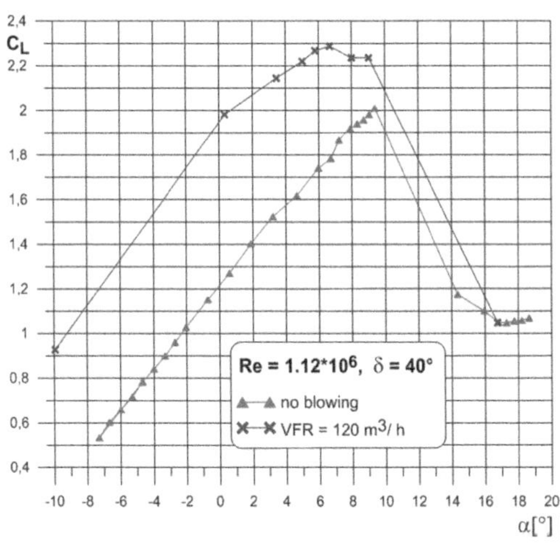

Figure 5. The influence of air blowing on lift coefficient.

The example of wind tunnel test results with the usage of CLC-System for active flow control on the airfoil flap is presented in Figure 6. This figure presents the changes in the pressure distribution on the upper and lower flap surface and rear part of the airfoil main body over time, during the CLC-System operation. Using a pulsed jets controlled by CLC System, the volume flow rate was diminished from VFR ≈ 120 m³/h (steady blowing) to VFR ≈ 68 m³/h with the similar effect of the airfoil lift increase.

The work was performed under the European project "ESTERA".

3.3 *Wind load control using blowing fluidic devices*

The wing load control systems are developed as the means to modify a distribution of aerodynamic load on the wing. Such systems are usually used in extraordinary, off-design flow conditions. Particularly, it concerns the reduction of bending

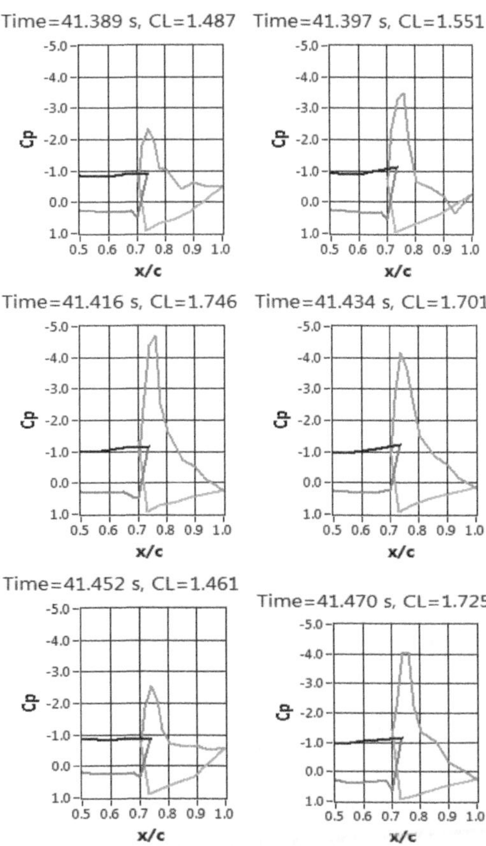

Figure 6. Changes in the pressure distribution over the time.

loads during accelerated flight manoeuvres or sudden gusts. In such situations, rising bending loads may lead to damage of the wing.

A new concept of active flow control system based on proposed blowing devices for the control of the aerodynamic load on aircraft wing was designed in Institute of Aviation. Two systems of fluidic control devices were tested. The first system (named "Fluidic Spoiler") was based on the nozzles blowing air in normal and in inclined directions, with respect to the upper wing surface. This system used the matrix of 540 mini nozzles arranged in nine rows, located at the 59 ÷ 92% of the wing span and 45 ÷ 65% of the wing chords (every 2.5%). The second system (named "DTEN") was based on specially designed nozzles located on a modified trailing edge surface. The fluidic control devices were supplied with air from external compressor, or alternatively from the high pressure area situated at lower wing surface, close to its leading edge.

The experimental tests were performed in IoA low speed wind tunnel T-3. For these tests the model of semi-span wing (2.4 m span), situated vertically on the endplate in wind tunnel test section, was used, Figure 7. The model was situated on the two aerodynamic wall balances. To measure the load distribution along the semi-span wing model, the 8 strain-gauge bridges were glued to the model front spar. Furthermore, recording and measurement system ESAM was used. Wind tunnel test were performed at Mach number $M = 0.1$.

The study was carried out in the framework of the European project STARLET.

The experimental studies showed, that using proposed blowing devices significant bending moment reduction can be achieved. In Figure 8 the achieved relative root bending moment reduction coefficient C_{BMA} versus air mass flow rate is presented.

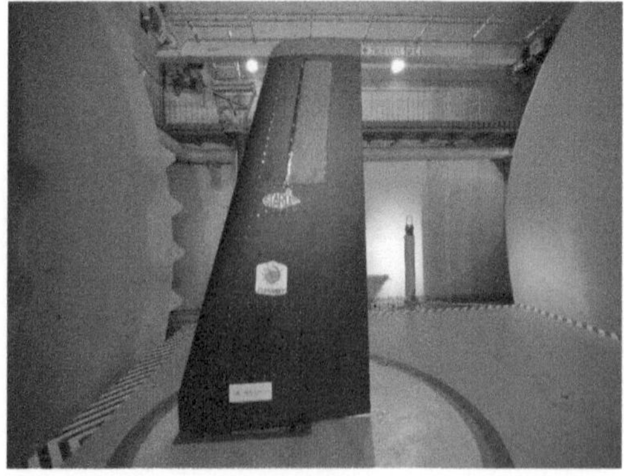

Figure 7. The semi-span wing in the wind tunnel T-3.

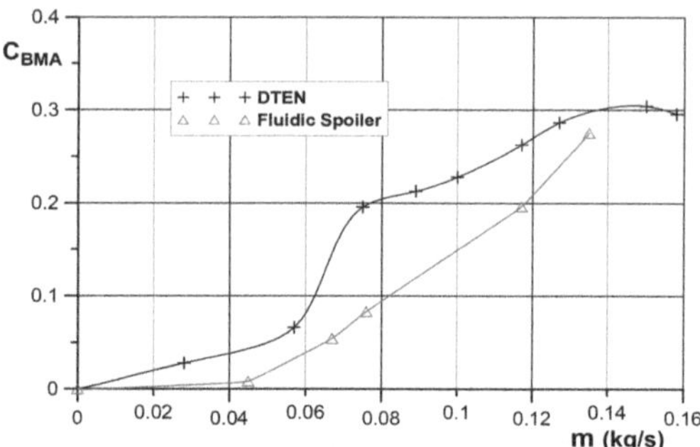

Figure 8. Relative root bending moment reduction coefficient versus air mass flow rate for tested fluidic device.

4 CONCLUSIONS

Experimental test results, described in the article related to the application of the three innovative flow control methods, lead to the following conclusions:

- Wind tunnel tests have confirmed the effectiveness of three proposed flow control methods, planned to be used on the flying objects lift surfaces.
- The presented flow control methods can be used both to the aerodynamic characteristics improvement of the flying objects (self-supplying AJVGs and CLC Blowing System) and to control them (Load Control Fluidic Devices).
- The proposed innovative flow control blowing devices may be an alternative for conventional, mechanical control systems.

REFERENCES

Alam, M. & Liu, W. 2006. Close-Loop Separation Control—An Analytic Approach. *Physics of Fluids* 18.

Bright, M. & Culley, D. 2005. Closed Loop Active Flow Separation Detection and Control in a Multistage Compressor. *NASA/TM* 213553.

Gad-el-Hak, M. 2000. *Flow Control—Passive, Active, and Reactive Flow Management.* Cambridge University Press.

Gad-el-Hak, M. 2001. Flow control—The future. *Journal of Aircraft* 38: 402–418.

Ikeda, M. & Yoshida, K. 2008. A Flow Control Technique Utilizing Air Blowing to Modify the Aerodynamic Characteristics of Pantograph for High-Speed Train. *Jour. of Mechanical Systems for Transportation and Logistics* 01: 264–271.

Wilson, J. & Schatzman, D. 2013. Suction and Pulsed-Blowing Flow Control Applied to an Axisymmetric Body. *AIAA Journal* 51: 2432–2446.

Advances in Mechanics: Theoretical, Computational and Interdisciplinary Issues – Kleiber et al. (Eds)
© 2016 Taylor & Francis Group, London, ISBN 978-1-138-02906-4

Influence of aluminum layer thickness on the fatigue properties of super-nickel alloy and crack detection by non-destructive techniques

D. Kukla & Z.L. Kowalewski
Institute of Fundamental Technological Research, Warsaw, Poland

ABSTRACT: The paper presents the results of fatigue tests performed on the super-nickel alloy after Chemical Vapour Deposition (CVD). CVD process was carried out for the periods of 4 and 12 hours, and as a consequence, a layer of the aluminum of the thickness 20 μm and 40 μm, respectively, was obtained on the surface of nickel alloy. The specimens with layers were tested on the servo-hydraulic testing machine under dynamic loading at high temperature of 900°C. The profiles of micro hardness on the cross-sections of specimens enabled identification of hardness variations. The main aim of the paper was to evaluate the effect of a film thickness on the fatigue properties of the alloy. Additionally, an identification of crack propagation in the layer subjected to cyclic loading was analysed on the basis of deformation changes in the subsequent cycles. The trial for identification of damage localization with the use of Electronic Speckle Pattern Interferometry optical method (ESPI) was made. It allowed for detection of cracks forming during cyclic loading of the sample. This enabled evaluation of fatigue life of the layer, as the formation of crack precede sample decohesion.

1 INTRODUCTION

Surface treatment of heat-resistant nickel based alloys is commonly used for improving their heat resistance in the operating conditions of the turbine engines for example, where temperature can reach even 1650°C at the outlet of the combustion chamber. The protective layers improve the high-temperature corrosion and erosion resistance, and as a consequence, improve the ability to raise the operating temperature. However, improvement of the heat resistance of alloys by the aluminide layer covering may lead to reduction of the creep resistance and fatigue strength.

The goal of this research was to estimate an influence of diffusion layer on the selected mechanical properties during creep or fatigue. It is well known that observations of a crack initiation of the substrate due to damage of the layer and determination of a dynamics of damage evolution are difficult from engineering point of view. Evaluation of the fatigue life and crack propagation in the aluminide layer on nickel based alloys are usually limited to microstructural observations of samples after different number of loading cycles. Such methodology, requiring fatigue test interruption, suffers to some disadvantages leading to considerable errors. Each removal of a sample from the testing machine changes the loading conditions of subsequent cycles, which strongly affects the reliability of the result, Khajavi & Shariat, 2004. The non-destructive techniques application during fatigue tests improves significantly such situation since it enables a detection of the crack without removing of the sample from the testing machine throughout the experiment from the start to the end. In this research the ESPI method takes the role of the non-destructive technique. It enables full-field displacement distribution measurement on the sample surface considered which can be subsequently used for analysis of damage development.

2 METHODOLOGY

The films for tests were obtained basing on the nickel alloy MAR 247. The aluminizing process was carried out by the Chemical Vapour Deposition (CVD) method using $AlCl_3$ vapours in hydrogen atmosphere as the carrier gas at temperature of 1040°C for the periods of 4 h or 12 h, and the reduced pressure of 150 hPa. Depending on the duration of this process the aluminum layers of 20 μm or 40 μm thickness were obtained on the nickel alloy surface, Sitek R. et al. 2014.

The fatigue tests were carried out on the MTS 810 testing machine of the axial force capacity equal to +/– 25 kN and equipped with the FLEX digital controller. The specimen temperature of 900°C during tests was obtained by means of the induction heater mounted on the testing stand as it is shown in Figure 1a. The geometry of specimens applied in the experimental programme is presented in Figure 1b.

All fatigue tests were force controlled under the assumption of the zero mean level and constant stress amplitude for a given specimen tested. The frequency during fatigue tests was equal to 20 [Hz], whereas the magnitudes of stress amplitude varied from 380 [MPa] up to 520 [MPa].

Since it was assumed that the diffusion layer is more susceptible to brittle fracture, micro hardness profiles on the cross section of specimens were prepared in order to evaluate a hardness gradient. The micro hardness was measured using the Hysytron device, under load of 0.0005 [N] starting from the edge of specimen. Observation of the fatigue damage development was conducted by means of the light and scanning microscopes in order to confirm whether the layer cracks are generating before the specimen's decohesion.

In order to detect a place of the crack initiation in the aluminide diffusion layer using the ESPI method a short interruptions of the fatigue test are required after selected number of cycles. During these periods the specimen was monotonically loaded with several stops during which ESPI images were captured, Figure 2.

Analysis of strain distribution changes observed with the use of ESPI camera allowed determination of crack appearance in the aluminide layer during fatigue test. ESPI images were taken when fatigue test was interrupted after the fixed number of cycles. In order to keep the high resolution of ESPI images, it was necessary to use handy operated pump for loading the specimen, since the

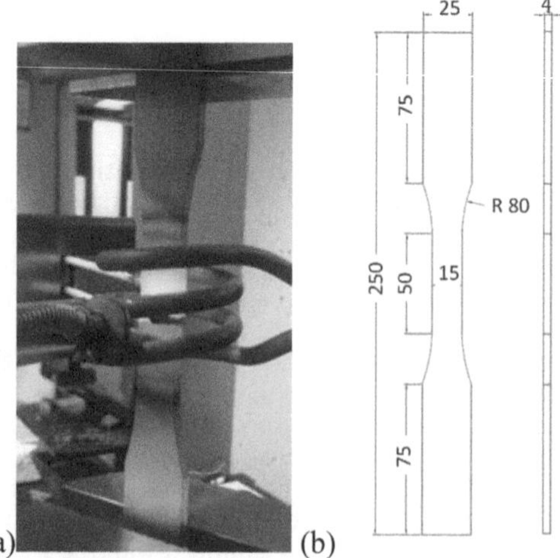

Figure 1. Specimen mounted in the gripping system of the testing machine with heater (a), and engineering drawing of the specimen (b).

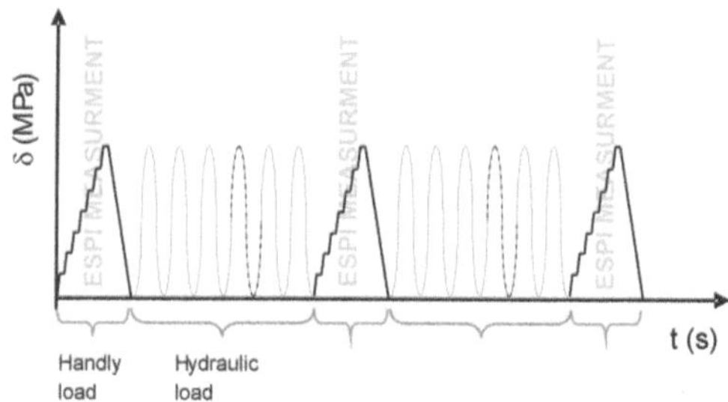

Figure 2. Diagram presenting sample load in fatigue cycles.

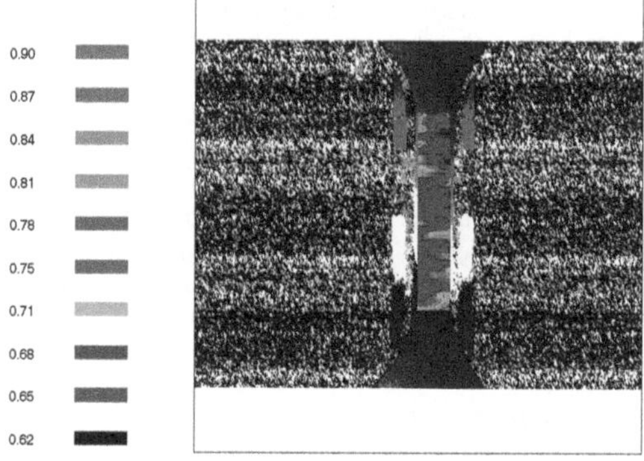

Figure 3. Interference image with superimposed map of specimen deformation generated by the ESPI.

hydraulic system of the testing machine interfered with the ESPI system leading to significant measurement errors. Measurements were repeated until a crack of the aluminide layer was detected prior to the sample fracture.

For each of tested samples from 5 to 12 displacement distribution images were recorded. They enabled effective analysis of deformation process evolution. Strain measurement using the ESPI camera requires specially prepared surface of sample. Displacement measurement is based on the analysis of the interference fringes due to the applied load. A typical example of image from displacement measurements performed by means of the ESPI camera is presented in Figure 3.

3 RESULTS

Experimental programme contained 14 fatigue tests (7 tests for each layer thickness considered). Microscopic measurements were carried out directly after fatigue tests. The results of fatigue investigations were elaborated in the form of Wöhler diagrams presented in Figure 4. The figure shows variation of the cycles number up to failure

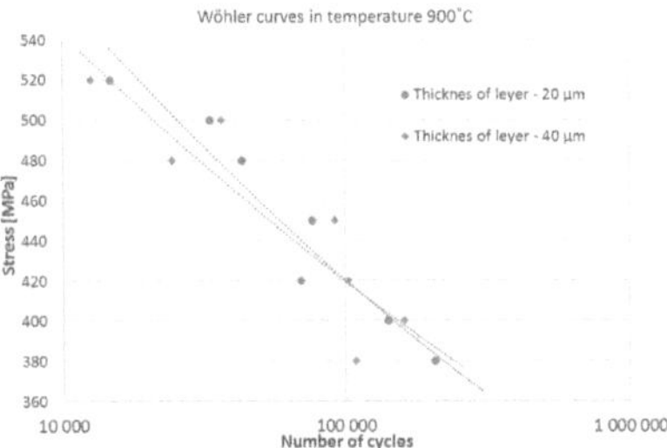

Figure 4. Wöhler characteristics for specimens with surface deposited by the aluminum layer, fatigue in temperature 900°C.

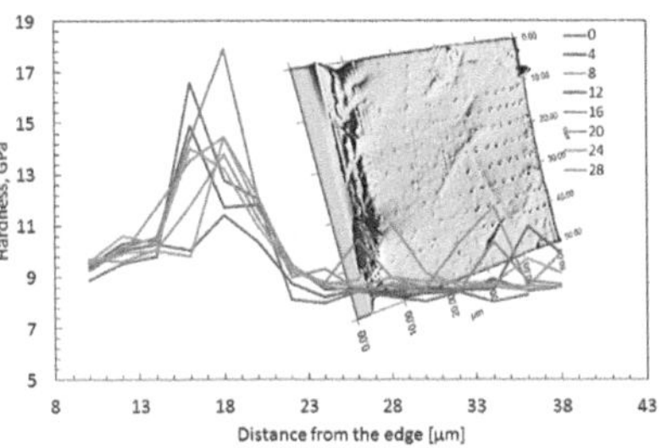

Figure 5. Micro hardness profiles and the map of indenter positions for micro hardness measurement.

depending on the thickness of the aluminum layer deposited on the surface of the super-nickel alloy specimen. It has to be noticed, that the fatigue results are not consistent taking into account the effect of the layer thickness. For three levels (of seven levels considered) of the stress amplitude a longer fatigue lifetimes were achieved for 20 µm layer thickness, for the remaining tests, however, longer lifetimes were obtained for thicker layer.

The profile of micro hardness variations is shown in Figure 5. It also contains a topography map of the surface tested where the spots of indentation are illustrated. As it is shown in this figure, the values of micro hardness of the aluminum layer are significantly higher than those for the nickel core determined.

Observation of the fatigue fracture using the light microscope, Figure 6, and SEM technique, Figure 7, enabled identification of the layer cracks generated just before the decohesion of the specimens tested. In the microscopic damage analysis of tests carrying under fatigue conditions usually such damage sensitive strain parameters as the accumulated inelastic strain or accumulated mean strain level well describe a damage development in the subsequent loading cycles. They can be easily determined on the basis of hysteresis loops, Socha (2003), Kukla et al. (2014).

In this case however, it has to be emphasized, that the effect observed during microscopic

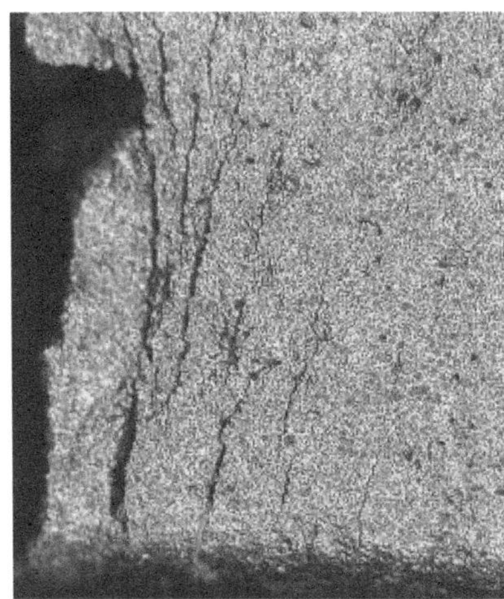

Figure 6. Illustration of cracks in the layer area close to the specimen failure—photo from light microscope.

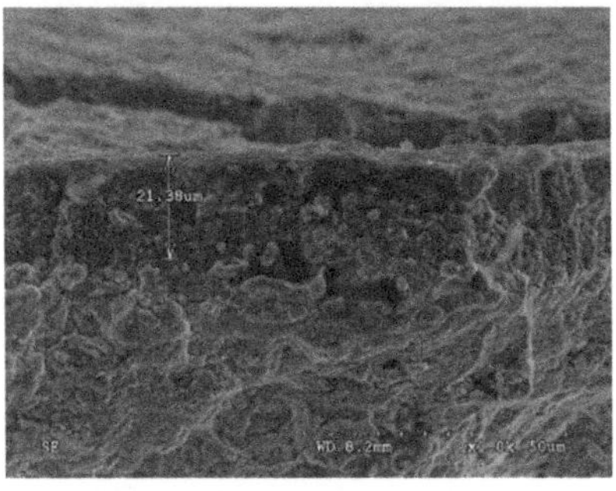

Figure 7. Illustration of cracks in the layer area close to the specimen failure—photo from SEM.

inspections could not be confirmed by the damage sensitive strain parameters mentioned above.

Looking at the hysteresis loops representing strain evolution during subsequent cycles of loading, a process of the layer cracking was completely invisible. Detailed analysis of the fatigue test under cyclic loading at stress amplitude of 600 MPa is presented in Figure 8. It shows the deformation maps worked out with the use of the ESPI technique. All images were recorded with the use of the procedure described above, starting from the image after first cycle and 5 selected numbers of cycles. The last measurement was made after 80000 cycles. The test was interrupted before fracture of the sample. Deformation values are expressed using dimensionless units [mm/mm] * 10^{-3} (Fig. 8). Analysis of deformation maps enabled identification of strain concentration areas (Fig. 8f) acted as crack growth initiators. One of them became to be dominant and led to the sample fracture.

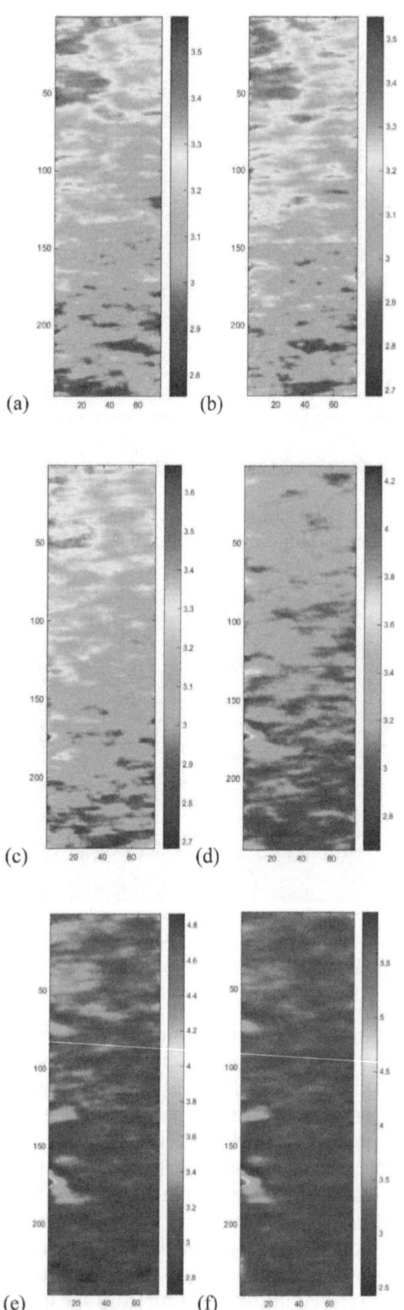

Figure 8. Deformation maps of sample surface after: (a) 1 cycle (b) 40001 cycles; (c) 60001 cycles; (d) 70001 cycles; (e) 75001 cycles (f) 80001 cycles of fatigue test.

4 CONCLUSIONS

The cracks of layer are responsible for damage initiation of the nickel based samples after CVD process. It has to be noted that the aluminum layer does not affect the fatigue strength of the nickel super-alloy. Also, a thickness of the layer covering this material is not important in this matter. The identification of the layer's cracking initiation is impossible by monitoring of the deformation level, only. Therefore, an application of the non-destructive methods capturing the layers cracking initiation and subsequent development is necessary in further studies.

The results confirmed applicability of the ESPI as the non-destructive technique in effective detection of fatigue cracks within the diffusion layer on the super-nickel alloy before the sample fracture.

Cracking of heat resistant layer reduces locally the heat resistance of a detail, and more importantly, leads to significant strength lowering due to the fact that areas of the reduced head resistance serve as notches.

Usage of the non–destructive technique described in this paper enables thorough investigation of cracking phenomena occurring during cycling loading of the coated samples, and estimation of heat resistance efficiency of the coating.

ACKNOWLEDGEMENTS

The authors gratefully acknowledge the funding by The National Centre for Research and Development, Poland, under Program for Applied Research, grant no 178781.

REFERENCES

Khajavi, M.R. & Shariat, M.H. 2004. Failure of first stage gas turbine blades. *Engineering Failure Analysis* 11: 589–597.

Kukla D. et al. 2014. Assessment of fatigue damage development in power engineering steel by local strain analysis. *Metallic Materials* 52: 269–277.

Sitek R. et al. 2014. Influence of High-Temperature Aluminizing on the Fatigue and Corrosion Resistance of Nickel Alloy Inconel 740, *Frontier of Applied Plasma Technology* 7(1).

Socha, G. 2003. Experimental investigations of fatigue cracks nucleation, growth and coalescence in structural steel. *International Journal of Fatigue* 25(2): 139–147.

Advances in Mechanics: Theoretical, Computational and Interdisciplinary Issues – Kleiber et al. (Eds)
© 2016 Taylor & Francis Group, London, ISBN 978-1-138-02906-4

Topology optimization as a tool for road pavement structure analysis

R. Kutyłowski & M. Szwechłowicz
Wrocław University of Technology, Wrocław, Poland

ABSTRACT: The paper presents the analysis of the road pavement structure from the point of view of topology optimization point of view. Minimum compliance approach was applied. As a result, the material distribution under the wheel is discussed. The representative stiffness of the road pavement structure is defined as a quantity characterizing the road pavement structure. The analysis of the material distribution is made taking into account the guideline requirements.

1 INTRODUCTION

In the paper variational formulation of the topology optimization problem is used (Bendsøe & Sigmund, 2003). A strain energy functional, being an equivalent of compliance, was minimized while constraints were imposed on the body mass:

$$F(\rho(x),\lambda) = \min \left\{ \begin{array}{l} \int_\Omega \mathbf{e}^{\mathrm{T}}\ \mathbf{C}(\rho(x))\ \mathbf{e}\ \mathrm{d}\Omega \\ +\lambda(\int_\Omega \rho(x)\ \mathrm{d}\Omega - m_0) \end{array} \right\} \tag{1}$$

where:

$$m_0 = \alpha m, \qquad 0 < \alpha < 1, \qquad m = V\rho \tag{2}$$

where ρ = is the design variable (density of the material); V = is the volume of the design domain.

During the optimization process in each j-th step the mass must be equal to the initial mass:

$$m_j = m_0 \tag{3}$$

This approach (a modified SIMP method) was a basis for preparing the numerical algorithm and the authors *Matlab* code, which was earlier tested and verified (Kutyłowski et al. 2011, Kutyłowski & Szwechłowicz 2008).

Pavement structures are designed in accordance with the proper guidelines taking the planned traffic load (traffic class) into account through the proper choice of the type of pavement structure (flexible, semi-rigid and rigid structures). The main goal of the paper is to discuss the material distribution from the topology optimization point of view and to define the representative stiffness of the road pavement structure.

2 ROAD PAVEMENT STRUCTURE ANALYSIS

In Figure 1a the considered, computational model of the road structure loaded by concentrated force (wheel) is shown.

A solution obtained for this model can be treated as a reliable basis for dimensioning the structure in the cross section and longitudinal section of the road. In order to maintain mutual correspondence between the recommended pavement structure and the adopted computational model the design domain was divided proportionally to the thickness of the particular road pavement structural courses.

At the left side of this figure the road pavement courses are marked from the top by colours, (they are reproduced by shades of grey): the yellow colour represents the bitumen wearing course, green colour—bitumen binder course,

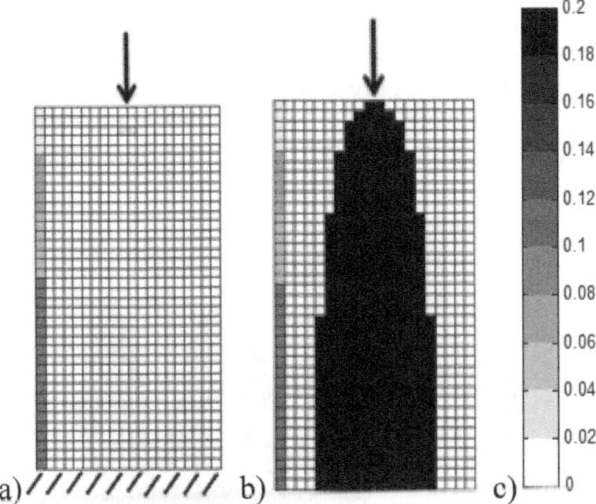

Figure 1. Computational model a), 0/1 topology b), grey scale assigned to interval of 0.0–0.2 c).

blue colour—asphalts concrete course, red colour—compacted aggregate or roadstone course. Typical material-void solution is presented in Figure 1b. An examination of the obtained solutions (Fig. 1b, and Fig 2) shows that the material is symmetrically distributed in the form of a triangle with a rather small base. The load is transferred to the ground through the base courses in such a way that the material needed to carry the load is distributed in the form of a rectangle. The computations distinguish areas in the design domain, in which material should be located when the load is applied to the design domain in the assumed place. Additionally, similar material amount is noticed in particular courses in considered cases and the distribution of material for one force (Fig. 1b) is similar to one particular force in three forces case (marked in box—Fig. 2) where the vehicle with 3 axis is shown: small vehicle on the left side and track vehicle on the right.

The question is considered what material should be distributed in the design domain, due to its properties. As previously, material density proportional to the strain energy accumulated in the particular areas, and so proportional to the strain of the areas under the prescribed load, is of interest here. Thus it will be possible to show what material is required in particular areas. It can be done by analyzing the strain energy distribution in the final optimization step (obtained from the boundary value problem solution—Fig. 3).

Then distributions will be considered of considerable share of shades of grey. In order to facilitate the analysis of such distributions a proper numerical tool was built. The tool enables to interactively analyze material strain in the particular subareas of the design domain owing to the fact that the scale in which the examined topology is viewed can be changed in real time during the analysis. The ten-interval grey scale covering the interval of 0–1, is applied to, e.g., an interval of 0–0.2, as shown in Figure 1c. As a result, all the elements with a density higher than 0.18, are assigned the black colour.

The viewing scale of 0–1 was adopted in Figure 1b and in Figure 3a—for the final topology

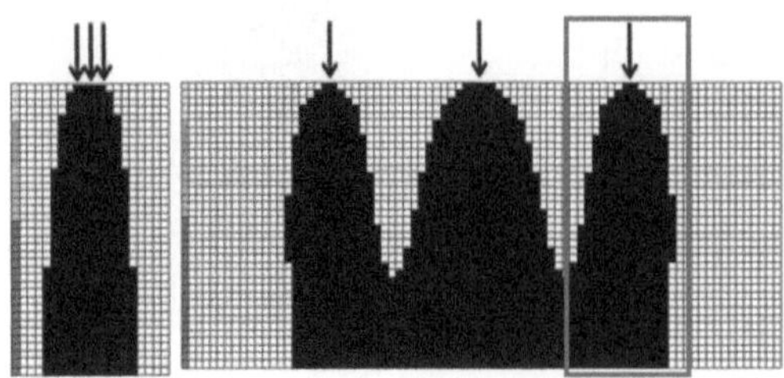

Figure 2. 0/1 topologies for various road loading.

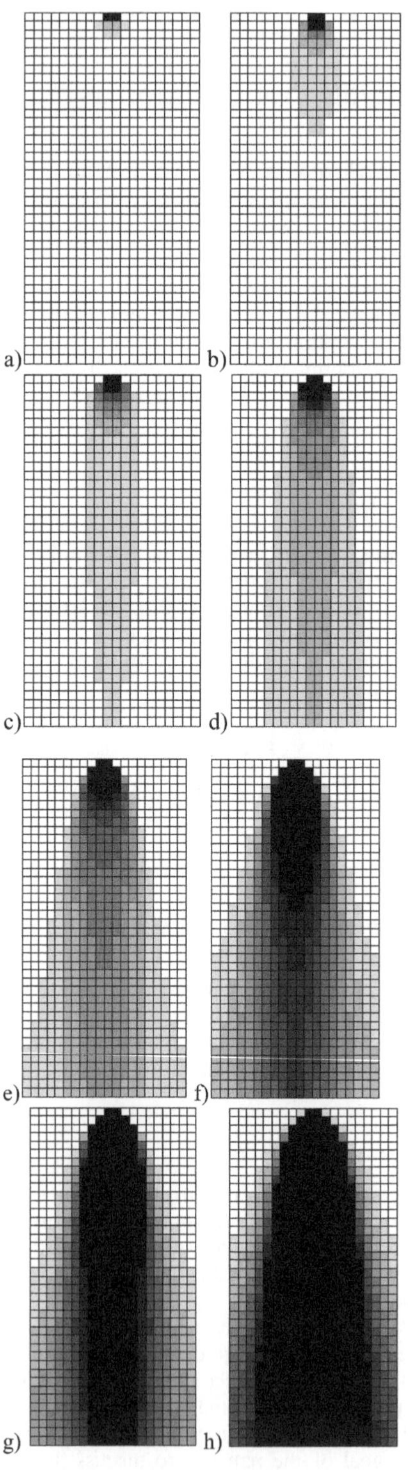

Figure 3. Density distribution topologies for α = 0.5 in scale intervals: a) 0–1; b) 0–0.3; c) 0–0.2; d) 0–0.1; e) 0–0.06; f) 0–0.03; g) 0–0.02; h) 0–0.01.

and for the strain energy distribution in this final step respectively. Sporadically occurring areas with a density intermediate between 0 and 1 (Fig. 3a) are marked with an appropriate shade of grey corresponding to the density value. In Figure 3a only in the place where the force is applied in the top wearing course two elements have a density of 1 (the black colour). Most of the other finite elements have a density in the interval of 0–0.1 (the white colour). The next solutions in Figures 3b–3g, and h have density distributions shown in appropriate scales in order to reveal the variation in material density distribution and the most strained places.

The scale adopted in the particular solutions is specified in the figure description. For example, Figure 3f shows a topology in the scale of 0–0.03, where densities in the interval of 0.027–1 are marked black and the lower density values are assigned shades of grey and the white colour.

The material with the highest density (Fig. 3d) is distributed in the top course (layer) of the analysed structure. The next courses should be made of a

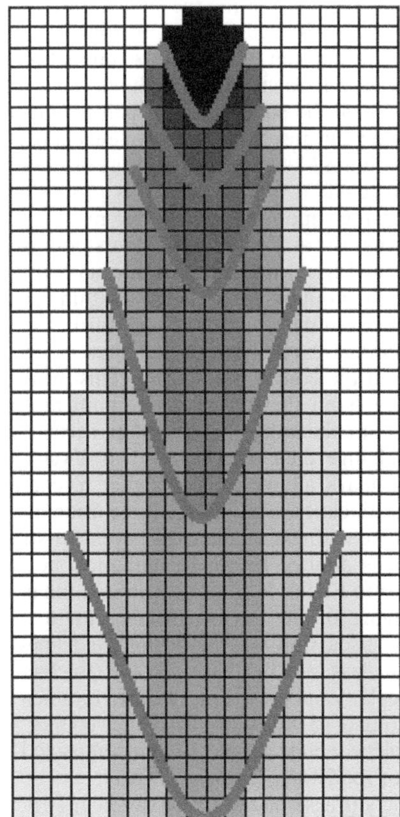

Figure 4. Density distribution topologies in scale interval of 0–0.06 with plotted isolines.

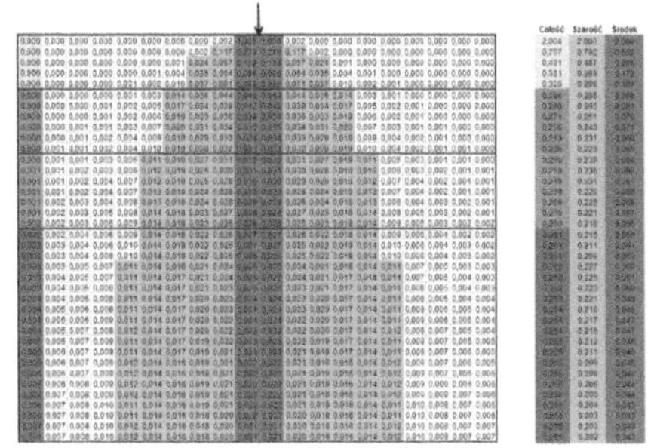

Figure 5. Density distribution topology in number notation (for topology showed in Fig. 3a).

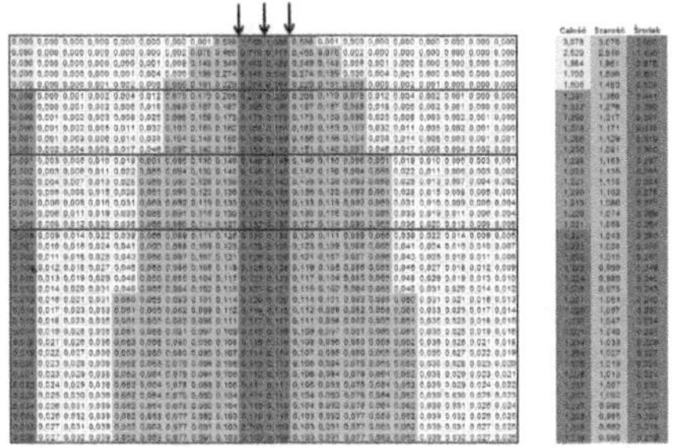

Figure 6. Numerical notation solution—small, 3 axis vehicle.

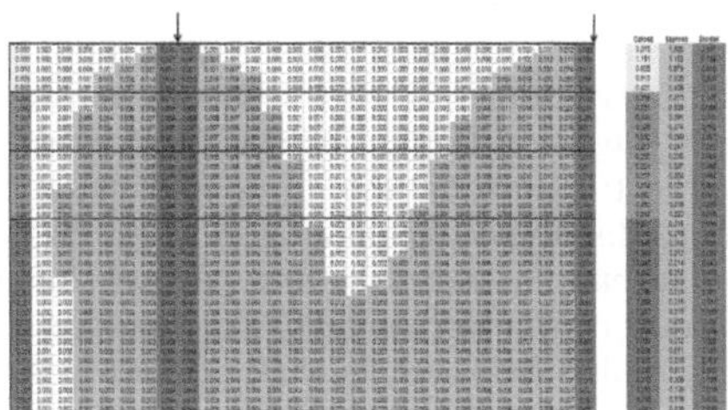

Figure 7. Numerical notation solution—truck, 3 axis vehicle (because of the symmetry—half of the solution).

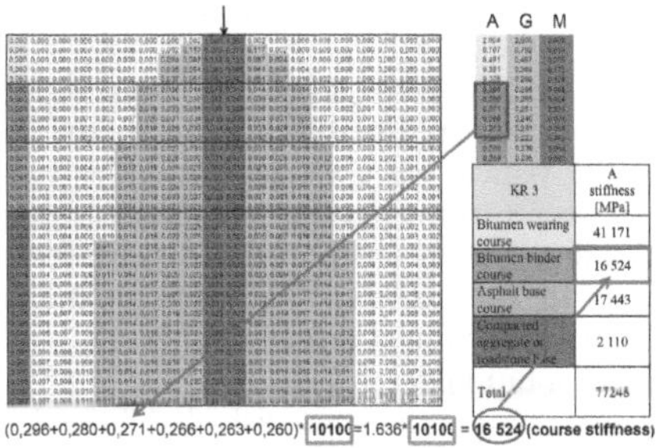

Figure 8. The example of calculating the course stiffness.

Table 1. The representative stiffness of the road pavement structure (KR3—traffic class).

KR3	A	G	M	*E* GPa
Bitumen wearing course	3.998	3.946	3.120	10.3
Bitumen binder course	1.636	1.375	0.462	10.1
Asphalt base course	1.816	1.598	0.416	9.6
Compacted aggregate or roadstone base	5.275	4.242	0.945	0.4
KR3	**A Stiffness MPa**	**G Stiffness MPa**	**M Stiffness MPa**	
Bitumen wearing course	41.171	40.643	32.131	
Bitumen binder course	16.542	15.103	4661	
Asphalt base course	17.443	15.342	3994	
Compacted aggregate or roadstone base	2110	1697	378	
Total	77.248	72.785	41.164	

relatively weaker (lower density) material (Fig. 3e). Less dense material is in a larger number of elements (the wider cross section) than more dense material (the narrower cross section). It should be also noticed that directly under the load (wheel) one can distinguish a material with a certain density, which transfers the load to the soil. Also a certain specific material distribution can be discerned (Figs 3d–3e) where the areas of similarly dense material have the shape of an ellipse.

The areas can be distinguished by means of isolines. Figure 4 shows such isolines drawn in Figure 3e. It is apparent that the load makes it necessary to distribute the material in a specific way. The distribution is laminar and parallel, but the layers do not run horizontally, which proves that directly under the wheel the strain of the material is relatively higher than in the neighbouring areas. In this way it has been demonstrated that optimal structures are laminar structures (Kutyłowski et al. 2011). As confirmed by computations, the larger the distance (downwards) from the load, the relatively weaker, but more widely distributed material is needed.

The above analysis should be treated as an analysis of the real distribution of material effort under the vehicle wheel. Since the latter moves, each consecutive cross section will require the same distribution. This means that the material requirements for the successive cross sections are the same as the ones determined directly under the force in the above solution.

Three domains are distinguished: All the design domain (A), the grey domain (G) separated as more strenuous domain and the black domain (M), situated just under the wheel. For detail analysis of the material distribution the number notation, normalized to 1 is used (Fig. 5). In Figure 5—on the right side one can find the rows summation of the material in following form: column A summation over the all the particular row, column G summation over the grey and black elements in particular row, column M summation over the black elements in particular row. The density is proportional to the Young's modulus of the real material and in this sense the analysis can be understood. As it can be seen (Fig. 5) the densities in obtained optimal solution should change fluently within each layer and they should decrease downwards. Analysing the courses it can be seen that there is no difference on the course boundaries. This means from the one hand that the guideline hints are proper, and from the other hand that particular course may change their thicknesses. It is a very important remark allowing the designer to change the course thicknesses. It may be done using the representative stiffness of the road pavement structure, which is defined below. In numerical notation—similar to Figure 5—mall vehicle solution is shown in Figure 6 and track vehicle in Figure 7. The above discussion was made for guidelines' KR3 traffic class and flexible structure (thicknesses of particular courses are as follow: bitumen wearing course—5 cm, bitumen binder course—6 cm, asphalts concrete course—7 cm, compacted aggregate or roadstone course—20 cm).

The stiffness of the road pavement structure is defined as the sum of courses stiffness of a considered structure. Particular course representative stiffness is defined as the sum of the material density of the course multiplying by the Young modulus of considered course. The example of the bitumen binder course is shown in Figure 8 where the Young's modulus of the course was assumed as equal to 10.100 MPa (autumn and spring season). In Table 1 the stiffnesses for particular domains are calculated. The stiffness of A domain (A—all the design domain—is equal to 77.248 MPa) and it is very similar to the stiffness of the grey domain (equal to 72.785 MPa). The difference in this case is equal to 5%. We propose to define the representative stiffness of the road pavement structure as the stiffness of the grey domain, which finally is the same as in the 0/1 solution black domain (Fig. 1b).

3 CONCLUSION

The topology optimization seems to be a relevant tool for the road pavement structure analysis. The real, material distribution under the wheel can be analysed. The guideline requirements were discussed. Based on the analysis of the material distribution the road pavement structure may be designed properly, according to the material effort. The representative stiffness of the road pavement structure was proposed as a quantity which characterized the road pavement structure. It may be useful in the design process of the road pavement structure.

REFERENCES

Bendsøe, M.P. & Sigmund, O. 2003. *Topology optimization, theory, methods and applications*. Berlin, Heidelberg, New York: Springer.

Kutyłowski, R., Rasiak, B. & Szwechłowicz, M. 2011. Topology optimization as a tool for obtaining a multimaterial structure. *Archives of Civil and Mechanical Engineering*, XI: 391–409.

Kutyłowski, R. & Szwechłowicz M. 2008. Analyse des Penalty-Faktors in SIMP-Methode in Bezug auf die Konvergenz der Lösung. *Proceedings in Applied Mathematics and Mechanics*. 8: 10803–10804.

Advances in Mechanics: Theoretical, Computational and Interdisciplinary Issues – Kleiber et al. (Eds)
 ISBN 978-1-138-02906-4

GPR simulations for diagnostics of a reinforced concrete beam

J. Lachowicz & M. Rucka
Faculty of Civil and Environmental Engineering, Gdańsk University of Technology, Gdańsk, Poland

ABSTRACT: The most popular technique for modelling of an electromagnetic field, The Finite Difference Time Domain (FDTD) method, has recently become a popular technique as an interpretation tool for Ground Penetrating Radar (GPR) measurements. The aim of this study is to detect the size and the position of damage in a reinforced concrete beam using GPR maps. Numerical simulations were carried out using the finite difference time domain method. Four different damage scenarios with different crack width and shape were investigated. The influence of the frequency of applied Electromagnetic (EM) waves on the possibility of damage detection was examined.

1 INTRODUCTION

The damage assessment of existing structures is of great importance to improve their reliability and safety. The actual state of a structure can be assessed in the process of diagnostics. Recently, various non-destructive testing techniques have been developed for evaluation of engineering structures, for example acoustic emission methods (e.g. Gołaski et al. 2012) ultrasonic methods (e.g. Rucka & Wilde 2013) or acoustic and vibroacoustic methods (e.g. Hoła et al. 2011). Particularly useful in diagnostics of reinforced concrete or masonry structures is the Ground Penetrating Radar (GPR) technique (e.g. Bęben et al. 2013, Binda et al. 2005, Cassidy et al. 2011, Hugenschmidt 2002, Lachowicz & Rucka 2015), which uses the phenomenon of Electromagnetic (EM) wave propagation.

A primary problem in the GPR technique is the interpretation of measurement data because of the variety of factors that can affect GPR signals. In order to make an appropriate analysis of GPR data, considerable amount of expertise in interpreting of experimental results is required. Moreover, different defects may cause similar types of patterns in GPR maps. Numerical modelling of electromagnetic wave propagation can provide a significant support in understanding the origin of reflections appearing in GPR experimental data. Moreover, numerical simulations allow performing a preliminary analysis, with different damage scenarios prior to the in-situ surveys.

2 PROBLEM FORMULATION

The aim of this study is to detect the size and the position of damage in a reinforced concrete beam using GPR maps. The beam had a length of 5 m and a rectangular cross-section area 30 cm by 40 cm. The scheme of the analyzed beam is illustrated in Figure 1. Damage was considered as an open crack of rectangle, triangle and trapezoid shape. Four different damage scenarios with different crack width were investigated. Numerical calculations were conducted using a software called gprMax 2D (Giannopoulos 2005). Calculations were performed on the middle part of the beam span over a distance of 2 m. The influence of different frequencies of excited waves (1 GHz, 1.6 GHz, 1.8 GHz and 2 GHz) of electromagnetic waves on resulting radargrams was examined.

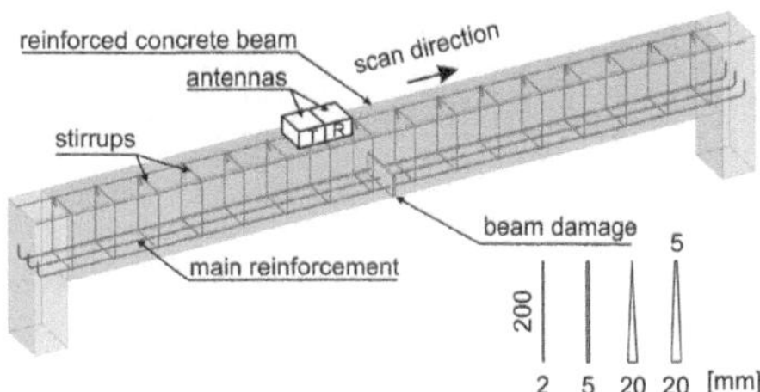

Figure 1. Scheme of the reinforced concrete beam with different damage scenarios.

3 NUMERICAL ANALYSIS BY FDTD

The Maxwell's equations form the theoretical basis of EM wave propagation. In lossy and anisotropic medium, the Maxwell's equations take the form (Giannopoulos 2005):

$$\nabla \times \mathbf{E} = \frac{-\partial \mathbf{B}}{\partial t}, \tag{1}$$

$$\nabla \times \mathbf{H} = \frac{-\partial \mathbf{D}}{\partial t} + \mathbf{J}_c + \mathbf{J}_s, \tag{2}$$

$$\nabla \cdot \mathbf{B} = 0, \tag{3}$$

$$\nabla \cdot \mathbf{D} = q_v, \tag{4}$$

where: **H** and **E**—magnetic and electric field strength vector, respectively, **B**—electric displacement vector, **D**—magnetic flux density vector, t—time, $\mathbf{J}_c$—conduction current density, $\mathbf{J}_s$—impressed current density, q_v—volume electric charge density. The most common numerical method of solving Maxwell's equations for electromagnetic wave propagation analysis is the FDTD method (Colla et al. 1998, Diamanti et al. 2008) developed by Yee (1966). The approach of this method is to discretize both time and space continua.

The numerical model used in this study was composed of Yee cells. To set a material, constitutive parameters (permeability μ, permittivity ε and conductivity σ) were assigned at nodes of Yee cells. At the border of computational space, absorbing boundary conditions (ABC) were applied.

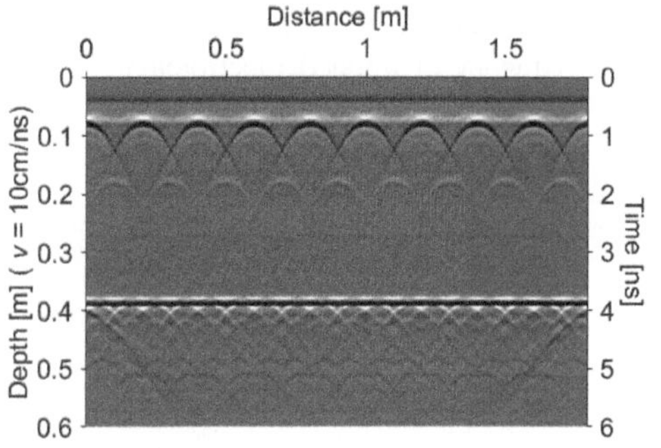

Figure 2. Radargram for the beam without damage using the antenna operating at a frequency of 2 GHz.

4 RESULTS OF DAMAGE DETECTION

Result of calculations for the intact beam using the antenna operating at a frequency of 2 GHz is shown in Figure 2. In the GPR map (also known as radargram) several hyperbolas are visible. These hyperbolas correspond to reflection from reinforcement bars. Moreover, a longitudinal reflection is visible which corresponds to the main reinforcement of the beam.

In Figure 3 results of calculations for the beam with damage in the form of 2 mm rectangular crack are presented. Radargrams were obtained using the antenna operating at a frequency of 1 GHz, 1.6 GHz, 1.8 GHz and 2 GHz. Regardless of the scanning frequency, no reflections from damage are visible. However, a strong reflection can be observed at the contact of air with the main reinforcement which remains continuous within the crack (Fig. 3d).

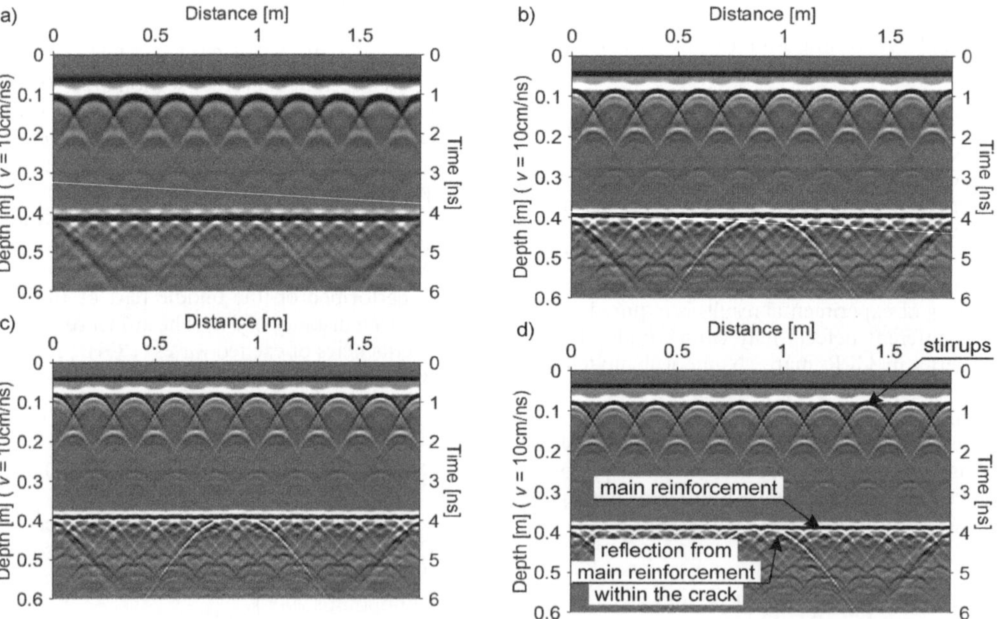

Figure 3. Radargram for the beam with 2 mm rectangular damage using the antenna operating at a frequency of: a) 1 GHz, b) 1.6 GHz, c) 1.8 GHz, d) 2 GHz.

Figures 4, 5 and 6 presents GPR maps calculated for the beam with 5 mm rectangular, triangular and trapezoid damage, respectively. For higher frequencies (1.6 GHz, 1.8 GHz and 2 GHz) a single hyperbolic reflection corresponding to damage is visible on the center part of radargrams. The vertex of this hyperbola indicates the upper part of the defect. On this basis the position of the defect can be found but there is no possibility to establish its width. Moreover, it can be observed that there is no difference between reflection from 5 mm rectangular and trapezoidal crack (Figs. 4 and 5), but for triangular damage weak multiple reflections appear on radargrams (Fig. 6).

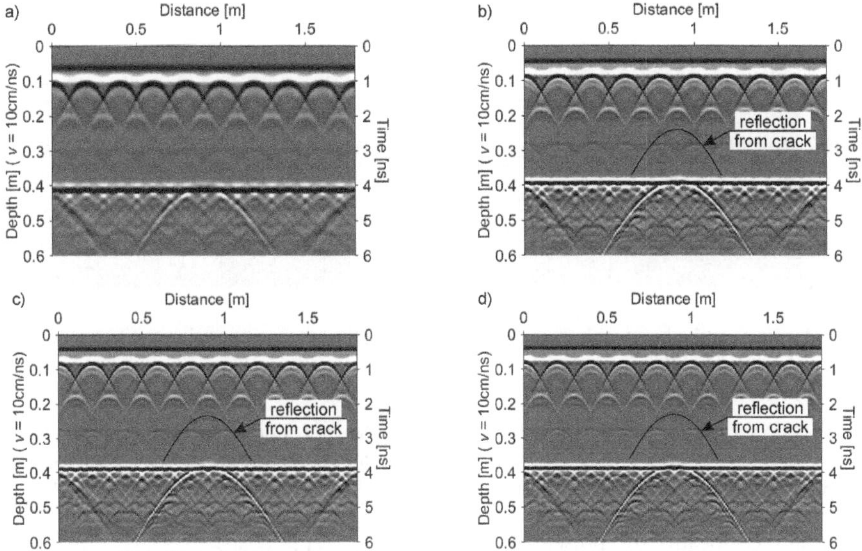

Figure 4. Radargram for the beam with 5 mm rectangular damage using the antenna operating at a frequency of: a) 1 GHz, b) 1.6 GHz, c) 1.8 GHz, d) 2 GHz.

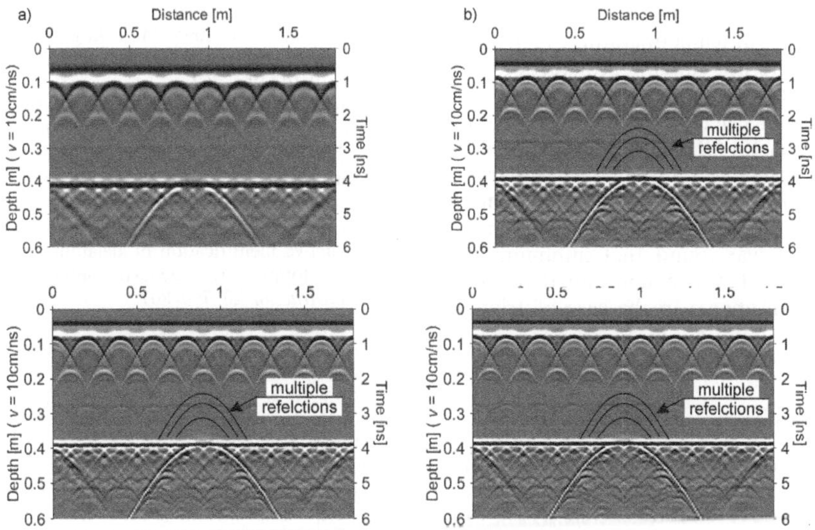

Figure 5. Radargram for the beam with triangular damage using the antenna operating at a frequency of: a) 1 GHz, b) 1.6 GHz, c) 1.8 GHz, d) 2 GHz.

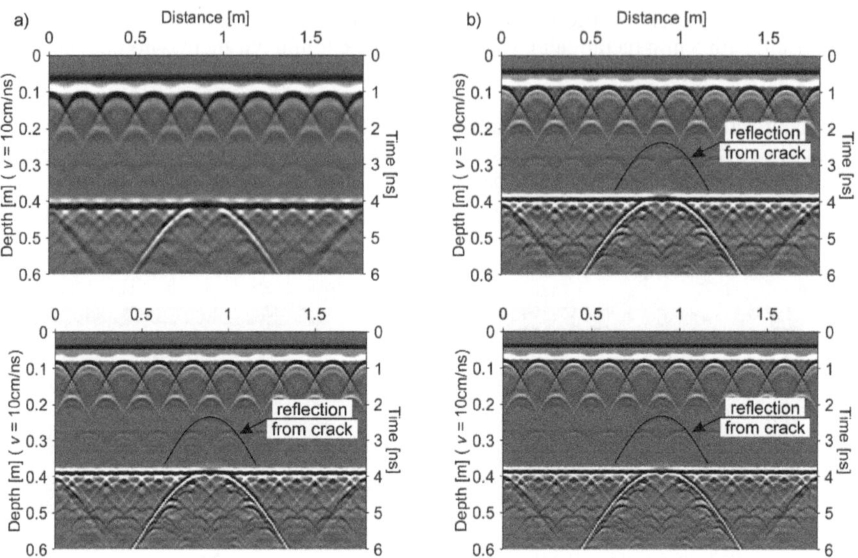

Figure 6. Radargram for the beam with trapezoid damage using the antenna operating at a frequency of: a) 1 GHz, b) 1.6 GHz, c) 1.8 GHz, d) 2 GHz.

5 CONCLUSIONS

In the paper results of numerical simulations of electromagnetic wave propagation in the concrete reinforced beam are presented. The FDTD calculations for various damage scenarios and various operating frequencies of the antenna allowed assessing the possibility of the application of the GPR method for damage diagnostics of reinforced concrete structures. It was found that electromagnetic waves excited and acquired by the antenna operating at a frequency of 1 GHz were not suitable in detection of any kind of considered damage types. On the other hand results of calculations for higher frequencies (1.6 GHz, 1.8 GHz and 2 GHz) revealed no significant differences in obtained radargrams. Numerical GPR simulations revealed some features of radargrams caused by particular types of defects. It was found that minimum delectable crack width was 5 mm and that multiple reflections can appear in the case of triangular damage.

REFERENCES

Bęben, D., Mordak, A. & Anigacz, W. 2013. Ground Penetrating Radar application to testing of reinforced concrete beams. *Procedia Engineering* 65: 242–247.

Binda, L., Zanzi, L., Lualdi, M. & Condoleo, P. 2005. The use of georadar to assess damage to a masonry Bell Tower in Cremona, Italy. *NDT&E International* 38: 171–179.

Cassidy, N.J., Eddies, R. & Dods, S. 2011. Void detection beneath reinforced concrete sections: The practical application of ground penetrating radar and ultrasonic techniques. *Journal of Applied Geophysics* 74: 263–276.

Colla, C., Burnside, C.D., Clark, M.R., Broughton, K.J. & Forde, M.C. 1998. Comparison of laboratory and simulated data for radar image interpretation. *NDT&E International* 31(6): 439–444.

Diamanti, N., Giannopoulos, A. & Forde, M.C. 2008. Numerical modelling and experimental verification of GPR to investigate ring separation in brick masonry arch bridges. *NDT&E International* 41: 354–363.

Giannopoulos, A. 2005. Modelling ground penetrating radar by gprMax. *Construction and Building Materials* 19(10): 755–762.

Gołaski, L., Goszczyńska, B., Świt, G. & Trąmpczyński, W. 2012. System for the global monitoring and evaluation of damage processes developing within concrete structure under service load. *The Baltic Journal of Road and Bridge Engineering* 7: 237–245.

Hoła, J., Sadowski, Ł. & Schabowicz, K. 2011. Nondestructive identification of delaminations in concrete floor toppings with acoustic methods. *Automation in Construction* 20: 799–807.

Hugenschmidt, J. 2002. Concrete bridge inspection with a mobile GPR system. *Construction and Building Materials* 16: 147–154.

Lachowicz, J. & Rucka, M. 2015. Application of GPR method in diagnostics of reinforced concrete structures. *Diagnostyka* 16(2): 31–36.

Rucka, M. & Wilde, K. 2013. Experimental study on ultrasonic monitoring of splitting failure in reinforced concrete. *Journal of Nondestructive Evaluation* 32: 372–383.

Yee, K.S. 1966. Numerical solution of initial boundary value problems involving Maxwell's equations in isotropic media. *IEEE Transactions on Antennas and Propagation* 14: 585–589.

Advances in Mechanics: Theoretical, Computational and Interdisciplinary Issues – Kleiber et al. (Eds)
© 2016 Taylor & Francis Group, London, ISBN 978-1-138-02906-4

Application of polymer element in reduction of temporary steel grandstand vibrations

N. Lasowicz & R. Jankowski
Gdańsk University of Technology, Gdańsk, Poland

ABSTRACT: The numerical analysis focused on reduction of vibrations of a temporary steel scaffolding grandstand has been conducted in this paper. These types of structures are regularly subjected to dynamic loads which, in conjunction with light and quite slender structural members, may induce dangerous vibrations. To increase their safety, temporary steel grandstands are usually strengthened with the diagonal stiffeners of tubular cross section. Another approach, using a diagonal element consisting of two L-shape steel members bonded with polymer mass, has been considered. Dynamic parameters, such as modes of free vibrations and the corresponding natural frequencies for both structural models, have been estimated and compared. Then, the dynamic transient analysis has been conducted and the behaviour of the structure under human-induced excitation due to jumping has been determined. The results of the study show that the response of the temporary steel grandstand equipped with the polymer element as well as with the typical stiffener are substantially different. The application of the polymer element leads to substantial reduction in the levels of peak acceleration and displacement values as the result of high damping properties of polymer applied.

1 INTRODUCTION

Civil engineering structures may be induced to vibrations as the result of ground motions (see, for example, Jankowski 2005, 2007; Mahmoud et al., 2011, 2012), large mechanical vibrating machines, vehicles using the nearby road or movement of people. The last excitation, if acting on relatively flexible structure, such as a temporary steel grandstand, may lead to substantial damages (De Brito & Pimentel 2009).

A large number of sporting, music and other entertainment events are nowadays organized around the world. Temporary steel grandstands erected with scaffolding system are often used during such events. These structures are regularly subjected to dynamic loads which, in conjunction with light and quite slender structural members, may induce dangerous vibrations (Reynolds et al., 2004). Such vibrations may cause serious problems, including damages, collapse of structures and panic among spectators (Nhleko et al., 2009; Jones et al., 2010). Previous numerical and experimental studies have confirmed that mass of the people significantly decreases values of natural frequencies making the structure more vulnerable (Lasowicz et al., 2014, 2015b).

One of the most common methods used to reduce the grandstand vibrations is the application of a bracing system. It consists of additional diagonal elements installed on the structure so as to increase its stiffness (Crick & Grondin 2008). In the paper, an alternative method of reduction of structural vibrations is considered. An additional element consisting of two L-shape steel members (50 × 50 × 5 mm) bonded with polymer mass of thickness 5 mm is analysed. The polymer mass consider in the study is a specially designed flexible two-component grout of high damping properties (Falborski et al., 2012, 2013; Lasowicz et al., 2015a). The aim of the paper is to analyse numerically the response of the temporary steel scaffolding grandstand equipped with two different diagonal elements, which has been subjected to dynamic loads due to jumping.

2 MODAL ANALYSIS

Two different numerical models of a part of a typical temporary steel grandstand have been generated using commercial programme MSC Marc. The first of them describes the structure with typical diagonal stiffener member of tubular cross-section (Model 1—see Fig. 1), while the second one concerns the structure equipped with polymer element (Model 2—see Fig. 2). Scaffolding system consisting of slender tubular structural member has been modelled by standard two-node beam-column elements, while platforms and benches have been generated as standard four-node shell elements available in the program. In the case of polymer,

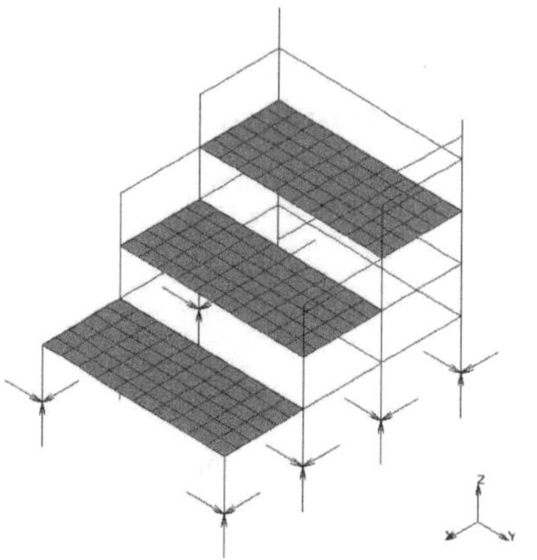

Figure 1. Numerical model of an empty temporary steel grandstand with tubular element (Model 1).

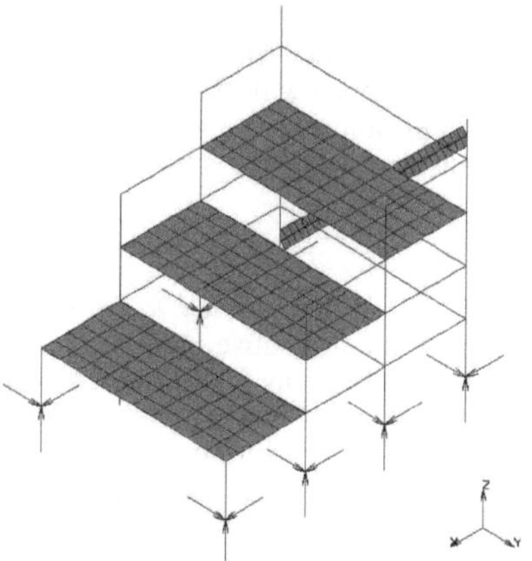

Figure 2. Numerical model of an empty temporary steel grandstand with polymer element (Model 2).

standard eight-node (six degrees of freedom for each node) solid elements have been used.

The behaviour of polymer has been simulated with the use of the Mooney-Rivlin material model that is the most frequently adopted method for modelling complex mechanical behaviour of elastomers and rubber-like solids. The following material constants for the five-parameter Mooney-Rivlin model have been applied: C_{10} = 889,490 kPa, C_{01} = −245,840 kPa, C_{20} = −155,310 kPa, C_{11} = 93,786 kPa, C_{30} = 11,148 kPa. The bulk modulus has been set to be 2.5 GPa, as commonly used for elastomers and rubber-like materials.

The bases of the models have been considered to be fixed only in the translational directions. The empty grandstand and the structure that takes into account mass of spectators have been considered in the study. It has been assumed that twelve people occupy the structure.

For both models and two load cases considered in the study, dynamic parameters, such as modes of free vibrations and corresponding natural frequencies, have been first determined. In the case of temporary and retractable grandstands, the left-to-right horizontal mode of free vibrations is usually the crucial one (Ellis et al., 2000). Such modes of free vibrations for the empty temporary steel grandstand equipped with tubular and polymer element are shown in Figure 3 and Figure 4, respectively. The estimated values of natural frequencies corresponding to the modes for the empty as well as for the occupied structure are summarized in Table 1.

It can be seen from Table 1 that mass of the spectator significantly decreases values of natural frequency (by as much as 41% in the case of Model 1).

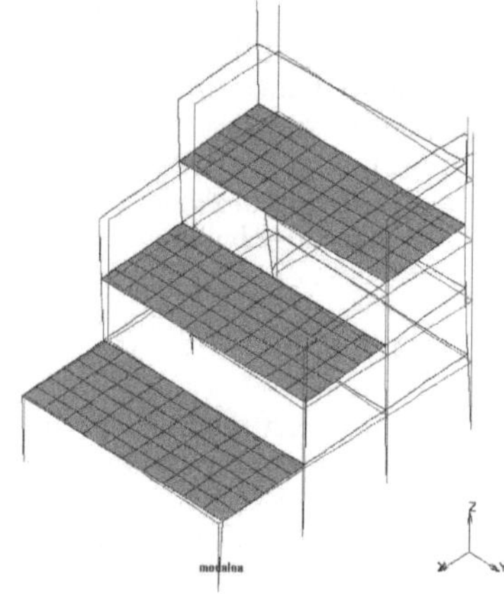

Figure 3. Horizontal mode of free vibrations for an empty temporary steel grandstand with tubular element.

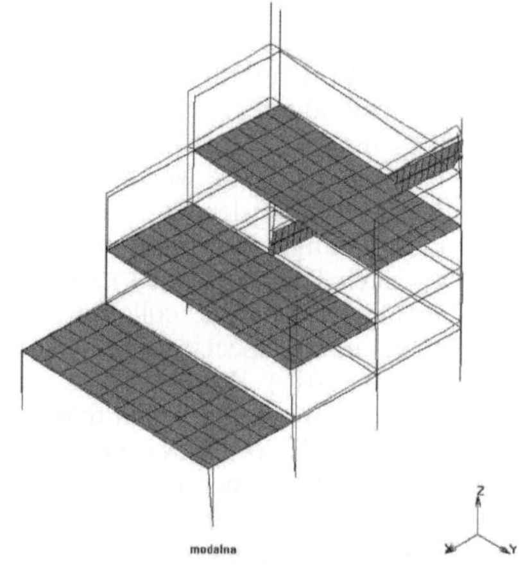

Figure 4. Horizontal mode of free vibrations for an empty temporary steel grandstand with polymer element.

Table 1. Natural frequencies for left-to-right horizontal mode of free vibrations for the empty and occupied structure.

	Natural frequency [Hz]	
	Empty structure	Occupied structure
Model 1	5.663	3.354
Model 2	5.711	3.465

Different variations of natural frequencies for Model 1 and Model 2 of grandstand equipped with typical diagonal stiffener and polymer element are connected with relatively high stiffness of the structure itself.

3 DYNAMIC TRANSIENT ANALYSIS

The second stage of the numerical study was focused on dynamic transient analysis. In the analysis, the dynamic load was assumed to consist of synchronous repetitive impacts, as expressed by Fourier series (Ellis et al., 2000):

$$F_s(t) = G_s\left[1+\sum_{n=1}^{\infty} r_n \sin\left(\frac{2n\pi}{T_p}t+\varphi_n\right)\right] \quad (1)$$

where $F_s(t)$ = dynamic load; G_s = weight; r_n = the Fourier coefficient (or dynamic load factor) of the n-th term; n = the number of Fourier terms; T_p = the period of the jumping load; and φ_n = the phase lag of the n-th term. According to Polish Standards, the full load has been applied in the vertical direction together with the additional load of 6% in its value acting in the horizontal (left-to-right) direction.

Dynamic transient analysis has been conducted so as to determine peak values of accelerations and displacements of the structure equipped with tubular and polymer element. Acceleration time histories for both models obtained at the level of the highest steel platform are presented in Figure 5 and Figure 6. On the other hand, the displacement time histories are presented in Figure 7 and Figure 8

The peak values of accelerations and displacements for left-to-right vibrations (Y direction) of the temporary steel scaffolding grandstand under human-induced excitation due to jumping have been determined and summarized in Table 2.

It can be seen from Table 2 that the peak value of 0.75 g (where g is the acceleration of gravity) has been obtained for the analysed grandstand with the typical stiffener member, while the value of 0.44 g has been reached in the case of the structure equipped with the polymer element. Both values are very high and can be even considered

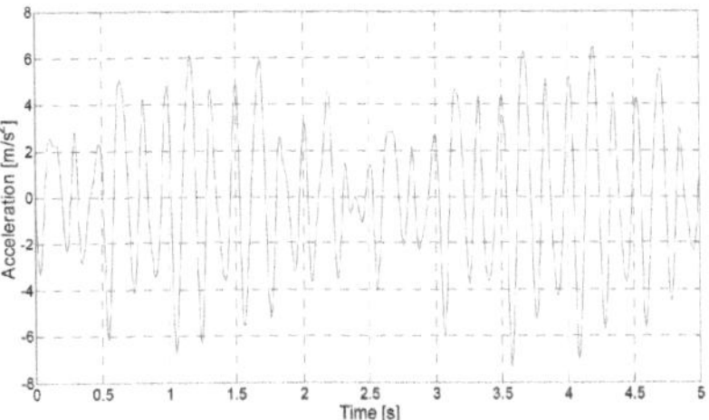

Figure 5. Acceleration time history for temporary steel grandstand equipped with tubular member (Y direction).

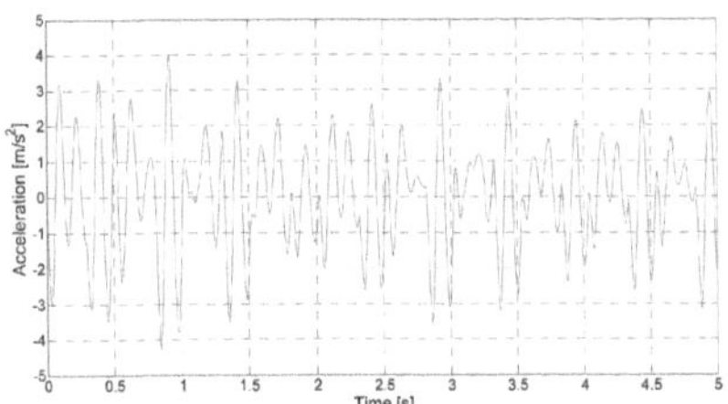

Figure 6. Acceleration time history for temporary steel grandstand equipped with polymer element (Y direction).

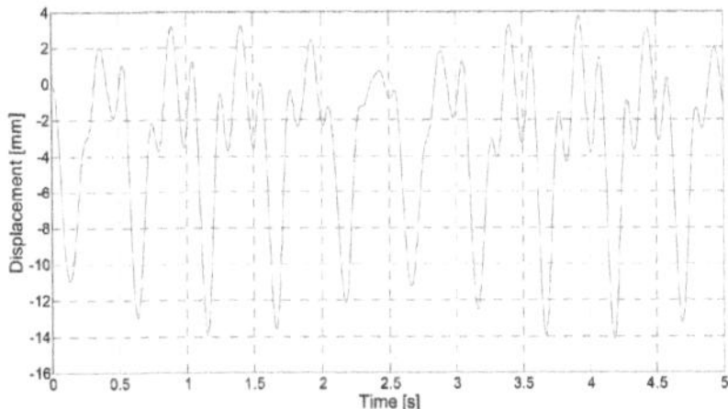

Figure 7. Displacement time history for temporary steel grandstand equipped with tubular member (Y direction).

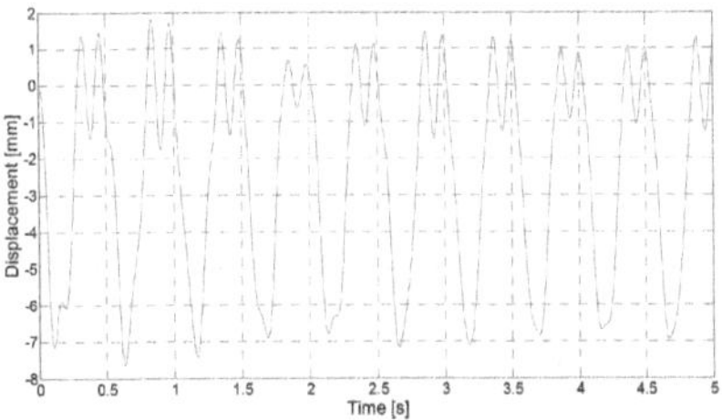

Figure 8. Displacement time history for temporary steel grandstand equipped with polymer element (Y direction).

as unacceptable (the lowest recommended value of acceleration, so as to avoid panic among spectators, is equal to 0.35 g—see Ellis et al., 2000). However, it should be underlined that the peak

Table 2. Peak left-to-right horizontal accelerations and displacements for Model 1 and Model 2.

	Acceleration [m/s^2]	Displacement [mm]
Model 1	7.317	14.13
Model 2	4.277	7.634

value of acceleration estimated for the structure equipped with the polymer element is almost two times lower than the value determined for grandstand with the typical tubular member. A similar conclusion concerns also peak displacements. The peak displacement obtained from the dynamic analysis for the structure with the polymer element is almost two times lower than the value determined for grandstand with the typical tubular member.

4 CONCLUSIONS

The numerical analysis focused on reduction of vibrations of a temporary steel scaffolding grandstand has been conducted in this paper. Typical solution of strengthening such a structure by the use of the diagonal tubular stiffener has been compared with the application of the polymer element consisting of two L-shape steel members bonded with polymer mass. In the first stage of the study, modal analysis has been conducted so as to determine dynamic parameters, such as modes of free vibrations and corresponding natural frequencies, of temporary steel scaffolding grandstand equipped with different types of diagonal elements. Then, the dynamic transient analysis has been conducted and the behaviour of the structure under human-induced excitation due to jumping has been determined.

The results of the study clearly show that the response of a temporary steel grandstand equipped with the polymer element as well as with the diagonal stiffener is substantially different. The application of the polymer element leads to substantial reduction in the values of peak accelerations (reduction by as much as 42%) and peak displacements (reduction by as much as 46%). This substantial decrease in the peak responses of the structure quipped with the diagonal polymer element is caused by the enhanced damping properties of polymer applied.

REFERENCES

Crick D. & Grondin G.Y. 2008. *Monitoring and analysis of a temporary grandstand.* Structural Engineering Report No.275. Alberta: Department of Civil and Environment Engineering. University of Alberta.

De Brito V. & Pimentel R. 2009. Cases of collapse of demountable grandstands. *Journal of Performance of Constructed Facilities* 23: 151–159.

Ellis B.R., Ji T. & Litter J.D. 2000. The response of grandstands to dynamic crowd loads. *Proceedings of the Institution of Civil Engineers, Structures and Buildings* 140: 355–365.

Falborski, T., Jankowski R. & Kwiecień A. 2012. Experimental study on polymer mass used to repair damaged structures. *Key Engineering Materials* 488–489: 347–350.

Falborski, T. & Jankowski, R. 2013. Polymeric bearings—a new base isolation system to reduce structural damage during earthquakes. *Key Engineering Materials* 569–570: 143–150.

Jankowski R. 2005. Impact force spectrum for damage assessment of earthquake-induced structural pounding. *Key Engineering Materials* 293–294: 711–718.

Jankowski R. 2007. Assessment of damage due to earthquake-induced pounding between the main building and the stairway tower. *Key Engineering Materials* 347: 339–344.

Jones C.A., Reynolds P. & Pavic A. 2010. Vibration service-ability of stadia structures subjected to dynamic crowd loads: A literature review. *Journal of Sound and Vibration* 330: 1531–1566.

Lasowicz N., Kwiecień A. & Jankowski R. 2015. Enhancing the seismic resistance of columns by GFRP confinement using flexible adhesive—experimental study. *Key Engineering Materials* 624: 478–485.

Lasowicz N. & Jankowski R. 2014. *Numerical analysis of a temporary steel grandstand.* Shell Structures: Theory and Applications Vol. 3—Proc. of the 10th SSTA Conference, Gdańsk, Poland 16–18 October 2013. Leiden, Netherlands: CRC/Balkema, 543–546.

Lasowicz N., Kwiecień A. & Jankowski R. 2015. *Experimental study on the effectiveness of polymer damper in damage reduction of temporary steel grandstand.* 11th International Conference on Damage Assessment of Structures DAMAS 2015, Journal of Physics: Conference Series 628, Article no. 012051.

Mahmoud, S. & Jankowski, R. 2011. Modified linear viscoelastic model of earthquake-induced structural pounding. *Iranian Journal of Science and Technology* 35(C1): 51–62.

Mahmoud S., Austrell P-E. & Jankowski R. 2012. Simulation of the response of base-isolated buildings under earthquake excitations considering soil flexibility. *Earthquake Engineering and Engineering Vibration* 11: 359–374.

Nhleko S.P., Williams M.S. & Blakeborough A. 2009. *Vibration perception and comfort level for an audience occupying a grandstand with perceivable motion.* Proc. of the IMAC-XXVII, Feb. 9–12, 2009, Orlando, Florida USA.

Reynolds P., Pavic A. & Ibrahim Z. 2004. *Changes of modal properties of a stadium structure occupied by a crowd.* Proc. of IMAC-XXII: Conference & Exposition on Structural Dynamics, Dearborn, USA 26–29 January 2004.

Advances in Mechanics: Theoretical, Computational and Interdisciplinary Issues – Kleiber et al. (Eds)
© 2016 Taylor & Francis Group, London, ISBN 978-1-138-02906-4

Frequency Response Function of structures with viscoelastic dampers

M. Łasecka-Plura & R. Lewandowski
Poznan University of Technology, Poznań, Poland

ABSTRACT: The Frequency Response Function (FRF) of planar frames with viscoelastic dampers mounted on them is considered in this paper. Two methods of the calculation of FRF are presented: the direct method and Adhikari's method. The second one is used for shear frames with viscoelastic dampers for the first time. Based on the obtained solution, the design sensitivity analysis of the FRF is presented. The direct differentiation method is used to determine the design sensitivities of the frequency response function.

1 INTRODUCTION

The frequency response function is one of the most important tools of evaluation of the dynamic response of structures. It is very important in many areas, such as model updating, vibration control, optimization, and other ones. The FRF can be calculated using the direct method by solving inversion of the dynamic stiffness matrix or the superposition method proposed by Adhikari where FRF is calculated by expressing them as a sum containing all modes. The method was described by Adhikari (2002) and Lewandowski (2014) in detail. Some other authors studied this approach to find the dynamic response for viscoelastic damping systems (Li et al. 2013).

The design sensitivity analysis of FRF has been studied by several authors. For example, the direct differentiation method and the adjoint variables method is described by Choi et al. (1997). The sensitivity of FRF of structures with viscoelastic dampers determined by the direct method is considered by Łasecka-Plura & Lewandowski (2014). In that paper, sensitivity of the first and the second order is considered.

In the paper, Adhikari's method and its application for the calculation of frequency response function of frame with viscoelastic dampers described by fractional derivatives is presented for the first time. The results are compared with those obtained by the direct method. A possibility of using Adhikari's method in the case in which only the lower modes are of interest is also presented. Then, the direct differentiation method is used for the design sensitivity analysis of FRF, also when only the lower modes are of interest.

2 THE EQUATION OF MOTION OF STRUCTURES WITH VISCOELASTIC DAMPERS

2.1 *Forces in dampers*

In the paper, the fractional Kelvin model of a damper is considered (Fig. 1a). The model consists of a spring of the stiffness k_0 connected parallel with a fractional dashpot described by the damping factor c_0 and the fractional parameter α ($0 < \alpha \leq 1$).

The equation of motion of the Kelvin damper is:

$$u_{0i}(t) = k_{0i}\Delta q_i(t) + c_{0i} D_t^{\alpha_i} \Delta q_i(t) \tag{1}$$

where: u_{0i} = is the force in the i-th damper; $\Delta q_i(t) = q_k - q_i$, q_k and q_j = denote nodal displacements of the considered damper model and i = is the damper number. $D_t^{\alpha}(\bullet)$ denotes the Riemann-Liouville fractional derivative of the order α with respect to time t (Podlubny 1999).

Applying the Laplace transform to (1) leads to:

$$\bar{u}_i(s) = G_i(s)\Delta\bar{q}_i(s) \tag{2}$$

where: $\bar{u}(s)$ and $\Delta\bar{q}_i(s)$ = denote the Laplace transforms of $u_i(s)$ and $\Delta q_i(s)$, respectively; s = is the

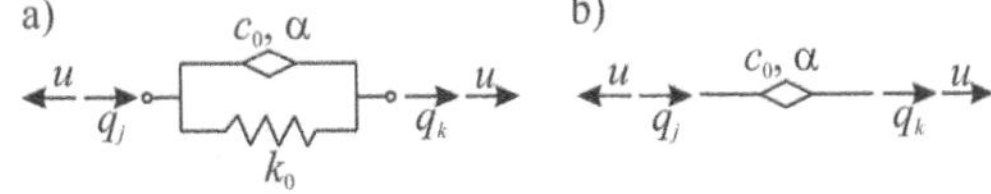

Figure 1. A model of dampers a) the fractional Kelvin model, b) the fractional spring-pot element.

Laplace variable and $G_i(s) = k_{0i} + c_{0i}s^{\alpha i}$. The considered damper model contains the fractional spring-pot element (Fig. 1b) when $k_{0i} = 0$ in (2). The classical models can be obtained by introducing $\alpha = 1$.

2.2 *Equation of motion of structures with dampers*

The equation of motion of a structure with dampers can be written in the following form (Lewandowski & Pawlak 2011):

$$\mathbf{M}_s\ddot{\mathbf{q}} + \mathbf{C}_s\dot{\mathbf{q}} + \mathbf{K}_s\mathbf{q} = \mathbf{p}(t) + \mathbf{f}(t) \tag{3}$$

where: $\mathbf{M}_s$, $\mathbf{C}_s$ and $\mathbf{K}_s$ = denote mass, damping, and stiffness matrices, respectively. $\mathbf{q}(t) = [q_1, ..., q_n]^T$ is the vector of displacements of the structure, $\mathbf{p}(t) = [p_1, ..., p_n]^T$ is the vector of the excitation forces and $\mathbf{f}(t) = [f_1, ..., f_n]^T$ is the vector of the interaction forces between the frame and the dampers.

Vector $\mathbf{f}(t)$ is a sum of vectors $\mathbf{f}_i(t)$. Each of them is formed if only the damper i is located on the frame, i.e.:

$$\mathbf{f}(t) = \sum_{i=1}^{r} \mathbf{f}_i(t). \tag{4}$$

After taking the Laplace transform, the equation of motion can be written in the form:

$$\left(s^2\mathbf{M}_s + s\mathbf{C}_s + \mathbf{K}_s\right)\bar{\mathbf{q}}(s) = \bar{\mathbf{p}}(s) + \bar{\mathbf{f}}(s) \tag{5}$$

where:

$$\bar{\mathbf{f}}(s) = \sum_{i=1}^{r} \bar{\mathbf{f}}_i(s) = \sum_{i=1}^{r} G_i(s)\mathbf{L}_i\bar{\mathbf{q}}(s) \tag{6}$$

and $\mathbf{L}_i$ = is the matrix of location of dampers; r = is the number of dampers. The above Equation 5 can be rewritten as:

$$\mathbf{D}(s)\bar{\mathbf{q}}(s) = \bar{\mathbf{p}}(s) \tag{7}$$

where:

$$\mathbf{D}(s) = s^2\mathbf{M}_s + s\mathbf{C}_s + \mathbf{K}_s + \sum_{i=1}^{r} \mathbf{G}_i \tag{8}$$

is the dynamic stiffness matrix, $\mathbf{G}_i = G_i(s)\mathbf{L}_i$.

If $\bar{\mathbf{p}}(s) = \mathbf{0}$, (7) leads to the following nonlinear eigenproblem:

$$\mathbf{D}(s)\,\bar{\mathbf{q}}(s) = \mathbf{0}. \tag{9}$$

The nonlinear eigenproblem was solved by the continuation method, described in detail by Pawlak & Lewandowski (2013).

3 FREQUENCY RESPONSE FUNCTION

3.1 *Direct Method (DM)*

From (7), the vector $\bar{\mathbf{q}}(s)$ can be obtained as:

$$\bar{\mathbf{q}}(s) = \mathbf{D}^{-1}(s)\,\bar{\mathbf{p}}(s) = \mathbf{H}(s)\,\bar{\mathbf{p}}(s) \tag{10}$$

where:

$$\mathbf{H}(s) = \mathbf{D}^{-1}(s) \tag{11}$$

is the frequency response matrix.

3.2 *Adhikari's Method (AM)*

From the residue theorem (Adhikari 2002, Riley et al. 1997) the frequency response matrix can be expressed in terms of poles and residues:

$$\mathbf{H}(s) = \sum_{j=1}^{m} \frac{\mathbf{R}_j}{s - s_j}, \tag{12}$$

where m = denotes the number of eigenvalues of eigenproblem (9); and $\mathbf{R}_j$ = is the residue of the transfer function at the pole s_j:

$$\mathbf{R}_j = \lim_{s \to s_j} \left(s - s_j\right)\,\mathbf{H}(s). \tag{13}$$

For a given s, it is necessary to solve the following eigenvalue problem associated with the dynamic stiffness matrix:

$$\mathbf{D}(s)\mathbf{x}_k(s) = \nu_k(s)\mathbf{x}_k(s) \tag{14}$$

where $\nu_k(s)$ and $\mathbf{x}_k(s)$ = are the eigenvalues and the eigenvectors, respectively. The above eigenproblem (14) has n eigenvalues and eigenvectors, where n denotes the number of degrees of freedom of structure. For distinct eigenvalues the following normalization condition of eigenvectors can be taken into account:

$$\mathbf{x}_j^T(s)\mathbf{x}_k(s) = \delta_{jk} \tag{15}$$

where δ_{jk} = is a Kronecker delta. Premultiplying Equation (14) by $\mathbf{x}_j^T(s)$ leads to the following formula in the matrix form:

$$\mathbf{X}^T(s)\mathbf{D}(s)\mathbf{X}(s) = \mathbf{v}(s) \tag{16}$$

where:

$$\mathbf{X}(s)=\left[\mathbf{x}_1(s),\, \mathbf{x}_2(s),\, \dots,\, \mathbf{x}_n(s)\right], \tag{17}$$

$$\mathbf{v}(s)=diag\left[\nu_1(s),\, \nu_2(s),\, \dots,\, \nu_n(s)\right]. \tag{18}$$

Now, it is necessary to establish the relationships between the eigenproblems defined by Equations (9) and (14). We assume that the r-th eigenvalue of eigenproblem (14) equals zero when $s \rightarrow s_j$. Hence, when $s = s_j$ then:

$$x_r(s_j)=\bar{q}_j(s_j). \tag{19}$$

From (16):

$$\mathbf{D}^{-1}(s)=\mathbf{X}(s)\mathbf{v}^{-1}(s)\mathbf{X}^T(s). \tag{20}$$

After using (11) and noting that $\mathbf{v}(s)$ is a diagonal matrix the following relationship can be obtained:

$$\mathbf{H}(s)=\mathbf{D}^{-1}(s)=\sum_{k=1}^{n}\frac{\mathbf{x}_k(s)\mathbf{x}_k^T(s)}{\nu_k(s)}. \tag{21}$$

By separating the *r-th* term we obtain:

$$\mathbf{H}(s)=\frac{\mathbf{x}_r(s)\mathbf{x}_r^T(s)}{\nu_r(s)}+\sum_{k=1,\,k\neq r}^{n}\frac{\mathbf{x}_k(s)\mathbf{x}_k^T(s)}{\nu_k(s)}. \tag{22}$$

The second term of the right-hand side of Equation (22) is an analytical function because $s \rightarrow s_j$ and $\nu_k(s_j) \neq 0$. From Equation (13), the residue at $s = s_j$ may be obtained as:

$$\mathbf{R}_j=\lim_{s\rightarrow s_j}\left(s-s_j\right)\frac{\mathbf{x}_r(s)\mathbf{x}_r^T(s)}{\nu_r(s)}. \tag{23}$$

Using l'Hospital rule and substituting relationship (19) leads to:

$$\mathbf{R}_j=\frac{\mathbf{q}_j(s)\mathbf{q}_j^T(s)}{\left.\dfrac{\partial \nu_r(s)}{\partial s}\right|_{s=s_j}}. \tag{24}$$

In order to obtain the unknown denominator in relationship (24) it is necessary to differentiate (14) for $k = r$ with respect to s. After using the normalization condition (15) we obtain:

$$\left.\frac{\partial \nu_r(s)}{\partial s}\right|_{s=s_j}=\mathbf{q}_j^T\left.\frac{\partial \mathbf{D}(s)}{\partial s}\right|_{s=s_j}\mathbf{q}_j=\mathbf{q}_j^T\frac{\partial \mathbf{D}\left(s_j\right)}{\partial s_j}\mathbf{q}_j. \tag{25}$$

Element $\partial\mathbf{D}(s_j)/\partial s_j$ can be determined by differentiating relationship (8). Equations (24) and (25) join the matrix of residues with eigenvectors determined for a structure with viscoelastic dampers. After taking into account these relationships and substituting $s = i\lambda$, where λ is the excitation frequency, the following frequency response matrix can be obtained:

$$\mathbf{H}(i\lambda)=\sum_{j=1}^{m}\frac{\gamma_j\bar{\mathbf{q}}_j\bar{\mathbf{q}}_j^T}{i\lambda-s_j} \tag{26}$$

where:

$$\gamma_j=\frac{1}{\bar{\mathbf{q}}_j^T\dfrac{\partial \mathbf{D}\left(s_j\right)}{\partial s_j}\bar{\mathbf{q}}_j}. \tag{27}$$

4 DESIGN SENSITIVITY ANALYSIS

In order to determine the relationship describing the sensitivity of FRF, it is necessary to use the following obvious equation:

$$\mathbf{H}(\lambda)\mathbf{H}(\lambda)^{-1}=\mathbf{I}. \tag{28}$$

Differentiating (28) with respect to the design parameter p leads to:

$$\frac{\partial \mathbf{H}(\lambda,p)}{\partial p}=-\mathbf{H}(\lambda,p)\frac{\partial \mathbf{D}(\lambda,p)}{\partial p}\mathbf{H}(\lambda,p) \tag{29}$$

where $\partial\mathbf{D}(\lambda, p)/\partial p$ can be obtained by differentiating (8).

5 NUMERICAL EXAMPLE

In order to compare the methods of the calculation of FRF presented in this paper, a ten-storey shear frame with dampers modeled by fractional springpot element situated on the first, second and third storey is considered. Data of the structure are taken from a paper by Pawlak & Lewandowski (2013): the mass of every floor m = 18,000 kg and the storey stiffness k = 51,600 kN/m. The damper's parameters are the following: $k_0 = 0$, c_0 = 400 Ns^{α}/m and $\alpha = 0.6$. Damping in the structure is neglected. The frequency response matrix $\mathbf{H}(\lambda)$ is determined for an excitation frequency in the range $\lambda \in$ (0, 120 rad/sec). The dynamic characteristics of the considered structure are presented in Table 1. The obtained eigenvalues are distinct and complex.

Table 1. Dynamic characteristics.

Modal number	Eigenvalues	Natural frequencies rad/s	Non-dimensional damping ratios
1	−0.0461±8.0365i	8.03665	0.005735
2	−0.1651±23.950i	23.9505	0.006895
3	−0.2306±39.299i	39.2998	0.005867
4	−0.4689±53.935i	53.9373	0.008693
5	−0.8193±67.416i	67.4213	0.012152
6	−0.8967±79.143i	79.1477	0.011329
7	−0.8069±89.210i	89.2136	0.009044
8	−0.9885±97.741i	97.746	0.010113
9	−1.9340±104.46i	104.479	0.018511
10	−1.3574±105.29i	105.297	0.012891

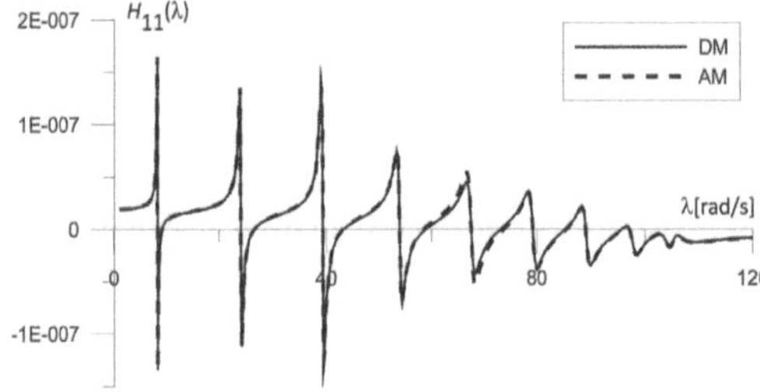

Figure 2. A comparison of frequency response function $H_{11}(\lambda)$.

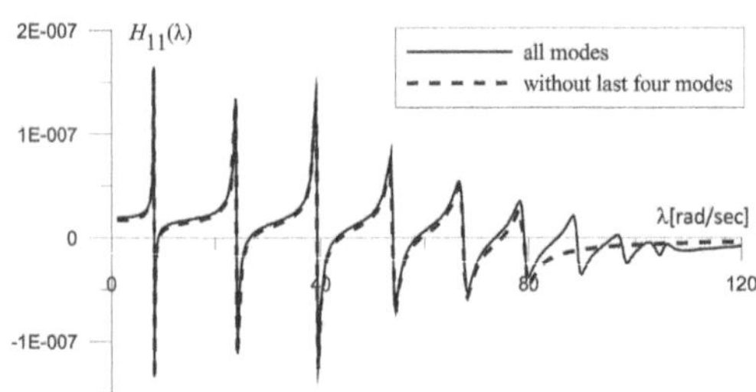

Figure 3. A comparison of frequency response function $H_{11}(\lambda)$.

The results of the calculation of the real part of the function $H_{11}(\lambda)$ are presented in Figure 2. The graphs obtained by both the direct method (solid line) and Adhikari's method (dashed line) overlap. Similar results are obtained for the imaginary part of the function $H_{11}(\lambda)$.

Sometimes only the first few modes are of interest. Then it is possible to take into account only the first few modes in Adhikari's method. In Figure 3, a comparison of the function $H_{11}(\lambda)$ calculated for all modes (solid line) and when the last four modes are neglected (dashed line) is presented. The graphs overlap in the case of the initial four resonance zones. This analysis proves that the results obtained by Adhikari's method are close to the exact solution if only the lower modes are taken into account.

Now, the design sensitivity of function $H_{11}(\lambda)$ with respect to the damping factor of the damper situated on the first floor is presented. The results are shown in Figure 4.

In order to verify the correctness of the presented calculation the values of FRF, after changing the design parameter, are determined based on the formula:

$$H_{11}\left(\lambda, c_{01}+\Delta c_{01}\right)=H_{11}\left(\lambda, c_{01}\right)+\frac{\partial H_{11}\left(\lambda, c_{01}\right)}{\partial c_{01}} \Delta c_{01} \tag{30}$$

where Δc_{01} = denotes a change of the damping factor. The obtained values are compared with the exact solution when the value of the design parameter changed by 5%. The calculation is carried out for selected frequencies and the obtained results are presented in Table 2. The results determined using the design sensitivity are very close to exact values. This proves the presented method is correct.

The design sensitivity analysis is also carried out when the last four modes are neglected. Sensitivity with respect to a change of the damping factor c_{01} is presented. The comparison is shown in Figure 5. The results calculated for all modes (solid line) are

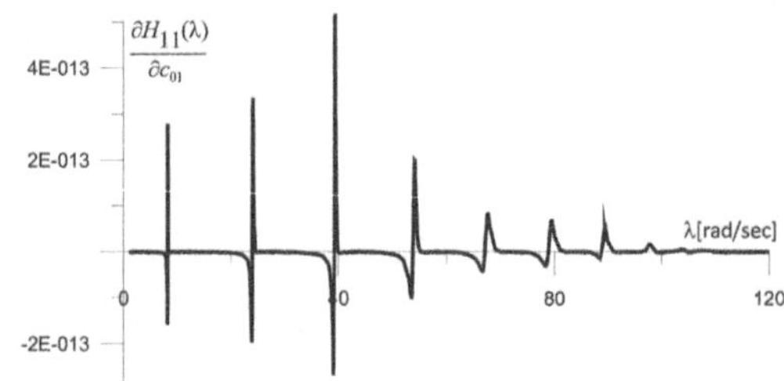

Figure 4. Sensitivity of frequency response function $H_{11}(\lambda)$.

Table 2. A comparison of $H_{11}(\lambda, c_{01})$.

λ rad/s	Values obtained using design sensitivity	Exact values
10	$1.008\cdot10^{-8}-3.602\cdot10^{-10}i$	$1.008\cdot10^{-8}-3.602\cdot10^{-10}i$
20	$2.441\cdot10^{-8}-1.239\cdot10^{-9}i$	$2.441\cdot10^{-8}-1.238\cdot10^{-9}i$
30	$1.211\cdot10^{-8}-9.353\cdot10^{-10}i$	$1.211\cdot10^{-8}-9.353\cdot10^{-10}i$
40	$-6.543\cdot10^{-8}-2.283\cdot10^{-8}i$	$-6.541\cdot10^{-8}-2.283\cdot10^{-8}i$

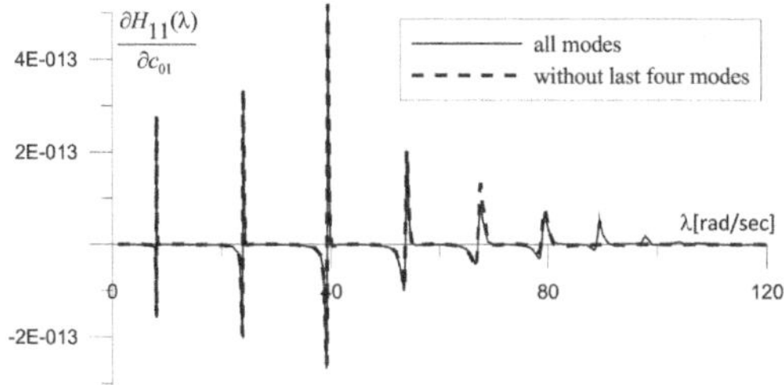

Figure 5. A comparison of sensitivity of frequency response function $H_{11}(\lambda)$.

compared with the results calculated for the last four modes neglected (dashed line). The obtained graphs overlap in the case of the initial four resonance zones.

6 CONCLUSIONS

In the paper two methods of calculation of FRF are compared: Adhikari's method and direct method. The obtained results confirmed the possibility of using the presented method for calculating the FRF for structures with viscoelastic dampers. The Adhikari's method gives results which are close to an exact solution if only lower modes are taken into account. Such analysis for shear frame with viscoelastic dampers described by fractional derivatives has been carried out for the first time. The formulae which enable determination of the sensitivity of FRF with respect to change of design parameters are also presented. The method used to calculate sensitivity is easy to formulate, systematic to apply, simple to code, and it agrees well with the exact results.

REFERENCES

Adhikari, S. 2002. Dynamics of nonviscously damped linear systems. *Journal of Engineering Mechanics* 128: 328–339.

Choi, K.K., Shim, I. & Wang, S. 1997. Design sensitivity analysis of structure-induced noise and vibration. *Journal of Vibration and Acoustics-ASME* 119: 173–179.

Łasecka-Plura, M. & Lewandowski, R. 2014. Design sensitivity analysis of frequency response functions and steady-state response for structures with viscoelastic dampers. *Vibrations in Physical Systems* 26: 129–136.

Lewandowski, R. 2014. *Reduction of vibration of building structures* (in Polish). Warsaw: PWN.

Lewandowski, R. & Pawlak, Z. 2011. Dynamic analysis of frames with viscoelastic dampers modeled by rheological models with fractional derivatives. *Journal of Sound and Vibration* 330: 923–936.

Li, L., Hu, Y. & Wang, X. 2013. Improved approximate methods for calculating frequency response function matrix and response of MDOF systems with viscoelastic hereditary terms. *Journal of Sound and Vibration* 332: 3945–3956.

Pawlak, Z. & Lewandowski, R. 2013. The continuation method for the eigenvalue problem of structures with viscoelastic dampers. *Computers and Structures* 125: 53–61.

Podlubny, I. 1999. *Fractional differential equations*. San Diego-Boston-New York-London-Tokyo-Toronto: Academic Press.

Riley, K.F., Hobson, K.F. & Bence, S.J. 1997. *Mathematical methods for physics and engineering*. Cambridge: Cambridge University Press.

Advances in Mechanics: Theoretical, Computational and Interdisciplinary Issues – Kleiber et al. (Eds)
 ISBN 978-1-138-02906-4

Influence of temperature on dynamic characteristics of structures with VE dampers

R. Lewandowski, M. Przychodzki & Z. Pawlak
Poznan University of Technology, Poznań, Poland

ABSTRACT: In the paper, the effect of changes of temperature in Visco Elastic (VE) dampers on the dynamic characteristics of structures having such dampers installed on them are presented. Both classical and fractional rheology models are used to describe the properties of VE dampers. The time-temperature superposition principle is used to describe changes in the properties of VE dampers due to temperature changes. The dynamic properties of structures with dampers are determined from the solution to an appropriately defined nonlinear eigenvalue problem.

1 INTRODUCTION

Viscoelastic dampers can effectively reduce the amplitudes of vibrations of building structures excited by the wind or earthquakes (Soong & Spencer 2002). Descriptions of the dynamic behavior of this type of dampers are complex because the properties of the Visco Elastic (VE) materials used in the dampers depend on excitation frequency and temperature (Nashif et al. 1985, Chang et al. 1992). In previous papers, only the influence of excitation frequency was taken into account (Singh & Chang 2009, Lewandowski et al. 2012).

The temperature of a VE damper can change because of changes of environmental temperature or due to the so-called self-heating, induced by dissipation of energy in the dampers. The influence of temperature on the properties of VE dampers has been described in several papers (see Chang et al. 1992, de Lima et al. 2015). However, the influence of changes of temperature in VE dampers on the dynamic characteristics of structures having such dampers installed on them has not been analyzed systematically. In this paper, a computation method which enables an analysis of the influence of temperature changes on natural frequencies, dimensionless damping ratios and eigenvectors (or, loosely speaking, modes of vibration) of structures with VE dampers is presented.

2 INFLUENCE OF TEMPERATURE ON PARAMETERS OF MODELS OF VE DAMPERS

In this work, the fractional Zener model is used to describe the VE damper. It enables the analysis of a number of simpler models, such as the simple Maxwell model and the classical Zener model.

In order to determine the VE material response to changes of temperature, the time-temperature superposition principle, as given by the following relationship (1), can be used:

$$K(t,T) = K(\alpha_T t_0, T_0) \tag{1}$$

where K = is the complex modulus; t_0 and T_0 = are the reference time and the reference temperature, respectively. The symbol α_T denotes the so-called shift factor. This principle can be applied to the frequency domain and sometimes named as the frequency-temperature correspondence principle (de Lima et al. 2015):

$$K(\lambda,T) = K(\alpha_T \lambda_0, T_0) \tag{2}$$

where λ_0 = is the reference frequency.

The shift factor is typically calculated from the Arrhenius formula or from the following William-Landel-Ferry formula:

$$\log \alpha_T = \frac{-C_1 \Delta T}{C_2 + \Delta T} \tag{3}$$

where C_1 and C_2 are constants and $\Delta T = T - T_0$.

According to relationship (2), the calculation of material response to any temperature T consists in shifting the argument of function $K(\lambda_0)$ for reference temperature T_0 in the following way:

$$\lambda_0 = \alpha_T \lambda. \tag{4}$$

3 EQUATION OF MOTION OF STRUCTURES WITH VE DAMPERS AND THE NONLINEAR EIGENPROBLEM

3.1 *Equation of motion of VE damper*

The equation of motion of the Zener model of damper is:

$$u(t) + \tau^{\alpha} D_t^{\alpha} u(t) = k_s \Delta q(t) + \tau^{\alpha}(k_s + k_d) D_t^{\alpha} \Delta q(t) \tag{5}$$

where u = is the force in the damper; q_j and q_i = denote the nodal displacements of the damper model and $\Delta q(t) = q_j(t) - q_i(t)$. $D_t^{\alpha}(\bullet)$ denotes the Riemann-Liouville fractional derivative of the order α with respect to time t. The element $\tau^{\alpha} = c_d / k_d$ could be understood as the relaxation time and the symbols k_s, k_d and c_d denote the stiffness coefficients and the damping factor of the damper, respectively (Fig. 1).

After using the Laplace transformation with zero initial conditions, the transform of force in the damper can be written as:

$$\bar{u}(s) = \frac{k_s + (k_s + k_d)(s\tau)^{\alpha}}{1 + (s\tau)^{\alpha}} \Delta\bar{q}(s) \tag{6}$$

where $\Delta\bar{\mathbf{q}}(s)$ = is the Laplace transform of $\Delta q(t)$ and s = denotes the Laplace variable. For $\alpha = 1$, (5–6) describe the classical Zener model.

Equation 6 can be simplified to the form:

$$\bar{u}(\lambda) = K(\lambda)\Delta\bar{q}(\lambda) \tag{7}$$

where λ fulfills the relationship:

$$s = \mathrm{i}\lambda. \tag{8}$$

The symbol $K(\lambda)$ denotes the so-called complex modulus which is described by the following formula:

$$K(\lambda) = K'(\lambda) + \mathrm{i}K''(\lambda) \tag{9}$$

where $K'(\lambda)$ = is the storage modulus and $K''(\lambda)$ = is the loss modulus. The storage and loss moduli can be written as follows:

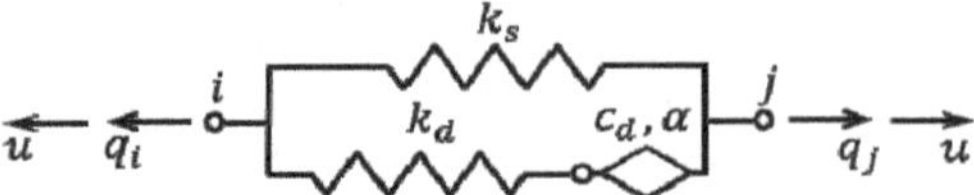

Figure 1. A diagram of the fractional Zener model of damper.

$$K'(\lambda) = \frac{k_0 + (\tau\lambda)^{\alpha}(k_0 + k_{\infty})\cos\frac{\alpha\pi}{2} + k_{\infty}(\tau\lambda)^{2\alpha}}{1 + 2(\tau\lambda)^{\alpha}\cos\frac{\alpha\pi}{2} + (\tau\lambda)^{2\alpha}} \tag{10}$$

$$K''(\lambda) = \frac{(\tau\lambda)^{\alpha}(k_{\infty} - k_0)\sin\frac{\alpha\pi}{2}}{1 + 2(\tau\lambda)^{\alpha}\cos\frac{\alpha\pi}{2} + (\tau\lambda)^{2\alpha}} \tag{11}$$

where $k_0 = k_s$ and $k_{\infty} = k_s + k_d$.

Under the assumption that the parameters k_0, k_{∞} and τ are related to the reference temperature T_0, (10) takes the form:

$$K'(\lambda_0, T_0) = \frac{k_0 + (\tau\lambda_0)^{\alpha}(k_0 + k_{\infty})\cos\frac{\alpha\pi}{2} + k_{\infty}(\tau\lambda_0)^{2\alpha}}{1 + 2(\tau\lambda_0)^{\alpha}\cos\frac{\alpha\pi}{2} + (\tau\lambda_0)^{2\alpha}}. \tag{12}$$

When temperature is different from the reference value the storage modulus can be described by the formula:

$$K'(\lambda, T) = \frac{\tilde{k}_0 + (\tilde{\tau}\lambda)^{\alpha}(\tilde{k}_0 + \tilde{k}_{\infty})\cos\frac{\alpha\pi}{2} + \tilde{k}_{\infty}(\tilde{\tau}\lambda)^{2\alpha}}{1 + 2(\tilde{\tau}\lambda)^{\alpha}\cos\frac{\alpha\pi}{2} + (\tilde{\tau}\lambda)^{2\alpha}} \tag{13}$$

Taking into account the frequency-temperature correspondence principle, expressed by (4), the above formula can be rewritten as follows:

$$K'(\lambda, T) = K'(\lambda_0 = \alpha_T\lambda, T_0)$$
$$= \frac{k_0 + (\tau\alpha_T\lambda)^{\alpha}(k_0 + k_{\infty})\cos\frac{\alpha\pi}{2} + k_{\infty}(\tau\alpha_T\lambda)^{2\alpha}}{1 + 2(\tau\alpha_T\lambda)^{\alpha}\cos\frac{\alpha\pi}{2} + (\tau\alpha_T\lambda)^{2\alpha}}. \tag{14}$$

The comparison (13–14) leads to the following results:

$$\tilde{k}_0 = k_0, \; \tilde{k}_{\infty} = k_{\infty}, \; \tilde{\tau} = \alpha_T\tau. \tag{15}$$

On the basis (15), it can be established, that only parameter c_d of the Zener model of damper will change with changes of temperature according to the relationship:

$$\tilde{c}_d = \alpha_T c_d. \tag{16}$$

3.2 *Equation of motion of structure with VE dampers*

The equation of motion of a structure with VE dampers can be written in the following form (compare Pawlak & Lewandowski 2013):

$$\mathbf{M}\ddot{\mathbf{q}}(t)+\mathbf{C}\dot{\mathbf{q}}(t)+\mathbf{K}\mathbf{q}(t)=\mathbf{p}(t)+\mathbf{f}(t) \tag{17}$$

where $\mathbf{M}$, $\mathbf{C}$ and $\mathbf{K}$ = denote the mass, damping, and stiffness matrices, respectively. Moreover, $\mathbf{q}(t)$ = is the vector of displacements of the structure; $\mathbf{p}(t)$ = is the vector of the excitation forces and $\mathbf{f}(t)$ = is the vector of the interaction forces between the frame and the dampers. After applying the Laplace transform, (8) can be written as follows (Pawlak & Lewandowski 2013):

$$(s^2\mathbf{M}+s\mathbf{C}+\mathbf{K})\,\bar{\mathbf{q}}(s)=\bar{\mathbf{p}}(s)+\bar{\mathbf{f}}(s). \tag{18}$$

The vector $\bar{\mathbf{f}}(s)$ is formed using Equation 6 (compare Pawlak & Lewandowski 2013).

After assuming that $\bar{\mathbf{p}}(s)=0$, the final form of the equation of motion in the frequency domain for the structure with VE dampers is as follows:

$$(s^2\mathbf{M}+s\mathbf{C}+\mathbf{K}+\mathbf{K}_d+\mathbf{G}_d(s))\,\bar{\mathbf{q}}(s)=\mathbf{0} \tag{19}$$

where:

$$\mathbf{K}_d=\sum_{r=1}^{m}k_{0r}\mathbf{L}_r,\quad \mathbf{G}_d(s)=\sum_{r=1}^{m}\frac{k_{dr}s^{\alpha}}{\nu_{dr}+s^{\alpha}}\mathbf{L}_r,\quad \nu_{dr}=\frac{k_{dr}}{c_{dr}}. \tag{20}$$

The symbol $\mathbf{L}_r$ denotes the location matrix of the r-th damper and m is the number of dampers in the structure.

In the considered case, the matrix of the frequency response functions could be written as follow:

$$\mathbf{H}(\lambda)=(-\lambda^2\mathbf{M}+\mathrm{i}\lambda\mathbf{C}+\mathbf{K}+\mathbf{K}_d+\mathbf{G}_d(\mathrm{i}\lambda))^{-1} \tag{21}$$

where i = is the imaginary unit.

3.3 *Nonlinear eigenproblem*

Equation 18 constitutes the nonlinear eigenproblem. In this research work, the problem is solved using the so-called continuation method (Pawlak & Lewandowski 2013). The calculated values of s are the eigenvalues and $\bar{\mathbf{q}}(s)$ are the eigenvectors. The knowledge of the eigenvalues $s_k=\mu_k+\mathrm{i}\eta_k$, enables the calculation of the natural frequencies of the structure ω_k and its dimensionless damping ratios γ_k. This can be done in the following way:

$$\omega_k^2=\mu_k^2+\eta_k^2,\quad \gamma_k=\mu_k/\omega_k. \tag{22}$$

4 NUMERICAL EXAMPLE

The numerical calculation was made for the shear frame model of an eight-storey building structure presented in Figure 2. The mass is lumped and same at every floor: m_s = 60,000 kg. The bending rigidity of the storeys is as follows:

$k_1=k_2=441{,}119$ kN/m,

$k_3=k_4=275{,}351$ kN/m,

$k_5=k_6=152{,}948$ kN/m,

$k_7=k_8=93{,}244$ kN/m.

The VE dampers are located on each storey. The parameters of the damper for the reference temperature $T_0=0.2$ °C take the values $k_s=10{,}856$ kN/m, $k_d=1{,}996{,}809$ kN/m, $c_d=22{,}963$ kNs$^{\alpha}$/m, $\alpha=0.609$. The shift factor α_T was computed according to the William-Landel-Ferry formula with the constants $C_1=19.5$ and $C_2=80.2$.

Figure 3 illustrates the influence of temperature on the first natural frequency of the model.

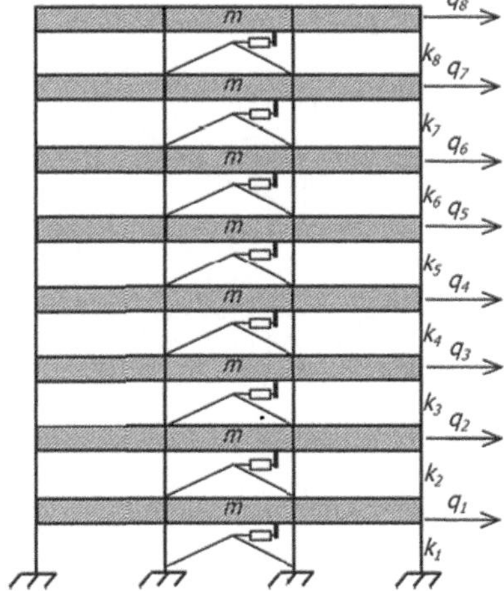

Figure 2. A diagram of the considered structure.

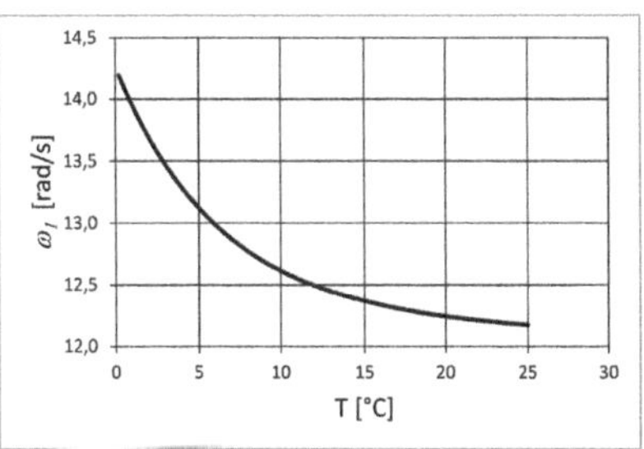

Figure 3. The first natural frequency function versus temperature.

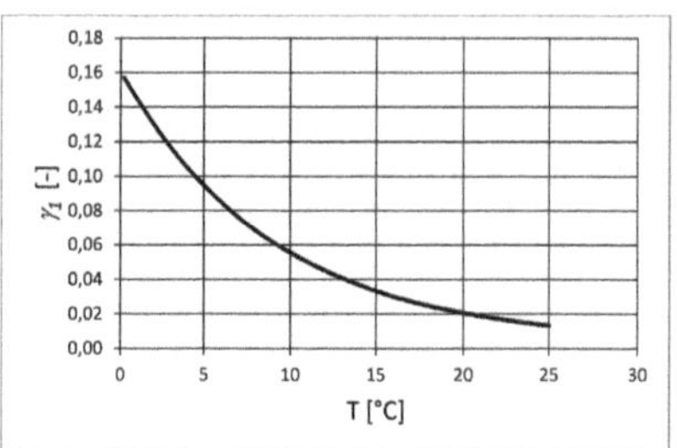

Figure 4. The dependence of the dimensionless damping ratio of the first mode of vibration on temperature.

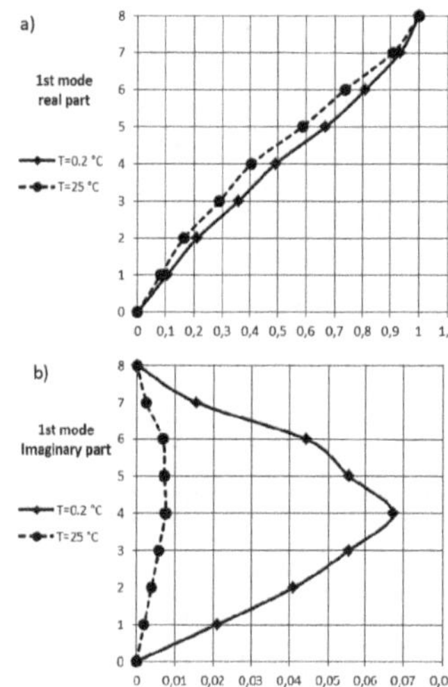

Figure 5. A comparison of the values of first eigenvectors for reference temperature T_0 and $T = 25$ °C: a—the real part,: b—the imaginary part.

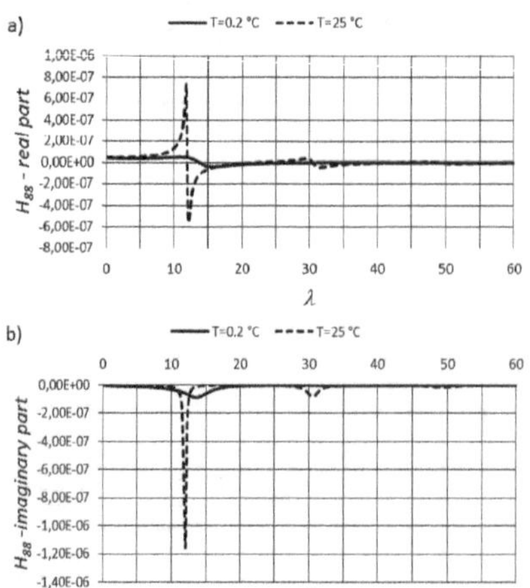

Figure 6. A comparison of the single element values of the frequency response function matrix for reference temperature T_0 and $T = 25$ °C: a—the real part,: b—the imaginary part.

It shows that the values of the first natural frequency decrease very fast with an increase in temperature. A similar effect is observed in the case of the dimensionless damping ratio of the first mode of vibration (Fig. 4). The significant influence of temperature on the values of eigenvector elements is visible in Figure 5. The eigenvector is normalized in such a way that $\overline{\mathbf{q}}(s)\,(8) = 1.0 + \mathrm{i}\,0.0$.

Moreover, the relationship between temperature and the frequency response function has been analyzed. Representative results are shown in Figure 6. Also in this case, the influence of temperature is very significant; especially in the resonance regions.

5 CONCLUSIONS

The presented analysis investigates the influence of temperature change on the dynamic characteristics of structures with VE dampers.

One of the most important conclusions is that the change of temperature has an influence on one parameter c_d of the Zener model of VE dampers only. The computations performed by the authors show a significant decrease in the natural frequencies and the dimensionless damping ratios of a structure with VE dampers with an increase in the temperature of the VE material in the dampers. The frequency response functions significantly differ with temperature as well.

Summing up, it should be emphasized that the change of temperature in VE dampers has an essential influence on all dynamic characteristics of structures.

REFERENCES

Chang, K.C., Soong, T.T., Oh, S.T. & Lai, M.L. 1992. Effect of ambient temperature on a viscoelastically damped structure. *Journal of Structural Engineering* 118: 1955–1973.

de Lima, A.M.G., Rade, D.A., Lacerda, H.B. & Araújo, C.A. 2015. An investigation of the self-heating phenomenon in viscoelastic materials subjected to cyclic loadings accounting for prestress. *Mechanical Systems and Signal Processing* 58–59: 115–127.

Lewandowski, R., Bartkowiak, A. & Maciejewski, H. 2012. Dynamic analysis of frames with viscoelastic dampers: a comparison of damper models. *Structural Engineering Mechanics* 41: 113–137.

Nashif, A.D., Jones, D.I.G. & Henderson, J.P. 1985. *Vibration Damping*. New York: John Wiley and Sons.

Pawlak, Z. & Lewandowski, R. 2013. The continuation method for the eigenvalue problem of structures with viscoelastic dampers. *Computers and Structures* 125: 53–61.

Singh, M.P. & Chang, T.S. 2009. Seismic analysis of structures with viscoelastic dampers. *Journal of Engineering Mechanics* 135: 571–580.

Soong, T.T. & Spencer, B.F. 2002. Supplemental energy dissipation: state-of-the-art and state-of-the-practice. *Engineering Structures* 24: 243–259.

Advances in Mechanics: Theoretical, Computational and Interdisciplinary Issues – Kleiber et al. (Eds)
© 2016 Taylor & Francis Group, London, ISBN 978-1-138-02906-4

Modified Hu-Washizu variational principle as a general basis for FEM plasticity equations

J. Lewandowski & K. Myślecki
Wrocław University of Technology, Wrocław, Poland

ABSTRACT: The modified Hu-Washizu variational principle is a universal approach to solve various plasticity problems. It enables to derive FEM equations for any yield criterion formulation (including a hardening model) and any element (regarding dimensions and shape functions). The procedure to obtain specific FEM equations is presented. Verification on several examples is done. Described FEM algorithm shows convergence, stability, validity and no return mapping error.

1 INTRODUCTION

The FEM equations may be derived from different variational principles. Each one has a specific application range and can be modified due to additional constraints. The simplest Hu-Washizu variational principle is used for elastostatics. It is also often modified to be applied to plasticity problems as well, but usually in terms of the deformation theory of plasticity (Washizu 1982). However, it has never been modified so that deriving FEM plasticity equations for any plasticity model and any FE definition is possible. The authors propose a modification which meets these aims, limiting "any plasticity model" to any plasticity criterion with, mostly applied, associated flow rule.

The universal approach has many advantages. It allows to use presented methods to derive FEM equations for a specific FE and yield criterion (including hardening). It is very simple, clear and direct way (implying from physical equations) to solve any plasticity problem. It also allows to omit problems with return mapping projection errors. Many tests also show that the incremental algorithm based on derived equations is convergent and stable irrespectively to iterations number.

2 MODIFIED PRINCIPLE

2.1 *Partial differential equations*

The final modified version of the principle arises from the original one (for elastostatics) by changing its form to incremental one and by applying the Lagrange multiplier method:

$$\Pi(\sigma_{ij},\varepsilon_{ij},u_i,\lambda)=\frac{1}{2}\int_V C_{ijkl}\left(\varepsilon_{ij}-\varepsilon_{ij}^n\right)\left(\varepsilon_{kl}-\varepsilon_{kl}^n\right)dV$$
$$-\int_V\left(\sigma_{ij}-\sigma_{ij}^n\right)\left[\varepsilon_{ij}+\bar{\varepsilon}_{ij}^n-\frac{1}{2}\left(u_{i,j}+u_{j,i}\right)\right]dV$$
$$-\int_V \Delta F_i u_i dV-\int_V \lambda\Phi(\sigma_{ij})dV+\cdots \tag{1}$$

in which the omitted part ("…") are boundary conditions, provided in FEM by stiffness matrix modification.

The functional stationarity satisfies complete set of incremental PDE equations (equilibrium equations, Hooke's law, geometrical relations including associated flow rule, yield criterion respectively):

$$\left.\begin{aligned}&\left(\sigma_{ij}-\sigma_{ij}^n\right)_{,j}+\Delta F_i=0\\&\left(\sigma_{ij}-\sigma_{ij}^n\right)=C_{ijkl}\left(\varepsilon_{kl}-\varepsilon_{kl}^n\right)\\&\varepsilon_{ij}+\bar{\varepsilon}_{ij}^n+\lambda\frac{\partial\Phi(\sigma_{ij})}{\partial\sigma_{ij}}=\frac{1}{2}\left(u_{i,j}+u_{j,i}\right)\\&\Phi(\sigma_{ij})=0\end{aligned}\right\}\text{ in volume } V \tag{2}$$

where respective symbols denote C_{ijkl} = elasticity tensor; σ_{ij}, ε_{ij} = stress and strain tensor; $\bar{\varepsilon}_{ij}^n$ = plastic strain tensor; u_i = displacement vector; ΔF_i = increment of body forces vector; Φ = yield function; λ = coefficient of plastic strain increment.

The index "n" corresponds to the values calculated in the previous incremental step whereas no index means the current value.

2.2 *FEM equations*

The functional (Eqn 1) may be used for FEM formulation. To simplify its discrete form, let's assume a constant deformation **B** matrix (as it often is for elements with linear shape functions). For any FE (regarding dimensions and linear shape functions) and any yield criterion (with or without hardening) one can obtain:

$$\Pi(\boldsymbol{\sigma},\boldsymbol{\varepsilon},\mathbf{q},\lambda)=\frac{1}{2}\left(\boldsymbol{\varepsilon}-\boldsymbol{\varepsilon}^n\right)^T\mathbf{C}\left(\boldsymbol{\varepsilon}-\boldsymbol{\varepsilon}^n\right)V-(\Delta\mathbf{f})^T\mathbf{q}+ \\ -\left(\boldsymbol{\sigma}-\boldsymbol{\sigma}^n\right)^T\left(\boldsymbol{\varepsilon}+\bar{\boldsymbol{\varepsilon}}^n-\mathbf{B}\mathbf{q}\right)V-\boxed{\lambda\Phi V} \tag{3}$$

where V = is measure of the element (length, area, volume). Symbols are analogous to these in Equation 1. Bold capital letters (**C**, **B**) are reserved for matrices and small letters ($\boldsymbol{\sigma}$, $\boldsymbol{\varepsilon}$, $\mathbf{q}$, $\Delta\mathbf{f}$) – for vectors.

The FEM equations are obtained by the stationarity of the functional:

$$\begin{bmatrix} 0 & 0 & \mathbf{B}^T V & 0 \\ 0 & \mathbf{C}V & -\mathbf{I}V & 0 \\ \mathbf{B}V & -\mathbf{I}V & \boxed{\mathbf{K}_S^{\sigma\sigma}} & \boxed{\mathbf{K}_S^{\sigma\lambda}} \\ 0 & 0 & \boxed{\mathbf{K}_S^{\lambda\sigma}} & \boxed{\mathbf{K}_S^{\lambda\lambda}} \end{bmatrix} \begin{Bmatrix} \mathbf{q} \\ \boldsymbol{\varepsilon} \\ \boldsymbol{\sigma} \\ \lambda \end{Bmatrix} = \begin{Bmatrix} \Delta\mathbf{f}+\mathbf{B}^T\boldsymbol{\sigma}^n V \\ \left(\mathbf{C}\boldsymbol{\varepsilon}^n-\boldsymbol{\sigma}^n\right)V \\ \bar{\boldsymbol{\varepsilon}}^n V \\ \boxed{b^\lambda} \end{Bmatrix} \tag{4}$$

where **I** = is the identity matrix. The framed elements are zero for elastic state (Eqs. 3–4). They are the difference between elasticity and plasticity equations and depend on a particular yield criterion.

3 FEM EQUATIONS FOR HMH CRITERION

3.1 *Assumptions for further examples*

Equation 4 is valid for various yield criteria. To show the particular application, Huber-Mises-Hencky (HMH) yield criterion for plane-stress state is used. The plane stress constitutive equation is:

$\boldsymbol{\sigma}=\mathbf{C}\boldsymbol{\varepsilon}$, where

$$\boldsymbol{\sigma}=\begin{Bmatrix}\sigma_x\\ \sigma_y\\ \tau_{xy}\end{Bmatrix}\quad \boldsymbol{\varepsilon}=\begin{Bmatrix}\varepsilon_x\\ \varepsilon_y\\ \gamma_{xy}\end{Bmatrix}\quad \mathbf{C}=\frac{E}{1-\nu^2}\begin{bmatrix}1 & \nu & 0\\ \nu & 1 & 0\\ 0 & 0 & \frac{1-\nu}{2}\end{bmatrix} \tag{5}$$

Two different hardening models are considered.

3.2 *Ideal plasticity*

The first one is ideal plasticity. In this case the yield function can be easily expressed as a function of stress vector:

$$\Phi(\boldsymbol{\sigma})=\frac{1}{2}\boldsymbol{\sigma}^T\psi\boldsymbol{\sigma}-2\sigma_0^2,\quad \psi=\begin{bmatrix}4 & -2 & 0\\ -2 & 4 & 0\\ 0 & 0 & 12\end{bmatrix} \tag{6}$$

where σ_0 = is a normal yield stress value. Then:

$$\begin{cases}\mathbf{K}_S^{\sigma\lambda}=-\boldsymbol{\psi}\boldsymbol{\sigma}V,\quad \mathbf{K}_S^{\lambda\sigma}=-\frac{1}{2}\boldsymbol{\sigma}^T\boldsymbol{\psi}V\\ \mathbf{K}_S^{\sigma\sigma}=\mathbf{K}_S^{\lambda\lambda}=0,\quad b^\lambda=-2\sigma_0^2 V\end{cases} \tag{7}$$

3.3 *Kinematic hardening*

The second model is kinematic hardening. Hardening process can be illustrated generally as shown in Figure 1.

Assuming a proportional material stiffness ($\mathbf{C}_1=\alpha\mathbf{C}$, $0<\alpha<1$) after yield stress exceeding, the yield function for the kinematic hardening is:

$$\Phi(\boldsymbol{\sigma})=\frac{1}{2}\boldsymbol{\sigma}^T\boldsymbol{\psi}\boldsymbol{\sigma}-2\sigma_0^2+ \\ -\frac{\alpha}{1-\alpha}\boldsymbol{\sigma}^T\boldsymbol{\psi}\mathbf{C}\bar{\boldsymbol{\varepsilon}}+\frac{1}{2}\left(\frac{\alpha}{1-\alpha}\right)^2\bar{\boldsymbol{\varepsilon}}^T\mathbf{C}\boldsymbol{\psi}\mathbf{C}\bar{\boldsymbol{\varepsilon}} \tag{8}$$

In this case:

$$\begin{cases}\mathbf{K}_S^{\sigma\sigma}=0,\ \mathbf{K}_S^{\sigma\lambda}=-\left(\boldsymbol{\psi}\boldsymbol{\sigma}-\frac{\alpha}{1-\alpha}\boldsymbol{\psi}\mathbf{C}\left[\bar{\boldsymbol{\varepsilon}}^n+\Delta\bar{\boldsymbol{\varepsilon}}\right]\right)V\\ \mathbf{K}_S^{\lambda\sigma}=-V\frac{1}{2}\boldsymbol{\sigma}^T\boldsymbol{\psi}+V\frac{\alpha}{1-\alpha}\left[\bar{\boldsymbol{\varepsilon}}^n+\Delta\bar{\boldsymbol{\varepsilon}}\right]^T\mathbf{C}\boldsymbol{\psi}\\ \mathbf{K}_S^{\lambda\lambda}=-V\left(\frac{\alpha}{1-\alpha}\right)^2\frac{\Delta\bar{\boldsymbol{\varepsilon}}^T}{\lambda}\mathbf{C}\boldsymbol{\psi}\mathbf{C}\left(\frac{1}{2}\Delta\bar{\boldsymbol{\varepsilon}}+\bar{\boldsymbol{\varepsilon}}^n\right)\\ \mathbf{b}_\lambda=V\frac{1}{2}\left(\frac{\alpha}{1-\alpha}\right)^2\bar{\boldsymbol{\varepsilon}}^{nT}\mathbf{C}\boldsymbol{\psi}\mathbf{C}\bar{\boldsymbol{\varepsilon}}^n-V2\sigma_0^2\end{cases} \tag{9}$$

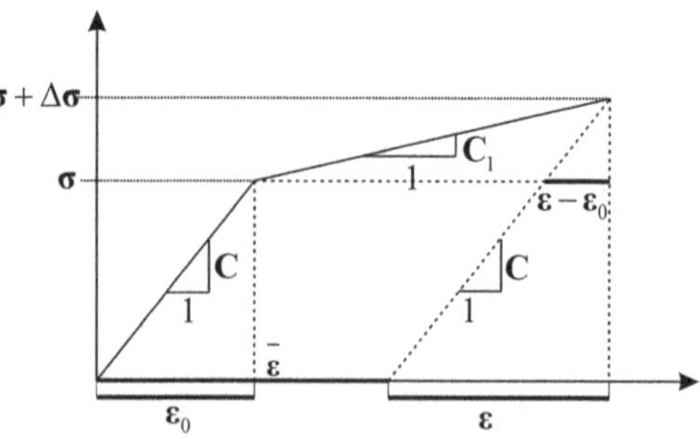

Figure 1. General hardening model.

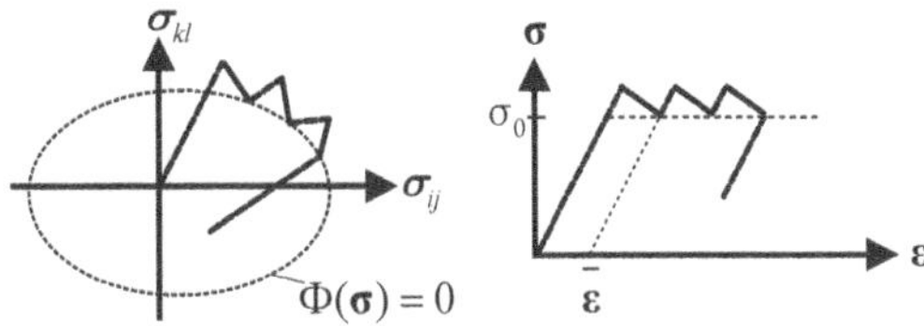

Figure 2. FEM algorithm.

4 FEM ALGORITHM

Equation 4 is nonlinear and solved with the Newton-Raphson method by authors' algorithm implemented in Mathematica. Considering its block form and zero submatrices, it is possible to reduce the FEM algorithm computation time by block elimination.

The big advantage of the algorithm is that it omits return mapping dilemmas. The return mapping method tentatively assumes elastic behavior of the element throughout a given time step. If the resulting stress violates the yield criterion, it is projected back to the plastic yield surface by enforcing Equation 4 where $\boldsymbol{\sigma}^n$ and $\boldsymbol{\varepsilon}^n$ are results from the "elastic prediction", as illustrated in Figure 2.

Therefore Equation 4 is satisfied without any error) as the projection to the plasticity surface respects all equations (equilibrium equations, Hooke's law, geometrical relations and yield criterion). The only inaccuracy implies from Newton-Raphson method imprecision.

The second advantage of the approach is that the solution is clear and simple as well as easy to implement if only Equation 4 is known. In particular, no elasticity matrix modification is needed to enforce the plastic flow, as often applied in other algorithms.

The algorithm does not have any problems with convergence and stability even if amount of FEs is small and if not many incremental steps are involved.

5 EXAMPLES

There are three examples presented in this paper. Results of first two examples are compared with analytical solutions. The outcome of the third is compared to another numerical result.

In each problem, common plane stress triangular FEs are used by Zienkiewicz & Taylor (2000). Their thickness is assumed as 1 m. In each case there are 3 iterations of Newton-Raphson method and 200 incremental steps. One of shown loading programs (Fig. 3) is applied.

5.1 *Ideal plasticity*

First verification is done on a simple example—membrane under uniaxial loading is shown in Figure 4.

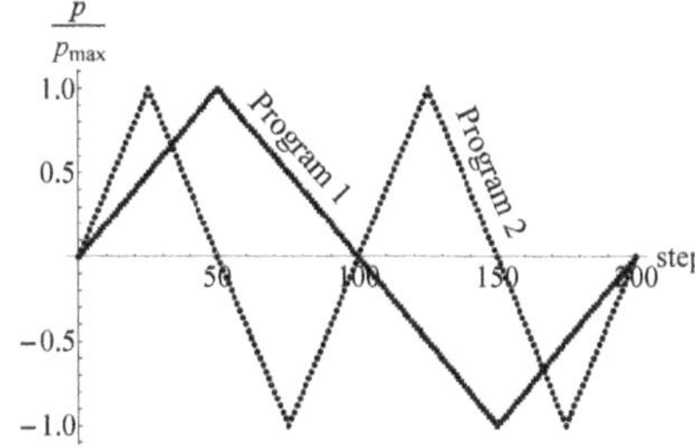

Figure 3. Loading programs.

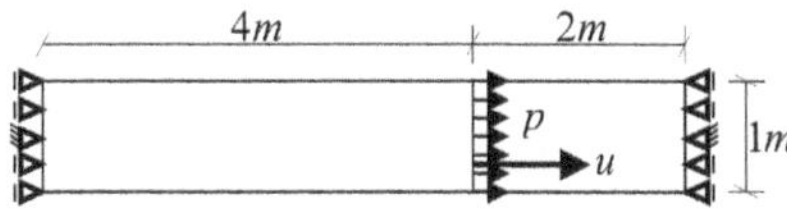

Figure 4. Membrane with two-side support.

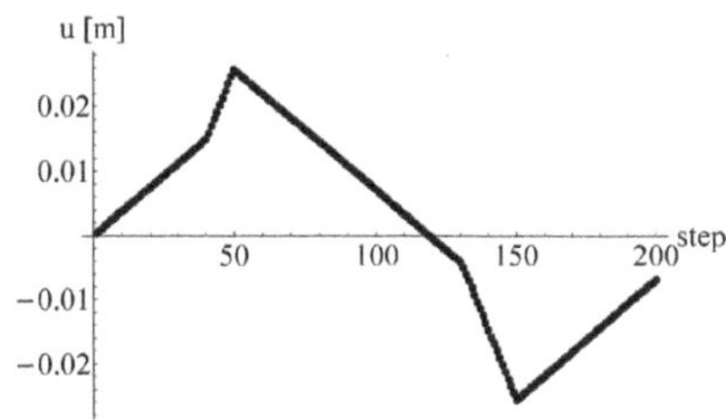

Figure 5. History of displacement u.

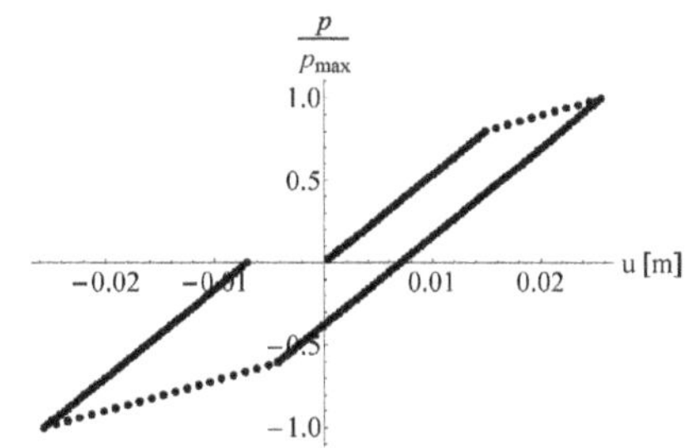

Figure 6. Loading—displacement relation.

Input data: E = 20 MPa, $\upsilon = 0$, $\sigma_0 = 150$ kPa, $p_{max} = 280$ kPa, Program 1, 96 FEs. Result produced by the algorithm is shown on Figures 5–6.

The analytical solution for equivalent 1D beam is exactly the same.

5.2 *Kinematic hardening*

In order to verify the kinematic hardening model, one-side supported membrane is employed (Fig. 7).

Input data: E = 20 MPa, $\upsilon = 0.2$, $\sigma_0 = 40$ kPa, $\alpha = 0.1$, $p_{max} = 60$ kPa, Program 2, 72 FEs. The algorithm outcomes are shown in Figure 8-9.

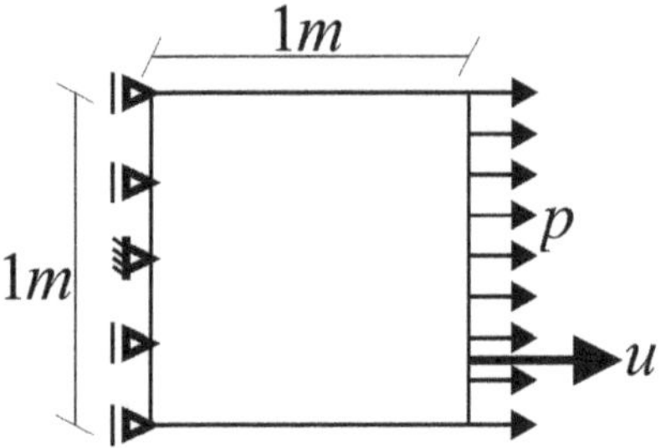

Figure 7. One-side supported membrane.

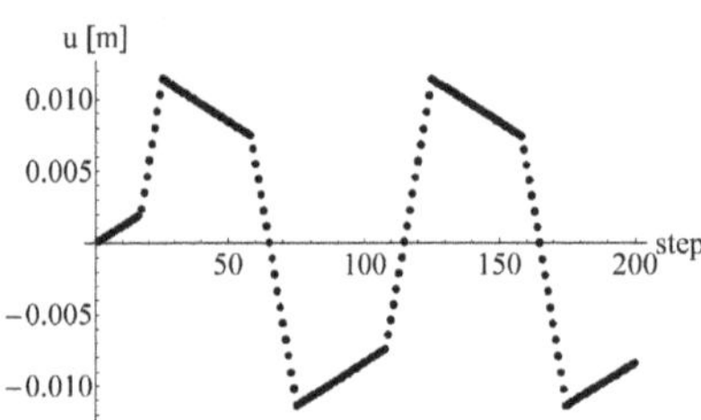

Figure 8. History of displacement u.

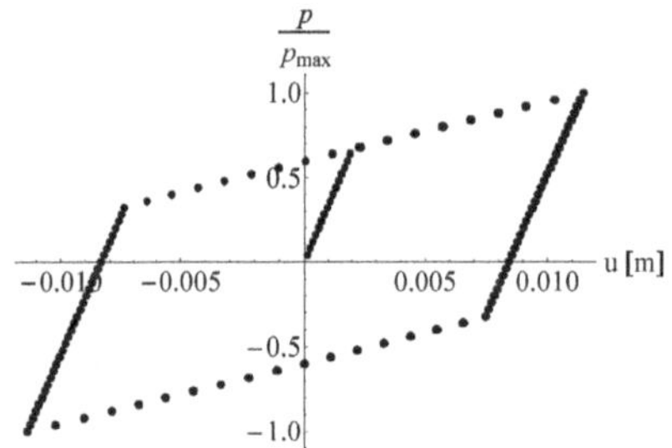

Figure 9. Loading—displacement relation.

The solution (Figs. 8–9) is almost the same as the analytical one for the equivalent 1D beam. Two loading cycles cover each other in the pressure-displacement plane. The p(u) line inclination (Fig. 9) in the plastic flow range is, 0.1 of the initial value, as expected.

5.3 *2D cantilever beam*

The last example – 2D cantilever beam (Fig. 10) is verified by comparing results of the authors' algorithm with Abaqus results. The kinematic hardening model is applied.

Input data: E = 20 MPa, $\upsilon = 0.2$, $\sigma_0 = 40$ kPa, $\alpha = 0.1$, $p_{max} = 6$ kPa, Program 2, 200 FEs. The authors applied the same data to Abaqus model, including the same FE mesh geometry and incremental iterations number. The results are presented in Figure 11.

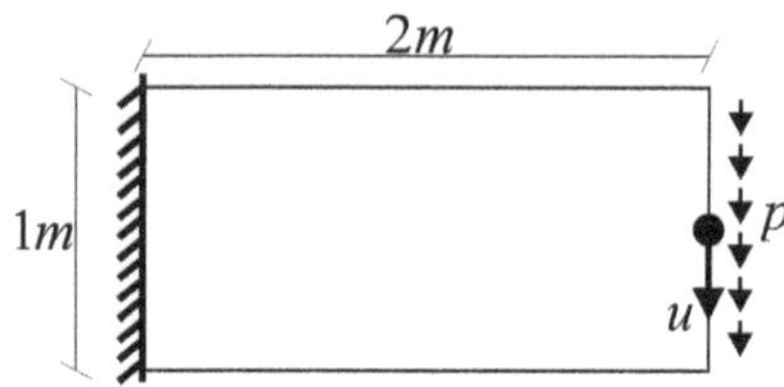

Figure 10. 2D cantilever beam.

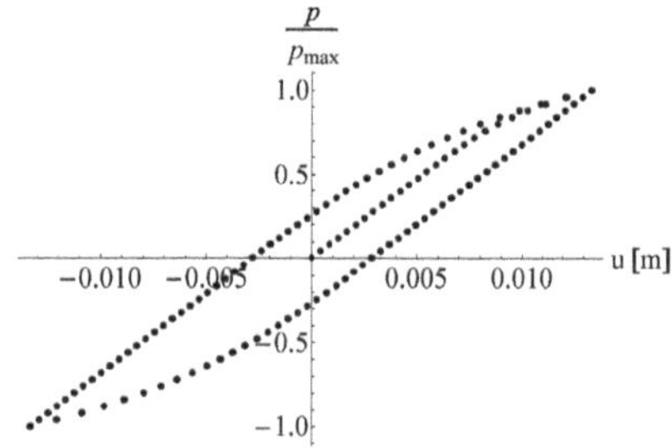

Figure 11. Loading—displacement relation.

Figure 11 contains both Abaqus and authors' algorithm results. They are virtually the same on each incremental step, despite the fact that 2 loading cycles were applied (Program 2).

6 CONCLUSION

The plane stress examples, as many others conducted by the authors, show that the functional (Eqn 1) can be successfully used to derive FEM plasticity equations for various yield criteria and finite elements. The presented simple algorithm allows to use it efficiently to solve plasticity problems without return mapping error.

Coming from physical plasticity equations, it is shown how to obtain FEM equations directly. Therefore the functional can be regarded as a unified FEM formulation for plasticity problems.

REFERENCES

Washizu, K. 1982. *Variational Methods in Elasticity and Plasticity*, third ed. New York: Pergamon Press.

Zienkiewicz, O.C. & Taylor, R.T. 2000. *The Finite Element Method, Vol. 1: The Basis*, fifth ed. Oxford: Butterworth-Heinemann.

Advances in Mechanics: Theoretical, Computational and Interdisciplinary Issues – Kleiber et al. (Eds)
© 2016 Taylor & Francis Group, London, ISBN 978-1-138-02906-4

Electro-mechanical multiple-point beam-to-beam contact

P. Litewka
Poznan University of Technology, Poznań, Poland

ABSTRACT: The electro-mechanical contact is analysed for beam-like electric conductors getting in touch at acute angles. The concept of multiple-point contact is used to better cover situations when contact cannot be considered as point-wise. Electric contact is introduced with constraints yielding from the Ohm's law and the phenomenon of long constriction is assumed, which is present in a situation when electric current flows through a contacting spot with dimensions much smaller than regular dimensions of the contacting conductors. The governing equations are written separately for electric and mechanical fields using voltages and displacements as principal unknowns, respectively. The consistent linearisation leads to a tangent stiffness matrix for the problem which features a semi-coupling nature of the problem with the electric field depending on the mechanical one only. Numerical example is presented to show advantages of the presented model.

1 INTRODUCTION

Contact between beams is a special case in 3D contact analysis. The topic was initiated by Wriggers & Zavarise (1997) and continued in (Zavarise & Wriggers, 2000, Litewka & Wriggers 2002a, 2002b) where contact without and with Coulomb friction for beams of circular and rectangular cross-sections was covered. The further development concerned inclusion of thermal and electric coupling (Boso et al. 2005). A rigorous approach to the question of point-wise contact was suggested by Konyukhov & Schweizerhof (2010). The authors focused their interest on the closest-point projection procedure, which for the beam-to-beam contact leads to the orthogonality conditions.

One of the key assumptions in the majority of those beam-to-beam analyses was, that the closest points of two curves can always be uniquely determined, at least locally. In other words, the point-wise contact between beams was assumed. However, such an approach may fail in some cases. Hence, the general approach should include possibilities when contacting beam-like objects form acute angles and are parallel or conforming.

The beam-to-beam contact finite element presented in this paper is a further enhancement of the multiple-point contact formulation (Litewka 2013, 2015) to include electro-mechanical coupling. For beams it is assumed that their circular cross-sections do not deform and are subjected to small strains but displacements are large. Moreover, it is necessary that the closed point projection for the curves representing beams axes has a unique solution, at least locally. Thus, the model is suited for the case of beams contacting at acute angles but not conforming ones.

Section 2 includes a description how the additional contact pairs are defined. In Section 3 the variables for electro-mechanical contact are introduced. Section 4 gives the equations of electro-mechanical contact and their linearisation is presented, too. Some details of the finite element discretisation are discussed in Section 5, while Section 6 includes numerical examples, illustrating advantages of the new approach, with some concluding remarks.

2 ADDITIONAL CONTACT POINTS

The multiple-point contact approach uses the orthogonality conditions to find the central (0) pair of points, C_{s0} and C_{m0}. Then on the beam s two additional points C_{sb} and C_{sf} with co-ordinates ξ_{sb} and ξ_{sf}, shifted back (b) and forward (f), are introduced (Litewka, 2013), see Figure 1.

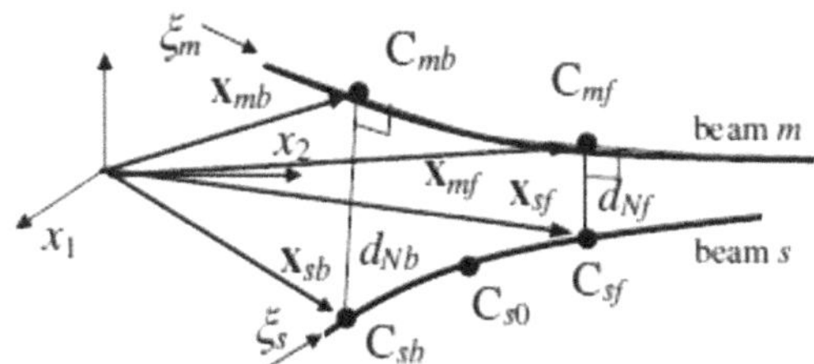

Figure 1. Additional contact pairs for the multiple-point contact.

Finally, the corresponding additional points C_{mb} and C_{mf} on the beam m are located using two separate orthogonality conditions. With three contact candidate pairs three normal gaps can be defined:

$$g_{Ni} = \left\| \mathbf{x}_{mi} - \mathbf{x}_{si} \right\| - r_m - r_s = d_{Ni} - r_m - r_s \tag{1}$$

where r_m and r_s are the cross-sectional radii and subscript i stands for b, 0 or f.

3 VARIABLES FOR ELECTRO-MECHANICAL CONTACT

To analyze the direct electric current flow through a contact spot its geometry is assumed using the Hertz classic solution. The case of two contacting cylinders features an elliptic contact spot (Boso et al. 2005). The semi-axes a_j of the ellipse in terms of the normal contact force F_N are given by:

$$a_j = \alpha_j \cdot \sqrt[3]{\frac{3}{2}\frac{F_N r_{av}}{E_{av}}} \tag{2}$$

where $j = 1, 2$ and equivalent radius and equivalent elasticity modulus are introduced:

$$r_{av} = \frac{r_m r_s}{r_m + r_s}, \quad E_{av} = \frac{E_m E_s}{E_s\left(1-\nu_m^{\ 2}\right) + E_m\left(1-\nu_s^{\ 2}\right)} \tag{3}$$

with E_m, E_s, ν_m, ν_s denoting the elasticity moduli and Poisson's ratios for the contacting cylinders (beams). The coefficients α_j in (2) depend on geometry of contacting cylinders.

To determine the electric conductance h the Concept of long constriction (Holm 1981) is used to get:

$$h = 2K_{av}a \tag{4}$$

where K_{av} is the equivalent electric conductivity:

$$K_{av} = \frac{K_m K_s}{K_s + K_m} \tag{5}$$

computed from conductivity values from both beams and a is the radius of a circular contact spot. The latter value is found using an equivalent of the ellipse defined in (2):

$$a = \sqrt{cd} = \sqrt{\alpha\beta} \cdot \sqrt[3]{\frac{3}{2}\frac{F_N r_{av}}{E_{av}}} \tag{6}$$

Substitution (6) into (4), use of penalty formulation for the contact force and Ohms law allow to define the electric current i through the contact spot. This derivation is repeated for all three contact pairs and the following relations are obtained (the subscript i stands for b, 0 or f):

$$i_i = 2K_{av}\sqrt{\alpha_{1i}\alpha_{2i}}\sqrt[3]{\frac{3}{2}\frac{r_{av}}{E_{av}}\varepsilon_N g_{Ni}} g_{Vi} \tag{7}$$

The values of voltage gaps g_V are found as differences of electric voltages V_m and V_s at contact points:

$$g_{Vi} = V_{mi} - V_{si} \tag{8}$$

The normal gaps and voltage gaps are the key factors in electro-mechanical contact. The variations of electric current with respect to them can be found in the form:

$$\frac{\partial i_i}{\partial g_{Ni}}\delta g_{Ni} = \frac{2}{3}K_{av}\sqrt{\alpha_{1i}\alpha_{2i}}\sqrt[3]{\frac{3}{2}\frac{\varepsilon_N r_{av}}{E_{av} g_{Ni}^{\ 2}}} g_{Vi}\delta g_{Ni} \tag{9}$$

$$\frac{\partial i_i}{\partial g_{Vi}}\delta g_{Vi} = \frac{2}{3}K_{av}\sqrt{\alpha_{1i}\alpha_{2i}}\sqrt[3]{\frac{3}{2}\frac{\varepsilon_N g_{Ni} r_{av}}{E_{av}^{\ 2}}}\delta g_{Vi} \tag{10}$$

4 GLOBAL EQUATIONS OF ELECTRO-MECHANICAL CONTACT

Solving a contact problem leads to minimization of a functional subjected to contact inequality constraints. In this mathematical problem the active set strategy can be used. A contact search routine yields active constraints which are treated as equality ones. In the presented three-point contact formulation there can be three active constraints—for the central pair and for two additional pairs.

The global set of equations is obtained by adding the mechanical virtual work δW for active pairs (Mc) and analogous quantities for the electric field (Ec) to the global virtual work expressions, mechanical (M) and electric (E), for two beams, m and s:

$$\begin{aligned} \delta_u W_M &= \delta_u W_{Mm} + \delta_u W_{Ms} + \bigcup_{active} \delta_u W_{Mc} = 0 \\ \delta_V W_E &= \delta_V W_{Em} + \delta_V W_{Es} + \bigcup_{active} \delta_V W_{Ec} = 0 \end{aligned} \tag{11}$$

The contact terms in (11) take the form:

$$\begin{aligned} \delta_u W_{Mc} &= \sum_{i=b,0,f} F_{Ni}\delta_u g_{Ni} \\ \delta_V W_{Ec} &= \sum_{i=b,0,f} i_i \delta_u g_{Vi} \end{aligned} \tag{12}$$

The Equations 11 are non-linear and to solve them the Newton-Raphson method is used. The variations of the contact terms are computed with respect to mechanical and electric variables, i.e. displacements u and voltages V, respectively:

$$\Delta_u \delta_u W_{Mc} = \sum_{i=b,0,f} (\Delta_u F_{Ni} \delta_u g_{Ni} + F_{Ni} \Delta_u \delta_u g_{Ni})$$
$$\Delta_u \delta_V W_{Ec} = \sum_{i=b,0,f} (\Delta_u i_i \delta_V g_{Vi} + i_i \Delta_u \delta_V g_{Vi})$$
$$\Delta_V \delta_u W_{Mc} = \sum_{i=b,0,f} (\Delta_V F_{Ni} \delta_u g_{Ni} + F_{Ni} \Delta_V \delta_u g_{Ni})$$
$$\Delta_V \delta_V W_{Ec} = \sum_{i=b,0,f} (\Delta_V i_i \delta_V g_{Vi} + i_i \Delta_V \delta_V g_{Vi}) \quad (13)$$

Substitution of the penalty formulation for the contact forces and (7) for the electric current allows to express the contact contributions in terms of voltages and displacements, which are ready for finite element discretisation.

The purely mechanical terms from (13_1) were discussed in Litewka (2013). The coupling terms coming from (13_3) vanish because the normal contact gaps do not depend on electric voltages. The coupling terms from (13_2) are:

$$\Delta_u \delta_V W_{Vc} = \sum_{i=b,0,f} \left(\frac{\partial i_i}{\partial g_{Vi}} \cdot \Delta_u g_{Vi} \cdot \delta_V g_{Vi} + \frac{\partial i_i}{\partial g_{Ni}} \cdot \Delta_u g_{Ni} \cdot \delta_V g_{Vi} + i_i \cdot \Delta_u \delta_V g_{Vi} \right) \quad (14)$$

while the purely electric terms can be expressed as:

$$\Delta_V \delta_V W_{Vc} = \sum_{i=b,0,f} \frac{\partial i_i}{\partial g_{Vb}} \cdot \Delta_V g_{Vi} \cdot \delta_V g_{Vi} \quad (15)$$

5 FINITE ELEMENT DISCRETISATION

The contact finite element for the electro-mechanical case has to possess displacements and voltages as nodal parameters. In the presented approach the method to smooth contact facets, denoted as inscribed-curve in (Litewka 2007), is used. The resulting element involves three nodes per each beam, m1, m2, m3 and s1, s2, s3, as shown in Figure 2.

The nodal degrees of freedom, displacements **u** and voltages **V** for each involved node from both beams, are assembled into the vector **q**:

$$\mathbf{q}_{[24\times1]} = \left\{ \mathbf{u}_M{}^T{}_{[9\times1]}, \mathbf{u}_S{}^T{}_{[9\times1]}, \mathbf{V}_M{}^T{}_{[3\times1]}, \mathbf{V}_S{}^T{}_{[3\times1]} \right\} \quad (16)$$

It should be noted that for clarity the sketch in Figure 2 shows the central pair contact points only.

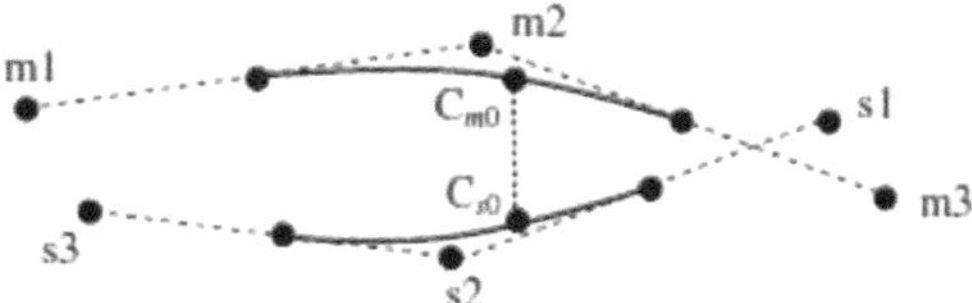

Figure 2. Inscribed-curve beam-to-beam contact finite element.

The presented element includes two additional pairs and in general all three particular points may not lie in the same contact facet on a beam. A thorough discussion of this issue was given in Litewka (2015). Here, for the sake of simplicity of the formulation it is assumed, that all three contact point per beam are within one contact facet. Yet, the computer program used to solve the example in the further section includes a provision for a more general case.

The details of the finite element discretisation of displacements for beams with the shape function matrices $\mathbf{N}_{umi}$ and $\mathbf{N}_{usi}$ used here, were given in Litewka (2013). For the voltages the linear shape functions are used (Boso et al. 2005):

$$V_{mi} = \frac{1}{2}\left[(1-\xi_{mi}) \cdot V_{m1} + (1+\xi_{mi}) \cdot V_{m2}\right] = \mathbf{N}_{Vmi}\mathbf{V}_M$$
$$V_{si} = \frac{1}{2}\left[(1-\xi_{si}) \cdot V_{s1} + (1+\xi_{si}) \cdot V_{s2}\right] = \mathbf{N}_{Vsi}\mathbf{V}_S \quad (17)$$

The residual vector yielding from the FE discretisation (12) takes the form:

$$\mathbf{R}_{[24\times1]} = \begin{bmatrix} \mathbf{R}_m \\ \mathbf{R}_e \end{bmatrix} \quad (18)$$

Its mechanical part $\mathbf{R}_m$ was given in Litewka (2013) and the electric part is given as:

$$\mathbf{R}_e = 2K_{av} \sqrt[3]{\frac{3\varepsilon_N r_{av}}{2E_{av}}} \times \sum_{i=b,0,f} g_{Vi} \sqrt{\alpha_{1i}\alpha_{2i}} \sqrt[3]{g_{Ni}} \begin{bmatrix} \mathbf{N}_{Vmi}{}^T \\ -\mathbf{N}_{Vsi}{}^T \end{bmatrix} \quad (19)$$

The tangent stiffness matrix yielding from FE discretisation (13) reflects the semi-coupled nature of the problem and takes the form:

$$\mathbf{K}_{[24\times24]} = \begin{bmatrix} \mathbf{K}_{mm} & \mathbf{0} \\ (\mathbf{K}_{me1} + \mathbf{K}_{me2} + \mathbf{K}_{me3}) & \mathbf{K}_{ee} \end{bmatrix} \quad (20)$$

For the purely mechanical part refer to Litewka (2013). The coupling part split into three components, as in (14), takes the following format:

$$\mathbf{K}_{me1} = 2K_{av}\sqrt[3]{\frac{3\varepsilon_N r_{av}}{2E_{av}}} \sum_{i=b,0,f} \left\{ \sqrt[3]{g_{Ni}} \sqrt{\alpha_{1i}\alpha_{2i}} \times \begin{bmatrix} \mathbf{N}_{Vmi}{}^T \\ -\mathbf{N}_{Vsi}{}^T \end{bmatrix} \left[V_{mi,m}\mathbf{F}_{mi} - V_{si,s}\mathbf{F}_{si} \right] \right\} \tag{21}$$

$$\mathbf{K}_{me2} = \frac{2K_{av}}{3}\sqrt[3]{\frac{3\varepsilon_N r_{av}}{2E_{av}}} \sum_{i=b,0,f} \left\{ g_{Vi} \sqrt{\alpha_{1i}\alpha_{2i}} \times (g_{Ni})^{-\frac{2}{3}} \begin{bmatrix} \mathbf{N}_{Vmi}{}^T \\ -\mathbf{N}_{Vsi}{}^T \end{bmatrix} \mathbf{n}_i{}^T \left[\mathbf{N}_{umi}, \ -\mathbf{N}_{usi} \right] \right\} \tag{22}$$

$$\mathbf{K}_{me3} = 2K_{av}\sqrt[3]{\frac{3\varepsilon_N r_{av}}{2E_{av}}} \times \sum_{i=b,0,f} \sqrt{\alpha_{1i}\alpha_{2i}}\, g_{Vi} \sqrt[3]{g_{Ni}} \begin{bmatrix} \mathbf{D}_{Vi}{}^T\mathbf{F}_{mi} \\ -\mathbf{D}_{Vi}{}^T\mathbf{F}_{si} \end{bmatrix} \tag{23}$$

where the matrices of **F** type result from the discretisation of variations of local co-ordinates ξ_m and ξ_s (Litewka 2013) and the matrices of **D** type are numeric matrices resulting from the differentiation of linear shape functions for voltages (17) with respect to local co-ordinates.

Finally, the purely electric terms from (15) lead to the following matrix:

$$\mathbf{K}_{ee} = 2K_{av}\sqrt[3]{\frac{3\varepsilon_N r_{av}}{2E_{av}}} \times \sum_{i=b,0,f} \sqrt{\alpha_{1i}\alpha_{2i}} \sqrt[3]{g_{Ni}} \begin{bmatrix} \mathbf{N}_{Vmi} \\ -\mathbf{N}_{Vsi} \end{bmatrix} \otimes \begin{bmatrix} \mathbf{N}_{Vmi} \\ -\mathbf{N}_{Vsi} \end{bmatrix} \tag{24}$$

6 NUMERICAL EXAMPLE AND CONCLUSIONS

To illustrate performance of the new approach to electro-mechanical contact a representative example is solved and discussed. Two cantilever beams getting in contact at an acute angle $\varphi = 9.53°$ are analyzed, as shown in Figure 3. The beams have the free ends subjected to displacements $\Delta = 0.4$ and there are external voltages applied at both ends of the beams, $V_1 = 0$ for Beam 1 and $V_2 = 10$ for Beam 2. These are applied simultaneously in 60 equal increments. The beams are identical with length 6.021, cross-section radii 0.1, Young's modulus $205{\cdot}10^5$, Poisson's ratio 0.3 and electric conductivity $8{\cdot}10^{-3}$, are separated initially by 0.01 and are divided into 10 finite elements of co-rotational type.

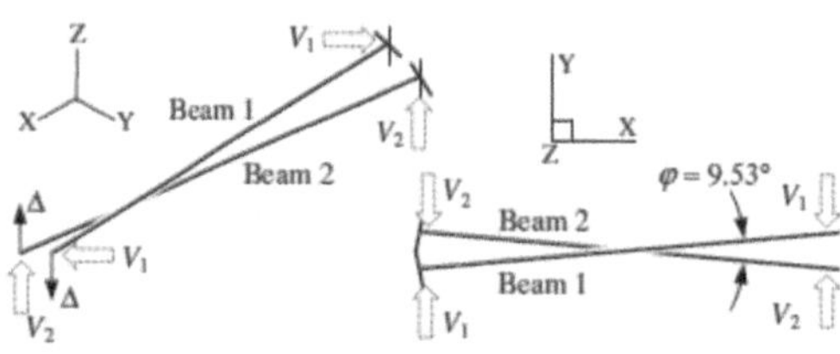

Figure 3. Beam axes layout for numerical example.

Figure 4a presents voltage distributions along both beams computed using the presented multiple-point contact model and compared with the solution coming from the point-wise approach.

The values of total electric current are: 1.01 for the point-wise vs 1.25 for the multiple-point. As can be seen in the distributions, the corresponding voltage gap for the point-wise (0.366) is larger than those for the multiple-point (0.220, 0.224, 0.228). Still, the latter case allows for a larger current flow.

An important advantage of the presented multiple-point model is a smaller sensitivity of results to the value of the penalty parameter, what can be verified in Figures 4b-4c, were the magnification of voltage distribution in the contact zone is shown. The results obtained for various values of the penalty parameter are less differing in the case of the multiple-point approach.

In general, it should be stated that the three-point model of contact for beams forming acute angles is more realistic. It features higher electric current flow taking into account in a better way the increased contact area. Simultaneously it substantially limits the disadvantages of penalty formulation making the results less dependent on penalty parameter values.

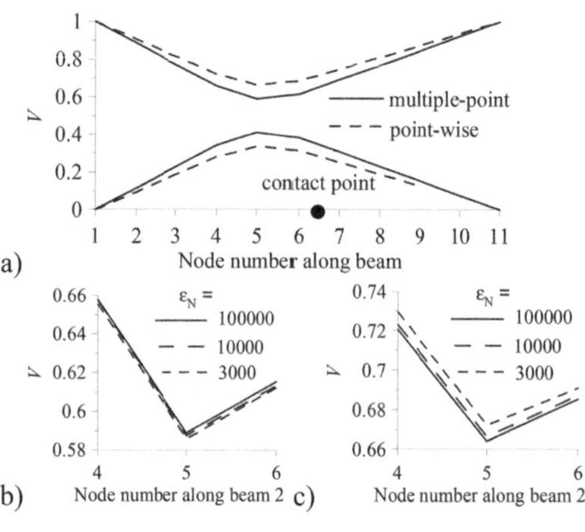

Figure 4. Results for Example: a) voltage distributions along beams, b) voltage distributions in contact zone for multiple-point model, c) voltage distributions in contact zone for point-wise model.

ACKNOWLEDGEMENTS

This research is done under the internal university grant 01/11/DSPB/400.

REFERENCES

Boso, D.P., Litewka, P., Schrefler, B.A. & Wriggers, P. 2005. A 3D beam-to-beam contact finite element for coupled electric-mechanical fields. *International Journal for Numerical Methods in Engineering* 64: 1800–1815.

Holm, R. 1981. *Electric Contacts: Theory and applications*, Springer-Verlag: Berlin, Heidelberg, New York.

Konyukhov, A. & Schweizerhof, K. 2010. Geometrically exact covariant approach for contact between curves. *Computer Methods in Applied Mechanics and Engineering* 199: 2510–2531.

Litewka, P. & Wriggers, P. 2002. Contact between 3D beams with rectangular cross-sections. *International Journal for Numerical Methods in Engineering* 53: 2019–2041.

Litewka, P. & Wriggers, P. 2002. Frictional contact between 3D beams. *Computational Mechanics* 28: 26–39.

Litewka, P. 2007. Hermite polynomial smoothing in beam-to-beam frictional contact problem. *Computational Mechanics* 40(6): 815–826.

Litewka, P. 2013. Enhanced multiple-point beam-to-beam frictionless contact finite element. *Computational Mechanics* 52(6): 1365–1380.

Litewka, P. 2015. Frictional beam-to-beam multiple-point contact finite element. *Computational Mechanics* 56(2): 243–264.

Wriggers, P. & Zavarise, G. 1997. On contact between three-dimensional beams undergoing large deflections. *Communications in Numerical Methods in Engineering* 13: 429–438.

Zavarise, G. & Wriggers, P. 2000. Contact with friction between beams in 3-D space. *International Journal for Numerical Methods in Engineering* 49: 977–1006.

Advances in Mechanics: Theoretical, Computational and Interdisciplinary Issues – Kleiber et al. (Eds)
© 2016 Taylor & Francis Group, London, ISBN 978-1-138-02906-4

HSρ—an isotropic material interpolation scheme based on Hashin-Shtrikman variational bounds

T. Łukasiak
Warsaw University of Technology, Warsaw, Poland

ABSTRACT: Topology optimization can be performed with several possible choices of design parametrizations. In particular the density method and the interpolation schemes as SIMP and RAMP are very popular due to their simplicity and efficiency. In the paper a new approach is put forward in which the element-wise pseudo-mass density of the material is treated as a decision variable and a new interpolation scheme controlling the effective material properties. This scheme is based on the variational Hashin-Shtrikman bounds of effective isotropic moduli for mixtures. A similar interpolation scheme for the relaxed optimization problem has been suggested by Ole Sigmund (2001), yet has not been widely applied in practice despite its theoretical advantages. The scheme of interpolation (called HSρ) proposed here is aimed at solving 2D classical topology optimization problems (compliance minimization) as well as 2D inverse-homogenization problems to determine topology of micro-structure exhibiting the prescribed effective material properties.

1 TOPOLOGY OPTIMIZATION AS RELAXED 0–1 PROBLEM

The modern topology or shape optimization problems which state the questions on the optimal layout of several materials in given proportions within a feasible domain Ω nowadays are solved by using the Finite Element Method. The conventional process of the optimization is carried out on usually the rectangular specimen (ground structure) divided into uniform n finite element mesh Ω_k ($k = 1..n$). In case of the two of component materials C_1 and C_2 (of Hooke tensors $\mathbf{C}_1$, $\mathbf{C}_2$ represented by the constitutive matrices $\mathbf{E}_1$, $\mathbf{E}_2$ respectively) and their volume fractions in the design domain expressed by $0 < \rho < 1$, for each element a variable ρ_k is assigned a number such that.

$$\rho_k = \begin{cases} 0 & \Omega_k \in C_1 \\ 1 & \Omega_k \in C_2 \end{cases} \quad \text{and} \quad \sum_{k=1}^{n} \rho_k = n\rho \tag{1}$$

Changing the value of $\{\rho_k\}$ – i.e. creating a new configuration of the placement of the components one can try get the expected result described by the objective function of the optimization problem. The optimization problem defined this way is a difficult, binary integer programming problem with a large number of variables. Accurate representation of structure requires a large number n of variables ρ_k i.e. a dense domain division into finite elements, and produce a huge number (at most) $n!/(m!\,(n-m)!)$ possible configurations to check, where $m = n\,\rho$. Note, that for $n = 100$ only and $\rho = 0.5$ there exists nearly 10^{29} different configurations of $\{\rho_k\}$. The optimization problems stated as pure 0–1-element wise problem may be solved using the so-called hard-kill methods e.g. Xie & Steven (1993) Evolutionary Structural Optimization (ESO) or Bidirectional ESO (BESO) proposed by Rozvany & Querin (2002) previously known as SERA i.e. Sequential Element Rejection and Admission, but the above-mentioned methods does not guarantee the achievement of the global or even a local optimum. This issue was discussed by Rozvany (2009). Note that for problem given by (1) the $\mathbf{E}_k$ of each element are clearly defined.

For a large n (> 10 000) to be effectively solved i.e. to reach the global optimum, the topology optimization problem must be relaxed, usually by allowing variation $\rho_k = \langle 0, 1\rangle$ and by making the material characteristics depended of the variables ρ_k (Fig. 1). Note, that the final proper solution of the optimization problem in case of two materials must be pure 0–1 solution i.e. contain only $\rho_k = \{0, 1\}$. The introduction of variable continuum ρ_k allows determining the gradient of the objective function and using the efficient gradient methods to search for the minimum.

The variable ρ_k determines the proportions by volume of materials in the element Ω_k. In fact such type of relaxation of the problem means admitting infinitely fine mixtures of the constituents in each element and requires proper calculating of its effective

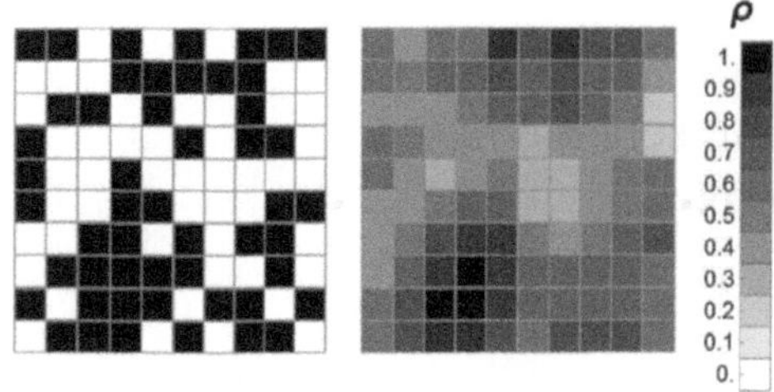

Figure 1. Distribution of $\{\rho_k\}$ for 01 and relaxed problem.

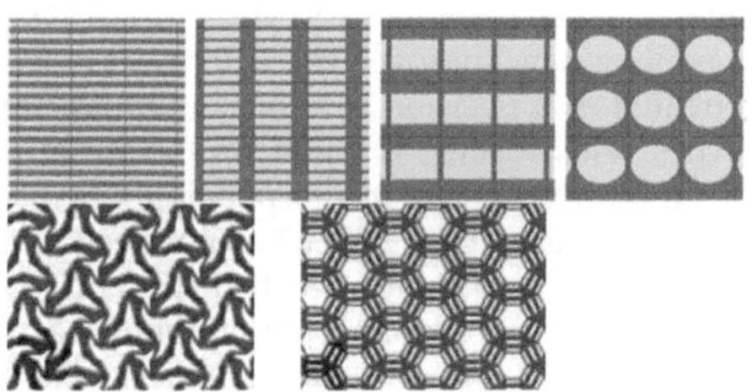

Figure 2. Miscellaneous underlying structures with the same $\rho_k = 0.5$ but different $\mathbf{E}(\rho_k)$. 1-rank or higher rank laminates, different shape or Vigdergauz inclusions, isotropic structures see Łukasiak (2013).

properties $\mathbf{E}(\rho_k)$. It was analyzed by Milton & Cherkaev (1995). This point is crucial for the relaxed topology optimization problems. A wrong approximation $\mathbf{E}(\rho_k)$ may not to lead to the pure 0–1 solution or even in case 0–1 solution the obtained result may be very far from the global optimum. Calculating $\mathbf{E}(\rho_k)$ can be done by homogenization approach. For the homogenization method some microscopic structure must be a priori assumed i.e. some type of the underlying periodic composite structure (Fig. 2): laminates or Vigdergauz (1989) cells etc., see also Grabovsky & Kohn (1995) or isotropic structures see Sigmund (2000).

This causes great limitations to the considered problem—the resulting solution by definitions is limited to a narrow class of the assumed microstructures. It is clear that the homogenization method involves orthotropic or anisotropic materials. Moreover, the main complications caused by homogenization approach are the additional design variables required to describe the structure of used composites i.e. for example: angle of orientation of layers, elliptic inclusion diameters and its orientations etc. Alternatively, a level-set method can be used, see Wang & Wang (2004) or the effective material properties for intermediate ρ_k can be determined by adopting so called material interpolation scheme for an artificial isotropic material. The latter method is very popular due to their simplicity. The comprehensive surveys regarding topology optimization methods was given among others by Rozvany (2001) and recently by Deaton & Grandhi (2014).

2 ISOTROPIC MATERIAL INTERPOLATION SCHEMES

The interpolation schemes usually assume that the element Ω_k contains an isotropic composite material of a "pseudo-density" ρ_k. The effective Young's modulus $E_k = E(\rho_k)$ can be approximated by the functions chosen arbitrary as in (2) e.g. SIMP scheme proposed by Bendsøe (1989) a.k.a power law or RAMP scheme utilized by Rietz (2001), Stolpe & Svanberg (2001) and others.

$$E_k^{SIMP} = E_1 + \rho_k^p \Delta E, \qquad E_k^{RAMP} = E_1 + \Delta E \frac{\rho_k}{1 + p(1 - \rho_k)} \tag{2}$$

where $\Delta E = E_2 - E_1$. Note, that classical SIMP or RAMP involve only on the Young's moduli E_1, E_2 while the Poisson's ratio $\nu_1 = \nu_2$ remains unperturbed. The proposal for case $\nu_1 \neq \nu_2$ known as GRAMP was proposed by Dzierżanowski (2012). The power law scheme (SIMP or RAMP) originated from solid-void problem ($E_1 = 0$) can be easily expanded to 2- or 3-materials case with more or less sophisticated dependencies between E_i and ρ_i, see Fujii et al. (2001). In (2) the "nature" of the additional parameter p is two-fold. Apart from ensuring the correct approximation Bendsøe & Sigmund (1999) it can be used as "penalization parameter" Rozvany (2001), aims to penalize the intermediate value of variables ρ_k.

From the micromechanical point of view the selected interpolation scheme and range of values of the penalty parameter p should produce effective

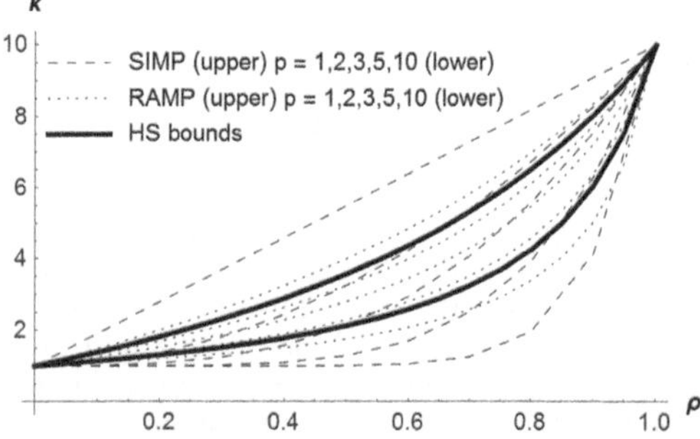

Figure 3. Comparison of SIMP and RAMP interpolation scheme of $\kappa(\rho)$ for $(\kappa_1, \mu_1) = (1.0, 0.5)$ and $(\kappa_2, \mu_2) = (10.0, 5.0)$. The similar nature of graph and curve approximation for $\mu(\rho)$.

E_k within the area bounded by variational bounds for effective material parameters of mixtures of material i.e. Hashin-Shtrikman (HS) bounds, see Hashin & Shtrikman (1963) and Walpole (1966) or more precise (CG) bounds found by Cherkaev & Gibiansky (1993). These bounds are expressed by bulk κ and shear μ moduli. The relationships in 2D between two sets of moduli (E, ν) and (κ, μ) are: $2\kappa = E/(1-\nu)$, $2\mu = E/(1+\nu)$.

For two materials and given fractions $\rho_1 + \rho_2 = 1$ (note that $\rho_2 \equiv \rho$) the HS uncoupled lower ($\#_{HS}$) and upper ($\#^{HS}$) bounds are given in form:

for case $\kappa_2 \geq \kappa_1$ and $\mu_2 \geq \mu_1$ (a.k.a well-ordered phase)

$$\kappa_{HS} = \kappa_1 + \frac{(\kappa_1+\mu_1)(\kappa_2-\kappa_1)}{(\kappa_1+\mu_1)+(1-\rho)(\kappa_2-\kappa_1)}\rho,$$

$$\mu_{HS} = \mu_1 + \frac{2\mu_1(\kappa_2+\mu_1)(\mu_2-\mu_1)}{2\mu_1(\kappa_2+\mu_1)+(1-\rho)(\kappa_2+2\mu_1)(\mu_2-\mu_1)}\rho,$$

$$\kappa^{HS} = \kappa_2 - \frac{(\kappa_2+\mu_2)(\kappa_2-\kappa_1)}{(\kappa_2+\mu_2)-\rho(\kappa_2-\kappa_1)}(1-\rho),$$

$$\mu^{HS} = \mu_2 - + \frac{2\mu_2(\kappa_1+\mu_2)(\mu_2-\mu_1)}{2\mu_2(\kappa_1+\mu_2)-\rho(\kappa_1+2\mu_2)(\mu_2-\mu_1)}(1-\rho) \tag{3}$$

hence, for any ρ_k and proper p the following relations must be fulfilled: $\#_{HS} \leq \#(\rho_k) \leq \#^{HS}$ where # is interchangeably κ or μ. Derivatives of Equations 3 for ρ, although somewhat complex, can be analytically determined and used in the gradient methods of optimization process.

Thus, by analogy to the SIMP, the completely acceptable thermo-dynamically new interpolation scheme HSρ (Fig. 5) can be formulated as:

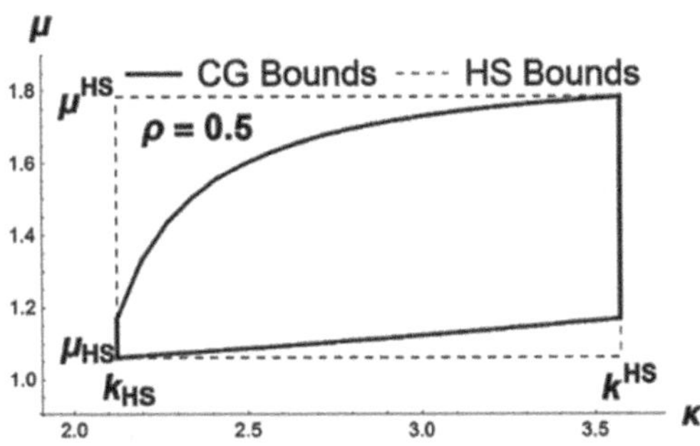

Figure 4. The bounds of isotropic structures effective properties for well-ordered (κ_1, μ_1) = (1.0, 0.5) and (κ_2, μ_2) = (10.0, 5.0) components.

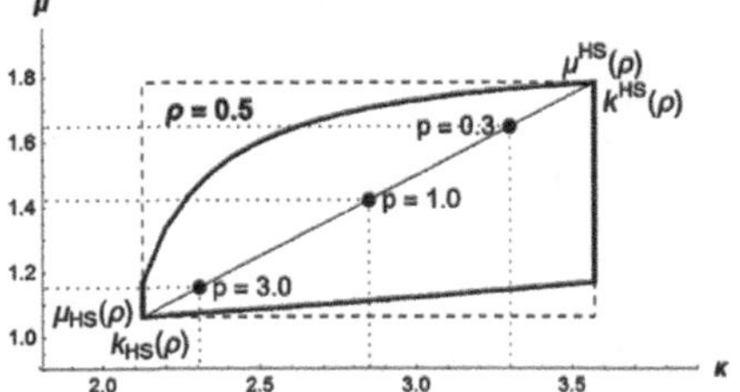

Figure 5. The HSρ interpolation scheme for (κ_1, μ_1) = (1.0, 0.5) and (κ_2, μ_2) = (10.0, 5.0). Results for different p with $\rho = 0.5$.

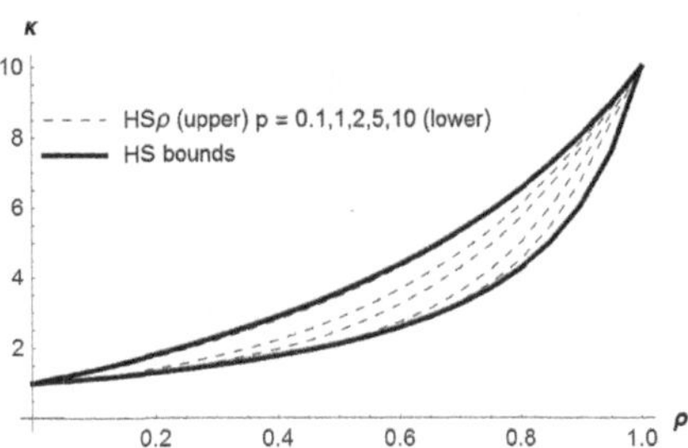

Figure 6. The HSρ interpolation scheme of $\kappa(\rho)$ for (κ_1, μ_1) = (1.0, 0.5) and (κ_2, μ_2) = (10.0, 5.0). The similar nature of graph and curve approximation for $\mu(\rho)$.

$$\kappa = (1-\rho^p)\kappa_{HS} + \rho^p\kappa^{HS}$$
$$\mu = (1-\rho^p)\mu_{HS} + \rho^p\mu^{HS} \tag{4}$$

Such kind of interpolation of the effective properties of isotropic composites fulfills not only HS-bounds but for all numbers of p will never violate more restrictive CG-bounds.

3 NUMERICAL EXAMPLES

The proposed method of interpolation of the material properties for different values of p has been tested in compliance optimization and inverse homogenization tasks. Shortly speaking, the inverse homogenization means reconstructing the layout of given materials (usually isotropic) within RVE to achieve prescribed effective properties (κ^*, μ^*) of the isotropic composite. The RVE is identified with a periodicity cell Y and here, the hexagonal repetitive cell with internal rotational symmetry of 120 degree was used. The objective function $P(\boldsymbol{\rho})$ for the tested inverse problem and $\boldsymbol{\rho} = \{\rho_k\}$ is chosen below:

$$P(\boldsymbol{\rho}) = \left(\frac{\kappa^* - \kappa}{\kappa^*}\right)^2 + \left(\frac{\mu^* - \mu}{\mu^*}\right)^2 \tag{5}$$

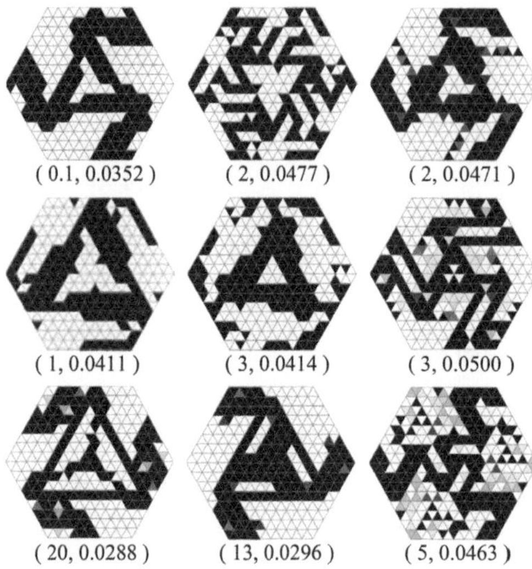

Figure 7. The results for HSρ, RAMP and SIMP (column-wise) material interpolation model. Adopted sets of p values well-nigh reflects the upper bound, middle area and the lower bound (row-wise respectively) of $\kappa(\rho)$ and $\mu(\rho)$, see Figures 3, 5, 6.

where the pair (κ, μ) is calculated from the homogenized constitutive matrix for the data values of $\boldsymbol{\rho}$, namely $\mathbf{E}^{\text{eff}} = \mathbf{E}(\kappa, \mu)$. The considered optimization problem is nonlinear and nonconvex and has been solved by the Sequential Linear Programming method. For details see Łukasiak (2013).

For all the presented cases adopted $(\kappa_1, \mu_1) = (5/7, 5/13)/20$, $(\kappa_2, \mu_2) = (5/7, 5/13)$ and $\rho_0 = 0.5$ for components (i.e. both with Poisson ratio $\nu = 0.3$). The selected $(\kappa^*, \mu^*) = (0.1231, 0.0958)$ gives almost the smallest possible effective $\nu = 0.1108$ for the components used. For all the presented results were the same fixed starting point, i.e. $\boldsymbol{\rho} = \boldsymbol{\rho}^{\text{start}}$, $\dim(\boldsymbol{\rho}^{\text{start}}) = 128$, $\langle\boldsymbol{\rho}^{\text{start}}\rangle = \rho_0$ and $0.45 < \boldsymbol{\rho}^{\text{start}} < 0.55$. The structure of the hexagonal representative cells was divided into 128×3 (due to symmetry) = 384 triangular grains of components approximated by 9 three-nodes linear finite elements each. Material for the ρ_k was approximated as SIMP, RAMP and HSρ with different values of p. The structure of grains of the representative cells recovered by this process and the final values of p and $P(\boldsymbol{\rho})$ are shown in Figure 7.

4 CONCLUSIONS

The paper puts forward a new interpolation scheme of material properties for the planar isotropic composites. The proposed in Equation 4 scheme of interpolation was used in the classical topology optimization problem i.e. the compliance minimization and to the inverse-homogenization problems to determine topology of micro-structure. It is noteworthy that for each p the results obtained using HSρ model have physical meaning even for intermediate values of ρ_k in the final mesh. For other models, the parameter p must be chosen carefully. The HSρ model gives for $p = 1$ at least as good solutions as RAMP with various p for problem tested above whilst the SIMP model produce significantly worse results than the HSρ and RAMP for wide range of penalization parameter p, especially in inverse homogenization problems. Moreover, many tests shows that the use the HSρ interpolation scheme is much easier to get the nearly 0–1 or even exact 0–1 solution to the optimization problem than in the case of the SIMP or RAMP interpolation schemes.

ACKNOWLEDGEMENTS

The paper was prepared within the Research Grant no 2013/11/B/ST8/04436 financed by the National Science Centre, entitled: Topology optimization of engineering structures. An approach synthesizing the methods of: free material design, composite design and Michell-like trusses.

REFERENCES

Bendsøe, M.P. 1989. Optimal shape design as a material distribution problem. *Struct. Optim.* 1: 193–202.

Bendsøe, M.P. & Sigmund, O. 1999. Material interpolation schemes in topology optimization. *Archive of Applied Mechanics* 69: 635–654.

Cherkaev, A.V. & Gibiansky, L.V. 1993. Coupled estimates for the bulk and shear moduli of a two-dimensional isotropic elastic composite. *J. Mech. Phys. of Solids* 41: 937–980.

Deaton, J.D. & Grandhi, R.V. 2014. A survey of structural and multidisciplinary continuum topology optimization: post 2000. *Struct. Multidisc. Optim.* 49: 1–38.

Dzierżanowski, G. 2012. On the comparison of material interpolation schemes and optimal composite properties in plane shape optimization. *Struct. Multidisc. Optim.* 46: 693–710.

Fujii, G., Chen, B.C. & Kikuchi, N. 2001. Composite material design of two-dimensional structures using the homogenization design method. *Int. J. Numer. Meth. Engng*, 50: 2031–2051.

Grabovsky, Y. & Kohn, R.V. 1995. Microstructures minimizing the energy of a two phase elastic composite in two space dimensions. II: the Vigdergauz microstructure. *J. Mech. Phys. Solids* 43: 949–972.

Hashin, Z. & Shtrikman, S. 1963. A variational approach to the theory of the elastic behaviour of multiphase materials. *J Mech Phys of Solids* 11: 127–140.

Łukasiak, T. 2013. Recovery of two-phase microstructures of planar isotropic elastic composites. WCSMO-10 http://www.issmo.net/wcsmo10/Papers/5367.pdf.

Milton, G.W. & Cherkaev, A.V. 1995. Which elasticity tensors are realizable? *J. Eng. Mat. Tech.* 117: 483–493.

Rietz, A. 2001. Sufficiency of a finite exponent in SIMP methods. *Struct. Multidisc. Optim.* 21: 159–163.

Rozvany, G.I.N. 2001. Aims, scope, methods, history and unified terminology of computer-aided topology optimization in structural mechanics. *Struct. Multidisc. Optim.* 21: 90–108.

Rozvany, G.I.N. 2009. A critical review of established methods of structural topology optimization. *Struct. Multidisc. Optim.* 37: 217–237.

Rozvany, G.I.N. & Querin O.M. 2002. Combining ESO with rigorous optimality criteria. *Int. J. Veh. Des.* 28: 294–299.

Sigmund, O. 2000. A new class of extremal composites. *Journal. Mech. Phys. of Solids* 48: 397–428.

Sigmund, O. 2001. Design of metaphysics actuators using topology optimisation—Part II: Two material structures. *Comput. Meth. Appl. Mech. Engrg.* 190: 6005–6627.

Stolpe, M. & Svanberg, K. 2001. An alternative interpolation scheme for minimum compliance topology optimization, *Struct. Multidisc. Optim.* 22: 116–124.

Vigdergauz, S.B. 1989. Regular structures with extremal elastic properties. *Mechanics of Solids* 24: 57–63.

Walpole, L.J. 1966. On bounds for the overall elastic moduli of inhomogeneous systems. *Journal. Mech. Phys. of Solids* 14: 151–162.

Wang, M.Y. & Wang X. 2004. "Color" level sets: a multiphase method for structural topology optimization with multiple materials. *Comput. Meth. Appl. Mech. Engrg.* 193: 469–496.

Xie, Y.M. & Steven, G.P. 1993. A simple evolutionary procedure for structural optimization. *Comput. Struct.* 49: 885–896.

Advances in Mechanics: Theoretical, Computational and Interdisciplinary Issues – Kleiber et al. (Eds)

 ISBN 978-1-138-02906-4

Modelling of the vibration reduction system used for protection of working machine operators

I. Maciejewski & T. Krzyżyński
Department of Mechatronics and Applied Mechanics, Faculty of Technology and Education, Koszalin University of Technology, Koszalin, Poland

ABSTRACT: In the paper a generalised model is proposed to determine the basic characteristics of non-linear visco-elastic elements used in typical vibration reduction systems. The proposed model takes into account the resonance of human body parts and organs in order to minimise the harmful vibrations affecting the machine operators during their work. Such a mathematical description of the vibration reduction systems determines the ability to use the presented model in a wide range of applications.

1 INTRODUCTION

The fundamental method of evaluating the effectiveness of vibration reduction system is to perform an experiment in the laboratory. The tested system should be excited by the signals that are representative of the different type of working machines (ISO-7096 2000). Based on measured signals, the vibro-isolation criteria of the tested system can be calculated. However, in many cases the test execution is difficult due to the technological restrictions and the high cost of the experiment. The duration of the test is also an important aspect, especially when the test must be performed repeatedly for different design parameters of the system. Taking into account the complexity of the research process, the authors of this paper recommend to carry out a simulation experiment based on a generalised mathematical model of vibration reduction system shown in Figure 1.

In order to analyse the dynamics of such system (see Fig. 1), models of the vibration reduction system and the isolated body have to be created. The bio-mechanical models are created in order to verify the vibration impact on individual parts and organs of the human body. Using these models, the vibration amplitudes and frequencies can be determined without the necessity of an experimental investigation. This kind of research may not be always reliable due to subjective feelings of the individuals. According to the paper (Nader 2001), the significant discrepancies of measurement results have occurred in the past, therefore the modelling studies are carried out at present.

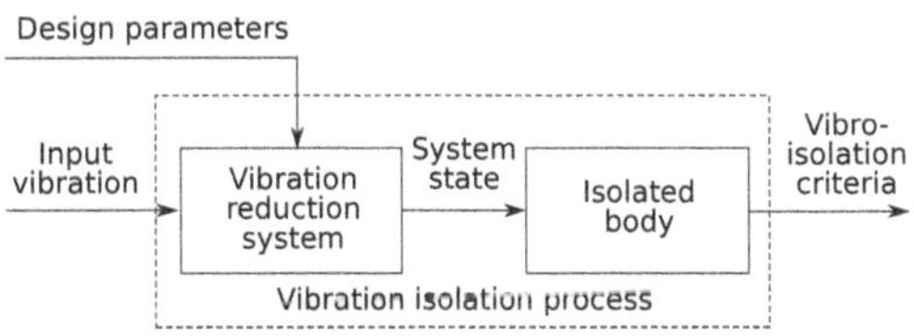

Figure 1. Concept model of the vibro-isolation process.

There are dozens of human body models presented in modern literature (Rutzel, Hinz, & Wolfel 2006). Usually, there are multi-degree of freedom lumped parameter models (Stein, Muka, Chmurny, Hinz, & Bluthner 2007) that consider seating and standing position (Toward & Griffin 2011). In this paper a generalisation of the well-known human body models is proposed that allows to use various bio-mechanical structure in the modelled vibration reduction system. An innovation of the model consists in a configurable mechanical system that can be used for the purpose of selecting the vibro-isolation properties of various suspension systems.

2 GENERALISED MODEL OF THE SYSTEM

The general model of vibration reduction system is presented in Figure 2. The suspended body is isolated against harmful vibrations in three orthogonal directions: longitudinal x, lateral y and vertical z. The passive visco-elastic elementsas well as the active force actuators are used in order to minimise of human exposure to whole-body vibration. Rotation vibrations around each axis of the Cartesian coordinate system (x, y, z) are neglected in this simplified model. The ongoing research (Directive-44/EC 2002) indicates that the exposure of workers to the risks arising from vibration are evaluated for translational axes, therefore

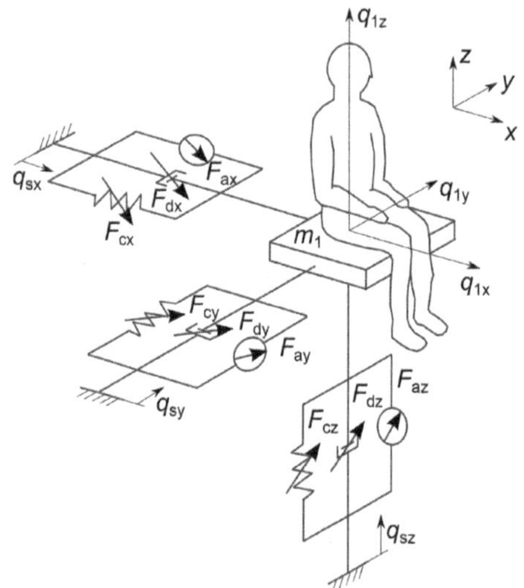

Figure 2. Generalised model of the non-linear vibration reduction system.

in the following paper the vibro-isolation properties are discussed for a specific direction.

A set of three independent equations of motion is formulated in the matrix form:

$$\mathbf{M}_i\ddot{\mathbf{q}}_i + \mathbf{D}_i\dot{\mathbf{q}}_i + \mathbf{C}_i\mathbf{q}_i = \mathbf{F}_{si} + \mathbf{F}_{ai}, \quad i = x,y,z \tag{1}$$

where: $\mathbf{q}_i$ is the displacement vector of an isolated body, $\mathbf{M}_i$, $\mathbf{D}_i$, $\mathbf{C}_i$ are the inertia, damping and stiffness matrices, $\mathbf{F}_{si}$ and $\mathbf{F}_{ai}$ are the vectors of exciting and active forces describing the non-linear vibration isolator.

The n-element vector represents the movement of elements contained in the bio-mechanical model of a human body:

$$\mathbf{q}_i = [q_{1i}, q_{2i}, ..., q_{ni}]^T, \quad i = x,y,z \tag{2}$$

where: $q_{1i}, q_{2i}, ..., q_{ni}$ are the displacements of different parts of the human body.

The n-element vectors of exciting and active forces are given by the following expressions:

$$\mathbf{F}_{si} = [F_{si}, 0, ..., 0]^T, \quad i = x,y,z \tag{3}$$

$$\mathbf{F}_{ai} = [F_{ai}, 0, ..., 0]^T, \quad i = x,y,z \tag{4}$$

The particular non-linear exciting forces F_{si} can be described in a general form as follows:

$$F_{si} = \sum_{j=1}^{k} F_{dij}(\dot{q}_{1i} - \dot{q}_{si}) + \sum_{j=1}^{k} F_{cij}(q_{1i} - q_{si}) \tag{5}$$

where: $F_{cij}(q_{1i} - q_{si})$ are the non-linear functions including force characteristics of the conservative elements as a function of the system relative displacement $q_{1i} - q_{si}$, $F_{dij}(\dot{q}_{1i} - \dot{q}_{si})$ are the non-linear functions including force characteristics of the dissipative elements as a function of the system relative velocity $\dot{q}_{1i} - \dot{q}_{si}$. The input displacement q_{si} and velocity $\dot{q}_{si}$ are modelled as excitation signals that are generated for the specific direction of vibration exposure x, y, z.

The non-linear active forces can be described as a function of the input signals u_i and the system relative displacement $q_{1i} - q_{si}$ or velocity $\dot{q}_{1i} - \dot{q}_{si}$ as follows:

$$F_{ai} = \sum_{j=1}^{k} F_{aij}(u_i, q_{1i} - q_{si}, \dot{q}_{1i} - \dot{q}_{si}) \tag{6}$$

where: F_{aij} defines dynamic characteristics of the actuators.

A generalised structure of the human body model proposed within the framework of this paper is shown in Figure 3. The human body is modelled as a discrete mechanical system with many degrees of freedom and its particular bodies are combined using visco-elastic elements.

The diagonal matrix $\mathbf{M}_i$ (size $n \times n$) represents the masses of each element contained in the human body as follows:

$$\mathbf{M}_i = \begin{bmatrix} m_1 & 0 & 0 & \cdots & 0 \\ 0 & m_2 & 0 & \cdots & 0 \\ 0 & 0 & m_3 & \cdots & 0 \\ \cdots & \cdots & \cdots & \ddots & \cdots \\ 0 & 0 & 0 & \cdots & m_n \end{bmatrix} \tag{7}$$

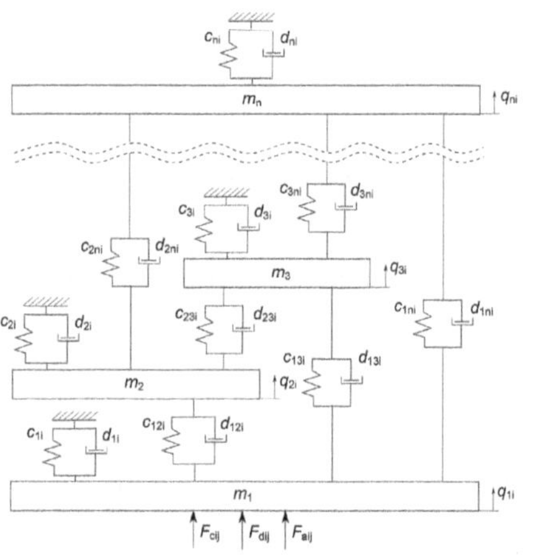

Figure 3. Generalised model of a human body for one direction of the vibration transmission ($i = x, y, z$).

The damping matrix $\mathbf{D}_i$ of size $n \times n$ is symmetric and should be defined for a specific direction of the vibration exposure ($i = x, y, z$) in the following form:

$$\mathbf{D}_i = \begin{bmatrix} d_{11i} & -d_{12i} & -d_{13i} & \cdots & -d_{1ni} \\ -d_{12i} & d_{22i} & -d_{23i} & \cdots & -d_{2ni} \\ -d_{13i} & -d_{23i} & d_{33i} & \cdots & -d_{3ni} \\ \cdots & \cdots & \cdots & \ddots & \cdots \\ -d_{1ni} & -d_{2ni} & -d_{3ni} & \cdots & d_{nni} \end{bmatrix} \tag{8}$$

where the principal diagonal terms are the sums of the corresponding damping coefficients:

$$\begin{aligned} d_{11i} &= d_{1i} + d_{12i} + d_{13i} + \ldots + d_{1ni} \\ d_{22i} &= d_{2i} + d_{12i} + d_{23i} + \ldots + d_{2ni} \\ d_{33i} &= d_{3i} + d_{13i} + d_{23i} + \ldots + d_{3ni} \\ &\ldots \\ d_{nni} &= d_{ni} + d_{1ni} + d_{2ni} + \ldots + d_{(n-1)ni} \end{aligned} \tag{9}$$

The stiffness matrix of the same size $n \times n$ can be determined in a similar way:

$$\mathbf{C}_i = \begin{bmatrix} c_{11i} & -c_{12i} & -c_{13i} & \cdots & -c_{1ni} \\ -c_{12i} & c_{22i} & -c_{23i} & \cdots & -c_{2ni} \\ -c_{13i} & -c_{23i} & c_{33i} & \cdots & -c_{3ni} \\ \cdots & \cdots & \cdots & \ddots & \cdots \\ -c_{1ni} & -c_{2ni} & -c_{3ni} & \cdots & c_{nni} \end{bmatrix} \tag{10}$$

wherein the following stiffness coefficients are located on the principal diagonal:

$$\begin{aligned} c_{11i} &= c_{1i} + c_{12i} + c_{13i} + \ldots + c_{1ni} \\ c_{22i} &= c_{2i} + c_{12i} + c_{23i} + \ldots + c_{2ni} \\ c_{33i} &= c_{3i} + c_{13i} + c_{23i} + \ldots + c_{3ni} \\ &\ldots \\ c_{nni} &= c_{ni} + c_{1ni} + c_{2ni} + \ldots + c_{(n-1)ni} \end{aligned} \tag{11}$$

In the section, the basic structure of a vibration reduction system is employed to formulate the generalised model of human body exposure. Such a model can be used to describe the general dynamic behaviour of many vibration reduction systems. However, their essential force characteristics (F_{cij}, F_{dij}, F_{aij}) have to be evaluated for the elements, e.g. mechanical, pneumatic, hydraulic, etc., applied in the suspension systems (Snamina, Kowal, & Orkisz 2013).

3 EXAMPLE: MODELLING AND SIMULATION OF A HORIZONTAL SEAT SUSPENSION

In Figure 4a the physical model of a horizontal seat suspension with seated human body is shown. The experimental set-up for laboratory evaluation of the operator seat vibration, which meets all requirements for the person posture on a seat (ISO-7096 2000), is presented in Figure 4b.

The system dynamics is determined in accordance to the modelling procedure presented in Section 2. The equations of motion are defined for the longitudinal x direction as follows:

$$\mathbf{M}_x \ddot{\mathbf{q}}_x + \mathbf{D}_x \dot{\mathbf{q}}_x + \mathbf{C}_x \mathbf{q}_x = \mathbf{F}_{sx} \tag{12}$$

where: $\mathbf{q}_x$ is the the displacement vectors of human body model, $\mathbf{M}_x$, $\mathbf{D}_x$, $\mathbf{C}_x$ are the corresponding inertia, damping and stiffness matrices, $\mathbf{F}_{sx}$ is the vector of applied forces.

The diagonal inertia matrix $\mathbf{M}_x$ includes the masses m_1, m_2, m_3 that exist in the human body model. The damping matrix $\mathbf{D}_x$ and $\mathbf{C}_x$ stiffness matrices take the following forms:

$$\mathbf{D}_x = \begin{bmatrix} d_{12x} & -d_{12x} & 0 \\ -d_{12x} & d_{2x} + d_{12x} + d_{23x} & -d_{23x} \\ 0 & -d_{23x} & d_{23x} \end{bmatrix} \tag{13}$$

$$\mathbf{C}_x = \begin{bmatrix} c_{12x} & -c_{12x} & 0 \\ -c_{12x} & c_{2x} + c_{12x} + c_{23x} & -c_{23x} \\ 0 & -c_{23x} & c_{23z} \end{bmatrix} \tag{14}$$

where: d_{12x}, d_{2x}, d_{23x} are the damping coefficients and c_{12x}, c_{2x}, c_{23x} are the stiffness coefficients of the human body model. The human body model parameters (Fig. 4a) are identified for the seated human body with cushioned seat system and backrest contact in the lumbar region (Fig. 4b). The numerical values established by identification of

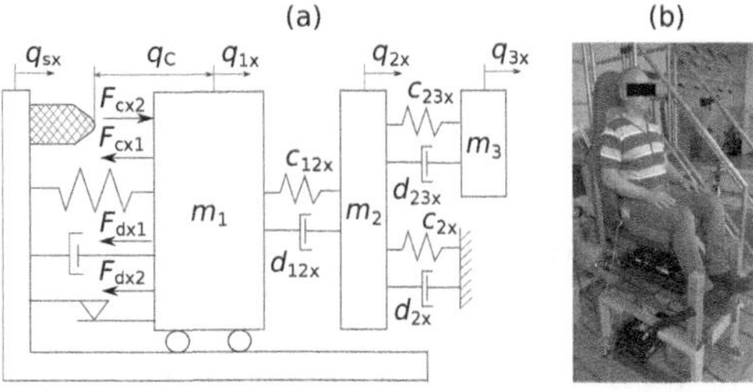

Figure 4. Physical model of the horizontal seat suspension with seated human body (a) and laboratory experimental set-up (b).

laboratory measurements under well-defined conditions are presented in the paper (Stein, Muka, Chmurny, Hinz, & Bluthner 2007).

Particular vectors describing the human body displacements and applied forces are set in the following order:

$$\mathbf{q}_x = \begin{bmatrix} q_{1x} \\ q_{2x} \\ q_{3x} \end{bmatrix} \tag{15}$$

$$\mathbf{F}_{sx} = \begin{bmatrix} -F_{cx1} + F_{cx2} - F_{dx1} - F_{dx2} \\ 0 \\ 0 \end{bmatrix} \tag{16}$$

where: q_{1x}, q_{2x}, q_{3x} are the displacement of the human body model, F_{cx1} is the mechanical spring force, F_{cx2} is the force from end-stop buffer, F_{dx1} is the force of hydraulic shock-absorber and F_{dx2} is the overall friction force of suspension system.

An evaluation of the force characteristics is determined with the help of the force-deflection measurement of the suspension system. The spring stiffness is evaluated by calculating the measured force per the measured suspension deflection (Fig. 5a). The stiffness characteristics of the end-stop buffers (Fig. 5b) are determined while the suspension system without mechanical spring is deflected using the same test conditions. The force of shock-absorber in velocity domain is evaluated by sinusoidal cycling of the damper at different velocities (Fig. 5c). The measured hysteresis of the friction force in displacement domain (Fig. 5d) is fitted to the simulation results by appropriate selection of the model parameters.

During the evaluation of the operator seat vibration in horizontal direction, an excitation signal similar to the white, band limited noise in the range of frequency 0.5–10 Hz is used at different excitation intensities: 1.02 m/s^2 (WN1x), 1.36 m/s^2 (WN2x), 1.91 m/s^2 (WN3x). The male subject with body mass 90 kg is chosen for the tests and such a subject is exposed to three vibration inputs in longitudinal direction.

The power spectral densities and transmissibility functions of the horizontal seat suspension with seated human body are shown in Figure 6. The simulation results of the system under examination correspond to the results measured experimentally with the error not exceeding 10%. The worst agreement between vibration amplitudes can be observed in the 3–6 Hz frequency range. In this range the cushioned seat back has a significant influence on the system dynamics. The particular

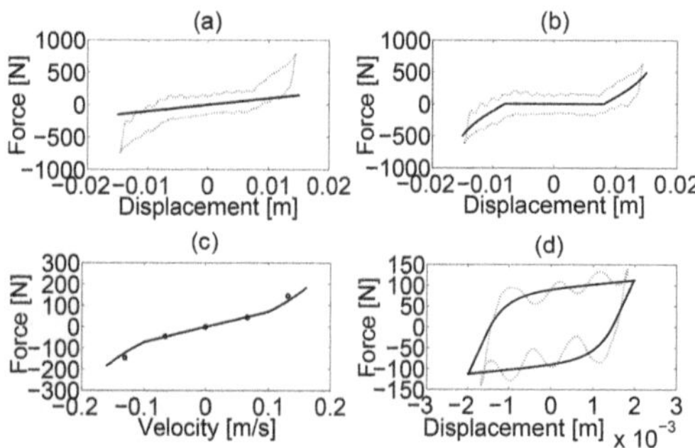

Figure 5. Stiffness characteristics of the mechanical spring (a), stiffness characteristics of the end-stop buffers (b), damping characteristics of the hydraulic shock-absorber (c), friction characteristics of the suspension system (d) obtained using simulation model (solid line) and measurement (dotted line).

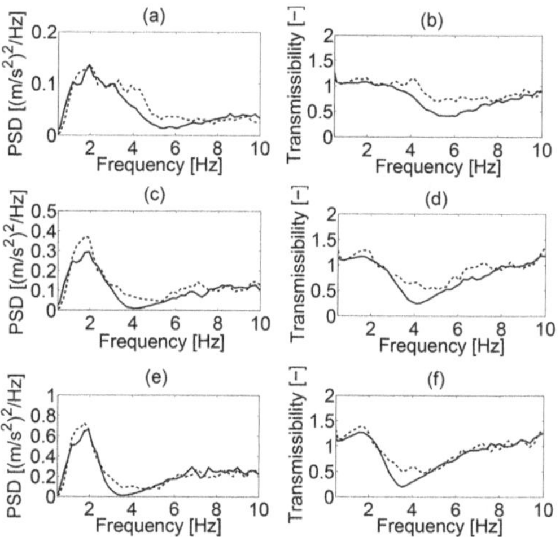

Figure 6. Simulated (solid line) and measured (dashed line) power spectral densities for different vibration inputs: WN1x (a), WN2x (c), WN3x (e) and transmissibility functions for different vibration inputs: WN1x (b), WN2x (d), WN3x (f).

Table 1. Simulated and measured transmissibility factors, suspension travels and relative errors of the seat suspension.

	Simulation		Measurement		Relative error	
Input vibration	TFE_x factor	s_{tx}, mm	TFE_x factor	s_{tx}, mm	δ_{TFE_x}, %	$\delta_{s_{tx}}$, %
WN1x	0.965	16.4	1.052	18.2	8.2	9.8
WN2x	0.965	23.2	1.051	25.1	8.2	7.5
WN3x	0.994	27.9	1.075	28.2	7.5	1.1

transmissibility factors TFE_x, suspension travels s_{tx} and their relative errors: δ_{TFE_x}, $\delta_{s_{tx}}$ are presented in Table 1.

4 CONCLUSIONS

The presented model provides the basis for using a mechanical analogue of the human body. Such a mechanical system should assists an appropriate selection of the vibro-isolating properties in view of the conflicted requirements for modern vibration reduction systems. Such a selection of the dynamic characteristics can be conducted for the vibration isolators of various designs that are activated using the excitation signals representing the work of diverse types of the machinery.

The paper is a part of the research project "Methods and procedures of selecting vibro-isolation properties of vibration reduction systems" founded by the National Science Center of Poland under the contract No. UMO-2013/11/B/ST8/03881.

REFERENCES

Directive-44/EC (2002). *On the minimum health and safety requirements regarding the exposure of workers to the risks arising from physical agents (vibration)*. The European Parliament and of the Council: Official Journal of the European Communities.

ISO-7096 (2000). *Earth-moving machinery—Laboratory evaluation of operator seat vibration*. Genewa: International Organization for Standardization.

Nader, M. (2001). *Modelling and simulation of vehicle vibration influence on human body (in Polish)*. Warsaw: Publishing House of Warsaw University of Technology.

Rutzel, S., B. Hinz, & H. Wolfel (2006). Modal description—a better way of characterizing human vibration behavior. *Journal of Sound and Vibration* 298, 810–823.

Snamina, J., J. Kowal, & P. Orkisz (2013). Active suspension based on low dynamic stiffness. *Acta Physica Polonica A* 123(6), 1118–1122.

Stein, G., P. Muka, R. Chmurny, B. Hinz, & R. Bluthner (2007). Measurement and modelling of x-direction apparent mass of the seated human body—cushioned seat system. *Journal of Biomechanics* 40, 1493–1503.

Toward, M. & J. Griffin (2011). The transmission of vertical vibration through seats: Influence of the characteristics of the human body. *Journal of Sound and Vibration* 330, 6526–6543.

Advances in Mechanics: Theoretical, Computational and Interdisciplinary Issues – Kleiber et al. (Eds)
© 2016 Taylor & Francis Group, London, ISBN 978-1-138-02906-4

Modelling of phase changes in thin metal film subjected to ultrafast laser heating using the two-temperature model

E. Majchrzak & J. Dziatkiewicz
Silesian University of Technology, Gliwice, Poland

ABSTRACT: Thin metal film subjected to an ultrafast laser heating is considered. The thermal processes which take place in the domain considered are described by two-temperature hyperbolic model supplemented by appropriate boundary and initial conditions. The model presented takes into account the phase changes of material (solid-liquid and liquid-vapour) occurring at the constant temperature. At the stage of numerical computations the finite difference method with staggered grid is used. In the final part the results of computations are shown.

1 INTRODUCTION

During the heating of thin metal films by a laser of high power and ultra-short pulse the phase transitions can occur, this means the melting and the evaporation. In order to analyze the process, the two-temperature hyperbolic model can be used (Tzou 1997, Zhang 2007, Chen et al. 2005, 2006). This model consists of equations describing the temporal and spatial evolution of the lattice and electrons temperatures, lattice and electron heat fluxes and also the isothermal solid-liquid (or liquid-vapour) phase change. In the paper the system of equations is solved using the explicit scheme of finite difference method with staggered grid (Dai &Nassar 2001, Dziatkiewicz et al. 2014). In the final part the results concerning the heating of gold film are presented.

2 GOVERNING EQUATIONS

The one-dimensional two-temperature model describing the temporal and spatial evolution of the lattice and electrons temperatures (T_l and T_e) in the irradiated thin metal film together with the isothermal solid–liquid and liquid-vapor phase changes is of the form (Chen et al. 2005, 2006):

$$C_e \frac{\partial T_e}{\partial t} = -\frac{\partial q_e}{\partial x} - G\left(T_e - T_l\right) + Q(x, t) \tag{1}$$

$$q_e + \tau_e \frac{\partial q_e}{\partial t} = -\lambda_e \frac{\partial T_e}{\partial x} \tag{2}$$

$$C_l \frac{\partial T_l}{\partial t} = -\frac{\partial q_l}{\partial x} + G\left(T_e - T_l\right), \ T_l \neq T_m, \ T_l \neq T_b \tag{3}$$

$$\int_{t_{mb}}^{t_{me}} \left[-\frac{\partial q_l}{\partial x} + G\left(T_e - T_l\right) \right] \mathrm{d}t = Q_m, \ T_l = T_m \tag{4}$$

$$\int_{t_{bb}}^{t_{be}} \left[-\frac{\partial q_l}{\partial x} + G\left(T_e - T_l\right) \right] \mathrm{d}t = Q_{ev}, \ T_l = T_b \tag{5}$$

$$q_l + \tau_l \frac{\partial q_l}{\partial t} = -\lambda_l \frac{\partial T_l}{\partial x} \tag{6}$$

where C_e, C_l = are the volumetric specific heats; G = is the electron-phonon coupling factor; $Q(x, t)$ = is the source function associated with the irradiation; x = is the spatial coordinate; t = is the time; q_e, q_l = are the heat fluxes; λ_e, λ_l = are the thermal conductivities; τ_e = is the relaxation time of free electrons in metals; τ_l = is the relaxation time in phonon collisions; Q_m = is the volumetric latent heat of fusion; T_m = is the melting temperature; t_{mb} and t_{me} = are the times corresponding to the start and the end of the solid–liquid phase transformation; Q_{ev} = is the volumetric latent heat of evaporation; T_b = is the boiling temperature; t_{evb} and t_{eve} = are the times corresponding to the start and the end of the liquid-vapor phase transformation.

The laser irradiation is described by source term introduced in (1):

$$Q(x, t) = \sqrt{\frac{\beta}{\pi}} \frac{1-R}{t_p \delta} I_0 \exp\left[-\frac{x}{\delta} - \beta \frac{(t-t_p)^2}{t_p^2} \right] \tag{7}$$

where I_0 = is the laser intensity; t_p = is the characteristic time of laser pulse; δ = is the optical penetration depth; R = is the reflectivity of the irradiated surface and $\beta = 4 \ln 2$ (Chen et al. 2005).

The mathematical model is supplemented by boundary conditions:

$$q_e(0,t)=q_e(L,t)=q_l(0,t)=q_l(L,t)=0 \tag{8}$$

and the initial ones

$$t=0:\ \ T_e(x,0)=T_l(x,0)=T_p \tag{9}$$

where T_p is the known constant temperature.

For high laser intensity the following formula describing the temperature-dependent volumetric specific of lattice is proposed (Chen et al. 2006):

$$C_e(T_e)=\begin{cases} AT_e, & T_e<T_g \\ AT_g+\dfrac{Nk_B-AT_g}{2T_F/\pi^2}\left(T_e-T_g\right), & T_g\le T_e<3T_g \\ Nk_B+\dfrac{Nk_B/2}{T_F-3T_g}\left(T_e-3T_g\right), & 3T_g\le T_e<T_F \\ 3Nk_B/2, & T_e>T_F \end{cases} \tag{10}$$

where $T_g = T_F/\pi^2$, T_F = is the Fermi temperature; N = is the density of electrons; k_B = is the Boltzmann constant and $A = \pi^2 Nk_B/(2T_F)$.

Electrons thermal conductivity and the coupling factor are dependent on the temperature of electrons and lattice (Chen et al. 2006), namely:

$$\lambda_e=\chi\frac{\left[\left(T_e/T_F\right)^2+0.16\right]^{5/4}\left[\left(T_e/T_F\right)^2+0.44\right]\left(T_e/T_F\right)}{\left[\left(T_e/T_F\right)^2+0.092\right]^{1/2}\left[\left(T_e/T_F\right)^2+\eta\left(T_l/T_F\right)\right]} \tag{11}$$

and

$$G=G(T_e,T_l)=G_{rt}\left[\frac{A_e}{B_l}\left(T_e+T_l\right)+1\right] \tag{12}$$

where χ, η, A_e, B_l are the constants and G_{rt} = is the coupling factor at room temperature (Chen et al. 2006).

3 METHOD OF SOLUTION

To solve the problem formulated the explicit scheme of the finite difference method presented in details in (Dziatkiewicz et al. 2014) is used. Here, only the modeling of phase changes is discussed.

In the differential grid the temperature nodes i = 0, 2, 4, ..., N and the heat fluxes nodes i = 1, 3, ..., N–1 are distinguished—Figure 1 (Dai & Nassar 2001). Let $T_{li}^{\,f} = T_l\,(ih, f\Delta t)$, $T_{ei}^{\,f} = T_e\,(ih, f\Delta t)$, i = 0, 2, ... N, and $q_{l\,i}^{\,f} = q_l\,(ih, f\Delta t)$, $q_{e\,i}^{\,f} = q_e\,(ih, f\Delta t)$, i = 1, 3, ... N–1, where h is the mesh step and Δt is the time step, f = 0, 1, ..., F.

If at the node x_i for time t^f the lattice temperature $T_{li}^{\,f}$ is higher or equal to T_m ($T_{li}^{\,f} \ge T_m$) then in the sub-domain $[x_{i-1}, x_{i+1}]$ the melting process starts (Fig. 1). For this node the value:

$$M_i^f=\left[-\frac{q_{l\,i+1}^f-q_{l\,i-1}^f}{2h}+G_i^f\left(T_{ei}^f-T_{li}^f\right)\right]\Delta t \tag{13}$$

is calculated and it is assumed that $T_{li}^{\,f} = T_m$. For next transition $t^f \to t^{f+1}$ the value M_i^{f+1} is determined. If $M_i^f + M_i^{f+1} < Q_m$ then $T_{li}^{f+1} = T_m$ and the temperature field for transition $t^{f+1} \to t^{f+2}$ is found. The calculations are continued until:

$$\sum_{k=f}^{K} M_i^k \ge Q_m \tag{14}$$

In a similar way the phase change liquid-vapor is modeled. Thus, if at the node x_i for time t^f the lattice temperature T_{li}^f is higher or equal to T_b ($T_{li}^f \ge T_b$) then in the sub-domain $[x_{i-1}, x_{i+1}]$ the evaporation process starts. For this node the value:

$$N_i^f=\left[-\frac{q_{l\,i+1}^f-q_{l\,i-1}^f}{2h}+G_i^f\left(T_{ei}^f-T_{li}^f\right)\right]\Delta t \tag{15}$$

is calculated and it is assumed that $T_{li}^f = T$. For next transition $t^f \to t^{f+1}$ the value N_i^{f+1} is determined. If $N_i^f + N_i^{f+1} < Q_{ev}$ then $T_{li}^{f+1} = T_b$ and the temperature field for transition $t^{f+1} \to t^{f+2}$ is found. The calculations are continued until:

$$\sum_{k=f}^{K} N_i^k \ge Q_{ev} \tag{16}$$

If the condition (16) is fulfilled, it means that the sub-domain $[x_{i-1}, x_{i+1}]$ is removed (ablation).

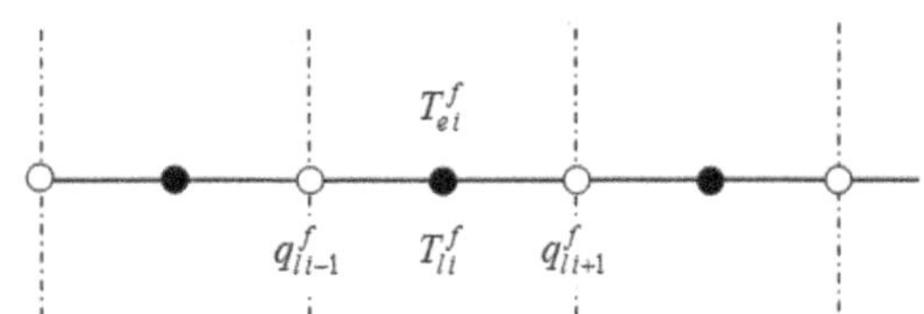

Figure 1. Fragment of differential mesh.

4 RESULTS OF COMPUTATIONS

The gold film of thickness $L = 100$ nm is considered. Initial temperature is equal to $T_p = 300$ K. The constants in equations (11–12) are the following: $\chi = 353$ W/(mK), $\eta = 0.16$, $A_e = 1.2 \cdot 10^7$ 1/(K²s), $B_l = 1.23 \cdot 10^{11}$ 1/(Ks) and $G_{rt} = 2.2 \cdot 10^{16}$ W/(m³K) (Chen et al. 2006). The Fermi temperature is equal to $T_F = 64.200$ K and the density of electrons is equal to $N = 5.9 \cdot 10^{28}$ 1/m (Chen et al. 2006). The other parameters are as follows: thermal conductivity of lattice $\lambda_l = 315$ W/(mK), volumetric specific heat of lattice $C_l = 2.5$ MJ/(m³ K), electrons relaxation time $\tau_e = 0.04$ ps, phonons relaxation time $\tau_l = 0.8$ ps (Chen et al. 2005), reflectivity $R = 0.93$, optical penetration depth $\delta = 15.3$ nm. The melting temperature is equal to $T_m = 1336$ K and the latent heat of fusion $Q_m = 1229.989$ MW/m³, while the boiling temperature is equal to $T_b = 3127$ K and the latent heat of evaporation $Q_{ev} = 32771.4$ MW/(m³K). As mentioned before, the problem is solved using finite difference method (Dziatkiewicz et al. 2014) under the assumption that $\Delta t = 0.001$ ps and $h = 0.5$ nm.

In Figures 2–3 the electrons and lattice temperatures at the irradiated surface $x = 0$ for the laser intensity $I_0 = 4182$ W/m² (Fig. 2) and $I_0 = 40000$ W/m² (Fig. 3) are shown. In both cases the characteristic time of laser pulse is equal to $t_p = 100$ ps.

For laser intensity $I_0 = 4182$ W/m² the solid-liquid phase change is observed, while for $I_0 = 40000$ W/m² the solid-liquid and the liquid-vapor phase changes are visible.

In Figure 4 the liquid state fraction M_i^f/Q_m at the neighbourhood of the points 1 ($x_4 = 2$ nm), 2 ($x_{50} = 25$ nm) and 3 ($x_{100} = 50$ nm) under the assumption that the laser intensity equals $I_0 = 4182$ W/m² is presented. As can be seen, in the sub-domains corresponding to the selected nodes the liquid state fraction does not exceed 0.21 (21%). Figure 5 illustrates the liquid state fraction in all nodes after the time 400 ps. The melting process takes place in all thin film, but is not finished.

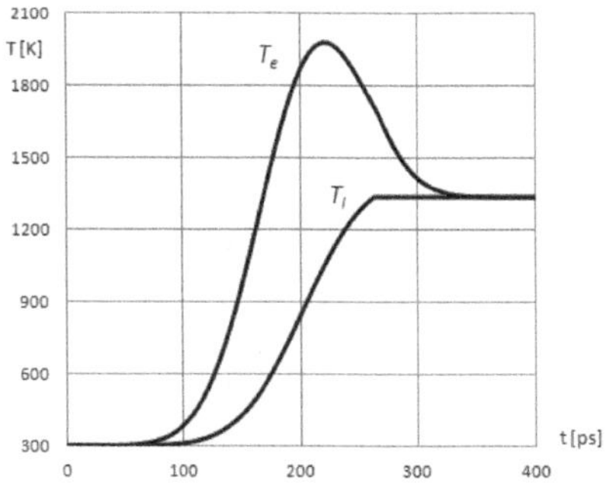

Figure 2. Electron and lattice temperature at the irradiated surface ($I_0 = 4182$ W/m²).

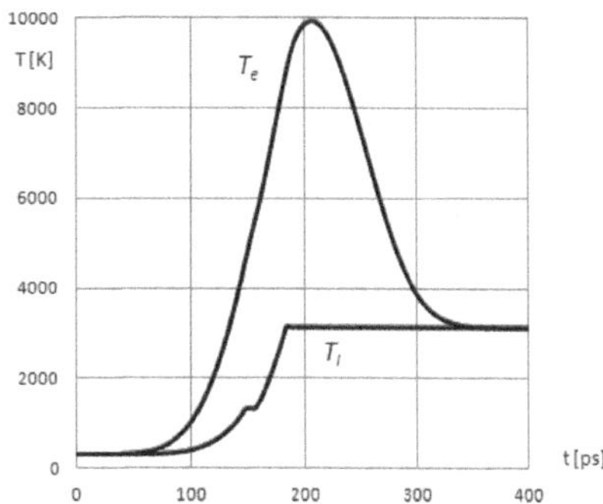

Figure 3. Electron and lattice temperature at the irradiated surface ($I_0 = 40000$ W/m²).

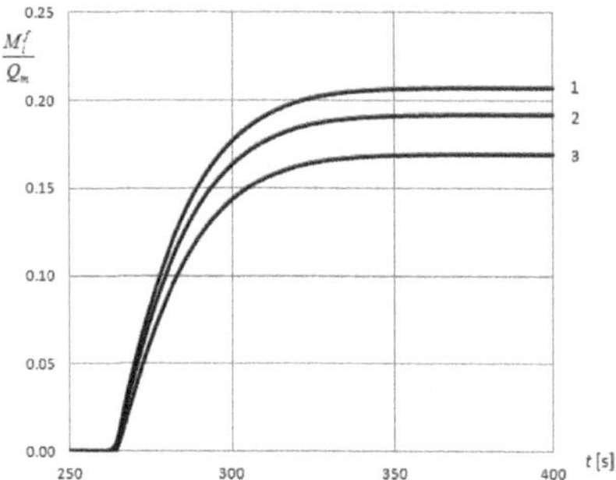

Figure 4. Liquid state fraction at the neighbourhood of the points 1–2 nm, 2–25 nm, 3–50 nm ($I_0 = 4182$ W/m²).

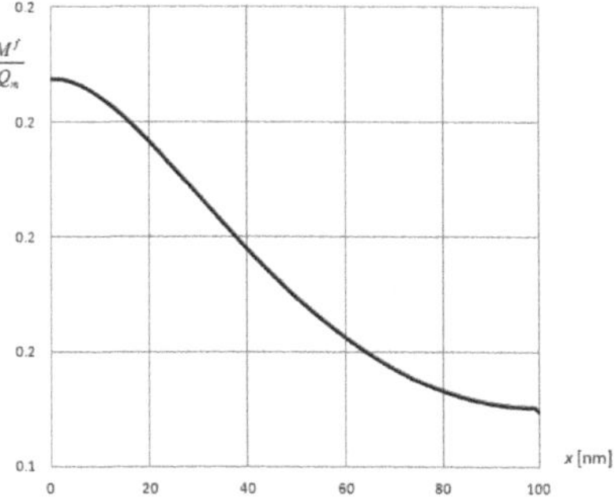

Figure 5. Liquid state fraction in thin film after the 400 ps ($I_0 = 4182$ W/m²).

Figures 6–8 concern the calculations for the laser intensity $I_0 = 40000$ W/m². As previously mentioned, in this case the temperature at the irradiated surface not only exceeded the melting temperature but also the boiling point (Fig. 3). These temperatures are also exceeded at the points 1, 2 and 3. In Figure 6 the liquid state fraction at the neighbourhood of these points is presented, while Figure 7 illustrates the volumetric fraction of gaseous phase. As can be seen, the phase change liquid-vapor is not finished.

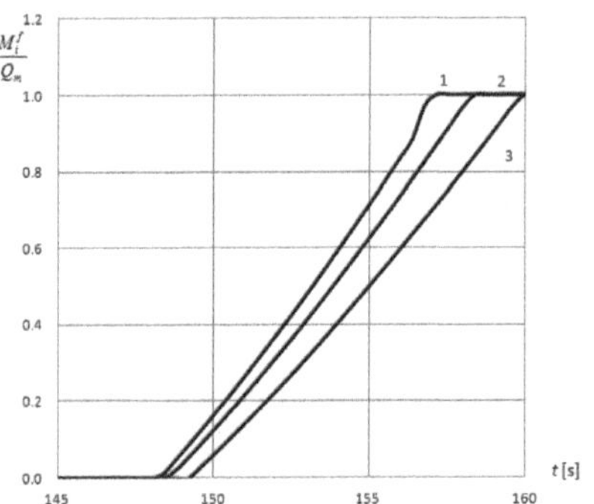

Figure 6. Liquid state fraction at the neighbourhood of the points 1, 2, 3 ($I_0 = 40000$ W/m^2).

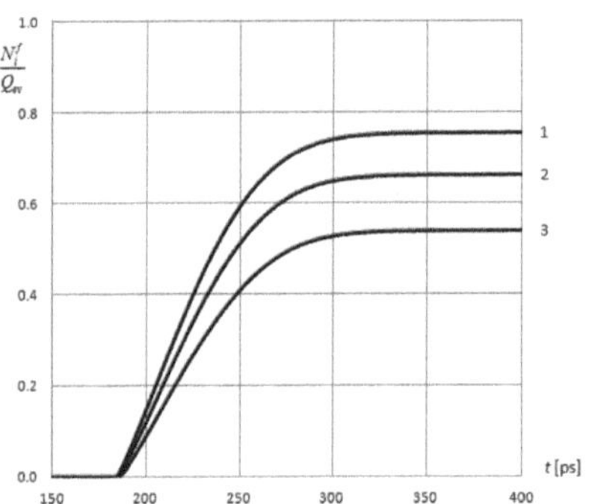

Figure 7. Volumetric fraction of gaseous phase at the neighbourhood of the points 1, 2, 3 ($I_0 = 40000$ W/m^2).

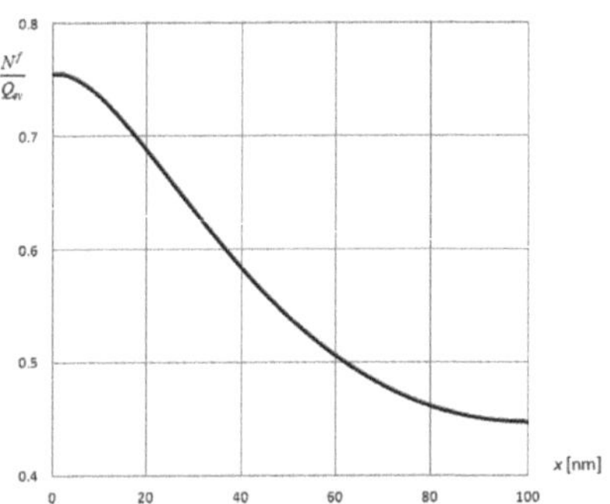

Figure 8. Volumetric fraction of gaseous phase in thin film after the 400 ps ($I_0 = 40000$ W/m^2).

5 FINAL REMARKS

The mathematical model of thermal processes which take place in a thin metal film subjected to an ultrafast laser heating has been considered. The formulation of the problem based on the two-temperature model supplemented by equations describing the phase transformations. The problem is solved using the explicit scheme of finite difference method. The examples illustrating the solid-liquid and liquid-vapour phase changes are shown. At present, the authors are preparing the similar computer program concerning the axially-symmetrical tasks. It should be noted that modeling of phase transformations may also be performed using other approaches, e.g. dual-phase lag equation (Tzou 1997, Mochnacki & Ciesielski 2012, Mochnacki & Paruch 2014).

ACKNOWLEDGEMENTS

This work is supported by the project No. 2012/05/B/ST8/01477 sponsored by the PNSC.

REFERENCES

Chen, J.K., Latham, W.P. & Beraun, J.E. 2005. The role of electron–phonon coupling in ultrafast laser heating. *Journal of Laser Applications* 17(1): 63–68.

Chen, J.K., Tzou, D.Y. & Beraun, J.E. 2006. A semiclassical two-temperature model for ultrafast laser heating. *Intern. Journal of Heat and Mass Transfer* 49: 307–316.

Dai, W. & Nassar, R. 2001. A compact finite difference scheme for solving a one-dimensional heat transport equation at the microscale. *Journal of Computational and Applied Mathematics* 132: 431–441.

Dziatkiewicz, J., Kuś, W., Majchrzak, E., Burczyński, T. & Turchan, Ł. 2014. Bioinspired identification of parameters in microscale heat transfer. *International Journal for Multiscale Computational Engineering* 12(1): 79–89.

Majchrzak, E. & Dziatkiewicz, J. 2012. Identification of electron-phonon coupling factor in a thin metal film subjected to an ultrashort laser pulse. *Computer Assisted Methods in Engineering and Science* 19: 383–392.

Majchrzak, E., Dziatkiewicz, J. & Turchan, Ł. 2015. Sensitivity analysis and inverse problems in microscale heat transfer. *Defect and Diffusion Forum* 362: 209–223.

Mochnacki, B. & Ciesielski, M. 2012. Numerical model of thermal processes in domain of thin film subjected to a cyclic external heat flux. *Materials Science Forum* 706–709: 1460–1465.

Mochnacki, B. & Paruch, M. 2013. Estimation of relaxation and thermalization times in in microscale heat transfer model. *Journal of Theoretical and Applied Mechanics* 51(4): 837–845.

Tzou, D.Y. 1997. *Macro- to Microscale Heat Transfer: The lagging behaviour,* Taylor and Francis.

Zhang, Z.M. 2007. *Nano/microscale heat transfer*. McGraw-Hill.

Advances in Mechanics: Theoretical, Computational and Interdisciplinary Issues – Kleiber et al. (Eds)
 ISBN 978-1-138-02906-4

Sensitivity analysis of temperature field in the heated tissue with respect to the dual-phase-lag model parameters

E. Majchrzak, Ł. Turchan & G. Kałuża
Silesian University of Technology, Gliwice, Poland

ABSTRACT: Heat transfer processes proceeding in axially symmetrical domain of heated tissue are analyzed. The problem is described by Dual-Phase Lag Equation (DPLE) supplemented by appropriate boundary and initial conditions. The aim of this research is to estimate the changes in the transient temperature field due to the perturbations of the DPLE parameters. For this purpose, the methods of sensitivity analysis are applied. The direct problem and additional ones connected with the sensitivity analysis are solved using the explicit scheme of finite difference method. In the final part of the paper the examples of computations are presented.

1 INTRODUCTION

Heat transfer processes taking place in the heated tissues can be described by different models (Stańczyk & Telega 2002). In this paper the dual-phase-lag equation is considered (Zhou et al. 2009). This model allows one to take into account the non-homogeneous structure of biological tissue. Thermophysical parameters occurring in the dual-phase-lag equation differ significantly from person to person. Thus, it is important to assess the impact of thermophysical parameters perturbations on the course of thermal processes proceeding in the domain considered. For this purpose, the method of sensitivity analysis (Dems 1996, Kleiber 1997) is used here. Governing equations are differentiated with respect to the parameters and in this way the additional problems are formulated. The basic problem and the additional ones are solved using the finite difference method. In the final part the results of computations are shown.

2 GOVERNING EQUATIONS

Dual-phase-lag equation in the cylindrical coordinate system has the following form (Zhou et al. 2009):

$$c\left[\frac{\partial T(r,z,t)}{\partial t}+\tau_q\frac{\partial^2 T(r,z,t)}{\partial t^2}\right] = \lambda\nabla^2 T(r,z,t)+\lambda\tau_T\frac{\partial \nabla^2 T(r,z,t)}{\partial t} + Q(r,z,t)+\tau_q\frac{\partial Q(r,z,t)}{\partial t} \tag{1}$$

where c = is the volumetric specific heat of tissue, λ = is the thermal conductivity of tissue, T is the temperature, (r, z) = are the spatial co-ordinates, t is the time and $Q(r, z, t)$ = is the source term due to metabolism and blood perfusion. In (1) τ_q = is the relaxation time and τ_T is the thermalization time.

The source term $Q(r, z, t)$ can be written in the form:

$$Q(r,z,t)=w_B c_B[T_B - T(r,z,t)]+Q_m \tag{2}$$

where w_B = is the blood perfusion rate, c_B = is the specific heat of blood, T_B = is the blood temperature and Q_m = is the metabolic heat source.

The equation (1) is supplemented by the boundary conditions (Fig. 1):

$$\begin{aligned}(r,z)\in\Gamma_1 &: \quad T=T_b\\ (r,z)\in\Gamma_2 &: \quad q(r,z,t)=q_b(r,z,t)\\ (r,z)\in\Gamma_3 &: \quad q(r,z,t)=0\end{aligned} \tag{3}$$

and initial ones

$$t=0:\quad T=T_p,\quad \left.\frac{\partial T}{\partial t}\right|_{t=0}=0 \tag{4}$$

where T_b = is the known boundary temperature, q_b = is the known boundary heat flux and T_p = is a constant initial tissue temperature. For $r \le r_D$, $z = 0$ and $t \le t_e$, where t_e = is the exposure time, the Neumann boundary condition is accepted (Mochnacki & Piasecka-Belkhayat 2013)

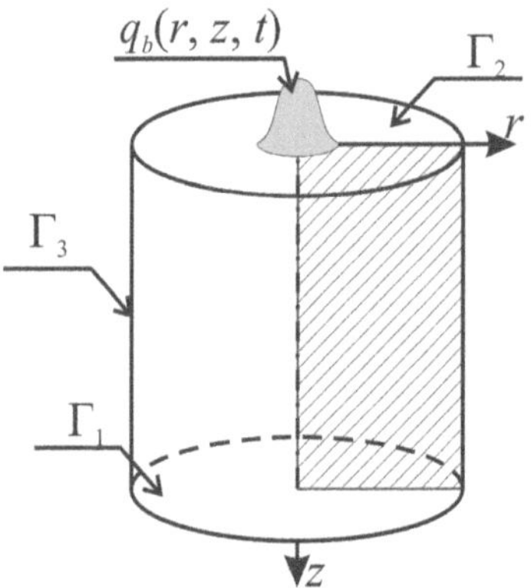

Figure 1. Domain considered.

$$q_b(r,0,t) = q_0 \frac{t}{t_e}\left(1-\frac{t}{t_e}\right)\exp\left(-\frac{r^2}{r_D^2}\right) \tag{5}$$

where q_0 = is the constant value, for $t > t_e$: $q_b(r, 0, t) = 0$.

It should be pointed out that using the dual phase lag model the following form of second type boundary condition should be considered:

$$(r,z)\in\Gamma_2:\ q_b + \tau_q \frac{\partial q_b}{\partial t} = -\lambda\left[\frac{\partial T}{\partial n} + \tau_T \frac{\partial}{\partial t}\left(\frac{\partial T}{\partial n}\right)\right] \tag{6}$$

where n = is the normal outward vector and $\partial(\cdot)/\partial n$ = is the normal derivative.

3 SENSITIVITY ANALYSIS

Sensitivity analysis of temperature distribution in the heated tissue can be done using the direct approach (Dems 1996, Mochnacki & Majchrzak 2003, Jasiński 2014, Majchrzak et al. 2013, Majchrzak & Mochnacki 2014, Dziewoński et al. 2011).

Let $p_1 = c$, $p_2 = \lambda$, $p_3 = \tau_q$, $p_4 = \tau_T$, $p_5 = w_B$, $p_6 = Q_m$. Equation (1) is differentiated with respect to the parameter p_s, $s = 1, 2, \ldots 6$ and then

$$\begin{aligned}
&(c+\tau_q w_B c_B)\frac{\partial U_s}{\partial t} + c\tau_q \frac{\partial^2 U_s}{\partial t^2} \\
&= \lambda\nabla^2 U_s + \lambda\tau_T \frac{\partial \nabla^2 U_s}{\partial t} + \\
&- w_B c_B U_s + \left[\frac{\partial \lambda}{\partial p_s}\frac{c}{\lambda} + \frac{\partial \lambda}{\partial p_s}\frac{\tau_q}{\lambda} w_B c_B - \right. \\
&\left. + \frac{\partial \tau_q}{\partial p_s} w_B c_B - \tau_q \frac{\partial w_B}{\partial p_s} c_B - \frac{\partial c}{\partial p_s}\right]\frac{\partial T}{\partial t} \\
&+ \left[\frac{\partial \lambda}{\partial p_s}\frac{c}{\lambda}\tau_q - \frac{\partial c}{\partial p_s}\tau_q - c\frac{\partial \tau_q}{\partial p_s}\right]\frac{\partial^2 T}{\partial t^2} + \lambda\frac{\partial \tau_T}{\partial p_s}\frac{\partial \nabla^2 T}{\partial t} \\
&+ \frac{\partial w_B}{\partial p_s} c_B (T_B - T) + \frac{\partial Q_m}{\partial p_s} \\
&- \frac{1}{\lambda}\frac{\partial \lambda}{\partial p_s}\left[w_B c_B (T_B - T) + Q_m\right]
\end{aligned} \tag{7}$$

where $U_s = \partial T/\partial p_s$ is the sensitivity function. The boundary conditions (3) and initial ones (4) are also differentiated with respect to p_s. Thus:

$$(r,z)\in\Gamma_1:\ U_s = 0 \tag{8}$$

$$\begin{aligned}
&(r,z)\in\Gamma_2:\ -\frac{q_b}{\lambda}\frac{\partial \lambda}{\partial p_s} + \left[\frac{\partial \tau_q}{\partial p_s} - \frac{\tau_q}{\lambda}\frac{\partial \lambda}{\partial p_s}\right]\frac{\partial q_b}{\partial t} \\
&+ \lambda\frac{\partial \tau_T}{\partial p_s}\frac{\partial}{\partial t}\left(\frac{\partial T}{\partial n}\right) = -\lambda\left[\frac{\partial U_s}{\partial n} + \tau_T\frac{\partial}{\partial t}\left(\frac{\partial U_s}{\partial n}\right)\right]
\end{aligned} \tag{9}$$

$$\begin{aligned}
&(r,z)\in\Gamma_3:\ \lambda\frac{\partial \tau_T}{\partial p_S}\frac{\partial}{\partial t}\left(\frac{\partial T}{\partial n}\right) \\
&= -\lambda\left[\frac{\partial U_S}{\partial n} + \tau_T\frac{\partial}{\partial t}\left(\frac{\partial U_S}{\partial n}\right)\right]
\end{aligned} \tag{10}$$

and

$$t = 0:\ U_s = 0,\quad \left.\frac{\partial U_s}{\partial t}\right|_{t=0} = 0 \tag{11}$$

In this way the additional problems connected with the sensitivity functions U_s are formulated.

4 RESULTS OF COMPUTATIONS

The cylindrical tissue domain (R = 0.02 m, Z = 0.02 m) of the initial temperature T_p = 37 °C is considered.

The following values of parameters are assumed: thermal conductivity of tissue λ = 0.5 W/(mK), volumetric specific heat c = 4 MJ/(m^{3}K), blood perfusion rate w_B = 0.53 kg/(m^{3}s), specific heat of blood c_B = 3770 J/(kgK), blood temperature T_B = 37 °C, metabolic heat source Q_m = 250 W/m^3, relaxation time τ_q = 15 s, thermalization time τ_T = 10 s. In boundary condition (5): q_0 = 10000 W/m^2, $r_D = R/4$, t_e = 100 s.

To solve the direct problem (1–5) and the additional ones (7–11) the explicit scheme of the finite difference method is used (Majchrzak &

Mochnacki 2014, Majchrzak & Turchan 2014). The time step is equal to $\Delta t = 0.0005$ s and the grid step equals $h = 0.0002$ m.

In Figure 2 the temperature history at the points $N_1(0, h)$, $N_2(0, 10\,h)$, $N_3(0, 20\,h)$ and $N_4(0, 50\,h)$ is shown. These points are located along the axis of cylinder, at the distances h, $10\,h$, $20\,h$, $50\,h$ from the upper base. One can see that the source function (5) achieves the maximum for $t = 50$ s, while the maximum value of temperature at these points appears later (about 65 s). This is due to the relaxation time and the thermalization time that occur in the considered model. Figure 3 illustrates the temperature distribution in the domain considered after 100 seconds. Figures 4–9 refer to the sensitivity functions distributions.

The sensitivity functions allow to estimate the changes of temperature distribution due to the perturbations of the parameters. The influence of

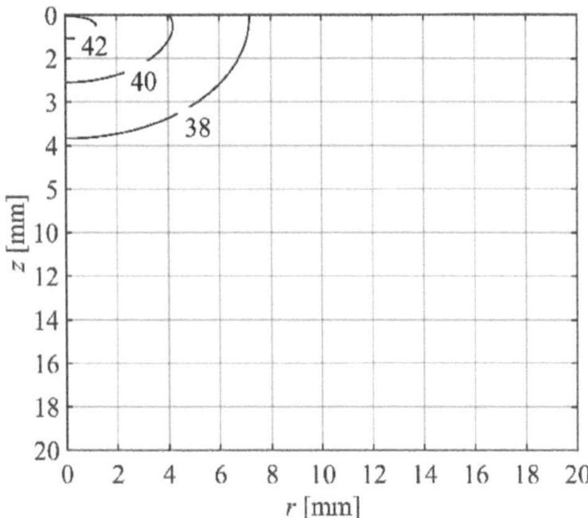

Figure 2. Temperature distribution after 100s.

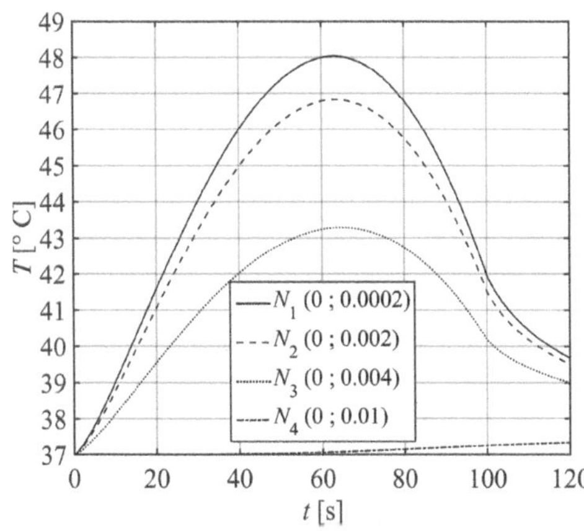

Figure 3. Temperature history at selected points.

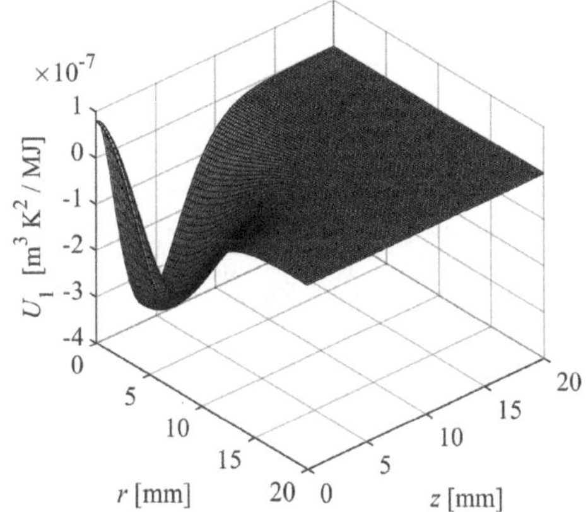

Figure 4. Distribution of function $U_1 = \partial T/\partial c$ after 100s.

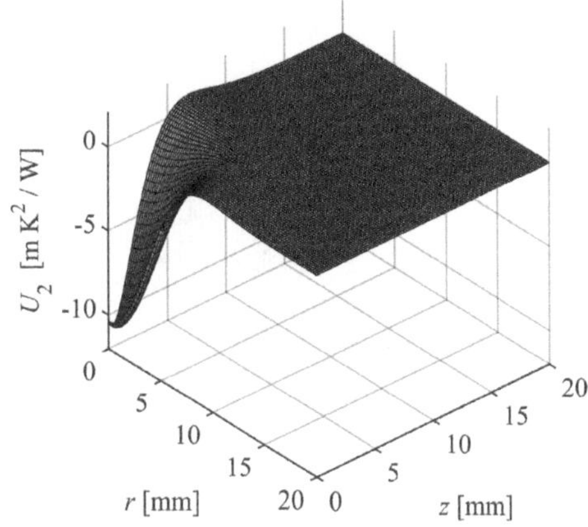

Figure 5. Distribution of function $U_2 = \partial T/\partial \lambda$ after 100s.

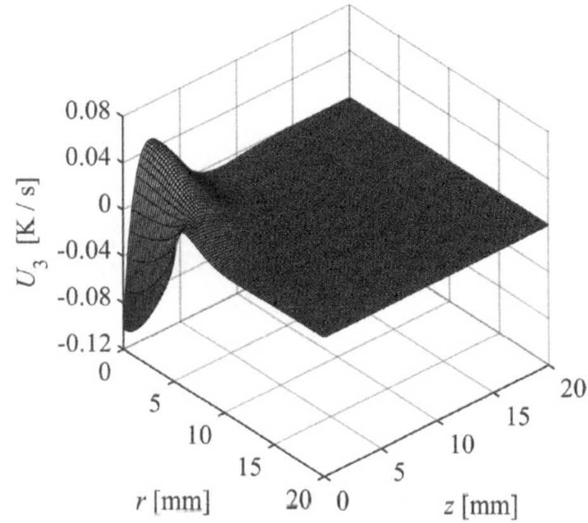

Figure 6. Distribution of function $U_3 = \partial T/\partial \tau_q$ after 100s.

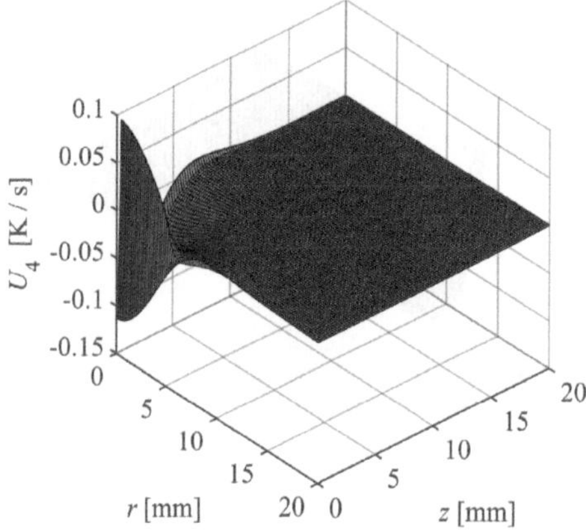

Figure 7. Distribution of function $U_4 = \partial T/\partial \tau_T$ after 100s.

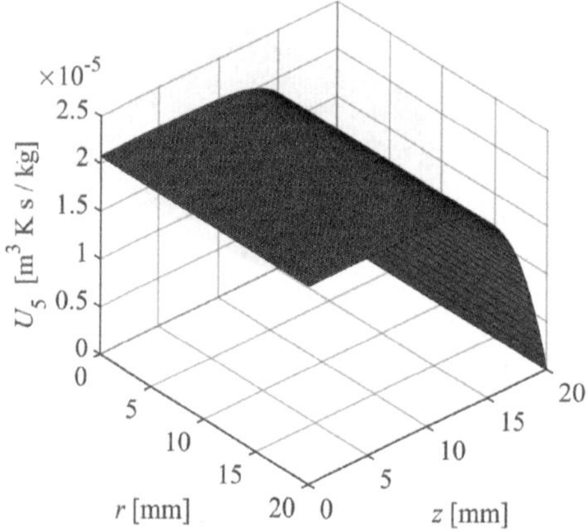

Figure 8. Distribution of function $U_5 = \partial T/\partial w_B$ after 100s.

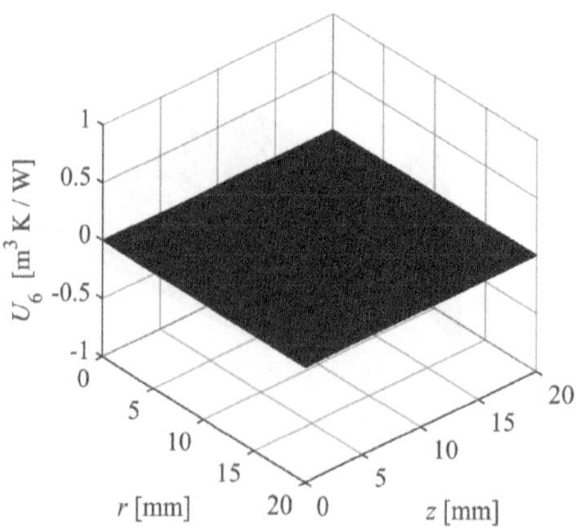

Figure 9. Distribution of function $U_6 = \partial T/\partial Q_m$ after 100s.

change of a single parameter on the temperature distribution can be analyzed using the formula

$$Z_s T = U_s \, \Delta p_s, \quad s = 1, 2, ..., 6 \tag{12}$$

where Δp_s = are the assumed perturbations of parameters.

In Figures 10 and 11 the temperature changes at the points N_1 and N_4 corresponding to 10% perturbations of successive parameters are shown. The calculations show that perturbation of the metabolic heat source and the blood perfusion rate almost does not change the temperature.

The greatest temperature changes are related to the changes of thermal conductivity (maximum value of temperature change is about 1.6 K) and volumetric specific heat of tissue (maximum value of temperature change is about 0.7 K). Perturbation of the relaxation and thermalization times

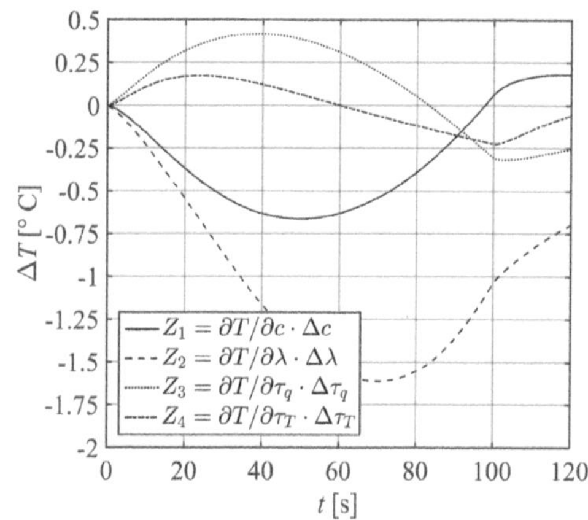

Figure 10. Temperature perturbation in node N_1.

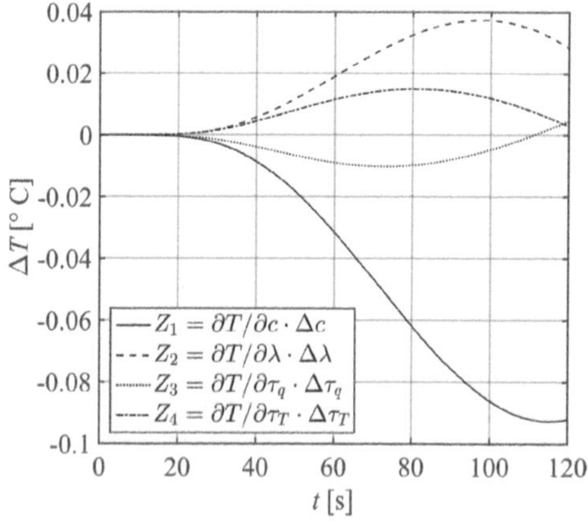

Figure 11. Temperature perturbation in node N_4.

causes the maximum change in temperature of 0.4 and 0.25 K, respectively.

It should be pointed out that it is also possible to estimate the changes of temperature due to the simultaneous perturbations of all parameters (e.g. Majchrzak & Mochnacki 2014).

5 CONCLUSIONS

The analysis of thermal processes proceeding in axially symmetrical domain of heated tissue is presented. The methods of sensitivity analysis showed that the greatest temperature changes are caused by perturbations in thermal conductivity and volumetric specific heat. The temperature changes due to the perturbations in the metabolic source and the blood perfusion rate are very small.

ACKNOWLEDGEMENTS

The research is financed by Grants No. 10/040/BK_15/0006 and 10/990/BK_15/0013.

REFERENCES

Dems, K. 1996. Sensitivity analysis in thermal problems, 1: Variation of material parameters within fixed domain. *Journal of Thermal Stresses* 9: 303–324.

Dziewoński, M., Mochnacki, B. & Szopa, R. 2011. *Sensitivity of biological tissue freezing process on the changes of cryoprobe cooling rate*, Mechanika 2011: Proceedings of the 16th International Conference, Book Series: Mechanika Kaunas University of Technology, 82–87.

Jasiński, M. 2014. Modelling of tissue thermal injury formation process with application of direct sensitivity method. *Journal of Theoretical and Applied Mechanics* 52 (4): 947–957.

Kleiber, M. 1997. *Parameter sensitivity and nonlinear mechanics*, Chichester: J. Wiley & Sons Ltd.

Majchrzak, E., Dziatkiewicz, J. & Kałuża, G. 2013. Application of sensitivity analysis in microscale heat transfer. *Computer Assisted Methods in Engineering and Science* 20: 113–121.

Majchrzak, E. & Mochnacki, B. 2014. Sensitivity analysis of transient temperature field in microdomains with respect to the dual phase lag model parameters. *International Journal for Multiscale Computational Engineering* 12 (1): 65–77.

Majchrzak, E. & Turchan, Ł. 2014. Numerical analysis of tissue heating using the generalized dual phase lag model. In T. Łodygowski, J. Rakowski & P. Litewka (eds), *Recent Advances in Computational Mechanics*: 355–362. London: CRC Press.

Mochnacki, B. & Piasecka-Belkhayat, A. 2013. Numerical modeling of skin tissue heating using the interval finite difference method. *MCB: Molecular & Cellular Biomechanics* 10 (3): 133–144.

Mochnacki, B. & Majchrzak, E. 2003. Sensitivity of the skin tissue on the activity of external heat sources. *CMES: Computer Modeling in Engineering and Sciences* 4 (3–4): 431–438.

Stańczyk M. & Telega, J.J. 2002. Modelling of heat transfer in biomechanics a review. Part I. Soft tissues. *Acta of Bioengineering and Biomechanics* 4 (1): 31–61.

Zhou, J., Zhang, Y. & Chen, J.K. 2009. An axisymmetric dual-phase-lag bioheat model for laser heating of living tissues. *International Journal of Thermal Sciences*, 48: 1477–1485.

Advances in Mechanics: Theoretical, Computational and Interdisciplinary Issues – Kleiber et al. (Eds)
 ISBN 978-1-138-02906-4

Multiscale model of the proximal femur with implanted bone scaffold

P. Makowski & W. Kuś
Institute of Computational Mechanics and Engineering, Silesian University of Technology, Gliwice, Poland

ABSTRACT: This paper is focused on multiscale modeling of the proximal femur with implanted bone scaffold. The Finite Element Method (FEM) macro-scale model of human proximal femur is built on the basis of Computed Tomography (CT) data. Micro-scale FEM model (RVE) of trabecular (cancellous) bone microstructure is built on the basis of micro-computed tomography (μCT) data. Effective anisotropic material parameters of trabecular bone are calculated using micro model with numerical homogenization algorithm and then are applied in the macro model. Homogenized anisotropic material parameters of the personalized bone scaffold structure are applied in the implantation area of the macro model and numerical analysis of proximal femur is conducted. Stain and stress state in the bone and implant microstructures are calculated using localization.

1 INTRODUCTION

1.1 *Trabecular bone and osteonecrosis*

Trabecular bone is a porous and heterogeneous bone tissue which is located mainly in ends of long bones e.g. end of the human femur. Osteonecrosis of femoral head is a bone tissue disease caused by abnormal blood supply to the bone. The affected bone tissue needs to be removed during surgical procedure. Bone exhibits the ability of self-regeneration, but in case of large size bone defects the healing process cannot be accomplished by itself. In such cases the bone scaffolds can be used to provide a temporary support for bone tissue regeneration.

1.2 *Bone scaffolds*

Bone scaffold is a biodegradable bone tissue implant. During the bone regeneration scaffold is gradually replaced by newly formed healthy bone tissue. Trabecular bone microstructure is adapted to the loads occurring in human skeletal system, so it is not random. The structure of bone scaffold also must be appropriate and specified in order to help induce bone healing and regeneration. Bone scaffold should exhibit open pore structure and at least 60% porosity (Karageorgiou & Kaplan 2005). Scaffold should be also characterized by anisotropic material parameters close to material parameters of bone—the biomechanical compatibility between tissue and implant (Lin et al. 2004). Material parameters of trabecular tissue are individual and can vary between different locations of bone. To achieve the biomechanical compatibility requirement the bone scaffold must be a personalized implant. Such an implant can be manufactured from biocompatible and biodegradable polymer (e.g. PLGA) using additive method like Fused Deposition Modeling (FDM). The anisotropic material parameters of bone and scaffolds can be calculated using multiscale modeling methods like numerical homogenization.

1.3 *Multiscale modeling of heterogeneous materials*

Multiscale modeling can be performed numerical homogenization method (Makowski et al. 2013a). Homogenization relies on replacement of heterogeneous material with a homogeneous material. The effective material parameters, which are applied in the homogeneous macro model, are calculated on the basis of micro model analysis—the Representative Volume Element (RVE) of considered heterogeneous medium (Fig. 1). Analyses of RVE model can be performed in parallel way (Kuś 2007).

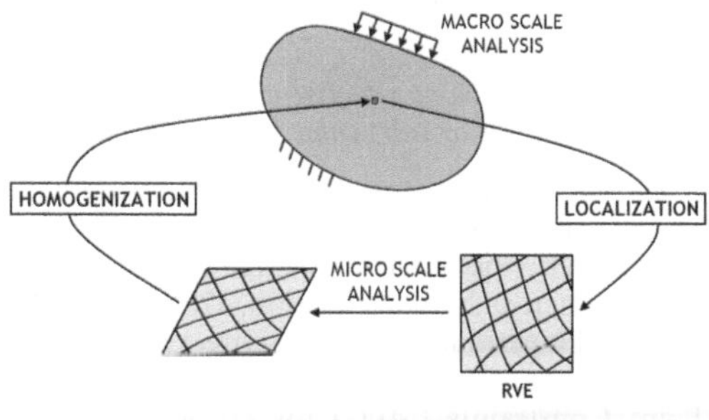

Figure 1. Idea of multiscale modeling using RVE model.

During the micro-scale analyses the strains (1) and stresses (2) from RVE model are averaged and effective anisotropic material parameters of heterogeneous material are calculated.

$$<\varepsilon_{ij}> = \frac{1}{V_{RVE}} \int_{V_{RVE}} \varepsilon_{ij} \, dV_{RVE} \tag{1}$$

where $<\varepsilon_{ij}>$ = averaged strain at macro scale; ε_{ij} = strain at micro scale; V_{RVE} = micro model (RVE) volume.

$$<\sigma_{ij}> = \frac{1}{V_{RVE}} \int_{V_{RVE}} \sigma_{ij} \, dV_{RVE} \tag{2}$$

where $<\sigma_{ij}>$ = averaged stress at macro scale; σ_{ij} = stress at micro scale; V_{RVE} = micro model (RVE) volume.

The constitutive relationship for linear elastic anisotropic material is then formulated (3).

$$<\sigma_{ij}> = \boldsymbol{C} \; <\varepsilon_{ij}> \tag{3}$$

where $\boldsymbol{C}$ is an elasticity matrix of equivalent homogeneous material (effective material parameters).

After performing the macro-scale analysis, the localization can be conducted for a given point of macro model in order to calculate the stress and strain state in the considered microstructure.

2 DETERMINATION OF TRABECULAR BONE MATERIAL PARAMETERS

2.1 *Identification of trabeluar bone material parameters at micro scale*

The microstructure of trabecular bone is composed of interconnected bone trabeculas. The FEM numerical model of bone microstructure (Fig. 2) was created on the basis of micro computed tomography (μCT) data segmentation. The evolutionary identification of trabeculas elastic material parameters using experimental and numerical data was considered in Makowski et al. 2013b.

2.2 *Computation of effective anisotropic material parameters of trabecular bone at macro scale*

After identification of material parameters of bone tissue at micro scale, the computation of macro-scale material parameters of bone was performed using numerical homogenization. Trabecular bone is a locally periodic structure. Direct applying of periodic boundary conditions with use of Multi Point Constraints (MPC) for porous numerical models with non-periodic opposite boundary meshes is not possible. The buffer zone based on the Embedded Cell Approach (Böhm 2015) was created for the microstructure model and the effective material parameters (matrix $\boldsymbol{C}$) of trabecular bone were calculated. More detailed description of numerical procedures and the results of computations can be found in Makowski & Kuś 2014.

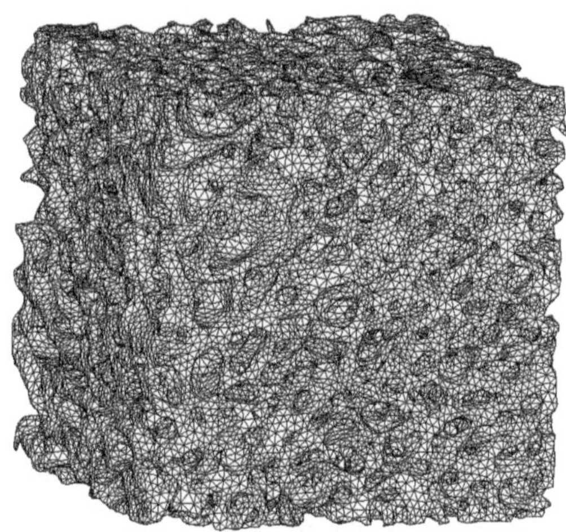

Figure 2. Numerical model of trabecular bone microstructure composed of bone trabeculas.

3 OPTIMIZATION OF PERSONALIZED BONE SCAFFOLD STRUCTURE

The topology optimization of personalized bone scaffold was performed using evolutionary algorithm (Makowski & Kuś 2015). The goal of optimization was obtaining the bone scaffold structure with macro-scale material parameters close to the anisotropic material parameters of bone—the biomechanical compatibility requirement. Homogenized effective anisotropic material parameters of trabecular bone sample (Fig. 2) were used. The three-scale numerical model of bone implant taking into account FDM additive manufacturing process (filament geometry, used raster) was built. Result of optimization is the periodic structure of bone scaffold (Fig. 3). Scaffold RVE can be multiplied in three directions to fit irregular shape of surgically removed bone tissue.

The numerical models shown in the Figure 2 and Figure 3 are used as an micro models (RVE) of bone tissue and bone implant in the multiscale analysis of proximal femur described in the following chapter.

Figure 3. Numerical model of optimized personalized bone scaffold structure.

4 BONE—IMPLANT MULTISCALE ANALYSIS

Macro-scale model of human proximal femur was created on the basis of Computed Tomography (CT) data segmentation. The boundary conditions for the analysis are the force $\boldsymbol{F}$ (Table 1) resulting from the Body Weight (BW) of the patients upper body part during walking (Bergman et al. 2001, Madej 2008). Force is applied in contact region between head of the femur and the pelvis bone, the displacements $\boldsymbol{u}$ for bottom region of model are fixed (Fig. 4).

Local coordinate system was defined and anisotropic material parameters of bone tissue and bone scaffold were applied in the macro model with accordance to main anatomical axis in the head of human proximal femur (Fig. 5).

The modeled region of implantation is shown in the Figure 6 using darker color. The volume of bone scaffold is 5.9 cm^3.

The resulting multiscale model is simplified due to fixed displacements in the bottom region and the homogeneous distribution of anisotropic material parameters of bone tissue in whole proximal femur. Creation of more detailed model is possible and will require higher number of bone RVE models with different local coordinate systems.

After defining the boundary conditions and material parameters with local coordinate system, the macro-scale analysis was conducted.

Table 1. Components of force F.

Components of force F	F_x	F_y	F_z	F
Load of femoral head [% of BW]	49	17	216	222
Load of femoral head [N]	421,4	146,2	1857,6	1909,2

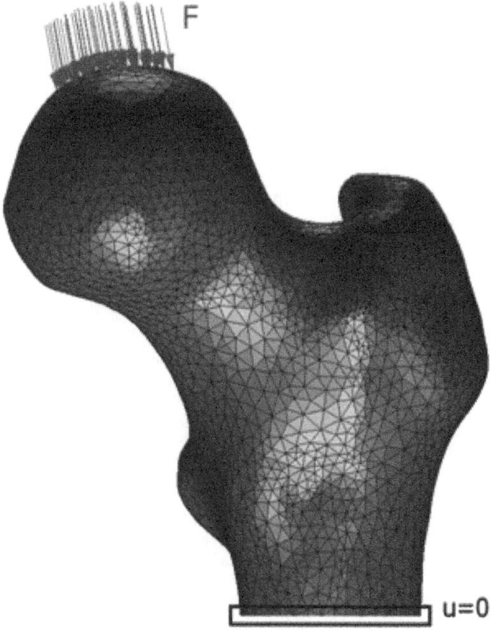

Figure 4. Macro-scale model of proximal femur with boundary conditions.

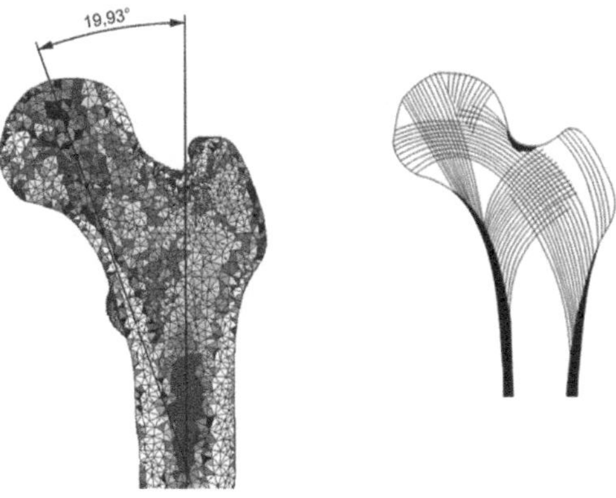

Figure 5. Heterogeneous distribution of material parameters and main anatomical axis in femoral head (Phillips 2012).

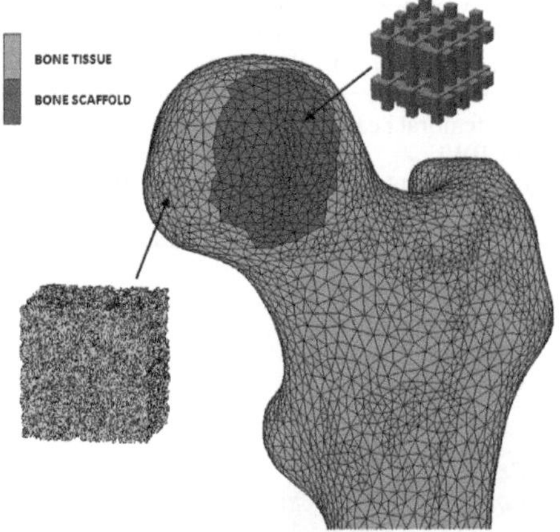

Figure 6. Regions of bone tissue and bone scaffold in proximal femur.

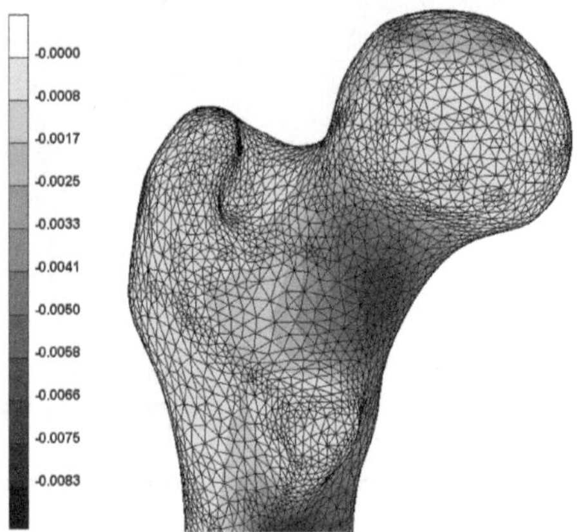

Figure 7. Distribution of principal strains ε_3 in proximal femur.

5 RESULTS OF MULTISCALE ANALYSIS

The macro-scale results of multiscale analysis are the principal strains in the proximal femur. The distribution of principal strains ε_3, which absolute value is highest due to compression character of force F, is shown in the Figure 7. More detailed results are summarized in Table 2.

Localization was performed for a point inside the implantation region and for a point of the tissue region in the surroundings of implantation region. The micro-scale result of multiscale analysis are the principal strains in the tissue (Fig. 8) and the implant (Fig. 9). More detailed data concerning micro-scale results is summarized in Table 2.

Calculated values of strains at both macro and micro scales are in the range of strains reported as typical for human bone tissue during daily activity (Duncan & Turner 1995).

Table 2. Extreme values of calculated principal strains in macro and micro models.

Principal strain	ε_1	ε_2	ε_3
Macro scale (proximal femur)			
max	0.59e-2	0.15e-2	0
min	0.1e-3	–0.9e-3	–0.83e-2
Micro scale (bone tissue)			
max	1.463e-3	2.607e-4	1.055e-5
min	1.355e-6	–4.634e-4	–9.516e-4
Micro scale (bone scaffold)			
max	4.544e-4	1.279e-4	–2.493e-5
min	2.523e-5	–1.286e-4	–5.429e-4

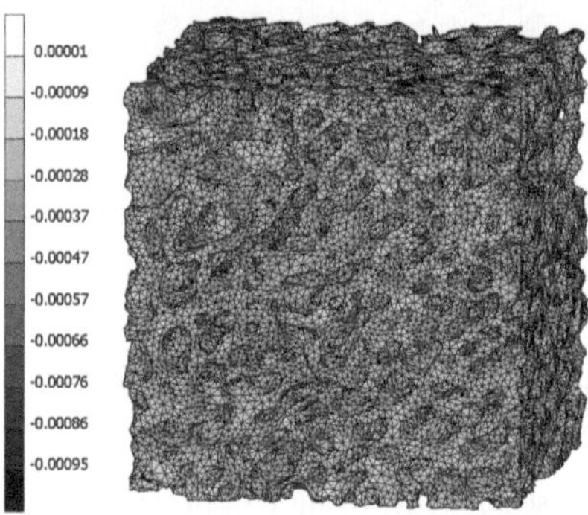

Figure 8. Distribution of principal strains ε_3 in trabecular bone microstructure.

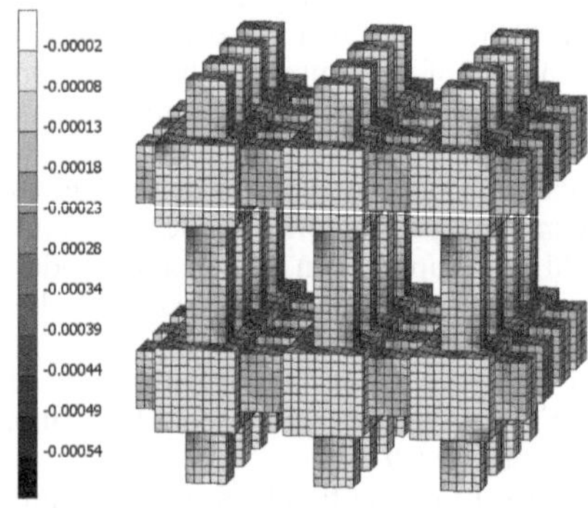

Figure 9. Distribution of principal strains ε_3 in bone scaffold microstructure.

6 CONCLUSIONS

Multiscale modeling methods were used to model bone—implant structure. Calculated values of strains at micro at macro scale are correct despite the simplification of the model. The ability to compute the stain fields in the microstructure of bone tissue and implant can be used to predict the occurrence of remodeling and further overgrowth of scaffold with new bone tissue. Bone—implant multiscale models can be further used to identify macro-scale loads of skeletal system which will induce bone remodeling, speed up the healing process and shorten rehabilitation of the patient.

ACKNOWLEDGEMENTS

The research is partially funded from the projects Silesian University of Technology, Faculty of Mechanical Engineering 10/990/BK_15/0013, 10/040/BK_15/0006, 10/040/BKM_14/0004.

REFERENCES

Bergmann, G., Deuretzbacher, G., Heller, M., Graichen, F., Rohlmann, A., Strauss, J. & Duda, G.N. 2001. Hip contact forces and gait patterns from routine activities. *Journal of Biomechanics* 34: 859–871.

Böhm, H.J. 2015. A Short Introduction to Basic Aspects of Continuum Micromechanics. Tech. *rep., Institute of Lightweight Design and Structural Biomechanics (ILSB), Vienna University of Technology*. Vienna.

Duncan, R.L. & Turner, C.H. 1995. Mechanotransduction and the functional response of bone to mechanical strain. *Calcified Tissue International* 57: 344–358.

Karageorgiou, V. & Kaplan, D. 2005. Porosity of 3D biomaterial scaffolds and osteogenesis. *Biomaterials* 26: 5474–5491.

Kuś, W. 2007. Grid-enabled evolutionary algorithm application in the mechanical optimization problems. *Engineering Applications of Artificial Intelligence* 20(5): 629–636.

Lin, C.Y., Kikuchi, N. & Hollister, S.J. 2004. A novel method for biomaterial scaffold internal architecture design to match bone elastic properties with desired porosity. *Journal of Biomechanics* 37: 623–636.

Madej, T. 2008. Modelling of the movement zone of endoprosthesis of hip joint in the aspect of biomaterials (In Polish). *PhD Thesis*. AGH University of Science and Technology.

Makowski, P., John, A., Kuś, W. & Kokot G. 2013a. Multiscale modeling of the simplified trabecular bone structure. *Mechanika, Kaunas*: 156–161.

Makowski, P., Kuś, W. & Kokot, G. 2013b. Evolutionary identification of trabecular bone properties. *International Conference on Inverse Problems in Mechanics of Structures and Materials IPM 2013*. Rzeszów.

Makowski, P & Kuś, W. 2014. Trabecular bone numerical homogenization with the use of buffer zone. *Computer Assisted Methods in Engineering and Science* 21(2): 113–121.

Makowski, P. & Kuś W. 2015. Optimization of bone scaffold structures using experimental and numerical data. *Acta Mechanica*. DOI:10.1007/s00707-015-1421-4.

Phillips, A.T. 2012. Structural Optimisation: Biomechanics of the Femur. *Proceedings of the ICE. Engineering and Computational Mechanics* 165.

Advances in Mechanics: Theoretical, Computational and Interdisciplinary Issues – Kleiber et al. (Eds)
 ISBN 978-1-138-02906-4

Reference FEM model for SHM system of cable-stayed bridge in Rzeszów

A. Mariak, M. Miśkiewicz, B. Meronk, Ł. Pyrzowski & K. Wilde
Gdańsk University of Technology, Gdańsk, Poland

ABSTRACT: The paper presents the reference model for a Structural Health Monitoring system (SHM) of cable-stayed bridge recently constructed in Rzeszów over Wisłok River. The SHM system is design to provide on-line information on the structure state and facilitate its maintenance procedures. The main feature of the SHM system is permanent observation of the dynamic behavior of the bridge with focus on cable vibrations. The paper discusses the process of creating a reference FEM model of the bridge with use of data from in situ static and dynamic test as well as the preliminary measurements obtained from the SHM system.

1 INTRODUCTION

In standard bridge inspection, deterioration and damage are discovered by a visual observation of the signs they exhibit on structural elements. In case of long-span bridges, the effectiveness of visual inspection is questionable since accesses to all critical locations and ability for comprehensive inspection are limited. Therefore, many long-span bridge owners have already adopted or are currently in the process of design and installation of various on-line monitoring techniques (Magalhaes et al. 2008).

The SHM system is a kind of continuous non-destructive diagnostic system. Sensors and actuators of the SHM system are permanently mounted on the object or even built into it. The aim of SHM system application is to determine technical condition of an engineering object by means of objective methods and to increase structure durability, reliability and effectiveness. Information on the object technical condition is a base for detection of the existence of damage, detection of damage location, identification of damage type and extent as well as prediction of damage development.

The advanced SHM systems use theoretical and numerical models for precise definition of the object technical state. One of a possible solution for the advanced SHM system is application of reference FEM models (Wilde 2013). The block diagram of this solution is shown in Figure 1. In this approach the diagnosis is obtained through the analysis of changes between the results from the numerical models in reference state, current state and damaged state with respect to data collected from the SHM system sensors. In this paper the reference FEM model of the cable-stayed bridge in Rzeszów is discussed.

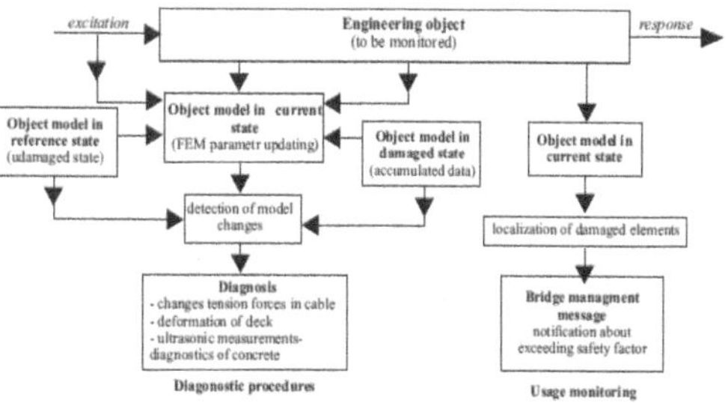

Figure 1. The block diagram of SHM system with FEM reference model.

2 DESCRIPTION OF THE CABLE-STAYED BRIDGE OVER WISŁOK RIVER

The bridge is 5-span cable-stayed structure with length 3 × 30.00 + 150.00 + 240.00 m (Nadolny 2014). The bridge tower in shape of an inverted Y has the total high equal to 108.50 m of which about 95.00 meters is located above the grade line. Center spacing of cable anchorage in the tower equals 1.70 m. The inner part of the tower has been intended for communications to the tower top. The superstructure of the bridge is concrete deck slab with steel box girder jointed by floor beam and hung to tower structures with cable spaced by 12.00 m. The bridge is designed for class load A according to PN-85/S-10030 code and STANAG 2021 class 150.

3 SHM SYSTEM OF CABLE-STAYED BRIDGE

The main hardware element of the SHM system is the central management unit, which controls the operation of the system, collects measurement data and performs numerical simulations and relevant analysis (Wilde et al. 2015). The monitoring system consists of three functional modules: a measurement, expert and the notification. The first module collects and stores data from sensors and measuring devices. The expert module analyses the environmental impacts and response structure in time. It also includes a segment numerical simulations and the algorithm for generating alert. Notification module informs about exceeding safety factors.

The measurement module consists of the following segments:

- acceleration measurements: tower (2 measurement channels—top of the tower), cables (78 channels), deck (24 channels—section B-B, D-D);
- force in cable measurement (direct measurements of changes in tension forces in 2 reference cables;
- static inclination measurements: top of the tower (2 channels), deck (12 channels, C-C, D-D, E-E);
- strain and temperature measurements: bridge girders (24 channels);
- normal force in cables by vibrational method: (78 measurement channels);
- meteorological: temperature, wind direction and speed (top of the tower).

The view of the bridge with SHM system during its installation is shown in Figure 2. The wiring of the acceleration segment of the SHM system is

Figure 2. View of the bridge with zoom on accelerometer/gyroscope sensor (6 degrees of freedom measurements).

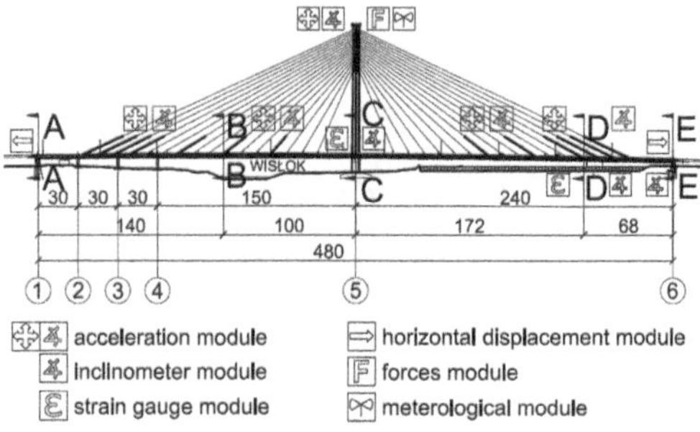

Figure 3. Sketch of the SHM system installed on the bridge.

visible on the first right cable. Figure 2 shows also zoom of the WE1013 sensor which allows on-line measurements of all 6 degrees of freedom of a rigid body (3 translations and 3 rotational angles). The location of the elements of the measurement segments are given in Figure 3.

4 REFERENCES FEM MODEL OF THE BRIDGE

Numerical simulations of the structure have been carried out with use of a linear-elastic finite element model. First, a detailed three-dimensional FE model of the as-built bridge has been developed in Sofistik software. Then, a group of numerical models for direct implementation in SHM system has been defined. The geometry of the bridge and construction materials parameters have been adopted on the basis of the object's documentation (Nadolny 2014).

The initial reference model consists of the following finite elements, marked as (R), (P) (K):

- (R)—1—dimensional, 2—nodal spatial rod, Timoshenko beam finite elements, taking into account shear effect and eccentricity of the beam axis;
- (P)—2—dimensional, 4—nodal quadrilateral, Timoshenko—Reissner finite elements, class C° of bilinear shape functions witch elimination locking effect. They take into account the shear effect and varied position of the reference surface;
- (K)—1—dimensional, 2—nodal spatial truss, class C° for linear shape functions.

The concrete deck and tower support zone is represented using shell elements, the girders and tower are defined by beam elements and cables are modelled with truss element. The visualization of the FE model grid is shown in Figure 4. The total number of degrees of freedom equals 268008.

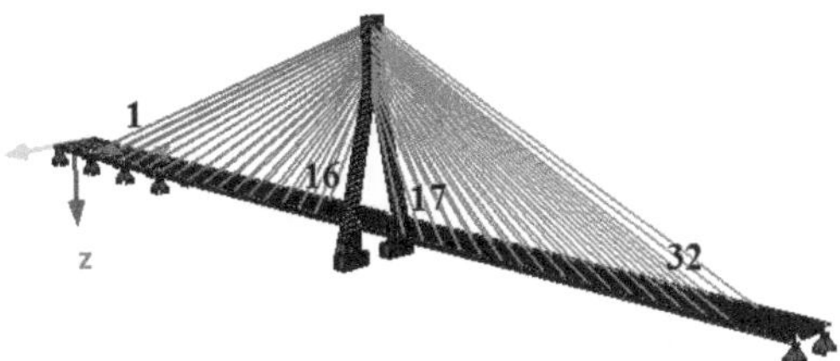

Figure 4. The visualization of the FE model.

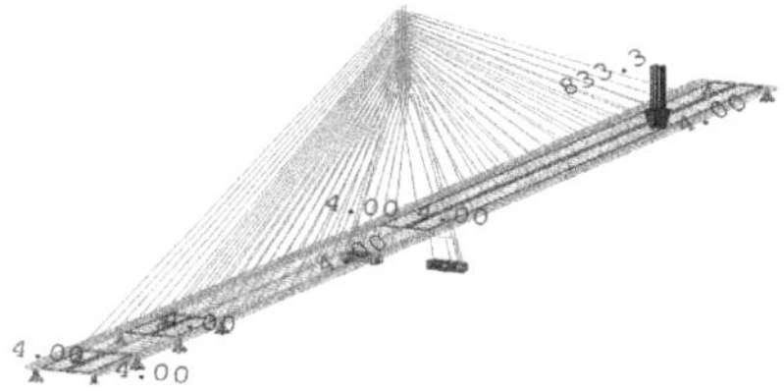

Figure 5. The K+q load combination for maximum normal forces in cables.

5 SELECTED RESULTS

5.1 *Static analysis*

The aim of the numerical analysis is to conduct the reference model parameter updating with respect to the in-situ measurement data so that the reference FEM model of the SHM system properly describes static and dynamic bridge responses. First, a standard preliminary calculation for SHM system design has been conducted. Static analysis has been performed for load K = 800 kN and distributed load q = 4.0 kN/m^2. The most unfavorable combination of external loads for maximum forces in cables is shown in Figure 5. This specific load combination produces significant bending of the bridge tower and the longest span at a distance of 68 m from the support no. 6 (section D-D). Figure 6 presents distribution of bending moments, M_y, in the beam elements under the K+q load combination.

The second stage of the reference FE model improvement has been performed on the measurement data from the in-situ static load tests conducted in September 2015 (Filar et al. 2015). The load is subjected by trucks sets placed in the selected locations in order to induce from 75% to 100% values of internal forces with respect to the characteristic load class A. Three locations (center of span 1–2, B-B sect., D-D sect.) of 4-axle trucks have been considered. Basic parameters of used Mercedes Benz truck are: the total weight N_c = 348.6 kN, the front axle load $N_p = 2 \times 76.65$ kN and rear axle load $N_t = 2 \times 97.65$ kN. Vehicles location in D-D section consist of 16 trucks is shown in Figures 7, 8. Sectional internal forces and moments have been computed and summarized in Table 1.

During bridge in situ static tests stress changes in box girders and traverse have been recorded by the SHM system. The location of stress points are presented in Figure 9 and the selected values of stress are summarized in Table 2.

5.2 *Dynamic analysis*

The final improvement of the reference FE model of the SHM system has been conducted on

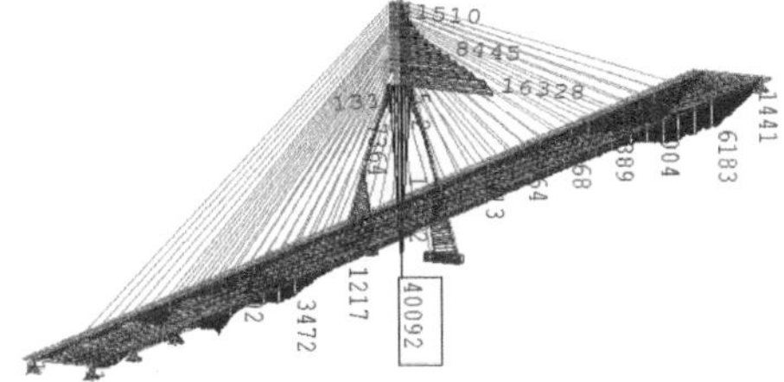

Figure 6. Maximum bending moments M_y under load K+q.

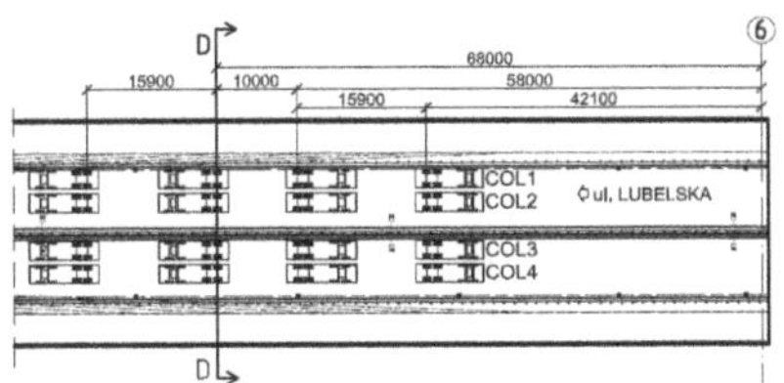

Figure 7. Vehicles location in D-D section.

Figure 8. View of the static load tests from the tower.

Table 1. Sectional D-D bending moment.

Load case	M_y (kNm)
1+2 COL	14567
1+2+3+4 COL	29139
3+4 COL	14572
K+q NORM	30748
Extreme [%]	95

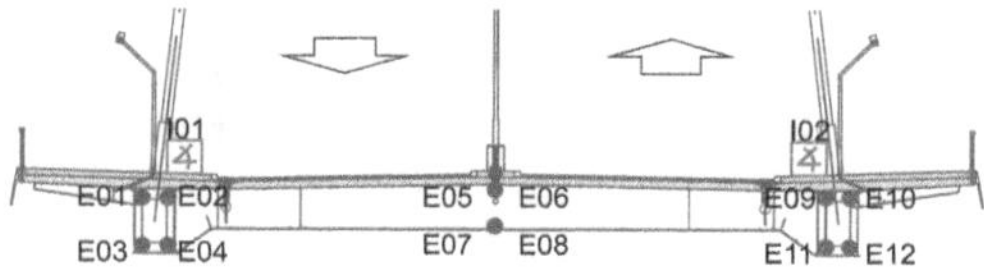

Figure 9. Cross section of the bridge witch stress measurement points.

Table 2. Values of stress in selected points of D-D section under vehicle load 1+2+3+4 COL.

Point	Test load (MPa)	Reference model (MPa)	Relative error (%)
E03	107.3	103.0	4.0
E04	100.5	100.7	–0.2
E07	44.1	43.8	0.7
E08	50.4	48.3	4.2
E11	86.1	91.0	–5.7
E12	89.5	93.8	–4.8

Table 3. Cable eigenfrequencies.

Eigenfrequency	SHM system (Hz)	Numerical (Hz)	Relative error (%)
f_1	0.586	0.581	0.9
f_2	1.158	1.162	–0.3
f_3	1.738	1.743	–0.3
f_4	2.314	2.324	–0.4
f_5	2.892	2.905	–0.4
f_6	3.465	3.486	–0.6
f_7	4.034	4.066	–0.8
f_8	4.612	4.647	–0.8
f_9	5.184	5.228	–0.8
f_{10}	5.754	5.809	–1.0
f_{11}	6.322	6.389	–1.1
f_{12}	6.906	6.970	–0.9
f_{13}	7.470	7.550	–1.1
f_{14}	8.036	8.131	–1.2
f_{15}	8.625	8.711	–1.0
f_{16}	9.208	9.291	–0.9

dynamic data. The dynamic load tests were carried out using vehicles moving along the object at different speeds. Time histories of vertical displacements and accelerations of several points located at all bridge spans have been measured. In addition the measurements were taken by the sensors of acceleration segment of the SHM system.

The first global theoretical frequency of the bridge, vertical bending mode equals 0.349 Hz and the first eigenfrequency of cable amounts 0.581 Hz (Table 3). Measured values are 0.350 Hz and

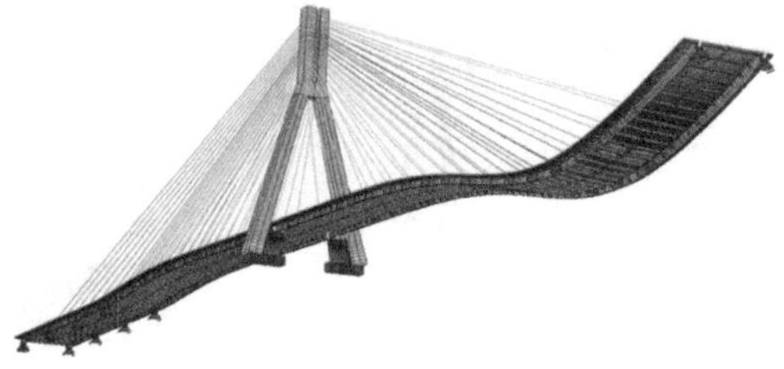

Figure 10. The first eigenmode of the bridge ($f_1 = 0.349$ Hz).

Figure 11. The first eigenmode of the longest cable ($f_1 = 0{,}581$ Hz).

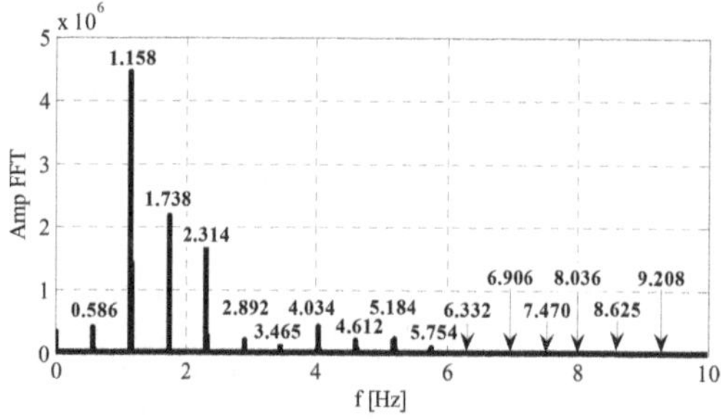

Figure 12. The frequency spectra of the longest cable oscillations recorded by the SHM system.

0.586 Hz respectively. The mode shapes corresponding to the above mentioned first frequencies are given in Figures 10–11.

Since the SHM system uses very sensitive sensors the identification of as much as 16 cable natural frequencies has been possible. The relative errors between the numerical and experimental data do not exceed 1.2% (Table 3). The updating of the FE model parameters of the bridge adopted the solutions developed for the SHM system of the roof structure of Sports Arena "Olivia" (Wilde et al. 2013).

6 CONCLUDING REMARKS

As a part of an ongoing long-term monitoring research project, the finite element reference model has been developed to monitor the behavior of the road bridge over the Wisłok River in Rzeszów. The presented validation and model parameter updating procedure allowed to develop the reference model of the bridge which properly describes static and dynamic properties of the bridge.

REFERENCES

Filar, Ł., Kałuża, J. & Wazowski, M. 2015. *Load test report of road bridge over Wisłok river.* Aspekt Sp. z o.o. (in Polish).

Magalhaes, F., Cunha, A. & Caetano, E. 2008. Dynamic monitoring of a long span arch bridge. *Engineering Structures* 30: 3034–3044.

Nadolny, A. 2014. *Architectural and construction design of road bridge over Wisłok River*. Mosty Gdańsk Sp. z o.o., (in Polish).

Wilde, K. 2013. *Structural health monitoring systems supported by numerical models for civil engineering structures. Proceedings of Computer Methods in Mechanics,* Poznań, Poland.

Wilde, K., Miśkiewicz, M. & Chróścielewski, J. 2013. SHM System of the Roof Structure of Sports Arena "Olivia". *Structural Health Monitoring 2013. – Vol. II/ ed. Fu-Kuo Chang, Alfredo Guemes Lancaster, Pennsylvania 17602 U.S.A.: DEStech Publications, Inc.*: 1745–1752.

Wilde, K., Chróścielewski, J., Miśkiewicz, M., Pyrzowski, Ł. & Mariak, A. 2015. *Conceptual project. Structural health monitoring of the bridge over Wisłok River in Rzeszów*. Wilde Engineering Sp. z o.o., (in Polish).

Advances in Mechanics: Theoretical, Computational and Interdisciplinary Issues – Kleiber et al. (Eds)
© 2016 Taylor & Francis Group, London, ISBN 978-1-138-02906-4

Numerically-based quantification of internal forces generated in a steel frame structure with flexible joints when exposed to a fire

M. Maślak & M. Pazdanowski
Cracow University of Technology, Cracow, Poland

M. Snela
Lublin University of Technology, Lublin, Poland

ABSTRACT: The distribution of internal forces generated in a steel frame structure subjected to fire is determined by the flexibility of joints connecting the structural members. Such flexibility, increasing with growing member temperature, may be modelled by the generalized component method, in which a special finite element is introduced in each joint, with properties corresponding to the structural response of the particular joint components. For small size frames the analogous analysis may be performed with the 3D finite elements applied to discretize the whole considered structure. Thus on a small scale a more precise approach may be used to calibrate and verify the parameters of a simplified model, which later shall be applied to analyse more complex structural systems. Identification of behaviour of a steel frame structure with flexible beam-to-column joints, subjected to a fire exposure, is the objective of the paper.

1 INTRODUCTION

Behaviour of steel frames subjected to fire is usually analyzed with beam elements applied to model the structural components (beams and columns). The generalized component method (Simoes da Silva et al. 2001) is then recommended for application to determine the nonlinear relationships of the $\mathbf{M} - \varphi$ type, accounting for the flexibility of joints increasing with the member temperature. However, more reliable results should be available with more precise numerical model, based either on the 3D modeling of only a joint, with remaining parts of the structure modeled using the classical beam elements, or on the full 3D modeling of the whole analyzed frame (Diaz et al. 2011). This latter approach should yield the most accurate results, but it may be applied only for the structures of limited size (Chen & Wang 2012). Identification of the behaviour of a steel frame with flexible beam-to-column joints subjected to a fire exposure, numerically modeled using the 3D analysis, is the objective of the paper.

2 MECHANICAL MODEL

In order to correctly estimate the fire resistance of a steel frame one should account for the joint rigidity decreasing with a developing potential fire. This means, that the joints designed initially, for persistent design situation, as nominally rigid, may become flexible during the fire exposure. Introduction of the statement of this type to the conventional computational procedure, oriented on the evaluation of both the fire resistance and the stability of the analyzed frame, results in the need to specify a special nodal element, such that its stiffness matrix $\mathbf{K}_c$ would constitute a solution of the equation:

$$\mathbf{F} = \mathbf{K}_c \cdot \mathbf{u} \tag{1}$$

where $\mathbf{F}$ = denotes the vector of the generalized internal forces generated in the joint and $\mathbf{u}$ = a vector of the generalized displacements identified for this joint. Due to the strong nonlinearity of the analysis associated with thermal action of a fire on the considered structure the equation (1) has to be solved iteratively, and the matrix $\mathbf{K}_c$ shall be interpreted as the tangent stiffness matrix $\mathbf{K}_c^*$. In such a case this equation is reduced to the following:

$$\Delta\mathbf{F} = \mathbf{K}_c^* \cdot \Delta\mathbf{u} \tag{2}$$

where, respectively:

$$\Delta\mathbf{F}^{\mathbf{T}} = [\Delta\mathbf{N}_{xj}, \Delta\mathbf{N}_{yj}, \Delta\mathbf{N}_{zj}, \Delta\mathbf{M}_{xj}, \Delta\mathbf{M}_{yj}, \Delta\mathrm{M}_{zj}, \Delta\mathbf{V}_{x,j}, \Delta\mathbf{V}_{y,j}, \Delta\mathbf{V}_{z,j}, \Delta\mathbf{M}_{x,j}, \Delta\mathbf{M}_{y,j}, \Delta\mathbf{M}_{z,j}] \tag{3}$$

$$\Delta\mathbf{u}^{\mathbf{T}} = [\Delta\mathbf{u}_i, \Delta\mathbf{v}_i, \Delta\mathbf{w}_i, \Delta\varphi_{x,i}, \Delta\varphi_{y,i}, \Delta\varphi_{z,i}, \Delta\mathbf{u}_j, \Delta\mathbf{v}_j, \Delta\mathbf{w}_j, \Delta\varphi_{x,j}, \Delta\varphi_{y,j}, \Delta\varphi_{z,j}] \tag{4}$$

Identification of the components of the matrix $\mathbf{K}_c^*$ is usually performed based on the classical component method generalized to the case of a fire. For this purpose the behavior of the joint, and especially its flexibility, is modeled with an appropriately selected set of springs, each of which corresponds to the conceptually isolated component of the considered joint. Each of such components is at the beginning assigned to the respective type of work, i.e. bending, compression, tension or shear. A scheme of a typical flexible end-plate joint connecting the beam with the column of a frame considered in the example is depicted in Figure 1, with a listing of the components determining its effective stiffness and the accompanying simplified mechanical model.

With the application of the formal model presented above to the persistent design situation, when no impact of a fire is considered, one obtains a single nonlinear $\mathbf{M} - \varphi$ relationship quantifying the dependence between the bending moment $\mathbf{M}$ applied to the joint and the resultant rotation φ in such a joint of the frame beam with respect to the frame column. In order to account for the temperature of the joint components increasing in fire, one should specify a whole family of such relationships, each of which would have to be determined as a solution of an isothermal problem at a set steel temperature (Maślak & Litwin 2010). It was shown previously (Maślak et al. 2015), that the $\mathbf{M} - \varphi$ curves relating to the fire situation may be generated in a simplified way, through the transformation of an a priori known reference curve determined for the room temperature. Such a method may, however, result in an overestimated joint rigidity, thus yielding an overestimated bearing capacity of the whole frame (Maślak & Snela 2014). A more accurate approach should be applied in this field, based on the generalization of the well-known component method. The first step is to identify the potentially possible failure modes of the joint components subjected to the foreseen fire, the second step—to estimate the bearing capacity of each component related to the considered failure mode (Spyrou et al. 2004). Such bearing capacity is later transformed into the parameters $\mathbf{k}_i$ characterizing the stiffness of the equivalent springs, modelling the behaviour of the *i*-th joint component under fire conditions. This parameter is, in general, a nonlinear function of the steel temperature Θ. Since the components of the stiffness matrix $\mathbf{K}_c^*$ are specified based on the characteristics $\mathbf{k}_i$ related to the previously identified joint components, they will change with increasing temperature of such components. However, the form of this matrix in the qualitative sense will not change.

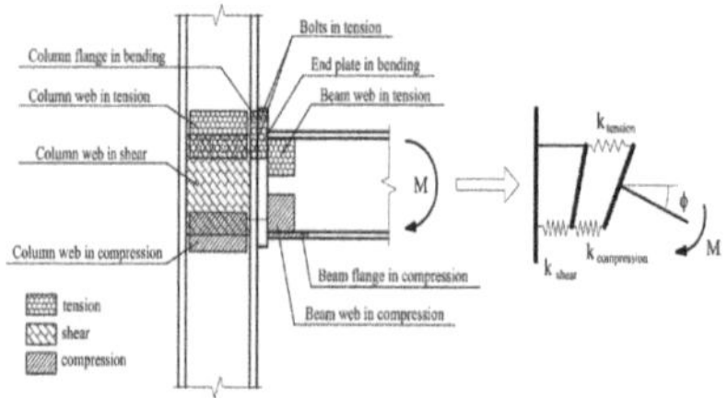

Figure 1. A typical steel end-plate beam-to-column joint, with components determining its rigidity—listed at left, an equivalent mechanical model of such the joint, prepared according to the rules of the component method—shown at right.

3 DESCRIPTION OF THE FRAME ANALYSED IN THE EXAMPLE

A structural response of a typical steel frame subcomponent, dimensioned and loaded as depicted in Figure 2, when exposed to a fire, was analyzed in detail. Two variants of the frame scheme, denoted with the symbols A and B, were analyzed separately. These variants differ in the way the column tops are supported. In the variant A (Fig. 2a) the top nodes of both frame columns may move freely in the plane defined by the undeformed structure. Thus both the horizontal translation and the vertical movement of these nodes in this plane are possible, as well as any combination of these displacements. However, both these nodes remain effectively constrained in the perpendicular direction, so as a result of this constraint they cannot displace out of the plane. Of course, the constraints imposed on the displacements of these nodes only do not prevent the potential risk of out of the plane buckling of the whole columns, as well as of other 3D forms of their instability. In the variant B (Fig. 2b) the top nodes of the frame columns may displace only in the vertical direction, while the displacements of these nodes in the horizontal direction are blocked. As in the case A, both nodes remain effectively constrained in out of the plane direction.

In both variants the connections between the beam and the columns are implemented via a flexible end-plate joint with four rows of bolts having the scheme and the dimensions as depicted in Figure 3.

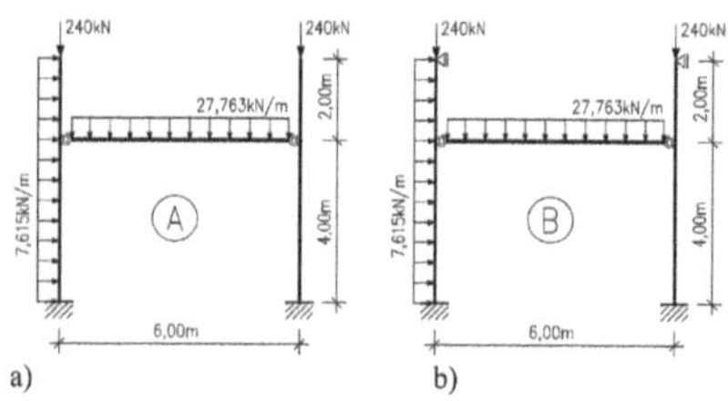

Figure 2. Variants of a steel frame substructure considered in the example (detailed description is given in the text).

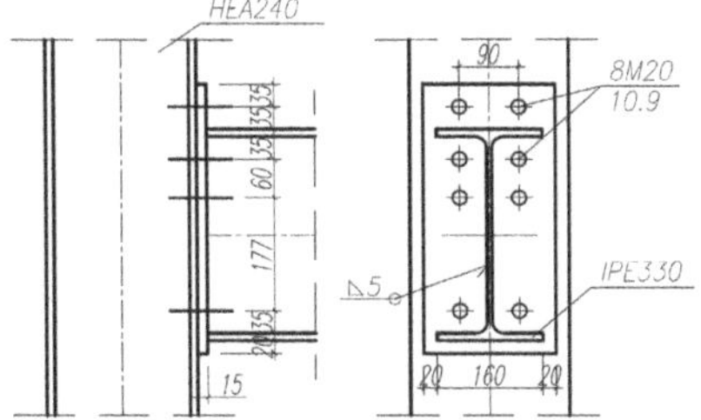

Figure 3. The end-plate beam-to-column joint used in both variants of a steel frame substructure considered in the example.

After application of external loads to the structural components the columns as well as the beam are subjected to a direct fire exposure. This exposure is conducted in such a manner, that in any given moment during the fire the steel temperature in all the components is constant and equal along the whole length as well as in any cross section of the considered structural members. However, the steel temperature increases monotonically, following the development of fire. One has to be aware, that the assumption of a proportional temperature growth in the steel structural elements constitutes a significant simplification of the analysis, since in the real fires the beams and the columns get heated with different speed because of having different cross sections.

4 BASICS OF THE NUMERICAL MODELING OF THE CONSIDERED FRAME

Detailed numerical analysis was performed to investigate the behaviour under fire conditions of the substructures depicted in Figure 2. The analysed frame was automatically subdivided into the 64025 three-dimensional finite elements (bricks and tetrahedra) without the mid-side nodes. Moreover, the average element size was separately adjusted to the dimensions of each meshed structural component, resulting in the components subdivided into between 300 (in the case of the bolts and the nuts) and 12800 (in the case of the lower part of a right frame column) elements. Regarding the material properties, the thermoplastic model of structural steel was applied to the analysis with basic parameters specified for the S235 low carbon steel and relating to the room temperature equal to 20°C. It was assumed that these steel properties, such as the yield limit, the longitudinal elasticity modulus as well as the coefficient of thermal expansion, change with temperature according to the standard EN 1993-1-2 (CEN 2005). Due to the expected large deformations during the thermal part of the analysis the total Lagrangian formulation was used for the incremental and the nonlinear computations. The loads were applied in two steps, i.e. at first—the whole static load was applied in 80 consecutive increments at the reference temperature equal to 20°C, resulting in the initial configuration of the deformed frame, to which the thermal load was applied in the second step, again in 80 identical increments, ending at evenly distributed steel temperature of the whole structure equal to 500°C. The iteration termination criteria at each load increment, both in the static and in the thermal part of the analysis, were assumed as the displacement only with the threshold value of 0.0005. In order to limit the calculation time, only the connections between the beam end-plates and the column flanges were assumed as the contact ones, allowing for the separation and for the relative displacement, while all the remaining connections between the other structural components were assumed as the bonded ones.

5 DISCUSSION OF THE OBTAINED RESULTS

Let us consider at first the situation before initiation of fire. If only the vertical loads were to act on the frame, then the beam ends would rotate around the bottom edge of the endplate (or around the axis of the bottom flange of the beam, depending on the geometrical properties of the joint components) as a result of the joint flexibility, and such joint would "open" at the top. On the other hand, if only the horizontal loads were to act on the column, then the beam ends would rotate around the top edge of the endplate (or around the axis of the top flange of the beam), with the joint "opening" at the bottom. The final rotation was then the combination of both loading modes mentioned above. Gradual heating of the frame components resulted in the diminishing rotation of this type, in consequence changing the direction. The knowledge of the horizontal displacements of the columns in the areas adjacent to the joints is important for the analysis of the redistribution of internal forces in the considered substructure. These displacements increase monotonically with the increasing steel temperature as a result of the thermal expansion of the beam pushing the both columns outwards. The rate of this expansion is high only in the initial stages of a fire. At higher values of the steel temperature the increase of the force pushing out is much less pronounced and, finally, the displacements of the beam are so large, that they compensate for this pushing out effect, pulling the joints together (Maślak & Snela 2013). The magnitude of the horizontal displacements in the column at the node determines the magnitude of the bending moment in such a column (Figs. 4–6).

Due to the specifics of the numerical analysis performed in search of the generalized internal

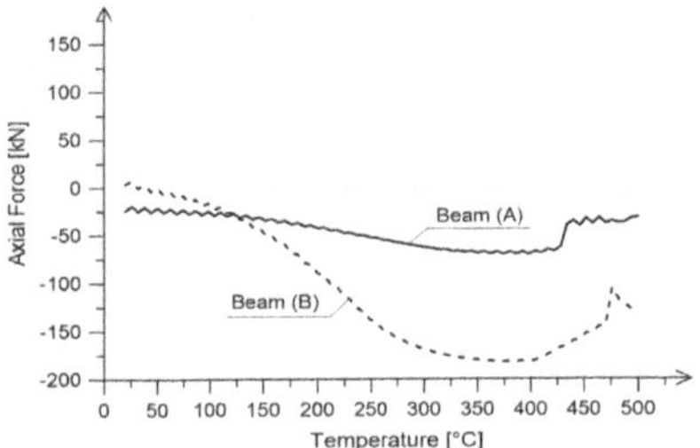

Figure 4. The axial force induced in the beam of the considered substructure.

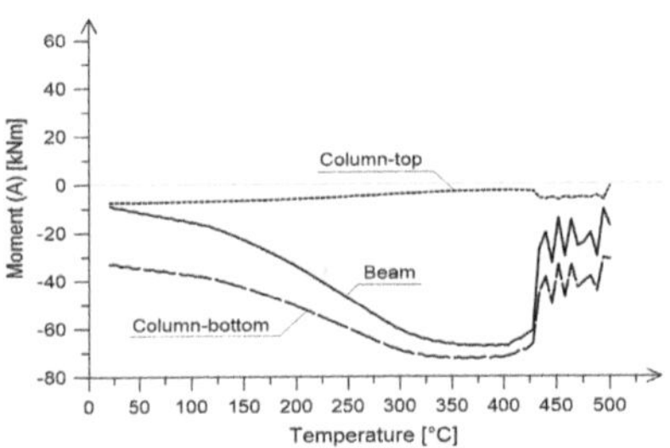

Figure 5. Redistribution of the bending moments in the cross sections adjacent to the joint, specified for the left joint of the substructure in the variant A.

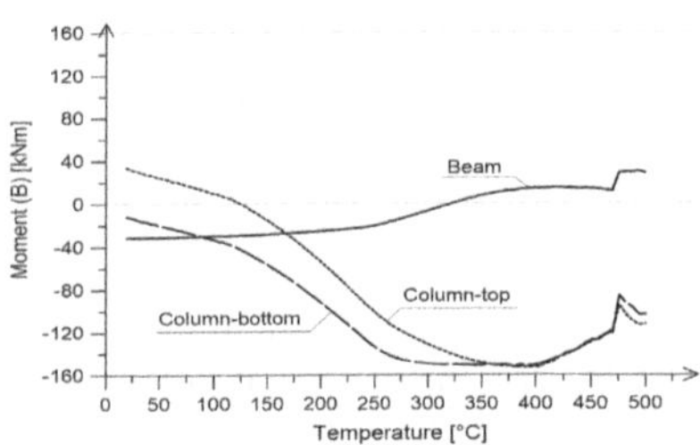

Figure 6. Redistribution of the bending moments in the cross sections adjacent to the joint, specified for the left joint of the substructure in the variant B.

forces, three transverse cross sections located respectively:

- for the beam—to the right of the left frame endplate joint,
- for the left bottom column—immediately below the joint,
- for the left top column—immediately above the joint,

are considered. Of course, the cross sections indicated above are offset by a small distance (approximately 25 cm) from the theoretical center of the joint, and thus the values of the generalized internal forces obtained in these cross sections may not be used directly to verify the equilibrium of the whole joint.

6 CONCLUDING REMARKS

The assessment of fire resistance of a steel frame will be more accurate if in the analysis pertaining to the fire situation the analyzed structure is considered as a whole, without isolating of not necessarily functionally or structurally homogenous subassemblies. Only such an approach allows for an identification of the phenomena which could not be observed in the analysis conducted in a conventional manner, with the application of only the simplified analytical models. Most of these phenomena are the inevitable consequence of the redistribution of internal forces, being specific for the accidental fire situation. As one can see in the example presented above, in the variant B the degree of the constraint imposed on the potential displacements of the frame components, generated in a fire due to the thermal expansion of structural steel, is substantially larger than in the variant A. This difference in an obvious manner results in the much larger amplitudes of internal forces induced in the considered structure when subjected to fire.

REFERENCES

Chen, L. & Wang, Y.C. 2012. Efficient modeling of large deflection behavior of restrained steel structures with realistic endplate beam/column connections in fire. *Engineering Structures* 43: 194–209.

Diaz, C., Marti, P., Victoria, M. & Querin, O.M. 2011. Review on the modeling of joint behavior in steel frames. *Journal of Constructional Steel Research* 67: 741–758.

EN 1993-1-2. 2005. Eurocode 3: Design of steel structures. Part 1–2: General rules. Structural fire design.

Maślak, M. & Litwin (Snela), M. 2010. Flexibility of beam-to-column steel joint under fire temperature (in Polish). *Inżynieria i Budownictwo* 8: 441–445.

Maślak, M., Pazdanowski, M. & Snela, M. 2015. Moment—rotation characteristics for flexible beam-to-column steel joint exposed to fire. *Journal of Civil Engineering and Architecture* 9(89): 257–261.

Maślak, M. & Snela, M. 2014. Alternative methods of identifying the moment—rotation characteristics related to fire conditions. *Journal of Civil Engineering, Environment and Architecture* 31(64): 135–145.

Maślak, M. & Snela, M. 2013. Axial force redistribution in a steel beam with thermal expansion capability increasing in fire. *Journal of Civil Engineering, Environment and Architecture* 30(60): 189–202.

Simoes da Silva, L., Santiago, A. & Vila Real, P. 2001. A component model for the behaviour of steel joints at elevated temperatures. *Journal of Constructional Steel Research* 57: 1169–1195.

Spyrou, S., Davison, J.B., Burges, I.W. & Plank, R.J. 2004. Experimental and analytical investigation of the "compression/tension zone" component within a steel joint at elevated temperatures. *Journal of Constructional Steel Research* 60: 841–896.

Advances in Mechanics: Theoretical, Computational and Interdisciplinary Issues – Kleiber et al. (Eds)

Numerical modelling of detonation in mining face cut-holes

L. Mazurkiewicz, J. Małachowski, P. Baranowski & K. Damaziak
Military University of Technology, Warsaw, Poland

W. Pytel & P. Mertuszka
KGHM Cuprum R&D Centre, Wrocław, Poland

ABSTRACT: One of the crucial factors determining the effective burden is an appropriate selection of geometry of cut-holes. Furthermore, leaving empty (uncharged) blasting holes may lead to enlargement of the fractured zone, and thus—improvement of the mining faces blasting effectiveness by reducing the number of blasting holes and/or extending the burden. The aim of the paper is to present a theoretical basis and preliminary results of numerical analyses carried out to determine the fractured zone caused by detonation of explosives. The authors used Arbitrary Lagrangian—Eulerian formulation for modelling the pressure wave propagation and its interaction with the mining face.

1 INTRODUCTION

The most common method for excavating copper ore in Polish mining industry is a blasting technique, which involves drilling holes in the mining face. An actual number of holes, their relative position and even charging sequence are the factors influencing effectiveness of the process. Any optimization of this process is incredibly difficult since acquisition of any field data different than an amount of the burden is nearly impossible. Therefore, an idea to implement numerical methods to simulate mining face fragmentation emerged. In the article, the authors present an approach to simulate blast wave propagation in the mining face. During investigations different patterns of holes were selected and their influence on the mining face damage, and consequently, fragmentation characteristic was analyzed. Furthermore, a detonation sequence implementation was also discussed.

2 NUMERICAL ASSUMPTIONS FOR PROBLEM REALIZATION

In the simulations the following assumptions were taken into account:

- The blast was modelled using Arbitrary Lagrangian-Eulerian formulation (ALE),
- Simulated model was a square with a side length of 3000 mm discretized using 360 000 elements,
- 2D plain strain was assumed, therefore simultaneous detonation of the entire depth was analyzed,
- Non-reflecting boundaries were applied (on the edges of the model),
- Riedel–Hiermaier–Thoma (RHT) material model was used for rock.

In the carried out study four different holes patterns were considered for selecting the most effective one. The first two patterns (Fig. 1a and Fig. 1b) are similar: they consist of 10 holes with 8 holes filled with explosive material. The third one (Fig. 1c) has 13 holes with 4 empty ones. The fourth (Fig. 1d) is different than others: it has 7 filled and 4 empty holes and they are distributed differently.

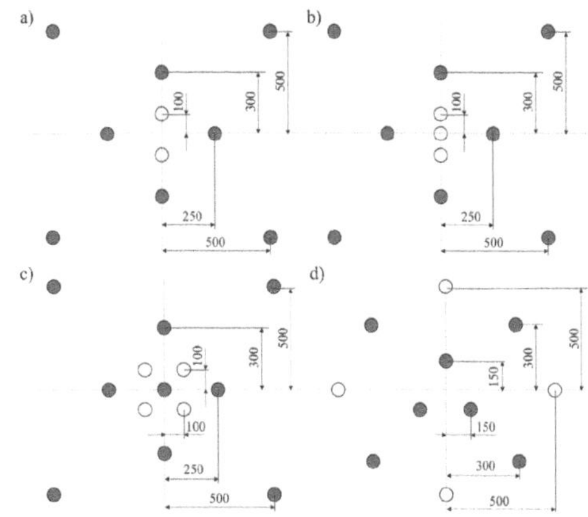

Figure 1. A pattern of holes located at the center of the mining face. White holes denote the empty ones (not filled with an explosive).

2.1 *Numerical simulations definition*

Numerical analyses were performed using explicit solver with central difference scheme and with the implementation of modified equation of motion time integration occur (Hallquist 2006). In this method acceleration and velocity at t time are given by:

$$\ddot{q}_t = \frac{1}{2\Delta t}\left[q_{t-\Delta t} - 2q_t + q_{t+\Delta t}\right], \tag{1}$$

$$\dot{q}_t = \frac{1}{2\Delta_t}\left[q_{t+\Delta t} - q_{t+\Delta t}\right]. \tag{2}$$

In the analyses carried out the stability of computations was achieved by Courant-Friedrichs-Lewy (CFL) condition, which can be described as follows (Hallquist 2006):

$$C = \frac{u_x \Delta t}{\Delta x} + \frac{u_y \Delta t}{\Delta y} + \frac{u_z \Delta t}{\Delta z} \leq C_{\max} \tag{3}$$

where u_x, u_y, u_z = velocities; Δt = time step; Δx, Δy, Δz = length intervals; C_{max} = varies with the method used (for highly dynamic cases this value should be set 0.66).

2.2 *Numerical blast modelling*

Out of many techniques of pressure wave propagation analysis, the authors focus on those involving discrete models. These methods are based on the Finite Element Method (FEM), which uses several algorithms for determination of dynamical loading of the structure, e.g. ConWep blast wave function, Smoothed Particle Hydrodynamics (SPH) method and Arbitrary Lagrangian-Eulerian formulation (ALE). In fact, the latter is adopted in the presented study. Such a choice is based on the previous study of several test models and the authors' experience in this research area (Mazurkiewicz et al. 2013). Exemplary investigations of the detonation process, pressure wave propagation and its interaction with structures can be found in many references (Ma & An 2008, Zhu et al. 2008, Malachowski 2010).

The ALE procedure contains two steps: a classical Lagrangian step and an advection step, with the following sub-steps included: relocation of nodes, recalculation of all variables as referred to elements and revaluation of the momentum and velocity updating. The advection step is carried out under the assumption that changes in the positioning of nodes are only slight (very small) in comparison to element size.

Detonation of the explosive is simulated using a burn model based on the function that determines the degree of burn of an explosive in the given finite element. This function consists of two parts. The first one—F_1 describing burning with known predetermined speed D_{CJ} and the second one—F_2, applicable where detonation velocity exceeds D_{CJ}. The bigger value of F_1 and F_2 is the final pressure multiplier resulting from the equation of state of the detonation products (Hallquist 2006):

$$F_1 = \begin{cases} \dfrac{2(t-t_1)D_{CJ}}{3L} & for\ t > t_1 \\ 0 & for\ t \leq t_1 \end{cases} \quad \text{and} \quad F_2 = \frac{1-V_r}{1-V_{CJ}} \tag{4}$$

where t = time; $t_1 = R_{det}/D_{CJ}$ moment of detonation; R_{det} = distance from the element centre to the point of detonation initiation; D_{CJ} = detonation velocity; L = characteristic length of element; V_{CJ} = volume of Capman-Jouguet point.

The blast wave generation was simulated describing the behavior of highly compressed detonation product after reaching successive locations by the detonation wave front. The Jones Wilkins Lee (JWL) equation of state with the following form was implemented (Hallquist 2006):

$$p = A\left(1 - \frac{\omega}{R_1\bar{\rho}}\right)\exp\left(-R_1\bar{\rho}\right) + B\left(1 - \frac{\omega}{R_2\bar{\rho}}\right)$$
$$\exp\left(-R_2\bar{\rho}\right) + \frac{\omega\bar{e}}{\bar{\rho}} \tag{5}$$

where $\bar{\rho} = \rho_{HE}\,\rho$; $\bar{e} = \rho_{HE} \cdot e$; ρ_{HE} = density of the high explosive; p = pressure; e = specific internal energy; ρ = density of detonation product; A, B, R_1, R_2, ω = empirical constants determined for specific type of an explosive material. All required constants (Table 1) were taken from literature Włodarczyk (1995).

2.3 *Rock modelling*

The mining face was modelled using a material model proposed by Riedel et al. (Riedel et al. 1999, Borrvall & Riedel 2011), known as RHT model. It is a macro-scale material model that incorporates features necessary for a correct dynamic strength description of concrete at impact relevant strain rates and pressures. The shear strength is described by means of three limit surfaces; an inelastic yield

Table 1. Material properties for HE with EOS (Włodarczyk 1995).

ρ (kg/m³)	D (m/s)	P_{CJ} (GPa)	A (GPa)	B (GPa)
1630	6930	21	371.2	3.23
R_1 (–)	R_2 (–)	ω (–)	E_0 (J/mm³)	
4.15	0.95	0.30	7	

surface, a failure surface and a residual surface, all dependent on the pressure value. The post-yield and post-failure behaviors are characterized by strain hardening and damage, respectively. Furthermore, the pressure is governed by the Mie-Gruneisen equation of state together with a p-α model (Equation 6) to describe the pore compaction hardening effects and thus give a realistic response in the high pressure regime (Hallquist 2006):

$$\alpha(t) = \max\left(1, \min\left(\begin{matrix}\alpha_0, \min \\ 1+(\alpha_0-1)\left[\frac{p_{comp}-p(t)}{p_{comp}-p_{el}}\right]^N\end{matrix}\right)\right) \quad (6)$$

In the above equation α represents porosity. The current pore crush pressure is given by the following formula (Riedel et al. 1999):

$$p_c = p_{comp} - \left(p_{comp} - p_{el}\right)\left[\frac{\alpha-1}{\alpha_0-1}\right]^{1/N} \quad (7)$$

where p_{comp} = compaction pressure; p_{el} = initial pore crush pressure; N = porosity exponent.

In Table 2 material parameters are presented for the RHT. Note, that only few are listed, due to automatic generation via ONEMPA card of other parameters for 35 MPa strength concrete (Riedler et al. 1999). These parameters (which are left as 0) are generated by interpolating between the 35 MPa and 140 MPa strength concretes. However, any of the automatically generated parameters can be overridden by the user, if necessary.

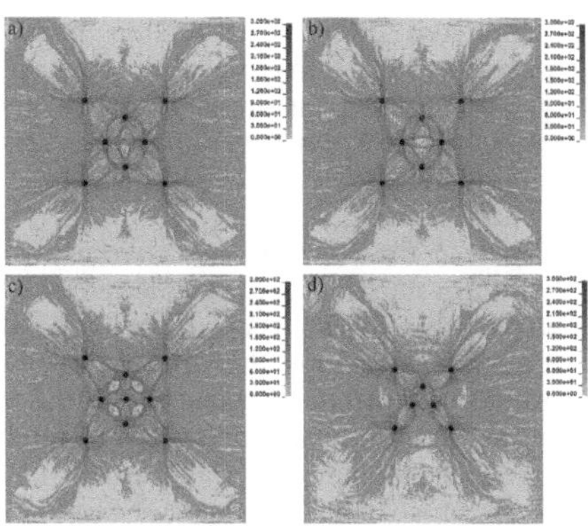

Figure 2. Pressure distribution [MPa]: a) pattern no 1, b) pattern no 2, c) pattern no 3, d) pattern no 4.

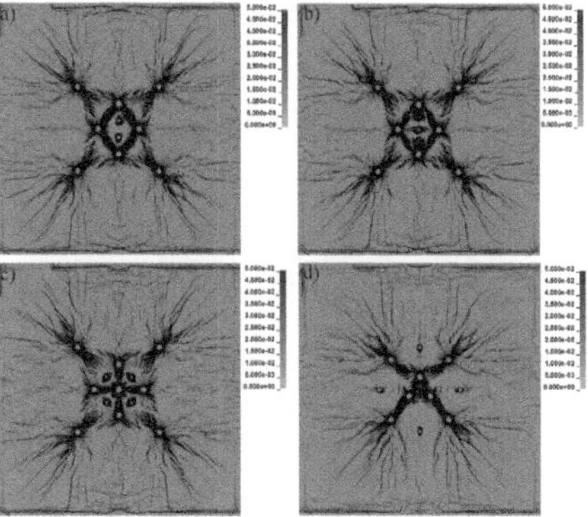

Figure 3. RHT model damage index: a) pattern no 1, b) pattern no 2, c) pattern no 3, d) pattern no 4.

3 RESULTS AND DISCUSSION

During the numerical investigations the following were analyzed:

- Pressure characteristic within the mining face,
- Damage characteristic within the mining face.

In Figure 2 pressure characteristic is presented for all four simulated cases. One can noticed similar behavior of its distribution over the mining face in three first patterns 1–3 (Fig. 2a–2c). The only difference in within the central area where various empty holes distribution was used. In Figure 2d results from pattern no 4 are shown. Here, the pressure characteristic is slightly different, generally due compaction of the holes in the central part of model. With increasing a distance from the center of the model pressure distribution becomes more similar to first three patterns.

In Figure 3 damage characteristic is presented also for all four simulated cases. The same conclusions rise as in the case of pressure. Three first patterns resulted in similar behavior of damage distribution over the mining face, whereas the fourth one gave slightly different results.

Table 2. Material properties for RHT material (Riedler et al. 1999).

ρ (kg/m³)	ONEMPA (-)	EPSF (%)	F_c (MPa)	F_s^* (MPa)	PTF (%)
2400	6930	21	371.2	3.23	0.001

3.1 *Simultaneous detonation vs. sequential detonation*

In order to improve the effectiveness of the blasting the authors decided to apply a detonation sequence in the model. For the simulation hole pattern no 1 was taken into account. The same initial—boundary conditions were applied, but with one major difference: different detonation times were used for holes with HE material. As previously, the RHT material model was implemented to model the rock behavior. Detonation sequence was as follows (Fig. 4). Time interval between the blast initiations was set to 0.05 s.

In Figure 5 the comparison between pressure distribution in the model is presented. The difference between two models is clearly seen: by applying sequential detonation the pressure as well as damage index characteristic is much more different. The authors assumptions were correct: by applying the sequential detonation it was possible to dramatically improve the effectiveness of blasting.

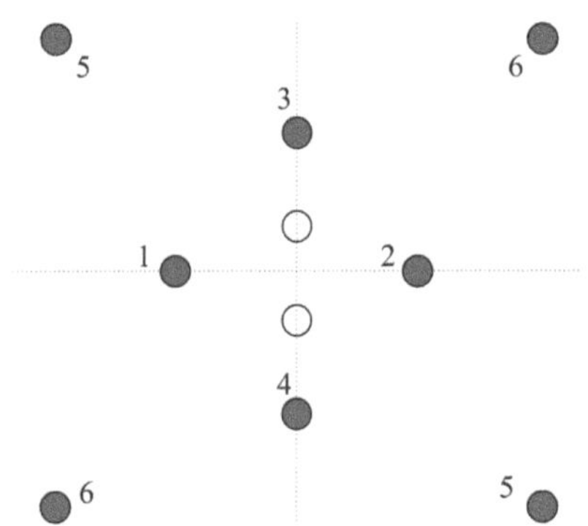

Figure 4. Detonation sequence applied in the model with pattern no 1: 1. Hole no 1 (left), 2. Hole no 2 (right), 3. Hole no 3 (upper), 4. Hole no 4 (bottom), 5. Holes no 5 (upper—left and bottom—right), 6. Holes no 6 (upper—right and bottom—left).

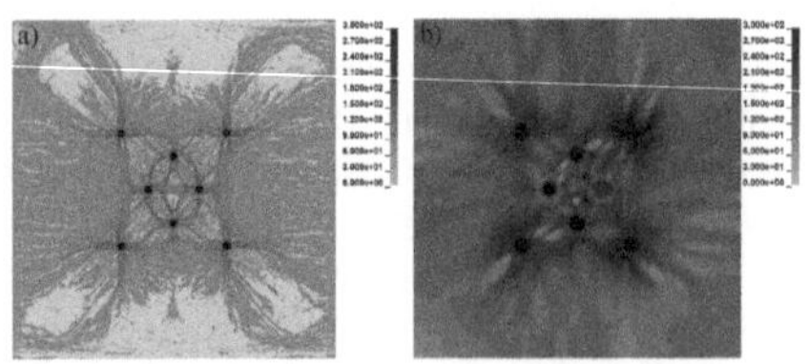

Figure 5. Pressure distribution comparison [MPa]: a) simultaneous detonation, b) sequential detonation.

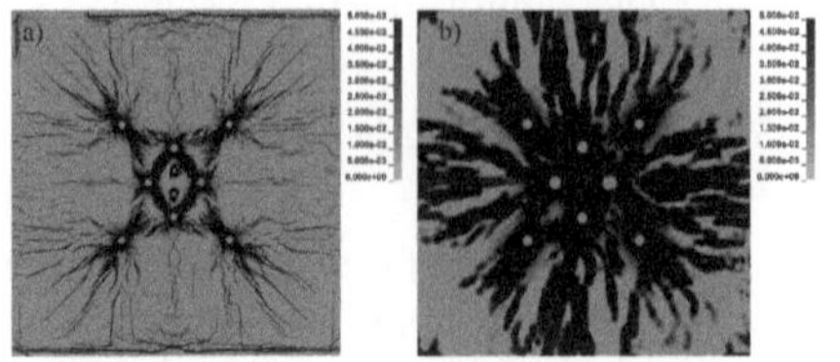

Figure 6. RHT model damage index comparison: a) simultaneous detonation, b) sequential detonation.

4 CONCLUSIONS

Preliminary results show that even a simplified numerical analysis can be very helpful in the selection of the most effective holes pattern. ALE procedure has proved its effectiveness and reliability in simulating the detonation in mining face cut-holes. However, the selected method is computationally expensive. Therefore, other methods of blast simulation must be tested and verified.

Currently, different methods (FEM, DEM, SPH) for rock modelling are tested in terms of effectiveness, reasonable computational time and correctness in simulating the fracture mechanism in material. Moreover, different material models for rock are also tested.

Additionally, the authors have undertaken further steps towards the acquisition of reliable material parameters that will guarantee a proper representation of rock material behavior during dynamic simulations. Quasi-static and dynamic tests will be carried out for different rock materials. For the dynamic testing Split Hopkinson pressure bar will be applied, which was used for other author's investigations (Baranowski et al. 2014).

REFERENCES

Baranowski P., Janiszewski J. & Małachowski J. 2014. Study on computational methods applied to modelling of pulse shaper in split—Hopkinson bar. *Archives of Mechanics* 66 (6): 1–14.

Borrvall, T. & Riedel, W. 2011. The RHT Concrete Model in LS-Dyna, *8th European LS-DYNA Users Conference, Proc. intern. conf., 23–24 May 2011.* Strasbourg.

Hallquist, J.O. 2006. *LS-Dyna. Theory manual.* California, Livermore: Livermore Software Technology Corporation.

Ma, G.W. & An, X.M. 2008. Numerical simulation of blasting-induced rock fractures. *International Journal of Rock Mechanics & Mining Sciences* 45: 966–975.

Małachowski, J. 2010. Modelowanie i badania interakcji ciało stałe-gaz przy oddziaływaniu impulsu ciśnienia na elementy konstrukcji rurociągu (in polish). Warsaw: BEL Studio.

Mazurkiewicz, Ł., Małachowski, J., Baranowski, P. & Damaziak, K. 2013. Comparison of numerical testing methods in terms of Impulse loading applied to structural elements. *Journal of Theoretical and Applied Mechanics* 51 (3): 615–625.

Riedel, W., Thoma, K., Hiermaier, S. & Schmolinske, E. 1999. Penetration of Reinforced Concrete by BETA-B-500, Numerical Analysis using a New Macroscopic Concrete Model for Hydrocodes. *Interaction of the Effects of Munitions with Structures; Proc. intern. symp., 3 July 1999:* 315–332. Berlin Strasberg.

Włodarczyk, E. 1995. *Fundamentals of detonation.* Warsaw: Military University of Technology.

Zhu, Z., Xie H. & Mohanty, B. 2008. Numerical investigation of blasting-induced damage in cylindrical rocks. *International Journal of Rock Mechanics & Mining Sciences* 45: 111–121.

Advances in Mechanics: Theoretical, Computational and Interdisciplinary Issues – Kleiber et al. (Eds)
© 2016 Taylor & Francis Group, London, ISBN 978-1-138-02906-4

Stress convergence in adaptive resolution of boundary layers in the case of 3D-based first- and higher-order shell models

Ł. Miazio
University of Warmia and Mazury, Olsztyn, Poland

G. Zboiński
Institute of Fluid Flow Machinery, Polish Academy of Sciences, Gdańsk, Poland
University of Warmia and Mazury, Olsztyn, Poland

ABSTRACT: The presented paper concerns stress convergence analysis in the case of the boundary layers within thin-walled structures. The structures under consideration are modeled with the 3D-based first-order Reissner-Mindlin model and the 3D-based higher-order shell models as well. We consider both parametric and adaptive convergence of various boundary stress components. The main objective of this research is to demonstrate that the appearance of elongated elements in the vicinity of the boundary does not destroy numerical stability of the solution of the problem. The elongated elements are generated in the second, modification step of the four-step adaptive strategy being our extension of the so-called Texas three-step strategy. The appearance of the elongated elements is a result of application of our numerical tools for a posteriori detection of boundary layers. Significance, intensity and division ratio assessment of boundary layers is made by means of the same tool. The application of the tools is followed by an operation of the corresponding mesh modification algorithms.

1 INTRODUCTION

In the research we investigate local convergence properties of the boundary stresses in the case of the edge effect occurring in thick- or thin-walled elastic structures described with 3D-based shell models. The 2D, both classical and hierarchical shell models were analyzed in this context in (Schwab & Suri 1996) and (Oden & Cho 1997), respectively.

The global convergence of the solution of the considered 3D-based models (Zboiński 2010) in the strain energy norm was investigated by the authors in Zboiński (2001) and Miazio (2013).

2 DETECTION AND RESOLUTION OF BOUNDARY LAYERS

2.1 *The detection and assessment tools*

In order to detect the boundary layer phenomenon a kind of sensitivity analysis is performed. The analysis is based on comparison of two local solutions obtained from the local meshes suitable for the cases of existence and lack of the phenomenon. The local problems are generated through the division of a couple of prismatic elements (Fig. 1) adjacent to the boundary. The division into four smaller elements is performed either arithmetically (Fig. 2) or exponentially (Fig. 3).

The solution of two four-element problems is performed within the domain V_c. The domain consists of two prismatic elements (marked 1 and 2 in Fig. 1). The pair of elements is adjacent to the boundary S of an analysed elastic body. The values of stress vector $\boldsymbol{r}$ on the whole boundary S_c of the domain V_c are taken from the global solution $\boldsymbol{u}^{hpq}$ corresponding to the initial mesh. In order to improve the quality of the calculation of stresses we calculate the inter-element stresses $\langle \boldsymbol{r} \rangle$ between the elements of 1 and 2 their external neighbours. The inter-element stresses are equilibrated as they are taken from the equilibrated residual method

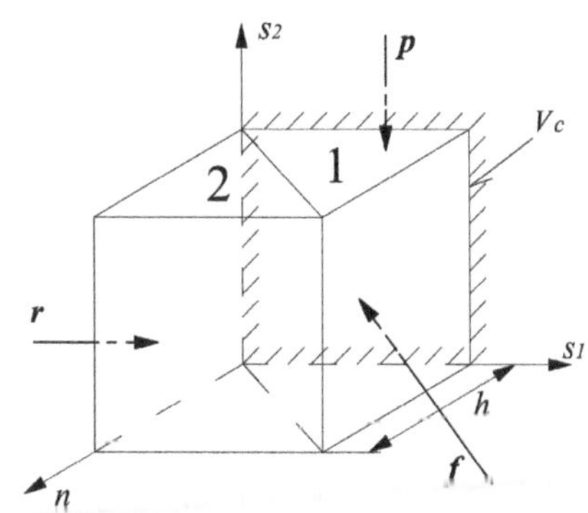

Figure 1. Mesh of two elements before the division.

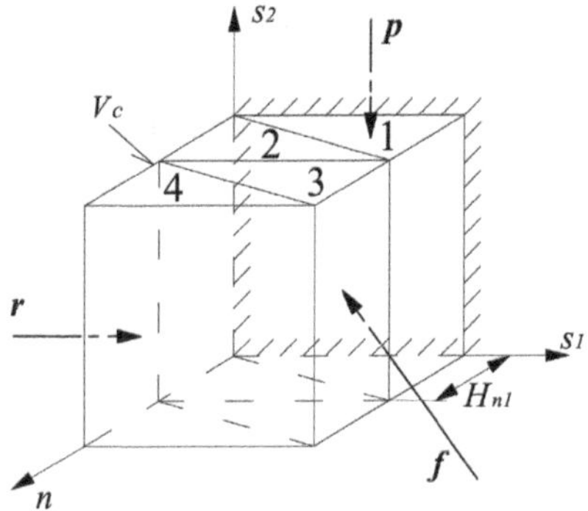

Figure 2. Arithmetic division mesh of the local problem.

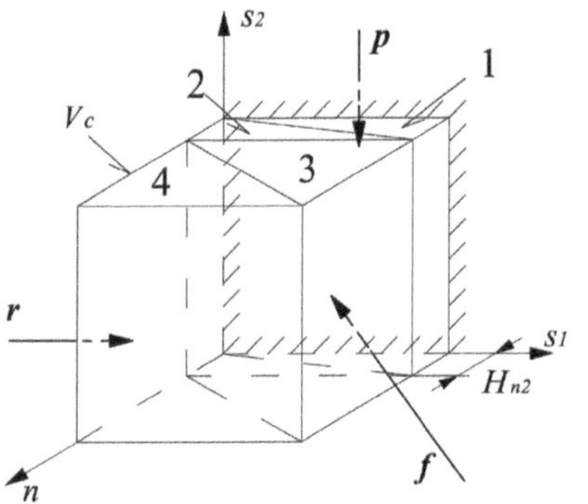

Figure 3. Exponential division mesh of the local problem.

of error estimation. The equilibration algorithm for the first- and higher order 3D-based elements can be found in Zboinski (2013). The equilibration means that the internal (inter-element) and external loadings acting in the two-element volume V_c and on the supported P_c or loaded Q_c parts of the two-element structure boundary are in force equilibrium.

The solution of two local problems is performed on the algebraic (Fig.2) and exponential (Fig. 3) four-element meshes. The division of the two elements into four is carried out in the direction n normal to the boundary S, without any changes in the tangential directions s_1 and s_2. Thus, the division numbers in these directions are $M_n = 2$ and $M_{s1} = M_{s2} = 1$. The division leads to two sub-domains $f = 1, 2$ with two elements in each sub-domain ($i = 1, 2$). The problems under consideration can be stated as:

$$\sum_{f=1}^{2}\sum_{i=1}^{M_n}\Bigg[B_{f_i}\left(\boldsymbol{u}_{f_i}^{HPQ}, \delta\boldsymbol{u}_{f_i}^{HPQ}\right) - L_{f_i}\left(\delta\boldsymbol{u}_{f_i}^{HPQ}\right) + \\ - \int_{S_c\cap S_{f_i}/P\cup Q} \delta\boldsymbol{u}_{f_i}^{HPQ^T}\left\langle \boldsymbol{r}_{f_i}\left(\boldsymbol{u}_{f_i}^{hpq}\right)\right\rangle dS_{f_i}\Bigg] = 0 \qquad (1)$$

Above, B represents the bilinear form corresponding to the virtual strain energy, L denotes the virtual work of external forces, and the vector $\boldsymbol{u}^{HPQ}$ stands for the element displacements. The local problem indices H, P and Q are the element size, and the element horizontal and vertical orders of approximation.

In the case of detection and significance assessment of the phenomenon only one solution for the exponential mesh is taken into account, while in the case of intensity and optimal division ratio assessments the proper sequences of the exponential meshes are considered (Zboiński 2001, Miazio & Zboiński 2011, 2012a, 2012b, 2014, Miazio 2013).

The above relation (1) can be also expressed in the language of finite elements. Using such a notation two four-element ($e = 1, 2, 3, 4$) local problems can be written in the form

$$\sum_e k^e q^e = \sum_e (f_M^e + f_S^e + f_R^e), \qquad (2)$$

with:

$$\boldsymbol{k}^e = \int_{V_e} \boldsymbol{B}^{e^T}\boldsymbol{D}\boldsymbol{B}^e dV_e, \qquad (3)$$

$$f_M^e = \int_{V_e} N^{e^T} f\, dV_e, \qquad (4)$$

$$f_S^e = \int_{S_e\cap P_e} N^{e^T} p\, dV_e, \qquad (5)$$

$$f_R^e = \int_{S_e\setminus(P_e\cap Q_e)} N^{e^T}\langle r\rangle\, dV_e. \qquad (6)$$

In the above equations $\boldsymbol{k}$ = the stiffness matrix of the element e, $\boldsymbol{q}$ = denotes the element displacement vector, $\boldsymbol{f}_M$ and $\boldsymbol{f}_S$ = stand for the element force vectors due to volume and surface loadings $\boldsymbol{f}$ and $\boldsymbol{p}$, while $\boldsymbol{f}_R$ = is the force vector from the inter-element stresses $\langle\boldsymbol{r}\rangle$. The symbols V_e and S_e denote the element volume and boundary surface, while P_e and Q_e are the loaded and supported parts of S_e.

It is worth emphasizing that the static equilibrium of the four elements of the local problem is analogous to their static state in the global problem.

2.2 *Our adaptive modification of the mesh*

The modification (second) step of the four-step adaptive procedure consists in exponential subdivisions of the layers of elements adjacent to the boundaries affected by the edge effect. Each such a boundary can be modified independently. Additionally, the number of subdivisions and the exponential subdivision ratio can be different for each boundary. The details of the mesh modification procedures can be found in Miazio (2013), where the standard Texas three-step, *hp*-adaptive strategy of Oden (1993), shown in Figure 4, was completed with the modification step (from A to B). This step (Fig. 5) consists of the edge effect detection and the initial mesh modification (only in the case of phenomenon detection).

Additionally, it should be mentioned that the boundary layer phenomenon may cause a loss of the asymptotic convergence during the *hp*-adaptation. The main task of our modification step is to remove the mentioned loss and to restore the asymptotic convergence before the *h*- and *p*-steps of the adaptive procedure are performed. As the convergence theories, relating the error level to the discretization parameters *h* and *p*, exist for the asymptotic range it is important to have the solution in such a range before the error-controlled adaptation starts.

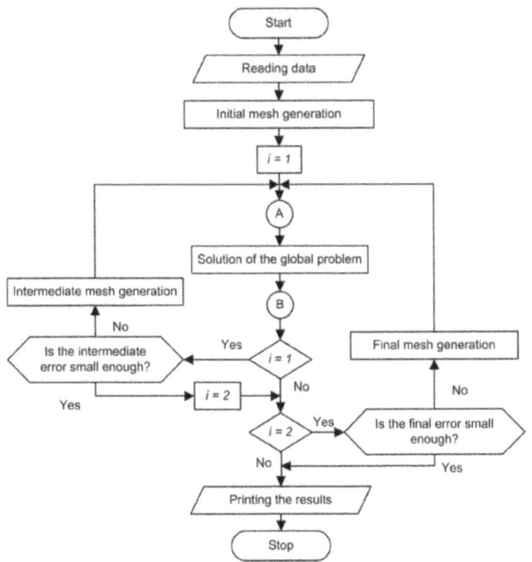

Figure 4. Block diagram of the standard *hp*-adaptive method.

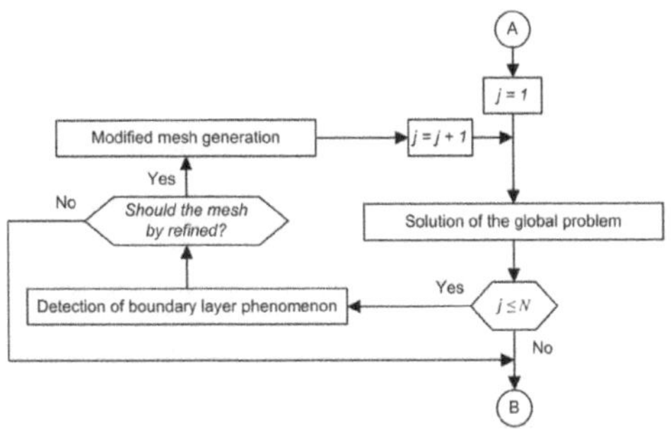

Figure 5. Block diagram of the improved *hp*-adaptive method with the additional modification step.

3 NUMERICAL TESTS

Our numerical experiments concern the model problem of a square plate (Fig. 5). Because of the symmetry of the plate geometry, loading and boundary conditions only a quarter of the plate is analyzed. The plate length is $l = 1.57075 \cdot 10^{-2}$ m, while its thickness is equal to $t = 0.03 \cdot 10^{-2}$ m. The Young modulus is $E = 2.11 \cdot 10^{11}$ N/m^2, the Poisson's ratio equals $\nu = 0.3$. The plate is loaded with the uniform normal traction $p = -4.0 \cdot 10^6$ N/m^2 and clamped around its lateral boundaries. We apply the first-order and hierarchical shell models of the transverse approximation orders q equal to 1 and 2, respectively.

3.1 *Parametric studies*

In the presented tests we apply initial, regular, 3 × 3 mesh shown in Figure 7. Numbering of the finite elements of this mesh are presented in the Figure 7.

In the tests concerning local convergence of the boundary stress components we modify the starting mesh by dividing it exponentially towards the outer boundaries of the plate. Such modification is performed in the layers of elements adjacent to two outer edges of the plate. The division concerns all elements in these layers. The dimensionless division ratio ρ describes the size ratio of the elements obtained through the modification. The divisions are shown in Figures 2–3. In the case $\rho = 1$ division into halves occurs (Fig. 2), while $\rho = 10$ and $\rho = 100$ correspond to the exponential divisions (Fig. 3). For each of three cases we perform *p*-convergence analysis, where *p* stands for the longitudinal order of approximation. The values of this order were set to 4, 6 and 8. We checked how all six stress components, and the effective stress as well, change at the centroid of the element lateral face adjacent to the outer boundary of the plate. The elements under consideration were numbered as 8 and 16 (Fig. 7). The presented results concern the hierarchical shell model and are presented in Figures 8–9 for the cases of the effective σ_{ef} and normal σ_{nn} stresses. We skip the 3D-based first-order model results as they are qualitatively the same.

The results presented in the figures show good convergence properties of the displayed stress

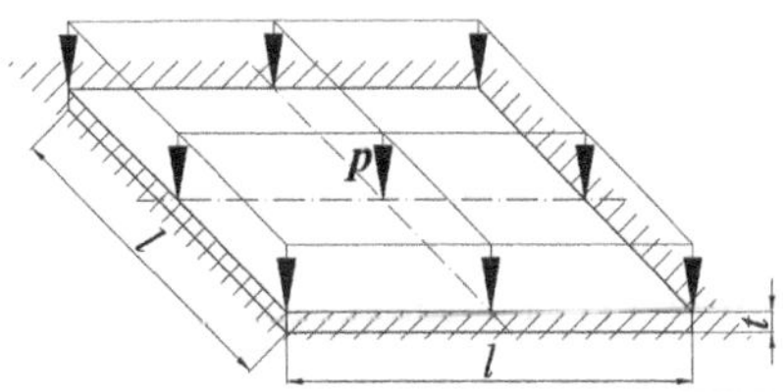

Figure 6. A load and support of a square plate.

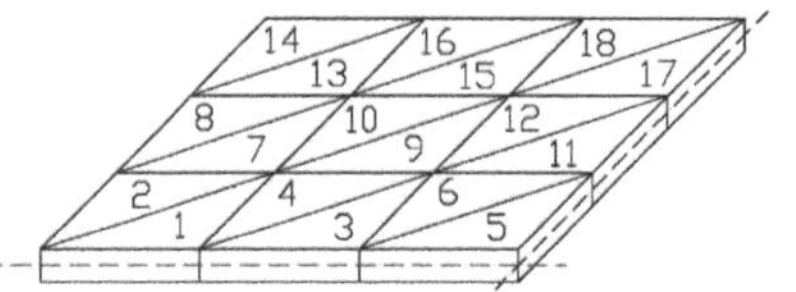

Figure 7. The global numbering of elements (3 × 3 regular mesh).

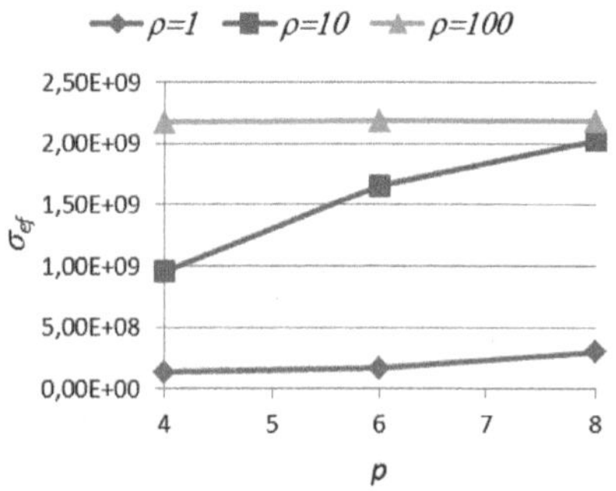

Figure 8. Effective stress convergence at the outer boundary.

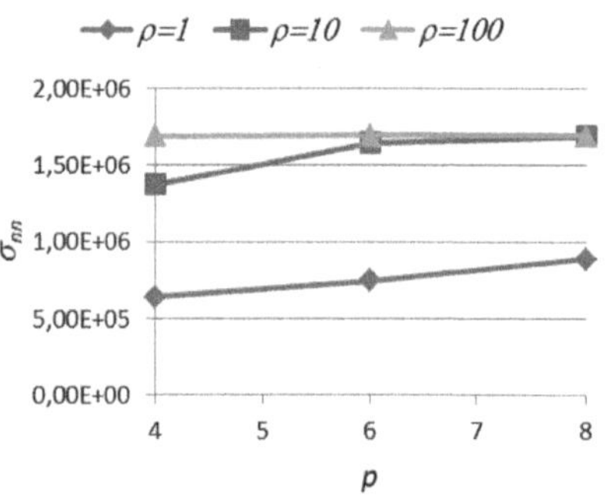

Figure 9. Normal stress convergence at the outer boundary.

components with the increasing values of ρ and p. The other stress components exhibit analogous behavior.

3.2 *Adaptive analysis*

We have performed the *hp*-adaptive analysis of the same model plate problem, so as to check the convergence of stresses in the case of four-step strategy. The strategy has included as a second step exponential modification of the mesh in the vicinity of the boundaries. The obtained results confirmed stable convergence properties of each individual stress component and the effective stress as well.

4 CONCLUSIONS

The results obtained in the convergence studies of stresses on the boundaries affected by the edge effect confirmed the anticipated numerical stability of the solution.

The presence of the elongated elements in the vicinity of the boundaries does not destroy the solution convergence both in the global sense (in terms of the strain energy of the plate) and in the local, boundary sense (in terms of the boundary stress values).

Our results are in good agreement with the analogous results obtained by other researchers in the case of the 2D classical and hierarchical shell models.

REFERENCES

Cho, J.R. & Oden, J.T. 1997. Locking and boundary layer in hierarchical models for thin elastic structures. *Comput. Methods Appl. Mech. Engrg*. 149: 33–48.

Miazio, Ł. 2013. *An Analysis of Elastic Structures with Boundary Layers by Means of hp-Adaptive Finite Element Method* (in Polish). Ph.D. Thesis (supervisior—G. Zboiński), University of Warmia and Mazury, Olsztyn.

Miazio, Ł. & Zboiński, G. 2011. The numerical tool for a posteriori detection of boundary layers in *hp*-adaptive analysis, *Short Papers of CMM 2011*: 351–352.

Miazio, Ł. & Zboiński, G. 2012a. A posteriori detection of boundary layers in the numerical problems of bending-dominated shells (in Polish). *Modelowanie Inżynierskie* 43: 177–184.

Miazio, Ł. & Zboiński, G. 2012b. Application of the numerical tools for boundary layer detection in adaptive plate and shell analysis (in Polish). *Mechanik* 7: 451–458.

Miazio, Ł. & Zboiński, G. 2014. *hp*-Adaptive finite element analysis of thin-walled structures with use of the numerical tools for detection and range assessment of boundary layers. *Recent Advances in Computational Mechanics*: 57–62.

Oden, J.T. 1993. Error estimation and control in computational fluid dynamics, The O.C. Zienkiewicz Lecture. *Proc. Math. of Finite Elements*—MEFELAP VII: 1–36, Brunnel Univ., Uxbridge.

Schwab, C. & Suri, M. 1996. The *p* and *hp* versions of the finite element method for problems with boundary layers. *Math. Comput.*, 65: 1403–1429.

Zboiński, G. 2001. *Hierarchical Modelling and Finite Element Method for Adaptive Analysis of Complex Structures* (in Polish). Sci. Instal. of IFFM, No. 520/1479, IFFM, Polish Academy of Sciences, Gdańsk.

Zboiński, G. 2010. Adaptive *hpq* finite element methods for the analysis of 3D-based models of complex structures. Part 1. Hierarchical modeling and approximations. *Comput. Methods Appl. Mech. Engrg*. 199: 2913–2940.

Zboiński, G. 2013. Adaptive *hpq* finite element methods for the analysis of 3D-based models of complex structures. Part 2. A posterior error estimation. *Comput. Methods Appl. Mech. Engrg*. 267: 531–565.

Advances in Mechanics: Theoretical, Computational and Interdisciplinary Issues – Kleiber et al. (Eds)

Experimental tests and numerical simulations of full scale composite sandwich segment of a foot- and cycle-bridge

M. Miśkiewicz, K. Daszkiewicz, T. Ferenc, W. Witkowski & J. Chróścielewski
Faculty of Civil and Environmental Engineering, Gdańsk University of Technology, Gdańsk, Poland

ABSTRACT: In the paper experimental tests and numerical simulations of a full cross-section segment of ultimately designed foot- and cycle-bridge are presented. The experimental tests were conducted on an element with length reduced to 3 m and unchanged (target) cross-sectional dimensions. The external skin of structure is GFRP laminate while internal core is PET foam. Several quasi-static tests were performed using hydraulic cylinder to generate vertical loads, and using specially designed system of rods to generate horizontal loads applied to the structure transversally and longitudinally. Results of the numerical simulations conducted in the FEM environment are shown and serve as the basis for the design process of target foot- and cycle-bridge.

1 INTRODUCTION

The purpose of the project *FOBRIDGE* is to design and manufacture a single span foot- and cycle-bridge (Fig. 1) made of composite materials and manufactured as one element in infusion process. The project was conducted in consortium by Gdańsk University of Technology, Military University of Technology in Warsaw and private company ROMA Sp. z o. o (Chróścielewski et al. 2014a).

In order to design innovative structures some experiments needs to be done. These can be divided into two groups: experiments that determine and identify material parameters for laminates and foams and experiments that validate the correctness of assumptions made during numerical simulations. In the paper validation tests of a numerical model of full cross-section segment are presented.

2 VALIDATION SEGMENT

Validation segment was constructed as a three-layer sandwich structure with external skin of GFRP laminate (0,663 mm) and internal core of PET foam (100 kg/m^3, 100 mm thick). The segment was made without any external protective layers and with specially reinforced support area. Preparation of the segment served also as a test stage of the infusion process assumed in the project. The cross-section for segment and target foot- and cycle-bridge are the same. Total length of the segment is 3 m, while theoretical span length was established as 2,5 m. The usable width is 2,59 m, the height of handrails is 1,3 m, which was assumed according to Polish law regulations for crowd and cycle traffic. The cross-section is not symmetric—the wall (a) has an extra foam core (10 mm) additionally inserted into parapet. The segment was placed on four rubber bearings of dimensions 300 × 300 × 30 mm, Figure 2. Material parameters of rubber were determined in separate experiments.

3 CONDUCTED EXPERIMENTS

Several quasi-static tests were performed for the composite sandwich segment (Chróścielewski et al. 2014b). The load schemes were classified into four categories, the example draft of each is

Figure 1. Visualization of foot- and cycle-bridge.

Figure 2. The validation segment—front view, A4 test.

presented in Figure 3: group A (Fig. 3a)—vertical load applied by hydraulic cylinder on platform or the top of walls (A1–A7), group B (Fig. 3b)—horizontal load applied on the top of walls transversally simulating walls bending over support (B1) or support (B2), group C (Fig. 3c)—horizontal load applied on the top of walls longitudinally simulating compression of wall (a) (Ca) or wall (b) (Cb), group D (Figs. 3d and 4)—combination of above loads (i.e. D3b = A3 + Cb). From the number of the test conducted, we present here only one dynamic scheme S1 with crowd loading (Fig. 5).

The load schemes B, C and D were conducted using specially design system of fixing elements and rods. On the base of preliminary FEM numerical simulations the load value for linear range was estimated. In calculations the failure index according to Tsai-Wu hypothesis (Chróścielewski et al. 2013, Chróścielewski et al. 2014c) for skin and compression principal stress for foam were controlled.

Several measurement points were used during experiments: 60 strain gauges (location of selected gauges are presented in Fig. 6), 18 displacements sensors (Fig. 7) and 4 acoustic sensors.

The load value was applied to segment in many steps (see example of load scheme A4 in Figs. 2 and 8), in order to verify whether the creep of sandwich structure appears.

During validation tests of the numerical model (FEM), the values in strain gauges were monitored (see Fig. 9) and compared with preliminary numerical results.

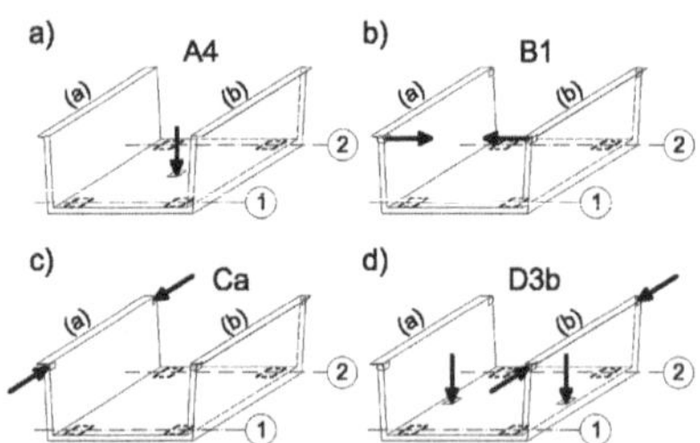

Figure 3. Draft of conducted load schemes.

Figure 4. The example of combined load designated as D3b.

Figure 5. Load scheme S1 caused by crowd.

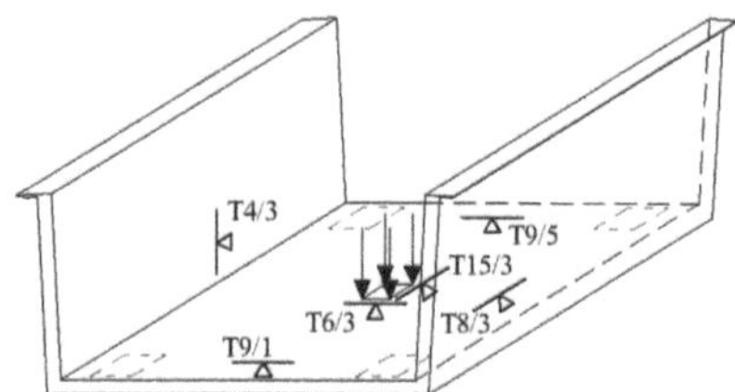

Figure 6. Location of selected strain gauges, scheme A4.

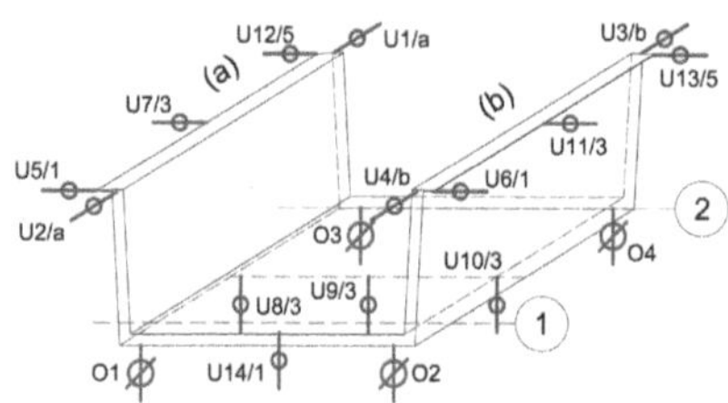

Figure 7. Location of displacements sensors.

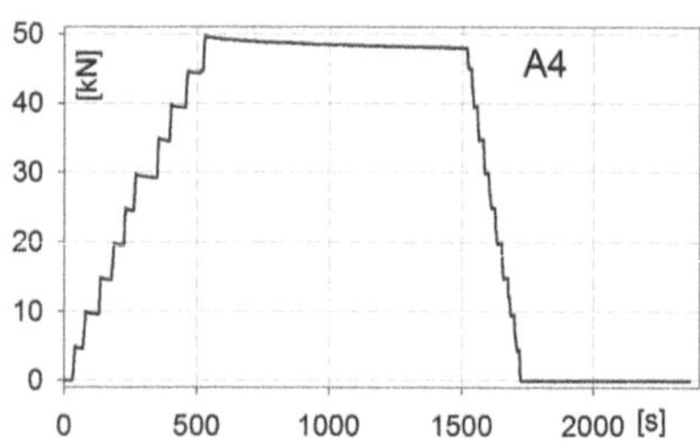

Figure 8. Force-time function, scheme A4.

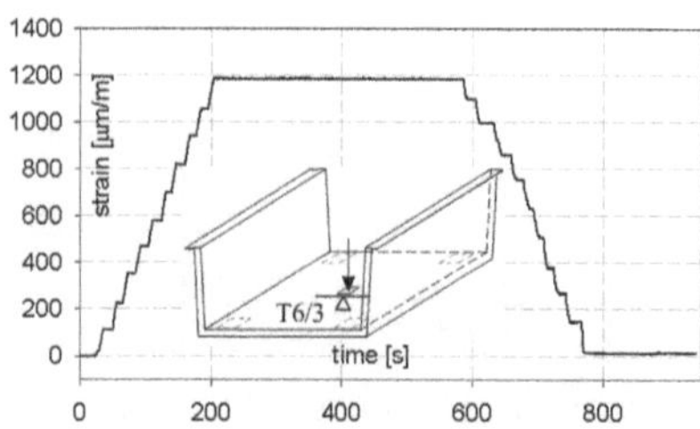

Figure 9. Strain-time function in strain gauge T6/3, scheme A4.

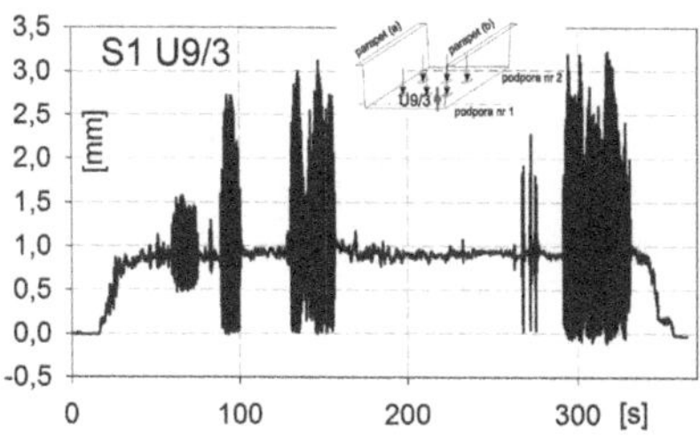

Figure 10. Displacement-time function in inductive sensor U9/3, scheme S1.

The example of displacement occurred during the load scheme S1, generated by crowd loading, is shown in Figure 10. The presented graph shows that static displacement caused by load is about 1 mm, while the crowd jumping displacement increases three times, to 3 mm.

4 NUMERICAL SIMULATIONS

The numerical simulations were performed in ABAQUS 6.14. The external laminate was assumed as the shell using Equivalent Single Layer approach. S4 ABAQUS shell element was used. In the range of the deformation for which the structure is designed for the inner foam core was assumed as linear elastic isotropic material and was meshed using C3D8 continuum elements. To account for incompressible behavior of the rubber bearings the corresponding mesh regions were meshed using C3D8H continuum finite elements. All the calculations were carried out in the geometrically non-linear analysis. Throughout the model, the characteristic element size of 20 mm was used yielding 5 finite elements across foam core thickness in the deck and 6 finite elements in the vertical walls. This amounted to 293887 nodes and 343428 elements. The ABAQUS skin feature was used to connect the laminate with the foam, which allowed for conforming mesh of shell and continuum elements. Material parameters determined in the previous experiments (Klasztorny et al. 2013) or values given by producers were assumed in the numerical analyses. In order to assure correspondence between experiment and the numerical simulation the load was applied as uniformly distributed pressure (Fig. 11) or through a modeled steel beam (Fig. 12). Hard contact condition in normal direction was assumed in the simulations. In tangential direction 0.5 friction coefficient was used.

Here detailed analysis of a load scheme A4 (Figs. 3a, 6, 11) is presented. In order to obtain more precise results for the numerical model, after validation tests the support zones (i.e. material parameters of the bearings) were modified (updated). The comparison of numerical and experimental values of vertical displacements at 4 sensors O1, O2, O3, O4 (Fig. 7) is presented in Table 1.

For the maximum load valuc 50 kN/ 300 × 300 mm numerical and experimental values of strains and displacements are compared in Tables 2 and 3, respectively. In addition

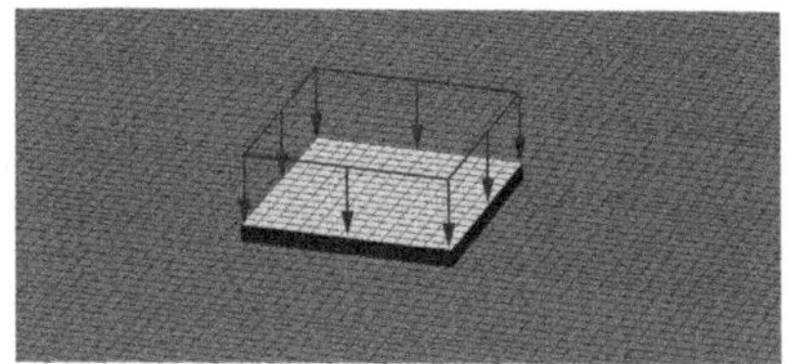

Figure 11. Detail of load application method in A4 load case.

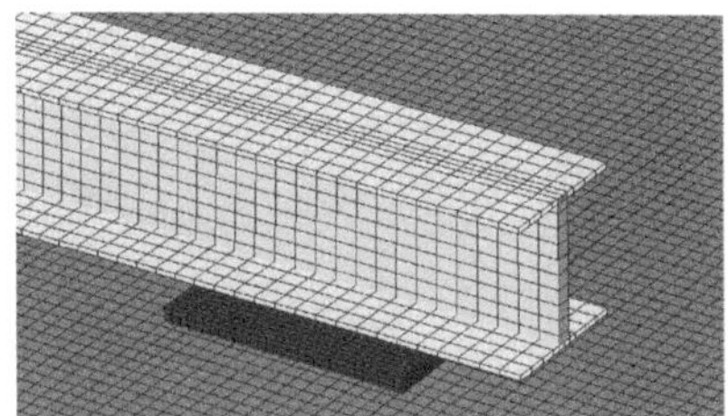

Figure 12. Detail of steel beam transmitting the load in A6 load case.

Table 1. Comparison of numerical and experimental vertical displacements at the support zones.

Displacement sensor	Experimental (mm)	Numerical (mm)
O1	1.36	1.35
O2	1.40	1.40
O3	1.16	1.16
O4	0.99	0.98

Table 2. Comparison of numerical and experimental values at displacement sensors.

Displacement sensor	Experimental (mm)	Numerical (mm)
U6/1	11.19	12.90
U7/3	11.10	12.92
U8/3	1.96	1.20
U9/3	14.60	16.68
U10/3	1.42	1.16
U11/3	11.21	13.12
U13/5	11.49	13.30

Table 3. Comparison of numerical and experimental strains.

Strain gauge	Experimental (μm/m)	Numerical (μm/m)
T9/1	668	635
T4/3	137	160
T6/3	1240	1589
T8/3	167	184
T15/3	835	760
T9/5	585	631

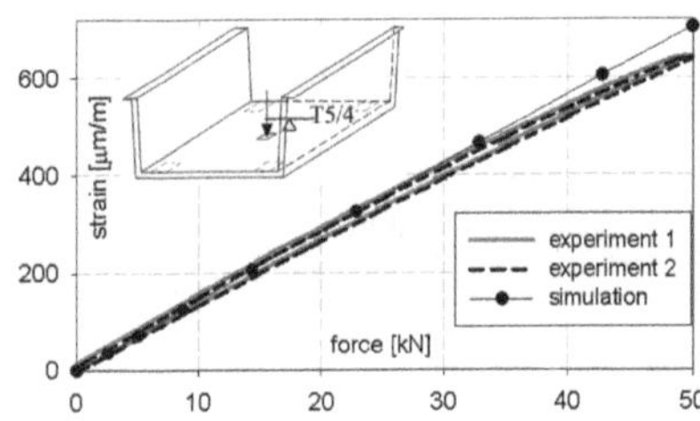

Figure 13. Strain-force relation in strain gauge T5/4, scheme A4.

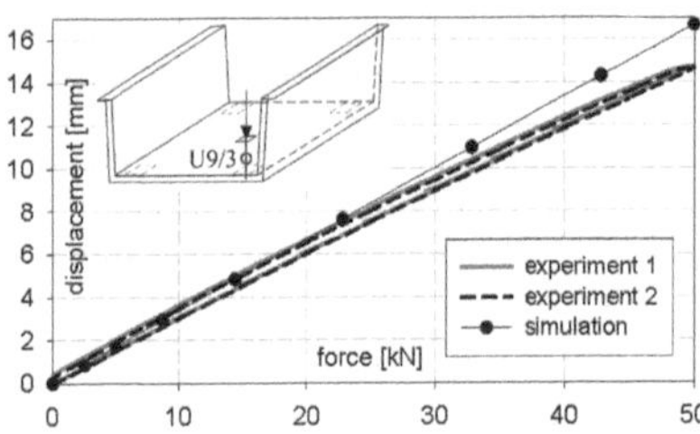

Figure 14. Displacement-force relation in inductive sensor U9/3, scheme A4.

Figures 13 and 14 show the comparison of two experimental curves with numerical curve for strain gauge T5/4 and inductive sensor U9/3, respectively.

Once the tests were conducted the segment was cut to investigate the structure of the sandwich, especially in the support zones. No signs of delamination or debonding were found.

5 SUMMARY

In the paper some experimental tests and numerical simulations of full scale composite sandwich segment of a foot- and cycle-bridge have been presented. The validation of numerical model is a difficult task, due to complexity of structure and great importance of the support and load zones in load-carrying process. The experimental values of strains and displacement are in good agreement with numerical results. Some discrepancy between results may be caused by nonlinearly varying contact surface between external faces of the segment and rubber bearings (Figs. 13 and 14). In April 2015 at Gdańsk University of Technology a full scale bridge with span length of 14 m was erected. During 6 months it is tested (Fig. 15) in many ways and monitored by SHM system (Wilde et al. 2013).

Figure 15. Target foot- and cycle-bridge on Gdańsk University of Technology during load tests.

ACKNOWLEDGEMENTS

The study is supported by the National Centre for Research and Development, Poland, grant no. PBS1/B2/6/2013. Abaqus calculations were carried out at the Academic Computer Center in Gdańsk.

REFERENCES

Abaqus 6.14-2 Documentation.

Chróścielewski, J., Kreja, I., Sabik, A., Sobczyk, B. Witkowski, W. 2013. Failure analysis of footbridge made of composite materials. Proceedings of 10th SSTA Conference, Gdańsk, Poland, 16–19 Oct. 2013, CRC Press, Taylor & Francis Group, Balkema, 389–392.

Chróścielewski, J., Klasztorny, M., Wilde, K., Miśkiewicz, M., Romanowski, R. 2014a. Composite foot- and cycle-bridge with sandwich construction, *Materiały Budowlane*, 7/2014, pp. 1–2 (in Polish).

Chróścielewski, J., Miśkiewicz, M., Pyrzowski, Ł., Rucka, M., Ferenc, T. 2014b. Experimental tests on validation element of composite foot- and cycle- bridge, *Materiały Budowlane,* 4/2015, pp. 72–73 (in Polish).

Chróścielewski, J., Klasztorny, M., Nycz, D., Sobczyk, B. 2014c. Load capacity and serviceability conditions for footbridges made of fiber-reinforced polymer laminates. *Roads and Bridges* 13, pp. 189–202.

Klasztorny, M., Gotowicki, P., Nycz, D. 2013. Identification tests of new soaked composite BG/F in temperature 20°C, Report 6-2013v4 of project Fobridge, (in Polish).

Wilde, K., Miśkiewicz, M., Chróścielewski, J. 2013. SHM System of the Roof Structure of Sports Arena "Olivia", *Structural Health Monitoring 2013*, Fu-Kuo C. Ed., DEStech Publications, Inc., Vol. II, pp. 1745–1752.

Advances in Mechanics: Theoretical, Computational and Interdisciplinary Issues – Kleiber et al. (Eds)
© 2016 Taylor & Francis Group, London, ISBN 978-1-138-02906-4

Numerical analysis and in situ tests of the Grot-Rowecki Bridge in Warsaw

M. Miśkiewicz, Ł. Pyrzowski, K. Wilde & J. Chróścielewski
Faculty of Civil and Environmental Engineering, Gdańsk University of Technology, Gdańsk, Poland

ABSTRACT: The paper presents the FEM analysis of the reconstructed the Grot-Rowecki Bridge over Vistula river in Warsaw. The bridge has seven spans and consists of two independent structures with the longest spans of 120 m. After reconstruction the bridge is over 10 m wider. The numerical, nonlinear analysis has been conducted on a global and local FEM models. These models are defined so as they strictly comply with global structural bridge behaviour and satisfy fundamental rules of mechanics. The analysis on detailed models makes it possible to have insight into the issues like the local accumulation and concentration of stresses, local stability or local dynamic factors possible. The overview of the in situ tests conducted on the bridge is presented.

1 INTRODUCTION

The Grot-Rowecki Bridge (Fig. 1) over Vistula River is one of the most important bridge in Warsaw since it is located within the road connecting the east with the west of Poland.

The bridge was built in 1981 and designed in accordance to the code PN-66/B-02015. Before reconstruction the bridge had 2 × 4 lanes and it allowed to cross Vistula river hundreds of thousands of cars every day. The new bridge consists of two independent steel 7-span structures with static scheme of continuous beam and total length of 646 m ($L_t = 75 + 3 \times 90 + 2 \times 120 + 60$ m). The longest spans with length of 120 m are designed as box superstructure. Other spans have steel girders.

According to the modernization plans of entire S-8 express road, the bridge was not wide enough and had insufficient strength, therefore it had to be rebuilt. The major structural change, in the new bridge, is addition of pre-stressed cables that allowed to widen each structure by truss girders with additional lane and a footpath (Fig. 2).

Figure 1. The Grot-Rowecki Bridge over Vistula river in Warsaw after modernization.

2 FEM MODELS

In order to verify a new bridge capacity and its agreement with design assumptions the FEM calculations were performed using two levels of the structure description—global and local ones (Wilde et al. 2014). Moreover, due to the exact description of tendons and associated raised forces some analyses were performed in geometrically nonlinear format. The purpose of the analysis at the global level was to check the overall behaviour of the entire structure and to determine an envelope of generalized internal forces. In the global model, where full length of the structure has been mapped, shell finite elements have been used to describe bridge plate, webs of girders and the bottom plate. Other structural elements were modelled by beam elements (Fig. 3). The local simulations provide detailed information concerning the stress distribution in the selected areas of the structure. The local approach is focused on the elements in question i.e. orthotropic deck plate with a detailed description of ribs and nodes of newly added truss members.

On the basis of the global description of the bridge three zones of orthotropic deck plate have been selected and analysed locally. In these models shell finite elements were used to describe bridge plate, top plate ribs in the impact zone of load, webs of girders and traverses and the bottom plate. Other structural elements were described with beam elements. The most complex detailed model of the bridge section consists of about 1.5 million degrees of freedom (Fig. 4).

In order to validate the agreement of the local formulations with global approach a number of numerical tests have been carried out for

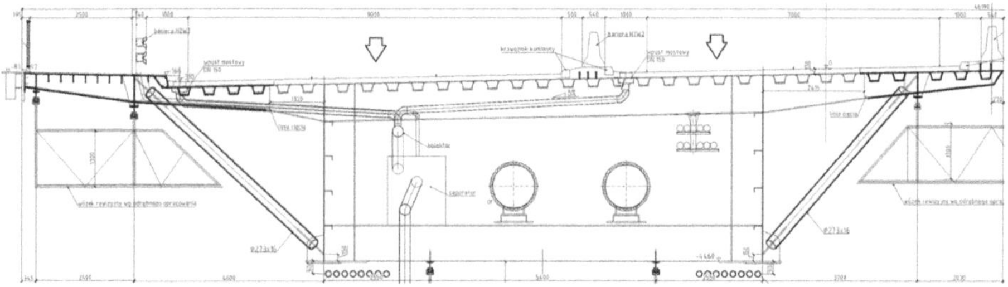

Figure 2. Redesigned cross section of the bridge. Old part of the structure in grey. New parts in black.

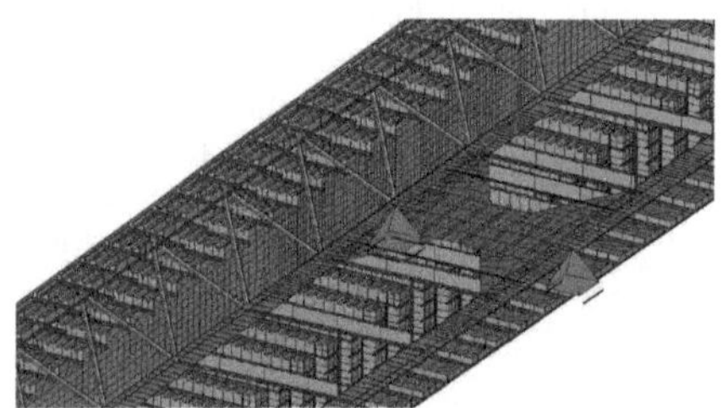

Figure 3. Visualization of a fragment of the global FEM model.

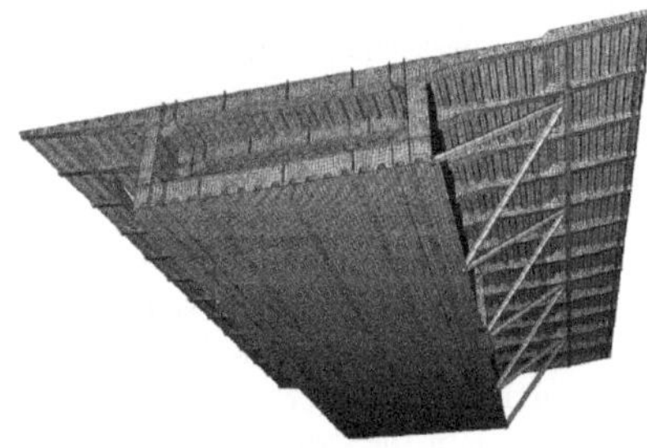

Figure 4. Visualization of one of the local FEM model.

Table 1. Cross-sectional internal bending moment in span 6–7.

Load	Global model (kNm)	Local model (kNm)	Global/local (%)
Dead weight	76606	77029	99
Structural accessories	91813	95430	96
Distributed live load q	111727	114404	98
Vehicle live load K	23145	23231	100

each detailed model. They aimed at verification of compliance of resultant cross-sectional internal forces as well as the verification of the stress distribution in the regions described using the same FE types in global and local approach

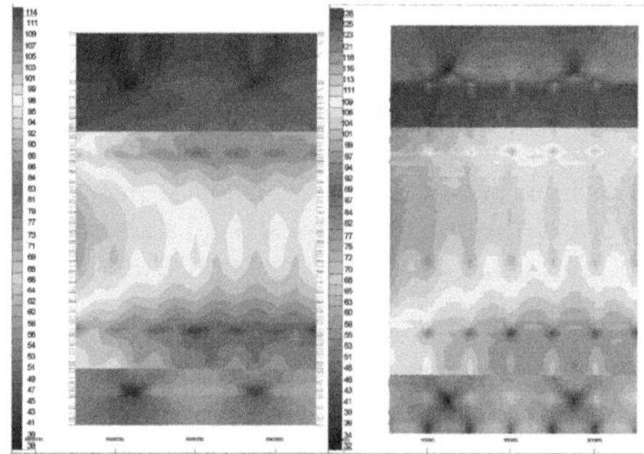

Figure 5. Plate stress distribution in unserviceable state obtained in global FEM model (left) and local one (right).

(Chróścielewski et al. 2004). The example verification of compliance for internal bending moment in half-length of span 6–7 is shown in Table 1 and for plate stress distribution in unserviceable state in Figure 5. The results show a very good agreement between the data obtained from the global and local models.

3 SELECTED RESULTS FOR BRIDGE DECK

The detailed deck analysis addresses the local stress distributions and their agreement with design assumptions. There has been a concern that the loads caused by the wheels of the code vehicle K may lead to dangerous, local increase of stresses in deck plate and in ribs located directly below the vehicle wheel. The extreme values of stress for the two types of FEM models are presented in Table 2.

The obtained results indicate redistribution of stresses in the bridge deck plate. The significant decrease of peak stress in the ribs with respect to the results obtained in the global approach is observed. Finally it is noted that in some critical configuration of loads the local stresses in steel may exceed the limit values by 6.7% for the plate

Table 2. Numerical results.

Part of the deck	Global model (MPa)	Local model (MPa)
Plate	186	239
Rib	390	244
Traverse	115	161

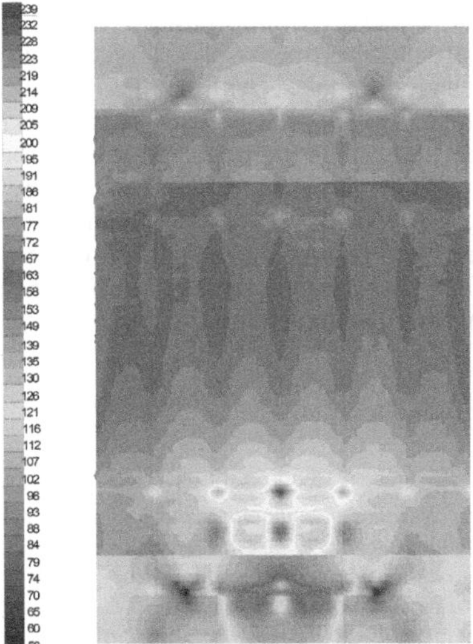

Figure 6. Plate stress distribution in serviceable state obtained in local FEM model.

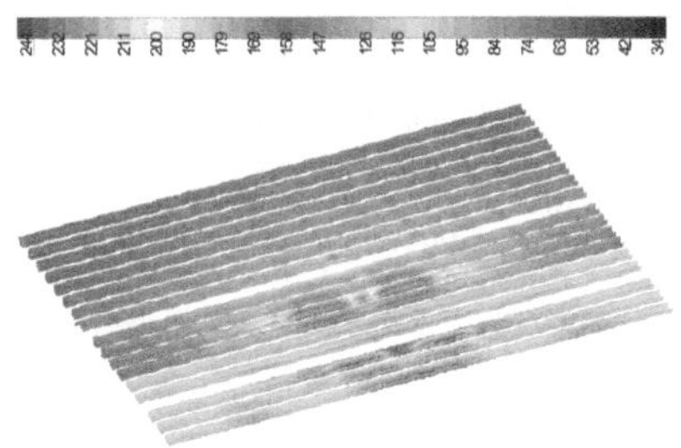

Figure 7. Ribs stress distribution in serviceable state obtained in local FEM model.

and 8.9% for the ribs. On the other hand, it should be noted that the stress concentrations are local effects (Figs. 6 and 7) due to the code vehicle K and they will never occur during bridge operation under real loads. Finally it has been decided that the deck load capacity is sufficient and the bridge is correctly designed. It was suggested that the safety level of the bridge might be increased by installation of Structural Health Monitoring system (SHM) Wilde et al. (2013).

4 IN SITU LOAD TEST

According to the Polish bridge design code road bridges with spans over 20 m and all railway bridges, before they are given to public use, require in situ load tests. The goal of the experimental work is to check the agreement of the true structural bridge behaviour (static and dynamic) with the design assumptions. The analysis is based on comparison of the measurement data with FEM results (Miśkiewicz 2008). The numerical model should faithfully reflect all important structural elements, material selections, structural member joints, proposed construction phases and structure sensitivity to external factors. In case of the Grot-Rowecki Bridge both global and local models are used to verify its theoretical behaviour on the experimental data.

Figures 8 and 9 show one of the in situ load case conducted with use of four axle trucks of weight 332 kN each. First, numerical analysis of bridge behaviour under the load of 30 trucks has been conducted. In the next step field tests were carried out and the real response of the structure was measured. The strains and displacements of the bridge under static and dynamic loading were recorded. Global response of the north part of the bridge presented by displacements in the middle of the measured spans and its comparison to the computed values is presented in Table 3.

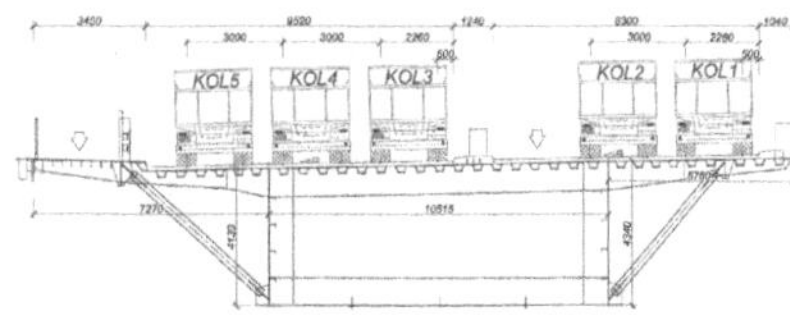

Figure 8. Cross-section view on one of the static load case of the Grot-Rowecki Bridge.

Figure 9. Static load tests of the Grot-Rowecki Bridge.

Table 3. The bridge maximum displacements in the middle of the spans loaded during tests.

Span between axes	Measured displacement (mm)	Calculated displacement (mm)	Measured/ computed (%)
4–5	96.4	111.0	87
5–6	128.2	143.5	89
6–7	119.8	136.5	88

The relative errors between the measured and numerical results are about 10%. During the dynamic tests accelerations and displacements at representative points on the bridge deck were recorded. On the basis of recorded signals and FFT the eigenfrequencies were estimated. The comparison of theoretical and measured values for south lane of the bridge is presented in Table 4.

The measured values of displacements, stresses and accelerations are in a very good agreement with the corresponding numerical values.

In addition to the global measurements also the detailed local measurements of the stress distribution in the nodal truss plates and orthotropic slab were conducted (Figs. 10 and 11).

Table 4. The comparison of theoretical and measured values of eigenfrequencies for south lane of the bridge.

Eigenfrequency	Theoretical (Hz)	Measured (Hz)
1	0.81	0.85
2	1.06	1.16
3	1.23	1.52
4	1.42	1.85
5	2.16	2.15
6	2.44	2.80

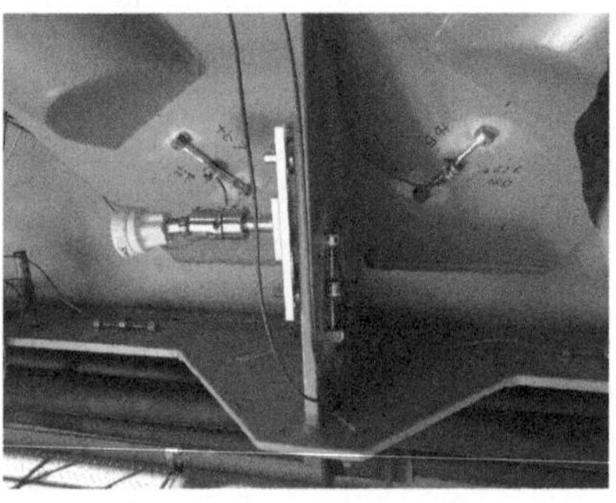

Figure 10. Points of strain measurements (vibrating wire sensors) located on the truss girder node.

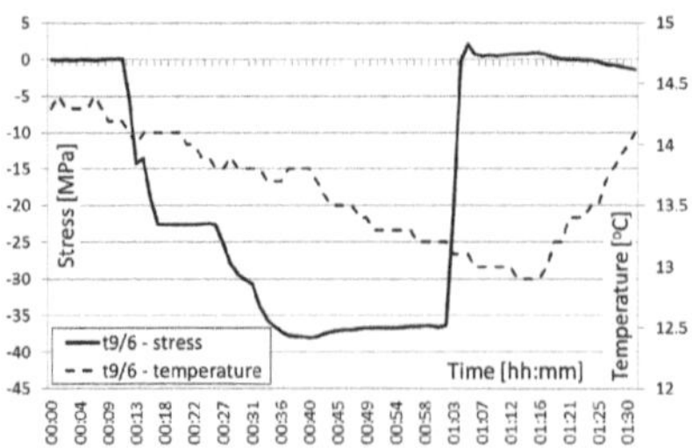

Figure 11. Stress and temperature change during U3 load setting (span 6–7) in measurement point located on the rib of orthotropic plate. Extreme theoretical value is –40.5 MPa giving agreement between in situ test and calculations about 94%.

5 CONCLUSIONS

In bridge design practice, in order to perform the numerical analysis of the structure, the beam grid scheme is commonly used. However, the results from the theoretical analysis and the in situ measurements of the Grot-Rowecki Bridge in Warsaw highlight the need of applying more advanced numerical models for getting better understanding of the real behaviour of the structure. This statement is particularly important for the analysis of local structural elements of a large bridge structure. The local analysis with detailed FEM models must take into account need to satisfy the fundamental rules of the mechanics in relation between the local and global model. In case of the Grot-Rowecki Bridge, a very good agreement between the numerical results, for both the global and local approach, and the data recorded during the in situ tests has been obtained.

REFERENCES

Chróścielewski, J., Makowski, J., Pietraszkiewicz, W. 2004. *Statyka i dynamika powłok wielopłatowych: Nieliniowa teoria i metoda elementów skończonych.* Biblioteka Mechaniki Stosowanej, IPPT PAN Warszawa (in Polish).

Miśkiewicz, M. 2008. Test loadings of arch bridges. *Current scientific challenges in concrete and steel structures and material technology.* Kaiserslautern Technische Univ.: 57–64.

Wilde, K., Miśkiewicz, M., Pyrzowski, Ł., Chróścielewski, J., Nadolny, A. 2014. Most im. gen. Grota-Roweckiego w Warszawie—analizy pomostu techniką "trepanacji". *Archiwum Instytutu Inżynierii Lądowej* 18: 231–240 (in Polish).

Wilde, K., Miśkiewicz, M., Chróścielewski, J. 2013. SHM System of the Roof Structure of Sports Arena "Olivia". *Structural Health Monitoring 2013.—Vol. II/ ed. Fu-Kuo Chang, Alfredo Guemes Lancaster, Pennsylvania 17602 U.S.A.: DEStech Publications, Inc.*: 1745–1752.

Advances in Mechanics: Theoretical, Computational and Interdisciplinary Issues – Kleiber et al. (Eds)
 ISBN 978-1-138-02906-4

Numerical analysis of the carpentry joints applied in the traditional wooden structures

A. Mleczek & P. Kłosowski
Gdańsk University of Technology, Gdańsk, Poland

ABSTRACT: The paper concerns the numerical analysis of the carpentry joints made of spruce wood. The material parameters of the wood have been calculated on the basis of nine independent material constants (an orthotropic material). The contact zone between the individual elements of the connection has been determined using Contact Tables in the MSC.Marc software. Five types of carpentry joints have been analyzed. The main aim of the research is selection of the areas of the greatest stress values in the planes of contact, and thus the determination of vulnerable damage zones.

1 INTRODUCTION

Nowadays solid timber structures are not so common as steel or concrete. This is essentially due to the properties of material and the price of wooden constructions. Nevertheless, there are many wooden buildings, mostly historical, requiring maintenance, renovation and the reinforcement of existing elements. This paper is focused on carpentry joints used in traditional, mainly historical wooden log walls. Unfortunately, the lack of the literature in the field mechanical behaviour of such connections is still observed. The engineering studies of carpentry joints are carried out primarily during historical analyses and roof constructions using photoelasticity (Gogolin 2006, Jasieńko et al. 2005, Jasieńko et al. 2006). These analyses are related to the connections with the use of a wooden mandrel, but these studies are different examples of the solution compared to traditional carpentry joints, which are the subject of the current paper.

The topic of carpentry joints is also extremely important from the point of view of growing interest of architects and engineers in wood structures or wooden components.

2 GEOMETRY

Five types of carpentry joints have been analyzed: long-corner dovetail connection, short-corner dovetail connection, cross connection, tooth lock connection and hidden-tooth lock connection. The geometry of each joint has been presented in Figures 1–5.

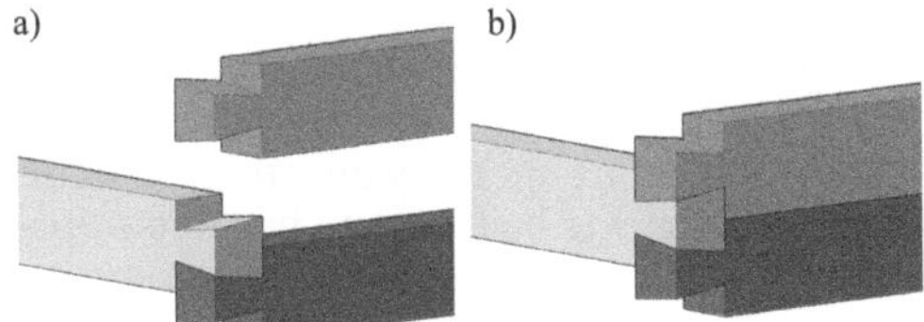

Figure 1. Long-corner dovetail: a) notch—visibility of one log geometry, b) connection.

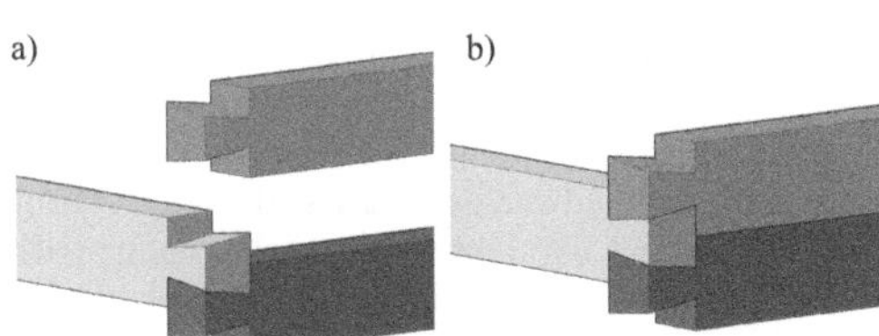

Figure 2. Short-corner dovetail: a) notch—visibility of one log geometry, b) connection.

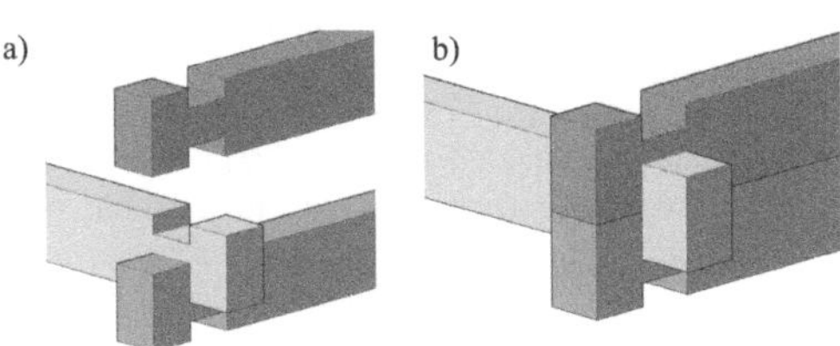

Figure 3. Cross: a) notch—visibility of one log geometry, b) connection.

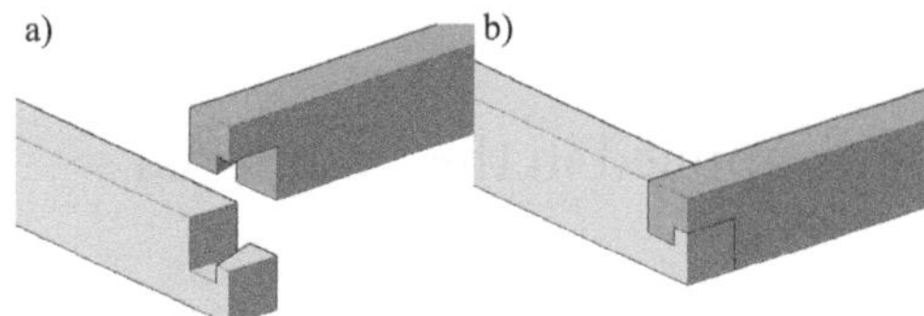

Figure 4. Tooth lock: a) notch—geometry, b) connection.

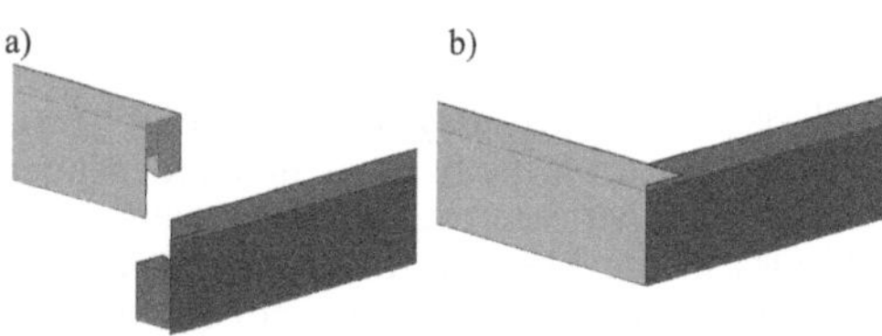

Figure 5. Hidden-tooth lock: a) notch—geometry, b) connection.

3 BOUNDARY CONDITIONS AND LOAD

Due to the nature of the work, the connection numerical model is built using three-dimensional tetrahedral solid elements. Each log consist of about 10,000 finite elements and the mesh is about three times more dense in the region of the analysed joint than for the rest of the log. The following boundary conditions have been adopted:

1. in the case of three logs in the analysed carpentry joint: at the end of two parallel logs—blocked translations in three directions (*X*, *Y*, *Z*), at the end of a single perpendicular log—blocked translations in two directions (*X* and *Z*) and the applied surface load at the *Y* direction,
2. in the case of two logs in the analysed carpentry joint: at the end of one log—blocked translations in three directions (*X*, *Y*, *Z*), at the end of second perpendicular log—blocked translations in two directions (*X* and *Z*) and the applied surface load at the *Y* direction.

In the MSC.Marc software the contact zones have been defined by using the Contact Tables. It has been necessary to indicate the interaction between the logs of the carpentry joints. In the case of two logs—first beam has been affected the other and inversely. In turn, in the case of three logs—each beam has been affected two remaining beams. The cross-sectional dimensions of each analyzed log are 80 × 120 mm.

4 MATERIAL PARAMETERS OF WOOD

The anatomical directions in wood are distinguished depending on the grain of the annual increment (Kyzioł 2005). In the literature concerning the properties of the wood is applied the literal notation, where *R*—radial direction, *T*—tangential direction, *L*—longitudinal direction to the surface of each layer of the fibres. In turn, in the theory of elasticity is used index notation, where x_1—radial direction, x_2—tangential direction, x_3—longitudinal direction.

The anatomical directions of wood are presented in Figure 6.

Wood is an orthotropic material, described by twelve material constants. Due to symmetry with respect to the main diagonal of the constitutive matrix, three reciprocal relations must be satisfied:

$$\frac{\nu_{ij}}{E_i} = \frac{\nu_{ji}}{E_j}, \quad i, j = 1, 2, 3 \tag{1}$$

According to this, nine independent material constants can be distinguished (Jones 1999, Kyzioł 2005). The analysed carpentry joint is made of spruce wood, which material parameters for the applied constitutive model are presented in Table 1 (Green 1999). In order to define the orthotropic material in the MSC.Marc software all the components of the constitutive matrix have been introduced individually according to the data presented in Table 1.

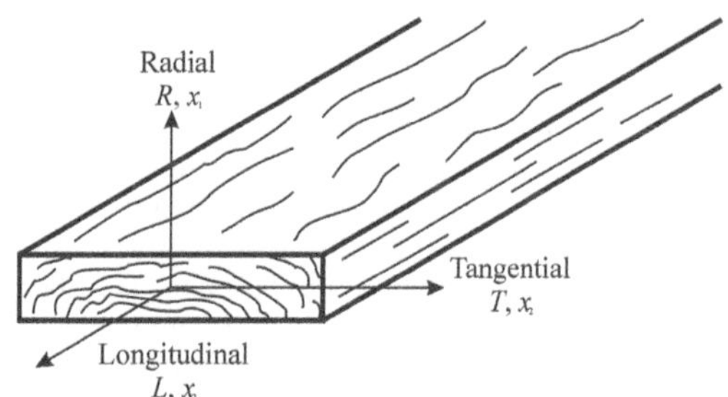

Figure 6. Anatomical directions of wood.

Table 1. Material parameters of spruce wood.

Material constants	Value	Unit
$E_R = E_1$	849	MPa
$E_T = E_2$	468	MPa
$E_L = E_3$	10,890	MPa
$G_{RT} = G_{12}$	33	MPa
$G_{TL} = G_{23}$	664	MPa
$G_{LR} = G_{31}$	697	MPa
$\nu_{RT} = \nu_{12}$	0.372	–
$\nu_{TL} = \nu_{23}$	0.435	–
$\nu_{LR} = \nu_{31}$	0.025	–

5 RESULTS

The geometrically nonlinear static analysis of the carpentry joint has been carried out taking into account the contact phenomenon between the individual elements. It is not possible to compare simultaneously all types of carpentry joints, because the tooth lock and hidden-tooth lock connections consist of two elements only while remaining connections of three parts (additional linking elements are used to link each section). Comparison of the obtained results has been carried out for the long-corner connections, for the dovetail connections and for the tooth lock and the hidden-tooth lock connections.

5.1 *Comparison of two long-corner notches—dovetail and cross connections*

The values of maximal displacement at the *Y* direction and the extreme principal stress in the connection corner have been presented in Table 2.

On the basis of data from Table 2, it can be noticed that values of displacement at the Y direction are almost identical for two types of the carpentry joints. In turn, the values of the extreme principal stress for the dovetail connection are about 6 MPa greater than in the cross one. The major principal stress distributions in the connection corner for both types of notches have been presented in Figures 7–8 (for a better visibility one log is hidden).

In the dovetail connections, the highest stress values are concentrated along the edges of the log. For the cross joint the maximum stress values are smaller, but the high stress area is larger than in the dovetail case.

Table 2. Values of displacements and stresses.

Maximum displacement on the Y direction		Extreme principal stress in the connection corner	
Dovetail (cm)	Cross (cm)	Dovetail (MPa)	Cross (MPa)
2.8	2.9	–45.8	–39.6

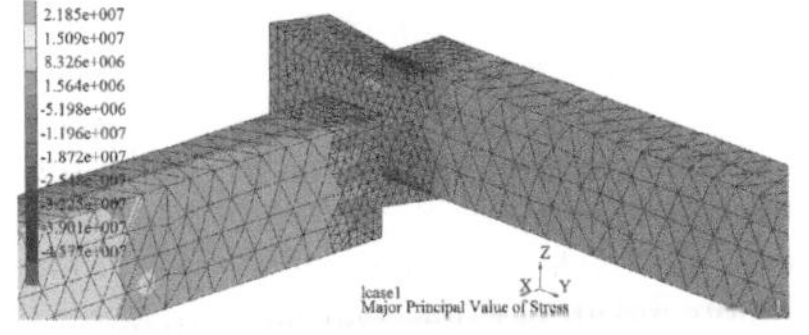

Figure 7. Major principal stress distribution in the connection corner for the dovetail notch.

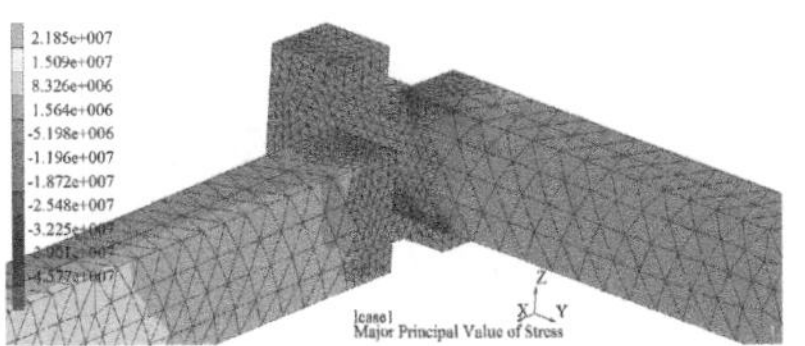

Figure 8. Major principal stress distribution in the connection corner for the cross notch.

Table 3. Values of displacements and stresses.

Maximum displacement on the Y direction		Extreme principal stress in the connection corner	
Long-corner dovetail [cm]	Short-corner dovetail (cm)	Long-corner dovetail (MPa)	Short-corner dovetail MPa
2.8	2.9	–45.8	–57.7

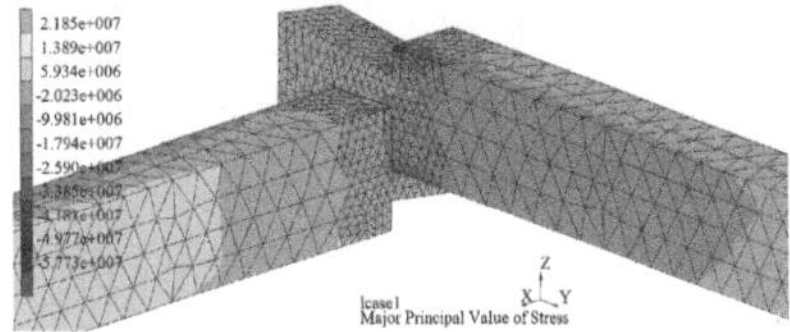

Figure 9. Major principal stress distribution in the connection corner—long-corner dovetail notch.

5.2 *Comparison of two types of dovetail notches—long-corner dovetail connection and short-corner dovetail connection*

The values of maximum displacement and extreme principal stress for two types of dovetail notches have been compared in Table 3.

On the basis of After analyzing Table 3, it can be noticed that the values of the extreme principal stress in short-corner dovetail connection are about 1,3 times greater than in the long-corner dovetail connection. In turn, for the applied load scheme the values of displacement are almost identical. Also the high stress areas are very similar (Figs. 9–10, for better visibility one log is hidden).

5.3 *Comparison the tooth lock connection and hidden-tooth lock connection*

The obtained results of maximum displacement at the *Y* direction and extreme principal stress in the connection corner have been presented in Table 4.

Analysing Table 4, it has been noticed that values of displacement at the *Y* direction of two types are almost identical. In turn, values of principal stress in the tooth lock notch are about two times greater than in the hidden-tooth lock notch.

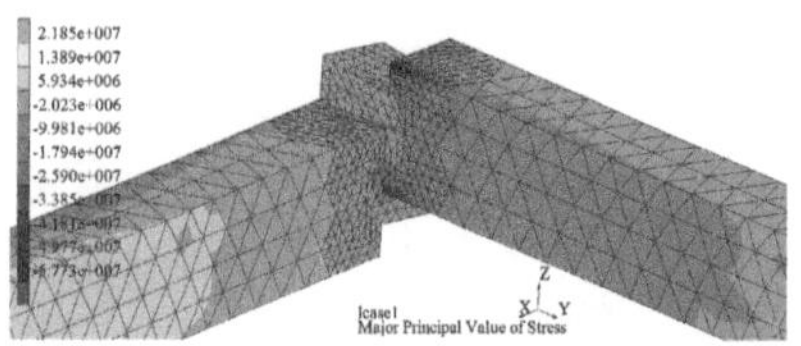

Figure 10. Major principal stress distribution in the connection corner—short-corner dovetail notch.

Table 4. Values of displacements and stresses.

Maximum displacement on the Y direction		Extreme principal stress in the connection corner	
Tooth lock (cm)	Hidden-tooth lock (cm)	Tooth lock (MPa)	Hidden-tooth lock (MPa)
2.8	2.9	–81.0	–43.4

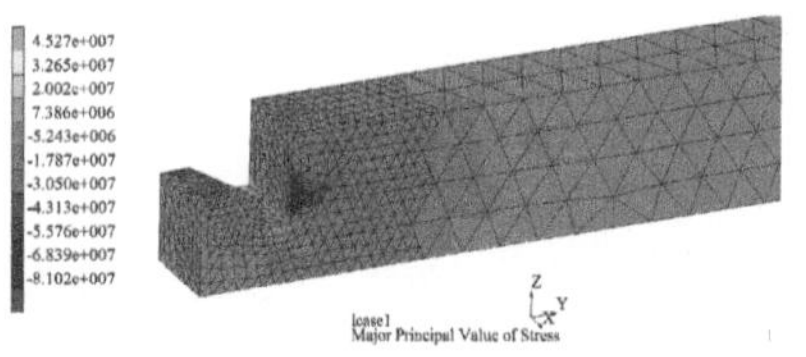

Figure 11. Major principal stress distribution in the connection corner—tooth lock notch.

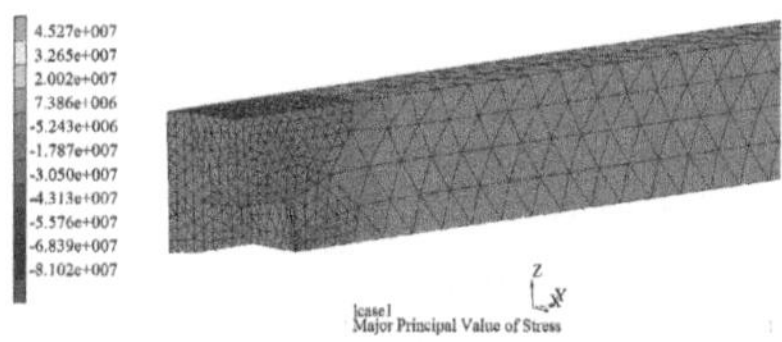

Figure 12. Major principal stress distribution in the connection corner—hidden-tooth lock notch.

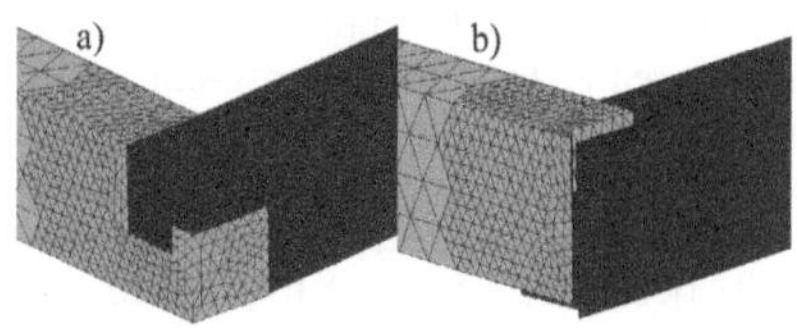

Figure 13. Deformation: a) tooth lock notch, b) hidden-tooth lock notch.

Major principal stress distributions in the connection corner for one log for tooth lock notch and hidden-tooth lock notch have been presented in Figures 11–12.

Deformations for two types of carpentry joints have been shown in Figures 13. For applied type of loading the tooth lock notch produces higher values of stresses, but the deformation is significantly smaller.

6 CONCLUSIONS

Five types of the carpentry joints, commonly used in the currently preserved structures of the wooden architecture, have been analysed. The areas of greatest stress in the planes of contact have been determined. On this basis, vulnerable damage areas have been assigned. The probable damage of the joints under cyclic loading should be expected in places of the greatest stress values. In the case of the long-corner notches, the principal stress at the corner in the cross notch is smaller than in the dovetail one. In turn, in the dovetail notch the stress values are about 1,2 times greater and occur only along the edges of the log. It has been certified small differences of the major principal stress (about 10 MPa) between the long-corner dovetail connection and short-corner dovetail connection, but for the analysed load scheme the high stress areas are very similar.

Two recent types of connections: the tooth lock and hidden-tooth lock connections constitute a separate group in comparison with typical weld in joints. There is a noticeable difference in the values of stresses observed in the connections corner. The maximum value of the principal stress in the tooth lock notch is almost twice the value in the hidden-tooth lock notch.

For all analysed groups of carpentry joints, it has been noted that the maximum displacements at the *Y* direction are almost identical, what means that the type of the connection has no influence of the structure deformation, if the force is applied along logs.

REFERENCES

Goglin, M.R. 2006. Durability of carpenter's joints in historic timberwork on the example of roof structures from the period of 13th–15th centuries. *Ed. Poznań University of Life Science* 37: 55–65.

Green, D.W., Winandy, J.E. & Kretschmann, D.E. 1999. *Wood handbook—Wood as an Engineering Material. Chapter 4: Mechanical Properties of Wood.* Madison: General Technical Report FPL-GTR 190.

Jasieńko, J., Engel, L. & Gospodarek, T. 2005. *Elastooptische und numerische Undersuchungen ausgewählter Verbindungen an hölzerner Denkmalkonstruktionen.* Wismar: Antikon.

Jasieńsko L., Engel, L. & Rapp, P. 2006. Study of stresses in historical carpentry joints by photoelasticity modelling. In Lourenço (ed.), *Structural Analysis of Historical Constructions, Possibilities of Numerical and Experimental Techniques, Proc. intern. conf., New Delhi, 6–8 November 2006*. New Delhi: Macmillan India Ltd.

Jones, R.M. 1999. *Mechanics of composite materials.* London: Taylor & Francis Group.

Kyzioł, L. 2005. Substitute material constants of construction wood with surface modification PMM (in Polish). *Scientific Notebooks of Polish Naval Academy* 1(160): 69–82.

Advances in Mechanics: Theoretical, Computational and Interdisciplinary Issues – Kleiber et al. (Eds)

 ISBN 978-1-138-02906-4

Numerical modeling of biological tissue freezing process using the Dual-Phase-Lag Equation

B. Mochnacki
University of Occupational Safety Management in Katowice, Poland

E. Majchrzak
Silesian University of Technology, Gliwice, Poland

ABSTRACT: Heat transfer processes proceeding during the biological tissue freezing are discussed. The problem is described by the Dual-Phase Lag Equation (DPLE) supplemented by the appropriate boundary and initial conditions. The freezing models so far presented based on the well-known Pennes equation, but taking into account the physical properties of tissue the DPLE seems to be the better approximation of thermal processes taking place in the sub-domain of a frozen tissue. The characteristic feature of DPLE is the introduction of two lag times called the relaxation and thermalization times (τ_q and τ_T). The cryosurgery treatment using the spherical internal cryoprobe is considered. At the stage of numerical computations the explicit scheme of FDM for nonlinear problems is applied. The example of simulation is also presented.

1 INTRODUCTION

The typical description of the bioheat transfer processes (Mochnacki & Piasecka Belkhayat 2013, Paruch 2014, Rabin & Schitzer 1998, Stańczyk & Telega 2002) is based on the well-known Pennes equation (Pennes 1948) supplemented by the appropriate boundary and initial conditions, but recently starts to dominate the view that taking into account the tissue properties, the model using DPLE is better. The specific, non-homogeneous structure of the tissue causes that the introduction of lag times is entirely justified (Zhou et al. 2009, Singh & Kumar 2014).

The biological tissue freezing process proceeds in the temperature interval (from −1°C to −8°C). The evolution of the freezing heat and the cooling process in the intermediate region can be taken into account by the introduction of parameter called an effective thermal capacity (Comini & Del Giudice 1976, Majchrzak et al. 2009, Rabin & Schitzer 1998). The governing equation obtained this way corresponds to the *one domain method* (Mochnacki & Suchy 1996), because only one equation describes the thermal processes in the whole, conventionally homogeneous tissue domain. The problem has been solved using the variant of the finite difference method presented in (Majchrzak & Mochnacki 2014). In the final part the results of computations are shown.

2 GOVERNING EQUATIONS

The following energy equation is considered (Zhou et al. 2009):

$$c(T)\left(\frac{\partial T}{\partial t}+\tau_q\frac{\partial^2 T}{\partial t^2}\right)=\nabla\left(\lambda(T)\nabla T\right) + \tau_T\frac{\partial}{\partial t}\{\nabla[\lambda(T)\nabla T]\}+Q+\tau_q\frac{\partial Q}{\partial t} \tag{1}$$

where $c(T)$ = is the volumetric specific heat of tissue; $\lambda(T)$ = is the thermal conductivity of tissue; τ_q = is the relaxation time and τ_T = is the thermalization time, $Q = Q(x, t)$ is the source term due to metabolism, blood perfusion and freezing process; $T = T(x, t)$ is the temperature, x, t = denote the spatial co-ordinates and time.

Thus, the source term $Q\,(x, t)$ is the sum of the following components:

$$Q(x,t)=w_B(T)c_B[T_B-T(x,t)]+Q_m(T)+L\frac{\partial S(x,t)}{\partial t} \tag{2}$$

where $w_B(T)$ = is the blood perfusion rate; c_B = is the specific heat of blood; T_B = is the blood temperature, $Q_m(T)$ = is the metabolic heat source; L = is the volumetric freezing heat and $S(x, t)$ = is the frozen state fraction at the neighborhood of the point considered.

Differentiation of source function (2) with respect to time gives:

$$\frac{\partial Q(x,t)}{\partial t}=\frac{\mathrm{d}w_B(T)}{\mathrm{d}T}c_B[T_B-T(x,t)]-$$
$$+w_B(T)c_B\frac{\partial T(x,t)}{\partial t}+\frac{\mathrm{d}Q_m(T)}{\mathrm{d}T}\frac{\partial T(x,t)}{\partial t}$$
$$+L\left\{\frac{\mathrm{d}^2 S(T)}{\mathrm{d}T^2}\left[\frac{\partial T(x,t)}{\partial t}\right]^2+\frac{\mathrm{d}S(T)}{\mathrm{d}T}\frac{\partial^2 T(x,t)}{\partial t^2}\right\} \tag{3}$$

Introducing formulas (2) and (3) to the equation (1) one obtains:

$$c(T)\left(\frac{\partial T}{\partial t}+\tau_q\frac{\partial^2 T}{\partial t^2}\right)=\nabla(\lambda(T)\nabla T)$$
$$+\tau_T\frac{\partial}{\partial t}\{\nabla[\lambda(T)\nabla T]\}+w_B(T)c_B(T_B-T)+Q_m(T)$$
$$+L\frac{\mathrm{d}S(T)}{\mathrm{d}T}\frac{\partial T}{\partial t}+\tau_q c_B\left\{\frac{\mathrm{d}w_B(T)}{\mathrm{d}T}(T_B-T)-w_B(T)\frac{\partial T}{\partial t}\right\}$$
$$+\tau_q\frac{\mathrm{d}Q_m(T)}{\mathrm{d}T}\frac{\partial T}{\partial t}$$
$$+\tau_q L\left\{\frac{\mathrm{d}^2 S(T)}{\mathrm{d}T^2}\left(\frac{\partial T}{\partial t}\right)^2+\frac{\mathrm{d}S(T)}{\mathrm{d}T}\frac{\partial^2 T}{\partial t^2}\right\} \tag{4}$$

or:

$$C(T)\left(\frac{\partial T}{\partial t}+\tau_q\frac{\partial^2 T}{\partial t^2}\right)=\nabla(\lambda(T)\nabla T)$$
$$+\tau_T\frac{\partial}{\partial t}\{\nabla[\lambda(T)\nabla T]\}+w_B(T)c_B(T_B-T)$$
$$+Q_m(T)+\tau_q v_B(T)c_B(T_B-T)$$
$$+\tau_q[P_m(T)-w_B(T)c_B]\frac{\partial T}{\partial t}+\tau_q LZ(T)\left(\frac{\partial T}{\partial t}\right)^2 \tag{5}$$

where:

$$C(T)=c(T)-L\frac{\mathrm{d}S(T)}{\mathrm{d}T} \tag{6}$$

is the substitute thermal capacity (Mochnacki & Suchy 1996), while $v_B(T)=\mathrm{d}w_B(T)/\mathrm{d}T$, $P_m(T)=\mathrm{d}Q_m(T)/\mathrm{d}T$ and $Z(T)=\mathrm{d}^2S(T)/\mathrm{d}T^2$.

For the biological tissue the function $C(T)$ can be defined as follows:

$$C(T)=\begin{cases}c_N, & T>T_1\\ c_1+c_2T+c_3T^2+c_4T^3+c_5T^4, & T_2\le T\le T_1\\ c_F, & T<T_2\end{cases} \tag{7}$$

where c_N, c_F = are the volumetric specific heats of natural and frozen state and (Majchrzak et al. 2007):

$$c_1=\frac{c_FT_1-c_NT_2}{T_1-T_2}+\frac{(c_N-c_F)T_1T_2(T_1+T_2)}{(T_1-T_2)^3}+\frac{30T_1^2T_2^2L}{(T_1-T_2)^5}$$
$$c_2=-\frac{6(c_N-c_F)T_1T_2}{(T_1-T_2)^3}-\frac{60T_1T_2(T_1+T_2)L}{(T_1-T_2)^5}$$
$$c_3=\frac{3(c_N-c_F)(T_1+T_2)}{(T_1-T_2)^3}+\frac{30(T_1^2+4T_1T_2+T_2^2)L}{(T_1-T_2)^5}$$
$$c_4=-\frac{2(c_N-c_F)}{(T_1-T_2)^3}-\frac{60(T_1+T_2)L}{(T_1-T_2)^5}$$
$$c_5=\frac{30L}{(T_1-T_2)^5} \tag{8}$$

The course of function $C(T)$ is shown in Figure 1.

To calculate $Z(T)$ (4) the formula (6) is used and then:

$$Z(T)=\frac{\mathrm{d}^2S(T)}{\mathrm{d}T^2}=\frac{1}{L}\left(\frac{\mathrm{d}c(T)}{\mathrm{d}T}-\frac{\mathrm{d}C(T)}{\mathrm{d}T}\right) \tag{9}$$

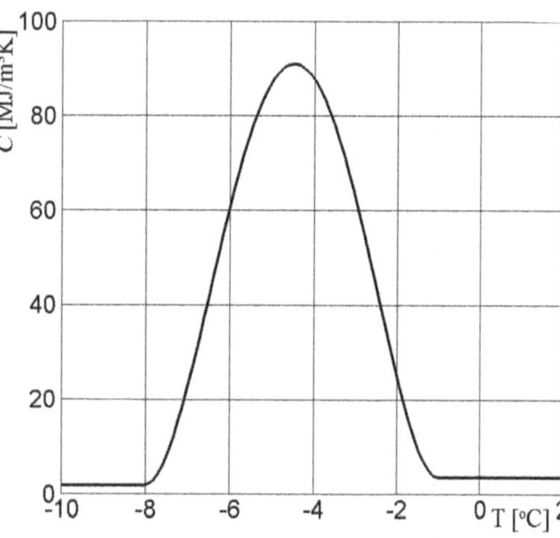

Figure 1. Substitute thermal capacity.

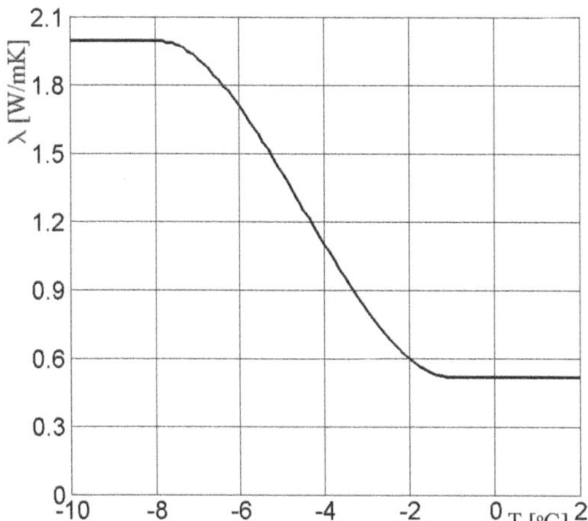

Figure 2. Thermal conductivity.

The course of thermal conductivity was approximated by the continuous and differentiable function created by the constant values and the polynomial of the third degree (Fig. 2):

$$\lambda(T)=\begin{cases}\lambda_N, & T>T_1\\ b_1+b_2T+b_3T^2+b_4T^3, & T_2\le T\le T_1\\ \lambda_F, & T<T_2\end{cases} \tag{10}$$

The coefficients b_1, b_2, b_3, b_4 are determined using the conditions: $\lambda(T_1)=\lambda_N$, $\lambda(T_2)=\lambda_F$, $d\lambda(T)/dT=0$ for $T=T_1$ and $T=T_2$.

It should be pointed out that the blood perfusion rate and the metabolic heat source are equal to zero for the frozen region ($T<T_2=-8°C$), while for the intermediate one the linear changes starting from the point $T=T_1=-1°C$ (corresponding to the natural state of tissue) are assumed:

$$\begin{aligned} w_B(T)&=\frac{w_{B0}}{T_1-T_2}[T(x,t)-T_2]\\ Q_m(T)&=\frac{Q_{m0}}{T_1-T_2}[T(x,t)-T_2]\end{aligned} \tag{11}$$

where index 0 corresponds to the natural state of tissue.

3 FORMULATION OF THE PROBLEM AND THE METHOD OF SOLUTION

As an example of freezing process modeling the action of spherical cryoprobe is considered. The cryoprobe tip causes the generation of time-dependent ice ball and the problem has been treated as 1D one for the object oriented in the spherical co-ordinate system and then (5):

$$\nabla(\lambda(T)\nabla T)=\frac{1}{x^2}\frac{\partial}{\partial r}\left(\lambda(T)x^2\frac{\partial T}{\partial x}\right) \tag{12}$$

On the conventionally assumed external surface of the system $x=R_2$ the no-flux condition is accepted, while on the contact surface between cryoprobe tip and tissue ($x=R_1$) the Dirichlet condition is introduced. In particular during the first 60 seconds the cryoprobe temperature changes linearly from 37°C to −90°C, next during the same time the temperature changes linearly from −90°C to 37°C. Additionally, for $t=0$:

$$t=0:\ T=T_p=37°\mathrm{C},\quad \left.\frac{\partial T}{\partial t}\right|_{t=0}=0 \tag{13}$$

Let us denote the successive time levels by $f-2$, $f-1$, f and the nodes creating 3-points star by $i-1$, i, $i+1$ ($x_i=R_1+ih/2$, h is the mesh step).

The FDM equation proposed by the authors (explicit differential scheme) is of the form:

$$\begin{aligned} &C_i^{f-1}\left(\frac{T_i^f-T_i^{f-1}}{\Delta t}+\tau_q\frac{T_i^f-2T_i^{f-1}+T_i^{f-2}}{(\Delta t)^2}\right)\\ &=\frac{\Delta t+\tau_T}{\Delta t}\left[\frac{1}{x^2}\frac{\partial}{\partial x}\left(\lambda x^2\frac{\partial T}{\partial x}\right)\right]_i^{f-1}-\\ &+\frac{\tau_T}{\Delta t}\left[\frac{1}{x^2}\frac{\partial}{\partial x}\left(\lambda x^2\frac{\partial T}{\partial x}\right)\right]_i^{f-2}\\ &+\left[w_B(T_i^{f-1})+\tau_q v_B(T_i^{f-1})\right]c_B\left(T_B-T_i^{f-1}\right)\\ &+Q_m(T_i^{f-1})\\ &+\tau_q\left[P_m(T_{i,j}^{f-1})-c_Bw_B(T_{i,j}^{f-1})\right]\frac{T_i^{f-1}-T_i^{f-2}}{\Delta t}\\ &+\tau_q LZ(T_i^{f-1})\left(\frac{T_i^{f-1}-T_i^{f-2}}{\Delta t}\right)^2\end{aligned} \tag{14}$$

where (Mochnacki & Suchy 1996)

$$\left[\frac{1}{x^2}\frac{\partial}{\partial r}\left(\lambda x^2\frac{\partial T}{\partial x}\right)\right]_i^s=\Phi_{i+1}\frac{T_{i+1}^s-T_i^s}{R_{i+1}^s}+\Phi_{i-1}\frac{T_{i-1}^s-T_i^s}{R_{i-1}^s} \tag{15}$$

while

$$\Phi_{i+1} = \frac{(x_i + 0.5h)^2}{x_i^2 h}, \quad \Phi_{i-1} = \frac{(x_i - 0.5h)^2}{x_i^2 h}$$
$$R_{i+1}^s = \frac{h}{2}\left(\frac{1}{\lambda_i^s} + \frac{1}{\lambda_{i+1}^s}\right), \quad R_{i-1}^s = \frac{h}{2}\left(\frac{1}{\lambda_i^s} + \frac{1}{\lambda_{i-1}^s}\right) \tag{16}$$

In the formulas (16) $s = f - 1$ or $s = f - 2$. The equation (14) allows to find the value T_i^f. The 'attaching' of boundary conditions for the case considered is presented in Mochnacki & Suchy (1996).

The problem of stability criterion formulation for equation (14) is very complex, perhaps even unfeasible. In this connection the critical time step has been determined 'experimentally'.

4 RESULTS OF COMPUTATIONS

The following input data have been taken into account: $R_1 = 0.005$ m, $R_2 = 0.02$ m, number of nodes $n = 200$, time step $\Delta t = 0.0002$s. The lag times are equal to $\tau_q = 3$ s, $\tau_T = 0.1$ s (Singh & Kumar 2014). Thermophysical parameters of tissue: $c_N = 3600000$ J/(m³ K), $c_F = 1930000$ (J/m³ K), $L = 330000 \cdot 10^3$ J/m³, $\lambda_N = 0.52$ W/(mK), $\lambda_F = 2$ W/(mK), $Q_{m0} = 250$ W/m³. $w_{B0} = 0.53$ kg/(m³ s), $c_B = 3770$ J/(kgK), $T_B = 37$ °C.

Part of the results is shown in Figures 3–4. In Figure 3 the temperature profiles for selected times are shown (the stage of cooling). The next Figure presents the temperature history (cooling/heating curves) at the distance 1, 2, and 3 mm to the cryoprobe surface (points 1, 2 and 3, respectively). The solution corresponding to the Pennes model is also shown (the dashed lines).

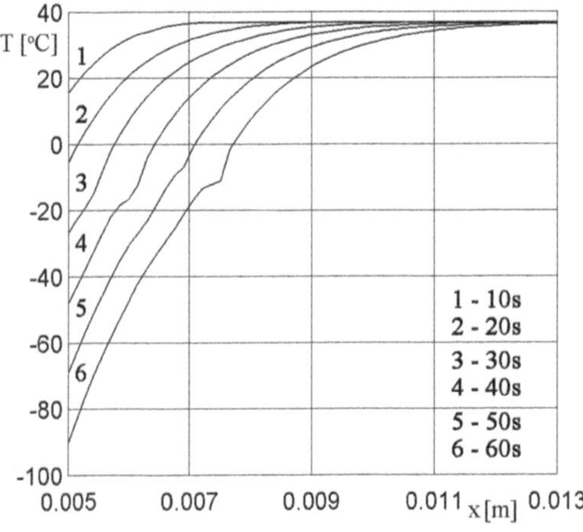

Figure 3. Temperature profiles.

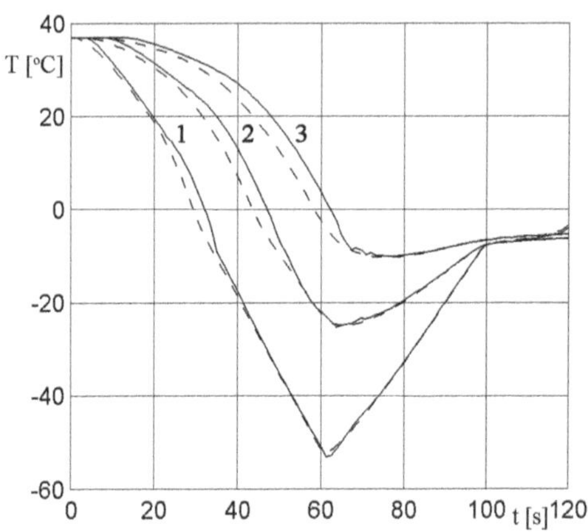

Figure 4. Cooling/heating curves.

5 CONCLUSIONS

Analysis of freezing process proceeding in the 1D spherical domain is presented. The mathematical model is based on the DPL equation supplemented by the appropriate boundary—initial conditions. The numerical solution has been obtained using the authorial variant of the FDM for hyperbolic equations. The results obtained appear to be entirely satisfactory. Interesting thing is the comparison of the DPLM solution with the solution basing on the Pennes model. It turned out that both solutions are close, but the differences are visible. The greater differences should appear for the bigger values of lag times. In literature (Zhang 2009) can be found the information that for the some of the tissue types the delay times may be of the order of several seconds. The influence of delay times changes on the course of temperature distribution in the tissue domain may be tested using the methods of sensitivity analysis (Dziewoński et al. 2011, Mochnacki & Majchrzak 2003, Jasiński 2014) and these problems in the future will be the subject of authors interest.

ACKNOWLEDGEMENTS

The research is a part of Project PB3/2013 sponsored by WSZOP in Katowice.

REFERENCES

Comini, G. & Del Giudice, L. 1976. Thermal aspects of cryosurgery. *Journal of Heat Transfer* 98: 543–549.

Dziewoński, M., Mochnacki, B. & Szopa, R. 2011. *Sensitivity of biological tissue freezing process on the changes of cryoprobe cooling rate*. Mechanika 2011: Proceedings of the 16th International Conference, Book Series: Mechanika Kaunas University of Technology, 82–87.

Jasiński, M. 2014. Modeling of tissue thermal injury formation process with application of direct sensitivity method. *Journal of Theoretical and Applied Mechanics* 52, 4: 947–957.

Majchrzak, E. Dziewoński, M. & Kałuża, G. 2007. Numerical algorithm of casting latent heat identification. *Journal of Achievement of Materials and Manufacturing Engineering* 22 (1): 61–64.

Majchrzak, E. & Mochnacki, B. 2014. Sensitivity analysis of transient temperature field in microdomains with respect to the dual phase lag model parameters. *International Journal for Multiscale Computational Engineering* 12 (1): 65–77.

Mochnacki, B. & Majchrzak, E. 2003. Sensitivity of the skin tissue on the activity of external heat sources. *CMES: Computer Modeling in Engineering and Sciences* 4 (3–4): 431–438.

Mochnacki, B. & Piasecka Belkhayat, A. 2013. Numerical modeling of skin tissue heating using the interval finite difference method. *MCB: Molecular & Cellular Biomechanics* 10 (3): 133–144.

Mochnacki, B. & Suchy, J. S. 1996. Numerical methods in computations of foundry processes, PFTA, Cracow.

Paruch, M. 2014: Hyperthermia process control induced by the electric field in order to destroy cancer. *Acta of Bioengineering and Biomechanics* 16 (4): 123–130.

Pennes, H.H. 1948. Analysis of tissue and arterial blood temperatures in the resting human forearm. *Journal of Applied Physiology* 1: 93–122.

Rabin, Y. & Schitzer, A. 1998. Numerical solution of the multidimensional freezing problem during cryosurgery. *ASME Journal of Biomechanical Engineering* 120: 32–37.

Singh, S. & Kumar, S. 2014. Numerical study on triple layer skin tissue freezing using dual phase lag bio-heat model. *International Journal of Thermal Sciences* 86: 12–20.

Stańczyk, M. & Telega, J.J. 2002. Modelling of heat transfer in biomechanics a review. Part I. Soft tissues. *Acta of Bioengineering and Biomechanics* 4 (1): 31–61.

Zhou, J., Zhang, Y. & Chen, J.K. 2009. An axisymmetric dual-phase-lag bioheat model for laser heating of living tissues. *International Journal of Thermal Sciences* 48: 1477–1485.

Zhang, Y. 2009. Generalized dual-phase lag bioheat equations based on nonequilibrium heat transfer in living biological tissues. *International Journal of Heat and Mass Transfer* 52: 4829–4834.

Advances in Mechanics: Theoretical, Computational and Interdisciplinary Issues – Kleiber et al. (Eds)

Real-time hybrid simulation using materials testing machine and FEM

W. Mucha
Institute of Computational Mechanics and Engineering, Gliwice, Poland

ABSTRACT: The paper presents an algorithm for real-time hybrid testing using Finite Element Method (FEM). Hybrid testing is a method that allows to investigate dynamic properties of complex mechanical structures or systems by creating two models: the experimental model (a physical component of the tested structure or system) and the virtual model (a numerical model of the rest of the tested structure or system). The paper describes an algorithm for hybrid testing where the experimental model is mounted in a dynamic materials testing machine which is controlled by an advanced microcontroller with the virtual model implemented in its algorithm. The virtual substructure is modeled using FEM. An example is given where the tested mechanical structure is a truss and one of its elements is mounted in the testing machine.

1 INTRODUCTION

Hybrid testing (commonly referred as hardware-in-the-loop simulation) is a method that allows to investigate dynamic properties of mechanical structures or systems. The method consists of creating two models strictly related to each other: the experimental model which is a physical component of the tested structure and the virtual model which is a numerical representation of the rest of the tested structure. The popularity of hybrid testing technique has grown rapidly since its concept (in the turn of 1960s and 1970s) because of its undeniable advantages like the ability to test complex structures containing elements difficult or even impossible to model numerically, a possibility to experimentally test only a component of a structure whose sizes makes it impossible to test it as a whole, or significantly lower costs compared to shake table tests (Bursi 2008, Nakashima et al. 2008).

When building the experimental model, the physical component of the structure is usually constrained and attached to one or more dynamic actuators that load it. There is also a necessity to connect some measurement devices. Therefore, in the classical approach, actuator control system, measurement system, computational system, A/D and D/A conversion system are often separate but cooperating devices (Bursi 2008, Nakashima et al. 2008).

The principal idea of the author is to mount the physical component of the tested system to a dynamic materials testing machine. The machine would be controlled by an advanced microcontroller with the virtual model of the Hardware-in-the-Loop (HIL) simulation implemented in its algorithm. The virtual subsystem of the tested object can be modeled using Finite Element Method (FEM). The tasks of the microcontroller is to generate signals controlling the testing machine, perform computations on the virtual model and acquire data from built-in measurement systems of the testing machine. The proposed solution has some important advantages. First of all, it is much simpler than the classical approach because there are only two devices: the testing machine and the microcontroller. The second advantage is that the measurement of displacements and forces of the physical substructure is quite easy with the embedded measurement systems of the machine. It is also worth mentioning that dynamic materials testing machines often have wide force ranges and are able to perform high-speed tests.

2 THE ALGORITHM

2.1 *General FEM algorithm for hybrid testing*

The algorithm of hybrid simulation using FEM is based on a modified FEM matrix equation of motion:

$$\mathbf{M}\ddot{\mathbf{u}} + \mathbf{C}\dot{\mathbf{u}} + \mathbf{K}\mathbf{u} + \mathbf{r}^{E} = \mathbf{f} \tag{1}$$

where **M**, **C** and **K** are mass, damping and stiffness matrix of the virtual model, respectively, $\mathbf{r}^E$ is the vector of forces developed in the experimental model (depending on structural displacement, velocity and acceleration), **u** is the displacement vector and **f** is the force vector (Shing 2008, Chandrupatla & Belegundu 2012, Bursi 2008).

Figure 1 presents the general FEM algorithm for hybrid testing of mechanical structures and systems. The algorithm requires iterative approach. In each time step the displacement for the next

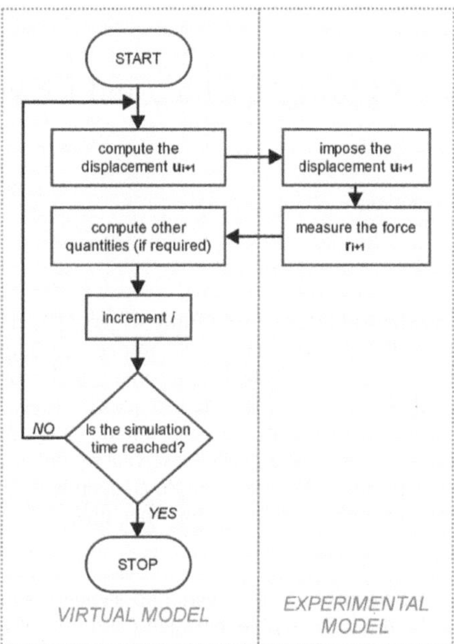

Figure 1. General hybrid testing algorithm.

time step needs to be computed, imposed on the physical component and the developed force in it measured (Nakashima 2008, Shing 2008).

2.2 *Explicit and implicit algorithms*

The modified FEM equation (1) can be solved step by step using explicit or implicit methods. In explicit algorithms the next state is calculated only from the current state and in implicit methods the next state is calculated from the data of the current state and the next state. Therefore implicit methods are computationally more demanding but on the other hand they are often unconditionally stable while explicit methods are only conditionally stable (the stability condition defines the maximum time step) (Zienkiewicz & Taylor 2000, Shing 2008).

2.3 *Real time computations*

In order to enable the control of the material testing machine in a closed loop, the FEM computations have to be performed in real time.

Real time is understood as the ability to reliably and, without fail, respond to an event by performing a set of operations in a guaranteed time period (Buttazo 2011, Kopetz 2011).

As the Finite Element Method is, in general, very computationally demanding method, it is very difficult however possible to use in real time applications (Kuś & Mucha 2014).

Before performing a hybrid simulation one should consider which algorithm is more suitable for the tested mechanical system, explicit or implicit. For models with many Degrees of Freedom (DOFs) and high frequency modes, the maximum time step in explicit methods is often shorter than the time of computations that must be performed in one time step. Such situation makes real time impossible for explicit methods. On the other hand, in implicit methods more operations need to be performed in a time step, but one has the convenience of adjusting the time step to be only slightly longer than the computation time.

2.4 *Newmark algorithm for hybrid simulation*

An unconditionally stable implicit Newmark algorithm (Zienkiewicz & Taylor 2000, Shing 2008) has been adapted to hybrid simulation and is presented below.

In each time step the equation (1) has to be solved in the form:

$$\mathbf{K}_{\mathbf{eff}}\,\Delta\mathbf{u}_{i+1} = \mathbf{r} \tag{2}$$

where $\Delta\mathbf{u}_{i+1}$ is the difference between the predicted and real displacement in time step i+1, $\mathbf{K}_{\mathbf{eff}}$ is the effective stiffness matrix of the virtual model and $\mathbf{r}$ is the right-hand side vector. The elements of equation (2) can be expressed as follows:

$$\mathbf{u}_{i+1} = \hat{\mathbf{u}}_{i+1} + \Delta\mathbf{u}_{i+1} \tag{3}$$

$$\mathbf{K}_{\mathbf{eff}} = \mathbf{K} + \frac{4}{\Delta t^2}\mathbf{M} + \frac{2}{\Delta t}\mathbf{C} \tag{4}$$

$$\mathbf{r} = \mathbf{f} - \mathbf{M}\left[\frac{4}{\Delta t^2}(\hat{\mathbf{u}}_{i+1} - \mathbf{u}_i) - \frac{4}{\Delta t}\dot{\mathbf{u}}_i - \ddot{\mathbf{u}}_i\right] - \mathbf{C}\left[\frac{2}{\Delta t}(\hat{\mathbf{u}}_{i+1} - \mathbf{u}_i) - \dot{\mathbf{u}}_i\right] - \mathbf{K}\hat{\mathbf{u}}_{i+1} - \hat{\mathbf{r}}^{\mathbf{E}}_{i+1} \tag{5}$$

where Δt is the constant time of a single step.

Probably the most difficult part in the algorithm is to predict the forces that will be developed in the experimental model in the next time step:

$$\hat{\mathbf{r}}^{\mathbf{E}}_{i+1} = \mathbf{r}^{\mathbf{E}}_i + \mathbf{K}^{\mathbf{E}}_{\mathbf{eff}}(\hat{\mathbf{u}}_{i+1} - \mathbf{u}_i) \tag{6}$$

where $\mathbf{r}^{\mathbf{E}}_i$ are the measured forces in the current iteration and $\mathbf{K}^{\mathbf{E}}_{\mathbf{eff}}$ is the effective stiffness matrix of the experimental model (calculated analogously as the effective stiffness matrix of the virtual model using Equation 4). The experimental model may consist physical elements that are very difficult to model numerically, but the stiffness, damping and mass matrices of the experimental model have to be at least roughly estimated in order to calculate $\mathbf{K}^{\mathbf{E}}_{\mathbf{eff}}$.

The whole procedure described above may have to be repeated a few times in a single time step (for nonlinear structures and systems). When repeating the procedure the calculated displacement using Equation 3 after solving the system of Equations 2 becomes the predicted displacement in the next iteration. In the first iteration in a time step, the displacement for the next time step can be predicted by extrapolation.

After performing the last iteration in a single time step, when the displacement for the next time step is already known, the velocity and acceleration are to be calculated using the following formulas:

$$\dot{\mathbf{u}}_{i+1} = \frac{2}{\Delta t}(\mathbf{u}_{i+1} - \mathbf{u}_i) - \dot{\mathbf{u}}_i \tag{7}$$

$$\ddot{\mathbf{u}}_{i+1} = \frac{4}{\Delta t^2}(\mathbf{u}_{i+1} - \mathbf{u}_i) - \frac{4}{\Delta t}\dot{\mathbf{u}}_i - \ddot{\mathbf{u}}_i \tag{8}$$

2.5 *Multi-threaded calculations*

In order to ensure continuous and smooth path of the actuators of the materials testing machine, sending a displacement command every integration time step may be not sufficient (especially when using implicit algorithms). Therefore performing some multi-threaded calculations in real time is necessary.

The author's idea is to utilize two threads, where the first thread is performing FEM computations for the next time step in a loop timed at the integration time step and the second thread is generating displacement signal for the machine (in a loop that is timed significantly faster than the FEM loop).

Figure 2 presents the idea of generating displacement signal by the second thread. When the current time step is *i* and calculations for the time step i+1 on the first thread are ongoing (the displacement for the time step i+1 is not known yet), an extrapolation must be performed and in every iteration of the loop executing on the second thread a displacement command of value read from the extrapolation curve must be sent to the testing machine. The moment that the first thread computes the exact displacement for step i+1, an interpolation must be performed in order to replace the extrapolation curve with the interpolation curve.

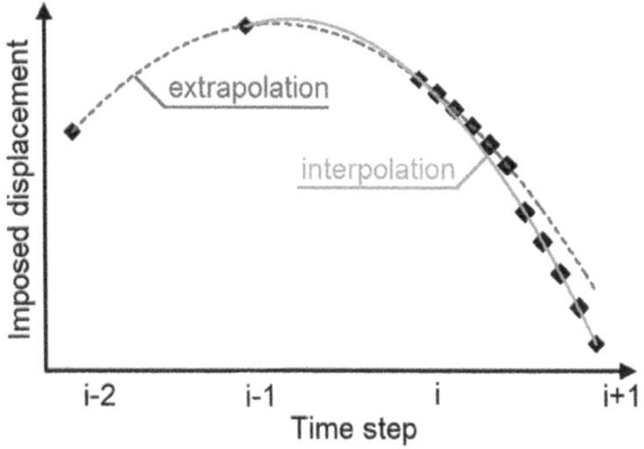

Figure 2. Idea of generating displacement signal.

3 EXAMPLE

3.1 *Tested structure*

The tested structure is a duralumin plane truss with constant Young's modulus 71 GPa and cross-sectional area 0.22 cm^2, presented in Figure 3. Element 11 of the truss is tested experimentally in the material testing machine and the rest is modeled using FEM.

3.2 *Utilized hardware*

Dynamic materials testing machine *Instron ElectroPuls E10000* has been used in the example. The machine is equipped in de-coupled linear and rotary actuators (for the test only the first was used), has got high dynamic performance (is capable of performing to more than 100 Hz) and ±10 kN dynamic linear load capacity.

As the microcontroller, National Instruments myRIO has been used. It is equipped in 2-cored Xilinx Z-7010 processor, clocked at 667 MHz, 512 MB DDR3 memory and real time operating system (*NI Linux Real-Time*).

3.3 *Obtained results*

The waveform of the applied force F(t) and obtained results are presented in Figure 4.

The FEM loop was clocked at 5 ms and signal generation loop at 1 ms.

The displacement presented in Figure 4b has been verified analytically (where the whole truss was modeled using FEM) and the obtained results have confirmed that the described Newmark algorithm for hybrid simulation works correctly.

As one can see in Figure 4c, the computed elongation of the experimental model (grey line) has not always been achieved by the machine (white line).

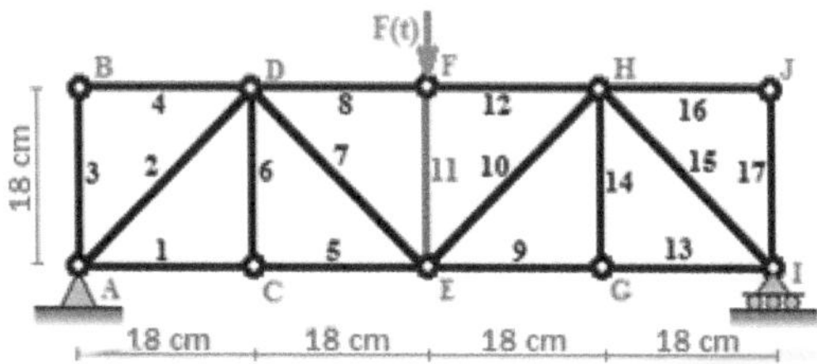

Figure 3. Tested truss.

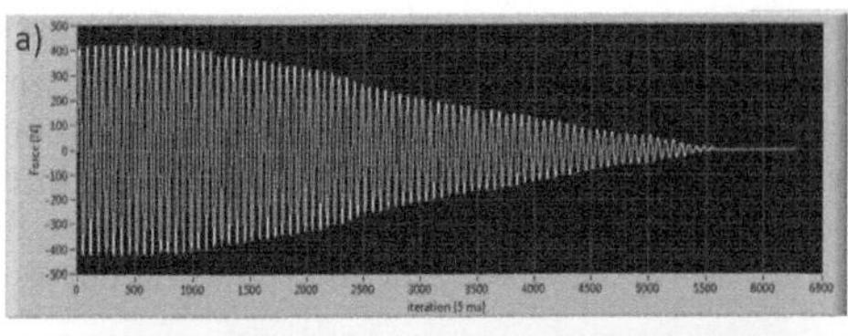

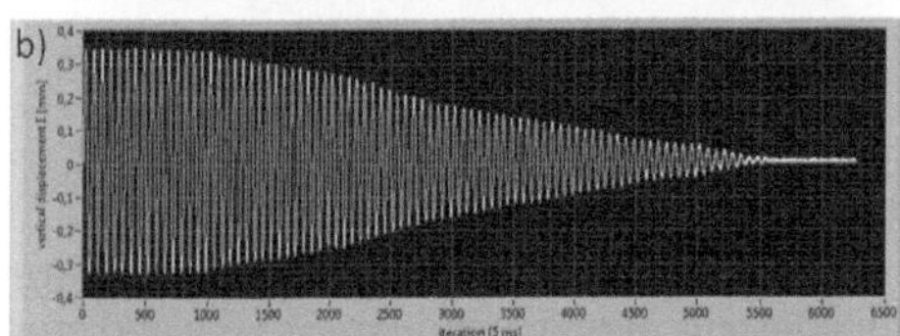

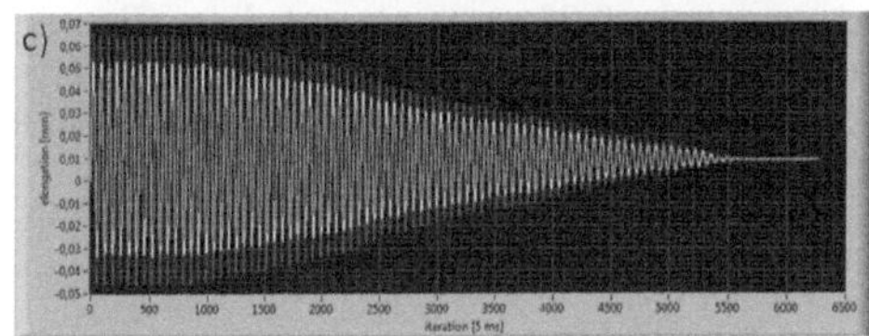

Figure 4. Waveforms of: a) applied force F(t), b) vertical displacement of point E, c) computed and real elongation of the experimental model.

The differences are not large however, they required introducing some extra correction in Equation 6 of the described Newmark algorithm:

$$\hat{\mathbf{r}}^{E}_{i+1} = \mathbf{r}^{E}_{i} + \mathbf{K}^{E}_{eff}\left(\hat{\mathbf{u}}_{i+1} - \frac{\Delta l_i^m}{\Delta l_i}\mathbf{u}_i\right) \tag{9}$$

where Δl_i^m and Δl_i are the measured and computed value of the elongation of the experimental model, respectively.

4 CONCLUSIONS

Utilizing materials testing machines and Finite Element Method in hybrid simulation is possible and may be useful for testing certain mechanical structures (of large sizes part by part or containing elements difficult to model numerically).

Microcontroller controlling the machine has to be equipped in real time operating system and fast, multithreaded processor.

The aim of the presented example was only to verify and validate the Newmark algorithm for hybrid simulation.

Improving the algorithm and examples of more complex structures are planned in author's future work. Author plans to utilize FPGA (field-programmable gate array) in which the micro-controller is equipped in order to speed-up FEM computations. Some arithmetic operations would be executed on FPGA instead on microprocessor. Author plans to test mobile robot components using hybrid simulation (similar as FEM simulations presented in Mucha & Kuś 2015).

ACKNOWLEDGEMENTS

The research is partially financed from project BKM-504/RMT-4/2014.

REFERENCES

Bursi, O.S. 2008. Computational Techniques for Simulation of Monolithic and Heterogeneous Structural Dynamic Systems. In O.S. Bursi & D. Wagg (eds), *Modern Testing Techniques for Structural Systems*: 1–96. Udine: SpringerWienNewYork.

Buttazzo, G. 2011. *Hard Real-Time Computing Systems, Predictable Scheduling Algorithms and Applications*. New York: Springer.

Chandrupatla, T.R. & Belegundu, A.D. 2012. *Introduction to Finite Elements In Engineering.* Upper Saddle River: Prentice-Hall.

Kopetz, H. 2011. *Real Time Systems, Design Principles for Distributed Embedded Applications*. New York: Springer.

Kuś, W. & Mucha, W. 2014. Real Time Computations Using FEM—Review of Applications and Possibilities. In *Mechanika 2014. Proceeding of the 19th International conference*: 153–156. Kaunas: Technologija.

Mucha, W. & Kuś, W. 2015. Exploration mobile robot, project and prototype. In J. Arwejcewicz, R. Szewczyk, M. Trojnacki, M. Kaliczyńska (eds), *Mechatronics: ideas for industrial application*: 231–238. Cham: Springer.

Nakashima, M., McCormick, J., Wang, T. 2008. Hybrid simulation: A historical perspective. In V. Saouma & M. Sevaselan (eds), *Hybrid Simulation: Theory, Implementation and Applications*: 3–14. London: Taylor & Francis Group.

Shing, P.B. 2008. Integration schemes for real-time hybrid testing. In V. Saouma & M. Sevaselan (eds), *Hybrid Simulation: Theory, Implementation and Applications*: 25–34. London: Taylor & Francis Group.

Zienkiewicz, O.C. & Taylor, R.L. 2000. *The Finite Element Method, Vol. 1: The Basis*, fifth ed. Oxford: Butterworth-Heinemann.

Advances in Mechanics: Theoretical, Computational and Interdisciplinary Issues – Kleiber et al. (Eds)
© 2016 Taylor & Francis Group, London, ISBN 978-1-138-02906-4

Rolling contact problems for plastically graded materials

A. Myśliński
Systems Research Institute, Warsaw, Poland
Faculty of Manufacturing Engineering, Warsaw University of Technology, Warsaw, Poland

A. Chudzikiewicz
Faculty of Transport, Warsaw University of Technology, Warsaw, Poland

ABSTRACT: The paper deals with the numerical solution of thermo-elastoplastic wheel-rail rolling contact problems including Coulomb friction, frictional heat generation and heat transport across the contact surface. The rail surface is assumed to be covered with a coating layer made from the functionally graded material. The mechanical and thermal properties of this layer are changing with its depth. This layer and the rail can undergo elastoplastic deformations. The contact phenomenon is governed by the coupled system of elastoplastic and heat conductive equations. Finite difference method and finite element methods are used as the discretization methods. The discretized contact problem is solved numerically using the semi-smooth Newton method. Numerical examples are provided and discussed.

1 INTRODUCTION

The paper concerns numerical solution of the wheel-rail contact problems including friction and frictional heat generation. The contact of a rigid wheel with a rail lying on a rigid foundation is considered. The friction between the bodies is governed by the Coulomb law (Han et al. 2013, Shillor et al. 2004, Wriggers 2006). The elastic or elastoplastic rolling contact problems for homogeneous materials were considered by many authors. For details see references in (Chudzikiewicz et al. 2011, Shillor et al. 2004, Wu et al. 2011, Wriggers 2006).

Numerous laboratory or numerical experiments (Ahn et al. 2010, Choi et al. 2008, Nikbakht et al. 2013, Prasad et al. 2009, Yuan et al. 2012) indicate that the use of a coating functionally graded material attached to the conventional steel body reduce the magnitude of residual or thermal stresses and rolling contact fatigue. Functionally graded materials are multiphase composites consisted mainly from a ceramic and a metal.

Therefore in this paper we solve numerically the wheel-rail contact problem with friction assuming plastically graded model of the coating layer rather than elastic as in (Chudzikiewicz et al. 2011). In the paper the time-dependent model of this rolling contact problem is formulated. The elastic and plastic responses are approximated by the Hooke's law and von Mises yield criterion with an isotropic power law of hardening, respectively (Hager 2009, Han et al. 2013). Finite difference and finite element methods are used as discretization methods. It is well known that the application of the classical finite element method, where material properties are constant, to solution of problems with the functionally graded materials may lead to large numerical errors. A proper approach to solve such problems requires application of nonhomogeneous finite element method containing additional approximation functions in order to interpolate material properties at the level of each finite element. This idea is implemented in the framework of the graded (Kugler et al. 2013) or multi-scale (Efendiev 2009) finite element methods. The discretized contact problem is solved numerically using the semi-smooth Newton method (Hager 2009). The distribution of stresses including the normal and tangent contact stresses as well as the distribution of the temperature are numerically calculated. The provided results are discussed.

2 ROLLING CONTACT PROBLEM

Consider deformations of a suitably long two-dimensional strip $\Omega \subset R^2$ lying on a rigid foundation (see Fig. 1). A wheel rolls along the upper surface of the strip. The axis of the wheel is moving along a straight line at a constant altitude and the wheel is pressed in the elastoplastic strip. Moreover it is assumed, that there are no mass forces in the strip. The strip consists of two layers $\Omega_c \cup \Omega_s = \Omega$. The mechanical and thermal properties of the surface layer are assumed to vary throughout the material

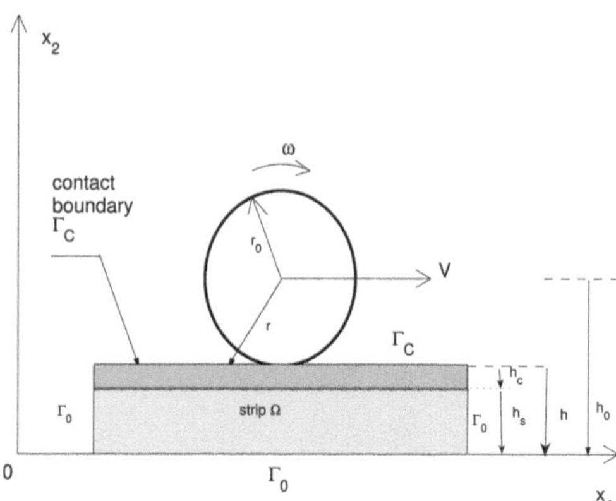

Figure 1. Wheel rolling over the two-layered rail.

according to the power law (Ahmadi et al. 2011, Ahn et al. 2010):

$$P = P(x) = P_c + (P_s - P_c)\left(\frac{h_s + h_c - x_2}{h_c}\right)^n, \tag{1}$$

for $h_s \le x_2 \le h_s + h_c$, $x = (x_1, x_2) \in \Omega$. $P = P(x)$, denotes the material property dependent on height x_2 and P_c, P_s are the ceramic property and the steel property, respectively. n denotes the nonhomogeneity parameter of the graded material. The symbol P may be used for the elastic modulus, the thermal expansion, the thermal conductivity or the thermal diffusivity. The Poisson's ratio is assumed constant. Conventional model of isotropic plasticity is assumed. The yield function depends on effective Mises stress, the yield stress on the upper surface and the isotropic hardening function.

2.1 *Thermo-elastoplastic model*

Consider wheel-rail contact described by the coupled thermo-elastoplastic system of equations. Let us denote by $u = (u_1, u_2)$, $u = u(x, t)$, $x \in \Omega$, $t \in (0, T)$, $T > 0$, a displacement of the strip and by $\theta = \theta(x, t)$ the absolute temperature of the strip. Since the rail consists from the coating layer Ω_c and the substrate layer Ω_s we denote by u_c and u_s as well as by θ_c and θ_s the displacement and temperature of these layers, respectively. Therefore $u = u_c$ in $\Omega_c \times (0, T)$ and $u = u_s$ in $\Omega_s \times (0,T)$. A similar relation holds for θ.

Assume wheel and rail brought into contact under the action of the static wheel load. We use two-dimensional elastoplastic model and Coulomb friction model to determine contact area and stress distribution. The displacement u of the strip Ω is governed by the system of the following equations (Ahmadi et al. 2011, Ahn et al. 2010, Chudzikiewicz et al. 2011, Wu et al. 2011)

$$\rho_c(x_2)\ddot{u}_c = \operatorname{div}\sigma_c - \alpha_c(x_2)(3\lambda_c(x_2) + 2\gamma_c(x_2))\nabla\theta_c \quad \text{in } \Omega_c \times (0,T), \tag{2}$$

$$\rho_s\ddot{u}_s = \operatorname{div}\sigma_s - \alpha_s(3\lambda_s + 2\gamma_s)\nabla\theta_s, \tag{3}$$

in $\Omega_s \times (0,T)$. Here $\ddot{u}_c = \frac{\partial^2 u_c}{\partial t^2}$. The gradient $\nabla\theta_j = \left[\frac{\partial\theta_j}{\partial x1}, \frac{\partial\theta_j}{\partial x2}\right]^T$ of temperature θ_j for $j =$ "c" or $j =$ "s" appears in equations (2) and (3). Moreover for $j =$ "c" or $j =$ "s" ρ_j, α_j, λ_j and γ_j denote mass density coefficient, thermal expansion coefficient, Lamé coefficients Lamé coefficients (Han et al. 2013, Shillor et al. 2004, Wriggers 2006), respectively. For $j =$ "c" these coefficients depend on x_2 according to (1). For $j =$ "s" these coefficients are constant.

The strain $\epsilon(u)$ is additively decomposed (Hager 2009, Han et al. 2013) into its elastic and plastic parts $\epsilon = \epsilon^e + \epsilon^p$ where $\epsilon(u) = \frac{1}{2}(\nabla u + \nabla u^T)$, stress tensor in domain Ω is $\sigma = \sigma(u) = \{\sigma_{ij}\}_{i=1}^{2} = A^e\, \epsilon^e = A^e(\epsilon(u) - \epsilon^p)$, $\operatorname{tr}\epsilon^p = 0$ on Ω. ∇u^T denotes a transpose of ∇u, A^e denotes Hooke's tensor and $\operatorname{tr}\varepsilon^p$ is the trace of ϵ^p. Let us define the plasticity conditions (Hager 2009, Han et al. 2013). Denote by η and f^p relative deviatoric stress and the yieldfunction, respectively, defined by:

$$\eta = \operatorname{dev}\sigma - K\,\epsilon^p, \quad \operatorname{dev}\sigma = \sigma - \operatorname{tr}\sigma Id, \tag{4}$$

$$f^p(\varepsilon,\eta) = \|\eta\| - Y^p(\varepsilon), \quad Y^p(\varepsilon) = \sigma_0 + H\varepsilon, \tag{5}$$

where Id is the identity operator, K, $H \ge 0$ are hardening parameters, $\sigma_0 > 0$ denotes the yield stress and ε is equivalent plastic strain. The plastic strain $\varepsilon^p : \Omega \times (0,T] \to \{D \in R^{2\times2} : D = D^T,\ \operatorname{tr}D = 0\}$ and the relative deviatoric stress η satisfy the flow rule in $\Omega \times (0,T]$ (Hager 2009, Han et al. 2013)

$$\dot{\epsilon}^p = \begin{cases} \gamma^p \dfrac{\eta}{\|\eta\|} & \text{for } \|\eta\| \ge 0, \\ 0 & \text{for } \|\eta\| = 0, \end{cases} \tag{6}$$

with $\epsilon^p(0) = \epsilon^p(x,0) = 0$. The yield function f^p and consistency parameter γ^p satisfy the complementarity conditions in $\Omega \times (0,T]$

$$\gamma^p \ge 0, \ f^p(\varepsilon,\eta) \le 0, \ \gamma^p f^p(\varepsilon,\eta) = 0. \tag{7}$$

The equivalent plastic strain ε satisfies the evolution law $\dot{\varepsilon} = \gamma^p$ and $\varepsilon(x,0) = \varepsilon(0) = 0$. From it follows the plasticity law is associative, i.e., the inner variable (z, ϵ^p) with $z = -\varepsilon H$ satisfies $\dot{z} = \gamma^p \frac{\partial f^p}{\partial \varepsilon}$ and $\dot{\epsilon}^p = \gamma^p \frac{\partial f^p}{\partial \eta}$.

Along the boundary $\Gamma_0 = \Gamma \setminus \Gamma_C$ the strip is assumed to be clamped, i.e., $u_c = u_s = 0$ on $\Gamma_0 \times (0,T)$. Displacement continuity condition is assumed on the interface between layers $\Gamma_c \cap \Gamma_s$.

In the contact zone the surface traction vector F is determined by:

$$-\sigma n = F \quad \text{on } \Gamma_C \times (0,T), \tag{8}$$

where $n=(n_1,n_2)$ denotes outward normal versor to the boundary Γ of the domain Ω. At the initial moment $t = 0$ the displacement and velocity of points of the rail domain are given, i.e., $u_c(0)=\bar{u}_{0c}$ $\dot{u}_c(0)=\bar{u}_{1c}$ in Ω_c and $u_s(0)=\bar{u}_{0s}$ $\dot{u}_s(0)=\bar{u}_{1s}$ in Ω_s where for $j="c"$ or $j="s"$ we denote $\dot{u}=du_j/dt$, $\bar{u}_{0j}$, $\bar{u}_{1j}$ are given functions. The surface traction vector $F=(F_1,F_2)=(F_t,F_n)$ on the boundary Γ_C is a priori unknown and is given by conditions of contact and friction. Assumed that the strip displacement is suitably small the contact conditions on $\Gamma_C \times (0,T)$ take a form (Chudzikiewicz et al. 2011, Hager 2009, Han et al. 2013, Wriggers 2006):

$$u_{c2}+g_r \le 0, \quad F_n \le 0, \quad (u_{c2}+g_r)F_n = 0, \tag{9}$$

$$|F_t| \le \mu |F_n|, \quad F_t \dot{u}_1 \le 0, \quad (|F_t| - \mu |F_n|)\dot{u}_1 = 0, \tag{10}$$

where $\dot{u}_1 = \frac{du_1}{dt}$, μ is a constant friction coefficient and u_{c2} denotes vertical component of u_c. Under suitable assumptions the gap between the bodies is equal to $g_r = h - h_0 + \sqrt{r_0^2 - (u_1+x_1)^2}$ (Chudzikiewicz et al. 2011, Shillor et al. 2004, Wriggers 2006).

The frictional heating within the contact patch is generally a time dependent heat source. The heat governing equation at each layer of the strip Ω becomes (Ahmadi et al. 2011, Ahn et al. 2010, Chudzikiewicz et al. 2011, Prasad et al. 2009, Sebes et al. 2014):

$$\rho_c(x_2)c_c(x_2)\dot{\theta}_c = \bar{\kappa}_c(x_2)\left(\frac{\partial^2\theta_c}{\partial x_1^2} + \frac{\partial^2\theta_c}{\partial x_2^2}\right) + \frac{\partial \bar{\kappa}_c(x_2)}{\partial x_2}\frac{\partial \theta_c}{\partial x_2} \quad \text{in } \Omega_c \times (0,T), \tag{11}$$

$$\rho_s c_s \dot{\theta}_s = \bar{\kappa}_s\left(\frac{\partial^2\theta_s}{\partial x_1} + \frac{\partial^2\theta_s}{\partial x_2}\right) \quad \text{in } \Omega_s \times (0,T), \tag{12}$$

where $\dot{\theta}_c = \frac{\partial\theta_c}{\partial t}$. For $j="c"$ or $j="s"$ $\bar{\kappa}_j$ and c_j denote a thermal conductivity coefficient and a heat capacity coefficient respectively. These coefficients in Ω_c $(j="c")$ are the functions of the height x_2 governed by (1) and in Ω_s $(j="s")$ are assumed to be constant. The boundary conditions associated with the heat equation are dependent on contact between the bodies and surrounding environment. The first case is when the surfaces of the wheel and the rail are in contact. It causes heat generation due to friction. This heat flow is expressed as boundary condition

$$-\bar{\kappa}\frac{\partial\theta_c}{\partial n}(x,t) = \bar{\alpha}\mu V F_n, \quad \text{on } \Gamma_C \times (0,T), \tag{13}$$

where $\bar{\alpha}$ represents the fraction of frictional heat flow rate entering the rail. When the contact surfaces are separated this condition takes the form $-\bar{\kappa}\frac{\partial\theta_c}{\partial n}(x,t)=0$. Moreover $\theta_s=\theta_c$ holds on $\Gamma_c \cap \Gamma_s$. Along the boundary Γ_0 temperature $\theta_c=\theta_s=\theta_g$ where θ_g denotes the given temperature of the surrounding air. At the initial moment $t=0$ holds $\theta_c(0,x)=\theta_s(0,x)=\theta_g$ in $\Omega_c \cup \Omega_s$. We assume problem (1)–(13) possesses a solution.

3 NUMERICAL ALGORITHM

The finite difference and finite element methods (Efendiev 2009, Han et al. 2013, Kugler et al. 2013, Wriggers 2006) have been used to discretize problem (1)–(13). Inequality plasticity and contact conditions (6)–(7) and (9)–(10) have been reformulated in the form of equations using suitable nonlinear complementary functions (Hager 2009). The coupled system of equations have been solved using operator splitting method combined with semi-smooth Newton and Cholesky methods.

4 NUMERICAL RESULTS

The numerical results indicate that the graded layer can reduce the values of the normal contact stress and the maximal temperature in the contact zone. Figure 2 compares the normal contact pressure distributions on the contact surface for different

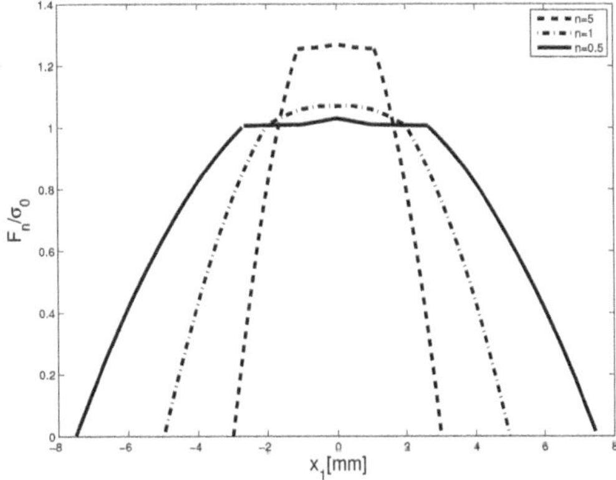

Figure 2. Normal contact distribution for different nonhomogeneity parameters n.

values of the nonhomogeneity gradient n. Three values of n have been employed $n = 0.5, 1, 5$. In all cases the maximum contact pressure of F_n occurs at the center of the contact zone. As n increases and the coated surface becomes more flexible the maximal contact pressure value is slightly increasing at a cost of shrinking the contact zone length. The obtained maximal values are lower than in pure homogeneous case, i.e., $n = 0$. In elastoplastic model the contact zone is generally greater than in elastic model. The obtained contact patches are characterized by longer zones and lower stress intensity than in the elastic case. Figure 3 displays the distribution of temperature in the strip in x_2 (i.e. normal) direction along the line $x_1 = 0$. Points $x_2 = 0$ and $x_1 = 1$ are lying on the interface between the rail and the rigid foundation and on the contact surface respectively. The maximum value of the temperature appears at in the contact area

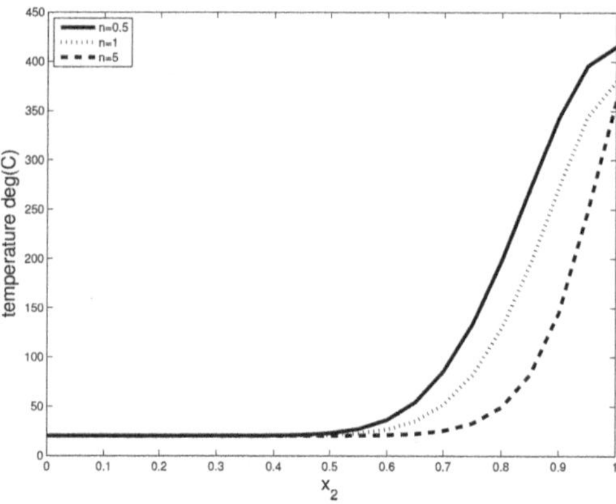

Figure 3. Rail temperature distribution along x_2 direction at $x_1 = 0$.

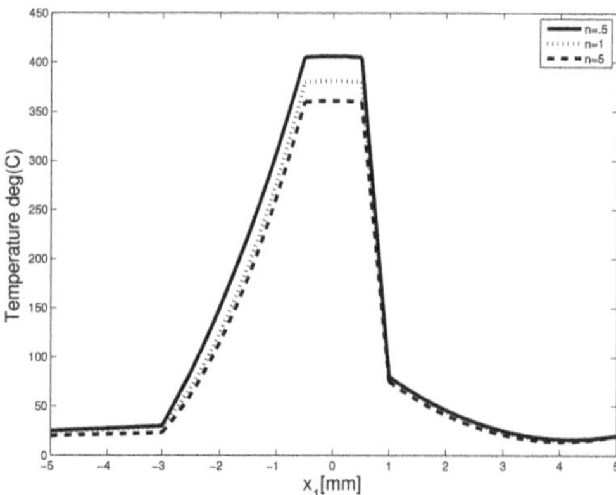

Figure 4. Temperature distribution along x_1 direction on the contact interface at $x_2 = 1$.

and in adjacent area. As n increases the temperature maximum value decreases. Moving from the contact surface inside the rail to the rigid foundation the temperature rapidly decays to the ambient temperature. Temperature distribution along the contact boundary (i.e. $x_2 = 1$) in x_1 direction (i.e. tangent) displays Figure 4. In this figure contact occurs at the point $x_1 = 0$. As nonhomogeneity gradient n increases from 0.5 to 5 the temperature maximum value decreases. The obtained temperature maximum values are lower than in pure homogeneous material case, i.e., $n = 0$.

5 CONCLUSIONS

The thermo-elastoplastic rolling contact problem where the properties of the elastoplastic upper coating layer of the rail surface are dependent on its depth is solved numerically using the operator splitting method combined with the semi-smooth Newton and Cholesky methods. The material properties of the graded layer are governed by the power law. The numerical results indicate that the obtained contact patches are characterized by longer zones and lower stress intensity than in the elastic case. The obtained stress and temperature distributions are dependent on the nonhomogeneity index. For its higher values one can obtain the higher differences in the maximum contact stress and temperature compared to the homogeneous material case. Application of the coating material reduces the contact pressure and the rolling contact fatigue. The analysis of two–material rails indicates that even for coated rails there are cases where the shakedown limit is exceeded. It implies that the choice of the rail profile or its material properties including the depth of the coating layer may be based on the optimization approach to ensure the best admissible reduction of the generated normal contact stress or temperatures. For the optimization results concerning contact systems see (Myśliński 2015).

REFERENCES

Ahmadi, S., Sheikhlou, M., & Gharebagh, V.M. (2011). Thermo-elastic/plastic semi-analytical solution of incompressible functionally graded spherical pressure vessel under thermo-mechanical loading, *Acta Mech.*, 222, 161–173.

Ahn, S., Jang, Y.H. (2010). Frictionally excited thermo-elastoplastic instability. *Tribology International*, 43, 779–784.

Branch, N., Arakere, N., Subhash, G., Klecka, M. (2011). Determination of constitutive response of plastically graded materials. *International Journal of Plasticity*, 27(5), 728–738.

Choi, I.S., Dao, M., Suresh, S. (2008). Mechanics of indentation of plastically graded materials–I: Analysis. *Journal of the Mechanics and Physics of Solids*, 56, 157171.

Chudzikiewicz, A., Myśliński, A. (2011). Thermoelastic Wheel–Rail Contact Problem with Elastic Graded Materials. *Wear*, 271, 417–425.

Efendiev, Y., Hou, T. (2009). *Multiscale Finite Element Methods: Theory and Applications*. Springer.

Hager, C., Wohlmuth B.I. (2009). Nonlinear complementarity functions for plasticity problems with frictional contact. *Comput. Methods Appl. Mech. Engrg.* 198, 3411–3427.

Han, W., Reddy, B.D. (2013). *Plasticity. Mathematical Theory and Numerical Analysis*. 2nd edition, Springer, New York.

Kugler, S., Fotiu, P.A., Murin, J. (2013). Enhanced functionally graded material shell finite elements. *ZAMM Journal of applied mathematics and mechanics*, 94(1–2), 72–84.

Myśliński A. Piecewise constant level set method for topology optimization of unilateral contact problems, *Advances In Engineering Software*, 2015, Vol. 80, pp. 25–32.

Nikbakht, A., Sadighi, M., Arezoodar, Fallahi A., Zucchelli, A. (2013). Elastic–plastic frictionless indentation analysis of a functionally graded vitreous enameled steel plate by a rigid spherical indenter. *Materials Science & Engineering A*, 564, 242–254.

Prasad, A., Dao, M., Suresh, S. (2009). Steady-state frictional sliding contact on surfaces of plastically-graded materials. *Acta Materialia*, 57, 511–524.

Sebes, M., Chollet, H., Ayasse, J.B., Chevalier, L. (2014). A multi-Hertzian model considering plasticity. *Wear*, 314, 118–124.

Shillor, M., Sofonea, M., Telega, J.J. (2004). *Models and Analysis of Quasistatic Contact: Variational Methods*. Springer, Berlin.

Yuan, F., Jiang, P., Xie, J., Wu, X., Analysis of spherical indentation of materials with plastically graded surface layer, *International Journal of Solids and Structures*, 49(3–4), pp. 527–536, 2012.

Wu, L., Wen, Z., Li, W., Jin, X. (2011). Thermo-elastic-plastic finite element analysis of wheel-rail sliding contact. *Wear*, 271, 437–443.

Wriggers, P. (2006). *Computational contact mechanics*. Second Edition, Springer, Berlin.

Advances in Mechanics: Theoretical, Computational and Interdisciplinary Issues – Kleiber et al. (Eds)
© 2016 Taylor & Francis Group, London, ISBN 978-1-138-02906-4

Structural optimization of contact problems using piecewise constant level set method

A. Myśliński & M. Wróblewski
Systems Research Institute, Warsaw, Poland

ABSTRACT: The paper deals with the topology optimization of the elastic contact problems using the level set approach. A piecewise constant level set method is used to follow the evolution of design domain interfaces rather than the standard level set method. The piecewise constant level set function takes distinct constant values in each subdomain of a whole design domain. Using a two-phase approximation the original optimization problem is reformulated as an equivalent constrained optimization problem in terms of the piecewise constant level set function. Necessary optimality condition is formulated. Finite difference and finite element methods are applied as the approximation methods. Numerical examples are provided and discussed.

1 INTRODUCTION

The paper deals with the numerical solution of a structural optimization problem for contact problems between an elastic body and a rigid foundation. This contact phenomenon is governed by an elliptic variational inequality. The structural optimization problem for the elastic body in unilateral contact consists in finding such topology of the domain occupied by the body and/or the shape of its boundary that the normal contact stress along the boundary of the body is minimized. The volume of the body is assumed bounded.

The standard level set method (Allaire et al. 2004, Osher et al. 2003) is employed in structural optimization for numerical tracking the evolution of the domain boundary on a fixed mesh and finding an optimal domain. This method is based on an implicit representation of the boundaries of the optimized structure. Recently, different modifications (De Cezaro et al. 2012, Yamada et al. 2010) of the standard level set method are developed to increase its effectiveness. Among others an arbitrary number of subdomains can be identified using only one discontinuous piecewise constant level set function taking distinct constant values on each subdomain (De Cezaro et al. 2012, Lie et al. 2005, Myśliński 2015, Wei et al. 2009, Zhu et al. 2011).

In the paper the original structural optimization problem is approximated by a two-phase optimization problem using weak and strong phases (Allaire et al. 2004). Using the piecewise constant level set method (De Cezaro et al. 2012) this approximated problem is reformulated as an equivalent constrained optimization problem in terms of the piecewise constant level set function only. Therefore neither shape nor topological sensitivity analysis is required. During the evolution of the piecewise constant level set function small holes can be created without use of the topological derivatives. The paper extends results contained in (Myśliński 2015). Necessary optimality condition is formulated. The finite difference and finite element methods are used as the approximation methods. This discretized optimization problem is solved numerically using the augmented Lagrangian method. Numerical examples are provided and discussed.

2 PROBLEM FORMULATION

Consider deformations of an elastic body occupying two-dimensional domain Ω with the smooth boundary Γ. The elastic body obeying Hooke's law is subject to body forces $f(x) = (f_1(x), f_2(x))$, $x \in \Omega$. Moreover, surface tractions $p(x) = (p_1(x), p_2(x))$, $x \in \Gamma$, are applied to a portion Γ_1 of the boundary Γ. Assume the body is clamped along the portion Γ_0 of the boundary Γ, and that the contact conditions are prescribed on the portion Γ_2, where $\Gamma_i \cap \Gamma_j = \emptyset$, $i \neq j$, $i, j = 0, 1, 2$, $\Gamma = \overline{\Gamma}_0 \cup \overline{\Gamma}_1 \cup \overline{\Gamma}_2$. We denote by $u = (u_1, u_2)$, $u = u(x)$, $x \in \Omega$, the displacement of the body and by $e(x) = \{e_{ij}(u(x))\}$ as well as by $\sigma(x) = \{\sigma_{ij}(u(x))\}$, $i, j = 1, 2$, the strain field and stress field in the body, respectively.

Let us formulate a contact problem in variational form. Denote by V_{sp} and K the space and set of kinematically admissible displacements given by

$V_{sp} = \{z \in [H^1(\Omega)]^2 = H^1(\Omega) \times H^1(\Omega) : z_i = 0 \text{ on } \Gamma_0, i = 1,2\}$ and $K = \{z \in V_{sp} : z_N \leq 0 \text{ on } \Gamma_2\}$. Let Λ denotes the set of Lagrange multipliers $\Lambda = \{\zeta \in L^2(\Gamma_2) : |\zeta| \leq 1\}$. Variational formulation of contact problem has the form: *find a pair* $(u, \lambda) \in K \times \Lambda$ *satisfying*

$$\int_\Omega a_{ijkl} e_{ij}(u) e_{kl}(\varphi - u) dx - \int_\Omega f_i(\varphi_i - u_i) dx - \tag{1}$$

$$\int_{\Gamma_1} p_i(\varphi_i - u_i) ds + \int_{\Gamma_2} \lambda(\varphi_T - u_T) ds \geq 0,$$

$$\int_{\Gamma_2} (\zeta - \lambda) u_T ds \leq 0, \tag{2}$$

for all $(\varphi, \zeta) \in K \times \Lambda$ $i, j, k, l = 1, 2$. The elasticity tensor satisfying usual requirements is denoted by $\{a_{ijkl}\}$ and the tangential (normal) displacement by $u_T (u_N)$. We use here and throughout the paper the summation convention over repeated indices (Myśliński 2015).

Before formulating a structural optimization problem for (1)–(2) let us introduce first the set U_{ad} of admissible domains in the form $U_{ad} = \{\Omega : \Omega$ is Lipschitz continuous, $Vol(\Omega) - Vol^{giv} \leq 0, Per(\Omega) \leq const_1\}$ where $Vol(\Omega) \overset{def}{=} \int_\Omega dx$ and $Per(\Omega) \overset{def}{=} \int_\Gamma dx$. The set U_{ad} is assumed to be nonempty. In order to define a cost functional we shall also need the following set M^{st} of auxiliary functions $M^{st} = \{\eta = (\eta_1, \eta_2) \in [H^1(D)]^2 : \eta_i \leq 0 \text{ on } D, i = 1,2, \|\eta\|_{[H^1(D)]^2} \leq 1\}$ where the norm $\|\eta\|_{[H^1(D)]^2} = (\Sigma_{i=1}^2 \|\eta_i\|^2_{H^1(D)})^{1/2}$. Recall from (Myśliński 2015) the cost functional approximating the normal contact stress on the contact boundary

$$J_\eta(u(\Omega)) = \int_{\Gamma_2} \sigma_N(u) \eta_N(x) ds, \tag{3}$$

depending on the auxiliary given bounded function $\eta(x) \in M^{st}$. σ_N and ϕ_N are the normal components of the stress field σ corresponding to a solution u satisfying system (1)–(2) and the function η, respectively.

We shall consider the following structural optimization problem: *for a given function* $\eta \in M^{st}$, *find a domain* $\Omega^\star \in U_{ad}$ *such that*

$$J_\eta(u(\Omega^\star)) = \min_{\Omega \in U_{ad}} J_\eta(u(\Omega)) \tag{4}$$

3 PIECEWISE CONSTANT LEVEL SET APPROACH

In (Allaire et al. 2004) the standard level set method (Osher et al. 2003) is employed to govern the evolution of domains and to solve numerically problem (4). Denote by $t > 0$ the artificial time variable and consider the evolution of a domain Ω under a velocityfield V. Under the suitable regular mapping $T(t, V)$ we have $\Omega_t = T(t,V)(\Omega) = (I + tV)(\Omega)$, $t > 0$. By Ω_t^- (resp. Ω_t^+) we denote the interior (resp. outside) of the domain Ω_t. The domain Ω_t and its boundary $\partial\Omega_t$ are determined by a function $\phi = \phi(x,t) : R^2 \times [0, t_0) \to R$ satisfying: $\phi(x,t) = 0$, if $x \in \Omega_t$, $\phi(x,t) < 0$, if $x \in \Omega_t^-$, $\phi(x,t) > 0$, if $x \in \Omega_t^+$. Function ϕ is called the level set function (Osher et al. 2003).

Let us reformulate problem (4) in terms of a piecewise constant level set function. For hold-all domain $D \subset R^2$ partitioned into N subdomains $\{\Omega_i\}_{i=1}^N$ such that $D = \bigcup_{i=1}^N (\Omega_i \cup \partial\Omega_i)$ where N is a given integer and $\partial\Omega_i$ denotes the boundary of the subdomain Ω_i a piecewise constant level set function $\phi : D \to R$ is defined as (Yamada et al. 2010, Wei et al. 2009)

$$\phi = i \ \text{ in } \Omega_i, \quad i = 1, 2, ..., N. \tag{5}$$

Consider piecewise constant density function $\rho : D \to R^2$ defined as

$$\rho(x) = \begin{cases} \epsilon & \text{if } x \in D \setminus \bar{\Omega}, \\ 1 & \text{if } x \in \Omega, \end{cases} \tag{6}$$

where $\epsilon > 0$ is a small constant. We confine to consider a two-phase problem in the domain D where the characteristic functions of the subdomains are $\chi_1(x) = 2 - \phi(x)$ and $\chi_2(x) = \phi(x) - 1$. Therefore $\rho(x) = \rho_1 \chi_1(x) + \rho_2 \chi_2(x) = (1 - \epsilon)\phi(x) + 2\epsilon - 1$.

Using it as well as (5) the structural optimization problem (4) can be transformed into the following one: *find* $\phi \in U_{ad}^\phi$ *such that*

$$\min_{\phi \in U_{ad}^\phi} J_\eta(\phi) = \int_{\Gamma_2} \rho(\phi) \sigma_N(u_\epsilon) \eta_N ds \tag{7}$$

where the set U_{ad}^ϕ of the admissible functions is given as

$$U_{ad}^\phi = \{\phi \in H^1(D) : Vol(\phi) - Vol^{giv} \leq 0, \tag{8}$$

$$Vol(\phi) \overset{def}{=} \int_\Omega \rho(\phi) dx,$$
$$W(\phi) \overset{def}{=} (\phi - 1)(\phi - 2) = 0, \tag{9}$$
$$Per(\phi) \overset{def}{=} \int_\Omega |\nabla \phi| dx \leq const_1\}.$$

The element $(u_\epsilon, \lambda_\epsilon) \in K \times \Lambda$ satisfies the state system (1)–(2) in the domain D rather than Ω.

4 NECESSARY OPTIMALITY CONDITION

Using a two-phase approximation the original structural optimization problem (4) is reformulated as an equivalent constrained optimization problem (7) in terms of the piecewise constant level set function $\phi(x)$. Using the Augmented Lagrangian associated to the problem (7) the derivative of the cost functional is calculated and a necessary optimality condition is formulated. Let us formulate the necessary optimality condition for the optimization problem (7)–(9). We denote by $\tilde{\mu}=\{\tilde{\mu}_i\}_{i=1}^3$ Lagrange multiplier associated with constraints in the set (8). Let us introduce the Augmented Lagrangian $L(\phi,\tilde{\mu})$ associated with this optimization problem:

$$\begin{aligned}L(\phi,\tilde{\mu}) &= L(\phi,u_\epsilon,\lambda_\epsilon,p^a,q^a,\tilde{\mu})\\ &= J_\eta(\phi)+\int_D \rho(\phi)a_{ijkl}e_{ij}(u_\epsilon)e_{kl}(p^a)dx\\ &\quad -\int_D \rho(\phi)f_ip_i^a dx-\int_{\Gamma_1} p_ip_i^a ds+\int_{\Gamma_2}\lambda_\epsilon p_T^a ds\\ &\quad +\int_{\Gamma_2} q^a u_{\epsilon T} ds+\tilde{\mu}c(\phi)+\sum_{i=1}^3\frac{1}{2\beta_i}c_i^2(\phi),\end{aligned} \tag{10}$$

where $i, j, k, l = 1,2$. Moreover $c(\phi)\overset{def}{=}\{c_i(\phi)\}_{i=1}^3 = [\boldsymbol{Vol}(\phi),\boldsymbol{Per}(\phi),\boldsymbol{W}(\phi)]^T$, $c^T(\phi)$ denotes a transpose of $c(\phi)$ and $\beta_m>0$, $m=1,2,3$, are a given real numbers. The pair $(\boldsymbol{p}^a,\boldsymbol{q}^a)\in K_1\times\Lambda_1$ denotes an adjoint state defined as the solution to the following variational inequality (Myśliński 2015):

$$\begin{aligned}&\int_D \rho(\phi)a_{ijkl}e_{ij}(\eta+p^a)e_{kl}(\varphi)dx\\ &\quad+\int_{\Gamma_2} q^a\varphi_T ds=0 \quad \forall\varphi\in K_1,\end{aligned} \tag{11}$$

$$\int_{\Gamma_2}\zeta(p_T^a+\eta_T)ds=0 \quad \forall\zeta\in\Lambda_1. \tag{12}$$

The sets K_1 and Λ_1 are given by

$$K_1=\{\xi\in V_{sp}:\xi_N=0 \text{ on } A^{st}\}, \tag{13}$$

and by

$$\Lambda_1=\{\zeta\in\Lambda:\zeta(x)=0 \text{ on } \boldsymbol{B}^{st}\}, \tag{14}$$

while the coincidence set $A^{st}=\{x\in\Gamma_2:\ u_N+v=0\}$ and $B^{st}=B_1\cup B_2\cup B_1^+\cup B_2^+$. Moreover $B_1=\{x\in\Gamma_2:\lambda(x)=-1\}$, $B_2=\{x\in\Gamma_2:\ \lambda(x)=+1\}$, $\tilde{B}_i=\{x\in B_i:u_N(x)+v=0\}$, $i=1,2$, $B_i^+=B_i\setminus\tilde{B}_i$, $i=1,2$. For interpretation of these sets see (Myśliński 2015).

The derivatives of functions $\rho(\phi)$, $c(\phi)$ with respect to ϕ are equal to (Myśliński 2015, Zhu et al. 2011) $\rho'(\phi)=1-\epsilon$, $c'(\phi)=[\boldsymbol{Vol'}(\phi),\boldsymbol{W'}(\phi),\boldsymbol{Per'}(\phi)]$, respectively. Moreover $Vol'(\phi)=1$, $W'(\phi)=2\phi-3$ and

$$Per'(\phi)=\chi_{\{\partial\Omega=const_0\}}\max\{0,$$

$$\left.-\nabla\cdot\left(\frac{\nabla\phi}{|\nabla\phi|}\right)\right\}-\chi_{\{\partial\Omega>const_0\}}\nabla\cdot\left(\frac{\nabla\phi}{|\nabla\phi|}\right). \tag{15}$$

Therefore the derivative of the Lagrangian L with respect to ϕ has the form:

$$\begin{aligned}\frac{\partial L}{\partial\phi}(\phi,\tilde{\lambda}) &= \int_D \rho'(\phi)[a_{ijkl}e_{ij}(u_\epsilon)e_{kl}(p^a+\eta)\\ &\quad -f_i(p_i^a+\eta)]dx+\tilde{\mu}c'(\phi)+\sum_{i=1}^3\frac{1}{\beta_i}c(\phi)c'(\phi).\end{aligned} \tag{16}$$

Using (15)–(16) we can formulate the necessary optimality condition for topology optimization problem (7)–(9). It takes the form (Myśliński 2015): if $\hat{\phi}\in U_{ad}^{\phi}$ is an optimal solution to the problem (7)–(9) than there exists Lagrange multiplier $\tilde{\mu}^\star\in R^3$ such that $\tilde{\mu}_1^\star,\tilde{\mu}_2^\star\geq 0$ and satisfying for all $\phi\in U_{ad}^{\phi}$ and $\tilde{\mu}\in R^3$ the inequalities

$$L(\hat{\phi},\tilde{\mu})\leq L(\hat{\phi},\tilde{\mu}^\star)\leq L(\phi,\tilde{\mu}^\star). \tag{17}$$

Condition (17) implies (Myśliński 2015) that for all $\phi\in U_{ad}^{\phi}$ and $\tilde{\mu}\in R^3$

$$\frac{\partial L(\hat{\phi},\mu^\star)}{\partial\phi}\geq 0 \text{ and } \frac{\partial L(\hat{\phi},\tilde{\mu}^\star)}{\partial\tilde{\mu}}\leq 0, \tag{18}$$

hold at the optimal point $(\hat{\phi},\tilde{\mu}^\star)\in U_{ad}^{\phi}\times R^3$.

5 NUMERICAL IMPLEMENTATION

The optimization problem (7) is discretized using the finite difference and the finite element methods. The discretized optimization problem is numerically solved using Uzawa type iterative algorithm. The minimization of the Augmented Lagrangianwith respect to function ϕ is realized using the gradient flow equation. For details see (Myśliński 2015).

6 NUMERICAL EXPERIMENTS

The discretized topology optimization problem (7)–(9) has been solved numerically in Matlab environment. The elastic body in unilateral contact with the rigid foundation is assumed to occupy two-dimensional domain Ω given by

$$\Omega = \{(x_1, x_2) \in R^2 : 0 \le x_1 \le 8 \\ \wedge 0 < v(x_1) \le x_2 \le 4\}, \quad (19)$$

with the function $v(x_1) = 0.125 * (x_1 - 4)^2$. The boundary Γ of the domain Ω is divided into three pieces

$$\Gamma_0 = \{(x_1, x_2) \in R^2 : x_1 = 0, 8 \\ \wedge 0 < v(x_1) \le x_2 \le 4\}, \quad (20)$$

$$\Gamma_1 = \{(x_1, x_2) \in R^2 : 0 \le x_1 \le 8 \wedge x_2 = 4\},$$

$$\Gamma_2 = \{(x_1, x_2) \in R^2 : 0 \le x_1 \le 8 \wedge x_2 = v(x_1)\}.$$

The computations are carried out for the elastic body characterized by the Poisson's ratio $\nu = 0.29$ and strong material Young's modulus $E = 2.1 \cdot 10^{11} N/m^2$. The weak material phase parameter $\epsilon = 10^{-3}$. The body is loaded by the boundary traction $p_1 = 0$, $p_2 = -5.6 \cdot 10^6 N$ along the boundary Γ_1. The body forces $f_i = 0$, $i = 1,2$. The computational domain $D = [0,8] \times [0,4]$ is selected. Domain D is discretized with a fixed rectangular mesh into 3200 elements. Auxiliary function η in (3) is selected as a piecewise linear on computational domain D. Material volume fraction $r_{fr} = 0.5$ is prescribed. The other computational parameters are equal to: the tolerance parameter $\varepsilon_1 = 10^{-4}$, the smoothness parameter $\varepsilon_2 = 10^{-6}$, the penalty parameter $\beta_i = 10^{-6}, i = 1,2,3$. The computations have been performed for the initial level set function $\phi^0 = 1.5$.

Figure 1 presents the optimal topology domain of structural optimization problem (7)–(9). The big area with low values of density function, i.e., filled with the weaker material, appear in the central part of the domain Ω and is symmetrically distributed. Two smaller such areas appear near the fixed edges. They are slightly unsymmetrically distributed due to numeric errors (Myśliński 2015). Figure 2 presents the distribution of the normal stress along the contact boundary for the initial and the optimal topology domains. The peak of the normal contact stress at the initial topology domain has been smeared out and significantly reduced. For the optimal topology domain the obtained normal contact stress along the contact boundary is almost constant.

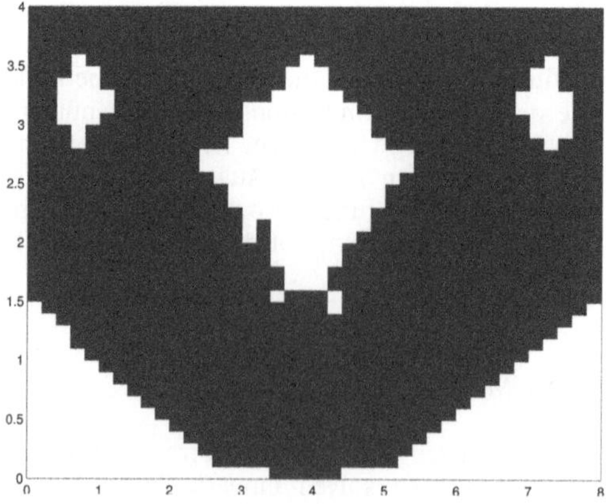

Figure 1. Optimal topology domain $\Omega^\star$. $\phi^0 = 1.5$.

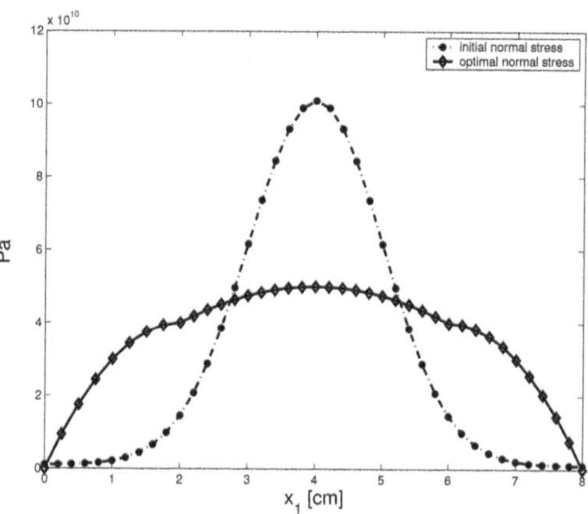

Figure 2. Normal contact stress distributions for initial and optimal domains. $\phi^0 = 1.5$.

7 CONCLUSIONS

A new method has been proposed for solving topology optimization problems for elastic unilateral contact problems with a given friction based on piecewise constant level set functions. The original topology optimization problem is approximated by the two-phase optimization problem and is transformed into the constrained optimization problem in terms of the piecewise constant level set function. The proposed method does not require to solve Hamilton—Jacobi equation and to perform the reinitialization process of the signed distance function as in standard level set approach. Moreover the proposed method has also voids nucleation capabilities as topological derivative based methods. It can be viewed as combining the elements of SIMP and topology derivative approaches.

The obtained numerical results that the optimal domains contain the areas with low values of density function in the central part of the body and near the fixed edges. The normal contact stress is almost constant along the optimal topology domain boundary and has been significantly reduced comparing to the initial one. The proposed algorithm is robust and finds optimal topologies which seems to be in accordance with the physical reasoning and engineering experience. Since the elliptic inequalities constrained topology optimization problems are generally nonconvex it is well known that their numerically obtained solutions are dependent on the initial design. Therefore the obtained optimal topology domains are likely of local character.

REFERENCES

Allaire, G., Jouve, F., Toader, A. (2004). Structural Optimization Using Sensitivity Analysis and a Level Set Method. *J. Comp. Phys.*, 194, 363–393.

De Cezaro, A., Leitão, A. (2012). Level-set approaches of L^2-type for recovering shape and contrast in ill-posed problems. *Inv. Prob. Sci. Eng.*, 20(4), 571–587.

Lie, J., Lysaker, M., Tai, X.C. (2005). A piecewise constant level set framework. *Int. J. Num. Anal. Model.*, 2(4), 422–438.

Myśliński, A. (2015). Piecewise constant level set method for topology optimization of unilateral contact problems. *Adv. Eng. Soft.*, 80, 25–32.

Osher, S., Fedkiw, R. (2003). *Level Set Methods and Dynamic Implicit Surfaces*. Springer, New York, New York.

Wei, P., Wang, M.Y. (2009). Piecewise constant level set method for structural topology optimization. *Int. J. Num. Meth. Eng.*, 78, 379–402.

Yamada, T., Izui, K., Nishiwaki, S., Takezawa, A. (2010). A topology optimization method based on the level set method incorporating a fictitious interface energy. *Comp. Meth. Appl. Mech. Eng.*., 199(45–48), 2876–2891.

Zhu, S., Wu, Q., Liu, C. (2011). Shape and topology optimization for elliptic boundary value problems using a piecewise constant level set method. *App. Num. Math.*, 61, 752–767.

Advances in Mechanics: Theoretical, Computational and Interdisciplinary Issues – Kleiber et al. (Eds)

 ISBN 978-1-138-02906-4

Multiscale analysis of metal matrix composite with ceramic reinforcement using Johnson-Holmquist constitutive model

W. Ogierman & G. Kokot
Silesian University of Technology, Gliwice, Poland

ABSTRACT: This paper deals with computational analysis of composite material reinforced with ceramic particles. The research is devoted to investigation of material behavior under different strain rates. In particular aluminum matrix composite reinforced with 20% of SiC particles is taken into consideration. During this study particle damage, that is often predominant in case of particle reinforced composites, is accounted by using Johnson-Holmquist material model. Matrix is modeled by elastic-plastic material model with yield stress scaling dependent on strain rate. Numerical simulation of dynamic compression test is performed. To analyze the material at micro level the embedded cell approach is applied and the boundary value problem is solved by the finite element method using LS-Dyna code. The obtained results are averaged stress-strain curves for different strain rates and micro stress and damage parameter fields in the particles.

1 INTRODUCTION

Modern metal matrix composites often contain reinforcement made of ceramic material providing high strength and stiffness. Strength and damage evolution in particle reinforced composites depend on a variety of factors such as particle characteristics, matrix strength and work hardening, strength and nature of the interface between the phases. Therefore modeling the behavior of mentioned materials is a complex issue form the computational point of view. In general three damage modes can be distinguished: particle damage, interfacial debounding and voids growth within matrix (Chawla 2013). During the study only particle damage is investigated, however experimental results report that in the case of many particle reinforced composites this is a predominant damage mechanism (Böhm et al. 2004). Computational study of the damage of particles in quasi-static condition is presented in work of Segurardo & Llorca (2006) where damage of particles is determined by using Weibull probability description. Similar approach is presented in work of Eckschlager et al. (2001). Mishnaevsky (2004) modeled damage of the particles by element weakening technique. In this case finite elements in which maximum principal stress exceeded a critical value were considered as failed and Young's modulus of these elements was set to a relatively low value. In order to simulate behavior of ceramic material under high strain rates the post-yield ceramic material response is significant. Therefore, during this study usage of Johnson-Holmquist damage model is proposed. The Johnson-Holmquist model incorporates the effect of damage on residual material strength and resulting bulking during the compressive failure of ceramic material (Cronin et al. 2003). This model is described in detail in section 2.2. The response of particle reinforced composite on dynamic load was previously modeled by using Johnson-Holmquist model by Zhang et al. (2007) however aforementioned investigation is restricted to assumption of plane strain condition. During this work numerical simulation of dynamic compression of 3D unit cell is performed in framework of finite element method. The main goals of this study is investigation of particle damage and strain rate sensitivity of a considered composite.

2 NUMERICAL MODEL

2.1 *Model description, boundary conditions*

The material considered is 6061-T6 aluminum matrix composite reinforced with 20% of SiC particles. To investigate the material response virtual dynamic compression test is performed. In this case material microstructure is represented by a unit cell containing randomly, non-periodically distributed particles. In order to set up dynamic compression simulation the embedded cell approach is used. The embedding models typically contains a local heterogeneous region, in which the microstructure is well resolved geometrically, that is embedded in an outer region that serves mainly for transmitting far-field loads (Böhm et al. 2004, Trias et al. 2006). This approach is particularly appropriate when material damage is studied at the micro-level (Böhm et al. 2004). Moreover the embedding region, which has the averaged properties of the composite, is necessary to avoid the boundary effects on the particle failure (Mishnaevsky 2004). Figure 1 shows geometry of the considered model. The cell contains 48 particles of this same diameter equal 30 μm. Geometry is discretized by tetrahedral finite elements, the whole model contains approximately 656000 finite elements.

Figure 2 shows dimensions of the model and applied boundary conditions. The Velocity that provides an appropriate strain rate and contact between embedding region and rigid bodies have been applied.

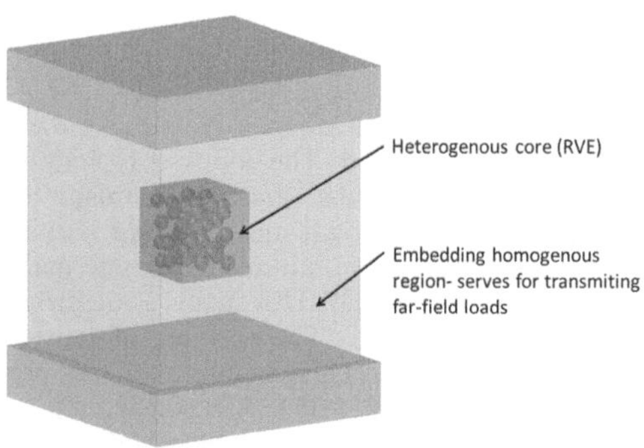

Figure 1. Considered embedded cell model.

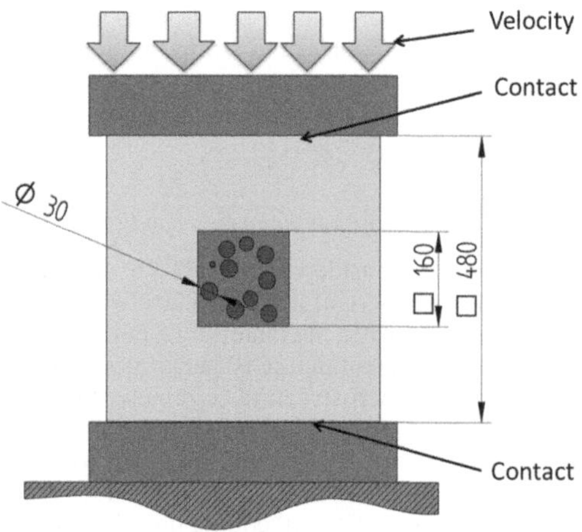

Figure 2. Model in section view with dimensions expressed in μm and applied boundary conditions.

2.2 *Material modeling*

Material properties of the embedding region are determined in the preliminary homogenization based on Mori-Tanaka mean field method. In this case particles are described by elastic model while matrix is described by elastic-plastic model with yield stress scaling by Cowper-Symonds equation (Hallquist 2006):

$$\sigma_y = \left(1 + \left(\frac{\dot{\varepsilon}}{C}\right)^{\frac{1}{p}}\right)\sigma_0 \tag{1}$$

The mean field homogenization is performed using Digimat-MF software that was previously applied in the case of quasi-static analysis by Ogierman & Kokot (2013). The result of the homogenization in the form of stress-strain curve for composite as well as curves representing composite constituents properties are presented in Figure 3. Moreover the analysis of strain rate effect on yield stress of composite is carried out. In this case Cowper-Symonds constants for matrix material have been prescribed as follows: $C = 1288000$ 1/s and $p = 4$. The particles were assumed strain rate insensitive. An effective Cowper-Symonds parameters for the composite have been calculated: $C_{EF} = 1612212$ 1/s and $p_{EF} = 4.72$. Figure 4 shows computed yield stresses for different strain rates and curve obtained by fitting the effective Cowper-Symonds parameters.

At the cell level ceramic reinforcement material behavior is described by Johnson-Holmquist constitutive model implemented in LS-Dyna code (Hallquist 2006). This model initially accounts an elastic behavior of the material until yield occurs. After yielding damage starts to accumulate, the material weakens and material behaves in accordance

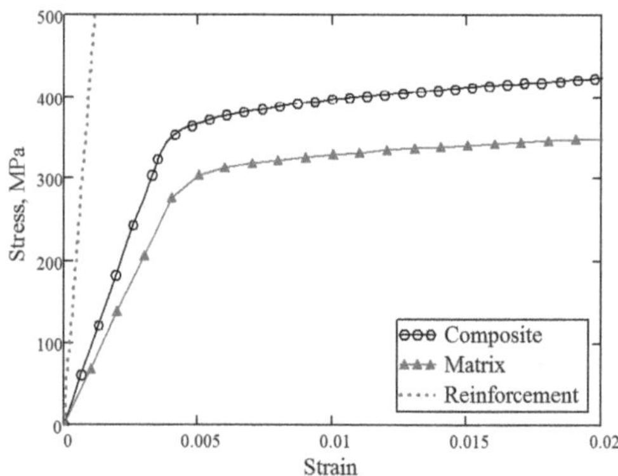

Figure 3. Stress-strain curves representing behavior of reinforcement, matrix and composite.

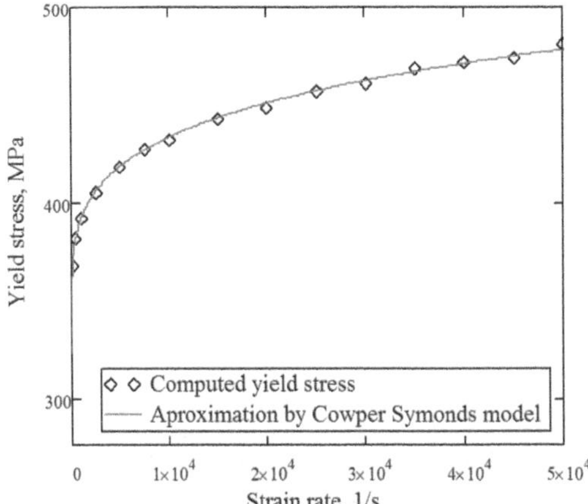

Figure 4. Yield stress in function of strain rate.

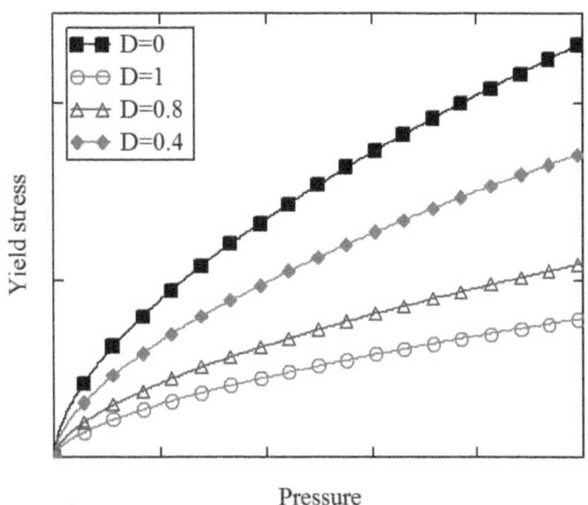

Figure 5. Yield stress in function of pressure for different damage parameter values.

with weakened material curve. Equivalent stress is given in terms of the damage parameter D by:

$$\sigma^* = \sigma_i^* - D\left(\sigma_i^* - \sigma_f^*\right) \tag{2}$$

where:

$$\sigma_i^* = a\left(p^* + t^*\right)^n \left(1 + c\ln\dot{\varepsilon}^*\right) \tag{3}$$

represents undamaged material behaviour and:

$$\sigma_f^* = b\left(p^*\right)^m (1 + c\ln\dot{\varepsilon}) \tag{4}$$

represents the damaged behavior. The superscript "*" indicates a normalized quantity. In particular the stresses σ are normalized by the equivalent stresses at the Hugoniot elastic limit, pressures p are normalized by the pressure at the Hugoniot elastic limit and the strain rates $\dot{\varepsilon}$ are normalized by the reference strain rate. Parameters a, b, c, n, m are model constants and

$$t^* = \frac{T}{p_{HEL}} \tag{5}$$

where T = is the maximum tensile strength and p_{hel} = is the pressure at the Hugoniot elastic limit (Hallquist 2006).

Figure 5 shows yield stress in function of pressure for exemplary damage parameters.

The damage parameter D represents the accumulated damage accounting the increase in the plastic strain per computational cycle and the plastic strain to fracture defined as follows:

$$D = \sum \frac{\Delta\varepsilon^p}{\varepsilon_f^p} \tag{6}$$

$$\varepsilon_f^p = d_1 (p + t)^{d_2} \tag{7}$$

where d_1, d_2 = are damage constants. Johnson-Holmquist constitutive model constants for SiC are presented in Table 1 (Cronin et al. 2003).

Table 1. Johnson-Holmquist material model constants for SiC.

Density	kg/m³	3163
Bulk modulus	GPa	204.78
Shear modulus	GPa	183
Tensile strength	GPa	0.37
Hugoniot Elastic Limit (HEL)	GPa	14.57
Pressure at HEL	GPa	5.9
a		0.96
b		0.35
c		0.00
m		1
N		0.65
D_1		0.48
D_2		0.48

3 RESULTS

Finite element computations were conducted using explicit time integration and LS-Dyna solver.

The results are collected in the form of effective stress-strain curves computed by volume averaging stresses and strains at each post-processed time step in the following way:

$$\sigma_x^{EF} = -\frac{1}{V_{CELL}} \int V_{CELL}\, \sigma_x dV_{CELL} \tag{8}$$

$$\varepsilon_x^{EF} = -\frac{1}{V_{CELL}} \int V_{CELL}\, \varepsilon_x dV_{CELL} \tag{9}$$

where V_{CELL} = is the cell volume; x = is the axis of loading. The Figure 6 shows effective stress-strain curves that represent the composite behavior at different strain rates. The Figure 7 shows effective stress-strain curves that represent the particles behavior taking into account elastic model and Johnson-Holmquist model.

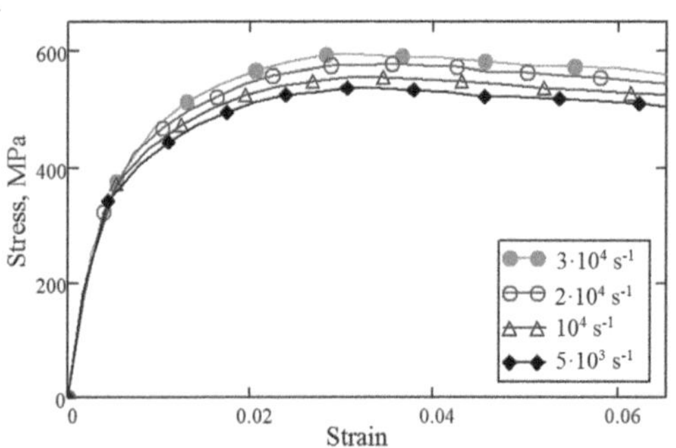

Figure 6. Effective stress-strain curves representing the composite behavior at different strain rates.

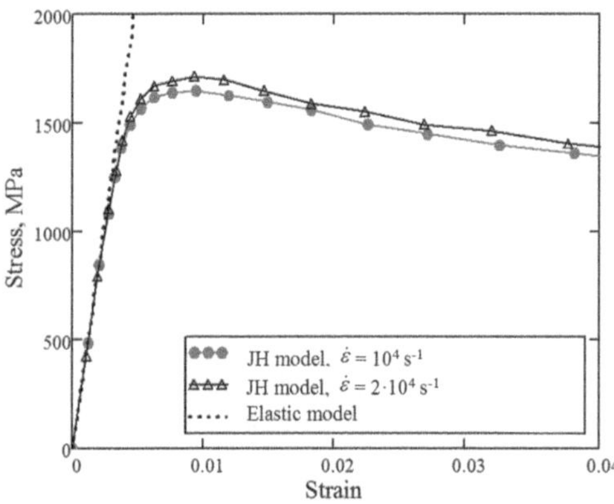

Figure 7. Effective stress-strain curves representing particles behavior at different strain rates computed by using Johnosn-Holmquist model (JH) and elastic model.

Figure 8 shows colored maps representing equivalent stress distribution in the particles modeled with Johnson-Holmquist model and elastic model. Figure 9 shows the damage parameter distribution in the particles.

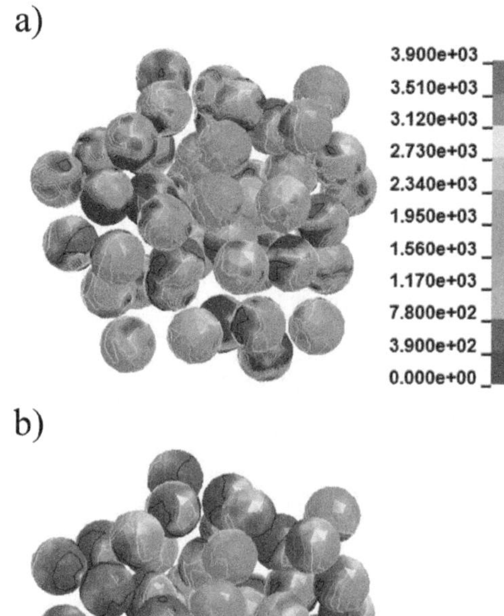

Figure 8. Equivalent stress (MPa) corresponding to macrostrain 0.03, and strain rate 10^{-4} s^{-1}:a) Johnson-Holmquist model, b) elastic model (displayed value threshold is 3900 MPa).

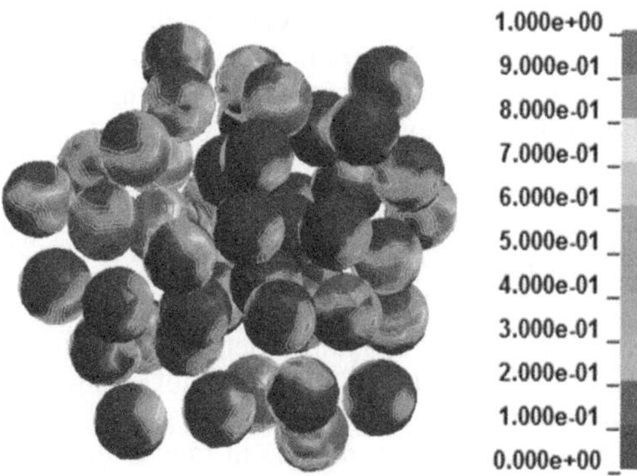

Figure 9. Damage parameter corresponding to macrostrain 0.03, and strain rate 10^{-4} s^{-1}.

4 CONCLUSIONS

Strain rate sensitivity of particle reinforced composite was successfully modeled and studied by means of performing numerical simulation of dynamic compression test. The Johnson-Holmquist model allowed to account the damage of the ceramic reinforcement. Presented results are preliminary and concerns simple, spherical geometry of particles. In further work a study on particle shape and their spatial distribution influence on the damage will be carried out.

ACKNOWLEDGEMENTS

The research is partially funded from the projects Silesian University of Technology, Faculty of Mechanical Engineering 10/990/BK_15/0013, 10/040/BK_15/0006, 10/040/BKM_14/0004.

REFERENCES

Böhm, H.J. 2004. *Mechanics of Microstructured Materials*, Springer-Verlag.

Chawla, N. & Chawla, K.K. 2013. *Metal Matrix Composites*, second ed. Springer.

Cronin, D., Khahn, B., Kaufmann, K., McIntosh, G. & Berstad. T. 2003. Implementation and Validation pf the Johnson-Holmquist Ceramic Material Model in LS-Dyna. *4th European LS-DYNA Users Conference*, Ulm, 2003.

Eckschlager, A., Han, W. & Böhm, H. 2001. Modeling of Brittle Particle Fracture in Particle Reinforced Ductile Matrix Composites by 3D-Unit Cells, *Proceedings of the 13th Inter. Conference on Composite Materials, Beijing, 2001*.

Hallquist, J.O. 2006. *LS-DYNA Theory Manual*, Livermore: Livermore Software Technology Corporation.

Mishnaevsky, L. 2004. Three-dimensional numerical testing of microstructures of particle reinforced composites. *Acta Materialia* 52(14): 4177–4188.

Ogierman, W., Kokot, G. 2013. Mean field homogenization in multi-scale modelling of composite materials. *Journal of Achievements in Materials and Manufacturing Engineering* 61(2): 343–348.

Segurado, J. & Llorca, J. 2006. Computational micromechanics of composites: The effect of particle spatial distribution. *Mechanics of Materials* 38: 873–883.

Trias, D., Costa, J., Mayugo, J.A. & Hurtado, J.E. 2006. Random models versus periodic models for fibre reinforced composites. *Computational materials science* 38: 316–324.

Zhang, J.-T., Liu, L.-S., Zhai, P-C., Fu, Z.-Y. & Zhang, Q.J. 2007. The prediction of the dynamic responses of ceramic particle reinforced MMCs by using multiparticle computational micro-mechanical method. *Composites Science and Technology* 67: 2775–2785.

Advances in Mechanics: Theoretical, Computational and Interdisciplinary Issues – Kleiber et al. (Eds)
© 2016 Taylor & Francis Group, London, ISBN 978-1-138-02906-4

Influence of geometrical and physical irregularities on dynamic characteristics of a passively damped structure

A. Olszewska & Z. Pawlak
Poznan University of Technology, Poznan, Poland

ABSTRACT: The focus of the paper is determination of dynamic parameters for structural systems with Viscoelastic (VE) dampers, when a building is irregular or simply non-symmetric. The structure is considered an elastic linear system. The dynamic characteristics of a structure with VE dampers are determined from a solution of an appropriately defined eigenvalue problem. The solution of the equations of motion is given in the frequency domain. The problem of qualitative differences between the solutions derived for regular and irregular structures is discussed. Several conclusions concerning torsional modes of vibration are formulated on the base of numerical test results.

1 INTRODUCTION

In civil engineering passive damping systems are mounted on structures in order to reduce excessive vibrations caused by wind or earthquake (Park 2001, Takewaki 2009). Different kinds of mechanical devices, such as viscous dampers, Viscoelastic dampers (VE), tuned mass dampers or base isolation systems, are used in the passive systems. The force—displacement characteristics of VE dampers are represented by rheological models (i.e. the Kelvin or Maxwell model).

The equations of motion for multi-degree-of-freedom structures with installed VE dampers are derived. The analytical solution of these equations is given in the frequency domain. For regular structures we assume the shear frame model with flexible columns and rigid beams (Fig. 1a).

For irregular structures we adopt a 3D model with additional masses concentrated at floor levels (Fig. 1b). Both, mass and stiffness irregularities in the floors planes are considered. An uneven weight distribution at floor level leads to different location of the centre of mass (C_M) for each storey (Fig. 2). Similarly, arrangement of stiffening elements in a considered storey determines the position of the centre of stiffness (C_S).

Dynamic characteristics of the structural system with VE dampers (i.e. the natural frequencies and non-dimensional damping ratios) are determined after solving the appropriate nonlinear eigenproblem (Daya & Potier-Ferry 2001, Adhikari & Pascual 2011).

The results of selected numerical tests obtained using the proposed method for structures with irregularities exhibit a certain relation between the stiffness eccentricity and modes of vibration. Basing on these results one can classify the considered structure as torsionally rigid or torsionally flexible (Kyrkos & Anagnostopoulos 2013).

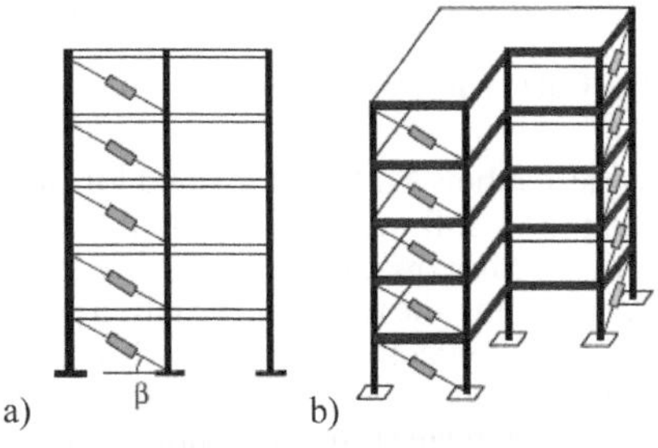

a) β b)

Figure 1. The model of structure: a) the plane shear frame model, b) the spatial frame model.

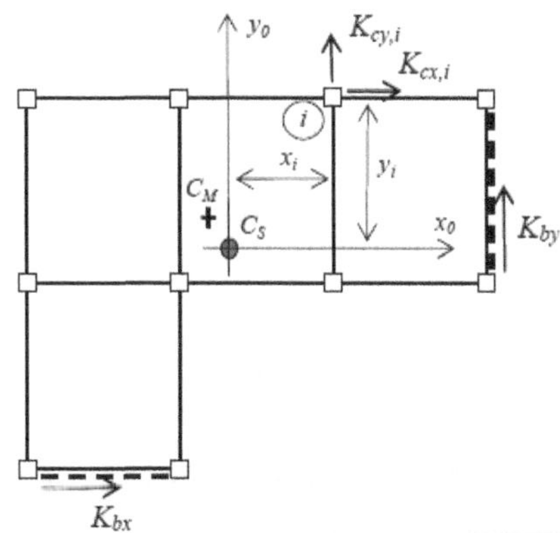

Figure 2. Plan view of a selected structure floor.

2 IRREGULAR STRUCTURES

2.1 *The structure with VE-dampers*

The non-symmetric floor plan or irregular live load distribution may generate the mass eccentricity varying between storeys. The uneven mass distribution leads to different forces and corresponding cross-sections of the structure members. Thus, the building is both, mass and stiffness eccentric. Further irregularities in structure stiffness are caused by VE dampers installed by means of a system of bracing.

Properties of VE dampers could be captured using Kelvin or Maxwell rheological models, where a dashpot with the constant c_d is connected with a spring of stiffness k_d in parallel or in series, respectively.

The force $u(t)$ which acts in the Kelvin model can be given as:

$$u(t) = k_d \Delta q(t) + c_d \Delta \dot{q}(t) \tag{1}$$

while the force in the Maxwell model of damper is governed by the equation:

$$\nu_d u(t) + \dot{u}(t) = k_d \Delta \dot{q}(t) \tag{2}$$

where:

$$\nu_d = \frac{k_d}{c_d}, \quad \Delta q(t) = q_j(t) - q_i(t).$$

The dot over a symbol denotes the derivative with respect to time, t.

2.2 *The centre of stiffness*

The bending stiffness of a column i for x and y axis, respectively, and the axial stiffness of a bracing j are:

$$K_{cx,i} = \frac{12EI_y}{h^3}, \quad K_{cy,i} = \frac{12EI_x}{h^3},$$
$$K_{b,j} = \left[\frac{L}{A_b E} + \frac{1}{k_d}\right]^{-1} \cos^2 \beta \tag{3}$$

where E = is the modulus of elasticity; I_x, I_y = are the column cross-section second moments of area; h = is the storey height; L = is the brace length; A_b = is the area of the brace cross-section; β = is the angle between the bracing member and the horizontal plane (Fig. 1) and k_d = is the stiffness parameter of the damper.

One can derive an approximate position of the centre of stiffness (e_{Sx}, e_{Sy}) for a considered storey in the auxiliary co-ordinate system in the following way (Kyrkos & Anagnostopoulos 2013):

$$e_{Sx} = \frac{\sum_{i=1}^{m} K_{cy,i} x_i + \sum_{j=1}^{n} K_{by,j} x_j}{\sum_{i=1}^{m} K_{cy,i} + \sum_{j=1}^{n} K_{by,j}},$$
$$e_{Sy} = \frac{\sum_{i=1}^{m} K_{cx,i} y_i + \sum_{j=1}^{p} K_{bx,j} y_j}{\sum_{i=1}^{m} K_{cx,i} + \sum_{j=1}^{p} K_j}, \tag{4}$$

where m = is the number of columns; n and p = the number of bracing along axis y and x, respectively.

2.3 *The stiffness factors for a storey*

In this approach, we propose a formula for estimation of the stiffness factor of a considered storey, as an appropriate sum of column and bracing stiffness values:

$$K_{Xo} = \sum_{i=1}^{m} K_{cx,i} \cdot y_i^2 + \sum_{j=1}^{p} K_{bx,j} \cdot y_j^2,$$
$$K_{Yo} = \sum_{i=1}^{m} K_{cy,i} \cdot x_i^2 + \sum_{j=1}^{n} K_{by,j} \cdot x_j^2,$$
$$K_{XYo} = \sum_{i=1}^{m} K_{cy,i} \cdot x_i y_i + \sum_{j=1}^{n} K_{by,j} \cdot x_j y_i$$
$$+ \sum_{i=1}^{m} K_{cx,i} \cdot y_i x_i + \sum_{j=1}^{p} K_{bx,j} \cdot y_i x_j. \tag{5}$$

The comparison of the values of these quantities allows one to predict the form of vibration at the first mode, i.e. if they are transverse or torsional.

3 THE EQUATION OF MOTION

The equation of motion of a structure with VE dampers could be written in the following form:

$$\mathbf{M}\ddot{\mathbf{q}}(t) + \mathbf{C}\dot{\mathbf{q}}(t) + \mathbf{K}\mathbf{q}(t) = \mathbf{f}(t) + \mathbf{p}(t) \tag{6}$$

where the symbols **M**, **C** and **K** stand for the mass, damping, and stiffness matrices of the structure, respectively. Moreover, $\mathbf{q}(t)$, $\mathbf{p}(t)$, $\mathbf{f}(t)$ are the vector of displacements, the vector of excitation forces and the vector of interaction forces, which act between the structure and the dampers, respectively.

Using the Laplace transformation with zero initial conditions to the Equations 1 and 2, the transform of the damper force can be written as:

$$\bar{u}(s) = k_d \Delta\bar{q}(s) + sc_d \Delta\bar{q}(s), \quad \bar{u}(t) = \frac{k_d s}{\nu_d + s} \Delta\bar{q}(s) \quad (7)$$

where quantities $u(s)$ and $\Delta q(s)$ with bars denote the Laplace transforms of the force $u(t)$ and the displacement $\Delta q(t)$, respectively, and s is the Laplace variable.

The equation of motion (6) after Laplace transformation could be written in a general form, which enables to take into account various models of dampers in one structure (Pawlak & Lewandowski 2013):

$$\left(s^2\mathbf{M} + s\mathbf{C} + \mathbf{C}_d(s) + \mathbf{K} + \mathbf{K}_d + \mathbf{G}_d(s)\right)\bar{\mathbf{q}}(s) = \mathbf{0} \quad (8)$$

where:

$$\bar{\mathbf{p}}(s) = \mathbf{0}, \quad \mathbf{K}_d = \sum_{r=1}^{m} K_r \mathbf{L}_r, \quad \mathbf{C}_d(s) = \sum_{r=1}^{m} C_r(s)\mathbf{L}_r,$$

$$\mathbf{G}_d(s) = \sum_{r=1}^{m} G_r(s)\mathbf{L}_r, \quad \mathbf{L}_r = \mathbf{e}_r \mathbf{e}_r^T$$

and $\mathbf{L}_r$ is the location matrix. The functions K_r, $C_r(s)$, $G_r(s)$ must be defined for each damper r, depending on the adopted model:

- for Kelvin model:

$$K_r = k_d, \quad C_r(s) = sc_d, \quad G_r(s) = 0,$$

- for Maxwell model:

$$K_r = 0, \quad C_r(s) = 0, \quad G_r(s) = \frac{sk_d}{\nu_d + s}.$$

Equation 8 constitutes the nonlinear eigenvalue problem. In the case of small damping, eigenvalues and eigenvectors can be obtained as complex conjugate numbers and vectors, respectively. The natural frequencies ω_i and non-dimensional damping ratios γ_i are determined from the eigenvalues in a similar way as for the viscous damping case:

$$\omega_i^2 = \mu_i^2 + \eta_i^2, \quad \gamma_i = \frac{-\mu_i}{\omega_i}, \quad (9)$$

where $\mu_i = \mathrm{Re}(s_i)$, $\eta_i = \mathrm{Im}(s_i)$.

4 NUMERICAL TEST

4.1 *Structure consisting of one segment*

In order to validate the effectiveness of the proposed approach we derived the dynamic characteristics for a structure with VE dampers. A regular five-storey steel frame was tested (Fig. 3).

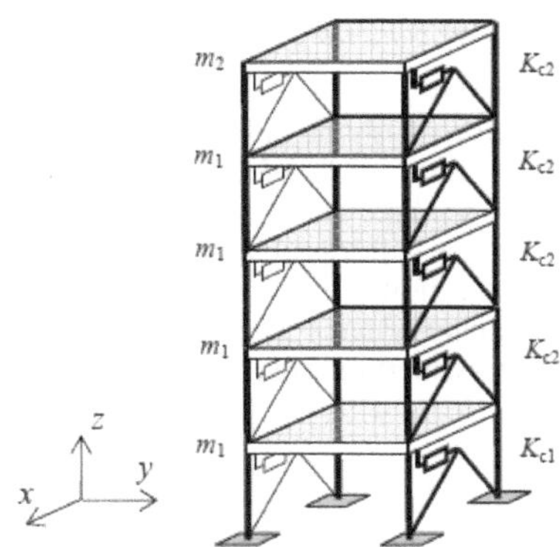

Figure 3. The model of regular five-storey frame.

Table 1. The first three natural frequency for regular five-storey frame.

	Natural frequency ω_i [Hz]		
Mode number	Shear frame	Spatial model*	Experimental (Chang et al. 1995)
1	3.173	3.007	3.17
2	3.634	3.236	3.61
3	–	5.524	–

*Abaqus Software Dassault Systèmes Simulia Corp. Providence, RI, USA.

The calculations were carried out basing on the data given in Chang et al. (1995). Applying Equation 3 one can derive the stiffness of the lowest storey column, as $K_{cx,1} = K_{cy,1} = 2927.5$ kN/m. For the upper column we obtained the stiffness $K_{cx,2} = K_{cy,2} = 1336.7$ kN/m. The mass of the four lower floors was the same $m_1 = 1132$ kg, the fifth floor mass was different and amounted to $m_2 = 1168$ kg.

There is a system of bracing with dampers in two opposite walls at each storey. The dampers are characterised using the Kelvin model, with the stiffness and damping parameters $k_d = 800$ kN/m, $c_d = 16$ s·kN/m, respectively. The axial stiffness of a bracing member together with a damper at the first storey is $K_{b,1} = 542.0$ kN/m, at the upper stories it is $K_{b,2} = 441.0$ kN/m.

The experimental results from the shaking-table test described in Chang et al. (1995) coincide with the analytical ones obtained using the proposed method (see Table 1).

The stiffness factors of upper stories of the considered structure derived from Equations 5 are as follows: $K_{Xo} = 2713.3$ kNm, $K_{Yo} = 2329.0$ kNm, $K_{XYo} = 0.0$ kNm. The above values indicate that the first mode of vibrations has a form of transverse displacements along the y axis.

4.2 *Structure consisting of three segments*

The dynamic behaviour of an irregular steel frame building with a L-shaped plan view was examined (see Fig. 1b and Fig. 4). We analyzed the structure consisting of three identical units presented in Figure 3, so the stiffness of column and bracing are taken from Section 4.1. The results were derived for various bracing systems. The first three natural frequencies presented in Table 2, are related to three cases: frame without bracing (Fig. 4a), frame with symmetric (Fig. 4b) and non-symmetric (Fig. 4c) system of bracing.

For each case the fundamental frequencies are similar, but the stiffness factors for the considered systems have a different value (see Table 3).

The forms of first mode of vibration for three considered structures are presented in Figure 4. For the symmetric systems ($K_{Xo} = K_{Yo}$) the displacements take place along the symmetry axis or perpendicularly to it.

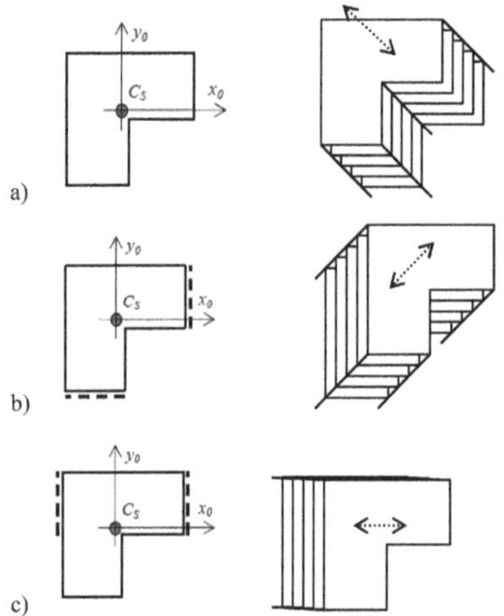

Figure 4. The first mode of vibration for structure: a) without bracing, b) with symmetric bracing, c) with non-symmetric bracing.

Table 2. The first three natural frequencies for regular five-storey frame.

Mode number	Natural frequency ω_i [Hz]		
	Without bracing	Bracing system: Symmetric	Bracing system: Non-symmetric
1	2.575	2.620	2.581
2	2.587	2.623	2.680
3	3.688	3.857	3.810

Table 3. The stiffness factors for a typical storey of considered structures.

Mode number	The stiffness factor [kNm] for structure		
	Without bracing	Bracing system: Symmetric	Bracing system: Non-symmetric
K_{Xo}	11,631	13,011	11,631
K_{Yo}	11,631	13,011	13,976
K_{XYo}	5368	5061	5594

5 CONCLUSIONS

The first mode of vibration associated with the fundamental natural frequency has a significant impact on the dynamic behavior of structures exposed to a dynamic load. Thus, the form of the first mode of vibration is important in the design of building structures.

The results of sample numerical calculations confirm that the estimation of the stiffness centre and some stiffness factors for storeys enables to predict the dynamic behavior of an irregular structure with dampers. It is possible to determine which form of vibration is fundamental, and whether the torsional movement dominates. Moreover, one can design appropriate dampers and a bracing system distribution to improve the torsional stiffness of a building structure.

REFERENCES

Adhikari S. & Pascual B. 2011. Iterative Methods for Eigenvalues of Viscoelastic Systems. *Journal of Vibration and Acoustics* 133(2): 021002.

Chang, K.C., Soong, T.T., Oh, S.T. & Lai, M.L. 1995. Seismic behaviour of steel frame with added viscoelastic dampers. *Jour. of Structural Engineering* 121: 1418–1426.

Daya, E.M. & Potier-Ferry, M. 2001. A numerical method for nonlinear problems application to vibrations of viscoelastic structures. *Computers and Structures* 79: 533–541.

Kyrkos, M.T. & Anagnostopoulos, S.A. 2013. Eccentric Steel Buildings Designed for Uniform Ductility Demands under Earthquake Actions, In M. Papadrakakis, V. Papadopoulos, V. Plevris (eds.), *4th ECCOMAS Thematic Conference on Computational Methods in Structural Dynamics and Earthquake Engineering, COMPDYN 2013, Kos Island, Greece, 12–14 June 2013.*

Park, S.W. 2001. Analytical modeling of viscoelastic dampers for structural and vibration control. *International Journal of Solids and Structures* 38: 8065–8092.

Pawlak, Z. & Lewandowski, R. 2013. The continuation method for the eigenvalue problem of structures with viscoelastic damper. *Computer and Structures* 125: 53–61.

Takewaki, I. 2009. *Building control with passive dampers, Optimal performance-based design for earthquakes.* Singapore: Wiley and Sons (Asia).

Advances in Mechanics: Theoretical, Computational and Interdisciplinary Issues – Kleiber et al. (Eds)
 ISBN 978-1-138-02906-4

Impact of an unsecured excavation on an underground pipeline

R. Ossowski & K. Szarf
Faculty of Civil and Environmental Engineering, Gdańsk University of Technology, Gdańsk, Poland

ABSTRACT: The paper presents a numerical analysis of the impact of an unsecured excavation on an underground pipeline in selected soil conditions. The research was inspired by a real-life failure of a water pipeline which was caused by a nearby unsecured excavation (Sikora et al. 2015). The failure was triggered by displacement of soil mass in the vicinity of the pipeline. The study conducted in the framework of Finite Element Method is focused on analyzing different soil conditions and pipe rigidity combinations and calculating the resulting displacements. The FEM parametric study aims at investigation of the soil model parameters sensitivity on the possible magnitude of horizontal and vertical deformations. The research and analysis can be incorporated into recommendations for engineers performing earthworks in urban areas.

1 INTRODUCTION

Urban areas with long history, like most large city centres in Europe, are often equipped with old infrastructure—water lines, gas lines, district heating etc.—dating back to the 1950s or 1960s (or older) but still operational and in use. Its presence can lead to numerous problems and complications and cannot be neglected. Moreover, these dense urban areas are being intensively renovated nowadays. Among the typical problems encountered when dealing with such infrastructure are:

- uncertainty of pipe localisation, often linked to the archive data being incomplete or destroyed,
- uncertainty and low quality of backfill material surrounding the pipes,
- presence of old and sometimes outdated technologies,
- shallow placement of the pipelines,
- condition of the pipeline material uncertainty,
- many old buildings in the vicinity of the pipe,
- work conditions that were not planned during the construction, e.g. heavy traffic loads on the surface, additional constructions in the area.

An example of such a situation and resulting problems is a pipe failure that occurred in December 2013 at one of the intersections in Gdańsk, Poland (Sikora et al. 2015). A rigid, gray cast iron water pipe with bell and spigot joints was still operational while several constructions were being performed in the area. Additionally the street was used as a nearby construction yard access road, with heavy trucks passing through it. When an unsecured excavation was dug in order to construct a manhole for a planned sewage line, one of the joints in the existing water pipeline got damaged and a large failure occurred.

Apparently an excavation made in the vicinity of the existing pipelines, especially ones made of stiff, brittle materials, can damage these pipelines due to large soil deformations around the excavation (Makar et al. 2001, Makar 2000, ORourke and Ahmed 1985). This problem is both current nowadays and will remain so in the future.

In the paper we describe numerical simulations of the pipe and soil using the Finite Element Method (FEM) performed in order to assess how an unsecured excavation can impact the soil and an underground pipeline in a scenario described above.

1.1 *Problem geometry*

Although typically pipe–soil interactions are modeled in the plane perpendicular to the pipeline axis, in this case it was decided to investigate a 2D cross section along the pipeline, perpendicular to an excavation, see Figure 1. Where as the top layer material

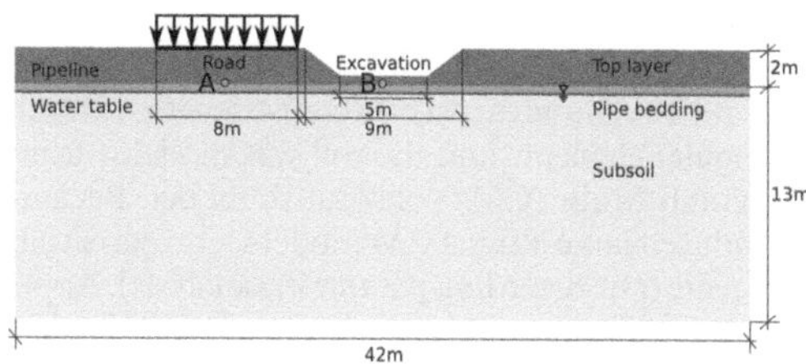

Figure 1. General geometry of the 2D problem. Point A denotes the road centerline. Point B denotes excavation centerline.

Table 1. Mechanical parameters of the soil materials used.

Soil symbol	M_0 (MPa)	γ_{sr} (kN/m^3)	γ_d (kN/m^3)	ν (–)	$\varphi'(°)$	$c'(kPa)$
Or—Peat	0.5	14.0	11.0	0.40	5.0	5.0
Or—Compacted mud/Peat	1.0	14.0	11.0	0.40	5.0	5.0
Or—Compacted mud	2.0	14.0	11.0	0.40	8.0	5.0
saOr	3.0	16.0	13.0	0.35	10.0	3.0
saclSi	5.0	18.0	14.0	0.35	15.0	10.0
orsiSa	10.0	18.0	14.0	0.35	20.0	3.0
siCl	20.0	21.0	17.0	0.35	15.0	30.0
FSa	30.0	20.0	16.0	0.30	30.0	0.0
MSa	50.0	20.0	16.0	0.30	35.0	0.0
Top layer and bedding	50.0	19.0	16.0	0.30	30.0	10.0

remained the same in every simulation, a range of subsoils and its mechanical parameters was used, see Table 1. The soil types and its mechanical parameters were based on the real soils in the Gdańsk area.

The pipe was assumed to correspond to typical DN400 pipe connected in 6 *m* long pieces with shallow bell and spigot joints. The applied traffic load corresponding to heavy loaded trucks was assumed to be equal 35 *kPa*. Dynamic load was neglected. The pipe was placed on a 0,5 *m* thick bedding with the water table assumed at the bottom of the bedding layer.

1.2 *Pipe loading and failure mechanism*

According to (Makar et al. 2001) pipe failure can be caused by numerous factors and can occur in different ways. In the paper we focused on a joint failure that could occur due to axial pullout of a pipe element, excessive rotation at the joint or bell material crack. These three mechanisms were supposed to be linked with large displacements of the soil surrounding the pipe.

2 FEM ANALYSIS—SOLUTION STRATEGY

The FEM analysis was focused on calculating soil displacement maps surrounding the pipeline caused by a shallow unsecured excavation nearby. The analysis was performed in simplified plane strain conditions, with use of PLAXIS FEM code (PLAXIS BV 2013). The area was discretised with 15-node triangular elements and the soil was modeled using Coulomb-Mohr (CM) constitutive model. Because of induced large strains CM model seems physically adequate (esp. regarding parametric analysis). Application of a more sophisticated model, e.g. Hardening Soil (HS), would not lead to significantly more precise results (Viggiani and Tamagnini 2000).

A 2D solution can be used as a time effective alternative to a 3D model. One important factor to consider is the influence of pipe rigidity and its zone of influence in the perpendicular plane. In this paper the perpendicular thickness of the model was equal to 1 *m* while the pipe diameter was 0.4 *m*. It was taken into account by reducing the pipe material stiffness values.

The phenomenon of differential displacements must be studied systematically, therefore the parametric analysis is a proper approach. The resulting vertical and horizontal displacements were calculated using different sets of constitutive model parameters (Young's modulus, Poisson's ratio, effective internal friction angle and effective cohesion of soil), see Table 1 representing different soil conditions.

Simulations were done in two series. The first series was used to establish the upper-bound solution, that is one with a pipe with zero stiffness. In this case the displacements of soil were largest. This approach was used to select the worst soil conditions that would lead to largest settlements. Second series of simulations were performed with calculated pipe stiffness values assumed as for grey cast iron pipe with diameter D = 0.4 *m*, elastic modulus E = 100 *GPa* and Poisson's ratio ν = 0.3.

In order to achieve reliable results soil stress history was emulated. The following steps were performed in every simulation: 1) loading and unloading of the road surface with traffic; 2) soil displacement values reset; 3) excavation; 4) traffic loading. The results presented in this paper refer to the two most essential steps of the simulations: steps 3) and 4).

3 RESULTS

First series of simulations show the upper bound of resulting displacements. The calculated differential displacements between the road center line and excavation center line (points A and B in the Fig. 1), $|U_x^A - U_x^B|$ and $|U_y^A - U_y^B|$, are shown in Figures 2 and 3.

It is no surprise that the analysis indicated the oedometric modulus M_0 was the most important

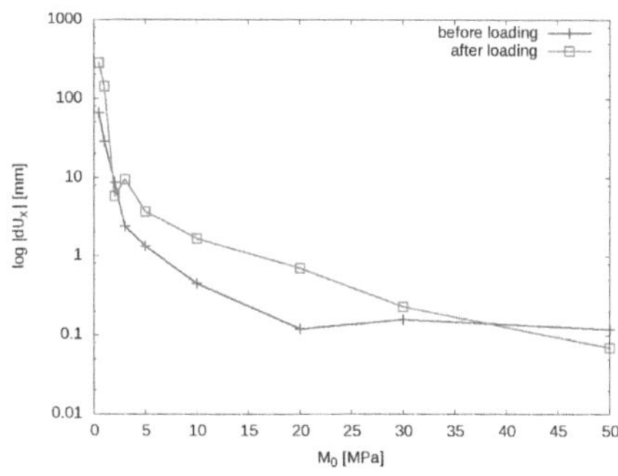

Figure 2. Upper bound case (pipe with no stiffness) horizontal differential displacements plotted against subsoil oedometric modulus.

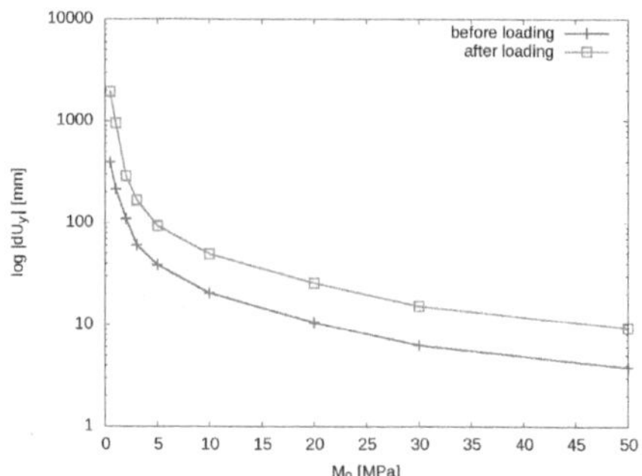

Figure 3. Upper bound case (pipe with no stiffness) vertical differential displacements plotted against subsoil oedometric modulus.

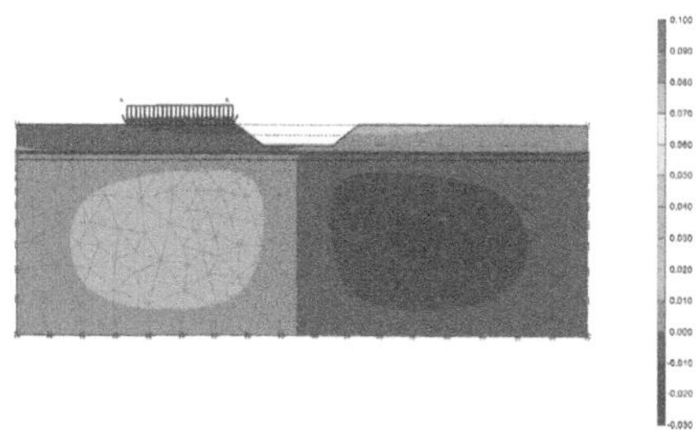

Figure 4. Horizontal soil displacements map before traffic loading. Subsoil type: Or—Compacted mud ($M_0 = 2$ *MPa*). Pipe with real stiffness.

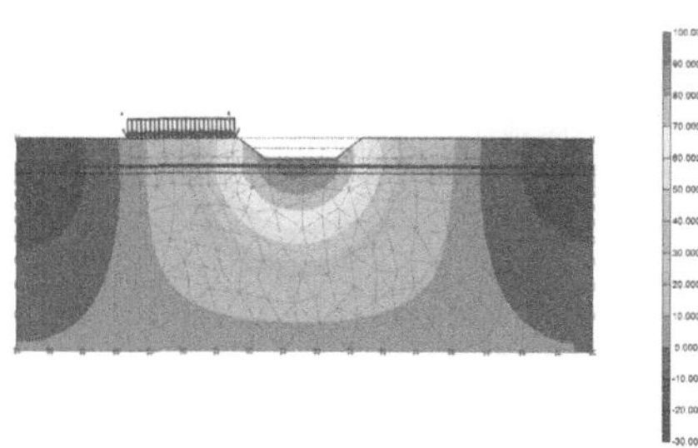

Figure 5. Vertical soil displacements map before traffic loading. Subsoil type: Or—Compacted mud ($M_0 = 2$ *MPa*) Pipe with real stiffness.

parameter for soil settlements. Moreover, its value is correlated to values of the other mechanical parameters. Influence of mechanical soil parameters other than M_0 was much smaller and less relevant.

Based on the upper-bound solution it can be concluded that presence of soils with $M_0 \le 5MPa$ can lead to differential settlements greater than 100 *mm*. Such values of displacements could be dangerous to bell and spigot pipe joints therefore subsequent simulations focused on those weak soils.

The second series of simulations (with real pipe stiffness) were in accordance with former calculations, showing—as expected—lower values of displacements. As an example, some results for Or-Compacted mud (M0 = 2 *MPa*) soil are presented below. Figures 4 and 5 show the horizontal and vertical components of displacement field caused by an excavation but without traffic loading. Soil displacements maps for the same soil but after traffic loading are presented in the Figures 6 and 7.

In Figure 7 (vertical displacements U_y) one can notice large difference of settlements: more than 90 *mm* below the road centerline, virtually 0 *mm* at the edge of the excavation nearest to the road and uplift of more than 60 *mm* below the opposite excavation edge. Such intensive differences result in high stresses that can cause pipe material failure by bell fracturing. Figure 6 (horizontal displacements U_x) show little soil displacements, in the range of 10 *mm*. Horizontal soil displacements around the pipe could lead to pipe joint failure by disjoining of pipe elements even though soil would obviously displace more than the pipe itself. That kind of failure was reported in Gdańsk in 2013 (Sikora et al. 2015). Soil displacements in the vicinity of stiff and brittle pipes, especially the ones that are aged and weakened (e.g. by corrosion), in the range of centimeters are sufficient to cause a pipeline failure (Kuliczkowski 2004).

Additionally combined vertical and horizontal soil displacements acting on pipe can also lead to deflection of each pipe element out of the axis of pipeline. It is reported that modern, less brittle spheroidal cast iron pipes connected with deeper spigot joints can bear such deflections smaller than 4° (Kuliczkowski 2004). It is safe to assume that the deflection limit for the old-type pipes, made with more brittle material and shallower joints is significantly lower. Our estimations indicate rotations in the range of 1–2°

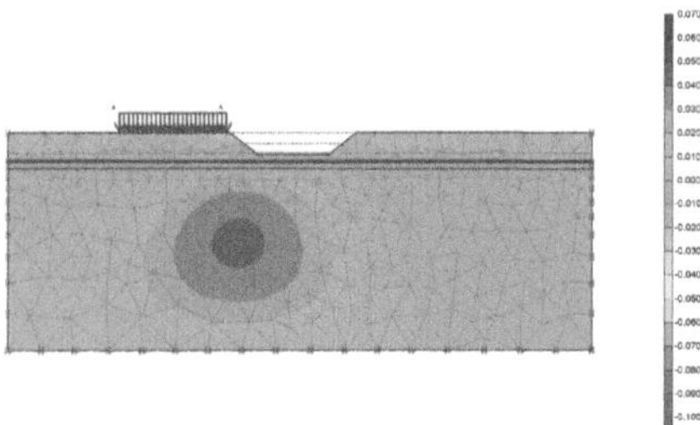

Figure 6. Horizontal soil displacements map after traffic loading. Subsoil type: Or—Compacted mud ($M_0 = 2$ MPa). Pipe with real stiffness.

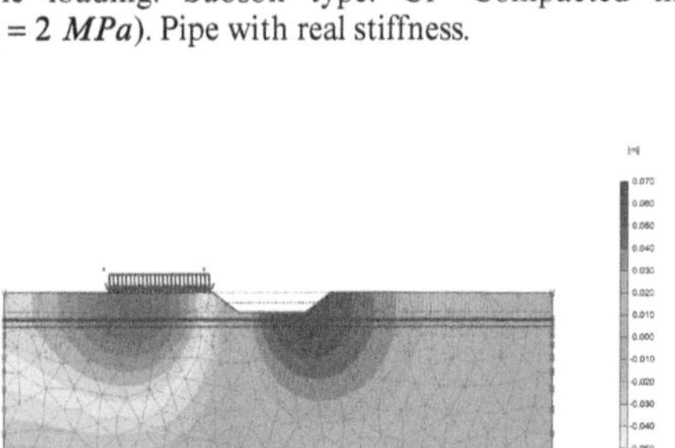

Figure 7. Vertical soil displacements map after traffic loading. Subsoil type: Or—Compacted mud ($M_0 = 2$ MPa) Pipe with real stiffness.

Figure 8. General view of Gdańsk 2013 real-life case of a pipe failure in the vicinity of an unsecured excavation that inspired this study. Red lines indicate road edges while blue line indicates the axis of the pipeline.

Figure 9. Close-up of the Gdańsk 2013 pipe failure area. Red lines indicate road edges while blue line indicates the axis of the pipeline.

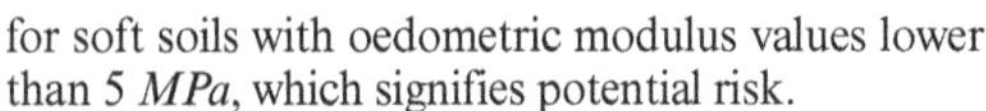

for soft soils with oedometric modulus values lower than 5 MPa, which signifies potential risk.

4 CONCLUSIONS

The results of works performed in an unsecured excavation in urban areas can be disastrous for nearby existing infrastructure due to large deformations of soil mass, see Figures 8 and 9. This article was focused on study of soil mass movement in the vicinity of underground pipeline as a result of unsecured excavation and shows the influence of soil model parameters on the resulting displacement of soil mass surrounding the pipeline, indicating the most prone conditions to pipe damage. Impact of traffic loading was pointed as well. The most dangerous soils were the most soft ones, with values of oedometric modulus lower than 5 MPa.

Further research of these problems will be focused on full 3D FEM analysis. That analysis can be enhanced with use of Discrete Element modeling to study the effects of soil friction on pipeline shaft in case of large displacements. That approach could also shed light on the problem of rigid body movements of pipe elements leading to bell and spigot joint failures.

REFERENCES

Kuliczkowski, A. (2004). *Rury kanalizacyjne, t. II.* Wydawnictwo Politechniki Świętokrzyskiej.

Makar, J. (2000). A preliminary analysis of failures in grey cast iron water pipes. *Engineering failure analysis* 7(1), 43–53.

Makar, J., R. Desnoyers, & S. McDonald (2001). Failure modes and mechanisms in gray cast iron pipe. *Underground Infrastructure Research*, 1–10.

ORourke, T. & I. Ahmed (1985). Effect of shallow trench construction on cast iron pipelines. In *Advances in Underground Pipeline Engineering*, pp. 21–31. ASCE.

PLAXIS BV (2013). PLAXIS FEM. [Online; accessed 15-09-2015].

Sikora, Z., M. Wyroślak, & R. Ossowski (2015). Przebudowa infrastruktury podziemnej w zwartej zabudowie studium przypadku (rebuilding of underground infrastructure in urban areas—case study). In *Awarie Budowlane 2015*, Międzyzdroje, Poland.

Viggiani, G. & C. Tamagnini (2000). Ground movements around excavations in granular soils: a few remarks on the influence of the constitutive assumptions on fe predictions. *Mechanics of Cohesive-Frictional Materials* 5(5), 399–423.

Advances in Mechanics: Theoretical, Computational and Interdisciplinary Issues – Kleiber et al. (Eds)

Computable two-sided a posteriori error estimates for *h*-adaptive Finite Element Method

O.Yu. Ostapov & O.V. Vovk
Department of Information Systems, Ivan Franko National University of Lviv, Lviv, Ukraine

H.A. Shynkarenko
Department of Mathematics and Applied Informatics, Opole University of Technology, Opole, Poland

ABSTRACT: We describe the simple element-wise a posteriori error estimators (AEEs) that are able to provide two-sided error estimates for Finite Element Method (FEM) approximations of the solutions to the elliptic boundary value problems. We name these estimators "Dirichlet and Neumann estimators" because the theoretical background of their two-sided estimations lays in the analysis of the solutions to the variational formulations of the residual problems with homogeneous Dirichlet and Neumann boundary conditions. Here we show how to construct Dirichlet and Neumann AEEs for the linear finite element approximations on the triangular meshes and prove theoretically their abilities to yield lower and upper bounds of finite element approximation errors for linear problems. The numerical results demonstrate and confirm the properties of Dirichlet and Neumann AEEs in the process of solving singularly perturbed and nonlinear diffusion-advection-reaction problems on the uniform or *h*-adaptively refined meshes.

1 INTRODUCTION

A posteriori error estimators have already become a standard supplement to the finite element method (Ainsworth & Oden 2000), (Babuška, Whiteman, & Strouboulis 2011), (Bank & Weiser 1985). Especially it is important for singularly perturbed problems. Here we continue the research (Kvasnytsia, Ostapov, Shynkarenko, & Vovk 2014), (Ostapov, Shynkarenko, & Vovk 2014), (Shynkarenko & Vovk 2013), (Ostapov, Shynkarenko, & Vovk 2013), (Ostapov, Shynkarenko, & Vovk 2015) by constructing Dirichlet and Neumann AEEs for the two-sided error estimation of FEM approximations. The ability to compute the element-wise approximation of the lower and upper FEM error bounds expands the control and correction options of the boundary value problems solving with the aim to increase its efficiency. Here the "correction" consists in *h*-adaptivity that is widely used for solving singularly perturbed problems, see (Deuflhard & Weiser 2012). And for "control" the possibility of error approximation refinement is especially important.

In section 2 we state a boundary value problem. FEM approximations, their AEEs, *h*-FEM and numerical results are described in sections 3–6.

2 FORMULATION OF THE PROBLEM

We consider an elliptic boundary value problem with a weak non-linearity in the right-hand side:

$$\begin{cases} -\nabla\cdot(\mu\nabla u)+\beta\cdot\nabla u+\sigma u=f[u] \text{ in } \Omega \\ u=0 \text{ on } \Gamma_D, mes(\Gamma_D)>0, \\ -(\mu\nabla u)\cdot\nu=g \text{ on } \Gamma_N=\partial\Omega\backslash\Gamma_D, \end{cases} \quad (1)$$

where $\Omega\subset R^2$ is bounded polygonal domain with Lipschitz boundary $\Gamma=\partial\Omega$, ν is an outward normal unit vector to Γ. We also assume that the given functions $\mu=\mu(x)$, $\beta=\{\beta_i(x)\}_{i=1}^2$, $\sigma=\sigma(x)$, $f[u]:=f(x,u)$ are regular enough for the problem (1) to have a unique solution $u=u(x)$, and allows the following variational formulation

$$\begin{cases} \text{find } u\in V \text{ such that} \\ a_\Omega(u,v)=n_\Omega(u;v) \quad \forall v\in V, \end{cases} \quad (2)$$

where $a_\Omega(w,v):=\int_\Omega[\mu\nabla w.\nabla v+v\beta.\nabla w+\sigma wv]dx$, $n_\Omega(w;v):=(f[w],v)_\Omega+(g,v)_\Gamma=\int_\Omega f[w]vdx-\int_{\Gamma_N}gvd\gamma$ $\forall v, w\in V$, $V:=\{v\in H^1(\Omega): v=0 \text{ on } \Gamma_D\}$. The conditions of well-posedness for the problem (2) one can find in (Kvasnytsia, Ostapov, Shynkarenko, & Vovk 2014).

3 FINITE ELEMENT APPROXIMATIONS

Let $\Im_h=\{K\}$ be the regular conforming triangulation of Ω, $h=max_{K\in\Im_h}h_K$, h_K $diamK$, K be a triangle, $I_h=\{A_i\}_{i=1}^N$ be a set of all vertices $A_i(x_i,y_i)\notin\Gamma_D$ in $\Im_h$, $P_1(K):=span[L_1,L_2,L_3]$ be the space of linear polynomials on K in barycentric coordinate system

$\{L_i\}_{i=1,2,3}$. We construct the finite dimensional subspace $V_h \in V$ for the triangulation $\Im_h$ with the basis that consist of the following piecewise functions

$$\begin{cases} \operatorname{supp}\phi_i := \Omega_i = \{K \in \Im_h : A_i \in \overline{K}\}, \\ \phi_i \in C(\Omega_i),\ \phi_i \in P_1(K) \quad \forall K \in \Omega_i, \\ \phi_i(A_j) = \delta_{ij} \quad \forall A_i, A_j \in I_h. \end{cases}$$

Then we approximate the solution $u(x) \in V$ of problem (2) by the expansion $u_h(x) = \sum_{i=1}^{N} q_i \phi_i(x) \in V_h \subset V \quad \forall x \in \Omega$. The coefficients $q := \{q_i\}_{i=1}^{N} \in \mathbb{R}^N$ are calculated from the following nonlinear system

$$\begin{cases} \sum_{i=1}^{N} S_{ki} q_i = F_k[q], \ k = 1, \dots, N, \\ S_{ki} = a_{\Omega_k \cap \Omega_i}(\phi_i, \phi_k), \\ F_k[q] = (f[u_h], \phi_k)_{\Omega_k} + (g, v)_{\Gamma_q \cap \partial\Omega_k}. \end{cases} \tag{3}$$

To linearize the system (4) we use the Newton method, which produces the sequence of systems of linear equations that are solved by restarted generalized minimal residual method (GMRES) with precondition (Ostapov, Shynkarenko, & Vovk 2014).

4 TWO-SIDED ERROR ESTIMATION

In order to estimate the quality of the obtained approximation u_h we consider the following linearized error problem:

$$\begin{cases} \text{given } \Im_h, u_h \in V_h, V = V_h \oplus E; \\ \text{find } e = e_h^{Ext} := u - u_h \in E \text{ such that} \\ b_\Omega(u_h; e, v) = \rho_\Omega(u_h; v) \quad \forall e \in E \ \forall v \in V, \end{cases} \tag{4}$$

where for all $w, z, v \in V$ we have

$$\begin{cases} b_\Omega(w; z, v) := a_\Omega(z, v) - \int_\Omega f_u[w] z v dx \\ \rho_\Omega(w; v) := n_\Omega(w; v) - a_\Omega(w, v). \end{cases} \tag{5}$$

To solve the problems (4), (5) approximately, we apply the Galerkin method and in the finite dimensional subspace $E_h \subset E$ we find the approximation $e_h = e_h(x)$ such that $e_h(x) = \sum_{K \in \Im_h} e_K(x) = \sum_{K \in \Im_h} \lambda_K \varphi_K(x) \ \forall x \in \Omega$. The functions $\phi_K = \phi_K(x)$ generate an orthogonal basis of the subspace $E_h \subset E$ and have the properties: $\operatorname{supp} \phi_K = \overline{K} \ \forall K \in \Im_h$, $\phi_K(A_i) = 0 \ \forall A_i \in K$, where A_i are vertices of K.

Consequently, for each finite element $K \in \Im_h$ we obtain a posteriori error estimator in the form

$$e_K(x) = \lambda_K \phi_K(x) = \frac{\rho_K(u_h; \phi_K)}{b_K(u_h; \phi_K, \phi_K)} \phi_K(x).$$

Throughout this paper we use the following basis functions

$$\begin{cases} \phi_K^{Dir} := 27 L_i L_j L_k, \\ \phi_K^{Neu} := 3[L_i L_j + L_j L_k + L_k L_i], \ \forall K \in \Im_h, \end{cases}$$

to construct Dirichlet e_h^{Dir} and Neumann e_h^{New} AEEs of linear finite element approximations.

Theorem 1. *Efficiency and reliability of AEE.*

Let $u \in V$ be the solution of linear case of (2), $u_h \in V_h$ be its approximation, $\| w \|_\Omega^2 := a_\Omega(w, w)$, and $\overline{u}_h \in W_h$ be the solution of the following problem

$$\begin{cases} \text{find } \overline{u}_h \in W_h := V_h \oplus E_h \text{ such that} \\ a_\Omega(\overline{u}_h, v) = \langle l_\Omega, v \rangle \quad \forall v \in W_h, \\ \langle l_\Omega, v \rangle := (f, v)_\Omega + (g, v)_\Gamma \quad v \in W_h. \end{cases}$$

Moreover let us assume that the saturation condition and strengthened Cauchy inequality holds

$$\| u - \overline{u}_h \|_\Omega \le \mathcal{X} \| u - u_h \|_\Omega, \mathcal{X} \in (0,1) \quad \forall h > 0$$
$$| a_\Omega(v, w) | \le \gamma \| v \|_\Omega \| w \|_\Omega, \gamma \in (0,1) \quad \forall v, w \in V.$$

Then AEE $e_h = \overline{u}_h - u_h \in E_h$ is efficient and reliable with constant $C = (1-\mathcal{X})^{-1}(1-\gamma^2)^{-1/2}$,

$$\| e_h \|_\Omega \le \| u - u_h \|_\Omega \le C \| e_h \|_\Omega .$$

Proof. See (Vovk, Kvasnytzia, Ostapov, & Shynkarenko 2014).

Theorem 2. *Two-sided error estimations.*

We assume e_K^{Dir}, e_K^{Neu} and $e = u - u_h$ are the solutions of the following local Dirichlet, Neumann error problems, and linear case of (4)

$$\begin{cases} \text{find } e_K^{Dir} \in E_h^{Dir}(K) \subset H_0^1(K) \text{ such that} \\ a_K\left(e_K^{Dir}, v\right) = \langle r_K(u_h), v \rangle \quad \forall v \in H_0^1(K), \end{cases} \tag{6}$$

$$\begin{cases} \text{find } e_K^{Neu} \in E_h^{Neu}(K) \subset H^1(K) \text{ such that} \\ a_K\left(e_K^{Neu}, v\right) = \langle r_K(u_h), v \rangle \quad \forall v \in H^1(K), \end{cases} \tag{7}$$

$$\begin{cases} \text{find } e \in E,\ V = V_h \oplus E \text{ such that} \\ a_K(e, v) = \langle r_K(u_h), v \rangle \quad \forall v \in V, \end{cases} \tag{8}$$

$\forall K \in \Im_h$ and $\langle r_K(u_h), v \rangle = \langle l_K, v \rangle - a_K(u_h, v)$.

$$\text{Then } \| e_K^{Neu} \|_K^2 = \| e_K^{Dir} \|_K^2 + \| e_K^{Neu} - e_K^{Dir} \|_K^2, \tag{9}$$

$$\| e_K^{Dir} \|_K \le \| e \|_K \le \| e_K^{Neu} \|_K \quad \forall K \in \Im_h, \tag{10}$$

$$\| e_K^{Dir} \|_\Omega \le \| e \|_\Omega \le ||| e_K^{Neu} |||_\Omega, \tag{11}$$

where $\| v \|_K^2 := a_K(v, v)$, $||| w |||_\Omega^2 := \sum_{K \in \Im_h} \| w \|_K^2$.

Proof. We start from the following equation

$$a_K\left(e_K^{Dir} - e_K^{Neu}, v\right) = 0 \quad \forall v \in H_0^1(K),$$

and obtain

$$\begin{aligned}
&\left\|e_K^{Neu} - e_K^{Dir}\right\|_K^2 \\
&\quad = a_K\left(e_K^{Neu} - e_K^{Dir}, e_K^{Neu} - e_K^{Dir}\right) \\
&\quad = \left\|e_K^{Neu}\right\|_K^2 - a_K\left(e_K^{Dir}, e_K^{Neu}\right) \quad \forall K \in \Im_h. \qquad (12)
\end{aligned}$$

On the other hand

$$\begin{aligned}
a_K\left(e_K^{Neu}, e_K^{Dir}\right) &= a_K\left(e_K^{Dir}, e_K^{Neu}\right) \\
&= \left\|e_K^{Dir}\right\|_K^2 \quad \forall K \in \Im_h. \qquad (13)
\end{aligned}$$

The combination of (12) and (13) yields inequality (9) that describes the subordination between Dirichlet and Neumann AEEs. Note if $e_K^{Dir} \neq e_K^{Neu}$ than $\| e_K^{Dir} \|_\Omega < \| e_K^{Neu} \|_\Omega \quad \forall K \in \Im_h$.

From the residual problems (6), (7) and (8) we obtain

$$\begin{cases}
a_K\left(e_K^{Dir}\right) = a_K(e,v) \quad \forall v \in H_0^1(K), \\
a_K\left(e_K^{Neu}\right) = a_K(e,v) \quad \forall v \in H^1(K).
\end{cases}$$

Finally, inequality (17) and then (18) are obtained from

$$\begin{cases}
\left\|e_K^{Dir}\right\|_K^2 = a_K\left(e, e_K^{Dir}\right) \leq \| e \|_K \left\|e_K^{Dir}\right\|_K, \\
\left\|e_h\right\|_K^2 = a_K\left(e_K^{Neu}, e\right) \leq \| e \|_K \left\|e_K^{Neu}\right\|_K.
\end{cases}$$

5 H-ADAPTIVE FEM

The aim of the adaptation process is to calculate more exact approximation $u_H \in V_H$ such that $\| u - u_H \|_\Omega \leq \| u - u_h \|_\Omega$, $u_h \in V_h$, $u \in V$. For this purpose the initial triangulation $\Im_h$ is refined to $\Im_H$ and simultaneously construct subspace $V_H \subset V$. Bisection method for dividing the triangles guarantees both $V_h \subset V_H$ and, as consequence, convergence of h-adaptive FEM, see (Mekchay & Nochetto 2002). At first, we create the set $D \in \Im_h$ of triangles such that the following error indicator

$$\eta_K = \sqrt{N} \left\|e_K\right\|_K \left\|u_h + e_h\right\|_\Omega^{-1} 100\%,$$

exceeds the given admissible error level δ. Then we divide all triangles of D by bisections (Ostapov, Shynkarenko, & Vovk 2014), gain triangulation $\Im_H$, calculate u_H. We repeat adaptation procedure for u_H if the refinement is needed. For other details of the h-adaptive FEM construction see (Kvasnytsia, Ostapov, Shynkarenko, & Vovk 2014).

6 NUMERICAL RESULTS

In the numerical analysis of the constructed Dirichlet and Neumann AEEs we solve the singularly perturbed and semi-linear boundary value problems with explicit exact solutions by using the uniform and local adaptive mesh refinement for comparison. We use the Dirichlet estimator in adaptive criterion and the bisection method for the local mesh refinement, see (Ostapov, Shynkarenko, & Vovk 2014).

The singularly perturbed boundary value problem with internal layer is stated as follows

$$\begin{cases}
-10^{-3}\Delta u + (\beta_1, \beta_2) \cdot \nabla u = 0 \;\text{ in } \Omega = [0,1]^2, \\
\qquad\qquad\qquad u = u_{ext} \;\text{ on } \Gamma_D = \Gamma,
\end{cases} \qquad (14)$$

where $\beta_1(x) = x_1 - 0.6$, $\beta_2(x_2) = x_2 - 0.3$, the solution $u^{ext}(x,y) = G[m\beta_1(x_1) + v\beta_2(x_2)]G[m\beta_2(x_2) - v\beta_1(x_1)]$, $m = \cos(\pi/6)$, $v = \sin(\pi/6)$, $G(z) = \frac{1}{2}[1 + erf(z/\sqrt{2\mu})]$.

We solve problem (14) by using linear FEM approximation on the uniform (see Table 1) and locally refined (see Table 2) meshes, see also Figure 1.

Above k denotes the step number of the mesh refinement, $Nod\Im_h$ the number of the nodes in the mesh $\Im_h$, $\varepsilon_h^a = \| e_h^a \| \| u_h \|^{-1} 100\%$ the relative errors, $p^a = 2\frac{\ln\|e_{h,k}^a\| - \ln\|e_{h,k+1}^a\|}{\ln N_{k+1} - \ln N_k}$ the convergence rates, where $a = \{Dir, ext, Neu\}$, $N_k = Nod\Im_h$ at k step, $\| \cdot \| := \| \cdot \|_{H^1(\Omega)}$.

Finally, we consider the following semi-linear problem

$$\begin{cases}
-\Delta u = a(r)u^3 + b(r)u^2 \;\text{ in } \Omega = [0,1]^2, \\
\quad u = u_{ext} \;\text{ on } \Gamma_D,\; \nu \cdot \nabla u = 0 \;\text{ on } \Gamma_N,
\end{cases} \qquad (15)$$

where $\Gamma_D = \{x \in \Gamma : x_1 = 1\} \cup \{x \in \Gamma : x_2 = 1\}$, $\Gamma_N = \{x \in \Gamma : x_1 = 0\} \cup \{x \in \Gamma : x_2 = 0\}$, $a(r) = -8l^2r^2\cos^2 r^2$, $b(r) = 4l^2[\cos r^2 - r^2 \sin r^2]$, the solution $u_{ext} = [\sin r^2 + 2]^{-1}$, $r^2 := x_1^2 + x_2^2$, $l^2 = 3\pi$.

The numerical results for the problem (15) are presented in Tables 3, 4. Note that in Tables 2, 4 we show only certain rows.

Tables 1–4 and Figure 1 show that Dirichlet ε_h^{Dir} and and Neumann AEEs ε_h^{Neu} provide the lower and upper bounds of exact error ε_h^{ext} in a sufficiently narrow interval correspondingly. The convergence rates p^{Dir} and p^{Neu} are close to the expected theoretical rate 1.0 which is numerically calculated as p^{ext}. All the above conclusions are true for the sufficiently fine meshes and, consequently, small approximation errors.

Table 1. The convergence of linear approximations and their AEEs for the problem (14) on the uniform triangular meshes.

k	Nod $\Im_h$	ε_h^{Dir}	ε_h^{ext}	ε_h^{Neu}	p^{Dir}	p^{ext}	p^{Neu}
1	221	64.704	50.087	92.233	–	–	–
2	841	22.202	24.658	62.533	1.9	1.0	1.5
3	3 281	5.769	11.344	21.320	2.0	1.1	1.9
4	12 961	2.371	5.620	9.089	1.3	1.0	1.3
5	51 521	1.124	2.809	4.343	1.1	1.0	1.1
6	205 441	0.554	1.404	2.145	1.0	1.0	1.0
7	820 481	0.276	0.702	1.069	1.0	1.0	1.0
8	3 279 361	0.138	0.351	0.534	1.0	1.0	1.0

Table 2. The convergence of linear AFEM approximations and their AEEs for the problem (14) on the bisection meshes with the tolerance 1%.

k	Nod $\Im_h$	ε_h^{Dir}	ε_h^{ext}	ε_h^{Neu}	p^{Dir}	p^{Neu}
2	49	93.584	89.817	97.141	–	–
4	169	76.431	61.281	95.008	1.1	0.4
6	573	36.187	32.019	79.723	1.9	1.6
8	1 215	8.873	14.823	31.612	2.9	2.7
10	2 841	3.104	7.366	11.836	1.5	1.5
11	4 758	1.953	4.727	7.511	1.8	1.8
12	8 072	1.386	3.745	5.348	1.3	1.3
13	13 693	0.973	2.512	3.761	1.3	1.3

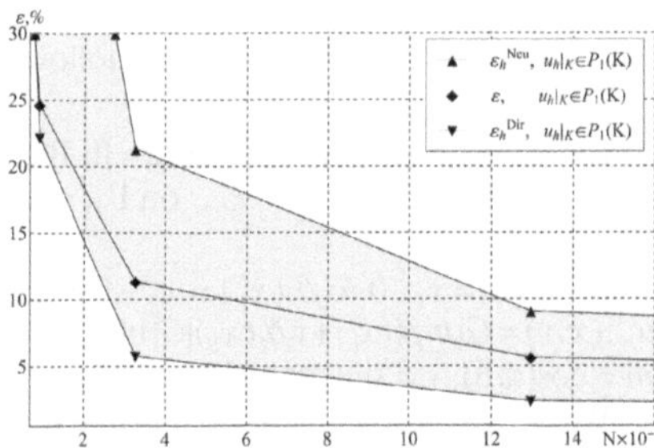

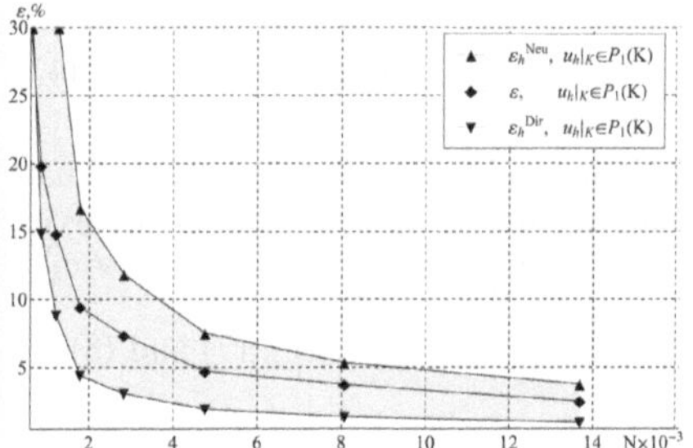

Figure 1. The convergence of AEEs Dirichlet, Neumann and exact error of finite element approximations on uniform meshes (left) and h-adaptive FEM (right) for problem (14).

Table 3. The convergence of linear FEM approximations and their AEEs for the problem (15) on the uniform triangular meshes.

k	Nod $\Im_h$	ε_h^{Dir}	ε_h^{ext}	ε_h^{Neu}	p^{Dir}	p^{ext}	p^{Neu}
1	221	17.252	58.702	70.145	–	–	–
2	841	9.506	36.180	37.016	0.6	0.7	1.1
3	3 281	4.935	17.063	19.127	0.7	1.1	0.8
4	12 961	2.509	7.332	9.721	0.9	1.2	0.9
5	51 521	1.262	3.328	4.888	1.0	1.1	1.0
6	205 441	0.632	1.609	2.448	1.0	1.1	1.0
7	820 481	0.316	0.797	1.224	1.0	1.0	1.0
8	3 279 361	0.158	0.397	0.612	1.0	1.0	1.0

Table 4. The convergence of linear AFEM approximations and their AEEs for the problem (15) on the bisection meshes with the tolerance 1%.

k	Nod $\Im_h$	ε_h^{Dir}	ε_h^{ext}	ε_h^{Neu}	p^{Dir}	p^{Neu}
2	49	30.462	91.929	71.412	–	–
4	165	19.175	68.771	73.593	0.4	−0.3
6	598	10.854	44.104	42.707	0.6	0.6
8	2135	5.742	23.838	22.246	0.9	0.9
10	7623	2.946	10.599	11.411	1.0	1.0
12	21420	1.541	4.780	5.965	1.3	1.3
13	32781	1.155	3.091	4.471	1.3	1.3
14	44874	0.921	2.623	3.562	1.4	1.4

7 CONCLUSIONS

The constructed Dirichlet and Neumann AEEs are suitable for solving the singularly perturbed and semi-linear boundary value problems with the established tolerance. The theoretically established efficiency, reliability, and two-sided error estimations of these AEEs in theorems 1,2 are confirmed by the numerical results.

Finally, the suggested Dirichlet and Neumann AEEs can be extended to both the bilinear FEM approximations, see (Shynkarenko & Vovk 2013), and the quadratic serendipity FEM approximations, see (Vovk, Kvasnytzia, Ostapov, & Shynkarenko 2014).

REFERENCES

Ainsworth, M. & J.T. Oden (2000). *A Posteriori Error Estimation in Finite Element Analysis*. New York: Wiley.

Babuška, I., J. Whiteman, & T. Strouboulis (2011). *Finite Elements: An Introduction to the Method and Error Estimation*. Oxford University Press.

Bank, R.E. & A. Weiser (1985). Some a posteriori error estimators for elliptic partial differential equations. *Mathematics of Computation* 44, 283–301.

Deuflhard, P. & M. Weiser (2012). *Adaptive Numerical Solutions Of PDEs*. Berlin: Walter de Gruyter GmbH & Co. KG.

Kvasnytsia, H.A., O.Y. Ostapov, H.A. Shynkarenko, & O. Vovk (2014). Computable double-sided a posteriori error estimates and h-adaptive finite element approximations. *Manufacturing Processes: Actual Problems-2014*. 1, 87–102.

Mekchay, K. & R. Nochetto (2002). Convergence of adaptive finite element methods for general second order linear elliptic pde. *SIAM J. Numer. Analysis*. 43, 1803–1827.

Ostapov, O.Y., H.A. Shynkarenko, & O. Vovk (2013). A posteriori error estimator and h-adaptive finite element method for diffusion-advection-reaction problems. In *20th Int. Conf. on Comp. Meth. is Mech. (CMM 2013)*, Poznan University of Technology, Poznan, Poland, pp. MS10:3–4.

Ostapov, O.Y., H.A. Shynkarenko, & O. Vovk (2014). A posteriori error estimator and h-adaptive finite element method for diffusion-advection-reaction problems. *Recent Advances in Comput. Mechanics*. Taylor & Francis Group, 329–337.

Ostapov, O.Y., H.A. Shynkarenko, & O. Vovk (2015). Computable double-sided a posteriori error estimates for hadaptive finite element method. In *21st Int. Conf. on Comp. Meth. is Mech. (CMM 2015), 3rd Polish Congress of Mechanics (PCM 2015)*, Gdańsk University of Technology, Gdańsk, Poland, pp. 45–46.

Shynkarenko, H.A. & O. Vovk (2013). A posteriori error estimations for finite element approximations on quadrilateral meshes. *J. Comp. & Appl. Math.* 3 (113), 107–118.

Vovk, O., G.A. Kvasnytzia, O.Y. Ostapov, & H.A. Shynkarenko (2014). A posteriori error estimators of quadratic serendipity finite element approximations for elliptic boundary value problems. *Visnyk of the Lviv University. Series Appl. Math. and Informatics* 21, 67–78. In Ukrainian.

Advances in Mechanics: Theoretical, Computational and Interdisciplinary Issues – Kleiber et al. (Eds)

Cancer ablation during RF Hyperthermia using internal electrode

M. Paruch
Silesian University of Technology, Gliwice, Poland

ABSTRACT: The paper presents numerical modeling of RF (Radiofrequency) hyperthermia caused by the introduction of internal electrode to the tumor. The main purpose of the publication is to analyze the relationship between electrode voltage and the duration of hyperthermia treatment necessary to achieve adequate temperature in the tumor region. Mathematical modeling based on the coupling of two problems: electrical—to generate additional heat and thermal—to estimate the change and rise of the temperature in the tumor. The distribution of electric potential in domain considered is described by the Laplace equation, while the temperature field is described by the Pennes equation. At the stage of numerical simulation the Boundary Element Method (BEM) is used. In the final part of the paper the results of numerical computations are shown.

1 INTRODUCTION

Artificial hyperthermia is a method of treating cancer, wherein the cancer tissues are subjected to external (Paruch 2014) or internal (Ng & Jamil 2014) impacts to high temperatures—more than 45°C. Up till now, other various heating methods for hyperthermia have been also used, such as microwaves (Gas 2015), ultrasound (Marmor et al. 1979) or lasers (Kim et al. 1996). Hyperthermia, or overheating of the tumor, becomes one of the esteemed methods of dealing with cancer. It is a kind of radiotherapy, which can significantly increase the effectiveness of cancer treatment. During hyperthermia, the tissues are subjected to the temperatures of 40 to 45°C, because at temperatures above 42°C begins the process of necrosis of living cells and the temperatures above 45°C are known as thermoablative (Gasselhuber et al. 2010). Prolonged exposure at such high temperature destroys the cells via coagulation necrosis. Research have shown that high temperatures can damage and kill cancer cells, usually accompanied by a low risk of damage to healthy tissue (Ng & Jamil 2014).

2 MATHEMATICAL MODELLING

Proper prediction of the temperature distribution in the tumor and surrounding tissue is a key element for ensuring the effectiveness of the artificial hyperthermia treatment. For this purpose, the methods of mathematical modeling may be used. A mathematical model of the process of heating by the electric field consists of two parts. The electric part concerns the system of two Laplace equations (tumor region and healthy tissue) supplemented by boundary conditions to obtain the electric field distribution. The thermal part is connected with the system of two bioheat transfer equations also supplemented by boundary-initial conditions to obtain the temperature distribution (here the transient problem is considered). In the bioheat transfer equations the additional source terms associated with the heat generation caused by electric field distribution appear.

The RF Hyperthermia (RFH) represents coupled electro-thermal problems, and is applicable to use of electromagnetic energy to heat a specific area. Therefore it is important to correctly describe how the electromagnetic field interacts with biological tissue. In the full Maxwell's equations, electric and magnetic fields are coupled, so it requires complex numerical calculations. Because the RFH uses low frequencies, in order to determine the intensity of the electric field the simplification known as the quasi-static approach can be taken into account.

Using the quasi-static formulation, the electric field intensity **E** (V/m) inside the biological tissue can be calculated as follows (Andreuccetti & Zoppetti 2006, Paruch 2014):

$$\mathbf{E}_e(X)=-\nabla\varphi_e(X)=-[\partial\varphi_e(X)/\partial x_1 \quad \partial\varphi_e(X)/\partial x_2] \tag{1}$$

where e = 1, 2 denotes the healthy tissue and tumor, respectively, $X = \{x_1, x_2\}$ and φ_e [V] = is an electric potential, while the heat generation

$Q_e^E(X)$ due to the electromagnetic heating is defined as follows:

$$Q_e^E(X) = \left[\sigma_e |\mathbf{E}_e(X)|^2\right]/2 = (\sigma_e/2) \sum_{i=1}^{2} (\partial \varphi_e(X)/\partial x_i)^2 \quad (2)$$

where σ_e [S/m] = is an electrical conductivity.

The electric potential $\varphi_e(X)$ inside the healthy tissue Ω_1 and cancer Ω_2 (Fig. 1) is described by the system of Laplace equations:

$$X \in \Omega_e: \ \nabla[\varepsilon_e(X)\nabla\varphi_e(X)] = 0 \quad (3)$$

where ε_e [C²/(Nm²)] = is a dielectric permittivity of tissue.

For a heat transfer process in biological tissue the Pennes model has been proposed (Pennes 1948):

$$X \in \Omega_e: \ c_e\rho_e \frac{\partial T_e(X)}{\partial t} = \lambda_e \nabla^2 T_e(X) + k_e\left[T_B - T_e(X)\right] + Q_{mete} + Q_e^E \quad (4)$$

where t = denotes time; ρ_e [kg/m³] = is the density; c_e [J/(kgK)] = is the specific heat; λ_e [W/(mK)] = is the thermal conductivity; T_e [K] = is the temperature, $k_e = G_{Be}c_B$ [W/(m³K)] is the perfusion rate (G_{Be} [1/s] = is the perfusion coefficient, c_B [J/(m³K)] = is the volumetric specific heat of blood); T_B = is the supplying arterial blood temperature and Q_{mete} [W/m³] = is the metabolic heat source.

Differential equations which describe the electric (3) and temperature (4) fields are supplemented by appropriate boundary-initial conditions. Constant voltage boundary condition $\varphi = U$ (U[V] is an electric potential of the electrode) is applied on the active part of the electrode (tip length: 1.5 cm), whereas the boundaries Γ_1, Γ_2 and Γ_3 are considered as the ground ($U = 0$). On the remaining boundaries electric insulation is assumed. In the case of temperature field, boundaries Γ_1–Γ_4, the body core temperature T_b = 37°C is assumed, while on the electrode surface the thermal insulation is applied. On the contact surface between healthy tissue and tumor the ideal electric and thermal contacts were assumed. Equation (4) is also supplemented by the initial condition $T_0(X) = T_{init}$.

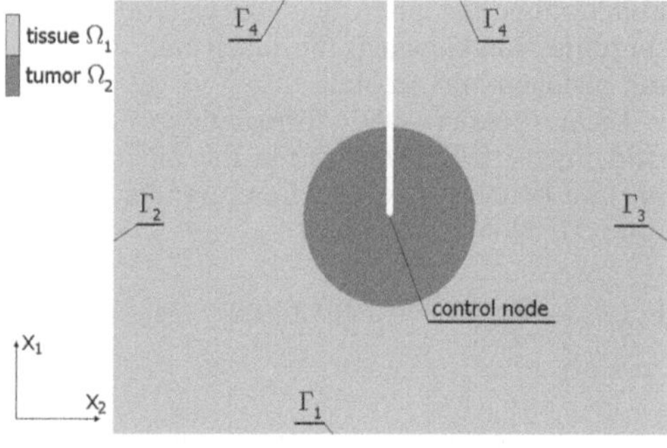

Figure 1. The healthy and tumor tissue including the action of electric field.

3 BOUNDARY ELEMENT METHOD

In order to solve the equations describing the potential of electric field and the temperature field in the considered domains the boundary element method has been applied (Majchrzak 2013).

The boundary integral equations corresponding to (3) can be expressed as follows:

$$B_e(\xi,\eta)\varphi_e(\xi,\eta) + \int_\Gamma \psi_e(X)\varphi_e^*(\xi,\eta,X)\,d\Gamma = \int_\Gamma \varphi_e(X)\psi_e^*(\xi,\eta,X)\,d\Gamma \quad (5)$$

where:

$$\varphi_e^*(\xi,\eta,X) = \frac{1}{4\pi\varepsilon_e r} \quad (6)$$

is a fundamental solution, while for temperature field (4):

$$\begin{aligned} &B_e(\xi,\eta)T_e(\xi,\eta,t^f) + \frac{1}{c_e\rho_e}\int_{t^{f-1}}^{t^f}\int_\Gamma T_e^*(\xi,\eta,X,t^f,t)q(X,t)\,d\Gamma\,dt \\ &= \frac{1}{c_e\rho_e}\int_{t^{f-1}}^{t^f}\int_\Gamma q_e^*(\xi,\eta,X,t^f,t)T(X,t)\,d\Gamma\,dt + \iint_\Omega T_e^*(\xi,\eta,X,t^f,t^{f-1})T(X,t^{f-1})\,d\Omega \\ &+ \frac{1}{c_e\rho_e}\int_{t^{f-1}}^{t^f}\iint_\Omega Q_e(X,t)T_e^*(\xi,\eta,X,t^f,t)\,d\Omega\,dt \end{aligned} \quad (7)$$

where:

$$Q_e = Q_{mete} + Q_e^E \quad (8)$$

while a fundamental solution is defined as follows:

$$T_e^*(\xi,\eta,X,t^f,t) = \frac{1}{4\pi a_e(t^f - t)}\exp\left[-\frac{r^2}{4a_e(t^f - t)}\right] \quad (9)$$

where $a_e = \lambda_e/(c_e\rho_e)$ is a diffusion coefficient.

The above boundary-integral equations are solved using the method described in Majchrzak (2013).

4 RESULTS OF COMPUTATIONS

The 2D domain of dimensions 0.16 m × 0.12 m has been considered. The tumor region which is located at the center of healthy tissue (Fig. 1) is approximated by circle (radius: 0.025 m). The electrical and thermophysical parameters are collected in Tables 1 and 2, respectively.

The initial temperature T_{init} = 37°C, time step Δt = 5s and maximum analysis time is assumed as 1 h.

Table 3 contains the results of numerical calculations for various values of voltage on the electrode together with the information about the time in the center of tumor the temperature increase to the value 45°C was observed.

In Figure 2 the temperature distribution for healthy tissue with tumor without action of electrode is shown, it is visible that the presence of the tumor disturbs the temperature distribution inside the tissue, but this is not a hyperthermia state.

Table 1. The electrical parameters (2–3).

Parameter	Healthy tissue	Tumor
Dielectric permittivity	$8.85 \cdot 10^{-10}$	$1.3275 \cdot 10^{-9}$
Electrical conductivity	0.2	0.3

Table 2. The thermophysical parameters (4).

Parameter	Healthy tissue	Tumor
Thermal conductivity	0.5	0.75
Perfusion rate	1998.1	7992.4
Metabolic heat source	420	4200
Arterial blood temperature	37°C	
Density	1090	
Specific heat	3421	

Table 3. Results of computations.

U (V)	Exposure time (s)	Temp. at the control node (Fig. 1) (°C)
10	3600	40.17
15	3600	43.56
17	1500	45.01
20	340	45.01
25	130	45.09

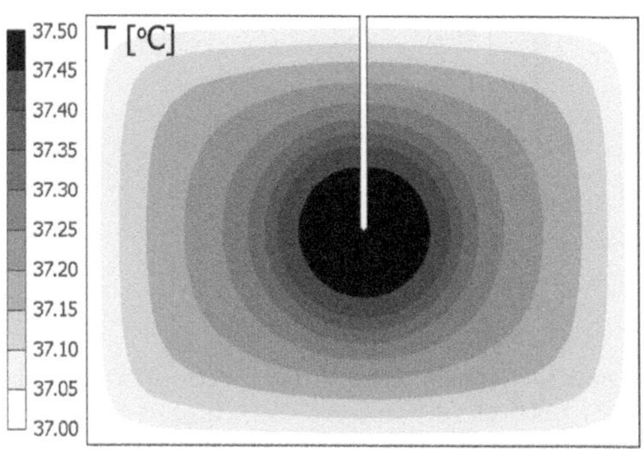

Figure 2. Temperature distribution (without electric field).

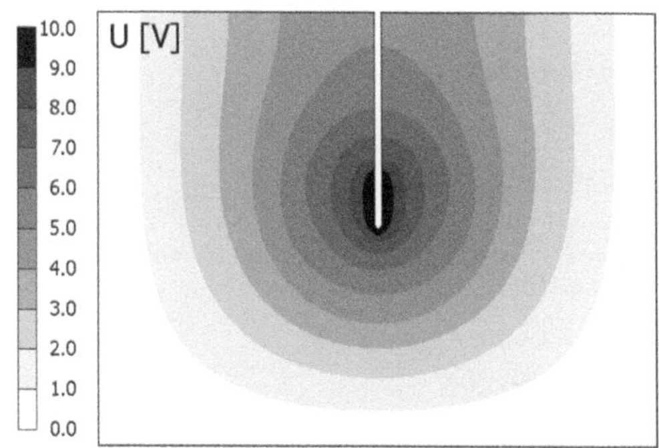

Figure 3. Electric potential distribution (U = 10V).

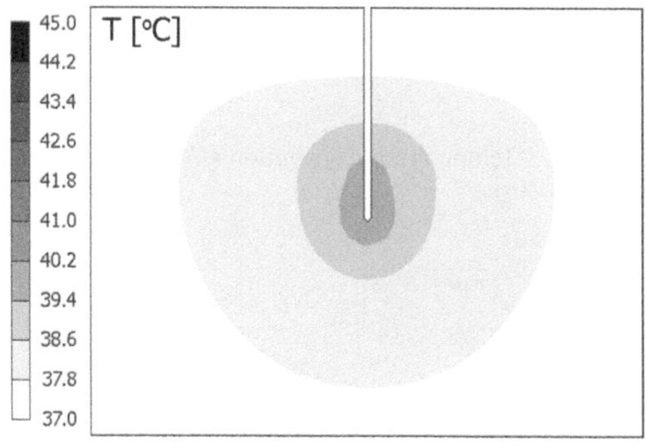

Figure 4. Temperature distribution (U = 10V, exposure time t = 3600s).

In Figure 3 the electric potential distribution for U = 10V is shown. The maximum value of the electric potential, of course, is observed in the neighbourhood of the active part of the electrode (in each cases collected in Table 3).

In Figures 4–8 the temperature field obtained after the action of the internal electrode are shown.

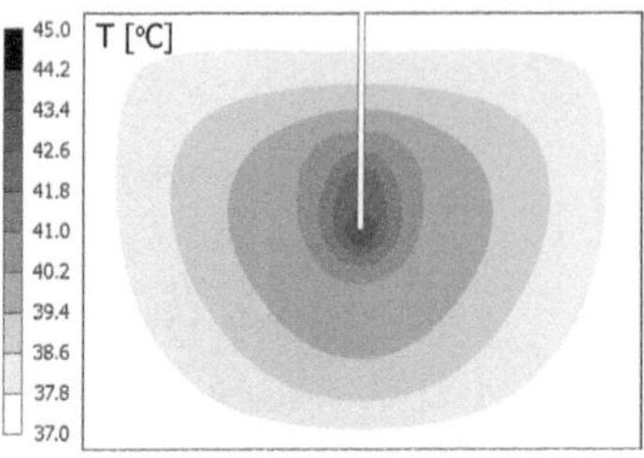

Figure 5. Temperature distribution ($U = 15$V, exposure time $t = 3600$s).

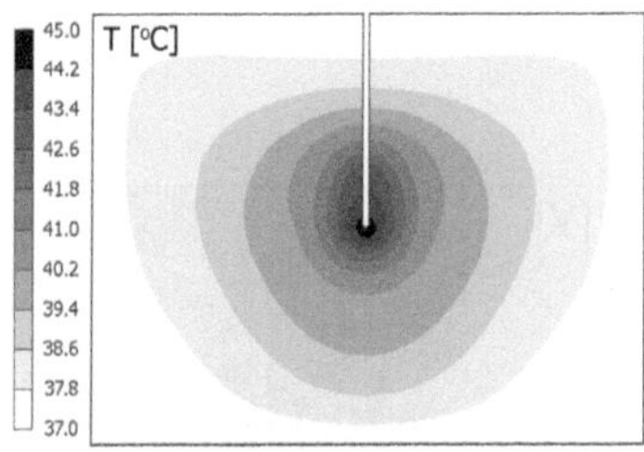

Figure 6. Temperature distribution ($U = 17$V, exposure time $t = 1500$s).

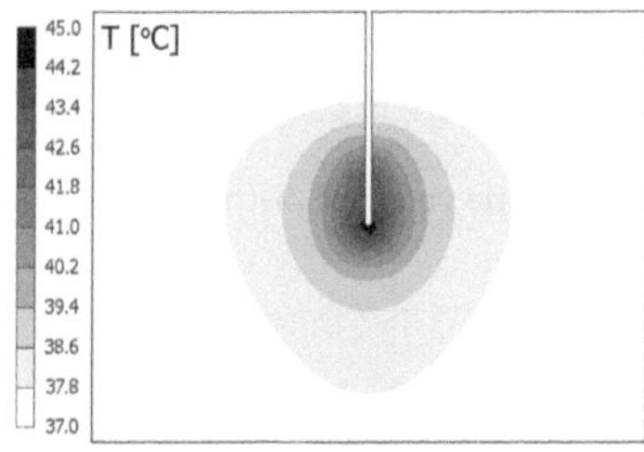

Figure 7. Temperature distribution ($U = 20$V, exposure time $t = 340$s).

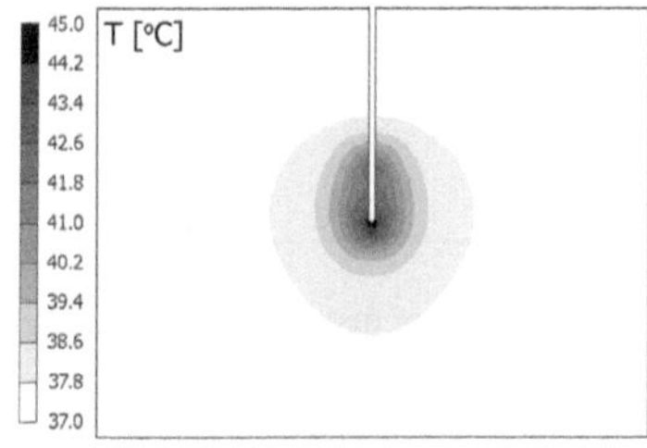

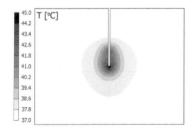

Figure 8. Temperature distribution ($U = 25$V, exposure time $t = 130$s).

5 CONCLUSIONS

The approach presented in this paper offers a methodology of finding the solution of radiofrequency hyperthermia treatment. Causing the phenomenon of RFH using electric field requires the proper selection of the potential on the electrode. The numerical simulations demonstrate that the changes of the voltage of electrode and exposure time have a significant influence to the changes of temperature in the entire domain considered, the higher the voltage, the shorter of the exposure time (Figs. 6–8), but the hyperthermia state is not obtained when the potential on the electrode is too low (Figs. 4–5). Since the interaction of the electric field causes heating of healthy tissue (Fig. 6), to concentrate the thermal energy exactly in the area of the tumor, the introducing of the magnetic nanoparticles is necessary. That problem will be under consideration in the future work.

ACKNOWLEDGEMENTS

This work was supported by statutory grant No 10/990/BK_15/0013 from Faculty of Mechanical Engineering of the Silesian University of Technology and grant No BKM-504/RMT-4/2014.

REFERENCES

Andreuccetti, D. & Zoppetti, N. 2006. Quasi-static electromagnetic dosimetry: from basic principles to examples of applications. *International Journal of Occupational Safety and Ergonomics* 12: 201–215.

Gas, P. 2015. Multi–frequency analysis for interstitial microwave hyperthermia using multi–slot coaxial antenna. *Journal of Electrical Engineering* 66(1): 26–33.

Gasselhuber, A., Dreher, M.R., Negussie, A., Wood, B.J., Rattay, F. & Haemmerich, D. 2010. Mathematical spatio-temporal model of drug delivery from low temperature sensitive liposomes during radiofrequency tumour ablation. *International Journal of Hyperthermia* 26: 499–513.

Kim, B.M., Jacques, S.L., Rastegar, S., Thomsen, S. & Motamedi, M. 1996: Nonlinear finite-element analysis of the role of dynamic changes in blood perfusion and optical properties in laser coagulation of tissue. *IEEE Journal of Selected Topics in Quantum Electronics* 2(4): 922–933.

Majchrzak, E. 2013. Application of different variants of the BEM in numerical modeling of bioheat transfer processes. *MCB: Molecular & Cellular Biomechanics* 10(3): 201–232.

Marmor, J.B., Pounds, D., Postic, T.B. & Hahn, G.M. 1979. Treatment of superficial human neoplasms by hyperthermic induced by ultrasound. *Cancer* 43: 196–200.

Ng, E.Y.K. & Jamil, M. 2014. Parametric sensitivity analysis of radiofrequency ablation with efficient experimental design. *International Journal of Thermal Science* 80: 41–47.

Paruch, M. 2014. Hyperthermia process control induced by the electric field in order to destroy cancer. *Acta of Bioengineering and Biomechanics* 16(4): 121–128.

Pennes, H.H. 1948: Analysis of tissue and arterial blood temperature in the resting human forearm. *Journal of Applied Physiology* 1: 93–122.

Advances in Mechanics: Theoretical, Computational and Interdisciplinary Issues – Kleiber et al. (Eds)
© 2016 Taylor & Francis Group, London, ISBN 978-1-138-02906-4

Self-sustained oscillations of the axisymmetric free jet at low and moderate Reynolds number

A. Pawłowska, S. Drobniak & P. Domagała
Institute of Thermal Machinery, Częstochowa University of Technology, Częstochowa, Poland

ABSTRACT: The paper presents the experimental verification of the self-ustained oscillations found numerically by the team of the Institute of Thermal Machinery. The special design of the experimental rig ensured the relevance between experiment and LES calculations. The measurements of both mean and fluctuating velocity fields confirmed the existence of self-sustained oscillations, however the phenomenon analyzed seems to be Reynolds number dependent.

1 MOTIVATION AND EXPERIMENTAL CONDITIONS

The research is focused on the experimental verification of the self-sustained oscillations in axisymmetric jet, which were found by Bogusławski et al. (2013) from LES calculations and received only limited experimental evidence. These results confirm, that despite a great deal of attention that the round jet received in the last century there remain some significant open questions, and the round jet is still a matter of interest as it is exemplified by the recent experimental work of Mi et al. (2013). The present paper was therefore intended as the experiment verification of the hypothesis contained in Bogusławski et al. (2013).

The main problem in comparison of experimental and numerical investigations in free jets is the identity of initial conditions, which in the flow considered should be a matter of extreme care. As it was stated by Ball et al. (2012), the round jet is extremely sensitive to inlet conditions, which change not only the near flow field but this influence may propagate even to the far field. LES computations performed by Bogusławski et al. (2013) revealed, that the regime of self-sustained oscillations appeared when the initial level of turbulence was very low (well below 1%) and the shear layer at the jet exit was sufficiently thin.

For the sake of stability of computations a moderate rate of co-flow was applied in Bogusławski et al. (2013), that changed the initial conditions from the free-slip ones (usually encountered in experiments) to no-slip initial conditions, where the entraining air flowed coaxially with the jet. The experimental research performed by Pawłowska et al. (2014) revealed, that no significant changes occurred in the flow when the free-slip initial conditions were replaced by no-slip ones. The existence of self-sustained oscillations was found at LES computations for the relative shear layer thickness confined in the range $R/\theta = 16$–32 (where R—nozzle outlet radius, θ—momentum thickness of the shear layer at the nozzle outlet). It appeared however, that the design of the nozzle applied in the Pawłowska et al. (2014) experiment gave the thickness of the exit shear layer that was in the range $R/\theta = 39$–86, that in turn did not provide the relevance between the computations and experiment.

The contraction ratio defined by the square of inlet and outlet nozzle diameter $(D/d)^2$ applied in the experiment of Pawłowska et al. (2014) was $(D/d)^2 = 144$ and in order to obtain the thinner shear layer it was decided to go for contraction ratio $(D/d)^2 = 225$ in the present experiment.

The sketch of the test rig is presented at Figure 1, the air is supplied at the bottom, the fan, filter, ducts and acoustic silencers are located outside the measuring area and are not shown at the sketch.

The measurements were performed with hot wire CTA type 55 M DISA and LDA DANTEC 55X System. The spectral and correlation analysis of CTA signals was performed with standard MATLAB software while for LDA signals with randomly spaced samples it was performed by BSA software. Detailed descriptions of the test rig and measurements methods may be found in Pawłowska et al. (2014).

The substantially higher contraction ratio of the nozzle should provide not only the thinner shear layer but also the lower value of turbulence intensity at the outlet. As the measurements had to be performed for sufficiently wide range of exit shear layer thickness, the thickness of exit shear layer was varied by the cylindrical extensions with various lengths L, the L/d being the ratio of the extension length to jet exit diameter d and this ratio

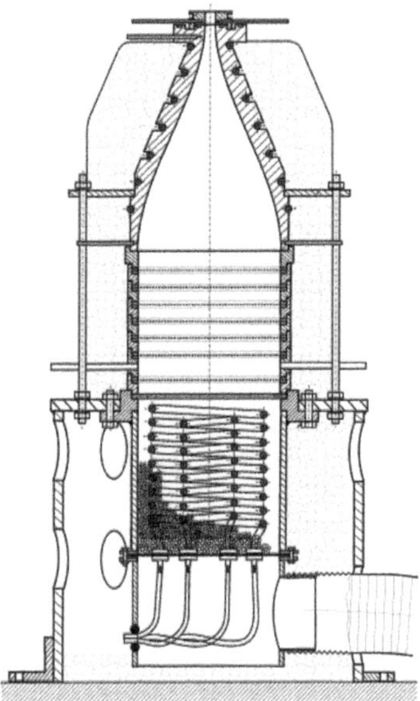

Figure 1. The sketch of the test rig.

was an important parameter of the present experiment. Measurements were performed for three values of Re number equal 5×10^3, 10×10^3 and 20×10^3 respectively, the four lengths L of cylindrical extensions were applied equal L/d = 1, 7, 15 and 25. The results of measurements performed at the jet outlet plane summarized in Table 1 reveal, that shear layer thickness could be varied in the range R/θ = 12–81, that provided the relevance between experiment and LES computations. The value of shape parameter H is very close to theoretical value H = 2.59 given by Blasius, that means that the laminar shear layer was present at the jet exit in all investigated cases. The turbulence intensity obtained at the jet exit was in the range Tu = 0.1–0.2% as can be seen at Table 1, for all the cases considered the turbulence intensity was substantially lower in the present experiment in comparison with the investigations described by Pawłowska et al. (2014). The sample distributions of mean velocity and turbulence intensity profiles at the jet exit (x/d = 0—where x is the distance along the jet axis measured from the exit is presented at Figure 2 for Re = 10 000 and the longest extension tube L/d = 25.

Both mean velocity and turbulence intensity profiles obtained from the present experiment are compared with the results from Pawłowska et al. (2014). The mean velocity profiles normalized by the centerline velocity U_{max} presented at Figure 2a reveal the existence of potential core with uniform distribution of velocity and the thick shear layer spanning from the jet wall (r/d = 0.5) to as far as r/d = 0.25 for both experiments. Contrary to mean velocity distributions which do not reveal any significant difference between particular experiments, the turbulence intensity profiles shown at Figure 2b indicate, that much lower turbulence intensity was obtained at the jet exit when compared with previous jet nozzle design from Pawłowska et al. (2014). One may expect therefore that all preliminary conditions outlined by Bogusławski et al. (2013) are fully met by the present design of experimental test rig and

Table 1. Parameters of exit shear layer.

Re	R/θ	H	Tu[%]
5×10^3	35.7	2.591	0.16
	22.3	2.486	0.21
	15.3	2.555	0.22
	12.3	2.528	0.26
104	54.3	2.567	0.13
	31.3	2.533	0.14
	20.1	2.540	0.14
	15.3	2.646	0.11
2×10^4	80.9	2.617	0.05
	39.0	2.700	0.07
	31.0	2.541	0.07
	22.6	2.661	0.09

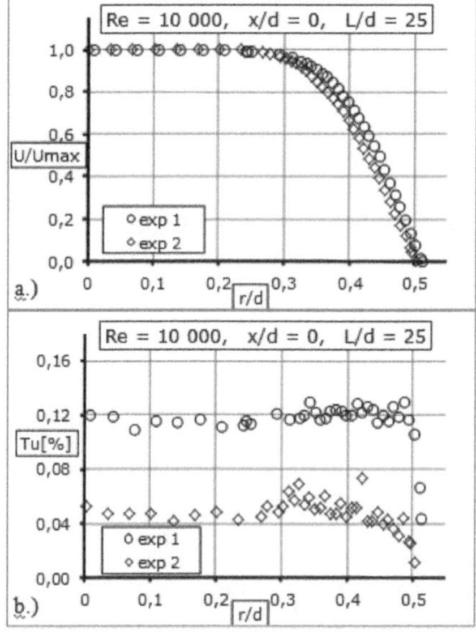

Figure 2. Radial profiles of mean velocity (a) and turbulence intensity (b) at the exit plane for Re = 10 000 for both test rigs, exp1—present experiment, exp2—Pawłowska et al. (2014).

furthermore the relevance between LES computations and experiment was also achieved.

2 EXPERIMENTAL RESULTS AND DISCUSSION

The sample evolution of mean velocity U and turbulence intensity Tu along the jet axis is presented at Figure 3 for all values of exit shear layer thickness obtained due to cylindrical extensions. Data presented at the Figure 3a reveal a typical jet behaviour i.e. the thicker is the shear layer at the exit (that corresponds to larger values of L/d ratio) the longer becomes the potential core, that means the thicker shear layer reduces the spreading angle of the jet. The turbulence intensity evolution shown at Figure 3b reveals the qualitative difference between the behaviour of jets with thin and thick exit shear layers. The jet with the thinnest shear layer (L/d = 1) reaches the highest value of Tu closer to jet exit, while the thickening of shear layer (larger L/d values) moves the maximum of Tu away from the jet exit. One should also notice that for the thinnest exit shear layer the jet develops a characteristic "plateau", which is commonly regarded as the behaviour related to the "pairing" of coherent structures, as it was first shown by Crow and Champagne (1971). In order to verify whether this statement is true for all values of shear layer thickness a spectral analysis of velocity fluctuations was performed along the jet axis. The sample results of spectral analysis are presented at Figure 4 for three points located at the jet axis at the distances x/d = 3, 4 and 6 from the jet exit plane for the thinnest shear (L/d = 1), where the plateau is visible at the axial Tu profile (see Fig. 3b).

For each value of Re four cylindrical extension tubes of various lengths were applied, the summary of exit shear layers are presented at Table 1, where R/θ is the non—dimensional shear layer thickness.

It may be noticed that at all axial locations the same dominant frequency $St_D \approx 0.45$ is observed, that contradicts the statements of Crow and Champagne (1971) and may suggest that the coherent structures observed for the thinnest shear layer do not reveal "vortex pairing" and may probably be related to self-sustained oscillations described in Bogusławski et al. (2013).

The next step was the spectral analysis of velocity fluctuations for thicker shear layer (obtained with longer L/d = 7 extension tube) and the results of this analysis are presented at Figure 5 for the same three axial locations x/d = 3, 4 and 6 from the jet exit plane. As can be seen at Figure 5a at the x/d = 3 distance the dominant frequency corresponding to $St_D = 0.6$ develops and then, when one moves along the jet axis (see Fig. 5b and c) the characteristic "halving of dominant frequency is observed. This behaviour is typical for vortex pairing attributed to "plateau" in axial Tu distribution by Crow and Champagne (1971), even if no "plateau" was observed at Figure 3b. One may conclude therefore, that the classical behaviour of coherent structures may only be observed above a certain value of shear layer thickness, while for sufficiently thin shear layer another mechanism develops.

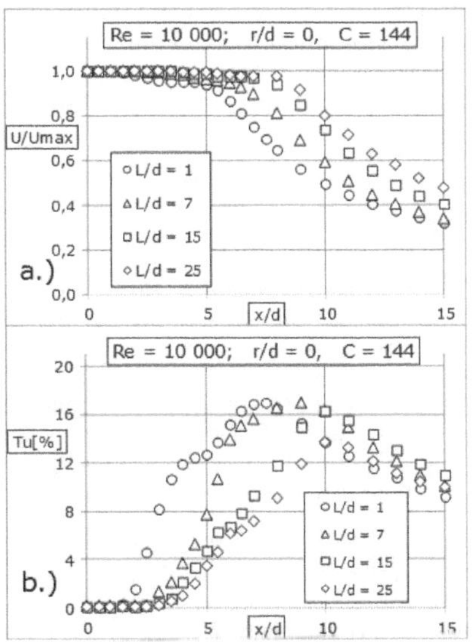

Figure 3. Mean velocity (a) and turbulence intensity (b) at the jet axis for Re = 10 000.

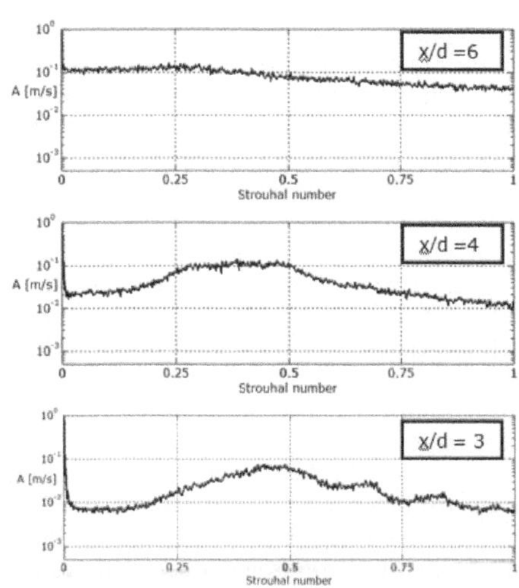

Figure 4. Spectral analysis of velocity fluctuations at the jet axis for Re = 10 000, extension tube L/d = 1.

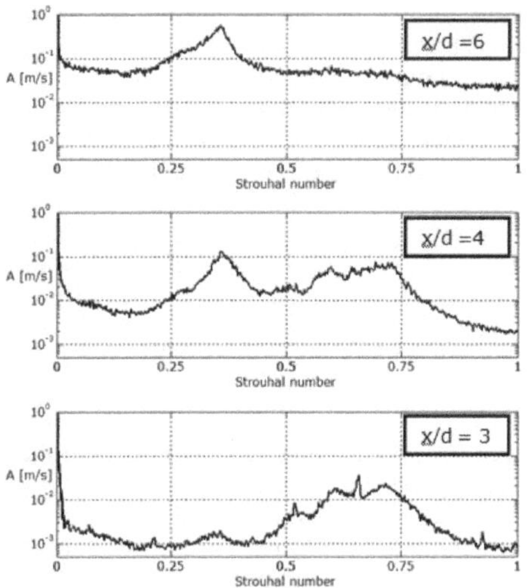

Figure 5. Spectral analysis of velocity fluctuations at the jet axis for Re = 10 000, extension tube L/d = 7.

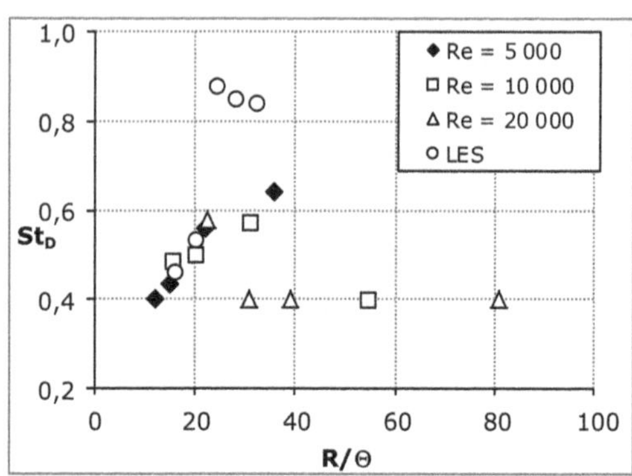

Figure 6. Non-dimensional frequency (Strouhal number) of the most amplified mode versus shear layer thickness obtained from LES calculations (Bogusławski et al. 2013) versus present experiment.

The amplitudes of dominant peaks from all spectra recorded for various Re and various ratios of R/θ were presented at Figure 6 as a relation between Strouhal number based on jet exit diameter and the relative thickness of exit shear layer. Furthermore, the results of LES calculations taken from Bogusławski et al. (2013) were also presented here. The relevance between LES and present experiment is qualitative only but the most important is the same tendency that was observed here and which was also present in the previous research reported in Bogusławski et al. (2013). For Re = 10^4 and 2×10^4 one may observe that above the certain value of R/θ the constant St_D is obtained that probably results from the presence of self-sustained oscillations.

The results obtained with the present experimental set-up confirm the existence of self-sustained oscillations, however the phenomenon analyzed seems to be Reynolds number dependent. In particular, one may notice that for the lowest value of Re the jet behavior resembles the results of linear theory calculations. The increase of Re leads to the behavior suggested by LES simulations, which are characteristic for large amplitude self—excited flow oscillations. It should also be noticed that the results obtained in the present experiment appear to be very similar to those typical for externally forced jets, further studies should be directed towards the comparison with LES numerical investigations and concentrated on explaining this unusual jet behavior. In particular seems necessary to explain whether these self-sustained oscillations, despite that they are growing locally in time up to high amplitudes limited by non-linear interactions (being in this feature similar to the self-sustained absolutely unstable mode present in hot jets) or if they reveal the convective character typical for the classical Kelvin-Helmholtz instability. Since the self-sustained regime seems to be a new finding as a consequence there is a need to explain how this regime reacts to density variations in heated jets and how this self-sustained mode interacts with absolutely unstable mode, when the density ratio is lower than the critical one for absolutely unstable regime.

ACKNOWLEDGEMENTS

The investigations presented in this paper have been performed with funding from Polish National Centre of Science within the grant: DEC 2011/03/B/ST8/06401 and Statutory Funds BS/PB-1-103-3010/2011/P.

REFERENCES

Ball C.G., Fellouah H., Pollard A. 2012. The Flow Field of a Turbulent Round Jet, *Progress in Aerospace Sciences*, 50: 1–26.

Bogusławski A., Tyliszczak A., Drobniak S. & Asendrych D. 2013. Self-sustained oscillations in a homogeneous-density round jet, *Journal of Turbulence*, 14, 4: 25–52.

Crow S.C., Champagne F.H. 1971. Orderly structure in jet turbulence, *Journal of Fluid Mechanics*, 48, 3: 547–592.

Mi J., Xu M., Zhou T. 2013. Reynolds number influence on statistical behaviors of turbulence in a circular free jet, *Physics of Fluids*, 25: 11–24.

Pawłowska A., Domagała P., Wysocki M. 2014. Hot-wire study on the axisymmetric free jet dependence on the nozzle shape, *Journal of Physics: Conference Series*, 530: 1–5.

Advances in Mechanics: Theoretical, Computational and Interdisciplinary Issues – Kleiber et al. (Eds)
© 2016 Taylor & Francis Group, London, ISBN 978-1-138-02906-4

On decreasing the optimization problem size

M. Pazdanowski
Cracow University of Technology, Cracow, Poland

ABSTRACT: Solution time of nonlinear constrained minimization problem depends on the number of constraints and the number of decision variables. While the number of constraints is external to the solution procedure itself, one may try to speed up the solution process by the proper conditioning of the decision variables space within the self-contained minimization routine, if possible accompanied by the elimination of the decision variables least affecting the objective function. Singular Value Decomposition (SVD) of the objective function may serve for this purpose. An idea to limit the number of the decision variables in case of quadratic objective function, using SVD, is presented in this paper. Results of computer tests performed on a problem having quadratic objective function and subjected to quadratic constraints are enclosed and discussed.

1 INTRODUCTION

Nonlinear minimization problems may be solved with an iterative method falling into one of two broad classes, namely deterministic and stochastic based ones (Nemhauser 1989, Spall 2003). In general stochastic based methods are efficient when one deals with objective functions having multiple local extremes, while gradient based search procedures prove themselves in the case of convex objective functions spanned over convex feasible domains (Nocedal & Wright 2006).

Solution time of a minimization problem by gradient based search algorithm depends on the number of decision variables, constraints, and conditioning of the decision variables space. While the number of constraints may not be changed without affecting the final solution, one may try to influence the calculation time by proper conditioning and limiting the size of the decision variables space. Thus let us consider a quadratic objective function $F(\boldsymbol{x})$:

$$F(\boldsymbol{x}) = \frac{1}{2}\boldsymbol{x}^T \mathbf{A}\boldsymbol{x}, \quad \mathbf{A} = \mathbf{A}^T, \tag{1}$$

where $\mathbf{A}$ = denotes a square, symmetrical matrix and $\boldsymbol{x}$ a vector of decision variables. In the case of such objective function, or any objective function locally approximated by a quadratic one, the conditioning factor may be defined (Golub & Van Loan 1996):

$$\kappa(\mathbf{A}) = \frac{|\lambda_{\max}(\mathbf{A})|}{|\lambda_{\min}(\mathbf{A})|}, \tag{2}$$

where $\lambda_{\max}$ and $\lambda_{\min}$ are defined as the largest and the smallest (by moduli) eigenvalues of the matrix $\mathbf{A}$ in (1). With such a definition the best possible conditioning factor κ for the considered minimization problem is equal to 1. Shall the objective function be of the type (1), the SVD may also be applied in order to decrease the number of decision variables in $\boldsymbol{x}$, if the application of SVD indicates, that there exists at least one singular eigenvalue in $\mathbf{A}$.

2 SOLUTION APPROACH

An application of SVD algorithm to a square, symmetric matrix $\mathbf{A}$ yields (Press et al. 1992):

$$\mathbf{A} = \mathbf{U}^T\mathbf{D}\mathbf{U}, \tag{3}$$

where $\mathbf{U}$ = is an orthonormal matrix ($\mathbf{U}^{-1} = \mathbf{U}^T$) and $\mathbf{D}$ = is a diagonal matrix, containing moduli of eigenvalues of $\mathbf{A}$, ordered by magnitude. In the case of rank deficient $\mathbf{A}$, at least one diagonal element of $\mathbf{D}$ is singular, i.e. equal to zero. Matrix $\mathbf{D}$ may be decomposed into a product of two diagonal matrices:

$$\mathbf{D} = \mathbf{S}\cdot\mathbf{J}\cdot\mathbf{S}, \tag{4}$$

where $\mathbf{S}$ = is a diagonal matrix, containing square roots of respective diagonal elements of $\mathbf{D}$, with the exception of the singular values, which are replaced by 1's, while $\mathbf{J}$ = is an unit matrix, again with an exception of the singular value locations in $\mathbf{D}$, at which 1's are replaced by 0's. In this way the singularity present in matrix $\mathbf{A}$ is transferred into matrix **J**. Finally, the objective function (1) is replaced by the following one:

$$F(y) = \frac{1}{2} y^T \mathbf{J} y, \tag{5}$$

accompanied by an appropriate change of variables:

$$y = \mathbf{S}\mathbf{U}x. \tag{6}$$

Of course an inverse relationship to (6) holds as well:

$$x = \mathbf{S}^{-1}\mathbf{U}^{-1}y = \mathbf{S}^{-1}U^T y. \tag{7}$$

It comes directly from formulas (5) and (6), that 0's in **J**, corresponding to singular eigenvalues of **A** denote, that certain linear combinations of original decision variables x do not affect the objective function, and thus may be disregarded during analysis. This, depending on the solved problem, may lead to substantial savings in calculation time and will be a focus of the following analysis.

The minimization problem presented by the objective function (1), and subjected to quadratic constraints $G_i(x)$:

$$G_i(x) = \frac{1}{2} x^T \mathbf{M}_i x + \mathbf{N}_i x + O_i \le 0,$$
$$\mathbf{M}_i = \mathbf{M}_i^T, \quad i = 1, 2, \ldots, m \tag{8}$$

where $\mathbf{M}_i$ = denote square symmetric matrices; N_i N_i = vectors; O_i = constants, while index i = enumerates all these constraints; may be transformed into the following problem, using the new decision variables y: find a minimum of $F(y)$ given in (5) while subjected to:

$$G_i(y) = \frac{1}{2} y^T \mathbf{U}\mathbf{S}^{-1}\mathbf{M}_i\mathbf{S}^{-1}\mathbf{U}^T y$$
$$+ N_i\mathbf{S}^{-1}\mathbf{U}^T y + O_i \le 0, \quad i = 1, 2, \ldots, m \tag{9}$$

Gradients of the objective function F and constraints G_i needed to find the search direction, and expressed in the new decision variables, may be calculated using the following formulas:

$$\nabla F(y) = \mathbf{J}y$$
$$\nabla G_i(y) = y^T \mathbf{U}\mathbf{S}^{-1}\mathbf{M}_i\mathbf{S}^{-1}\mathbf{U}^T + N_i\mathbf{S}^{-1}\mathbf{U}^T \tag{10}$$

If needed, the gradients of constraints ∇G_i may be expressed using the original decision variables:

$$\nabla G_i(y) = x^T \mathbf{M}_i\mathbf{S}^{-1}\mathbf{U}^T + N_i\mathbf{S}^{-1}\mathbf{U}^T. \tag{11}$$

This may offer some additional time advantage, should one prefer to operate on both the original (x) and new, orthogonal (y) decision variables spaces.

3 COMPUTATIONAL APPROACH

In each iteration, when new decision variables vector y is found, a corresponding vector x is calculated using (7), then values and gradients of constraints are computed using original decision variables, and transformed to the conditioned variables space as needed, using (11), to execute next iteration in search of the minimum. The search termination criteria are checked after each iteration in the new (y) decision variables space.

In this approach, at the expense of having to perform the transformations (7) and (11) during each iteration, one avoids the need to perform the matrix operations present in (9). The information flow logic and the sequence of operations are depicted in the Figure 1. If the singular or nearly singular values are found during the decomposition

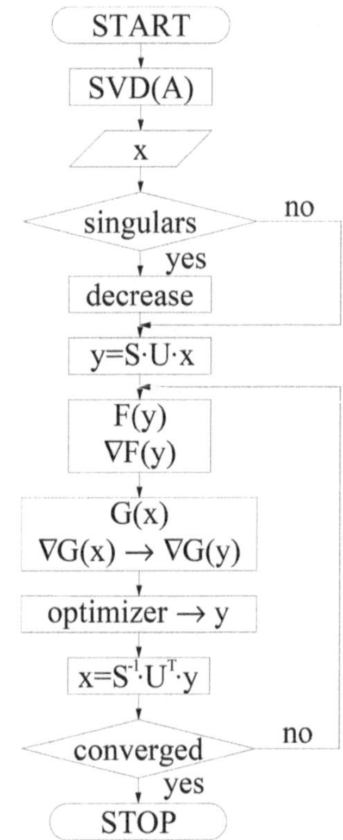

Figure 1. Information flow logic applied in the code.

of matrix **A** the user has to decide whether to decrease the number of the new decision variables (i.e. the dimension of the new decision variables vector $\boldsymbol{y}$ is smaller than the dimension of the original vector $\boldsymbol{x}$) or keep it unchanged (dimensions of both vectors remain the same).

One should note, that the SVD on matrix **A**, the single most time consuming step in the whole procedure outlined in the preceding section, is performed only once at the beginning of calculations.

4 NUMERICAL IMPLEMENTATION

A set of computer routines in FORTRAN programming language has been prepared to implement the extension, supplemented by an externally developed SVD routine (Press et al. 1992).

The implementation is fully transparent to the end user of the FDM optimization package (Orkisz & Pazdanowski 1991), i.e. user needs only to supply the matrix **A** to the code at the beginning of calculations. During subsequent iterations the values of indicated constraints and constraint gradients expressed in terms of original decision variables $\boldsymbol{x}$ must be supplied, exactly the same as if the problem was solved without using the extension.

5 TEST PROBLEM

A search for an estimate of residual stress distribution in an elastic perfectly plastic body subjected to cyclic loads exceeding its elastic bearing capacity may be formulated as a two step nonlinear minimization procedure (Cecot & Orkisz 1997). In the first step one must find a minimum of a quadratic objective function subjected to linear equality constraints (Pazdanowski 2010), while in the second step a solution to the minimization problem of another quadratic objective function subjected to quadratic inequality constraints is sought (Pazdanowski 2010).

It is worth noting that in this case, based on the general assumptions of the theory of plasticity (Martin 1975), matrix **A** is square, but singular, with substantial rank deficiency.

6 NUMERICAL RESULTS AND CONCLUSIONS

The test problem (Cecot & Orkisz 1997, Pazdanowski 2010) was solved several times for input data selected so, as to get minimization problems differing in size (number of decision variables and constraints) as well as the decision variables to constraints ratio. During calculations the Kuhn-Tucker optimality conditions (Bod & Vandenberghie 2004) were used as termination criterion. Due to the practical considerations (Pazdanowski 2010) two levels of this criterion were applied. In order to get a 'rough' approximate solution the termination criterion was set to 0.001, while for accurate 'smooth' final results the value thousand times smaller was used. This was accompanied by a setting of 0.0001 for constraint thickness for the first case and a thousand times smaller value for the second case. This was done, since previous experience (Pazdanowski 2010) has shown, that the final solution of the minimization problem (Cecot & Orkisz 1997) has to be found in at least two consecutive steps of seeking the volume, in which the constraints (8) are active. Thus the 'rough' solution is used only to estimate the volume, while the 'smooth' one is used to find the accurate final solution in the considered domain.

During all the tests, at first it had to be checked, how the inherent inaccuracy of all applied numerical procedures (discrete method itself, other numerical errors) does affect the singular eigenvalues of **A**. In each case the spectrum of eigenvalues similar to those presented on Figure 2 was observed. This clearly indicates that, though strictly speaking all the singular eigenvalues in **A** vanished due to numerical errors, the computed eigenvalues split up into two distinct groups of values, separated by at least an order of magnitude. Therefore in all tests this separation was assumed as the dividing line between significant (to the left of the vertical jump) and insignificant (to the right of it) conditioned decision

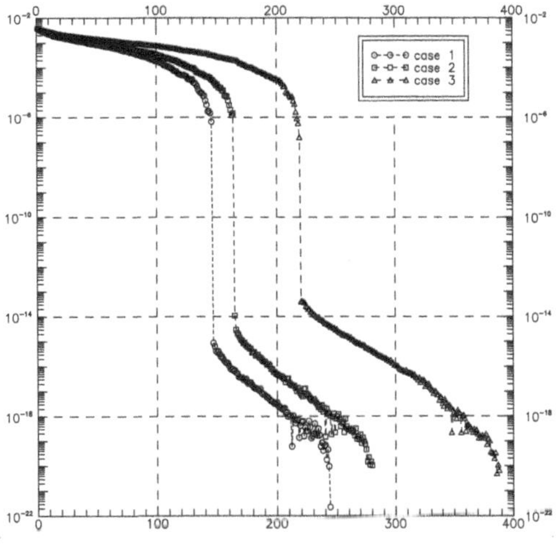

Figure 2. Spectrum of eigenvalues for **A** matrix—sample case.

variables y. The insignificant values were candidates for subsequent elimination.

For comparative purposes, and to ensure, that the presented above method of selecting the conditioned variables is acceptable, all tests presented here were computed five times.

For the first time, (denoted as STD) no conditioning was applied, for the second time (denoted as SVD) the conditioning was applied, but with all y variables kept during calculations, for the remaining three times (denoted as SVD0, SVD+ and SVD—respectively) the conditioning accompanied by elimination of insignificant variables y was applied, while the number of insignificant variables was kept exactly the same as determined by the examination of spectrum graph (SVD0), decreased by 10% (SVD+), and increased by 10% (SVD–).

The tests performed may be arranged in six different groups based on the ratio of decision variables to constraints, and setting of termination criteria. Selected results of these tests are presented in tables (Table 1—'rough' preliminary solutions, Table 2—'smooth' final solutions).

One should note, that all these results were arrived at using automated, heuristics based, procedure applied to control the minimization process itself through changes in the values of governing parameters, such as constraint thickness and Kuhn-Tucker criterion limit values. This procedure, designed to limit the user interaction during minimization process, usually works fine (Pazdanowski

Table 1. Solution calculation times (in seconds)—low quality results.

No.	Case	Iter.	Total time	Speed up	Time per iter.	Ratio
	STD	1340	16.72	–	0.0125	–
	STD	111	7.44	2.247	0.0670	5.36
1	STD+	100	8.16	2.049	0.0816	6.53
	STD0	100	7.89	2.119	0.0789	6.31
	STD–	157	9.82	1.703	0.0625	5.00
	STD	1076	28.33	–	0.0263	–
	STD	133	9.83	2.882	0.0739	2.81
2	STD+	85	9.96	2.844	0.1172	4.46
	STD0	85	7.52	3.767	0.0885	3.37
	STD–	113	10.49	2.701	0.0928	3.53
	STD	19,859	502.76	–	0.0253	–
	STD	955	112.80	4.457	0.1175	4.64
3	STD+	438	49.94	10.067	0.1140	4.51
	STD0	352	31.98	15.721	0.0909	3.59
	STD–	700	61.82	8.133	0.0883	3.49

Table 2. Solution calculation times (in seconds)—high quality results.

No.	Case	Iter.	Total time	Speed up	Time per iter.	Ratio
	STD	4503	96.41	–	0.0214	–
	STD	295	20.99	4.593	0.0712	3.33
1	STD+	552	28.39	3.396	0.0514	2.40
	STD0	426	19.82	4.864	0.0465	2.17
	STD-	425	22.59	4.268	0.0532	2.49
	STD	3682	90.79	–	0.0247	–
	STD	267	23.24	3.907	0.0870	3.52
2	STD+*	474	29.07	3.123	0.0576	2.33
	STD0	296	20.07	4.524	0.0678	2.75
	STD–*	503	28.97	3.134	0.0613	2.48
	STD	15,261	628.44	–	0.0412	–
	STD	686	90.41	6.951	0.1318	3.20
3	STD+*	1365	112.03	5.610	0.0821	1.99
	STD0	746	59.47	10.567	0.0797	1.93
	STD–*	915	83.59	7.518	0.0914	2.22

2010), but under rare circumstances may result in extended calculation time, as has happened in cases denoted by asterisk in Table 2.

Based on the results of the tests performed so far it may be stated, that though the singular values of **A** vanish as a result of the inherent computational inaccuracy, the calculated eigenvalues of this matrix, in each case may be separated into two distinct groups, differing by at least an order of magnitude. This phenomenon facilitates a quick and accurate decision on how many of the original decision variables $\boldsymbol{x}$ are linearly independent, and thus how many conditioned decision variables $\boldsymbol{y}$ must be preserved during optimization. This is clearly indicated by the abrupt jump visible on each spectral plot of the eigenvalues (Fig. 1). In all cases considered, application of the proposed procedure resulted in the substantial time savings during calculations, varying between 40% and 980%, problem size depending, with accuracy of results fully preserved when compared to the solutions obtained using the unmodified algorithm. It was observed, that the time gains tend to increase with increasing problem size (as measured by the number of decision variables and constraints). This may be considered a very fortunate occurrence, as the calculation times tend to grow excessively with increasing problem size. It may be also interesting, that even a substantially underestimated number of conditioned decision variables is sufficient, when only extreme values of the solution are sought. Current application of the proposed procedure is limited to quadratic functions, but the proposed idea is quite general and the SVD decomposition may be applied to any objective function, which may be locally approximated by a quadratic one. Unfortunately, this may be achieved only at the expense of efficiency, as the SVD decomposition, the single most time consuming step of the proposed procedure, would have to be applied after every change of the local approximation. This does require additional testing.

REFERENCES

Bod, S.P. & Vandenberghie, L. 2004. Convex Optimization. Cambridge University Press.

Cecot, W. & Orkisz, J. 1997. Prediction of Actual Residual Stresses Resulting from Cyclic Loading in Kinematic Hardening Material. Proc. COMPLAS V: 1879–1891. Springer.

Golub, G.H. & Van Loan, C.F. 1996. Matrix Computations, 3rd ed., Johns Hopkins University Press.

Martin, J.B. 1975. Plasticity—fundamentals and general results. MIT Press.

Nemhauser, G.L., Rinooy Kan, A.H.G. & Todd, M.J. 1989. Optimization. North Holland.

Nocedal, J. & Wright, S.J. 2006. Numerical optimization. Springer.

Orkisz, J. & Pazdanowski, M. 1991. On a new Feasible Directions solution approach in constrained optimization. In: E. Onate, J. Periaux, A. Samuelsson, (eds), The Finite Element Method in the 1990's: 621–632. Springer.

Pazdanowski, M. 2010. On Estimation of Residual Stresses in Rails Using Shake-Down Based Method. Archives of Transport XXII(3): 319–336.

Pazdanowski, M. 2014. SVD as a preconditioner in nonlinear optimization, CAMES 21(2): 141–150.

Press, W.H., Teukolsky, S.A., Vetterling, W.T. & Flannery, B.P. 1992. Numerical Recipes in Fortran, Cambridge University Press.

Spall, J.C. 2003. Introduction to stochastic search and optimization: Estimation, Simulation and Control. Wiley.

Advances in Mechanics: Theoretical, Computational and Interdisciplinary Issues – Kleiber et al. (Eds)
© 2016 Taylor & Francis Group, London, ISBN 978-1-138-02906-4

Gum Metal—unique properties and results of initial investigation of a new titanium alloy—extended paper

E.A. Pieczyska & M. Maj
Institute of Fundamental Technological Research, Warsaw, Poland

T. Furuta
Toyota Central Research and Development Laboratories, Inc., Nagakute, Aichi, Japan

S. Kuramoto
Ibaraki University, Nakanarusawa, Hitachi, Japan

ABSTRACT: Initial results of effects of thermomechanical couplings occurring in a new titanium alloy, called Gum Metal, subjected to tension loading are presented. The mechanical characteristics obtained by a testing machine confirm an ultra-low elastic modulus and high strength of the alloy. The temperature changes accompanying the alloy deformation process, obtained in a contactless manner by an infrared camera, related to the mechanical results were used to estimate yield limit basing on thermoelastic effect. It was found that the maximal drop in temperature, denoting the Yield point in solid materials, is related to significantly lower strain value in contrast to such a high limit of the reversible deformation, which characterizes the Gum Metal alloy. Furthermore, the temperature distribution on the specimen surface allowed to investigate development of the strain localization leading to the specimen necking and rupture.

1 INTRODUCTION—EXPERIMENTAL DETAILS

The aim of the research is investigation of the effects of thermomechanical couplings in a new multifunctional titanium alloy called Gum Metal, developed in Japan in the beginning of the 21st century. It combines high elasticity and flexibility of rubber and strength of metal. It is a beta-type titanium alloy with a simple body-centered-cubic (bcc) crystal structure and fundamentally expressed composition as Ti_3 (Ta, Nb, V)+(Zr, Hf, 0); prepared by a powder sintering process. Experimental set-up designed for the Gum Metal tension tests, the MTS Testing Machine and Flir Co Phoenix Infrared System is presented in Figure 1.

The Gum Metal is characterized by an ultra-low elastic modulus with very high strength, superelastic nature—one digit higher in nonlinear elastic deformation (≈ 2.5%) compared to other metallic materials, super-plastic nature allowing cold plastic working without hardening up to (≈ 90%), very low linear coefficient of thermal expansion (similar to Invar) and a constant elastic modulus in the temperature range from –200°C up to +250°C; similarly to Elinvar (Saito et al. 2003, Kuramoto

Figure 1. Photograph of the experimental set-up.

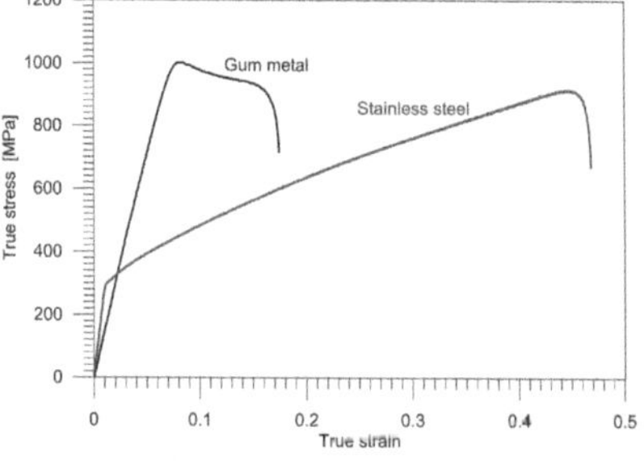

Figure 2. Comparison of stress vs. strain curves for Gum Metal and stainless steel until the specimen rupture.

et al. 2006). Such unique properties create huge potential for applications (Miyazaki. 2005). A comparison of the mechanical curves obtained for Gum Metal and Stainless steel during tension at a strain rate of 2×10^{-3} s^{-1} is presented in Figure 2.

2 INVESTIGATION OF GUM METAL YIELD LIMIT BASING ON THERMOMECHANICAL COUPLINGS

During the loading, the mechanical parameters and infrared radiation from the specimen surface have been simultaneously recorded. Stress-strain curves and the temperature changes of the specimen during the deformation process have been elaborated. The stress and the strain quantities are related to the current (instantaneous) values of the specimen cross-section values, obtaining in this way the so-called true stress (σ_{true}) and true strain (ε_{true}) values (Pieczyska et al. 2015), shown in the diagrams presented in Figures 2–4.

Moreover, a mean temperature of the specimen during the deformation process has been calculated, using the infrared measurement methodology, elaborated in IPPT PAN (Maj et al. 2012). The stress and average temperature changes vs. strain obtained for the Gum Metal loading until rupture for the strain rate of 10^{-3} s^{-1} is shown in Figure 3 and for the strain rate of 10^{-2} s^{-1} in Figure 4, respectively.

The specimen temperature referred to the deformation parameters allows indicating a limit of the alloy reversible deformation with high accuracy, according to the thermodynamic laws (Pieczyska. 1999). The temperature change of the specimen subjected to adiabatic uniaxial elastic deformation is called a thermoelastic effect and can be described by Lord Kelvin (Thomson W. 1853) theory (1):

$$\Delta T_{el} = -\frac{\alpha T \, \Delta\sigma_s}{c_p \, \rho} \quad (1)$$

where α denotes the coefficient of the material linear thermal expansion, T—the specimen absolute initial temperature, $\Delta\sigma_s$—the isentropic change of stress, c_p—the specific heat at constant pressure, ρ—the material density. Thus, the discrepancy between the linear dependence of the change in temperature vs. stress (≈ strain for small values) or the value of maximal drop in the specimen temperature (for the simplicity) can be used for evaluating a limit of the reversible material deformation with a high accuracy.

Comparison of the stress vs. strain and changes in the specimen average temperatures vs. strain curves obtained for Gum Metal subjected to tension at two strain rates of: 10^{-2} s^{-1} and 10^{-3} s^{-1} is presented in Figure 5a–b, respectively.

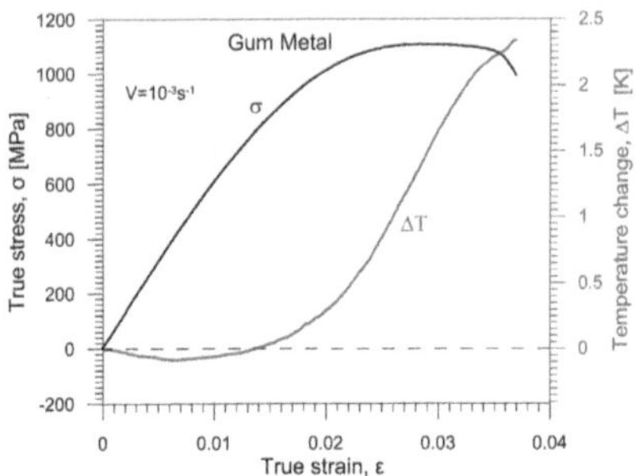

Figure 3. Stress and maximum temperature vs. strain of Gum Metal specimen subjected to tension at a strain rate of 10^{-3} s^{-1}.

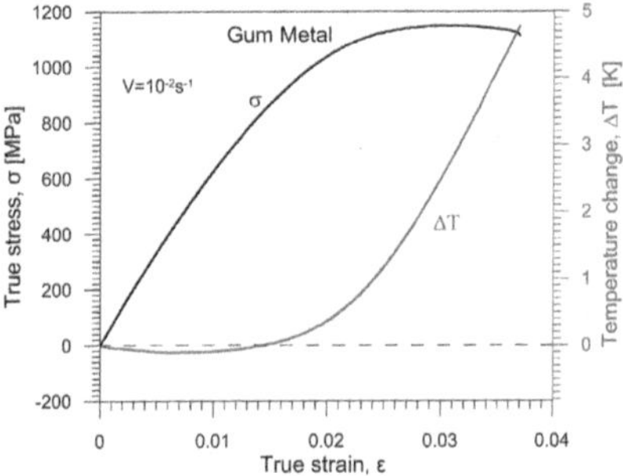

Figure 4. Stress and maximum temperature vs. strain of Gum Metal specimen subjected to tension at a strain rate of 10^{-2} s^{-1}.

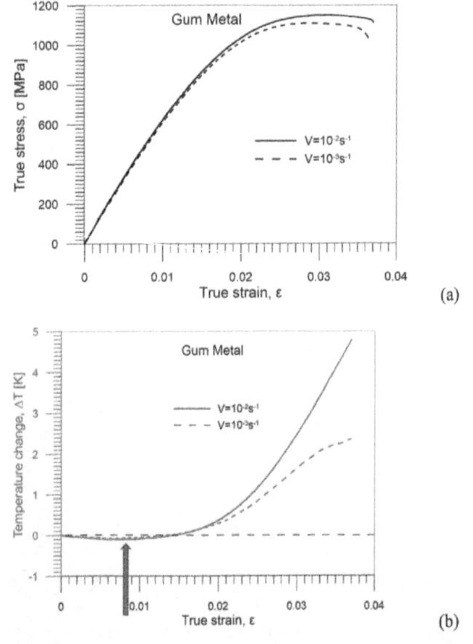

Figure 5. (a) Stress vs. strain and (b) the specimen average temperature vs. strain curves obtained for Gum Metal subjected to tension at two strain rates of: 10^{-2} s^{-1} and 10^{-3} s^{-1}.

Looking at the mechanical and thermal curves presented in Figure 5a–b one can notice that a maximal drop in the Gum Metal specimen temperature occurs at significantly lower strains than the limit of the reversible deformation, macroscopically estimated. It means that such a large limit of the alloy reversible elastic deformation (nonlinear) stressed as the Gum Metal "super" property (Saito et al. 2003, Furuta et al. 2012), originates from other deformation mechanisms and probably cannot be described by the Lord Kelvin formula. These results are quite different from those observed until now for other titanium alloys, stainless steel, alumina and polymers and will be a subject of our further research.

3 GUM METAL NECKING AND RUPTURE PHENOMENA ANALYSIS BY INFRARED CAMERA

Thermograms, showing the temperature distribution on the specimen surface, obtained by a fast and sensitive infrared camera, enabled to investigate the Gum Metal necking and rupture phenomena at higher strains. A photograph of a Gum Metal specimen in grips of the testing machine just after rupture is presented in Figure 6, whereas thermograms of a Gum Metal specimen in grips of testing machine just before and after rupture are shown in Figure 7 a and 7 b respectively. A specially designed holder was used to fit the Gum Metal specimen in grips of testing machine. Moreover, a thin layer of carbon powder black ink was put on the sample surface in order to obtain higher and well known emissivity, resulting in calculating the temperature changes with a higher accuracy.

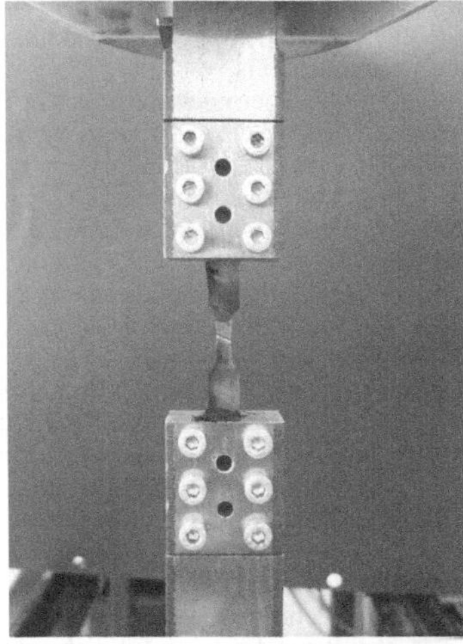

Figure 6. Photograph of Gum Metal specimen in grips of testing machine just after rupture.

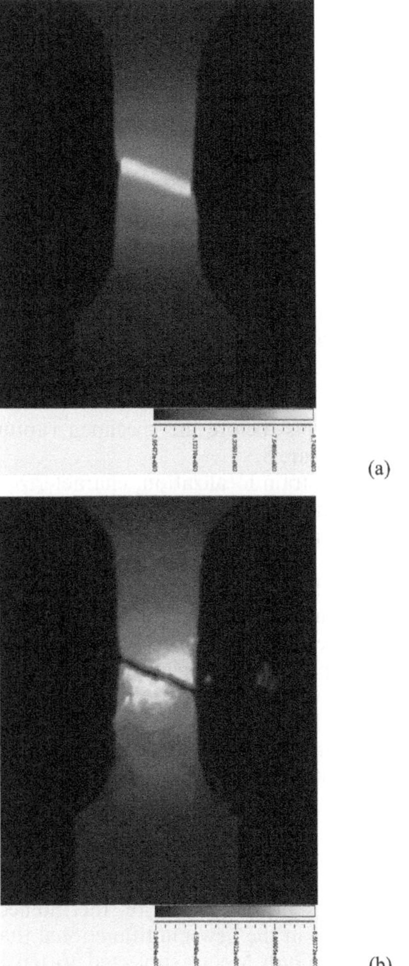

Figure 7. (a) Thermogram of Gum Metal specimen in grips of testing machine just before rupture; (b) thermogram of Gum Metal specimen in grips of testing machine just after rupture.

The thermogram clearly depicts that maximum temperature values are concentrated in the centre of the specimen where the rupture process occurs. The further one goes from the localization, the lower the temperature values get. Such a temperature gradient evolution can be monitored by the infrared camera and the recorded results will be simulated by mathematical model. The simulations might enable to predict the necking and rupture phenomena occurring in the alloys under tension, for potential applications of Gum Metal as a material used for mechanical components.

An example of the temperature distribution on the Gum Metal specimen and the temperature profile along the specimen center obtained for the

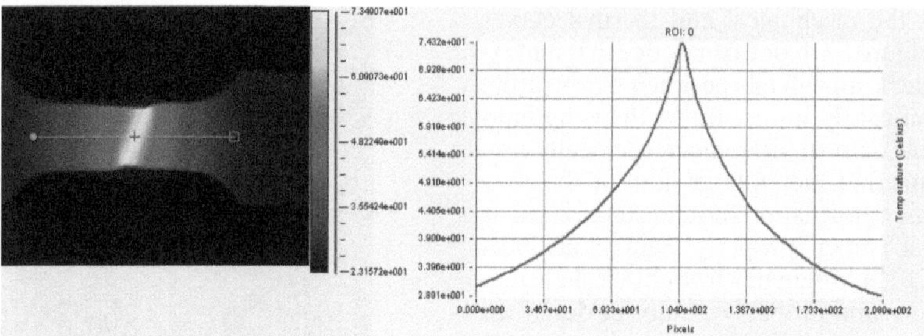

Figure 8. In left—temperature distribution on the Gum Metal sample during tension just before fracture; in right—the temperature change profile, indicated along the sample center denoted by a white line.

tension test just before the specimen rupture are shown in Figure 8.

A strong strain localization, characterized by a high temperature increase, was captured in the presented thermogram. Such thermograms and temperature profiles demonstrate a possibility to study the more advanced deformation process of Gum Metal, leading to necking and rupture. The maximum temperature, measured in the localization and the rupture area, depends on the strain rate applied, since at the higher strain rates the test conditions are closer to adiabatic. The calculated value of the maximum temperature change obtained for the strain rate of $2.4 \times 10^{-2}\ s^{-1}$ is equal to 96°C.

4 CONCLUDING REMARKS

Initial results of the effects of thermomechanical couplings in the new multifunctional titanium alloy, called Gum Metal, subjected to tension at various strain rates have been presented.

The obtained mechanical characteristics confirm an ultra-low elastic modulus and very high strength of the new titanium alloy.

The temperature distribution obtained with fast and sensitive infrared camera showed effects of strain localization, characterized by a significantly higher temperature, leading to the specimen necking and rupture.

The temperature changes accompanying the material initial deformation stage indicate thermoplastic effect, usually related to the material Yield point. However, it has been noticed that the maximal drop in the temperature obtained for Gum Metal occurs earlier than limit of its reversible deformation.

All of these aspects as well as a microstructure analysis, single crystal deformation tests and mathematical modeling will be subjects of our future research on Gum Metal.

ACKNOWLEDGMENTS

The research has been carried out with the support of the Polish National Centre of Science under Grant No. 2014/13/B/ST8/04280. Authors are also grateful to Leszek Urbański, Maria Staszczak and Karol Golasiński for obtaining mechanical data, elaborating diagrams and many useful experimental remarks.

REFERENCES

Furuta, T., Kuramoto S., Morris J.W., Nagasako N., Withey E., Chrzan D.C. 2013. The mechanism of strength and deformation in Gum Metal. *Scripta Materialia* 68: 767–772.

Kuramoto S., Furuta T., Hwang J., Nishino K., Saito T. 2006. Elastic properties of Gum Metal. *Materials Science and Engineering A* 442: 454–457.

Maj M. & Oliferuk W. 2012. Analysis of plastic strain localization on the basis of strain and temperature fields. *Archives of Metallurgy and Materials* 57: 1111–6.

Miyazaki S. 2005–2014. University of Tsukuba, Japan—Private Communications.

Pieczyska E.A. 1999. Thermoelastic effect in austenitic steel referred to its hardening, *Journal of Theoretical and Applied Mech*anics 2, 37 (Ph. D. thesis).

Pieczyska E.A., Maj M., Kowalczyk-Gajewska K., Staszczak M., Gradys A., Majewski M., Cristea M., Tobushi H., Hayashi S. 2015. Thermomechanical properties of polyurethane shape memory polymer—experiment and modelling, *Smart Materials and Structures* 24 045043-1-16.

Saito T., Furuta T., Hwang J.H., Kuramoto S., Nishino K., Suzuki N., Chen R., Yamada A., Ito K., Seno Y., Nonaka T., Ikehata H., Nagasako N., Iwamoto C., Ikuhara Y., Sakuma T. 2003. Multifunctional Alloys obtained via a dislocation-free plastic deformation mechanism. *Science* 300: 464–467.

Thomson W. (Lord Kelvin) 1853. On the thermoelastic and thermomagnetic properties of matter, *Transactions of the Royal Society of Edinburgh* 20(161): 57–77.

Advances in Mechanics: Theoretical, Computational and Interdisciplinary Issues – Kleiber et al. (Eds)
© 2016 Taylor & Francis Group, London, ISBN 978-1-138-02906-4

On the resultant six-field linear theory of elastic shells

W. Pietraszkiewicz
Institute of Fluid-Flow Machinery, Polish Academy of Science, Gdańsk, Poland

ABSTRACT: As compared with the classical linear shell models of Kirchhoff-Love and Timoshenko-Reissner type, the six-field linear shell model contains the drilling rotation as an independent kinematic variable as well as two surface drilling couples with two work-conjugate surface drilling bending measures are present in description of the shell stress-strain state. Within the six-field linear theory of elastic shells the following new results are presented: 1) the extended static-geometric analogy; 2) the complex BVP for complex independent variables; 3) description of deformation of the shell boundary element; 4) the Cesáro type formulas and expressions for the vectors of stress functions along the shell boundary contour; 5) estimates of surface gradients of 2D stress and/or strain measures to be of 2nd order small in the 2nd approximation to the complementary energy density of the elastic shell. These theoretical results should be of interest to specialists of the linear theory of shells and those developing computer FEM software for analyses of irregular multi-shell structures.

1 INTRODUCTION

The resultant non-linear theory of elastic shells was proposed by Reissner (1974), developed in a number of papers and summarised in several monographs, for example by Libai & Simmonds (1998) and Chróścielewski et al. (2004). In this formulation the 2D non-linear shell equilibrium equations are derived by the *exact* through-the-thickness integration of equilibrium conditions of non-linear elasticity. Then the 2D virtual work identity allows one to construct *uniquely* the 2D shell kinematics consisting of the translation vector $\boldsymbol{u}$ and rotation tensor $\boldsymbol{Q}$ (or an equivalent finite rotation vector $\boldsymbol{\psi}$) fields (six independent scalar variables) defined on the shell base surface. The 2D surface stretch and bending measures follow then again *uniquely* as direct consequence of the exact resultant equilibrium equations. When such a resultant shell model is linearized for infinitesimal translations, rotations, stretches and bending measures, the linearized drilling rotation remains as the independent kinematic variable, as well as two linearized drilling couples and two work-conjugate drilling bending measures remain in the description of 2D stress and strain state. The latter features contradict all classical shell formulations of the Kirchhoff-Love and Timoshenko-Reissner type following from linear elasticity by any 3D-to-2D reduction technique.

In this note we wish to investigate what the resultant six-field linear theory of shells brings to the classical linear shell models.

2 EXTENDED STATIC-GEOMATRIC ANALOGY

For the general system of notation let me refer to the recent paper by Pietraszkiewicz & Konopińska (2014), where the exact resultant 2D relations of the six-field non-linear theory of shells are briefly recalled. When translations and rotations of the shell base surface M are assumed to be small, the corresponding vectors $\boldsymbol{u}$, $\boldsymbol{\psi}$ on M are given by:

$$\boldsymbol{u} = u_\alpha \boldsymbol{a}^\alpha + w\boldsymbol{n}, \quad \boldsymbol{\psi} = \boldsymbol{n} \times (\psi_\alpha \boldsymbol{a}^\alpha) + \psi \boldsymbol{n} \tag{1}$$

where $\boldsymbol{a}^\alpha$, $\boldsymbol{n}$ are the contravariant base vectors of M. The above relations contain the drilling rotation $\boldsymbol{\psi}$ which is not present in the classical shell models.

The corresponding vectorial surface strain measures are:

$$\begin{aligned} \boldsymbol{\varepsilon}_\alpha &= \boldsymbol{u},_\alpha - \boldsymbol{\psi} \times \boldsymbol{a}_\alpha = E_{\alpha\beta}\boldsymbol{a}^\beta + E_\alpha \boldsymbol{n} \\ \boldsymbol{\kappa}_\alpha &= \boldsymbol{\psi},_\alpha = \varepsilon_{\lambda\beta} K_\alpha^\lambda \boldsymbol{a}^\beta + K_\alpha \boldsymbol{n} \end{aligned} \tag{2}$$

The resultant vectorial surface stress measures are:

$$\boldsymbol{n}^\alpha = N^{\alpha\beta}\boldsymbol{a}_\beta + Q^\alpha \boldsymbol{n}, \quad \boldsymbol{m}^\alpha = \varepsilon_{\lambda\beta} M^{\alpha\lambda} \boldsymbol{a}^\beta + M^\alpha \boldsymbol{n} \tag{3}$$

Linearization of the component form of exact equilibrium equations gives:

$$N^{\alpha\beta}\Big|_{\alpha} - b^{\beta}_{\alpha}Q^{\alpha} + f^{\beta} = 0, \quad Q^{\alpha}\Big|_{\alpha} + b_{\alpha\beta}N^{\alpha\beta} + f = 0$$
$$M^{\alpha\beta}\Big|_{\alpha} - Q^{\beta} + \varepsilon^{\lambda\beta}b_{\alpha\lambda}M^{\alpha} + m^{\beta} = 0$$
$$M^{\alpha}\Big|_{\alpha} + \varepsilon_{\alpha\beta}(N^{\alpha\beta} - b^{\alpha}_{\lambda}M^{\lambda\beta}) + m = 0 \tag{4}$$

These twelve PDEs involve the drilling stress couples M^{α} as well, which are not present in analogous PDEs of the classical linear shell models of the K-L and T-R type. The corresponding exact 2D compatibility conditions of the six-field non-linear theory of shells follow from integrability conditions $\varepsilon^{\alpha\beta}\boldsymbol{u}_{|\alpha\beta} = \mathbf{0}$ for the surface translation vector and $\varepsilon^{\alpha\beta}\boldsymbol{Q}_{|\alpha\beta} = \mathbf{0}$ for the surface rotation tensor fields. For small stretch and bending surface measures the component form of non-linear compatibility conditions can be linearized:

$$\varepsilon^{\alpha\beta}(E_{\alpha\lambda|\beta} - E_{\alpha}b_{\beta\lambda} + \varepsilon_{\alpha\lambda}K_{\beta}) = 0$$
$$\varepsilon^{\alpha\beta}(E_{\alpha|\beta} + E_{\alpha\lambda}b^{\lambda}_{\beta} + K_{\alpha\beta}) = 0$$
$$\varepsilon^{\alpha\beta}(\varepsilon^{\rho\lambda}K_{\alpha\rho|\beta} + b^{\lambda}_{\alpha}K_{\beta})$$
$$= 0, \varepsilon^{\alpha\beta}(K_{\alpha|\beta} + \varepsilon^{\lambda\rho}K_{\alpha\lambda}b_{\beta\rho}) = 0 \tag{5}$$

These twelve PDEs involve the 2D surface drilling bending measures K_{α}. Again, the fields K_{α} are not present in compatibility conditions of any classical linear shell models. Between the homogeneous equilibrium equations (4) and the compatibility conditions (5) there exists the following correspondence:

$$N^{\alpha\beta} \Leftrightarrow \varepsilon^{\alpha\rho}\varepsilon^{\beta\lambda}K_{\rho\lambda}, \qquad Q^{\alpha} \Leftrightarrow -\varepsilon^{\alpha\rho}K_{\rho}$$
$$M^{\alpha\beta} \Leftrightarrow -\varepsilon^{\alpha\rho}\varepsilon^{\beta\lambda}E_{\rho\lambda}, \quad M^{\alpha} \Leftrightarrow -\varepsilon^{\alpha\rho}E_{\rho} \tag{6}$$

When the resultant stress measures in (4) are replaced by the 2D strain measures according to (6), the homogeneous equilibrium equations (4) are converted exactly to the compatibility conditions (5). The correspondence (6) can be called *the extended static-geometric analogy* in the resultant six-field linear theory of shells.The property (6) allows one to introduce the 2D complex stress and strain measures in analogy to those used by Novozhilov (1964) in the classical linear shell theory of K-L type:

$$\tilde{N}^{\alpha\beta} = N^{\alpha\beta}_{*} - i\,Ehc\,\varepsilon^{\alpha\rho}\varepsilon^{\beta\lambda}\tilde{K}_{\rho\lambda}$$
$$\tilde{Q}^{\alpha} = Q^{\alpha}_{*} + i\,Ehc\,\varepsilon^{\alpha\rho}\tilde{K}_{\rho}$$
$$\tilde{M}^{\alpha\beta} = M^{\alpha\beta}_{*} + i\,Ehc\,\varepsilon^{\alpha\rho}\varepsilon^{\beta\lambda}\tilde{E}_{\rho\lambda}$$
$$\tilde{M}^{\alpha} = M^{\alpha}_{*} + i\,Ehc\,\varepsilon^{\alpha\rho}\tilde{E}_{\rho} \tag{7}$$

where $i = \sqrt{-1}$, $c = h/\sqrt{12(1-\nu\cdot\nu)}$, the star means a particular solution of inhomogeneous PDEs (4), and $\tilde{K}_{\alpha\lambda}$, $\tilde{K}_{\alpha}$, $\tilde{E}_{\alpha\lambda}$, $\tilde{E}_{\alpha}$ are the 2D strain measures but constructed from the complex translations and rotations:

$$\tilde{u}_{\alpha} = u_{\alpha} + i\bar{u}_{\alpha}, \ \tilde{w} = w + i\bar{w}$$
$$\tilde{\psi}_{\alpha} = \psi_{\alpha} + i\bar{\psi}_{\alpha}, \ \tilde{\psi} = \psi + i\bar{\psi} \tag{8}$$

where by overbar we denote the stress functions. This leads to the following complex linear equilibrium equations:

na nowo wpisać

$$\tilde{N}^{\alpha\beta}|_{\alpha} - b^{\beta}_{\alpha}\tilde{Q}^{\alpha} + f^{\beta} = 0, \ \tilde{Q}^{\alpha}|_{\alpha} + b_{\alpha\beta}\tilde{N}^{\alpha\beta} + f = 0$$
$$\tilde{M}^{\alpha\beta}|_{\alpha} - \tilde{Q}^{\beta} + \varepsilon^{\lambda\beta}b_{\alpha\lambda}\tilde{M}^{\alpha} + m^{\beta} = 0$$
$$\tilde{M}^{\alpha}|_{\alpha} + \varepsilon_{\alpha\beta}(\tilde{N}^{\alpha\beta} - b^{\alpha}_{\lambda}\tilde{M}^{\lambda\beta}) + m = 0 \tag{9}$$

The system (9) of PDEs is of 6th order in the complex domain as compared with PDEs (4) which are of 12th order in the real domain. One can apply the complex shell equations (9) with hope to obtain more accurate analytical results than those based on K-L shell model presented in Novozhilov (1964) and Novozhilov et al. (1991) and those based on T-R type shell model presented in Pelekh (1978).

3 RELATION ON THE SHELL BOUNDARY

In the resultant shell theory, the undeformed rectilinear shell lateral boundary surface ∂B^{*} deforms into the deformed lateral boundary surface $\chi(\partial B^{*})$, which is not the rectilinear one anymore. To describe the linearized deformation of ∂B^{*} into $\chi(\partial B^{*})$, let the boundary contour ∂M of the shell base surface be parametrized by the length coordinate s. Along ∂M we can introduce the triad of orthonormal vectors: the tangent $\boldsymbol{\tau}$, the $\boldsymbol{n}$ normal to M, and the exterior normal $\boldsymbol{\nu} = \boldsymbol{\tau}\times\boldsymbol{n}$. In the base $\boldsymbol{\nu}$, $\boldsymbol{\tau}$, $\boldsymbol{n}$ the translation $\boldsymbol{u}$ and rotation $\boldsymbol{\psi}$ vectors are represented by:

$$\boldsymbol{u} = u_{\nu}\boldsymbol{\nu} + u_{\tau}\boldsymbol{\tau} + w\boldsymbol{n}, \quad \boldsymbol{\psi} = -\psi_{\tau}\boldsymbol{\nu} + \psi_{\nu}\boldsymbol{\tau} + \psi\boldsymbol{n} \tag{10}$$

According to the modified polar decomposition theorem for shells, the material fiber $\boldsymbol{\tau}ds$ tangent to ∂M should be rotated by a total rotation vector $\boldsymbol{\psi}_{\tau}$ into the fiber $\hat{\tau}ds$ tangent to $\chi(\partial M)$, where $\hat{\tau}$ denotes the unit vector tangent to the deformed boundary contour. This can be achieved by two rotations: the global rotation vector $\boldsymbol{\psi}$ of the neighborhood of $x \in \partial M$ followed by an additional rotation $\boldsymbol{\psi}_{\alpha}$ associated with the stretch in the direction $\boldsymbol{\tau}$. After transformations given in detail by Pietraszkiewicz (2015) the linearized total rotation vector is given by:

$$\psi_\tau = \psi + \psi_a = (E_\tau - \psi_\tau)\boldsymbol{\nu} + \psi_\nu \tau - (E_{\tau\nu} - \psi)\boldsymbol{n} \quad (11)$$

Differentiation of the translation and rotation fields along ∂M gives:

$$\frac{d\boldsymbol{u}}{ds} = E_{\tau\tau}\boldsymbol{\tau} + \boldsymbol{\psi}_\tau \times \boldsymbol{\tau}, \boldsymbol{k}_\tau = \frac{d\boldsymbol{\psi}_\tau}{ds} = -k_{\tau\tau}\boldsymbol{\nu} + k_{\tau\nu}\boldsymbol{\tau} + k_\tau \boldsymbol{n} \quad (12)$$

where the components $k_{\tau\tau}$, $k_{\tau\nu}$, k_τ = are expressed entirely through the surface strain measures and geometry of ∂M. These relations can be integrated along ∂M to obtain:

$$\boldsymbol{u} = \boldsymbol{u}^0 + \int_{s_0}^{s} (E_{\tau\tau}\boldsymbol{\tau}' + \boldsymbol{\psi}_\tau \times \boldsymbol{\tau}')ds', \ \boldsymbol{\psi}_\tau = \boldsymbol{\psi}_\tau^0 + \int_{s_0}^{s} \boldsymbol{k}_\tau ds' \quad (13)$$

Using some transformations the relation $(13)_1$ can also be given in the form:

$$\boldsymbol{u} = \boldsymbol{u}^0 + \psi_\tau^0 \times (\boldsymbol{x} - \boldsymbol{x}_0) + \int_{s_0}^{s} \left[\left(\int_{s_0}^{s} \boldsymbol{k}_\tau ds'' \right) \times \tau' + E_{\tau\tau}\tau' \right] ds' \quad (14)$$

The relation above and $(13)_2$, derived here along the boundary contour ∂M, are also valid for any regular curve C on the shell base surface. Since for the simply connected base surface M the resulting translations and rotations do not depend on the type of surface curve connecting the initial and current points on M, these relations are just the Cesáro type formulas in the resultant linear six-field theory of shells. When solving shell problems in terms of stress functions, it is of importance to know how the stress function vectors $\bar{u}$ and $\bar{\psi}$ are related to the resulting force and moment vectors of loads acting along the boundary contour ∂M_f.

The boundary loads $\boldsymbol{n}_\nu = \boldsymbol{n}^\alpha \boldsymbol{\nu}_\alpha$ and $\boldsymbol{m}_\nu = \boldsymbol{m}^\alpha \boldsymbol{\nu}_\alpha$ per unit length of ∂M_f are:

$$\boldsymbol{n}_\nu = \boldsymbol{n}_\nu^* - Ehc\bar{\mathbf{k}}_\tau, \ \boldsymbol{m}_\nu = \boldsymbol{m}_\nu^* - Ehc\left(\bar{E}_{\tau\tau}\boldsymbol{\tau} + \bar{\psi}_a \times \boldsymbol{\tau}\right) \quad (15)$$

where the first terms $\boldsymbol{n}_\nu^*$, $\boldsymbol{m}_\nu^*$ correspond to the chosen particular solution of the inhomogeneous equilibrium equations (4).

The resulting force vector $\mathbf{F}$ of the loads $\boldsymbol{n}_\nu$ acting along the part $s_0 s$ of ∂M_f is:

$$\mathbf{F} = \int_{s_0}^{s} \boldsymbol{n}_\nu ds' = \int_{s_0}^{s} \left(\boldsymbol{n}_\nu^* - Ehc\bar{\mathbf{k}}_\tau \right) ds' \quad (16)$$

The resulting moment vector $\mathbf{B}$ of the loads $\boldsymbol{n}_\nu$ and $\boldsymbol{m}_\nu$ acting along the part $s_0 s$ of ∂M_f, taken with regard to the current point $x \in \partial M_f$ with the coordinate s, is (Pietraszkiewicz 2015):

$$\mathbf{B} = \mathbf{B}^* - Ehc\left[\bar{\boldsymbol{u}} - \bar{\boldsymbol{u}}^0 + (\boldsymbol{x} - \boldsymbol{x}_0) \times \bar{\boldsymbol{\psi}}^0 \right] \quad (17)$$

Solving the two relations given above, for the stress function vectors $\bar{\boldsymbol{u}}$ and $\bar{\boldsymbol{\psi}}$ we obtain:

$$\bar{\boldsymbol{\psi}} = \bar{\boldsymbol{\psi}}^0 - \frac{\mathbf{F} - \mathbf{F}^*}{Ehc}, \bar{\boldsymbol{u}} = \bar{\boldsymbol{u}}^0 + \bar{\boldsymbol{\psi}}^0 \times (\boldsymbol{x} - \boldsymbol{x}_0) - \frac{\mathbf{B} - \mathbf{B}^*}{Ehc} \quad (18)$$

The relations (18) along ∂M_f are also valid for any regular curve C on the shell base surface M. In the case of simply connected M and the closed curve C we can assume $\bar{\boldsymbol{u}}^0 = \bar{\boldsymbol{\psi}}^0 = 0$. This means the change of stress function vectors by the term $\bar{\boldsymbol{u}}^0 + \bar{\boldsymbol{\psi}}^0 \times (\boldsymbol{x} - \boldsymbol{x}_0)$ of "rigid-body motion" type which does not influence the stress-deformation state of the shell. In the case of shell problems with non-vanishing differences $\mathbf{F} - \mathbf{F}^*$ and $\mathbf{B} - \mathbf{B}^*$ as well as with a multi-connected M the relations (18) should allow to construct multivalued parts of the stress functions in analogy to procedures developed by Chernykh (1968) and Pietraszkiewicz (1968) for the linear K-L type elastic shells.

4 GRADIENTS OF 2D SHELL MEASURES IN THE RESULTANT STRESS WORKING

From the resultant thermomechanics of shells worked out by Pietraszkiewicz (2011) it follows that even if thermal fields are neglected and kinematic fields are linearized, the constitutive equations of elastic shells are still allowed to depend upon surface gradients of 2D strain measures. We are not aware of any discussion of this problem in the literature.

To have some insight into the problem, let us remind that distribution of translations through the shell thickness is non-linear, in general. For an arbitrary deformation of the shell space, we have introduced in Chróścielewski et al. (2004) the intrinsic deviation vector $\mathbf{e}(\theta^\alpha, \xi)$ defined by:

$$\mathbf{e} = Q^T\boldsymbol{\zeta} - \xi\boldsymbol{n} = e^\rho(\xi)\mathbf{g}_\rho(\xi) + e(\xi)\boldsymbol{n} \quad (19)$$

where $Q\mathbf{e}$ is a measure of deviation of the deformed curved material fiber from its approximately linear rotated shape $\xi Q\boldsymbol{n}$, see Figure 4 in Pietraszkiewicz (2015).

The 3D stress power density is defined by $\Sigma = (\mathbf{FS}) : \dot{\mathbf{F}}$, where $\mathbf{F}$ is the 3D deformation gradient, the overdot means the material time derivative, $\mathbf{S} = \mathbf{s}^i \otimes \mathbf{g}_i$ is the 2nd Piola-Kirchhoff stress tensor, and: denotes the double-dot (scalar) product in the tensor space. Using the modified polar decomposition $\mathbf{F} = \boldsymbol{Q}\boldsymbol{\Lambda}$, where $\boldsymbol{\Lambda} \neq \boldsymbol{\Lambda}^T$ is the modified stretch tensor, we can calculate exactly the resultant 2D stress power density in the form:

$$\mathcal{\Sigma} = \int_{-}^{+} \Sigma \mu d\xi = N^{\alpha\beta}\dot{E}_{\alpha\beta} + Q^{\alpha}\dot{E}_{\alpha} + M^{\alpha\beta}\dot{K}_{\alpha\beta} + M^{\alpha}\dot{K}_{\alpha} + \int_{-}^{+}(\dot{\mathbf{e}} \times \boldsymbol{\Lambda}\mathbf{s}^{\alpha})\mu d\xi \cdot \boldsymbol{\kappa}_{\alpha} + \int_{-}^{+}(\boldsymbol{\Lambda}\mathbf{S}) : \nabla\dot{\mathbf{e}}\mu d\xi \tag{20}$$

where ∇ means the 3D gradient performed in the undeformed shell space. The last two integrals in (20) represent that part of $\mathcal{\Sigma}$ which is not expressible through the 2D shell stress and strain measures alone. In case of linear equilibrium problems, material time derivatives in (20) become linear increments of the fields from the undeformed state and the 2D stress power (20) reduces to the resultant 2D stress working, i.e. to the relation (20) without overdots. We have thoroughly analyzed the principal terms of the last two integrals in (20) within the accuracy of equilibrium problems of the linear theory of isotropic elastic shells. In particular, we have used some results obtained Rychter (1988) for kinematically admissible components of 3D displacement field of the Reissner type linear shell theory and postulate approximate through-the-thickness distributions of the intrinsic deviation vector:

$$e_{\rho} \simeq k(\xi)q_{\rho} + g(\xi)c_{\rho}, \quad e \simeq r(\xi)q + s(\xi)d$$

$$q = -\frac{1}{40}h^2 D^{\alpha\beta}K_{(\alpha\beta)}, \; d = -\frac{h^3}{48}D^{\lambda\mu}D^{\alpha\beta}E_{(\alpha\beta)|\lambda\mu}$$

$$r(\xi) = \frac{20\xi^2}{h^2} - 1, \quad s(\xi) = \frac{2}{h}\left(\frac{4\xi^3}{h^2} - \xi\right) \tag{21}$$

while $k(\xi)$, $g(\xi)$, q_0, c_0 and $D^{\alpha\beta}$ are defined in Pietraszkiewicz & Konopińska (2014). Then for the principal terms of $(20)_2$ after appropriate transformations we are able to obtain the following consistently estimated relations:

$$\int_{-}^{+}(\mathbf{e} \times \boldsymbol{\Lambda}\mathbf{s}^{\alpha})\mu d\xi \cdot \boldsymbol{\kappa}_{\alpha} \simeq \frac{2}{3}qN^{\alpha\beta}K_{(\alpha\beta)} \sim Eh\eta^2 \cdot \left(\frac{1}{60}\nu\eta\right)$$

$$\int_{-}^{+}(\boldsymbol{\Lambda}\mathbf{s}^{\alpha}) \cdot \mathbf{e}_{,\alpha}\mu d\xi \simeq \frac{3}{4}Q^{\alpha}q_{,\alpha} \sim Eh\eta^2 \cdot \left(\frac{3}{160}\nu\theta\frac{h}{\lambda}\right)$$

$$\int_{-}^{+}(\boldsymbol{\Lambda}\mathbf{s}^{3}) \cdot \mathbf{e}_{,3}\mu d\xi \simeq -9M^{\alpha\beta}{}_{|\beta\alpha}q - \frac{25}{3}b_{\alpha\beta}N^{\alpha\beta}q \sim Eh\eta^2 \cdot \left(\nu\theta^2, \nu\frac{h^2}{\lambda^2}\right) \tag{22}$$

where θ = is a small parameter defined in Pietraszkiewicz & Konopińska (2014); the large parameter $\lambda = \min_{x \in M}(L, b, \sqrt{hR}, h/\sqrt{\eta})$, $(\cdot)_{,\alpha} \sim (\cdot)/\lambda$ and $h/\lambda \sim \theta$.

Within the first-approximation linear theory of elastic shells, all terms in (22) can be neglected as compared with the main terms $\sim Eh\eta^2$ in (20). However, in the Timoshenko-Reissner type and the resultant six-field linear shell models the terms (22) are of the same order (aside from small numerical factors) as those following from Q^{α}. Their role should still be discussed if these linear models are to be regarded as energetically consistent ones.

5 CONCLUSIONS

Within the resultant six-field linear theory of elastic shells we have formulated several results which are not available elsewhere. Among the new results let us point out the following:

- Formulation of the extended static-geometric analogy and derivation of complex shell relations for the complex displacements (section 4).
- Description of infinitesimal deformation of the shell boundary element (section 5) and derivation of corresponding Cesáro type formulas.
- Expressions (18) for the vectors of stress functions in terms of the resulting force and moment vectors acting along the boundary contour and along an arbitrary curve on the shell base surface.
- Estimation that surface gradients of 2D stress and/or strain measures may be of 2nd order small in the 2nd approximation to the complementary energy density.

These theoretical results should be of interest to specialists of the linear theory of shells and to those developing computer FEM software for analyses of irregular multi-shell structures.

ACKNOWLEDGEMENTS

The research reported in this note was supported by the National Centre of Science of Poland with the grant DEC—2012/05/D/ST8/02298.

REFERENCES

Chernykh, K.F. 1968. *Linear Theory of Shells.* NASA-TT-F-II 562.

Chróścielewski, J., Makowski, J. & Pietraszkiewicz, W. 2004. *Statics and Dynamics of Multifold Shells: Nonlinear Theory and Finite Element Method.* Warsaw: Institute of Fundamental Technological Research Press (in Polish).

Libai, A. & Simmonds, J.G. 1998. *The Nonlinear Theory of Elastic Shells*, second ed. Cambridge, UK: Cambridge Univ. Press.

Novozhilov, V.V. 1964. *Thin Shell Theory.* Groningen: P. Noordhoff.

Novozhilov, V.V., Chernykh, K.F. & Mikhailovskii, E.I. 1991. *The Linear Theory of Thin Shells.* Leningrad: Politekhnika (in Russian).

Pelekh, B.L. 1978. *Generalized Theory of Shells.* L'vov: Vishcha Shkola (in Russian).

Pietraszkiewicz, W. 1968. Multivalued stress functions in the linear theory of shells. *Arch. Mech. Stos.* 20(1): 37–45.

Pietraszkiewicz, W. 2011. Refined resultant thermomechanics of shells. *Inter. Jour. of Eng. Science* 49: 1112–1124.

Pietraszkiewicz, W. 2015. The resultant linear six-field theory of elastic shells: What it brings to the classical linear shell models? *Journal of Applied Mathematics and Mechanics—ZAMM* (submitted).

Pietraszkiewicz, W. & Konopińska, V. 2014. Drilling couples and refined constitutive equations in the resultant geometrically non-linear theory of elastic shells. *International Journal of Solids and Structures* 51: 2133–2143.

Reissner, E. 1974. Linear and nonlinear theory of shells. In Y.C. Fung & E.E. Sechler (eds), *Thin Shell Structures*: 29–44. Englewood Cliffs, NJ: Prentice Hall.

Rychter, Z. 1988. Global error estimates in Reissner theory of thin elastic shells. *International Journal of Engineering Science* 26(8): 787–795.

Advances in Mechanics: Theoretical, Computational and Interdisciplinary Issues – Kleiber et al. (Eds)
© 2016 Taylor & Francis Group, London, ISBN 978-1-138-02906-4

Explanation of the mechanism of destruction of the cylindrical sample in the Brazilian test

J. Podgórski & J. Gontarz
Lublin University of Technology, Lublin, Poland

ABSTRACT: The paper presents the analysis of the so-called Brazilian compression tests of the cylinder loaded by the two linearly distributed balanced forces, in terms of possibility of determining the proper tensile strength of the material. These analyses contain the precisely determined stress field, without the singularity at the point of force application, the determination of the critical stress from the point of view of the classical and contemporary failure criteria for the brittle materials, and the position of the point at which the destructive crack may start to destroy the sample. The classical mechanism of the destruction relies on the analysis of the plane stress state which requires revision and its replacement with a 3D model. These analyses are supported by experimental data from the authors' laboratory tests.

1 INTRODUCTION

The paper presents the study of the indirect method of determining the tensile strength of brittle materials such as concrete and rock. Most often this type of testing is done using the "Brazilian test" by compressing the cylindrical sample by the two linear, balanced loads. The simplicity of this test, and the convenience of using drilling cores as laboratory samples made the "Brazilian test" the dominant method for determining the tensile strength of natural rock and concrete. The tensile strength (f_t) of the sample material is usually determined by taking the maximum tensile stress (σ_t), reached at the moment of destruction. The stress value is usually determined on the basis of the consideration of 2D elasticity problem for the circular shield compressed by two balancing forces acting along a diameter, which gives:

$$f_t = \frac{2P_{\max}}{\pi dh}, \qquad (1)$$

where $P_{\max}$ = is a destructive force of sample, d = diameter and h = height of the tested cylinder.

Tensile strength, determined in this way, has values lower than that determined on a direct tensile test. The reason for this are excessive simplifications of determining the stress field and neglecting the impact of compressive stress, which have a significant impact on the effort of material.

The problem of determining the material strength and modulus of elasticity basing on the results of the measurements made during the Brazilian test is still an interesting topic of research, which can be observed in many studies appearing in scientific journals dedicated to the problems of rock mechanics and concrete (Jianhong et al. 2009, Aréoglu et al. 2006, Chen & Hsu 2001).

The authors of the study have attempted to accurately determine the tensile strength obtained in Brazilian test by analyzing the material effort basing on modern and conventional failure conditions applicable for concrete and natural rocks. The conditions of Lame-Rankine, Mohr-Coulomb, Drucker-Prager, Ottosen-Podgorski were analyzed.

2 STRESS FIELD

Determination of stress values in compression along the circular disc diameter (Fig. 1a) is a classic issue, solved at the end of XIX century by Flamant and Hertz (Timoshenko & Goodier 1962). The equations of the stress tensor components in the Cartesian coordinate system can be expressed as follows (Timoshenko & Goodier 1962, Mushelishvili 1949):

$$\sigma_x = \frac{-2P}{\pi dh}\left[\frac{(1-\zeta)\xi^2}{((1-\zeta)^2+\xi^2)^2} + \frac{(1+\zeta)\xi^2}{((1+\zeta)^2+\xi^2)^2} - \frac{1}{2}\right]$$
$$\sigma_y = \frac{-2P}{\pi dh}\left[\frac{(1-\zeta)\xi^3}{((1-\zeta)^2+\xi^2)^2} + \frac{(1+\zeta)\xi^3}{((1+\zeta)^2+\xi^2)^2} - \frac{1}{2}\right]$$
$$\tau_{xy} = \frac{-2P}{\pi dh}\left[\frac{(1-\zeta)\xi^2}{((1-\zeta)^2+\xi^2)^2} + \frac{(1+\zeta)\xi^2}{((1+\zeta)^2+\xi^2)^2}\right] \qquad (2)$$

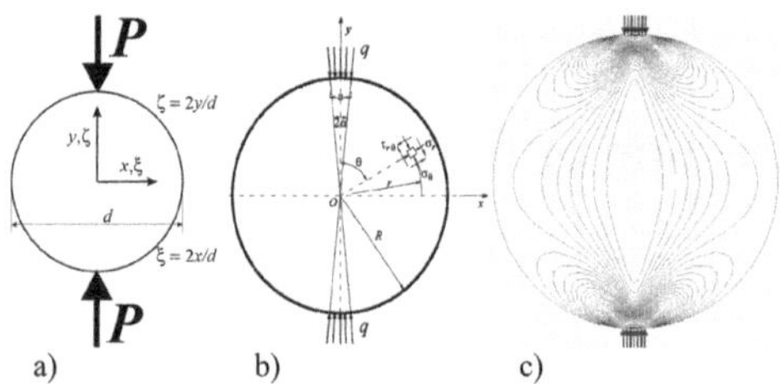

Figure 1. The issue of circular disc compression—Brazilian test, a) concentrated force, b) distributed load by. Hondros, c) distributed load considered in FEM analysis.

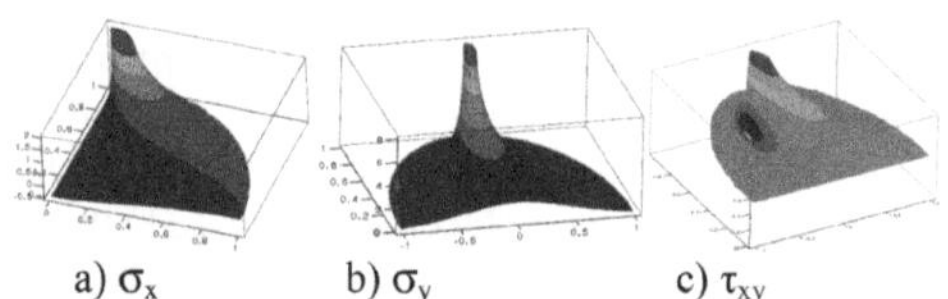

Figure 2. Stress distribution in the circular disc compressed with concentrated forces.

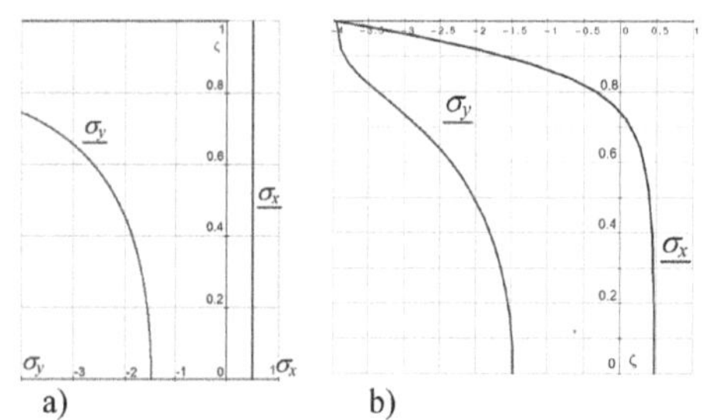

Figure 3. Charts of stresses σ_x and σ_y in section $x = 0$, a) for the concentrated load, b) distributed load.

Figure 2 shows a map of stress fields in areas corresponding to quarter of the compressed cylinder in the case of the graph σ_x, and half of the cylinder in the case of compressive stress σ_y and tangential τ_{xy}.

As it can be seen, in the middle section ($\xi = 0$, $\zeta = 0$) compressive stresses σ_y are in absolute value three times greater than the tensile stresses σ_x ($\sigma_y/\sigma_x = -3$). In applying point of concentrated force, there are singularities of both the stress fields because $\sigma_y \to -\infty$, $\sigma_x \to \infty$. Graphs of the stresses in section where $x = 0$ are shown in Figure 3a.

In the real task for a cylinder, the stress field singularities do not exist, because the force always have to be applied as a pressure on a small area of the lateral surface (Fig. 3b). Field of the stresses in this case has been designated by Hondros (Jianhong et al. 2009, Hondros 1959).

Figure 3b is a graph of stresses σ_x and σ_y in section $x = 0$ of circular disc loaded by pressure p, similar to the diagram of stresses induced by the concentrated force shown in Figure 3a. A lack of singularity at the point of load application is visible here. In the case of a small width "a" of load pressure strip, the stress distribution in the center region of the disc is not significantly different from distribution σ_x and σ_y shown in Figure 3a.

3 CRITERIA FOR MATERIAL DAMAGE

3.1 *Lame-Rankine criterion*

Lame and Rankine assumed that the excision of the biggest strength of the main stress determines the failure of the material:

$$f_c \le \sigma_1 \le f_t, \quad f_c \le \sigma_2 \le f_t \tag{3}$$

The border envelope of this condition in the area of "compression-stretching" is shown in Figure 4.

3.2 *Coulomb-Mohr criterion*

According to this hypothesis the maximum shear stress that exceeds the cohesive value enlarged by the friction force determines the failure of the material:

$$\tau_{\max} \le c - \sigma_n \text{tg}\phi, \tag{4}$$

In this equation σ_n = is the stress normal to the plane $\tau_{\max}$; c = is cohesive and ϕ = is the angle of internal friction. The values of parameters c and $\tan\phi$ = can be determined from known values of compressive strength—f_c and tensile strength—f_t.

$$c = \frac{f_c}{2(\eta+1)}. \tag{5}$$

The boundary envelope of Coulomb-Mohr condition in the area $\sigma_1 \le 0$, $\sigma_2 \ge 0$ is shown in Figure 4.

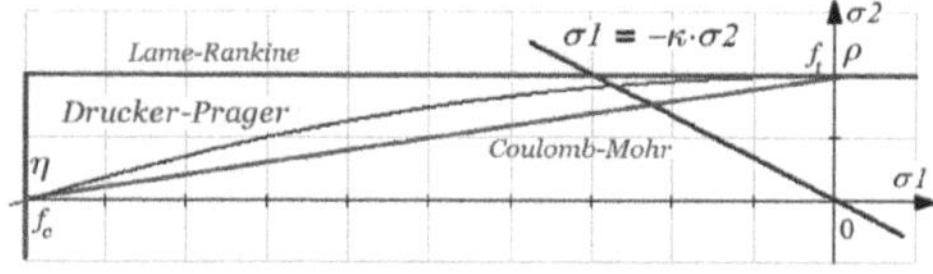

Figure 4. The envelope of Lame-Rankine, Coulomb-Mohr and Drucker-Prager conditions.

3.3 *Drucker-Prager criterion*

Drucker-Prager condition (Podgórski 1984) is based on similar to Coulomb-Mohr dependence expressed in the form of invariants in an octahedral plane:

$$\tau_0 \le c - b\sigma_0,$$
$$\sigma_0 = (\sigma_1 + \sigma_2 + \sigma_3)/3, \qquad (6)$$
$$\tau_0 = \sqrt{\frac{2J_2}{3}}$$

In this equation σ_0 = is the average normal stress; τ_0 = is the tangential octahedral stress; J_2 = is the second invariant of stress deviator; c and b = are constants that can be determined from known values of compressive strength—f_c and tensile strength—f_t.

$$c = f_c \frac{2\sqrt{2}}{3(\eta+1)}, \qquad (7)$$

The envelope boundary of Drucker-Prager condition in the area $\sigma_1 \le 0$, $\sigma_2 \ge 0$ is shown in Figure 4.

3.4 *Ottosen-Podgórski (JP) criterion*

Ottosen-Podgórski condition (Podgórski 1984a, 1985) has been proposed in the form which expresses the relationship of three alternative stress tensor invariants:

$$\sigma_0 - C_0 + C_1 P(J)\tau_0 + C_2\tau_0^2 = 0, \quad \sigma_0 = \frac{1}{3}I_1,$$
$$\tau_0 = \sqrt{\frac{2}{3}J_2}, \qquad (8)$$

where $P(J)$ = is a function describing the cross-section of limit state surface by the deviatoric plane, proposed by Podgórski in the form:

$$P(J) = \cos\left(\frac{1}{3}\arccos(\alpha J) - \beta\right) \qquad (9)$$

where α, β, C_0, C_1, C_2 = are constants dependent on the material.

3.5 *Determination of tensile strength*

Comparing the ordinate of the point of the load path intersection in the Brazilian test ($\sigma_1 = \kappa \cdot \sigma_2$) with the envelope of the material failure condition (Figs. 4–5), it can be seen that only in the case of the simplest Lame-Rankine criterion the maximum tensile stress obtained with Brazilian test (1) can be regarded as tensile strength, $f_t = \sigma_{max}$, in other cases there is always $f_t > \sigma_{max}$, which means $\rho < 1$.

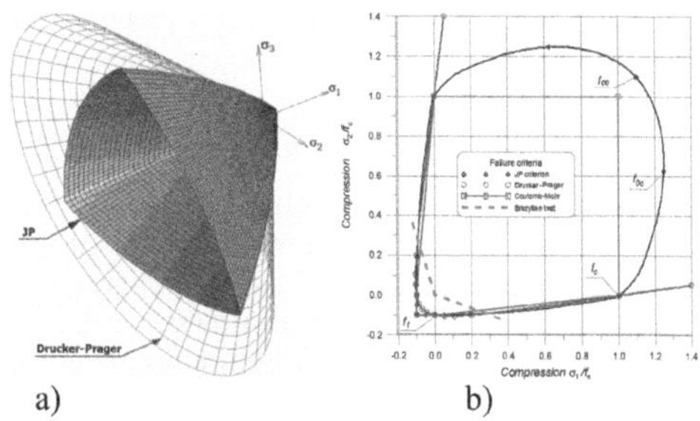

Figure 5. a) Boundary surface for J. Podgórski criterion b) its envelope for 2D stress state $\sigma_3 = 0$.

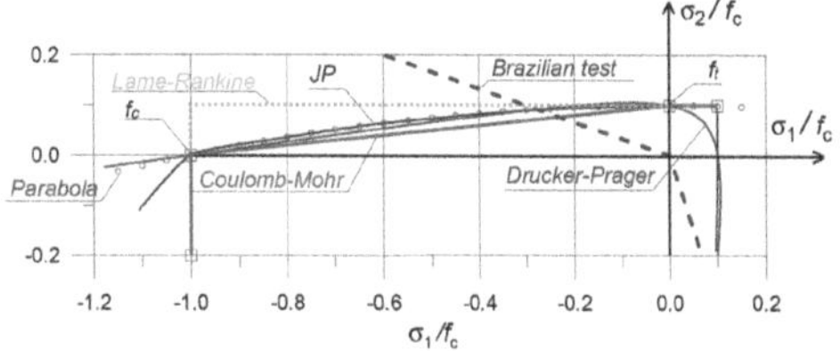

Figure 6. The envelope of the *JP*, Drucker-Prager and Coulomb-Mohr conditions on the plane σ_1–σ_2.

In case of other criteria, to obtain such a simple relationship between σ_{max} and f_t as in equation (9) is not possible, but observing a small difference (Fig. 6) between the position of the envelope described by Drucker-Prager and *JP* conditions and the parabola of the equation (10) it can be assume with sufficient accuracy that the intersection of the parabola with the load path in Brazilian test gives the wanted value for σ_{max}.

$$\frac{\sigma_2}{f_t} = 1 - \left(\frac{\sigma_1}{f_c}\right)^2. \qquad (10)$$

The breaking stress now can be received by using a simple relationship: $\sigma_{max} = \rho \cdot f$, where:

$$\rho = \frac{\sqrt{1+4\gamma^2} - 1}{2\gamma^2}, \quad \gamma = \frac{\kappa}{\eta}. \qquad (11)$$

Taking $\kappa = 3{,}0$ and $\eta = 10$ we get $\rho = 0{,}92328$, and the exact calculations for the Drucker-Prager condition give value $\rho = 0{,}92332$, which deviates less than $4 \cdot 10^{-5}$ between that received from equation (11). The tensile strength calculated using the *JP* and Drucker-Prager conditions is therefore approx. 8% higher than the σ_{max}, which means that:

$$f_t = \frac{\sigma_{max}}{\rho} \approx\sim 1.083\sigma_{max}. \qquad (12)$$

4 THE FORMATION OF A DESTRUCTIVE CRACK

4.1 *The criterion for crack initiation*

As a criterion for crack initiation which starts the damaging process of the material we adopted the achievement of effort, which corresponds to the point in the stress space situated at the border surface defined by the equations (4), (6) or (8) depending on the taken condition of the failure. It was assumed that the loading of the sample increases monotonically and the failure process starts at a point where the effort reaches the maximum value. Effort (μ) of the material is defined as the ratio of modules of vectors r_σ and r_f:

$$\mu = \frac{|\mathbf{r}_\sigma|}{|\mathbf{r}_f|} = \frac{r_\sigma}{r_f}, \tag{13}$$

where r_σ = is a vector indicating the point (in the stress space) according to the stress state in the analyzed place of the sample, and r_f = is a vector indicating a point on the boundary surface, which also belongs to the path of monotonic load (Fig. 7).

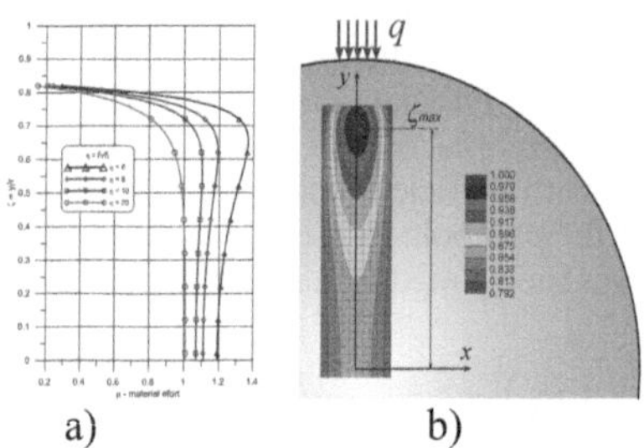

Figure 7. Definition of the material effort coefficient μ, a) values of this coefficient on the axis of the sample compressed with a distributed load depending on the ratio of strength η, b) map of the effort coefficient μ according to *PJ* criterion, determined on the FEA numerical analysis.

4.2 *Location of crack initiation*

The effort of the sample material determined in accordance with equation (13), after adopting the stress field described by Hondros equations (2), in points lying on the axis $x = 0$ is shown in Figure 7a. Here is used a parabolic approximation, setting effort curves for different ratios $\eta = f_c/f_t$. The displacement is seen of the point of maximum effort from the center of the sample to the loaded edge of the sample with η values lower than 15. The location of this point is defined with a dimensionless coordinate ζ_{max} for different effort hypotheses for a fixed $\eta = 10$, and is shown in Table 1 (11).

5 SUMMARY AND CONCLUSIONS

The paper presents a study of the indirect method of determining the tensile strength of brittle materials known as the "Brazilian" method. The strength of the material in the sample is usually determined in standards (PN-EN 12390, Test method T222, ASTM Standards) by taking the maximum tensile stress at the time of failure, as the value of this characteristic. As it was shown, this stress is much lower than the breaking stress of the sample. These differences can be estimated on the assumed condition of failure in a 3D state of stress. We can assume that this difference is approx. 8%, while taking the Drucker-Prager and *PJ* conditions, and can reach up to 30%.

An interesting issue is also the location of the place of initiation of a destructive crack. As shown in the graphs of material effort, the place of maximum effort depends on the ratio $\eta = f_c/f_t$. When $\eta < 17$, a point of crack initiation moves toward the loaded edge of the sample and when $\eta \geq 17$ the maximum effort point is located in the center of the sample. The real mechanism of destruction of the sample is different and its explanation requires taking take into account the spatial work of the sample material. The corresponding analysis performed by finite element method taking into account the JP fracture criterion and triaxial stress state, explains the mechanism of destruction and indicates the origin of the destructive crack (Fig. 8).

Table 1. Comparison of efforts depending on the criterion of failure, the $\eta = 10$.

	Concentrated force			Distributed load		
Failure criterion	ζ_{max}	$\kappa = \sigma_1/\sigma_2$	ρ	ζ_{max}	$\kappa = \sigma_1/\sigma_2$	ρ
Coulomb-Mohr	0	3	0.769	0.593	5.167	0.659
Drucker-Prager	0	3	0.923	0.566	4.885	0.834
Ottosen-Podgórski	0	3	0.923	0.566	4.885	0.834
Parabolic envelope	0	3	0.923	0.566	4.885	0.834
Lame-Rankine	0	3	1.0	0	3.112	1.0

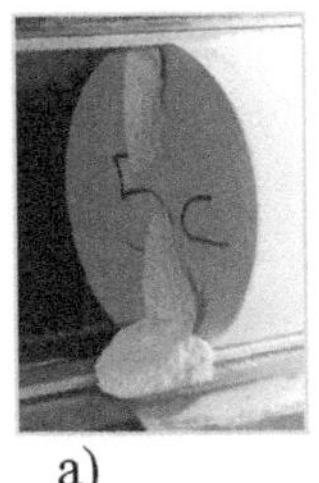

a)

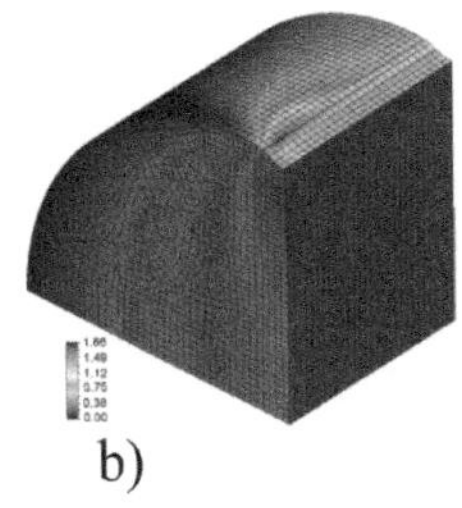

b)

Figure 8. View of destroyed sample after the Brazilian test (a) and map of effort coefficient calculated for 3D FEM model (b).

Table 2. Position of the point of maximum effort, depending on the relative of strength $\eta = f_c/f_t$.

η	5	7	10	13	15	16	17	20
ζ_{max}	0.668	0.638	0.566	0.439	0.293	0.170	0.0	0.0

REFERENCES

Aréoglu, N., Girgin, Z.C. & Aréoglu, E. 2006. Evaluation of Ratio between Splitting Tensile Strength. *ACI Materials Journal* 103 (1): 18–24.

Chen, C.S. & Hsu, S.C. 2001. Measurement of Indirect Tensile Strength of Anisotropic Rocks by the Ring Test. *Rock Mech. Rock Engng.* 34 (4): 293–321.

Hondros, G. 1959. The evaluation of Poisson's ratio and the modulus of mat .... *Aust. J. Appl. Sci.* 10: 243–268.

Jianhong, Ye, Wu, F.Q. & Sun, J.Z. 2009. Estimation of the tensile elastic modulus using Brazilian disc .... *Int. Journ. Rock Mechanics & Mining Sciences* 46: 568–576.

Mushelishvili, N.I. 1949. *Nekotorye Osnovnye Zadachi Matematicheskoj Teorii Uprugosti*. Moskwa: Nauka.

Norma PN-EN 12390-6:2011. Wytrzymałość na rozciąganie przy rozłupywaniu próbek do badań, Polish Standard.

Ottosen, N.S. 1977. A Failure Criterion for Concrete. *Journ. Eng. Mech. ASCE* 103 (EM4): 527–535.

Podgórski, J. 1984a. Limit state condition and the dissipation function for isotropic materials. *Archives of Mechanics* 36: 323–342.

Podgórski, J. 1984b. Limit state condition and the dissipation function ... *Archives of Mechanics* 36: 323–342.

Podgórski, J. 1985. General Failure Criterion for Isotropic Media. *Journ. Eng. Mech. ASCE* 111: 188–201.

Podgórski, J. 2011. Criterion for angle prediction for the crack ... *Mechanics and Control* 30 (4): 229–233.

Standard Test Method for splitting Tensile Strength of Intact Rock Core Specimens, ASTM Standards, USA.

Test method T222. Indirect tensile strength of rock drill core, NSW Roads and Maritime Services, Australia.

Timoshenko, S. & Goodier, J.N. 1962. *Theory of elasticity.* Warszawa: Arkady.

Advances in Mechanics: Theoretical, Computational and Interdisciplinary Issues – Kleiber et al. (Eds)
© 2016 Taylor & Francis Group, London, ISBN 978-1-138-02906-4

Stress redistribution at the support of a transversely loaded sandwich panel

Z. Pozorski
Institute of Structural Engineering, Poznan University of Technology, Poznań, Poland

J. Pozorska
Institute of Mathematics, Częstochowa University of Technology, Częstochowa, Poland

ABSTRACT: The paper concerns the problem of static structural behavior of sandwich panels. The local stress concentration and distribution at the supports are discussed. The 3-D numerical model with the definition of damage initiation and evolution is applied. Different failure modes are taken into account. The numerical results are compared with real experiments and engineering simplifications. The assumption concerning uniformity of stress distribution is verified. The evaluation of the support capacity of the sandwich panel is discussed.

1 INTRODUCTION

The paper considers sandwich structures, which consist of two thin external steel faces and a thick and soft core. Sandwich elements are very attractive because of high load-bearing capacity at low self-weight and excellent thermal insulation. Unfortunately, in sandwich structures various failure mechanisms may occur. The failure mode depends on many geometrical and mechanical parameters, also on boundary conditions (Daniel et al. 2002). The following failure mechanisms can be distinguished: face yielding, global and local instability, debonding (usually caused by impact loads), shear and indentation of the core.

An important failure mode of the sandwich panel, which can occur at the point loads, is local indentation. A concentrated load spreads at different area of the sandwich core depending on the form of excitation, the flexural stiffness of a sandwich face and the stiffness of the core. The indentation loads were studied for various materials (Lolive and Berthelot 2002, Wu and Sun 1996) and the great significance of an indenter shape was reported (Wen et al. 1998). Experimental studies of the indentation phenomenon are dominating, because of the complexity of the problem (Flores-Johnson and Li 2011). Theoretical models of indentation were started with a sheet treated as an elastic beam located on the ideal elastic foundation (Hetenyi 1946). Currently, there are more complex models, which assume plastic yielding of the core (Triantafillou and Gibson 1987), rigid and ideally plastic behavior of the core and face material or elastic deformation of the faces and compressive yield of the core (Ashby et al. 2000). It is also a challenge to take into account the anisotropy of the core (Chuda-Kowalska et al. 2015).

This paper discusses an engineering problem of compressive stress redistribution at supports of the sandwich panel. Sandwich panels applied in civil engineering are installed to the substructure (purlins, side rails). Interaction between the substructure and the panel induces compression of the sandwich core. Stress concentration can be particularly important in the case of multi-span systems loaded by temperature gradient. The support conditions strongly influence the scope of the application of the sandwich panels (Studziński et al. 2015). From a practical point of view, it is essential to spread out the impact of the support on a relatively large area of the panel. The paper takes up the problem of estimation of the stress redistribution at the supports. The influence of geometrical and mechanical parameters of the system on the redistribution level is also discussed.

2 PROBLEM FORMULATION

To analyze the problem of core compression and stress redistribution at the support, the system shown in Figure 1 was examined. A short panel is subjected to a one-line loading to provide high stress concentration at the support. The system corresponds to the assumptions of the standard EN 14509. The support reaction is calculated using the relation:

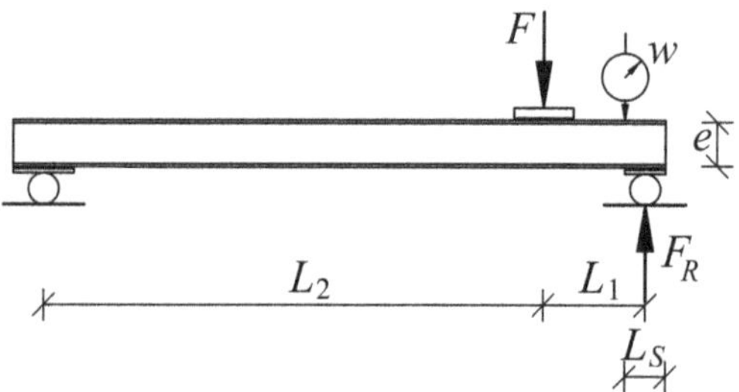

Figure 1. The system for determination of stress redistribution at the support.

$$F_R = \frac{L_2}{L_1 + L_2} F. \qquad (1)$$

Looking for a parameter describing the distribution of the support reaction, we can define the support capacity F_R using the distribution parameter k:

$$F_R = b\left(L_S + 0.5ke\right) f_{Cc}, \qquad (2)$$

where b, L_S, e and f_{Cc} denote the width of the panel, the width of the end support, the distance between the centroids of faces and the compression strength of the core, respectively. Theoretically, the parameter k determines the effective width of the support (and thus the support reaction redistribution) related to the mid-plane of the panel.

In contrast to the theoretical assumptions, the observations made during the laboratory tests indicate that the support capacity does not depend on the support width L_S. In addition, the calculated stress level corresponding to the actual support reaction capacity (F_R/bL_S) is much higher (2–3 times) than the value of compressive strength fCc obtained from the classical compression test of the cuboid specimen. Both observations call the formula (2) into question and prompt for a detailed analysis of the problem of stress distribution at the support. The question is what determines the destruction of the sandwich panel at the support and how to find the proper value of the support reaction capacity. The issue is analysed numerically.

3 THE NUMERICAL MODEL

In numerical simulations, the parameters of the system correspond to the previously carried out laboratory tests. The numerical model was prepared in the ABAQUS system. Only 3-D models were applied. Sandwich panel of the length 2.20 m is located on two supports ($L_1 = 0.34$ m, $L_2 = 1.80$ m). The width of the left support is 0.08 m, whereas the width of the right support is 0.04 m, see Figure 2. The loading is applied as a pressure distributed over a width of 0.08 m. Core compression at the right support is analyzed. The total depth of the panel is D (initially 98.43 mm) and the thickness of each of the faces is t_F (base value 0.471 mm). Note that the distance between centroids of faces $e = D - t_F$.

Steel, flat facings were assumed as isotropic, elastic-plastic material. The elasticity modulus $E_F = 195$ GPa and Poisson ratio $\nu_F = 0.3$ were assumed. The actual relationship between stress and strain was introduced. The yield strength was 360 MPa, and the ultimate strength reached 436 MPa. Facings were modeled using a four node doubly curved, thin or thick shell, reduced integration, hourglass control, finite membrane strain elements S4R. The core of the panel was modeled using eight node linear brick elements C3D8R, though for comparison elements C3D8 were also used. The core was considered isotropic, elastic-plastic (yield strength 0.09 MPa, ultimate strength 1.0 MPa). To ensure shear modulus $G_C = 4.22$ MPa, the following initial core parameters were assumed: elasticity modulus $E_C = 8.61$ MPa and Poisson's ratio $\nu_C = 0.02$. In the parametric studies, constant value G_C was assumed and only ν_C and E_C were changed respectively.

To include the core failure by shear or tension, the interface layer of the thickness 0.5 mm was introduced between the core and the faces. The quadratic nominal stress criterion of damage initiation and the displacement type of damage evolution were defined. The interface was modeled using COH3D8 8-node 3D cohesive elements. The uncoupled elasticity law for cohesive material was used. The quadratic nominal stress criteria of damage initiation and the linear softening damage evolution were defined. Interaction between all parts was assumed as TIE type, which makes equal displacements of nodes. Between the right support and the sandwich structure, the surface-to-surface contact was also examined.

The main aim of this study was the observation of phenomena occurring at the support, whereby

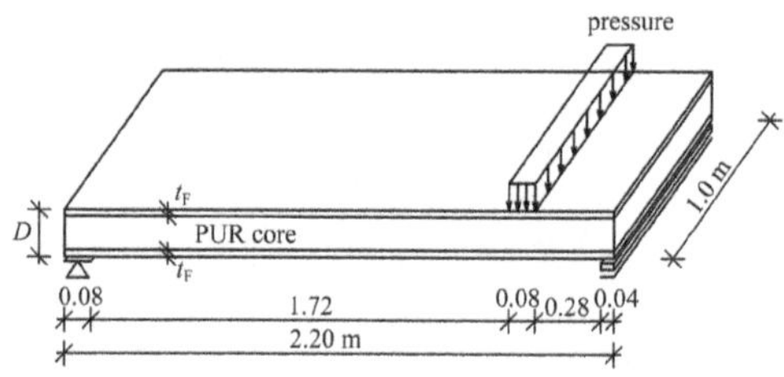

Figure 2. The geometry of the analyzed numerical model.

the general conclusions regarding the nature of these phenomena were the most important. For this reason, for the simplicity the model a constant density of the FEM mesh was used. The mesh sizes 0.02 m and 0.01 m were applied. The analyzed problem is certainly mesh-dependent, though the reduction in a mesh size alters the values of extreme stress not more than 10%. Therefore, most of the results were compared for the mesh size 0.02 m.

4 DISCUSSION OF THE RESULTS

The sandwich panel with the total thickness D = 98.43 mm and the thickness of facings t_F = 0.471 mm was taken as a reference. Let us also keep in mind that the Poisson's ratio $\nu_C = 0.02$ (at the shear modulus $G_C = 4.22$ MPa). The model of the system has been marked as *Sup_2_1*. The distribution of the normal stress σ_z (perpendicular to the facings), the shear stress τ_{xz} and the equivalent plastic strain at the integration points in the core of the sandwich are presented in Figure 3. The stress values correspond to the load pressure $q = 112$ kPa, while the ultimate pressure is 120 kPa.

As expected, normal stresses reach extreme values at the support, and then decrease in the thickness of the panel. It is surprising, however, that these stresses spread to the relatively narrow width of the core (about 60 mm). Moreover, the higher core layer, the narrower the normal stress concentration zone. On the other hand, the shear stress reaches the proper value only 20 mm from the edge of the support. This effect is even more visible in Figure 4, which shows the normal and shear stresses in the nodes of the core, at two different levels. The presented results show that at a very narrow space of the core, normal stresses are transferred into the shear stresses. Of course, normal stresses in facings consequently increase.

To assess the influence of the facings thickness, Poisson's ratio and the panel depth on the stress level in the core, more numerical analyses have been done. Only one selected parameter has been changed in relation to the reference model *Sup_2_1*. The facing thickness was equal to 0.6 mm and 0.7 mm in models *Sup_2_1a* and *Sup_2_1b*, respectively. In model *Sup_2_5* the Poisson's ratio ν_C = 0.10 was introduced leaving G_C = 4.22 MPa. In model *Sup_2_6* is ν_C = 0.30. Models *Sup_2_7* and *Sup_2_8* differ

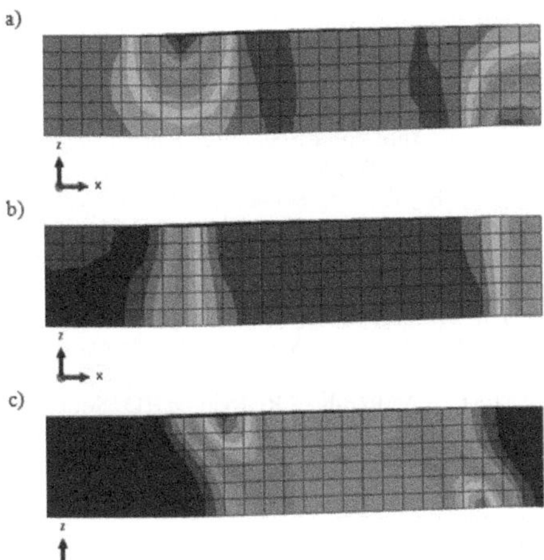

Figure 3. The stress and strain distribution in the core part: a) normal stress σ_z, b) shear stress τ_{xz}, c) equivalent plastic strain at the integration points.

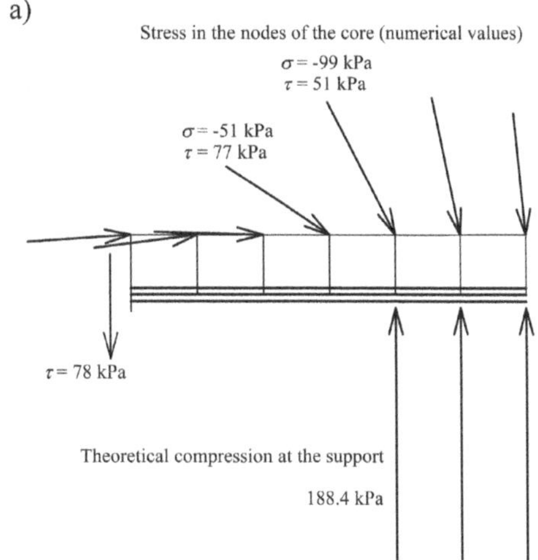

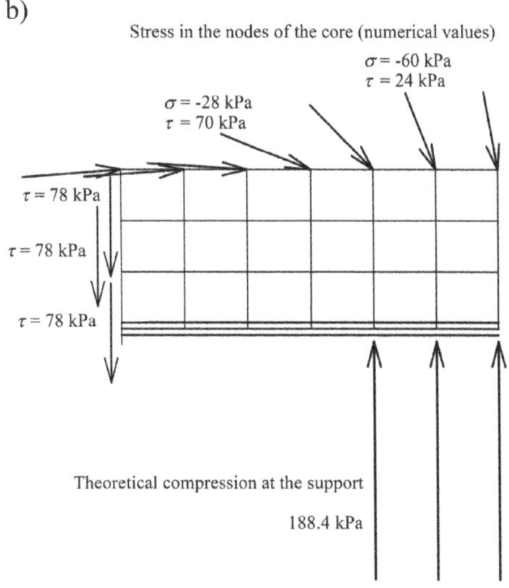

Figure 4. Normal and shear stress values in the nodes of the numerical model: a) the first, b) the third core layer.

from *Sup_2_1* only the depth of the panel (core). In *Sup_2_7* the depth was increased by 20 mm, and in *Sup_2_8* it was decreased by 20 mm. In Table 1 are shown the values of the extreme normal stress at the support σ_z and the extreme shear stress τ_{xz} obtained in the core for the load pressure $q = 112$ kPa.

The results presented in Table 1 show that increase in the thickness of facings (*Sup_2_1a*, *Sup_2_1b*) has reduced the extreme normal stress in the core only about 6%. It is interesting that increasing the Poisson's ratio (*Sup_2_5*, *Sup_2_6*) results in an increase of extreme normal stress. For a thicker core (*Sup_2_7*) the value of extreme normal stress agreed numerically with the value obtained in the model *Sup_2_1*. Extreme values of shear stresses in each model are arranged in a certain regularity, but it should be clearly noted that differences between models are slight. The exception is of course the case of a thicker core. It should be noted that Table 1 does not include *Sup_2_8* model (core thinner 20 mm) because the panel is damaged before reaching the assumed pressure. For the sake of completeness, it is noted that at a pressure of 112 kPa support reaction is 8.96 kN, which gives a theoretical compression on the support 188.4 kPa. This level of stress in the core would cause strain of 20%. However, both in reality and in the numerical models, a change in the panel depth at the support is much smaller.

5 CONCLUSIONS

The proposed numerical model is useful in the analysis of the problem of stress redistribution at the supports of sandwich panels. The 3-D model allows to assess the level of stress concentration. The introduced interface layers enable observation of different failure mechanisms.

The method of support capacity estimation proposed in the standard EN 14509 is inappropriate. The distribution parameter k determined according to Equation 2 does not reflect adequately the phenomena occurring at the support. Numerical simulations, preceded by laboratory tests show that in all practical cases shear of the core determines the failure of the sandwich panel at the support. Therefore shear stresses determine the support reaction capacity. Stress distribution in the case of axial compression of specimens is quite different than in the case of compression at the support of the panel. This is because support compressive stresses are transferred into the shear stresses at a very narrow space of the core. Degradation of foam cells caused by compression is very local (only at the surface of the facing). For this reason, an acceptable level of the foam local degradation should be determined.

Table 1. Extreme stresses in the core part in the case of pressure $q = 112$ kPa, for different numerical models.

Model	Normal stress at the support (kPa)	Shear stress (kPa)
Sup_2_1	−115.9	83.9
Sup_2_1a	−111.9	83.2
Sup_2_1b	−109.2	82.3
Sup_2_5	−117.4	83.4
Sup_2_6	−123.3	82.2
Sup_2_7	−115.9	73.6

REFERENCES

Ashby, M.F., A.G. Evans, N.A. Fleck, L.J. Gibson, J.W. Hutchinson & H.N.G. Wadley. 2000. *Metal Foams: A Design Guide*. Butterworth Heinemann.

Chuda-Kowalska, M., T. Gajewski & T. Garbowski. 2015. Mechanical characterization of orthotropic elastic parameters of a foam by the mixed experimental-numerical analysis. *Journal of Theoretical and Applied Mechanics 53(2)*, 383–394.

Daniel, I.M., E.E. Gdoutos, K.-A. Wang & J.L. Abot. 2002. Failure modes of composite sandwich beams. *International Journal of Damage Mechanics 11(4)*, 309–334.

Flores-Johnson, E.A. & Q.M. Li. 2011. Experimental study of the indentation of sandwich panels with carbon fibre-reinforced polymer face sheets and polymeric foam core. *Composites Part B: Engineering 42(5)*, 1212–1219.

Hetenyi, M. 1946. *Beams on elastic foundation: theory with applications in the fields of civil and mechanical engineering.* Ann Arbor: University of Michigan Press.

Lolive, E. & J.-M. Berthelot. 2002. Non-linear behaviour of foam cores and sandwich materials. Part 2: indentation and three-point bending. *Journal of Sandwich Structures and Materials 4(4)*, 297–352.

Studziński, R., Z. Pozorski & J. Błaszczuk. 2015. Optimal support system of sandwich panel. *Journal of Engineering Mechanics 141(3)*, 04014133 1-8.

Triantafillou, T. & L. Gibson. 1987. Minimum weight design of foam core sandwich panels for a given strength, *Materials Science and Engineering 95*, 55–62.

Wen, H.M., T.Y. Reddy, S.R. Reid & P.D. Soden. 1998. Indentation, penetration and perforation of composite laminate and sandwich panels under quasi-static and projectile loading. *Key Engineering Materials 141–143*, 501–552.

Wu, C.L. & C.T. Sun. 1996. Low velocity impact damage in composite sandwich beams. *Composite Structures 34(1)*, 21–27.

Advances in Mechanics: Theoretical, Computational and Interdisciplinary Issues – Kleiber et al. (Eds)
© 2016 Taylor & Francis Group, London, ISBN 978-1-138-02906-4

Three-point bending test of sandwich beams supporting the GFRP footbridge design process—validation

Ł. Pyrzowski, B. Sobczyk, W. Witkowski & J. Chróścielewski
Faculty of Civil and Environmental Engineering, Gdańsk University of Technology, Gdańsk, Poland

ABSTRACT: Some selected aspects concerning material and construction design issues for pedestrian footbridge made of GFRP composite materials are elaborated in this paper. The analysis is focused on validation tests, which are particularly important because of the advanced technology and materials that are used for this innovative bridge. The considered footbridge is a sandwich-type shell structure comprising of PET foam core and outer skins made of glass fibre reinforced polymer laminate. A proper connection of foam core and laminate skins is crucial, regarding its load bearing capacity. The aim of this paper is to compare experimental and Finite Element Analysis (FEA) results of sandwich beams subjected to bending loading in three point bending tests, which are afterwards used to validate the stiffness properties of the full structure.

1 INTRODUCTION

The general subject of the study is to elaborate architectural, material and construction design aspects, concerning the process of pedestrian footbridge made of composite materials design. A typical span of the bridge has an U-shape sandwich type cross section comprising of Polyethylene Terephthalate (PET) foam core and skins made of Glass Fibre Reinforced Polymer (GFRP) laminate. The simply supported 14 m long spans are intended to be applied for instance over two lane roadways. The design architectural concept of the bridge is presented in Figure 1. More detailed information about the project can be found e.g. in (Chróścielewski et al. 2014a).

The standard design process includes the development of concepts and material selections, identification of material properties, numerical simulations, strength calculations and serviceability analyses. In the case of innovative structures, especially made of non-classical materials, which are not included in the existing design codes (Chróścielewski et al. 2014b), it is necessary to perform additional validating analyses (Miśkiewicz et al. 2015). Material parameters identification as well as experimental and numerical investigations are also required during the process of the innovative bridge design. Some selected above mentioned aspects are described in detail in the paper. Comparison of the experimental and FEA results of three-point bending tests carried out on sandwich beams is shown.

2 EXPERIMENTAL INVESTIGATIONS

2.1 *Description of specimens*

Many sandwich beams (laminate-core-laminate) with different dimension and layups have been investigated in the project. Here we discuss a case of $90 \times 74 \times 1220$ mm beam (Fig. 2).

The main components of GFRP laminate skins are fire retardant vinylester resin and four layers (4×0.663 mm) of bi-directional balanced sewn E-glass woven fabrics, with [0/90] and [45/–45] orientation (Fig. 2). The core (70 mm) is made of 100 kg/m^3 PET foam. The samples, such as the intended footbridge, are formed using the infusion technology.

2.2 *Experimental tests*

The sandwich specimens considered in this work are subjected to three-point bending tests. The distance between supports is set to 1000 mm, the force is applied in the middle of the span via steel cross-head with a radius of 40 mm. There are no additional pads put between the head and the beam. The experiments are carried out by the Zwick/Roell Z400 testing machine using the displacement control technique with the speed of the cross-head of 2 mm/min. The test setup is shown in Figure 3.

During the experiments the following data is recorded: cross-head motion, force, vertical displacement in the span middle (inductive sensor) and longitudinal skin strains (resistance strain gauges), see Figure 4.

Figure 1. Design concept of the composite footbridge.

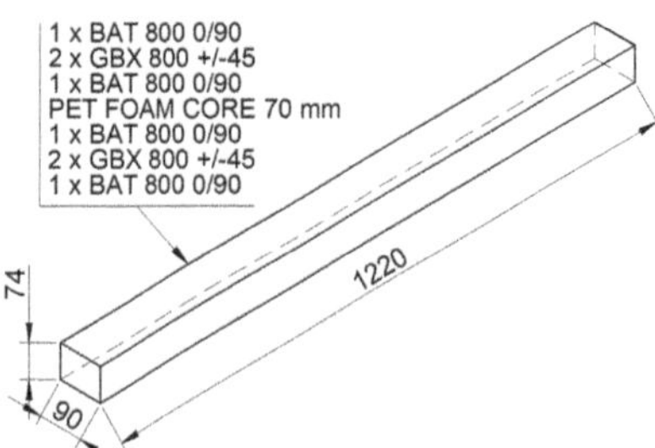

Figure 2. Geometry of composite beam.

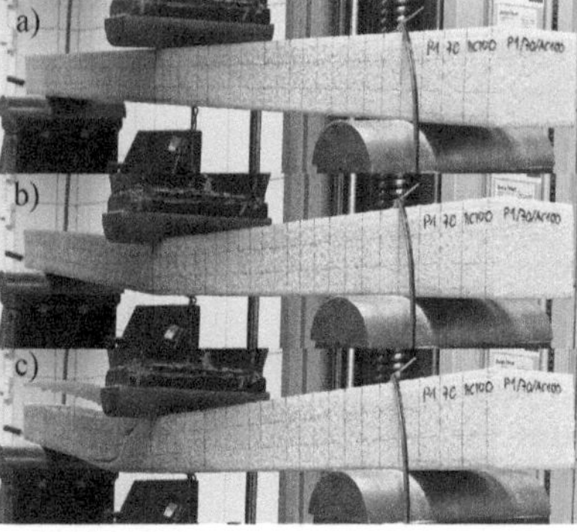

Figure 3. Three-point bending test: a) beginning, b) maximum force point, c) beam failure moment—debonding between skin and foam.

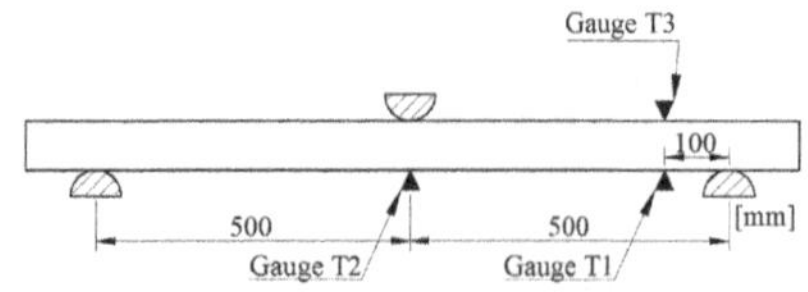

Figure 4. Location of strain gauges.

The results of measured force vs. vertical displacement is shown in Figure 5. The displacement registered by inductive sensor is lower than the cross-head one. It is caused by foam crushing effect directly under the cross-head half roller. The measured strains are presented in Figure 6. The failure process of the beam is started in the area of cross-head roller action, see Figure 3. The crushed foam initiates shear damage of the core, which results in debonding between skin and foam.

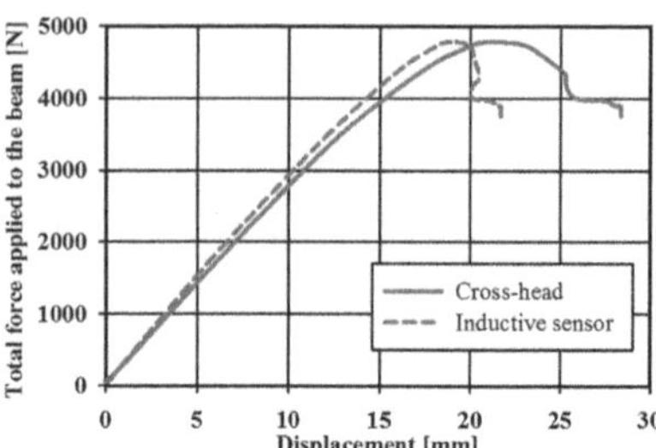

Figure 5. Experimental result—total force vs. vertical displacement.

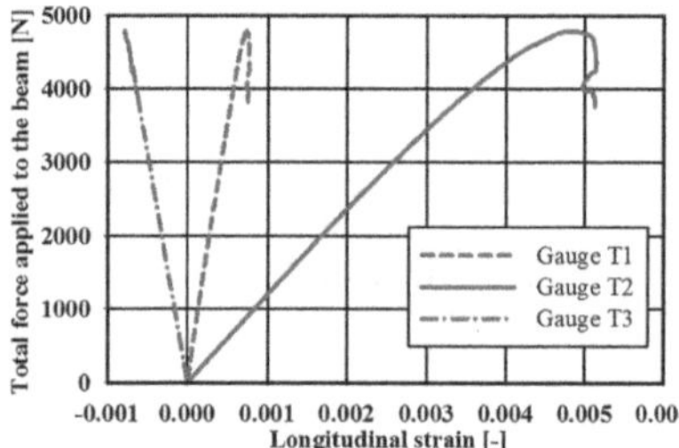

Figure 6. Experimental result—total force vs. longitudinal strain.

3 NUMERICAL MODEL

Some numerical simulations of three-point bending test of sandwich beams (described in chapter 2) are carried out in order to assess whether the design assumptions, concerning material models selection and material constant experimental determination are correct. The discrete models are built in Abaqus 6.14-2 Finite Element Method (FEM) environment. Two model variants are considered, both based on the dimensions given in Figures 2 and 4. Because of the geometry, applied material solutions, and observed failure mechanism we decided to model the beam using symmetry with respect to the beam width. The first model represents pure laminate-foam-laminate structure. Additional resin columns are included in the second one, which are a consequence of the infusion technology. It requires application of small holes in the foam (~3 mm), which allow resin to flow through it and form both laminate skins. It is believed that their consideration will improve results accuracy of obtained response of the beam. The columns height is the same as height of the foam core. The distance between adjacent columns is equal to 60 mm. Two columns are observed in the analysed specimen width direction.

Following finite elements are used to create discrete beam structure. The laminate skins are modelled with S4 shell elements with full integration scheme. Equivalent single layer technique and first order shear deformation kinematics theory, see e.g.

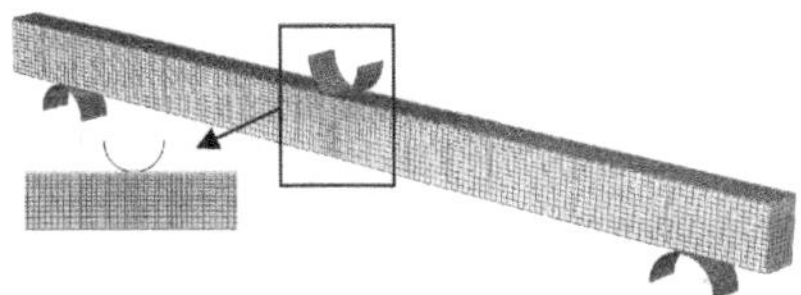

Figure 7. Visualisation of numerical model, red lines represent additional resin columns, included in the 2nd beam model.

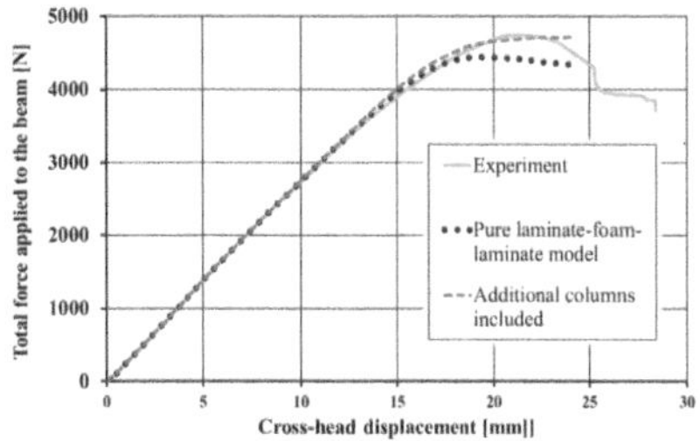

Figure 8. Total force vs. cross-head displacement.

(Reddy 2004) are utilised to represent behaviour of this multi-layered medium. C3D8 brick finite elements with full integration represent the foam. We assume that the steel cross-head and supports are rigid. Additional resin columns, used in the 2nd model variant, are built with truss T3D2 finite elements. The structural mesh of finite elements consists circa 20 000 nodes (see) Figure 7.

Linear elastic, orthotropic material law under plane stress is used to represent behaviour of lamina stiffness and strength properties. The elastic material constants, describing the material law, are determined on the basis of experiments performed by (Klasztorny et al. 2013). Isotropic constitutive relation is used for foam material, basing on the data provided by the material distributor. Crushable foam plastic behaviour with hardening (Abaqus) is included for the core description, in which the material parameters determined by means of additional experiments, namely, uniaxial compression and hydrostatic compression test. The laminate skins are bonded to the foam. A friction contact is assumed between: the cross-head and laminate upper skin and between laminate lower skin and rigid supports interfaces. The resin columns are included in the second analysis variant using the embedded region Abaqus technique. Isotropic material model is used to represent their behaviour. Analysis of internal laminate skins failure is not here, as well as debonding at the foam-skin interface. However, despite no pads were applied between the head and the laminate skins, we did not observe any laminate damage during the experiment. Moreover, the rapid failure of beam, connected with debonding happened after the maximum force was achieved. The aim of our numerical analysis is to recreate beam behaviour to the point when maximum measured force for the head is achieved, without analysis of failure mechanism.

The three-point bending experiment is recreated by means of static calculations in which the cross head is rigidly moving downwards. Geometrical and material nonlinearities (only for foam core) are considered.

4 COMPARISON OF RESULTS

The following results were obtained during the FEA. Figure 8 presents comparison of total force vs. cross-head displacement relations for the experiment and numerical simulations. Figure 9 depicts relation of total force vs. longitudinal strain, measured in strain gauges.

On the basis of the data given in Figures 8 and 9 it can be stated that both numerical models are able to describe the range of linear response of the sandwich beam properly. Because laminate skin failure is not considered here only the nonlinear foam destruction process is initiated in numerical simulations. Therefore, the load carried by beam does not decrease rapidly in FEA. The total beam load capacity, predicted numerically, is more accurate, when the additional resins are included. This effect also arises, because one row of resin columns is located almost directly under the cross-head. However, the range of structural response determined to be used during the design process of such a structure must be reduced in our opinion, because of the behaviour of FRP skins. Uniaxial tension tests performed on multilayered laminates, composed of biaxial fabrics, reveal that the stress-strain curve slightly decreases at the level of ~1/3 of their ultimate load capacity. This phenomenon is named as the knee effect, see e.g. (Królikowski 2012), and is caused by some micro cracks, developing parallel to the direction of tension. However, in the authors opinion, the knee effect should indicate the upper limit of the design loads. This is due to the required bridge durability in different environmental conditions. This limitation is introduced despite the fact that from the materials strength point of view the laminate retains its full load carrying capabilities. The pure laminate-foam-laminate model yields the same response as the model with additional columns before the knee effect occurs. Hence, it will be utilised in the design process, since it is the simpler one.

Some differences between the obtained numerical and experimental strains are observed. The strain is a local parameter measured on the resin surface, which is also the outer laminate surface. Moreover, the adopted material constants of the composite are defined as some statistically mean values, resulting from analysis of many test coupons. The difference between experimental and numerical values is believed to be caused by the above mentioned

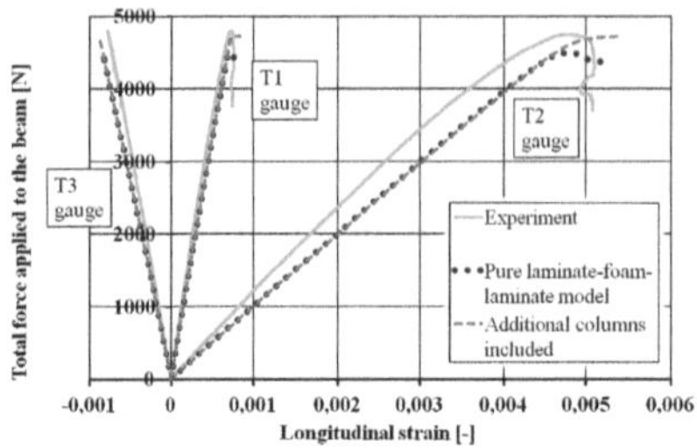

Figure 9. Total force vs. measured strain.

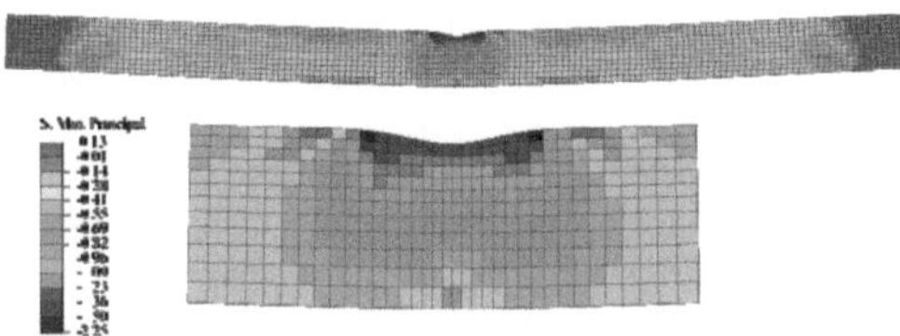

Figure 10. Contours (quilt values) of minimal principal stresses at the end of the analysis for the 2nd numerical model.

aspects. However, it is acceptable in an engineering sense in the design process. Even though the stresses or strains are slightly underestimated, the procedure for bridges, that include e.g. load partial factors that increase action values, see e.g. (Chróścielewski et al. 2014b), allows for a safe design.

Contours (quilt values) of minimal principal stresses in foam core are shown in Figure 10 in deformed configuration at the end of the analysis for the numerical model with additional resin columns. The values exceeding foam compression strength (1.5 MPa) are coloured in grey. The stress pattern shown in Figure 10 corresponds with the failure mode of the beam, shown in Figure 4c.

5 CONCLUSIONS

The design process of an innovative footbridge is very complex. Many additional calculations, simulations and experiments are required to be carried out, before design of full scale structure can be started. In this work two finite element simulations of the sandwich beam (being one of many others studied and analysed in the project) three-point bending test were performed and compared with experimental data. These calculations revealed that the design variables as material parameter and constitutive relations are correctly determined. Both numerical models, namely, pure laminate-foam-laminate and the model with additional resin columns, were able to describe global response of the beam. The pure laminate-foam-laminate variant is chosen for further application to design of full scale structure. This is because the range of structural response is by assumption limited only to the linear elastic response of undamaged material, in FRP skins in this case. On the basis of the analyses described here, it can be concluded that it is possible to design a full scale footbridge with use of information determined from the presented research.

ACKNOWLEDGEMENTS

The study has been supported by the National Centre for Research and Development, Poland, as a part of a research project No. PBS1/B2/6/2013, realized in the period 2013–2015. B. Sobczyk is supported under Gdansk University of Technology (Poland), Faculty of Civil and Environmental Engineering, Young Scientist Support Programme. The support is gratefully acknowledged. Abaqus calculations were carried out at the Academic Computer Centre in Gdansk.

REFERENCES

Abaqus 6.14-2 Documentation.

Chróścielewski, J., Klasztorny, M., Miśkiewicz, M., Romanowski, R., Wilde, K. 2014. Innovative design of GFRP sandwich footbridge. Footbridge 2014, *5th International Conference Footbridges: Past, present & future*. London.

Chróścielewski, J., Klasztorny, M., Nycz, D., Sobczyk, B. 2014. Load capacity and serviceability conditions for footbridges made of fibre reinforced polymer laminates. *Drogi i Mosty* 13(3): 189–202 (www.rabdim.pl).

Klasztorny, M., Gotowicki P., Nycz D. 2013. Badania identyfikacyjne kompozytu BG/F nowego wygrzewanego (NS) w temperaturze 20°C. *Foobridge project report* 6-2013v2.

Królikowski, W. 2012. *Polimerowe kompozyty konstrukcyjne*. PWN, Warszawa.

Miśkiewicz, M., Daszkiewicz, K., Ferenc, T., Witkowski, W., Chróścielewski J. 2015. Experimental tests and numerical simulations of a full scale composite sandwich segment. *3rd Polish Congress of Mechanics and 21st International Conference on Computer Methods in Mechanics Short Papers*. Gdańsk: 413–414.

Reddy, J.N. 2004. Mechanics of Laminated Composite Plates and Shells: Theory and Analysis. Second Edition. CRC Press.

Advances in Mechanics: Theoretical, Computational and Interdisciplinary Issues – Kleiber et al. (Eds)
 ISBN 978-1-138-02906-4

An adaptive finite element method for contact problems in finite elasticity

W. Rachowicz & W. Cecot
Institute of Computer Science, Institute for Computational Civil Engineering, Cracow University of Technology, Cracow, Poland

A. Zdunek
Aeronautics and System Integration, Swedish Defence Research Agency FOI, Stockholm, Sweden

ABSTRACT: We consider a mortar finite element method for 3D finite deformation frictionless contact problems proposed by Popp et al. (Popp, Gitterle, Gee, & Wall 2010) in a version enabling error estimation and mesh adaptivity. Contact of elastic bodies made of hyperelastic nearly incompressible materials is considered. Mixed formulation of finite elasticity with contact is approximated with hexahedral elements. The contact problem is transformed into an equality constrained nonlinear problem which is solved by the Newton-Raphson method. Error estimation and adaptivity of finite element meshes is used to reduce the error of approximation.

1 INTRODUCTION

Contact problems play an important role in modeling of many phenomena in solid mechanics. Numerical simulation of such problems is still a challenging task especially in the context of finite elasticity. This paper is devoted to the so-called mortar technique with Lagrange multipliers for modeling contact between elastic bodies (Puso & Laursen 2004). It is also called a Segment-To-Segment (STS) method and it competes with a very popular Node-To-Segment (NTS) algorithm (Wriggers 2002, Laursen 2002). The major difference between the two approaches is that in the node-to-segment method one yields the no-penetration condition between the nodes of the discretised surface of one body and the finite elements covering the surface of the second body while in the mortar technique no-penetrability is enforced weakly between elements of both surfaces. The STS method is superior over the NTS approach. It satisfies the so-called contact patch test and it is more robust in the context of problems with large deformation and large sliding.

Superiority of the mortar technique can also be observed if one applies an *h*-adaptive FEM for disretization. In this method we subdivide elements in subdomains of large estimated errors to obtain optimal distribution of elements leading to fast reduction of the error with possibly smallest increase of the size of the discrete problem. This may result in substantial difference of densities of meshes on surfaces in contact. In such a case enforcing of no-penetration of nodes of the coarse mesh into the elements of the fine mesh resembles enforcing of pointwise Dirichlet boundary conditions on the elastic body and thus it frequently results in instability of solutions. As far as the mortar method is concerned, it uses smoothly distributed contact forces: the Lagrange multipliers to enforce impenetrability. For this reason one has practically unlimited freedom to choose densities of elements on surfaces in contact. The advantages of the STS approach do not come, however, without a price. This algorithm is essentially more complicated than the NTS procedure. The major reason is the need for a sophisticated integration procedure over intersections of elements of surfaces in contact. It also applies a complex multi-step linearization procedure.

2 MIXED APPROXIMATION OF FINITE ELASTICITY WITH A NEARLY INCOMPRESSIBLE MATERIAL

We apply the material description of finite elasticity. We assume that the density of strain energy takes form:

$$\psi = \frac{K}{4}(J-1)^2 + \frac{\mu}{2}(\bar{I}_1 - 3) \tag{1}$$

where $J = \det \boldsymbol{F}$, with $\boldsymbol{F}$ being the deformation gradient, $\bar{I}_1 = J^{-2/3}\mathrm{tr}[\boldsymbol{C}]$, $\boldsymbol{C} = \boldsymbol{F}^T\boldsymbol{F}$. We assume that the bulk stiffness module K exceeds the shear stiffness module μ by a factor of the order of 1000 which implies that the material is nearly incompressible. The second Piola-Kirchhoff stress tensor $\boldsymbol{S}$ corresponding to ψ takes form:

$$\boldsymbol{S} = 2\frac{\partial \psi}{\partial C} = -pJC^{-1} + 2J^{-2/3}\{\psi_1 \mathrm{Dev}[\boldsymbol{I}]\} \tag{2}$$

where the pressure $p = \partial\psi/\partial J = K/2(J-1)$. Near-incompressibility suggests the mixed formulation of the problem with displacements u and the pressure p as the unknowns, and with a principle of variational displacements expressing equilibrium and weak enforcement of the constitutive expression for the pressure: find $u \in V + \hat{u}$ and $p \in Q$ such that

$$\int_\Omega \boldsymbol{S} : \boldsymbol{E}'(\phi,v)dX = \int_\Omega Jb \cdot v dX + \int_{\Gamma^N} (\boldsymbol{F}^T v)\cdot \hat{t} dS,$$

$$\int_\Omega -\left(\frac{2}{K}p - (J-1)\right) q\, dX,$$

for all $v \in V$ and $p \in Q$, with $V = \{v \in H^1(\boldsymbol{\Omega}) : v = 0$ on $\Gamma_D\}$, $Q = L^2(\Omega)$. Here Ω denotes the domain occupied by the body in the reference configuration, Γ_D is the part of the boundary with the Dirichlet boundary condition $u = \hat{u}$, and Γ_N is the boundary with the Neumann boundary condition $\boldsymbol{SN} = \hat{\boldsymbol{t}}$, $\boldsymbol{N}$ denoting the unit normal to Γ_N. $\boldsymbol{E}'(\phi,v) = \mathrm{sym}(F^T \boldsymbol{\nabla}_X v)$ and b are volume forces. The above formulation is subject to linearization leading to the saddle point problem of the form:

$$\begin{aligned} &a(u,v) + b(p,v) = l_1(v) \ \ \forall v \in V, \\ &b(q,u) + c(p,q) = l_2(q) \ \ \forall q \in Q, \end{aligned} \tag{3}$$

where a,b,c are the appropriate bilinear and l_1,l_2 the linear forms (Holzapfel 2000). We discretise this problem with a $Q^p(K)/P^{p-1}(K)$ family of hexahedral elements with $p = 1$.

3 MORTAR APPROACH

We adopt a popular in contact mechanics terminology calling surface $\Gamma^{(1)}$ of the first body in contact as the slave and the surface $\Gamma^{(2)}$ of the second body as the master surface. We denote by $x^{(i)}$ current locations of points on $\Gamma^{(i)}$. They are functions of reference locations and time: $x^{(i)} = x^{(i)}(\boldsymbol{X}^{(i)},t)$. The gap function $g(\boldsymbol{X}^{(1)},t)$ between the two bodies is defined as follows:

$$g(\boldsymbol{X}^{(1)},t) = -n \cdot (x^{(1)} - \hat{x}^{(2)}) \tag{4}$$

where $\hat{x}^{(2)}$ is a projection of the slave point $x^{(1)}$ onto the master surface along the unit vector normal to $\Gamma^{(1)} : \boldsymbol{n}(x^{(1)}(\boldsymbol{X}^{(1)},t))$. The negative contact tractions z on the slave surface $\Gamma^{(1)}$ are expressed in the natural orthogonal basis of unit vectors $(\boldsymbol{n},\boldsymbol{\tau}^\xi,\boldsymbol{\tau}^\eta) : z = z_n n + z_\tau^\xi \boldsymbol{\tau}^\xi + z_\tau^\eta \boldsymbol{\tau}^\eta$. We state the frictionless contact problem as the following set of conditions:

$$\begin{cases} g & \geq 0, \\ z_n & \geq 0, \\ z_n g & = 0, \\ z_\tau^\xi & = 0, \\ z_\tau^\eta & = 0. \end{cases} \tag{5}$$

Inequality $(5)_1$ expresses impenetrability of the surfaces, $(5)_2$ positiveness of the contact pressure, $(5)_{4,5}$ the lack of friction. Equation $(5)_3$ manifests the fact that the contact force appears only if the bodies are in contact. In the mortar approach impenetrability is formulated in the weak form:

$$\tilde{g} := \int_{\Gamma^{(1)}} g\, \delta z_n\, dS \geq 0; \quad \forall \delta z_n \geq 0.$$

After applying the bilinear approximation for the displacements and forces on the contact surfaces we can write the contact conditions for nodes x_i and the corresponding shape functions $N_j(x)$ as follows:

$$\begin{cases} \tilde{g}_j & \geq 0, \\ (z_n)_j & \geq 0, \\ (z_n)_j \tilde{g}_j & = 0, \\ (z_\tau^\xi)_j & = 0, \\ (z_\tau^\eta)_j & = 0, \end{cases} \tag{6}$$

with $\tilde{g}_j = \tilde{g}(N_j), (z_n)_j = z_j \cdot n, (z_\tau^\xi)_j = z_j \cdot \boldsymbol{\tau}^\xi, (z_\tau^\eta)_j = z_j \cdot \boldsymbol{\tau}^\eta$, with z_j being a vector-valued degree-of-freedom of the contact force at node x_j. According to the idea of Primal-Dual Active Set Strategy (PDASS) of Popp et al. (Popp, Gitterle, Gee, & Wall 2010) we can formulate the involving inequalities conditions $(6)_{1-3}$ by a single equality condition being expressed by a specially designed function $C_j := [(z_n)_j - \max(0,(z_n)_j - c_n \tilde{g}_j)]$, $c_n > 0$:

$$\begin{cases} \tilde{g}_j & \geq 0, \\ (z_n)_j & \geq 0, \Longleftrightarrow C_j((z_n)_j, \tilde{g}_j) = 0, \\ (z_n)_j \tilde{g}_j & = 0 \end{cases} \tag{7}$$

which allows to express linearization of the contact problem in a standard way by derivatives of C_j. Such linearization together with saddle point problem (3) constitute the final statement of the discretized contact problem. Contact forces are included as additional stress load on active contact surfaces.

4 INTEGRATION OVER CONTACT SURFACES

Evaluation of the virtual work of contact forces and the weak statement of impenetrability involves integration over the contact surface. In practice we must consider pairs of elements of $\Gamma^{(1)}$ and $\Gamma^{(2)}$ whose parts do not coincide exactly (as contact is imposed only in a weak form) but they are only close one to another. We define the procedure of integration over a pair of such elements as follows. We select a plane defined by a unit normal vector n coinciding with such a vector at the center of the slave element. Next we perform the orthogonal projection along n of both elements on the plane and we find the polygonal intersection of their images on the plane. We subdivide it into triangles and perform the Gauss integration over each of the triangles. The procedure is illustrated in Figure 1.

5 ERROR ESTIMATION FOR THE MIXED FORMULATION

We follow the idea of Rüter and Stein (Rüter & Stein 2000) in designing the procedure for error estimation of FE approximation of the mixed formulation of finite elasticity. In this approach the total error is understood as a sum of the H^1-seminorm of the error of displacements and the L^2-norm of the pressure, and it is expressed as follows:

$$|u-u_h|_1+\|p-p_h\|_0 \le C\left(|\psi|_1+\left\|(J-1)-\frac{2}{K}p_h\right\|_0\right),$$

where ψ is the solution of the auxiliary problem:

$$\psi\in V: \int_\Omega \nabla\psi:\nabla v\,dx = R(v)\;\forall v\in V, \qquad (8)$$

with $R(v)$ denoting the known residual of the principle of virtual displacements corresponding to FE solution. The H^1-seminorm of ψ can be estimated without a necessity of solving (8) by applying a residual technique of estimating the energy norm (H^1-norm in our case) of the solution of the elliptic problem. We apply here the element residual technique of Bank and Weiser (Bank & Weiser 1985).

Adaptivity is based on the standard approach of refining the elements with the largest estimated errors (Demkowicz, Kurtz, Pardo, Paszynski, Rachowicz, & Zdunek 2008). Typically one breaks the elements with errors exceeding a prescribed percentage of the maximum error.

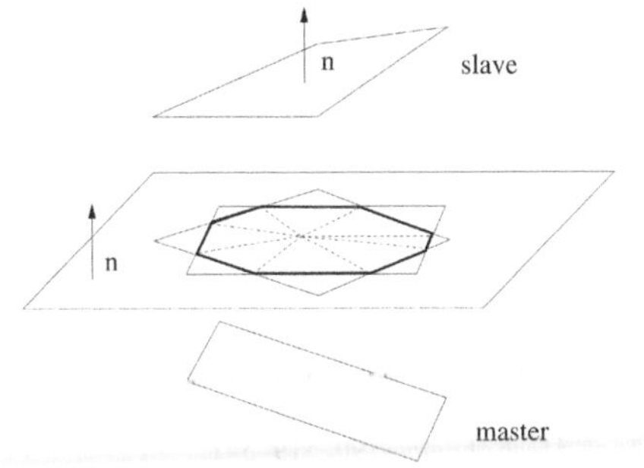

Figure 1. Integration over a pair of elements.

6 NUMERICAL EXAMPLES

We test the mortar contact algorithm on three example problems. In the first of them a half-cylindrical die of the hyperelastic almost incompressible material with the shear modulus $\mu_1 = 1.0$, and with the internal and external radii $r_1 = 2.8$, $r_2 = 3.0$ is intruding into an elastic block of dimensions $4.0 \times 6.0 \times 3.0$ with the shear modulus $\mu = 0.002$. The vertical displacement of the endpoints of the die is 1.4. We performed 3 levels of h-adaptation of the mesh. Figure 2 shows the initial mesh and distribution of the effective stress on the deformed configuration.

In the second test we consider a thin-walled tube being squeezed between two plates. The total length of the tube is $l = 250$, its internal and external radii are $r_1 = 28$ and $r_2 = 30$, respectively. The dimensions of the plates are $60 \times 50 \times 2$. The shear modulus $\mu = 0.002$. The plates were loaded with the vertical

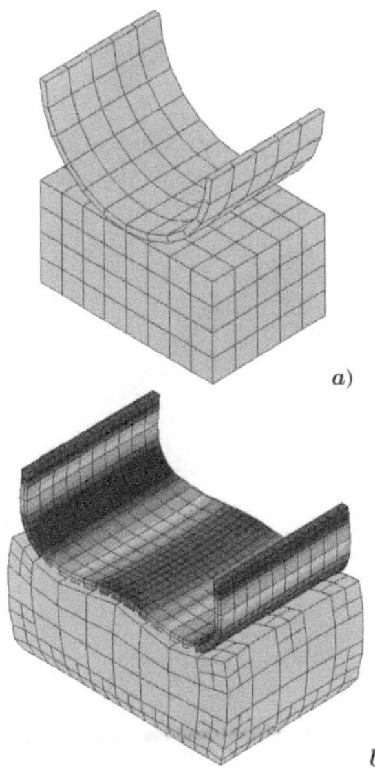

Figure 2. A half-cylindrical die intruding into an elastic block: a) the initial mesh, b) contours of the effective stress σ_0.

stress $t_z = 2.10^{-6}$. Due to symmetries of the problem we solved only 1/8 of the structure imposing appropriate no-penetration (symmetry) boundary conditions on symmetry planes, Figure 3a. We performed 3 levels of h-refinements of the mesh. Figure 3 presents the initial mesh on the reference configuration and distribution of the effective stress on the deformed configuration.

In the third test we use the same geometry as in the second one. This time, however, we push the plate by a prescribed distance of 28 units downwards. The solution is presented in Figure 4. Due to the Dirichlet boundary condition imposed on the plate it is very stiff so that it does not align with the tube. Contrary to the previous example the contact area is now limited to points very close to the edges of the plate. The rest of the plate does not touch the tube.

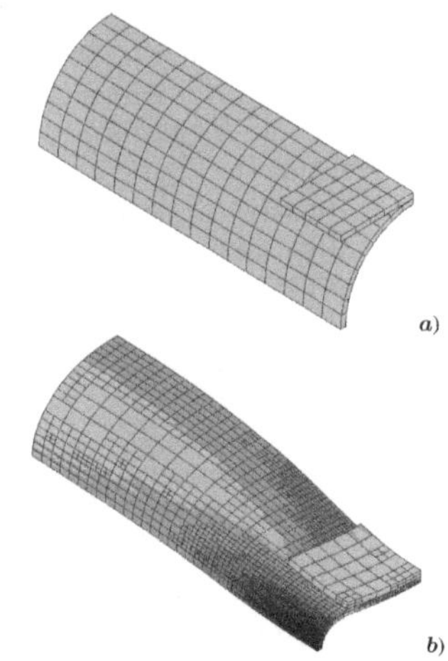

Figure 3. An elastic plate pressing on a thin-walled tube: a) an initial mesh, b) contours of the effective stress σ_0.

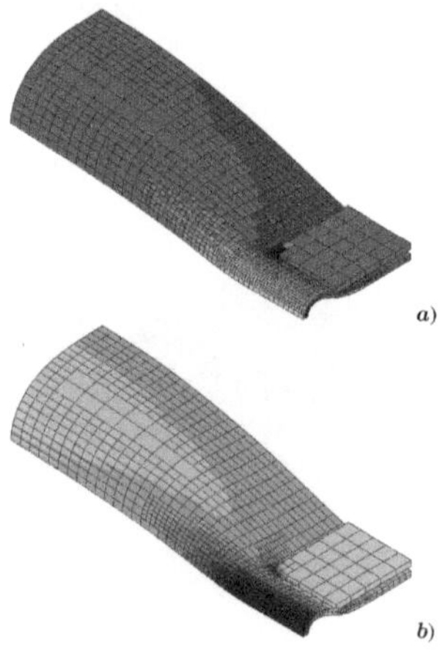

Figure 4. A stiff plate pressing on a thin-walled tube: a) contours of pressure, b) contours of the effective stress σ_0.

7 CONCLUSIONS AND FUTURE WORK

We demonstrated the possibility of applying the typical adaptive procedures to finite element simulation of contact problems in finite elasticity. This includes error estimation and subdividing the elements in areas of large error. We extended the mortar approach to h-adaptive meshes. Adaptivity helps to resolve irregular behavior of the solution which is clearly seen in the third example where the plate seems to "cut" the tube with its sharp rigid corner.

The possibility of using adaptivity in contact problems is very promising as many solutions of problems of this kind are characterized by large gradients and concentration of stresses. Accurate recovery of these characteristics is an essential goal of the numerical simulation.

What is clearly missing in the present approach is the (at least) quadratic approximation of displacements. This involves developing of the version of the mortar contact algorithm with quadratic elements. The present algoritm is almost ready to include friction in modeling. This is a matter of developing a time integration scheme and adding the tangential friction forces to the formulation.

ACKNOWLEDGEMENTS

W. Rachowicz and W. Cecot acknowledge the financial support via grant No. UMO-2011/01/B/ST6/07306 received from the Polish National Center of Science.

REFERENCES

Bank, R. & A. Weiser (1985). Some a posteriori error estimates for elliptic partial differential equations. *Math. Comp. 44*, 283–301.

Demkowicz, L., J. Kurtz, D. Pardo, M. Paszynski, W. Rachowicz, & A. Zdunek (2008). *Computing with hp-adaptive finite elements, vol 2, Frontiers: Three Dimensional Elliptic and Maxwell Problems with Applications*. Boca Raton, London, New York: Chapman and Hall/CRC.

Holzapfel, G. (2000). *Nonlinear solid mechanics*. Chichester, Weinheim, New York: Wiley.

Laursen, T. (2002). *Computational Contact and Impact Mechanics*. Berlin, Heildelberg: Springer.

Popp, A., M. Gitterle, M. Gee, & W. Wall (2010). A dual mortar approach for 3d finite deformation contact with consistent linearization. *Int. J. Numer. Meth. Engng*. 83, 1428–1465.

Puso, M. & T. Laursen (2004). A mortar segment-to-segment contact method for large deformation solid mechanics. *Comp. Meth. Appl. Mech. Engng*. 193, 601–629.

Rüter, M. & E. Stein (2000). Analysis, finite element computation and error estimation in transversely isotropic nearly incompressible finite elasticity. *Comput. Meth. Appl. Mech. Engng*. 190, 519–541.

Wriggers, P. (2002). *Computational Contact Mechanics*. New York: Wiley.

Advances in Mechanics: Theoretical, Computational and Interdisciplinary Issues – Kleiber et al. (Eds)
© 2016 Taylor & Francis Group, London, ISBN 978-1-138-02906-4

Vibrations of a double-beam complex system subjected to a moving force

J. Rusin
University of Zielona Góra, Zielona Góra, Poland

ABSTRACT: In the paper dynamic response of a complex double-beam system traversed by moving load is considered. The conflation is represented by a set of linear springs in Winkler model. The classical solution of the response of complex systems subjected to a force moving with a constant velocity has a form of an infinite series. The main goal of the paper is to show that in the considered case part of the solution can be presented in a closed, analytical form instead of an infinite series. The presented method to search for a solution in a closed, analytical form is based on the observation that the solution of the system of partial differential equations in the form of an infinite series is also a solution of an appropriate system of ordinary differential equations. The dynamic influence lines of complex systems may be used for the analysis the complex models of moving load.

1 INTRODUCTION

Composite beams are very important in aeronautical, civil and mechanical engineering as structural members with high strength to weight ratios. Dynamics response of complex beam systems has been studied by many authors in the recent decades. The problem of a dynamic response of a structure subjected to moving loads is interesting and important. Different models of moving loads have been assumed by Fryba (1999). This problem occurs in dynamics of bridges, aircrafts, missiles as well as railways and motorways. The problem of vibration of the complex beams and strings has been considered in the reference list, for instance, Abu-Hilal (2006), Oniszczuk (2000, 2003) and Rusin et al. (2006).

It would be interesting to study the problem of dynamic response of a composite beam to moving loads. The paper includes the study of a dynamic response of a finite, simply supported double-beam complex system subject to a concentrated force moving with a constant velocity. The beams are prismatic, parallel one upon the other and continuously coupled by a linear Winkler elastic element. The classical solution has a form of an infinite series. The main goal of the paper is to show that the aperiodic part of the solution can be presented in a closed form instead of an infinite series. Using the method, of superposed deflections Kączkowski (1963) has shown for a simply supported Euler-Bernoulli beam that, in the case of undamped vibration, the aperiodic part of the solution can be presented in a closed-form. Next, a closed-form solutions were obtained for a beam with arbitrary boundary conditions by Reipert (1969) and for a frame by Reipert (1970). Based on this method, the study was conducted by Misiurek et al. (2013), Rusin et al. (2011) and Śniady (2008). In these works they were considered a sandwich beam, a complex system of strings and a Timoshenko beam sequentially. The presented method to finding a solution in a closed-form is based on the observation that the solution of the system of partial differential equations in the form of an infinite series is also a solution of an appropriate system of ordinary differential equations. The solution for the dynamic response of the composite beam under moving force is important because it can be used also in order to find the solution for other types of moving loads. The double beam connected in parallel by a linear elastic elements, can be studied as a theoretical model of composite beam in which bending and coupling effects are taken into account.

2 MATHEMATICAL MODEL AND GOVERNING EQUATION

Let us consider the vibration of complex system consist of couple Euler-Bernoulli beams interface by a linear springs in Winkler model under axial compression N_1 and N_2 excited by a force $p(x,t)$ moving with a constant velocity v as on Figure 1. The solution of the system in the classical forms is investigated it is possible to find the closed-forms of the deflection function and internal forces.

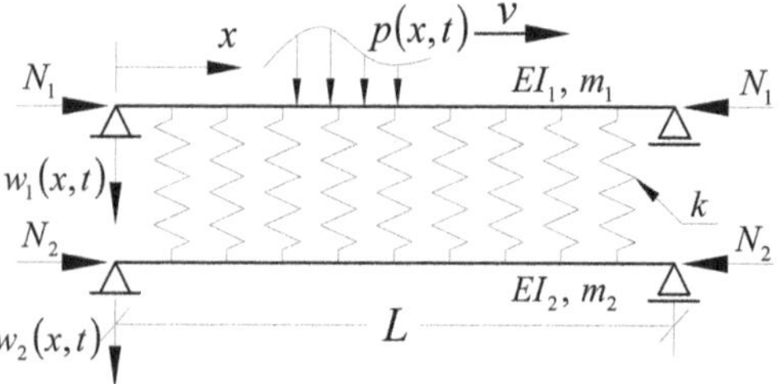

Figure 1. Double-beam system under a moving force.

The transverse vibration of double-beam system is governed by two conjugate partial differential equations:

$$EI_1 w_1^{IV}(x,t) + N_1 w_1^{II}(x,t) + m_1 \ddot{w}_1^{II}(x,t) + k\left[w_1(x,t) - w_2(x,t)\right] = p(x,t), \tag{1}$$

$$EI_2 w_2^{IV}(x,t) + N_2 w_2^{II}(x,t) + m_2 \ddot{w}_2^{II}(x,t) + k\left[w_2(x,t) - w_1(x,t)\right] = 0, \tag{2}$$

where EI_i = flexural rigidity of the i-th beam; E = Young's modulus of elasticity; I = moment of inertia of the cross-section area; m_i = mass spread over the length of the i-th beam; k = stiffness modulus of a Winkler elastic element; and $i = \{1, 2\}$ denote the first and the second beam. The load function in expression (1) and from Figure 1 has the form:

$$p(x,t) = P\,\delta(x - vt), \tag{3}$$

where P = constant load value; $\delta(.)$ = Dirac delta. After introducing the dimensionless variables:

$$\xi = x/L, \quad T = vt/L, \quad \xi \in [0,1], \quad T \in [0,1], \tag{4}$$

the differential equations of motion of the beams system have the form:

$$w_1^{IV}(\xi,T) + N_{01} w_1^{II}(\xi,T) + \sigma_1^2 \ddot{w}_1(\xi,T) + k_1\left[w_1(\xi,T) - w_2(\xi,T)\right] = P_o\,\delta(\xi - T), \tag{5}$$

$$w_2^{IV}(\xi,T) + N_{02} w_2^{II}(\xi,T) + \sigma_2^2 \ddot{w}_2(\xi,T) + k_2\left[w_2(\xi,T) - w_1(\xi,T)\right] = 0. \tag{6}$$

The parameters from equations (5) and (6) have the following designations:

$$N_{0i} = N_i L^2/EI_i, \quad \sigma_i^2 = m_i v^2 L^2/EI_i = v^2/v_{si}^2, \quad k_i = k L^4/EI_i, \quad P_o = P L^3/EI_1, \tag{7}$$

where v_{si} = shear wave velocity propagated in the first and second beam respectively. Therefore boundary conditions expressions (5) and (6) have the form:

$$w_i(0,T) = w_i(1,T) = 0, \quad w_i^{II}(0,T) = w_i^{II}(1,T) = 0. \tag{8}$$

Let the initial conditions have the form:

$$w_i(\xi,0) = 0, \quad \dot{w}_i(\xi,0) = 0. \tag{9}$$

The solutions (5) and (6) for boundary conditions (8) are assumed to be in the form of sine series:

$$w_i(\xi,T) = \sum_{n=1}^{\infty} y_{in} \sin n\pi\xi. \tag{10}$$

By substituting expression (10) into (5) and (6) and using the orthogonalization method one obtains set of uncoupled ordinary differential equations. Eventually, the solution of the system (5) and (6) are sums of the particular integrals $w_i^A(\xi, T)$ and general integrals $w_i^S(\xi, T)$ take the form:

$$w_i(\xi,T) = w_i^A(\xi,T) + w_i^S(\xi,T). \tag{11}$$

3 CLASSICAL-FORM SOLUTIONS

The classical solution has a form of an infinite series, so classical part of the particular solution presents itself as:

$$w_1^A(\xi,T) = \sum_{n=1}^{\infty} \frac{2P_o\left[k_2 - b(n\pi)^2 + (n\pi)^4\right] \sin n\pi T \sin n\pi\xi}{-a_1(n\pi)^2 + a_2(n\pi)^4 - a_3(n\pi)^6 + (n\pi)^8}, \tag{12}$$

$$w_2^A(\xi,T) = \sum_{n=1}^{\infty} \frac{2P_o k_2 \sin n\pi T \sin n\pi\xi}{-a_1(n\pi)^2 + a_2(n\pi)^4 - a_3(n\pi)^6 + (n\pi)^8}, \tag{13}$$

where:

$$a_1 = k_1\left(N_{02} + \sigma_2^2\right) + k_2\left(N_{01} + \sigma_1^2\right), \quad a_3 = N_{01} + N_{02} + \sigma_1^2 + \sigma_2^2, \quad b = N_{02} + \sigma_2^2, \quad a_2 = k_1 + k_2 + \left(N_{01} + \sigma_1^2\right)\left(N_{02} + \sigma_2^2\right). \tag{14}$$

The general solution depends on the roots of the uncoupled ordinary differential equations. So, it determines the solution as trigonometric functions or hyperbolic functions. For example, a positive roots d_{3n}^2 and d_{4n}^2 puts series:

$$w_1^S(T,\xi)=\sum_{n=1}^{\infty}\frac{n\pi v\sin n\pi\xi}{d_{3n}d_{4n}\left(-d_{3n}^2+d_{4n}^2\right)L}$$
$$\cdot\left[d_{4n}\left(C_{1n}\left(A_{2n}^2+d_{3n}^2\right)+B_1C_{2n}\right)\sinh d_{3n}T+\right.$$
$$\left.-d_{3n}\left(C_{1n}\left(A_{2n}+d_{4n}^2\right)+B_1C_{2n}\right)\sinh d_{4n}T\right],\tag{15}$$

$$w_2^S(T,\xi)=\sum_{n=1}^{\infty}\frac{n\pi v\sin n\pi\xi}{d_{3n}d_{4n}\left(-d_{3n}^2+d_{4n}^2\right)L}$$
$$\cdot\left[d_{4n}\left(C_{2n}\left(A_{1n}+d_{3n}^2\right)+B_2C_{1n}\right)\sinh d_{3n}T+\right.$$
$$\left.-d_{3n}\left(C_{2n}\left(A_{1n}+d_{4n}^2\right)+B_2C_{1n}\right)\sinh d_{4n}T\right],\tag{16}$$

where:

$$A_{1n}=\left[(n\pi)^4-N_{01}(n\pi)^2+k_1\right]/\sigma_1^2,\quad B_1=k_1/\sigma_1^2,$$
$$A_{2n}=\left[(n\pi)^4-N_{02}(n\pi)^2+k_2\right]/\sigma_2^2,\quad B_2=k_2/\sigma_2^2,\tag{17}$$

$$C_{1n}=\frac{2P_o\left[k_2-b(n\pi)^2+(n\pi)^4\right]}{-a_1(n\pi)^2+a_2(n\pi)^4-a_3(n\pi)^6+(n\pi)^8},$$
$$C_{2n}=\frac{2P_ok_2}{-a_1(n\pi)^2+a_2(n\pi)^4-a_3(n\pi)^6+(n\pi)^8},\tag{18}$$

and roots:

$$d_{3n}^2=\frac{1}{2}\left[-A_{1n}-A_{2n}-\sqrt{\left(A_{1n}-A_{2n}\right)^2+4B_1B_2}\right],$$
$$d_{4n}^2=\frac{1}{2}\left[-A_{1n}-A_{2n}+\sqrt{\left(A_{1n}-A_{2n}\right)^2+4B_1B_2}\right].\tag{19}$$

4 CLOSED-FORM SOLUTIONS

Below we present a method for finding the solution also in a closed analytic form. Let the function $f(\xi, T)$ for $\xi\in[0, 1]$, $T\in[0, 1]$ be given by the series:

$$f(\xi,T)=\sum_{n=1}^{\infty}\frac{b_1(n\pi)^{2r}\sin n\pi T\sin n\pi\xi+\ldots}{a_o(n\pi)^{2k}+a_1(n\pi)^{2(k-1)}+\ldots}$$
$$\frac{\ldots+b_2(n\pi)^{2s-1}\cos n\pi T\sin n\pi\xi}{\ldots+a_{k-1}(n\pi)^2+a_k},\tag{20}$$

where k, r, $s\in N_0$, quantities a_i $(i = 0, 2, \ldots, k)$ are real numbers such that $a_o\neq 0$. To function (20) one can associate the following partial differential equation:

$$a_o(-1)^k\frac{d^{2k}f(\xi,T)}{d\xi^{2k}}+a_1(-1)^{k-1}\frac{d^{2(k-1)}f(\xi,T)}{d\xi^{2(k-1)}}+\ldots$$
$$\ldots+a_{k-1}(-1)\frac{d^2f(\xi,T)}{d\xi^2}+a_kf(\xi,T)=$$
$$=b_1(-1)^r\frac{d^{2r}\delta(\xi-T)}{d\xi^{2r}}+b_1(-1)^{s-1}\frac{d^{2s-1}\delta(\xi-T)}{d\xi^{2s-1}}.\tag{21}$$

Expression (21) has a solution with the boundary conditions:

$$[f(0,T)]^{(2,j)}=[f(1,T)]^{(2,j)}=0\tag{22}$$

where $[f(\xi, T)]^{(2j)} = d^{2j}f(\xi, T)/d\xi^{2j}$ and $j = 0, 1,\ldots, k-1$. On the background of algorithm (20–22) it is easy to notice that functions (12) and (13) are solution not only to the system of partial differential system expressions (5) and (6) but also to the system of ordinary system equations:

$$w_{A1}^{VIII}(\xi,T)+a_3w_{A1}^{VI}(\xi,T)+a_2w_{A1}^{IV}(\xi,T)+a_1w_{A1}^{II}(\xi,T)$$
$$=P_o\left[k_2\delta(\xi-T)+b\delta^{II}(\xi-T)+\delta^{IV}(\xi-T)\right]\tag{23}$$

$$w_{A2}^{VIII}(\xi,T)+a_3w_{A2}^{VI}(\xi,T)+a_2w_{A2}^{IV}(\xi,T)$$
$$+a_1w_{A2}^{II}(\xi,T)=P_ok_2\delta(\xi-T)\tag{24}$$

which conforms with the following boundary conditions:

$$w_{Ai}^{2j}(0,T)=w_{Ai}^{2j}(1,T)=0,\tag{25}$$

where order derivative $2j = \{0, II, IV, VI, VIII\}$. This can be verified by solving (23) and (24) using finite Fourier sine transform. After solving (23) and (24) by, for example, Laplace transform and taking into account the boundary conditions (25) we get the function $w_1^A(\xi, T)$ in a closed-form with a very extensive form of symbolic.

5 NUMERICAL RESULTS

Figure 2 presents deflections $w_i(\xi,T)$ of double-beam complex system under the concentrated moving force. Figure 3 shows the bending moment. The following dimensionless values of the parameters are used in the numerical calculations: $N_1 = 2$,

$N_2 = 3$, $\sigma_1 = 0.1$, $\sigma_2 = 0.2$, $k_1 = 20$, $k_2 = 10$, $P_o = 20$ and $T = \{0.25, 0.50, 0.75\}$. The results for different location of the moving point force are presented in graphical form in Figures 2 and 3. The continuous line represents the functions of the loaded beam. The dashed line shows the functions of the second beam for which the load is transferred with the coupling.

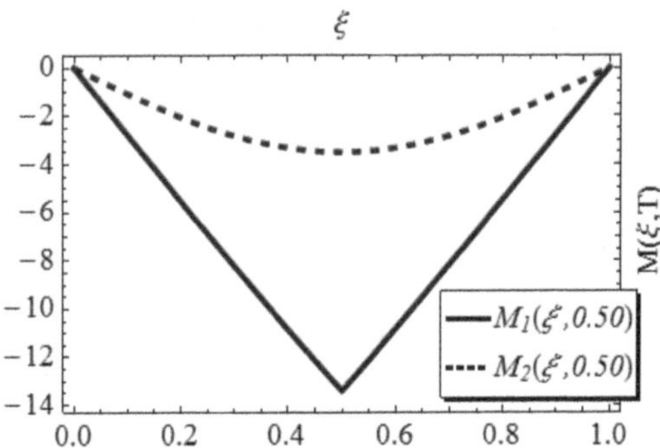

Figure 3. Bending moment at the location of the moving force with regard to time parameter $T = 0.50$.

6 CONCLUSION

The dynamics response of an elastically connected double-beam complex system loaded by a concentrated moving force, moving with a constant velocity has been studied. The motion of the system is described by a non-homogeneous conjugate set of two partial differential equations and aperiodic vibrations can be described by ordinary system equations. The classical solution for transverse displacement function has a form of a sum of two infinite series. It has been shown that aperiodic vibrations of the beam of the series can be presented in a closed, analytical form. The closed solutions take different forms depending if the velocity v of a moving force is smaller or bigger than the shear wave velocity v_{is} of the beams. This follows from the fact that in string wave phenomena may occur. The closed solutions improve the preciseness of the classical sine series expansion of the complex system of beams response by considering the aperiodic part as the solution not only to partial of the differential equation but also to appropriate ordinary differential equation. So the closed-form allows analyzing the vibration phenomena due to moving loads without performing numerical calculations. Having determined the closed-form allows us to assess a quantity of necessary function approximation to determine the classical solution. It is possible to conduct an analysis the impact of more complex load models.

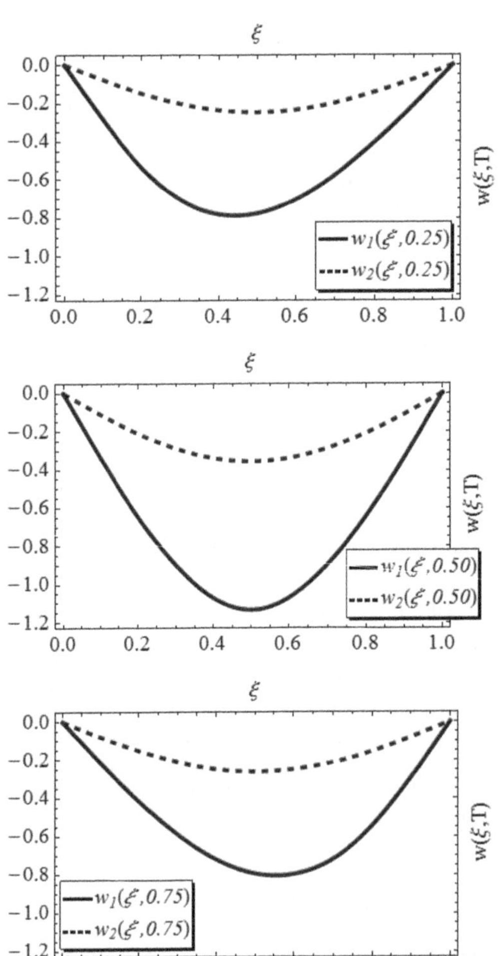

Figure 2. Displacements at the location of the moving force with regard to time parameter $T = \{0.25, 0.50, 0.75\}$.

REFERENCES

Abu-Hilal, M. 2006. Dynamic response of a double Euler-Bernoulli beam due to a moving constant load. *Journal of Sound and Vibration* 297: 477–491.

Fryba, L. 1999. *Vibration of Solids and Structures under Moving Load.* London: Telford.

Hodges, D.H. & Pierce, G.A. 2002. *Introduction to Structural Dynamics and Aeroelasticity.* Cambridge: Cambridge University Press.

Kączkowski, Z. 1963. Vibration of a beam under a moving load. *Proceedings of Vibration Problems* 4 (4): 357–373.

Kukla, S. 1994. Free vibration of the system of two beams connected by many translational springs. *Journal of Sound and Vibration* 172: 130–135.

Kumar, A.V. & Ganguli, R. 2012. Analogy between rotating Euler-Bernoulli and Timoshenko beams and stiff strings. *Computer Modeling in Engineering & Science* 88:443–474.

Misiurek, K. & Śniady, P. 2013. Vibrations of sandwich beam due to a moving force. *Composite Structures* 104: 85–93.

Oniszczuk, Z. 2000. Free transverse vibrations of an elastically connected simply supported double-beam complex system. *Journal of Sound and Vibration* 232 (2): 387–403.

Oniszczuk, Z. 2003. Forced transverse vibrations of an elastically connected complex simply supported double-beam system. *Journal of Sound and Vibration* 264:273–286.

Reipert, Z. 1969. Vibration of a beam arbitrarily supported on its edges under moving load. *Proceedings of Vibration Problems* 2 (10): 249–260.

Reipert, Z. 1970. Vibration of frames under moving load. *Archiwum Inżynierii Lądowej* 16 (3): 419–447.

Rusin, J. & Śniady, P. 2006. Vibration of a Complex Strings System under a Moving Force. *Proceedings in Applied Mathematics and Mechanics* 6 (1): 839–840.

Rusin, J., Śniady, P. & Śniady, P. 2011. Vibrations of double-string complex system under moving forces. Closed solutions. *Journal of Sound and Vibration* 330: 404–415.

Śniady, P. 2008. Dynamic response of a Timoshenko beam to a moving force. *Journal of Applied Mechanics*, Transactions of the ASME, 75, March, 024503-1-024503-4.

Vu, H.V., Ordonez, A.M. & Karnopp, B.H. 2000. Vibration of a double-beam system. *Journal of Sound and Vibration* 229 (4): 807–822.

Zhang, Y.Q., Lu, Y., Wang, S.L. & Liu, X. 2008. Vibration and buckling of a double-beam system under compressive axial loading. *Jour. of Sound and Vibration* 318: 341–352.

Advances in Mechanics: Theoretical, Computational and Interdisciplinary Issues – Kleiber et al. (Eds)
© 2016 Taylor & Francis Group, London, ISBN 978-1-138-02906-4

Dynamic behavior of Zymne Monastery Cathedral on soil base with consideration of non-linear deformation of materials

V.O. Sakharov
Kyiv National University of Construction and Architecture, Kyiv, Ukraine

ABSTRACT: The study contains results of the numerical analysis of the seismic impact on massive constructions using Zymne Monastery Cathedral building as an example. The cathedral has been founded in X-XI century AD using mostly brickwork. Due to the numerous destructions, reconstructions, change of ownership through history between Poland, Soviet Union and Ukraine, the cathedral has a complex configuration and is now in the immediate need of strengthening its foundation. Simulation has been performed using FEM with modified explicit central difference method in the nonlinear formulation. The calculations involved a soil model that takes into account visco-elastic-plastic deformation properties, structural strength and water pressure in soil pores. The study also contains an analysis of the influence of nonlinear orthotropic properties of the brick masonry on the stress-strain state of the building and the estimation of the dynamic behaviour of the cathedral structures. It demonstrates the distribution of stress concentration zones and locations of the localized structural damage. Finally, it contains an analysis of the accumulated plastic deformations in the soil base that leads to an uneven foundation settlement and the estimations how water pressure in the soil pores affected the above-mentioned processes.

1 INTRODUCTION

Due to the increase of the seismic activity in Ukraine, architecturally- and historically—valuable buildings need to be inspected for the seismic resistance. Existing buildings have different design and subjected to the different soil conditions. During this process, it is important to consider not only the specific soil conditions and material parameters, but also the real nonlinear properties of the environment. Today this can be realized only by utilizing numerical modeling using advanced algorithms and technologies.

2 INTERACTION OF CATHEDRAL BUILDING WITH SOIL BASE

2.1 *Problem definition*

The goal is to ensure reliable operation of buildings and structures under dynamic load, including seismic activity. Due to the recent general increase of the seismic activities in the Ukraine, it became necessary to review seismic resistance of existing buildings and structures. About 12% of the Ukrainian territories are currently classified as seismically dangerous (Kendzera 2011). A numerical simulation has been performed in order to evaluate Stress-Strain State (SSS) of the massive, predominantly brick-based cathedral building under the seismic load. Numerous studies have shown large influence of the irreversible deformation of the soil mass and structures on the dynamic behaviour of the structures (Boyko 2013).

2.2 *Structural features of the cathedral*

This study presents results of the analysis of the design of the massive building based on the monument of architecture that belongs to the XI–XII century AD—Uspenskij Cathedral of the Zymne Monastery in Volyn, Ukraine. Founded by the Vladimir The Great, it experienced numerous destructions, reconstructions and architectural changes over its more than 1000-year long history, all of which affected its present state. The main load-bearing structures of the cathedral, including its foundation are made using brick masonry, that was partly was laid before 1495 AD. The Cathedral is located on the slope with the retaining wall built in the upper part of the slope.

The geological structure consists of layers of loam and sandy loam, underlain by Cretaceous sediments. On the surface, natural soil is covered with filled ground with depths from 0.5 to 5.0 m. Seismicity of the territory is considered to be 6 points by MSK-64 scale.

2.3 *Finite element model*

The simulation was performed using actual geographic relief, enclosing structures and the

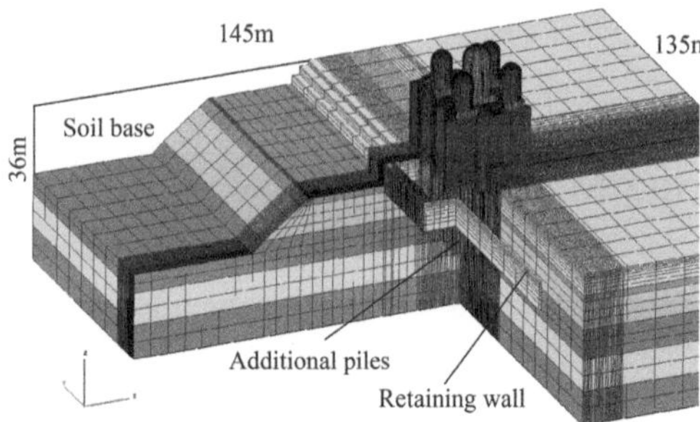

Figure 1. Finite element model of the Cathedral as a "soil base—foundation—building" system.

strengthening elements of the foundation. The study was conducted using developed Finite Element Model (FEM), which included a soil array, cathedral building and the retaining wall as a "soil base—foundation—building" system (see Fig. 1).

Soil base, the Cathedral itself and other constructional elements were modelled using volumetric finite elements. Finite elements used by the model were utilizing FEM moment scheme (Altenbach 1982 & Bazhenov 2002), which takes into account rigid offsets when computing deformations and provides better convergence (confirmed by multiple studies). This has greatly reduced the number of required finite elements while preserving sufficient accuracy. Dimensions of the modeled soil mass, the density of the FE grid and the boundary conditions were selected based on the preliminary calculations and the impact assessment of these parameters on the strength-stress state bearing structures of the Cathedral. The minimum size of the soil mass was selected to ensure the at least 10x vibration absorption for the given geological conditions. Final model is described by the system of algebraic equations with 940,530 variables. Modeling of the seismic action was carried out on the basis of an Automated System for Scientific Research (ASSR) «VESNA-DYN» in spatial statement by the modified method with the explicit nonlinear deformation of materials (Sakharov 2014a). Integration step has been set at 0.025 ms with analysis taking 43 h to complete.

2.4 *Soil model*

The model used to describe soil behaviour takes into account the structural strength, visco-elastic-plastic volume and shape deformation processes (Boyko 2013 & Sakharov 2014b) and processes associated with the pore pressure. This allows for the simulation of real processes of interaction between the soil base and the foundation under the seismic loads. Structural and rheological model of soil is shown in Figure 2.

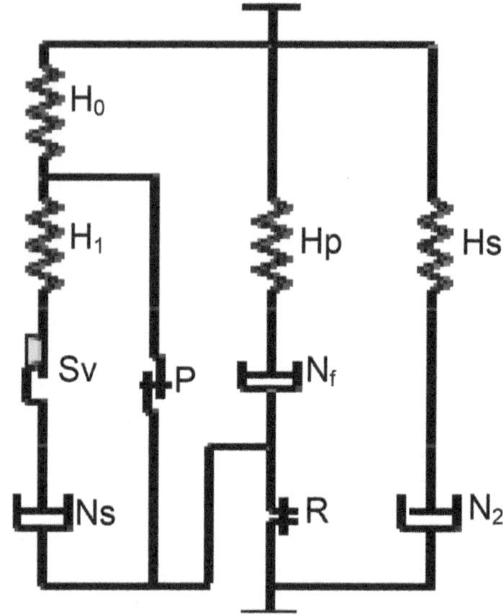

Figure 2. Structural-rheological model of visco-elastic-plastic deformation of the soil under dynamic load.

The model is based on the ideas of the Kelvin viscoelastic model and has a three branches. First branch includes elements H_0 and H_1, representing an elastic modulus, the element Sv of the plastic strength. In generic case the model should also contain viscous element Ns to account for the creep process. However such processes require significant time and are not manifested during seismic load modeling. Brittle element P models elastic behavior of the initially rigid soil until its structural strength is exceeded. Second branch simulates a high-speed of viscoelastic properties of the soil. Finally, third branch is responsible for the simulation of the pore pressure and water filtration process. In addition there is a unilateral constraint R simulating features of ground tension. Also, the model takes into account the viscous energy dissipation from the interaction with the environment.

Relying on the time independence of the elastic-plastic deformations, it is possible to apply standard methods of determining the physical and mechanical properties of the soil used by the static research. In this model, the volume change is described in accordance with the theory by Terzaghi and shape change—by using modified law Mises—Schleicher—Botkin and accounting for dilatancy.

To describe the viscoelastic properties of the soil (branch 2), special method for determination of dynamic stiffness and toughness over a wide load range has been developed and implemented (Sakharov V.O., 2014c). The technique is based on the analysis of soil the deformation at different deformation rates while performing odometric (and, in particular, stabilometric) testing of samples (1):

$$\eta = \Delta_1/\Delta, \quad \mu_V = \Delta_2/\Delta, \quad E_V = \mu_V/\eta \tag{1}$$

where

$$\Delta = a_{11}a_{22} - a_{12}a_{21};\ \Delta_1 = q_1a_{22} - q_2a_{12};$$
$$\Delta_2 = q_2a_{11} - q_1a_{21}$$
$$a_{11} = E_{H1}\dot{\varepsilon}_1 + E'_{H1}\varepsilon_1\dot{\varepsilon}_1 - \dot{\sigma}_1;\ a_{11} = \dot{\varepsilon}_1;$$
$$q_1 = \sigma_1 - E_{H1}\varepsilon_1$$
$$a_{21} = E_{H2}\dot{\varepsilon}_2 + E'_{H2}\varepsilon_2\dot{\varepsilon}_2 - \dot{\sigma}_2;\ a_{22} = \dot{\varepsilon}_2;$$
$$q_2 = \sigma_2 - E_{H2}\varepsilon_2$$

where first index: 1—slow deformation (static test); 2—fast (dynamic) test; EH = elastic modulus; σ = stress, ε = deformation, q—value of current load.

Influence of the pore water filtration on the mechanical properties of the soil was described using Darcy-Gersenvanov's law (Sakharov 2013). In accordance with the parameters of the Maxwell body (see Fig. 2) there are two components: an elastic θ_e, which is proportional to the speed of the hydrostatic stress (Hooke element Hs) and θ_f, that has properties corresponding to the Newton element (N2), but models a more complex movement of a viscous fluid in a porous medium while taking into account the filtration coefficient k_f (Eq. 2).

$$\dot{\Theta} = \dot{\Theta}_e + \dot{\Theta}_f;\ \dot{\Theta}_e = \dot{\sigma}_w/\tilde{K}_w$$
$$\begin{cases} \dot{\Theta}_f = -n\vec{\nabla}(k_f\vec{I}) \\ \dot{\Theta}_f = -\dfrac{k'_f}{\gamma_w}\left[\nabla^2\sigma_w - \dfrac{\rho_w}{\rho}\vec{\nabla}\cdot(\vec{\nabla}\cdot\hat{\sigma})\right] \\ when\ [I] > I'_0 \end{cases} \tag{2}$$

where ρ, ρ_w = density of soil and water, I = hydraulic gradient.

If either vector I or filtration coefficient are equal to zero, it means an absence of the filtration—thus branch 3 on Figure 2 will contain only elastic Hooke element. This case happens under short-lived dynamic (e.g. seismic) load when filtration does not have time to occur.

This model permits us to adequately describe the deformation processes of sandy and clay soils under dynamic and static loads.

2.5 *Brick masonry model*

Modelling of the deformation of brick masonry walls and piers of the Cathedral was based on piecewise linear stress-strain dependency diagram (Fig. 3). The model is based on the research by Ivanov (2011), Onishchik (1957) and G. Kashevarova & Zobachev (2010). Orthotropic properties appear during the deformation and accumulation of the local damage along the respective directions. Orthotropic axes are oriented in accordance with the masonry joints. At the initial stage the masonry is considered to be an isotropic material.

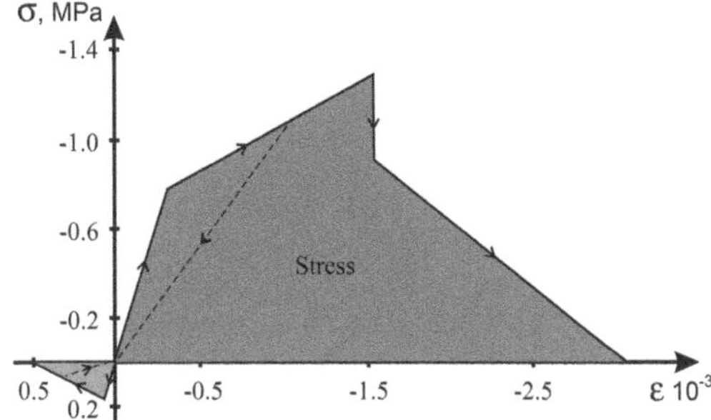

Figure 3. Stress-strain dependency graph for the masonry.

Elastic modulus of masonry (as a homogeneous material) is determined based on the condition of the same pliability of real and modeled masonry while accounting for different rigidity of the brick and mortar:

$$\frac{1}{E_m} = \left(\frac{h_{br}}{E_{br}} + \frac{h_{mr}}{E_{mr}}\right)\frac{1}{h_{br} + h_{mr}} \tag{3}$$

where E_{br}, E_{mr}, E_m—elastic moduli of brick and mortar, and reduced modulus of masonry as a homogeneous material;

h_{br}, h_{mr}—the total thickness of the brick and mortar, corresponding to the masonry fragment in the desired direction.

During compression further analysis of the masonry shear along the seams is performed using the shear stress value on the orthotropic axes. If threshold value is exceeded then stress value is corrected stress in accordance with the Prandtl diagram of the elastic-plastic behavior of the solution.

If tensile strength of the brick masonry is exceeded, the cracks are formed in a plane perpendicular to the direction of the stress while corresponding elastic modulus and Poisson's ratio become equal to zero. The moduli of elasticity in other directions remain unchanged. There is also a possibility that cracks will close in the future—then compressive strength of the masonry is restored and only shear modulus is corrected. During the formation of cracks in the two directions it is assumed that the masonry material is crushed and will not be able to sustain any load.

3 RESULTS

Simulation has shown that due to the seismic load unfortified soil has accumulated plastic deformation over time. This has resulted in an uneven settlement and caused the structure to lean. The maximum amplitude of the upper part of the Cathedral during oscillation has reached 10 cm.

Accounting for the water in the porous layer increased calculated foundation settlement up to 30%. This confirms significant influence of water in pores on the soil deformation process.

In the bottom level of the Cathedral stress reaches 0,45–1,2 MPa, which may damage constructions. At the column bases, some wall sections, as well as at the arch locations, the stress level reached 1.3–1.5 MPa, sufficient to crush masonry material. Such stress distribution can be explained by design of the central columns which sit on separate foundations. During oscillation this can lead to the movement that does not match movement of the surrounding supporting structures. Predicted stress state (MPa) and damage zones closely match existing cracks in the Cathedral. This approach allows to identify potential problem areas and address them during the reconstruction phase.

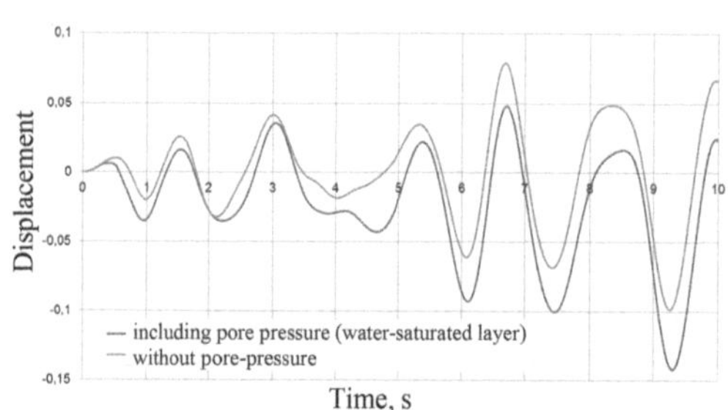

Figure 4. Settlement of foundation with water in the soil pores.

Crushing the bottom part of the column (south nave)

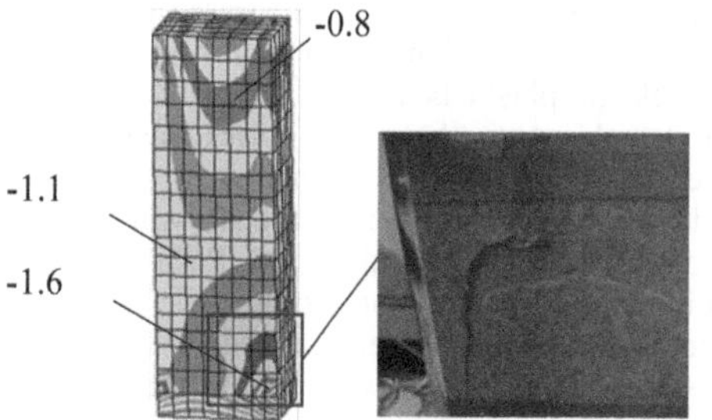

Figure 5. Location of stresses (MPa) and damaged areas.

4 CONCLUSIONS

The simulation has shown that the nonlinear properties have a significant impact on the structural behaviour of the cathedral. It identified localized damage zones caused by the tension and compression, predominantly located under squinches, at the base of pylons and buttresses and at several places near the entrance. Analysis also revealed the formation of the non-uniform sediment, causing tilt of the structure. Accounting for the pore pressure in the water-saturated layer increased calculated foundation settlement values. However, it should be noted that the discovered damaged zones are not sufficient to fundamentally affect work of the load-bearing structures of the cathedral and will not cause an emergency situation.

REFERENCES

Altenbach, J., Sacharov, A.S. 1982. Die Methode der finiten Elemente in der Festkörpermechanik, VEB Fachbuchverlag: Leipzig, 1982.

Bazhenov, V.A., Sakharov, A.S., Tsykhanovskii, V.K. 2002. The Moment Finite-Element Scheme in Problems of Nonlinear Continuum Mechanics. International Applied Mechanics Vol. 38. 2002.

Boyko, I.P., Sakharov, O.S., Sakharov, V.O. 2013. Behaviour of the multi-story building under seismic loads with the account of the viscoplasticity of the soil base. Proceedings International Conference on Soil Mechanics and Geotechnical Engineering, Paris, Vol. 2, 2013.

Kendzera, A.V., Pigulevskij, P.I., Shherbina, S.V. 2011. Prognozirovanie opasnyh sejsmicheskih sobytij kak obobshhennyj metod sejsmicheskogo rajonirovanija (in rus.). Available at: http://www.seismo.kiev.ua/.

Sakharov, V.O. 2013. Model of deformation of water-saturated soil base under seismic and long-term dynamic loads (in ukr). Osnovy ta fundamenty [Base and foundation], Vol. 33, 2013.

Sakharov, V.O. 2014a. Modyfikatsiia iavnoho metodu dlia efektyvnoho rozv'iazannia nelinijnykh zadach heotekhniky (in ukr). [Modification of an explicit method for the efficient solution of nonlinear geotechnic's problems] Osnovy ta fundamenty [Base and foundation], Vol. 35, 2014.

Sakharov, V. 2014b. An investigation of system "soil base-foundation-structure" response to seismic forces with provision for nonlinear properties of materials. Proceedings of the X Conference "Konstrukcje zespolone", Poland, Zielona Góra, 2014.

Sakharov, V. 2014c. An investigation of the dynamic soil parameters in laboratory conditions (in ukr). Visnyk of Prydniprovsk State Academy of Civil Engineering and Architecture, Vol.: Dnipropetrovsk, 2014.

Advances in Mechanics: Theoretical, Computational and Interdisciplinary Issues – Kleiber et al. (Eds)
© 2016 Taylor & Francis Group, London, ISBN 978-1-138-02906-4

Loadings in rail vehicle due to bulk material

H. Sanecki
Institute of Rail Vehicles, Faculty of Mechanical Engineering, Cracow University of Technology, Poland

ABSTRACT: The paper concerns loadings appearing during transport of bulk and aggregate materials such as sand, gravel, crushed coal etc. Nowadays, designers use very general recommendations and then they solve problems of stress analysis in a simplified way. The author derived and solved numerically equations similar to the ones taken from soil mechanics allowing for a better description of bulk material pressure on retaining walls of containers. A Significant achievement was to solve the problem of the bulk material pressure on the container walls of a running vehicle, when vertical and lateral accelerations occur. After application of the method, numerous results were obtained, which were next used for sensitivity analysis of the pressure force for various factors. The topic was undertaken after author's experience with experimental and theoretical testing of freight wagons as well as crash tests made on samples of cabs.

1 INTRODUCTION

The author undertook the subject of loadings arisen during transport of granular and aggregate materials, after experiences with design and testing of wagons (Sanecki 2005). Designers usually apply to very general guidelines (EN 2010) concerning statics and dynamics of a vehicle and then they solve the problem of the pressure distribution on the container walls in some simplified ways. Admittedly, some of the numerical methods were also used to solve the problem but the results were difficult for application. Recently a Smoothed Particle Hydrodynamics (SPH), a meshless method described for example in (Liu & Liu 2010) was adopted for the numerical simulation of pressure caused by the granular material. However, studies such as (Schema 2008) show that the results obtained by the SPH method are not accurate enough and even for static case give too high pressure acting on the walls of the container.

The author of the paper derived and numerically solved equations similar to ones taken from soil mechanics (Das 2002) but also for the running vehicle, when vertical and lateral accelerations occur. The model of pressure distribution, presented here, caused by bulk payload is required for better description of vehicle dynamics and is also necessary in order to optimize the vehicle structure.

2 STATIC LOADINGS CAUSED BY BULK MATERIAL

The paper presents a method for determining the static and dynamic loads originating from the bulk cargo transported in wagons, whose containers have elongated and prismatic shape.

The key problem is to determine the real value of thrust force and a pressure on the walls of the tank.

The averaged pressure acting on the horizontal floor of the tank (vertical component, p_v) can be calculated according to the formula $p_v = \gamma h$, where γ—average specific gravity of the bulk material, h—height of the pile.

According to the theory applicable to bulk materials presented e.g. in (Das 2002) unit pressure p acting on the side wall of the tank should be determined as

$$p(\zeta) = \gamma K \cdot \zeta \tag{1}$$

where K = lateral pressure coefficient (acc. EN 2010, Halicka & Franczak 2011), ζ = vertical coordinate measured down from the top of the pile.

While using formulas adopted from soil mechanics it should be noted that the bulk material pressure on the retaining wall depends on whether it is in active state—if a small movement of the wall outward from the granular material is expected, passive state—if it is possible substantial movement of the wall towards the pile, or the wall is at rest. There is a relationship

$$K_a < K_o < K_p \tag{2}$$

where K_a = active lateral pressure coefficient, K_o = lateral pressure coefficient at rest, K_p = passive lateral pressure coefficient.

In order to determine the coefficient K_a and K_p one can use known Coulomb or Rankine formulas. The latter include only unilateral embankment slope with angle β, but they neglect friction between the wall and the backfill. They also assume that the wall is positioned vertically.

Now, consider the container, which is commonly used to transport bulk material, filled with the pile with two-sided embankment. Since the above mentioned models do not include important parameters of the embankment, the author attempts to determine the pressure coefficient K also for more general situations. Presented approach consists in examining the balance of forces acting on faction OABCO (wedge) of the bulk material, as shown in Figure 1.

The thrust force of the wall loaded by the bulk material is given by formula

$$P = W \cdot \frac{\sin(\theta-\varphi)}{\cos(\theta-\varphi-\omega-\delta)} = \gamma AL \cdot \frac{\sin(\theta-\varphi)}{\cos(\theta-\varphi-\omega-\delta)} \tag{3}$$

where W = weight of the faction, A = area of the faction, L = length of the container box, θ = angle of quasi-rupture lines behind the retaining wall, ϕ = average angle of internal friction of bulk material, ω = angle of inclination of the wall face, δ = an angle of friction between the wall and the bulk material.

The formula (3) can be rewritten as following function

$$P[\hat{\theta}(\zeta),\zeta] = \frac{1}{2}\gamma L \cdot \zeta^2 \cdot K_a[\hat{\theta}(\zeta),\zeta] \tag{4}$$

where

$$K_a[\hat{\theta}(\zeta),\zeta] = \frac{\cos(\omega-\beta_1)}{\cos^2\omega} \cdot k(\hat{\theta}) \cdot \left[1-\lambda(\hat{\theta},\zeta)\chi(\hat{\theta})\right] \tag{5}$$

$$k(\theta) = \frac{1}{2} \cdot \frac{\sin(\theta-\varphi)\cos(\theta-\omega)}{\sin(\theta-\beta_1)\cos(\theta-\varphi-\omega-\delta)},$$

$$\chi(\theta) = \frac{\cos(\theta-\omega)}{\sin(\theta+\beta_2)},$$

$$\lambda(\theta,\zeta) = \left[1-\frac{\zeta_o(\theta)}{\zeta}\right]^2 \cdot \frac{\sin(\beta_1+\beta_2)}{\cos(\omega-\beta_1)},$$

$$\zeta_o(\theta) = b_{o1} \cdot \frac{\mathrm{tg}\theta - \mathrm{tg}\beta_1}{1+\mathrm{tg}\theta\,\mathrm{tg}\omega},$$

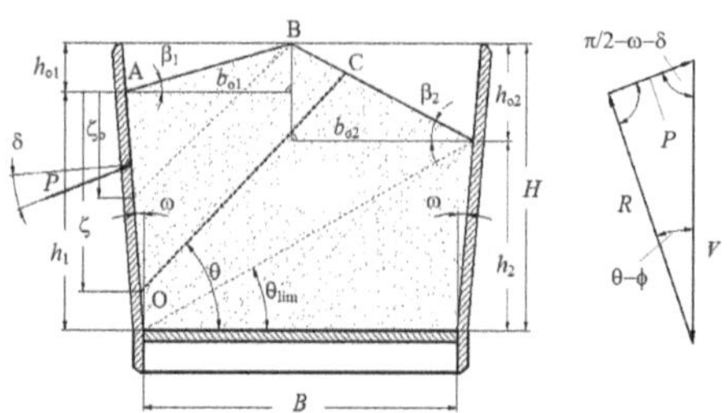

Figure 1. Cross section of container filled with bulk material and forces acting on faction—polygon OABCO.

β_1, β_2 = the embankment slope angles, b_{o1} = lateral position of the top of the embankment relative to one of the wall, $\hat{\theta}$ = an angle of quasi-rupture lines for maximal force P (Fig. 1). The angle $\hat{\theta}$ ought to be determined from a condition $P'(\theta) = 0$ for a fixed value of the variable ζ. This equation is suitable for obtaining a solution for $\hat{\theta} = \hat{\theta}(\zeta)$ with the use of one of the numerical methods. Next, it can be used for determining the pressure function

$$p(\zeta) = \frac{\cos\omega}{L} \cdot \frac{\mathrm{d}P}{\mathrm{d}\zeta} \tag{6}$$

which leads to equation

$$p(\zeta) = \gamma\zeta \cdot \frac{\cos(\omega-\beta_1)}{\cos\omega} \cdot k(\hat{\theta}) \cdot \left[1-\frac{\chi(\hat{\theta})\cdot\lambda(\hat{\theta},\zeta)}{\eta(\zeta)}\right] \tag{7}$$

where (H—Heaviside function)

$$\eta(\zeta) = \left(1-\frac{\zeta_o}{\zeta}\right)H(\zeta,\zeta_o).$$

A series of results was achieved after application of the method described above in order to get the sensitivity analysis on various factors. The distributions of the function $p(z)$ (where $z = h-\zeta$) for the cargo containers filled with the bulk material with different shapes of the embankment were investigated. The slope angles β_1, β_2 as well as b_{o1} – lateral position of the top of the embankment relative to one of the two walls were main variables of the analysis. The side walls were arranged vertically or were inclined relatively to the vertical at different angles ω. It was assumed that the internal friction angle of the bulk material was $\phi = 31°$ (for crushed coal) and the coefficient of friction on the side walls was $\mu = 0.3$. Figure 2 shows an example of

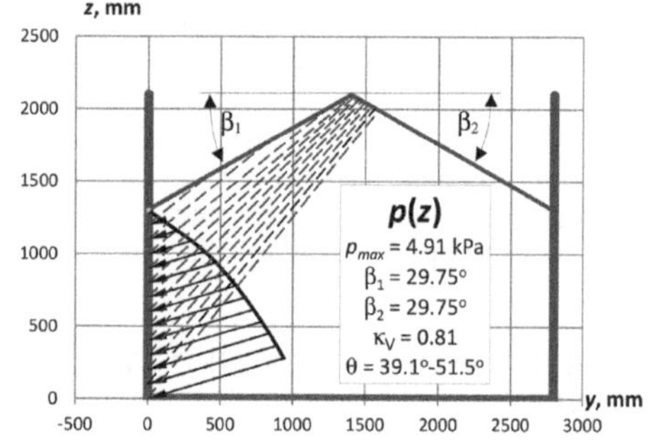

Figure 2. Exemplary distribution of pressure $p(z)$ acting on vertical wall of container filled with bulk material with embankment slope angles $\beta_1 = \beta_2 = 29.75°$.

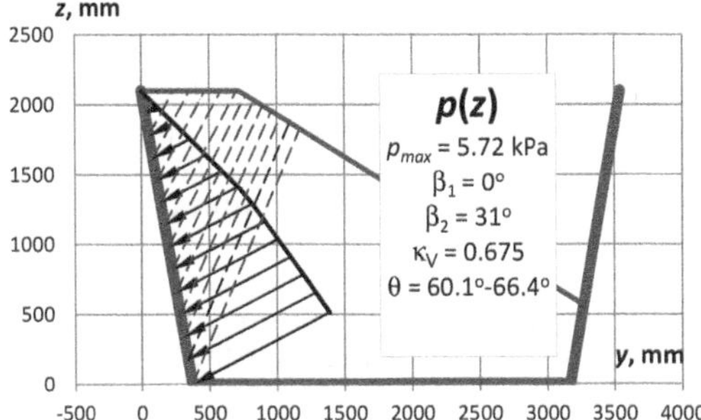

Figure 3. Exemplary distribution of pressure $p(z)$ acting on slanted wall of container filled with bulk material with embankment slope angles $\beta_1 = 0$ and $\beta_2 = 31°$.

the pressure distribution acting on a vertical wall of container filled with bulk material. The quasi-rupture lines inclined at the angle θ are marked with dashed lines. Next Figure shows another distribution of pressure acting on a slanted wall of container which has asymmetric embankment slopes. In order to provide the sensitivity analysis so called cargo volume utilization factor κ_V was introduced

$$\kappa_V = 1 - \frac{(b_{o1} + h_{o1}\mathrm{tg}\omega)h_{o1} + (b_{o2} + h_{o2}\mathrm{tg}\omega)h_{o2}}{2(B + H\mathrm{tg}\omega)H} \tag{8}$$

which allows controlling the size of the bulk cargo.

3 DYNAMIC LOADINGS CAUSED BY BULK MATERIAL

A much more difficult problem is to define the impact of the bulk material on the walls of the tank of the running vehicle, when dynamic loads occur. In such cases usually various kinds of simplifications have to be made. In this section it is proposed to adopt the approach used to analyze the pressure effects occurring in the retaining walls during the earthquakes. Taking into account the vertical and lateral accelerations caused by seismic waves was proposed for example in (Das 2002).

Pressure of the bulk material on the walls of the typical freight vehicle may be considered as the active state because the walls tend to deform outside the container while being under the load. Therefore, the weight of the faction of the bulk material makes the friction force acting down.

Figure 4 presents a frequent case of the symmetrical embankment of the bulk material, which is subjected to a vertical acceleration $a_Z = (1 + \alpha_Z)g$ and a lateral one $a_Y = \alpha_Y g$, where α_Z, α_Y = coefficients acc. EN, 2010. Due to these accelerations, a force of inertia is defined as

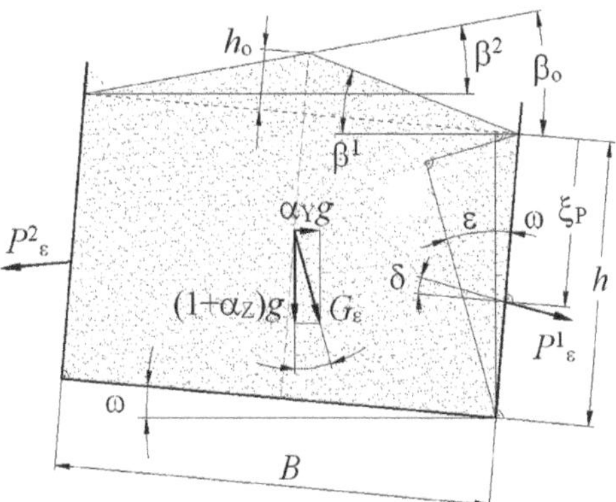

Figure 4. Nonstationary container filled with bulk material loaded with accelerations a_Y and a_Z, ω—rotation angle of container caused by rail cant.

$$G_\varepsilon = \sqrt{\alpha_Y^2 + (1+\alpha_Z)^2}\, g m_3 \tag{9}$$

(m_3 = mass of the bulk material, EN 2010) and acts instead of the weight of the bulk material. The force G_ε is inclined from the vertical direction with an angle

$$\varepsilon = \mathrm{arctg}\frac{\alpha_Y}{1+\alpha_Z} \tag{10}$$

Therefore, the walls of the container are loaded with the forces P^s_ε, $s = 1, 2$ determined by taking into account the changed direction of the vector of the force of inertia G_ε with the angle ε. Moreover, modified specific gravity of the bulk material should be changed from γ into

$$\gamma_\varepsilon = \gamma\sqrt{\alpha_Y^2 + (1+\alpha_Z)^2} \tag{11}$$

so it leads to formula

$$P^s_\varepsilon = \frac{1}{2}\gamma L\sqrt{\alpha_Y^2 + (1+\alpha_Z)^2}\, h^2 \cos^2(\omega+\varepsilon) K^s_{a\varepsilon}, \tag{12}$$

The coefficients $K^s_{a\varepsilon}$ can be calculated with the following analytical formulas

$$K^1_{a\varepsilon} = \frac{\cos^2(\varphi-\omega-\varepsilon)}{\cos^2(\omega+\varepsilon)\cos(\delta+\omega+\varepsilon)} \cdot \frac{1}{\left[1+\sqrt{\frac{\sin(\varphi+\delta)\sin(\varphi-\beta^1-\varepsilon)}{\cos(\delta+\omega+\varepsilon)\cos(\beta^1-\omega)}}\right]^2} \tag{13}$$

$$K_{a\varepsilon}^2 = \frac{\cos^2(\varphi+\omega+\varepsilon)}{\cos^2(\omega+\varepsilon)\cos(\delta-\omega-\varepsilon)} \cdot \frac{1}{\left[1+\sqrt{\frac{\sin(\varphi+\delta)\sin(\varphi-\beta^2+\varepsilon)}{\cos(\delta-\omega-\varepsilon)\cos(\beta^2+\omega)}}\right]^2} \tag{14}$$

or with the method presented in Section 2 (5).

The formulas are valid if the following inequalities are satisfied:

$\varepsilon-\omega \le \phi-\beta_o$ for the 1st wall (13) and

$\omega-\varepsilon \le \phi-\beta_o$ for the 2nd wall (14), Figure 4.

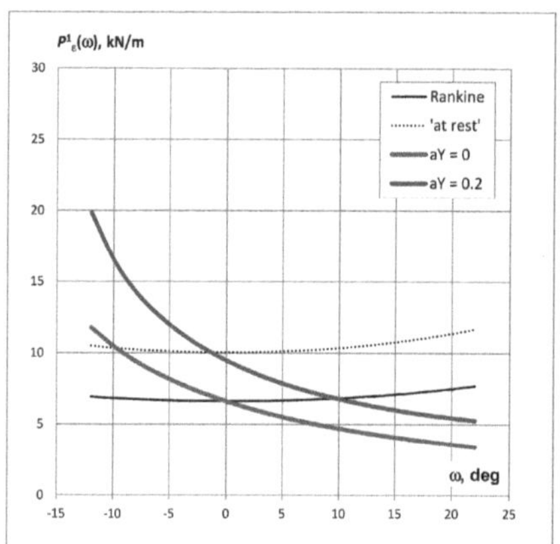

Figure 5. Total pressure force $P^1_\varepsilon(\omega)$ acting on right hand side wall (as in Fig. 4) of container filled with bulk material (for L = 1 m).

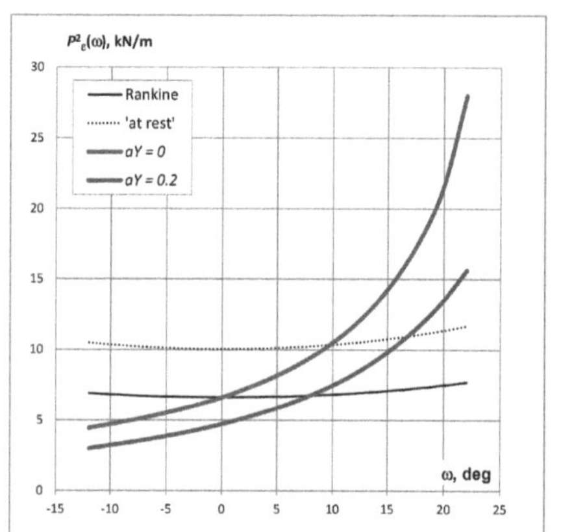

Figure 6. Total pressure forces $P^2_\varepsilon(\omega)$ acting on left hand side wall of container (Fig. 4).

More general case with an asymmetrical embankment of the bulk material can be also solved by this method. However, only the approach presented in Section 2 (5) would be useful for obtaining the coefficients $K^s_{a\varepsilon}$ and then the solutions for the pressure forces (12) as well as the pressure distributions according to (7).

In Figures 5 and 6 the total pressure forces $P^1_\varepsilon(\omega)$ and $P^2_\varepsilon(\omega)$ were plotted as functions of the angle ω. Their values were calculated using the Equations 13 and 14. These results could be compared with the Rankine and 'at rest' solutions also presented in the charts. The figures below represent an analysis concerning the sensitivity of the pressure forces on the angle ω, nevertheless investigations of effect of other parameters (above mentioned, e.g. ϕ, δ, β^s) are also very useful and possible.

4 CONCLUSIONS

Significant achievement of the paper was the solution of the problem of the pressure acting on the container walls of a running vehicle caused by the granular material. Having such information the designer can predict loadings occurring while the vehicle is operating in real conditions. It could be achieved by changing the rotation angle of the container ω caused by the rail cant as well as changing the coefficients of the vertical and lateral accelerations α_Z and α_Y. Such simulation seems to be a very good tool for static as well as for fatigue analysis of some freight wagons.

REFERENCES

Das, B.M. 2002. *Principles of geotechnical engineering*. Chapter 12. *Lateral Earth Pressure: At-Rest, Rankine, and Coulomb, 5th Edition*.

EN 12663-1,2. 2010. *Railway applications. Structural requirements of railway vehicle bodies—Part 1: Locomotives and passenger rolling stock (and alternative method for freight wagons); Part 2: Freight wagons*.

Halicka, A. & Franczak, D. 2011. *Design of reinforced concrete tanks. V. 1, Tanks for bulk materials* (in Polish), Warszawa PWN.

Liu, M.B. & Liu G.R. 2010. Smoothed Particle Hydrodynamics (SPH): an Overview and Recent Developments. *Archives of Computational Methods in Engineering* 17: 25–76.

Sanecki, H. 2005. Experience in Full Scale Tests of Passive Safety Assessment of Rolling Stock, *SAMNET Workshop on Railway Safety Management Systems*, Warsaw, 30.06.2005.

Schena, A. 2008. Manufacturing Simulations: Simulation of the Actions in Silo and Tanks with the Smooth Particles Hydrodynamics Method. *EHTC 2008*, Strasbourg.

Advances in Mechanics: Theoretical, Computational and Interdisciplinary Issues – Kleiber et al. (Eds)

Dynamics of bodies under symmetric and asymmetric orthotropic friction forces

O.A. Silantyeva & N.N. Dmitriev
Department of Mathematics and Mechanics, Saint-Petersburg State University, Saint-Petersburg, Russia

ABSTRACT: Recent studies show that the anisotropic behavior of frictional forces is an important factor in contact problems. This research provides the detailed explanation of dynamic behavior of a mass point and thin plates with circular and elliptical contact areas on the rough surface. Symmetric and asymmetric orthotropic frictional forces were mainly studied. The presented model describes terminal movement of bodies in the horizontal plane. Some situations are analyzed numerically for regular pressure distribution. Results for the asymmetric orthotropic friction for elliptical and circular contact areas are provided. An example illustrates the influence of anisotropic effects on the evolution of dynamical characteristics of the elliptical and circular thin plates. Some remarks dedicated to the dynamic behavior of the mass point are presented.

1 INTRODUCTION

Anisotropy of friction forces appears in many cases. Wear, special surface textures, plasticity, changes in structure of surface layer, all lead to a situation when friction coefficients depend on the direction of sliding. Experimental and theoretical investigations of the phenomenon are done in the papers (Zmitrowicz 2005, Zmitrowicz 1992, Antoni, Ligier, Saffre, & Pastor 2007, Konyukhov, Vielsack, & Schweizerhof 2008) and others. The effect of frictional anisotropy influences significantly the dynamic characteristics of the body motion on the rough surface.

The objective of the study is to provide a complete theory to analyze anisotropic effects in different cases. We point out main features dedicated to orthotropic symmetric and orthotropic asymmetric friction.

We investigated a circular area with orthotropic friction and regular pressure distribution in (Dmitriev 2009), an elliptical area with orthotropic friction, regular and linear pressure distribution in (Dmitriev & Silantyeva 2014), a mass point with asymmetric friction in (Dmitriev 2013) and a ring with asymmetric friction in (Dmitriev 2015).

In the paper we present description, regarding the developed theory for circular and elliptical thin plates with asymmetric orthotropic friction forces. For a mass point some remarks dedicated to asymmetric friction were mentioned in (Dmitriev & Silantyeva 2015).

2 FORMULATION OF THE PROBLEM

Let us consider terminal movement (Waidman & Malhotra 2007) of thin elliptical plate on the horizontal plane with anisotropic friction forces. Anisotropic friction force T on the point M of a moving body according to (Zmitrowicz 1989) can be written in the following form:

$$\mathbf{T} = -p_M \mathcal{F}(M)\frac{\mathbf{v}}{|\mathbf{v}|}, \quad \mathcal{F}(M) = \begin{pmatrix} f_x & f \\ -f & f_y \end{pmatrix}. \tag{1}$$

here p_M is a normal pressure at a point M, $\mathcal{F}(M)$ is a friction matrix written in the coordinate system Oxy (Dmitriev 2002), $\mathbf{v}$ is a velocity vector of the point M.

Friction is *symmetric orthotropic* in case the friction matrix $\mathcal{F}(M)$ is a tensor which components are constant and do not depend on the orientation of contacting areas. This approximation is possible for the case when hardness of one plane is greater than hardness of another one or one of the contact bodies has isotropic frictional properties and $f = 0$. If hardness of each material of the contacting pair is similar we should use a more complicated law for friction force (Dmitriev 2002).

Friction is *asymmetric orthotropic* if in the contact pair friction matrix components differ in negative and positive directions with $f_{x+} \geq f_{x-}, f_{y+} \geq f_{y-}$ and $f = 0$. Thus, in (1) we take:

$$f_x = \begin{cases} f_{x+}, & v_x \geq 0 \\ f_{x-}, & v_x < 0 \end{cases} f_y = \begin{cases} f_{y+}, & v_y \geq 0 \\ f_{y-}, & v_y < 0 \end{cases},$$

where v_x, v_y are projections of velocity vector in Oxy.

2.1 *Equations of motion*

Let us introduce a moving coordinate system $C\xi\eta\zeta$, associated with the plate. Axis $C\zeta$ is perpendicular to the plane of sliding. Stationary coordinate system Oxy is selected thus, that the friction matrix has the form (1) and axes Ox and Oy are in the sliding plane. Let φ be an angle between Ox and $C\xi$, ϑ—an angle between axis Ox and $\mathbf{v}_C$—a velocity of a mass center of the plate.

$$\mathbf{v_C} = v_C(\cos\vartheta\mathbf{i} + \sin\vartheta\mathbf{j}), \qquad (2)$$

where v_C is a velocity value, **i**, **j** are unit vectors of axes Ox, Oy. Vector of an angular velocity is $\omega = \omega\mathbf{k}$, where $\omega = \dot{\varphi}$, **k** is a unit vector of axis Oz.

Euler equation $\mathbf{v_M} = \mathbf{v_C} + \omega \times \mathbf{C_M}$ allows us to write the following statements:

$$\begin{aligned} &v_x = v_C\cos\vartheta - \omega y', \quad v_y = v_C\sin\vartheta + \omega x', \\ &x' = \xi\cos\varphi - \eta\sin\varphi, \quad y' = \xi\sin\varphi + \eta\cos\varphi, \\ &h = \eta\cos(\vartheta-\varphi) - \xi\sin(\vartheta-\varphi), \\ &v_M = \sqrt{v_C^2 + \omega^2(\xi^2+\eta^2) - 2v_C\omega h}. \end{aligned} \qquad (3)$$

With the anisotropic friction law (1) and equations (3) we can write projections of the main friction force vector **T** and the friction moment **M** in the form:

$$\begin{aligned} &T_x = \iint_\Omega \tau_x d\xi d\eta, \quad T_y = \iint_\Omega \tau_y d\xi d\eta, \\ &M_{C\zeta} = \iint_\Omega (\tau_y x' - \tau_x y') d\xi d\eta, \\ &\tau_x = -f_x p(\xi,\eta)\frac{v_x(\xi,\eta)}{v_M(\xi,\eta)}, \quad \tau_y = -f_y p(\xi,\eta)\frac{v_y(\xi,\eta)}{v_M(\xi,\eta)}, \end{aligned} \qquad (4)$$

where Ω is an integration area.

Equations of motion in the Frennet-Serret frame are:

$$\begin{aligned} &m\dot{v}_C = T_\tau = T_x\cos\vartheta + T_y\sin\vartheta, \\ &mv_C\dot{\vartheta} = T_n = -T_x\sin\vartheta + T_y\cos\vartheta, \\ &I\dot{\omega} = M_{C\zeta}, \end{aligned} \qquad (5)$$

where m is a mass of the plate, I is a plate inertia moment around $C\zeta$, T_τ and T_n are projections of a friction force vector **T** on tangential and normal axes, $M_{C\zeta}$ is a friction moment around axis $C\zeta$.

Let us rewrite the system (5) in the dimensionless form using following relations:

$$I = ma^2I^*, \quad \xi = a\xi^*, \quad \eta = a\eta^*, \quad v_C = v_C^*\sqrt{ag},$$

$$\omega = \omega^*\sqrt{\frac{g}{a}}, \quad t = t^*\sqrt{\frac{a}{g}}, \quad \dot{\vartheta} = \frac{d\vartheta}{dt^*}\sqrt{\frac{g}{a}}, \quad p = p^*\frac{mg}{S}$$

and let us introduce a variable $\beta = \frac{v_C}{\omega} = a\beta^*$ and a parameter $\mu = f_y - fx$. In these equations parameter a is measurable, it is the length of the largest line from point C to the area contour, S—contact area.

Equations (5) in dimensionless form (asterisks are omitted later):

$$\begin{aligned} &\frac{dv}{dt} = -\iint_\Omega p(\xi,\eta)I_1 d\xi d\eta, \\ &v_C\frac{d\vartheta}{dt} = -\iint_\Omega p(\xi,\eta)I_2 d\xi d\eta, \\ &\frac{d\omega}{dt} = -\iint_\Omega \frac{p(\xi,\eta)}{I}I_3 d\xi d\eta, \\ &I_1 = \left[\frac{\beta(f_x + \mu\sin^2\vartheta) + f_x s_1 + \mu s_3}{s}\right], \\ &I_2 = \left[\frac{\beta\mu\sin\vartheta\cos\vartheta + f_x s_2 + \mu s_4}{s}\right], \\ &I_3 = \left[\frac{\beta(f_x s_1 + \mu s_3) + f_x(\xi^2+\eta^2) + \mu s_0^2}{s}\right], \\ &s = \sqrt{\beta^2 + \xi^2 + \eta^2 + 2\beta s_1}), \\ &s_0 = \xi\cos\varphi - \eta\sin\varphi, \\ &s_1 = \xi\sin(\vartheta-\varphi) - \eta\cos(\vartheta-\varphi), \\ &s_2 = \xi\cos(\vartheta-\varphi) + \eta\sin(\vartheta-\varphi), \\ &s_3 = \xi\cos\varphi\sin\vartheta - \eta\sin\varphi\sin\vartheta, \\ &s_4 = \xi\cos\varphi\cos\vartheta - \eta\sin\varphi\cos\vartheta. \end{aligned} \qquad (6)$$

System of equations (6) is general. It is possible to numerically evaluate this system directly. However, in most cases it is better to integrate forces in the right part of the system at least onc—it accelerates calculations and simplifies analysis. It is possible to integrate the system for regular pressure distribution using a method described in (Dmitriev & Silantyeva 2014) or with the method developed by A.I. Lurye (Lurye 2002).

3 SELECTED RESULTS

Let us divide the first equation of the system (6) by the third and introduce $\Phi(\beta,\vartheta)$—derived right part:

$$\begin{aligned} &\frac{dv}{d\omega} = \Phi(\beta,\vartheta), \\ &v_C\frac{d\vartheta}{dt} = T_n(\beta,\vartheta). \end{aligned} \qquad (7)$$

A friction force has a negative power, so the motion with non-zero initial conditions terminates.

Thus, the second equation in the system (7) allows us to write a relation:

$$T_n(\beta,\vartheta) \underset{\substack{t\to t_* \\ \vartheta\to\vartheta_* \\ \beta\to\beta_*}}{\longrightarrow} 0 \tag{8}$$

where ϑ_* and β_* are limit values of corresponding parameters, t_* is a terminal moment. Integrating first equation of the system (7) we achieve:

$$\omega = \omega_0 \exp\left[-\int_{\beta_0}^{\beta_1} \frac{d\beta}{\beta - \Phi(\beta,\vartheta)} \right]. \tag{9}$$

It is important to mention, that function $\Phi(\beta, \vartheta)$ depends not only from β and ϑ but also from the form of the contact area, pressure distribution law $p(\xi, \eta)$, components of the friction matrix f_x, f_y and the angle φ (orientation of the body on the surface). Thus, value of β_1, when integral in (9) becomes improper and seeks $-\infty$, depends on parameters of the mechanical system:

$$\beta_1 \underset{\substack{t\to t_* \\ \vartheta\to\vartheta_* \\ \varphi\to\varphi_*}}{\longrightarrow} \beta_*(\vartheta_*, \varphi_*, \Omega, f_x, f_y, p(\xi,\eta)). \tag{10}$$

Summarizing, let us mention, that by the time t_* relation (8) and

$$\beta - \Phi(\beta,\vartheta) \underset{\substack{t\to t_* \\ \vartheta\to\vartheta_* \\ \beta\to\beta_*}}{\longrightarrow} 0 \tag{11}$$

should be achieved.

Furthermore, with fixed values of $\beta = \tilde{\beta}$ equations $T_n(\tilde{\beta},\vartheta) = 0$ and (11) may have several solutions. However, both conditions (8) and (11) are achieved with singular ϑ_*, β_* (Dmitriev 2002), which depend on initial conditions.

Figures 1 and 2 show evolution of parameters β and ϑ during sliding in asymmetric orthotropic conditions. For all cases parameter β for elliptical plate is lower than for circular contact area and ϑ changes more—the form of the sliding body influences the sliding process.

Table 1 shows parameters β_* and ϑ_* and area, where simultaneous velocity center is located at the terminal moment. Results for elliptical plate are generally less accurate than for circular in asymmetric friction case, because of numerical problems, which typically appear at the most final period of the sliding when both linear and angular velocities are close to zero. Furthermore, interrelations between the inertia moment of the plate and coefficients of friction influences the solution of

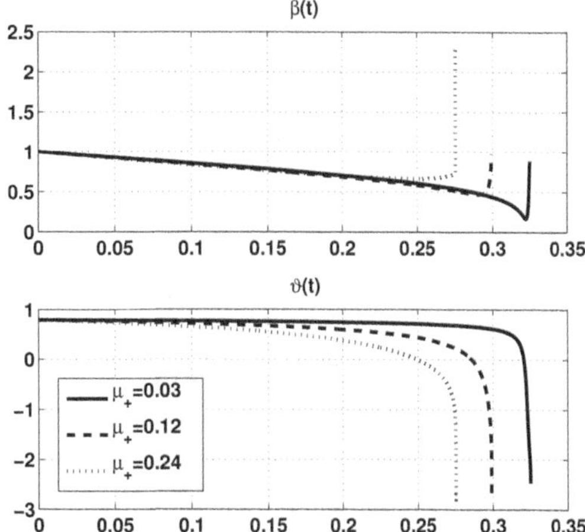

Figure 1. Parameters β and ϑ evolution for the circle. Asymmetric orthotropic friction case.

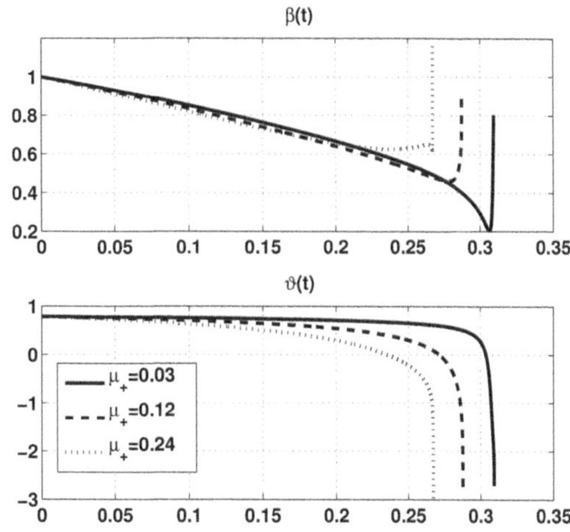

Figure 2. Parameters β and ϑ evolution for the ellipse ($e = 0.6$). Asymmetric orthotropic friction case.

Table 1. Parameters β_*, ϑ_* for circular and elliptical plates for asymmetric orthotropic friction. ($\vartheta_0 = \frac{\pi}{4}, \varphi_0 = \frac{\pi}{3}$)

	Circle			Ellipse ($e = 0.6$)		
μ_+	β_*	ϑ_*	Area	β_*	ϑ_*	Area
0.03	0.887	−2.46	13	0.81	−2.71	13
0.06	0.908	−2.57	13	0.83	−2.77	13
0.09	0.937	−2.65	13	0.86	−2.82	13
0.12	0.976	−2.71	13	0.89	−2.86	13
0.15	1.042	−2.78	12	0.91	−2.88	12
0.18	1.197	−2.86	19	0.99	−2.93	19

the equations (6). Area column describes position of the simultaneous velocity center at the terminal point. In Figure 3 the way of splitting the full integration area is shown. Each area corresponds to different velocity vector orientation and friction coefficient (Dmitriev 2015).

Comparing with symmetric orthotropic friction case studied in (Dmitriev & Silantyeva 2014) one can see significant influence of even very low asymmetry on the evolution of parameters during sliding.

In our work (Dmitriev 2015) some results regarding narrow ring with asymmetric friction are presented. In the case of zero initial linear velocity of the mass center and non-zero initial angular velocity, mass center gets acceleration directed to the third quadrant, if coefficients of friction along axes are two times higher than versus axes. For circular contact area results for these cases are the same.

In the case of non-zero initial linear velocity and non-zero initial angular velocity, velocity vector at the stopping point will be directed to the quadrant with minimal values of friction coefficients (to the third quadrant for the case described here). However, if the initial motion is translational, it stays translational up to the final point.

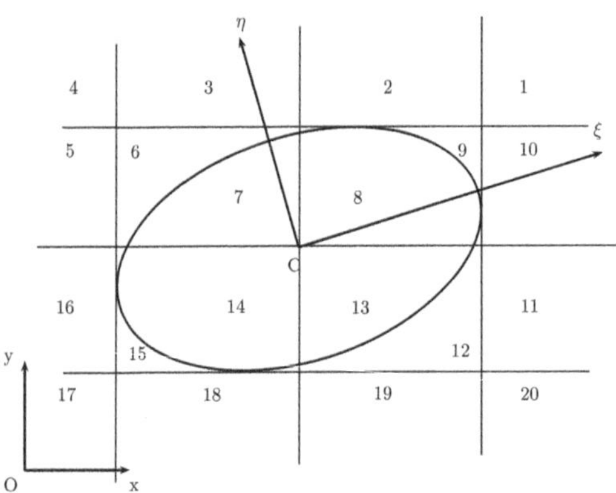

Figure 3. Full integration area is partitioned according to the directions of velocity vector.

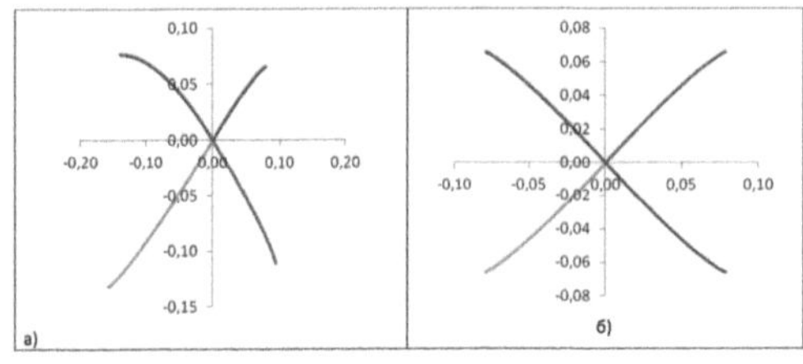

Figure 4. Trajectories of a mass point.

Table 1 shows that for elliptical and circular thin plates velocity vector is also oriented to the third quadrant.

Figure 4 shows trajectories of the movement of a material point under symmetric (Fig. 4b) and asymmetric (Fig. 4a) orthotropic friction. The full description of the motion is presented in (Dmitriev 2013). On the figure we observe deviation of the trajectories for asymmetric case over symmetric. It is worth mentioned here that we should take into account directions of velocities in each quadrant (Dmitriev & Silantyeva 2015).

4 CONCLUSIONS

We sequentially develop methods for considering effects of anisotropy of friction forces for a number of cases. Our research tends to define exact way of calculating friction forces and predicting bodies movement. This may be useful to understand better the basis of the process. Furthermore, the more information we get about frictional behavior, the more carefully we predict wear and fatigue in contact pair.

We achieved differential equations describing dynamical behavior of material point, circular plate, elliptical plate in case of frictional anisotropy. System of equations was solved numerically for symmetric orthotropic and asymmetric orthotropic friction with various initial conditions.

For symmetric orthotropic friction both for circular and elliptic plates, we postulate that spinning and sliding finish simultaneously. For elliptic and circular contact areas sliding regimes depend on the interrelations between inertia moment and frictional coefficients. Initial orientation of elliptic area also influences the dynamic behavior.

For asymmetric orthotropic friction the contact area should be partitioned due to relations between friction coefficients. Direction of the velocity vector at the stopping point results from the interrelations between frictional coefficients.

REFERENCES

Antoni, N., J.-L. Ligier, P. Saffre, & J. Pastor (2007). Asymmetric friction: Modeling and experiments. *Int. J. Eng. Sci.* 45, 587–600.

Dmitriev, N. (2002). Movement of the disk and the ring over the plane with anisotropic friction. *J. Fric. Wear* 23, 10–15.

Dmitriev, N. (2009). Sliding of a solid body supported by a round platform on a horizontal plane with orthotropic friction. part 1. regular load distribution. *J. Fric. Wear* 30(4), 227–234.

Dmitriev, N. (2013). Motion of material point and equilibrium of two-mass system under asymmetric orthotropic friction. *J. Fric. Wear* 34, 429–437.

Dmitriev, N. (2015). Motion of a narrow ring on a plane with asymmetric orthotropic friction. *J. Fric. Wear* 36, 80–88.

Dmitriev, N. & O. Silantyeva (2014). About the movement of a solid body on a plane surface in accordance with elliptic contact area and anisotropic friction force. In *Proc. of jointly organised WCCM XI, ECCM V, ECFD VI. Vol. IV*, Volume 4, Barcelona, Spain, pp. 4440–4452.

Dmitriev, N. & O. Silantyeva (2015). Dynamics of bodies under symmetric and asymmetric orthotropic friction forces. In *Short Papers of PCM-CMM-2015*, Volume 1, Gdansk, Poland, pp. 135–136.

Konyukhov, A., P. Vielsack, & K. Schweizerhof (2008). On coupled models of anisotropic contact surfaces and their experimental validation. *Wear* 264, 579–588.

Lurye, A. (2002). *Analytical Mechanics*. Berlin: Springer-Verlag Berlin Heidelberg.

Waidman, P. & C. Malhotra (2007). On the terminal motion of sliding spinning disks with uniform coulomb friction. *Phys. D* 233(1), 1–13.

Zmitrowicz, A. (1989). Mathematical descriptions of anisotropic friction. *Int. J. Solids Struct.* 25(8), 837–862.

Zmitrowicz, A. (1992). A constitutive modeling of centrosymmetric and non-centrosymmetric anisotropic friction. Int. J. Solids Struct. 29, 3025–3043.

Zmitrowicz, A. (2005). Models of kinematic dependent anisotropic and heterogeneous friction. *Int. J. Solids Struct.* 43, 4407–4451.

Advances in Mechanics: Theoretical, Computational and Interdisciplinary Issues – Kleiber et al. (Eds)
© 2016 Taylor & Francis Group, London, ISBN 978-1-138-02906-4

Dynamic and positioning analysis of the feed drive of the rope threading lathe

P. Sitarz, B. Powałka & A. Parus
West Pomeranian University of Technology in Szczecin, Szczecin, Poland

ABSTRACT: The paper presents analysis of dynamics and feed drive positioning of a lathe while machining rope threads defined by ISO 10208, with an innovative rope threading method. This highly efficient method, requires high acceleration of the cutting tool in the X axis. The kinematics of the tool was based on the geometry of the thread. It serves as the input signal in the servo drive model for the lathe. The effect of spindle rotational speed on the inertia force and the error of tool position was investigated as it directly affects the precision of machining. The tool, the turret and the support were considered flexible system components. While conducting FEM calculations, the transmission parameters between the nut and the cutting edge were determined which enabled to take into consideration vibration occurring in the investigated components and its effect on the accuracy of machining.

1 INTRODUCTION

Rope threads (Fig. 1) manufactured in accordance with ISO 10208:1991(E) are widely used in mining industry. The aim of the paper is to develop a lathe capable of machining rope threads using an innovative rope threading method.

At first, rope threads were made on copying lathes. Dedicated master patterns, however, had to be made beforehand. Currently, copying lathes are being replaced by CNC machine tools. Standard rope thread processing method is conducted on CNC machines at a constant X axis position of the cutting edge. It is possible owing to an appropriate synchronisation of the spindle rotational speed n with tool feed rate in the Z axis. The drawback of the process is that it involves several machine motion operations.

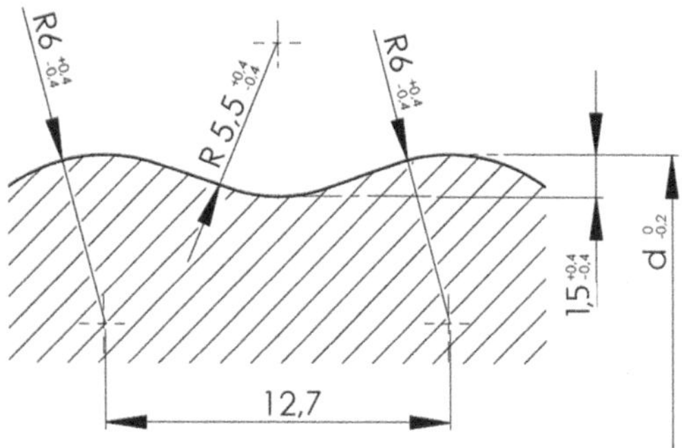

Figure 1. Definition of rope thread crest geometry.

The rope threading method has been proposed as an alternative to the above process. Although the rope threading method requires high dynamics of the moving tool components in the X axis, it enables machining threads and other workpieces with non-circular diameters with a single toolpath. High efficiency of the rope threading method requires, though, high acceleration of the cutting tool in the X axis, changing with rotational frequency. These tough requirements must be met by the dynamic components of a lathe and its fast control system.

Given the complexity of the task, the constant striving for better efficiency standards and the motivation to develop models that better reflect the real world, much research was conducted on the feed drives (Jeong et al. 2010, Lee et al. 2015, Tounsi et al. 2003). Many papers focus on control and tool position error correction issues (Erkorkmaz & Altintas 2001, Pislaru et al. 2004). Their findings usually apply to typical or theoretical cases, e.g. for lathe feed drives—the issues of trajectory tracking for ring beam corners or circles.

2 SERVOMECHANISM MODEL

A servomechanism model is presented below (Fig. 2). The course of the cutting force in time and the preset tool position are fed into the model as the input. The model uses a cascade control system, with a feed forward PIV controller with experimentally tuned feedback loop gain of the position, velocity and current.

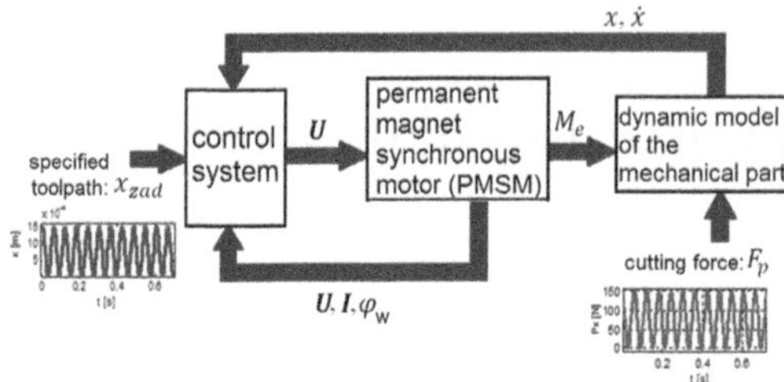

Figure 2. Servomechanism model scheme.

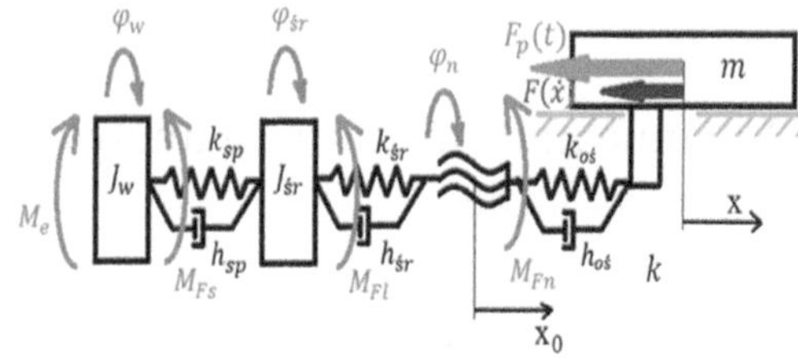

Figure 3. Dynamic rigid body model of table feed in the X axis.

3 THE MECHANICAL PART OF THE MODEL

3.1 *Rigid body model*

Dynamic rigid body model of table feed in the X axis is described by the following equations:

$$M_e - J_r\varphi_r - k_{co}(\varphi_r - \varphi_{sc}) - h_{co}(\dot{\varphi}_r - \dot{\varphi}_{sc}) + \\ -M_{Fs} sign^o \dot{\varphi}_r = 0 \tag{1}$$

$$k_{co}(\varphi_r - \varphi_{sc}) + h_{co}(\dot{\varphi}_r - \dot{\varphi}_{sc}) - k_{sc}(\varphi_{sc} - \varphi_n) + \\ - h_{sc}(\dot{\varphi}_{sc} - \dot{\varphi}_n) - J_{sc}\ddot{\varphi}_{sc} - M_{Fl} sign\dot{\varphi}_{sc} = 0 \tag{2}$$

$$k_{sc}(\varphi_{sc} - \varphi_n) + h_{sc}(\dot{\varphi}_{sc} - \dot{\varphi}_n) - h_{ax}\frac{b}{2\pi}(\dot{x}_A - \dot{x}) + \\ -\frac{b}{2\pi}k_{ax}(x_A - x) - M_{Fn} sign\dot{\varphi}_n = 0 \tag{3}$$

$$h_{ax}(\dot{x}_A - \dot{x}) + k_{ax}(x_A - x) - m\ddot{x} - F_p - F(\dot{x}) = 0 \tag{4}$$

$$x_A = \frac{b}{2\pi}\varphi_n \tag{5}$$

where M_e = the torque of the electromagnetic motor; x = the translational displacement of the feed assembly; x_A = the translational displacement of the nut resulting from the turn of the screw; b = the screw pitch; J_r = the inertia moment of the rotor; J_{sc} = the inertia moment of the drive screw and the coupling; φ_r = the rotation angle of the rotor; φ_{sc} = the rotation angle of the coupling relative to the screw; φ_n = the rotation angle of the screw relative to the nut; k_{co} = the torsional stiffness coefficient of the connection between the rotor and the coupling and that of the coupling; k_{sc} = the torsional stiffness coefficient of the connection between the coupling and the screw and that of the screw; k_{ax} = the resultant longitudinal stiffness of the screw, the bearings, the nut, the bearing housings and the nut with the motor shaft; h_{co} = the torsional dumping coefficient of the viscous coupling unit between the rotor and the coupling and that of the coupling; h_{sc} = the torsional dumping coefficient of the viscous coupling unit between the coupling and the screw: k_{ax} = the longitudinal viscous damping coefficient of the screw M_{Fs} = the resistance torque of the rotor; M_{Fl} = the friction torque in the screw bearings; M_{Fn} = the friction torque in the nut.

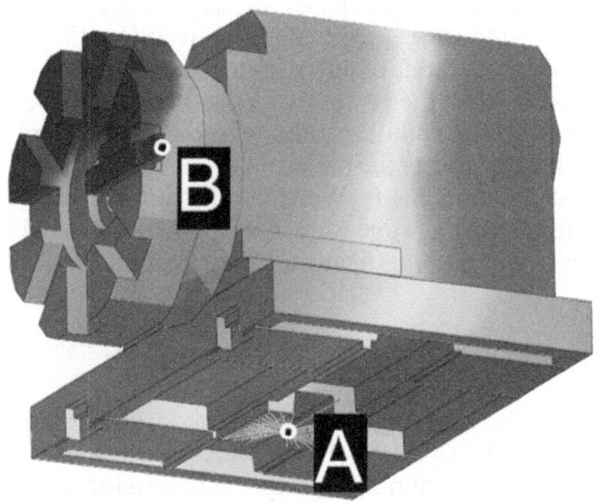

Figure 4. Components treated as flexible.

3.2 *Flexible turret*

The model (1–5) was extended to include a deformable turret, a support and a tool (Fig. 4). Point A denotes a place where traverse motion is obtained from a turning screw x_A.

Given the deformable character of the components, the motion of the cutting edge x_B, can be given by:

$$x_B(\omega) = \frac{H_{BA}(\omega)}{H_A(\omega)}x_A(\omega) \tag{6}$$

where $H_{BA}(\omega)$ = is the receptance between points A and B; $H_A(\omega)$ = is the receptance in point A.

The receptances $H_A(\omega)$ and $H_{BA}(\omega)$ were obtained by Finite Element Method. Simulation results obtained for the rigid model and those found for the model with flexible elements are discussed in more detail below.

4 NUMERICAL CALCULATIONS

Numerical experiments were conducted for three spindle rotational speeds to investigate the effect of spindle speed on error values, to measure the effect of vibration while taking into consideration the deformable nature of components. The spindle rotational speed of $n_1 = 400$ rpm was preset as the standard speed. The aim of tests was to determine how machining accuracy change when the spindle speed increased to $n_3 = 1000$ rpm. The spindle speed of $n_2 = 600$ rpm was predefined as the intermediate speed. The feed rate in the Z axis was $f_z = 0.1$ mm and the cutting resistance coefficient was $k_p = 1000$ N/mm^2.

5 RESULTS

The figure below present displacement and transmissivity, which is the quotient of dynamic flexibility from (3–5). Frequency is the domain of both functions. Therefore, it is possible to compare both parameters for the investigated spindle rotational speeds.

The displacement of the cutting edge is the quotient of transmissivity and nut motion in the frequency domain. Transmissivity is constant for given deformable components and it does not depend on the spindle rotational speed. In the investigated case, tool tip displacement, at the frequency of approximately 1250 Hz, is almost 50 times larger than the nut displacement at point A (Fig. 4).

The spindle rotational speed affects the spectrum of nut translational signal. That is why it is advisable to avoid high value harmonics in the resonance interval of transmissivity.

A clear example in which the recommendation is not met is machining at the speed of $n_2 = 600$ rpm (Fig. 5b). High amplitudes of the harmonics overlap transmissivity resonance frequencies, which excites vibration with the highest amplitude of all investigated cases (Fig. 6b).

The smallest vibration with the amplitude not exceeding 0.1 µm occur at the spindle speed of $n_1 = 400$ rpm. For the speed of $n_3 = 1000$ rpm there is a series of x_A signal harmonics with high amplitudes at the whole width of investigated frequencies (Fig. 5c). They are not well pronounced in the vicinity of resonance frequency and that is why vibration amplitude was approximately 2 µm. For machining at the speed of $n_2 = 600$ rpm, the structure of the components should be changed, to shift the normal modes from the area of 1200–1300 Hz, where signal harmonics are relatively large. It is not easy because of the need to interfere in the lathe construction. The vibration amplitude while machining at the speed was 3 µm.

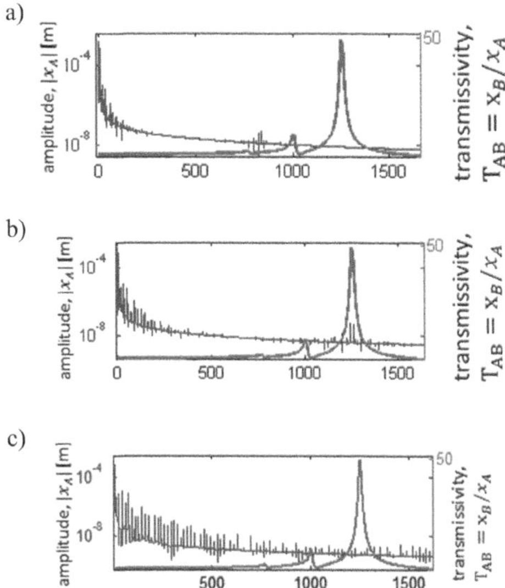

Figure 5. FFT of the x_A nut amplitudes—blue line; flexible elements transmissivity function in frequency domain—dark green line, for rotational spindle speeds: (a) $n_1 = 400$ rpm; (b) $n_2 = 600$ rpm; (c) $n_3 = 1000$ rpm.

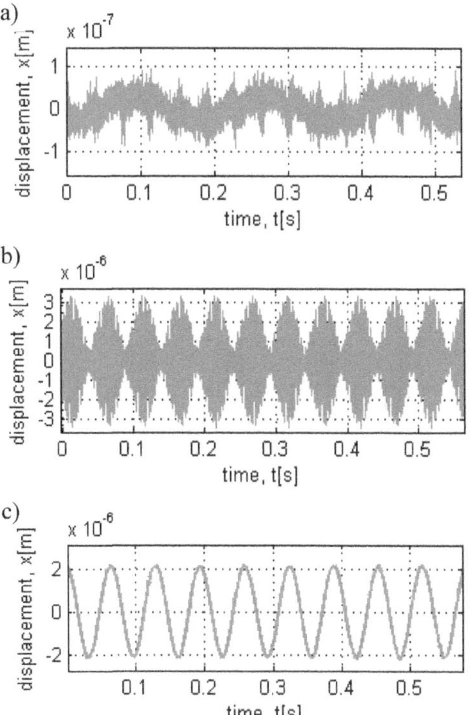

Figure 6. Displacements of cutting edge vibration in time domain, for rotational spindle speeds: (a) $n_1 = 400$ rpm; (b) $n_2 = 600$ rpm; (c) $n_3 = 1000$ rpm.

The error values obtained in the simulation of machining with the rigid body model and with the model with a flexible turret are compared below (Fig. 7).

When machining is conducted at the spindle rotational speed of $n_1 = 400$ rpm, deformation of components will have little effect on the accuracy of machining. The error values obtained in simulations conducted for the rigid and flexible models, for the above spindle speed, almost overlap (Fig. 7a). The maximum error does not exceed 4 μm. For the spindle speed of $n_2 = 600$ rpm differences are much larger. In the flexible model, the error value increased by 3 μm compared to the rigid model, and amounted to 10 μm (Fig. 7b). For the spindle speed of $n_3 = 1000$ rpm the vibration amplitude was slightly over 2 μm (Fig. 6c), which is still only approximately 10% of the total error value, which for this speed was over 20 μm.

a)

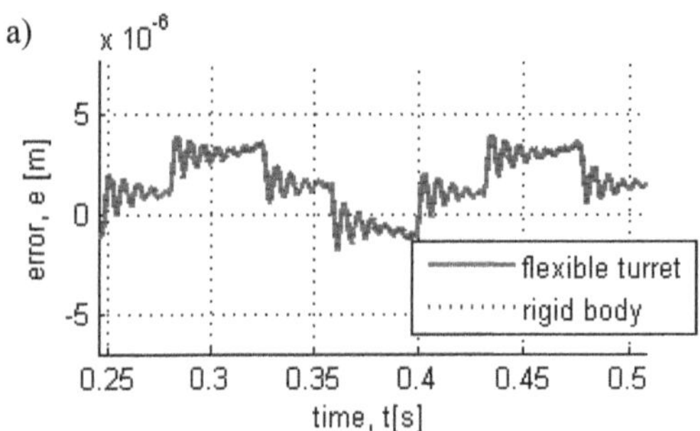

b)

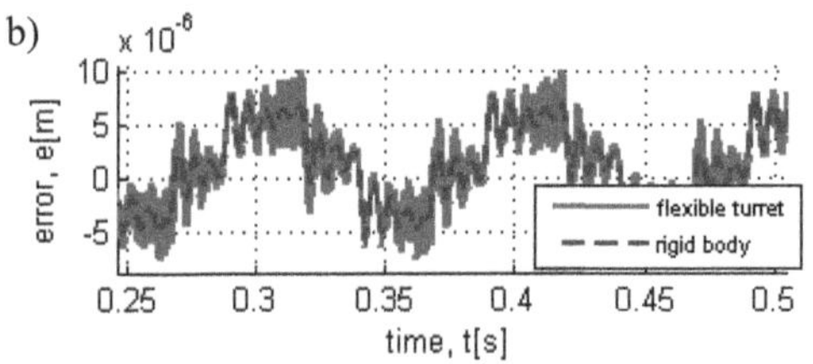

c)

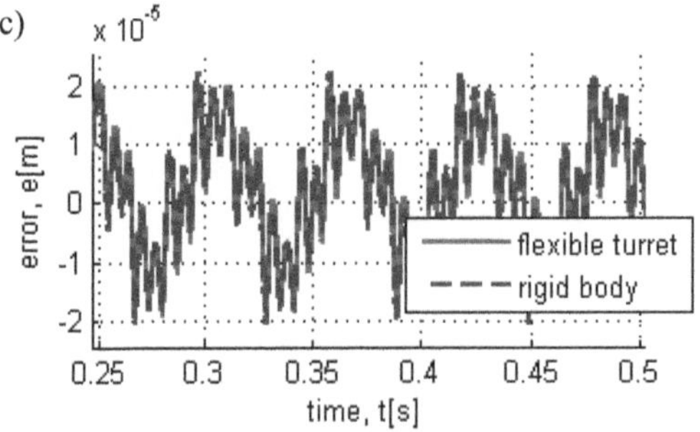

Figure 7. Processing error values achieved for both models, for rotational spindle speeds: (a) $n_1 = 400$ rpm; (b) $n_2 = 600$ rpm; (c) $n_3 = 1000$ rpm.

6 SUMMARY

The developed model enables to gain information about error values and vibration occurring in the drive system of the lathe depending on the spindle rotational speed. At higher spindle speeds, error values are larger. Larger are also the amplitudes of successive harmonics of the input defining the nut displacement, which increases the chance of exciting vibration of larger amplitude at the cutting edge. Vibration with a larger amplitude, at higher rotational speeds of the spindle, is exception rather than the rule, which is evidenced by the case of $n_2 = 600$ rpm. It is advisable to avoid harmonics of nut motion with high amplitude values in the vicinity of turret normal modes. It can be achieved either by changing the spindle rotational speed or by modifying the structure in order to shift its resonance frequency.

REFERENCES

Erkorkmaz, K. & Altintas, Y. 2001. High speed CNC system design. Part III: high speed tracking and contouring control of feed drives. *International Journal of Machine Tools and Manufacture* 41(11): 1637–1658.

ISO 10208:1991(E): Rock drilling equipment—Left-hand rope threads. *ISO 1991*.

Jeong, Y.H., Min, B.K., Cho, D.W., & Lee, S.J. 2010. Motor current prediction of a machine tool feed drive using a component-based simulation model. *International Journal of Precision Engineering and Manufacturing* 11(4): 597–606.

Lee, C.H., Yang, M.Y., Oh, C.W., Gim, T.W. & Ha, J.Y. 2015. An integrated prediction model including the cutting process for virtual product development of machine tools. *International Journal of Machine Tools and Manufacture* 90: 29–43.

Pislaru, C., Ford, D.G. & Holroyd, G. 2004. Hybrid modelling and simulation of a computer numerical control machine tool feed drive. Proceedings of the Institution of Mechanical Engineers. *Part I: Journal of Systems and Control Engineering* 218(2): 111–120.

Tounsi, N., Bailey, T. & Elbestawi, M.A. 2003. Identification of acceleration deceleration profile of feed drive systems in CNC machines. *International Journal of machine tools and manufacture* 43(5): 441–451.

Advances in Mechanics: Theoretical, Computational and Interdisciplinary Issues – Kleiber et al. (Eds)
© 2016 Taylor & Francis Group, London, ISBN 978-1-138-02906-4

Experimental and numerical evaluation of mechanical behaviour of composite structural insulated wall panels under edgewise compression

Ł. Smakosz & I. Kreja
Faculty of Civil and Environmental Engineering, Gdansk University of Technology, Gdańsk, Poland

ABSTRACT: A Composite Structural Insulated sandwich Panel (CSIP) is a quite novel approach to the idea of sandwich structures. A series of natural-scale experimental test is required each time a change in panel's geometry is planned and a reliable computational tool is required to precede actual laboratory testing with virtual simulations. An attempt of creating such a tool has been made with the use of a commercial FEM code ABAQUS, in order to predict the behaviour of a specific kind of CSIPs with magnesium-oxide board facings and expanded polystyrene core. The results obtained from simulations taking into account geometrical nonlinearity as well as material nonlinearity of both core and facing materials are presented and compared with experimental data from natural-scale edgewise compression tests of wall CSIPs.

1 INTRODUCTION

Composite structural insulated sandwich panel is an advanced approach to the idea of sandwich structures. It is based on the main principle of combining two materials with diametrically different properties—a light-weight, thick core sandwiched between two high-strength, durable and thin facings—joined together by an adhesive layer. The CSIP facings are made of composite materials, making a light and easy to handle prefabricated element considerably stronger, immune to biological corrosion and more durable to weather conditions. Such improvements make CSIPs very attractive alternative to classical SIPs with OSB facings and broaden their field of application (FAS 2009, Lstiburek 2008).

High facing-core elastic moduli ratio causes CSIPs to be challenging in design. Due to a large number of possible failure modes, natural-scale experimental tests are required with each change in panel's cross-section. A reliable computational tool is required to precede laboratory tests with numerical simulations. Abaqus (2010) software was used to create a suitable computational model for the CSIPs with magnesium-oxide board (MgO board) facings and Expanded Polystyrene (EPS) core. To observe behaviour and failure modes of this kind of CSIPs a series of natural- and small-scale tests was performed on samples with a specific cross-section (Fig. 1). A comparable FEA was performed and a comparison of computational results with experiments was presented in (Smakosz & Kreja 2014, Smakosz & Tejchman 2014).

The paper focuses on behaviour prediction of a selected type of CSIPs under an edgewise compression with the use of mentioned FEM model.

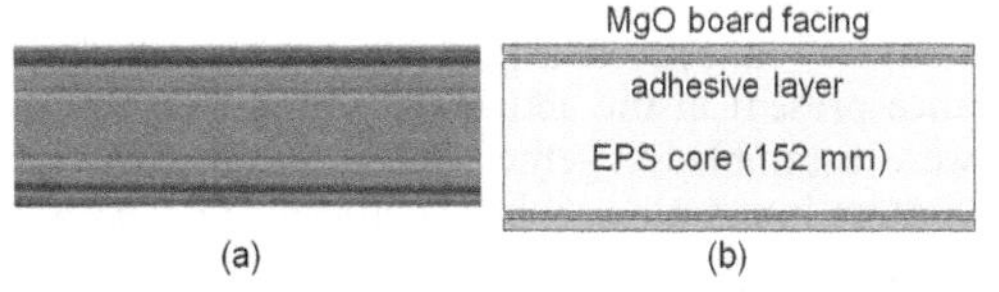

Figure 1. Analysed CSIP's cross-section: (a) view and (b) scheme.

2 MATERIAL PROPERTIES

The analysed panel's cross-section is shown in Figure 1. It consists of two 11 mm thick MgO boards, reinforced with a fiberglass mesh on top and bottom surfaces, a 152 mm thick EPS core layer with volumetric density of 19.5 kg/m^3 and a thin layer of polyurethane adhesive joining the layers together, adding up to a total thickness of 174 mm.

Material properties of EPS are well recognized and strongly dependent on its density, while, the available characteristics of the MgO board are very limited. To determine the mechanical properties of both materials the Authors performed a series of experimental tests on small-scale samples described in Smakosz & Tejchman (2014).

In order to improve the process of parameter identification, small-scale bending and compression tests of CSIP samples composed of core and both facings were performed. Based on results presented in Figure 6 and in Smakosz & Tejchman (2014), a set of values for the application in FEM analysis was established.

3 EDGEWISE COMPRESSION TEST

3.1 *Experimental set-up*

Wall panel edgewise compression tests under an eccentric load were carried out on two test samples in natural scale. The dimensions of the tested panels were 2750 mm × 1000 mm × 174 mm and two values of the compressive force eccentricity were taken as $e = d/6$ and $e = d/3$, where $d = 174$ mm is the panel's total thickness. Due to technical limitations, the tests were performed in a horizontal position (Fig. 2a) with both ends supported on hinges (Fig. 2b, c). In order to obtain relatively uniform distribution of stresses at the edges, the panels were inserted into steel profiles and sealed with concrete in first case and polyurethane foam in the other. The steel profile depth was 165 mm and the total distance between the support points was 3080 mm. The complete experimental set-up is shown in Figure 2.

3.2 *Test results*

In both cases the load eccentricity was positioned in the lower half of the cross-section which resulted in a flexural deformation with an arch tip directed upwards, opposite to the eccentricity's direction.

The samples failed by a facing cracking in the contact zone between sample and steel profiles. In both cases, damage initiated as a horizontal crack appearing suddenly in the bottom facing (Fig. 3a), signalled with a barely audible snapping sound and a clearly visible drop of mid-span deflection. Further loading led to a gap forming between the damaged sample edge and the steel profile (Fig. 3d)—this shift in sample's position in relation to the loading profile eventually caused a vertical crack in the top facing (Fig. 3b). Slow, progressive deterioration of bottom facing's material (Fig. 3c) and increase of the gap size followed. No buckling behaviour—neither global nor local—was observed throughout the test, and no debonding occurred prior to the damage initiation.

The applied load and the horizontal movement of the piston were recorded and presented as force-displacement relationships in Figure 7.

Figure 2. Experimental set-up for wall panel compression tests: (a) general view, (b) loading point and left hinge, (c) right hinge.

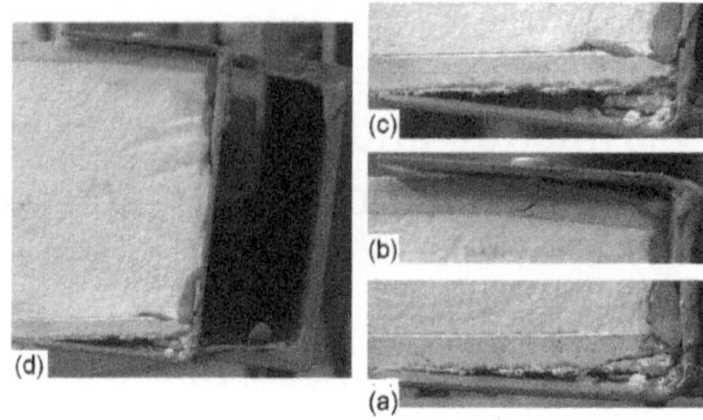

Figure 3. Failure development: (a) bottom facing horizontal crack—damage initiation, (b) top facing vertical crack—damage propagation, (c) deterioration of damaged facing, (d) a gap between the steel profile and the sample.

4 NUMERICAL SIMULATIONS

4.1 *Assembly's description*

A commercial FEM software—Abaqus (2010)—was used to build a numerical model of two tests described in previous section. In order to capture the nonlinear response of tested panels a fine mesh of solid elements in two-dimensional space was used to discretise full-length samples.

Both samples were represented as plane stress sections divided into partitions corresponding to core and facing as shown in Figures 4–5. A single FE mesh divided in subsections with different material properties was employed (Figs. 5–6) because no pre-failure deboning was observed in the compression tests so no additional contact interactions between the sample layers were required. Steel mounting profiles were substituted with a rigid body planar wire sections. A gap of 1 mm was introduced between the profiles and the sample (Fig. 5) to take into account a small clearance present in the actual test setup. These parts were assembled together and a penalty friction contact interaction with a coefficient of 0.05 was used between the rigid supporting parts and the sample. Pinned boundary conditions were created at a single reference point assigned to each rigid profile. One half of the numerical assembly is

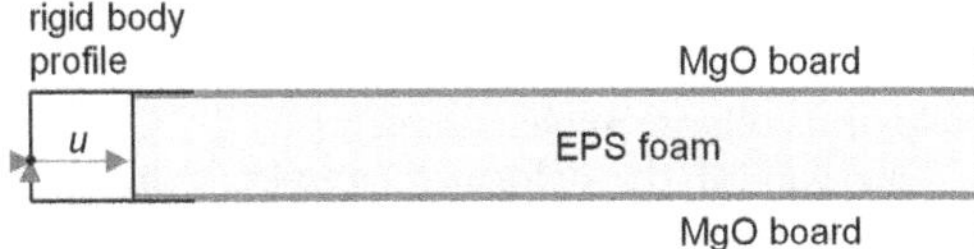

Figure 4. Half of the FEM model assembly: boundary conditions and section assignments.

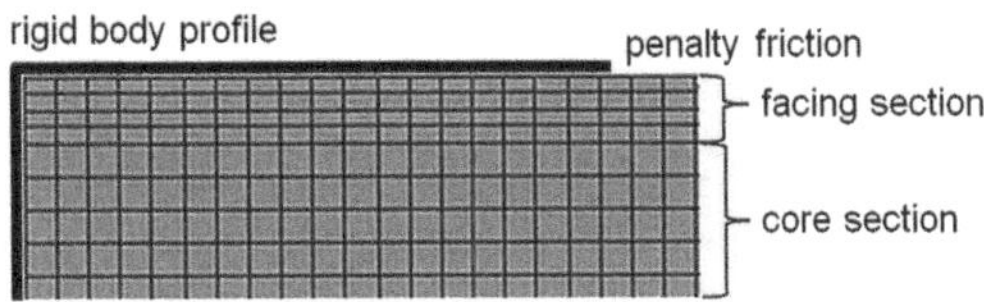

Figure 5. FEM assembly vertex zoom—mesh size comparison, section property assignment and contact zone.

presented in Figure 4 and a magnified view of one of the sample's vertices is shown in Figure 5.

The assembly was loaded in two steps—first the dead weight was taken into, next a horizontal displacement of one of the supports was forced. Geometric nonlinearity effects were taken into account.

A fine mesh of 4-node plane stress elements with reduced integration and hourglass control was used. Global mesh size of 5.5 mm resulted in 28 elements of 1:1 proportions through the core thickness. In the facing area the mesh was refined in one direction resulting in 4 elements of 1:2 proportions through the facing thickness. An enlarged view of a mesh section is presented in Figure 5.

4.2 *Material models*

Both, the MgO facings and the EPS core were described with an elastic-plastic Drucker-Prager material model. Since both layers show different behaviour in compression and tension, two sets of properties were defined. A user-defined field procedure was created to generate an additional Field Variable (*FV*) in each integration point in such a way that it takes values from –1 (when both principal stresses are negative) to 1 (when both principal stresses are positive). Elastic properties and a hardening behaviour for compression are used for both materials when $FV = -1$, while the tensile properties are assigned in EPS, and flexural ones in MgO board for $FV = 1$. For all the other values of *FV* the adequate material properties are linearly interpolated between these two data sets.

Elastic properties used to describe tensile and compressive properties of both materials were obtained in a two-step parameter identification process. First mean values obtained from compression, tension and bending tests (Table 1) were used directly and

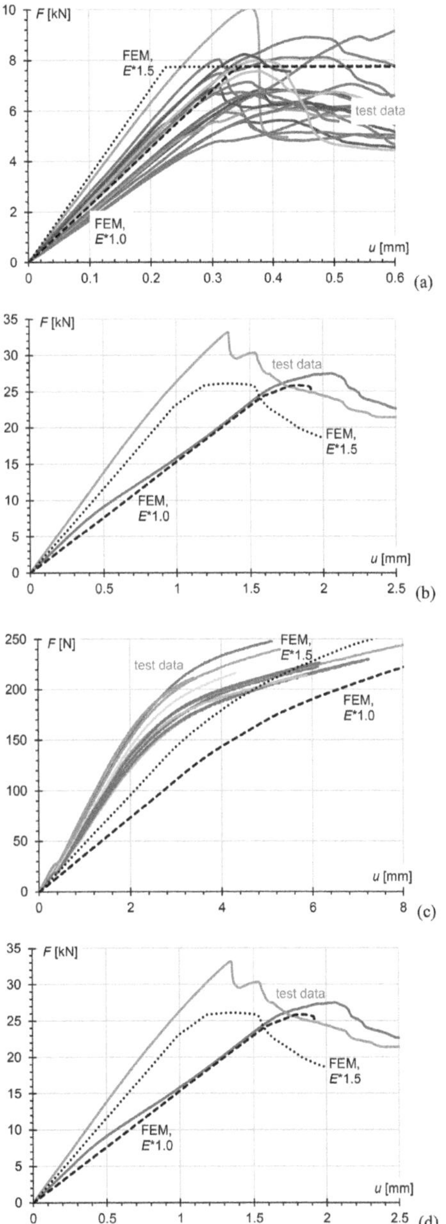

Figure 6. Comparison of FEA with small-scale tests used for parameter identification: (a) MgO board uniaxial compression, (b) CSIP sample edgewise compression, (c) MgO board bending, (d) CSIP beam three point bending.

afterwards additional simulations of analogical tests performed on small composite CSIP samples (core and both facings) were carried out to verify them (Fig. 6). These additional simulations indicated that MgO board's averaged modulus of elasticity obtained

Table 1. Average properties obtained from small scale tests on specific materials: E—elastic modulus, σ_y—yield stress.

	EPS		MgO board	
	Compr. (MPa)	Tension (MPa)	Compr. (MPa)	Flexure (MPa)
E	6.1	10.4	1964	7043
σ_y	0.09	0.11	13.3	10.7

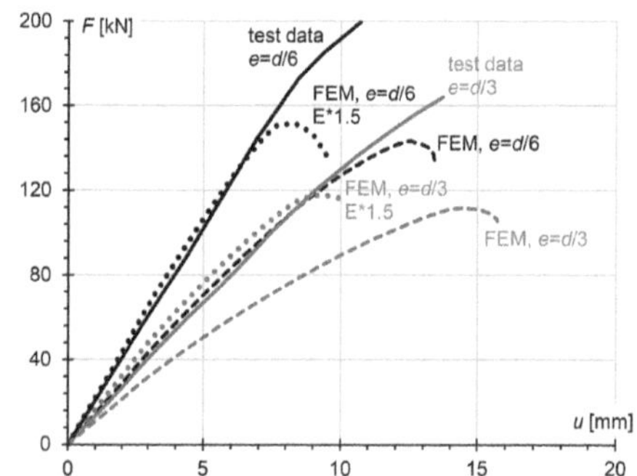

Figure 7. Comparison of force-displacement curves from laboratory tests and numerical simulations.

Figure 8. FEM simulation: deformed shape and initial yield zone signalling failure initiation.

from uniaxial compression tests was considerably underestimated. A parametric study showed that by increasing this value by a multiplier equal to 1.5 one obtains results being more accurate to the ones recorded in small-scale CSIP sample tests. A value obtained this way is slightly greater than a maximum value obtained in compression of separate MgO board samples (Fig. 6a). The following values of the Poisson's ratios were assumed: 0.11 for EPS and 0.18 for the MgO board.

4.3 *FEA results*

The comparison of numerical and experimental results with natural-scale experimental data (force vs. horizontal displacement curves) is presented in Figure 7. Two sets of numerical results are presented: with MgO board's $E_{MgO}^{(FV=-1)} = 1964$ MPa and $E_{MgO}^{(FV=-1)} = 3000$ MPa. Results computed from FEM models with increased facing Young's modulus in compression are in better agreement with experimental data. In both cases the simulations reflect elastic behaviour of panels well but yield and collapse appear sooner.

One half of the deformed sample is shown in Figure 8 with a yield zone signalising the sample failure. Since bearing capacity of a panel is exceeded after damage initiation, the following progression process was not analysed.

5 CONCLUSIONS

The comparison illustrated in Figure 7 shows clearly that the numerical model with MgO board's modulus of elasticity in compression entered as a mean value of small-scale uniaxial tests generally underestimates the performance of an actual panel. By using strength properties estimated in an additional parametric study one can obtain results that are much closer to the experimental ones. There are two possible explanations to this situation: 1) the performed experimental tests showed that MgO board's behaviour in compression can vary significantly from batch to batch, but it is also possible that 2) material properties obtained in the uniaxial edgewise compression of small board samples might not correspond well to conditions in the actual panel facing.

The lower strength obtained in both simulations can be explained the same way. Since significant variation in yield strength was observed during the small-scale laboratory tests, an average value of compressive yield stress and hardening curve were used which placed the numerical results on the safe side.

The behaviour of simulated samples gives satisfactory results in terms of deformed shapes and failure initiation zone—they both match the ones observed in experiments.

REFERENCES

Abaqus 2010. *Standard theory and user's manuals*. Hibbit, Karlsson & Sorensen Inc. Dassault Systèmes.

Federation of American Scientists 2009. *Report on Expanding the Scope and Market of SIP Technologies: a history of SIPs and CSIP manufacturing, construction and market issues*.

Lstiburek J. 2008. *Builder's Guide to Structural Insulated Panels (SIPs) for all Climates*. Building Science Press.

Smakosz Ł. & Kreja I. 2014. Experimental and numerical evaluation of mechanical behaviour of composite structural insulated panels. In T. Łodygowski, J. Rakowski & P. Litewka (eds), *Recent Advances in Computational Mechanics*: 269–276. London: CRC Press/Balkema.

Smakosz Ł. & Tejchman J. 2014. Evaluation of strength, deformability and failure mode of composite structural insulated panels. *Materials & Design* 54: 1068–1082.

Advances in Mechanics: Theoretical, Computational and Interdisciplinary Issues – Kleiber et al. (Eds)
© 2016 Taylor & Francis Group, London, ISBN 978-1-138-02906-4

A new adaptive ground structure method for multi-load spatial Michell structures

T. Sokół
Warsaw University of Technology, Warsaw, Poland

ABSTRACT: In the paper a new method of solving large-scale linear programming problems related to Michell trusses, generalized to multiple load conditions and three-dimensional domains, is proposed. The method can be regarded as an extension of the adaptive ground structure methods developed recently by the author. In the present version both bars and nodes can be switched between active and inactive states in subsequent iterations allowing significant reduction of the problem size. Thus, the numerical results can be attained for denser ground structures giving better approximation of exact solutions to be found. The examples of such exact solutions (new 3D Michell structures), motivated by the layouts predicted numerically, are also reported and can serve as benchmark tests for future methods of numerical optimization of structural topology in 3D space.

1 INTRODUCTION

In spite of some inherent limitations (e.g. neglecting stability requirements for compression bars), the classical theory of Michell structures plays an important role in structural topology optimization, by enabling the derivation of exact analytical solutions for the least-weight trusses capable of transmitting the applied loads to the given supports within limits on stresses in tension and compression. Thus the exact solutions derived by means of this theory may serve as valuable benchmarks for any structural topology optimization method.

In general, the exact solutions are very hard to obtain since they require in advance a good prediction of the optimal layouts. Fortunately, this difficult task of predicting the optimal layout can effectively be carried out numerically using the adaptive *ground structure* method developed recently by the author. In this method the solution is achieved iteratively using a small number of properly chosen active bars from the huge set of possible connections. The problem of enormous number of potential bars becomes particularly hard for space trusses subjected to multiple load conditions because the optimization problem grows very rapidly and becomes too large even for supercomputers.

In the paper a new and more advanced solution method for large-scale optimization problems is proposed. It is based on adaptive activating of new bars in the ground structure and eliminating large number of unnecessary bars. Moreover, in this new method the nodes can also be switched between active and inactive states. This is particularly important for 3D problems because the optimal 3D trusses tend to assume forms of lattice surfaces (shell-like structures) while most of design space becomes empty. As a result, the size of the problem can be significantly reduced. The method makes it possible to obtain new numerical solutions for deriving new optimal topologies for 3D Michell structures. A class of new 3D exact solution inspired by numerical results is also presented. The method proposed in the present paper is a natural extension of the adaptive ground structure methods developed by Gilbert & Tyas (2003), Pritchard et al. (2005) and Sokół (2011a, b, 2013, 2014).

Concluding, the aim of this paper is two-fold: a) to develop a reliable and efficient optimization method based on the adaptive ground structure approach, and b) to obtain substantially new exact solutions of spatial Michell trusses subjected to multiple load cases. Note that many new exact solutions of multi-load but plane problems have recently been discussed by Rozvany et al. (2014).

2 PRIMAL AND DUAL FORMULATIONS OF MULTI-LOAD PLASTIC DESIGN OPTIMIZATION PROBLEMS

According to well-known duality principles, the plastic design optimization problem can be written in either primal or in dual form. Both of them play an important role in the proposed method and should be considered together.

The most concise formulation of plastic design optimization problem for multi-load cases can be written as follows:

$$\begin{aligned}
&\min_{\mathbf{A},\mathbf{S}_{(l)}\in\mathbf{R}^M} V=\mathbf{L}^T\mathbf{A}\\
&\text{s.t.}\quad \mathbf{B}^T\mathbf{S}_{(l)}=\mathbf{P}_{(l)}\\
&\qquad -\mathbf{A}\sigma_C\le\mathbf{S}_{(l)}\le\mathbf{A}\sigma_T\\
&\qquad \text{for } l=1:K
\end{aligned}\tag{1}$$

where V is the total volume of the structural material in the truss of M potential bars; $\mathbf{L}$ is the vector of lengths of bars; $\mathbf{B}$ is the geometric matrix; vectors $\mathbf{P}_{(l)}$ define nodal forces for the given load cases $l=1, 2, \ldots, K$, where K is the number of load cases; $\mathbf{S}_{(l)}$ is the vector of member forces for load case l; $\mathbf{A}$ is the vector of cross-section areas (the main design variables); and σ_T and σ_C denote the permissible stresses in tension and compression, respectively.

The primal form (Eq. 1) is not convenient for direct application of *simplex* or *interior point method* and it is recommended to convert it to a more applicable form (see Sokół 2014 for details). The inequalities (Eq. $1)_3$ can be converted to equality constraints using properly adjusted slack variables $\mathbf{c}_{(l)}$ and $\mathbf{t}_{(l)}$:

$$\begin{aligned}
&\sigma_T\mathbf{A}-\mathbf{S}_{(l)}-\frac{\sigma_T+\sigma_C}{\sigma_C}\mathbf{c}_{(l)}=\mathbf{0},\quad \mathbf{c}_{(l)}\ge\mathbf{0}\\
&\sigma_C\mathbf{A}+\mathbf{S}_{(l)}-\frac{\sigma_T+\sigma_C}{\sigma_T}\mathbf{t}_{(l)}=\mathbf{0},\quad \mathbf{t}_{(l)}\ge\mathbf{0}
\end{aligned}\tag{2}$$

which then allow elimination of original design variables $\mathbf{A}$ and $\mathbf{S}_{(l)}$

$$\begin{aligned}
&\mathbf{A}=\frac{\mathbf{t}_{(l)}}{\sigma_T}+\frac{\mathbf{c}_{(l)}}{\sigma_C}\\
&\mathbf{S}_{(l)}=\mathbf{t}_{(l)}-\mathbf{c}_{(l)}
\end{aligned}\tag{3}$$

Note that $\mathbf{c}_{(l)}$ and $\mathbf{t}_{(l)}$ are the vectors of slack variables which can be interpreted as the additional forces which can be added without violating the restrictions of permissible stresses (Eq. $1)_3$ (i.e. they denote not forces itself but unused reserves of internal forces).

Using (Eq. 3) the original problem (Eq. 1) can be converted to the standard linear programming problem

$$\begin{aligned}
&\min_{\mathbf{t}_{(l)},\mathbf{c}_{(l)}\in\mathbf{R}^M} V=\mathbf{L}^T\left(\frac{\mathbf{t}_{(1)}}{\sigma_T}+\frac{\mathbf{c}_{(1)}}{\sigma_C}\right)\\
&\text{s.t.}\quad \mathbf{B}^T(\mathbf{t}_{(l)}-\mathbf{c}_{(l)})=\mathbf{P}_{(l)}\\
&\qquad \frac{\mathbf{t}_{(l+1)}}{\sigma_T}+\frac{\mathbf{c}_{(l+1)}}{\sigma_C}=\frac{\mathbf{t}_{(l)}}{\sigma_T}+\frac{\mathbf{c}_{(l)}}{\sigma_C}\\
&\qquad \mathbf{t}_{(l)}\ge\mathbf{0},\ \mathbf{c}_{(l)}\ge\mathbf{0}
\end{aligned}\tag{4}$$

which is more complex but in fact two times smaller. Note that the form (Eq. 4) is new in the literature and more economical than the formulations proposed in (Pritchard et al. 2005 or Sokół 2014). Moreover, for $K=1$ it automatically reduces to a well-known form used for one-load case problem (see Sokół 2011a, 2013) thus no additional separate code is needed for this special case.

For activating new bars in the adaptive ground structure method we need also dual variables but they are calculated automatically and for free using the primal-dual version of the interior point method. The convenient dual form of multi-load case problem was derived in (Sokół 2014) and is given by

$$\begin{aligned}
&\max_{\mathbf{u}_{(l)}\in\mathbf{R}^N,\boldsymbol{\varepsilon}^+_{(l)},\boldsymbol{\varepsilon}^-_{(l)}\in\mathbf{R}^M} W=\sum_l\mathbf{P}^T_{(l)}\mathbf{u}_{(l)}\\
&\text{s.t.}\quad \sum_l(\sigma_T\varepsilon^+_{(l),i}+\sigma_C\varepsilon^-_{(l),i})\le 1\\
&\qquad \varepsilon^+_{(l),i}\ge\max(\mathbf{B}_i\,\mathbf{u}_{(l)}/L_i,0)\\
&\qquad \varepsilon^-_{(l),i}\ge\max(-\mathbf{B}_i\,\mathbf{u}_{(l)}/L_i,0)\\
&\qquad \text{for } l=1:K,\ i=1:M
\end{aligned}\tag{5}$$

where $\mathbf{B}_i$ denotes the i-th row of matrix $\mathbf{B}$ and $\mathbf{u}$, $\boldsymbol{\varepsilon}^+$, $\boldsymbol{\varepsilon}^-$ are Lagrange multipliers, called adjoint nodal displacements and adjoint strains for tension and compression, respectively. They are independent variables for every load condition (l) but constrained together by (Eq. $5)_2$ which enables to derive the generalized optimality criteria for multi-load trusses (see Sokół 2014 for details).

3 THE ADAPTIVE GROUND STRUCTURE METHOD WITH SELECTIVE SUBSETS OF ACTIVE BARS AND NODES

Due to limited space of the paper the new method can be described only briefly. The main idea of activating new bars is the same as before (Sokół 2014) but now after each iteration the nodes are split into two subsets: active and inactive nodes. Then, in the subsequent iteration the adjoint displacements are updated only for active nodes. Inactive nodes appear in empty regions where no material is needed. Thus before starting the next iteration all bars connected with inactive nodes have to be eliminated too (temporary for the current iteration). Consequently, the size of the coefficient matrix of the problem (Eq. 4) is much smaller in terms of the number of rows and columns. It should be noted that inactive nodes are not removed forever from the ground structure and can be activated if necessary. Moreover, the adjoint displacements of these nodes have to be preserved for subsequent iterations for checking the optimality criteria. Of course this complicates the code but is necessary and crucial to assure convergence to a globally optimal solution. The step by step procedure for the proposed method can be described as follows:

First iteration:

1. Set $iter = 1$, $d = 1$ and generate the initial ground structure $N_x \times N_y \times N_z$:1×1×1 with bars connecting only the neighbouring nodes. Contrary to the previous versions these bars can also be deactivated in subsequent iterations.
2. Solve the problem (4) for this initial ground structure and get the dual variables $\mathbf{u}_{(l)}^{(1)}$.

Next iterations:

3. Increment the number of iteration *++iter*.
4. Increment the distance of connections $d := \max(d_{max}, d+1)$, together with d_x, d_y, d_z.
5. Select the new set of active bars in the ground structure $N_x \times N_y \times N_z : d_x \times d_y \times d_z$:
   - for every bar compute normalized strain using the displacement fields from the previous iteration:

   $$\mathbf{u}_{(l)}^{(iter-1)} \Rightarrow \hat{\varepsilon}_i = \sum_l (\sigma_T \varepsilon_{(l),i}^{+} + \sigma_C \varepsilon_{(l),i}^{-}), \text{ (see (5))}$$

   - if $\hat{\varepsilon}_i \geq 1 - tol$, then activate (add) i-th bar,
   - otherwise, if $\hat{\varepsilon}_i < 0.3$ and $d < d_{max}$ then deactivate (remove) bar,
   - if $d < d_{max}$ and the number of added bars is too small then go to step 4.
6. Check the stopping criterion:
   - if $d = d_{max}$ and there are no new bars added then finish (we approach the optimum solution because for all potential bars $i = 1{:}M$ the constraints $(5)_2$ are satisfied and the solution cannot be further improved).
7. Calculate the volumes of material connected to nodes; if the volume of a chosen node is equal or close to zero and no any new bar is added to this node, then deactivate this node together with all connecting bars; then remove the appropriate degrees of freedom from the system but keep adjoint displacements of inactive nodes 'frozen' for the next iteration.
8. Solve primal problem (4) for reduced system of active bars and nodes and get dual variables $\mathbf{u}_{(l)}^{(iter)}$ (combine the updated adjoint displacements of active nodes with frozen displacements of inactive nodes from previous iteration).
9. Repeat from step 3.

The program implementing the above algorithm has been written in Wolfram Mathematica using parallel computing and sparse arrays.

4 EXAMPLES OF TWO-LOAD CASE PROBLEMS WITH SPATIAL MICHELL TRUSSES

In both examples presented in this section let us assume: a) equal permissible stresses in tension and compression $\sigma_T = \sigma_C = \sigma_0$; b) equal magnitudes of applied point forces $\|\mathbf{P}_{(1)}\| = \|\mathbf{P}_{(2)}\| = P$.

As the first example, consider the two-load case problem presented in Figure 1, in which the two independent point loads are applied in the centre of the upper square of the bounding cuboidal domain $d/\sqrt{2} \times d/\sqrt{2} \times 3d$ and directed along x and y axes, while the continuous full support is applied on the whole bottom square.

Figure 1a presents the exact optimal solution obtained using the superposition principles and the concept of component loads (Nagtegaal & Prager 1973, Rozvany & Hill 1978). The optimal structure is composed of two orthogonal long cantilevers lying in diagonal planes. The exact volume of the structure can be calculated using the formulae derived by Lewiński et al. (1994). The numerical confirmation of this analytical prediction is presented in Figure 1b and was performed for the ground structure with 20 × 20 × 60 cells, 26 901 nodes and more than 300 mln bars. The solution was obtained in less than 2 hours using classical computer with Intel i7 processor which clearly indicates a good efficiency of the proposed method. Note that 'numerical' volume is only 0.3% worse than the exact analytical solution.

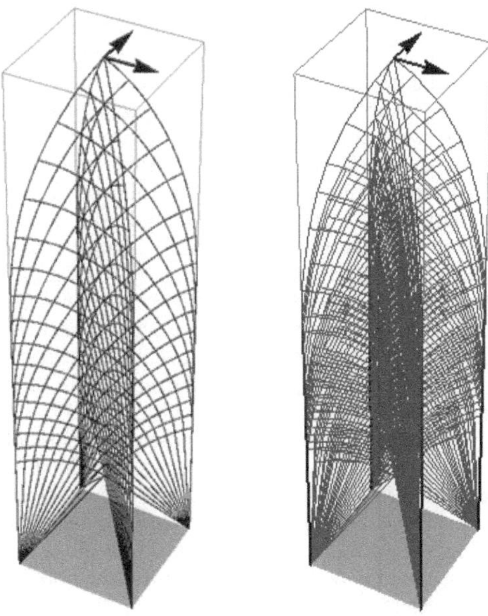

Figure 1. Example of a 3D Michell structure for two loading conditions: a) an exact analytical prediction using the concept of component loads, b) numerical confirmation using the ground structure with 300mln potential bars.

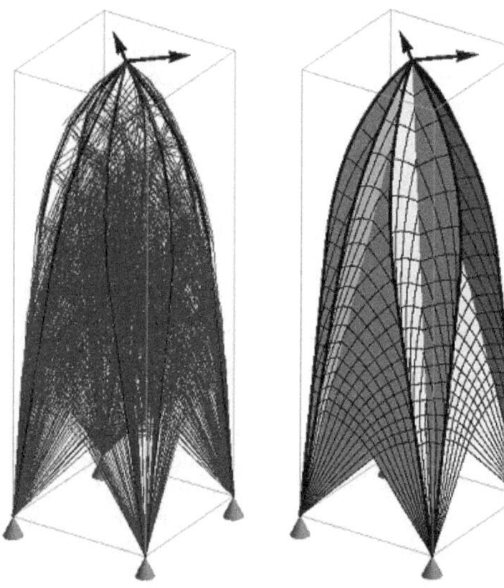

Figure 2. Example of 3D Michell truss for two independent load cases: a) numerical recognition of the optimal layout; b) exact analytical solution.

The second example is presented in Figure 2 and is a subtle modification of the previous problem. Now the independent point loads are directed along the diagonals of the upper square of the bounding cuboidal domain $d \times d \times 3d$. Moreover, these loads have to be optimally transmitted to four fixed supports in the corners of bottom square. The exact analytical solution of this modified problem is harder to predict even using the concept of component loads. Hence in this example the optimal layout was discovered numerically using the same density of the ground structure as before (i.e. more than 300 mln bars). The numerical result presented in Figure 2a suggests that the exact solution consists of four plane Michell trusses forming a specific hip roof (it is not clearly visible from Figure 2a but is evident after rotating this structure in 3D space). Then, employing the layout predicted numerically, the new exact analytical solution was obtained and presented in Figure 2b. As before, the exact volume of this complex structure can be calculated using the formulae from Lewiński et al. (1994), and is about 0.3% better than the volume obtained numerically.

5 CONCLUDING REMARKS

Note that the optimal solutions presented in Figures 1 and 2 form shell-like structures composed of lattice surfaces with Michell trusses inside. Hence most of design space is empty. It was the main motivation for developing an improved method in which both bars and nodes can be eliminated if unnecessary. As before, despite the iterative nature of the method the convergence to a global optimum is guaranteed.

ACKNOWLEDGEMENTS

The paper was prepared within the Research Grant no 2013/11/B/ST8/04436 financed by the National Science Centre, entitled: *Topology optimization of engineering structures. An approach synthesizing the methods of: free material design, composite design and Michell-like trusses.*

REFERENCES

Gilbert, M. & Tyas, A. 2003. Layout optimization of large-scale pin-jointed frames, *Engineering Computations* 20(8): 1044–1064.

Lewiński, T., Zhou, M. & Rozvany, G.I.N. 1994. Extended exact solutions for least-weight truss layouts–part I: cantilever with a horizontal axis of symmetry, *Int. J. Mech. Sci.*, 36(5): 375–398.

Nagtegaal, J.C. & Prager, W. 1973. Optimal layout of a truss for alternative loads. *International Journal of Mechanical Sciences* 15: 583–592.

Pritchard, T.J., Gilbert, M. & Tyas, A. 2005. Plastic layout optimization of large-scale frameworks subject to multiple load cases, member self-weight and with joint length penalties, *WCSMO-6, Rio de Janeiro, Brazil, 30 May–03 June*.

Rozvany, G.I.N. & Hill, R.H. 1978. Optimal plastic design: superposition principles and bounds on the minimum cost, *Computer Methods in Applied Mechanics and Engineering* 13(2): 151–173.

Rozvany, G.I.N., Sokół, T. & Pomezanski, V. 2014. Fundamentals of exact multi-load topology optimization—stress-based least-volume trusses (generalized Michell structures)—Part I: Plastic design. *Structural and Multidisciplinary Optimization* 50(6): 1051–1078.

Sokół, T. 2011a. A 99 line code for discretized Michell truss optimization written in Mathematica, *Structural and Multidisciplinary Optimization* 43(2): 181–190.

Sokół, T. 2011b. Topology optimization of large-scale trusses using ground structure approach with selective subsets of active bars. In A. Borkowski, T. Lewiński, G. Dzierżanowski (eds), *19th International Conference on Computer Methods in Mechanics, Warsaw, 9–12 May*.

Sokół, T. 2013. Numerical approximations of exact Michell solutions using the adaptive ground structure approach. In S. Jemioło, M. Lutomirska (eds), *Mechanics and materials*, Ch. 6: 87–98, Warsaw University of Technology.

Sokół, T. 2014. Multi-load truss topology optimization using the adaptive ground structure approach. In T. Łodygowski, J. Rakowski, P. Litewka, *Recent Advances in Computational Mechanics*, Ch. 2: 9–16, CRC Press, London.

Advances in Mechanics: Theoretical, Computational and Interdisciplinary Issues – Kleiber et al. (Eds)
 ISBN 978-1-138-02906-4

Stochastic Finite Element Method SORM study of the corrugated web steel plate girder

D. Sokołowski & M. Kamiński

Department of Structural Mechanics, Faculty of Civil Engineering, Architecture and Environmental Engineering, Lodz University of Technology, Łódź, Poland

ABSTRACT: The main purpose of this paper is to present a second order reliability, stress and stability analysis of the steel plate girder with the corrugated web subjected to the Gaussian random fluctuations in its web thickness. Such an analysis is carried out using the Stochastic Finite Element Method (SFEM) based on the generalized stochastic perturbation technique and discretization of structure with the volumetric finite elements. It is numerically implemented using the FEM system ABAQUS and the symbolic algebra system MAPLE. We compare the perturbation-based results with these obtained from the traditional Monte-Carlo simulation and, separately, an analytical solution calculated by a symbolic integration carried out in MAPLE. The reliability index is calculated according to the First Order Reliability Method (FORM) or Second Order Reliability Method (SORM) and can be further used in durability prediction of such structures. A stochastic variable is web thickness, whose randomness comes from an extensive corrosion or manufacturing imperfection.

1 INTRODUCTION

The corrugated web beams increase their influence in civil engineering practice since their first appearance principally due to their high transverse rigidity and small weight. They tend to substitute the classical I-beams and columns and are extensively used as supports, roof trusses and columns in hall constructions, homogeneous steel bridge girders with large spans, as well as in composite box bridge girders, where the flanges are made of concrete. Contrary to the beliefs of the producers and practitioners, theoretical foundations and numerous computational issues connected with the corrugated, and particularly SIN web, remain still unresolved and may contribute to their failure. Furthermore, the current state of the art is insufficient to distinguish a response of the corrugated web beams to the elevated temperature and reliability of such girders is still questionable. This is principally due to longitudinal waviness and vertical slenderness (and extreme thinness) of their webs, which additionally makes them exceedingly predisposed to the stochastic processes like corrosion. That is the reason why the corrugated-web beams recently constitute a topic of the scientific interest on numerical and experimental basis, especially in terms of its stability, reliability, response to the elevated temperature and behavior with web openings.

2 NUMERICAL ANALYSIS

A numerical analysis concerns the simply supported girder with a span of 40 m, the web height of 2,5 m, flange width of 1,4 m and loaded by the uniform pressure of 107.14 kN/m^2 on its upper flange and by the self-weight.

2.1 *Details of the FEM model*

An analysis is made in the FEM system ABAQUS with 573 043 linear hexahedra (C3D8R) and 152 460 second-order tetrahedral finite elements (C3D10). Geometry includes the web and flanges, as well as the transverse ribs (hexahedral FEs) and welds, which are discretized by tetrahedral elements. A constitutive model used to simulate both, steel and welds is linear with a plastic limit of 460 MPa and connections between these materials are considered as a full contact and defined as ties of a total amount of 1392. Such an assumption is principally adopted to reduce the computation time and additionally because the authors are not interested in the breakage of bonds between the weld and steel. Details of the girder geometry and discretization are provided in Figure 1.

2.2 *von Mises stress*

The numerical results include von Mises stresses, whose distribution is presented on Figure 2.

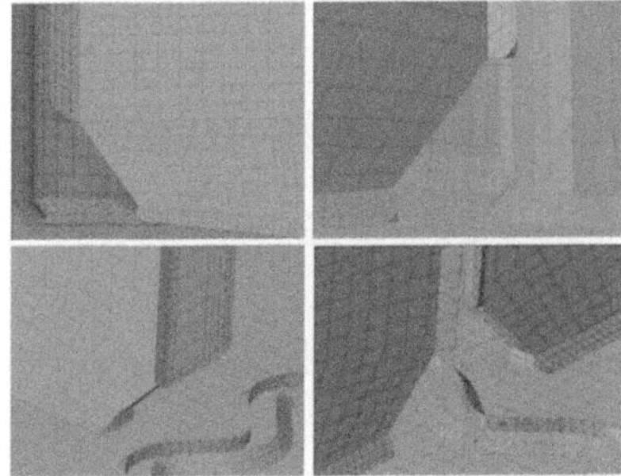

Figure 1. A discretization of the FEM model.

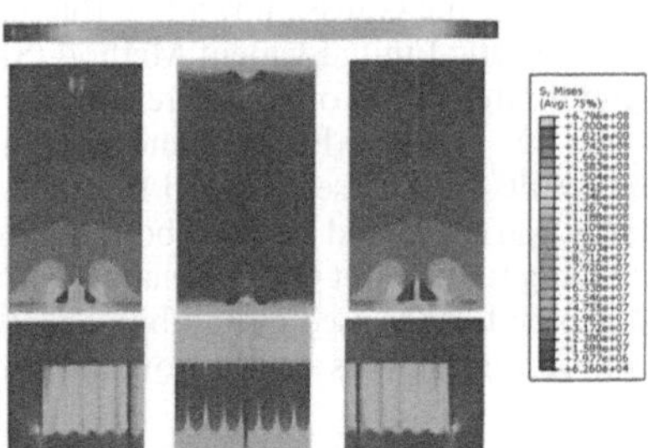

Figure 2. A von Mises stress of the volumetric model with ribs and welds.

In addition to the stress distribution revealed by the beam theory, some substantial periodic stress peaks are placed on the weld and nearby the support. Interestingly, the highest stress is reported in the middle of the girder span on the weld, and the (shear) stress pattern nearby the support is nowhere near to proportional with the main axis of the girder as the beam theory states. An additional remark is, that the transverse ribs on the support and, particularly, in the middle of the girder span are majorly unused. It is also worth to mentioning, that some of the stress peaks located in the weld may be strengthened by the geometrical complexity of contact between the weld and the web (a commonly known numerical problem of the commercial FEM software). Nevertheless, such a behaviour is evident for other (shell) models simultaneously made for the same geometry and therefore should be taken into consideration during the design process. Some additional insight in this matter is available in (Sokołowski et al. 2014).

2.3 *Stability*

Considering the stability of such a girder, its mode is strictly related to the web thickness. When web is thick, a girder fails for the moment

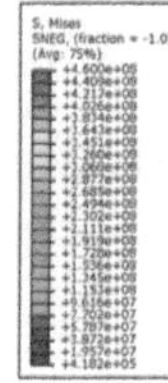

Figure 3. A von Mises stress pattern at the critical load for $t_w = 5.0$ mm.

in the middle of its span and the web is perfectly stable. Only when the web becomes 5 mm and less (ratio of web thickness to its height reaches 1:500), it starts to behave locally unstable on the support. This behaviour is presented on Figure 3, showing reach of a girder plastic limit nearby the support. This scheme is obtained with the use of an additional, shell model with 77 422 rectangular 4-noded shell elements discretizing the above reported geometry and under the load equivalent to its first critical value revealed by the buckling FEM analysis.

3 RELIABILITY ANALYSIS

A reliability analysis is made for the three different kinds of weighting scheme of the Least Squares Method (LSM), Dirac, uniform and triangular with polynomial response function and according to the FORM and SORM. This is done for the two principal limit functions of deflection and normal stress describing the girder Ultimate and Serviceability Limit States (ULS and SLS).

3.1 *Theoretical background*

General formulas used for reliability index determination are the following ones (after Kamiński 2013, Melchers & Horwood 1987 and Waarts & Vrouwenvelder 1999):

$$\beta_{FORM} = \frac{E[g]}{\sigma(g)}, \tag{1}$$

$$\beta_{SORM} = -\Phi^{-1}\left(P_{f2}\right), \tag{2}$$

where $E[g]$ denotes the limit state function, $\sigma(g)$ is its variance and P_{f2} means the probability of failure according to the chosen probability distribution, which can be calculated according in the following manner:

$$P_{f2} = \frac{\Phi(\beta_{FORM})}{\sqrt{1+\beta_{FORM}\kappa}}, \tag{3}$$

where κ denotes the curvature approximating the primary surface for the given probability distribution.

3.2 *Reliability index*

The resulting graphs of reliability given on Figures 4–6 firstly show, that the index is non-linearly decreasing with an increase of the input uncertainty for both, FORM and SORM and is basically independent of the probabilistic method type. An additional observation is that its value is much higher than the limit one (3.3–4.3) up to a high input uncertainty of more than $\alpha = 0.20$ for the limit function of displacement (Figs. 4 and 5) and $\alpha = 0.12$ for the limit function of the normal stress (Figs. 5 and 6). Furthermore, the type of the weighting scheme selected for the analysis has a noticeable influence on both the character and magnitude of the index, especially for the moderate input uncertainty for displacements and low input uncertainty for normal stress. Interestingly, the reliability for the latter limit function is converging only for the vanishing or high input uncertainty, revealing a big dependence of the resulting reliability for the realistic input Coefficients of Variation (CoVs).

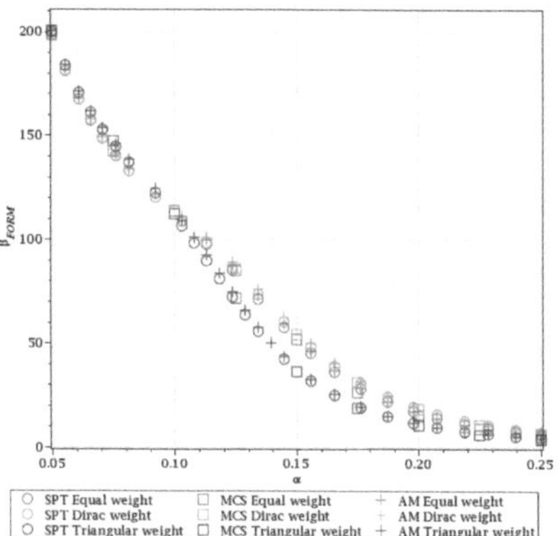

Figure 4. A reliability index of the corrugated I-beam for the different types of a weighting scheme according to the FORM and based on the limit function of displacement.

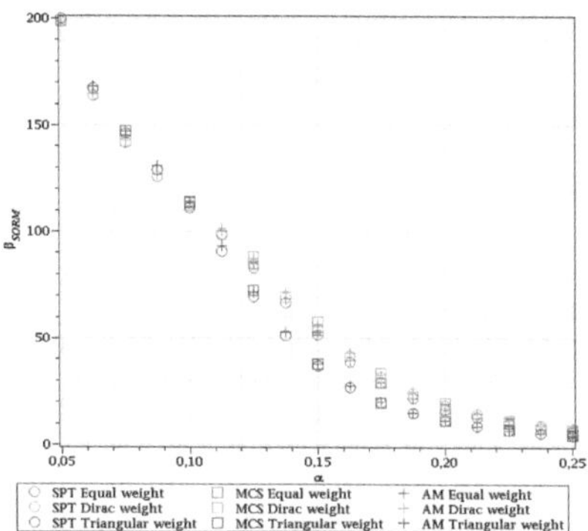

Figure 5. A reliability index of the corrugated I-beam for the different types of a weighting scheme according to the SORM and based on the limit function of displacement.

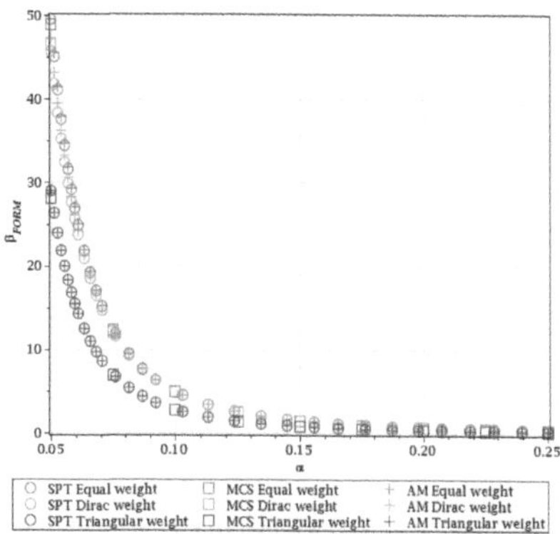

Figure 6. A reliability index of the corrugated I-beam for the different types of a weighting scheme according to the FORM and based on the limit function of the ultimate normal stress.

A fundamental remark is, that the reliability index may substantially differ for various limit functions and, consequently, a check of all of these functions applicable for the studied structure should be supplemented in reliability analysis (such as the ultimate displacement, shear, normal, deviatoric and Huber Mises stress, stability, eigenfrequency, etc). The overall reliability in that case can be defined as the lowest of all the studied ones, which in the case of the girder is the one for the normal stress.

The difference between the FORM and SORM approach is principally visible for the high input CoVs, for which the consequent approaches tend to produce higher (SORM) or lower (FORM) results.

3.3 *Skewness and kurtosis for the ultimate normal stress*

The skewness and kurtosis of the ultimate normal stress limit function (Figs. 8 and 9) show a high dependence of the trend, magnitude and sign on the type of the weighting scheme, while are much less sensitive to the change of the stochastic method applied. This is not valid for the kurtosis of the SPT method, which produces an opposite sign to the other methods, yet still having a comparable magnitude of results. This problem is common for the SPT method, but it does not have any influence on the resulting reliability index of the girder (see Figs. 4–7) and can, consequently, be disregarded in the analysis.

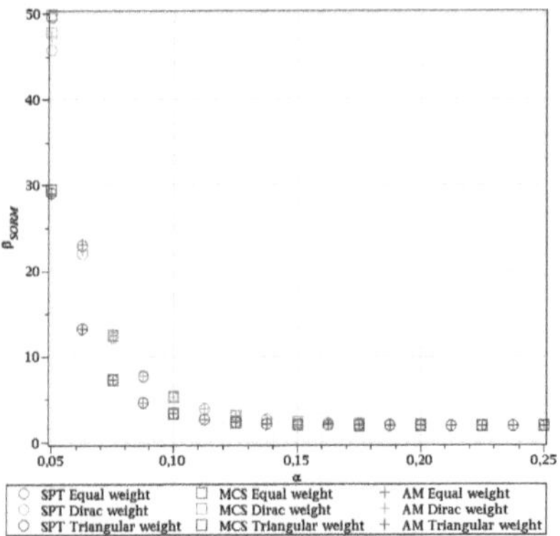

Figure 7. A reliability index of the corrugated I-beam for the different types of a weighting scheme according to the SORM and based on the limit function of the ultimate normal stress.

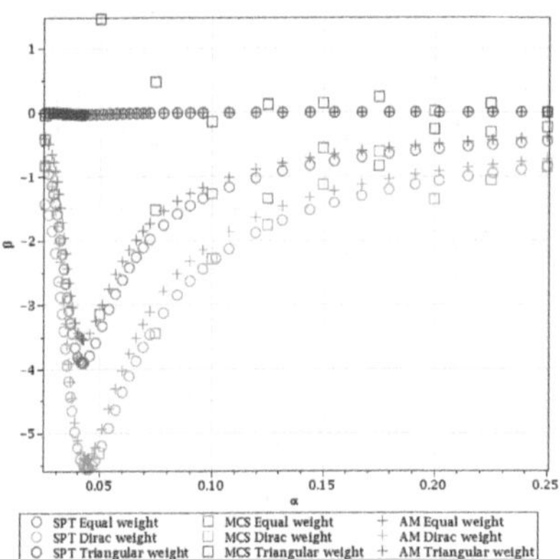

Figure 8. A skewness of the corrugated I-beam for the different types of a weighting scheme based on the limit function of the ultimate normal stress.

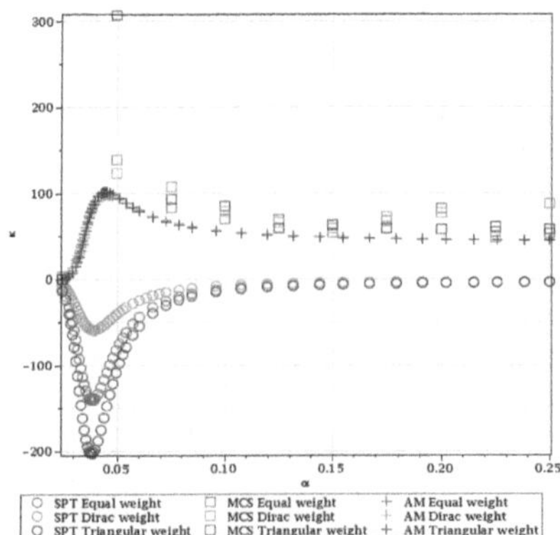

Figure 9. A kurtosis of the corrugated I-beam for the different types of a weighting scheme based on the limit function of the ultimate normal stress.

4 CONCLUSIONS

The results show, that the analysed SIN web girder is not exactly following the beam theory, especially for the stress field in the web. Nevertheless, the ultimate normal stress on the flange together with the ultimate deflection of the girder in the middle of its span can be calculated with this theory accurately enough. A second observable is, that the web corrugation brings an excellent stability of te web allowing its slenderness up to 500 and by this can substantially reduce the amount steel used in the I-beam section.

The reliability of such girder is sufficiently high for the low and moderate input CoV of the web thickness (up to $\alpha = 0.12$) and the girder is much more sensitive for the normal stress (ULS) than the displacement (SLS). The weighting scheme of the WLSM is not a leading factor in computations of the reliability index, but has a small influence on the result and all the stochastic methods show a common result of the index. This is also valid for the skewness, but questionable for the kurtosis, where the stochastic perturbation method shows an opposite sign with comparable magnitude to other methods. This divergence, however, has no influence on the reliability index itself, which currently is recommended for checking in the binding European standards (Eurocode 0 2002).

Further design attention is planned for design and analysis of the hybrid girders, where the flanges are made of a higher plastic limit steel than the web and, separately, modelling of imperfections, geometrical sensitivity and corrosion of such systems.

ACKNOWLEDGEMENTS

This article has been written to commemorate the 60th anniversary of the Faculty of Civil Engineering, Architecture and Environmental Engineering and 70th anniversary of Łódź University of Technology, Alma Mater of both Authors.

REFERENCES

Eurocode 0, 2002, Basis of Structural Design. *European Committee for standardization*. Brussels.

Kamiński, M. 2013, The Stochastic Perturbation Method for Computational Mechanics, Wiley, Chichester.

Melchers, R.E., Horwood, E., 1987, Structural reliability. Analysis and prediction. Wiley, Chichester.

Sokołowski, D., Kamiński, M., Strąkowski, M., 2014, Stochastic Finite Element Method reliability analysis of the corrugated I-beam girder. CMES: *Computer Modeling in Engineering & Sciences* 99(3): 209–231.

Waarts, P.H., Vrouwenvelder, A.C.W.M., 1999, Stochastic finite element analysis of steel structures, *Journal of Construction Steel Research* 52(1): 21–32.

Advances in Mechanics: Theoretical, Computational and Interdisciplinary Issues – Kleiber et al. (Eds)
© 2016 Taylor & Francis Group, London, ISBN 978-1-138-02906-4

Influence of separation gap on the response of colliding models of steel structures under seismic and paraseismic excitations

B. Sołtysik & R. Jankowski
Gdańsk University of Technology, Gdańsk, Poland

ABSTRACT: As a result of high urbanization, the need to erect closely-spaced buildings forces designers to consider collisions between structures taking place during ground motions. Experimental and numerical studies have confirmed that such collisions (often referred to as structural pounding) may cause serious damages to the structural elements and may even lead to the total collapse of colliding structures. The aim of the paper is to show the results of shaking table experimental investigation focused on dynamic behaviour of closely-separated steel structures under seismic and paraseismic excitations. To perform this study, models of three steel towers, with different dynamic parameters, have been constructed and mounted to the platform of a shaking table. The influence of the size of separation gap ($\Delta_1 = 0$ mm; $\Delta_2 = 30$ mm; $\Delta_3 = 60$ mm) on the response of colliding structures has been investigated. Experimental study confirms that pounding may lead to significant changes in the structural behaviour under seismic and paraseismic excitations. Generally speaking, the smallest structural response has been observed for the case of zero gap size as well as when the gap size is large enough to prevent pounding.

1 INTRODUCTION

Seismic as well as paraseismic excitations are among the most unpredictable loads acting on civil engineering structures. The safety and reliability of closely-spaced buildings under ground motions are among a number of different aims during the design stage. This is due to the fact that interactions between insufficiently separated structures may cause serious damages, finally leading to their total collapse. Collisions between adjacent buildings have been repeatedly observed under moderate to strong ground motions. Interactions between two parts of a building during the earthquake of 2004 in the north-eastern Poland resulted in local damage at the places of impacts (Zembaty et al. 2005). After the Athens earthquake of 1999, a significant damage due to pounding between adjacent wings of school buildings was observed (Vasiliadis & Elenas 2002). Structural collisions resulted in total collapse of a number of structures during the Loma Prieta earthquake in 1989 (Kasai & Maison 1997). Also, the San Fernando earthquake in 1971 caused serious damage at the places of interactions between the main building of the Olive View Hospital and the stairway tower (Mahin et al. 1976). The observations after the Mexico earthquake in 1985 indicate that one of the main reasons of structural damage was related to pounding between adjacent buildings (Rosenblueth & Meli 1986). Besides the inadequate gap distance between neighbouring buildings, a major reason leading to structural pounding results from the differences in dynamic parameters of the structures (Anagnostopoulos 1994, Jankowski 2005, Mahmoud et al. 2013). These differences lead to the out-of-phase vibrations and induce interactions during the time of ground motion.

Earthquake-induced pounding is a complicated phenomenon which involves deformations at contact points, local cracking or crushing, fracturing due to impact, friction, etc. Therefore, the phenomenon is very difficult for a mathematical analysis. In spite of its complexity, the earthquake-induced structural pounding has recently been intensively analysed using various structural models and applying different models of collisions (Goldsmith 1960, Anagnostopoulos 1988, Jankowski 2007b, Mahmoud & Jankowski 2011). The fundamental investigation of collisions between adjacent buildings, modelled as single degree-of-freedom systems, was conducted by Anagnostopoulos (1988). More advanced analyses were also performed, using multi degree-of-freedom lumped-mass models (Maison & Kasai 1992, Anagnostopoulos & Spiliopoulus 1992, Karayannis & Favvata 2005ab, Cole et al. 2011, Efraimiadou et al. 2013) and applying the Finite Element Method (Jankowski 2007a, Sołtysik & Jankowski 2013).

The research on structural pounding between buildings has recently been much advanced. However, most of the research works has concerned reinforced concrete buildings and studies on steel structures (especially the experimental ones) are very limited (Sołtysik & Jankowski 2013, 2015). Therefore, the aim of the paper is to present the results of experimental investigation focused on dynamic behaviour of closely-separated models of steel structures under seismic and paraseismic excitations. The influence of the size of separation gap on the response of colliding model structures has been investigated by conducting tests on a shaking table.

2 EXPERIMENTAL STUDY

2.1 *Experimental model*

To perform the experimental study, models of three steel towers, with height of 1000 mm and different dynamic parameters, were constructed and mounted to the platform of a shaking table (see Fig. 1). Towers were made out of four steel columns (rectangular box section 15 × 15 × 1.5 mm) with spacing of 480 mm in the longitudinal direction (corresponding to the load direction) and 571 mm in the transverse direction. Additional skew bracings (also rectangular box section 15 × 15 × 1.5 mm) were used to prevent transverse and torsional vibrations. In order to obtain different dynamic characteristics of structures, additional mass (concrete plates with the dimensions of 500 × 500 × 70 mm and 42.2 kg weight) were installed at the top of model structures. In the configuration analysed, two concrete plates were mounted at the top of the middle tower (tower no. 2), only one concrete plate was mounted at the top of the external towers (tower no. 1 and 3)—see Figure 1.

In order to conduct the experimental study, a unidirectional shaking table, located at the Faculty of Civil and Environmental Engineering, Gdańsk University of Technology, was used. This device has a platform of 2000 × 2000 mm, which allows us to test models of the maximum weight of 1000 kg. The linear actuator, which may induce movement with maximum acceleration of 10 m/s^2 and a maximum force of 44.5 kN is connected to the platform. The following equipment was used to carry out the measurements:

- four single-axis accelerometers,
- amplifier with low pass filter of 100 Hz,
- analogue-digital card.

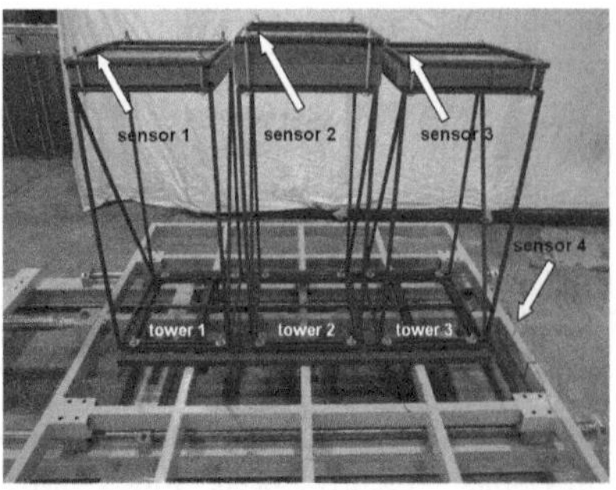

Figure 1. Setup of the shaking table experiment.

2.2 *Free vibration tests*

The first stage of the study has been focused on identification of the dynamic properties of each structure by conducting the free vibration tests. Four accelerometers have been used during each test to measure structural vibrations. Three of the sensors have been located at the top of the towers and one has been placed at the platform to control its movement (see Fig. 1). Values of natural frequency for each tower, as obtained from the free vibration tests, are summarized in Table 1. As it can be seen in the table, the natural frequencies of the models are very similar to the parameters of small, few-storey buildings. This fact justifies the acceptance of the scaled structural models and allows us to draw more general conclusions related to real structures.

2.3 *Dynamic tests*

After conducting the free vibration tests, the dynamic tests have been performed under different seismic and paraseismic excitations. In this paper, the examples of the results for the following excitations are presented:

- Northridge earthquake of January 17, 1994 (Santa Monica station, EW component),
- Polkowice mining tremor of February 20, 2002 (NS component).

Additionally, different gap size values between the towers have been taken into consideration. In this paper, the results for the gap size of $\Delta_1 = 0$ mm, $\Delta_2 = 30$ mm and $\Delta_3 = 60$ mm are presented.

The examples of the results, in the form of acceleration time histories for tower no. 1 under the Northridge earthquake and the Polkowice mining tremor for the gap distance equal 0 mm are shown in Figures 2 and 3, respectively. The peak structural

Table 1. Natural frequency values of towers.

Tower number	Natural frequency value Hz
Tower no. 1	3.190
Tower no. 2	2.450
Tower no. 3	3.120

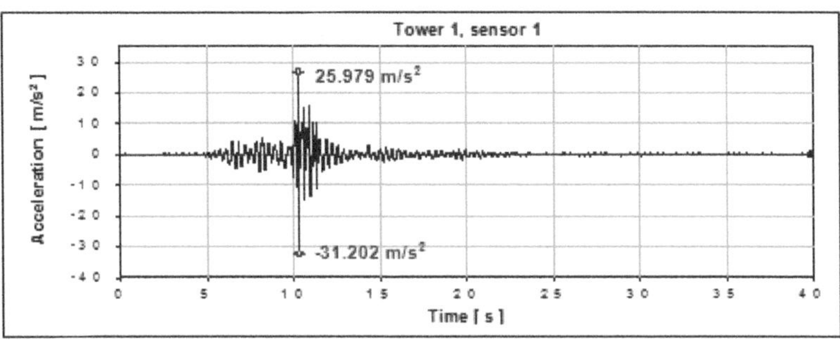

Figure 2. Acceleration time history for tower no. 1 under the Northridge earthquake for gap distance equal 0 mm.

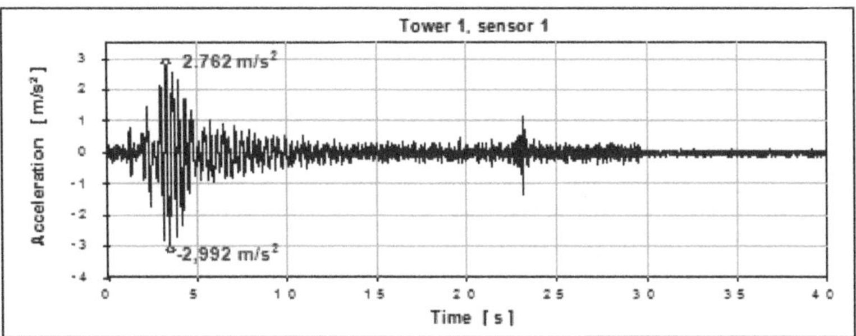

Figure 3. Acceleration time history for tower no. 1 under the Polkowice mining tremor for gap distance equal 0 mm.

Table 2. Peak acceleration values for the Northridge earthquake and the Polkowice mining tremor.

	Peak acceleration (m/s^2)		
	Separation gap 0 mm	Separation gap 30 mm	Separation gap 60 mm
	Northridge earthquake (1994)		
Tower no. 1	31.202	56.439	12.394
Tower no. 2	26.174	66.384	31.228
Tower no. 3	22.820	143.776	71.080
	Polkowice mining tremor (2002)		
Tower no. 1	2.992	3.151	3.838
Tower no. 2	3.562	2.105	2.565
Tower no. 3	3.395	3.166	3.148

response accelerations for all towers and gap size values considered in the analysis are summarized in Table 2.

It can be seen from Table 2 that, in the case of the Northridge earthquake, the change in value of separation gap from 0 mm to 30 mm results in the increase in the peak value of acceleration for tower no. 1 from 31.202 m/s^2 to 56.439 m/s^2 (increase by 80.9%). On the other hand, when the gap size is further enlarged to 60 mm, which is large enough to prevent pounding, the response of this tower decreases by about 78.0%. Similar situation takes place for tower no. 2, where the change of the gap distance from 0 mm to 30 mm results in the increase in the structural response (by 153.6%), but also when the gap size is equal 60 mm, the response is decreased (by 53.0%) comparing to the 30 mm distance. This trend is repeated for tower no. 3. When the gap size is enlarged from 0 mm to 30 mm, the peak response of this structure increases from 22.820 m/s^2 to 143.776 m/s^2 (increase by as much as 530.0%). Then, the value falls down to 71.080 m/s^2 (decrease by 50.6%) when the gap size is enlarged to 60 mm.

For the Polkowice mining tremor, the increase in the gap size leads to the uniform increase in the peak acceleration of tower no.1 (increase by 28.3%

for the change in value of separation gap from 0 mm to 60 mm). However, different trends concern tower no. 2 and 3. In the case of tower no. 2, the peak response of the structure first decreases (by 40.9%) and then increases (by 21.9%) when the gap size value is increased to 30 mm and 60 mm, respectively. On the other hand, the peak acceleration of tower no. 3 shows the uniform decrease trend when the separation gap changes from 0 mm to 60 mm (decrease by 7.3%).

3 CONCLUDING REMARKS

The results of the shaking table experimental study focused on pounding between models of steel structures with different dynamic parameters have been presented in this paper. The first stage of the study has been focused on identification of the dynamic properties of each structure by conducting the free vibration tests. Then, the dynamic tests have been conducted to observe the structural response of colliding structures under different seismic and paraseismic excitations.

The results of the study show that structural pounding during ground motions, due to insufficient in-between separation, may lead to significant changes in the responses of colliding structures. In the case of the behaviour of three structures analysed in this paper, interactions have resulted in the decrease as well as in the increase in the structural behaviour. Generally speaking, the smallest structural response has been observed for the case of zero gap size as well as when the gap size is large enough to prevent pounding.

The results of the study clearly indicate that pounding may have a significant influence on the response of colliding structures under seismic and paraseismic excitations. Further research is therefore required so as to fully identify the threats related to structural interactions during different ground motions.

REFERENCES

Anagnostopoulos, S.A. 1988. Pounding of buildings in series during earthquakes. *Earthquake Engineering and Structural Dynamics* 16: 443–456.

Anagnostopoulos, S.A. 1994. Earthquake induced pounding: State of the art. *Proceedng of the 10th European Conference on Earthquake Engineering*: 897–905.

Anagnostopoulos, S.A. & Spiliopoulus, K.V. 1992. An investigation of earthquake induced pounding between adjacent buildings. *Earthquake Engineering and Structural Dynamics* 21: 289–302.

Cole, G., Dhakal, R., Carr, A. & Bull, D. 2011. An investigation of the effects of mass distribution on pounding structures. *Earthquake Engineering and Structural Dynamics* 40: 641–659.

Efraimiadou, S., Hatzigeorgiou, G.D. & Beskos, D.E. 2013. Structural pounding between adjacent buildings subjected to strong ground motions. Part I: The effect of different structures arrangement. *Earthquake Engineering and Structural Dynamics* 42: 1509–1528.

Goldsmith, W. 1960. *Impact: The theory and physical behavior of colliding solids*. Edward Arnold Ltd, London, UK.

Jankowski, R. 2005. Impact force spectrum for damage assessment of earthquake-induced structural pounding. *Key Engineering Materials* 293–294: 711–718.

Jankowski, R. 2007a. Assessment of damage due to earthquake-induced pounding between the main building and the stairway tower. *Key Engineering Materials* 347: 339–344.

Jankowski, R. 2007b. Theoretical and experimental assessment of parameters for the non-linear viscoelastic model of structural pounding. *Journal of Theoretical and Applied Mechanics* 45(4): 931–942.

Karayannis, C.G. & Favvata, M.J. 2005a. Earthquake-induced interaction between `adjacent reinforced concrete structures with non-equal heights. *Earthquake Engineering and Structural Dynamics* 34: 1–20.

Karayannis, C.G. & Favvata, M.J. 2005b. Inter-story pounding between multistory reinforced concrete structures. *Structural Engineering and Mechanics* 20: 505–526.

Kasai, K. & Maison, B. 1997. Building pounding damage during the 1989 Loma Prieta earthquake. *Engineering Structures* 19: 195–207.

Mahin, S.A., Bertero, V.V., Chopra, A.K. & Collins, R.G. 1976. Response of the Olive View Hospital main building during the San Fernando earthquake. *Report No. EERC 76–22, Earthquake Engineering Research Center, University of California* Berkeley, USA.

Mahmoud, S., Abd-Elhamed, A. & Jankowski, R. 2013. Earthquake-induced pounding between equal height multi-storey buildings considering soil-structure interaction. *Bulleting of Earthquake Engineering* 11(4): 1021–1048.

Mahmoud, S. & Jankowski, R. 2011. Modified linear viscoelastic model of earthquake-induced structural pounding. *Iranian Jour. of Science and Technology* 35(C1): 51–62.

Maison, B. & Kasai, K. 1992. Dynamics of pounding when two buildings collide. *Earthquake Engineering and Structural Dynamics* 21: 771–786.

Rosenblueth, R. & Meli, R. 1986. The 1985 earthquake: causes and effects in Mexico City. *Concrete International, American Concrete Institute* 8: 23–34.

Sołtysik, B. & Jankowski, R. 2013. Non-linear strain rate analysis of earthquake-induced pounding between steel building. *International Journal of Earth Sciences and Engineering* 6(3): 429–433.

Sołtysik, B. & Jankowski, R. 2015. Building damage due to structural pounding during earthquakes. *11th International Conference on Damage Assessment of Structures DAMAS 2015, Journal of Physics: Conference Series* 628: 1–10.

Vasiliadis, L. & Elenas, A. 2002. Performance of school buildings during the Athens earthquake of September 7 1999. *12th European Conference on Earthquake Engineering*, paper ref. 264: 429–433.

Zembaty, Z., Cholewicki, R., Jankowski, R. & Szulc J. 2004. Trzęsienia ziemi 21 września 2004 r. w Polsce północno-wschodniej oraz ich wpływ na obiekty budowlane. *Inżynieria i Budownictwo* 61(1): 3–9 (in Polish).

Advances in Mechanics: Theoretical, Computational and Interdisciplinary Issues – Kleiber et al. (Eds)
© 2016 Taylor & Francis Group, London, ISBN 978-1-138-02906-4

Discrete random variables in reliability calculations of a reticulated shell

P. Sorn, J. Górski & K. Winkelmann
Faculty of Civil and Environmental Engineering, Gdańsk University of Technology, Gdańsk, Poland

J. Przewłócki
Faculty of Architecture, Gdańsk University of Technology, Gdańsk, Poland

ABSTRACT: Implementation of the Point Estimation Method (PEM) in the reliability analysis of a three-dimensional truss structure is presented in the paper. The influence of geometric and material random parameters on the truss load-carrying capacity was investigated. The analysis was performed for different combinations of basic variables. Symmetric and asymmetric cases of snow load were taken to assess the structural reliability. Sensitivity of the structural response to cross-sectional area was also taken into account. The obtained results were compared with the Monte Carlo and Response Surface Method calculations.

1 INTRODUCTION

The problem of cost minimization in construction industry is always important and up–to–date. The cost reduction can be obtained by a careful structural design, employing sensitivity analysis or optimization methods. A number of key factors should be taken to solve the considered task. A proper selection of structural element cross–sections is such a high importance issue.

Engineering calculations are usually conducted for an idealized structural model and deterministic loads. However, many structures are highly sensitive to variations of their material and geometric parameters. These fluctuations are usually random, so a probabilistic description should be used in engineering design. Applicable methods and related software usually take a solid theoretical background on probability theory to deal with the random variables used and their moments. The Monte Carlo simulation Method (MCM) is crucial in the field, usually used for the result verification. A number of the so–called variance reduction techniques, e.g. Stratified Sampling (SSMC) and Latin Hypercube Sampling (LHS) were formulated to reduce the computational effort.

In the paper, a reticulated shell structure described by random parameters is analyzed. The Point Estimate Method (PEM) is applied (Rosenblueth, 1975). The method is easy in its use, requiring little theoretical probabilistic background. The main advantage of PEM, in contrast to other probabilistic routines, is a small number of calculations required for a relevant solution. The PEM is sufficiently accurate to estimate probabilistic moments of engineering importance. It was widely applied in geotechnical problems, only a few papers took other structural reliability problems into account.

The obtained results were also compared with calculations utilizing the Response Surface Method (RSM) using an authors' dedicated software RSM–Win, described in Winkelmann & Górski (2014).

2 THE POINT ESTIMATE METHOD (PEM)

The PEM maps the continuous random variable of a given probability density function into a discrete variable of a probability mass function containing N points.

The following formula holds

$$p_X(x) = \sum_{i=1}^{N} \delta(x - x_i) p(x_i) \tag{1}$$

where $\delta(x - x_i)$ is Dirac delta, and $p(x_i)$ is a probability corresponding to a fixed value of random variable x_i.

The number of points N selected for the estimation depends on the order of probabilistic moments to be analyzed. The standard PEM is inefficient in case of a large number of random variables to be considered. In this case n^2 samples should be used. Thus many modified versions of the classical approach exist. The Rosenblueth method is one of the simplest and the most effective procedures, reducing the number of samples only to $2n + 1$. The algorithm proposed by Hong (1998) reduces substantially the number of evaluation points to $2n$.

Although this method is dedicated only to uncorrelated random variables, their dispersion may be assessed by means of skewness.

3 RETICULATED SHELL

A typical example of a reticulated structure described among others by Hensley and Azar (1968), was taken into consideration (Fig. 1). The radius of a three-dimensional truss was 50 m, the truss was 8.216 m high. Tubular sections RO 647.8 × 20 were designed for the structural elements. All elements were made of S355 steel. The elements were connected by means of ball joints.

The critical load for the ideal structure was calculated λ_{ideal} = 0.26123. The limit state function (performance function) was defined as the value of the admissible load multiplier, its non–exceedance standing for a safe state. Several preliminary examples made using the MSC Nastran code (2001) proved that even small changes in geometric description (node displacements) resulted in considerable changes of the load multiplier. For example, displacement of the highest structure point (no. 13 in Fig. 1) only 0.10 m down led to a 14.3% load multiplier drop (Sorn et al. 2015).

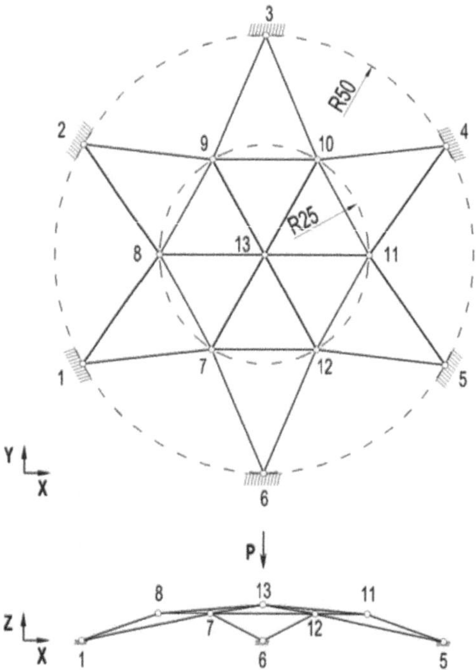

Figure 1. Three dimensional truss structure (reticulated shell).

4 RELIABILITY CALCULATIONS

4.1 *Variants of one and two random variables*

In the first step of the probabilistic analysis a random geometric discrepancy was represented by its mean value m_u = 0.0 m and the standard deviation σ_u = 0.08 m. Thus the majority of generated imperfections covered the interval (–0.13, 0.13 m). Initial vertical displacements of nodes 7–12 were computed proportionally to node 13 discrepancies (Fig. 1). The Direct Monte Carlo method (DMC) was applied. The convergence analysis proved that the results obtained for 55 realizations can be a proper reference level. It is worth noting that only two realizations are required using PEM. In this case the errors of computed mean values and standard deviations are 0.3% and 17%, respectively (Sorn et al. 2015).

Similar analysis was performed using two random variables, the geometric and material parameters. The random node displacements were identical to the case of one-dimensional analysis. The second random parameter, Young's modulus, was also assumed Gaussian (m_E = 210.0 Gpa, σ_E = 4.0 GPa).

The results were compared with the outcomes of the SSMC. The number of subsets was adjusted in order to obtain the number of samples equal to the DMC variant i.e. 6 × 6, 5 × 5, 4 × 4 and 3 × 3. In the case of the PEM computations were only made for four truss models. Relative errors, with respect to the DMC are 1.2% and 1.4%. The results may be considered satisfactory.

The main results from all calculations are summarized in Table 1.

4.2 *Multidimensional random variables*

Following the satisfactory conclusion of the PEM applications, an advanced analysis was then

Table 1. Results of limit load analysis.

Method used	NR—number of realisations	Mean value of resistance *R*	Stan. deviation of resistance *R*
DMC	60	2653.21 kN	232.54 kN
DMC	36	2658.71 kN	229.21 kN
DMC	25	2651.48 kN	227.16 kN
DMC	16	2602.15 kN	253.78 kN
DMC	9	2728.93 kN	309.69 kN
SSMC 6 × 6	36	2637.43 kN	388.28 kN
SSMC 5 × 5	25	2645.16 kN	410.19 kN
SSMC 4 × 4	16	2647.99 kN	389.89 kN
SSMC 3 × 3	9	2633.06 kN	348.67 kN
PEM	4	2620.48 kN	229.38 kN

performed. The space truss model was defined using the following random variables (mean value and standard deviations are given in the brackets):

case a—initial vertical displacement of node 13 (0.0, 0.08 m), Young's modulus (210 GPa, 4.0 GPa),

case b—initial vertical displacements of nodes 7–12 proportional to the displacement of node 13, cross-sectional area (394.458 cm^2, 2.000 cm^2),

case c—Young's modulus (210 GPa, 4.0 GPa), cross-sectional area (394.458 cm^2, 2.000 cm^2),

case d—initial displacement of node 7–13 calculated proportionally to the displacement of node 13, Young's modulus, cross-sectional area,

case e—initial vertical displacement of node 13, initial displacement of node 7–12 calculated proportionally to the displacement of node 12 (0.0 m, 0.03 m), Young's modulus,

The Rosenblueth and Hong versions of the PEM were taken into consideration (Table 2). Each case brought about the mean value and standard deviation of structural limit load-carrying capacity.

The Hong approach (8 realizations) gave almost the same estimators of the limit load and the standard deviations as provided by the classical Rosenblueth method (16 realizations, see Sorn et al. 2015). Note that a greater number of random variables makes the results diverge from the true solutions. That is because the values of random variables defined by the Hong method are not realistic. There is another advantage of the Rosenblueth method in comparison to the Hong method: an additional random variable $n + 1$ leaves the remaining n results accessible. That is because the sample calculations are always performed at points defined by their mean values and standard deviations $m_{x_n} \pm \sigma_{x_n}$.

The results of various methods regarding the mean values (see Table 2) are similar. The dispersion described by standard deviations change when various sets of random variables are analyzed.

Table 2. Results of limit load analysis.

	Rosenblueth version			Hong version		
Case	NR	Mean value	Stan. dev.	NR	Mean value	Stan. dev.
a	5	2605.94	246.09	4	2620.70	228.94
b	5	2603.39	241.03	4	2620.69	223.85
c	5	2603.39	51.00	4	2613.94	51.78
d	7	2605.76	246.35	6	2606.38	245.94
e	9	2658.55	341.40	8	2628.09	363.64

Generally, a greater number of random variables produces an increment of standard deviations in the results. The standard deviation values in case *c* suggest that the specific random variables defined in this set do not act considerably upon the results.

4.3 *Snow load*

A problem of snow–loaded space truss reliability was investigated further on. Two variants of snow load: a uniformly distributed load and an asymmetric load are provided in the PN-EN 1991-1-3 code (Fig. 2). The snow load s [kN/m^2] is described using the formula

$$s = \mu_i C_e C_t s_k \tag{2}$$

where

s_k—characteristic values of ground snow load at the relevant location [kN/m^2],

μ_i—snow load shape coefficient (roof shape factor),

C_e—exposure coefficient (in the analysis it is assumed that $C_e = 1.0$),

C_t—thermal coefficient (in the analysis, $C_t = 1.0$).

Figure 2 presents the distribution of the maximum values of snow load over the plane area along the diameter of the structure. Appropriate shape factors correspond to the model geometry. The snow load is imposed in the form of point forces at the model nodes, to reflect the reactions of the roofing structure on the truss nodes.

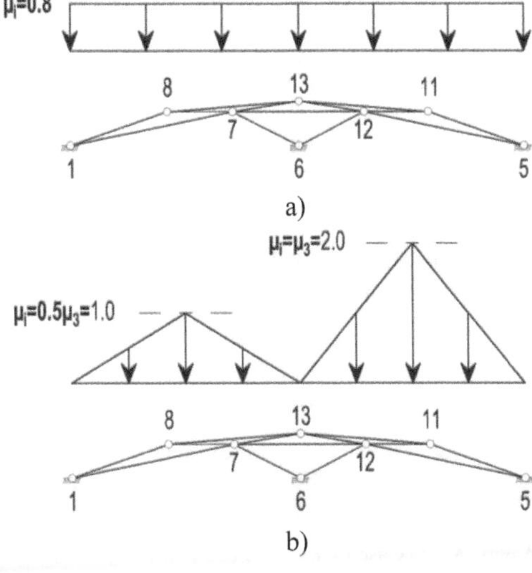

Figure 2. Cases of snow loads.

The space truss parameters were described in two variants, using four and seven random values:

case f – four variables—initial vertical displacement of node 13 and displacements of nodes 7–12 generated as one random variable, Young's modulus and a cross-sectional area,

case g – seven variables—initial vertical displacement of nodes 7–13 generated independently.

The Hong version of the PEM (1998) was applied. The obtained results (Table 3) show a considerable impact of a load class (symmetric, asymmetric) on the probabilistic response parameters.

The obtained results show a considerable impact of a load class (symmetric, asymmetric) on the probabilistic response parameters. The symmetric load acts stronger, in terms of load effects and the geometric response, on nodes 7–12 compared with node 13. It is also worth noting the results for the symmetric snow load when random variables are described by the independent deflections of nodes 7–13 (*case g* in Table 3). This case results in 4 iterations only, due to axisymmetry of load and structural geometry.

Significant difference may be observed between the results obtained for a single force loading (Table 1) and the realistic snow load (Table 3). In the latter case nodes 7–12 are loaded, so their initial geometric displacement has a greater effect on the limit load value. The mean values and the standard deviations of the limit loads decrease significantly under an asymmetric snow load. Small values of standard deviations are detected in this case (Table 3).

4.4 *Reliability estimation*

In order to estimate the reliability of the truss the distribution of the snow load should be considered. According to the data given in (Gwóźdź and Machowski, 2011) the Gumbel distribution is:

$$f(x) = 4.95 \cdot \exp\left[-4.95(x-0.259)\right] \cdot \exp\left[-\exp\left(-4.95(x-0.259)\right)\right] \quad (3)$$

The sensitivity of the truss reliability to the variation of cross-sectional parameters was estimated.

Table 3. Results of limit load analysis—snow load.

Load type	Var. dim.	Var. type	No.	Mean value of limit load	Standard dev. of limit load
Sym.	4	Case *f*	8	23.106 kN/m²	5.43 kN/m²
Asym.	4	Case *f*	8	4.605 kN/m²	0.668 kN/m²
Sym.	7	Case *g*	4	6.024 kN/m²	3.599 kN/m²

According to the Standard PN-EN-1990 the level of structural reliability depends on the assumed lifetime of the structure. A classical Hasofer–Lind approach to reliability index was applied. The results in the case of asymmetric snow load are given in Figure 3.

The truss failure probability can be estimated by the following functions:

– symmetric snow $p_f^s = 4.2965 \cdot 10^{10} A^{-7.5}$
– asymmetric snow $p_f^a = 4.5319 \cdot 10^{4} A^{-3.4351}$

Assuming a 50-year service and the RC3 reliability class the minimum reliability index is $\beta = 4.3$. The results in Figure 3 indicate that the elements' cross–sectional area cannot be less than 300 cm².

4.5 *Verification using the response surface method*

Subsequently, PEM reliability estimation was verified using the RSM, which applies an approximate function (in this case, the 1st order polynomial) to represent the response of the structure.

On the basis of the obtained surface slope factors again, the Hasofer–Lind approach reliability index is calculated, assuming the same cross–sectional area for which the PEM calculations were performed. The RSM results are also shown in Figure 3.

It can be shown, that both approximations are slightly divergent, however their trend lines remain similar, which confirms the accuracy of both approaches. The RS method, due to an insufficient number of approximation points used, seems to be erroneously overestimating the probability of failure.

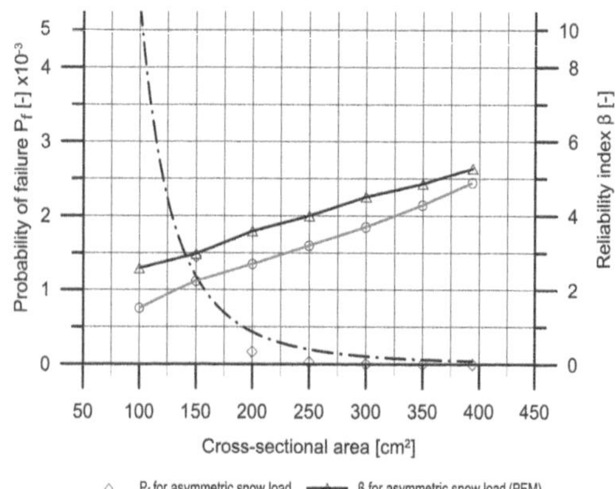

Figure 3. Probability of failure and reliability index for asymmetric snow load, calculated with PEM (dark) and RSM (grey).

5 CONCLUSIONS

The PEM has proved to be an efficient tool for probabilistic engineering applications. The method does not strictly require probability density functions of particular random variables in a reliability analysis. Unlike the FOSM or SORM methods, in PEM it is not necessary to carry on laborious differentiations or iterations. In contrary to the simulation methods only a small number of deterministic realizations is required.

In the case of a low number of random variables (up to three) standard analyses should be applied. When greater numbers of structural random variables are used the PEM can significantly reduce the sample space, not losing the accuracy of the results. A greater number of random variables makes it possible to easily implement the Rosenblueth and Hong methods. On the other hand the field of applications of the above methods is limited. The direct Monte Carlo method and the stratified sampling method can be applied in the case of multivariate random variables. Attention should be paid to the accuracy of the results and specific distribution types, e.g. bimodal.

The investigations have proved that the probabilistic methods can be applied with ease to standard engineering problems. The results of a random analysis may have a substantial influence on the design process.

REFERENCES

Gwóźdź, M., Machowski, A. 2011. Selected studies and calculation of structures using probabilistic methods. Cracow University of Technology.

Hensley, C.C., Azar, J.J. 1968. Computer analysis of nonlinear truss-structures. Journal of Structural Division ASCE, 94, 1437–1439.

Hong, H.P. 1998. An efficient point estimate method for probabilistic analysis. *Reliability Engineering and System Safety* 59(3): 261–267.

MSC Nastran for Windows. Version 2001. MSC Software Corporation. Los Angeles. USA.

PN-EN 1991-1-3 Eurocode 1—Actions on structures—Part 1–3: General actions—Snow loads.

Rosenblueth, E. 1975. Point estimates for probability moments. *Proc. Nat. Acad. of Sci.* 72(10): 3812–3814.

Sorn P., Górski J., Przewłócki J. 2015. Probabilistic analysis of a space truss by means of a multidimensional variable description. Archives of Civil Engineering, 61 (2): 99–124.

Winkelmann, K., Górski J. 2014. The use of response surface methodology for reliability estimation of composite engineering structures. *Journal of Theoretical and Applied Mechanics* 52(4): 1019–1032.

Advances in Mechanics: Theoretical, Computational and Interdisciplinary Issues – Kleiber et al. (Eds)

The influence of utilizing different materials and their configurations on ballistic panels blast resistance

S. Stanisławek & G. Sławiński
Military University of Technology, Warsaw, Poland

ABSTRACT: The article is focused on materials and their configurations to provide protection against explosive blasts of lightly armored vehicles. The authors consider elastomeric material, aluminum and polyurethane foams which are easily available on the local market at a reasonable cost. A computer simulation method was chosen to solve the problem in an efficient way. The Finite Element Method (FEM) implemented in the LS-DYNA commercial code was used with an explicit (central difference) time integration algorithm. Space discretization for each option was made with three dimensional elements ensuring satisfying accuracy of the calculations. The obtained results indicate that the discussed ballistic panels may be utilized as vehicle crew protection. The investigation proves that panels made of energy consuming materials limit the maximum level of reaction force. The highest reaction is observed when polyurethane foam interacts with a blast wave. In other variants both the forces maximum values and curve shapes are similar. Simulations show that ballistic panel supports rebounds from the massive block quickly after the blast reaches its target, which is crucial for the obtained results.

1 INTRODUCTION

The article is focused on materials and their configurations to provide protection against explosive blasts of light armored vehicles. Traditionally, the term "armor" connotes a thick layer, typically steel which can protect an object regardless the nature of a threat. Such images accurately depict traditional protection systems (Gamaa 2001) utilized in past decades but, according to present vehicle mobility requirements, they are not suitable for the modern army. Versatility of an armor is strongly recommended, however, is often hard to achieve. Threats may be divided into two main groups. The first group includes all types of projectiles or blast debris the energy of which is concentrated on a relatively small area during impact. The other danger, which is a subject of this paper, is a blast wave which, while interacting with a large vehicle surface, can injure a crew due to high acceleration of their bodies (Lemonick 2011).

Typical materials for blast absorbing are fiber composites, foams, magneto-rheological fluids, rubbers and porous materials (Uddin 2010). Blast mitigation systems based on such materials demonstrate the capability to absorb large portions of the energy of a blast as well as extend the interval of the blast significantly reducing the peak shock wave.

In the research, the authors focused on a few solutions which are easily available on a local market at a reasonable cost. The first one is elastomeric material in which energy consumption is caused by its hysteresis behavior (Fana 2015). Moreover, aluminum and polyurethane foams, which are presently one of the most commercially available foams, are used. Open-pore foams appear to be particularly suitable for the construction of blast absorption components due to their unique cellular structure (Shim 2012).The geometry and dimensions of the ballistic panels and the test stand were chosen to be easily tested experimentally.

2 INVESTIGATION

A steel stand (Fig. 1) was subjected to a blast wave from a 120 g TNT charge placed 430 mm above the ballistic panel support. Such a value was chosen because it is a distance between the ground and the chassis of KTO Rosomak vehicle (armored personnel carrier) and is considered to be a good example of a light armored vehicle. The height of the ballistic panel was set to 50 mm.

Four kinds of ballistic panels have been tested numerically. Each calculation variant included the same type of a charge mass and a different material or materials structure. In the reference variant (V1), no ballistic panel was used. Two kinds of initial boundary conditions for a massive block were used. In the first one, it is simply placed in the space (V1a),while in the second one its rear

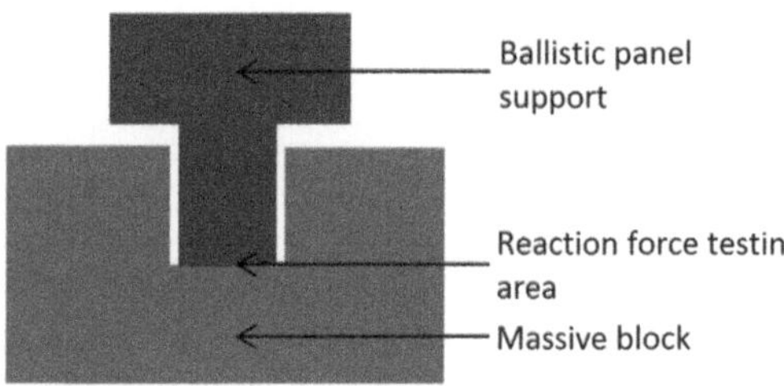

Figure 1. Physical model of a blast test.

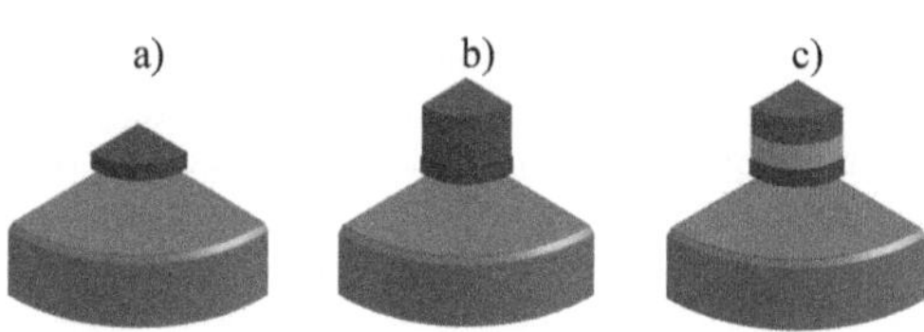

Figure 2. Physical model of a blast test: a) reference variant (V1a, V1b), b) one-layer panel (V2, V3, V4), c) two-layer panel (V5, V6).

Table 1. Simulation tests description.

No.	Name	Description	TNT mass
1	V1a	Reference variant fixed	120 g
2	V1b	Reference variant not fixed	120 g
3	V2	Elastomer	120 g
4	V3	Aluminum foam	120 g
5	V4	Polyurethane foam	120 g
6	V5	Aluminum foam + elastomer	120 g
7	V6	Elastomer + aluminum foam	120 g
8	V3a	Aluminum foam	60 g
9	V3b	Aluminum foam	180 g
10	V3c	Aluminum foam	240 g

wall is fixed (V1b). In V2, V3, V4 variants, rubber material, aluminum foam and polyurethane foam were used respectively. In V5 and V6 variants, two-layer sandwich structures containing rubber and aluminum foams were utilized. Furthermore, V3 variant was tested under different ballistic loads in order to check an influence of a charge mass on panel behavior. All the examined structures are depicted in Figure 2 and all kinds of simulation tests in Table 1. The most important factor deciding about ballistic panel protection ability is force transferred to the massive steel block.

3 CONSTITUTIVE MODELS AND NUMERICAL METHOD DESCRIPTION

A computer simulation method was chosen to solve the problem in an efficient way. The Finite Element Method (FEM) implemented in the LS-DYNA (Hallquist 2006) commercial code was used with an explicit (central difference) time integration algorithm. The boundary conditions were defined by supporting the test stand at its back wall. In order to minimize computation time, only a quarter of the model was analysed and the symmetry was defined.

To describe the contact between the ballistic panel components and the projectile core, a penalty method was utilized. To estimate a wave blast impulse, a widely known method, ConWep, developed by Kingery & Bulmash (1984), is used. All components in the ballistic panels were modeled using hexagonal elements only. The elements size was the same for the whole volume, of each component. In Figure 3, a mesh for a chosen V5 variant is depicted.

Different materials require models which are able to describe their behavior properly. Steel material utilized a simple elastic model as the strain range was very little. The elastomeric material may change its properties in a huge range under tensile, shear or compression load, so Ogden material model (Dubois 2003) was used. Its behavior is based on an energy density equation as follows:

$$W(\lambda_1,\lambda_2,\lambda_3)=\sum_{p=1}^{N}\frac{\mu_p}{\alpha_p}(\lambda_1^{\alpha_p}+\lambda_2^{\alpha_p}+\lambda_3^{\alpha_p}-3) \quad (1)$$

where λ_1, λ_2, λ_3 = principal strains; N, μ_p, α_p = material constants.

A material characteristic was determined based on the laboratory tensile and compression tests and is depicted in Figure 4.

For both aluminum and polyurethane foams, a material model crushable foam (Hallquist 2006)

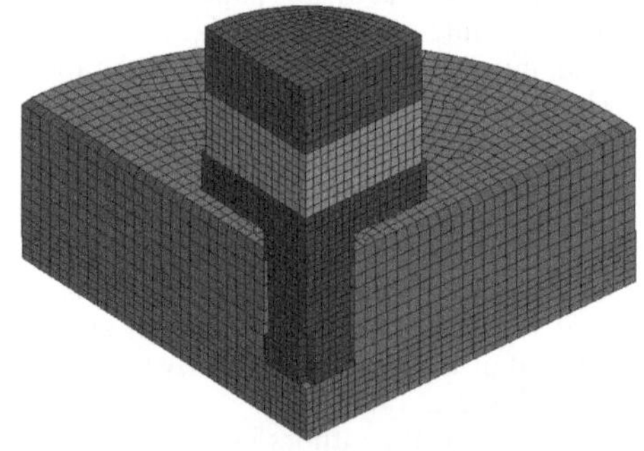

Figure 3. Mesh of two-layer panel (V5).

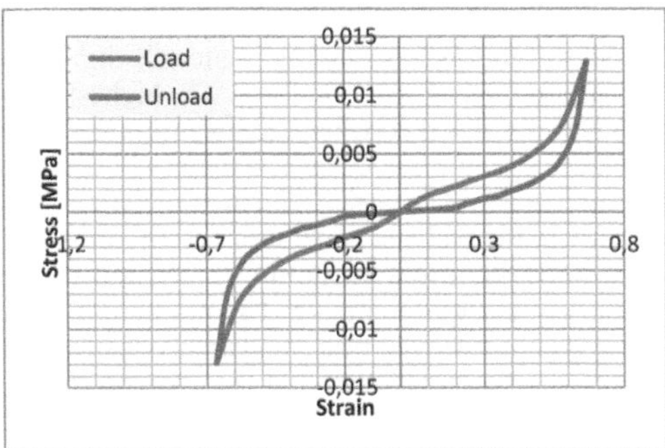

Figure 4. Load and unload stress-strain curves for rub type material.

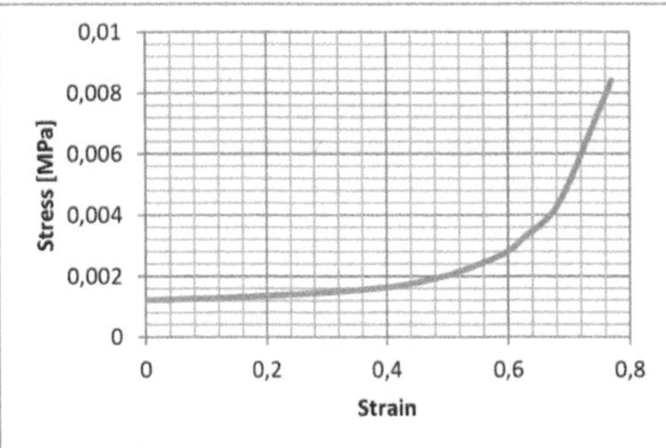

Figure 5. Stress quantity versus volumetric strain for aluminum foam.

was utilized. It predicts that foam can exist in two forms: compressed and uncompressed, in which its properties differ substantially. Young modulus is described by a formula:

$$E_{aa} = E_{aau} + \beta(E - E_{aau}) \tag{2}$$

where E_{aa} = actual elastic modulus; E_{aau} = initial elastic modulus; E = elastic modulus of fully compacted material.

$$G_{ab} = G_{abu} + \beta(E - E_{abu}) \tag{3}$$

where G_{ab} = actual elastic shear modulus; G_{abu} = initial elastic shear modulus; G = elastic shear modulus of fully compacted material.

$$\beta = \max\left[\min\left(\frac{1-V}{1-V_f}, 1\right), 0\right] \tag{4}$$

where V = relative volume; V_f = fully compacted volume.

4 SIMULATION RESULTS

The main purpose of this investigation is an assessment of ballistic panels ability to absorb blast energy. Their layer structure and material used make them easy to manufacture. The assessment is based on the analysis of reaction force which acts on the ballistic panel support. Such force is dangerous for a vehicle crew who suffers from the blast load. It is especially harmful when it increases rapidly.

The results presented in Figure 6 show that maximum forces and work which is performed by those forces during the process differ for each variant. The comparison of V1a and V1b variants indicates the importance of a proper boundary conditions description. Although the mass of the test stand is much bigger than the mass of the panels, it is crucial to represent the way it is fixed in the space or, more adequately, to model the supporting material. In the simulation, the massive block seams to remain steady during the whole process, nevertheless, it is placed freely or its bottom wall is fixed. However, the comparison of the reaction forces (Fig. 6) proves that there is a small movement of the massive block. That fact changes the results substantially. In reality, such a test stand would probably be placed on the ground. Therefore, it is crucial to use a proper soil material model.

The highest reaction force level is observed in the case of V4 variant, where the polyurethane foam interacts with a blast wave (Fig. 7). It is quickly compressed and after this process is finished, the reaction force increases rapidly. The similar results are achieved for V2, V3 and V6 variants. However, in the panel where a blast interacts with the aluminum foam supported by elastomeric material, relatively higher forces can be observed. Moreover, in V3 and V6 variants, the reaction force is reduced mainly due to a foam internal structure rather than inertia of the panel. In both V3 and V6 variants, it can be observed that interaction between the ballistic panel support and the massive block is the longest.

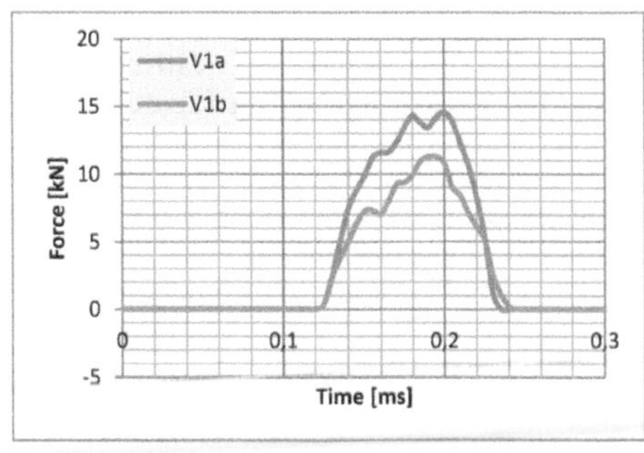

Figure 6. Force versus time for both ways of modeling.

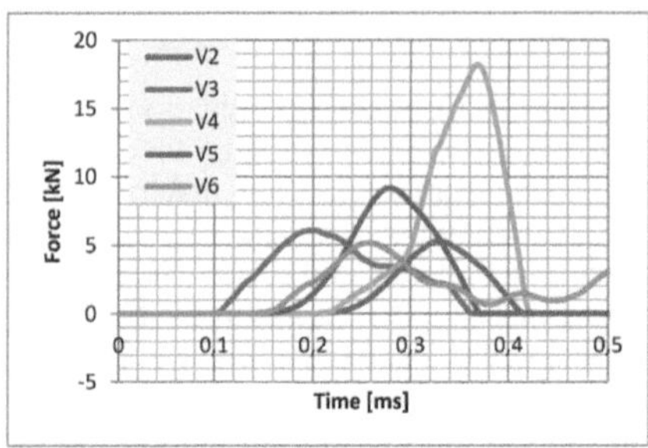

Figure 7. Force versus time for each analyzed panel.

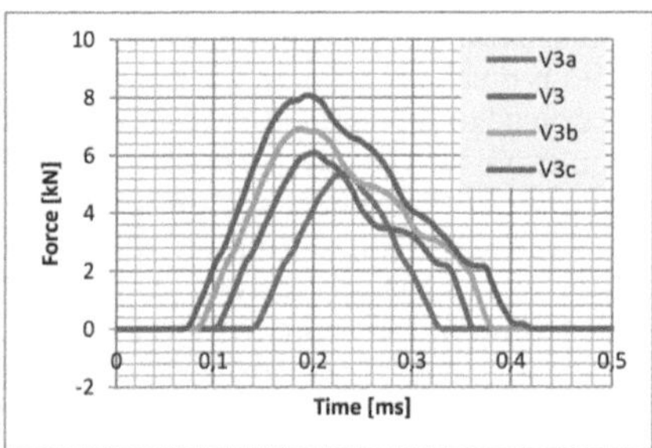

Figure 8. Force versus time for different charge mass.

It is important to notice two significant facts. Firstly, short reaction impulse duration is caused by the fact that the ballistic panel support rebounds from the massive block. If those two parts were connected with each other, the simulation results would be substantially different. Secondly, each test was carried out for different areal density. In those simulation tests, this fact is not considered according to prior assumptions. However, in certain conditions, the mass of the ballistic panel may be found even more important than geometric limitations.

All the above results were achieved for 120 g of TNT material. Is important to verify an influence of different mass of an explosive on the ballistic panels response. Such a simulation was performed for the aluminum foam which showed good properties previously (Fig. 8) and has a relatively low areal density. The results depicted in Figure 6 confirm good energy absorption properties of the aluminum foam. Despite the fact that the biggest mass is 240 g and the mass of the lightest one is only 60 g, the reaction force increases only by 36%. The shape of the curve is similar for all the examined variants.

5 CONCLUSIONS

The obtained results indicate that the discussed ballistic panels may be utilized as vehicle crew protection. The investigation proves that panels made of energy consuming materials limit the maximum level of reaction force. The highest reaction force is observed when the polyurethane foam interacts with a blast wave. The material is quickly compressed and in a such form transfers energy directly to the ballistic panel support. In other variants, both the forces maximum values and shapes of curves are quite similar. The aluminum foam panel is characterized by a small areal density comparing to others. The fact that the reaction force is reduced is caused mainly by its internal structure. In both V3 and V6 variants, it can be observed that interaction between the ballistic panel support and the massive block is the longest. Moreover, the simulations show that the ballistic panel support rebounds from the massive block quickly after the blast reaches its target. The achieved values of reaction forces depend on numerical aspects, especially on properly defined boundary conditions. Although the mass of the test stand is much bigger than the mass of the panels, it is important to represent the way it is fixed in the space. The results achieved for different explosive mass confirm good energy absorption properties of aluminum foam.

ACKNOWLEDGMENTS

This investigation was supported by a research grant, as a part of the project DOBR-BIO/22/13149/2013 financed by NCBiR.

REFERENCES

Du Bois, P.A. 2003. A simplified approach to the simulation of rubber-like materials under dynamic loading, *4th European LS-DYNA Users Conference*, Ulm.

Fana, J.T., Weerheijma, J. & Sluysa, L.J. 2015. High-strain-rate tensile mechanical response of a polyurethane elastomeric material. *Polymer* 65: 72–80.

Gamaa, B.A., Bogettid, T.A., Finkd, B.K., Yue, C.J., Claare, T.D., Eiferte, H.H. & Gillespie, J.W. 2001. Aluminum foam integral armor: a new dimension in armor design. *Composite Structures* 52: 381–395.

Hallquist, J.O. 2006. LS_DYNA Theoretical Manual. Livermore Software Technology Corporation.

Kingery, C.N. & Bulmash, G. 1984. *Air-Blast Parameters from TNT Spherical Airburst and Hemispherical Surface Burst*. Maryland: US Army Ballistic Research laboratory.

Lemonick, D.M. 2011. Bombings and blast injuries: A primer for physicians. *Am. J. Clin. Med.* 8(3): 134–140.

Shim, C., Yun N., Yu, R. & Byun, D. 2012. Mitigation of Blast Effects on Protective Structures by Aluminum Foam Panels. *Metals* 2: 170–177.

Uddin, N. 2010. *Blast Protection of Civil Infrastructures and Vehicles Using Composites*. Oxford: Woodhead Publishing.

Advances in Mechanics: Theoretical, Computational and Interdisciplinary Issues – Kleiber et al. (Eds)

Numerical modeling of thermopiezoelectricity steady state forced vibrations problem using adaptive Finite Element Method

V.V. Stelmashchuk
Ivan Franko National University of Lviv, Lviv, Ukraine

H.A. Shynkarenko
Opole University of Technology, Opole, Poland

ABSTRACT: The purpose of our research is to construct h-adaptive Finite Element Method (FEM) scheme for solving a particular kind of thermopiezoelectricity problems. We use the linear mathematical model of thermopiezoelectricity and then construct the initial boundary value problem. Then a case of steady state forced vibrations of pyroelectrics is considered. After formulation of the corresponding variational problem, variable separation and applying Galerkin classic procedure, we obtain the numerical scheme for solving the problem. Finally, a posteriori error estimators and h-adaptive scheme for thermopiezoelectricity forced vibration problem is built. Numerical experiments are set for PZT-4 pyroelectric bar under influence of different types of loadings.

1 INTRODUCTION

Many modern engineering applications and devices utilize the piezoelectric and pyroelectric materials (Preumont 2011, Tichy et al. 2010, Yang 2006). In our research we consider only the case of thermopiezoelectric steady state forced vibrations. Our aim is also to achieve solution accuracy of some preset level. For this purpose we will construct numerical scheme based on h-adaptive FEM.

2 THERMOPIEZOELECTRICITY STEADY STATE FORCED VIBRATIONS PROBLEM

2.1 *Initial boundary value problem of linear thermopiezoelectricity*

One of the first works on linear mathematical models of thermopiezoelectricity was Nowacki (1978). A comprehensive review of the studies on piezoelectricity can be seen in Benjeddou (2000).

Let us suppose that a pyroelectric specimen occupies the domain Ω in euclidean space R^n, where $n=1,2$ or 3. In accordance with Shynkarenko (1993), Shynkarenko (1994), Chaban & Shynkarenko (2009), Chaban et al. (2012) and Stelmashchuk & Shynkarenko (2014), we assume that thermopiezoelectric behaviour of a specimen can be quite fully described by elastic displacement vector $\mathbf{u} = \{u_i(x, t)\}^n_{i=1}$, electric potential $p = p(x, t)$ and temperature increment $\theta = \theta(x, t)$, which satisfy the system of coupled partial differential equations of motion, electrodynamics and heat conduction. The whole system can be seen in Shynkarenko (1993).

2.2 *Variational form of thermopiezoelectricity initial boundary value problem*

We are introducing the spaces V, Q and Z of admissible elastic displacements, electric potentials and temperature increments respectively and using notations $\Phi = V \times Q \times Z$, $H = [L^2(\Omega)]^n$. Then, after applying virtual work principle, the initial boundary value problem of thermopiezoelectricity can be rewritten in the following variational form:

$$\begin{cases} \text{given } \psi_0 = (\mathbf{u}_0, p_0, \theta_0) \in \Phi,\ v_0 \in H \\ \text{and } (l, r, \mu) \in L^2(0,T;\Phi'); \\ \text{find triple } \psi = \{\mathbf{u}, p, \theta\} \in L^2(0,T;\Phi) \text{ such that} \\ m(\mathbf{u}''(t),\mathbf{v}) + a(\mathbf{u}'(t),\mathbf{v}) + c(\mathbf{u}(t),\mathbf{v}) \\ \qquad - e(p(t),\mathbf{v}) - \gamma(\theta(t),\mathbf{v}) = < l(t),\mathbf{v} >, \\ g(p'(t),q) + e(q,\mathbf{u}'(t)) + z(p(t),q) \\ \qquad + \pi(\theta'(t),q) = < r(t),q >, \\ s(\theta'(t),\xi) + k(\theta(t),\xi) + \pi(\xi,p'(t)) \\ \qquad + \gamma(\xi,\mathbf{u}'(t)) = < \mu(t),\xi > \ \forall t \in (0,T], \\ m(\mathbf{u}'(0) - \mathbf{v}_0,\mathbf{v}) = 0, \\ c(\mathbf{u}(0) - \mathbf{u}_0,\mathbf{v}) = 0 \quad \forall \mathbf{v} \in V, \\ g(p(0) - p_0,q) = 0 \quad \forall q \in Q, \\ s(\theta(0) - \theta_0,\xi) = 0 \quad \forall \xi \in Z. \end{cases} \tag{1}$$

Here Φ' is a dual space of space Φ. The question of well-posedness of the variational problem (1) is sufficiently studied in Shynkarenko (1993) and Shynkarenko (1994).

2.3 *Variational form of thermopiezoelectricity steady state forced vibrations problem*

A particular kind of the problem (1) is the case, when a piezoelectric specimen is subjected to some harmonic loadings with angular frequency $\omega = \text{const} > 0$, that is, the linear forms in the right part of (1) can be represented in the following way:

$$\begin{aligned} &l(t) = (l_1 + il_2)e^{-i\omega t},\ l_1, l_2 \in V', \\ &r(t) = (r_1 + ir_2)e^{-i\omega t},\ r_1, r_2 \in Q', \\ &\mu(t) = (\mu_1 + i\mu_2)e^{-i\omega t},\ \mu_1, \mu_2 \in Z'\ \forall t \in (0,T], \end{aligned} \tag{2}$$

where $i = \sqrt{-1}$.

In this case it is natural to look for approximate solution of the problem (1) as follows:

$$\begin{aligned} &\mathbf{u}(t) = (\mathbf{u}_1 + i\mathbf{u}_2)e^{-i\omega t},\ \mathbf{u}_1, \mathbf{u}_2 \in V, \\ &p(t) = (p_1 + ip_2)e^{-i\omega t},\ p_1, p_2 \in Q, \\ &\theta(t) = (\theta_1 + i\theta_2)e^{-i\omega t},\ \theta_1, \theta_2 \in Z\ \forall t \in (0,T], \end{aligned} \tag{3}$$

where the vector of amplitudes $y = (\mathbf{u}_1, \mathbf{u}_2, p_1, p_2, \theta_1, \theta_2) \in W$, $W = \Phi \times \Phi$ is unknown.

After substituting (2) and (3) into (1) and performing simplifications, we obtain the following variational problem for steady state forced vibrations of piezoelectric specimen:

$$\begin{cases} \text{given angular frequency } \omega = const > 0, \\ \text{the loadings } \ell = (l_1, r_1, l_2, r_2, \mu_1, \mu_2) \in W', \\ \text{find the amplitude vector} \\ \psi = (\mathbf{u}_1, p_1, \theta_1, \mathbf{u}_2, p_2, \theta_2) \in W \text{ such that} \\ -\omega^2 m(\mathbf{u}_1, \mathbf{v}_1) + \omega a(\mathbf{u}_2, \mathbf{v}_1) + c(\mathbf{u}_1, \mathbf{v}_1) \\ -e(p_1, \mathbf{v}_1) - \gamma(\theta_2, \mathbf{v}_1) = < l_1, \mathbf{v}_1 > \forall\ \mathbf{v}_1 \in V, \\ \omega g(p_2, q_1) + \omega e(q_1, \mathbf{u}_2) + z(p_1, q_1) \\ \qquad + \pi(\theta_2, q_1) = < r_1, q_1 > \forall\ q_1 \in Q, \\ k(\theta_1, \xi_1) + s(\theta_2, \xi_1) + \omega\gamma(\xi_1, \mathbf{u}_2) \\ \qquad + \omega\pi(\xi_1, p_2) = < \mu_1, \xi_1 > \forall\ \xi_1 \in Z, \\ -\omega^2 m(\mathbf{u}_2, \mathbf{v}_2) - \omega a(\mathbf{u}_1, \mathbf{v}_2) + c(\mathbf{u}_2, \mathbf{v}_2) - \\ -e(p_2, \mathbf{v}_2) - \gamma(\theta_1, \mathbf{v}_2) = < l_2, \mathbf{v}_2 > \forall\ v_2 \in V, \\ -\omega g(p_1, q_2) - \omega e(q_2, \mathbf{u}_1) + z(p_2, q_2) \\ \qquad - \pi(\theta_1, q_2) = < r_2, q_2 > \forall\ q_2 \in Q, \\ k(\theta_2, \xi_2) - s(\theta_1, \xi_2) - \omega\gamma(\xi_2, \mathbf{u}_1) \\ \qquad - \omega\pi(\xi_2, p_1) = < \mu_2, \xi_2 > \forall\ \xi_2 \in Z. \end{cases} \tag{4}$$

After adding up all the equations from (4), we can introduce the bilinear form $\Pi_\omega(\cdot,\cdot)$: $W \times W \to R$ and linear functional χ_ω: $W \to R$. Then we can rewrite the variational form as shown below:

$$\begin{cases} \text{given angular frequency } \omega = const > 0, \\ \chi_\omega = (l_1, r_1, l_2, r_2, \mu_1, \mu_2) \in W' = \Phi' \times \Phi', \\ \text{find the vector of amplitudes} \\ \psi = (\mathbf{u}_1, p_1, \theta_1, \mathbf{u}_2, p_2, \theta_2) \in W = \Phi \times \Phi \text{ such that} \\ \Pi_\omega(\psi, w) = <\chi_\omega, w>\ \forall\ w \in W. \end{cases} \tag{5}$$

The detailed expression for bilinear form Π_ω and linear functional χ_ω is given in Shynkarenko (1993).

3 GALERKIN FINITE ELEMENT DISCRETIZATION

Let $\Im_h$ be a partition of domain Ω into non-overlapping shape regular elements, and W_h is a space of finite element approximations based on that partition. Then after applying standard Galerkin procedure, we get the following the system of linear algebraic equations for calculation of nodal values of finite element approximation for the unknown amplitudes:

$$\begin{bmatrix} \omega\mathbf{A} & -\mathbf{D} & 0 & \mathbf{E}^T & 0 & \mathbf{Y}^T \\ \mathbf{D} & \omega\mathbf{A} & -\mathbf{E}^T & 0 & -\mathbf{Y}^T & 0 \\ 0 & \omega\mathbf{E} & \mathbf{Z} & \omega\mathbf{G} & 0 & \omega\mathbf{B}^T \\ -\omega\mathbf{E} & 0 & -\omega\mathbf{G} & \mathbf{Z} & -\omega\mathbf{B}^T & 0 \\ 0 & \omega\mathbf{Y} & 0 & \omega\mathbf{B} & \mathbf{K} & \omega\mathbf{S} \\ -\omega\mathbf{Y} & 0 & -\omega\mathbf{B} & 0 & -\omega\mathbf{S} & \mathbf{K} \end{bmatrix} \times \left[\mathbf{U}_1, \mathbf{U}_2, \mathbf{P}_1, \mathbf{P}_2, \Theta_1, \Theta_2\right]^T = \left[\mathbf{L}_1, \mathbf{L}_2, \mathbf{R}_1, \mathbf{R}_2, \mathbf{F}_1, \mathbf{F}_2\right]^T. \tag{6}$$

Here the unknown vectors $\mathbf{U}_1$, $\mathbf{U}_2$, $\mathbf{P}_1$, $\mathbf{P}_2$, Θ_1, Θ_2 are the nodal values of amplitudes (real and imaginary parts) of elastic displacement, electic potential and temperature respectively. Matrices **M**, **C**, **A**, **E**, **Y**, **Z**, **G**, **B**, **K**, **S** are mass, stiffness, viscosity, piezoelectricity, thermal expansion, electric conductivity, dielectric susceptibility, pyroelectricity, thermal conduction and heat capacity matrices respectively, while matrix $\mathbf{D} = \omega^2\mathbf{M} + \mathbf{C}$. Vectors $\mathbf{L}_1$, $\mathbf{L}_2$, $\mathbf{R}_1$, $\mathbf{R}_2$, $\mathbf{F}_1$, $\mathbf{F}_2$ represent amplitudes of mechanical, electrical and heat loadings correspondingly. Matrices **A**, **Z**, **K** are positively defined, which provides us the existence and uniqueness of solution.

4 A POSTERIORI ERROR ESTIMATOR

Similarly to Chaban & Shynkarenko (2009), with respect to variational problem (5) and its discretized form, we can easily formulate the variational problem for finding approximation error:

$$\begin{cases} \text{given } \omega = const > 0, \psi_h \in W_h; \\ \text{find error } e = \psi - \psi_h \in E = W \setminus W_h \\ \text{such that} \\ \Pi_\omega(e,w) = < \rho_\omega(\psi_h), w > := \\ := < \chi, w > - \Pi_\omega(\psi_h, w) \ \forall w \in E. \end{cases} \quad (7)$$

After introducing finite-dimensional subspace E_h of space of admissible errors E, we obtain the discretized variational problem for finding the error estimator e_h:

$$\begin{cases} \text{given } \omega = const > 0, \psi_h \in W_h; \\ \text{find error estimator } e_h \in E_h \text{ such that} \\ \Pi_\omega(e_h, w) = < \rho_\omega(\psi_h), w > \forall w \in E_h. \end{cases} \quad (8)$$

The subspace E_h is then constructed in the way to provide calculation of a posteriori error estimator e_h on each finite element $K \in \Im_h$ without dependence on other elements in the mesh. For that purpose, element-wise defined non-negative bubble functions b_K are taken as basis of subspace E_h:

$$\begin{cases} b|_K := b_K \in H_0^1(K), \\ b_K(x_K) = 1 \ \forall K \in \Im_h \ \forall h > 0, \end{cases} \quad (9)$$

where $x_K \in K$ is, for example, the mass-center of finite element $K \in \Im_h$.

A posteriori error estimator $\hat{\varepsilon}_K = (\hat{\varepsilon}_K^1, \hat{\varepsilon}_K^2)$ of elastic displacements approximations $\mathbf{u}_h$ = $(\mathbf{u}_{1h}, \mathbf{u}_{2h})$ is then looked for as the following linear combination:

$$\hat{\varepsilon}_K^\alpha(x) := \varepsilon_{1K}^\alpha \begin{pmatrix} b_K(x) \\ 0 \\ \vdots \\ 0 \end{pmatrix} + \ldots + \varepsilon_{nK}^\alpha \begin{pmatrix} 0 \\ 0 \\ \vdots \\ b_K(x) \end{pmatrix} \quad (10)$$
$$\alpha = 1, 2, \ \forall x \in K$$

with unknown coefficients $\hat{\varepsilon}_{1K}^\alpha, \ldots, \hat{\varepsilon}_{nK}^\alpha$.

Similarly, a posteriori error estimator $\hat{\xi}K = (\hat{\xi}^1 K, \hat{\xi}^2 K)$ of electric potential approximations $p_h = (p_{1h}, p_{2h})$ can be written in the following way:

$$\hat{\xi}_K^\alpha(x) := \xi_K^\alpha b_K(x), \alpha = 1, 2, \ \forall x \in K. \quad (11)$$

Finally, a posteriori error estimator $\hat{\mathbf{v}}K = (\hat{\mathbf{v}}^1 K, \hat{\mathbf{v}}^2 K)$ of temperature increment approximations $\theta_h = (\theta_{1h}, \theta_{2h})$ can be written as the expansion:

$$\hat{v}_K^\alpha(x) := v_K^\alpha b_K(x), \alpha = 1, 2, \ \forall x \in K. \quad (12)$$

The total a posteriori error estimator $e_h = (\hat{\boldsymbol{\varepsilon}}_K^1, \hat{\boldsymbol{\varepsilon}}_K^2, \hat{\boldsymbol{\xi}}_K^1 \hat{\boldsymbol{\xi}}_K^2 \hat{\mathbf{v}}_K^1, \hat{\mathbf{v}}_K^2)$. After performing standard Galekin procedure with basic bubble functions b_K and using the expansions (10–12), we obtain the following system of linear algebraic equations for finding a posteriori error estimator coefficients on each finite element $K \in \Im_h$:

find vectors $\hat{\varepsilon}_K = (\varepsilon_K^1, \varepsilon_K^2) \in R^{2n}$,
$\hat{\xi}_K = (\xi_K^1, \xi_K^2) \in R^2$,
$\hat{v}_K = (v_K^1, v_K^2) \in R^2$ such that

$$\begin{bmatrix} \omega\mathbf{A} & -\mathbf{D} & 0 & \mathbf{E}^T & 0 & \mathbf{Y}^T \\ \mathbf{D} & \omega\mathbf{A} & -\mathbf{E}^T & 0 & -\mathbf{Y}^T & 0 \\ 0 & \omega\mathbf{E} & Z & \omega G & 0 & \omega B^T \\ -\omega\mathbf{E} & 0 & -\omega G & Z & -\omega B^T & 0 \\ 0 & \omega\mathbf{Y} & 0 & \omega B & K & \omega S \\ -\omega\mathbf{Y} & 0 & -\omega B & 0 & -\omega S & K \end{bmatrix}$$
$$\times [\varepsilon_K^1, \varepsilon_K^2, \xi_K^1, \xi_K^2, v_K^1, v_K^2]^T$$
$$= [\mathbf{L}^1, \mathbf{L}^2, R^1, R^2, F^1, F^2]^T. \quad (13)$$

Here matrices **M**, **C**, **A**, **E**, **Y**, **D** and numbers Z, G, B, K, S are calculated using corresponding bilinear forms of bubble functions b_K. The numbers, participating in the right part of the system (13) are calculated by the following expressions:

$$\begin{cases} L_m^1 = < \rho_\omega(\psi_h), (\hat{\mathbf{b}}_K^m, 0, 0, 0, 0, 0) >, \\ L_m^2 = < \rho_\omega(\psi_h), (0, \hat{\mathbf{b}}_K^m, 0, 0, 0, 0) >, m = 1, 2, \ldots, n, \\ \hat{\mathbf{b}}_K^1 = (b_K, 0, \ldots, 0), \ldots, \hat{\mathbf{b}}_K^n = (0, \ldots, 0, b_K), \\ R^1 = < \rho_\omega(\psi_h), (0, 0, b_K, 0, 0, 0) >, \\ R^2 = < \rho_\omega(\psi_h), (0, 0, 0, b_K, 0, 0) >, \\ F^1 = < \rho_\omega(\psi_h), (0, 0, 0, 0, b_K, 0) >, \\ F^2 = < \rho_\omega(\psi_h), (0, 0, 0, 0, 0, b_K) >. \end{cases} \quad (14)$$

5 ADAPTIVE FEM

On each iteration of adaptive scheme an error indicator η_K is calculated for each finite element K of partition $\Im_h$ by the following rule:

$$\eta_K = \frac{\sqrt{N} \| e_h \|_{1,K}}{\| e_h \|_{1,\Omega}} \quad \forall K \in \Im_h, \quad (15)$$

where $N = card\, \Im_h$ is the number of finite elements in conforming partition $\Im_h$. If on some finite element $\eta_K > 1$ (the error is greater than average on partition), the element K is then refined. If total error indicator:

$$\eta = \frac{\| e_h \|_{1,\Omega}}{\| \psi_h + e_h \|_{1,\Omega}} \cdot 100\% \tag{16}$$

is less or equal to a given admissible error level δ, the algorithm stops.

6 NUMERICAL EXPERIMENT

Various singularities are often seen when modeling heat loadings applied to the pyroelectric specimen. Therefore, we consider the following numerical experiment. PZT-4 (physical properties can be seen in Stelmashchuk & Shynkarenko 2014, Yang 2006) ceramic bar with length $L = 0.01$ m, loaded on the right edge with a heat flux $\overline{q}(x, t) = 100 \cos\omega t$ $\mathrm{Jm^{-2}s^{-1}}$, where angular frequency $\omega = 3 \cdot 10^6$ $\mathrm{rad \cdot s^{-1}}$. The left edge is fixed, grounded and with constant temperature. We start with the uniform mesh of $N = 256$ finite elements with piecewise linear approximation of solution. The solutions contain oscillations (Fig. 1), and the relative error is equal to 32.48%.

The adaptive scheme with a posteriori error estimator, which uses quadratic bubble-function, is then applied. The preset level of accuracy is $\delta = 0.01\%$. It took $I = 12$ iterations of adaptation process to achieve the goal. The final mesh consists of 410 finite elements.

Figure 2 demonstrates the relative error convergence.

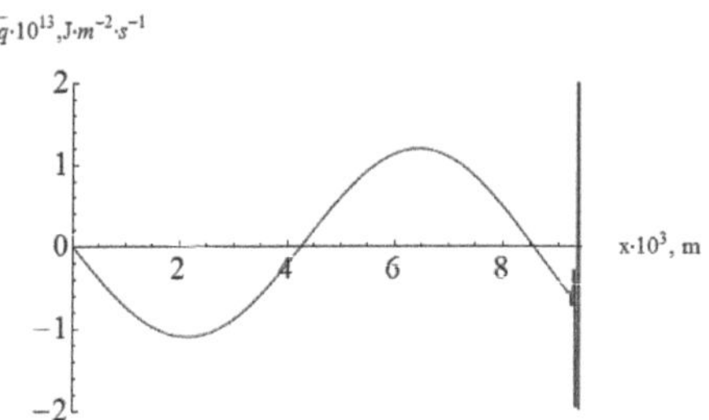

Figure 1. Heat flux amplitude $\overline{q}(x)$.

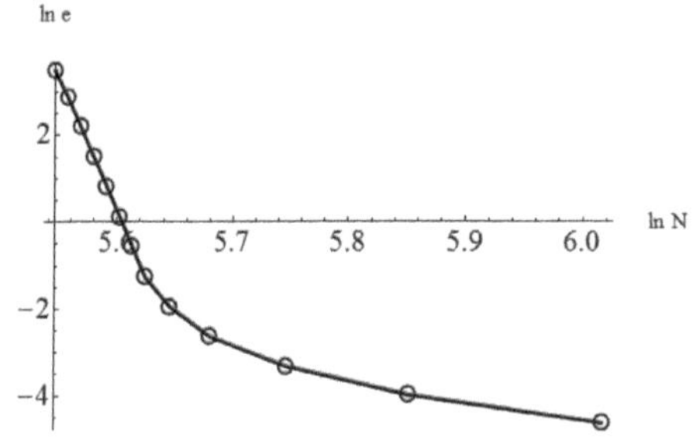

Figure 2. Relative error convergence.

7 CONCLUSIONS

The constructed h-adaptive scheme allows to solve the thermopiezoelectricity forced vibrations problem with a preset level of accuracy. In the case of mechanical or electrical loadings to piezoelectric specimen (not described in this paper), all the solutions are quasi-harmonic smooth functions. Therefore, under these conditions the appliance of h-adaptive scheme is not very useful. Actually, the standard uniform mesh refining allows us to obtain solutions with small relative error. On the other hand, since velocity of heat transfer and mechanical impulse propagation speed are of different scales, singularities are often noticed, when heat loadings are applied, and it is impossible to obtain sufficient solutions without using h-adaptive scheme. The described numerical experiment shows the scheme in action.

REFERENCES

Benjeddou, A. 2000. Advances in piezoelectric finite element modeling of adaptive structural elements: a survey. *Computers and Structures* 76: 347–363.

Chaban, F. & Shynkarenko, H. 2009. The construction and analysis of a posteriori error estimators for piezoelectricity stationary problems. *Operator Theory: Advances and Applications* 191: 291–304.

Chaban, F., Shynkarenko, H., Stelmashchuk, V. & Rosinska, S. 2012. Numerical modeling of mechanical and electric fields interaction in piezoelectric. *In M. Gajek, O. Hachkevych and A. Stanik-Besler (ed.), Manufacturing Processes. Some Problems*: 107–118. Opole: Politechnika Opolska.

Jalili, N. 2010. *Piezoelectric-Based Vibration Control. From Macro to Micro/Nano Scale Systems*. New York: Springer.

Nowacki, W. 1978. Some theorems of thermopiezoelectricity. *J. Thermal Stresses* 1: 171–182.

Preumont, A. 2011. *Vibration Control of Active Structures. An Introduction*, third ed. Berlin: Springer.

Schwartz, M. 2003. *Encyclopedia of Smart Materials*. New York: John Wiley & Sons, Inc.

Shynkarenko, H. 1993. Projection-mesh approximations for pyroelectricity variational problems. I. Problems statement and analysis of forced vibrations. *Differential equations* 29 (7): 1252–1260 (in Russian).

Shynkarenko, H. 1994. Projection-mesh approximations for pyroelectricity variational problems. II. Discretization and solvability of non-stationary problems. *Differential equations* 39 (2): 317–326 (in Russian).

Stelmashchuk, V. & Shynkarenko, H. 2014. Numerical modeling of dynamical pyroelectricity problems. Visnyk of the Lviv University. *Series Applied Mathematics and Computer Science* 22: 92–107 (in Ukrainian).

Tichy, J., Erhart, J., Kittinger, E. & Privratska, J. 2010. *Fundamentals of Piezoelectric Sensorics, Mechanical, Dielectric and Thermodynamical Properties of Piezoelectric Materials*. Berlin: Springer.

Yang, J. 2006. *The Mechanics of Piezoelectric Structures*. Singapore: World Scientific Publishing Co. Pte. Ltd.

Advances in Mechanics: Theoretical, Computational and Interdisciplinary Issues – Kleiber et al. (Eds)
© 2016 Taylor & Francis Group, London, ISBN 978-1-138-02906-4

Wind tunnel tests of the development and demise of Vortex Ring State of the rotor

K. Surmacz, P. Ruchała & W. Stryczniewicz
Institute of Aviation, Warsaw, Poland

ABSTRACT: The Vortex Ring State (VRS) was the reason of several crashes of helicopters. This state of flight appears when a helicopter descents vertically (or with slight forward speed), and descent speed is nearly equal to induced velocity of the helicopter. However, this state may occur during hovering or slow horizontal forward flight, if a strong vertical gust appears. In the VRS a toroidal vortex is created around the rotor disc, which significantly reduces the thrust force generated by the helicopter's rotor. The process of development and demise of the VRS phenomenon was investigated in a wind tunnel using 6-component strain-gage balance and an Particle Image Velocimetry (PIV) system. The investigated object was Remotely Controlled (RC) helicopter with rotor diameter of 0.71 m. Results show the decrease of thrust related to the vortex ring state occurrence a visualization of the development and demise of the VRS.

1 INTRODUCTION

The development of VRS was responsible for accidents involving many different types of helicopters (Hughes 269C (2000), Robinson R-22 (2002), MH-60 (2011)) and tiltrotors (V-22 Osprey (2000)). The vortex ring state in a descent flight is prone to form in case when the upward flow velocity w and the velocity of the induced flow v_i have similar values. The theoretical range of the VRS occurrence is in a range $w = (0{,}5 \div 1{,}5)v_i$. The development of recirculation zone in a vortex ring state causes severe loss of lift of the main rotor (Leishman J. G., 2000). The settling with power condition is especially dangerous during landing of the helicopter. Therefore, the Vortex Ring State has been investigated for decades because of its threat for helicopters and their crews. A detailed investigation of this phenomenon was conducted by experimental tests (Dress & Hendal 1951, Szumański 2006, Wielgus 1963, Wiśniowski 2014), and numerical simulations (Stalewski & Dziubiński 2006, Florczuk 2009, Grzegorczyk 2012). However, most of these investigations were focused on the visualization of the streamlines thrust of the rotor during flight tests or wind tunnel tests in steady conditions. Only few research concerns simulations in transient conditions (Grzegorczyk 2013).

A development of measurement techniques used in wind tunnel investigations allow performing the unsteady experimental tests of the VRS phenomenon, including measurement of the rotor thrust as a function of the rate of descent. In the presented paper a wind tunnel tests of the development and demise of Vortex Ring State of the helicopter rotor are described. The process of development and demise of the VRS phenomenon was investigated in a wind tunnel using 6-component strain-gage balance. As a result a variation of thrust versus the upward flow velocity was determined. Application of Particle Image Velocimetry (PIV) method allowed to measure the velocity field around the rotor disc and to obtain a shape, size and location of the vortex. These results may be used for better understanding the process of development of the VRS, which should improve the safety of helicopter flight.

2 METHODOLOGY OF INVESTIGATION

In order to provide the necessary conditions for development of VRS, the rotation axis of the main rotor of a helicopter (RC model) was set in the axis of the wind tunnel's test section. The flow induced by the rotor and the flow generated by the wind tunnel was oppositely directed. The measurements were performed during increment or decrement of the velocity. The speed of the freestream was continuously reduced from value $w > v_i$ to zero in a decrement conditions in order to provide transient conditions for the VRS occurrence.

It must be noted that the acceleration phase lasted significantly shorter than the deceleration phase. It was caused by the design of the wind tunnel propulsion and its control system.

2.1 *Object of investigation*

The object of investigation of the VRS was a Remotely Controlled (RC) helicopter T-REX 450 PRO Super Combo, with rotor diameter of 0.71 m and weight of 0.78 kg. The pitch angle of the blades can be changed during the test in range from 0° to 10°. The maximum rotational speed of the rotor is 2400 rpm and the maximum induced velocity of the main rotor in hover is approximately 7 m/s. The helicopter was mounted in the wind tunnel via 6-component strain-gage balance. The axis of the rotor was parallel to the axis of the wind tunnel (Fig. 1).

The investigation was performed in the T-1 wind tunnel in the Institute of Aviation in Warsaw. It is a closed-circuit, open test section wind tunnel (with the test section diameter of 1.50 m), powered by 55 kW electric motor and 4-bladed constant-speed fan. The maximum airflow velocity in this wind tunnel is over 40 m/s and minimum steady velocity of the airflow is about 11 m/s.

2.2 *Strain-gage balance investigation*

The strain-gage balance was used to measure aerodynamic loads acting on the helicopter, especially the rotor thrust T. The loads were calculated basing on the voltages of Wheatstone strain-gage bridges, measured by a computer with the NI USB-6259 I/O card. In the postprocessing phase timings of the aerodynamic loads were filtered by a zero-phase Butterworth lowpass filter.

The airflow velocity was calculated from Equation 1:

$$V = \sqrt{\frac{2q}{\rho}} = \sqrt{\frac{2(p_0 - p_S)}{\rho}} \tag{1}$$

where $\rho = 1.225$ kg/m^3 is air density, q means a dynamic pressure of the airflow, p_0 means total pressure of the airflow and p_S—its static pressure.

Both total and static pressure were measured by two Druck pressure sensors, connected with a Prandtl tube. The electrical outputs of these sensors were measured by the measurement & control system of the T-1 wind tunnel (Ruchała 2013).

The timings of the airflow velocity were approximated as a polynomial of the fifth order. The timings of the thrust were filtered with the zero-phase digital filter. This approach simplified an analysis of results, presented as T = T(V) plots.

2.3 *Particle image velocimetry investigation*

The PIV is a modern technique of measurement and visualization of the flow velocity field (Adrian 2005, Stryczniewicz 2012). To make the measurement, seeding particles (the DEHC oil in this case) must be atomized in the flow. Droplets of the seeding are illuminated by a lightsheet (i.e. a laser light, formed in a thin sheet by the lenses) and photographed by a camera. The diameter of a seeding droplet is few microns. For every measurement the camera grabs two frames. The time interval between frames in presented measurements was Δt = 80 μs. In the post-processing phase the measured velocity field is obtained by determination of the particles displacements (Adaptive Correlation scheme, with integration windows size of 64 × 64 pixels with 50% window overlap). The displacements are divided by the time interval Δt to calculate the velocity field. The outlier vectors and missing data was removed in post-processing with use of average and median filtering. The measurement area of the PIV system included the area around the tip of the blades. Measurements were performed with three sizes of the PIV test region (380 × 380, 400 × 400, 450 × 450 mm).

A scheme of the PIV test stand has been presented in Figure 2. More details can be found in (Stryczniewicz & Surmacz 2014).

The PIV system setup was validated in two tests. First test was designed to validate the settings of the PIV system for measurements of the speed of air in the wind tunnel. The freestream velocity was

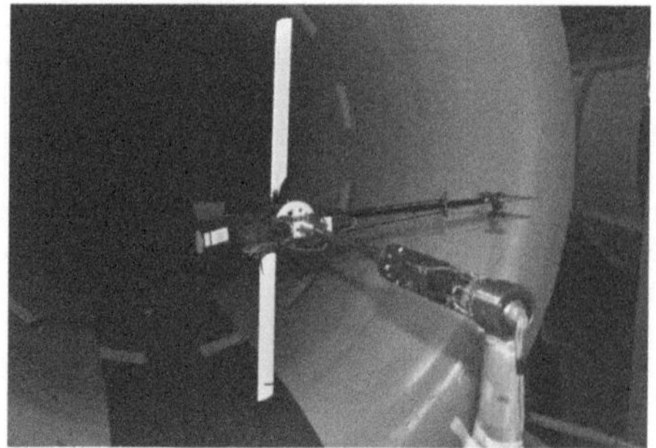

Figure 1. The investigated helicopter in the wind tunnel.

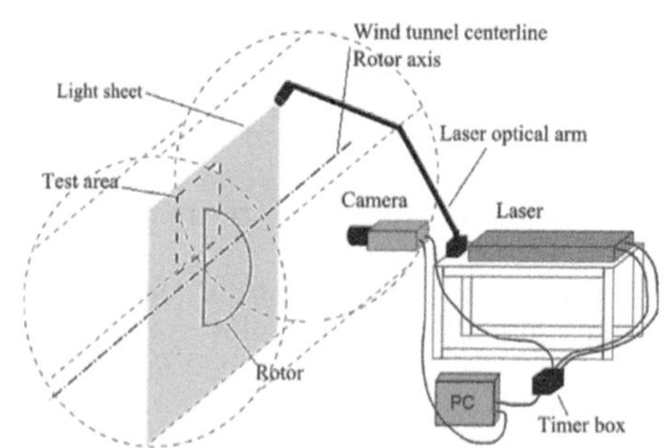

Figure 2. Particle Image Velocimetry test stand.

Table 1. Comparison of the inducted velocity measured with PIV and vane anemometer.

	Velocity v_i	
Pitch angle of the blades [°]	Measured by vane anemometer [m/s]	Measured by PIV [m/s]
0	0.75	0.8
2	0.59	–
4	2.0	2.0
6	3.4	3.5
8	5.9	5.9
10	7	7.1

measured by the PIV system and compared the results of the velocity measured by the Pitot tube in the wind tunnel test section. The second test of the PIV system setup involved the measurement of the induced flow velocity. In this test the induced flow velocity was measured by the PIV system and a vane anemometer. Both tests proved that the PIV system settings were proper (see Table 1).

3 RESULTS

In Figure 3 a sample relation between the thrust of the rotor and the descent velocity has been presented. The range of VRS appearance (theoretical and observed) has been marked as well. In this case the thrust decreases from 12.5 N to 9.2 N, thus the difference caused by the VRS phenomenon is about 26%.

Figure 3 presents results obtained for increasing the descent velocity and decreasing it. It should be noted that in the first case the curve T = T(V) is smooth. It is caused by a relatively short time when the airflow velocity was appropriate for the VRS phenomenon (about 3 seconds). For decreasing the descent velocity, when this time was extended, the plot is more ragged—which means that the vortex around the rotor is more deployed and the flow becomes turbulent. It confirms a conclusion made during the RC helicopter flight tests of the VRS (Szumański 2006). It also should be noted that the VRS for increasing descent rate has been observed for slightly higher velocity than for decreasing. It may be related with longer time of deceleration phase.

In Figure 4 the streamlines in two consecutive frames have been presented. It should also be noted that the core of the vortex moves towards the rotor disc. Typical path of the vortex ring center determined from PIV measurements is presented in Figure 5. The black vertical line indicates the position of the blade's tip. Firstly the position of the vortex core travels form region above the rotor's plane to region

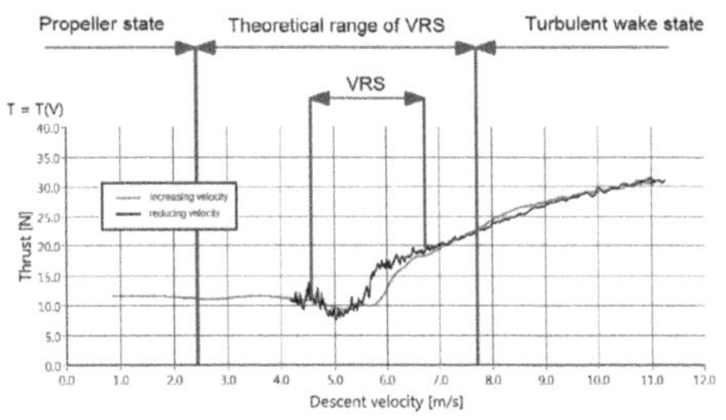

Figure 3. VRS marked on the plot of thrust vs. descent velocity.

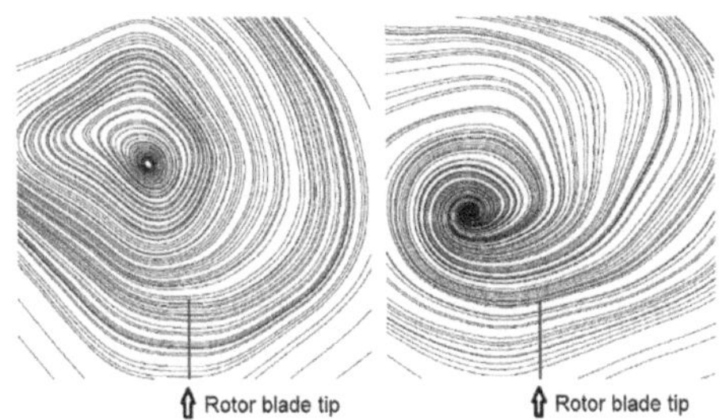

Figure 4. Sample streamlines in the VRS.

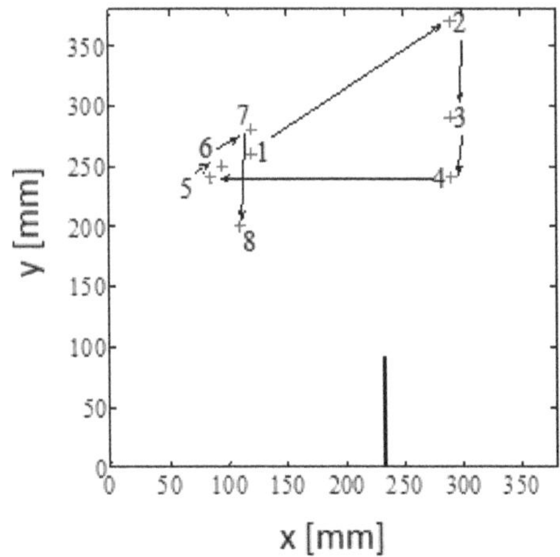

Figure 5. Typical path of the vortex ring center determined from PIV measurements (Stryczniewicz & Surmacz 2015).

underneath. Secondly reverse transition is observed. The measurement frequency is 7 Hz.

4 SUMMARY

In the recent years, the use of helicopters increased. Determination of limits on the use provides acceptable levels of safety. One of the restrictions for use of the helicopter is the vortex ring state boundary. Research on the issue of the VRS is intended to increase the safety and reliability of helicopters. The results of experimental and numerical analyses show that the real area of occurrence of VRS can be a little different than the theoretical area.

The experimental results presented in this paper are part of the work carried out on the vortex ring state phenomenon. The conditions of vertical descent of a helicopter (in the vicinity and inside the area of the VRS) were considered.

Results of the investigation described in the paper illustrate a process of development and demise of the Vortex Ring State on the rotor of the helicopter. During the strain-gage balance investigation the loss of thrust caused by the VRS has been obtained, as well as the rate of descent when this phenomenon appears.

The PIV measurement allowed to investigate size and location of the vortex created around the rotor disc. The visualization of a flow field provided information about the changing nature of the flow in vortex ring state conditions. The 2D vector velocity field of the investigated flow was obtained. In order to get insight on development of VRS, the vortex ring center was traced on consecutive frames. The velocity of the vortex ring center was in range 0,5 to 1,5 m/s. The time of the VRS occurrence for the experimental conditions was in range 0,5 to 2 seconds. The position of the vortex ring center changed between frames. It is worth to notice that the region identified as a vortex core of the VRS structure was observed in a region above and underneath the rotor's plane and it was traversing freely during development and demise vortex (Fig. 5). Although the sampling frequency of the PIV system was relatively low (7 Hz) it was possible to observe the development and demise of VRS in controlled laboratory conditions and to characterise the movement of the vertex ring canter. The settings of the PIV system was validated and results of PIV velocity measurements was compared with the vane anemometer indications. Good agreement of the results in most cases was obtained (see Table 1). Due to the technical limitations of the laser the sampling frequency was 7 Hz. Higher sampling frequency would provide more accurate information of the vortex core trajectory. The field of view of the PIV system cameras do not allowed to visualise the whole investigated structure. Tests proved that PIV method is a reliable tool for an investigation of flow around the rotor of a helicopter in a controlled wind tunnel conditions.

The presented experiment permitted an quantitative and qualitative analysis of VRS phenomena. The research will be continued with use of synchronized a strain gauge balance and a stereo-PIV system and in order to acquire more detailed understanding of VRS phenomenon development. Also the field of view will be enlarged due to application of special optics. In the future, experimental results will be compared with results of CFD calculations.

REFERENCES

Adrian R.J. Twenty Years Of Particle Image Velocimetry, 2005, *Experiments in Fluids* 39: 159–169.

Dress J., and Hendal W.P., 1951, Airflow Patterns In The Neighbourhood Of Helicopter Rotors, *Aircraft engineering*, vol. 23(266): 107–111.

Florczuk W., 2009, CFD Hover Analysis Of W-3 Falcon Rotorcraft For Different Upstream Air Conditions, *Transactions of the Institute of Aviation*, Vol. 201: 67–80.

Grzegorczyk K., 2009, An Analysis of Vortex Ring State on the Rotorcraft, *Transactions of the Institute of Aviation*, Vol. 201, pp. 52–66.

Grzegorczyk K., Helicopter flight simulation close to safety limits conditions, Mechanics in Aviation no. ML-XV, 2012

Grzegorczyk K., 2013, Analysis of influence of helicopter descent velocity changes on the phenomena of vortex ring state, *Postępy Nauki i Techniki*, Vol. 7:35–41.

Leishman J.G., 2000, *Principles of helicopter aerodynamics*, Cambridge University Press.

Raffel M., Willert C.E., Wereley S.T., Kompenhans J., 2007, *Particle Image Velocimetry—A practical Guide*, 2nd ed., Springer.

Ruchała P., 2013, The Measurement and Control System in the T-1 Wind Tunnel, *Transactions of the Institute of Aviation*, Vol. 232: 63–78.

Stalewski W. and Dziubiński A., Symulacja zjawiska pierścienia wirowego wokół wirnika śmigłowca w oparciu o rozwiązanie równań N-S z uproszczonym modelem wirnika w postaci powierzchni skoku ciśnienia, *Prace Instytutu Lotnictwa*, nr 184–185: 65–71, 2006.

Stryczniewicz W., 2012, Development of Particle Image Velocimetry Algorithm, *Problems of Mechatronics*, Vol. 9,: 41–54.

Stryczniewicz W., Surmacz K., 2014, Investigations of the Vortex Ring State of the Main Rotor of a Helicopter, *Transactions of the Institute of Aviation*, Vol. 235: 17–27.

Stryczniewicz W., Surmacz K., 2015, PIV measurements of the Vortex Ring State of the Main Rotor of a Helicopter, 15 th Conference on Modelling Fluid Flow, Budapest.

Szumański K., Przelot dynamiczny śmigłowca przez strefę występowania strumienia wirowego, *Prace Instytutu Lotnictwa*, nr 184–185: 110–118, 2006

Wielgus S., 1963, "Sprawozdanie z prób w locie śmigłowca SM-1 w stanie pierścienia wirowego", *Technical Report*, Institute of Aviation.

Wiśniowski W., 2014, Twenty Years of Light Aircraft and Safety Program, *Transactions of the Institute of Aviation*, Vol. 236: 7–25.

Advances in Mechanics: Theoretical, Computational and Interdisciplinary Issues – Kleiber et al. (Eds)
© 2016 Taylor & Francis Group, London, ISBN 978-1-138-02906-4

Assessment of the biomechanical parameters of the disabled driver during frontal crash

K. Sybilski, J. Małachowski & P. Baranowski
Faculty of Mechanical Engineering, Military University of Technology, Warsaw, Poland

P. Płatek
Faculty of Mechatronics and Aerospace, Military University of Technology, Warsaw, Poland

ABSTRACT: The presented paper is related to safety issues of disabled drivers. Based on results of numerical analyses the authors have described behaviour of disabled drivers who use special equipment during frontal crash. To perform analysis numerical model of MPV car (Multi-Purpose Vehicle) using 3D scanning was built. Adaptive devices in the form of knobs on the steering wheel and control device for acceleration and braking pedals were added. Driver with disabilities was modelled using the modified Dummy Hybrid III 50th model, developed by Humanetic company, which was seated using gravity load before simulations. As the result of performed numerical analysis the charts of dummy's mass centre movement, angle of body rotation, head accelerations, forces between dummy and seatbelts are presented. Based on those charts several injury criteria factors were estimated (e.g. HIC and NIC).

1 INTRODUCTION

The world population surpassed seven billion. Nearly 15% of them are disabled persons, among which are a large group of people with movement disabilities. These people are often well-educated and have unique skills that could be successfully used in a professional or social life. Unfortunately, lack of mobility is the huge obstacle preventing participation in the normal life. This is particularly noticeable in smaller towns, where there is no well-organized public transport with the fleet of low-floor buses or trams.

One of many possibilities to increase the level of mobility of people with disabilities is to adapt cars to their needs. In Poland there is an active association Spinka (Spinka 2014), whose goal is to create a nationwide network of specialized Disabled Driver Training Centres, and financial support programs aiming to help in the purchase and adaptation of vehicles. These activities results in growing availability of both used and new cars that can be driven by disabled.

There is a lot of adaptive equipment for people with disabilities available on the market (Manocelli 2009, Gil 2013). Choice of a specific solution depends on the type of disability. However, mounting of any type of the equipment in the car changes the position of the driver and kinematics of his movements compared to that anticipated by car manufacturer. In particular, such devices affect behaviour of the driver during collision, and thus its safety. Therefore, each new type of equipment for disabled driver (including a combination of existing solution and new vehicle) should be tested before passing to an user. One of the method to verify the safety of the mechanical system, which is the additional adaptive equipment, are numerical simulations (Golman 2014).

The aim of the article is to determine the behaviour of the disabled driver during frontal crash, and to estimate biomechanical parameters and criteria characterizing his safety. The paper presents a methodology of car geometry data collection, acquisition of material data, numerical model preparation and numerical analysis algorithm.

The assessment of a disabled driver safety it the presence of additional equipment was carried out for the MPV (Multi-Purpose Vehicle) car type, which is currently being produced. Its interior geometry for the numerical analysis was obtained using the 3D F5 Mantis Vision scanner (Małachowski 2014). As a result of the scanning process the cloud of point was obtained, which was further processed to obtain set of triangles (mesh). In the next stage of the work these triangles were converted into four-node shell elements forming surfaces of the car interior.

2 NUMERICAL MODEL DEVELOPMENT

During the numerical simulations three initial velocities were analysed: 30, 40 and 50 km/h.

Curves defining velocity vs. time during crash were taken from additional numerical analyses aimed at projection of the frontal car impact into the flat rigid wall (Marzuogui 2012). The obtained curves (Fig. 1) were used to control the body of the car only. All other parts of the model have been associated with the initial velocity and movement was implemented thanks to the interaction with a car body.

In the numerical analyses the Hybrid III 50th dummy model was used (Fig. 2). This is a very accurate model of Anthropomorphic Test Device (ATD), that gives results complying with the experimental test in about 95% (Pradeep 2010). Preparation of the numerical model included settling of the dummy under gravity load and fastening using the seatbelts. Seatbelt model featured tensioner, retractor and scrolling through the lugs.

During numerical analyses the driver with paralyzed legs was considered. In such cases the driver has to use a specialized knob on the steering wheel and hand control for accelerator and brake (Fig. 3).

All the above simulations were able to perform thanks to numerical investigations which were

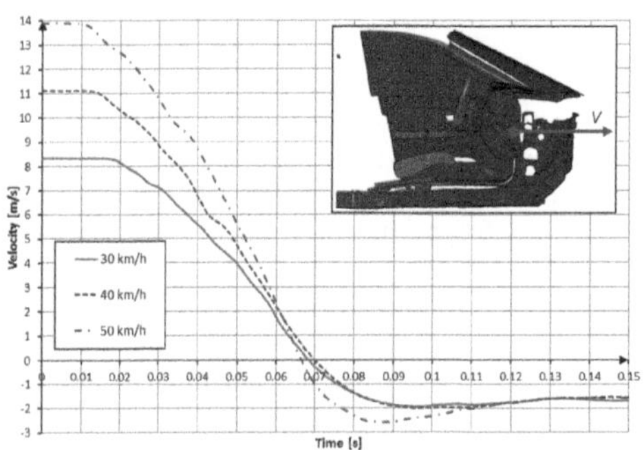

Figure 1. Velocity of car body during numerical analyses.

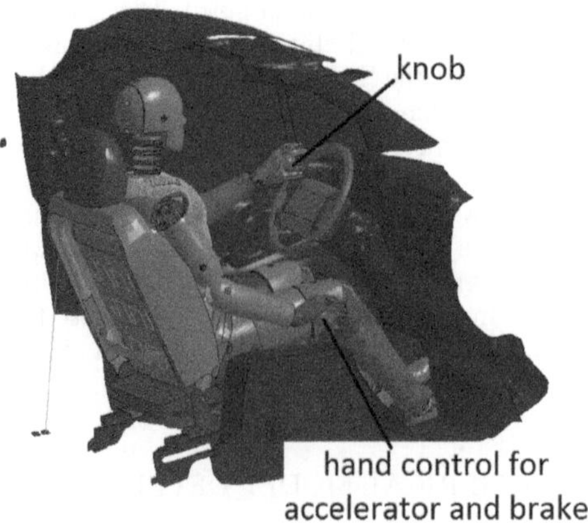

Figure 2. Numerical model used in simulations.

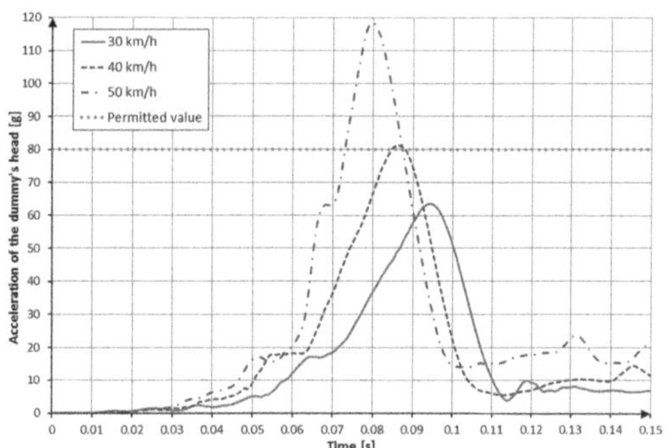

Figure 3. Acceleration of dummy's head.

carried out using software with explicit time integration scheme. This scheme was chosen because of fast-changing and highly nonlinear character of analysed crash process. During the investigations, the central difference method for integration over time was adopted, which is one of the variant of the finite difference method. The major advantage of this method is lack of time-consuming operation involving the stiffness matrix inversion. Instead, only a diagonal matrix of mass is inverted. However, the main disadvantages is that the method is conditionally stable requiring a time step to be limited according to the Courant-Friedrichs-Lewy (CFL) stability condition.

One of the most important issues of the paper is an accurate numerical description of the two interacting bodies. The contact process played a significant role due to its direct influence on the interaction process between contacting/interacting bodies. Thus, it was necessary to simulate the interaction between the segments of collaborating parts as accurately as possible. In explicit software, the interaction between two or more bodies is define using a so called penalty function approach. For elements with different stiffness, the contact stiffness is determined by the following formula (Hallquist 2006):

$$k_{cs}(t) = 0.5 \cdot SLSFA \cdot \begin{Bmatrix} SFS \\ or \\ SFM \end{Bmatrix} \cdot \left(\frac{m_1 m_2}{m_1 + m_2} \right) \cdot \left(\frac{1}{\Delta t_c(t)} \right)^2 \tag{1}$$

where $SLSFAC$—scaling factor, SFS and SFM—scaling factor for slave and master side, m_1 and m_2—masses of master and slave nodes, $\Delta t_c(t)$—initial time step dependent on the contact procedure (if the solution time step grows, Δt_c is reset to the current time step to prevent unstable behavior of the simulation).

Another challenging problem was to properly describe behavior of materials subjected to large deformations. One of them were seat cushions. To describe it low density foam model was adopted. In tension foam behaves in a linear fashion until tearing occurs. In compression material behavior was determined by stress vs. strain curve.

The next material subjected to large deformations were seatbelts. In order to describe 2D elements fabric material model was implemented, in which the main role played stress vs. strain curve. For 1D elements stiffness was calculated based on force vs. strain characteristic (Hallquist 2006).

Numerical analyses were carried out in two stages. In the first stage a seating phase of dummy under gravity load was simulated. In the second stage seatbelts and initial velocity were added and then frontal crash was simulated.

3 OBTAINED RESULTS

During simulations several similar stages can be observed in each cases. In the first step seatbelts were rolled up, what minimized clearances between them and the dummy. Next the car started to slow down sharply, what caused triggering of the sensor and thereby activating pretensioners and filling airbag. In the next stage the dummy started to move on the seat which was accompanied by increasing reaction between the dummy and the seatbelts. At the same time the airbag was activated and the dummy's chest started pressing it. The filled airbag also rapidly pushed away dummy's hand towards the door. In the last stage the car changed moving direction and dummy's head hit the airbag.

Process described above, especially change of car direction, results in occurrence of peak on dummy's head acceleration chart (Fig. 3). The maximum allowed acceleration is 80 g. For initial velocity of 30 km/h and 40 km/h this value was not reached but for 50 km/h the maximum acceleration was 120 g so it exceeded the allowed value one and a half times. It had significant impact on Head Injury Criterion (HIC). For the lowest initial velocity HIC15 was 380, for 40 km/h it reached respectively the limit 700 and for the highest initial velocity it was 1597.

The HIC informs about probability of trauma which the driver may suffer. Prasad & Mertz curves give the information how serious this trauma could be. HIC values referenced to this curves (Fig. 4) show, that for 30 km/h driver has only minor and moderate injuries. For 40 km/h driver has minor, moderate injuries and has 30% chances to get serious injury. For the highest initial velocity driver has 65% and 25% chances to get severe and critical injuries.

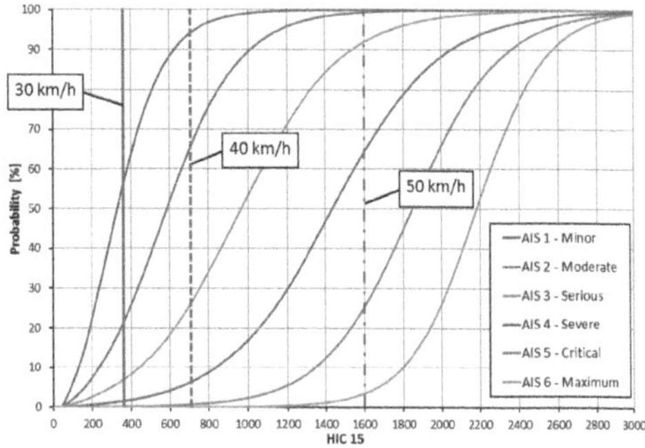

Figure 4. Obtained HIC15 on Prasad & Mertz curves.

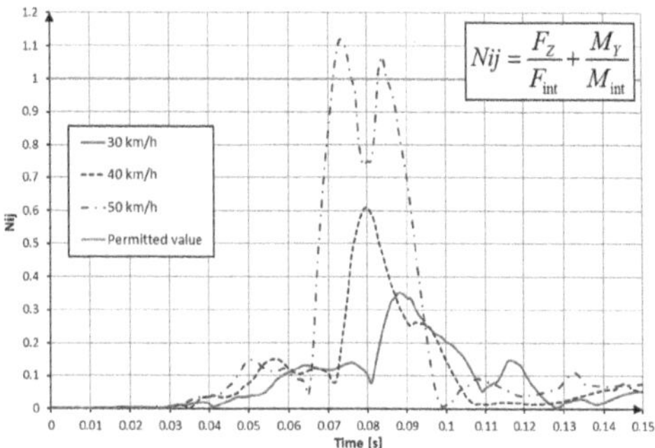

Figure 5. Neck Injury Criterion (Nij) calculated for obtained results.

High accelerations acting on dummy's head had an impact on forces in neck. In Figure 5 Neck Injury Criterion (Nij) is shown. The permitted value of this criterion is 1 and only for analysis with the initial velocity of 50 km/h this value was exceeded (1.1). For other cases it was 0.6 (40 km/h) and 0.35 (30 km/h).

The equipment mounted on steering wheel and above the central tunnel in the car results in a significant change in the behavior of the body in the crash phase. This is due to a change in body position, which is very asymmetrical relative to a plane passing through the steering wheel. In Figure 6 the angle of rotation of the arms for the analysed cases is shown.

In the first stage when the car was moving forward the lowest value was obtained for the highest initial velocity (2°). In other cases angle of rotation of the dummy's arms was twice time higher (4°). In the last stage when the car was moving backward the smallest angle (approx. 11°) occurs for lowest initial velocity. The highest angle was noticed for 50 km/h (18°).

In Figure 7 the impact of the initial velocity on the reaction force between the seatbelts and the

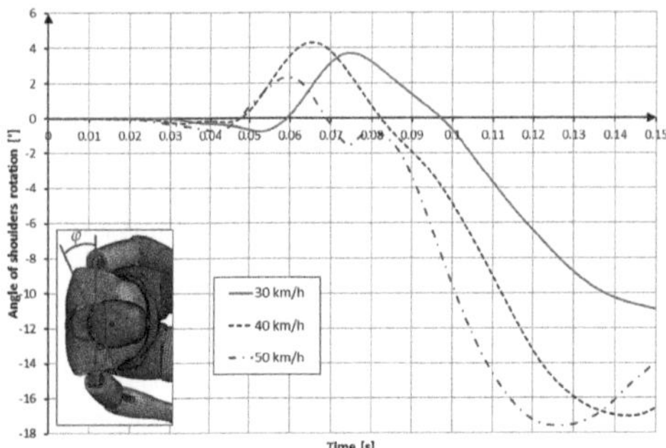

Figure 6. Angle of rotation of dummy's arms.

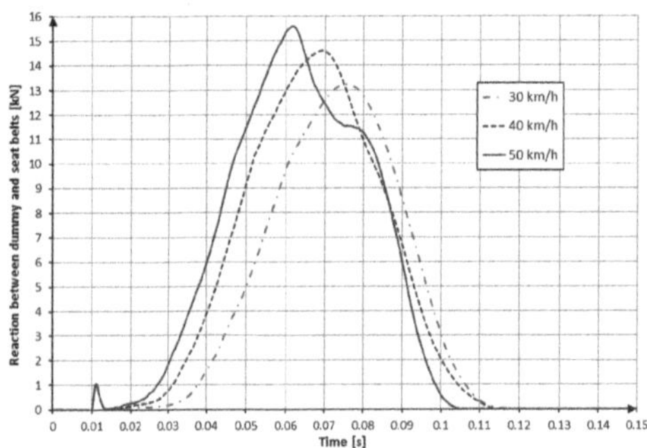

Figure 7. Reaction between dummy and seatbelts.

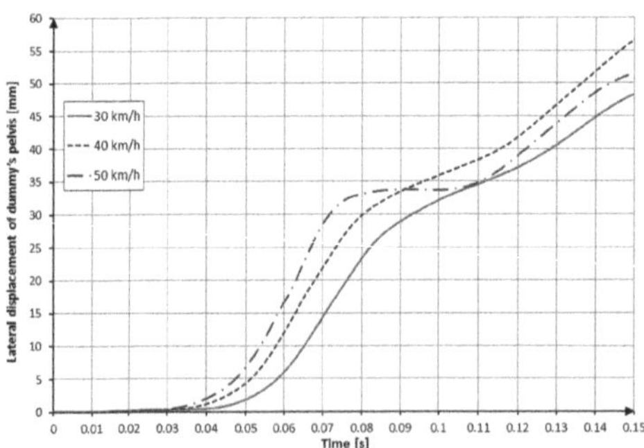

Figure 8. Lateral displacement of dummy's pelvis.

dummy is shown. It can be noticed that the increase of the initial velocity results in the increase of the reaction force. The maximum obtained value was 15.5 kN, whereas the lowest was 13 kN.

Asymmetric dummy position and asymmetric operation of the seatbelts causes lateral displacement of the centre of driver mass (Fig. 8). During performed simulations the dummy was moving toward the left door pillar, which is very dangerous. During the crash a driver can hit the door pillar which additionally increases the probability of serious injury. He may also miss the headrest and suffer serious neck injuries while moving backwards.

4 CONCLUSIONS

Testing of the solutions for disabled drivers is a very complex process. It should include operation of all safety systems and it should be performed for each combination of equipment and car separately.

Additional equipment mounted in the car changes significantly driver ergonomics and kinematics which also has impact on his safety.

It also forces a driver to sit closer to the steering wheel. It results in shortening the distance the action of the seatbelts and increasing impact of the airbag, and thereby increasing forces in the neck and the head.

The combination of asymmetric operation of the seatbelts and the equipment for the disabled, causes movement of the driver's body toward the door pillar, which, as previously discussed, can be very dangerous in terms of the driver's safety.

REFERENCES

Gil J. & Diaz I. & Ciaurriz P. & Echaverria M. New driving control system with haptic feedback: Design and preliminary validation tests, *Transportation Research Part C*, 33, 2013.

Golman A. & Danelson K. & Miller L. & Stitzel J. Injury prediction in a side impact crash using human body model simulation, *Accident Analysis and Prevention*, 64, 2014.

Hallquist J. LS-Dyna theory manual, Livemore Software Technology Corporation, 2006.

Małachowski J. & Sybilski K. & Muszyński A. & Baranowski P. Problematyka bezpieczeństwa kierowcy wykorzystującego dodatkowe oprzyrządowanie na kierownicy w warunkach zderzenia czołowego, *Logistyka*, 3, 2014.

Małachowski J. & Sybilski K. & Szafrańska A. & Baranowski P. Analiza kinematyki kierowcy wykorzystującego oprzyrządowania dla osoby niepełnosprawnej na podstawie skanowania 4D, *Transport Przemysłowy i Maszyny Robocze*, 2, 2014.

Marzougui D. & Samaha R.R. & Cui C. & Kan C.D. & Opiela K.S. Extended validation of the finite element model for the 2010 Toytota Yaris Passenger Sedan, *NCAC The National Crash Analysis Center*, 2012.

Monacelli E. & Dupin F. & Dumas C. & Wagstaff P. A review of the current situation and some future developments to aid disabled and senior drivers in France, *IRBM*, 30, 2009.

Pradeep M. & Chung-Kyu P. & Dhafer M. & Cing-Dao K. & Guha S. & Maurath C. & Bhalsod D. LSTC/NCAC Dummy Model Development, *11th International LS-Dyna Users Conference*, 2010.

Stowarzyszenie Pomocy Niepełnosprawnym Kierowcom SPiNKa http://spinka.org.pl/, 2014.

Advances in Mechanics: Theoretical, Computational and Interdisciplinary Issues – Kleiber et al. (Eds)
 ISBN 978-1-138-02906-4

Dynamic response of the steel chimney by the stochastic perturbation-based Finite Element Method

J. Szafran, K. Juszczyk & M. Kamiński
Łódź University of Technology, Łódź, Poland

ABSTRACT: The main aim of the paper is a presentation of the results obtained for the stochastic analysis concerned steel chimney with the height equal to 40 meters and subjected to the random wind loading. Analysis was performed using generalized stochastic perturbation technique based on a Taylor expansion and traditional Finite Element Method. Polynomial response functions of the observed design parameters such as deflections of the top of the structure, bending moments or transverse forces are numerically determined via symbolic algebra system MAPLE with the Least Square Method embedded in this system. The presented computational procedure including both dynamic and stochastic analysis can serve as a tool for the reliability estimation of slender, lightweight structures.

1 INTRODUCTION

Steel chimneys (Rykaluk 2005) are used widely in industry as structures that allow remove the exhaust into the higher layers of the atmosphere. Their main advantages comparing to concrete structures are weight, strength (resulting as a high value of the strength/weight ratio), simplicity of the assembly and dismantling. Considering bearing capacities of the chimneys, a very important factor is the influence of the main load, namely wind in the context of the structural responses. A case has been studied—observing the deflections of the top of the structures, transverse forces and bending moments with reference to the wind loading. The main attention was focused to changes of the wind loading as the most natural source of the randomness in performance of this kind of structures. In the context of stochastic analysis, large random variations of the input uncertainty sources and a necessity of third and fourth probabilistic characteristics determination, the Second Order Second Moment (SOSM) perturbation-based method (Kleiber 1985, Kleiber & Hien 1997) is still extended to include as long stochastic Taylor expansions as necessary. The second issue here is replacement of the Direct Differentiation Method (DDM) consisting of successive formation of the increasing order equilibrium equations obtained via partial differentiation with respect to the random input by the Least Square Method (LSM) (Björck 1996). The series of iterative deterministic solutions to the given boundary-initial problem using FEM with the given random input varying around its expectation enable to get the local polynomial approximation at each degree of freedom separately thanks to the weighted version of the LSM itself (Kamiński & Szafran 2012). All the calculations are carried out thanks to the interoperability of the computer algebra program MAPLE with commercial FEM system ROBOT. We determine up to the fourth probabilistic characteristics of the structural random response.

2 GOVERNING EQUATIONS

Stochastic perturbation method (Kamiński 2013) is based on an expansion of the all random functions into the Taylor series of the required order. In this particular formulation, assumptions of the Gaussian probability density function is not necessary, because we can implement such an approach to the non—symmetrical density distribution as well. Expansion for the random deflection u with respect to the random wind velocity $v(\omega)$ can be expressed as:

$$u(v(\omega)) = u^0(v^0(\omega)) + \varepsilon \left. \frac{\partial u(v(\omega))}{\partial v} \right|_{v=v^0} \Delta v + \dots + \frac{\varepsilon^n}{n!} \left. \frac{\partial^n u(v(\omega))}{\partial v^n} \right|_{v=v^0} \Delta v^n \quad (1)$$

and classical integral definitions like for the Mth central probabilistic moments:

$$\mu_M(u(v)) = \int_{-\infty}^{+\infty} (u(v) - E(u(v)))^M p_{u(v)}(x)dx \quad (2)$$

A full symbolic approach guarantees the expansion with *a priori* given length and its *a posteriori* modifications according to the numerical errors obtained for higher than the second probabilistic moments of the structural deformations. The method in the structural context is based on the iterative solution of the discrete equations of motion

$$M_{\alpha\beta}^{(i)}\ddot{q}_{\beta}^{(i)}(t)+C_{\alpha\beta}^{(i)}\dot{q}_{\beta}^{(i)}(t)+K_{\alpha\beta}^{(i)}q_{\beta}^{(i)}(t)=Q_{\alpha}^{(i)}(t) \qquad (3)$$

using the Hilber-Hughes-Taylor α-method (Hilber et al. 1977, Hughes 2010) to get the series of deterministic nodal responses $q_B^{(i)}(t)$, $i = 1, ..., n$, where n is the total number of the LSM tests. It enables for a recovery via the Least Squares Method of the polynomial response functions $q_\beta(t, b)$, where for the given $\tau \in [0, \infty)$ one may get:

$$q_{\beta}(\tau,b)=A_{\beta j}(\tau)\,b^{j}, \quad j=1,...,m \qquad (4)$$

where m is the approximation order such that $m \leq n$. Further determination of the probabilistic moments proceeds from their integral definitions extracted from (2) thanks to the symbolic derivation of all partial derivatives with respect to the given b.

3 COMPUTATIONAL ANALYSIS

The main aim of this work is to apply higher order Stochastic perturbation-based Finite Element Method. Motivation of this work comes from the necessity of the reliability index determination for the thin-walled steel structures exhibited to the dynamic excitation of the wind blows that quite naturally have stochastic character. The steel chimney presented in Figure 1 is modeled with the use of 4-noded thin shell finite elements in the FEM system ROBOT and the dynamic response functions are found thanks to 11 tests with varying basic velocity of the wind acting on the chimney surface, while the dynamic spectrum of the wind blow is adopted after (Flaga 2008) and shown below in Figure 2.

In first numerical approach (as it is denoted in Fig. 2) we have tested this basic velocity by taking plus minus 10% of its initial value with an increment equal to 1%. The FEM discretization of the chimney consists of 12 × 320 shell elements (with the height equal to 40,0 m and having 1,20 m diameter) together with the variations of its thickness is given on the figure above. The entire structure is fully restrained at its bottom and is homogenous having a thickness equal 6 mm (at its top) to 12 mm (at its bottom). The wind spectrum is given in Figure 2 and it corresponds to time interval of 600 s, while time step in the FEM equals Δt = 1 s. Each time we calculate the basic first four probabilistic characteristics of the observed parameters and this is done with the use of the 16th order stochastic perturbation method. This is done thanks to the polynomial response function, whose

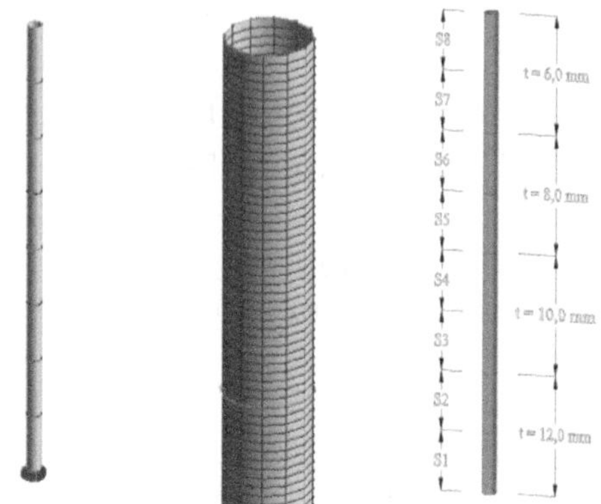

Figure 1. FEM discretization of the chimney (left and middle) and its thickness variations (right).

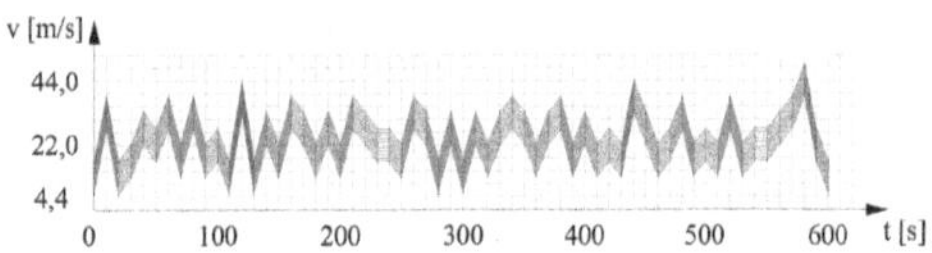

Figure 2. Dynamic excitation of the steel chimney—first numerical approach.

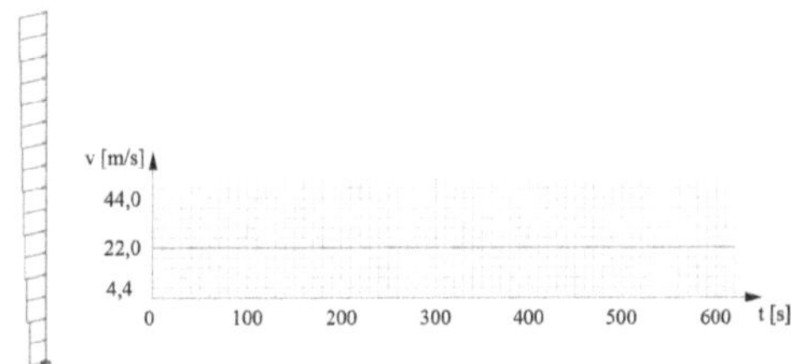

Figure 3. FEM discretization of the chimney and the typical standard load state—second numerical approach.

argument is the input Gaussian random variable of the wind speed v (adopted after Wittmann & Schneider 1974) and the ninth order polynomial is been adopted after initial FEM tests with 3D cantilever model of this chimney.

A comparative structural analysis has also been performed for the chimney modelled as the vertical cantilever beam (by using of 8 2-noded 3D beam finite elements), where the wind velocity value is time-independent (Fig. 3) and its expectation E[v] equals to 22 m/s. It is worth to emphasize that this type of simplified computational approach is most frequently used in steel chimneys design procedure. Numerical results obtained for this reduced computational model are presented in the graphs by blue solid line. The average wind pressure in both models is adopted as 0.3 kN/m², while the aerodynamic resistance of the structures has been determined based on geometric parameters of the object as well as the additional formulas provided by the standard—Eurocode 3.

4 NUMERICAL RESULTS

Expected values of a deflection of the top of the chimney E[u] (in [cm]) obtained for both computational models (full time analysis and shell model *versus* beam model and constant wind velocity as a function of the coefficient of variation α assumed to be equal 0.05, 0.10, 0.15 have been shown on Figure 4, correspondingly. All these results have been obtained by the use of the eighth order perturbation method.

Execution of the full time analysis allows for an observation of probabilistic parameters fluctuations—in considered case these variations are significant, especially in the context of the serviceability limit state. It can be also noticed that the differences in expected values for different time moments are not too great (comparing the values determined for the same time moments and for different input coefficients α). A comparison of the results of full SFEM dynamic analysis and the analysis performed for the fixed value E[v] looks completely different. The expected values of displacements are apparently larger for the full SFEM dynamic analysis model in most of the analyzed discrete time moments, which clearly shows the benefits of using of this approach in analysis of slender and low-weight skeletal steel structures.

The coefficients of variation $\alpha(u)$ of horizontal displacements at the top of the chimney have been presented in Figure 4 as the discrete functions of time and of an input coefficient of variation.

We can observe in this particular analysis that time responses of the displacements in addition to input random wind velocity are nonlinear in some time moments. Some significant differences in the results in-between full dynamic and simplified static approaches are noticed for the output coefficients of variation; they are two times larger for this second analysis type. These differences are less remarkable for the dynamic approach despite the fact that the relationship between input and output coefficient of variation is also nonlinear for certain time moments. As it was expected, the output coefficients of variation increase together with an increasing standard deviation of the uncertain wind velocity (Fig. 5).

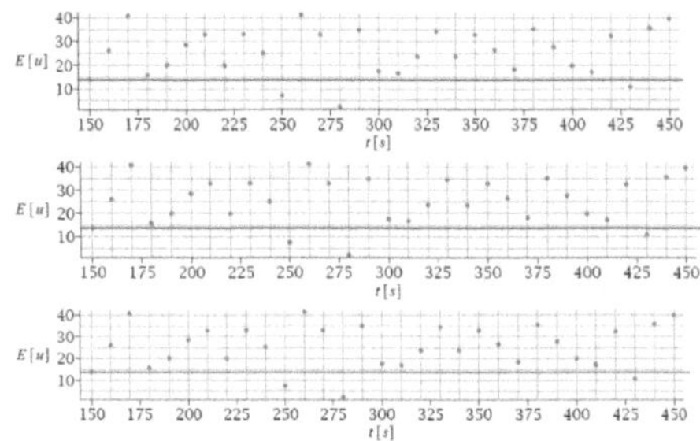

Figure 4. Expected values of the top of the chimney displacements [cm] for input coefficient of variation $\alpha(v)$ equal to 0.05 (top), 0.10 (middle) and 0.15 (bottom).

Figure 6 contains time fluctuations of the skewness for horizontal displacements at the chimney top, namely $\beta(u)$, for three different input coefficients of variation $\alpha(v)$—0.05, 0.10 and 0.15, correspondingly.

We can conclude from these plots that complexity of the structural response is rather high, especially in the context of the displacements skewness. It also confirms that full dynamic time analysis is appropriate for that class of the design issues. The coefficients of skewness for the given set of discrete time moments reaches the value of about 6 (for $\alpha(v) = 0.15$). It should be also emphasized that a necessity of execution of the full SFEM dynamic analysis is in this case also clearly visible. The results obtained for the simplified static model are significantly different, so that the output probability distribution may be affected by a remarkable numerical error.

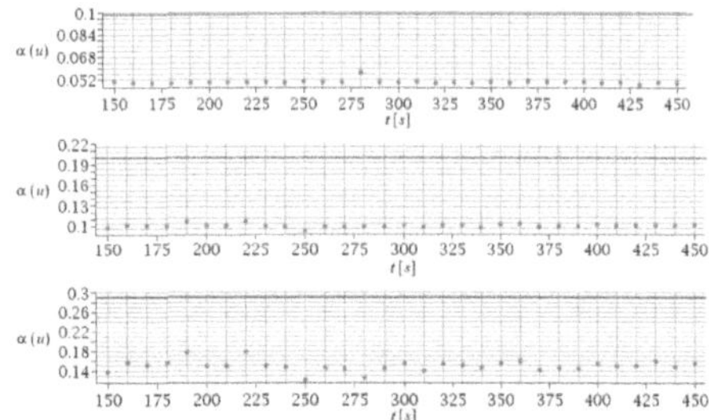

Figure 5. Coefficient of variation of the top of the chimney displacements for input coefficient of variation $\alpha(v)$ equal to 0.05 (top), 0.10 (middle) and 0.15 (bottom).

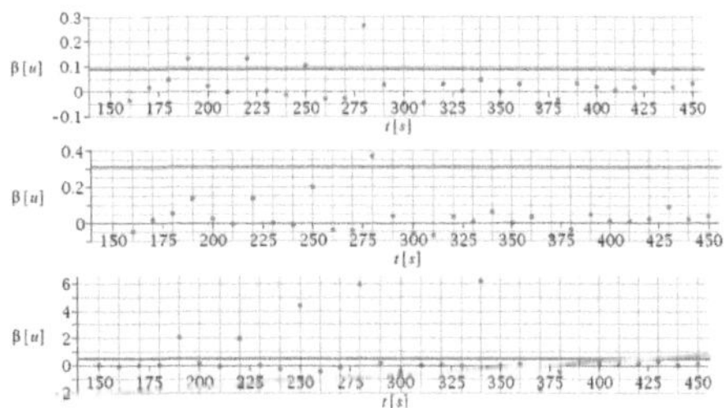

Figure 6. Skewness of the top of the chimney displacements for input coefficient of variation $\alpha(v)$ equal to 0.05 (top), 0.10 (middle) and 0.15 (bottom).

Figure 7 includes the resulting values of kurtosis for the horizontal displacements of the top of the chimney. As one may see, the dispersion of kurtosis values increases together with an increase of the standard deviation for the wind velocity. It is interesting that for some discrete time moments, the values obtained for full SFEM dynamic analysis are higher than these estimated for static analysis ($\alpha(v) = 0.05$ and $\alpha(v) = 0.10$). However, an increasing value of the input coefficient of variation of the wind velocity—as 0.15, for instance, may change this trend. The second approach of a static nature gives larger results in this case. It is worth to notice that kurtosis exhibits also negative values in certain period of time domain, for example in the interval [250, 310] sec. Such a situation in the reduced static model does not occur.

Figure 8 shows the basic probabilistic parameters of the transverse force H at the clamped edge of the chimney. This parameter, in a form of the rotating moment, is especially important in stability analysis of the chimney foundations. The graphs including time fluctuations of random parameters such as expectation value E[H], coefficient of variation $\alpha[H]$, and also kurtosis $\kappa[H]$ are presented below.

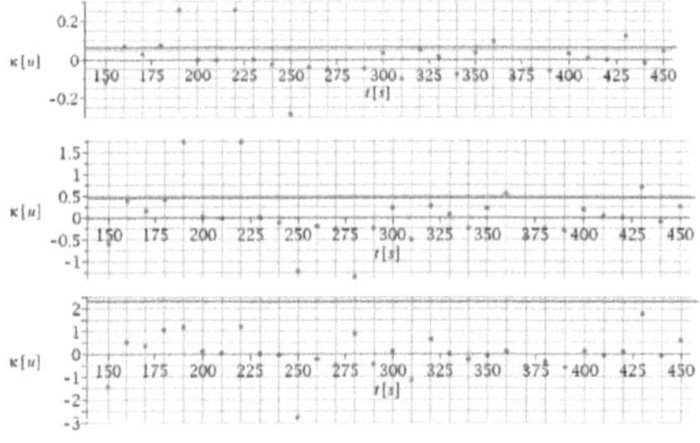

Figure 7. Kurtosis of the top of the chimney displacements for input coefficient of variation $\alpha(v)$ equal to 0.05 (top), 0.10 (middle) and 0.15 (bottom).

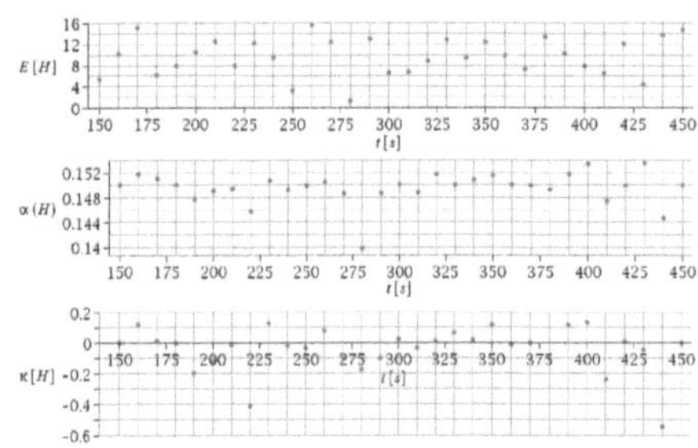

Figure 8. Expected values E[H], coefficient of variation $\alpha[H]$, kurtosis $\kappa[H]$ of the horizontal reaction for input coefficient of variation $\alpha(v)$ equal to 0.15.

5 CONCLUSIONS

Computational analysis performed in this work allows to observe that the scattered dispersion of the displacements expectations leads to a conclusion that precise full SFEM dynamic study needs further the entire wind excitation spectrum. Execution of the full time analysis allows for an observation of probabilistic parameters fluctuations—these variations are significant, especially in the context of the serviceability limit state. Higher order statistics allow to validate the probability distribution of the random output as apparently non-Gaussian. Introduction of the full-dynamic analysis to the design process of the steel chimneys allow for more efficient, economical and safe specification of parameters like: foundation size, thickness of the structural members and etc.

The future tasks on this subject should be focused on an improvement of the SFEM algorithm to increase time discretization as well as to use several wind velocity functions following directly meteorological research.

REFERENCES

Björck, A. 1996. *Numerical methods for least squares problems*. SIAM, Philadelphia.

Eurocode 3 2006. Design of steel structures—Part 3–2: Towers, masts and chimneys—Chimneys. *European Committee for standardization*. Brussels.

Flaga, A. 2008. *Wind Engineering. Base and applications* (in Polish), Warsaw, Poland.

Hilber, H.M., Hughes, T.J.R. & Taylor, R.L. 1977. Improved numerical dissipation for time integration algorithms in structural dynamics. *Earthquake Engineering and Structural Dynamics* 5 (3): 283–292.

Hughes, T.J.R. 2010. *The Finite Element Method: Linear Static and Dynamic Finite Element Analysis*. Dover: Chichester.

Kamiński M. 2013, *The Stochastic Perturbation Method for Computational Mechanics*. Wiley: Chichester.

Kamiński, M. & Szafran, J. 2012. Stochastic finite element analysis and reliability of steel telecommunication towers. *CMES: Computer Modelling in Engineering and Sciences* 83 (2): 143–168.

Kleiber M. 1985. *Finite Element Method in Nonlinear Continuum Mechanics* (in Polish). – Warsaw-Poznań: Polish Scientific Publishers.

Kleiber, M. & Hien T.D. 1997: Parameter sensitivity of inelastic buckling and post-buckling response. *Computer Methods in Applied Mechanics and Engineering* 145: 239–262.

Rykaluk, K. 2005, *Steel Structures. Chimneys, Towers, Masts* (in Polish), Wroclaw, Poland.

Wittmann, F.H. & Schneider, F.X. 1974. *Wind and vibrations measurements at the Munich Television Tower*. SFSM, Canada.

Advances in Mechanics: Theoretical, Computational and Interdisciplinary Issues – Kleiber et al. (Eds)
© 2016 Taylor & Francis Group, London, ISBN 978-1-138-02906-4

Welding deformation in a structure strengthened under load in an empirical-numerical study

P. Szewczyk
Faculty of Civil Engineering and Architecture, West Pomeranian University of Technology, Szczecin, Poland

M. Szumigała
Institute of Structural Engineering, Poznan University of Technology, Poznań, Poland

ABSTRACT: The paper presents the numerical and experimental results of an analysis conducted on welding deformations occurring in an element loaded with force during welding. The effect of the level of structural effort on the weakening of an element while welding and on welding shrinkage were determined. Initial investigations were conducted on overlay welded flat bars at various load levels. Then, full-size steel-concrete composite beams strengthened with welded sheet metal elements under load were investigated. The same tests were carried out numerically in Abaqus environment. The problem of strengthening structures is the focus of the authors' research interests.

1 INTRODUCTION

Welding shrinkage is a commonly known phenomenon. Many methods are used to determine its effects depending on the technology, method and parameters of welding. Most available papers enable calculation of welding shrinkage effects for elements that are not loaded with any forces, e.g. during the production of steel structures. However, sometimes partially loaded elements must be welded, e.g. to strengthen a structure. Few papers dealing with the issue of structureal behaviour following welding under load have been published to date. Augustyn & Skotny (1991) as well as Bródka (1995) used a simplified approach, assuming that the effect of shrinkage does not depend on the level of loading.

The study attempts to determine whether or not and to what extent the level of loading, for a particulate element being welded, affects welding deformations at the moment when the welded element reaches its maximum temperature and once it has cooled down. The study is a part of a larger project aiming at determining the effectiveness of strengthening steel-concrete composite beams under load (Szumigała & Szewczyk, 2014a).

2 PRELIMINARY TESTS

To determine the scale of the problem, initial tests were conducted on steel flat bars, with dimensions of 40 × 10 mm and the length of 1,000 mm. Samples were taken from the flat bars and static tensile testing was conducted. The yield point Re = 260 MPa, ultimate tensile strength Rm = 375 MPa and Young's modulus E = 206 GPa were determined. A numerical model of the flat bar was developed in Abaqus/CAE environment to take into account the above elastic and plastic parameters of the material. A sample view of the numerical and physical models of a longitudinal weld is presented in Figure 1.

Two tests were conducted. First, a transverse weld was placed perpendicularly to the axis of the element, along all its width (40 mm). Second, a longitudinal weld, 100 mm in length, was placed along the axis of the element. In both cases a 3.25 mm rutile coated electrode was used and the electric current was 140A.

Each measurement cycle consisted of:

1. the introduction of initial load: 10, 30 or 50 kN
2. welding and measurement of the force
3. cooling and measurement of the force.

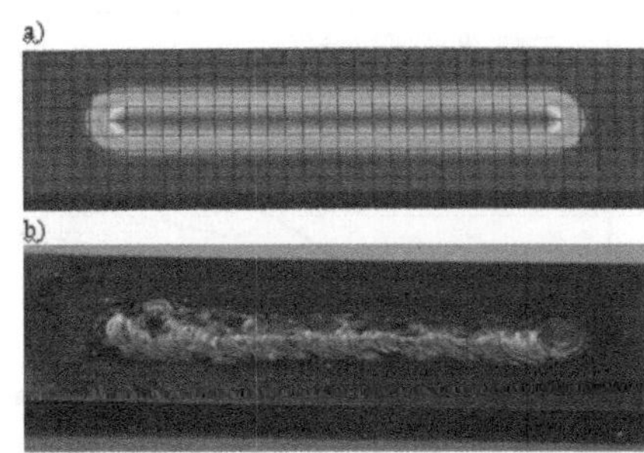

Figure 1. Steel flat bars, a) numerical model, b) physical model.

Since load was introduced in the experiment as a kinematic input, the effect of welding can be seen as a change of the force occurring in the examined element. For example, the measurement cycle for a longitudinal weld placed under 30 kN load is shown in Figure 2.

As can be seen in Figure 2, satisfactory consistency between experimental and numerical results was achieved. Similar consistency was achieved in other tests. Numerical analysis was complemented with a whole range of load, from zero to the yield point. Figures 3 and 4 below show the curves for the longitudinal and transverse welds, respectively. Results are shown as force—stress combination in the element before welding.

The continuous line denotes the load introduced before the test. The dashed line stands for the force in the element shortly after welding, at its largest decrease (approximately 5–10 seconds after welding). The dotted line represents the force measured once the rod has completely cooled down.

The curves show that the effort of the element during welding markedly affects welding deformation. For the longitudinal weld, in the investigated case, the welding shrinkage at 80 MPa is so small that it is not able to exceed the level of initial stress.

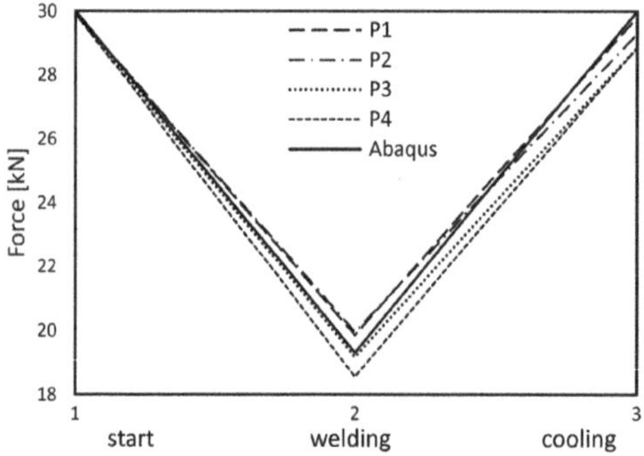

Figure 2. The exemplary measurement cycle.

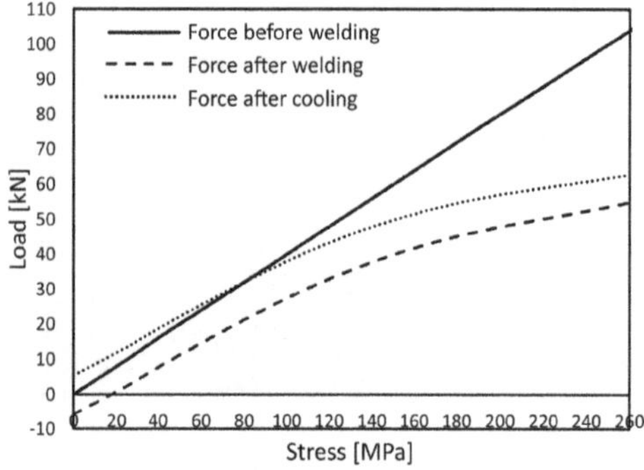

Figure 3. The results for the longitudinal weld.

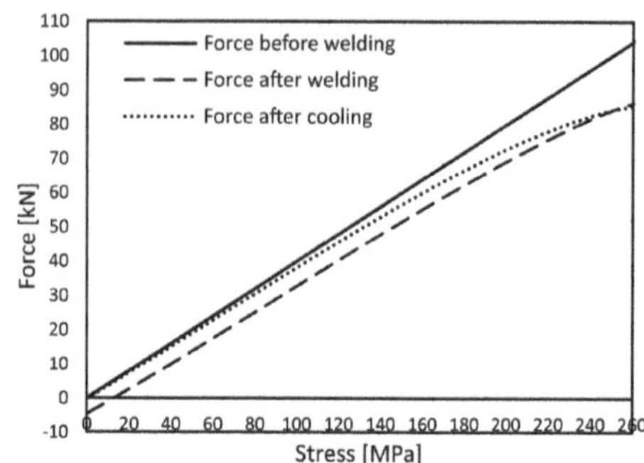

Figure 4. The results for the transverse weld.

Above 150 MPa, a significant decrease of the force was recorded, due to welding.

Research of welds placed transversely to the direction of main stress very clearly shows their negative effect on a structure due to the heating of the whole width of a cross-section and provides ample evidence that such welding techniques should be avoided.

3 MAIN TESTS

The aim of tests conducted on simple elements and described in Section 2 was to calibrate numerical models. The main objective of the study was to perform an experiment simulating the process of strengthening, through welding a steel-concrete composite beam under load.

Welding parameters, such as the method, type and diameter of an electrode, electric current or even the experience of a welder, have an enormous effect on results. Therefore, exactly the same materials, parameters and welding technique were used in strengthening tests compared to those applied in the tests described in Section 2. Welding operations were performed by the same welder. The beam cross-section and the location of the sheet metal used for strengthening purposes are shown in Figure 5.

The cross-section was selected so that the elastic neutral axis was at the level of connection between steel and concrete. Given the yielding, the neutral axis is translated inside the reinforced concrete zone, which justifies the strengthening operations through the enlargement of the steel part of the cross-section. Owing to the technological considerations (i.e. enabling welding in a flat position), the constant width of sheet metal was set at 120 mm. The optimal choice of the length of strengthening sheet metal has been the focus of much previous research (Szumigała & Szewczyk, 2014a). The final dimensions of sheet metal used in the study were the thickness of 10 mm and the length of 3200 mm which was supposed to increase the load bearing

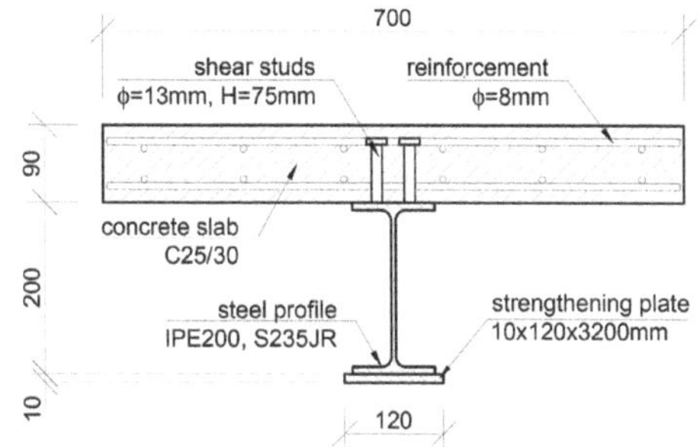

Figure 5. Cross-section of composite beam.

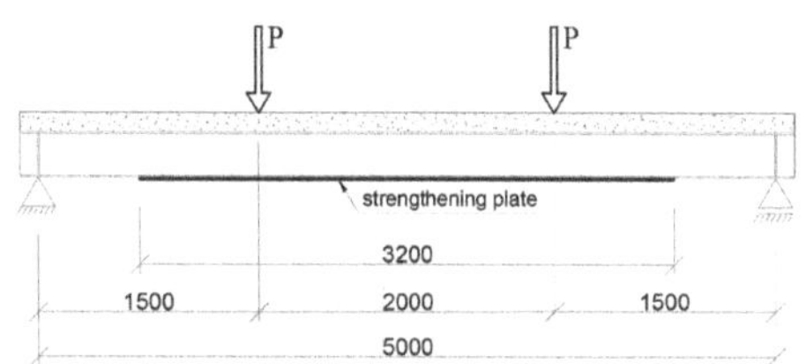

Figure 6. The testing beam with strengthening plate.

capacity of the beam by 50%. Figure 6 shows the strengthened beam and the load schematic for a 4-point bending test.

Three experimental models of the above geometry and a 3D numerical model accounting for material and geometrical non-linearities developed with Abaqus/CAE were developed for the tests. Figure 7 shows the experimental model during a welding attempt.

The load during strengthening operations was preset so that the stress in the bottom flange, i.e. the location of welding, was approximately 70% of the allowable stress (for S235JR steel). The introduction of large amounts of heat during welding reduces the cross-sectional stiffness of an element and increases dislocation. The effect of welding shrinkage was to be observed during cooling and as a result of that dislocation directed opposite to that which arises during welding occurs. Load was introduced in experiments with a hydraulic actuator. Its regulator was supposed to maintain a constant force, therefore during welding the piston enabled the beam to deform by changing its position while maintaining the preset load at all times.

Three identical tests were carried out. Figure 8 shows the recorded results of displacement in the middle of the beam in the function of time. First, the initial load simulating the partial loading of the structure before strengthening was introduced. Next, C-clamps were used to fasten the strengthening sheet metal to the bottom flange of the beam so that it assumed the curvature identical with that of the strengthened element. Then, according to

Figure 7. The experimental model during a welding attempt.

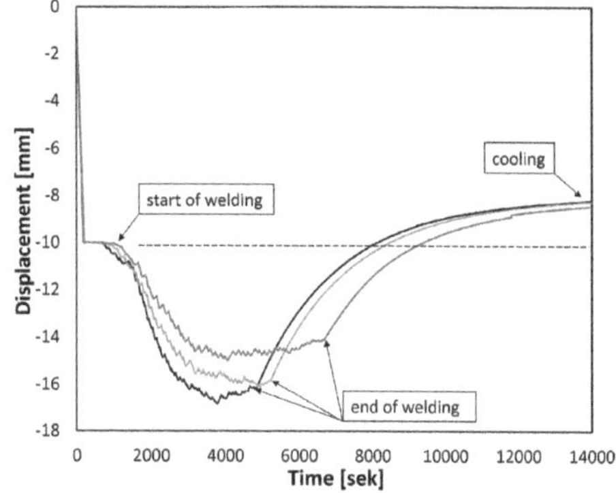

Figure 8. Experimental results for three tests.

the prepared technology longitudinal welds were made to connect the strengthened element with the strengthening one. First, the ends of sheet metal were welded to prevent its dislocation. Next, subsequent welds were placed alternately starting from the middle towards the end of sheet metal. The welding process of the first beam lasted approximately 1.5 h while the other two beam were welded faster, approximately 1 h each. When the welding was finished, each structure was left to cool (approximately 2 h).

As can be seen from Figure 7, all the 3 beams showed the same bending values before welding. The welding time, and following on from that the amount of heat introduced into the structure in a unit of time, markedly affects beam deflection during the strengthening operations. It can be clearly observed while comparing the graphs for the beam welded for 1.5 h and the other two welded for approximately 1 h. The observation of beams during cooling provides important information. Regardless of bending observed during welding, the deflection of all the beams after cooling stabilised at a very similar level. The same experiment was numerically conducted in Abaqus environment, which is presented in Figure 9. The numerical simulations did not use the notion of real time

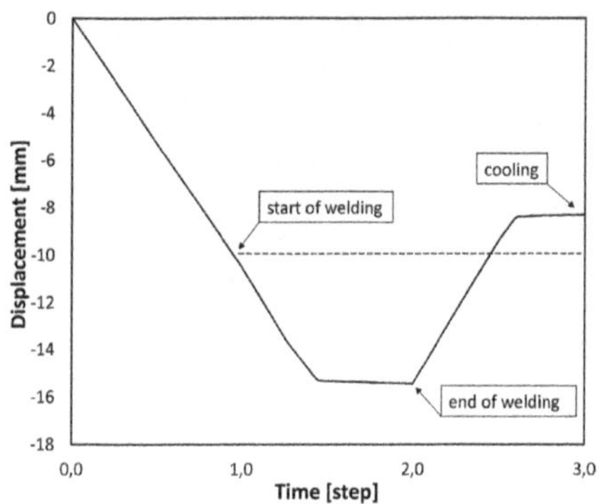

Figure 9. Results of numerical model.

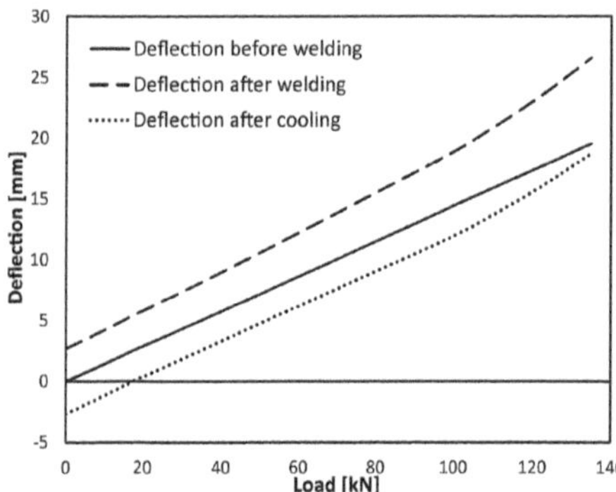

Figure 10. Collection of results for full range of load.

expressed, for example, in seconds. Instead, they were measured using the successive steps of analysis, whose course was expressed in the range from 0 to 1. The first step was the introduction of the initial load, the second step was welding and the last was cooling. A comparison of results, shown in Figures 8 and 9, demonstrates good consistency between the numerical and real models.

Based on that, similarly to the procedure described in section 2, the scope of numerical analysis was broadened, ranging from zero initial load to load close to the yield point. Results are presented in Figure 10 as a graph showing the relationship between load and deflection. Three characteristic deflection parameters were determined for each load value: the first—denoting deflection before welding—shown with the continuous line, the second—denoting deflection after welding—shown with the dashed line and the last—standing for deflection after cooling—shown with the dotted line.

Our analysis revealed an additional, positive effect of strengthening through welding. The shrinkage of long fillet welds placed in the lower part of the cross-section results in the prestressing of the bottom flange, i.e. in the reduction of beam deflection after welding. Needless to say, it is a very favourable effect in view of the ultimate limit state. Unfortunately, when the load was increased, the effect of shrinkage was getting smaller and smaller with the beam deflection gaining substantial values during welding.

The strengthening of structures through welding also has some negative effects. The introduction of large amounts of heat irreversibly changes the mechanical properties of steel and introduces an additional stress, which is difficult to estimate. The latter can be inferred through stiffness decrease in strengthened elements, which was discussed at length elsewhere (Szumigała, 2002; Szumigała & Szewczyk, 2014b). If residual stress is taken account of in a numerical model, it significantly improves the consistency between numerical and experimental results.

4 CONCLUSIONS

The presented results were obtained for one type of weld only (3.25 mm rutile coated electrode). Given the variety of materials and welding techniques used in the industry, our findings should be treated cautiously. However, based on the results obtained in the study, several conclusions can be drawn:

- welding deformations markedly depend on the effort of a structure during welding
- welding shrinkage can have a positive effect on the performance of a structure, causing for example lower deflection values. However, as load increases, the positive effect is replaced with excessive deflection occurring when the structure is heated up
- the placement of welds transverse to the main stress has an adverse effect and is not encouraged to be used in technological practice (Ziółko, 1991).

REFERENCES

Augustyn J. & Skotny J. 1991. *Tymczasowe wytyczne wzmacniania elementów konstrukcji stalowych przy pomocy spawania pod obciążeniem,* Warszawa, Izba Projektowania Budowlanego.

Bródka J. 1995. *Przebudowa i utrzymanie konstrukcji stalowych,* Warszawa-Łódź, Mostostal-Projekt S.A.

Szumigała M. 2002. *Model prętowy i przestrzenny stalowego elementu obetonowanego w świetle wyników ekspetrymentu.* XLVIII Konferencja Naukowa KILiW PAN i KN PZITB "Krynica 2002".

Szumigała M. & Szewczyk P. 2014a. *The trial of optimal strengthening of composite beams,* Recent Advances in Computational Mechanics, Taylor&Francis Group, London.

Szumigała M. & Szewczyk P. 2014b. *Analiza numeryczna wzmacniania belki zespolonej,* X Konferencja Naukowa "Konstrukcje Zespolone", Zielona Góra.

Ziółko J. 1991. *Utrzymanie i modernizacja konstrukcji stalowych,* Warszawa, Arkady.

Advances in Mechanics: Theoretical, Computational and Interdisciplinary Issues – Kleiber et al. (Eds)
© 2016 Taylor & Francis Group, London, ISBN 978-1-138-02906-4

An influence of notch type on material behaviour under monotonic tension

T. Szymczak
Motor Transport Institute, Warsaw, Poland

Z.L. Kowalewski
Institute of Fundamental Technological Research, Warsaw, Poland

A. Brodecki
Motor Transport Institute, Warsaw, Poland

ABSTRACT: The paper presents results of numerical and experimental investigations conducted in order to determine an influence of notch type on material behaviour under monotonic tension. Two kinds of specimens having U and V notches were applied. Digital Image Correlation (DIC) system was used to detect variations of strain/stress state components from beginning of the test up to fracture of the material in question. Assessments of the results enabled identification of stress concentrations in tips of notches. A comparison of tensile characteristics of unnotched and multi-notched specimens exhibited a great reduction of the yield point independently of the geometrical discontinuities applied. In the case of V-notched specimen 50% lowering of elongation was observed.

1 INTRODUCTION

Geometrical discontinuities in a form of notches modify magnitudes of stress state components. It is related with their geometrical sizes such as radius, angle or depth. Their influence on a material behaviour is usually examined either by theoretical or experimental analysis. The theory enables to assess possible fracture on the basis of the stress concentration factor and to illustrate variations of this parameter as a notch geometry function. In the case of experiment, an influence of geometrical discontinuities on fracture may be determined by the application of various specimen types. Among of them one can distinguish tubular or flat specimens with notches to be cut in the way reflecting special cases of interest. The notches may have various sizes, usually classified as: small (Fatemi et al. 2002, Qian et al. 2010), medium (Fatemi et al. 2002) or large (Fatemi et al. 2004). It has been found experimentally, that an increase of the notch radius from 0 to 6.35 mm reduces by 50% the number of cycles necessary for an initiation of the crack (Bennett & Weinberg 1954). The same effect was noticed for an increasing stress concentration factor taking values within a range of 1÷2.833 (Fig. 1) (Fatemi et al. 2002, 2004). The narrow notches may change the S-N curve significantly (Fig. 2) (Qian et al. 2010). For smooth specimens, a stepwise S-N curve was obtained representing different places of the crack initiation. For lower number of cycles the surface initiated fracture was observed whereas for higher ones the internally initiated fracture was discovered. In the case of notched specimens, a continuously decreasing curve occurred with fatigue crack started from the surface. These results do not reflect sufficiently geometrical discontinuities effect in a zone close to the notch.

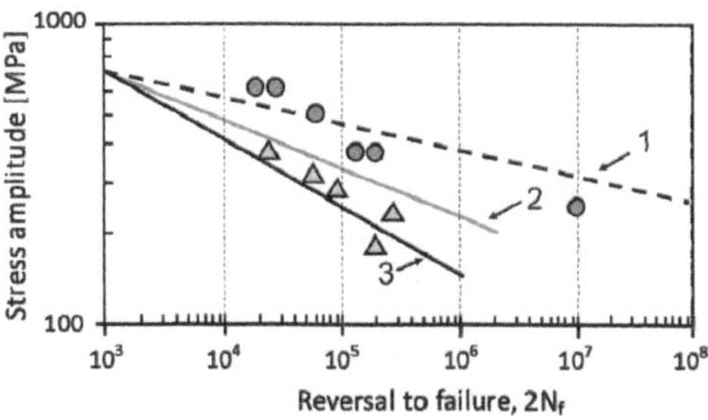

Figure 1. An influence of the SCF values: 1.0 (1), 1.787 (2), 2.833 (3) on the number of cycles to fracture of the 1141 medium carbon steel micro-alloyed with Vanadium, (Fatemi et al. 2002, 2004).

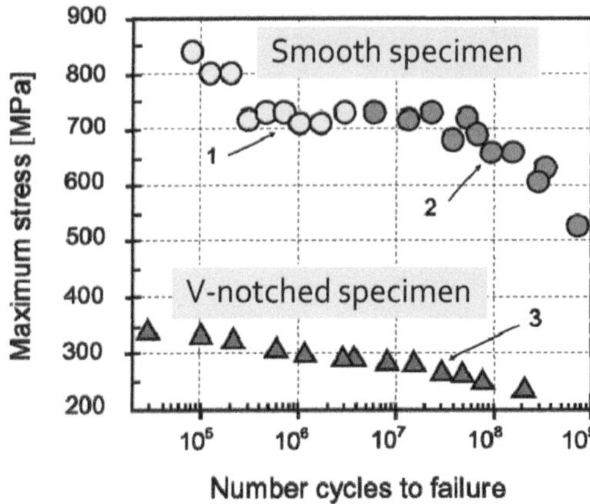

Figure 2. Influence of notches on the S-N curve of 40Cr steel, numbers identifying the results for specimens in which fracture initiated on their external surfaces (1, 3) or inside the material of specimen (2), (Qian et al. 2010).

This problem may be solved using such modern techniques as Digital Image Correlation (DIC) for example. Therefore, the main aim of the paper was focused on application of DIC system for investigation of notches influence on material behaviour during tensile tests carrying out up to the fracture.

2 EXPERIMENTAL PROCEDURE

An influence of notch radius on the crack initiation and its subsequent propagation was analysed on the basis of results obtained using Finite Element Method (FEM) and tests performed by means of DIC technique. Flat specimens of 3 mm thickness with three U and V notches having radius equal to 0.25, 1.5 and 2.5 mm and angle 30, 60, 90° respectively were tested, Figure 3a, c. A depth of the notches was equal to 1.3 mm. Specimen thickness was equal to 3 mm.

In FEM calculations the specimen geometry was reflected by network of 3D Solid Hex Elements of 0.3 mm high. Calculations were performed for the perfectly elastic material. Before the main tests a geometry of discontinuities was checked by the profilometric measurements. Monotonic tensile tests were carried out at room temperature on the servo-hydraulic testing machine at constant displacement velocity equal to 0.5 mm/s. Distribution of the strain components was determined by means of the 4M Aramis Digital Image Correlation system.

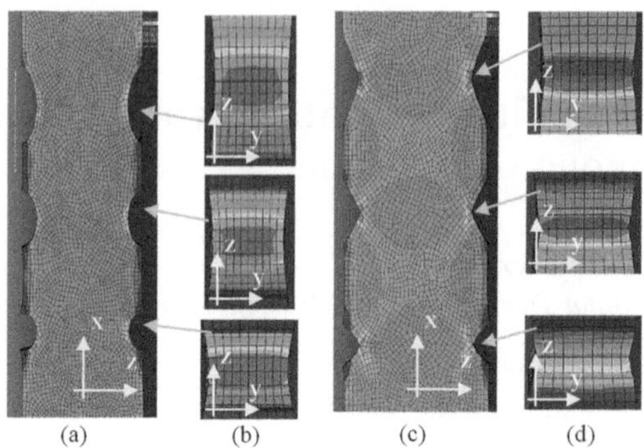

Figure 3. Results of FEM analysis for U and V multi-notched specimens in form of the HMH effective stress distribution on 0XZ (a, c) and 0YZ (b, d) planes, respectively.

3 RESULTS

3.1 *Numerical analysis*

In order to determine a role of radius and angle of notches, the HMH effective stress distribution in 3D coordinate system was considered, Figure 3. The results presented on the 0XZ plane did not show any significant differences in the stress distribution due to application of the notches, Figure 3a, c.

An opposite result was achieved for the HMH effective stress distribution presented on the 0YZ plane, Figure 3b, d. In this case a notch effect is expressed by differences between zones of the maximum stress. The largest area of the stress, expressing the crack occurrence, was obtained for the smallest radius (U specimen) and biggest angle (V specimen), Figure 3b, d, respectively. Moreover, one can notice that a radius increase did not cause variations of the stress level, however, it led to reduction of the maximum stress zone area along the y axis and to its expansion along the z axis. In the case of V specimen the stress reduction was observed with lowering of the notch angle, Figure 3d. The effect of radius and angle of notches on the stress level was also evaluated on the basis of calculation performed using the equations (1÷8) recommended for notched specimens (Pilkey, 1997).

$$\sigma_{\max} = K_t \sigma_{nom} \tag{1}$$

$$K_t = \frac{\sigma_{\max}}{\sigma_{nom}} \tag{2}$$

$$K_{tU} = C_1 + C_2\left(\frac{2r}{L}\right) + C_3\left(\frac{2r}{L}\right)^2 + C_4\left(\frac{2r}{L}\right)^3 \tag{3}$$

$$C_1 = 3.1055 - 3.4287\left(\frac{2r}{D}\right) + 0.8522\left(\frac{2r}{D}\right)^2 \quad (4)$$

$$C_2 = -1.4370 - 10.5053\left(\frac{2r}{D}\right) - 8.7547\left(\frac{2r}{D}\right)^2 - 19.6237\left(\frac{2r}{D}\right)^3 \quad (5)$$

$$C_3 = -1.6753 - 14.0851\left(\frac{2r}{D}\right) + 43.6575\left(\frac{2r}{D}\right)^2 \quad (6)$$

$$C_4 = 1.7207 + 5.7974\left(\frac{2r}{D}\right) - 27.7463\left(\frac{2r}{D}\right)^2 + 6.0444\left(\frac{2r}{D}\right)^3 \quad (7)$$

$$K_{tV} = C_{1V} + C_{2V} + \sqrt{K_{tU}} + C_3 K_{tU} \quad (8)$$

As presented in Figure 4, the stress concentration factor and maximum stress decrease linearly with the notch radius increase. In the case of the V-notched specimen the largest values of these parameters were obtained for the biggest angle considered, Figure 5. These effects are consistent with the results obtained by means of FEM analysis. The representative results for this type analysis are illustrated in Figure 3.

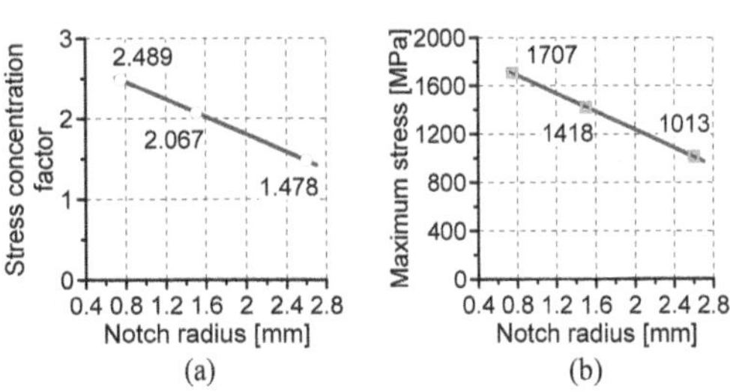

Figure 4. Diagrams representing variations of: (a) stress concentration factor; (b) maximum stress versus notch radius for U-notched specimen.

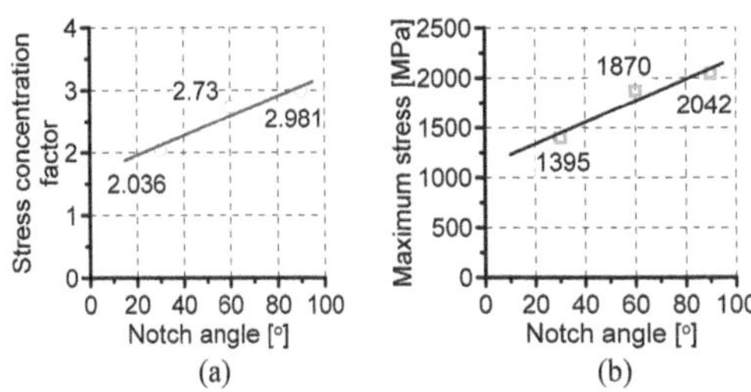

Figure 5. Diagrams representing variation of: (a) stress concentration factor; (b) maximum stress versus notch angle for V-notched specimen.

3.2 *Experimental results*

The 41Cr (1.7035) steel, commonly operated in automotive and power plant branches of industry, was selected for testing. All investigations were carried out under monotonic tension at room temperature using the 8802 Instron servo-hydraulic testing machine and two devices recommended for determination of strain components, i.e.: 2620 Instron static extensometer and Digital Image Correlation (DIC) system called 4M Aramis. Application of the both measurement independently working systems enabled validation of the DIC technique. It was attained by comparison of stress-strain curves.

As it is presented in Figure 6, a discrepancy between the results for extensometer and DIC system was negligible. Selected properties of the steel captured using both techniques are summarized in Table 1. Differences are very small and do not exceed 2%.

An effect of the notches is well reflected by variations of the HMH effective strain isolines. At the

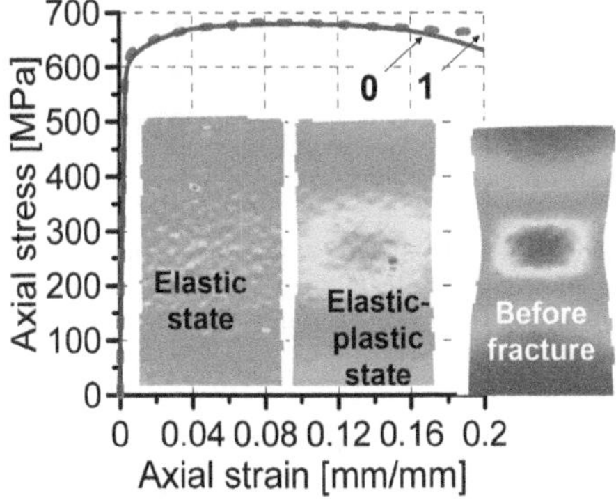

Figure 6. Tensile characteristics of the 41Cr steel determined using: extensometer (0); DIC system 4M Aramis (1).

Table 1. Selected mechanical properties of the 41Cr steel: E—Young's modulus, $\sigma_{0.2}$—yield point, σ_m—ultimate tensile stress.

Measurement technique	E (MPa)	$\sigma_{0.2}$ (MPa)	σ_m (MPa)
Extensometer	17 4396	610	677
4M Aramis	17 1548	616	684

beginning of tension, represented by 500th stage, two dominant cracks appeared and served as the significant stress/strain concentrators, Figures 7a, 8a. In the next step of the U-notch specimen test, the stress/strain components were increasing in the middle notch leading, as a consequence, to specimen fracture (Fig. 7b, c). A different fracture mode was observed for the V-notch specimen. Here the first crack appeared in the middle notch like for the U-notch specimen, but in contrast to that case also another crack initiated in the largest notch (Fig. 8b).

Comparison of the tensile curves determined for smooth and notched specimens enables identification of their essential differences, Figures 9, 10. They are expressed by a clear drop of the yield point observed for test performed on the notched specimens, Figures 9b, 10b. In the case of U-notched

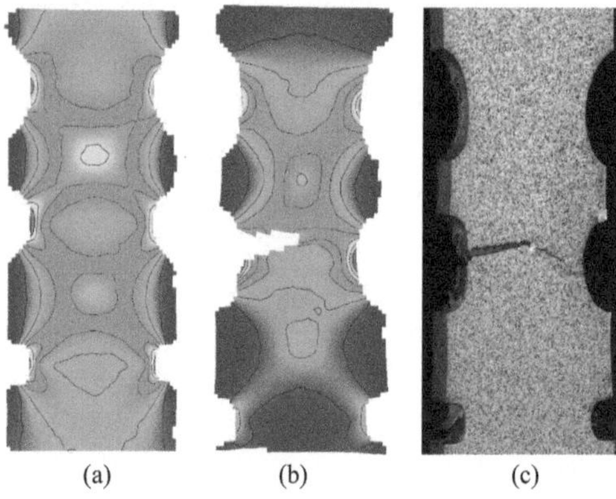

Figure 7. Results of DIC analysis for U-notched specimen showing: the HMH effective strain distribution at stage of: 500 (a), 1449 (b); and general view of specimen fracture at 1449th stage (c).

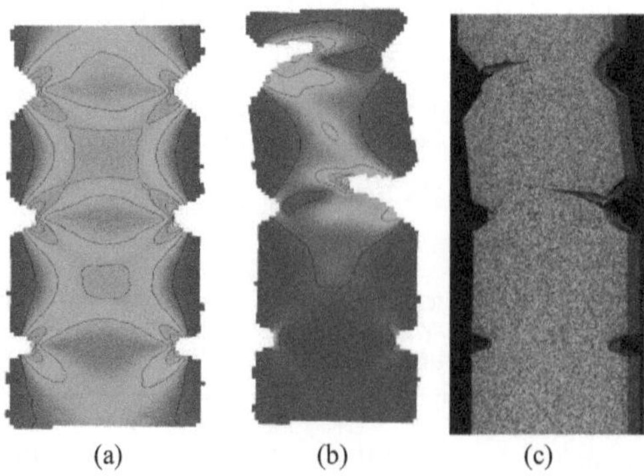

Figure 8. Results of DIC analysis for V-notched specimen showing: the HMH effective strain distribution at stage of: 500 (a), 750 (b); and general view of specimen fracture at 750th stage (c).

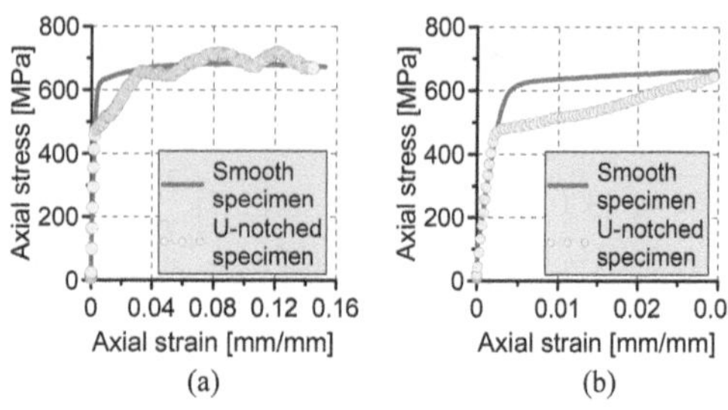

Figure 9. Comparison of tensile characteristics of the 41Cr4 steel for smooth and U-notched specimens: (a) entire curves, (b) initial parts of the curves.

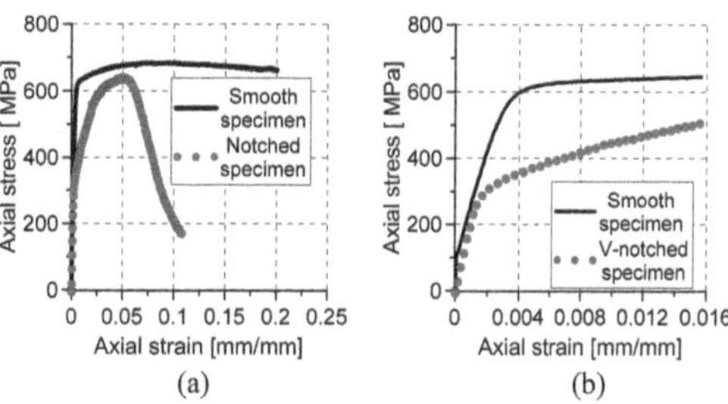

Figure 10. Comparison of tensile characteristics of the 41Cr4 steel for smooth and V-notched specimens: (a) entire curves, (b) initial parts of the curves.

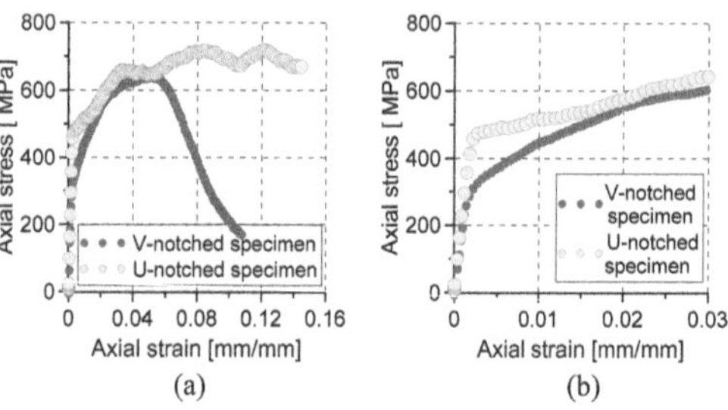

Figure 11. Stress—strain curves determined using specimens with artificial defects: (a) entire curves, (b) initial parts of the curves.

specimen a periodically varying effects of softening and hardening were visible up to material fracture, Figure 9a. Considering a level of the ultimate tensile stress achieved for all specimens tested one can notice that it is almost the same if the results for smooth and U-notched specimens are compared, Figure 9a. Similar comparison for smooth and V-notched specimen exhibits slightly lower ultimate tensile stress for the notched specimen, Figure 10a. Additionally, tensile curves presented in Figure 10a differ significantly for deformation range higher than 5%.

Comparison of the results for U- and V-notched specimens enables identification of an influence of the notch shape on mechanical properties of the steel tested, Figure 11. The higher values of the yield point and ultimate stress were obtained for the U-notched specimens. Also in terms of the elongation better result was achieved for such type of specimen.

4 REMARKS

A reduction of the notch radius for the U-notched specimen leads to the linear increase of the stress concentration factor and maximum stress. The same effect can be observed for the V-notched specimen when the notch angle takes higher values.

The main difference in the shape of tensile curves for the smooth and notched specimens takes place in the range of stresses higher than the yield point.

Comparison the yield point values for both types of notch considered showed 40% difference. Higher value was obtained for U-notched specimen. In the case of elongation of the V-notched specimen the 27% reduction was identified with respect to the U-notched specimen.

REFERENCES

Bennett, J.A. & Weinberg, J.G. 1954. Fatigue notch sensitivity of some aluminium alloys, *Journal of Research of the National Bureau of Standards*, 52 (5): 235–245.

Fatemi, A. et al., 2002. Notched fatigue behaviour under axial and torsion loads: Experiment and Predictions, *8th International Fatigue Congress*, Stockholm, Sweden, 3: 1905–1914.

Fatemi, A. et al., 2004. Fatigue behavior and life predictions of notched specimens made of QT and forged microalloyed steels, *International Journal of Fatigue*, 26: 663–672.

Pilkey, W.D. 1997. *Peterson's Stress Concentration Factors*, Wiley, New York.

Qian, G. et al. 2010. Investigation of high cycle and very-high-cycle fatigue behaviors for a structural steel with smooth and notched specimens, *Engineering Failure*, Analysis, 17 (7–8): 1517–1525.

Advances in Mechanics: Theoretical, Computational and Interdisciplinary Issues – Kleiber et al. (Eds)
 ISBN 978-1-138-02906-4

Fire performance of composite Concrete Filled Tubular columns exposed to localized fire

W. Szymkuć, A. Glema & M. Malendowski
Faculty of Civil and Environmental Engineering, Poznań University of Technology, Poznań, Poland

ABSTRACT: The main goal of the paper is to analyse the difference in the structural behaviour of Concrete Filled Tubular (CFT) columns in case of standard and localised fire. A script for coupling FDS and Abaqus is developed and used to transfer output from CFD analysis to heat transfer in solid followed by mechanical analysis performed in Abaqus. A set of concrete filled tubular columns is analysed in terms of their response to mechanical loading when exposed to fire. Heat transfer from the gas into solid is described with the use of Adiabatic Surface Temperature (AST). The structural behaviour of elements is obtained taking into account both material and geometrical non-linearities and non-uniformly distributed temperature along length and circumference. Comparison of the structural behaviour of the columns in terms of temperature distribution, vertical and horizontal displacements and fire resistance time is provided.

1 INTRODUCTION

Two usual approaches are commonly employed when the effect of fire on structures is considered, i.e. heating conditions according to standard temperature-time curve (3) or parametric fire described by EN 1991-1-2. A different approach can be applied for large compartments, where a fully developed fire is rather unlikely to occur. Heating conditions according to standard temperature-time curve assume identical exposure to high temperatures both along height and circumference of the element whereas for localized fire temperature distribution in the compartment will be varying in Cartesian space. This led to a study concerning fire performance of Concrete Filled Tubular (CFT) columns exposed to localized fire. For calculating heat transfer to locally exposed structures, Adiabatic Surface Temperature introduced by Wickstrom et al. (2007) is used. Study of the structural behaviour of CFT columns exposed to localised fire is based on a previously validated mechanical model by Szymkuc et al. (2015) able to reflect the structural behaviour of CFT columns exposed to heating conditions according to standard temperature-time curve. In order to perform mechanical analysis in the natural fire conditions Fire Dynamics Simulator software (FDS) and Abaqus are used.

2 HEAT TRANSFER

Heat transfer to solid can in general be described by two entities, radiative (1) and convective (2) heat transfer:

$$\dot{q}''_{net,r} = \varepsilon\sigma(T_r^4 - T_s^4), \quad (1)$$

$$\dot{q}''_{net,c} = h(T_g - T_s), \quad (2)$$

The total heat transfer is a sum of (1) and (2).

2.1 *Standard fire*

Standard temperature-time curve is used in majority of fire resistance tests. It defines a monotonic temperature rise in the compartment with only one variable *time*. Its purpose is harmonization of test conditions for structural elements in case of fire. It is expressed by the equation:

$$T_g = 293 + 345\log_{10}(8t + 1), \quad (3)$$

where t is time in minutes and T_g is gas temperature in Kelvin in the compartment or furnace.

2.2 *Localized fire*

As a concept of coupling numerical simulations concerning fire and structural model Wickström et al. (2007) proposed a concept of Adiabatic Surface Temperature (AST):

$$\dot{q}''_{tot,SM} = \varepsilon\sigma(T_{AST}^4 - T_{s,SM}^4) + h(T_{AST} - T_{s,SM}), \quad (4)$$

where $\dot{q}''_{tot,SM}$ is the total net heat flux to the surface, ε is the surface emissivity, σ is Stefan-Boltzmann constant, T_{AST} is the Adiabatic Surface Temperature, $T_{s,SM}$ is the surface temperature in structural model and h is the heat transfer coefficient.

The AST can be measured directly in experiment using Plate Thermometers used in the furnaces to control the temperature. It can also be an output of FDS simulation further used in FE structural model as an effective black body radiation temperature for calculating the incident radiation as well as gas temperature for calculating the convective heat flux.

2.3 *Abaqus-FDS coupling*

The coupling between FDS and Abaqus is done in sequential way, meaning the simulation of fire is performed apriori the heat transfer and mechanical analyses. After that the results of fire simulation are transferred to the finite element code as boundary conditions in form of nodal temperature amplitudes of dummy surface adjacent to analysed column. Then heat transfer is governed by equation (4). The cross-sectional dimensions of analysed columns are great enough to be able to reproduce the column in fire simulation performed in FDS. That means, the boundary data needed for the heat transfer analysis can be transferred directly from the selected surface in fire simulation model prepared in FDS. Nonetheless, because of the rectangular mesh in FDS, the actual shape of circular columns cannot be reflected. The approach similar to the one proposed by Paajanen et al. (2013) is implemented to deal with this particular problem. Heat transfer from the gas into solid is described with the use of AST (Wickström et al. 2007). In practice this allows the reduction of the output from CFD computations without loose of accuracy (Sandstrom 2013). Adiabatic surface temperature isthe one quantity which can be used to describe the actual exposure of the surface to the energy coming from the convection and radiation phenomena.

Due to the incompatible mesh size in FDS (10 × 10 × 10 cm) and Abaqus (order of magnitude less) a simple k-nearest neighbours algorithm is used to prescribe correct temperatures to nodes in the structural model. Weight of *i*th nearest node is described as:

$$w(x,p_i) = \frac{exp(-D(x,p_i)}{\sum_{i=1}^{n} exp(-D(x,p_i))} \quad (5)$$

where $D(x,p_i)$ is the distance between FEM node in Abaqus (x) and *i*th node in FDS (p_i). The same procedure was applied to every node of dummy surface in FE domain taking into account 3 nearest neighbours. Comparison of AST obtained directly from FDS and AST used as a boundary condition in structural model is presented in Figure 1. Highest temperature is in the upper layer of the compartment, with the maximum of about 830°. At the same time maximum temperatures on other faces are approximately 550, 700 and 900°.

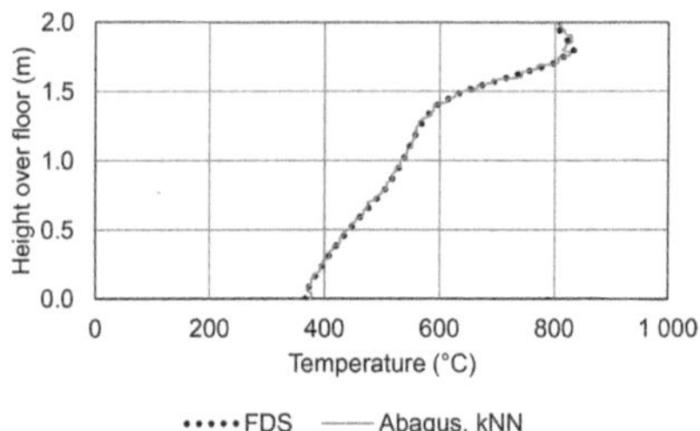

Figure 1. Verification of developed interface between Abaqus and FDS in terms of Adiabatic Surface Temperature distribution over column length on one of the surfaces.

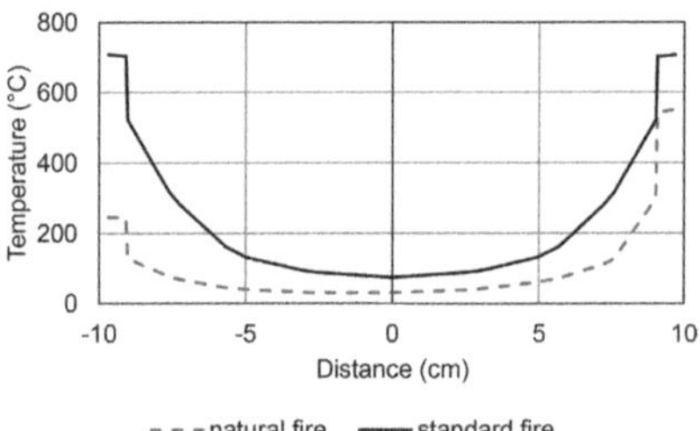

Figure 2. Temperature profile inside the cross-section after 28 minutes of standard and natural fire exposure.

Figure 2 shows a difference between temperature distribution inside the cross-section after 28 minutes of standard and natural fire exposure at selected height over the floor. Column is nearly 20 cm in diameter. Outer parts with higher temperature in both natural and standard fire represent temperature of steel tube, which is directly exposed to fire. A vertical part of both lines is due to thermal gap conductance incorporated in the model. This property allows to better describe temperatures inside the cross-section that are affected by non-ideal heat transfer between steel tube and concrete core. It is anticipated that non-symmetrical temperature distribution shown in Figure 2 that is an effect of natural fire exposure will result in additional thermally-induced bending of the column.

3 CFT COLUMNS ON FIRE

3.1 *Standard fire*

Analysed columns consist of steel tubes filled with plain concrete. The typical response presented in terms of evolution of vertical displacement in time is shown in Figure 3. The response can be divided into three separate stages, the first one is the thermally induced expansion of steel tube. During this stage tube separates form the concrete core and all loads are carried by the tube. Steel and concrete have rather similar thermal expansion, but due to temperatures differences (see Fig. 2) in the first stage they might act independently. After that, loss of stability and a noticeable shortening of the column occurs. From the point when loading plate comes in contact with the concrete core, both steel tube and concrete core serve their load-bearing role. Failure occurs after a certain time due to further degradation of mechanical properties.

Results of a detailed study on the prediction of fire resistance time are presented in Szymkuc et al. (2015), where measured and calculated fire resistance times for 41 analysed specimens are compared. The above-mentioned model served as a base for study presented in this paper. Specimens included in the validation process were taken from research conducted in the recent years (1992–2013) by various researchers (Han et al. 2003, Kim et al. 2005, Lie and Chabot 1992, Moliner et al. 2013, Romero et al. 2011, Wainman and Toner 1992). The sequentially coupled thermal mechanical analysis was applied to the models with initial imperfections. 8-node solid elements are used. The utilisation ratio of the analysed columns varied from 20 to 77%, while the length varied from 3180 to 3810 mm. The results are obtained with a linear Drucker-Prager yield criterion for concrete. It is combined with temperature-dependent stress-strain relationship for concrete based on siliceous or calcareous aggregates given by Eurocode 1992-1-2 with a non-linear descending branch after reaching the peak strength. Concrete fracture is not taken into account in the presented model. The mechanical properties of steel and thermal properties of both concrete and steel were taken from Eurocodes. In terms of normalised resistance the mean ratio between calculated and experimental results is 1.00 with standard deviation equal to 0.26. Detailed results are given by Szymkuc et al. (2015).

3.2 *Localized fire*

Non-standard fire exposition both along the length of a column and its circumference causes additional effects on the element. A non-symmetrical temperature distribution inside the cross-section such as one that might be seen in Figure 2 might result in significant thermal bowing along with additional eccentricities combined with second order effects. These effects have to be taken into account as far as the structural behaviour and fire resistance time of the columns are concerned.

The model prepared in FDS is meant to reflect possible conditions during a car park fire with a burning vehicle close to the structural element. Modelled compartment was 9 × 9 × 2.2 meters. The analyzed column was 2.7 m long with 2.0 m directly exposed to elevated temperature (0.5 m under the floor surface and 0.2 m above ceiling surface). Burning area was 7.2 m^2 with maximum Heat Release Rate per Unit Area (HRRPUA) = 1180 kW/m^2. Maximum HRRPUA was reached after 23 minutes of fire duration. Its development is shown in Figure 4.

3.3 *Standard vs. Localized fire*

The analysed CFT column consists of a steel circular tube with diameter of 193.7 mm and thickness equal to 6 mm. Structural steel with 235 MPa yield strength and concrete with 40 MPa compressive strength are used. The model allowed for a separation between steel tube and concrete core in order to capture behaviour similar to experiments (see Fig. 3). The column is loaded to approximately 36% (600 kN) of its load-bearing capacity in normal conditions. Column is fixed at the bottom and has rotational degrees of freedom at the top left unrestrained. The length of column directly exposed to elevated temperature is 2.0 meters with the total length of 2.7 meters. For the comparison all abovementioned properties are the same in both natural and standard fire. Figure 5 shows development of temperature in one of most heated part of steel tube along with transverse displacement in the middle of

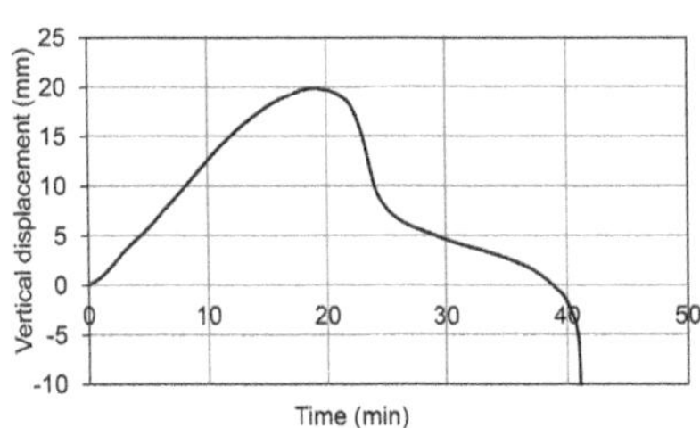

Figure 3. Typical CFT column response under elevated temperatures, Romero et al. (2011).

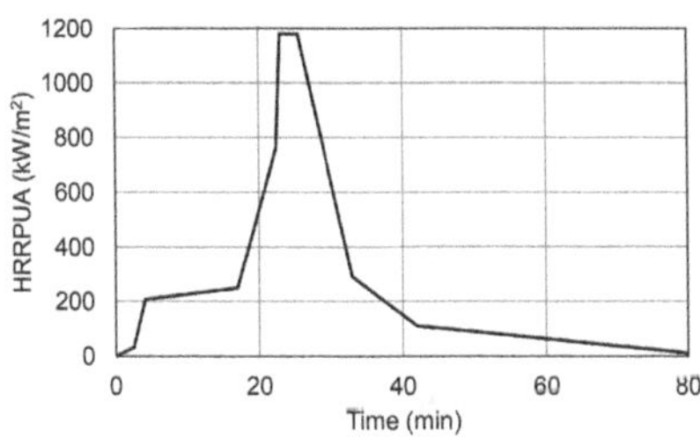

Figure 4. Heat Release Rate per Unit Area (kW/m^2) used in natural fire scenario.

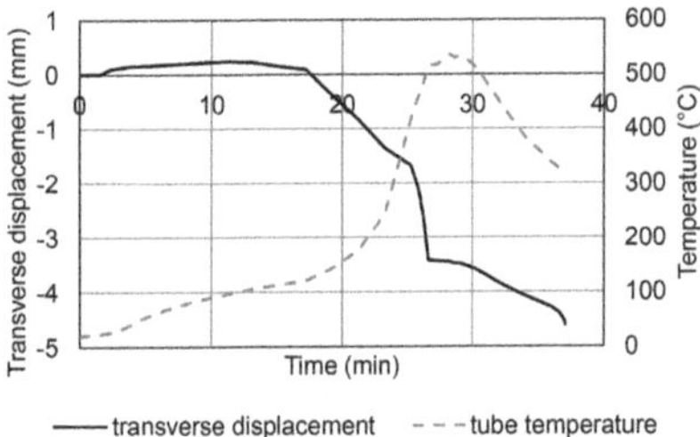

Figure 5. Transverse displacement of CFT column and temperature of selected point on the steel tube surface.

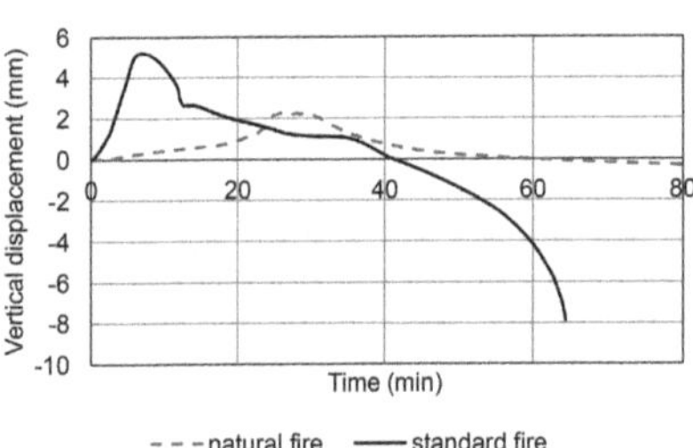

Figure 6. Vertical displacement of CFT column, loaded at 36% of its load-bearing capacity (600 kN).

the column. During the initial phase, column shifts towards the fire due to more significant thermal expansion on this side. Later on, due to degradation of mechanical properties, it shifts away from the fire. Figure 5 refers to specimen loaded with 1500 kN, with a fire resistance time equal to 37 minutes. The structural behaviour at elevated temperatures in standard (3) and localized fire (Fig. 4) is presented in Figure 6. Obviously natural fire represents only one of many possible cases of fire development hence the comparison cannot be ultimate. Nevertheless, the recorded behaviour shows a fire resistance time of CFT column exposed to standard fire limited to 65 minutes, while column exposed to natural fire succeeded in serving its load-bearing role for the whole time of exposure. For a given natural fire scenario, failure occurred for loads higher than 1200 kN. For smaller values columns were able to carry the load for the whole time. In the real case more fire scenarios should be used, including multiple vehicles and not just one. However the goal of this study was to connect a previously developed model with analyses in case of natural fire and not to analyse all possible cases. Applying post-fire stress-strain relationships, especially for concrete, would also be beneficial.

4 CONCLUSIONS

The results presented in the paper show a possible way of connecting Computational Fluid Dynamics models used to calculate temperature development in natural fires with Finite Element structural models. Presented approach successfully used a concept of adiabatic surface temperature that was transferred into finite element model and further used to study the structural behaviour of columns and calculate their fire resistance time. Approach shown in the paper can be used for performance-based design of structures in cases where it is possible to limit the number of fire scenarios.

REFERENCES

Han, L.H., X.L. Zhao, Y.F. Yang, & J.B. Feng (2003). Experimental study and calculation of fire resistance of concretefilled hollow steel columns. *Journal of Structural Engineering 129*, 346–356.

Kim, D.K., S.M. Choiz, J.H. Kim, K.S. Chung, & S.H. Park (2005). Experimental study on fire resistance of concretefilled steel tube column under constant axial loads. *Steel Structures 5*, 305–313.

Lie, T.T. & M. Chabot (1992). Experimental studies on the fire resistance of hollow steel columns filled with plain concrete. Internal report No. 611. Technical report, Institute for Research in Construction, National Research Council of Canada (NRCC), Ottawa, Canada.

McGrattan, K., S. Hostikka, R. McDermott, J. Floyd, C. Weinschenk, & K. Overholt (2014). Fire Dynamics Simulator (Version 6), Technical Reference Guide. Technical report, National Institute of Standards and Technology.

Moliner, V., A. Espinos, M. Romero, & A. Hospitaler (2013). Fire behavior of eccentrically loaded slender high strength concrete-filled tubular columns. *Journal of Constructional Steel Research 83*, 137–146.

Paajanen, A., T. Korhonen, M. Sippola, S. Hostikka, M. Malendowski, & R. Gutkin (2013). FDS2FEM a tool for coupling fire and structural analyses. In *Proceedings of the IABSE Workshop: Safety, Failures and Robustness of Large Structures*, Volume 100, pp. 218–224. International Association for Bridge and Structural Engineering.

Romero, M., V. Moliner, A. Espinos, C. Ibaez, & A. Hospitaler (2011). Fire behavior of axially loaded slender high strength concrete-filled tubular columns. *Journal of Constructional Steel Research 67*, 1953–65.

Sandström, J. (2013). *Thermal Boundary Conditions Based on Field Modelling of Fires*. Lule°a University of Technology, Finland.

Szymkuc, W., A. Glema, & M. Malendowski (2015). Fire performance of steel tubular columns filled with normal strength concrete. In *Proceedings of 5th International Workshop on Performance, Protection & Strengthening of Structures under Extreme Loading (PROTECT 2015)*, pp. 865–872. DEStech Publications, Inc.

Wainman, D.E. & R.P. Toner (1992). BS476:Part 21 Fire Resistance Tests. The construction and testing of three loaded CHS columns filled with concrete, British Steel Report No. SL/HED/R/S2139/1/92/D. Technical report, Swinden Laboratories, Rotherham, UK.

Wickström, U., D. Duthinh, & K. McGrattan (2007). Adiabatic surface temperature for calculating heat transfer to fire exposed structures. In *Proceedings of the 11th International Interflam Conference (Interflam '07)*, pp. 943–953. Interscience Communications Ltd.

Advances in Mechanics: Theoretical, Computational and Interdisciplinary Issues – Kleiber et al. (Eds)
© 2016 Taylor & Francis Group, London, ISBN 978-1-138-02906-4

Topological optimization of formworks' meshes for free-form surfaces

R. Tarczewski & M. Święciak
Faculty of Architecture, Wroclaw University of Technology, Wroclaw, Poland

ABSTRACT: Application of free-forms in architecture became very popular as CAD tools like Rhino 3D gave the ability to shape forms by curved surfaces. However, if the designed surface is to be constructed as a reinforced concrete shell casted on site, there is lack of sufficient methods for formwork construction. An alternative approach described in the paper is based on structural properties of 3-valent polyhedra. The suggested method presumes a discretization of double curved surfaces into plate stable polyhedra feasible to be realized as formwork on construction site. Also, to provide better performance of formwork stability and its closer approximation to base surface—optimization of a face distribution is described.

1 INTRODUCTION

1.1 *Assumptions*

Discretization is a necessary step in shaping free-form surfaces, which are to be constructed as a reinforced concrete wall or shell cast on-site. It provides division of surfaces into a set of planar panels, which can be easily manufactured, e.g. cut from plywood. Since the discretization causes inaccuracies in approximation of base surface, therefore optimization of faces distribution is required, also for structural purposes. The topological properties of obtained polyhedra should provide formwork stability during concrete casting and generally, the formwork should be easily assembled on site.

2 FORMAT

2.1 *Plate stability of polyhedral*

Ture Wester in his work "Structural Order in Space" states, that any general (i.e. abstract in topological meaning) polyhedron materialized in a form of hinge connected faces (plates) is structurally stable, only if it's 3-valent (vertex valency). In 3-valent polyhedron all its vertices are incident with 3 edges and 3 faces only. An example of such polyhedron is dodecahedron whose form is fixed even when its faces are rigid, but connections between them are hinged.

For spatial structures where rigid plates are hinge-connected and such connections are able to transmit shear forces, Ture Wester states a condition:

$$e + s \geq 3f \tag{1}$$

where: e = number of connected edges, s = number of fulcrums, f = number of plates.

2.2 *Panelization*

The discretization of free-formed surface into such polyhedral surface is a demanding task. While triangular discretization is well developed and there are several methods of such division (majority of free-formed glazed canopies are triangulated), it cannot satisfy condition of 3-valency. Compared to triangular meshes, where the incident triplets of vertices lie on the same plane for each face (moreover, these can also lie on the discretized surface), in 3-valent meshes sets of vertices incident with the face will unlikely lie on the same plane if they are placed on the discretized surface. Thus the panels have to be designated by points on free-formed surface, where faces will be tangent to that surface. The edges and vertices are determined by intersections of pairs (edges) and triplets (vertices) of neighboring faces.

A free-form surface can be discretized uniformly into a triangular mesh, whose dual graph indicates the topology of a newly created 3-valent mesh. In that process vertices of triangular mesh are proxies for plate tangency and the topology of point connections is now the topology of neighboring plate intersections.

Discretization of surface with positive Gaussian curvature only results with concave out-lines of plates (hexagons in that case), while for surfaces with negative Gaussian curvatures the plate outlines are concave only, see Figure 1.

As long as the given surfaces have no transitions between positive and negative Gaussian curvatures—the outlines of plates are proper, while at the transitions outlines tend to intersect themselves. Therefore, additional optimization is required.

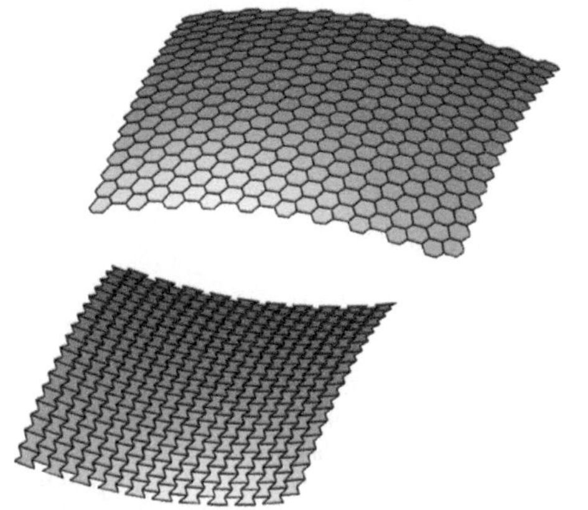

Figure 1. Plate stable discretizations of positive (upper) and negative (lower) Gaussian curvatures.

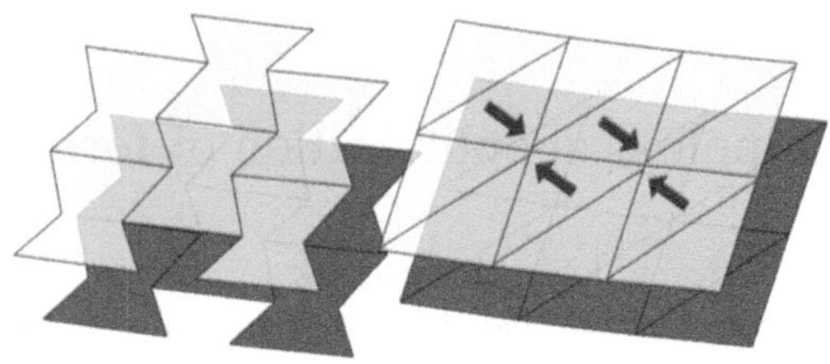

Figure 2. In both offsets the distances between corresponding faces are constant. In the right offset however, the intersections between newly created faces result in creating new vertices.

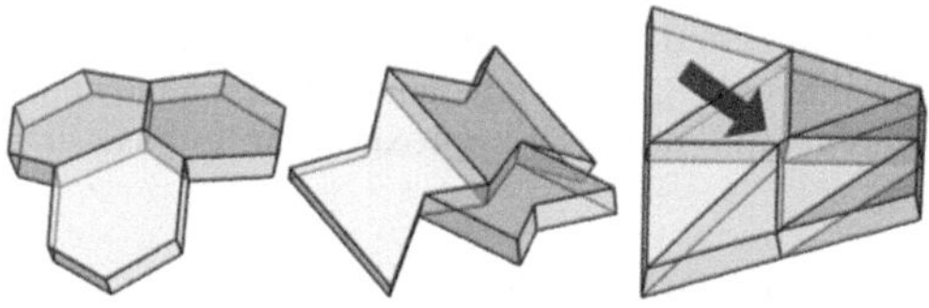

Figure 3. Necessity of additional carvings on the right examples.

2.3 *Offsetability*

Offset is a transformation of a mesh that creates another mesh which faces are parallel to the corresponding faces on the first mesh. Also the distances between corresponding faces have to coincide. The offsets, of the distances between corresponding vertices or edges are equal are also possible. Anyway, to build a formwork for a wall of constant thickness, the first type of offset is best suited.

The chosen type of mesh offsets may produce some shifts in non 3-valent mesh topologies if some requirements are not fulfilled. Fortunately, these shifts do not occur in offsets of 3-valent meshes, if the distance of offset is reasonable compared to faces dimensions, Figure 2.

The ability of preserving topology by 3-valent meshes after offset benefits in designing second layer of formwork, it does not have to be computed n the same way the first was, it can be obtained by offset transformation of the first mesh. Additionally, that property of preserving topology allows saving additional carvings at apexes of panels, Figure 3.

Each panel, however, needs to be chamfered along each of its edges according to angles of connections with corresponding neighboring panels, thus offsetability gives an advantage even for designating panel thickness.

2.4 *Assembly process*

Assuming the 3-valent system, all plates can be manufactured by means of CNC Cutting Machine, e.g. in plywood, providing the correct angles of all edges. For the simplification of assembly process, the assumed topology of connections have to be marked on each plate, that is to say each plate have its number and each edge have its number of neighboring plate. As described above in section 2.1, 3-valent topologies guarantee plate-stability. It also means that if properly assembled, formworks will recreate proper geometry without additional survey works. For each triple of connected plates, the angle deficiency (according to Descartes equation, sum of angles on faces incident with one vertex is less than 2π) determines convexity of an apex at the intersection of these connections. That also means, that connections along edges do not necessarily need to counteract against bending between plates. Consequently hinge connectors can be applied and one type of connector is sufficient for any angle of connection.

3 THEORETICAL BASE

3.1 *Variational shape approximation*

The discretization of free-form surface refers to Variational Shape Approximation process. The VSA algorithm which was introduced in computer graphics for simplification of complex triangular meshes has already been used for architectural design purposes in the past. Generally the algorithm distributes proxy points over the free-form surface, where faces of final polyhedra are tangent with that surface. The distribution refers to surface curvature. In areas with higher curvature the proxies are distributed denser. The distribution also refers to the directions of curvatures—for example—if the Gaussian curvature is close to zero, but surface is curved in one direction, then the proxies will be distributed along the curved direction, but the distribution along direction of zero principal curvature will be minimized. Not only the algorithm helps to minimize the errors of discretization, but also maintains the angles between resulted panels within reasonable values—high angles would result in harsh surface of final wall, while low angles would cause buckling at the connections.

Precisely, the algorithm is an iterative process of assigning faces of dense triangular surface discretization into clusters. The process starts with random proxy distribution, the number of proxies refers to final polyhedron's face amount. Then, a process called Distortion-minimizing Flooding sorts proxy neighboring triangles by an error metric which is given by formula:

$$\|n_i - N_i\|^2 |T_i| \tag{2}$$

where n_i is triangle normal vector, T_i is its area and N_i is proxy normal vector.

Each triangle error may be calculated correspondingly to multiple proxies if it neighbors with their clusters. The least error indicates triangle which will be assigned to proper cluster. After assigning all triangles to clusters, the proxies are redistributed within their clusters. The best fitting triangle for new proxy is the one which normal vector is closest to a vector given by a formula:

$$\sum_{T_i \in R_i} |T_i| n_i \tag{3}$$

where R_i is a set of triangles of one cluster. The process of Distortion-minimizing Flooding and redistributing proxies repeats until finding the best spots of plate tangency.

3.2 *Enhancement*

During the tests on VSA algorithm implemented in Rhino3D software, an issue occurred. In areas of surface with negative Gaussian curvature, faces of discretized polyhedron are concave polygons. In defined conditions these polygons can be self-intersecting.

Our observations, as well as conclusions from other works suggest that it is caused by exclusion of the distance between the planes consisting two neighboring proxies in the formula (2). Hence our proposal for enhanced error metric is:

$$\|n_i - N_i\|^2 |T_i| \left(\frac{d_1 \pm d_2}{2} \right) \tag{4}$$

where d_1 and d_2 are distances between proxies and their projections on the plane given by neighboring proxy or triangle.

In the Figure 4, the geometry of surface being discretized tends for proxies to omit left slope which result in the intersection between two faces on the wrong side of middle proxy. To bypass this issue, the introduction of the distance factor is suggested. While the distance between planes rises, the error metric forces a new proxy to occupy space between two clusters causing problematic intersections.

3.3 *Approximation of polyhedral mesh*

After distributing proxies over the surface, Figure 5, the plates need to be defined. The method adopted by the authors assumes to obtain vertices by intersection of plane triplets given by the proxies and normal vector of free-form surface at those proxies. Each plate is described by a set of those plane triplets giving vertex—a set of vertices connected in proper order gives outline of polygon. Each triplet consists of an origin plane (the plane where the plate belongs) and two other that belong to clusters neighboring each other and the cluster of origin plane.

In the first case, the discretization was made automatically and results in several self-intersecting panels, Figure 6. Some manual corrections are also

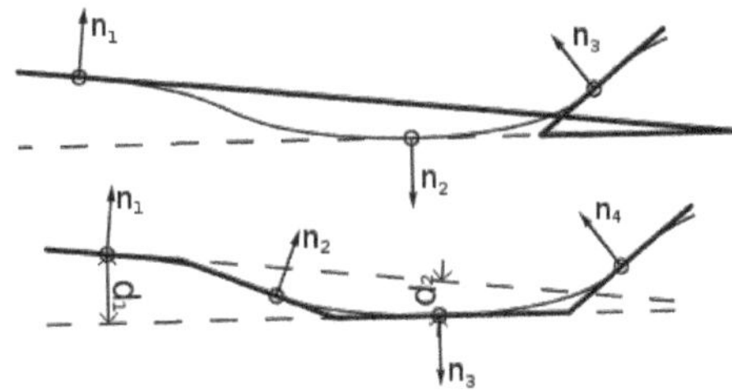

Figure 4. Section of discretized surface: two cases of plane intersections—with (top) and without (bottom) consideration of distances between proxies and planes.

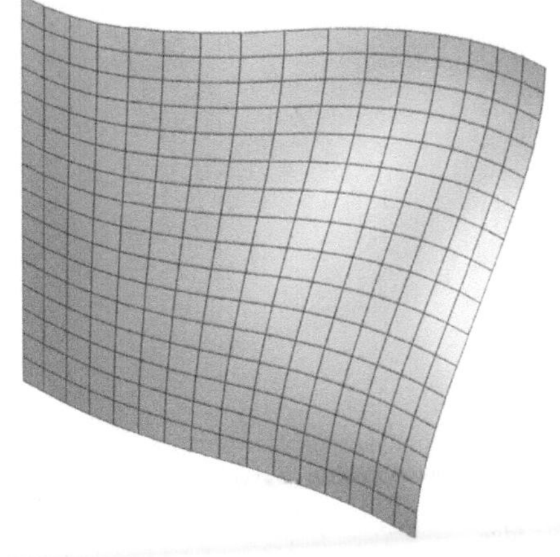

Figure 5. Free-form surface consisting zero, positive and negative Gaussian curvatures used for discretizations.

required in plate placement in order to obtain proper intersections.

In the second attempt, Figure 7, the issue, however still occurring, was significantly reduced, but resulted in very small lengths of some edges—possibly considered as vertices making the polyhedron 4-valent at those locations, which in fact are double 3-valent nodes connected by a very short edge. These edges do not allow for hinge to be attached and even if conditions of formula (1) would still be satisfied, it would affect the stability of formwork.

The last step of designing of the formwork presupposes offset of resulted mesh by a factor of assumed wall thickness, Figure 8.

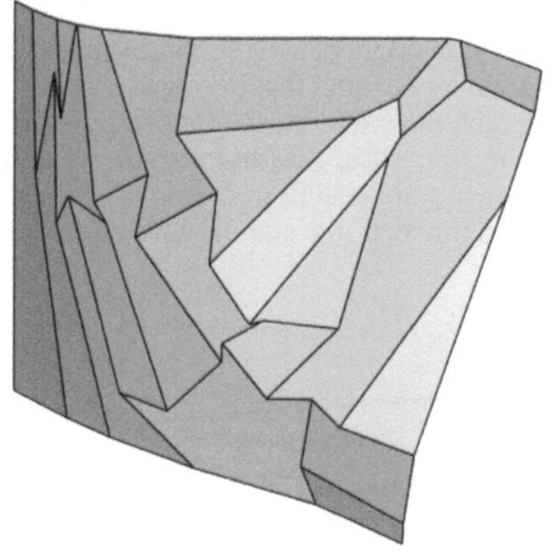

Figure 6. First attempt of discretization according to original VSA algorithm. Some middle panels had to be fixed manually.

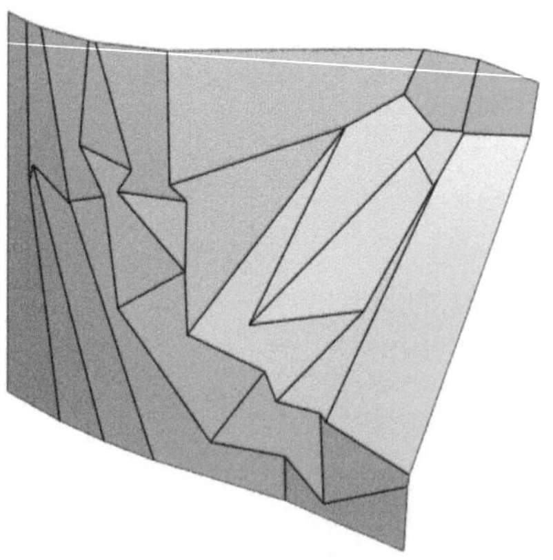

Figure 7. Second attempt of discretization according to enhanced VSA algorithm. Several 4-valent spots occur, actually consist of two 3-valent nodes connected by very short edge.

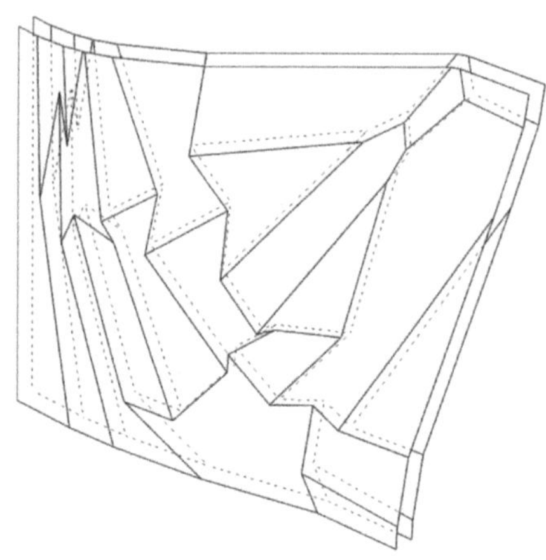

Figure 8. Two layers of polyhedron ready to be further processed into formwork design.

4 CONCLUSIONS

Further development of the discretization method considers optimizing the algorithm due to boundary dimensions of possible panels according e.g. to manufacturing capabilities. In comparison to state of the art methods of constructing free-form shaped structures, the suggested method allows avoiding the use of the additional supporting construction (scaffolding) e.g. for free-formed cladding—consequently the wall thickness can be reduced. Also a lot of survey work is avoided and the amount of connector types is reduced to only one type hinge. After all, the resulted structure requires finishing works to obtain a continuously smooth surface.

REFERENCES

Cohen-Steiner D., Alliez P., Desbrun M. 2004. Variational shape approximation. *ACM Transactions on Graphics* 23(3): 905–914.

Cutler B. & Whiting E. 2007. Constrained planar remeshing for architecture. *Proceedings of the Graphics Interface 2007 Conference*: 11–18. Healey C.G., Lank E. (eds) ACM Press New York, NY, USA.

Ferenc M. & Żmijewska A. *Museum of the History of Polish Jews*. http://architektura.nimoz.pl/2013/03/09/...

Pottman H. et al. 2007. Geometry of multi-layer freeform structures for architecture. *ACM Trans. Graphics* 26(3), Proc. SIGGRAPH.

Tarczewski R. & Święciak M. 2015. Methods for creating structural meshes on free-form surfaces. *Proceedings of the IASS 2015 Symposium,* Amsterdam.

Wester T. 1983. *Structural Order in Space—The Plate-Lattice Dualism*. Royal Academy of Fine Arts, School of Architecture, Copenhagen.

Advances in Mechanics: Theoretical, Computational and Interdisciplinary Issues – Kleiber et al. (Eds)
© 2016 Taylor & Francis Group, London, ISBN 978-1-138-02906-4

A new combined model of dynamic problems for thin uniperiodic cylindrical shells

B. Tomczyk
Lodz University of Technology, Łódź, Poland

ABSTRACT: The objects of consideration are thin linearly elastic Kirchhoff-Love-type cylindrical shells having a periodically micro-heterogeneous structure in circumferential direction (*uniperiodic shells*). The aim of this note is to formulate a new non-asymptotic averaged model for the analysis of selected dynamic problems for these shells. This, so-called, *general combined model* is derived by applying the combined modelling technique which includes both the asymptotic and tolerance non-asymptotic procedures. The combined model equations have constant coefficients depending also on a cell size. An important advantage of this model is that it makes it possible to study micro-dynamics of periodic shells independently of their macro-dynamics.

1 INTRODUCTION

Thin linearly elastic Kirchhoff-Love-type cylindrical shells with a periodically micro-inhomogeneous structure in circumferential direction (*uniperiodic shells*) are analysed (Fig. 1). At the same time, the shells have constant structure in axial direction.

The properties of such shells are described by highly oscillating and non-continuous periodic functions, so the exact equations of the shell theory are too complicated to apply to investigations of engineering problems. To obtain averaged equations with constant coefficients, various approximate modelling methods for shells of this kind have been proposed. Periodic cylindrical shells (plates) are usually described using *homogenized models* derived by means of *asymptotic methods*

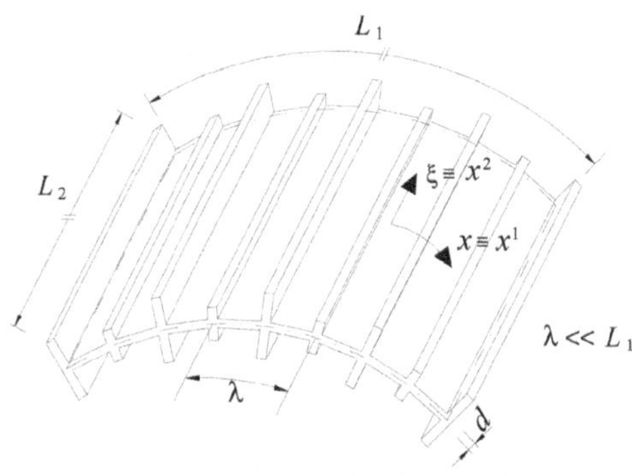

Figure 1. Example of uniperiodic cylindrical shell.

(Lewiński & Telega 2000). Unfortunately, in the models of this kind *the effect of a cell size* (called *the length-scale effect*) on the overall shell behaviour is neglected.

In order to analyse the length-scale effect in dynamic or/and stability problems, the new averaged non-asymptotic models of thin cylindrical shells with a periodic micro-heterogeneity either along two directions tangent to the shell midsurface (*biperiodic structure*) or along one direction (*uniperiodic structure*) have been proposed and discussed by Tomczyk in a series of publications and summarized as well as extended in Tomczyk (2014). These, so-called, *tolerance models* have been obtained by applying *the non-asymptotic tolerance averaging technique* (Woźniak et al. 2010). *Governing equations of the tolerance models have coefficients which are constant or slowly varying and depend on a cell size.*

The aim of this note is to formulate a new *general combined model* for the analysis of dynamic problems for the shells under consideration. This model will be derived by applying the combined modelling which includes two techniques: *the consistent asymptotic modelling procedure* given by Woźniak (2010) and the extended *tolerance non-asymptotic modelling technique* proposed by Tomczyk & Woźniak (2012). An important advantage of the combined model is that it makes it possible to study micro-dynamics of periodic shells independently of their macro-dynamics. Some applications of the tolerance averaging technique to the modelling of various periodic structure are shown in Woźniak et al. (2008, 2010), Ostrowski & Michalak (2015).

2 STARTING EQUATIONS

We assume that x^1 and x^2 are coordinates parametrizing the shell midsurface M in circumferential and axial directions, respectively (Fig. 1). Define $\tilde{\Omega} = (0, L_1) \times (0, L_2)$ as a set of midsurface parameters $\boldsymbol{x} \equiv (x_1, x_2)$ in R^2, where L_1, L_2 are length dimensions of M. It is assumed that on M the orthonormal parametrization is introduced. Sub- and superscripts α, β... run over 1, 2 and are related to x^1, x^2, summation convention holds. Partial differentiation related to x^α is represented by ∂_α. Moreover, it is denoted $\partial_{\alpha,...\delta} = \partial_\alpha ... \partial_\delta$. Let $a^{\alpha\beta}$ stand for the midsurface first metric tensor. The time coordinate is denoted by t. We define a bounded domain $\Omega \times \Xi \times \mathrm{I}$ by means of $\Omega = (0, L_1)$, $\Xi = (0, L_2)$ and $\mathrm{I} = [t_0, t_1]$ as well as we denote $x \equiv x_1 \in \Omega$ and $\xi \equiv x^2 \in \Xi$. Let $d(x)$ and r stand for the shell thickness and the midsurface curvature radius, respectively.

We define λ as a period of the shell structure in $x \equiv x^1$-direction, Figure 1. The period λ satisfies conditions: $\lambda/d_{max} >> 1$, $\lambda/r << 1$ and $\lambda/L_1 << 1$. *The basic cell* Δ is defined by: $\Delta \equiv [-\lambda/2, \lambda/2]$. Denote by $u_\alpha = u_\alpha(x, \xi, t)$, $w = w(x, \xi, t)$, $x \in \Omega$, $\xi \in \Xi$, $t \in \mathrm{I}$, the shell displacements in directions tangent and normal to M, respectively. Elastic properties of the shell are described by shell stiffness tensors $D^{\alpha\beta\gamma\delta}(x)$, $B^{\alpha\beta\gamma\delta}(x)$. Let $\mu(x)$ stand for a shell mass density per midsurface unit area. The external forces will be neglected.

It is assumed that the behaviour of the shell under consideration is described by the action functional determined by Lagrange function L being a highly oscillating function with respect to x and having the well-known form:

$$L = -\frac{1}{2}\Big(D^{\alpha\beta\gamma\delta}\partial_\beta u_\alpha \partial_\delta u_\gamma + \frac{2}{r} D^{\alpha\beta 11} w \partial_\beta u_\alpha + \frac{1}{r^2} D^{1111} ww + B^{\alpha\beta\gamma\delta}\partial_{\alpha\beta} w \partial_{\gamma\delta} w - \mu a^{\alpha\beta}\dot{u}_\alpha \dot{u}_\beta - \mu \dot{w}^2 \Big). \tag{1}$$

Applying the principle of stationary action we arrive at the system of Euler-Lagrange equations, which can be written in explicit form as:

$$\partial_\beta\left(D^{\alpha\beta\gamma\delta}\partial_\delta u_\gamma\right) + r^{-1}\partial_\beta\left(D^{\alpha\beta 11} w\right) - \mu a^{\alpha\beta}\ddot{u}_\beta = 0,$$
$$r^{-1} D^{\alpha\beta 11}\partial_\beta u_\alpha + \partial_{\alpha\beta}\left(B^{\alpha\beta\gamma\delta}\partial_{\gamma\delta} w\right) + r^{-2} D^{1111} w + \mu \ddot{w} = 0. \tag{2}$$

It can be observed that equations (2) coincide with the well-known governing equations of Kirchhoff-Love theory of thin elastic shells. For periodic shells, coefficients $D^{\alpha\beta\gamma\delta}(x)$, $B^{\alpha\beta\gamma\delta}(x)$, $\mu(x)$ of (1) and (2) are highly oscillating non-continuous functions depending on x with a period λ. Applying the combined modelling technique to lagrangian (1), we will derive the averaged model equations with constant coefficients depending also on a cell size. The combined modelling under consideration includes two techniques: *the consistent asymptotic modelling procedure* given by Woźniak et al. (2010) and *an extended version of the known tolerance non-asymptotic modelling technique* based on a new notion of *weakly slowly-varying function* proposed by Tomczyk & Woźniak (2012).

3 MODELLING CONCEPTS AND ASSUMPTIONS

The fundamental concepts of *the extended tolerance modelling approach* are those of *two tolerance relations, weakly slowly-varying function, fluctuation shape function and the averaging operation.*

Tolerance between points: let λ be a positive real number; points x, y belonging to $\Omega = (0, L_1)$ are said to be in tolerance determined by λ, if and only if the distance between points x, y does not exceed λ.

Tolerance between real numbers: let $\bar{\delta}$ be a positive real number; real numbers μ, v are said to be in tolerance determined by $\bar{\delta}$, if and only if $|u - v| \leq \bar{\delta}$.

The above relations are denoted by: $x \approx^\lambda y$, $\mu \approx^{\bar{\delta}} v$. Parameters λ, $\bar{\delta}$ are called *tolerance parameters*. Let $F \in C^R(\bar{\Omega})$, where $C^R(\bar{\Omega})$ is a space of functions being continuous, bounded and differentiable in $\bar{\Omega} \equiv [0, L_1]$. Nonnegative integer R is assumed to be specified in every problem under consideration. Let us denote by $\delta \equiv (\lambda, \delta_0, \delta_1, ..., \delta_R)$ the set of tolerance parameters. The first of them is related to the distance between points in Ω, the second one is related to the distance between values of function $F(\cdot)$ and the *k-th* one to the distance between values of the *k-th* derivative of $F(\cdot)$, $k = 1, ..., R$. Denote by ∂_1^k the *k-th* derivative of $F(\cdot)$ in Ω, $k = 1, ..., R$.

A function $F(\cdot)$ will be referred to as *weakly slowly-varying*, $F = WSV_\delta^R(\Omega, \Delta)$, if and only if:

$$\left(\forall (x,y) \in \Omega^2\right)\Big[(x \overset{\lambda}{\approx} y) \Rightarrow F(x) \overset{\delta_0}{\approx} F(y) \quad \partial_1^k F(x) \overset{\delta_k}{\approx} \partial_1^k F(y), \quad k = 1, 2, ..., R \Big]. \tag{3}$$

Let us also recall that the known *slowly-varying function* $F(\cdot)$, $F = SV_\delta^R(\Omega, \Delta) \subset WSV_\delta^R(\Omega, \Delta)$, satisfies not only condition (3) but also the extra restriction $(\forall x \in \Omega)[\lambda|\partial_1(\partial_1^{(k-1)}F(x)| \approx^{\delta(k-1)} 0$, $k = 1, 2, ..., R]$, $\partial_1^0 F \equiv F$. A function $h(\cdot)$ being λ-periodic, rapidly oscillating and continuous in Ω together with derivatives $\partial_1^k h$, $k = 1, ..., R-1$ and having a piecewise continuous bounded derivative ∂_1^R is called *the fluctuation shape function*, $h(\cdot) \in FS^R(\Omega, \Delta)$, if it depends on λ and satisfies certain conditions (Tomczyk 2014). Let $f(\cdot)$ be a function defined in Ω, which is integrable and bounded in every cell. *The averaging operation of* $f(\cdot)$ is defined by:

$$<f>(x) \equiv \frac{1}{\lambda} \int_{x-\lambda/2}^{x+\lambda/2} f(z)dz, \quad z \in \Delta(x), \quad x \in \Omega_\Delta. \tag{4}$$

If $f(\cdot)$ is λ-periodic function then $<f>$ is constant. The tolerance modelling is based on two assumptions. The first of them is called *the tolerance averaging approximation*. Let $b(\cdot)$, $c(\cdot)$, $e(\cdot)$ be arbitrary integrable λ-periodic functions defined in $\Omega = (0, L_1)$ and let $F = WSV_\delta^1(\Omega, \Delta)$, $G = WSV_\delta^2(\Omega, \Delta)$. *The tolerance averaging approximation has the form:*

$$\begin{aligned}&<bF + c\partial_1 F>(x) = <b> F(x) + <c> \partial_1 F(x)\\&\quad + O(\delta),\ \delta \equiv (\lambda, \delta_0, \delta_1),\\&<bG + c\partial_1 G + e\partial_{11} G>(x) = <b> G(x) + <c> \partial_1 G(x)\\&\quad + <e> \partial_{11} G(x) + O(\delta),\ \delta \equiv (\lambda, \delta_0, \delta_1, \delta_2).\end{aligned} \tag{5}$$

In the course of modelling, terms $O(\delta)$ will be neglected. The second fundamental assumption, called *the micro-macro decomposition*, states that the displacements occurring in the lagrangian under consideration can be decomposed into *unknown averaged displacements* being *weakly slowly-varying functions* in x and *fluctuations* represented by *known highly oscillating λ-periodic fluctuation shape functions* multiplied by *unknown fluctuation amplitudes* being *weakly slowly-varying* in x.

On passing from tolerance averaging to *the consistent asymptotic averaging*, we retain only the concept of *fluctuation shape function*. The notion of weakly slowly-varying function is not introduced. The fundamental assumption imposed on the lagrangian under consideration in the framework of the asymptotic approach is called *the consistent asymptotic decomposition*. It states that the displacement fields occurring in the lagrangian have to be replaced by families of fields depending on small parameter $\varepsilon = 1/m$, $m = 1, 2, \ldots$ and defined in an arbitrary cell. These families of displacements are decomposed into averaged part independent of ε and highly-oscillating part depending on ε.

4 EQUATIONS OF COMBINED MODEL

The combined modelling technique used to the starting lagrangian is realized in two steps.

The first step is based on *the consistent asymptotic averaging* of lagrangian (1) under *the consistent asymptotic decomposition*:

$$\begin{aligned}u_{\varepsilon\alpha}(z,\xi,t) &\equiv u_\alpha(z/\varepsilon,\xi,t)\\&= u_\alpha^0(z,\xi,t) + \varepsilon h_\varepsilon(z) U_\alpha(z,\xi,t),\\w_\varepsilon(z,\xi,t) &\equiv w(z/\varepsilon,\xi,t)\\&= w^0(z,\xi,t) + \varepsilon^2 g_\varepsilon(z) W(z,\xi,t),\end{aligned} \tag{6}$$

where $\varepsilon = 1/m$, $m = 1, 2, \ldots$, $z \in \Delta_\varepsilon(x)$, $\Delta_\varepsilon \equiv (-\varepsilon\lambda_1/\varepsilon\lambda_1/2)$, $(\xi, t) \in \Xi \times I$. Unknown functions u_α^0, U_α in (6) are assumed to be continuous and bounded in Ω together with their first derivatives. Unknown functions w^0, W in (6) are assumed to be continuous and bounded in Ω together with their derivatives up to the second order. Unknowns u_α^0, w^0 and U_α, W are called macrodisplacements and fluctuation amplitudes, respectively. They are independent of ε.

By $h_\varepsilon(z) = h(z, \varepsilon) \in FS^1(\Omega, \Delta)$, $g_\varepsilon(z) = g(z, \varepsilon) \in FS^2(\Omega, \Delta)$, in (6) are denoted highly oscillating λ-periodic known *fluctuation shape functions* depending on ε and satisfying conditions: $h \in O(\lambda)$, $\lambda\partial_1 h \in O(\lambda)$, $g \in O(\lambda^2)$, $\lambda\partial_1 g \in O(\lambda^2)$, $\lambda^2\partial_{11} g \in O(\lambda^2)$, $<\mu h> = <\mu g> = 0$, where $\mu(\cdot)$ is the shell mass density being a λ-periodic function.

Introducing decomposition (6) into (1), under limit passage $\varepsilon \to 0$ we obtained the averaged form of lagrangian (1). Then, applying the principle of stationary action we obtain *the governing equations of consistent asymptotic model for the uniperiodic shells under consideration*. Eliminating fluctuation amplitudes from these equations by means of:

$$\begin{aligned}&U_\gamma = -(G^{-1})_{\gamma\eta}\big[<\partial_1 h D^{1\eta\mu\vartheta}> \partial_\vartheta u_\mu^0 + r^{-1}\\&\quad \cdot <\partial_1 h D^{1\eta 11}> w^0\big],\ W = -E^{-1} <\partial_{11} g B^{11\gamma\delta}> \partial_{\gamma\delta} w^0,\end{aligned} \tag{7}$$

we arrive finally at the following form of the asymptotic model equations:

$$\begin{aligned}&D_h^{\alpha\beta\gamma\delta} \partial_{\beta\delta} u_\gamma^0 + r^{-1} D_h^{\alpha\beta 11} \partial_\beta w^0 - <\mu> a^{\alpha\beta} \ddot{u}_\beta^0 = 0,\\&B_h^{\alpha\beta\gamma\delta} \partial_{\alpha\beta\gamma\delta} w^0 + r^{-1} D_h^{11\gamma\delta} \partial_\delta u_\gamma^0 + r^{-2} D_h^{1111} w^0\\&\quad + <\mu> \ddot{w}^0 = 0,\end{aligned} \tag{8}$$

where:

$$\begin{aligned}&D_h^{\alpha\beta\gamma\delta} \equiv <D^{\alpha\beta\gamma\delta}> -\\&\quad + <D^{\alpha\beta\eta 1} \partial_1 h> (G^{-1})_{\eta\zeta} <\partial_1 h D^{1\zeta\gamma\delta}>\\&B_h^{\alpha\beta\gamma\delta} \equiv <B^{\alpha\beta\gamma\delta}> - <B^{\alpha\beta 11} \partial_{11} g> (E^{-1}) <\partial_{11} g B^{11\gamma\delta}>\end{aligned} \tag{9}$$

with $G = <\partial_1 h D^{\alpha 1\gamma 1} \partial_1 h>$ and $E = <\partial_{11} g B^{1111} \partial_{11} g>$.

Coefficients of equations (8) are constant and independent of a cell size. Hence, the model obtained in the first step is referred to as *the macroscopic model*. It is assumed that within this model, solutions u_α^0, w^0, U_α, W to the problems under consideration are known. Hence, there are also known functions $u_{0\alpha} = u_\alpha^0 + hU_\alpha$ and $w_0 = w^0 + gW$. The second step of combined modelling is based on *the tolerance averaging* of lagrangian (1) under so-called *superimposed decomposition*. In this step, the extended version of the known tolerance modelling technique proposed in Tomczyk & Woźniak (2012) is applied. We introduce *the extra micro-macro decomposition* superimposed on the known solutions $u_{0\alpha}$, w_0 obtained within the macroscopic model:

$$u_{c\alpha}(x,\xi,t) = u_{0\alpha}(x,\xi,t) + c(x)Q_\alpha(x,\xi,t),$$
$$w_b(x,\xi,t) = w_0(x,\xi,t) + b(x)V(x,\xi,t), \tag{10}$$

where $Q_\alpha \in WSV_\delta^1(\Omega, \Delta)$, $V \in WSV_\delta^1(\Omega, \Delta)$ are *new unknown weakly slowly-varying fluctuation amplitudes* and $c(\cdot) \in FS^1(\Omega, \Delta)$, $b(\cdot) \in FS^1(\Omega, \Delta)$ are new known periodic highly oscillating fluctuation shape functions. We substitute the right-hand sides of (10) into (1). The resulting lagrangian is denoted by L_{cb}. Then, we average L_{cb} over cell Δ using averaging formula (4) and applying *the tolerance averaging approximation* (5). As a result we obtain function $<L_{cb}>$ called *the tolerance averaging of starting lagrangian* (1) *in* Δ *under superimposed decomposition* (10). Then, applying the principle of stationary action, we obtain the system of Euler-Lagrange equations for Q_α, V, which can be written in explicit form as:

$$< D^{\alpha\beta\lambda\delta}c^2 > \partial_{\beta\gamma}Q_\delta - < D^{\alpha 1\gamma 1}\partial_1 c\, \partial_1 c > Q_\gamma -$$
$$+ < \mu c^2 > a^{\alpha\beta}\ddot{Q}_\beta = r^{-1} < D^{\alpha 111}\partial_1 c\, w_0 >$$
$$+ < D^{\alpha\beta\gamma 1}\partial_1 c\, \partial_\beta u_{0\gamma} >, \tag{11}$$

$$< B^{\alpha\beta\gamma\delta}b^2 > \partial_{\alpha\beta\gamma\delta}V + [2 < B^{\alpha\beta 11}b\partial_{11}b > +$$
$$- 4 < B^{\alpha 1\beta 1}(\partial_1 b)^2 >]\partial_{\alpha\beta}V + < B^{1111}(\partial_{11}b)^2 > V$$
$$+ < \mu b^2 > \ddot{V} = - < B^{\alpha\beta 11}\partial_{11}b\,\partial_{\alpha\beta}w_0 >. \tag{12}$$

Equations. (11–12) together with *the micro-macro decomposition* (10) constitute *the superimposed microscopic model*. Coefficients of the derived model equations are *constant* and some of them (i.e. averages including $c(\cdot)$, $b(\cdot)$, $\partial_1 b(\cdot)$ *depend on a cell size* λ. The right-hand sides of (11–12) are known under assumption that $u_{0\alpha}$, w_0 were determined in the first step of combined modelling. The basic unknowns Q_α, V of the model equations must be *the weakly slowly-varying functions in periodicity direction*. This requirement can be verified only *a posteriori and it determines the range of the physical applicability of the model*. It can be shown, that under assumption that fluctuation shape functions h, g of macroscopic model coincide with those of microscopic model we can obtain microscopic model equations, which are independent of the solutions obtained in the framework of the macroscopic model

$$< D^{\alpha\beta\lambda\delta}h^2 > \partial_{\beta\gamma}Q_\delta - < D^{\alpha 1\gamma 1}\partial_1 h\, \partial_1 h > Q_\gamma -$$
$$+ < \mu h^2 > a^{\alpha\beta}\ddot{Q}_\beta = 0, \tag{13}$$

$$< B^{\alpha\beta\gamma\delta}g^2 > \partial_{\alpha\beta\gamma\delta}V + [2 < B^{\alpha\beta 11}g\partial_{11}g > +$$
$$- 4 < B^{\alpha 1\beta 1}(\partial_1 g)^2 >]\partial_{\alpha\beta}V + < B^{1111}(\partial_{11}g)^2 > V$$
$$+ < \mu g^2 > \ddot{V} = 0. \tag{14}$$

It means, that an important advantage of the combined model is that it makes it possible to describe selected problems of the shell micro-dynamics independently of the shell macro-dynamics.

5 REMARKS AND CONCLUSION

Summing up our considerations, the following remarks and conclusions can be formulated:

- *The general combined model equations consist of macroscopic model* (8) formulated by means of *the consistent asymptotic procedure* which are combined with *superimposed microscopic model* (11–12) derived by applying *an extended version of the tolerance modelling technique* (Tomczyk & Woźniak, 2012) and under assumption that in the framework of the macroscopic model the solution to the problem under consideration is known.
- In contrast to exact shell equations (2) with discontinuous highly oscillating periodic coefficients, the combined model equations *have constant coefficients*. Moreover, *some coefficients depend on a cell size* λ. Hence, this model makes it possible to analyse *the length-scale effect*.
- It can be shown, that under special conditions imposed on the fluctuation shape functions we can obtain superimposed microscopic model equations, which are independent of the solutions obtained in the framework of the macroscopic model, (13–14). It means, that *an important advantage of the combined model is that it makes it possible to separate the macroscopic description of some special problems from the microscopic description of these problems*.
- The general combined model equations proposed here, which were derived under assumption that the unknown functions Q_α, V in the micro-macro decomposition (10) are *weakly slowly-varying*, definition (3), contain a bigger number of terms

depending on the microstructure size then the standard combined model equations presented in Tomczyk (2014), which were derived under assumption that the unknown functions Q_α, V in (10) are *slowly-varying*. So, *the general combined model allows us to investigate the length-scale effect in more detail.*

REFERENCES

Lewiński, T. & Telega, J.J. 2000. *Plates, Laminates and Shells. Asymptotic Analysis and homogenization.* Singapore: Word Scientific Publishing Company.

Ostrowski, P. & Michalak, B. 2015. The combined asymptotic-tolerance model of heat conduction in a skeletal micro-heterogeneous hollow cylinder. *Comp. Struc.* 134: 343–352.

Tomczyk, B. & Woźniak, C. 2012. Tolerance models in elastodynamics of certain reinforced thin-walled structures. In Z. Kołakowski & K. Kowal-Michalska (eds), *Static, Dynamics and Stability of Structures*: 123–153, Lodz: Lodz Technical University Press.

Tomczyk, B. 2014. *Length-scale effect in dynamics and stability of thin periodic cylindrical shells.* Bulletin of the Lodz University of Technology, No. 1166, series: Scientific Dissertations. Lodz: Lodz Technical University Press.

Woźniak, C., Michalak, B & Jędrysiak, J. 2008. *Thermomechanics of Microheterogeneous Solids and Structures. Tolerance Averaging Approach.* Lodz: Lodz Technical University Press.

Woźniak, C., et al. (eds) 2010. *Mathematical modelling and analysis in continuum mechanics of microstructured media.* Gliwice: Silesian Technical University Press.

Advances in Mechanics: Theoretical, Computational and Interdisciplinary Issues – Kleiber et al. (Eds)
© 2016 Taylor & Francis Group, London, ISBN 978-1-138-02906-4

Taylor-Couette flow with radial temperature gradient

E. Tuliszka-Sznitko & K. Kiełczewski
Institute of Thermal Engineering, Poznan University of Technology, Poznań, Poland

ABSTRACT: The authors investigate the Taylor-Couette flow with and without heat transfer using 3D Direct Numerical Simulations and Spectral Vanishing Viscosity methods (DNS/SVV). In the analyzed flow cases the inner rotating cylinder is cooled and the stationary outer one is heated. The authors study the influence of curvature parameter and the radial temperature gradient on the flow structure and on the statistical parameters. The Boussinesq approximation is used to take into account the buoyancy effects induced by the involved body forces. The authors additionally study the influence of the periodic modulation superimposed on the inner cylinder rotational speed on the instability critical parameters and on the flow structure. The distributions of the local Nusselt numbers along disks and cylinders are investigated, as well as, the axial and radial profiles of many structural parameters. The λ_2 criterion is used for numerical visualization.

1 INTRODUCTION

The Taylor-Couette flow (the shear flow between a rotating inner cylinder and a stationary outer cylinder) is one of paradigmatical systems in hydrodynamics, widely used for studying the primary instability, pattern formation, transitional flows and fully turbulent flows. Simultaneously, the flows in rotating cavities appear in numerous machines in the field of mechanics and chemistry, e.g., in ventilation, desalination tanks and waste water tanks, in cooling systems, in gas turbines and axial compressors. The results of the Taylor-Couette numerical investigations can be useful to estimate the strengths and limitations of the RANS models which are presently used to predict transitional and turbulent flows with heat transfer in more complicated configurations. Additionally, the Taylor-Couette flow is highly suitable for predicting phenomena in geophysics and astrophysics.

The review of literature shows that one aspect of the Taylor-Couette flow, which has been insufficiently researched so far, is the laminar-turbulent transition and fully turbulent flow in the varying temperature field which occurs in most industrial applications. Gaining insight into laminar-turbulent transition in the Taylor-Couette flow (with temperature gradient) requires very detailed experimental and numerical investigations.

In the paper the authors investigate the influence of curvature parameter and radial temperature gradient on the Taylor-Couette flow structure. The results are obtained for different aspect ratios $\Gamma = H/(R_2 - R_1) = 3.76, 11.75, 20.0$, for curvature parameters from the range $Rm = (R_2 + R_1)/(R_2 - R_1) = 2.2$–$19.0$, for the thermal Rossby number $B = \beta (T_2 - T_1) = 0.1$ (where T_2 is temperature of warmed walls and T_1 is temperature of cooled walls, β is a thermal expansion coefficient), and for different Reynolds numbers $Re = R_1(R_2 - R_1)\Omega/\nu$. The authors also investigate the influence of the harmonically modulated rotation of the inner cylinder on the Taylor-Couette flow instability. The angular velocity of the inner cylinder oscillates around some mean value $\bar{\Omega}$ with a given amplitude $\hat{\Omega}$, which is described as:

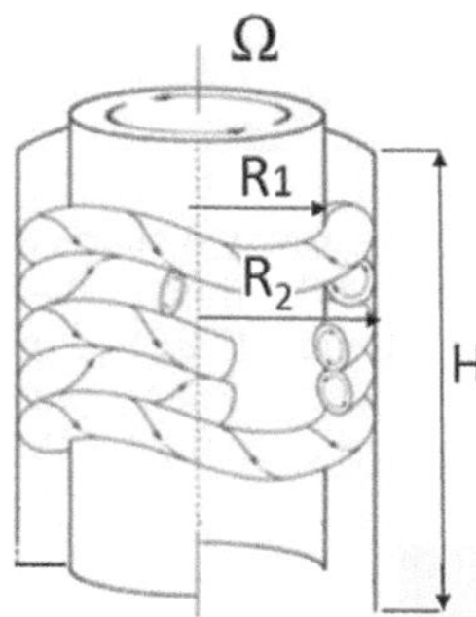

Figure 1. Schematic picture of the Taylor-Couette flow.

$$\Omega(t) = \bar{\Omega} + \hat{\Omega}\cos(\omega t), \quad \mathrm{Re}(t) = \overline{\mathrm{Re}} + \hat{\mathrm{Re}}\cos(\omega t) \quad (1)$$

where ω is the frequency of modulation.

2 MATHEMATICAL AND NUMERICAL MODELS

The flow is described by continuity, Navier-Stokes and energy equations. The equations are written in a cylindrical coordinate system (R, φ, Z) with respect to a rotating frame of reference. Numerical investigations are performed using Direct Numerical Simulations (DNS/SVV). To stabilize the discretization scheme for higher Reynolds number, the authors use the Spectral Vanishing Viscosity (SVV) method, (Kiełczewski & Tuliszka-Sznitko 2013). Numerical simulations are based on a pseudo-spectral Chebyshev-Fourier-Galerkin collocation approximation. In the time approximation the authors use a second-order semi-implicit scheme, which combines an implicit treatment of the diffusive terms and an explicit Adams-Bashforth extrapolation for the non-linear convective terms. The dimensionless axial and radial coordinates are: $z = Z/(H/2)$, $r = (2R(R_2 + R_1))/(R_2 - R_1)$. The velocity components are normalized by ΩR_2; the dimensionless components of the velocity vector in radial, azimuthal and axial directions are denoted by $u = U/\Omega R_2$, $v = V/\Omega R_2$, $w = W/\Omega R_2$. Time is normalized by $1/\Omega$. The dimensionless temperature is defined in the following way: $\Theta = (T - T_1)/(T_2 - T_1)$. The flow in rotating cavity is characterized by four main parameters: aspect ratio Γ, curvature parameter Rm, Reynolds number Re and thermal Rossby number B, which create a large parameter space, very rich in physical phenomena. Prandtl number is constant Pr = 0.

3 SELECTED RESULTS

3.1 *Influence of curvature parameter on the flow structure*

To check the influence of curvature parameter Rm on the flow structure and on statistics the authors perform computations for Γ = 3.76 (Rm = 2.2, 3.2, 4.2, 7.2 and 10.2) and for Γ = 11.75 (Rm = 19.0) with the asymmetric end-wall boundary conditions. This investigation allows them to observe different laminar-turbulent transition scenarios depending on Rm. For Rm = 2.2 and small Re the authors observe the appearance of three-cell structures (typical for T-C flow with the asymmetric end-walls boundary conditions), then, beyond critical Reynolds number Re = 281 the transition from the three-cell structure to the one cell-structure. The process of transition from the three-cell structure to the one cell-structure is described, among others, in (Mullin & Blohm 2001, Tuliszka-Sznitko & Kiełczewski 2015, Seelig et al. 2015). For further increase of Re the transition to unsteadiness takes place followed by the gradual development of the boundary layers along the inner and outer cylinder. The transition process from the three-cell structure to the one cell structure also takes place for Rm = 3.2. For Rm = 4.2, 7.2 and 10.2 the classic T-C series of consecutive bifurcations is observed: Beyond a certain critical Reynolds number the Taylor-Couette flow structure appears (pairs of counter-rotating, axisymmetric vortices fill in the annulus). For higher Reynolds numbers a supercritical Hopf bifurcation leads to a wavy vortex flow. After this transition the Taylor-Couette vortical structure is retained but vortices are modified i.e. the characteristic feature of the wavy vortex flow is the azimuthal waviness of vortices. With further increase of Re the modulated waves state and the turbulent Taylor vortex flows appear.

The example of the modulated wave vortex structure obtained for Γ = 3.76, Rm = 10.2 and Re = 1570 is presented in Figure 2a–c. In Figure 2a the 3D structure (λ_2 iso-surfaces) is presented in the vicinity of the rotating inner cylinder. For the flow case Γ = 3.76, Rm = 10.2

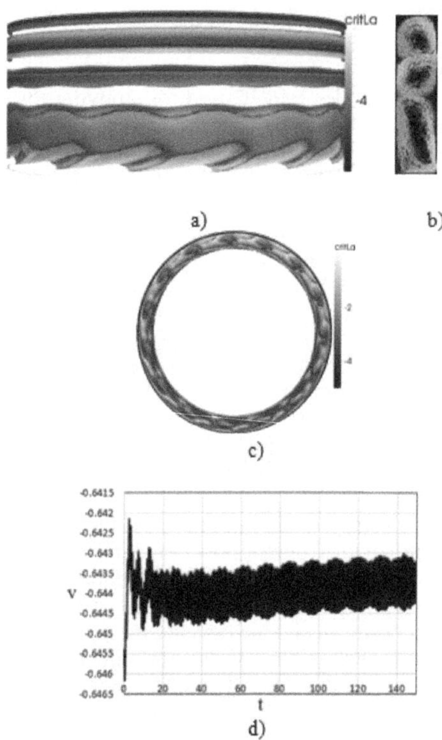

Figure 2. a) The 3D structure (λ_2 iso-surfaces) in the vicinity of the rotating inner cylinder. b) The meridian flow. c) The flow structure in the azimuthal section near the rotating disk (λ_2 iso-surfaces). d) The instability characteristics v = f(t). Γ = 3.76, Rm = 10.2, Re = 1570.

the critical Reynolds number of transition to unsteadiness takes place at Re = 1523. In Figure 2a it is visible that in vicinity of the rotating bottom disk there are 15 regular vortices (see also Fig. 3c where the azimuthal section of the flow near the bottom disk is presented). The instability characteristic is showed in Figure 2d; it is visible that on the fundamental wave another wave of lower frequency is superimposed. The number of spirals depends strongly on Rm: for Rm = 2.2 the authors observe 6 spirals, whereas for Rm = 10.2 there are 15 spirals.

The authors also perform computations for higher curvature parameter Rm = 19 (Γ = 11.75, asymmetric end-wall boundary conditions) and as a result report 9 spirals. For this particular flow case the computations are performed for Re up to 2500, which allows the authors to analyze statistical data. Among others, the authors analyze the local transverse angular momentum current which is defined in the following way, (Brauckmann, Eckhardt 2013):

$$j^{\omega}(R,\varphi,Z,t) = R^3(U\omega - \nu\partial\omega/\partial R); \quad \omega = V/R \quad (2)$$

The averaged value of the local transverse angular momentum current over any cylindrical surface gives the mean current:

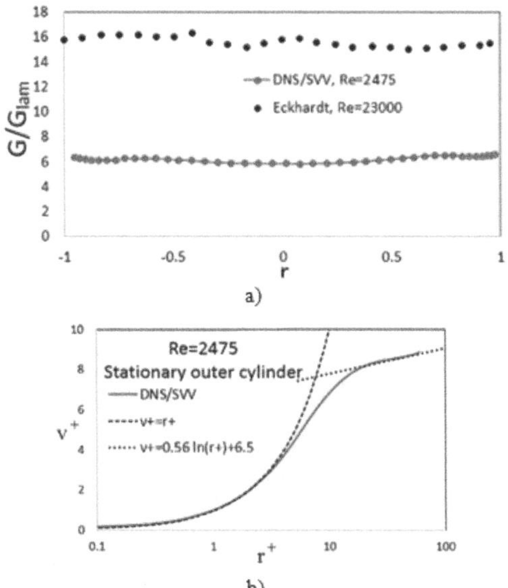

Figure 3. a) The rescaled value of the mean torque by its laminar value as a function of r. b) The azimuthal velocity component in wall value v^+ as a function of r^+ in the stationary outer cylinder boundary layer. Data is averaged over time, azimuthal direction and cylindrical surfaces. Γ = 11.75, Rm = 19.0, Re = 2475.

$$J^{\omega} = \langle j^{\omega}(R,\varphi,Z,t)\rangle_{A(R),t} = R^3\left[\langle U\omega\rangle_{A(R),t} - \nu\partial(\langle\omega\rangle_{A(R),t})/\partial R\right] \quad (3)$$

which is related to the dimensionless torque in the following way:

$$G = T/[2\pi(2h)\rho\nu^2] = J^{\omega}/\nu^2 \quad (4)$$

T is the torque required to drive the cylinders. The rescaled value of the mean torque by its laminar value as a function of radius r obtained for Re = 2475 (Γ = 11.75, Rm = 19) is presented in Figure 3a. The results are compared to data obtained for Re = 23000 in Brauckmann, Eckhardt (2013). Figure 3b shows the azimuthal velocity components in wall units v^+ as a function of r^+ obtained for the stationary outer cylinder boundary layer (Γ = 11.75, Rm = 19, data is averaged over the azimuthal direction and over any cylindrical surface). In Figure 3b agreement with the wall law $v^+ = r^+$ in the area of viscous layer is visible. The authors approximate their solution in the logarithmic layer by the following formula: $v^+ = 0.56\ln(r^+) + 6.5$.

3.2 *Influence of temperature gradient in radial direction on the flow structure*

The authors analyze the influence of thermal boundary conditions (superimposed on the inner and outer cylinders) on the Taylor-Couette flow structure. The computations are performed for the non-isothermal fluid in the cavity of aspect ratio Γ = 3.76 with a asymmetric boundary conditions (Rm = 10.2, the bottom rotating disk and stationary outer cylinder are heated) and in the cavity of Γ = 20 with symmetric end-walls boundary conditions (Rm = 3.0, the outer stationary cylinder and two stationary end-walls are heated). The difference in temperature between cylinders causes extensive heat transfer between them.

The exemplary distributions of the local Nusselt number along the outer heated and inner cooled cylinders obtained for Re = 550 (Γ = 3.76, Rm = 10.2) are presented in Figure 4b and c. Figure 4a shows the meridian flow structure. The maximum values of the local Nusselt numbers are observed in the jet areas. The numerical investigations performed for the flow case of Γ = 20, Rm = 3 show, among others, the existence of 24 vortices in the cavity of Γ = 20. This result shows that the wavelength in axial direction of the non-isothermal fluid is shorter than in the isothermal flow, where 20 vortices appear. This result confirms the finding published in paper by Peres (1995).

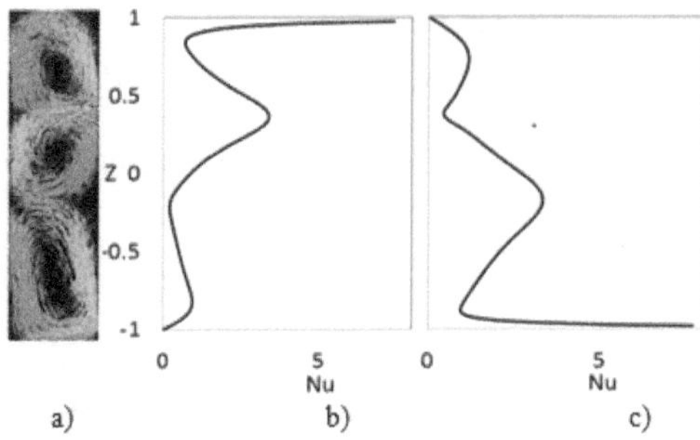

Figure 4. a) The meridian flow structure. b) The Nusselt number distribution along the outer cylinder. c) The Nusselt number distribution along the inner cylinder. Γ = 3.76, Rm = 10.2, Re = 550.

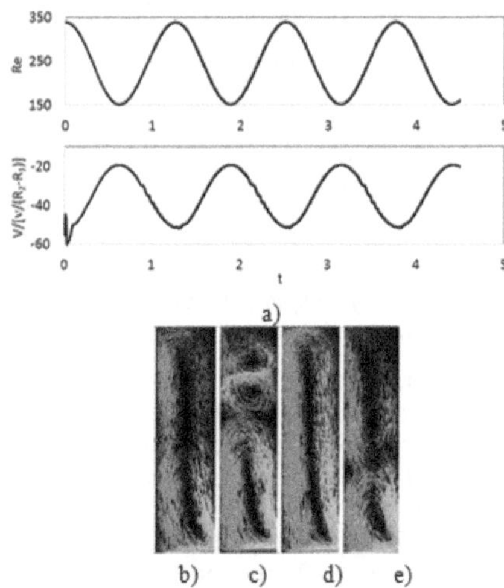

Figure 5. a) The flow response to the forcing function. b)–e) The meridian flow obtained in particular time sections: t = 3.05, 3.53, 4.14, 4.51, respectively. Γ = 3.76, Rm = 2.2.

3.3 *Influence of superimposed fluctuations on the rotational speed on the flow structure*

The authors also investigate the flow cases in which the outer cylinder is stationary and the non-stationary inner one does not rotate with the constant angular velocity (as was considered in previous sections) but oscillates as it is described in equation (1). This type of the Taylor–Couette flow has been the subject of a number of investigations, Youd (2005). The question arises, whether the modulation of rotational speed makes the flow more or less stable with respect to the onset of the Taylor-Couette vortices. Temporally forced Taylor-Couette flow is a paradigm issue for the parametric control of the instability by temporal forcing. The oscillations induce a damped viscous wave which penetrates into fluid to a depth of the Stokes layer $(2\nu/\omega)^{1/2}$.

Figure 5a presents the exemplary response of the flow obtained numerically to the forcing function (with $\hat{R} = 93.8$, $\overline{\text{Re}} = 246$, $\omega = 5$) i.e. $V/[\nu/(R_2 - R_1)]$ as a function of dimensionless time t (time normalized by $(R_2 - R_1)^2/\nu$). For parameters $\hat{R} = 93.8$, $\overline{\text{Re}} = 246$, $\omega = 5$ the period of forcing $T = 2\pi/\omega$ and the period of the flow response are identical. Figure 5b–e present the meridian flow structures obtained in different time sections. Different flow structures are observed as a function of Re(t).

4 CONCLUSIONS

In the paper the authors report on their numerical investigations of the phenomena occurring in the Taylor-Couette flows. The authors focus on the influence of the curvature parameters, radial temperature gradients and the influence of the modulated rotational speed (superimposed on the inner cylinder) on the flow structure. The obtained results can be useful for researches dealing with the RANS models and, among others, for engineers dealing with cooling systems in gas turbines.

ACKNOWLEDGMENTS

The authors are grateful to the Poznań Supercomputing and Networking Center, where the computations were performed.

REFERENCES

Brauckmann, H. & Eckhardt, B. 2013. Direct numerical simulations of local and global torque in Taylor-Couette flow up to Re D 30000. *J. Fluid Mech.* 718: 398.

Kielczewski, K. & Tuliszka-Sznitko, E. 2013. Numerical study of the flow structure and heat transfer in rotating cavity with and without jet. *Arch. Mech.* 65: 527.

Mullin, T. & Blohm, C. 2001. Bifurcation phenomena in a Taylor–Couette flow with asymmetric boundary conditions. *Phys. Fluids.* 13: 136–140.

Peres, I. 1995. Contribution a lanalyse de lecoulement et des transferts convectifs dans un espace annulaire lisse ou encoche par voie de simulations numeriques. *PhD Thesis, Universite de Poitiers.*

Seelig, T., Kiełczewski, K., Tuliszka-Sznitko, E., Harlander, U., Egbers, Ch. & Bontoux, P. 2015. Taylor-Couette flow with asymmetric end-walls boundary conditions. *Proc. of 15th European Turbulence Conference, Delft, The Netherlands.*

Tuliszka-Sznitko, E. & Kiełczewski, K. 2015. Numerical investigations of Taylor-Couette flow using DNS/SVV method. *J. CMST, accepted.*

Youd, A.J. 2005. Bifurcations in Forced Taylor–Couette Flow. *PhD Thesis, Newcastle University.*

Advances in Mechanics: Theoretical, Computational and Interdisciplinary Issues – Kleiber et al. (Eds)
© 2016 Taylor & Francis Group, London, ISBN 978-1-138-02906-4

Application of artificial neural networks in man's gait recognition

T. Walczak, J.K. Grabski, M. Grajewska & M. Michałowska
Institute of Applied Mechanics, Poznań University of Technology, Poznań, Poland

ABSTRACT: In the paper the method of human gait recognition based on artificial neural network is presented. The method is based on observation, that each individual has a different, unique way of basic movement activity. Hence, parameters representing typical characteristics of gait of each person can be defined. In the classical approach the video analysis is used to obtain the set of sequential pictures of a moving person. In the paper, all considered gait parameters are calculated from data obtained from dynamometric platforms used to measure the acting forces during the gait. Those gait characteristics are unique for each person, and some of geometrical qualities of curve which represent vertical and anterior-posterior components of force are taken into the account. To implement the recognition process the back-propagation neural network algorithm is used. In experiments, higher gait recognition performances have been achieved.

1 INTRODUCTION

The gait could be defined as the coordinated, cyclic combination of movements, that results in human locomotion (Inman et al. 1981). Human gait has common patterns of movements, which are coordinated in the sense, that they must occur with a specific temporal pattern for the gait to occur. These movements are repeated as a man cycles between steps with alternating feet. It means, that gait could be described as a periodic motion.

A period of the gait cycle exists between the successive heel strikes, and the gait motion in space and time, satisfies spatial and temporal symmetry. In similar way, one can describe other kind of motions like running, jogging or climbing stairs. Moreover, human motion analysis has many challenging issues, because of a highly flexible structure of the human body. Human gait is known to be one of the most universal and complex of all human activities. For that reason, the study of human gait has been increased extensive interests in various fields such as clinical analysis, biomechanics, robotics, and biometrics (Bartlett 2006; Inman et al. 1981; Michalski et al. 2011; Oh et al. 2013; Shukla et al. 2012; Xiao 2012).

The human gait is usually described by kinetic or kinematic parameters, which are obtained from simple mechanical model, where all considered human parts are assumed as pendulums, the motion of which is characterized by simple harmonic motion. Those parameters describe some gait patterns. One can observe, that each individual appears to have different and unique characteristics of gait. It means, that it is possible to identify a person on the basis of gait parameters (Narasimhulu & Jilani 2012; Nixon et al. 2010; Shukla et al. 2012; Su & Wu 2000; Tafazzoli & Safabakhsh 2010; Yoo et al. 2008).

That kind of recognition of individuals could be also used in clinical practice to identify improper or pathological gaits.

In this paper, the method for recognizing humans by their gait using Artificial Neural Network (ANN) is presented. In order to describe and recognize characteristics of gait for each considered individual, new approach was proposed. In classical way, the pattern of gait description is based on video analysis, where obtained sequences of pictures from camera are processed to 2D stick human body model. Next some necessary kinematic parameters, like trajectories and velocities of characteristic points or characteristic angles, are determined. That kind of analysis is complex and requires the use of many time-consuming algorithms.

In the paper, the method of human gait recognition based only on parameters of force characteristic is presented. Important data are obtained from dynamometric platforms, commonly used in biomechanical research. Typical shape of vertical component of force acting on feet during gait is widely described in literature (Inman et al. 1981). However, one can observe, that the character of curve, representing that force, seems to be different for every individual. Therefore, the main aim of the paper is to examine possibility of the gait recognition, that is based on dynamometric measurement data.

2 NEURAL NETWORK

The artificial neural networks are widely used to solve many problems of recognition, prediction or control processes (Antonini et al. 2006; Bartlett 2006; Perl 2004; Scholnhorn 2004; Sorci et al. 2010). Algorithms based on ANN are effective and could process a lot of data in the relatively short time. For that reason they are commonly used in recognition problems, where usually one can find the pattern to classify. The neural network methods facilitate gait analysis because of their highly flexible, inductive, non-linear modeling ability, unlike any other approaches. The nonlinear property of multi-layered neural networks is useful for analysis of complicated gait parameters, which traditionally have been difficult to model in analytical way.

To recognize the gait, back-propagation algorithm for training multi-layered neural network, based on selective retraining and a dynamic adaptation of learning rate and momentum, is implemented. The pattern recognition system contains an input subsystem, that accepts sample pattern vectors and a decision-maker subsystem, that decides the classes to which an input pattern vector belongs (Narasimhulu & Jilani 2012). The scheme of multi-layered neural network structure used to gait recognitions is presented on Figure 1.

Network architecture presented on the Figure 1 consists of input layer, one or two inner layers and output layer. The function **f** is defined for every neuron in the system as neuron activation function with S-shaped characteristic, vector **a** contains parameters, which describe gait, and vector **s** is an answer vector, which contains response of the network in recognition process, where s_i component is a network response responsible for i-th recognizing individual. That architecture should allow not only to recognize of the individuals, but also to show similarity of gait of the considered persons.

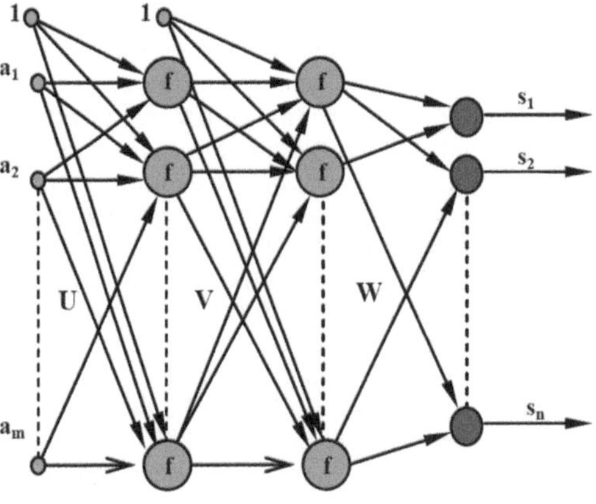

Figure 1. The scheme of used neural network.

The Lavenberg-Marquardt method will be applied to train network, and no more than 30% of data collected for each individual will be used during the learning process of the network.

3 GAIT REGISTRATION

Ten volunteers, five females and five males, took part in this study, in the Biomechanical Laboratory at Poznan University of Technology. Average ± SD values and ranges (min ÷ max) for selected characteristics of women amounted to:

- Age: 25 ± 4.2 (21 ÷ 33) years
- Height: 170 ± 3.6 (165 ÷ 176) cm
- Body mass: 61.6 ± 7.6 (54.8 ÷ 76) kg.

Male subjects information:

- Age: 28 ± 11.6 (22 ÷ 51) years
- Height: 183.4 ± 3.6 (181 ÷ 190) cm
- Body mass: 82.2 ± 15.9 (69 ÷ 113) kg.

The group comprised the subjects who did not suffer from any lower extremity pain or musculoskeletal disorders. Participants were asked to walk barefoot with self-chosen velocity and step frequency, through 10-m-long walkway with two built in dynamometric platforms. During this study subject had to step one foot on the one platform (always in the same configuration to differentiate left and right foot) to record ground reaction forces during one cycle of walk. The motion was repeated at least one hundred times and recordings that did not meet the requirements were rejected.

Three-dimensional GRF were measured by AMTI BP400600 force plates (600 × 400 × 82,5 mm) with the 400 Hz sampling rate. Signals were recorded directly to the computer with SMART Capture software which enable to export measured data into text files. These files were used to calculate dynamic parameters of gait for vertical and fore-aft components of force. The variables were normalized to body weight and stance phase time.

4 GAIT IDENTIFICATION PARAMETERS

The proposed method of human gait recognition is based on data obtained from measurement of forces of which feet acting on dynamometric platforms. High sensitivity and accuracy of those measurements allows to define parameters used to describe gait of each individual. Typical shapes of curves, that represent vertical component of those forces together with characteristic curve parameters are presented in Figure 2.

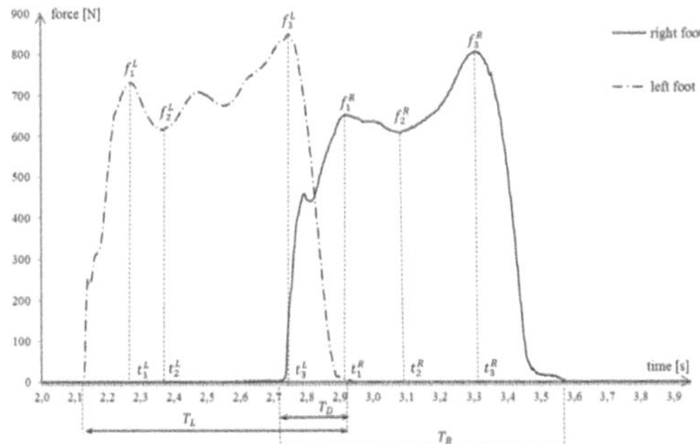

Figure 2. Vertical component of force for both feet with important time and force parameters.

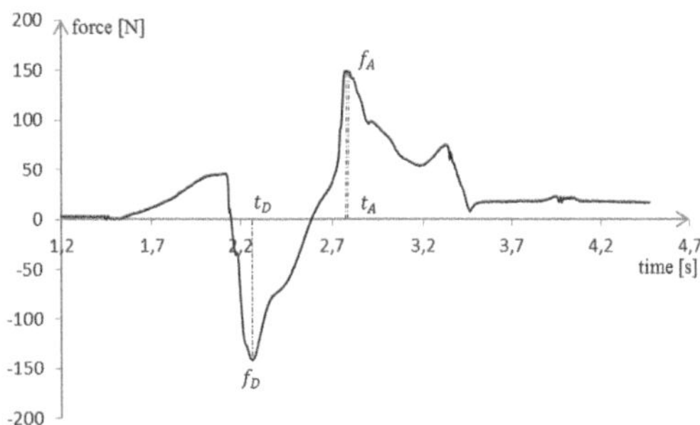

Figure 3. Anterior-posterior component of ground reaction force with parameters.

It should be noticed, that for most cases the curves representing vertical component of forces have always two maxima and one local minimum. However, some curve geometric parameters could be completely different for each considered individual. In order to examine the mentioned parameters, data should be normalized. We have scaled the values of force curves by the total weight of recognized person F, and time domain by the time of nonzero signal registration of left (T_L) and right (T_R) foot (Fig. 2) respectively. It allows us to determine two groups of recognition parameters: parameters obtained in forces and in the time domain.

Vertical force parameters include:

- Maxima of loading response phase for both feet f_1^L, f_1^R.
- Minima of midstance phase for both feet f_2^L, f_2^R.
- Maxima of terminal stance phase for both feet f_3^L, f_3^R.
- f_1^L/f_2^L, f_1^L/f_3^L, f_3^L/f_2^L and f_1^R/f_2^R, f_1^R/f_3^R, f_3^R/f_2^R ratios.
- The areas determined between curves and time axis for three gait phases: from start to maximum of loading response—p_1^L, p_1^R, between maxima—p_2^L, p_2^R and from maximum of terminal phase to the end of stance phase—p_3^L, p_3^R.
- The ratios of those areas.
- Coordinates of the geometrical curve center for both feet—x_C^L, y_C^L, x_C^R, y_C^R where:

$$x_C = \frac{\sum_{i=0}^{n} l_i * \frac{y_i + y_{i+1}}{2}}{\sum_{i=0}^{n} l_i}, \tag{1}$$

$$y_C = \frac{\sum_{i=0}^{n} l_i * \frac{x_i + x_{i+1}}{2}}{\sum_{i=0}^{n} l_i}, \tag{2}$$

where: n is number of samples.

Time parameters:

- Time of contact with floor of both feet T_L, T_R.
- Time of gait cycle T.

Occurrence time of characteristic vertical force parameters: $t_1^L, t_1^R, t_2^L, t_2^R, t_3^L, t_3^R$.

- Duration time measured for both feet in stance phase: $t_2^L - t_1^L, t_3^L - t_1^L, t_2^R - t_1^R, t_3^R - t_1^R$.
- Duration of double support phase: $T_D = T_R + T_L - T$.

Generally the anterior-posterior component of ground reaction force is irregular and has not smooth shape. One can only observe that there always exists one visible maximum and one minimum. Typical shape of this component of the reaction force together with marked important parameters is presented in the Figure 3.

Anterior-posterior force parameters:

- Maximum of acceleration force f_A^L, f_A^R and minimum of braking force f_D^L, f_D^R, measured for both feet (Fig. 3).

Time parameters include:

- Time of occurrence of the maximum of acceleration force t_A^L, t_A^R, measured for both feet.
- Time of occurrence of minimum of braking force t_D^L, t_D^R, measured for both feet.

To reduce the size of neural network, one parameter responsible for identify foot (1 if left and 0 if right) has been added. The total number of 25 parameters were used in the recognition processes. That set allows to determine patterns for each person and leads to effective recognition with use of neural network.

5 RESULTS AND CONCLUSIONS

The experiments were performed for 10 persons, where number of gaits registration for each

Table 1. The obtained RMSE of human recognition for each considered person.

Person	RMSE of recognition
1	0,0050
2	0,0033
3	0,0029
4	0,0020
5	0,0061
6	0,0032
7	0,0048
8	0,0043
9	0,0033
10	0,0057

individual was more than 90. Exactly 30 registrations of each person were taken to learn ANN. The rest of the collected data were used to validate the efficiency of learning process of the ANN. The results are presented in the form of the Root Mean Square Errors (RMSE) for each person in Table 1.

Analyzing the data presented in Table 1, one can observe the really small error of recognition calculated for every individual. It should also be mentioned that all unused in learning process data, was used to examine the efficiency of recognition of neural network. Hence, one can conclude that average recognition error of each considered individual in our research, is less than one percent. It means that every human person acts on a floor in really unique way during gait and recognition of characteristic patterns of gait could be considered not only with use of kinematic indicators.

The dynamic parameters proposed in the paper are based on characteristic shapes of the curves which represent reaction forces of ground, acting on feet during gait. Together with kinematic indicators, commonly used in biomechanical research of gait, they could provide complete information about a human gait. This approach could be also used in clinical practice for example to identification of improper gait or as a diagnostic tool.

It was proved that it is possible to recognize a person, based on characteristic parameters of gait, obtained from dynamometric platforms. The RMSE of human recognition was very small and that leads to conclusion that the possibility of human recognition which is based on the data from force plates and the ANN is very high.

ACKNOWLEDGEMENTS

The presented work is supported by grant 02/21/DSPB/3463.

REFERENCES

Antonini G., Sorci M., Bierlaire M., Tiran J. 2006. Discrete choice models for static facial expression recognition., *8th International Conference on Advanced Concepts for Intelligent Vision Systems* 4179: 710–721.

Bartlett R., 2006. Artificial intelligence in sports biomechanics: New dawn or false hope?, *Journal of Sports Science and Medicine* 5(4): 474–479.

Inman V.T., Ralston H.J., Todd F., Lieberman J.C. 1981. *Human Walking*, Baltimore: Williams & Wilkins.

Michalski R., Wit A., Gajewski J. 2011. Use of artificial neural networks for assessing parameters of gait symmetry, *Acta of Bioengineering and Biomechanics* 13(4): 65–70.

Narasimhulu V.G., Jilani S.A.K. 2012. Back Propagation Neural Network based Gait Recognition, *International Journal of Computer Science and Information Technologies* 3(5): 5025–5030.

Nixon M.S., Tan T., Chellappa R. 2010. *Human Identification based on Gait*, New York: Springer.

Oh S.E., Choi A., Mun J.H. 2013. Prediction of ground reaction forces during gait based on kinematics and a neural network model, *Journal of Biomechanics* 16(14): 2372–2380.

Perl J. 2004. Artificial neural networks in motor control research, *Clinical Biomechanics* 19(9): 873–875.

Scholnhorn W.I. 2004. Application of artificial neural nets in clinical biomechanics, *Clinical Biomechanics* 19(9): 876–898.

Shukla R., Shukla R., Shukla A., Sharma S., Tiwari N. 2012. Gender Identification in Human Gait Using Neural Network, *I.J. Modern Education and Computer Science* 11: 70–75.

Sorci M., Antonini G., Cruz J., Robin T., Bierlaire M., Thiran J.P. 2010. Modelling human perception of static facial expressions., *Image and Vision Computing* 28(5): 790–806.

Su F.Ch., Wu W.L. 2000. Design and testing of a genetic algorithm neural network in the assessment of gait patterns, *Medical Engineering and Physics* 22(1): 67–74.

Tafazzoli F., Safabakhsh R. 2010. Model-based human gait recognition using leg and arm movements, *Engineering Applications of Artificial Intelligence* 23(8): 1237–1246.

Xiao Q. 2012. A note on computational intelligence methods in biometrics., *International Journal of Biometrics* 4(2): 180–188.

Yoo J.H., Hwang D., Moon K.Y., Nixon M.S. 2008. Automated human recognition by gait using neural network, *First Workshop on Image Processing Theory, Tools and Application* 1–6.

Advances in Mechanics: Theoretical, Computational and Interdisciplinary Issues – Kleiber et al. (Eds)
© 2016 Taylor & Francis Group, London, ISBN 978-1-138-02906-4

Reliability assessment of truss towers using Monte Carlo Method, PEM and RSM

K. Winkelmann, M. Oziębło & K. Rybaczyk
Faculty of Civil and Environmental Engineering, Gdańsk University of Technology, Gdańsk, Poland

ABSTRACT: The paper discusses reliability assessment of simple random truss structures using three different probabilistic methodologies: the Monte Carlo Method (MCM), the Point Estimate Method (PEM) and the Response Surface Method (RSM). A benchmark truss structure and a simplified full-size engineering tower are both analyzed. A set of ultimate load numerical calculations is performed and the results are taken as the basis of advanced probabilistic calculations. Using the samples, numerical efficiency and accuracy of these methods are compared and the advantages and drawbacks of PEM and RSM in relation to MC are discussed.

1 INTRODUCTION

Nowadays, the simulation-based methods of reliability assessment through repetitive numerical model calculations regain their importance, mainly due to a swift development of computer-aided design tools performance.

Probabilistic structural reliability analysis is conducted to determine the impact of relevant input parameters on the change of model safety. An appropriate identification of methodologies to deal with simple engineering problems may be helpful both in design and optimization of the new and improving of the existing constructions.

This paper critically discusses the behaviour of three most popular probabilistic techniques used for reliability assessment, on two examples of simple truss towers—a benchmark example and a slightly simplified full-size engineering structure.

2 ANALYSIS OF A BENCHMARK EXAMPLE

2.1 *Subject of the analysis*

In the first part, the paper deals with a 25-member truss, proposed in e.g. Kleiber et al. (1989), presented in Figure 1.

The supports of the tower have a spacing of 508 cm. The width of the intermediate and upper storeys connectors is equal to 190.5 cm. The height of each storey is 254 cm. The tower is made of aluminum ($E = 69$ GPa) and the following starting set of loads is taken in each of the top two nodes: $P = (P_x, P_y, P_z) = (19.2$ kN, 19.2 kN, 2400 kN). The initial cross section area of the members is $A = 3.225$ cm^2.

It was assumed that the random variables will be associated with the possible variation of the cross sectional area of bars during design process, similarly to an optimization approach.

It was noted, that the bars can be steel profiles only. For 32 steel sections found in the catalogues with a sectional area close to that of the initial value, a histogram of the cross-sectional area availability is shown in Figure 2. It fits Gaussian PDF more than the uniform type. Based on this observation, a PDF of the cross sectional area was proposed, resulting in two parameters: $\mu_A = 3.246$ cm^2 and $\sigma_A = 0,334$ cm^2.

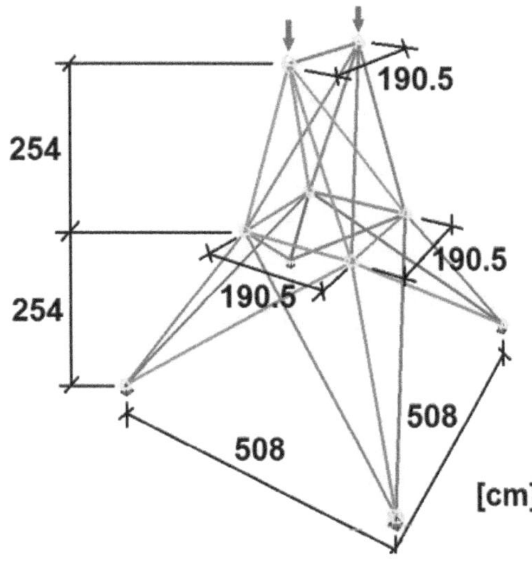

Figure 1. Geometry of the considered simple truss tower.

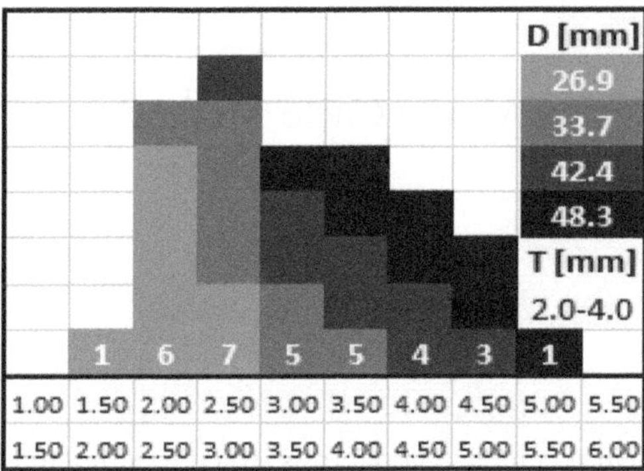

Figure 2. A histogram based on the bar's cross-sectional area availability in steel catalogues.

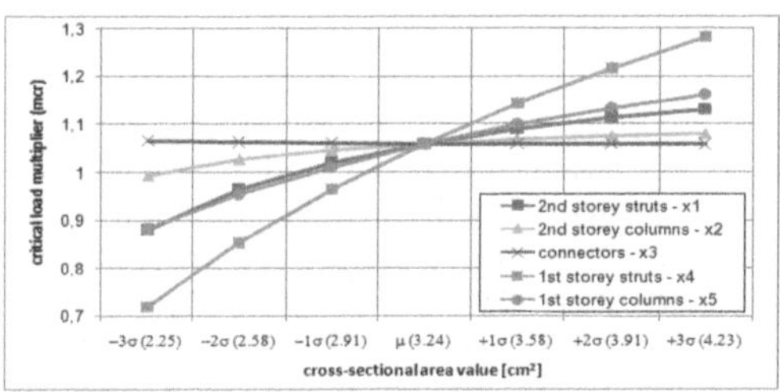

Figure 3. The sensitivity of structural response (m_{cr}) to the change of random variables (cross-sectional area of bars A_i).

However, considering the 25 cross–sectional area values of each element random is suboptimal and inconsistent with the methodology of truss tower erection process. Five distinct random variables were eventually proposed—2nd storey struts (x_1), 2nd storey columns (x_2), connectors (x_3), 1st storey struts (x_4), and 1st storey columns (x_5). This number of variables was also selected because the reduction of the numerical problem is strongly desirable.

The sensitivity of structural response to the change of cross–sectional area is shown in Figure 3.

2.2 *Formulation of the SLS requirement*

In the paper the critical multiplier m_{cr} of the applied set of loads at the stability loss of the tower is analyzed. Assuming all cross–sectional mean values (the proposed PDF mean value), a multiplier equal to $m_{cr,id} = 1.066$ was obtained.

Afterwards, three discrete values, representing the cross–sectional area variable PDF were chosen specifically for the task—the mean value ($\mu_A = A_{mean} = 3.246$ cm^2) and the boundary values related to $+3\sigma_A$ and $-3\sigma_A$ (giving $A_{min} = 2.252$ cm^2 and $A_{max} = 4.240$ cm^2). Due to this choice, a set of $3^5 = 243$ samples is created and then computed by means of SOFiSTiK.

A minimum value of the multiplier, equal to $m_{min} = 0.738$ is found for the design case of maximum connectors and minimal columns and struts. The maximum value of the multiplier, equal to $m_{max} = 1.399$ was found for the opposite design case. It was decided to define structural failure as the occurrence of buckling under a load less than 80% of the initially applied set of loads.

2.3 *Reliability assessment*

2.3.1 *Monte Carlo Method calculations*

In the first approach, slightly resembling Stratified Sampling Monte Carlo (SS–MC) methodology (where only one sample from a strictly defined stratum of a subspace of the problem is chosen), all 243 samples generated are taken into consideration.

This leads to the probability of failure equal to $P_{f,SS} = 0.05761$ and a resultant value of reliability index equal to $\beta_{SS} = 1.57513$. The SS–MC results were taken as reference values for further calculations.

Due to engineering processes a number of 243 samples may be too great, another MC variance reduction technique, Directional Sampling, is used.

The Directional Sampling (DS–MC) technique is based on selecting crucial samples from the polar ends of the sample space—in this task, it is done for the combinations of $\pm 3\sigma_A$ values due to each random variable of the task, giving $2^5 = 32$ points of interest.

This results in a probability of failure $P_{f,DS} = 0.12500$, and a reliability index $\beta_{DC} = 1.15048$, with an 26,96% error (in β), compared to the MC results.

2.3.2 *Point estimate method calculations*

The second approach incorporated the Rosenblueth form of Point Estimate Method (PEM).

In this methodology, each variable is assigned to be either $+3\sigma_A$ or $-3\sigma_A$, consistently with the weights approach, as noted in Christian & Baecher (1999), whereas the remaining variables are set at their mean values. This approach gives 10 or 11 computational samples, the latter when the sample of every variable set at its mean value is also taken into account (such an approach is named extended PEM or XPEM).

The probability of failure is equal to $P_{f,PEM} = 0.05983$ in both cases, giving a resultant value of reliability index equal to $\beta_{PEM} = 1.55620$ and an 1,21% error (in β), compared to the MC results.

Table 1. Numerical results of the RSM calculations.

RSM approximation	SS–MC (243 samples)	DS–MC (32 samples)	PEM (10 samples)	XPEM (11 samples)
1st order				
$P_{f,RSM}$	0.04527	0.05705	0.06173	0.06173
β_{RSM}	1.69258	1.58003	1.54042	1.54042
ε_β (error)	7.46%	0.31%	2.20%	2.20%
2nd order				
$P_{f,RSM}$	0.05761	0.11836	N/A	N/A
β_{RSM}	1.57513	1.18323	N/A	N/A
ε_β (error)	Ref. value	24.88%	N/A	N/A

2.3.3 *Response surface method calculations*

Subsequently, reliability was also assessed using the Response Surface Method, which applies an approximate function (in this case, the 1st and 2nd order polynomials) to represent the structural response.

The approximations are performed by an author's software RSM–Win, described in Winkelmann & Górski (2014), with four sets of sample points, taken directly from previous calculations, namely the SS–MC, DS–MC, PEM, and XPEM.

On the basis of the obtained response surfaces slope factors, safety measures (based on the Hasofer–Lind–Rackwitz–Fiessler 2nd order approach, most fitting for the undertaken task) are calculated. The results are summarized in Table 1.

Basing on the presented calculations, a real-life tower is to be considered. As both PEM and RSM proved efficient, trustworthy and accurate, crude MC analysis will be omitted.

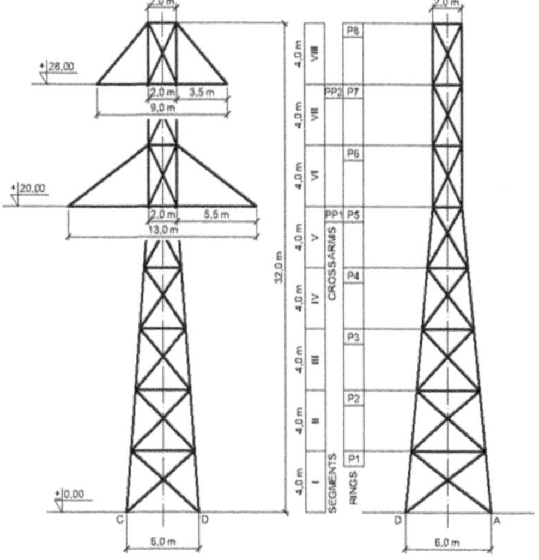

Figure 4. Geometry of the considered QVH line support tower.

3 ANALYSIS OF A SIMPLIFIED FULL–SIZE ENGINEERING STRUCTURE

3.1 *Subject of the analysis*

Next, a 160–member truss of geometry, size and cross–sectional dimensions similar to those of real–life overhead power lines supporting tower, is considered. The structure is presented in Figure 4.

The tower comprises of 8 segments, each 4 m high, giving 32 m of elevation in total. The tower is made of S275 steel (E = 210 GPa, R_e = 275 MPa) and the following loads are taken in each top node of the four crossbars: $P = (P_x, P_y, P_z)$ = (0.01 kN, 0.01 kN, 100 kN), bound to represent an excessive icing of the cables. All the members are made of equal leg angles, in three different types: L 80 × 80 × 8 ($A_{80\times8}$ = 12.37 cm^2) for columns, L 60 × 60 × 6 ($A_{60\times6}$ = 6.91 cm^2) for bracings and rings of segments I to V and L 60 × 60 × 4 ($A_{60\times4}$ = 4.71 cm^2) for cross–arms and rings of segments VI to VIII.

3.2 *Sensitivity analysis*

As in the previous example, it was assumed that the random variables are related to changes in cross–sectional area values. Because of the fact, that the areas of the elements are radically different it was decided adopt the angle profiles with cross–sectional areas approximately 2 and 4 times smaller and larger than the initial area of the bar, to improve the readability of the sensitivity analysis.

An important point of sensitivity analysis for such a complex structure is to choose the number of design variables optimally—to try to get results of satisfactory accuracy and creditability using minimal computational effort.

Despite consisting of only three different profiles, the structure is subdivided into 10 groups of elements, in accordance with the design specifications. A test solution with such a large number of

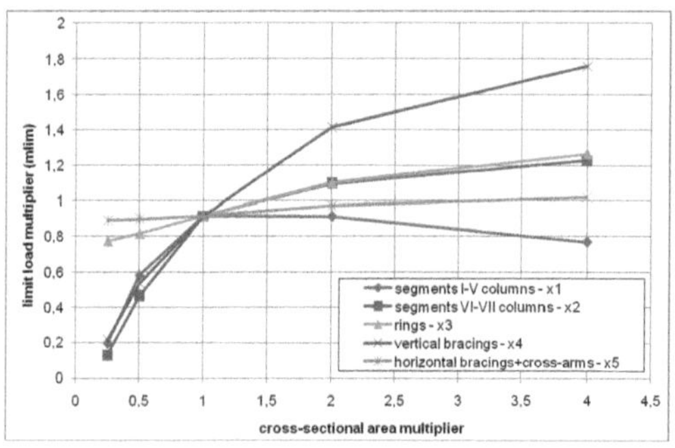

Figure 5. The sensitivity of structural response (m_{lim}) to the change of random variables (cross-sectional area of bars A_i).

variables showed that the structure is poorly sensitive to modifications of certain groups of elements. Based on this conclusion, the focus was only set, as previously, on only five crucial variables: segments I–V columns (x_1), segments VI–VIII columns (x_2), rings (x_3) vertical bracings (x_4), and horizontal bracings in connection with cross-arms (x_5), a most optimal subdivision of the elements into groups.

The sensitivity of structural response to the change of cross–sectional area is shown in Figure 5.

It can be noted that the structure is very sensitive to changes in cross-sections of individual elements. This example clearly shows this analysis type is reasonable for this type of structures, rationalizing their shape is one of the most basic engineering problems to be solved in the design process. Strengthening of the cross bracing over the entire height of the shaft is suggested as optimal. It is also advantageous to enlarge the sections of the upper segments and of rings.

3.3 *Formulation of the SLS requirement*

In the analysis, the load multiplier m_{lim} of the applied set of loads at the bearing capacity loss of the tower is analyzed. Assuming all sections as L 80 × 80 × 8, an initial multiplier equal to $m_{lim,i} = 1.535$ was obtained.

It was decided to define structural failure as total failure under a load less than 75% of the initially applied set of loads (about a half of the above mentioned multiplier).

3.4 *Reliability assessment*

3.4.1 *Monte Carlo Method calculations*

Making use of the cross-sectional area PDF (a uniform type distribution was used this time), 100 samples of the variable vector were generated and then computed in SOFiSTiK. The limit multiplier values were calculated for every sample, together with their mean values. Computations were to stop if a desired low error rate (3%) between successive samples was achieved and this was done after analyzing 45 of them, with the mean value equal to $m_{lim,mean} = 0.919$.

The probability of failure according to MC was computed as $P_f = 0.46667$, mostly because of samples consisting of only low values of sectional areas.

3.4.2 *Directional simulation calculations*

Afterwards, a Monte Carlo Directional Simulation approach was proposed, and 7 discrete values for each variable were chosen for the task—the mean value of each member (μ_A) and the intermediate values related to 0.25, 0.5, 2 and $4 \times \mu_A$. Due to this choice, a set of 3 × 5 = 35 samples is created and then computed by means of SOFiSTiK software.

A minimum value of the multiplier, equal to $m_{lim,min} = 0.535$, and the maximum value, equal to $m_{lim,max} = 2.778$ was found for the design case of boundary values of segments VI–VIII columns (x_1).

The DS probability of failure was calculated as equal to $P_{f,DS} = 0.05714$.

3.4.3 *Point estimate method calculations*

Due to the distribution of the area PDF, the asymmetry factor = 0.45 and the weights: $x_+ = 16.31$ (closest to L 100 × 100 × 8 with $A_{100\times8} = 15.51$ cm^2) and $x_- = 6.04$ (closest to L 60 × 60 × 5 with $A_{60\times5} = 5.82$ cm^2) were calculated in the first place.

Consistently with the Rosenblueth approach, a set of 10 (PEM) or 11 (XPEM) computational samples were created and then computed in SOFiSTiK.

The PEM probabilities of failure were equal to $P_{f,PEM} = 0.01000$ and $P_{f,PEM} = 0.09091$.

3.4.4 *Response surface method calculations*

Finally, the probabilities of failure were also assessed using the Response Surface Method, by an author's software RSM–Win, again with all four sets of sample points, taken directly from previous generations and computations.

On the basis of the obtained response surfaces slope factors, probabilities of failure and reliability indices (based on the Hasofer–Lind approach this time, to show the versatility of RSM–Win) are calculated. The results are summarized in Table 2.

Only the 1st order approximation is conducted this time, for in the previous example it was proven to be more trustworthy and accurate for such a task.

Table 2. Numerical results of the RSM calculations.

RSM approximation	MC (45 samples)	DS–MC (35 samples)	PEM (10 samples)	XPEM (11 samples)
1st order				
$P_{f,RSM}$	0.12974	0.12256	0.16358	0.16425
β_{RSM}	1.12764	1.16227	0.97984	0.97714
ε_β (error)	Ref. value	3.07%	11.69%	11.94%

4 CONCLUSIONS

Probabilistic reliability assessment of simple truss structures, using the standard probabilistic techniques is clearly possible. However it should be noted, that due to overstated levels of initial loading, the trusses analyzed in the paper have a relatively high probability of failure, significantly higher than those that have to be of real–life engineering structures, a fact previously noted for the benchmark example also in Thoft–Christensen & Murotsu (1999).

Moreover, PEM and RSM proved to be a good alternative for the biased MC calculations of such simple engineering structures.

The RSM approximation is an effective way to reduce the MC sample population, combined with either MC variance reduction techniques or a PEM–based sampling points choice is an efficient tool to assess reliability of geometrically nonlinear models.

However, it should be pointed out, that such a small number of samples as used in RSM with PEM approach gives completely erroneous results for the quadratic approximation (an error of 78,3% for the second truss, thus not shown in Table 2), indicating the fact, that for more advanced and accurate approximations, the number of points of approximation should also be adequately greater.

As mentioned in the previous studies made by the authors in Winkelmann & Górski (2014), the Directional Sampling approach, a derivative of Importance Sampling technique, is not intended for the analysis of simple models. While the DS-based 1st order approximation is satisfactory (presumably circumstantially), the 2nd order approximation is showing a large stand-off error, nearly ten times greater than the average error result.

As prompted by the sensitivity trend lines, the best approximation is obtained for the 1st order polynomial regardless of the used method which seems to be a curious phenomenon.

Finally, it is shown, that such an analysis can be an excellent starting point for sensitivity and reliability analysis of real-life engineering structures.

REFERENCES

Christian, J.T., Baecher, G.B. 1999. Point Estimate Method as numerical quadrature. *Journal of Geotechnical and Geoenvironmental Engineering* 125(9): 779–786.

Kleiber, M., Antúnez, H., Hien, T.D., Kowalczyk, P. 1997. *Parameter sensitivity in nonlinear mechanics*. Chichester: Wiley & Sons.

Thoft–Christensen, P., Murotsu, Y. 1986. *Application of structural systems reliability theory*. Heidelberg: Springer–Verlag.

Winkelmann, K., Górski, J. 2014. The use of response surface methodology for reliability estimation of composite engineering structures. *Journal of Theoretical and Applied Mechanics* 52(4): 1019–1032.

Advances in Mechanics: Theoretical, Computational and Interdisciplinary Issues – Kleiber et al. (Eds)
 ISBN 978-1-138-02906-4

The influence of presumed border conditions on FEM thermal analysis results based on the example of an LNG tank support saddle

J. Zapłata & M. Pajor
Faculty of Mechanical Engineering and Mechatronics, West Pomeranian University of Technology, Szczecin, Poland

ABSTRACT: The growing accessibility of CAE environments has made FEM more egalitarian, calling into question the reliability of some analyses. This paper aims to investigate the influence of some popularly taken assumptions on results of FEM thermal analyses on the example of a LNG tank support saddle. The temperature distribution of that structure is of interest since the classification societies require that the temperature of construction materials should not exceed allowed service conditions.

1 INTRODUCTION

The growing accessibility of CAE environments has made FEM more egalitarian, calling into question the relevance of some analyses results. Except for the well know rules that have to be followed to ensure the FEM analysis accurateness, such as fine mesh or the convergence criteria, one of the most rudimentary requirement is the proper boundary condition formulation. The last requirement often causes problems when thermal analysis of machines or technical structures is to be performed. This paper aims to investigate the influence of some popularly made assumptions on thermal FEM analyses results on the example of thermal analysis of a LNG tank support saddle.

In case of a LNG tank foundation the temperature field of the gas tanker hull is of interest since the classification societies (e.g. Polish Registry of Shipping) require that the temperature of construction materials of a ship does not exceed allowed service conditions. The LNG is transported in cryogenic tanks, in a temperature below −162°C. The regulations require appropriate calculations, confirming that the temperature of the hull does not exceed the allowed temperature limits in any condition, including a leakage of the inner LPG tank (PRS 2014). The upper mentioned regulations state, that that during those calculations the temperature of surrounding air is to be presumed at level of 5°C, whereas the temperature of the sea water at level of 0°C.

Reliability of such calculations is important since the increase of safety margins means that additional expenditures have to be made. This paper considers the influence of often accepted simplifications on the results of thermal FEM analysis.

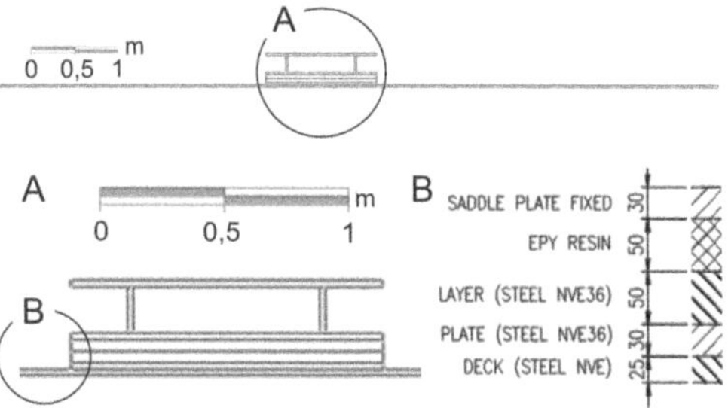

Figure 1. Geometry of the tank support saddle including a segment of the deck.

2 THE GEOMETRY

Geometric model of the tank support saddle is shown at Figure 1. In order to build a cost effective structure, the tank saddle includes a polymer layer (Grudziński 2014) used as an insulation. The order and thickness of layers is shown at detail view B (Fig. 1). The overall dimensions of the tank support saddle, not including the deck, are 0.4 m × 1.4 m × 4.6 m.

3 PRESUMING A TWO DIMENSIONAL TEMPERATURE FIELD

Two-dimensional models are often used in order to decrease the calculation time. Such a simplification is allowable if one of the overall dimensions of the concerned body is an order of magnitude smaller than the other dimensions, or if two opposite border surfaces are insulated. In other cases, a distortion in analysis results occur. A difference between 3D and 2D, evaluation is shown at Figures 2 and 3 on the example of a LNG tank support saddle.

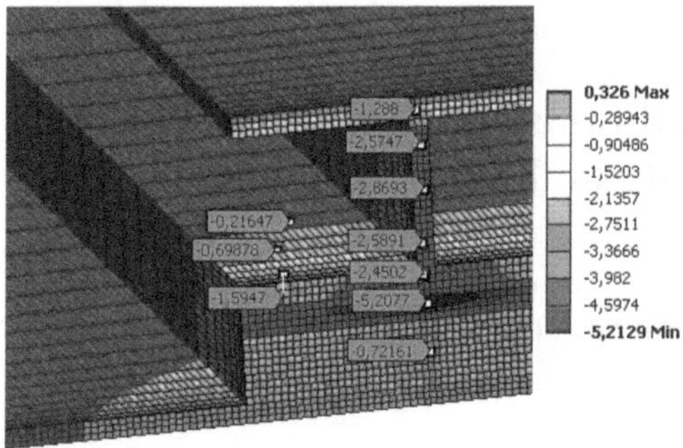

Figure 2. The error made as a result of using a 2D model instead of a 3D model [°C].

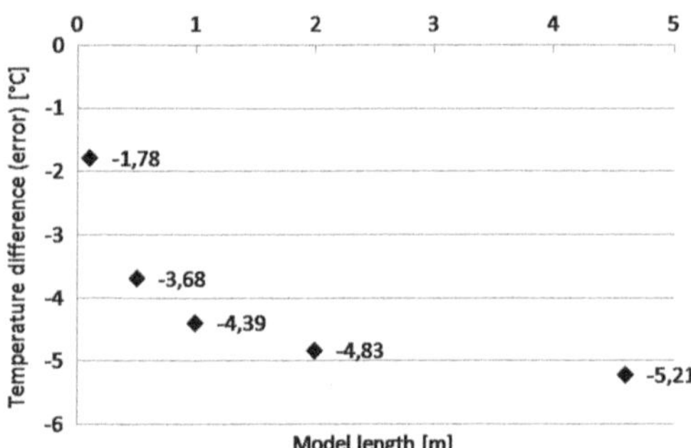

Figure 3. The maximal temperature difference in results between 2D and 3D model, as a function of model's length.

In the flowing case, the convection coefficients were assumed 5°C/W/m^2, the surrounding air temperature was equal to 5°C and the upper surface temperature was equal to −165°C. The convection boundary conditions were set at all other surfaces, including the front and the rear surface of the 2D model.

As shown in Figure 2 the value of the error is not uniform, increasing towards the exterior. The relation between the depth of the body and the 2D approximation exactness is shown at Figure 3.

4 THE CONVECTION COEFFICIENTS

4.1 *The heat convection coefficients for simple shapes*

One of the thermal FEM analysis inaccuracy sources lies in imprecisely estimated convection coefficients. The value of these coefficients depends on the surrounding fluid temperature, the physical properties of the surrounding fluid, the type of flow and the shape of the heat exchanging body. One of the most convenient method for estimating those coefficients is using literature formulas (e.g. Holman 1986, Incopera 2006, Kreith 2011), which are defined for some basically shaped surfaces. Accuracy of such estimated coefficients usually does not exceed 15% (Kreith 2011). As a result, many literature formulas concerning similar geometrical shapes can be found. Figure 4 shows the comparison of the mean Nusselt number for a vertical plane surface in the natural convection, estimated on the basis of various literature sources.

The advantage of literature formulas lies in its ease of use and its quickness in comparison to the FEM CFD method. The results are consentient, but the literature methods are only adequate for simple, already examined shapes.

4.2 *Convection coefficients for unrestricted shapes*

The formulas mentioned above are estimated for simple shapes, e.g. planes, solids of revolution, etc. Though attempts were made to generalize the formulas, still there are no simple empirical ones applicable for every complex surface (Holman 1986). Convection coefficients for complicated surfaces can be estimated by means of CFD method, or by means of an experiment. Both methods are not frequently used to enhance the accuracy of thermal FEM analysis, since they are both cost— and time-consuming. Many authors of papers dealing with the subject of machine temperature analysis or CNC machine thermal error calculation use a arbitrary chosen values or literature formulas appropriate for basic shapes in order to estimate the coefficients for complex surfaces.

Table 1 shows comparison of convection coefficients calculated on grounds of semi adequate literature formulas (Kreith 2011, Sparrow & Ansari 1983) and CFD analysis. The geometry used for

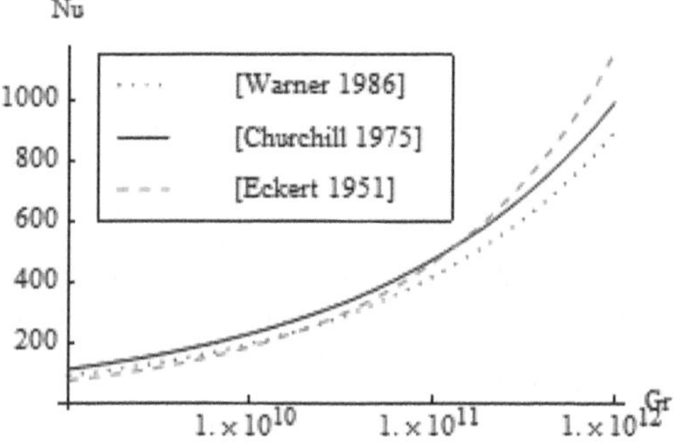

Figure 4. The mean Nusselt number for vertical plane, natural convection, calculated on the basis of various literature sources (Warner & Arpaci 1968, Churchill & Chu 1975, Eckert & Jackson 1951).

Table 1. Values of the heat convection coefficients estimated by means of miscellaneous methods.

	Kreith 2011 (W/m/°C)	Sparrow 1983 (W/m/°C)	CFD (W/m/°C)
Upper surface	1.7	–	2.3
Side surface	4.3	–	2.5
Lower surface	3.8	–	4.9
Weighted mean value	3.2	4.7	3.3

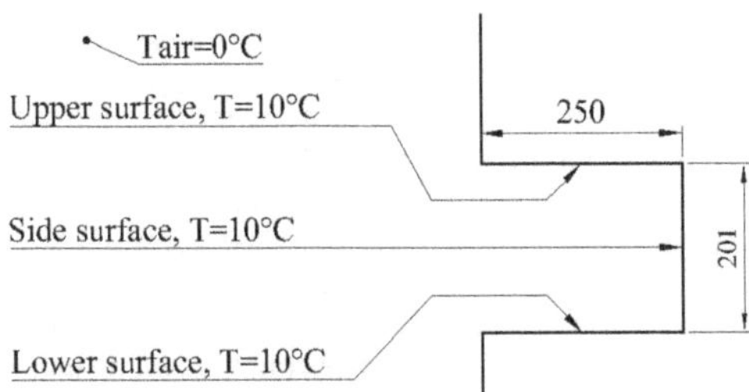

Figure 5. The geometry used in order to compare methods of heat coefficient estimation.

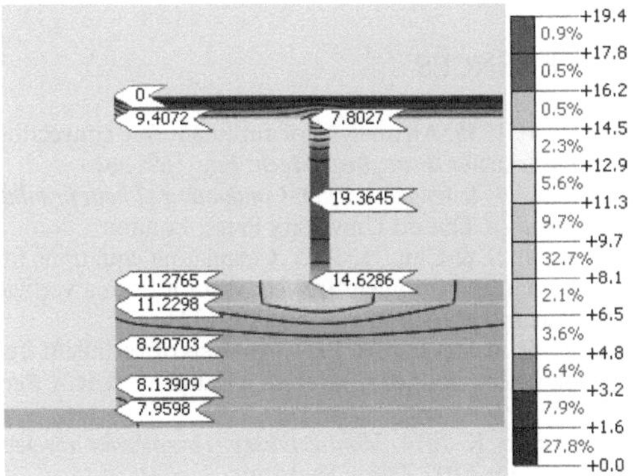

Figure 6. The difference in temperature distribution [°C] as a result of heat coefficient alteration increasing all heat transfer coefficients by 50%.

creating the comparison is shown in Figure 5. The temperature of selected surfaces was equal 10°C and the surrounding air temperature was equal 0°C.

If the CFD method is taken as a reference, the method proposed by Sparrow & Ansari (1983) seems to be a crude approximation. When studying Table 1, it is clear that the heat transfer coefficients, calculated independently for each wall using literature formulas, have values differing from the CFD reference. Although the mean values are quite similar.

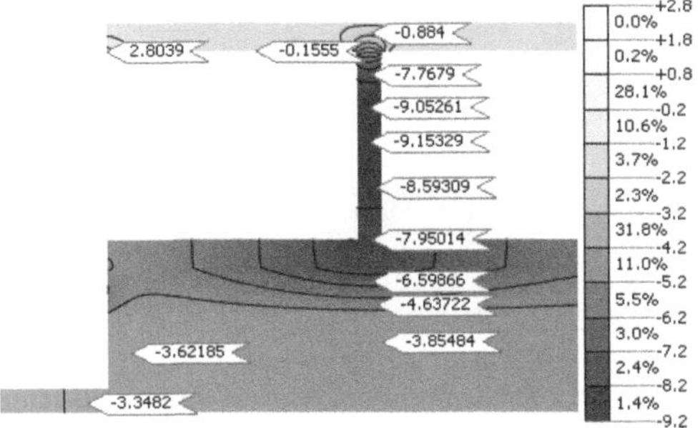

Figure 7. The change of temperature distribution [°C] as a result of forming of 10 mm thick ice layer on surfaces of the tank saddle.

Figure 6 shows how altering the value of heat coefficients influence the value of the temperature field of the LNG tank support saddle. In shown case, the change of the convection coefficient by 50% leads to change of the temperature results up to 20°C.

5 INFLUENCE OF FORMING OF AN ICE LAYER

If the temperature of a surface drop below the water freezing point and humidity of air is sufficient, an ice layer starts to grow. Ice layer thermal conductivity varies with its temperature from 1.9 W/m/°C at 0°C to 3 W/m/°C at −100°C (Raznjevic 1964), and is noticeably smaller than the thermal conductivity of steel. In result the ice forms an insulating layer.

Figure 7 shows the change in temperature distribution as a result of 10 mm thick ice layer formed on the surface of the tank support saddle. In presented case the additional insulation layer lead to the reduction of temperature value up to 9°C.

6 NONLINEARITY OF MATERIAL PROPERTIES

In case of large temperature value span over a body, it is often necessary to consider nonlinearity of heat conductivity of materials. If this is omitted an error in results occurs. The Figure 8 shows the error made as a result of assuming linear thermal properties of steel plate at level k = 14.4 W/m/°C, which is accurate for 0°C temperature.

Figure 9 shows an error made as a result of assuming linear thermal properties of material of steel plate at level 12.6 W/m/°C accurate for −80°C. This temperature is a mean of the LNG liquid temperature and ambient air temperature.

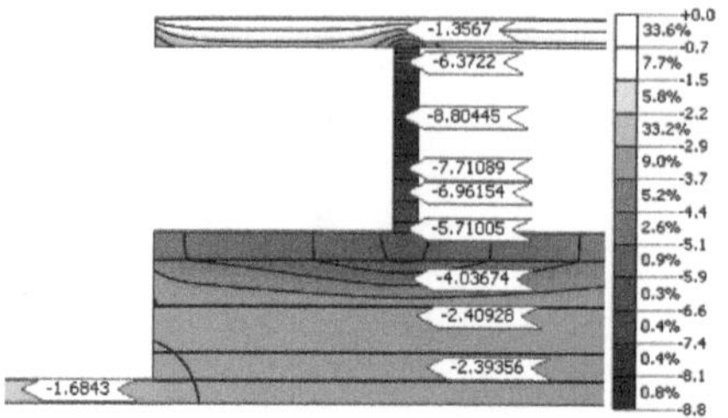

Figure 8. The temperature error as a result assuming linear thermal properties of steel k = 14.4 W/m/°C, as for t = 0°C.

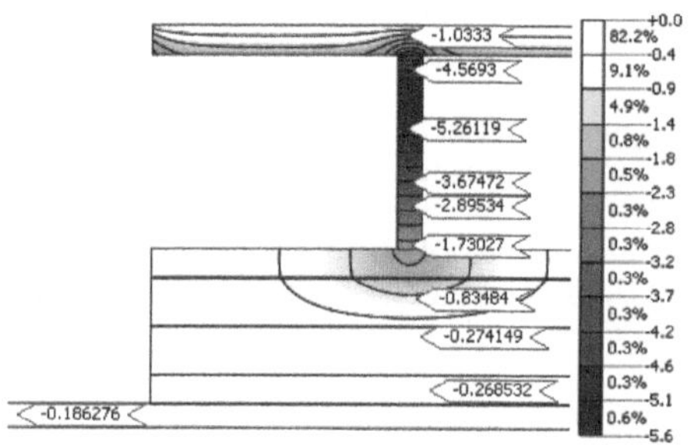

Figure 9. The temperature error as a result assuming linear thermal properties of steel k = 12.7 W/m/°C, as for t = −80°C.

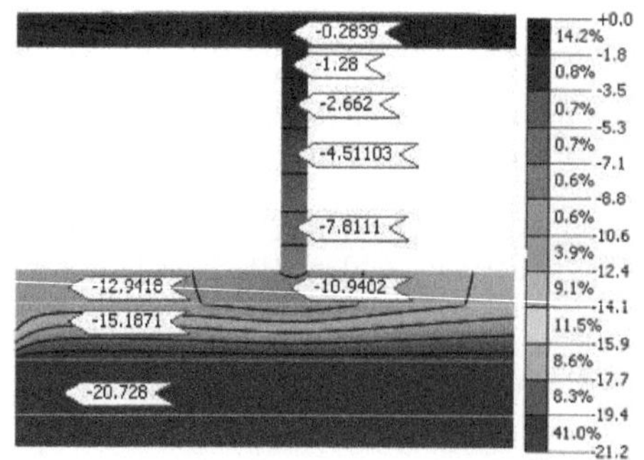

Figure 10. The difference in temperature distribution as a result of presuming false insulation border condition.

The comparison of Figures 8 and 9 shows that the second assumption is more precise.

7 PRESUMING INCORRECT INSULATIONS

The most serious distortion of FEM analysis result may be introduced by presuming inadequate insulation border conditions. Thought, it seems to be clear that such BCs are appropriate at a planes of symmetry of temperature distribution and in areas where thermal resistance is of such value that heat flux is close to null, one might however stumble across analyses where surfaces, at which heat flow is inevitable, are considered to be insulated. This leads to incorrect calculation result. Effects of such a crude mistake are presented at Figure 10, at which the lower surface of the tank saddle was presumed to be insulated from the steel deck plate. The resulting error at the exemplary case reaches up to 20°C.

8 CONCLUSIONS

The paper presents the influence of frequently made simplifications on the result of FEM thermal analyses, on basis of an LNG tank support foundation analysis. It has been shown, that the irrelevant method used for thermal coefficient calculation affects the analysis accuracy. In presented case presuming 2D model instead of 3D model gives moderate error situated mainly near exterior surfaces. In case of steel structures growing ice layer acts as an insulator, and such effect should not be neglected. Substituting nonlinear thermal conductivity with constant values is possible only if the temperature span is reasonable. The greatest error might be inducted if incorrect insulation borders are assumed.

REFERENCES

Bayley, F. 1955. An analysis of turbulent free convection heat transfer. *Proc. Inst. Mech. Eng.* 169: 361.

Carlslaw, H. & Jeager, J. 1959. *Conduction of heat in solid, second ed.* Oxford University Press: London.

Churchill, S. & Chu, H. 1975. Correlating equations for laminar and turbulent free convection from a vertical plate. *Int. J. Heat Mass Tran.* 18: 1323.

Eckert, E. & Jackson, W. 1951. Analysis of turbulent free convection boundary layer on a flat plate. *NACA Rep. 1015.*

Grudziński, K. 2014. *Montaż maszyn i urządzeń z użyciem tworzywa EPY*. Zapol: Szczecin

Holman, J.P. 1986. *Heat Transfer, sixth ed.* McGraw-Hill Book Company: Singapore.

Incopera, F.P. et al. 2006. *Fundamentals of heat and mass transfer, sixth ed.* John Wiley and Sons: New York.

Kreith, F. et al. 2011. *Principles of heat transfer, seventh ed.* Cengage Learning: Stamford.

PRS, 2014. *Publikacja NR 48/P—Wymagania dla gazowców—2014*. Polski Rejestr Statków: Gdańsk.

Raznjevic, K. 1964. *Tablice cieplne z wykresami: Dane liczbowe w układzie technicznym i międzynarodowym.* WNT: Warszawa.

Sparrow, E. & Ansari, M. 1983. Refutation of a King's rule for multi dimensional external natural convection. *Int. J. Heat Mass Tran.* 26: 1357.

Warner, C. & Arpaci, V. 1968. An experimental investigation of turbulent natural convection in air at low pressure along a vertical heated flat plate. *Int. J. Heat Mass Transfer* 11: 379.

Advances in Mechanics: Theoretical, Computational and Interdisciplinary Issues – Kleiber et al. (Eds)

Application of the Element Residual Methods to dielectric and piezoelectric problems

G. Zboiński
Institute of Fluid Flow Machinery, Polish Academy of Sciences, Gdańsk, Poland
Faculty of Technical Sciences, University of Warmia and Mazury, Olsztyn, Poland

ABSTRACT: This paper deals with some theoretical and implementation aspects of application of the Element Residual Methods (ERM) of error estimation to dielectric and piezoelectric problems. In particular the Residual Equilibration (or equilibrated residual) Method (REM) is a subject of our interest. The aspects presented in the paper are still under investigation. In the being conducted research we focus our attention on the upper bound properties of REM in the dielectric and piezoelectric cases. Furthermore the algorithms of the method for both cases are of our interest, especially the details concerning the equilibration process and the solution of REM element local problems. The convergence properties for both types of problems, and for the mechanical elasticity problem as well, are the base for determination of the algorithmic details in the piezoelectric case.

1 INTRODUCTION

This research is motivated by the present interest in application of the piezoelectric members as actuators and sensors in the so-called intelligent structures and structural health monitoring, respectively. Also, the questions of application of the adaptive finite element methods to dielectric and piezoelectric problems are in its early stage of theorisation and implementation.

In this paper we focus on error estimation in the adaptive modelling and adaptive finite element analysis of the dielectric and piezoelectric problems. The question we would like to answer in this research is if the element residual methods in general, and the residual equilibration method in particular, can be applied to two mentioned types of problems. Additionally, we would like to learn how the theory of the error estimation of the elliptic problems, and the specific elasticity problem as well, has to be modified in the case of the dielectricity and piezoelectricity. We are also interested in the modifications of the residual equilibration algorithms necessary for these two cases. Note that such algorithms were applied successfully to the adaptive modelling and analysis of complex elastic structures (Zboiński 2013) so far.

2 STATE OF THE ART

We address two issues in this brief survey. The first one is adaptive modelling and analysis of dielectric and piezoelectric media or systems. The second one concerns application of the element residual methods to error estimation in the electric and coupled electro-mechanical problems.

2.1 *Adaptivity in dielectricity and piezoelectric systems*

The adaptive analysis of dielectric and piezoelectric systems is not very popular and only a few papers address the issue. Many of the available papers on the topic has preliminary character and are published in conference proceedings. Among the papers on numerical analysis of the piezoelectric media one can mention the works (Lammering & Mesecke-Rischmann 2003) and (Carrera et al. 2007), which present the classical (non-adaptive) calculations via *h*-method and parametric studies with *p*-method, respectively. The rare example of the application of the adaptive approach to the analysis of piezoelectric problems can be found in (Vokas & Kasper 2010), where the *h*- and *p*-methods are applied. The suggestions concerning application of the *hp*-adaptivity to the dielectric and piezoelectric problems can be found in our conference and post-conference papers, i.e. (Zboiński 2012, Zboiński 2014b), where the *hp*-adaptive finite element method as well as the hierarchical dielectric and piezoelectric models are proposed, respectively. The related comparative *h*- and *p*-convergence studies for the model dielectric and reference elastic problems are presented in (Zboiński 2014a), together with the resultant conclusions on how the *hp*-mesh adaptation should be performed in the case of piezoelectrics.

2.2 *Error estimation for dielectricity and piezoelectricity*

No examples of the direct application of the Element Residual Methods (ERM) and the Residual Equilibration Method (REM) to the dielectric and piezoelectric problems can be found in the commonly available literature. The suggestion on the application of REM to dielectrics is presented in the conference abstract (Zboiński 2015) only. The analogous suggestion on piezoelectric systems is present in (Zboiński 2012).

Even though the mentioned methods are not common in the case of dielectricity and piezoelectricity, their application to elastic and other elliptic problems is a standard. The REM in the general context of elliptic problems was presented by Ainsworth and Oden (Ainsworth & Oden 1993a, Ainsworth & Oden 1993b). The 3D-elasticity case is described in (Ainsworth et al. 1994). The case of conventional shell-like and plate-like elastic structures was described in (Oden & Cho 1996), while the application of the residual equilibration method to 3D-based complex elastic structures was proposed in (Zboiński 2013). The mentioned works can be a direct hint and justification for the application of the REM to dielectric problems due to its similarity to elliptic problems in general, and the elastic problem in particular. The case of piezoelectricity is much more difficult and needs careful treatment due to its coupled character.

3 CONVERGENCE STUDIES

Here we check *h*- and *p*-convergence in three model problems: purely elastic, purely dielectric and coupled piezoelectric ones. The problems concern the square plate of length $2l$ and thickness t, clamped or/and grounded along its lateral edges. The mechanical and electric constants correspond to a typical PZT material. The plate is loaded with a uniform vertical traction or/and electric charge applied on its upper surface. The values of the electric and mechanical potential energies are of the same orders. We skip the corresponding figures here as we are interested only in a numerical qualitative comparison of the three problems. The *h*-convergence curves are presented in Figures 1–3, where N is the number of dofs, U, E and V are the strain, electric and electro-mechanical energies, while their reference values U_r, E_r and V_r correspond to the over-killed meshes. Below, h denotes the longitudinal size of the element, with $m = l / h$, while p and $pi \equiv \pi$ the element longitudinal orders

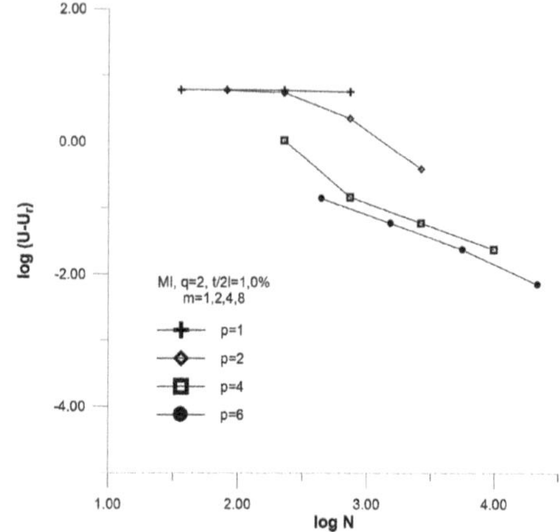

Figure 1. *h*-convergence—elasticity case.

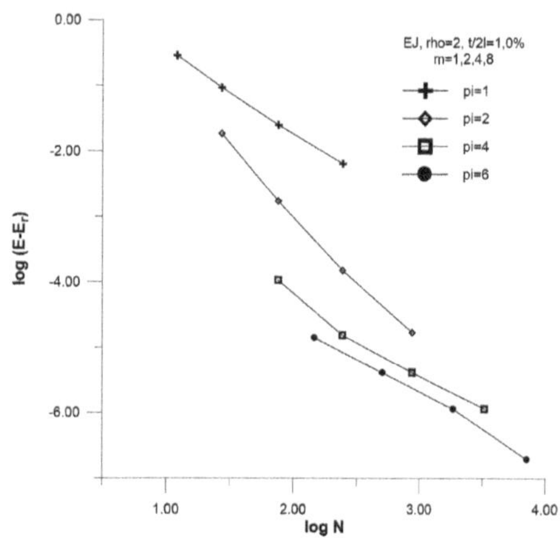

Figure 2. *h*-convergence—dielectricity case.

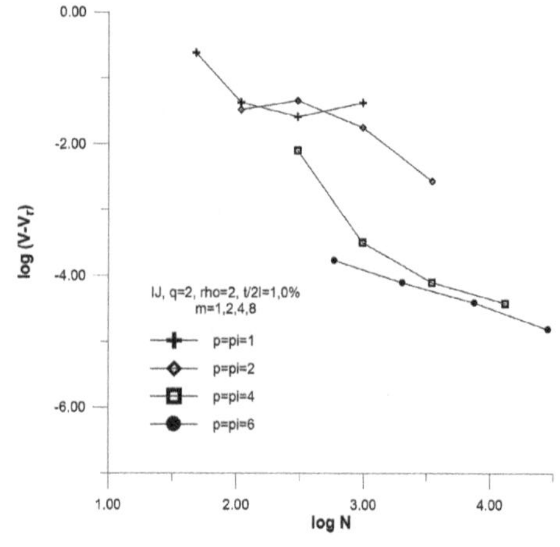

Figure 3. *h*-convergence—piezoelectricity case.

of approximation within the displacement and electric potential fields. The terms $q = I$ and $rho \equiv \rho = J$ are the transverse approximation orders equal to the orders of the applied hierarchical models MI and EJ, utilized within the mechanical and electric fields, respectively.

The main observation from the figures is that in the case of piezoelectricity the h-convergence may loose its monotonic character, observed in the cases of both pure problems. This refers also to the not included p-convergence curves.

4 REM IN THREE CLASSES OF PROBLEMS

It should be noticed that one deals with formal similarity of formulations of the problems of elasticity and dielectricity. This similarity is reflected by the analogy in the local, variational and finite element formulations of both problems. The available theoretical and implementation findings concerning the elasticity can be easily extended onto the dielectricity. Below we also consider extension of the residual equilibration method onto piezoelectric media.

4.1 *Bounding features of the estimation with REM*

It is well known that the residual equilibration method possesses the upper bound property of the global approximation error estimate, when applied to elliptic problems of the second order (Ainsworth & Oden 1993a, Ainsworth & Oden 1993b). This general case includes also the linear elasticity problems of various type (compare (Ainsworth, Oden, & Wu 1994), (Oden & Cho 1996) and (Zboiński 2013)). The upper boundedness for the elastic systems can be expressed as follows

$$\| \overset{MI}{\boldsymbol{e}} \|_U^2 \le \sum_e (\overset{MI}{\eta_e})^2 = (\overset{MI}{\eta})^2, \tag{1}$$

where

$$(\overset{MI}{\eta_e})^2 = B_e(\overset{MI}{\boldsymbol{u}_e} - \boldsymbol{u}^{hpq}, \overset{MI}{\boldsymbol{u}_e} - \boldsymbol{u}^{hpq}). \tag{2}$$

Above, $\| e \|_U^2$ represents the true value of the global approximation error in the strain energy norm U of the body modelled with the hierarchical mechanical model MI. The term $(\eta)^2$ represents the upper-bounding global approximation error estimator obtained be means of REM, i.e. as a sum of the corresponding element indicators $(\eta_e)^2$. The element indicators are defined by means of the quadratic form B_e equivalent to the strain energy defined on the difference of the global solution u^{hpq} and the REM local solution u_e obtained from the decoupled element local problem. The decoupled problem solution replaces the unknown exact or true global solution.

The above results can be extended onto dielectric problems, i.e.

$$\| \overset{EJ}{e} \|_E^2 \le \sum_e (\overset{EJ}{\eta_e})^2 = (\overset{EJ}{\eta})^2, \tag{3}$$

with

$$(\overset{EJ}{\eta_e})^2 = b_e(\overset{EJ}{\phi_e} - \phi^{h\pi p}, \overset{EJ}{\phi_e} - \phi^{h\pi p}), \tag{4}$$

where the element indicators contributing to the global approximation error estimator are defined by means of the quadratic norm b_e equal to the electric energy E defined on the point-wise differences of the ERM local problem solution ϕ_e and the known global problem solution $\phi^{h\pi p}$. In both problems the hierarchical electric model EJ is employed. The electric energy norm $\| e \|_E^2$ is defined on the scalar approximation error function e.

The proof of the upper bounding property of the global approximation error estimate for the general piezoelectric problem, in a sense applied to elasticity and dielectricity, is not possible due to a coupling of the potential energies of the electric and mechanical fields. We deal with subtraction of both energies, i.e. the mechanical one is converted into the electric one or vice versa. As a result the inequality sign of the below bounding relation cannot be determined, i.e.

$$\begin{aligned} \| \overset{IJ}{e} \|_V^2 \approx \left| \sum_e \overset{IJ}{\eta_e} \right| &= \left| \sum_e [(\overset{MI}{\eta_e})^2 - \overset{EM}{\eta_e} - (\overset{EJ}{\eta_e})^2] \right| \\ &= \left| (\overset{MI}{\eta})^2 - \overset{EM}{\eta} - (\overset{EJ}{\eta})^2 \right| = (\overset{IJ}{\eta})^2. \end{aligned} \tag{5}$$

In the above relation the global approximation error of the piezoelectric problem is defined on the error vector containing three displacement and one potential components in the electro-mechanical co-energy norm V. Above, the sign of the sum of the component element indicators is unknown. And so is the sign of the middle component. This component is defined as a bilinear form C_e:

$$\overset{EM}{\eta_e} = 2C_e(\overset{MI}{\boldsymbol{u}_e} - \boldsymbol{u}^{hpq}, \overset{EJ}{\phi_e} - \phi^{hpq}), \tag{6}$$

equal to the coupling part of the potential energy. This energy is defined on the point-wise differences of the local and global displacements and potentials.

Please note that the difference form of (5) explains the lack of monotonicity of the curves from Figure 3.

4.2 *The equilibration process of REM*

Equilibration of the residuals performed in the case of elasticity can be adapted to the dielectric case. This requires replacement of the vectors of the equilibrated stresses between elements *e* and *f*:

$$\langle \boldsymbol{r}_e(\boldsymbol{u}^{hpq})\rangle = \boldsymbol{\alpha}_f \boldsymbol{\sigma}_e(\boldsymbol{u}^{hpq})\boldsymbol{\nu}_e + \boldsymbol{\alpha}_e \boldsymbol{\sigma}_f(\boldsymbol{u}^{hpq})\boldsymbol{\nu}_e \tag{7}$$

with the equilibrated scalar electric charge densities:

$$\langle h_e(\phi^{h\pi\rho})\rangle = \beta_f \boldsymbol{d}_e(\phi^{h\pi\rho})\boldsymbol{\nu}_e + \beta_e \boldsymbol{d}_f(\phi^{h\pi\rho})\boldsymbol{\nu}_e, \tag{8}$$

where $\boldsymbol{\nu}_e$ represents the normal to the common face of the elements. The adaptation needs also replacement of the mechanical quantities (stresses $\boldsymbol{\sigma}_e$ and displacements u^{hpq}) with the electric ones (electric displacements $\boldsymbol{d}_e$ and electric potential $\phi^{h\pi\rho}$, respectively) and the stress equilibration function α_e with charge equilibration one β_e. Let us notice that the above two vectors (7) and (8) are necessary for the determination of the unknown local fields of displacements $\boldsymbol{u}_e$ and potentials ϕ_e from equations (2) and (4) via solution of the REM local problems.

There are no formal obstacles to obtain the equilibrated stresses and the equilibrated electric charge for the electro-mechanically coupled piezoelectric problems. This needs inclusion of the coupling terms (resulting from the coupling part of the potential energy) into the equilibration equations corresponding to the elasticity and dielectricity. From these equations the equilibrating functions $\boldsymbol{\alpha}_e$ and β_e are obtained by means of the standard approach, i.e. as shown in (Zboiński 2013), for example. The stresses and electric displacements are now obtained from the coupled field ($\boldsymbol{u}^{hpq}$, $\phi^{h\pi\rho}$), being the result of the solution of the global electro-mechanical problem. The equilibrating functions, stresses and electric displacements can then be substituted to (7) and (8).

4.3 *REM local problems*

The REM local problems of dielectricity can be defined in analogy to the elasticity case, i.e. the local element equations describing element mechanical equilibrium have to be replaced with the corresponding equations describing electric equilibrium of the element. In particular, the element stiffness $\boldsymbol{k}_M$ of the element *e*, the element external volume and surface load vectors $\boldsymbol{f}_V$ and $\boldsymbol{f}_P$, and the element load vector $\boldsymbol{f}_R$ due to the equilibrated stresses $\langle \boldsymbol{r}_e\rangle$ in the equation:

$$\overset{e}{\boldsymbol{k}}_M \boldsymbol{q}_e^{HPQ} = \overset{e}{\boldsymbol{f}}_V + \overset{e}{\boldsymbol{f}}_P + \overset{e}{\boldsymbol{f}}_R \tag{9}$$

have to be changed for the element dielectric matrix $\boldsymbol{k}_E$ and the element charge vector, composed of the external charges $\boldsymbol{f}_Q$ and the internal ones $\boldsymbol{f}_G$ due to the equilibrated charge density $\langle h_e\rangle$, in order to form:

$$\overset{e}{\boldsymbol{k}}_E \boldsymbol{\varphi}_e^{H\Pi\mathrm{R}} = \overset{e}{\boldsymbol{f}}_Q + \overset{e}{\boldsymbol{f}}_G. \tag{10}$$

Note that above the unknown fields $\boldsymbol{u}_e$ and ϕ_e from (2) and (4), respectively, are locally *HPQ*- and *H*ΠR-approximated by means of the dof vectors $\boldsymbol{q}_e^{HPQ}$ and $\varphi_e^{H\Pi\mathrm{R}}$ of the element displacements and electric potentials. Above *H*, *P*, *Q* and Π, R are the element discretization parameters in the element local problems. They replace their counterparts *h*, *p*, *q* and *π*, *ρ* from the global problem whose error is searched.

The coupled local problems can be formulated in the similar way, i.e. by addition of the coupling terms to the local equations corresponding to the elasticity and dielectricity cases:

$$\begin{cases} \overset{e}{\boldsymbol{k}}_M \boldsymbol{q}_e^{HPQ} - \overset{e}{\boldsymbol{k}}_C \boldsymbol{\varphi}_e^{H\Pi\mathrm{R}} = \overset{e}{\boldsymbol{f}}_V + \overset{e}{\boldsymbol{f}}_P + \overset{e}{\boldsymbol{f}}_R \\ \overset{e}{\boldsymbol{k}}{}_C^T \boldsymbol{q}_e^{HPQ} + \overset{e}{\boldsymbol{k}}_E \boldsymbol{\varphi}_e^{H\Pi\mathrm{R}} = \overset{e}{\boldsymbol{f}}_Q + \overset{e}{\boldsymbol{f}}_G \end{cases}. \tag{11}$$

Some simplifying decoupled options also exist. For example, the unknown vectors $\varphi_e^{H\Pi R}$ and $\boldsymbol{q}_e^{HPQ}$ can be replaced with their known global counterparts $\varphi_e^{h\pi\rho}$ and $\boldsymbol{q}_e^{hpq}$ in the mechanical and electric equilibrium equations, respectively:

$$\begin{aligned} &\overset{e}{\boldsymbol{k}}_M \boldsymbol{q}_e^{HPQ} - \overset{e}{\boldsymbol{k}}_C \boldsymbol{\varphi}_e^{h\pi\rho} = \overset{e}{\boldsymbol{f}}_V + \overset{e}{\boldsymbol{f}}_P + \overset{e}{\boldsymbol{f}}_R, \\ &\overset{e}{\boldsymbol{k}}{}_C^T \boldsymbol{q}_e^{hpq} + \overset{e}{\boldsymbol{k}}_E \boldsymbol{\varphi}_e^{H\Pi\mathrm{R}} = \overset{e}{\boldsymbol{f}}_Q + \overset{e}{\boldsymbol{f}}_G. \end{aligned} \tag{12}$$

Theoretical substantiation and numerical verification of the original and decoupled approaches (11) and (12) are underway.

5 CONCLUSIONS

Algorithmic extension of the residual equilibration method onto the dielectric and piezoelectric cases is possible. However, the upper bound property of the estimation can proved for the dielectricity case only.

REFERENCES

Ainsworth, M. & J. Oden (1993a). A posteriori error estimators for second order elliptic systems. Part 1: Theoretical foundations and a posteriori error analysis. *Computers Math. Applic. 25*, 101–113.

Ainsworth, M. & J. Oden (1993b). A posteriori error estimators for second order elliptic systems. Part 2: An optimal order process for calculating self-equilibrating fluxes. *Computers Math. Applic. 26*, 75–87.

Ainsworth, M., J. Oden, & W. Wu (1994). A posteriori error estimation for *hp* approximation in elastostatics. *Appl. Numer. Math. 14*, 23–55.

Carrera, E., M. Boscolo, & A. Robaldo (2007). Hierarchic multilayerd plate elements for coupled multifield problems of piezoelectric adaptive structures: Formulation and numerical assessment. *Arch. Comput. Methods Engng 14* (4), 384–430.

Lammering, R. & S. Mesecke-Rischmann (2003). Multifield variational formulation and related finite elements for piezoelectric shells. *Smart Mat. Struct. 12*, 904–913.

Oden, J. & J. Cho (1996). Adaptive *hpq* finite element methods of hierarchical models for plate- and shell-like structures. *Comput. Methods Appl. Mech. Engrg 136*, 317–345.

Vokas, C. & M. Kasper (2010). Adaptation in coupled problems. *Int. J. Comput. Math. Electr. Electronic Engng 29* (6), 1626–1641.

Zboiński, G. (2012). Adaptive finite element method for modelling and analysis of electro-mechanical systems. In *6th European Congress on Computational Methods in Applied Sciences and Engineering. Book of Abstracts*, Vienna, Austria, pp. 304.

Zboiński, G. (2013). Adaptive *hpq* finite element methods for the analysis of 3d-based models of complex structures. Part 2: A posteriori error estimation. *Comput. Methods Appl. Mech. Engng 267*, 531–565.

Zboiński, G. (2014a). Convergence properties of the hierarchical models in coupled electromechanical systems. In *11th World Congress on Computational Mechanics. Abstracts*, Barcelona, Spain, pp. 1–2. http://www.wccm-eccmecfd2014.org/admin/files/fileabstract/a999.pdf.

Zboiński, G. (2014b). Hierarchical models for adaptive modelling and analysis of coupled electro-mechanical systems. In *Recent Advances in Computational Mechanics*, London, Great Britain, pp. 339–334. CRC Press.

Zboiński, G. (2015). An algorithm of error estimation for adaptive analysis of electric problems with the finite element method (in Polish). In *52th Symposium on Modelling in Mechanics. Abstract Book*, Gliwice, Poland, pp. 135–136.

Advances in Mechanics: Theoretical, Computational and Interdisciplinary Issues – Kleiber et al. (Eds)

 ISBN 978-1-138-02906-4

Modelling of the viscoelastic properties of the technical fabric VALMEX

K. Żerdzicki & P. Kłosowski
Faculty of Civil and Environmental Engineering, Gdańsk University of Technology, Gdańsk, Poland

K. Woźnica
Institut National des Sciences Appliquées Centre Val de Loire, Bourges, France
Laboratoire PRISME, Orleans, Bourges, France

ABSTRACT: The analysis of the Burgers model used for the constitutive description of the technical fabric VALMEX viscoelastic behavior is presented. It has revealed that the correctness of the identification depends strongly on the level of the immediate strain taken for calculations. Moreover, it has been proven that the Burgers model can be used for modelling of the viscoelastic properties of the polyester reinforced PVC coated VALMEX fabric. Finally, the obtained results of the Burgers model parameters has been used for the influence evaluation of the service ageing on the viscoelastic properties of the VALMEX material.

1 INTRODUCTION

Based on research and engineering observation, the long-lasting behavior of technical fabrics fastened on a real construction can be defined as viscoelastic. For its identification, the rheological tests of creep or relaxation type are required. When modelling, the Argyris' and more advanced Schapery's non-linear approaches can be used (Ambroziak & Kłosowski 2011). However, for the numerical calculations of the large scale civil engineering structures simpler models are more practical. Therefore, the aim of the presented study is to analyze the Burgers model used for description of the viscoelastic behavior of the technical fabric VALMEX. Additional purpose is to evaluate the influence of the service ageing on the viscoelastic properties of the VALMEX material.

2 VISCOELASTIC BURGERS MODEL

The Burgers model is a four parameter model representing the linear viscoelasticity properties of a material using derivative relations. It is mainly used for the viscoelastic analysis of fluids (Siginer 2014), but it has been also successfully applied into the analysis of polymer composites (Chatree et al. 2014).

The Burgers model is a series combination of the Maxwell and Kelvin-Voigt models (Nowacki 1963). For the uniaxial case of creep tests it has the following formula:

$$\varepsilon(t) = \frac{\sigma_0}{E_1} + \frac{\sigma_0}{\eta_1} t + \frac{\sigma_0}{E_2}\left(1 - e^{\frac{-E_2}{\eta_2}t}\right) \tag{1}$$

where $\sigma = \sigma_0$ = constant is the stress level, E_1 is the instantaneous elastic modulus, E_2 is the delayed elastic modulus and η_1, η_2 are the viscous coefficients.

3 EXPERIMENTS

The VALMEX material is a polyester reinforced PVC coated architectural fabric that has been used as the canopy of the Forest Opera in Sopot (Poland). Two different kinds of the same VALMEX fabric have been examined in this study. The first one is the fabric used for 20 years as the canopy of the Forest Opera (called—the USED material). One should conclude that during exploitation for almost 20 years the first VALMEX material type underwent at least 20 cycles of stretching and unstretching, and additionally was exposed to environment influences all the time. It has confidently underwent ageing, in both weathering and mechanical aspects. The second kind of VALMEX fabric is the material from the same production part as the USED one, but it was kept as a spare material to repair the roof if necessary (called—the NOT USED material). It can be assumed that it has endured only natural, non-mechanical ageing process.

Figure 1. The strength machine for creep tests at the Gdańsk University of Technology.

In this research the creep tests with the stress level $\sigma = 31.7$ kN/m have been conducted separately for the warp and fill direction of the VALMEX fabric and for the USED and NOT USED kind of material. The tests have been carried out on the special testing machine dedicated to creep tests. This particular device has been designed at the Gdańsk University of Technology (Fig. 1).

4 PARAMETERS IDENTIFICATION

4.1 *Identification procedure*

The parameters identification has been realized according to the procedure described in Kłosowski (2011) and is presented in the graphic form in Figure 2.

The t_0, t_1, t_2 denote particular time points of the creep test in succession, while ε_0, ε_1, ε_2 stand for the corresponding strain levels. It is assumed that for the constant stress level σ_0 = constant the immediate strain ε_0 is reached at time t_0. With above mentioned values evaluated from the experimental curve (Table 1), the parameters E_1, E_2, ε_k, η_1 have been determined according to the below relations (Table 2):

$$E_1 = \frac{\sigma_0}{\varepsilon_0}, \qquad \eta_1 = \sigma_0 \frac{t_2 - t_1}{\varepsilon_2 - \varepsilon_1},$$
$$\varepsilon_k = \varepsilon_1 - \frac{\sigma_0}{\eta_1} t, \qquad E_2 = \frac{\sigma_0}{\varepsilon_k - \varepsilon_0}. \tag{2}$$

The viscous coefficient η_2 can be found by the approximation of the function (1) in time range $t_0 - t_1$ using the least square method and introducing the all already identified parameters. The typical regression with parameters E_1, E_2, ε_k, η_1 as constants has failed. Therefore already obtained parameters E_1, E_2, ε_k, η_1 have been treated together with η_2 as unknowns in the regression process. Their values obtained before have been used as the initial ones for the regression.

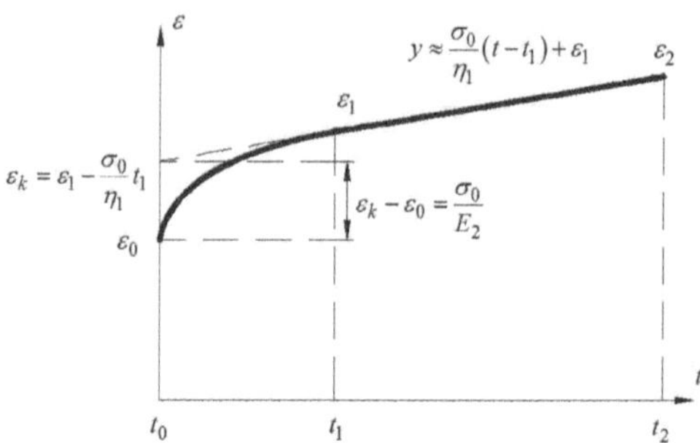

Figure 2. Methodology of Burgers model parameters identification (Kłosowski 2011).

Table 1. Burgers model parameters for the warp direction of the USED VALMEX fabric for t_1 = 2e+5 s: initial values taken from experimental curve.

Initial values						
t_0	t_1	t_2	ε_0	ε_1	ε_2	σ_0
s			–			kN/m
0	2e+5	12e+6	0.065	0.073	0.078	31.7

Table 2. Burgers model parameters for the warp direction of the USED VALMEX fabric for t_1 = 2e+5 s: calculated values based on Table 1.

Calculated values			
η_1	ε_k	E_1	E_2
–		kN/m	
6.8e+9 ± 3.1e+8	0.072 ± 0.001	490 ± 10	4520 ± 20

The final values of E_1, E_2, ε_k, η_1 have slightly changed but the value of η_2 has been successfully calculated (Table 3). Tables 1–3 include the example of calculations for stain level ε_0 = 0.065 and boundary time t_1 = 2e+5 s for the warp direction of the USED material. For the fill direction the same procedure has been executed. The full results juxtaposition can be found in Żerdzicki 2015.

4.2 *Analysis of the results*

Analysis of the obtained results has revealed that the correctness of the identification depends strongly on the value of the immediate strain ε_0 taken for calculations. Therefore the sensitivity analysis with respect to immediate strain ε_0.

Table 3. Burgers model parameters for the warp direction of the USED VALMEX fabric for t_1 = 2e+5 s: identified values based on Table 2.

Identification results			
η_1	η_2	E_1	E_2
–		kN/m	
6.8e+9 ± 3.1e+8	1.1e+8 ± 2.2e+6	500 ± 10	4460 ± 20

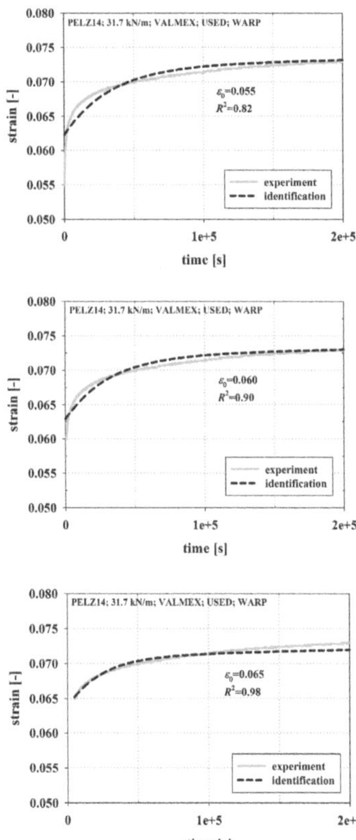

Figure 3. Burgers model parameters for the boundary time t_1 = 2e+5 s and different ε_0: identification of parameters.

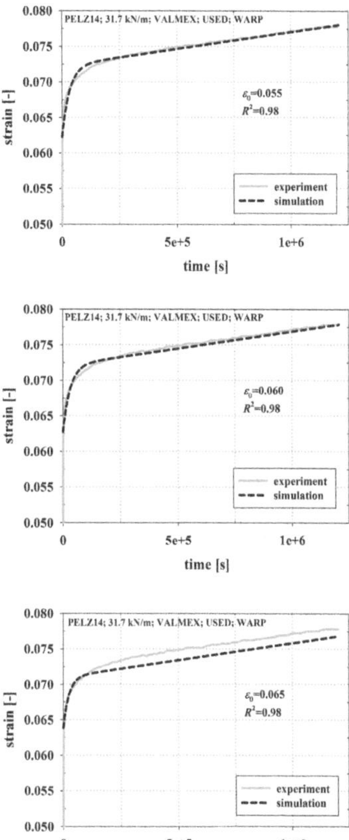

Figure 4. Burgers model parameters for the boundary time t_1 = 2e+5 s and different ε_0: verification by numerical simulation.

The results of the identification procedure for the warp direction of the USED VALMEX fabric (sample No. PELZ14) for three different levels of immediate strain: ε_0 = 0.055 [–], ε_0 = 0.060 [–], ε_0 = 0.065 [–] DHFDH and the limit time t_1 = 2e+5 s are presented in Figure 3. The verification has been accomplished by the calculation of the overall response of the material using the final values of the obtained parameters (Fig. 4). It can be noticed that for all presented cases the determination coefficients R^2 of both identification and verification have very high values.

However, the greater the immediate strain ε_0 is, the better correlation of identification in the range t = 0÷2e+5 s is observed. On the other hand, the higher the correlation coefficient is, the worse the fitting of the simulation for the whole creep test is, especially in the linear range $t_1 - t_2$. For the fill direction of the VALMEX fabric the identical identification and verification procedure has been performed. The obtained results have revealed the same tendencies as observed for the warp direction of the fabric. It proves that the value of the immediate strain ε_0 is of the greatest influence on the identification's results for both material directions.

5 INFLUENCE OF SERVICE AGEING

Taking above observations into account, it has been decided to compare the NOT USED and USED VALMEX fabric bearing in mind the parameters that have given the best fitting of the simulation curve in analyzed ranges separately: in the non-linear (high value of ε_0) and linear part of the creep relation (low value of ε_0). The boundary time taken for comparative analysis has been t_1 = 2e+5 s.

The relative change between the Burgers model parameters for the NOT USED and USED samples are collected in Tables 4 and 5, for the warp and fill direction, respectively. The relative change has been calculated according to the formula δ = [(NOT USED-USED)/NOT USED] × 100%.

The low and high values of the immediate strain ε_0 have produced almost the same variation of the identification results for the NOT USED and USED fabric. It is seen for each material direction, that the difference between the NOT USED and USED samples does not depend on the level of the immediate strain ε_0, that has been not the same for both types of the material. Analysis of each parameter separately has given the following observations:

- the change of the value E_1 is almost the same for the warp and fill direction;
- the changes of the values E_2 and η_1 are two times greater for the fill direction than for the warp one;
- the value of η_2 stays unchanged for each identification and direction of the material and its value is very low.

Table 4. Relative change between Burgers model parameters for the NOT USED and USED VALMEX fabric for the warp direction.

ε_0	Parameters			
NOT USED/USED	η_1	η_2	E_1	E_2
–	%			
0.055/0.0485	−11	−0.4	11	−5
0.065/0.0585	−11	−0.2	10	−4

Table 5. Relative change between Burgers model parameters for the NOT USED and USED VALMEX fabric for the fill direction.

ε_0	Parameters			
NOT USED/USED	η_1	η_2	E_1	E_2
–	%			
0.1325/0.115	−20	−0.8	−12	−12
0.1425/0.125	−20	−0.9	−12	−14

Summing up, the ageing process is more distinct when analyzing the fill direction of the fabric material. The environmental ageing has influenced mainly the parameters E_1, E_2, η_1 that are related to the linear (long-lasting) part of the creep test. The coefficient η_2 identified from the non-linear (initial) part of the creep curve can be neglected in ageing analysis of the VALMEX fabric. It proves that ageing has not affected the initial stage of creep, regarding the development of strains below t_1. Consequently, when using the polyester reinforced PVC coated membranes several times (like in the structure of the Forest Opera for 20 years), each time the material would adjust in the same way to the desired geometry, but with respect to the immediate elongation ε_0. Finally, for the long-term prediction of VALMEX performance the parameters E_1, E_2, η_1 are of the greatest importance.

6 CONCLUSIONS

The obtained results can be used for practical purposes. Using the Burgers model (with the high value of ε_0), it is possible to describe material behavior in the first period of inelastic deformation, that will be welcomed by engineers preparing the cut patterns of the fabric sheets. The Burgers model with parameters detected for the lower value of ε_0 can reflect better the long-lasting behavior under long-term constant loading. It would be of the primary importance for prediction of material performance built in a real construction and used for many years (e.g. the VALMEX fabric).

It has been observed, that the environmental ageing has great impact on the parameters E_1, E_2, η_1 of the Burgers model. No ageing influence on the coefficient η_2 has been recognized.

REFERENCES

Ambroziak, A. & Kłosowski, P. 2011. Review of constitutive models for technical woven fabrics in finite element analysis. *AATCC Review* 11(3): 58–67.

Chatree, H. & Thanate, R. & Wiriya, T. 2014. Time–temperature and stress dependent behaviors of composites made from recycled polypropylene and rubberwood flour. *Construction and. Builing Materials* 66: 98–104.

Kłosowski, P. 2011. Identyfikacja lepkosprężystego modelu Burgersa dla Poliwęglanu. Nowe kierunki rozwoju mechaniki. Hucisko.

Nowacki, W. 1963. *Teoria pełzania*. Warszawa: Arkady.

Siginer, D.A. 2014. *Stability of non-linear constitutive formulations for viscoelastic fluids*. New York: Springer.

Żerdzicki, K. 2015. *Durability evaluation of textile hanging roofs materials*. PhD thesis, Gdańsk, Orleans.

Advances in Mechanics: Theoretical, Computational and Interdisciplinary Issues – Kleiber et al. (Eds)
 ISBN 978-1-138-02906-4

Experimental and numerical analysis of wave propagation in ground anchors

B. Zima & M. Rucka
Faculty of Civil and Environmental Engineering, Gdańsk University of Technology, Gdańsk, Poland

ABSTRACT: The article focuses on wave propagation phenomenon in ground anchors. The main aim of the investigation is the non-destructive diagnostics and the assessment of the state of ground anchors, using the guided wave propagation method. Laboratory models of anchors with different lengths of the anchor body were tested and voltage signals of propagating waves were registered at several locations. For all tested specimens corresponding numerical models were created. Results of calculations were compared with experimental signals.

1 INTRODUCTION

Wave propagation testing has become an important subject in non-destructive evaluation (Rose 1999). The effectiveness of testing methods based on guided wave propagation has been proved in many previous pieces of research (Giurgiutiu 2007, Mallet et al. 2004). Wave propagation has a particular great potential in diagnostics of these structural elements, which state cannot be assessed on the basis of standard visual inspection, for example rock bolts or ground anchors.

Ground anchorages are geotechnical structures which provide effective support of stability of excavations by transferring tensile forces into surrounding ground layers. The ground anchor consists of a steel tendon, which can be made of a rod, a multi-wire strand or a cable and an anchor body, which is formed in the subsoil by injecting grouting mortar (Hobst & Zajic 1983). Despite very restrictive acceptance tests and high strength of reinforced concrete elements, the real problem of ground anchors is their durability. They are subjected to continuous deterioration of their state. The most common problem is creep of soil or steel, which results in stress release. One can try to avoid the danger by reducing the load supported by the anchorage (Habib 1983) or monitoring the level of stresses in the tendon. Another real threat is corrosion of the tendon because of a constant contact with ground water. Bad ground conditions or poor workmanship can also lead to deterioration of bond between the anchor body and the tendon. In contrast to the problem of stress release, the process of corrosion or bond deterioration cannot be easily monitored and controlled. Acceptance tests allow the evaluation the object directly after its realization and they are not carried out regularly. Hence, there is a need to develop diagnostic methods dedicated to constant monitoring of such type of structures. Non-destructive diagnostic techniques based on elastic wave propagation allowing the assessment of the state of elements, objects or parts of structures, which are not easily available have been recently the subject of growing interests (Cui & Zou 2012, Zou et al. 2010).

The study deals with the numerical and experimental research of guided wave propagation in laboratory models of ground anchors. The anchors with different lengths of the anchor body have been tested. The research focuses on the analysis of waveforms registered by piezoelectric transducers attached in several locations of anchors.

2 EXPERIMENTAL INVESTIGATION

The geometry of the examined model of the anchor is given in Figure 1. The specimens consist of a steel rod embedded in a concrete anchor body. The rod of a length of 1.5 m had a circular cross section with

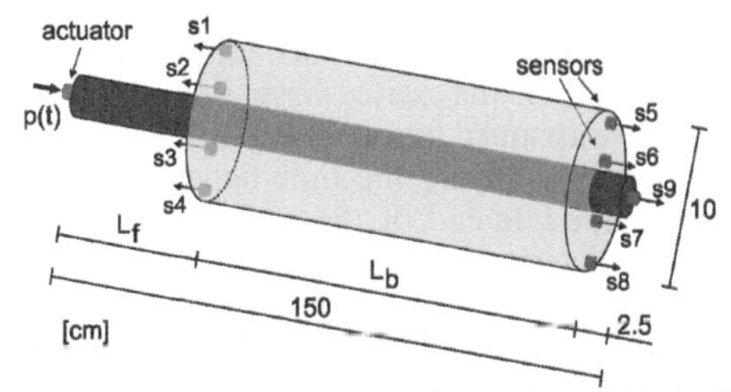

Figure 1. Geometry of ground anchor.

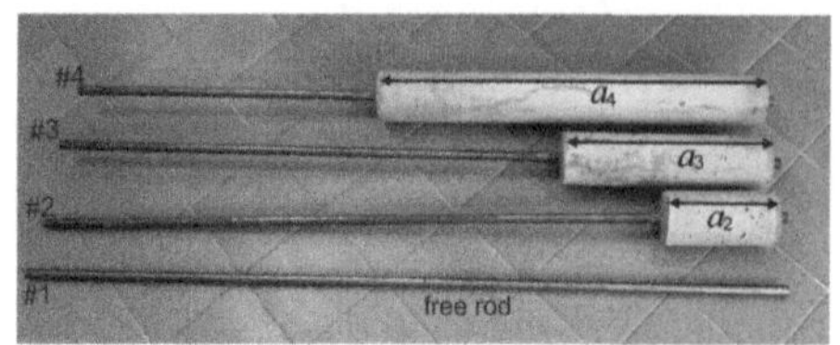

Figure 2. Photograph of experimentally tested specimens.

a diameter equal to 2 cm. Experimental investigations were conducted for different bonding lengths: $a_1 = 0$ cm, $a_2 = 20$ cm, $a_3 = 40$ cm and $a_4 = 80$ cm. Material parameters of steel and concrete were equal to $E = 210$ GPa, $\rho = 7830$ kg/m^3, $\nu = 0.3$ and $E = 26$ GPa, $\rho = 2084$ kg/m^3, $\nu = 0.2$, respectively. The thickness of the concrete cover was equal to 4 cm. In Figure 2 the photograph of experimentally tested models of anchors and free rod is presented.

Excitation and measurements of elastic waves were carried out by the device PAQ-16000D and piezoelectric transducers Noliac NAC2011. The actuator was located at the free end of the rod and receiving traducers were attached at different locations of the anchor. One of the sensors was located and the end of the anchor and the rest at the anchor body. The scheme of the arrangement of sensors and directions of measured signals are presented in Figure 1. Configuration of sensors was the same for all cases of anchors. In case of a free rod, only signals at the opposite end of the rod were registered by one sensor only. The excitation was obtained by multiplication of the sine wave with a frequency of 80 kHz and the Hanning window.

3 NUMERICAL CALCULATIONS

Numerical investigation of elastic wave propagation in models of ground anchors were conducted by the finite elements method code Abaqus/Explicit. The FEM models were performed with the assumption of the rotational symmetry.

Anchors were discretized using 4-node axisymmetric elements with dimensions 1 mm × 1 mm. The size of the finite elements was adopted according to the length of the excited wave. Boundary conditions were assumed free at all edges. Excitation was imposed by applying the time depending surface load on the free end of the anchor. In numerical models deterioration conditions of adhesion bonding between concrete and steel rod were not taken into account. The Rayleigh damping model was applied and the mass-proportional coefficients for steel and concrete were adopted on the basis of calibration with experimental models.

4 RESULTS

Comparison of experimental and numerical signals registered by sensor s#2 and the end of the anchor for all of test specimens is given in Figure 3. Numerical signals are plotted as envelopes created on the base of spectrum of Hilbert transform (Potamianos & Maragos 1994). The incident wave packet which is clearly visible in each case was identified and indicated. It can be observed that the value of the amplitude of this wave packet decrease with increasing length of the anchor body. This phenomenon, known as wave leakage, is caused by the transfer of wave energy into surrounding medium.

The intensity of the wave leakage depends on the relationship between values of the acoustic impedance of bordering media. The impact of the leakage on signals characteristics has been demonstrated in previous pieces of research (Rucka & Zima 2015, Zima & Rucka 2015). Another factor that influences the intensity of the wave energy transfer is the quality of the connection between steel and concrete. Good quality of the connection can lead to more intensive energy dissipation, while poor quality can cause complete reflection from the border of the media. For this reason amplitude

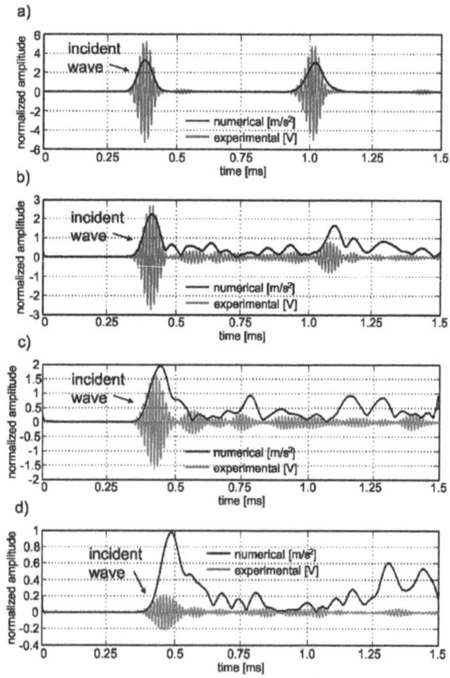

Figure 3. Experimental and numerical time signals registered by sensor s#2 for anchors with different bonding lengths: a) 0 cm; b) 20 cm c) 40 cm, d) 80 cm.

changes registered in signals can indicate well the quality of the cooperation between steel and surrounding composite material like concrete.

Comparisons of experimental and numerical signals registered by sensors s#3 and s#5 are given in Figures 4 and 5. Shading of the signal starts from the first registered wave packet. Registration time of the first peak value calculated numerically fits with experimental results well. In Figure 4 signals registered by sensor s#3 located at the anchor body are plotted. It is clearly visible that the increasing bonding length and decreasing distance between the sensor and the actuator, make the time-of-flight of the incident wave packet decrease. In Figure 5 signals registered by sensor s#5 located at the end of the anchor body are presented. One can notice, that distances between the actuator and the sensors located at the end of anchor body (s#4 and s#5) do not depend on the rod fixed length. For the same distances and for different bonding lengths, the time-of-flight of the incident wave changes. The longer concrete part, the later incident wave reaches the sensor. One can conclude that the presence of concrete cover affects the disturbance propagation velocity, even though wave was excited only in steel rod.

In all presented cases the registration time of the first peak is comparable for laboratory and numerical tests but the value of the amplitudes

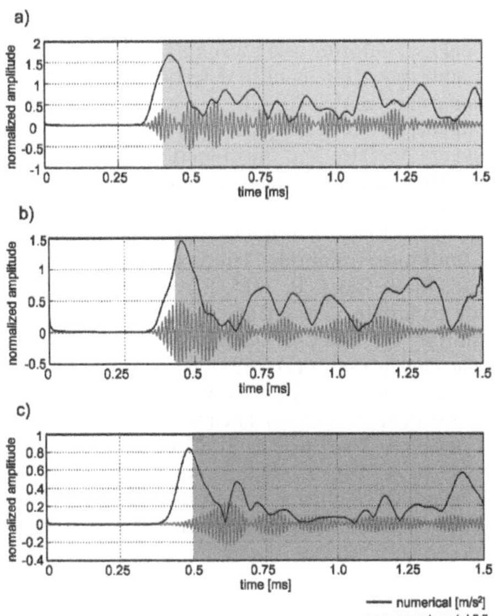

Figure 4. Experimental and numerical time signals registered by sensor s#3 located at the anchor body for anchors with different bonding lengths: a) 20 cm; b) 40 cm c) 80 cm.

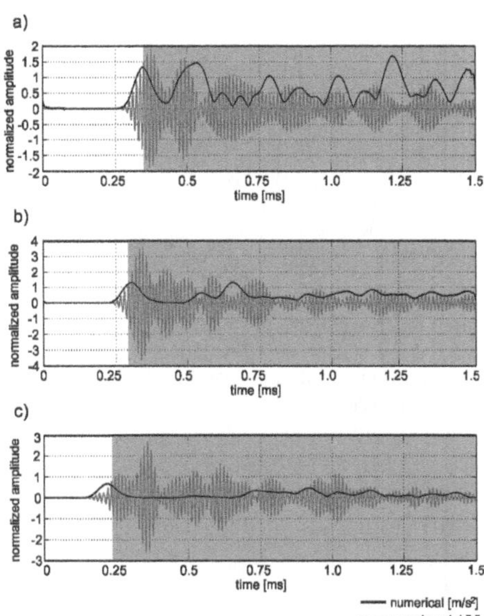

Figure 5. Experimental and numerical time signals registered by the sensor s#5 located at the anchor body for anchors with different bonding lengths: a) 20 cm; b) 40 cm c) 80 cm.

of the remaining part of the signal varies considerably. The difference between the results is the more apparent the greater bonding length (see Fig. 4). It is caused by the fact that performed numerical models do not reflect accurately the complexity of experimental model. Complex microstructure of concrete, deterioration of the contact conditions between reinforcement and concrete as well as cracks were not taken into account in calculations. All of these factors result in a multitude of reflections and affect the shape of signals, what cannot be obtained accurately in numerical signals.

The envelopes of experimental signals registered by sensors s#4 (Fig. 6) and s#6 (Fig. 7) which were closely attached to the outer edge of the anchor body are presented. The incident waves were identified and indicated. It can be observed in Figure 6 that the intensity of the energy dissipation is related with the length of propagation path. Even though, concrete part has a length equal to 20 cm, the greatest decrease of the amplitude was observed for this case, because of the longest distance between sensor s#4 and actuator.

In Figure 7 the inverse relation can be noticed: the shorter concrete part, the higher amplitude. In this case the wave energy leakage has greater influence on the amplitude value rather than the energy dissipation along the propagation path, which has the same length for all anchors.

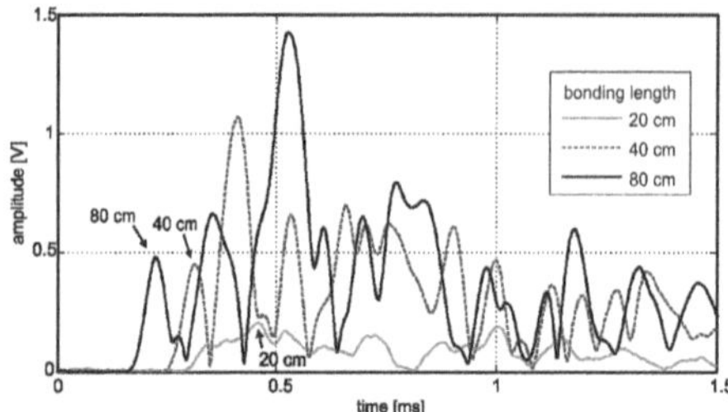

Figure 6. Envelopes of experimental time signals registered by sensor s#4 located at the anchor body for anchors with different bonding lengths: a) 20 cm; b) 40 cm c) 80 cm.

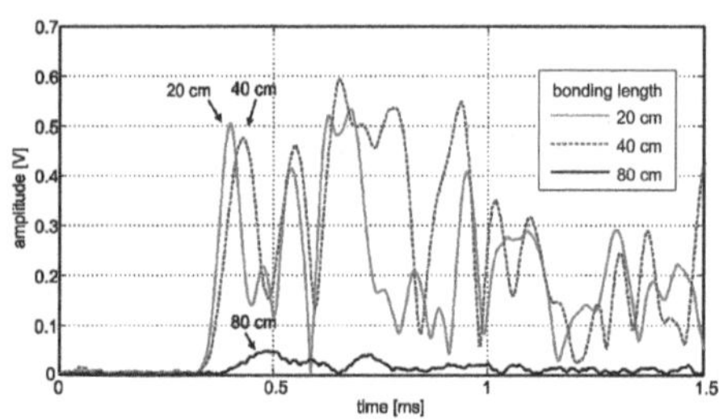

Figure 7. Envelopes of experimental time signals registered by sensor s#6 located at the anchor body for anchors with different bonding lengths: a) 20 cm; b) 40 cm c) 80 cm.

5 CONCLUSIONS

The paper presents results of experimental and numerical investigations on guided wave propagation in laboratory models of ground anchors. Investigations were conducted for steel rods embedded in a concrete body with different bonding length. Results of presented investigations indicate large capabilities of the wave propagation method in the assessment of bonding lengths. It is of great importance especially in the case of objects, which state cannot be checked during standard visual inspection. In all of presented cases, the relationship between signals characteristics and bonding lengths was visible. Even for relatively small laboratory models of anchors, energy dissipation of propagating wave was enough intensive to significantly complicate information extraction from signals.

Despite from the fact that interpretation of a single signal gives an opportunity to determine basic geometric parameters of tested element, it does not provide information about its state. Deterioration of bonding connection or the emergence of damage of tendon caused for example by corrosion can be detected only on the base of comparison of a set of signals registered at different times. For this reason the application of wave propagation methods in the context of diagnostics should assume the use of constant monitoring systems.

ACKNOWLEDGMENTS

Calculations were carried out at the Academic Computer Center in Gdańsk.

REFERENCES

Cui, Y. & Zou, D.H. 2012. Assessing the effects of insufficient rebar and missing grout in grouted rock bolt using guided ultrasonic waves. *Journal of Applied Geophysics* 79: 64–70.

Giurgiutiu, V. 2007. *Structural Health Monitoring with Piezoelectric Wafer Active Sensors*. Oxford: Academic Press.

Habib, P. 1983. *An Outline of Soil and Rock Mechanics*. Cambridge: Cambridge University Press.

Hobst, L. & Zajic, J. 1983. *Anchoring in Rock and Soil*. Amsterdam: Elsevier Science Publishing Company.

Mallet, L., Lee, B. & Staszewski, W.J. 2004. Structural health monitoring using scanning laser vibrometry: II. Lamb waves for damage detection. *Smart Materials and Structures* 13: 261–269.

Potamianos, A. & Maragos, P. 1994. A comparison of the energy operator and the Hilbert transform approach to signal and speech demodulation. *Signal Processing* 37: 95–120.

Rose, J.L. 1999. *Ultrasonic waves in Solid Media*. Cambridge: Cambridge University Press.

Rucka, M. & Zima, B. 2015. Elastic wave propagation for condition assessment of steel bar embedded in mortar. International *Journal of Applied Mechanics and Engineering* 20(1): 159–170.

Zima, B. & Rucka, M. 2015. Wave propagation in damage assessment of ground anchors. *Journal of Physics: Conference Series* 628(1): 012026.

Zou, D.H., Cheng, J., Yue, R. & Sun, X. 2010. Grout quality and its impact on guided waves in grouted rock bolts. *Journal of Applied Geophysics* 72: 102–106.

Advances in Mechanics: Theoretical, Computational and Interdisciplinary Issues – Kleiber et al. (Eds)
© 2016 Taylor & Francis Group, London, ISBN 978-1-138-02906-4

Effectiveness of damage detection in 3-D structures using discrete wavelet transformation

K. Ziopaja
Faculty of Civil Engineering, Poznan University of Technology, Poznań, Poland

ABSTRACT: The method of damage identification using 1-D discrete wavelet transformation is proposed in the paper. The 3-D structures subjected to static action are considered. The damage identification is based on multiresolution analysis and structural response registered in a single measurement point. The experiments are numerically simulated. Measurement errors are introduced by white noise simulation.

1 INTRODUCTION

The problem of structural damage identification is one of the most important engineering issues. From a wide range of identification methods, the non-destructive techniques are the most interesting and promising. In identification methods, the structural response is analysed. The non-destructive methods, based, among others, on optimization algorithms (Mróz & Garstecki 2005), on the dynamic response of the structure (Dems & Mróz 2001), artificial neural networks (Waszczyszyn & Ziemiański 2001) or thermal conductivity (Ziopaja et al. 2011). In identification methods, the structural response is analysed. One of powerful tool of signal analysis is wavelet transformation. It allows multiresolution analysis of 1-D or 2-D signals and due to its features is useful in detection of local disturbances of signals. Application of wavelet transformation to problem of damage identification was presented in (Gentile & Messina 2003, Knitter-Piątkowska et al. 2014, Rucka & Wilde 2006, Wang & Deng 1999).

The focus of this work is connected with the issue of the assessment of the technical conditions of bridges. Bridge constructions are of the three-dimensional complex beam and shell structures. Bridges are exposed to heavy variable loads of high amplitudes. The issue of damage detection in bridges is crucial due to high costs of abovementioned constructions. Some examples of damage identification are presented, amongst others, in (Czarnecki & Nowak 2008, Dutta & Talukdar 2004, Yeung & Smith 2005).

2 WAVELET TRANSFORMATION AS A TOOL IN SIGNAL ANALYSIS

Mathematical basis of wavelet analysis were created in 1980s. Newland's, Chui's, Daubechies's or Mallat's publications and monographs introduced theoretical basis and provided a broad range of applications connected with wavelet analysis. In 1980s and during the next decades wavelet transformation became an extremely effective and advanced tool of the signal analysis of a wide range of problems.

The unquestionable and fundamental advantage of wavelet transformation is the fact that the analyzing wavelet is characterized by a good localization in time and frequency (basic sine and cosine trigonometric functions applied by Fourier's analysis do not possess this quality). This key feature of wavelets is a great asset as for practical uses such as local singularities detection in signals, which is applied, amongst others, in the localization and detection of damage in engineering structures and defects in materials. In this article the Discrete Wavelet Transform (DWT) is applied.

2.1 *Discrete wavelet transform and multiresolution analysis*

Wavelet coefficients $d_{j,k}$ of the discrete wavelet transform W_ψ of a signal $f(t)$ are defined as the scalar product of the signal and the wavelet function

$$W_\psi f\left(\frac{1}{2^j}, \frac{k}{2^j}\right) = \left\langle f(t), \psi_{j,k}\right\rangle = d_{j,k}. \tag{1}$$

The basis function ψ (mother wavelet) creates the family of wavelets:

$$\psi_{j,k}(t) = 2^{\frac{j}{2}} \psi(2^j t - k), \tag{2}$$

where the scale parameter is defined as $1/2^j$, and translation parameter as $k/2^j$ for $j, k \in C$.

The advantage of the discrete wavelet transform is the possibility of the representation of the signal $f(t)$ as the sum of the smooth and detailed representation

$$f(t)=\sum_k a_j(k)\,2^{\frac{j}{2}}\,\varphi(2^j t-k)$$
$$+\sum_k d_j(k)\,2^{\frac{j}{2}}\,\psi(2^j t-k), \qquad (3)$$

where $\varphi(t)$ is called scaling function (father) describe by an equation similar to (2).

2.2 *One dimensional decomposition*

The efficient and simple Mallat's algorithm of one dimensional decomposition of the signal $f_J(t)$ in this article, is used:

$$f_J(t)=S_J+D_J+D_{J-1}+\cdots+D_m+\cdots+D_2+D_1, \quad (4)$$

where S_J means a smooth, constant part of the signal $f_J(t)$, and the symbols $D_{1,\ldots,J-1}$ reflect the details consisting the local information about the signal. Parameter $m = J - j$ indicates the level of the detail D_m. In order to identify the damage, we will use only the decomposition details. The most detailed, so i.e. carrying a lot of detailed and local information is, usually, the D_1 detail. The function f_J, must be approximated by $N = 2^J$ discrete values.

2.3 *Types of wavelet*

In the paper, orthogonal and biorthogonal wavelets family are applied:

- Daubechies (6) and (8)—orthogonal and far from symmetry wavelet
- Coiflet (18)—orthogonal and near from symmetry wavelet
- Bispline (3,1)—biorthogonal symmetrical wavelet.

The choice of wavelets with different properties will examine their usefulness in damage identification process.

3 PROBLEM FORMULATION

The main objective of this study is to present the ability to detect and locate damage in the numerical models of three-dimensional structures.

3.1 *Damage identification procedure*

Assuming that there is damage in the construction (its definition was given in 3.2.); the methodology of the identification procedure is as follows (see also Fig. 1):

- concentrated load is moving on the deck of the bridge (quasi static action)
- in selected point of the main girder, or other structural element of the supporting structure, the geometric parameters (deflection, angle of rotation) were monitored
- structural response is recorded in relationship with moving load (so we get an influence function of vertical displacement or angle of rotation)
- size of the signal depends on the load-shifting
- signal is analyzed using one-dimensional DWT
- DWT detects and localizes the damage.

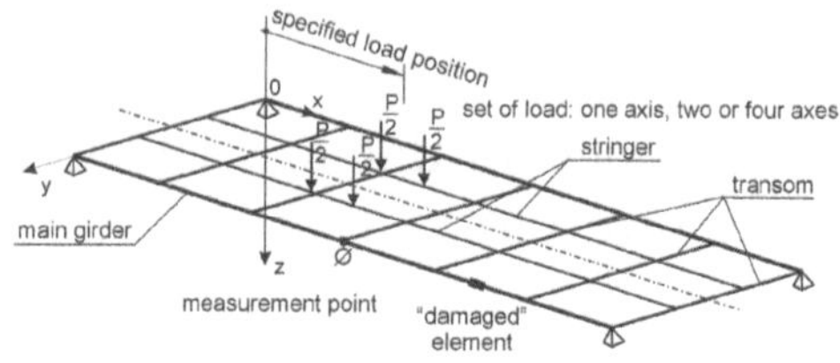

Figure 1. Model of a bridge structure.

3.2 *Definition and model of the damage*

Defects in structures (e.g. as a fatigue crack, corrosion) lead to stiffness reduction, increase of damping and decrease of natural frequencies. In the numerical examples a simple model of damage is used, most commonly frequent in the technical literature, which has the form of local stiffness reduction. Its formula $EJ_d = \alpha \cdot EJ$, where α is intensity of stiffness reduction dependable on the cross-section weakness and the value of moment of inertia J_d. It has been assumed that EJ_d includes two elements, each of 4.0 cm in length. In comparison to the construction size is seems very small. The model of damage used in analysis maps in a simple way the material discontinuity caused by crack in cover plate, crack in the belt plate girder and corrosion.

4 NUMERICAL ANALYSIS

Single-track railway bridge has been selected to the numerical analysis. Its a beam construction consisting of two main girders. The model of loading with the principles defined in PN-EN 1991-2, as for railway bridges, has been applied, with some simplification. The numerical model of bridge construction and all calculations have been done on the basis of the engineering RSTAB computer program.

4.1 *Numerical example*

A common two-girder railway bridge has been used in order to examine the damage detection effectiveness (Fig. 2). Rails coupling is located on the ties which are directly attached to the main girders. It is crucial due to the fact that the detection and localization of damage is directly connected with the main girders. The structure was modeled with the use of one-dimensional beam elements. The object and its model specification are presented in Table 1.

The discretization of the main girder involved dividing it into sixty-two elements of 0.35 m length and two elements of 0.175 m length which corresponds to 64 nodes. Numerical model and the location of damages is shown in Figure 3.

Damage "d1" corresponds to a significant loss material (discontinuity) in cover plate of the main girder belt plate 8.0 cm length and some loss in the belt plate, too. In the place of damage "d2" the thickness of the belt plate girder was reduced. Finally, in the position of damage "d3" a significant reduction of cross-section element was introduced or the linear elastic joint was applied.

4.1.1 *The influence of stiffness reduction*

The possibility of detection and localization of the damage depends mainly on its influence on the signal which is subjected to wavelet transform. Following the identification procedure (see 3.1) the influence function were marked. These functions include: the vertical displacement and an angle of rotation function for the chosen nodes (Fig. 3). The load, shifted across main girders, consisted of a single axis of 250 kN. The results appeared to be very interesting. Figure 4 shows the DWT graph of the detail D1, which allow to detect the "d1"

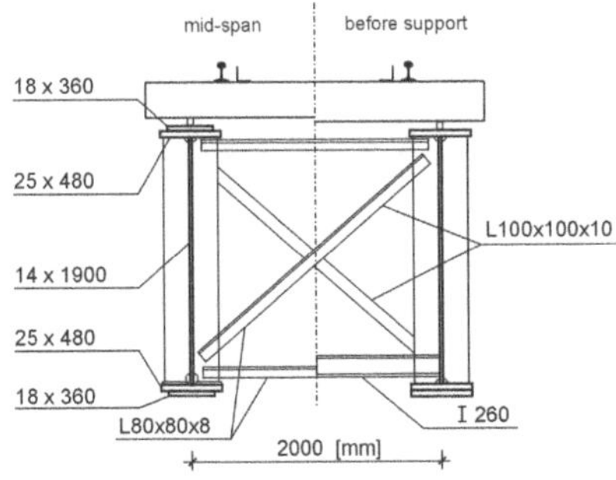

Figure 2. Bridge cross-section.

Table 1. Basic data, physical parameters and section properties of the bridge.

Span	20.05 m	Girder spacing	2.0 m
Young's Modulus	210 GPa	Unit mass	1300 kg/m
Region 1 (I-section with cover plate)			
Moment of inertia	J	4 309 310.0	cm^4
Cross-section	A	639.0	cm^2
Region 2 (I-section without cover plate)			
Moment of inertia	J	3 054 420,0	cm^4
Cross-section	A	509,4	cm^2

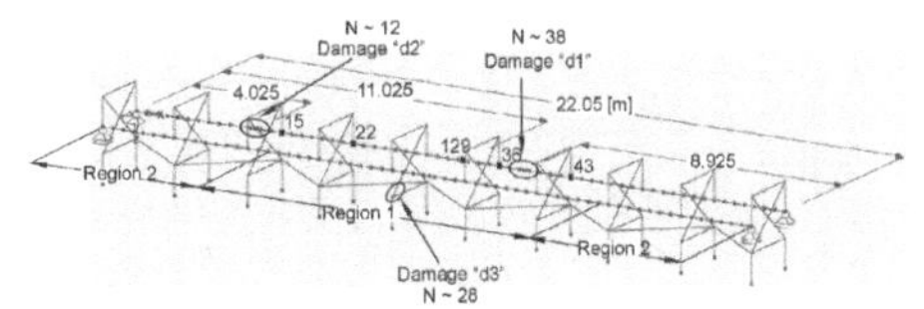

Figure 3. Numerical model of the bridge—damage zone, measurement points and changes in cross-section localization.

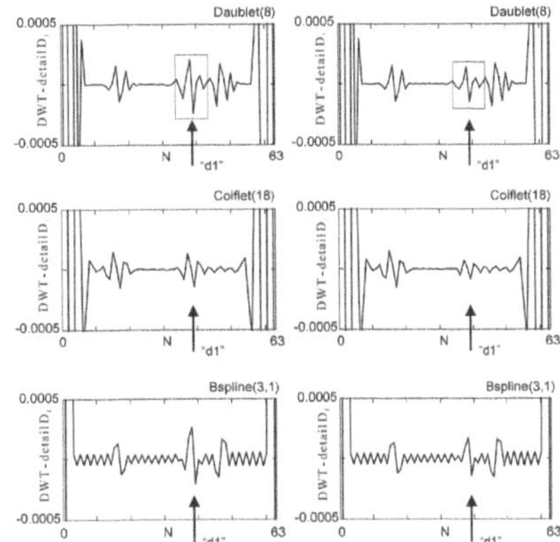

Figure 4. 1-D DWT (detail D_1) of the influence function of vertical displacement recorded in node 36 for $\alpha = 0.52$ (left) and $\alpha = 0.63$ (right), one axis of load $1P = 2 \times 125$ kN, three different types of wavelets.

damage in accordance with its spatial location in the model. Apart from the detection of changes in stiffness connected with the damage, we can also observed peaks on DWT graph located in the ends of cover plate of the main girder belt plate. The results representing the detection of damage "d2" (not shown) are promising too.

4.1.2 *Measurement point position*

Location of measurement is one of very important issues. The analyzes, especially those concerning the transform of the function of the rotation angle and also the influence function of the rotation angle function, indicate correctly only the location of measurement. Figure 5 shows how the peaks placed in position N~32 and N~43 exactly determine the location of the measurement, losing simultaneously information about the "d1"damage. It is important that the location of the measurement point on the nearby girder does not allow detect any damage zone. This follows from the nature of the bridge structure. The choice of the measurement point is absolutely vital. In the process of the damage identification its location should be determined in a lot of characteristic places such as the half of the span length, in L/4 and in L/8 of the span or in each connection between the main girder with a cross beam.

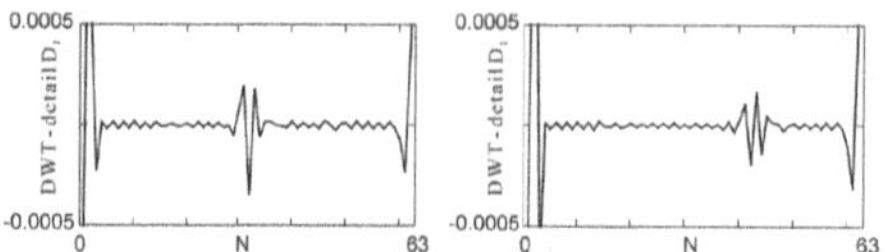

Figure 5. 1-D DWT (detail D_1) of and the influence function of the rotation angle function recorded in node 129 (left) and 43 (right) for $\alpha = 0.52$, one axis of load $1P = 2 \times 125$ kN, wavelet: Daubechies (6).

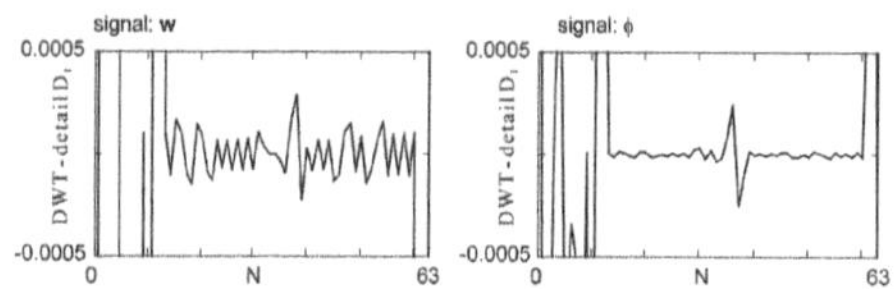

Figure 6. 1-D DWT (detail D_1) of the influence function of vertical displacement (left) and the influence function of the rotation angle function (right) recorded in node 129 for $\alpha = 0.52$, two load axes $2P = 4 \times 125$ kN, wavelet: Bspline (3,1).

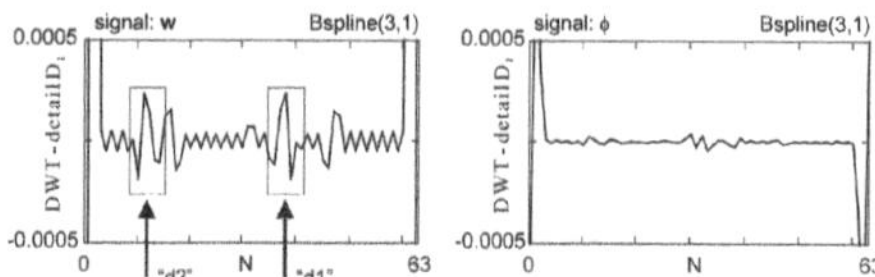

Figure 7. 1-D DWT (detail D_1) of the influence function of vertical displacement (left) and the influence function of the rotation angle function (right) recorded in node 129 for $\alpha = 0.52$ ("d1") and $\alpha = 0.41$ ("d2"), one axis of load $1P = 2 \times 125$ kN, wavelet: Bspline (3,1).

4.1.3 *Number of load axes*

Application of the load as a set of concentrated forces on two or four axes sought to better representation of the actual load. As shown in Figure 6 detection of damage "d1" is possible and easy for both response signals, but very big boundary disturbances appear on the left side. This can seriously hinder or even prevent the detection of defects in this area.

4.1.4 *Identification of more damages*

One of the advantages of the wavelet transformation is the fact that it can easily identify not only one, but also more defects.

4.1.5 *The addition of white noise*

Now, let us study the effectiveness of damage detection in real life situations, when inevitable errors in measurements appear. Several examples, similar to the case illustrated in Figure 4, but with variable intensity of noise were solved. These examples demonstrated that DWT detected the damage "d1" when the noise level was not greater than $\pm 1.24 \cdot 10^{-4}$ mm, when the maximum vertical displacement was 3.22482 mm. This means high sensitivity of the proposed method.

5 CONCLUSIONS

Numerical application of discrete wavelet transformation to the damage detection proved that the method is very efficient. The main advantage of presented approach of damage detection is exertion of the response signal of "damaged" structure and measurement realized in one selected point. The new approach presents the three-dimensional character of the structure work and an indirect loads action on the main analyzed structural elements.

On the other hand numerical analyses show that damage detection in real structures can be difficult due to: lack of possibility of direct loading of structural element, use of vehicles with several axes and, finally, the accuracy of the measurement method.

REFERENCES

Czarnecki, A.A. & Nowak, A.S. 2008. Time-variant reliability profiles for steel girder bridges. *Structural Safety* 30: 49–64.

Dems, K. & Mróz, Z. 2001. Identification of damage in beam and plate structures using parameter dependent frequency changes. *Engineering Computation* 18(1/2): 96–120.

Dutta, A. & Talukdar, S. 2004. Damage detection in bridges using accurate modal parameters. *Finite Elements in Analysis and Design* 40: 287–304.

Gentile, A. & Messina, A. 2003. On the continuous wavelet transform applied to discrete vibrational data for detecting open cracks in damaged beams. *Solids & Structures* 40.

Knitter-Piątkowska, A. & Guminiak, M. & Przychodzki, M. 2014. Damage detection in truss structures being the part of historic railway viaduct using wavelet transformation. In Łodygowski, T. & Rakowski, J. & Litewka, P. (ed.), *Recent Advances in Computational Mechanics*: 157–163. CRP Press/Balkema, Taylor and Francis Group.

Mróz, Z. & Garstecki, A. 2005. Optimal loading conditions in design and identification of structures. Part 1: Discrete formulation. *Structural and Multidisciplinary Optimization* 29: 11–18.

Rucka, M. & Wilde, K. 2006. Application of continuous wavelet transform in vibration based damage detection method for beams and plates. *Journal of Sound and Vibration* 297: 536–550.

Wang, Q. & Deng, X. 1999. Damage detection with spatial wavelets. *Journal of Solid and Structures* 36: 3443–3468.

Waszczyszyn, Z. & Ziemiański, L. 2001. Neural networks in mechanics of structures and materials—new results and prospects of application. *Computers and Structures* 79(22–25): 2261–2276.

Yeung, W.T. & Smith, J.W. 2005. Damage detection in bridges using neural networks for pattern recognition of vibration signatures. *Engineering Structures* 27: 685–698.

Ziopaja, K. & Pozorski, Z. & Garstecki, A. 2011. Damage detection using thermal experiments and wavelet transformation. *Inverse Problems in Science and Engineering* 19(1): 127–153.

Advances in Mechanics: Theoretical, Computational and Interdisciplinary Issues – Kleiber et al. (Eds)
© 2016 Taylor & Francis Group, London, ISBN 978-1-138-02906-4

Development of friction constitutive relations for polymers

A. Zmitrowicz
Institute of Fluid-Flow Machinery, Polish Academy of Sciences, Gdańsk, Poland

ABSTRACT: Variations of friction with respect to a sliding direction and a sliding path are observed at boundaries of solid polymers. In the contribution the author proposes several friction relations taking into account friction anisotropy, heterogeneity and fluctuating surface microstructures. Adhesion and hysteresis effects are included in friction models of elastomers. In melted polymers and polymer solutions the micro-elements composing the material move with a highly anisotropic friction. Continuum based models and microscopic-level models are used for the friction description inside the polymeric materials. Specific models of polymeric macromolecules (Rouse model, de Gennes model, rods and discs) and different modes of their kinematics (sliding, rolling and spinning) are taken into account.

1 INTRODUCTION

Friction models of polymers must be considered in terms of different scales and different friction processes at external boundaries of solids and inside the materials. In the case of polymeric materials, friction phenomena inside the materials are equally important as well as at external boundaries of solids. For example, an intrinsic structure of certain polymeric solids affects friction. Furthermore, a microstructure evolution at a solid boundary can induce variations of friction (Liu et al. 2007, Dunn et al. 2008), see Figure 1. Friction with anisotropy and heterogeneity effects at external boundaries of solids takes place in polymeric composites, semi-crystalline polymers, reinforced elastomers, self-lubricating polymers. Friction with anisotropy effects inside the polymeric materials occurs in polymeric solutions, melts and liquid crystalline polymers.

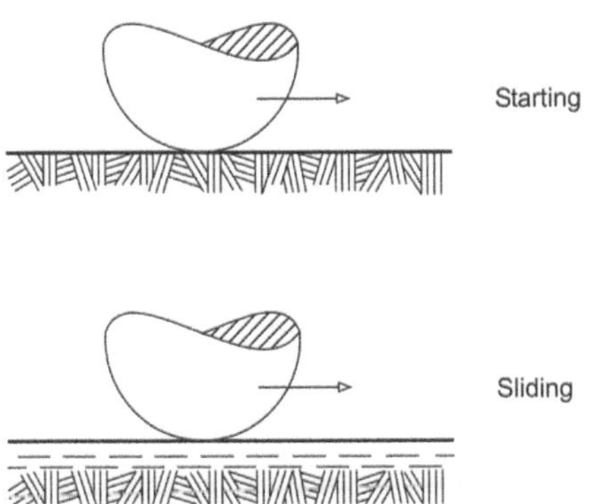

Figure 1. Sliding induced re-orientation of surface molecules in certain polymers.

Specific constitutive relations for friction of polymers are required with respect to their complex microstructure and evolving mechanical properties. The purpose of this contribution is to formulate several models of dry and viscous friction at external boundaries of solid polymers and inside the polymeric materials and to discuss their properties. Richard von Mises (1901–1908) collected main properties of dry and viscous friction models as follows:

1. Dry friction: (a) proportional to normal pressure, (b) independent on sliding velocity, (c) dependent on sliding surface roughness, (d) finite friction at zero sliding velocity.
2. Viscous friction: (a) independent on pressure, (b) proportional to velocity of slip, (c) independent on surface roughness, (d) zero friction at zero slip velocity.

2 EQUATIONS OF FRICTION AT EXTERNAL BOUNDARIES OF SOLID POLYMERS

There are various reasons of anisotropy of mechanical and frictional properties of polymeric solids. In composites consisting of components of different materials, external surfaces have mosaic structures, since different components are exposed at different points on the surface. Some polymers undergo morphological transformations, and they can transform from an amorphous phase into a semi-crystalline phase with regular intrinsic structure. In self-assembled and self-lubricating polymers, the sliding initiates microstructural changes in the surface and near-surface material (Liu et al. 2007, Dunn et al. 2008), see Figure 1. An oriented surface roughness is produced at solid external surfaces in the process of materials machining. Friction depends on the orientation of the surface roughness.

Models of friction at external boundaries of solid polymers describe friction of polymers one rubbing against another or against surfaces of different materials. In most cases, the model of dry friction or its contemporary modifications are used. The author have formulated constitutive equations for anisotropic and non-homogeneous friction force, i.e. for variations of friction with respect to a sliding direction and to a sliding path. Linear and nonlinear, centrosymmetric and non-centrosymmetric models have been developed, as well as models including a sliding path curvature (Zmitrowicz 2010). They are useful in the case of solid polymers since they include the influence of the microstructure on friction.

The dry friction linear equation with respect to the unit vector $\mathbf{v}$ of the sliding velocity is given by:

$$\mathbf{p}_t = -|p_n|\mathbf{C}_1\mathbf{v}, \quad \mathbf{v} = \frac{d\mathbf{u}_t}{dt}\left|\frac{d\mathbf{u}_t}{dt}\right|^{-1}, \quad \mathrm{C}_1 : \mathrm{R}^2 \to \mathrm{R}^2 \tag{1}$$

where $\mathbf{p}_t$ = friction force vector; p_n = normal pressure; $\mathbf{u}_t$ = tangential displacement vector; $\mathbf{C}_1$ = dry friction second-order tensor with simple cases of friction anisotropy included; R^2 = two-dimensional space of real numbers, see Figure 2.

The non-linear equation with respect to the unit vector $\mathbf{v}$ given as a polynomial has the following form:

$$\mathbf{p}_t = -|p_n|\,[\mathbf{C}_1\,\mathbf{v} + \mathbf{C}_2\,(\mathbf{v}\otimes\mathbf{v}\otimes\mathbf{v}) + \ldots] \tag{2}$$

where $\mathbf{C}_2$ = fourth-order friction tensor with complex cases of friction anisotropy included.

Dependence of the friction tensors on an oriented angle is described as follows:

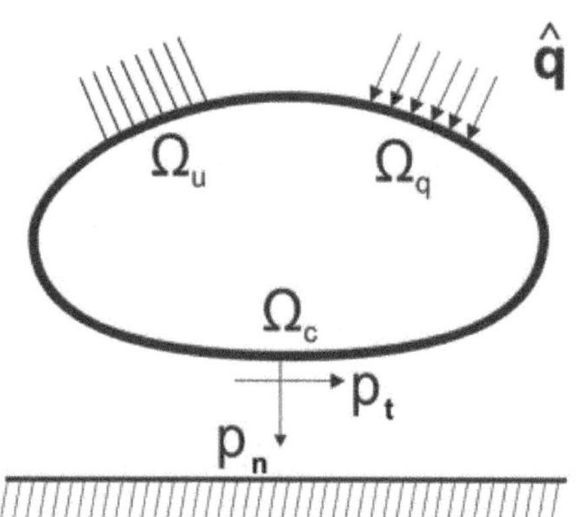

Figure 2. Boundary conditions of the contacting solid body.

$$\mathbf{C}_i = \mathbf{C}_i\,(\alpha_v), \quad \alpha_v \in \langle 0, 2\pi \rangle \quad i = 1, 2 \ldots \tag{3}$$

where α_v = measure of the oriented angle between the sliding direction $\mathbf{v}$ and a reference direction (e.g. Ox-axis), due to this non-centrosymmetric friction (i.e. asymmetry of friction) is included.

The friction equation can be defined as the sum of two single-term polynomials with respect to the unit vectors $\mathbf{v}$ and $\mathbf{n}$:

$$\mathbf{p}_t = -|p_n|\left[\mathbf{C}_1\mathbf{v} + \mathbf{E}_1\,\mathbf{n}\frac{1}{\rho}\right], \quad \mathbf{v}\cdot\mathbf{n} = 0, \quad |\mathbf{n}| = 1 \tag{4}$$

where $\mathbf{n}$ = unit vector normal to the sliding trajectory; ρ = curvature radius of the sliding trajectory; $\mathbf{E}_1$ = second-order tensor with friction heterogeneity effects included (Dunn et al. 2008).

The friction constitutive relations are restricted by thermodynamics (Zmitrowicz 2010, 2012): (a) variables of friction equations must transform as vectors, (b) the friction force power P_1 is non-positive in every case i.e.:

$$\mathrm{P}_1 = \mathbf{p}_t \cdot \frac{d\mathbf{u}_t}{dt} \le 0 \quad \forall \frac{d\mathbf{u}_t}{dt} \tag{5}$$

3 TWO-COMPONENT EQUATIONS OF FRICTION FOR ELASOMERS

Usually friction laws with respect to applications in elastomers contain two main components: the first deals with processes at the external boundaries of the material, the second includes an influence of deformations inside the material (Persson 1998). Due to this, at the external boundary of rubber the energy dissipation and friction are influenced by: (a) contact conditions (adhesion), (b) bulk properties (hysteresis effects). If two processes operate independently, the friction force of rubber can be expressed as the sum of two terms:

$$\mathbf{p}_t = \mathbf{p}_{adh} + \mathbf{p}_{hyst} \tag{6}$$

where $\mathbf{p}_{adh}$ = friction force due to adhesion; $\mathbf{p}_{hyst}$ = friction force due to hysteresis.

Adhesion occurs when a normal tensile force must be done to separate surfaces from contact, and it is defined by Frémond law:

$$p_{adh} = \beta|p_n|, \quad p_{adh} \ge 0, \quad \beta \in \langle 0, 1 \rangle \tag{7}$$

where β = adhesive intensity; p_n = initially applied compressive force (the normal pressure). The adhesion law is valid in normal and tangential directions.

Solid polymers (and melted polymers) may have simultaneously viscous properties typical in fluids and elastic properties as in solids (viscoelasticity). Time is the most important factor in models of polymers (time-derivatives appear). In elastomers, the energy dissipation accompanies every time-varying bulk deformation (e.g. under cyclic normal and tangential loadings at the external boundary). The hysteresis loop (i.e. stress versus strain or loading versus displacement) defines a dissipated work inside the material. We postulate that the dependence between the friction force at the external boundary and the hysteresis can be estimated by a dependence between the friction force and time. We describe the friction due to hysteresis with the aid of the relaxation (or creep) behaviour of friction as follows:

$$p_t(t) = |p_n| \left\{ \mu_0 + (\mu_\infty - \mu_0) \int_{\tau=0}^{\tau=t} \left(1 - e^{\frac{t-\tau}{\lambda}} \right) d\tau \right\} \tag{8}$$

where $p_t(t)$ = friction force with exponential time dependence; t = time of contact; λ = time of relaxation (or creep), this is the empirical factor; μ_0 = friction coefficient as $t \to 0$; μ_∞ = friction coefficient as $t \to \infty$.

4 CONTINUUM-BASED EQUATIONS OF FRICTION INSIDE POLYMERIC MATERIALS

Polymers are modelled as assemblies of a great number of isolated individual micro-elements composing the material. Initially the micro-particles are randomly oriented, but under large external loads or at high temperature the evolution of microstructure takes place in the following two forms: (a) unraveling of the molecular chains, (b) relative motion of the unraveled chains or their segments through the surrounding moving molecules. In this case, the polymeric macro-molecules are independent kinematical elements, they can slide against another. The process is liquid-like. The macromolecules have translational and rotational degrees of freedom.

Under sliding motion the molecular chains can be aligned in one direction, so that the structure may be very highly anisotropic. Due to this, the polymer chains move with the highly anisotropic friction, i.e. with low friction for motion parallel to the chain and high friction for transverse motion. Friction anisotropy is intrinsic property of these materials.

During the flow of objects in fluids or gases, they meet a resistance to motion (i.e. wet friction). In most cases, friction between molecular chains is described by viscous friction laws. The viscous friction force acting on the molecular chain is proportional to the relative velocity at the given point of the continuum (since the macromolecules move relative to one another). This is in accordance with thermodynamics i.e. with the principle of objectivity. The equation of translational viscous friction of the macromolecule, due to interactions with the surrounding fluid, is given by:

$$\begin{aligned} &\mathbf{p} = -\mathbf{B}\frac{d\mathbf{u}}{dt}, \quad \mathbf{B}: \mathrm{R}^3 \to \mathrm{R}^3 \\ &\mathbf{p} = \mathbf{p}_t + \mathbf{p}_n, \quad \frac{d\mathbf{u}}{dt} = \frac{d\mathbf{u}_t}{dt} + \frac{d\mathbf{u}_n}{dt} \end{aligned} \tag{9}$$

where $\mathbf{p}$ = force of translational viscous friction $\mathbf{u}$ = displacement vector; $\mathbf{B}$ = translational viscous friction second-order tensor with viscous friction anisotropy included; R^3 = three-dimensional space of real numbers.

The equation of rotational viscous friction of macromolecules (Cosserat-like molecules) is as follows:

$$\begin{aligned} &\mathbf{m} = -\mathbf{D}\boldsymbol{\omega}, \quad \boldsymbol{\omega} = \frac{d\boldsymbol{\Phi}}{dt} \\ &\mathbf{D} = \mathbf{R}\begin{bmatrix} \gamma & 0 & 0 \\ 0 & 0 & 0 \\ 0 & 0 & 0 \end{bmatrix}\mathbf{R}^T, \quad \mathbf{R} = \mathbf{R}(\boldsymbol{\Phi}) \end{aligned} \tag{10}$$

where $\mathbf{m}$ = couple of rotational viscous friction; $\boldsymbol{\omega}$ = angular velocity; $\boldsymbol{\Phi}$ = rotation vector; $\mathbf{D}$ = tensor of rotational viscous friction (given for cylindrical microstructures); $\mathbf{R}$ = orthonormal matrix; γ = friction coefficient to a rotation of the cylinder about its axis, see Cyron & Wall (2012). In accordance with thermodynamics, the power of the friction force couple P_2 is non-positive in every case:

$$P_2 = \mathbf{m} \cdot \boldsymbol{\omega} \leq 0 \quad \forall \boldsymbol{\omega} \tag{11}$$

5 MICROSCOPIC-LEVEL MODELS OF FRICTION INSIDE POLYMERIC MATERIALS

The macromolecules as independent kinematical elements have idealized simple shapes. Their individual motion modes consist of translations, rotations and spinning. These motions are of Brownian type. Two classical models of macromolecules are proposed in the literature, (a) Rouse model: the macromolecules are chains of spherical beads and springs (or rods), (b) De Gennes tube model: the long macromolecules move like a snake inside long and narrow tubes. Polymers in liquid crystal phase can form long rigid molecules similar to rods and rigid flat molecules similar to discs (Zmitrowicz 2013).

The energy dissipation takes part due to the interfacial friction between these objects. The frictional resistance between molecular chains must be considered with respect to the molecular structure and kinematical properties of the chains.

5.1 *Friction of beads in the Rouse model*

In the Rouse model, during the motion the beats are affected by the surrounding molecules and friction forces arise, e.g. viscous friction forces located at bead centres since the beads have no volume. The springs simulate elastic properties of the macromolecules.

5.2 *Friction of macromolecules in the De Gennes model*

The snake-like motion of long macromolecules within the tube describes dynamics of polymers in melts and solutions. The individual chain segment moves within a narrow tube around the axis of the motion path. Polymer chains move under a highly anisotropic friction due to entanglements with neighbouring chains. This anisotropy manifests itself as the snake-like motion.

5.3 *Friction of rod-like and disc-like macromolecules*

The rod-like and disc-like macromolecules can slide one against another. Friction couples are present, and dry friction models can be applied. Two particular cases are considered: (a) friction of rod-like molecules under sliding and rolling, (b) friction of disc-like molecules under sliding and spinning. Cyron & Wall (2012) modelled the rod-like microstructures with the aid of Cosserat continuum. Dmitriev (2008) investigated dynamics of the rigid disc including the presence of anisotropic friction in the sliding plane.

5.4 *Langevin model of polymer dynamics with anisotropic friction*

Models of friction inside the polymeric materials are used in Langevin equation of motion. This is the equation of stochastic dynamics, and it describes Brownian motions using the Newton equation of motion with an additional term including external random excitations (i.e. stochastically fluctuating forces). Usually this equation of motion is a stochastic ordinary differential equation with the stochastic white-noise term as the applied load. Motion of the i-th bead follows the Langevain equation:

$$m\frac{d^2\mathbf{r}_i}{dt^2}=\left(B^{11}\mathbf{k}_1\otimes\mathbf{k}_1+B^{22}\mathbf{k}_2\otimes\mathbf{k}_2\right)\frac{d\mathbf{r}_i}{dt}-\mathbf{f}_i-\mathbf{F} \quad (12)$$

$$\mathbf{r}_i=r_i^1\mathbf{k}_1+r_i^2\mathbf{k}_2$$

where m = mass of the bead; $\mathbf{r}_i$ = position vector of the bead; $\mathbf{f}_i$ = stochastically fluctuating force (e.g. Gaussian white noise); $\mathbf{F}$ = other external forces; $\mathbf{k}_1$, $\mathbf{k}_2$ = unit vectors parallel and normal to the chain; B^{11}, B^{22} = friction coefficients for motion parallel and normal to the chain respectively.

6 CONCLUSIONS

The relations for friction at the external boundaries of solids can be used to predict properties of unlubricated contacts in polymer-based systems e.g. dry sliding bearings, sliding seals, transmission belts, brakes, clutches, car tyres, prostheses of human joints, etc. The equations of friction inside the polymeric materials have practical applications in simulations of various production processes in the chemical industry.

REFERENCES

Cyron, C.J. & Wall, W.A. 2012. Numerical method for the simulation of the Brownian dynamics of rod like structures with three-dimensional nonlinear beam elements. *Int. J. Numer. Meth. Engng.* 90(8): 955–987.

Dmitriev, N.N. 2008. Movement of a disk with displaced center of masses over a plane with anisotropic friction. *Journal of Friction and Wear* 29(2): 86–91.

Dunn, A.C., Steffens, J.G., Burris, D.L., Banks, S.A. & Sawyer W.G. 2008. Spatial geometric effects on the friction coefficient of UHMWPe. *Wear* 264(7–8): 648–653.

Liu, X.X., Li, T.S., Liu, X.J., Lv, R.G. & Cong, P.H. 2007. An investigation on the friction of oriented polytetrafluoroetylene (PTFE). *Wear* 262(11–12): 1414–1418.

Persson, B.N.J. 1998. On the theory of rubber friction. *Surface Science* 401(3): 445–454.

Von Mises, R. 1901–1908. Experimentelle Untersuchungen der Reibung. In F. Klein & C. Müller (eds), *Encyklopädie der mathematische Wissenschaften: mit Einschluss ihrer Anwendungen*, Band IV/1–2, 196–218. Leipzig: Teubner.

Zmitrowicz, A. 2010. Contact stresses: a review of models and methods of computations. *Archive of Applied Mechanics* 80(12): 1407–1428.

Zmitrowicz, A. 2012. Models of friction and wear of polymers: thermodynamic considerations. In Michel Raous (ed.), *Euromech No. 514, New trends in contact mechanics, Abstracts, Cargése, 27–31 March 2012*. Marseille: Laboratoire de Mécanique et d'Acoustique CNRS.

Zmitrowicz, A. 2013. Friction models of polymers at external boundaries of solids and inside the materials. In T. Łodygowski et al. (eds), *CMM 2013, 20th International Conference on Computer Methods in Mechanics, Short Papers, Poznań, 27–31 August 2013*. Poznań: Poznań University of Technology.

Advances in Mechanics: Theoretical, Computational and Interdisciplinary Issues – Kleiber et al. (Eds)
© 2016 Taylor & Francis Group, London, ISBN 978-1-138-02906-4

Investigation of stability and limit load of a truss overhead opened bridge

Ł. Żmuda-Trzebiatowski, P. Iwicki & M. Krajewski
Gdańsk University of Technology, Gdańsk, Poland

ABSTRACT: The paper presents selected methods of determining stability and limit load of a truss top chord in opened bridges. These methods include linear buckling and non-linear static analysis based on the finite element method and algorithms based on design code procedures. The described methods were tested on an example of a steel footbridge situated in Straszyn. The results of stability analysis are compared. The results of geometrical and material non-linear analysis are close for the perfect structure, but very large (40%) for large imperfection amplitudes. Using first global free vibration mode as in initial imperfection resulted in greater load-carrying capacity than first global buckling mode. The critical forces is nearly equal to the results from non-linear analysis. A Polish standard procedure showed some difficulties during analysis so the final condition of stability was not satisfied, while a Eurocode procedure confirmed correctness of a footbridge design.

1 INTRODUCTION

Stability is an engineering problem which occurs in light elements due to compression and flexure. Large compressive forces appear in upper chords, diagonal posts of trusses, arches or columns. Renowned examples of such structures are truss and arch bridges (De Backer et al. 2014). A convenient tool to investigate stability and limit load-carrying capacity is a commercial software using different numerical methods based on the finite element method (Choong & Ramm, 1998). The most popular method of investigating structural stability is Linear Buckling Analysis (LBA). The methods leads to buckling modes and corresponding buckling factors to given loads. The LBA is often used in design offices. Critical forces are required in further analysis using standard procedures to compute a design buckling resistance which should be higher than the greatest compressive force.

Another approach does not apply any standards. The maximum value of the compressive forces can be obtained by means of a non-linear analysis, geometrically (GNA) or Geometrically And Materially (GMNA). These procedures are more time-consuming, but more precise.

The stability conditions can be checked according to standard procedures. The Polish standard—PN-82/S-10052 (1982) recommends stability verification of compressive truss chords sensitive to out-of-plane buckling, introducing stiffness of the so-called half-frame. It is created by posts and a rigidly connected transverse. The half-frame stiffness is determined by a horizontal force in its plane subjected to a compressed chord, which causes a horizontal translation equal to 1 cm. According to the standard the stiffness can be calculated as follows:

$$H = \frac{E}{\frac{h_s^3}{3I_s} + \frac{h^2 b}{2I_p}} \tag{1}$$

where E—Young's modulus; h_s—reduced height of a post measured from a centroid of a plate stiffening a connection between a post and a transverse to the axis of a compressed chord; I_s—moment of inertia of a post cross-section; h—theoretical height of a post; b—theoretical length of a transverse; I_p—moment of inertia of a transverse cross-section.

The determined half-frame stiffness should satisfy the following condition:

$$H \geq c_{1,2} \cdot H_0 \tag{2}$$

where

$$H_0 = \frac{3.2 P_{max} \gamma_s}{\mu_{sr}^2 d_{min}} \tag{3}$$

where $c_{1,2}$—coefficients dependent on the supporting of compressed chords of a truss girder; P_{max}—the maximum compressive force in the compressed chord; γ_s—material factor; $\mu_{śr}$—arithmetic mean of buckling length factors of all chord bars; d_{min}—the smallest length of a chord between two half-frames.

The European standard—Eurocode 3 (2006)—recommends to determine a design buckling resistance

Figure 1. A view of an investigated footbridge in Straszyn.

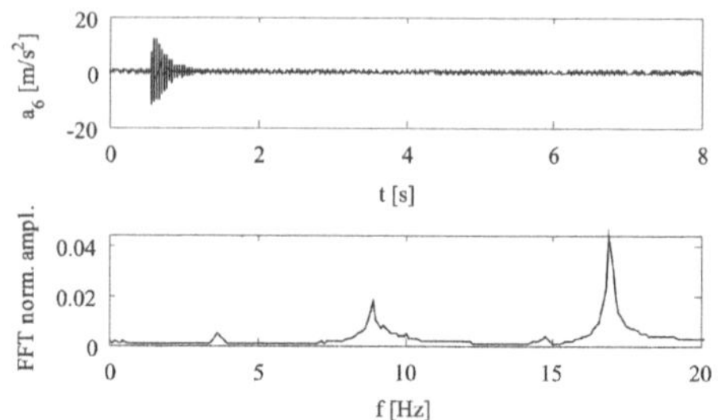

Figure 2. The measured signal of horizontal acceleration of an upper chord node and Fourier transform of the signal.

$N_{b,Rd}$ which should be higher or equal to a maximum compressive force in a compressed element:

$$N_{b,Rd} = \frac{\chi A f_y}{\gamma_{M1}} \tag{4}$$

where χ—reduction factor for the relevant buckling mode; A—cross-section area; f_y—yield strength; γ_{M1}—resistance factor for members due to instability.

The reduction factor is affected by shape and dimensions of a cross-section, buckling load and yield strength.

In the paper the abovementioned methods of investigating stability and limit load analysis were used for a real structure. The results obtained by means of LBA, GNA and GMNA were compared. Two standards—Polish and European—were used to check the stability conditions.

2 DESCRIPTION OF THE STRUCTURE

The abovementioned methods of investigating stability and limit load of structures were tested on a steel truss pedestrian bridge over Radunia river located on a hydroelectric power station field in Straszyn. Its view is presented in Figure 1.

The footbridge is a one-span truss structure of 21.0 m length. The width of the bridge in girders axes is 1.20 m and its overall width equals to 2.40 m. The upper truss chord consists of two isosceles angle sections L 80 × 80 × 8, while the bottom truss chord is made of two non-isosceles angle sections L 100 × 65 × 10. Angle sections L 50 × 50 × 6 are used for posts and L 90 × 90 × 11 for diagonals. Two strips of a balustrade lie between two sections of diagonals and posts in each truss girder and are welded to them, providing a joint work of the footbridge and the balustrade. Pedestrians walk on a deck made of truss plates. It is considered a non-cooperating element of the structure.

3 EXPERIMENTAL TESTS

The footbridge was subjected to a dynamic test. Vibration acceleration measurement was performed in 16 nodes of the structure by means of PCB 356A16 accelerometers (on the upper chord) and PCB 356B18 (on the bottom chord). At each node acceleration was measured in two directions: vertical and horizontal (perpendicular to the truss girder plane). Time range was recorded by 40-channel LMS SCADAS analyser. Modal hammer PCB 086D20 was used to excite vibrations.

The Fast Fourier Transform (FFT) was conducted on the measured signals of acceleration (Fig. 2). Three evident natural vibration frequencies were obtained—f_1 = 3.75 Hz, f_2 = 8.75 Hz and f_3 = 16.88 Hz. The first free vibration mode was used as an initial imperfection in further analysis. The detailed dynamic tests and results are described in Rucka & Żmuda-Trzebiatowski (2013).

4 NUMERICAL MODEL

Numerical calculations were conducted in a commercial program Abaqus (2010) applying the finite element method (Zienkiewicz & Taylor, 2000). A numerical model of the investigated footbridge was fully created with 4-node shell elements with reduced integration. Each node is assumed six degrees of freedom. All truss elements with gusset plates were included in the model. Joint work of the pedestrian bridge and the balustrade was considered as well. Although fasteners in the bridge were mostly rivets, the connections between the elements of the footbridge were modelled in a form of ties between the contacting surfaces of connected members. A steel class was assumed S235 due to the lack of any technical specifications. Geometry was set on a basis of stocktaking. The footbridge supports are bearings—rollers on one side of the bridge and fixed on the second side.

Loads with safety factors, acting on a structure were determined according to the Polish standard including dead load, load of equipment and live loads—wind and pedestrians. Analyses: static, modal, LBA, GNA and GMNA were executed.

Static analysis was necessary to obtain the compressive forces in the upper chord. In non-linear analysis an arc-length method by Riks was used to determine equilibrium paths. Initial arc-length increment was set as 0.05, minimum and maximum increments were default—10^{-5} and 0.1.

5 RESULTS

The compressive force in the upper chord of the footbridge was equal to 173.3 kN. The corresponding bending moments were very low, thus neglected in further analysis. The first five buckling modes were local deformations of the strip of the balustrade. They did not affect the entire structure. The sixth buckling mode was global buckling of the upper chord out of its plane (Fig. 3). The corresponding buckling factor was equal to 2.633, resulting in a critical force equal to 456.3 kN. Note that the first global buckling mode consists of two half-waves. This deformation was used as an initial imperfection in the non-linear analysis. The footbridge without any imperfections was also analysed.

The modal analysis was used to confirm a correctness of the numerical model. Natural frequencies and free vibration modes obtained from the experimental test were compared with the results of the modal analysis. The first fundamental frequency calculated in the model was equal to 4.03 Hz. The difference is about 7%. The real structure is old and deformed, thus the stiffness of the footbridge is lower than in the case of a perfect structure in the numerical model. This is the reason why the natural frequencies obtained from FFT are lower. The first free vibration mode consists of one half-wave and was used as an initial imperfection in non-linear analysis, too (Fig. 4).

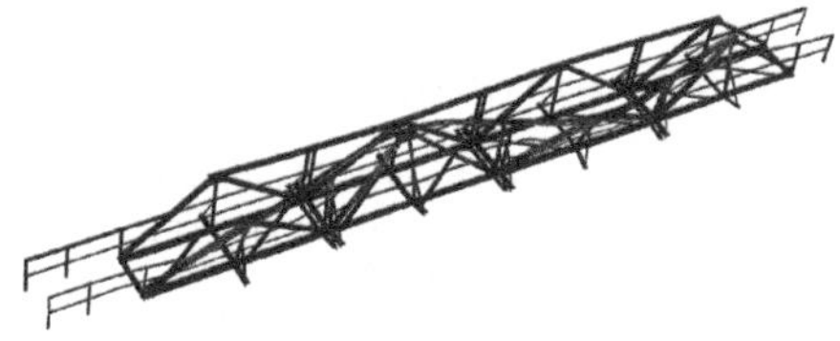

Figure 3. The first global buckling mode of the footbridge.

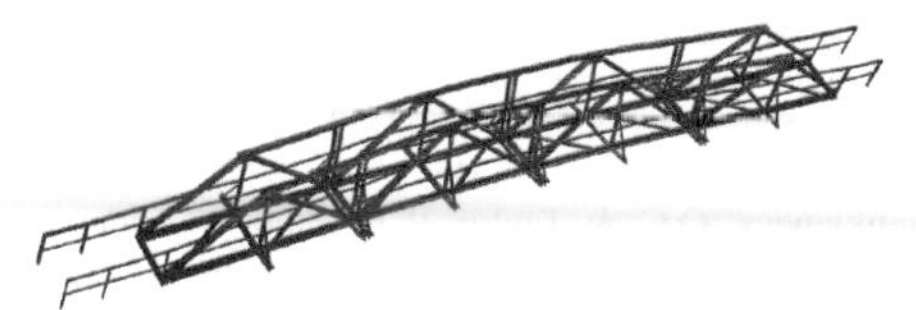

Figure 4. The first free vibration mode of the footbridge.

The relationship between maximum compressive forces in the upper chord and a horizontal displacement of a node in the middle of the chord was presented in the form of equilibrium paths obtained by means of a non-linear analysis. In the non-linear analyses three amplitudes of imperfections were used. They were obtained by dividing length by 250 (according to Eurocode). The length has three values: 15.45 m (the length of the upper chord), 4.025 m (the effective length of the upper chord) and 2.475 m (the distance between two nodes of the upper chord) and corresponding imperfection amplitudes: 61.8 mm, 16.1 mm and 9.9 mm, respectively.

The paths incorporating perfectly elastic and elastic-plastic models and different imperfection amplitudes were drawn in Figures 5 and 6. The comparison of the equilibrium paths from GNA and GMNA for different shapes of imperfections and constant amplitude $w = 16.1$ mm is given in Figure 7. The maximum compressive forces obtained from non-linear analyses were presented in Table 1.

It can be seen that elastic-plastic model results in much lower values of the maximum normal forces

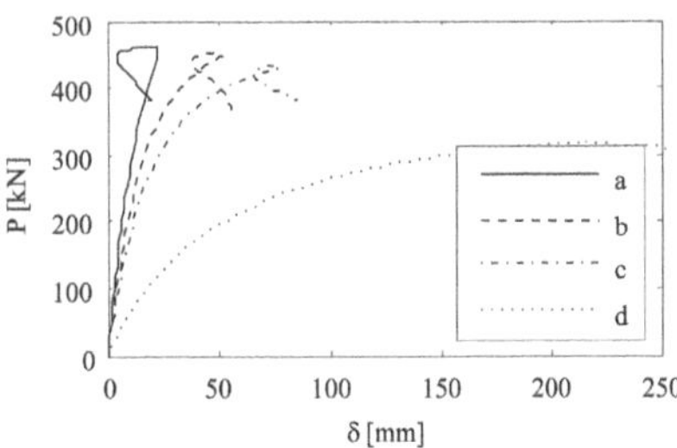

Figure 5. The evolution of maximum compressive forces and the horizontal displacement of the upper chord obtained from GNA with the imperfection based on the first buckling mode with amplitude: a) $w = 0$ mm, b) $w = 9.9$ mm, c) $w = 16.1$ mm, d) $w = 61.8$ mm.

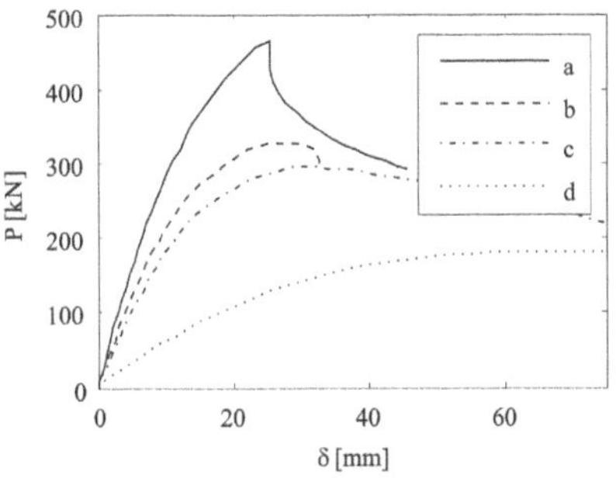

Figure 6. The evolution of maximum compressive forces and the horizontal displacement of the upper chord obtained from GMNA with the imperfection based on the first buckling mode with amplitude: a) $w = 0$ mm, b) $w = 9.9$ mm, c) $w = 16.1$ mm, d) $w = 61.8$ mm.

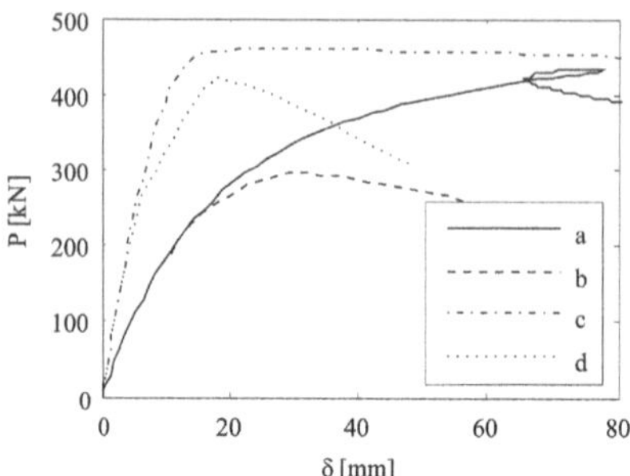

Figure 7. The evolution of maximum compressive forces and the horizontal displacement of the upper chord with the imperfection amplitude $w = 16.1$ mm for: a) GNA and buckling mode, b) GMNA and buckling mode, c) GNA and vibration mode, d) GMNA and vibration mode.

Table 1. Maximum normal forces [kN] for different models of the material, imperfection shapes and imperfection amplitudes.

Imperfection amplitude	buckling mode		vibration mode	
	GNA	GMNA	GNA	GMNA
0 mm	464.6	459.4	464.6	459.4
9.9 mm	451.9	328.1	459.7	425.0
16.1 mm	433.7	295.5	459.2	420.0
61.8 mm	320.4	182.7	458.1	377.9

in the upper chord of the truss girder than linearly elastic model. The difference is small (1.1%) for the footbridge with no imperfections and large (43%) for the structure with great imperfection amplitudes. The procedures in the standards contain some clues for determining buckling length in trusses. It is assumed as a distance between two inner nodes of the chord. Such length implies lower imperfection amplitude (9.9 mm) than in the case of effective length obtained from LBA (16.1 mm). The difference between maximum normal forces in the structure with imperfection amplitudes $w = 9.9$ mm and 16.1 mm is about 4–10% depending on the material model. Applying free vibration mode as the initial imperfection resulted in higher carrying capacity of the footbridge. The difference increased with the increase of the imperfection amplitude. It is worth noting that the buckling length determined on the basis of LBA is greater than assuming that the buckling length is equal to the distance between two nodes in the chord and smaller than the total length of the compressed upper chord. Thus the effective length should not be assumed according to the recommendations of the standards but determined on the basis of LBA.

The Polish standard procedure is difficult to use in the case of developed geometry of the bridges, such difficulties were evident in the footbridge considered here. The final stability condition according to the Polish standard was not satisfied ($H = 0.96$ kN/cm $< c_1H_0 = 1.06$ kN/cm). The European standard, Eurocode 3, was less restrictive and the design buckling resistance was higher than the compressive forces appearing in the model ($N_{Ed} = 173.3$ kN $< N_{b,Rd} = 300.6$ kN). It should be noted that the effective length was determined numerically, not assumed according to the standards.

6 CONCLUSIONS

Stability and limit load studies of the truss footbridge with different initial geometric imperfections allow to draw the following conclusions:

1. First global buckling mode as an initial imperfection results in lower load-carrying capacity of the structure than in the case of the first global free vibration mode.
2. Non-linear analyses are very effective in obtaining the full load-displacement relationship during the entire buckling process of the upper truss chord.
3. The procedure of checking stability conditions based on the so-called half-frames is much more restrictive and safer than Eurocode procedure.
4. The material non-linearity does not influence the load-bearing capacity of the perfect structure. The influence occurs with the increase of the imperfection amplitudes.
5. Assumption of the buckling length can lead to wrong results, thus it is better to determine it numerically.

REFERENCES

Abaqus 2010. *Standard theory and user's manuals*. Hibbit, Karlsson & Sorensen Inc.

Choong, K.K. & Ramm, E. 1998. Simulation of buckling process of shells by using the finite element method. *Thin-Walled Structures* 31: 39–72.

De Backer, H., Outtier, A. & Van Bogaert, P. 2014. Buckling design of steel tied-arch bridges. *Journal of Constructional Steel Research* 103: 159–167.

Eurocode 3, ENV 1993-2. 2006. Design of steel structures—Part 2: Steel Bridges.

PN-82/S-10052. 1982. Bridges. Steel Structures. Design.

Rucka, M. & Żmuda-Trzebiatowski, Ł. 2013. Identyfikacja parametrów modalnych kładki dla pieszych przy użyciu wzbudzenia impulsowego. *Inżynieria i Budownictwo* 7–8: 412–415.

Zienkiewicz, O.C. & Taylor, R.L. 2000. *The Finite Element Method, Vol. 1: The Basis, fifth ed.* Oxford: Butterworth-Heinemann.

Advances in Mechanics: Theoretical, Computational and Interdisciplinary Issues – Kleiber et al. (Eds)
© 2016 Taylor & Francis Group, London, ISBN 978-1-138-02906-4

Author index